CCS 2006

Proceedings of the 13th ACM Conference on Computer and Communications Security

October 30-November 3, 2006 • Alexandria, Virginia, USA

Sponsored by
ACM Special Interest Group on Security, Audit & Control

Rebecca N. Wright, Sabrina De Capitani di Vimercati, & Vitaly Shmatikov, *Editors*

The Association for Computing Machinery
2 Penn Plaza, Suite 701
New York, New York 10121-0710

Copyright © 2006 by the Association for Computing Machinery, Inc. (ACM). Permission to make digital or hard copies of portions of this work for personal or classroom use is granted without fee provided that copies are not made or distributed for profit or commercial advantage and that copies bear this notice and the full citation on the first page. Copyright for components of this work owned by others than ACM must be honored. Abstracting with credit is permitted. To copy otherwise, to republish, to post on servers or to redistribute to lists, requires prior specific permission and/or a fee. Request permission to republish from: Publications Dept., ACM, Inc. Fax +1 (212) 869-0481 or <permissions@acm.org>.

For other copying of articles that carry a code at the bottom of the first or last page, copying is permitted provided that the per-copy fee indicated in the code is paid through the Copyright Clearance Center, 222 Rosewood Drive, Danvers, MA 01923.

Notice to Past Authors of ACM-Published Articles

ACM intends to create a complete electronic archive of all articles and/or other material previously published by ACM. If you have written a work that has been previously published by ACM in any journal or conference proceedings prior to 1978, or any SIG Newsletter at any time, and you do NOT want this work to appear in the ACM Digital Library, please inform permissions@acm.org, stating the title of the work, the author(s), and where and when published.

ISBN: 1-59593-518-5

Additional copies may be ordered prepaid from:

ACM Order Department
General Post Office
P.O. Box 30777
New York, NY 10087-0777

Phone: 1-800-342-6626
(US and Canada)
+1-212-626-0500
(all other countries)
Fax: +1-212-944-1318
E-mail: acmhelp@acm.org

ACM Order Number 537060
Printed in the USA

Welcome to CCS 2006

It is a pleasure and honor to welcome you to the *13th ACM Conference on Computer and Communications Security*. This year's conference continues and extends its tradition as a premier forum for new data-security research. We have an excellent program comprising a research track, an industry track, three tutorials, and eleven workshops. The conference covers a strikingly broad spectrum of interests and disciplines in the area of computer and communications security.

The conference has benefited from many contributors to its success. I would like to thank Rebecca Wright and Sabrina De Capitani di Vimercati, the research-track program chair and co-chair respectively, along with the members of their program committee. The research program remains very highly selective, as evidenced by the caliber of the papers even more than acceptance statistics. My thanks also to the industry-track chair, Peter Dinsmore, and his program committee for arranging talks on applied topics to complement the other conference offerings.

I would like to extend my appreciation to Marianne Winslett, the workshops chair, for assembling a broad, strong program of workshops and ensuring their smooth execution. I would also like to thank all the workshop program-chairs, specifically: Kaoru Kurosawa and Rei Safavi-Naini (Workshop on Digital Rights Management); Donggang Liu and Sencun Zhu (Workshop on Security of Ad Hoc and Sensor Networks); Roger Dingledine and Ting Yu (Workshop on Privacy in Electronic Society); Guenter Karjoth and Fabio Massacci (Workshop on Quality of Protection); Ethan Miller and Erez Zadok (Workshop on Storage Security and Survivability); Farnam Jahanian (Workshop on Recurring Malcode); Andrew D. Gordon and David Sands (Workshop on Formal Methods in Security Engineering: From Specifications to Code); Stefan Axelsson, Kiran Lakkaraju, Soon Tee Teoh, and Bill Yurcik (Workshop on Visualization for Computer Security); Gene Tsudik, Shouhuai Xu, and Moti Yung (Workshop on Scalable Trusted Computing); Atsuhiro Goto (Workshop on Digital Identity Management); and Ernesto Damiani and Alban Gabillon (Workshop on Secure Web Services). In the past two years, the conference has expanded its program of satellite workshops from six to eleven. I would also like to express my thanks to Wenliang (Kevin) Du for a strong program of tutorials.

I am grateful to Peter Neumann for delivering the keynote address. My thanks as well to those who have offered their untiring administrative support, namely to Peng Ning for managing the budget, Vitaly Shmatikov for assisting with the preparation of the proceedings, and Michael Locasto and Angelos Keromytis their joint work as publicity chairs. I wish to express my appreciation to the CCS steering committee, Sushil Jajodia (Chair), Carl Gunter, Ravi Sandhu, and Pierangela Samarati, for their help with myriad logistical questions. I would also like to acknowledge the administrative staff of the ACM SIGSAC and George Mason University for their support, and Executive Events for its management of the registration process.

Finally, I wish to thank the following institutions for their generous sponsorship: The Defense Advanced Research Projects Agency, The Army Research Office, and IBM Research.

I hope that you, the conference and workshop attendees, will find this year's programs stimulating and beneficial for your research. Welcome—and enjoy.

Ari Juels
General Chair
ACM CCS 2006

Message from the Program Chairs

We are delighted to join Ari Juels in welcoming you to the *13th ACM Conference on Computer and Communications Security*, held October 30 to November 3, 2006 at the Hilton Alexandria Mark Center, Alexandria, Virginia, USA. These proceedings contain 38 papers presented in the research track and a paper contributed by our keynote speaker, Peter Neumann. The conference also comprises an industry track, tutorials, and workshops.

The 38 research track papers contained in this proceedings were selected from 256 received submissions. These submissions were carefully reviewed by at least three members of the program committee, or four in the case of papers authored by program committee members. Reviewing was double-blind, meaning that the program committee was not able to see the names and affiliations of the authors, and the authors were not told which committee members reviewed which papers. Program committee members were allowed to use external reviewers, but were responsible for the contents of the review and representing papers during the decision making. The review phase was followed by a three-week discussion phase in which each paper with at least one supporting review was discussed, outside experts were consulted where needed, and final decisions were made.

We thank the program committee for their hard work in selecting the program from these papers. We also thank the external referees that helped with the reviewing task and Cathy Meadows for providing helpful advice based on her experience as CCS'05 program chair. We thank the CCS steering committee and other CCS'06 organizers for their help, which allowed us to focus on putting together the research program. We also thank Thomas Herlea for running the WebReview system that was used for the electronic submission and review of the submitted papers.

Finally, we thank all authors who submitted papers and all conference attendees. Without them, there would be no conference.

Rebecca Wright
CCS 2006 Research Track
Program Chair

Sabrina De Capitani di Vimercati
CCS 2006 Research Track
Program Co-Chair

Table of Contents

CCS 2006 Conference Organization ix

Sponsor & Supporters x

Keynote Address

- **System and Network Trustworthiness in Perspective** 1
  P. G. Neumann *(SRI International)*

Session 1: Anonymity
Session Chair: V. Atluri *(Rutgers University)*

- **Providing Witness Anonymity in Peer-to-Peer Systems** 6
  B. Zhu, S. Setia, S. Jajodia *(George Mason University)*

- **Salsa: A Structured Approach to Large-Scale Anonymity** 17
  A. Nambiar, M. Wright *(University of Texas at Arlington)*

- **Hot or Not: Revealing Hidden Services by Their Clock Skew** 27
  S. J. Murdoch *(University of Cambridge)*

Session 2: Intrusion Detection
Session Chair: P. McDaniel *(Penn State University)*

- **Packet Vaccine: Black-box Exploit Detection and Signature Generation** 37
  XF. Wang, Z. Li *(Indiana University)*, J. Xu *(Google Inc. and North Carolina State University)*,
  M. K. Reiter *(Carnegie Mellon University)*, C. Kil *(North Carolina State University)*,
  J. Y. Choi *(Indiana University)*

- **Protomatching Network Traffic for High Throughput Network Intrusion Detection** 47
  S. Rubin, S. J. Jha, B. P. Miller *(University of Wisconsin, Madison)*

- **Evading Network Anomaly Detection Systems:
  Formal Reasoning and Practical Techniques** 59
  P. Fogla, W. Lee *(Georgia Institute of Technology)*

Session 3: Data Protection
Session Chair: N. Li *(Purdue University)*

- **Data Collection with Self-Enforcing Privacy** 69
  P. Golle *(Palo Alto Research Center)*, F. McSherry, I. Mironov *(Microsoft Research)*,

- **Searchable Symmetric Encryption: Improved Definitions and Efficient Constructions** 79
  R. Curtmola *(Johns Hopkins University)*, J. Garay *(Bell Labs - Lucent Technologies)*,
  S. Kamara *(Johns Hopkins University)*, R. Ostrovsky *(University of California at Los Angeles)*

- **Attribute-Based Encryption for Fine-Grained Access Control of Encrypted Data** 89
  V. Goyal, O. Pandey, A. Sahai *(University of California at Los Angeles)*,
  B. P. Waters *(SRI International)*

- **Secure Attribute-Based Systems** 99
  M. Pirretti, P. Traynor, P. McDaniel *(Pennsylvania State University)*, B. Waters *(SRI International)*

Session 4: Access Control
Session Chair: R. Sailer *(IBM T.J.Watson Research Center)*

- **Resiliency Policies in Access Control** 113
  N. Li *(Purdue University)*, M. V. Tripunitara *(Motorola Labs)*, Q. Wang *(Purdue University)*

- **Safety and Consistency in Policy-Based Authorization Systems** 124
  A. J. Lee, M. Winslett *(University of Illinois at Urbana-Champaign)*

- **On the Modeling and Analysis of Obligations** 134
  K. Irwin, T. Yu *(North Carolina State University)*,
  W. H. Winsborough *(University of Texas at San Antonio)*

- **RoleMiner: Mining Role using Subset Enumeration** 144
  J. Vaidya, V. Atluri, J. Warner *(Rutgers University)*

Session 5: Privacy and Authentication
Session Chair: R. Safavi-Naini *(University of Wollongong)*

- **Doppelganger: Better Browser Privacy Without the Bother** 154
  U. Shankar, C. Karlof *(University of California at Berkeley)*

- **Fourth-Factor Authentication: Somebody You Know** 168
  J. Brainard, A. Juels *(RSA Laboratories)*, R. L. Rivest *(Massachusetts Institute of Technology)*,
  M. Szydlo, M. Yung *(RSA Laboratories)*

- **An Effective Defense Against Email Spam Laundering** 179
  M. Xie, H. Yin, H. Wang *(The College of William and Mary)*

Session 6: Applied Cryptography I
Session Chair: M. Backes *(Saarland University)*

- **Forward-Secure Signatures with Untrusted Update** 191
  X. Boyen *(Voltage Security Inc.)*, H. Shacham *(Weizmann Institute of Science)*,
  E. Shen *(Stanford University)*, B. Waters *(SRI International)*

- **How to Win the Clone Wars: Efficient Periodic n-Times Anonymous Authentication** 201
  J. Camenisch, S. Hohenberger *(IBM Research)*, M. Kohlweiss *(Katholieke Universiteit Leuven)*,
  A. Lysyanskaya, M. Meyerovich *(Brown University)*

- **A Fully Collusion Resistant Broadcast, Trace, and Revoke System** 211
  D. Boneh *(Stanford University)*, B. Waters *(SRI International)*

Session 7: Attacks and Cryptanalysis
Session Chair: P. Vora *(George Washington University)*

- **Puppetnets: Misusing Web Browsers as a Distributed Attack Infrastructure** 221
  V. T. Lam, S. Antonatos, P. Akritidis, K. G. Anagnostakis *(Institute for Infocomm Research)*

- **A Natural Language Approach to Automated Cryptanalysis of Two-Time Pads** 235
  J. Mason, K. Watson, J. Eisner, A. Stubblefield *(Johns Hopkins University)*

- **Dictionary Attacks Using Keyboard Acoustic Emanations** 245
  Y. Berger, A. Wool, A. Yeredor *(Tel Aviv University)*

- **Inferring the Source of Encrypted HTTP Connections** 255
  M. Liberatore, B. N. Levine *(University of Massachusetts at Amherst)*

Session 8: Sensors and Networking
Session Chair: B. Levine *(University of Massachusetts at Amherst)*

- **TinySeRSync: Secure & Resilient Time Synchronization in Wireless Sensor Networks** 264
  K. Sun, P. Ning *(North Carolina State University)*, C. Wang *(Army Research Office)*,
  A. Liu, Y. Zhou *(North Carolina State University)*

- **Secure Hierarchical In-Network Aggregation in Sensor Networks** 278
  H. Chen, A. Perrig, D. Song *(Carnegie Mellon University)*

- **Provably-Secure Time-Bound Hierarchical Key Assignment Schemes** 288
  G. Ateniese *(Johns Hopkins University)*, A. De Santis, A. L. Ferrara, B. Masucci *(Università di Salerno)*

- **Optimizing BGP Security by Exploiting Path Stability** 298
  K. Butler, P. McDaniel *(The Pennsylvania State University)*, W. Aiello *(University of British Columbia)*

Session 9: Software and Network Exploits
Session Chair: S. De Capitani di Vimercati *(University of Milan)*

- **Replayer: Automatic Protocol Replay by Binary Analysis** 311
  J. Newsome, D. Brumley, J. Franklin, D. Song *(Carnegie Mellon University)*

- **EXE: Automatically Generating Inputs of Death** 322
  C. Cadar, V. Ganesh, P. M. Pawlowski, D. L. Dill, D. R. Engler *(Stanford University)*

- **A Scalable Approach to Attack Graph Generation** 336
  X. Ou *(Purdue University)*, W. F. Boyer, M. A. McQueen *(Idaho National Laboratory)*

Session 10: Formal Methods
Session Chair: T. Yu *(North Carolina State University)*

- **Formal Specification and Verification of Data Separation in a Separation Kernel for an Embedded System** 346
  C. L. Heitmeyer, M. Archer, E. I. Leonard, J. McLean *(Naval Research Laboratory)*

- **Beyond Separation of Duty: An Algebra for Specifying High-level Security Policies** 356
  N. Li, Q. Wang *(Purdue University)*

- **Computationally Sound Secrecy Proofs by Mechanized Flow Analysis** 370
  M. Backes *(Saarland University)*, P. Laud *(Tartu University)*

Session 11: Applied Cryptography II
Session Chair: M. Goodrich *(University of California at Irvine)*

- **Stateful Public-Key Cryptosystems: How to Encrypt with One 160-bit Exponentiation** 380
  M. Bellare *(University of California at San Diego)*,
  T. Kohno *(University of Washington)*, V. Shoup *(New York University)*

- **Multi-Signatures in the Plain Public-Key Model and a General Forking Lemma** 390
  M. Bellare *(University of California at San Diego)*,
  G. Neven *(Katholieke Universiteit Leuven)*

- **Deniable Authentication and Key Exchange** 400
  M. Di Raimondo *(Università di Catania)*,
  R. Gennaro, H. Krawczyk *(IBM T.J. Watson Research Center)*

- **Secure Function Evaluation with Ordered Binary Decision Diagrams** 410
  L. Kruger, S. Jha *(University of Wisconsin-Madison)*, E.-J. Goh, D. Boneh *(Stanford University)*

Author Index 421

CCS 2006 Conference Organization

General Chair: Ari Juels *(RSA Laboratories, USA)*

Program Chair: Rebecca Wright *(Stevens Institute of Technology, USA)*

Program Co-Chair: Sabrina De Capitani di Vimercati *(University of Milan, Italy)*

Industry and Government Track Chair: Peter Dinsmore *(Johns Hopkins University Applied Physics Lab, USA)*

Publicity Chairs: Angelos Keromytis *(Columbia University, USA)*
Michael E. Locasto *(Columbia University, USA)*

Publication Chair: Vitaly Shmatikov *(The University of Texas at Austin, USA)*

Tutorials Chair: Wenliang (Kevin) Du *(Syracuse University, USA)*

Treasurer: Peng Ning *(North Carolina State University, USA)*

Workshops Chair: Marianne Winslett *(UIUC, USA)*

Program Committee: Carlisle Adams *(University of Ottawa, Canada)*
Giuseppe Ateniese *(Johns Hopkins University, USA)*
Vijay Atluri *(Rutgers University, USA)*
Michael Backes *(Saarland University, Germany)*
Giampaolo Bella *(Università di Catania, Italy)*
John Black *(University of Colorado, USA)*
Nikita Borisov *(UIUC, USA)*
Jan Camenisch *(IBM Research, Switzerland)*
Rosario Gennaro *(IBM T. J. Watson Research Center, USA)*
Michael Goodrich *(UC Irvine, USA)*
Stefanos Gritzalis *(University of the Aegean, Greece)*
Trent Jaeger *(Penn State University, USA)*
Markus Jakobsson *(Indiana University, USA)*
Somesh Jha *(University of Wisconsin, USA)*
Trevor Jim *(AT&T Research, USA)*
Jonathan Katz *(University of Maryland, USA)*
Brian Levine *(University of Massachusetts, Amherst, USA)*
Ninghui Li *(Purdue University, USA)*
Peng Liu *(Penn State University, USA)*
Javier Lopez *(University of Malaga, Spain)*
Patrick McDaniel *(Penn State University, USA)*
Catherine Meadows *(Naval Research Laboratory, USA)*

Program Committee (continued): Nasir Memon *(Polytechnic University, USA)*
John Mitchell *(Stanford University, USA)*
Refik Molva *(Institut Eurecome, Sophia Antipolis, France)*
Eiji Okamoto *(University of Tsukuba, Japan)*
Phillip Porras *(SRI International, USA)*
Rei Safavi-Naini *(University of Wollongong, Australia)*
Reiner Sailer *(IBM T. J. Watson Research Center, USA)*
Pierangela Samarati *(Università degli Studi di Milano, Italy)*
Andre Scedrov *(University of Pennsylvania, USA)*
R. Sekar *(SUNY Stony Brook, USA)*
Shiuhpyng Shieh *(National Chiao Tung University, Taiwan)*
Sean Smith *(Dartmouth College, USA)*
Dawn Song *(Carnegie Mellon University, USA)*
Jessica Staddon *(Palo Alto Research Center, USA)*
Giovanni Vigna *(UC Santa Barbara, USA)*
Poorvi Vora *(George Washington University, USA)*
Susanne Wetzel *(Stevens Institute of Technology, USA)*
Shouhuai Xu *(University of Texas at San Antonio, USA)*
Alec Yasinsac *(Florida State University, USA)*
Ting Yu *(North Carolina State University, USA)*
Sheng Zhong *(SUNY Buffalo, USA)*

Sponsor:

Supporters: **IBM Research**

External Reviewers

Fenia Aivaloglou
Soon Ae Chun
Isaac Agudo
Mansour Alsaleh
Kun Bai
Theodoros Balopoulos
Davide Balzarotti
Greg Banks
Lujo Bauer
Steve Bellovin
Petros Belsis
Abhilasha Bhargav-Spantzel
Sandeep Bhatkar
Christophe Bidan
Olivier Billet
George Bissias
Marina Blanton
Damiano Bolzoni
Xavier Boyen
Linda Briesemeister
Arslan Broemme
Kevin Butler
Mike Burmester
Ji-Won Byun
Christian Cachin
Ran Canetti
Alvaro Cardenas
Dario Catalano
Iliano Cervesato
Hong Chen
Mihai Christodorescu
Jeremy Clark
Richard Clayton
Martin Cochran
Robert Cole
Carine Courbis
Lorrie Cranor
Stefano Crosta
Frederic Cuppens
Reza Curtmola
Breno de Medeiros

Drew Dean
Jing Deng
Mario Di Raimondo
Wenliang Du
Markus Duermuth
Glenn Durfee
Aleks Essex
Jianping Fan
Siamak Fayyaz-Shahandashti
Nelly Fazio
Nick Feamster
Gerardo Fernandez
Maria C. Fernandez
Luca Foschini
Martin Gagne
Vinod Ganapathy
Dimitris Geneiatakis
Jonathan Giffin
Matthew Green
Guofei Gu
Qijun Gu
Lazaros Gymnopoulos
Shai Halevi
Keith Harrison
Susan Hohenberger
Jason Holt
Jeffrey Horton
John Iliadis
Sotiris Ioannidis
Keith Irwin
Yoon-Chan Jhi
Lisa Johansen
Rob Johnson
Marc Joye
Christos Kalloniatis
Yogesh Kalyani
Seny Kamara
George Kambourakis
Costas Karafasoulis
Paul Karger
Erhan Kartaltepe

Maria Karyda
Aggelos Kiayias
Yunhua Koglin
Spyros Kokolakis
Kameswari Kotapati
Louis Kruger
Costas Lambrinoudakis
Adam Lee
Corrado Leita
Chris Lesniewski-Laas
Dave Levin
Marc Liberatore
Fengjun Li
Jiangtao Li
Lunquan Li
Qiming Li
Tiancheng Li
Zhenkai Liang
Jiqiang Liu
Liang Lu
Bo Luo
Matteo Maffei
Ziqing Mao
Boris Margolin
Brad Metz
Mira Meyerovich
Pietro Michiardi
Gerome Miklau
Sebastian Moedersheim
Frédéric Montagut
Razvan Musaloiu-E.
Antonio Nicolosi
Jesper Buus Nielsen
Melek Önen
Aabhas Paliwal
Chi-Chun Pan
Paul Parker
Bryan D. Payne
Adrian Perrig
Marinella Petrocchi
Van Hau Pham

Angela Piper
Richard Qing
Wei Qiu
Michael Rabinovich
Vijay Ramachandran
Rodrigo Roman
Kurt Rosenfeld
Yves Roudier
Farzad Salim
Taha Sencar
Hovav Shacham
Kulesh Shanmugasundaram
Raman Sharykin
abhi shelat
Nicholas Sheppard
Clay Shields
Abdullatif Shikfa
Heechang Shin
Radu Sion
Randy Smith
Miguel Soriano
Sid Stamm
Michael Steiner
Adam Stubblefield
Weiqing Sun
Paul Syverson
Gelareh Taban
Parisa Tabriz
Andreas Terzis
Ashish Thapliyal
Bhavani Thuraisingham
Alok Tongaonkar
Slim Trabelsi
Patrick Traynor
Mahesh V. Tripunitara
Theodoros Tzouramanis
Tri van Le
V.N. Venkatakrishnan
Jose L. Vivas
David Wagner
Hai Wang
Hao Wang
Qihua Wang
Xiaofeng Wang
Janice Warner
Brent Waters
William H. Winsborough
Hoeteck Wee
David Woodruff
Fan Wu
Qianhong Wu
Xintao Wu
Yuqing Wu
Jun Xu
Wei Xu
Zhiqiang Yang
Stefano Zanero
Zhijun Zhan
Chengqiang Zhang
Qing Zhang
Li Zhuang

System and Network Trustworthiness in Perspective

Peter G. Neumann
Computer Science Laboratory, Principled Systems Group
SRI International, Menlo Park CA 94025-3493

ABSTRACT

Characteristic problem areas experienced in the past are considered here, as well as some of the challenges that must be confronted in trying to achieve greater trustworthiness in computer systems and networks and in the overall environments in which they must operate. Some system development recommendations for the future are also discussed.

Category

D.4.6 Security and Protection

General Terms

Design, security, reliability, verification

Keywords

Computer systems, networks, trustworthiness, security, reliability, survivability, assurance, threats, vulnerabilities, risks

1. INTRODUCTION

In the context of this paper, *trustworthiness* means simply that something is worthy of being trusted to satisfy its specified requirements, typically relating to overall information system properties such as security, reliability, human safety, and survivability in the presence of a wide range of adversities, subject to some sort of assurance measures. In this context, *security* can be broadly thought of as encompassing everything that promotes the prevention, detection, and remediation of deleterious system behavior (for example, including actions of people, information technology, and physical environments).

Trustworthiness is relevant throughout system life cycles, from conception to requirements to detailed architectural designs to implementation, and then continuing into use, operation, maintenance, and recycling through revisions as appropriate. Trustworthiness is very difficult to achieve unless it is systematically integrated throughout the development cycle.

We begin with a few high-level remarks that might help motivate some of the problems that arise in designing, developing, operating, maintaining, and using complex highly distributed systems and networks — and especially those with extensive requirements for trustworthiness.

- "The future isn't what it used to be."[1] It is becoming ever more difficult to keep up with critical needs for trustworthy applications in the face of increasing system and network security vulnerabilities, increasing development complexity, multiple fault modes, sophisticated threats from insiders and outsiders, ubiquitously increasing societal dependence on information technology, increased dependence on the roles of administrators and users, and increasing costs and disruptions associated with the resulting risks. All these factors — and more — tend to exacerbate the problems of developing trustworthy intensively computer-based enterprises.

- "We need to go back to the future." Past research and development efforts present many important lessons for the future, although many of those lessons are either forgotten or never learned. Thus, it is useful to reconsider how we arrived at the present state of the art, and extrapolate on what might be needed in the future. We consider here both short-term and long-term approaches, and outline some of the hard problems that must be more systematically addressed.

- "Progress is always slower than you'd think." Proactive system development is often shunned, but has enormous potential payoffs. Mistakes of the past are frequently repeated, many of which could be easily surmounted — for example, using approaches such as evolvable system architectures, development that facilitates pervasive and predictable subsystem composability, intelligent interface design, sound system embeddings of strong cryptography, sound analysis tools, and up-front attention to assured trustworthiness.

2. PRINCIPLED SYSTEMS

For many years, the security community has been evolving various principles for the development of trustworthy systems, such as [14] and the Saltzer-Schroeder principles [28], both of which grew out of the Multics development.

An extended approach to principled development [16] also provides some analysis of how these principles and others can affect trustworthiness and assurance of systems as a whole. Several aspects of such an approach are highlighted here, notably the principle of least privilege; minimization of the extent to which portions of a system or network must be trusted, thereby limiting what is

---

[1] I first heard this almost-Yogi-Berra-like statement in 1969 in a keynote address by Arthur Clarke, who lamented that it was becoming ever more difficult to write good science fiction.

vulnerable to accidents and misuse; modular abstraction with encapsulation and complete mediation; and thorough nonbypassable auditing, In addition, various principles under the rubric of good software engineering practice can add significantly to assured trustworthiness, which — in addition to the aforementioned properties relating to security, reliability, safety, and survivability — can also encompass other concepts such as predictable composability, assured interoperability, and real-time performance.

Multics was perhaps the most principled of system developments. It broke some important new ground with hardware-software support for virtual memory, segmentation, paging, protection domains, stacks in which anything outside of the current stack frame was nonexecutable and nonaddressable, hierarchical directories, access control lists, dynamic linking, the use of a stark subset of PL/I that naturally prevented certain flaws, and so on. However, most instructive may have been its development discipline [5, 6].

Other principled efforts from the 1960s and 1970s include the T.H.E. system and its use of a deadlock-avoiding hierarchical locking strategy described by Dijkstra [7], and a seminal series of papers by Parnas (e.g., [20]). The 1977 Robinson–Levitt paper [22] on hierarchical formal specifications introduced the concept of formal mappings between different layers of functional specifications that represent abstract implementations of each layer as a function of the lower layers, as the foundation for the design of the Provably Secure Operating System (PSOS) [8, 17, 18].

The desire for multilevel security (MLS) prompted considerable research and development in the 1970s and 1980s, although those efforts did not bear much fruit with respect to commercially available MLS systems with strong assurance. Various efforts that were countercultural at the time (e.g., Rushby-Randell [27] and Proctor-Neumann [21]) are gaining new currency with respect to the concept of Multiple Independent Levels of Security (MILS) within the Secure Global Information Grid.

In addition, there is renewed interest in Rushby's 1981 notion of a separation kernel [26] and also in virtual machine monitors (such as VMware and Xen), which have a basic need for disciplined but limited interpartition communications [3].

Principles that are generally accepted *in principle* are often not widely used *in practice*. There are several possible explanations for this. For example, principles are not absolute; they are also not independent, and may sometimes interfere with one another; and their effective use requires appropriate educational stimuli, experiential feedback, acquired discipline, and (above all) foresight — which often runs counter to short-term motives such as profits and marketplace.

To illustrate the relevance of principled developments, we consider some particularly problematic real-world uses of computers and communications in the next two sections.

3. NETWORK DISRUPTIONS

Past networking difficulties offer lessons that need to be assimilated by researchers and system developers. This section revisits some old and new history relating to widespread network outages and other extreme network behavior. The earlier history may be of particular interest to younger researchers and developers who were not active in the olden days.

Various examples of widespread propagation effects are illustrative of some of the complexities arising in networked systems. Of particular interest are two cases in which total network failure modes resulted from local faults — namely the 1980 ARPANET collapse and the 1990 AT&T long-distance collapse — as well as numerous instances of domino-like propagating power outages.

- The 1980 ARPANET collapse: As a result of unchecked bits dropped in the highest-priority status messages in the memory of one ARPANET router (a so-called Interface Message Processor) and a garbage-collection algorithm that could subsequently not eliminate any nonrecent status messages, a memory overflow occurred in every node in the network. It took 4 hours to diagnose the unprecedented failure mode, telephone the administrator of every node and request that it be shut down manually, and finally, upon confirmation that all nodes had been shut down, manually initiate restarting every node. (See [24].)

- The 1990 AT&T long-lines collapse: An untested upgrade to the fault-tolerant recovery software had been installed in 114 ESS Number 4 telephone switches. (The C program contained a `break` statement within an `if` clause nested within a `switch` clause.) On 15 January 1990, one node crashed, and auto-recovered. However, as a result of the flaw, all the immediate neighbor nodes crashed, followed by iterated crashes of every node in the network repeatedly for something approaching half a day. Completing long-distance telephone calls became almost impossible. (See the ACM Risks Forum, vol 9, number 61, and *Telephony*, 22 January 1990, p. 11.)

In both of these cases, the developers of the networking technology seemed to have believed that network-wide outages could not possibly result from single-point failure modes.

Computer-controlled electrical-power networks have also been implicated in numerous widespread propagating power outages, evidently with some *a priori* disbelief on the part of developers and operators that widespread outages could result. (References for these and other cases noted below can be found in [13], with some of the older cases documented in [15].)

- Northeast U.S. blackout, November 1965: a threshold was exceeded that had been set too low, resulting in a 13-hour outage that affected New York, Pennsylvania, Vermont, Connecticut and Massachusetts.

- Lower New York State blackout, July 1977: recovery took in excess of 26 hours.

- Parts of ten Western U.S. states blacked out, October 1984: a faulty computer reading (the actual power loss was off by a factor of two) caused a chain reaction, although the triggering loss was reportedly an insignificant routine event that was supposed to have been handled seamlessly.

- Western U.S., July 1996: a heat-wave expanded power lines, which came in contact with a tree. Resulting power outages propagated over at least 12 states.

- Western U.S./Canada/Baja California (Mexico), August 1996: a summer heat-wave triggered cascading power outages, resulting in air-traffic slowdowns and many collateral problems.

- Northeast U.S., August 2003: a software race condition and an alarm-system failure caused a computer crash and widespread power outages; recovery took almost 2 days.

- Circuit breakers tripped in Maryland, in Queens, New York, and in Philadelphia on 25 May 2006, shutting down Amtrak, NJ Transit, and MARC trains in the morning rush-hour for the day. Five trains were stuck in tunnels.

- Queens, New York, July 2006: century-old wiring failed over a wide local region; the outage lasted a week.

What is perhaps most alarming about this list of outages is that such cases continue to occur. In some cases, this happens despite efforts to take remedial measures; in other cases, such measures are not even taken. Again, we often hear that a particular propagating outage is impossible because of the system design. Then, after it happens, we are typically told that the system or software or algorithms have been changed to prevent the problem from ever happening again. And then something similar happens again. Clearly, more proactive efforts are needed to analyze systems for potential widespread fault modes and ensuing risks, and to take appropriate timely actions. Of course, similar conclusions apply to proactive defenses against environmental disasters, as in the case of tidal waves, global warming, and anticipation of hurricanes (which was lacking in New Orleans before Katrina).

Leslie Lamport has remarked that a distributed system is a system in which you have to trust components whose existence is completely unknown to you. In the present context, distributed control of distributed systems with distributed sensors and distributed actuators is typically more vulnerable to propagated outages and other perverse failure modes such as deadlocks and other unrecognized hidden interdependencies (particularly on components that are untrustworthy and perhaps even hidden), race conditions and other timing quirks, coordinated denial-of-service attacks, and so on. Achieving reliability, fault tolerance, and system survivability in highly distributed systems is problematic; both formal analyses and system testing are highly desirable, but also potentially more complex than in nondistributed systems.

Furthermore, security, reliability, and system survivability are closely linked. A system is not likely to be secure if it is unreliable; similarly, a system is not likely to be reliable if it is not secure. A system is not likely to be survivable in the face of natural, accidental, and malicious adversities if it is not secure, reliable, fault tolerant, people tolerant, usable, and easily administered, with sufficient performance even under crises.

To illustrate this point, although the ARPANET outage and the AT&T long-lines collapse were triggered spontaneously by mechanisms that required no human intervention, each could have been equally well caused by (possibly remote) human activity, either intentionally (by anyone who knew about the fault mode) or accidentally. In the ARPANET case, the insertion of two bogus network status messages could have created the same effect as the dropped bits. In the AT&T case, inadvertent action by maintenance personnel or intruders on one telephone switch could have initiated the same sequence of propagating crashes. (There is some reason to suspect that this might actually have happened.) Similar reasoning also applies to the distribution of electrical power, where the control systems are heavily dependent on computer systems, with considerable functionality accessible from the Internet. Thus, security, reliability, and survivability need to be considered together rather than separately.

Any discussion of network disruptions must of course mention viruses, worms, propagating Trojan horses, and other forms of distributed malware. It is worth revisiting the 1988 Internet Worm [23, 29], which exploited four different system weaknesses — a buffer overflow in the `finger` daemon, systems configured with overly permissive `.rhosts` access, the `sendmail` debug option, and dictionary attacks on encrypted password files; however, it ran amok because of an overly aggressive algorithm for keeping the worm alive. Preventing malware might be aided by a system environment that provides a combination of sound computer system architectures with confined execution environments or virtual machines, rigorously controlled interactions among partitions (e.g., [3]), carefully configured fine-grained access control policies (e.g., [9]), good software engineering practice with safe uses of sensible programming languages and static analysis tools for software. With these and other approaches, the challenge of preventing exploitations of malware and security flaws such as buffer overflows could be much less problematic than it is today. (See [1, 10, 19, 31] for taxonomies of security flaws.)

With respect to confining the propagation of undesirable behavior in networks and distributed systems, the principles of security and good software development could contribute considerably. For example, the principle of least privilege suggests the development of architectures that enforce confined-domain or virtual-machine process isolation, with intercomponent interfaces and network protocols that are well defined, carefully controlled, and carefully analyzed. The principle of complete mediation suggests an architecture in which only a subset of the system needs to be highly trustworthy, and in which the integrity of that subset cannot be circumvented.

4. COMPUTERIZED ELECTIONS

The problems of trying to ensure the fairness, accuracy, and overall integrity of elections necessitate many requirements that span the entire election process. Various end-to-end requirements are needed to address system integrity, reliability, and survivability, vote integrity, overall accountability and oversight, nonsubvertible audit trails, impartial resolution of disputes, and uncompromised voter privacy (to identify just a few requirements). Simultaneously satisfying both vote integrity and voter privacy requires special care in design and implementation of operational procedures and relevant computer systems.

Since the mid-1980s, various efforts have been made to increase the automation of the voting process, particularly with the wider use of computer-based systems. (See Mercuri's doctoral thesis [11] for an extensive set of requirements casting the computer system requirements into the framework of the Common Criteria.) However, the need for trustworthy computing is only part of the problem. Trustworthiness is required throughout the end-to-end process.

Ideally, the total-process combination of computer system architectures and operational procedures should seek *strength in depth*, to prevent isolated failures or single-point attacks from compromising the end results. Unfortunately, the entire election process today represents *weakness in depth*: every step is a potential weak link — from registration to voter authentication to vote casting, vote counting, and transmission of the partial and final results (perhaps even via the Internet or wireless communications, perhaps unencrypted or weakly protected, and subject to denial-of-service attacks). Indeed, each stage in the voting process represents potential vulnerabilities that can be compromised in many ways — for example, accidentally or intentionally, detectably or undetectably, technologically or procedurally. The potential risks of faults, errors, failures, and misuse encompass human, environmental, and technological causes. Each step must be safeguarded from the outset with extensive cross-checks, and should be noncompromisibly audited. These requirements become especially important whenever election results are contested.

In today's all-electronic paperless voting machines, there is effectively no independent confirmation that the vote that is cast is the same as the vote that is counted, and that the results remain unchanged throughout. Furthermore, there is no independent audit trail that would enable an irrefutable recount. Today's unauditable proprietary touch-screen systems (e.g., [25]) are typically evaluated under a proprietary process that is commissioned by the vendors, relative to voluntary standards that are inherently weak. The result is fundamentally unconvincing from the perspective of trustworthiness.

One potential remedy for having to trust today's unauditable proprietary all-electronic voting systems is to add an independent voter-verified paper trail that is the vote of record [11], as some vendors are now attempting to do. However, this introduces further complexities, and appears to be only a palliative addition to an already overly expensive PC-based solution when compared with less expensive paper-based alternatives that are much less dependent on untrustworthiness of computer systems, such as optical scan technologies (but which of course have their own potential risks).

Of great potential interest to security researchers are possible voting systems and associated procedures that are designed, implemented, and rigorously evaluated openly to be demonstrably able to satisfy stringent end-to-end requirements. One very promising direction involves cryptographic approaches (e.g., [2, 4, 12]) that could be formally verified. Of course, a fundamental challenge for the cryptography community is to provide convincing evidence that the resulting total systems (that is, not just the cryptography in isolation) are demonstrably resistant to internal tampering, external manipulation, and undetected system errors, including compromise from any underlying operating systems (which do not even have to be considered under today's voluntary evaluation guidelines) or from compiler subversions (such as in [30]). Such evidence would need to be convincing to independent cryptographers, system security experts, and — to a considerable extent — the general public.

Although the problem of adding up a bunch of votes might intuitively seem easy, the complexity of the overall election process is considerable, particularly in operational practice. With respect to computerized elections, serious observance of the principles of Section 2 could (in principle) provide dramatic improvements in today's voting technology, but would need to be applied end-to-end to detect, prevent, and recover from software-hardware failures and power outages, as well as surreptitious manipulations and other unexpected irregularities in human procedures throughout the election process.

Of particular relevance here is the principle of least privilege. As an extension of that principle, what is needed are system architectures, network protocols, and operational procedures that minimize the areas of vulnerability, for reasons similar to those in Section 3, but with rather different threat models. Also needed are independent cross-checks that can ensure the integrity of the process despite accidental failures and intentional manipulations. In addition, the principle of pervasive nonbypassable auditing is critical, with nonalterable audit trails and oversight sufficient to enable incontrovertible reconstruction of all votes as cast.

5. CONCLUSIONS

If we take the principled advice of Section 2 and the examples of Section 3 and Section 4 seriously, several benefits can result.

- The development process could be dramatically improved. Principled system developments and selective uses of formal methods are increasingly needed. Greater up-front emphasis on requirements, sound architectures, and good software engineering practice can result in fewer cost overruns, fewer delivery delays, easier system operation, and much less need for recurring patch management. Systems should be designed to be predictably composable out of evaluated subsystems, with correspondingly predictable properties.

- Many characteristic system vulnerabilities can be systematically avoided or at least considerably diminished, especially in newly developed systems, but also through judicious use of well-conceived static analysis tools applied to existing systems.

- Although there are always many difficulties that arise in trying to retrofit wisdom into legacy systems, many of the principles can also be applied to incremental evolutions — for example, incorporating an architecturally sound combination of old and new subsystems, or attempting rather humbly to encapsulate or otherwise mask the weaknesses of older systems (albeit with somewhat lowered expectations of trustworthiness).

Trustworthiness (especially with respect to overall security, reliability, system survivability, and human safety) is inherently a weak-link phenomenon. In poorly designed systems, one flaw may be enough to result in compromise. Simplistic solutions are likely to be untrustworthy. Merely patching a few of the more obvious flaws is not likely to be very successful. Therefore, considerable pessimism should accompany any efforts to make silk purses out of sows' ears. However, trying to minimize what must be trustworthy, avoiding violations of the principle of least privilege, and wisely applying some of the other principles can help significantly.

This paper addresses just the tip of a huge iceberg that is lurking in our path. It should not be surprising. Its message is clearly not novel (for example, see [14, 28]), but nevertheless still fundamental and still underappreciated. In addition, more of the less visible portions of the iceberg are examined in [16], along with how to attain predictable composability of trustworthy systems with some meaningful measures of assurance.

The increasing complexity of modern systems, the critical dependence on computer-communication technology, and the ever greater need for trustworthiness make it imperative that we reflect on mistakes of the past and attempt to avoid them in the future. The road to disciplined system development is clearly paved with good intentions, and the suggested course is riddled with practical pitfalls. However, designing and implementing systems with trustworthiness as an integral fundamental goal is a very important challenge, and is well worth pursuing vigorously in research and development practice.

6. ACKNOWLEDGMENTS

This paper was prepared in part under National Science Foundation Grant Number 0524111. The author thanks Douglas Maughan, who sponsored the work culminating in [16] when he was a Program Manager in the U.S. Defense Advanced Research Projects Agency (DARPA). (Doug is now a Program Manager in the Science and Technology Directorate in the Department of Homeland Security, where he spearheads research and development relating to cybersecurity.) The author is grateful to Rebecca Wright for her very helpful feedback.

7. REFERENCES

[1] R.P. Abbott et al. Security analysis and enhancements of computer operating systems. Technical report, National Bureau of Standards, 1974. Order No. S-413558-74.

[2] B. Adida and C.A. Neff. Ballot casting assurance. In *Workshop on Electronic Voting Technology Workshop*, Vancouver, BC, Canada, August 2006. USENIX.

[3] Steven M. Bellovin. Virtual machines, virtual security? *Communications of the ACM*, 49(10), October 2006. *Inside Risks* column.

[4] J. Benaloh. Simple verifiable elections. In *Workshop on Electronic Voting Technology Workshop*, Vancouver, BC, Canada, August 2006. USENIX.

[5] F.J. Corbató. On building systems that will fail (1990 Turing Award Lecture, with a following interview by Karen Frenkel). *Communications of the ACM*, 34(9):72–90, September 1991.

[6] F.J. Corbató, J. Saltzer, and C.T. Clingen. Multics: The first seven years. In *Proceedings of the Spring Joint Computer Conference*, volume 40, Montvale, New Jersey, 1972. AFIPS Press.

[7] E.W. Dijkstra. The structure of the THE multiprogramming system. *Communications of the ACM*, 11(5), May 1968.

[8] R.J. Feiertag and P.G. Neumann. The foundations of a Provably Secure Operating System (PSOS). In *Proceedings of the National Computer Conference*, pages 329–334. AFIPS Press, 1979. http://www.csl.sri.com/neumann/psos.pdf.

[9] P.A. Karger. Limiting the damage potential of discretionary Trojan horses. In *Proceedings of the 1987 Symposium on Security and Privacy*, pages 32–37, Oakland, California, April 1987. IEEE Computer Society.

[10] C.E. Landwehr, A.R. Bull, J.P. McDermott, and W.S. Choi. A taxonomy of computer program security flaws, with examples. Technical report, Center for Secure Information Technology, Information Technology Division, Naval Research Laboratory, Washington, D.C., November 1993.

[11] R. Mercuri. *Electronic Vote Tabulation Checks and Balances*. PhD thesis, Department of Computer Science, University of Pennsylvania, 2001. http://www.notablesoftware.com/evote.html.

[12] C.A. Neff. A verifiable secret shuffle and its application to e-voting. In *Proceedings of the ACM Conference on Computer and Communications Security*, pages 116–125, Philadelphia, Pennsylvania, November 2001.

[13] P.G. Neumann. Illustrative risks to the public in the use of computer systems and related technology, index to RISKS cases. Technical report, Computer Science Laboratory, SRI International, Menlo Park, California. Updated regularly at http://www.csl.sri.com/neumann/illustrative.html; also in .ps and .pdf form for printing in a denser format.

[14] P.G. Neumann. The role of motherhood in the pop art of system programming. In *Proceedings of the ACM Second Symposium on Operating Systems Principles, Princeton, New Jersey*, pages 13–18. ACM, October 1969. http://www.multicians.org/pgn-motherhood.html.

[15] P.G. Neumann. *Computer-Related Risks*. ACM Press, New York, and Addison-Wesley, Reading, Massachusetts, 1995.

[16] P.G. Neumann. Principled assuredly trustworthy composable architectures. Technical report, Computer Science Laboratory, SRI International, Menlo Park, California, December 2004. http://www.csl.sri.com/neumann/chats4.html, .pdf, and .ps.

[17] P.G. Neumann, R.S. Boyer, R.J. Feiertag, K.N. Levitt, and L. Robinson. A Provably Secure Operating System: The system, its applications, and proofs. Technical report, Computer Science Laboratory, SRI International, Menlo Park, California, May 1980. 2nd edition, Report CSL-116.

[18] P.G. Neumann and R.J. Feiertag. PSOS revisited. In *Proceedings of the 19th Annual Computer Security Applications Conference (ACSAC 2003), Classic Papers section*, pages 208–216, Las Vegas, Nevada, December 2003. IEEE Computer Society. http://www.acsac.org/ and http://www.csl.sri.com/neumann/psos03.pdf.

[19] P.G. Neumann and D.B. Parker. A summary of computer misuse techniques. In *Proceedings of the Twelfth National Computer Security Conference*, pages 396–407, Baltimore, Maryland, 10–13 October 1989. NIST/NCSC.

[20] D.L. Parnas. On the criteria to be used in decomposing systems into modules. *Communications of the ACM*, 15(12), December 1972.

[21] N.E. Proctor and P.G. Neumann. Architectural implications of covert channels. In *Proceedings of the Fifteenth National Computer Security Conference*, pages 28–43, Baltimore, Maryland, 13–16 October 1992. (http://www.csl.sri.com/neumann/ncs92.html).

[22] L. Robinson and K.N. Levitt. Proof techniques for hierarchically structured programs. *Communications of the ACM*, 20(4):271–283, April 1977.

[23] J.A. Rochlis and M.W. Eichin. With microscope and tweezers: The Worm from MIT's perspective. *Communications of the ACM*, 32(6):689–698, June 1989.

[24] E. Rosen. Vulnerabilities of network control protocols. *ACM SIGSOFT Software Engineering Notes*, 6(1):6–8, January 1981.

[25] A. Rubin. *Brave New Ballot*. Random House, 2006.

[26] J.M. Rushby. The design and verification of secure systems. In *Proceedings of the Eighth ACM Symposium on Operating System Principles*, pages 12–21, Asilomar, California, December 1981. (ACM Operating Systems Review, 15(5)).

[27] J.M. Rushby and B. Randell. A distributed secure system (extended abstract). In *Proceedings of the 1983 IEEE Symposium on Security and Privacy*, pages 127–135, Oakland, California, April 1983. IEEE Computer Society.

[28] J.H. Saltzer and M.D. Schroeder. The protection of information in computer systems. *Proceedings of the IEEE*, 63(9):1278–1308, September 1975.

[29] E.H. Spafford. The Internet Worm: crisis and aftermath. *Communications of the ACM*, 32(6):678–687, June 1989.

[30] K.L. Thompson. Reflections on trusting trust. *Communications of the ACM*, 27(8):761–763, August 1984.

[31] K. Tsikpenyuk, B. Chess, and G. McGraw. Seven pernicious kingdoms: A taxonomy of software security errors. *IEEE Security and Privacy*, 3(6), November-December 2005.

Providing Witness Anonymity in Peer-to-Peer Systems

Bo Zhu
Center for Secure Information Systems
George Mason University
4400 University Drive
Fairfax, VA 22030-4444
bzhu@gmu.edu

Sanjeev Setia
Department of Computer Science
George Mason University
4400 University Drive
Fairfax, VA 22030-4444
setia@cs.gmu.edu

Sushil Jajodia
Center for Secure Information Systems
George Mason University
4400 University Drive
Fairfax, VA 22030-4444
jajodia@gmu.edu

ABSTRACT

In this paper, we introduce the concept of *witness anonymity* for peer-to-peer systems. Witness anonymity combines the seemingly conflicting requirements of anonymity (for honest peers who report on the misbehavior of other peers) and accountability (for malicious peers that attempt to misuse the anonymity feature to slander honest peers). We propose the *Secure Deep Throat* (**SDT**) protocol to provide anonymity for witnesses of malicious or selfish behavior to enable such peers to report on this behavior without fear of retaliation. On the other hand, in SDT the misuse of anonymity is restrained in such a way that any malicious peer that attempts to send multiple claims against the same innocent peer for the same reason (i.e., the same misbehavior type) can be identified. We also describe how SDT can be used in two modes. The active mode can be used in scenarios with real-time requirements, e.g., detecting and preventing the propagation of peer-to-peer worms, whereas the passive mode is suitable for scenarios without strict real-time requirements, e.g., query-based reputation systems. We analyze the security and overhead of SDT and present countermeasures that can be used to mitigate various attacks on the protocol. Our analysis shows that the communication, storage, and computation overheads of SDT are acceptable in peer-to-peer systems.

Categories and Subject Descriptors

K.4.1 [**Computers and Society**]: Public Policy Issues—*Privacy*; C.2.0 [**Computer-Communication Networks**]: General—*Security and protection (e.g., firewalls)*

General Terms

Security

Keywords

Privacy, Peer-to-Peer Systems, Witness Anonymity, k-Times Anonymous Authentication

Permission to make digital or hard copies of all or part of this work for personal or classroom use is granted without fee provided that copies are not made or distributed for profit or commercial advantage and that copies bear this notice and the full citation on the first page. To copy otherwise, to republish, to post on servers or to redistribute to lists, requires prior specific permission and/or a fee.
CCS'06, October 30–November 3, 2006, Alexandria, Virginia, USA.
Copyright 2006 ACM 1-59593-518-5/06/0010 ...$5.00.

1. INTRODUCTION

One of the fundamental challenges in peer-to-peer systems is how to build trust relationships between peers. In large-scale peer-to-peer systems, the chance that a given pair of peers will have repeated interactions with each other is small. Hence, two interacting peers may not have prior experience and knowledge of each other, and need a way to evaluate the risk involved with a transaction. To address this issue, several research studies [12, 13, 19, 33] have proposed mechanisms for building and using reputation-based trust models in peer-to-peer systems.

In these systems, a peer is assigned a trust value or reputation based on a trust metric. Although various systems differ in how this metric is defined, in general, the trust value associated with a peer is calculated based on the feedback provided by other peers. Peers rate the performance or behavior of another peer based on their previous interactions. When a peer encounters a new peer, it can query the network for trust ratings of that peer, and then based on the received feedback it can decide whether to proceed with the transaction.

There are three important requirements for these trust management systems:

- **Reliability:** A peer that issues a query for the trust ratings of another peer should be able to compute the *true* trust value despite the presence of malicious peers.

- **Anonymity:** It should not be possible to identify the peers who provide feedback in the form of their trust ratings for another peer. This is especially important when the feedback is negative in nature, since it could lead to retaliation.

- **Accountability:** It should be possible to identify malicious peers who attempt to misuse the anonymity property to manipulate the trust value computed for a peer. For example, without accountability, a malicious peer may submit multiple negative claims anonymously about the trustworthiness of another peer.

Most previous work on trust management in peer to peer systems has focused on the first requirement above, i.e., how to reliably compute trust values in the presence of malicious peers. However, the issues of anonymity and accountability have not received much attention. Although some of these systems [13, 29] make some provisions for peer anonymity, these provisions are easily circumvented as discussed in Section 9.

We observe that the issues of anonymity and accountability are closely coupled. Anonymity without accountability can be easily abused by malicious peers, so any system that enables peers to anonymously provide feedback on another peer must also include a mechanism for being able to identify peers that misuse the anonymity feature. We introduce the term *witness anonymity* to refer to this combination of seemingly conflicting requirements, i.e., identity anonymity for honest peers and accountability for misbehaving peers.

The major goal of our work is to show how peer-to-peer trust management systems can be extended to provide *witness anonymity*. The primary motivation for witness anonymity in peer-to-peer systems is similar to the need for whistleblower anonymity in real life. Without witness anonymity, peers that report on the misbehavior (e.g. false transactions) of other peers by submitting low trust ratings in response to a query, can become targets for retaliation [3]. This could take the form of tit-for-tat behavior, in which malicious peers intentionally lower their own trust rating for an honest peer. In extreme scenarios, peers giving a low rating to another peer may become targets for electronic attacks, e.g., denial of service, and even physical attacks.

Another important motivation for witness anonymity is simply to preserve the privacy of peers participating in the peer-to-peer trust management system. As a specific example, a peer A that responds to a query for the trust ratings of another peer B may not want to make public the fact that it has had previous interactions with B.

The third motivation for witness anonymity is to hide the trust topology of the peer-to-peer system from malicious parties. In particular, when the reputation system uses transitive trust (e.g. [19]) without witness anonymity the trust topology of the peer-to-peer system becomes public knowledge and can be exploited by malicious parties. For example, an adversary that wishes to launch an attack on a peer A may choose to compromise another peer B, if it knows that A has a high degree of trust in B. It can then exploit this trust by using B to launch an attack on A.

In this paper, we present a protocol called the *Secure Deep Throat* (**SDT**)[1] for providing witness anonymity in peer-to-peer systems. To the best of our knowledge, SDT is the first protocol that can support both aspects of witness anonymity, i.e., identity anonymity for honest peers, and accountability for peers that attempt to misuse the anonymity feature. SDT ensures the anonymity of a peer as long as she sends out only one feedback message per peer per malicious or selfish operation type. However, if a peer sends multiple claims against the same peer for the same reason, SDT includes a tracing mechanism to identify the peer.

The SDT protocol is based on the k-times anonymity authentication protocol proposed by Nguyen and Safavi-Naini [24]. We adapt their protocol to match the distributed and decentralized nature of peer-to-peer systems. In addition, we describe how SDT can be used in two modes: active mode and passive mode. The active mode is used in scenarios with real-time requirements for detecting malicious behavior, e.g. for detecting and preventing the propagation of peer-to-peer worms [34]. The passive mode is suitable for scenarios that do not have strict real-time requirements, e.g. query-based reputation systems [12].

The rest of the paper is organized as follows. In Section 2, we define the goals of SDT and analyze possible solutions based on available cryptographic techniques. In Section 3, we present the SDT framework including the system, adversary, and network models assumed in its design. In Section 4, we present the four procedures of the SDT protocol, i.e. *setup*, *registration*, *claim broadcasting*, and *public tracing*, under the active mode. In Section 5, we discuss two approaches for trading security for efficiency. We analyze the security and anonymity-related properties achieved in SDT and present the countermeasures to the collusion, Sybil and denial-of-service attacks in Section 6 and Section 7 respectively. In Section 8, we analyze the storage, communication, and computation costs of the SDT protocol. Related work is presented in Section 9. Finally, Section 10 contains our conclusions.

2. THE GOALS OF OUR WORK AND POSSIBLE SOLUTIONS

2.1 Design Goals

To achieve both anonymity and security in sending feedback (e.g., about the reputation of a peer) and event reports (e.g., about the misbehavior detected), a protocol that supports witness anonymity in peer-to-peer networks is expected to provide the following security and anonymity-related properties. In this paper, the terms *claim*, *feedback message*, and *report* are synonyms and used interchangeably. Similarly, the terms *user* and *peer* are synonyms and used interchangeably.

- **Identity Anonymity** Even if all the adversaries collude with each other, they will not be able to identify the source of an anonymous claim, i.e., the witness, as long as she sends the claim only once per adversary per type of malicious or selfish operation.

- **Backward Anonymity** The anonymity of a user that has acted as a witness is maintained even if other members in the network are compromised at a later time.

  Backward anonymity addresses the situation where users other than a witness are compromised. In this situation, adversaries can obtain the secrets known to the compromised users, e.g., the secret keys of the compromised users and the claims that were sent by witnesses and are stored by the compromised users. Due to backward anonymity, however, adversaries cannot use this information to deduce the identity of the user that has acted as a witness.

  Clearly, if a witness herself is compromised, her anonymity cannot be maintained. Using the secret key of the witness and the claims sent by the witness, adversaries can easily determine whether the compromised user has acted as a witness in the past.

- **Traceability** If a malicious or selfish user sends multiple claims against the same user for the same type of malicious or selfish operation, she will be identified.

---
[1] It is well known that the information from the anonymous source dubbed "Deep Throat" played an important role in helping unravel the Watergate scandals in the early 1970s.

- **Non-Slanderability** A good user can never be framed by adversaries, even if all of them collude with each other. This property is optional, and it is a must only when we assume there is a distributed trust mechanism in place so that the number of adversaries in the peer-to-peer system is lower than a threshold at any given time.

In addition, it is desirable that the new protocol can provide other properties such as *efficiency* (i.e., the storage, communication, and computation cost should be acceptable) and *decentralization* (i.e., an online central party is not necessary for the protocol).

2.2 Available Cryptographic Techniques for Providing Witness Anonymity

2.2.1 Blind Signature and Untraceable Electronic Cash

Blind signature and its applications to untraceable electronic cash cannot be used for our purpose. In untraceable electronic cash schemes [11, 7], the problem of detecting double-spending is similar to the problem of detecting multiple claims regarding the same peer in reputation systems. However, these schemes require the use of an online trusted party such as a bank. However, we cannot assume the existence of an online trusted party in peer-to-peer systems. Even if we assume that there is such a trusted party and it would be online periodically, a few problems still remain. One problem is that during the periods that the trusted party is off-line a group of innocent peers P might have been deleted from the friend lists of another group of peers O. This can occur if a single malicious peer has sent numerous claims to slander the members of P and these claim are accepted by the peers in O. Communication and computation overhead is another issue. The online trusted party could become the bottleneck of the verification. Most importantly, the content of the claim in a reputation system such as "the peer with the identity A is a malicious user because it has misbehavior of type I" cannot be predetermined. However, untraceable electronic cash schemes are not flexible enough to handle claims that are not predetermined.

2.2.2 Group and Ring Signatures

A typical group signature scheme must satisfy *unforgeability*, *anonymity*, *unlinkability*, *exculpability*, *traceability*[2], and *coalition-resistance* [2]. In group signature schemes [10, 2, 6], each group member can sign documents on behalf of the whole group. The receiver of a signed document can verify the signature to ensure that the document is signed by a group member. However, no one except the trusted group manager can recover the exact identity of the signer. The major weakness of group signature schemes is that in order to solve possible disputes at a later time, they give the group manager the unnecessary ability to trace any user even if there are no disputes. In other words, group signature schemes provide only partial anonymity to the signer, and are not suitable for scenarios where users have privacy concerns, e.g., peer-to-peer networks.

---

[2] Note that traceability as provided by group signatures and ring signatures is different from the one defined in Section 2.1.

Ring signatures[21, 8] can be viewed as a variant of group signatures without the traceability property. In ring signature, the group consists of only users without a group manager. Consequently, there is no way to revoke the anonymity of the signer, even if there is a dispute. That is to say, ring signatures can provide full anonymity, but fail to ensure accountability.

2.2.3 k-Times Anonymity Authentication

k-times anonymous authentication was first proposed by Teranishi, Furukawa, and Sako [31]. The entities in the k-times anonymous authentication scheme include a *group manager*, *users*, and *application providers*. This scheme provides stronger anonymity to a user in the sense that even the group manager cannot trace the identity of the user, as long as she follows the rule, i.e. authenticating herself and using the service provided by application providers (e.g., trial browsing of content) fewer than k times, where k is a predetermined number. In [31], the group membership is decided by the group manager, and application providers have no control over giving users access to their services. Nguyen and Safavi-Naini [24] proposed dynamic k-times anonymous authentication to enhance the privileges of application providers. In [24], application providers can select their user groups and grant or revoke access to users independently.

Compared to blind signature, group signature, and ring signature, k-times anonymous authentication is more suitable for achieving our design goals. It can provide *identity anonymity*, *backward anonymity*, and *traceability* at the same time.

3. THE FRAMEWORK OF THE SECURE DEEP THROAT PROTOCOL

In this section, we describe the system, adversary, and network models assumed in the design of the SDT protocol. We then provide an outline of the SDT protocol before presenting a detailed description in Section 4.

3.1 System Model

Our work assumes a peer-to-peer system where there is no online centralized *Trusted Third Party* (**TTP**). There are two types of entities that participate in SDT, namely the *Offline*[3] *Group Manager* (**OGM**) and *users*. We assume that there are a large number of users in the network and that a small fraction of the users are adversaries. The number of all the users and adversaries are denoted as n and t_a, respectively.

Unless explicitly specified, we assume the existence of a distributed trust mechanism, e.g., one of the protocols proposed in [22, 20, 35]. In these protocols, the number of adversaries in the system during a given time period, e.g., the key refresh period, is less than a threshold denoted as t. In addition, any group of t or more good users can cooperate together to provide certification services, e.g., assigning a certificate to a new user or revoking the certificate of an existing user when there are t or more claims against this user.

---

[3] The OGM is involved only in the first two procedures, which are assumed to be executed offline in SDT.

3.2 Adversary Model

We consider two types of adversaries – malicious users and selfish users. Malicious users attempt to disable the normal functionalities of the network. They may sniff, modify, or replay network communication messages. Selfish users take advantage of services provided by other peers without contributing back. For example, a selfish user may refuse to forward packets for other users.

For both types of adversaries, we assume that they will collude together to maximize the effectiveness of their attacks. We assume that it is possible to detect malicious and selfish operations, e.g., via worm detection software or by reviewing previous transactions with a specific peer. The details of how to detect such operations are beyond the scope of this paper.

3.3 Network Model

We assume the existence of a Mixnet-based anonymous communication system [27, 28, 15] so that an adversary cannot discover the identity-related information (e.g., the IP address) of the sender of a claim through traffic analysis. In this paper, we assume the existence of a mechanism for monitoring the number of claims sent by a given peer, irrespective of whether they are generated or forwarded by her.

3.4 Outline of The SDT Protocol

In both active and passive modes, the SDT protocol uses the following four procedures: *setup*, *registration*, *claim broadcasting*, and *public tracing*. We now present an outline of the operation of SDT in the active mode. (The passive mode of operation is described in Section 5.1. In this mode, the *claim broadcasting* and *public tracing* procedures operate differently.)

During the setup phase, the OGM generates a network-wide public/secret key pair, and publishes the public key. In addition, the OGM needs to publish the method of generating the tag bases, which correspond to the messages (or the meaningful content of the claim such as *user x executed a malicious behavior of type y*) to be anonymously authenticated in the *claim broadcasting* procedure, and other public information, e.g. the security parameters chosen.

Registration is done between the OGM and the user who wants to join the network. After this step, the user obtains a member public/secret key pair, and the OGM adds the user's identification and public key to an *identification list* (**LIST**). A user who has completed the *registration* procedure is called a member of the network.

Once a user detects any malicious or selfish behavior, and would like to act as a witness, i.e., broadcast a claim bearing witness to the misbehavior, she first calculates a tag base using the method published in the *setup* phase. Then she generates an anonymous claim using this tag base, and broadcasts the claim through a Mixnet-based anonymous communication system. The claim will be accepted by other users only if the sender (i.e., the witness) is a member of the network and the claim is generated by her secret key assigned in the *registration* procedure. Users that receive this claim store it into a local claim log, if they have not seen the same claim before. Otherwise, they simply drop the claim. Then the claim is forwarded again using the Mixnet-based anonymous communication system. Note that "the same claim" is different from "the claim against the same user for the same reason". In SDT, due to a random parameter, a witness can generate different claims against the same user for the same reason, and thus these claims are considered as distinct claims. (See Section 4.2 for more details.)

Using only the public information (i.e. LIST) and the claim log, anyone can do public tracing. This procedure outputs a user ID i or **NO-ONE**, which respectively mean "the user i has tried to misuse witness anonymity by sending multiple claims against the same user for the same reason" and "the public tracing procedure cannot find malicious entities misusing witness anonymity".

4. THE SECURE DEEP THROAT PROTOCOL

In this section, we present the details of the four procedures of the SDT protocol in the active mode of operation.

4.1 Preliminaries

4.1.1 Notation and Terminology

Let $\mathbb{N}$ and $\mathbb{Z}_p$ denote the set of natural integers and the set of natural integers in the range from 0 to $p-1$, respectively. A function $f: \mathbb{N} \to \mathbb{R}^+$ is called *negligible*, if for every positive number α, there exists a positive integer κ_0 such that for every integer $\kappa > \kappa_0$, it holds that $f(\kappa) < \kappa^{-\alpha}$. Let PT denote polynomial-time, and PPT denote probabilistic PT. For a set X, "$x \in_U X$" denotes that x is an element randomly and uniformly chosen from X. Let $\mathcal{H}_X$ denote a one-way hash function from the set of all finite binary strings $\{0,1\}^*$ onto the set X.

4.1.2 Bilinear Groups

Let $\mathbb{G}_1, \mathbb{G}_2$ be additive cyclic groups generated by P_1 and P_2, respectively, whose orders are a prime p, and $\mathbb{G}_T$ be a cyclic multiplicative group with the same order p. Suppose there is an isomorphism $\Psi: \mathbb{G}_2 \to \mathbb{G}_1$ such that $\Psi(P_2) = P_1$. Let $e: \mathbb{G}_1 \times \mathbb{G}_2 \to \mathbb{G}_T$ be a bilinear pairing with the following properties:

- **Bilinearity:** $e(aP, bQ) = e(P, Q)^{ab}$ for all $P \in \mathbb{G}_1$, $Q \in \mathbb{G}_2$, $a, b \in \mathbb{Z}_p$

- **Non-degeneracy:** $e(P_1, P_2) \neq 1$

- **Computability:** There is an efficient algorithm to compute $e(P, Q)$ for all $P \in \mathbb{G}_1, Q \in \mathbb{G}_2$

For simplicity, hereafter we set $\mathbb{G}_1 = \mathbb{G}_2$ and $P_1 = P_2$, but the proposed schemes can be easily modified for the general case when $\mathbb{G}_1 \neq \mathbb{G}_2$. In the rest of this paper, for a group $\mathbb{G}$ of prime order, we denote the set $\mathbb{G}^* = \mathbb{G}\setminus\{\mathcal{O}\}$, where $\mathcal{O}$ is the identity element of the group. We define a Bilinear Pairing Instance Generator as a PPT algorithm $\mathcal{G}$ that takes as input a security parameter 1^κ and returns a uniformly random tuple $t = (p, \mathbb{G}_1, \mathbb{G}_T, e, P)$ of bilinear pairing parameters, including a prime number p of size κ, a cyclic additive group $\mathbb{G}_1$ of order p, a multiplicative group $\mathbb{G}_T$ of order p, a bilinear map $e: \mathbb{G}_1 \times \mathbb{G}_1 \to \mathbb{G}_T$ and a generator P of $\mathbb{G}_1$.

4.2 Procedures of The Protocol

4.2.1 Setup

Given as input a security parameter 1^κ, the Bilinear Pairing Instance Generator generates a tuple $(p, \mathbb{G}_1, \mathbb{G}_T, e, P)$ as in Section 4.1.2. The OGM selects $P_0, H \in_U \mathbb{G}_1$, $\gamma \in_U \mathbb{Z}_p^*$, and sets $P_{pub} = \gamma P$ and $\Delta = e(P, P)$. The group public and secret keys are $gpk = (P, P_{pub}, P_0, H, \Delta)$ and $gsk = \gamma$, respectively. The identification list of group members denoted as LIST is initially empty.

An important task of the *setup* procedure is to define the way of generating the tag bases, which are used in the *claim broadcasting* procedure to create anonymous claims. Various methods can be employed to generate the tag bases. In this paper, we use the method defined in Equation (1).

$$(T_j, \check{T}_j) = \mathcal{H}_{\mathbb{G}_T \times \mathbb{G}_T}(Type_{ad}, ID_{ad}, MAX_{Claim}, j), \quad (1)$$

for $j = 1, \ldots, MAX_{Claim}$, where

- $Type_{ad}$ — denotes the type of event reported in the claim, e.g., accusing a user of refusing to forward a packet. $Type_{ad} \in TYPE$, where $TYPE$ is the set of all the event or claim types supported in SDT.

- ID_{ad} — denotes the identity of the user that the claim is accusing.

- MAX_{Claim} — the maximum number of claims that a user can send to accuse the same user for the same type of operations.

- j — The number of claims, including this one, that have been sent against a user with the identity ID_{ad} because of misbehavior of type $Type_{ad}$.

In this paper, we set $MAX_{Claim} = 1$. Namely, no one should send multiple claims against the same user for the same reason. Otherwise, she will be detected and identified. As such, there is only one tag base (i.e., $j = 1$) per type of misbehavior per user.

4.2.2 Registration

A user U_i can join the network as follows.

a. User U_i interacts with the OGM to determine her identity in the network. This can be done in various ways. In one possible approach, U_i selects an identity and forwards it to the OGM, who checks the availability of this identity. If it has been chosen by others, U_i will have to select another identity and repeat this process until the identity picked is available. Let the identity of U_i be denoted by i.

b. User U_i selects $x', r \in_U \mathbb{Z}_p^*$, and sends a commitment $C' = x'P + rH$ of x' to the OGM.

c. The OGM sends $y, y' \in_U \mathbb{Z}_p^*$ to U_i.

d. User U_i computes $x = y + x'y'$ and $(C, \beta) = (xP, \Delta^x)$. Next, U_i sends (C, β) to the OGM with a standard proof $Proof_1$ (please refer to [9] for this proof) to show that C is correctly computed from C', y, y', and that U_i knows the value of x satisfying $C = xP$.

e. The OGM verifies that the proof is valid, and that $e(C, P) = \beta$ is satisfied. If the verification succeeds, the OGM adds a new entry (i, β) to the LIST. The OGM then generates $a \in_U \mathbb{Z}_p^*$ different from all corresponding previously generated values, computes $S = \frac{1}{\gamma + a}(C + P_0)$, and sends (S, a) to user U_i. In addition, the OGM selects s items from the LIST (not including (i, β)), and forwards them to U_i. All the LIST items are signed using the OGM's secret key, and thus can be verified by any user.

f. User U_i confirms that $e(S, aP + P_{pub}) = e(C + P_0, P)$ is satisfied. The new member U_i's secret key is $usk = x$, and her public key is $upk = (a, S, C, \beta)$. U_i also verifies the s LIST items from the OGM.

4.2.3 Claim Broadcasting

Each user maintains two claim databases locally. One is a private claim database denoted as DB_{PC}. DB_{PC} records the claims that have been sent by the user herself. The other database is a common claim database denoted as DB_{CC}. DB_{CC} records the claims that she receives from other users.

Once a user detects any malicious or selfish operation that is a member of the set $TYPE$, she first examines DB_{PC} to check whether she has sent a claim against the malicious or selfish user for the same reason in the past. If the witness cannot find any entry with the same $(ID_{ad}, Type_{ad})$ pair, she generates a new anonymous claim and stores it in DB_{PC}. Otherwise, the witness will not generate another claim.

The anonymous claim is generated via the following steps. The witness first selects a random number $l \in_U \mathbb{Z}_p^*$. Then she computes the tag base $(T_1, \check{T}_1)$ following the method published in the *setup* procedure. Next, she calculates the tag $(\Gamma, \check{\Gamma}) = (T_1^x, (\Delta^l \check{T}_1)^x)$ using the tag base. Finally, the witness broadcasts $(\Gamma, \check{\Gamma})$ with a proof $Proof_2$ (please refer to [25] for this proof) using a Mixnet-based anonymous communication system. The format of the anonymous claim is $[Type_{ad}, ID_{ad}, \Gamma, \check{\Gamma}, l, Proof_2]$.

When a user receives the claim, she first checks whether there is an entry corresponding to the same claim in her DB_{PC} and DB_{CC}. If so, she simply drops the claim. Otherwise, the receiver computes T_1 according to equation (1), and checks whether $Proof_2$ is valid. If the proof is invalid, the user ignores the claim. Otherwise, she records the claim in DB_{CC}, and forwards the claim using the Mixnet-based anonymous communication system.

Once any user U_i finds t claims against the same user U_a for the same type of malicious or selfish operation (i.e., with the same ID_{ad} and $Type_{ad}$), she first checks whether all these claims are from distinct sources. Let the set of these t claims be denoted by V. Users can easily judge whether the claims in V are from distinct sources by comparing their Γ's. If any pair of claims in V has the same Γ, it means that a malicious user sent multiple claims against the same user for the same reason. She can be traced using the method presented in Section 4.2.4, and these claims are removed from V. If there are still t claims after this check, U_i can assert that U_a is an adversary who has executed the type of malicious or selfish operation indicated in all these claims. A message including the t claims collected is generated by U_i and broadcast to the network. Any user receiving the message can verify the claims included and agree with U_i's judgment on U_a if all the t claims are valid.

4.2.4 Public Tracing

Each member in the network can trace the identity of an adversary, if the adversary sends multiple claims against the same user for the same type of misbehavior. The *public tracing* procedure is as follows.

a. Look for two entries $[Type_{ad}, ID_{ad}, \Gamma, \check{\Gamma}, l, Proof_2]$ and $[Type'_{ad}, ID'_{ad}, \Gamma', \check{\Gamma}', l', Proof'_2]$ in the DB_{PC} and DB_{CC}, such that $Type_{ad} = Type'_{ad}$, $ID_{ad} = ID'_{ad}$, $\Gamma = \Gamma'$ and $l \neq l'$, and that both $Proof_2$ and $Proof'_2$ are valid. If no such entry can be found, output **NO-ONE**.

b. If such a pair of entries is found, compute β as $\beta = (\frac{\check{\Gamma}}{\check{\Gamma}'})^{\frac{1}{l-l'}} = [\frac{(\Delta^l \tilde{T}_1)^x}{(\Delta^{l'} \tilde{T}_1)^x}]^{\frac{1}{l-l'}} = \Delta^x$.

c. Search for a pair (i, β) in the part of LIST stored locally. If such a pair is found, broadcast a message including this pair together with the two entries as proofs to disclose the malicious user's identity.

d. If such a pair cannot be found locally, generate a message including β and the two entries and broadcast it to the network. If any member receiving this message finds a pair (i, β) in her local LIST, she repeats step a-c to verify the proofs and to disclose the malicious user's identity.

5. TRADEOFF BETWEEN SECURITY AND EFFICIENCY

In this section, we discuss how the security requirements of SDT can be relaxed in return for higher protocol efficiency, i.e., lower communication, computation, and storage costs.

5.1 Passive Mode Operation of SDT

The active mode of SDT is designed to support real-time detection of misbehavior by adversaries. In other words, a malicious or selfish peer is disclosed and expelled from the network as soon as t good peers detect her misbehavior, e.g. proliferating worms [34] or Trojans [23]. However, in situations where the real-time requirement is not critical, e.g. query-based reputation systems, SDT can operate in the passive mode to achieve better efficiency. More specifically, to reduce the communication, computation, and storage overheads, peers accept a delay in the disclosure and banishment of malicious or selfish peers.

To support the passive operation mode, modifications are required to the *claim broadcasting* and the *public tracing* procedures discussed in Section 4, as described below.

In the *claim broadcasting* procedure, when a peer detects a malicious or selfish operation performed by a peer A, instead of generating and sending an anonymous claim immediately, she keeps silent until she receives a query for the trust ratings of A. For example, another peer B who wants to know the trust ratings of A will broadcast a query to collect feedback from other peers. On receiving this query, each good peer who has witnessed A's misbehavior generates an anonymous claim in the same manner as in the active mode, and sends it through a Mixnet-based anonymous communication system. After collecting a sufficient number of feedback messages (e.g., t negative claims), B calculates the reputation of A based on the feedback received according to the trust metric defined. If B receives t or more claims from distinct peers accusing of A for the same reason, she knows that A is a malicious or selfish peer. As a result, B will refuse to participate in transactions with A, and inform other peers about this fact by broadcasting a message containing all these claims.

To prevent an adversary from sending multiple claims to slander or boost other peers, B needs to perform the *public tracing* procedure over all the claims received using the procedure described in Section 4.2.4.

In a peer-to-peer system, it is likely that there will be multiple peers who are interested in the trust ratings of A. Thus, when the witness receives a query on A for the second time, she should not generate a new claim. Otherwise, her anonymity will be compromised because of sending multiple claims against A. Instead, she can locate the claim regarding A in her DB_{PC}, which was generated in response to the first query process, and use it as the reply. Note that the same claim is counted only once even if any querying peer receives multiple copies of it.

The discussion so far has mainly focused on negative feedback. However, SDT can be used for positive feedback as well. For example, let N_p and N_n denote the numbers of positive and negative claims collected, respectively. A simple trust metric can be defined as $\frac{N_p}{N_p+N_n}$. However, we argue that positive feedback should not be considered in the active mode. Generally, the larger the number of positive claims received for a given peer, the higher the trust rating of the peer. However, a major difference between positive and negative feedback is that there is no threshold such that a peer is fully trusted if the number of positive claims regarding the peer is larger than this threshold. Suppose that such a threshold exists and is denoted as t'. An adversary A can cheat the system by first conducting t' transactions with distinct peers honestly. Thereafter A will be fully trusted and can cheat other peers in its subsequent transactions. This attack is a type of strategic oscillation attack [30]. Since we assume that most of peers in the network are benign and the claims are flooded in the active mode, counting positive feedback will result in large overheads. Fortunately, the goal of the active mode of operation is to detect malicious behavior in real-time, and thus it is sufficient to consider only negative feedback. In contrast, the goal of the passive mode of operation is to obtain an accurate trust value about the peer queried. Thus, we should consider both positive and negative feedback in the passive mode.

5.2 Probabilistic Forwarding

An approach for improving the efficiency of SDT in the active mode is for each peer receiving a claim to forward the claim with a probability p_f, instead of flooding claims to the whole network and storing them on each peer. Intuitively, the lower the probability p_f, the smaller the average number of peers storing a claim (denoted as n_s), and the lower the probability that at least one good peer stores t or more claims against the same adversary for the same type of misbehavior (denoted as p_r).

In this approach, the tradeoffs in security include: (i) a larger number of witnesses are needed before the adversary is disclosed (ii) given an upper bound on the number of witnesses needed, there is a non-zero probability that an adversary will escape disclosure. Let t_r denote the number of witnesses detecting an adversary's misbehavior and generating claims against her. Let p_d denote the security requirement, i.e., the lower bound of p_r. We first analyze the relationship between n_s and t_r while ensuring a high p_d, and then discuss how to select an appropriate p_f.

Assuming that an adversary engages in malicious or selfish behavior while interacting with peers that are uniformly distributed in the network, the claims against this adversary are generated, forwarded, and stored with a uniform probability. Therefore, the probability that a peer stores a given claim is n_s/n. Thus, we have $p_r = (n - t_a) \cdot (\frac{n_s}{n})^t \cdot C_{t_r}^t \geq (n - t + 1) \cdot (\frac{n_s}{n})^t \cdot C_{t_r}^t \geq p_d$.

Figure 1 shows the average number of peers required to store a claim (i.e. n_s) so that a high-level security (i.e. $p_d = 0.9999$) is achieved under different t_r's. We notice that n_s declines very fast when t_r increases. For example, when $t_r = 1.1 \cdot t$, i.e., requiring 10% more witnesses, n_s is around 24% to 27% lower than the case when $t_r = t$. Therefore depending on the level of security required and potential number of adversaries in the network, we can find an optimal n_s that achieves a good balance between security and efficiency. Let this optimal n_s be denoted by n_s^o.

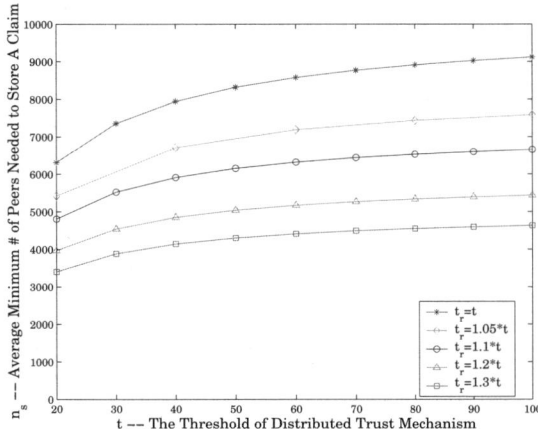

Figure 1: n_s **under different** t_r**'s** ($n = 10000$, $p_d = 0.9999$)

After determining n_s^o, we need to compute an appropriate probability of forwarding such that if every peer forwards the claims received with a probability p_f, the number of peers storing a claim is approximately equal to n_s^o. Let d denote the average degree/number of neighbors of a peer. When $d \cdot p_f < 1$ we have:

$$n_s < d + d^2 \cdot p_f + \cdots + d^{i+1} \cdot p_f^i + \cdots \approx \frac{d}{1 - d \cdot p_f} \quad (2)$$

Finally, we can compute the range of p_f as follows:

$$\frac{n_s^o - d}{n_s^o \cdot d} < p_f < \frac{1}{d} \quad (3)$$

In peer-to-peer systems, peers can join and leave the system at will. For both the active and passive modes of SDT, the trust value of a peer A is calculated based on the trust ratings collected from all the online good peers who have interactions with her. Therefore as long as there are at least t good peers online that detect A's misbehavior, A will be detected and expelled from the network. However, when we use the probabilistic forwarding approach, the dynamic nature of peer-to-peer networks does have an impact on the parameter setting. More specifically, let p_{on} denote the probability that a peer is online. Equations (2) and (3) should be modified by replacing d with $d \cdot p_{on}$.

6. SECURITY AND ANONYMITY-RELATED PROPERTIES ACHIEVED IN SDT

In this section, we examine how the security and anonymity-related goals defined in Section 2.1 are achieved in SDT.

6.1 Identity Anonymity

Generally, the possible methods of breaking identity anonymity as defined in Section 2.1 can be divided into two categories: traffic-based analysis and protocol-based analysis.

The idea behind traffic-based analysis is to detect common information among sniffed network traffic, assuming that any two packets are transferred along the same route if they have information in common. The "common information" could be either identical content (e.g., the same sequence number) in sniffed packets, or identical time consumed in handling sniffed packets (i.e., time analysis). In SDT, to prevent traffic analysis [26], the witness sends her claim through a Mixnet-based anonymous communication system [27, 28, 15] so that adversaries cannot discover the identity of the sender. Previous research [14, 16] shows that reputation systems can improve the performance of Mixnets. As such, we expect that the combination of SDT-based reputation systems and Mixnet-based anonymous communication systems is mutually beneficial and can enhance the effectiveness of both systems.

In protocol-based analysis, adversaries try to deduce the identity of the sender by analyzing the semantic context of messages. For example, to ensure that the receiver can verify the signature, the sender may include her public key in the packet. If that is the case, adversaries can easily discover the identity of the sender. In SDT, there is no public key or identity-related information in a claim, and the verification process is based on the zero knowledge proof. Given that the *Discrete Logarithm Problem* (**DLP**) is hard, SDT is robust against brute force attacks. More specifically, given T_1 and Γ, there is no PPT function which adversaries can use to find the secret of the witness (i.e., x) and thus deduce her identity.

6.2 Backward Anonymity

Most peer-to-peer systems can be viewed as overlay networks on top of another network, e.g. the Internet. The security of peer-to-peer systems will certainly be affected by the security of the underlying network. In addition, due to its open nature a peer-to-peer system is also vulnerable to attacks within the system, e.g. worms [34]. Therefore, our protocol should be resistant to possible compromises of peers. In SDT, a claim itself does not provide information disclosing the identity of the sender. Even if adversaries compromise multiple peers and collect additional claims, they cannot differentiate the claims sent by a peer from those sent by others. Thus, compromising more peers does not increase the probability of deducing the identity of the sender, i.e. the witness. Backward anonymity is ensured even when the OGM cooperates with adversaries, because in the *registration* procedure the OGM is convinced that a peer is holding the secret key corresponding to a given public key through zero knowledge proof and has no information about the secret key of that peer.

6.3 Traceability

In SDT, if an adversary sends multiple claims against a user for the same reason, she is identified by the *public tracing* procedure. To successfully disclose an adversary misusing witness anonymity, good peers need to find a valid record of this adversary, i.e., the LIST item containing the identity and public key of the adversary. In SDT, each item in the LIST has s copies distributed within the peer-to-peer system. As shown in Section 8.1, the probability that all the copies of a given item are controlled by the adversary group is tiny, even when s is very small. If we assume that such a probability is negligible, any adversary sending multiple claims against the same user for the same reason will be traced, and her identity will be disclosed.

6.4 Non-Slanderability

In this paper, we assume the existence of a distributed trust mechanism, e.g., one of the protocols proposed in [22, 20, 35]. Together with the *traceability* property provided by SDT, the maximum number of claims that the adversary group can send against the same good user for the same reason is the number of adversaries in that group, which is assumed to be less than t. In other words, adversaries cannot collect enough number of claims, i.e. t or more claims, to convince others that an innocent user is malicious or selfish.

7. COUNTERMEASURES TO VARIOUS ATTACKS

In this section, we consider several attacks on the SDT protocol and present possible countermeasures.

7.1 Collusion Attacks

The effectiveness of collusion attacks is limited by the combination of the *traceability* property provided by SDT and the distributed trust mechanism. The former ensures that any adversary sending multiple claims against the same user for the same reason will be detected and traced. As a result, an adversary can only send one claim per user per type of malicious or selfish behavior. Moreover, the existence of the latter guarantees that the number of adversaries in the peer-to-peer system is less than t. Consequently, even if all the adversaries collude with each other, they cannot generate t or more claims against an innocent user for the same reason.

7.2 Sybil Attacks

In a Sybil attack [17], malicious peers can assume multiple identities, and thus can control a substantial part of the system. If such an attack is possible, it would undermine the basis of trust schemes for peer-to-peer systems, i.e., most peers are honest and each peer can vote only once when assessing a given peer. According to [17], without a logically centralized authority, Sybil attacks are always possible except under certain assumptions regarding the resources possessed by the attacker.

A straightforward solution for SDT is to let the OGM assume the responsibility of the centralized authority. More specifically, during the *registration* procedure, the OGM is responsible for binding the user registering as a member to a real-world identity, and for limiting the number of the registrations per identity to one.

Assume that in a Sybil attack there is single real-world user assuming multiple identities. We refer to this real-world user as the *Sybil source* and the users corresponding to the assumed identities as *Sybil users*. To prevent Sybil attacks when there is no centralized authority, Douceur [17] proposed the use of a scheme in which the potential Sybil users are challenged to solve some resource-intensive task that can only be accomplished by multiple real-world users but will be impractical for a Sybil source. For example, suppose that a peer A receives t claims against another peer B. When employing the direct identity validation method proposed in [17] against Sybil attacks, A should first challenge all the t witnesses who generated the claims to perform some task that can only be completed by t real-world users before asserting that B is malicious or selfish. However, in SDT, the identities of the witnesses are unknown due to the requirement of witness anonymity. Thus the only way to deliver the task to the witnesses is via a system-wide broadcast. Although this method is relatively expensive, it may not be impractical given certain assumptions regarding resource parity and coordination among peers [17].

7.3 Denial-of-Service Attacks

In the SDT protocol, *Denial-of-Service* (**DoS**) attacks can be launched against the *claim broadcasting* procedure and the *public tracing* procedure.

The DoS attacks launched in the *claim broadcasting* procedure are similar to those against reputation systems, e.g., sending numerous fake claims [30]. In SDT, such attacks are restrained by the traceability property. However, a smart adversary may choose to send one claim per innocent peer per misbehavior type. Given that the size of the network is large, she can still flood the network with a number of fake claims, although it does not help her successfully slander any innocent peer.

When SDT is in the passive mode, this type of attack can be prevented by the following method. When a peer A receives an anonymous claim against another peer B, she first checks whether it has received a query regarding B within the previous T seconds. If there is no such query, A will drop the claim. If we set a small upper bound on the number of queries that each peer can launch every T seconds, the effectiveness of DoS attacks can be reduced.

In this paper, we assume the existence of a mechanism for monitoring the number of claims sent by a peer, irrespective of whether they are generated or forwarded by her. In the active mode, each valid claim is generated after an interaction between the sender and the peer accused. Suppose that an interaction between two peers takes time t_i. Assuming that each peer accused can have only one interaction at a time, since we know that only negative feedback is considered in the active mode and the number of adversaries is assumed to be less than t, any peer that sends out claims at a rate higher than $\frac{t}{t_i}$ per unit time is suspected of launching a DoS attack. For example, given that $t = 100$ and $t_i = 300$ seconds, a peer sends claims at a rate of one claim per second can be suspected to be launching a DoS attack.

Given that we can limit the number of claims that a peer can send within a certain time period as shown above, and since the computation requirements of SDT are relatively small (even on a low-powered Pentium III platform as shown in Section 8.4), we argue that most users in today's peer-to-peer systems have sufficient computational power to handle such DoS attacks. Similarly, considering the amount of storage available on computers [32] and the small size of each claim in SDT (as shown in Section 8.3), DoS attacks launched with the goal of exhausting the storage of a given peer are also ineffective.

To launch DoS attacks during the execution of the *public tracing* procedure, adversaries may generate fake broadcasting messages to exhaust the computational resources of other peers and the communication resources of the network. However, the effectiveness of these attacks is limited. In each message, the attacker needs to include the proofs that could be verified by other peers. Otherwise, the receiver will drop the message so that it cannot be propagated. The proofs in the pair of entries can be verified only if they are generated using the same secret key of a member in the network. As a result, the identity of the holder of this secret key will be disclosed in the *public tracing* procedure. In other words, in order to launch this attack successfully, the adversary has to expose herself or a peer that has been compromised by her.

8. THE COMMUNICATION, STORAGE, AND COMPUTATION COSTS OF SDT

8.1 Distributed Storage of the LIST

In the SDT protocol, the OGM maintains a complete identification list LIST offline. However, in most cases, e.g., to disclose the identity of malicious or selfish users during the *public tracing* procedure, users may want to access the LIST immediately instead of waiting for the OGM to be online. In addition, due the large size of the network, it would be too onerous to let each peer maintain a full copy of the LIST. Therefore, in SDT the LIST is stored in a distributed form. As discussed in Section 4.2.2, during the *registration* procedure the OGM assigns each member a portion of the LIST, which is signed by her secret key. The assignment is executed in such a way that each peer has no knowledge about the items of the LIST stored on another peer, and each item in the LIST has multiple copies stored on different members.

When there is no central party storing the LIST, it is possible that all the information about an adversary in the LIST is stored only by the adversary group. They can remove the public information of this adversary from the LIST. As a result, this adversary will not be identified even if submits multiple claims against the same peer for the same reason because the output of the *public tracing* procedure will always be always **NO-ONE**. However, we argue that, given that there are s copies of the public information for each member, and they are uniformly distributed within the network, the probability (denoted as p_{ad}) that the entry for an adversary in the LIST is stored only by the adversary group is vert small. The probability can be calculated as $p_{ad} = C_{t_a}^s / C_n^s$. In Table 1, we show the values of p_{ad} for different settings, when the network size is 10000. For example, when each member stores only around three items of the LIST, the probability that the public information of an adversary is stored only by the adversary group is less than 9.70×10^{-7}.

In fact, even if this low probability event does occur, this attack can be utilized by good peers to detect more adversaries with the help from the OGM, as illustrated by the following example. Assume that there are a group of adversaries denoted as $M = \{M_1, M_2, \ldots, M_6\}$, and s is set to

Table 1: p_{ad} under Different t_a and s ($n = 10000$)

	$t_a = 50$	$t_a = 75$	$t_a = 100$
$s=2$	2.45×10^{-5}	5.55×10^{-5}	9.90×10^{-5}
$s=3$	1.18×10^{-7}	4.05×10^{-7}	9.70×10^{-7}
$s=4$	5.53×10^{-10}	2.92×10^{-9}	9.41×10^{-9}
$s=5$	2.55×10^{-12}	2.07×10^{-11}	9.04×10^{-11}

5. M_1 sends multiple claims against a good peer denoted as Q_1. Another good peer denoted as Q_2 receives these claims, and thus detects the misuse of witness anonymity according to the *public tracing* procedure. Then Q_2 broadcasts a query, and asks for helps to disclose the identity of the senders (i.e. M_1) corresponding to β calculated in the *public tracing* procedure. Unfortunately all the five copies of M_1's LIST information are stored by $M_2, M_3, \ldots, M_6$, and they refuse to respond to the query. In such a case, Q_1 may ask for the help from the OGM, who will check the LIST stored, and then broadcast a signed message claiming that members of M are malicious and at the same time disclosing their public information.

8.2 Communication Costs

Assume that the SDT protocol is implemented by an elliptic curve or hyperelliptic curve over a finite field. To ensure the security of the protocol, the prime p should be at least 160 bits long, $\mathbb{G}_1$ should be a subgroup of an elliptic curve or a Jacobian of a hyperelliptic curve over a finite field of order p, and $\mathbb{G}_T$ should be a subgroup of a finite field of size at least approximately 2^{1024}. This will ensure that the *Decisional Bilinear Diffie-Hellman* (**DBDH**) problem [24] is sufficiently hard.

The major communication cost of the SDT protocol is for forwarding claims. By employing techniques in [18], elements of $\mathbb{G}_T$ can be compressed by a factor of three. As a result, the size of each claim is 405 bytes, excluding $Type_{ad}$ and ID_{ad}. Usually, it is sufficient to set the length of $Type_{ad}$ to one byte. As to the length of ID_{ad}, it should be set to at least $\lceil \log n \rceil$ bits. For a network with at most 10 million peers, we can set the length of ID_{ad} to three bytes, and thus the length of a claim is 409 bytes in total. Due to the relatively small size of the claim, given that we can limit the number of queries that a peer can launch and the number of claims that a peer can send within a certain time period as shown in Section 7.3, the communication overhead of SDT is acceptable in peer-to-peer systems.

If we store the LIST in a distributed manner, it leads to additional additional communication costs. Specifically, in the *public tracing* procedure, a peer may need to broadcast a message to search for pair (i, β) corresponding to a claim against a user. Intuitively, the smaller the number of LIST items that are stored on a peer, the higher is the probability that she may broadcast a message to ask for help to identify the adversary in the *public tracing* procedure. Fortunately, assuming that adversaries are rational in the sense that they may not launch attacks in which they are doomed to be caught, we argue that such cases may happen infrequently.

8.3 Storage Requirements

The storage requirements of SDT are due to (i) the cryptographic keys (ii) the items of the LIST, and (iii) the local claim databases (i.e. DB_{PC} and DB_{CC}). A user in the SDT protocol only needs to store her public/secret key pair and the public key of the OGM. With respect to the items of the LIST, as shown in Section 8.1, they can be stored in a distributed form, and the cost is only s LIST items per user. Each LIST item is a (i, β) pair, and its size is 46 bytes after compression, given that the identity of a peer is represented with three bytes.

With respect to the storage required for the local claim database, in the passive mode, the claims stored by each peer are actually the claims that were generated by her. She does not store the claims that are forwarded. In the active mode, however, to avoid the looping of the claim message during its broadcast, each peer needs to store the claims forwarded. As shown in Section 5.2, the probabilistic forwarding approach can help reduce the storage cost. Moreover, for both the active and passive modes, the storage cost can be reduced by deleting claims after a suitable time interval.

8.4 Computation Costs

The most expensive computational operation used in the SDT protocol is the calculation of bilinear pairing. If the bilinear mapping used is the well-known Tate Pairing, using MIRACL libraries [1] (optimized using Comba method for modular multiplication) on a Pentium III 1GHz desktop, it takes 20 milliseconds to calculate a 512-bit Tate pairing (for effective 1024-bit security) without pre-computation [4]. More recent results [5] show that the Tate pairing can be evaluated up to 10 times faster than in previously reported implementations by the use of various optimizations.

9. RELATED WORK

In [14, 16], Dingledine *et al* propose to use reputation systems to increase the reliability and efficiency of various Mixnet-based applications, e.g., remailer networks and anonymous publishing. Their protocols do not provide anonymity for the witnesses in reputation systems.

XRep [13] supports weak anonymity for the peer. The reputation is bound to a pseudonym or an opaque identifier, i.e. the digest associated with a servent. However, the real IP address of the peer is still required when it replies to a voting query.

In TrustMe [29], when a peer i joins the network, a bootstrap server randomly assigns a set of peers to be its *Trust-Holding Agent* (**THA**) peers. Subsequently, any peer having interactions with peer i will send out an encrypted report which can only be decrypted by the THA peers for peer i. As such, each THA peer stores all the reports related to peer i and can thus reply to queries from other peers regarding peer i. A major weakness of TrustMe is that if any THA peer randomly chosen at the bootstrapping stage is an adversary, the identities of the reporting peers are disclosed. In other words, TrustMe does not support, or at best partially supports witness anonymity.

10. CONCLUSIONS

This paper presents the *Secure Deep Throat* (**SDT**) protocol to provide witness anonymity to users that report on the misbehavior or trust ratings of other peers in peer-to-peer systems. SDT can be used in an active mode in scenarios with real-time requirements for detecting malicious behavior, e.g., for detecting and preventing the propagation of peer-to-peer worms [34], or it can be used in a passive mode in scenarios that do not have strict real-time requirements, e.g., query-based reputation systems [12]. We discuss the security and anonymity-related requirements of a protocol for providing witness anonymity, and show that these requirements are met by SDT. Further, we describe countermeasures that can be used to defend against collusion attacks, Sybil attacks, and denial-of-service attacks against SDT. Finally, we analyze the storage, communication, and computation costs of SDT and show that the overhead of the protocol is acceptable in peer-to-peer systems.

11. REFERENCES

[1] MIRACL library. http://indigo.ie/~mscott/.

[2] G. Ateniese and G. Tsudik. Some open issues and new directions in group signatures. In *Proceedings of The Third International Conference on Financial Cryptography (FC'99), LNCS 1648*, pages 196–211, 1999.

[3] AuctionBytes. Online auction feedback survey. Retrieved from http://www.auctionbytes.com/cab/pages/feedbacksurvey1105 on May 5, 2006.

[4] P. S. Barreto, H. Y. Kim, B. Lynn, and M. Scott. Efficient algorithms for pairing-based cryptosystems. In *Proceedings of Advances in Cryptology – CRYPTO 2002, LNCS 2442*, pages 354–368, 2002.

[5] P. S. L. M. Barreto, B. Lynn, and M. Scott. On the selection of pairing-friendly groups. In *Proceedings of Annual International Workshop on Selected Areas in Cryptography (SAC'03), LNCS 3006*, pages 17–25, 2003.

[6] M. Bellare, D. Micciancio, and B. Warinschi. Foundations of group signatures: Formal definitions, simplified requirements, and a construction based on general assumptions. In *Proceedings of Advances in Cryptology - EUROCRPYT 2003, LNCS 2656*, pages 614–629, 2003.

[7] S. Brands. Untraceable off-line cash in wallets with observers (extended abstract). In *Proceedings of the 13th Annual International Cryptology Conference on Advances in Cryptology, LNCS 773*, pages 302–318, 1993.

[8] E. Bresson, J. Stern, and M. Szydlo. Threshold ring signatures and applications to ad-hoc groups. In *Proceedings of Advances in Cryptology - CRYPTO 2002, LNCS 2442*, pages 465–480, 2002.

[9] J. Camenisch and M. Michels. A group signature scheme with improved efficiency (extended abstract). In *Proceedings of Advances in Cryptology - ASIACRYPT'98, LNCS 1514*, pages 160–174, 1998.

[10] D. Chaum and E. V. Heyst. Group signatures. In *Proceedings of Advances in Cryptology - EUROCRYPT '91, LNCS 547*, pages 257–265, 1991.

[11] D. L. Chaum, A. Fiat, and M. Naor. Untraceable electronic cash. In *CRYPTO88, Lecture Notes in Computer Science 403*, pages 319–327, 1989.

[12] F. Cornelli, E. Damiani, S. D. C. D. Vimercati, S. Paraboschi, and P. Samarati. Choosing reputable servents in a P2P network. In *Proceedings of the 11th International Conference on World Wide Web*, pages 376–386, 2002.

[13] E. Damiani, S. D. C. D. Vimercati, S. Paraboschi,

P. Samarati, and F. Violante. A reputation-based approach for choosing reliable resources in peer-to-peer networks. In *Proceedings of the 9th ACM Conference on Computer and Communications Security*, pages 207–216, 2002.

[14] R. Dingledine, M. J. Freedman, D. Hopwood, and D. Molnar. A reputation system to increase MIX-net reliability. In *Proceedings of The 4th International Workshop on Information Hiding (IHW'01), LNCS 2137*, pages 126–141, 2001.

[15] R. Dingledine, N. Mathewson, and P. Syverson. Tor: The second-generation onion router. In *Proceedings of the 13th USENIX Security Symposium*, 2004.

[16] R. Dingledine and P. Syverson. Reliable MIX cascade networks through reputation. In *Proceedings of Financial Cryptography (FC'02), LNCS 2357*, 2002.

[17] J. R. Douceur. The sybil attack. In *Proceedings of The First International Workshop on Peer-to-Peer Systems (IPTPS 2002)*, pages 251–260, 2002.

[18] R. Granger, D. Page, and M. Stam. A comparison of CEILIDH and XTR. In *Algorithmic Number Theory, 6th International Symposium, ANTS-VI*, pages 235–249, 2004.

[19] S. D. Kamvar, M. T. Schlosser, and H. Garcia-Molina. The eigentrust algorithm for reputation management in P2P networks. In *Proceedings of the 12th International Conference on World Wide Web (WWW 2003)*, pages 640–651, 2003.

[20] H. Luo, J. Kong, P. Zerfos, S. Lu, and L. Zhang. URSA: Ubiquitous and robust access control for mobile ad hoc networks. *IEEE/ACM Transactions on Networking*, 12(6):1049–1063, 2004.

[21] M. Naor. Deniable ring authentication. In *Proceedings of Advances in Cryptology - CRYPTO 2002, LNCS 2442*, pages 481–498, 2002.

[22] M. Narasimha, G. Tsudik, and J. H. Yi. On the utility of distributed cryptography in P2P and MANETs: The case of membership control. In *Proceedings of the 11th IEEE International Conference on Network Protocols (ICNP'03)*, pages 336–345, Nov. 2003.

[23] I. B. S. News. New peer-to-peer trojan worm attacks enterprises, Mar. 2006. Retrieved from http://www.justloadit.com/pr/6169 on May 5, 2006.

[24] L. Nguyen and R. Safavi-Naini. Dynamic k-times anonymous authentication. In *Proceedings of The Third International Conference on Applied Cryptography and Network Security (ACNS 2005)*, pages 318–333, 2005.

[25] L. Nguyen and R. Safavi-Naini. Dynamic k-times anonymous authentication. Full version, 2005.

[26] J.-F. Raymond. Traffic analysis: Protocols, attacks, design issues, and open problems. In *DIAU00, Lecture Notes in Computer Science 2009*, pages 10–29, 2000.

[27] M. K. Reiter and A. D. Rubin. Crowds: Anonymity for web transactions. *ACM Transactions on Information and System Security (TISSEC)*, 1(1):66–92, 1998.

[28] C. Shields and B. N. Levine. A protocol for anonymous communication over the internet. In *ACM Conference on Computer and Communications Security (CCS 2000)*, pages 33–42, 2000.

[29] A. Singh and L. Liu. TrustMe: Anonymous management of trust relationships in decentralized P2P systems. In *Proceedings of The Third International Conference on Peer-to-Peer Computing (P2P 2003)*, pages 142–149, 2003.

[30] M. Srivatsa, L. Xiong, and L. Liu. TrustGuard: Countering vulnerabilities in reputation management for decentralized overlay networks. In *Proceedings of the 14th International Conference on World Wide Web*, pages 422–431, 2005.

[31] I. Teranishi, J. Furukawa, and K. Sako. K-times anonymous authentication (extended abstract). In *Proceedings of ASIACRYPT 2004, LNCS 3329*, pages 308–322, 2004.

[32] Www.Programmersheaven.Com. Poll archive – how much storage capacity does your computer have?, June 2004. Available at http://www.programmersheaven.com/c/userpoll/Poll_archive.htm?PollID=148.

[33] L. Xiong and L. Liu. PeerTrust: Supporting reputation-based trust for peer-to-peer electronic communities. *IEEE Transactions on Knowledge and Data Engineering*, 16(7):843–857, July 2004.

[34] W. Yu, C. Boyer, S. Chellappan, and D. Xuan. Peer-to-peer system-based active worm attacks: Modeling and analysis. In *Proceedings of IEEE International Conference on Communications (ICC '05)*, pages 295–300, 2005.

[35] B. Zhu, F. Bao, R. H. Deng, M. S. Kankanhalli, and G. Wang. Efficient and robust key management for large mobile ad-hoc networks. *Computer Networks*, 48(4):657–682, July 2005.

Salsa: A Structured Approach to Large-Scale Anonymity

Arjun Nambiar
arjun.nambiar@uta.edu

Matthew Wright
mwright@cse.uta.edu

Dept. of Computer Science and Engineering
University of Texas at Arlington
Box 19015
Arlington, TX 76019-0015

Highly distributed anonymous communications systems have the promise of better distribution of trust and improved scalability over more centralized approaches. Existing distributed approaches, however, face security and scalability issues. Requiring nodes to have full knowledge of the other nodes in the system, as in Tor and Tarzan, limits scalability and leads to intersection attacks in peer-to-peer configurations. MorphMix avoids giving nodes complete system knowledge, but new research shows that a collaborating fraction of the peers can control the paths of many users.

To overcome these problems, we propose Salsa, a structured approach to organizing highly distributed anonymous communications systems for scalability and security. Salsa is designed to select nodes to be used in anonymous circuits randomly from the full set of nodes, even though each node has knowledge of only a small subset of the network. It uses a distributed hash table based on hashes of the nodes' IP addresses to organize the nodes into groups. With a virtual tree structure, limited knowledge of other nodes is enough to route node lookups throughout the system. We use redundancy and bounds checking when performing lookups to prevent malicious nodes from returning false information without detection. We show that our scheme prevents attackers from biasing path selection, while incurring moderate overheads, as long as the fraction of malicious nodes is less than 20%. Additionally, the system prevents attackers from obtaining a snapshot of the entire system until the number of attackers grows too large (e.g. 15% of 10000 peers, given 256 groups). The number of groups can be used as a tunable parameter in the system, depending on the number of peers, that can be used to balance performance and security.

Categories and Subject Descriptors

C.2.4 [**Computer-Communication Networks**]: Distributed Systems; C.2.0 [**Computer-Communication Networks**]: General—*Security and protection*

General Terms

Security

Keywords

Anonymous Communications, Peer-to-Peer Networks, Privacy

1. INTRODUCTION

Anonymous communications systems on the Internet provide protection against eavesdroppers and others that seek to link users with their communications. These systems have many important applications in areas such as law enforcement, intelligence gathering, business privacy, anonymous publishing, and personal privacy. Current systems, such as Tor [11], rely on a relatively small set of advertised servers to forward messages for the user. These systems can suffer from scalability problems, with potentially large bandwidth and system overhead costs, and the servers themselves can be targets of direct attacks.

Peer-to-peer anonymous communications systems, such as Tarzan [13] and MorphMix [23], have been proposed as a way to alleviate these problems with a large and dynamic set of peers acting as servers. This makes direct attacks less effective and increases scalability. Tarzan, however, requires that each peer know the identity of all other peers, which makes it highly vulnerable to intersection attacks [28] and does not scale beyond 10,000 nodes [13]. MorphMix does not have this requirement, but it requires that users allow untrusted peers to choose the proxies that will forward the user's messages; attacker-controlled peers will always select other colluding peers to be on the path. Although the authors of MorphMix propose a collusion detection scheme, recent work has shown that this scheme can be fooled while attackers continue to control many paths in the system [26]. The fundamental problem facing these systems is one of selecting peers independently at random from the set of peers to ensure unbiased path selection, while not requiring full knowledge of the set of available peers.

To solve this problem, we propose a new peer-to-peer anonymous communications system using distributed hash tables (DHTs). Similar to peer-to-peer file-sharing systems that use DHTs, like the Chord system [24], our system maps each IP address to a point on the ID space using consistent hashing. We further divide the ID space into groups, conceptually organized as a binary tree for purposes of node lookup. Each node has knowledge of all the nodes in its own group, as well as knowing a limited number of nodes in other groups. This knowledge is enough to effectively route lookups throughout the system. Nodes use redundancy and probabilistic checking when performing lookups to prevent malicious nodes from returning false information without detection.

We show that our scheme prevents attackers from biasing path selection, while incurring moderate overheads, as long as the fraction of malicious nodes is less than 20%. Additionally, the system prevents attackers from obtaining a snapshot of the entire system until the number of attackers grows too large (e.g. 15% for 10000 peers and 256 groups). The number of groups can be used as a tunable parameter in the system, depending on the number of peers, that can be used to balance performance and security.

In Section 2, we describe existing work in peer-to-peer anonymous communications systems, relevant attacks on these systems, and related work in structured peer-to-peer systems. Section 3 describes the design of Salsa, including our novel network architecture. We give an analysis of Salsa's security properties in Section 4. We describe the simulation methodology and our results in Section 5 and conclude with future directions in Section 6.

2. BACKGROUND

In this section, we cover three areas that are critical to our work. First, we make a case for highly distributed anonymous communications in light of a broad overview of work in anonymity. Second, we describe relevant attacks and challenges facing current proposed systems for highly distributed anonymous communications. Third, we look at other structured peer-to-peer overlays and security considerations that have been studied to date.

Before this, we give some concepts that we use throughout the paper. As we are considering systems in which a user's node may also be a proxy, we define the *initiator* as the node of the user who initiates an anonymous connection. We call the node that the initiator contacts, such as a Web site, the *responder*. Most systems are based on Chaum's idea of mixes [6], in which there is a path of proxies, or *circuit*, between the initiator and the responder that adds indirection to help hide their connection. A technique called *layered encryption* limits an attacker's ability to track packets passing through the circuit. However, most of these systems are vulnerable to attacks in which the first and last proxies on a circuit, or a well-placed eavesdropper, can collaborate and use the timings of packets to correlate the initiator with the responder [9, 17, 29].

2.1 Highly Distributed Anonymity has Strong Potential

Experts in anonymous communications do not have consensus on the issue of what types of mix-based systems are most secure. Some argue for *mix cascades*, such as Web Mixes [3], in which all messages pass through the same set of proxy servers [4]. Others argue for *mix networks*, such as Tor [11], in which users choose proxy servers randomly for each position on the path. More recently, starting with Crowds [22], the idea of peer-to-peer systems for anonymous communications has been developed. Rather than attempt to present complete arguments for these approaches, we describe here only the benefits of both the peer-to-peer approach and versions of mix networks with many servers. Our goal is only to demonstrate that this is a promising approach worth the present investigation – we expect that debate over which approach is best will remain open for some time.

Peer-to-peer anonymity systems provide a nice security property: the first proxy in the path does not know whether or not it follows the initiator or another peer serving as a proxy. This is the main argument for security against malicious peers in Crowds [22]. Although Crowds does not protect against what we believe to be reasonable attacker models, such as a substantial subset of the peers [27], this property also applies to systems that use layered encryption. A significant security property of both highly distributed mix networks and peer-to-peer networks with many servers comes from the large number of proxies. When there are fewer proxies, the relatively small number of them present a viable set of targets for direct and active attacks. Such attacks include eavesdropping, node corruption, node takeover, and legal subpeonas.

A particularly devastating attack for smaller, less distributed, systems was presented by Murdoch and Danezis in 2005 [20]. In this attack, which was demonstrated on the Tor network with 35 operating onion routers, a single corrupt client fills the available bandwidth of a server and watches for drops in the connection's bandwidth. If those drops in bandwidth correlate to the times when packets were received by the responder, the server was involved in forwarding the initiator's traffic. The attack succeeds by testing many servers and tracing back the initiator's path. In highly distributed systems, this and other direct attacks are significantly harder due to the sheer number of tests that must be conducted. Similarly, eavesdropping may require a truly global adversary with substantial capability to extract relevant data from vast amounts of traffic – placing eavesdropping equipment directly on the networks of even a fraction of the nodes is not practical for many attackers.

Another benefit of highly distributed systems is in the distribution of costs and the possibilities for beneficial incentive structures. Since some users are providing services as a cost of being in the system, they do not need to pay for service. This is important, because paying for anonymity can be challenging to do in a fair and anonymous way and may require special forms of digital cash [21]. Further, users who are particularly concerned about privacy have an incentive to provide service, which somewhat reduces the *freeriding* problem [14]. In particular, freeriders have weaker anonymity as they do not forward packets for other nodes, so all packets must be initiated by the freerider himself [1]. Other incentives issues may exist for peer-to-peer systems, but the current situation is promising.

2.2 Attacks on Highly Distributed Anonymous Communications

A significant issue with peer-to-peer systems, and any system that does not strongly verify the identity of the participants, is the Sybil attack [12]. This attack is essentially a recognition that an attacker can, at relatively low cost, construct many online identities and use them to control or attack the system.

The Sybil attack can be partially mitigated. For example, we can force the attacker to own many IP addresses by ensuring that each identity maps to a unique address; this has been used in the Tarzan system [13]. Another technique pioneered by Tarzan and adopted by MorphMix [23] is the use of *hierarchical address selection*. In this scheme, a user who wants to select a node at random first chooses a subnet at random, based on subnets represented in the peer's IP addresses, and then select a node from the chosen subnet at random. This ensures that an attacker who controls one subnet cannot flood the system with nodes from that subnet to gain control in the system. An attacker must then control nodes in a large number of subnets to succeed.

In today's Internet, the threat of *botnets*, in which the attacker controls a large population of corrupted nodes widely distributed in the network, continues to make the Sybil attack a dangerous threat. Botnets of 100,000 nodes have been reported, but botnets of 20,000 nodes or less are more common, partly because of the smaller chance of being caught [15]. Against a widely-distributed botnet, the only known defense is to have a large number of semi-honest nodes that are not collaborating with the attacker. Thus, scalability is critical for open anonymous communications systems to succeed.

One of the most challenging attacks to defend against in anonymous communications is the intersection attack. In this attack, a passive observer takes logs of the set of users participating in the system at regular intervals. By comparing these logs with logs of the responders contacted via the system at the same times, the attacker can profile different users and undermine the users' anonymity [8, 18, 28]. In most systems, this attack is difficult to conduct, as it requires complete knowledge of all the users in the system. An attacker controlling a large subset of the proxies could not be sure that they have the complete list. Note that an incomplete list means that the attacker would be profiling users based on unseen users' activity. The attackers do not, however, need to witness all outgoing activity. If the attacker is only interested in a single responder (e.g., the owner of the responder wants to know who is contacting it anonymously), then it need only get information on when the responder is contacted via the system.

In the Crowds and Tarzan systems this attack is particularly dangerous. When joining the system, the user obtains a list of all peers, which is used to select proxies. This is considered necessary for the security of the system, as a partial list might be biased with a large fraction of attackers [13]. However, the list of peers provides a list of all users. Thus, an attacker need only join a single peer to the system to obtain a great deal of the information needed to perform the intersection attack. This attack has been shown to be very efficient in narrowing the possible initiators to a small set (e.g., five or ten peers) [28].

Using complete lists of all peers is also a scalability issue. Tarzan's gossiping protocol for propagating peer information throughout the network only scales well to about 10,000 nodes [13]. The operators of Tor have discussed, for scalability reasons, reconfiguring the system so that each proxy only knows and connects to a subset of other proxies (R. Dingledine, personal communication).

The MorphMix peer-to-peer system similarly has each peer maintain connection information about a subset of the other peers [23]. While this helps to avoid the intersection attack and improves scalability, choosing a path securely becomes challenging. For example, MorphMix initiators choose the first node on the path from their list of known peers. For the second node, however, the first node makes the choice from among its list of known peers. If the first node is an attacker, it can select an attacker for the second node, and this can continue throughout the path. In this way, whenever an attacker is selected for the first node, the entire path can be easily compromised. MorphMix employs a *witness* process to detect when this kind of collusion is used, but recent work by Tabriz and Borisov has shown that the witnesses can be deceived by intelligent selection of nodes [26]. This allows the attackers to control many paths in a system indefinitely and without detection.

In general, limited knowledge of the network can be a significant issue for secure paths, as the initiator may be deceived. This is a key motivation of our work.

2.3 Structured P2P Systems and Anonymity

A number of works have considered security and privacy issues for structured peer-to-peer systems. Danezis *et al.* propose an alternative routing strategy, called *zig-zag routing* for DHT-based systems that helps defend against Sybil attacks [10]. Zig-zag routing aims to avoid Sybil groups by ensuring diversity while getting close to the target. Both Borisov and Ciaccio have proposed adding anonymity to structured peer-to-peer systems [5, 7]. Borisov proposes the use of random walks on de Bruijn networks to help provide anonymity with reasonably short paths. Ciaccio proposes the use of *imprecise routing* in which the construction of neighbor tables is done from a range rather than with a precise value (as in Chord). This makes it difficult for an attacker to work backwards to determine the source of a request.

These systems all have in common the idea of adding randomness or diversity to the structured overlay, while keeping the general approach of rapid reduction in the distance to the target. Salsa also shares this feature, but it uses a unique structure that is designed to satisfy the different system requirements of providing a structured overlay for large-scale anonymity.

3. DESIGN OF SALSA

In this section, we describe the Salsa system design. The purpose of Salsa is to organize a large anonymous communications network to enable random and unbiased node selection for users. The basic design combines a structured overlay with redundant lookups to ensure that randomly selected nodes are not biased towards selection of attackers. With such a system, we argue, the initiator can select the nodes in her circuit at random without significant bias and without knowledge of the entire set of proxies.

The overall anonymous communications system may either be a peer-to-peer anonymous communications system or a server-based system similar to Tor. We assume that the system creates a circuit of proxies for each user by selecting a set of nodes from the available peers or servers. This is the basic approach used in most anonymous communications systems, including Tor (and Onion Routing), MorphMix, Tarzan, and Freedom [11, 25, 23, 13, 2]. We have intended Salsa to focus on node organization and selection, independent of building and maintaining anonymous communication circuits, forwarding traffic, and returning replies.

The fundamental challenge that motivates the Salsa design is the selection of nodes at random from the set of available peers with only limited knowledge of the nodes. As explained in Section 2.2, allowing every node to have only limited knowledge of the system helps protect the system from the intersection attack (in peer-to-peer systems) and enhances the scalability of the system. It is also important to select nodes at random from the entire set of nodes. Selecting from subsets of the network would allow an attacker to focus an attack on any given user. However, selecting a node at random despite limited knowledge is not trivial.

3.1 A Naive Approach

We now describe a simple but naive approach to address the challenge of random node selection, based on the Chord system. Similar to Chord, we can give each node an identifier (ID) based on a cryptographic hash of the node's IP address [1]. The IDs are placed in sorted order around a circular *ID-space*, and each node N_i is said to *own* the fraction of the ID-space between itself and it's preceeding node N_{i-1}. A user makes a request for a specific ID, called the *target ID*, and a request for a target ID between N_{i-1} and N_i will be routed to N_i, as it owns that ID-space. In this case, we say that N_i is the *target owner*.

Also as with Chord, each node keeps a *finger table*, which stores routing information. The finger table includes the addresses of a series of nodes, called *fingers*, that are at varying distances away in the ID-space. Specifically, we can define the minimal distance between two adjacent nodes as one unit, and the fingers can then be said to at one unit, two units, four units, and further powers of 2 units away from the node, up to approximately one-half of the total ID-space (a more precise description may be found in the Chord paper [24]). This allows for fast routing of requests in the system through recursive propogation. If a node knows the address of the target owner, it will forward the request directly to that node. Otherwise it will forward the request to the node in its finger table that most closely preceeds the target ID. This scheme ensures that the number of nodes that need to be contacted is bounded by $O(\log n)$ [24]. Intuitively, this is because the distance is cut in half at each step until the owner is only one unit of ID-space away.

We now have what we need to select nodes at random with only limited knowledge of the system. First pick a random ID from the ID-space. Then send a request for the owner of that ID. This is not entirely the same as uniform random selection from the set of nodes, as each node may own different amounts of the ID-space, but the amounts are probabilistically bounded by $O((\log n)/n)$ and cannot be controlled by the owners [24]. In particular, an attacker who controls c nodes in the system will own, on expectation, c/n of the ID-space. For large values of c and n, the actual value will be close to expectation.

Due to the use of consistent hashing, the system has several other important features. Consistent hashing provides a weak form of authentication: every other proxy can quickly compare the hash of the IP address of the node it connects to with the node's claimed ID. As described in Tarzan, a simple two-way handshake is enough to ensure that the node controls the IP address (at least with respect to the connecter) [13]. Additionally, the attacker cannot attempt to place a node in a certain part of the address space without controlling an IP address that would map to the desired ID range. If the node density is high enough, this becomes difficult for most attackers. Another benefit of consistent

hashing is that when a node joins the system, the ID-space can be determined from its successor node. The successor node cannot lie to the joining node to gain more space for itself – the space between them is defined by the IP addresses that they use to communicate. When a node leaves the system, it can inform its successor that the address space now belongs to the successsor. With high probability, no node owns more than $O((\log n)/n)$ fraction of ID-space.

However, this approach has an obvious security problem: since the user must rely on other nodes to forward the requests, any node along the path of a request can modify the results. In particular, attackers may stop propogating the request and return the identity of another attacker. To alleviate this problem, we propose the use of simultaneous redundant requests. Rather than relying on a single request, the requesting node asks each of its fingers to make the request on it's behalf. Although redundancy increases overhead in the system, the lookups can be done in parallel to keep the delay bounded, and we demonstrate in Section 5 that, with a different network structure, the amount of redundancy is also bounded at what we believe to be a reasonable level.

A particularly nice property of consistent hashing in this regard is that the target owner will be closer than any other node. This means that only one of the redundant requests needs to return the correct result for the requesting node get the IP address of the true target owner – if multiple results are returned, the closest node can be selected. Here, the attacker also must take some risk of being discovered in modifying lookup results unless it can be sure that it has modified all of the redundant lookups. It may be possible to use incorrect lookup results as part of a reputation system, but we do not study this possibility here.

The issue with using this Chord-like structure for redundant requests is that many requests may go through a few of the same nodes, who are then in a position to modify more than one request if they are malicious. To see this, we first note that all requests flow in a single direction around the circular ID-space. Second, as requests get closer to the target, the distance between hops decreases. This ensures a greater density of the requests close to the target. If some of the nodes close to the target are malicious, the attacker is much more likely to be able to modify all the returned results. See Section 5 to see how using Crowds falls short.

3.2 The Salsa Network Architecture

To improve the value of redundant requests, we propose the Salsa network architecture, a novel structured overlay designed to aid the random selection of nodes for anonymous communications. The Salsa architecture improves upon the naive approach by ensuring that redundant requests proceed on random pathways and do not converge on a few nodes. We show in Section 5 that when the fraction of attacker nodes in the system is less than 20%, between four to six redundant lookups is sufficient to find the correct node a very high percentage of the time.

There are some key similarities between the Salsa architecture and the Chord-based approach. Salsa is based on a DHT and has an ID-space to which we map nodes by taking a hash of their IP-address. Each node has a list of a subset of the nodes in the system, its *contacts* (akin to fingers in Chord), that is used to route requests throughout the ID-space. Also, each lookup is resolved by contacting

---
[1] Recent advances in the birthday attack on cryptographic hashes such as MD5 do not affect our system, as the IPv4 address space is too small to yield many collisions.

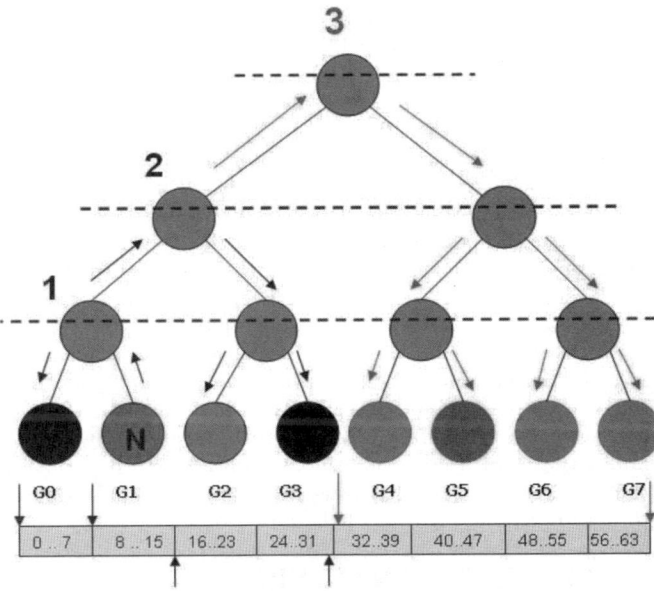

Figure 1: The binary tree structure of Salsa.

other nodes in a recursive manner until the target owner is reached. Each of the nodes contacted during a lookup is called a *lookup hop*.

However, rather than having a single circular ID-space, we divide the entire ID space into equal-sized groups that are defined by a contiguous portion of the ID-space. Each group's portion of the ID-space is cyclic, i.e., the ends of the ID-space within a group wrap around. Let us say that a node belongs to a group if its ID is in the group's ID-space; each node belongs to exactly one group. Within the group, each node *owns* the ID-space between it and the node preceeding it; the ID-space segment of the first node in the group loops around to the last node. A node knows the connection information and ID-space segment for all nodes in its own group, and these are the node's *local contacts*.

Groups are conceptually organized in the form of a binary tree, as illustrated in Fig 3.2. To facilitate lookups beyond a node's group, each node has a set of *global contacts* in different groups, according to the tree structure. In particular, a node has one global contact for each level of the tree. At each level, the global contact is selected at random from the subtree corresponding to the other child of the node's parent that level. For example, in Figure 3.2, a node in group 4 (black), will have one contact in group 3 (dark grey), one in either group 1 or group 2 (group 1 is chosen and shaded in grey), and one in any of groups 5 through 8 (group 8 is chosen and shaded light grey). To ensure that nodes cannot select arbitrary global contacts, the contact for a given segment of the address space is chosen by hashing the concatenation of the selecting node's IP address and the height of the tree. The selected node can easily verify that it should be a global contact for the requesting node.

With both global and local contacts, a lookup for any ID in the ID-space is possible. Lookups proceed recursively. The requesting node asks a global contact in the same subtree as the target owner to continue the lookup and return the results. The contacted node becomes the requesting node and repeats the request to a contact in a smaller subtree that includes the target owner. This continues until the target owner is in the requesting node's group, in which case the requesting node has the contact information and returns it along the path of the request.

Since each node has multiple local contacts, we utilize these for redundancy. The initiator asks a subset of the local contacts, chosen at random, to each find the owner of a given target ID. The result that is closest to the target ID is assumed to be the target owner and is selected. A key property of the Salsa architecture is that any two nodes, even from the same group, will share very few global contacts on average. Thus, redundant requests are likely to proceed along different paths on the way to the target owner's group. This property helps to reduce the possibility of a single node being on multiple paths, thereby reducing the chance that the attacker can modify all the results returned to the requesting node.

An additional security measure is a simple bounds check. Since the target owner should be close to the target ID, we can reject results that are too far from the target ID. This may lead to rejections of legitimate nodes, so we need to consider false positives as well as the false negatives in setting our boundary. We have found that when the number of attackers is less than 20%, it is possible to set a boundary with low error rates. Further, we can modify the boundary for different systems to balance the delay in finding an acceptable node with the security of finding only correct target owners. We show the effectiveness of this in Section 5.

3.3 Building a Circuit

An initiator must select a small set of nodes, e.g. three for Tor, to use as proxies in the circuit. Normally, the initiator would select these nodes privately. In Salsa, however, the node lookup can be linked to the initiator if any of the local contacts that is used for redundant look up is controlled by the attacker. If this happens, and the last node is an attacker, then it can link the initiator to her messages.

First, we calculate the chance of this attack succeeding and then provide a solution to decrease this chance. If there are n nodes in the system and c of them are controlled by the attacker, then any given node will be an attacker with probability c/n. Thus, the chance that at least one of k chosen nodes will be an attacker is $1-(1-c/n)^k$. For $k=3$, and $c = 0.1*n$, i.e. 10% attackers, this is approximately 27%. For the same $k=3$, but $c = 0.2*n$, i.e. 20% attackers, this is approximately 49%. The chance that the last node is an attacker is slightly more than the fraction of attackers in the system, as we show in Section 5. Let us assume that it is 11% and 22% for 10% and 20% attackers in the system, respectively. For $k=3$ and 10% attackers, there is a 3.0% chance of attacker success; for 20% attackers, it is 11%.

A more secure approach is to incorporate circuit-building into the redundant lookups. First, r nodes are looked up in the normal way. Keys are established with each of these nodes. Second, each of the first set of nodes does a lookup for r additional nodes. A Tor circuit is built from the initiator, through one of the first nodes, to one of the second nodes. The second set of nodes does a redundant lookup for a final node. One of the paths created between the first and second sets of nodes is selected, and the final final node is added to the end of the circuit. In this approach, the first node does not learn the identity of the final node and vice versa.

The attack against this last approach would require one attacker in the first set and one attacker in the second set as well as the final node. For $r = 3$ and 10% attackers, this means an approximately 0.8% chance of success. For $r = 3$ and 20% attackers, this means a 5.2% chance of success. Note that the chance of getting the first and last nodes as attackers on the circuit is 1.0% and 4.0% for 10% and 20% attackers in the system respectively. Since this scenario is considered sufficient to break the user's anonymity with a timing analysis attack in low-latency anonymity systems [11, 17, 9], further defenses are not likely to be worth the cost.

When $r = 3$ and a circuit length of three proxies, the added cost of this technique, over using redundant lookups with $k = 3$, is four additional key establishments and six additional lookups. The key establishments can be done in parallel with existing steps and likewise with the additional lookup, so there is relatively little additional delay.

3.4 Initialization and Network Dynamics

We now consider how a Salsa system could be built up securely and handle nodes entering and exiting the system. One issue for any peer-to-peer system is that an attacker who adds a large botnet's worth of nodes (e.g., 20,000 nodes) could dominate the system when it is relatively small and control most of the system's functions. We see little hope in stopping this through the design of the system – Captcha-inspired Turing tests designed to ensure that a human is using each connection might be tried, but it is beyond the scope of this work [16]. For Tor-like server-based systems, with presumably fewer nodes, we note that a higher barrier to entry is required to ensure that most nodes are at least associated with a unique owner. Again, the best method of doing this is beyond the scope of this work.

We propose that nodes should, if possible, join through trusted friends that already have nodes in the network. This is also the best way of building a small network into a larger one. The new node N can have the friend perform lookups to identify a subset of the nodes in N's group, as determined via a hash of N's IP address. N can then contact these nodes with a *join* message, to which the nodes respond with a full list of the group members. Again we face a performance and security tradeoff between the number of lookups performed by the trusted friend and the correct identification of the full group. However, mismatched lists would alert users to a possible attacker presence in the group, so the attacker must be confident that N only contacts attackers and not other nodes in the group. Otherwise, the attacker nodes must also send the full list.

In the event that new node N has no friends using the system, we propose that a subset of nodes could be advertised on, e.g., a website or a bulletin board. For example, a node may post its own IP and those of its global contacts. Since the global contacts are random and verifiable by hashing the IP with the level of the tree, N has some assurance that they are not an attacker's hand-picked selection. For $g = 256$ groups, $n = 10000$ nodes, and $c = 1000$ attacker nodes, the chance that a specific set of global contacts is all corrupt is approximately $(c/n)^{\log_2 g} = 0.1^8$. The chance that any of the 1000 attackers has a full set of attacker global contacts is:

$$1 - \left(1 - (c/n)^{\log_2 g}\right)^c = 0.00001$$

The user must be sure that nearly all of the contacts connect correctly, lest an attacker node have multiple attacker contacts and provide fake addresses for the rest. Once N connects to the local contacts, she can redundantly request multiple addresses within her new group. The security of redundant requests applies as with normal lookups, and the joining proceedure follows as in the trusted friend case.

Finally we consider what happens when a node leaves the system. The node should, ideally, notify the other group members and nodes that have recently connected to it as a global contact. This should be sufficent to create a smooth transition, as nodes can update their global contacts prior to the next round of requests. The disconnect message must include a brief handshake to prevent spoofed *leave messages* from becoming a denial of service attack or a way to redirect traffic to corrupt nodes. If a node abruptly disconnects, the nodes that had used it as a global contact will find out in the next round of requests and must issue a set of requests to determine the identity of the new global contact.

4. SECURITY ANALYSIS

We now discuss how well Salsa protects against a variety of attacks, including standard attacks against anonymity systems and attacks specific to the Salsa system. One type of attacker that it is critical to defend against is one with many compromised hosts to put into the system. Although this is a major issue for peer-to-peer systems, it can also be an issue in highly-distributed server-based systems in which keeping track of the identity of server operators may not be feasible. With compromised nodes in the system, possible attacks include intersection attacks, control of the initiator's circuit, end-to-end timing analysis, predecessor attacks, and denial of service attacks. We study each in turn.

Intersection Attacks. Intersection attacks are particularly dangerous when the attacker can learn the membership of the set of all users currently active in the system [28]. In Salsa, obtaining this information is significantly harder than in Tarzan or Crowds, but may be somewhat easier than in MorphMix. In particular, the attacker can own one node in each group, which ensures that he knows the membership of each group and therefore the entire system.

We model this attack as a balls-and-bins probability problem, as the placement of an attacker node into a group is random, based on a hash of the IP address [19]. Each bin is a group, and each ball is an attacker node. If there are g groups, the expected number of attacker nodes required to get one in each group is $g \cdot H_g$, where H_x is the x-th harmonic number (roughly $log(x)$). To ensure that there is an attacker in each group with high probability $1 - 1/g$, $O(2g \ln g)$ attacker nodes are required. When $g = 256$, the expected number of attackers needed is approximately 1570, and 2800 attackers will ensure that the attacker belongs to each group with high probability 0.996. When $g = 4096$, the expected number of attackers needed is approximately 36,400, with 68,100 attacker nodes needed to ensure attacker success with probability 0.9998. Clearly, the number of groups is critical to the security of the system. Consequently, the number of honest nodes is critical, as tens of thousands of nodes would be needed to support 4096 groups.

Unfortunately, the attacker can also learn about other nodes by using the lookup process. We note that the attacker must establish keys with each contact except for the final hop, increasing the cost of the attack. To slow the connection process, each lookup response could introduce a small delay before responding. This delay can keep the attacker from discovering every node quickly enough to be confident that older results still hold; new nodes may have joined in the meantime. A small delay can protect against high-volume attacks, while not greatly inconveniencing users. Secondly, the attacker can never be sure that he has discovered every node in the network without testing every point in the ID-space. Sometimes two nodes will be very close, meaning that the first node owns a very small segment of ID-space and is difficult to detect via random search. Critically, intersection attacks require a complete or nearly complete view of all users. Otherwise, the attacker could fail to find the initiator; even if the initator is observed later in the course of the attack, she will have been intersected out of the possible initiator set.

Control of the Initiator's Circuit. If the attacker can control all of the nodes on the initiator's circuit, he will be able to easily link the initiator with her communications. MorphMix is particularly vulnerable to this attack, as the initator cannot select nodes from the entire system. We claim that Salsa limits the initator's exposure to this attack.

In Salsa, the attacker may attempt to succeed by biasing the random selection of the proxies to get the initiator to select attacker-controlled nodes. Salsa mitigates this threat through redundancy and bounds checking. Despite these mechanisms, the attacker can still bias the selection; we show the extent of this in Section 5. If this bias is limited, the attack becomes similar to random selection. However, given that attackers selected early in the circuit have a greater chance to bias later results, even a moderate bias can lead to substantial increases in the attacker's success rate. We also study this effect in Section 5.

End-to-end timing analysis. The attacker can observe the timings of packets that enter and exit the system and find correlations between the two ends of the same stream [9, 29, 17]. If the attacker controls a subset of nodes, this can occur when the first and last nodes are controlled by the attacker. When each node is selected at random, with replacement, the chance of the attacker being in place is $(c/n)^2$; for a node to attacker ratio of 10.0, this is 0.01. Again, bias in the node selection can increase this chance, and it is critical to minimize this bias. When the attackers are in place, it is not guaranteed that timing analysis will work, but error rates for correlating streams without mixing or cover traffic are low [17]. Thus, we use this as a baseline for other attacks; if the attacker's chance of success in another attack is less than $(c/n)^2$ (and attacks can't be combined), the initiator's vulnerability has not been significantly increased.

Predecessor attacks. Crowds [22] introduced the idea of the *predecessor attack*, which was later extended in [27]. In this attack, the attacker takes observations of path order over time to link the initiator and the responder. The predecessor attack against mix-based systems relies on end-to-end timing or complete circuit control attacks. Thus, defenses against these attacks also inhibit the predecessor attack. In peer-to-peer mix-based systems, the predecessor attack requires $(1/S) \log n$ connections by the initiator to work with high probability, where S is the chance of attacker success in the placing nodes for the end-to-end timing attack [27, 17]. This further demonstrates the need to limit bias in the selection of nodes, as the length of the attack depends inversely on the attacker success rate.

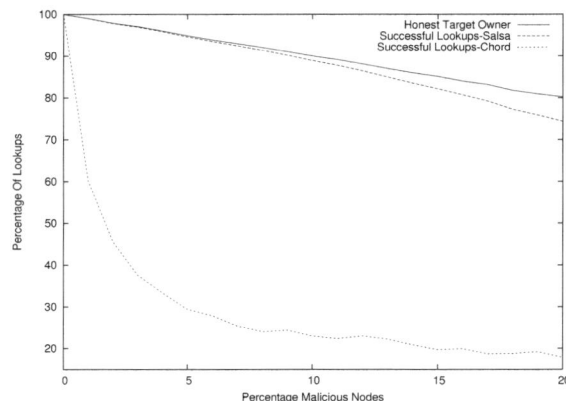

Figure 2: Lookup success: 1,000 nodes, 128 groups, Redundancy=5

Denial of Service Attacks. The last type of attack that we consider is a denial of service (DoS) attack. We do not intend to protect fully against such an attack, as determined attackers with enough resources can always flood the proxies. This is a strength of highly distributed anonymity systems, as there are many proxies to attack. However, we must be careful that our system design does not allow for attackers to easily take down the system without using substantial resources. One possibility for such an attack is to make excessive lookup requests in the network. Each request generates a series of up to $2 \log G + 2$ messages, which is good leverage for a DoS attack. Intermediate nodes can introduce additional delay for each request before forwarding it. This provides dual benefits of slowing any DoS attempts and keeping an attacker from quickly learning about the entire network.

Another possibility for attack is to create many computationally expensive key establishment requests. This can be mitigated by starting with a simple handshake process. Each requester initiates key establishment by first sending send a challenge, e.g. a nonce, that the proxy should return along with a second nonce. The second nonce should be returned in the key establishment request, providing weak authentication that the requesting node can receive at its advertised IP address. This is a simple extension of the handshake protocol from Tarzan [13]. Here it provides some assurance that both the requester and the proxy own their respective addresses, and it can be a vehicle for exchanging public keys. The proxy will only begin expensive key establishment with nodes that it has done this handshake with and can limit each requester to one request per fixed time unit. Thus, an attacker has more work to do per request, and has limited requests per machine per time unit, making the DoS attack more difficult.

5. SIMULATION RESULTS

We performed a number of experiments aimed at testing the security properties of the Salsa system, as described in Section 3. Specifically, we demonstrate the effectiveness of redundant node selection and bounds checking in limiting the attacker's ability to bias node selection. Additionally, we show the extent to which the levels of attacker bias in the system lead to compromise of the initiator's path.

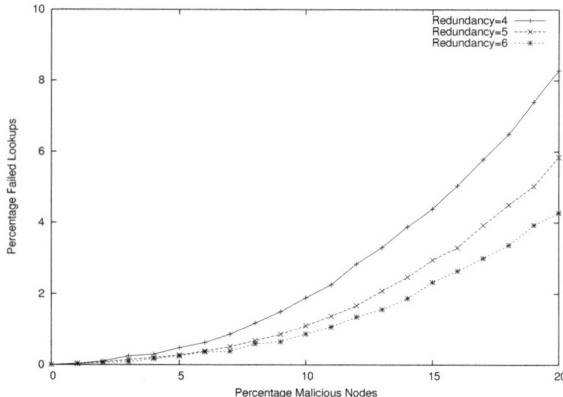

Figure 3: Percentage of Failed Lookups vs Redundancy: 1,000 nodes, 128 groups, 20% malicious nodes

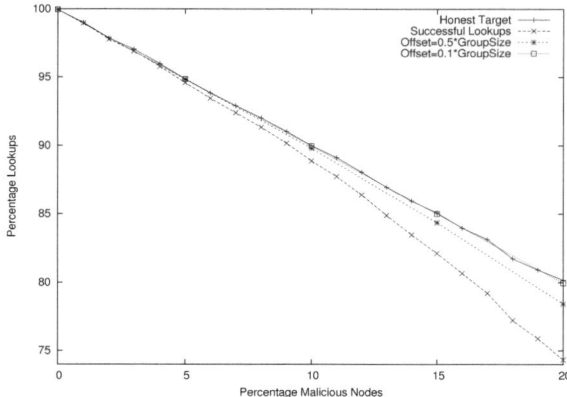

Figure 5: Lookup Success With Probabilistic Checking: 1,000 nodes, 128 groups

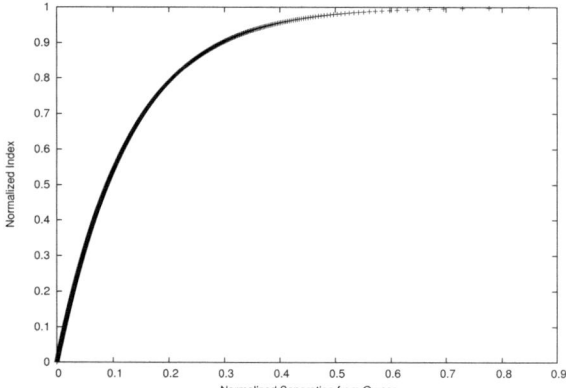

Figure 4: Probabilistic Checking: 1,000 nodes, 128 groups

For our tests, we used a 30-bit hash space and considered systems of 1,000 and 10,000 nodes. We used $G = 128$ and $G = 256$ groups for 1,000 nodes and $G = 256$, $G = 512$, and $G = 1024$ for 10,000 nodes. For each test, we simulated 1,000 separate systems and made 1,000 lookups per system. The system was simulated in Java; we do not simulate in detail the underlying network, encryption, or other system details. Rather, the experiments focus on the security of the system under varying degrees of attack. We tested with the percentage of malicious nodes ranging from 0% to 20%. Beyond 20% malicious nodes, greater than 4% of all paths are compromised due to end-to-end timing analysis attacks.

We claimed earlier that the id-space is organized such that it provides resilience to attacker-induced bias. Here we assume that once a node chooses a malicious node as a proxy on the circuit, that route fails. There are two ways to select a malicious node:

- When a node chooses an ID at random for one of the proxies on the circuit, that ID could be owned by a malicious node.

- While performing a lookup for an ID, the results of the lookup are modified by attackers to be a malicious node.

To perform a lookup we choose one node at random from the set of all nodes (assuming that it is honest for simplicity). Let us call this the *source node*. We then choose an ID at random from the entire ID-space, and we call this the *target ID*. The source node then performs a lookup for the target ID. The chance that the target ID is owned by a malicious node is approximately given by the fraction of malicious nodes in the system. As nothing can be done about the selection of these malicious nodes, we call these *abandoned lookups*. We aim to test the case where the target ID is owned by an honest node, as this is the chance for the attacker to bias the node selection to an attacker-controlled node. If we choose an honest target ID and the lookup result has been biased in this way, we say call it a *failed lookup*. Correspondingly, we define *successful lookups* as those in which the target ID was owned by an honest node and that node was returned. We checked the percentage of the total lookups that were successful, and compared the results with the number of lookups not abandoned. The difference–the percentage of failed lookups–shows the resiliency of the system to attackers randomly distributed over the ID-space.

We use redundancy to minimize the effect of running into malicious node while lookup up an ID. The level of redundancy, defined by the number of redundant lookup requests per lookup, is a tunable parameter to balance security with performance. The more redundant lookup requests the initiator uses, the greater the chance of getting a successful lookup. However, more lookups mean greater overhead in terms of numbers of messages. We show results for systems using between three and six redundant lookups.

To study the use of bounds checking, we evaluate the system using different bounds, with results for both false positives and false negatives. When a lookup result falls outside the bound, the initator must perform a lookup for a new ID. This increases the overhead of the system, but provides greater security, and we study this tradeoff as well.

5.1 Results

We now present the results of our simulations.

For the first set of experiments we use a redundancy of five, i.e., for groups of greater than five nodes, five of the local contacts were asked to do the lookup. As shown in Figure 2, about 20% of the target IDs were owned by ma-

R	Avg. min. group size	Avg max. group size	Avg. messages sent	Avg. max pathlength
4	1.68	15.85	39.63	8.92
5	1.74	15.94	48.53	8.94
6	1.68	15.98	56.42	8.94

Table 1: Group, Path, Message Statistics: System Details: 1000 nodes, 128 groups, 10% malicious nodes.

Max. offset	False Positives	False Negatives			
		5	10	15	20
0.5	2.0%	9.2%	17.0%	24.2%	29.9%
0.1	45.8%	0.95%	2.0%	2.9%	3.8%

Table 2: Probabilistic checking

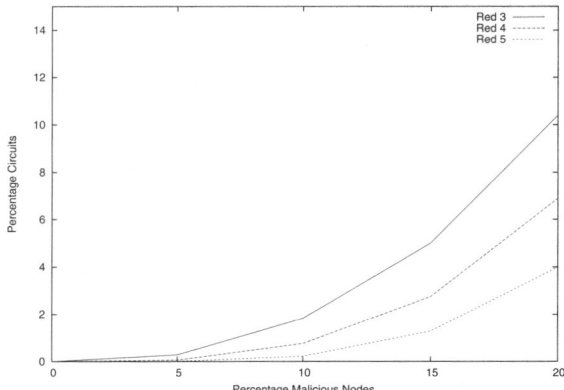

Figure 6: Lookup Success With Probabilistic Checking: 1,000 nodes, 128 groups

licious nodes when there were 20% malicious nodes in the system. The value is consistently close to the percentage of malicious nodes in the system, as expected. We see that even when there are 20% malicious nodes in the system, the number of failed lookups amounts to less than 6%. The naive system, using Crowds, fails badly. Figure 3 shows the results for varying levels of redundancy. As we increased the redundancy from four to five the percentage of failed lookups decreases by 2.44% and a further 1.57% as we increased the redundancy from five to six.

Next we did experiments to check general statistics of the lookups and systems. Table 1 shows information like the average minimum and maximum group population, the number of messages exchanged to perform lookups as well as the average maximum pathlength. The number of messages indicates the overhead the redundancy places on the system. The average maximum pathlength indicates the delay induced in the system. Since the lookups are similar to a binary search on the groups, we expect the number of lookup hops to be about $log_2(G)$. The results show that to be true. Here each of the group members asked to do a redundant lookup is considered the first lookup hop.

As is expected the number of messages exchanged during lookups increases as the redundancy is increased. Of course, the other values remain the same since the redundancy doesn't affect any of them.

We also show how well bounds checking works. Figure 4 shows how IDs and their corresponding owners are distributed. The values are normalized by the group sizes. Table 2 shows the results numerically. False positives mean that the honest pick was out-of-bounds. False negatives mean that the attacker pick was in-bounds. The false positive rate is high, which results in the requesting node needing to check multiple random IDs until she gets a good one. If the false positive rate is given by fp, the user would need to test $1/fp$ IDs on average and will succeed with probablity $1 - fp^k$ after testing k IDs. For example, with a 50% false positive rate, the user must test two IDs on average and no more than ten with 99.9% probability. As shown in Figure 2, at 20% malicious nodes in the system, only about 6% lookups turn false, so to pick the highest value of false negatives in the table i.e., 29.9% means 29.9% of those 6% or 1.8% of the total lookups returned malicious nodes and went un-

detected. This can be seen in Figure 5. With the offset set at $0.5 groupSize$ there is a non-negligible false negative rate, but we still get a substantial decrease in the attacker's ability to bias the lookup results. If we set the offset more strictly, e.g. to $0.1 groupSize$, we greatly reduce the false negative rate at the cost of more lookups due to more false positives.

Finally, we demonstrate that redundant checking continues to operate over the length of the path to prevent bias in node selection. In Figure 6, we show the chance that both the first and last nodes are compromised. As shown, the attacker does not gain much over systems with full random selection, such as Tarzan. For example, with 20% attackers, Tarzan would have 4% compromised paths, while for Salsa with redundancy of four, this chance is 6.9%. With redundancy of five, the chance is the same 4%.

6. CONCLUSIONS

In this paper we proposed a structured overlay architecture, based on a DHT, for securely and scalably organizing and selecting nodes in a highly distributed anonymous communications system. Nodes have limited knowledge of the system, while still being able to choose proxies at random from the entire system. This helps protect it from the intersection attack and enhances scalability. Due to the use of consistent hashing, attackers cannot influence where the nodes fall on the ID space. We have shown that the system organization helps its resilience to attack. The system uses redundant lookups to mitigate the risk of finding attackers on the path. This redundancy is a tunable parameter to balance performance and security. The system also uses distance checking to find malicious nodes returned by lookups. We show that we can reduce the false negatives to 3.8% in a system containing 20% malicious nodes. We also show that for an attacker to try to perform the intersection attack, he'd have to control a large fraction of the nodes in the system.

7. ACKNOWLEDGMENTS

We would like to thank the NSF for support of this project with grant CNS-549998. Additionally, thanks to Dawn Song for initial discussions about this project and Roger Dingledine for his input about Tor.

8. REFERENCES

[1] A. Acquisti, R. Dingledine, and P. Syverson. On the economics of anonymity. In *Proc. Financial Cryptography*, Jan 2003.

[2] A. Back, I. Goldberg, and A. Shostack. Freedom 2.0 security issues and analysis. Zero-Knowledge Systems, Inc. white paper, Nov 2000.

[3] O. Berthold, H. Federrath, and M. Kohntopp. Project anonymity and unobservability in the Internet. In *Proc. Computers Freedom and Privacy*, April 2000.

[4] O. Berthold, A. Pfitzmann, and R. Standtke. The disadvantages of free mix-routes and how to overcome them. In *Proc. Intl. Workshop on Design Issues in Anonymity and Unobservability*, July 2000.

[5] N. Borisov. *Anonymous Routing in Structured Peer-to-Peer Overlays*. University of California, Berkeley, CA, 2005. Ph.D Thesis.

[6] D. Chaum. Untraceable electronic mail, return addresses, and digital pseudonyms. *Communications of the ACM*, 24(2):84–88, Feb 1981.

[7] G. Ciaccio. Improving sender anonymity in a structured overlay with imprecise routing. In *Proc. Privacy Enhancing Technologies Workshop (PET)*, June 2006.

[8] G. Danezis. Statistical disclosure attacks: Traffic confirmation in open environments. In *Proc. Security and Privacy in the Age of Uncertainty (SEC 2003)*, pages 421–426, May 2003.

[9] G. Danezis. The traffic analysis of continuous-time mixes. In *Proc. Privacy Enhancing Technologies Workshop (PET)*, May 2004.

[10] G. Danezis, C. Lesniewski-Laas, M. F. Kaashoek, and R. Anderson. Sybil-resistant DHT routing. In *Proc. ESORICS*, Sep. 2005.

[11] R. Dingledine, N. Mathewson, and P. Syverson. Tor: The next-generation Onion Router. In *Proc. 13th USENIX Security Symposium*, August 2004.

[12] J. Douceur. The Sybil attack. In *Proc. IPTPS*, Mar. 2002.

[13] M. Freedman and R. Morris. Tarzan: A peer-to-peer anonymizing network layer. In *Proc. ACM CCS*, Nov. 2002.

[14] S. D. Kamvar, M. T. Schlosser, and H. Garcia-Molina. The eigentrust algorithm for reputation management in p2p networks. In *Proc. 12th International World Wide Web Conference*, 2003.

[15] J. Kirk. Botnets shrinking in size, harder to trace. http://tinyurl.com/nfxgk, Jan. 2006.

[16] L. von Ahn, M. Blum, N. J. Hopper, and J. Langford. CAPTCHA: Using hard AI problems for security. In *Proc. Eurocrypt*, 2003.

[17] B. N. Levine, M. Reiter, C. Wang, and M. Wright. Timing analysis in low-latency mix systems. In *Proc. Financial Cryptography*, February 2004.

[18] N. Mathewson and R. Dingledine. Practical traffic analysis: extending and resisting statistical disclosure. In *Proc. Privacy Enhancing Technologies workshop (PET 2004)*, May 2004.

[19] R. Motawani and P. Raghavan. *Randomized Algorithms*, chapter 3. Cambridge University Press, 1995.

[20] S. J. Murdoch and G. Danezis. Low-cost traffic analysis of Tor. In *Proceedings of the 2005 IEEE Symposium on Security and Privacy*, May 2005.

[21] M. Reiter, X. Weng, and M. Wright. Building reliable mix networks with fair exchange. In *Proc. 3rd Applied Cryptography and Network Security Conference (ACNS)*, June 2005.

[22] M. K. Reiter and A. D. Rubin. Crowds: Anonymity for Web Transactions. *ACM TISSEC*, 1(1):66–92, Nov 1998.

[23] M. Rennhard and B. Plattner. Practical anonymity for the masses with MorphMix. In *Proc. Financial Cryptography (FC '04)*, February 2004.

[24] I. Stoica, R. Morris, D. Karger, F. Kaashoek, and H. Balakrishnan. Chord: A scalable Peer-To-Peer lookup service for Internet applications. In *Proceedings of the 2001 ACM SIGCOMM Conference*, 2001.

[25] P. Syverson, G. Tsudik, M. Reed, and C. Landwehr. Towards an analysis of Onion Routing security. In *Workshop on Design Issues in Anonymity and Unobservability*, July 2000.

[26] P. Tabriz and N. Borisov. Breaking the collusion detection mechanism of MorphMix. In *Proc. Privacy Enhancing Technologies Workshop (PET)*, June 2006.

[27] M. Wright, M. Adler, B. Levine, and C. Shields. An analysis of the degradation of anonymous protocols. In *Proc. ISOC Sym. on Network and Distributed System Security*, Feb 2002.

[28] M. Wright, M. Adler, B. Levine, and C. Shields. Defending anonymous communications against passive logging attacks. In *Proc. IEEE Sym. on Security and Privacy*, May 2003.

[29] Y. Zhu, X. Fu, B. Graham, R. Bettati, and W. Zhao. On flow correlation attacks and countermeasures in mix networks. In *Proc. Privacy Enhancing Technologies (PET)*, May 2004.

Hot or Not: Revealing Hidden Services by their Clock Skew

Steven J. Murdoch
Computer Laboratory
University of Cambridge
15 JJ Thomson Avenue
Cambridge CB3 0FD, UK
http://www.cl.cam.ac.uk/users/sjm217/

ABSTRACT

Location-hidden services, as offered by anonymity systems such as Tor, allow servers to be operated under a pseudonym. As Tor is an overlay network, servers hosting hidden services are accessible both directly and over the anonymous channel. Traffic patterns through one channel have observable effects on the other, thus allowing a service's pseudonymous identity and IP address to be linked. One proposed solution to this vulnerability is for Tor nodes to provide fixed quality of service to each connection, regardless of other traffic, thus reducing capacity but resisting such interference attacks. However, even if each connection does not influence the others, total throughput would still affect the load on the CPU, and thus its heat output. Unfortunately for anonymity, the result of temperature on clock skew can be remotely detected through observing timestamps. This attack works because existing abstract models of anonymity-network nodes do not take into account the inevitable imperfections of the hardware they run on. Furthermore, we suggest the same technique could be exploited as a classical covert channel and can even provide geolocation.

Categories and Subject Descriptors

C.2.0 [**Computer-Communication Networks**]: General—*Security and protection*; D.4.6 [**Operating Systems**]: Security and Protection—*Information Flow Controls*; C.2.5 [**Computer-Communication Networks**]: Local and Wide-Area Networks—*Internet*; K.4.1 [**Computers and Society**]: Public Policy Issues—*Privacy*

General Terms

Security, Experimentation

Keywords

Anonymity, Clock Skew, Covert Channels, Fingerprinting, Mix Networks, Temperature, Tor

1. INTRODUCTION

Hidden services allow access to resources without the operator's identity being revealed. Not only does this protect the owner, but also the resource, as anonymity can help prevent selective denial of service attacks (DoS) [35, 36]. Tor [15], has offered hidden services since 2004, allowing users to run a TCP server under a pseudonym. At the time of writing, there are around 80 publicly advertised hidden services, offering access to resources that include chat, low and high latency anonymous email, remote login (SSH and VNC), websites and even gopher. The full list of hidden services is only known by the three Tor directory servers.

Systems to allow anonymous and censorship-resistant content distribution have been desired for some time, but recently, anonymous publication has been brought to the fore by several cases of blogs being taken down and/or their authors being punished, whether imprisoned by the state [43] or being fired by their employers [5]. In addition to blogs, Tor hidden websites include dissident and anti-globalisation news, censored or otherwise controversial documents, and a PGP keyserver. It is clear that, given the political and legal situation in many countries, the need for anonymous publishing will remain for some time.

Because of the credible threat faced by anonymous content providers, it is important to evaluate the security, not only of deployed systems, but also proposed changes believed to enhance the security or usability. Guaranteed quality of service (QoS) is one such defence, designed to protect against indirect traffic-analysis attacks that estimate the speed of one flow by observing the performance of other flows through the same machine [33].

QoS acts as a countermeasure by preventing flows on an anonymity-network node from interfering with each other. However, an inevitable result is that when a flow is running at less than its reserved capacity, CPU load on the node will be reduced. This induces a temperature decrease, which affects the frequency of the crystal oscillator driving the system clock. We measure this effect remotely by requesting timestamps and deriving clock skew.

We have tested this vulnerability hypothesis using the current Tor implementation (0.1.1.16-rc), although – for reasons explained later – using a private instance of the network. Tor was selected due to its popularity, but also because it is well documented and amenable to study. However, the attacks we present here are applicable to the design of other anonymity systems, particularly overlay networks.

In Section 2 we review how hidden services have evolved from previous work on anonymity, discuss the threat models used in their design and summarise existing attacks. Then in Section 3 we provide some background on clock skew, the phenomenon we exploit to link hidden service pseudonyms to the server's real identity. In Section 4 we present the results of our experiments on Tor and discuss the potential impact and defences. Finally, in Section 5 we suggest how the general technique (of creating covert channels and side channels which cross between the digital and physical world) might be applied in other scenarios.

2. HIDDEN SERVICES

Low latency anonymity networks allow services to be accessed anonymously, in real time. The lack of intentional delay at first glance decreases security, but by increasing utility, the anonymity set can increase [1]. The first such proposal was the ISDN mix [39], but it was designed for a circuit switched network where all participants transmit at continuous and equal data rates and is not well suited to the more dynamic packet switched Internet. PipeNet [10] attempted to apply the techniques of ISDN mixes to the Internet, but while providing good anonymity guarantees, it is not practical for most purposes because when one node shuts down the entire network must stop; also, the cost of the dummy traffic required is prohibitive.

The Anonymizer [3] and the Java Anon Proxy (JAP) [7] provide low-latency anonymous web browsing. The main difference between them is that while Anonymizer is controlled by a single entity, traffic flowing through JAP goes through several nodes arranged in a fixed cascade. However, in neither case do they obscure where data enters and leaves the network, so they cannot easily support hidden services. This paper will instead concentrate on free-route networks, such as Freedom [4, 8] and the Onion Routing Project [42], of which Tor is the latest incarnation.

2.1 Tor

The attacks presented in this paper are independent of the underlying anonymity system and hidden service architecture, and should apply to any overlay network. While there are differing proposals for anonymity systems supporting hidden services, e.g. the PIP Network [17], Tor is a popular, deployed system, suitable for experimentation, so initially we will focus on it. Section 5 will suggest other cases where our technique can be used.

Tor hidden services are built on the connection anonymity primitive the network provides. As neither our attack nor the Tor hidden service protocol relies on the underlying implementation, we defer to [12, 13, 14, 15] for the full details. All that is important to appreciate in the remaining discussion is that Tor can anonymously tunnel a TCP stream to a specified IP address and port number. It does this by relaying traffic through randomly selected nodes, wrapping data in multiple layers of encryption to maintain unlinkability. Unlike email mixes, it does not intentionally introduce any delay: typical latencies are in the 10–100 ms range.

There are five special roles in a hidden service connection and all links between them are anonymised by the Tor network. The *client* wishes to access a resource offered by a *hidden server*. To do so, the client contacts a *directory server* requesting the address of an *introduction point*, which acts as an intermediary for initial setup. Then, both nodes connect to a *rendezvous point*, which relays data between the client and hidden server.

For clarity, some details have been omitted from this summary; a more complete description is in Øverlier and Syverson [38]. In the remainder of the paper, we will deal only with an established data connection, from the client to the rendezvous point and from there to the hidden server.

2.2 Threat Model

The primary goal of our attacker is to link a pseudonym (under which a hidden service is being offered) to the operator's real identity, either directly or through some intermediate step (e.g. a physical location or IP address). For the moment, we will assume that identifying the IP address is the goal, but Section 5.3 will discuss what else can be discovered, and some particular cases in which an IP address is hard to link to an identity.

Low-latency anonymity networks without dummy traffic, like Tor, cannot defend against a global passive adversary. Such an attacker simply observes inputs and outputs of the network and correlates their timing patterns, so called *traffic-analysis*. For the same reason, they cannot protect against traffic confirmation attacks, where an attacker has guessed who is communicating with whom and can snoop individual network links in order to validate this suspicion.

It is also common to assume that an attacker controls some of the anonymity network, but not all. In cases like Tor, which is run by volunteers subjected to limited vetting, this is a valid concern, and previous work has made use of this [33, 38]. However, the attacks we present here do not require control of any node, so will apply even to anonymity networks where the attacker controls no nodes at all.

In summary, we do not assume that our attacker is part of the anonymity network, but can access hidden services exposed by it. We do assume that he has a reasonably limited number of candidate hosts for the hidden service (say, a few thousand). However, we differ from the traffic confirmation case excluded above in that our attacker cannot observe, inject, delete or modify any network traffic, other than that to or from his own computer.

2.3 Existing Attacks

The first documented attack on hidden servers was by Øverlier and Syverson [38]. It proposes and experimentally confirms that a hidden service can be located within a few minutes to hours if the attacker controls one, or preferably two, network nodes. It relies on the fact that a Tor hidden server selects nodes at random to build connections. The attacker repeatedly connects to the hidden service, and eventually a node he controls will be the one closest to the hidden server. Now, by correlating input and output traffic, the attacker can confirm that this is the case, and so he has found the hidden server's IP address.

Another attack against Tor, but not hidden services *per se*, is described by Murdoch and Danezis [33]. The victim visits an attacker controlled website, which induces traffic patterns on the circuit protecting the client. Simultaneously, the attacker probes the latency of all Tor nodes and looks for correlations between the induced pattern and observed latencies. The full list of Tor nodes is, necessarily, available in the public directories along with their IP addresses. When there is a match, the attacker knows that the node is on the target circuit and so can reconstruct the path, although not discover the end node. In a threat model where the attacker has a limited number of candidates for the hidden service, this attack could also reveal its identity. Many hidden servers are also publicly advertised Tor nodes, in order to mask hidden server traffic with other Tor traffic, so this scenario is plausible.

Several defences are proposed by Murdoch and Danezis, which if feasible, should provide strong assurances against the attack. One is non-interference – where each stream going through a node is isolated from the others. Here each Tor node has a given capacity, which is divided into several slots. Each circuit is assigned one slot and is given a guaranteed data rate, regardless of the others.

Our new observation, which underpins the attack presented, is that when circuits carried by a node become idle, its CPU will be less active, and so cool down. Temperature has a measurable effect on clock skew, and this can be observed remotely. We will show that an attacker can thus distinguish between a CPU in idle state and one that is busy. But first some background is required.

3. CLOCK SKEW AND TEMPERATURE

Kohno et al. [24] used timing information from a remote computer to fingerprint its physical identify. By examining timestamps from the machine they estimated its *clock skew*: the ratio between actual and nominal clock frequencies.

They found that a particular machine's clock skew deviates very little over time, around 1–2 parts per million (ppm), depending on operating system, but that there was a significant difference between the clock skews (up to 50 ppm) of different machines, even identical models. This allows a host's clock skew to act as a fingerprint, linking repeated observations of timing information. The paper estimates that, assuming a stability of 1 ppm, 4–6 bits of information on the host's identity can be extracted.

Two sources of timestamps were investigated by Kohno et al.: ICMP timestamp requests [40] and TCP timestamp options [21]. The former has the advantage of being of a fixed nominal frequency (1 kHz), but if a host is Network Time Protocol (NTP) [28] synchronised, the ICMP timestamp was found to be generated after skew adjustment, so defeating the fingerprinting attack. The nominal frequency of TCP timestamps depends on the operating system, and varies from 2 Hz (OpenBSD 3.5) to 1 kHz (Linux 2.6.11). However, it was found to be generated before NTP correction, so attacks relying on this source will work regardless of the NTP configuration. Additionally, in the special case of Linux, Murdoch and Lewis [34] showed how to extract timestamps from TCP sequence numbers.

We will primarily use TCP timestamps, which are enabled by default on most modern operating systems. They improve performance by providing better estimates of round trip times and protect against wrapped sequence numbers on fast networks. Because of their utility, TCP timestamps are commonly passed by firewalls, unlike ICMP packets and IP options, so are widely applicable. The alternative measurement techniques will be revisited in Section 5.4.

3.1 Background and Definitions

Let $T(t_\mathrm{s})$ be the timestamp sent at time t_s. Unless specified otherwise, all times are relative to the receiver clock. As we are interested in changes of clock frequency, we split skew into two components, the constant s_c and the time-varying part $s(t)$. Without loss of generality, we assume that the time-varying component is always negative.

Before a timestamp is sent, the internal value of time is converted to a number of *ticks* and rounded down. The nominal length of a tick is the clock's resolution and the reciprocal of this is its nominal frequency, h. The relationship between the timestamp and input parameters is thus:

$$T(t_\mathrm{s}) = \left\lfloor h \cdot \left(t_\mathrm{s} + s_\mathrm{c} t_\mathrm{s} + \int_0^{t_\mathrm{s}} s(t)\, dt\right)\right\rfloor \quad (1)$$

Now we sample timestamps T_i sent at times t_{s_i} chosen uniformly at random between consecutive ticks, for all i in $[1\ldots n]$, with $t_{\mathrm{s}_1} = 0$. The quantisation noise caused by the rounding can be modelled as subtracting a random variable c with uniform distribution over the range $[0,1)$. Also, by dividing by h, we can recover the time according to the sender in sample i:

$$\tilde{t}_i = T_i/h = t_{\mathrm{s}_i} + s_\mathrm{c} t_{\mathrm{s}_i} + \int_0^{t_{\mathrm{s}_i}} s(t)\, dt - c_i/h \quad (2)$$

We cannot directly measure the clock skew of a remote machine, but we can calculate the *offset*. This is the difference between a clock's notion of the time and that defined by the reference clock (receiver). The offset o_i can be found by subtracting t_{s_i} from $\tilde{t}_i$. However, the receiver only knows the time t_{r_i} when a packet was received.

Let d_i be the latency of a packet, from when it is timestamped to when it is received, then $t_{\mathrm{s}_i} = t_{\mathrm{r}_i} - d_i$. Skew is typically small ($< 50\,\mathrm{ppm}$) so the effect of latency to these terms will be dominated by the direct appearance of d_i and is ignored otherwise. The measured offset is thus:

$$\tilde{o}_i = \tilde{t}_i - t_{\mathrm{r}_i} = s_\mathrm{c} t_{\mathrm{r}_i} + \int_0^{t_{\mathrm{r}_i}} s(t)\, dt - c_i/h - d_i \quad (3)$$

Figure 1 shows a plot of the measured offset over packet receipt time. Were the sampling noise c/h, latency introduced noise d and variable skew $s(t)$ absent, the constant skew s_c would be the derivative of measured offset with respect to time. To form an estimate of the constant skew observed, $\hat{s}_\mathrm{c}$, in the presence of noise, we would like to remove these terms. Note that in (3) the noise contributions, as well as $s(t)$, are both negative.

Following the approach of Kohno et al., we remove the terms by fitting a line above all measurements while minimising the mean distance between each measurement and the point on the line above it. By applying the linear-programming based algorithm described by Moon et al. [29], we derive such a line. More formally this finds an estimate of the linear offset component $\hat{o}(t) = \hat{s}_\mathrm{c} \cdot t + \beta$ such that, for all samples, $\hat{o}(t_{\mathrm{r}_i}) > \tilde{o}_i$ and minimises the expression:

$$\frac{1}{n} \cdot \sum_{i=1}^{n} \left(\hat{o}(t_{\mathrm{r}_i}) - \tilde{o}_i\right) \quad (4)$$

The offset $\hat{o}(t)$ is also plotted on Figure 1. The band of offset samples below the line is due to the sampling noise c/h, as illustrated by the different width depending on h. Points are outside this band because of jitter in the network delay (any constant component will be eliminated), but latencies are tightly clustered below a minimum which remains fixed during the test. This is to be expected for an uncongested network where routes change rarely. The characteristics of these noise sources will be discussed further in Section 5.4.

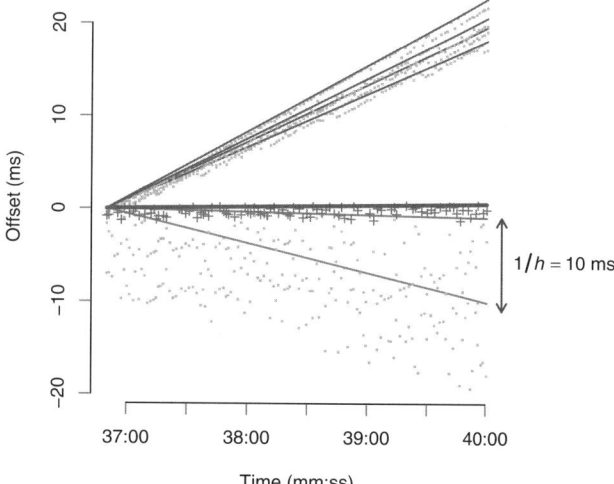

Figure 1: Offset between TCP timestamps of seven machines and the measurer's clock over time. The bottom two lines (—) show clocks with 100 Hz resolution and the others are 1 kHz. The range of the quantisation noise is $[0, 1/h)$, as indicated for the $h = 100\,\text{Hz}$ case. The time since the beginning of the experiment was subtracted from the measurer's clock and the first timestamp received was subtracted from all timestamps. All machines were on the same LAN, except one (+) which was accessed over a transatlantic link, through 14 hops.

3.2 Impact of Temperature

The effect of temperature on remote clock skew measurements has been well known to the NTP community since the early 1990s [25, 27] and was mentioned by Kohno et al. However, we believe our paper to be the first that proposes *inducing* temperature change and measuring the change in clock skew, in order to attack an anonymity system.

As shown in Figure 2, the frequency of clock crystals varies depending on temperature [9]. Exactly how depends on tradeoffs made during manufacture. The figure shows an AT-cut crystal common for PCs, whose skew is defined by a cubic function. BT-cut is more common for sub-megahertz crystals and is defined by a quadratic. The angle of cut alters the temperature response and some options are shown. It can be seen that improving accuracy in the temperature range between the two turning points degrades performance outside these values. Over the range of temperatures encountered in our experiments, skew response to temperature is almost linear, so for simplicity we will treat it as such.

The linear offset fit shown in Figure 1 matches almost perfectly, excluding noise. This indicates that although temperature varied during the sample period, the constant skew s_c dominates any temperature dependence $s(t)$ present. Nevertheless, the temperature dependent term $s(t)$ is present and is shown in Figure 3. Here, $\hat{o}(t_{r_i})$ has been subtracted from all $\tilde{o}_i$, removing our estimate of constant skew $\hat{s}_c$. To estimate the variable skew component $\hat{s}(t)$, the resulting offset is differentiated, after performing a sliding window line-fitting. We see that as the temperature in the room varied over the day, there is a correlated change in clock skew.

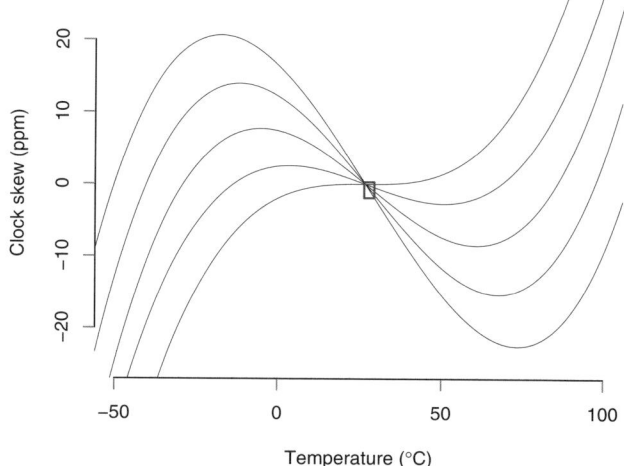

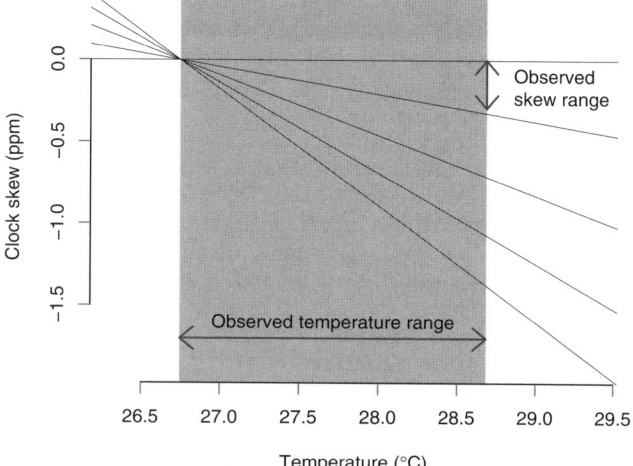

Figure 2: AT-cut crystal clock skew over two temperature ranges. Full operational range is shown above, with indicated zoomed-in area below. On the zoomed graph, the temperature and skew ranges found in Figure 5(a) are shown. As skews are relative, the curves have been shifted vertically so that skew is zero at the minimum observed temperature.

4. ATTACKING TOR

We aim to show that a hidden server will exhibit measurably different behaviour when a particular connection is active compared to when it is idle. Where the Murdoch and Danezis attacks [33] probed the latency of other connections going through the same node, we measure clock skew.

This is because when a connection is idle, the host will not be performing as many computations and so cool down. The temperature change will affect clock skew, and this can be observed remotely by requesting timestamps. The goal of our experiment is to verify this hypothesis.

Such an attack could be deployed in practice by an attacker using one machine to access the hidden service, varying traffic over time to cause the server to heat up or cool down. Simultaneously, he probes all candidate machines for

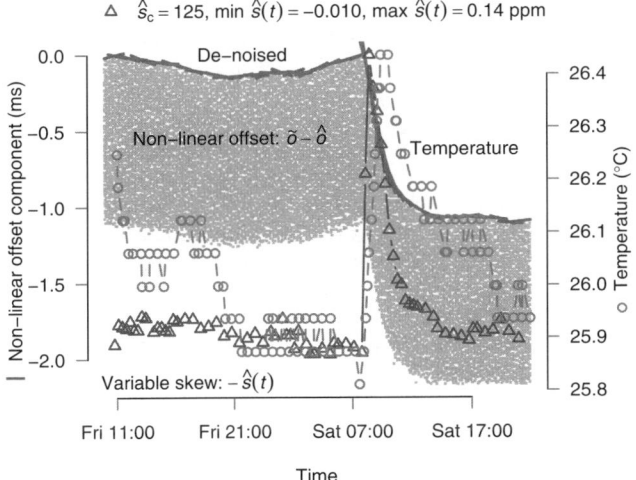

Figure 3: Offset after removing linear component (i.e. $\tilde{o} - \hat{o}$). The line (—) above is the de-noised version. The △ show the negated slope of each piece ($-\hat{s}(t)$) and ○ show the temperature. The maximum and minimum values of $\hat{s}(t)$ are shown, along with the constant skew $\hat{s}_c$. Results are from a mini-tower PC with ASUS A7M266 motherboard and 1.3 GHz AMD Athlon processor.

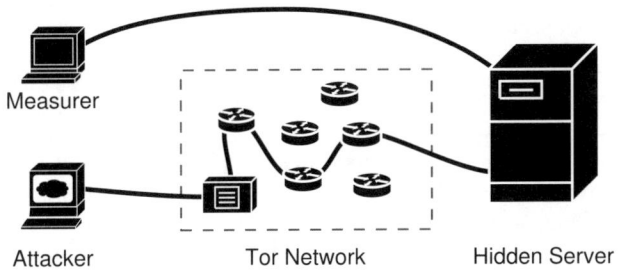

Figure 4: Experimental setup with four computers.

timestamps. From these the attacker infers clock skew estimates and when a correlation between skew and the induced load pattern is found, the hidden service is de-anonymised.

The reliability and performance of the Tor network for hidden servers is currently quite poor, so, to simplify obtaining results, our experiments were run on a private Tor network. We see no reason these results would not transfer to the real Tor network, even when it is made more reliable and resistant to the Murdoch and Danezis attacks.

The computers used in each test (shown in Figure 4) are:

Hidden Server: Tor client and webserver, hosting a 10 MB file; fitted with a temperature sensor.

Tor Network: Two Tor directory server processes and five normal servers, which can act as introduction and rendezvous points, all unmodified. While all processes are on the same machine, this does not invalidate our results as only the Hidden Server is being analysed.

Attacker: Runs the Tor client, repeatedly requesting the file hosted by Hidden Server, through the Tor Network. For performance, this is modified to connect directly to the rendezvous point.

Measurer: Connects directly to the Hidden Server's public IP address, requesting TCP timestamps, ICMP timestamps and TCP sequence numbers, although only the results for the first are shown.

For two hours the 10 MB file is repeatedly downloaded over the Tor network, with up to 10 requests proceeding in parallel. Then for another two hours no requests are made. During both periods timestamps are requested directly from the server hosting the hidden service at intervals of 1 s plus a random period between 0 s and 1 s. This is done to meet the assumption of Section 3.1, that samples are taken at random points during each tick. Otherwise, aliasing artifacts would be present in the results, perturbing the line-fitting.

Finally, the timestamps are processed as described in Section 3.2. That is, estimating the constant skew through the linear programming algorithm and removing it; dividing the trace into pieces and applying the linear-programming algorithm a second time to estimate the varying skew.

Were an attacker to deploy this attack, the next step would be to compare the clock skew measurements of all candidate servers with the load pattern induced on the hidden service. To avoid false-positives, multiple measurements are needed. The approach taken by Murdoch and Danezis [33] is to treat the transmission of the load pattern as a covert channel and send a pseudorandom binary sequence. Thus, after n bits are received, the probability of a false-positive is 2^{-n}. From inspection, we estimate the capacity of the covert channel to be around 2–8 bits per hour. An alternative taken by Fu et al. [16] is to induce a periodic load pattern which can be identified in the power spectrum of the Fourier transformed clock skew measurements. With either approach, the confidence could be increased to arbitrary levels by running the attack for longer.

4.1 Results

Overall throughput was limited by the CPU of the server hosting the private Tor network, so the fastest Hidden Server tested ran at around 70% CPU usage while requests were being made. CPU load on the Hidden Server was almost all due to the Tor process, we suspect due to it performing cryptographic operations. A 1–1.5 °C temperature difference was induced by this load modulation.

Ideally, the measuring machine would have a very accurate clock, to allow comparison of results between different experiments over time and with different equipment. This was not available for these experiments, however as we are interested only in relative skews, only a stable clock is needed, for which a normal PC sufficed. It would also be desirable to timestamp packets as near as possible to receipt, so while adding the timestamp at the network card would be preferable, the one inserted by the Linux kernel and exposed through the `pcap` [22] interface has proved to be adequate.

Figure 5 shows the results of two experimental runs, in the same style as Figure 3. Note that the top graph shows a relationship between clock skew and temperature opposite to expectations; namely when temperature increases, the clock has sped up. One possible explanation is that the PC is using a temperature compensated crystal oscillator (TCXO),

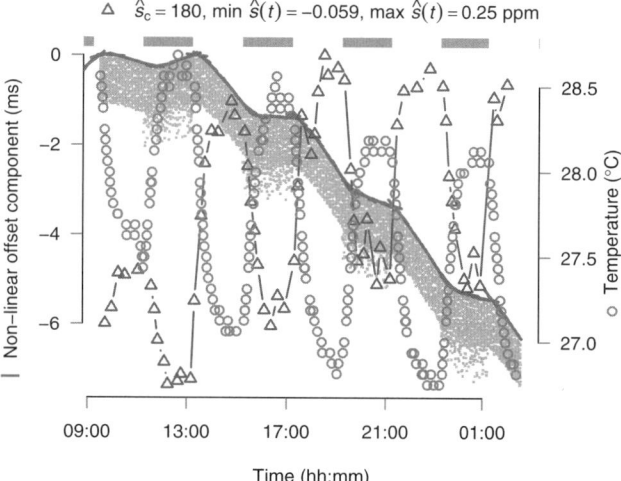

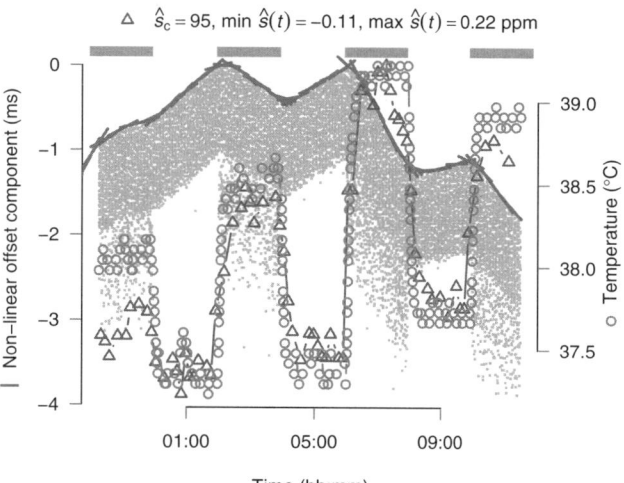

Figure 5: Clock skew measurements for two machines. The graph is as Figure 3, but the grey bars at the top indicate when the hidden server was being exercised. The top graph is from a mini-tower PC with Dell GX1MT motherboard and Intel Pentium II 400 MHz processor; the bottom is from a mini-tower PC with ASUS A7V133 motherboard and AMD Athlon 1.2 GHz processor.

but is over compensating; another is that the temperature curve for the crystal is different from Figure 2. In both cases there is a clear correlation between temperature and skew, despite only a modest temperature change.

While the CPU is under load, there is increased noise present in the results. This could be due to increased latency on the network, or more likely because the CPU is so busy, the operating system sometimes allocates a quantum to the Tor process in between adding a timestamp to a packet and dispatching it. However, note that the minimum latency is unchanged (and is often reached) so the linear programming algorithm still performs well. Were the minimum to change, then a step in the graph would be expected, rather than the smooth curve apparent.

4.2 Discussion

Murdoch and Danezis [33] proposed a defence to their flow interference attacks, that did not require dummy traffic. It was to ensure that no anonymous stream flowing through a hidden server should affect any other, whether it belonged to the anonymity service or not. All connections are thus given a fixed quality of service, and if the maximum number of connections is reached, further connections are refused.

Implementing this is non-trivial as QoS must not only be guaranteed by the host (e.g. CPU resources), but by its network too. Also, the impact on performance would likely be substantial, as many connections will spend much of their time idle. Whereas currently the idle time would be given to other streams, now the host carrying such a stream cannot reallocate any resources, thus opening a DoS vulnerability. However, there may be some suitable compromise, for example dynamic limits which change sufficiently slowly that they leak little information.

Even if such a defence were in place, our temperature attacks would still be effective. While changes in one network connection will not affect any other connections, clock skew is altered. This is because the CPU will remain idle during the slot allocated to a connection without pending data. Unless steps are taken to defend against our attacks, the reduced CPU load will lower temperature and hence affect clock skew. To stabilise temperature, computers could be modified to use expensive oven controlled crystal oscillators (OCXO), or always run at maximum CPU load. External access to timing information could be restricted or jittered, but unless all incoming connections were blocked, extensive changes would be required to hide low level information such as packet emission triggered by timer interrupts.

While the above experiments were on Tor, we stress that our techniques apply to any system that hides load through maintaining QoS guarantees. Also, there is no need for the anonymity service to be the cause of the load. For example, Dean and Stubblefield [11] show that because SSL allows the client to force a server to perform an RSA operation before doing any substantial work itself, DoS attacks can be mounted well before the connection is saturated. Such techniques could be used to attack hidden servers where the anonymity network cannot sustain high throughput.

Inducing clock skew and remotely measuring it can be seen as a *thermal covert channel* because attacking a hidden server could be modelled as violating an information flow control policy in a distributed system. The client accessing the hidden service over the anonymity network is using the link between between the server's pseudonym and its public IP address, which is information at a "high" confidentiality level. However, the client is prevented from leaking this information by the trusted computing base of the anonymity network. The user accessing the hidden server directly only has access to "low" information, the real IP address by itself, however if the "high" process can leak information to the "low" process, the server's anonymity is violated.

This scenario is analogous to covert channel attacks on the *-property of the BLP model [6]: that processes must not be able write to a process lower than its own privilege level. This approach to the analysis of anonymity systems was proposed by Moskowitz et al. [30, 31, 37], but we have shown here further evidence that past research in the field of covert channels can be usefully applied in enhancing the security of modern-day anonymity systems.

5. EXTENSIONS AND FUTURE WORK

The above experiments presented an example of how temperature induced clock skew can be a security risk, but we believe that this is a more general, and previously under-examined, technique which can be applied in other situations. In this section we shall explore some of these cases and propose some future directions for research.

5.1 Classical Covert Channels

The above section discussed an unconventional application of covert channels, that is within a distributed system where users can only send data but not execute arbitrary code. However, clock skew can also be used in conventional covert channels, where an operating system prevents two processes communicating which are on the same computer and can run arbitrary software.

CPU load channels have been extensively studied in the context of multilevel secure systems. Here, two processes share CPU time but the information flow control policy prohibits them from directly communicating. Each can still observe how much processing time it is getting, thus inferring the CPU usage of the other.

A process can thus signal to another by modulating load to encode information [26]. One defence against this attack is to distort the notion of time available to processes [19] but another is fixed scheduling and variations, ensuring that the CPU of one process cannot interfere with the resources of any at a conflicting security rating [20]. Temperature induced clock skew can circumvent the latter countermeasure. Covert channels are also relevant to recently deployed separation kernels such as MILS [2, 44].

Figure 6 shows one such example. In previous cases, the temperature in the measured machine has been modulated, but now we affect the clock skew of the measurer. This graph was plotted in the same way as before, but on the measurer machine the `CPUBurn` program [41] was used to induce load modulation, affecting the temperature as shown. Timestamps are collected from a remote machine and as we are calculating relative clock skew, we see the inverse of the measurer's clock skew, assuming the remote clock is stable.

Note that the temperature difference is greater than before (5 °C vs. 1–1.5 °C). This is because we are no longer constrained by the capacity of the Tor network, and can optimise our procedure to induce the maximum temperature differential. While this attack is effective, it requires fairly free access to network resources, which is not common in the general case of high-assurance systems where covert channels are a serious concern.

Where access to a remote timing source is blocked, the skew between multiple clock crystals within the same machine, due to their differing temperature responses and proximity to the heat source, could be used. For example, in a typical PC, the sound card has a separate crystal from the system clock. A process could time how long it takes (according to the system clock), to record a given number of samples from the sound card, thus estimating the skew between the two crystals.

5.2 Cross Computer Communication

Physical properties of shared hardware have previously been proposed as a method of creating covert channels. For example, hard disk seek time can be used to infer the previous position of the disk arm, which could have been affected by "high" processes [23]. However, with temperature, such effects can extend across "air-gap" security boundaries.

Our experiments so far have not shown evidence of one desktop computer being able to induce a significant temperature change in another which is in the same room, but the same may not be true of rack-mount machines. Here, a 3 °C temperature change in a rack-mount PC has been induced by increasing load on a neighbouring machine [18]. Blade servers, where multiple otherwise independent servers are mounted as cards in the same case, sharing ventilation and power, have even more potential for thermal coupling.

If two of these cards are processing data at different security levels, the tight environmental coupling could lead to a covert channel as above, even without the co-operation of the "low" card. For example, if a "low" webserver is hosted next to a "high" cryptographic co-processor which does not have Internet access, the latter could leak information to an external adversary by modulating temperature while the webserver clock-skew is measured. Side-channels are also conceivable, where someone probing one card could estimate the load of its siblings.

We simulated this case by periodically (2 hours on, 2 hours off) exposing a PC to an external heat source while a second computer measured the clock skew. The results showed that 3 °C temperature changes can be remotely received. Additionally, this confirmed that it is temperature causing the observed clock skew in the previous experiments, and not an OS scheduling artifact. The resulting graph was similar to Figure 5 except there is no increased noise during heating, as would be expected from the hypothesised interference attack resistant anonymity system.

5.3 Geolocation

In the attacks on anonymity systems so far, we have been inducing load through the anonymity system and measuring clock skew directly. An alternative is to measure clock skew through the anonymity network and let the environment alter the clock skew. This allows an attacker to observe temperature changes of a hidden server, so infer its location.

Clock skew does not allow measurement of absolute temperature, only changes. Nevertheless this still could be adequate for geolocation. Longitude could simply be found by finding the daily peak temperature to establish local time. To find latitude, the change in day length over a reasonably long period could be used.

It was apparent in our experiments when a door to the cooler corridor was left open, so national holidays or when daylight saving time comes into effect might be evident. Distortion caused by air-conditioning could be removed by inferring the temperature from the duty cycle (time on vs. time off) of thermostatically controlled appliances.

In this section we have assumed that we probe through the anonymity network. In the case of Tor, this will introduce significant jitter, and it is unclear how badly this will affect timing measurements. Alternatively, the attacker could connect directly to the external IP address.

This raises the question of utility – often IP addresses can easily be mapped to locations [32]. However, this is not always the case. For example, IP anycast and satellite connections are hard to track to a location; as are users who seek to hide by using long-distance dialup. While latency in the last two cases is high, jitter can be very low, lending itself to the clock skew attacks.

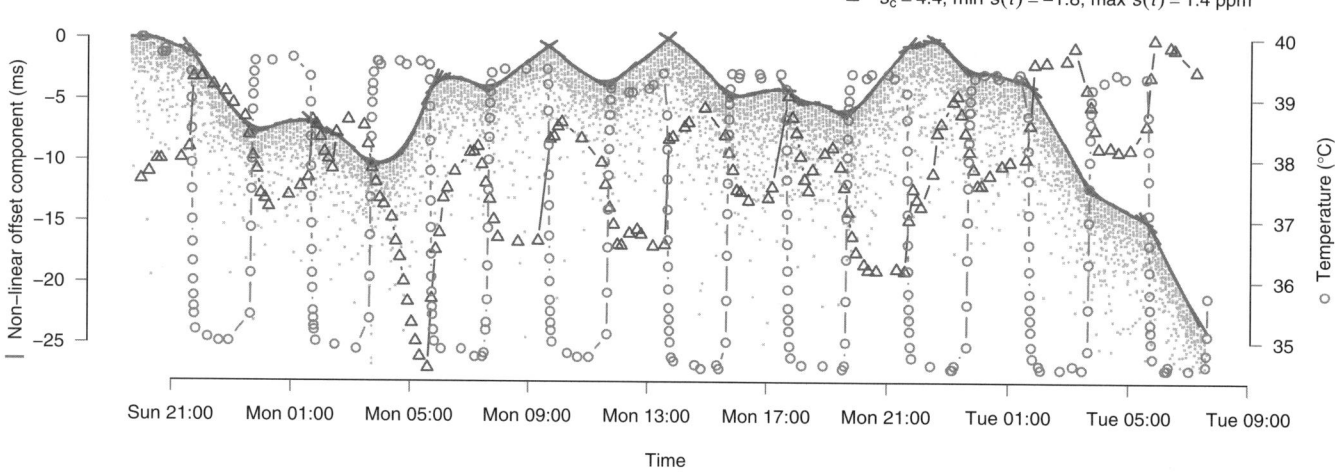

Figure 6: Clock skew measurements of a remote machine while modulating CPU load of the measurer (mini-tower, Intel D875 motherboard, Pentium 4 3.2 GHz CPU), for which temperature is also shown. The measurer and remote machine are separated by a transatlantic link, so the noise level is higher.

5.4 Noise Sources and Mitigation

In the above section, we proposed acquiring timing information from a hidden server through the anonymity network. Here, in addition to the problem of increased jitter, the timing sources we have used (ICMP/TCP timestamps and TCP sequence numbers) may not be available. For example, Tor operates at the TCP layer so these possibilities do not exist, whereas Freedom [4, 8] allows the transmission of arbitrary IP packets.

One option proposed by Kohno *et al.* is to use a Fourier transform to extract the frequency of a periodic event, for example, packet emission caused by a timer interrupt. Another possibility is to use application level timestamps. The most common Tor hidden service is a web server, typically using Apache, and by default this inserts a timestamp into HTTP headers. However, this only has a 1 Hz resolution, compared to the 1 kHz used in our experiments.

To improve performance in these adverse conditions by mitigating the effect of noise, we must first understand the source. The noise component of (3) is the sum of two independent parameters: quantisation noise c_i/h and latency d_i, although we only care about the variable component of the latter, *jitter* j_i. The quantisation noise is chosen uniformly at random from $[0, 1/h)$, and so is trivially modelled, but j_i can only be found experimentally.

The top graph of Figure 7 shows the smoothed probability density for round trip jitter (divided by two), which can be measured directly. If we assume that forward and return paths have independent and similar jitter, then j_i would be the same distribution. By convolving the estimated densities of the two noise sources, we can show the probability density of the sum, which matches the noise measurements of clock offset shown on the bottom of Figure 7.

The linear programming algorithm used for skew calculations is effective at removing j_i, because values are strongly skewed towards the minimum, but for c_i/h, it is possible to do better. One obvious technique is to increase h by selecting a higher resolution time source. We have used TCP timestamps in this paper, primarily with Linux 2.6, which have a nominal frequency of 1 kHz. Linux 2.4 has a 100 Hz TCP timestamp clock, so for this, ICMP timestamps may be a better option, as they increment at a nominal 1 kHz.

Unlike TCP timestamps, we found ICMP to be affected by NTP, but initial experiments show that while this is a problem for finding out absolute skew, the NTP controlled feedback loop in Linux intentionally does not react fast enough to hide the changes in skew this paper considers. Another option with Linux is to use TCP sequence numbers, which are the sum of a cryptographic result and a 1 MHz clock. Over short periods, the high h gives good results, but as the cryptographic function is re-keyed every 5 minutes, maintaining long term clock skew figures is non-trivial.

Note that to derive (2) from (1) we assumed that samples are taken at random points between ticks. This allows the floor operation ($\lfloor \ \rfloor$) to be modelled as uniformly distributed noise. Regular sampling introduces aliasing artifacts which interfere with the linear programming algorithm.

However, the points which contribute to the accuracy of the skew estimate, those near the top of the noise band, are from timestamps generated just after a tick. Here, the value of c_i is close to zero, and just before the tick, c_i is close to one and the timestamp is one less. An attacker could use the previous estimate of skew to target this point and identify which side of the transition the sample was taken. From this, he can estimate when the tick occurred and so refine the skew estimate.

This approach effectively removes the quantisation error. Rather than $1/h$ defining the noise band, it now only limits the sampling rate to h. Multiple measurements would still be needed to remove jitter, most likely by using the same linear programming algorithm as in the simple case, but perhaps also taking into consideration the round trip time. Adequate results can be achieved using naïve random sampling, but the improved technique would be particularly valuable for low resolution clocks, such as the 1 Hz Apache HTTP timestamp mentioned above.

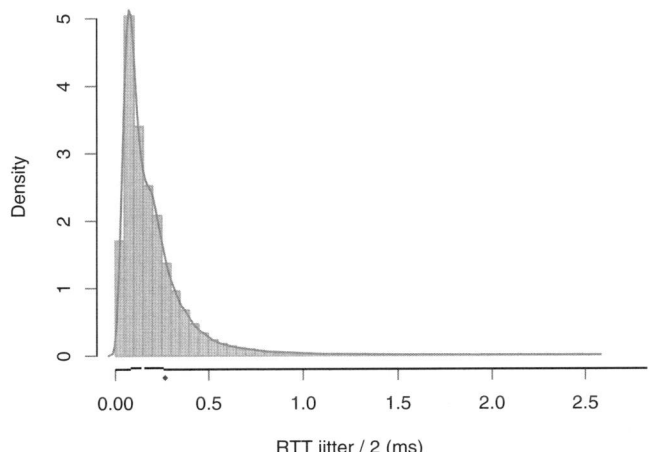

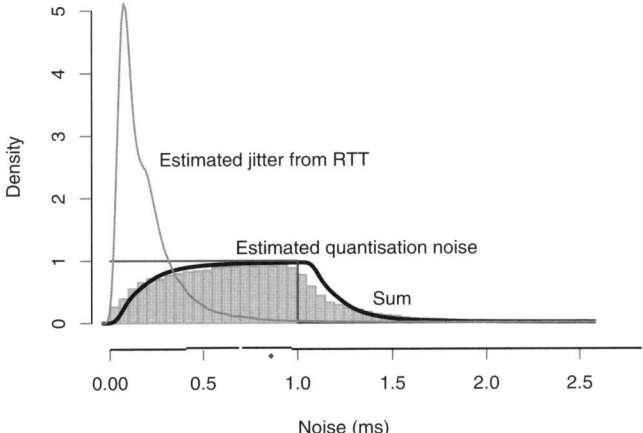

Figure 7: Top graph shows probability density of measured round trip time jitter (divided by two) with overlaid kernel density estimate (—). Bottom graph is density of measured offset noise, overlaid with the above density, uniform quantisation noise model (—) and the calculated sum of the two components (—). The breaks in the x axis indicate quartiles and the mean is shown as ⬥. Measurements were taken over a transatlantic link (14 hops).

6. CONCLUSION

We have shown that changes in clock skew, resulting from only modest changes in temperature, can be remotely detected even over tens of router hops. Our experiments show that environmental changes, as well as CPU load, can be inferred through these techniques. However, our primary contribution is to introduce an attack whereby CPU load induced through one communication channel affects clock skew measured through another. This can link a pseudonym to a real identity, even against a system that ensures perfect non-interference when considered in the abstract.

We have demonstrated how such attacks could be used against hidden services. We validated our expectations by testing them with the deployed Tor code, not simulations, although on a private network, rather than the publicly accessible one. Our results show that proposed defences against interference attacks which use quality of service guarantees, are not as effective as previously thought. We suggest that when designing such systems, considering only the abstract operating system behaviour is inadequate as their implementation on real hardware can substantially decrease security.

We proposed future directions for security research using thermal covert channels. These include allowing two computers which share the same environment, but are otherwise isolated, to communicate. Also, processes on the same computer, under an information-flow-control policy, can send information through temperature modulation, despite fixed scheduling preventing CPU load based covert channels.

Finally, we proposed how localised temperature changes might aid geolocation and suggested methods to deal with low resolution clocks.

7. ACKNOWLEDGEMENTS

Thanks are due to Richard Clayton and Roger Dingledine for assistance in running the experiments. We also thank Markus Kuhn, Nick Mathewson, Lasse Øverlier, and the anonymous reviewers for their valuable comments.

8. REFERENCES

[1] A. Acquisti, R. Dingledine, and P. F. Syverson. On the economics of anonymity. In R. N. Wright, editor, *Financial Cryptography*, volume 2742 of *LNCS*, pages 84–102. Springer-Verlag, 2003.

[2] J. Alves-Foss, C. Taylor, and P. Omanl. A multi-layered approach to security in high assurance systems. In *Proceedings of the 37th Hawaii International Conference on System Sciences*, Hawaii, January 2004. IEEE CS.

[3] Anonymizer, Inc. http://www.anonymizer.com/.

[4] A. Back, I. Goldberg, and A. Shostack. Freedom Systems 2.1 security issues and analysis. White paper, Zero Knowledge Systems, Inc., May 2001.

[5] BBC News. US blogger fired by her airline, November 2004. http://news.bbc.co.uk/1/technology/3974081.stm.

[6] D. E. Bell and L. J. LaPadula. Secure computer systems: Mathematical foundations. Technical Report 2547, Volume I, MITRE Corporation, March 1973.

[7] O. Berthold, H. Federrath, and S. Köpsell. Web MIXes: A system for anonymous and unobservable Internet access. In H. Federrath, editor, *Designing Privacy Enhancing Technologies*, volume 2009 of *LNCS*, pages 115–129. Springer-Verlag, July 2000.

[8] P. Boucher, A. Shostack, and I. Goldberg. Freedom Systems 2.0 architecture. White paper, Zero Knowledge Systems, Inc., December 2000.

[9] C-MAC MicroTechnology. HC49/4H SMX crystals datasheet, September 2004. http://www.cmac.com/mt/databook/crystals/smd/hc49_4h_smx.pdf.

[10] W. Dai. PipeNet 1.1, November 1998. http://www.eskimo.com/~weidai/pipenet.txt.

[11] D. Dean and A. Stubblefield. Using client puzzles to protect TLS. In *Proceedings of the 10th USENIX Security Symposium*, Aug. 2001.

[12] R. Dingledine and N. Mathewson. Tor protocol specification. Technical report, The Free Haven Project, October 2004. http://tor.eff.org/cvs/doc/tor-spec.txt.

[13] R. Dingledine and N. Mathewson. Tor path specification. Technical report, The Free Haven Project, April 2006. http://tor.eff.org/cvs/doc/path-spec.txt.

[14] R. Dingledine and N. Mathewson. Tor rendezvous specification. Technical report, The Free Haven Project, February 2006. http://tor.eff.org/cvs/doc/rend-spec.txt.

[15] R. Dingledine, N. Mathewson, and P. F. Syverson. Tor: The second-generation onion router. In *Proceedings of the 13th USENIX Security Symposium*, August 2004.

[16] X. Fu, Y. Zhu, B. Graham, R. Bettati, and W. Zhao. On flow marking attacks in wireless anonymous communication networks. In *Proceedings of the 25th IEEE International Conference on Distributed Computing Systems*, pages 493–503, Columbus, Ohio, USA, June 2005. IEEE CS.

[17] I. Goldberg. *A Pseudonymous Communications Infrastructure for the Internet*. PhD thesis, UC Berkeley, December 2000.

[18] H. Grundy. Personal communication.

[19] W.-M. Hu. Reducing timing channels with fuzzy time. In *1991 IEEE Symposium on Security and Privacy*, pages 8–20, Oakland, California, May 1991. IEEE CS.

[20] W.-M. Hu. Lattice scheduling and covert channels. In *1992 IEEE Symposium on Security and Privacy*, pages 52–61, Oakland, California, May 1992. IEEE CS.

[21] V. Jacobson, R. Braden, and D. Borman. TCP extensions for high performance. RFC 1323, IETF, May 1992.

[22] V. Jacobson, C. Leres, and S. McCanne. libpcap, March 2004. http://www.tcpdump.org/.

[23] P. A. Karger and J. C. Wray. Storage channels in disk arm optimization. In *1991 IEEE Symposium on Security and Privacy*, pages 52–63, Oakland, California, May 1991. IEEE CS.

[24] T. Kohno, A. Broido, and k. claffy. Remote physical device fingerprinting. In *2005 IEEE Symposium on Security and Privacy*, pages 211–225, Oakland, California, May 2005. IEEE CS.

[25] M. G. Kuhn. Personal communication.

[26] B. W. Lampson. A note on the confinement problem. *Communications of the ACM*, 16(10):613–615, 1973.

[27] M. Martinec. Temperature dependency of a quartz oscillator. http://www.ijs.si/time/#temp-dependency.

[28] D. L. Mills. Network time protocol (version 3) specification, implementation and analysis. RFC 1305, IETF, March 1992.

[29] S. B. Moon, P. Skelly, and D. Towsley. Estimation and removal of clock skew from network delay measurements. Technical Report 98–43, Department of Computer Science University of Massachusetts at Amherst, October 1998.

[30] I. S. Moskowitz, R. E. Newman, D. P. Crepeau, and A. R. Miller. Covert channels and anonymizing networks. In P. Samarati and P. F. Syverson, editors, *Workshop on Privacy in the Electronic Society*, pages 79–88, Washington, DC, USA, October 2003. ACM Press.

[31] I. S. Moskowitz, R. E. Newman, and P. F. Syverson. Quasi-anonymous channels. In M. Hamza, editor, *IASTED Communication, Network, and Information Security*, pages 126–131, New York, USA, December 2003. ACTAPress.

[32] J. A. Muir and P. C. van Oorschot. Internet geolocation and evasion. Technical Report TR-06-05, Carleton University – School of Computer Science, April 2006.

[33] S. J. Murdoch and G. Danezis. Low-cost traffic analysis of Tor. In *Proceedings of the 2005 IEEE Symposium on Security and Privacy*. IEEE CS, May 2005.

[34] S. J. Murdoch and S. Lewis. Embedding covert channels into TCP/IP. In M. Barni, J. Herrera-Joancomartí, S. Katzenbeisser, and F. Pérez-González, editors, *Information Hiding: 7th International Workshop*, volume 3727 of *LNCS*, pages 247–261, Barcelona, Catalonia (Spain), June 2005. Springer-Verlag.

[35] R. M. Needham. Denial of service. In *CCS '93: Proceedings of the 1st ACM conference on Computer and communications security*, pages 151–153, New York, NY, USA, 1993. ACM Press.

[36] R. M. Needham. Denial of service: an example. *Commun. ACM*, 37(11):42–46, 1994.

[37] R. E. Newman, V. R. Nalla, and I. S. Moskowitz. Anonymity and covert channels in simple timed mix-firewalls. In *Proceedings of Privacy Enhancing Technologies workshop (PET 2004)*, volume 3424 of *LNCS*. Springer-Verlag, May 2004.

[38] L. Øverlier and P. F. Syverson. Locating hidden servers. In *Proceedings of the 2006 IEEE Symposium on Security and Privacy*, Oakland, CA, May 2006. IEEE CS.

[39] A. Pfitzmann, B. Pfitzmann, and M. Waidner. ISDN-mixes: Untraceable communication with very small bandwidth overhead. In W. Effelsberg, H. W. Meuer, and G. Müller, editors, *GI/ITG Conference on Communication in Distributed Systems*, volume 267 of *Informatik-Fachberichte*, pages 451–463. Springer-Verlag, February 1991.

[40] J. Postel. Internet control message protocol. RFC 792, IETF, September 1981.

[41] R. Redelmeier. CPUBurn, June 2001. http://pages.sbcglobal.net/redelm/.

[42] M. G. Reed, P. F. Syverson, and D. M. Goldschlag. Anonymous connections and onion routing. *IEEE Journal on Selected Areas in Communications*, 16(4):482–494, May 1998.

[43] Reporters Without Borders. Blogger and documentary filmmaker held for the past month, March 2006. http://www.rsf.org/article.php3?id_article=16810.

[44] G. Uchenick. MILS middleware for secure distributed systems. *RTC magazine*, 15, June 2006 2006. http://www.rtcmagazine.com/home/article.php?id=100685.

Packet Vaccine: Black-box Exploit Detection and Signature Generation

XiaoFeng Wang, Zhuowei Li
Indiana University
{xw7,zholi}@indiana.edu

Jun Xu
Google Inc. & NCSU
jxu3@unity.ncsu.edu

Michael K. Reiter
Carnegie Mellon University
reiter@cmu.edu

Chongkyung Kil
North Carolina State University
ckil@ncsu.edu

Jong Youl Choi
Indiana University
jychoi@indiana.edu

ABSTRACT

In biology, a *vaccine* is a weakened strain of a virus or bacterium that is intentionally injected into the body for the purpose of stimulating antibody production. Inspired by this idea, we propose a *packet vaccine* mechanism that randomizes address-like strings in packet payloads to carry out fast exploit detection, vulnerability diagnosis and signature generation. An exploit with a randomized jump address behaves like a vaccine: it will likely cause an exception in a vulnerable program's process when attempting to hijack the control flow, and thereby expose itself. Taking that exploit as a template, our signature generator creates a set of new vaccines to probe the program, in an attempt to uncover the necessary conditions for the exploit to happen. A signature is built upon these conditions to shield the underlying vulnerability from further attacks. In this way, packet vaccine detects and filters exploits in a black-box fashion, i.e., avoiding the expense of tracking the program's execution flow. We present the design of the packet vaccine mechanism and an example of its application. We also describe our proof-of-concept implementation and the evaluation of our technique using real exploits.

Categories and Subject Descriptors: K.6.5 [Security and Protection]: Invasive software, Unauthorized access

General Terms: Security

Keywords: Black-Box Defense, Exploit Detection, Signature Generation, Worm, Vaccine Injection

1. INTRODUCTION

In biology, a *vaccine* is a living, weakened strain of a virus or bacterium that is intentionally injected into the body for the purpose of stimulating antibody production. That strain is weakened so as to prevent it from causing infection.

Similarly, a "weakened" exploit packet with important elements of its payload scrambled would quickly expose itself through the exception it causes in a vulnerable program. Forensic analysis of the exception could uncover the related program vulnerability and enable the generation of an "immunity", a signature for capturing future exploits on the same vulnerability.

The above intuition can be applied to exploit detection, vulnerability diagnosis and automatic signature generation. Design of such mechanisms has been impeded by the constraints of commodity software, for which access to source or binary recompilation is often prohibited. Existing approaches [23, 7, 5] have suggested tracking the input data as the program executes until the point at which control-flow hijacking happens. We call these approaches *gray-box analysis*, as they do not need source code (as a *white-box* approach would) but do have to monitor a program's execution flow closely (a *black-box* approach would not). Gray-box analysis is accurate and applicable to commodity software. However, it incurs significant runtime overheads, often slowing the system by an order of magnitude.

Inspired by the principle of vaccination, we develop a much faster *black-box* approach. Rather than using expensive dataflow tracking, it detects and analyzes an exploit using the outputs of a vulnerable program. Specifically, we first identify anomalous tokens in packet payloads, e.g., byte strings resembling injected jump addresses in a control-flow hijacking attack, and randomize the contents of these tokens to generate a *vaccine*. If the packets carrying these tokens indeed contain an exploit, the vaccine will likely cause an exception in the vulnerable software. When this happens, our approach will automatically generate a signature to protect the software using the forensic data gleaned from the exception and fault injection techniques [18]. We call this approach *packet vaccine*.

Compared to other techniques, packet vaccine offers some important benefits:

Fast, black-box exploit detection. Packet vaccine detects an exploit attempt by directly injecting vaccine packets into a program. Therefore, it performs as fast as a normal run of that program, and up to an order of magnitude faster than gray-box approaches. In addition, packet vaccine does not use source code or recompiled binaries and thereby works well with commodity software.

Effective signature generation. Packet vaccine generates signatures using host information, so it is immune to interference from Internet noise [28] and poisoning [25], which can mislead network-based signature generators (e.g., Early Bird [30], Polygraph [22], Nemean [41]) into generating false signatures. Moreover, the resulting signature tends to capture some key properties of a vulnerability such as the size of a vulnerable buffer, which can be used to detect a range of exploit mutations employed by polymorphic worms.

Using a confirmed exploit as a template, packet vaccine can generate a number of vaccines, i.e., variations of that exploit, to gain a better characterization of a software application's vulnerability. For instance, one type of our signatures uses a packet's field length as an attribute to identify a buffer-overflow attack; injection of vaccines with different field lengths allows us to accurately estimate the size of the underlying vulnerable buffer and thereby generate a more accurate signature (Section 2.3). Moreover, our technique can generate a signature without any information about an application or its protocol.

Some gray-box approaches perform static analysis [3, 21] over a vulnerable program's binary code and could generate signatures more accurate than our signatures. However, our black-box approach tends to be faster than those approaches and even works with obfuscated code [37, 19]. For many exploits, our black-box technique can produce signatures close to their signatures in quality, as we report in our experimental study. We argue that a rapidly-generated and reasonably accurate signature could be more useful in practice because such a signature is supposed to serve as a band-aid to a vulnerable application rather than a permanent fix [20], for use before a software manufacturer finishes developing its patch.

Low overhead and easy deployment. Packet vaccine is more lightweight and easier to deploy than many existing techniques. Exploit detection using our approach does not require installing *anything* on the host running vulnerable programs. Vulnerability diagnosis needs only a lightweight collector to gather forensic data from an exception, and even this requirement can be waived for operating systems which already offer error logging and debugging services. For example, Windows XP's event logs contain everything we need, such as corrupted pointer contents.

We present the design of the packet vaccine mechanism (Section 2) and the implementation of this technique in the paper. We evaluate it using real exploits and signatures generated by a gray-box approach (Section 3). Our study shows that packet vaccine can effectively detect exploits, and efficiently generate signatures of high quality. A problem of a vaccine is that it could modify a server's state, and interrupt its service. To apply this technique to protect an online service, we present an architecture which employs test servers to carry out exploit detection, and empirically evaluate its performance with a proof-of-concept implementation (Section 4). We also discuss the limitations of our approach (Section 5) and review related work (Section 6).

2. DESIGN

In this section, we present the design of the packet vaccine mechanism. Figure 1 illustrates the major steps of our approach: vaccine generation, exploit detection, vulnerability diagnosis and signature generation.

Vaccine generation is based upon detection of anomalous packet payloads, e.g., a byte sequence resembling a jump address, and randomization of selected contents. A vaccine generated in this way can detect an exploit attempt, since it should now trigger an exception in a vulnerable program. Vulnerability diagnosis correlates the exception with the vaccine to acquire information regarding the exploit, in particular the corrupted pointer content and its location in the exploit packet. Using this information, the signature generation engine creates variations of the original exploit to probe the vulnerable program, in an effort to identify necessary exploit conditions for generation of a signature.

2.1 Vaccine Generation

To generate a vaccine, we need to preserve the exploit semantics—i.e., its behavior that leads to an attempt to hijack control flow—while weakening it enough to prevent a control-flow hijacking from succeeding. Here, we describe a simple way to do that.

A key step in most exploits is to inject a jump address to redirect the control flow of a vulnerable program. Such an address points to somewhere in the stack or heap in a code-injection attack, or to a global library entry in an existing-code attack. Our approach is to check every 4-byte sequence (32-bit system) or 8-byte sequence (64-bit system) in a packet's application payload, and then randomize those which fall in the address range of the potential jump targets in a protected program. The vaccine generated in this way should cause an exception, segmentation fault (SEGV) or illegal instruction fault (ILL), to a vulnerable program's process if an exploit is indeed present in the original packet. A question here is how to determine the address range.

Address Range. A process's virtual memory layout is usually easy to obtain. On Linux and UNIX, the `proc` virtual filesystem maintains a file called `maps` under the directory `/proc/pid/` that offers the runtime memory layout for the process `pid`. From that file, we can obtain the base addresses for the stack (usually from 0xc0000000 downwards) and the entry for function libraries (in segment 0x40000000). The base address for heap is the end of the BSS segment, which can be determined by analyzing the binary executable using tools such as `objdump` or `readelf`. To find out the address range, we also need to know an application's stack and heap sizes. These can be estimated by monitoring stack and heap usage recorded in the `status` file of the application's process for a period of time. Using these data, we determine the address ranges as follows. Let b_s and u_s be the stack's base address and typical maximum usage, respectively. Stack addresses are estimated to range from $b_s - \alpha u_s$ to b_s, where $\alpha \geq 1$ is a ratio for keeping a safe margin. Similarly, the heap range is approximated as b_h to $b_h + \alpha u_h$, where b_h and u_h are the heap's base and typical maximum usage, respectively.[1] Address ranges can also be customized by the user. For example, one could restrict monitoring to the heap on an operating system with a nonexecutable stack.

---

[1] A process may have multiple heap regions, which can be observed from its memory maps. In this case, we can use the base addresses of these regions plus αu_h to estimate multiple heap address ranges.

Figure 1: The design of packet vaccine.

We can pinpoint the address range of the global libraries intensively used by exploits, e.g., `msvcrt.dll` or `libc.so`, and even the entry addresses of some "dangerous" functions, such as `system()` and `execve()`. These addresses can be easily acquired on Linux or UNIX using the `maps` file and the command `nm`. A Windows application's memory information can be collected using memory monitoring tools like Memview [16] or debugging tools such as CDB or NTSD [34]. The address range could also cover the global offset table (GOT), though this might not be necessary: an exploit usually changes a function pointer in the GOT to an address in the stack or heap, where the attack code lies. Again, it is at the user's discretion to decide the coverage of the address range. The larger the range becomes, the more packets must be checked and randomized.

Address ranges can also be approximated through an empirical study of known exploits, which could reveal 'hotspots' to which most exploits jump. In our research, we collected around 1000 jump addresses from known exploits and discovered that on Linux, most code-injection attacks use the jump addresses either in the range 0xbfff0000 to 0xbfffffff for the stack or 0x08040000 to 0x08ffffff for the heap. This treatment also works for existing-code attacks, as most of these exploits use a small set of libc (Linux or UNIX) or dll (Windows) functions as stepping stones.

Vaccine Generation Algorithm. Now we are ready to present the vaccine generation algorithm, which is formally described as follows.

- Gather data from the application being protected and build a *target address set* $T = [b_s - \alpha u_s, b_s] \cup [b_h, b_h + \alpha u_h] \cup S$, where S is a set containing the address ranges of objects other than the stack and heap, such as the entries for global library functions.
- Aggregate the application payloads of the packets in one session into a dataflow, carry out a proper decoding (e.g., Unicode decoding, URL decoding, etc.) if necessary and scan that dataflow to find all byte sequences $\tau \in T$.
- For every τ, replace its most significant byte with a byte randomly drawn from a scrambler set R to output a new dataflow.
- Construct vaccine packets using the new dataflow as application payloads.

In the above algorithm, the scrambler set R could be set to avoid introducing undesired symbols (such as syntax tokens) which could interrupt a protocol, and ensure a randomized byte sequence falls *outside* a process's memory map. An example of R is {A to Z, a to z, 0 to 9, '+' and '-'}.

For example, the payload of the Code Red II worm is presented in Figure 2. Our vaccine generator identifies multiple occurrences of the byte sequence 0x7801cbd3 from the payload after Unicode decoding. This sequence falls in the address range of `msvcrt.dll`, which is being monitored. Therefore, a vaccine is generated as illustrated in Figure 2,

The Orignal Packet of Code Red:
GET /default.ida?NNNNNNNNNNNNNNNNNNNNNNNNNNNNNNNN
NNNNNNNNNNNNNNNNNNNNNNNNNNNNNNNNNNNNNNNNNNNNNNNN
NNNNNNNNNNNNNNNNNNNNNNNNNNNNNNNNNNNNNNNNNNNNNNNN
NNNNNNNNNNNNNNNNNNNNNNNNNNNNNNNNNNNNNNNNNNNNNNNN
NNNNNNNNNNNNNNNNNNNNNNNNNN%u9090%u6858%ucbd3%u7801%u909
0%u6858%ucbd3%u7801%u9090%u6858%ucbd3%u7801%u9090%u9090%u8190
%u00c3%u0003%u8b00%u531b%u53ff%u0078%u0000%u00=a HTTP/1.0\r\n

A Vaccine Packet for Code Red:
GET /default.ida?NNNNNNNNNNNNNNNNNNNNNNNNNNNNNNNN
NNNNNNNNNNNNNNNNNNNNNNNNNNNNNNNNNNNNNNNNNNNNNNNN
NNNNNNNNNNNNNNNNNNNNNNNNNNNNNNNNNNNNNNNNNNNNNNNN
NNNNNNNNNNNNNNNNNNNNNNNNNNNNNNNNNNNNNNNNNNNNNNNN
NNNNNNNNNNNNNNNNNNNNNNNNNN%u9090%u6858%ucbd3%ua001%u9090
%u6858%ucbd3%u0401%u9090%u6858%ucbd3%u8c01%u9090%u9090%u8190%
u00c3%u0003%u8b00%u531b%u53ff%u0078%u0000%u00=a HTTP/1.0\r\n

Figure 2: A vaccine generated from Code Red II worm.

in which the most significant bytes of the sequence have been scrambled.

Discussion. A central question here is whether the vaccine generated above is effective in detecting an exploit if it is indeed present. Exploits tend to be fragile—a random perturbation could cause them to vanish. For example, randomization of protocol syntax tokens, such as the keyword 'GET' in the above example, renders the vaccine impossible to parse; modification of other exploit tokens can modify the exploit semantics, i.e., interfere with the exploit's attempt to hijack control flow. We address these concerns as follows.

Our approach is very unlikely to modify a protocol's syntax tokens, which usually look quite different from a suspicious jump address. We checked the most frequently used syntax tokens in HTTP, FTP and SMTP, and found none of them coincide with a typical Linux stack segment (0xbfff) and heap segment (0x08). To make the break of protocol syntax even less likely to happen, we can use a *whitelist* to guide vaccine generation. The whitelist contains all syntax tokens of a protocol, which can be either collected from the protocol's RFC or extracted from users' normal traffic. In our research, we were able to extract all important HTTP syntax tokens from one million HTTP traces. When generating vaccines, the generator checks a byte sequence τ against that whitelist. If it contains a syntax token, or it is a substring of such a token, the generator will refrain from scrambling it.

Our approach can also preserve exploit semantics in most cases. Exploits typically provide certain protocol parameters in the payload, in order to drive the target program's state to a "breakpoint" where exploit payload can be injected [3, 7]. Theoretically, it is possible for these parameters to coincide with addresses in T. However, this seems to be rare in practice, especially for protocols with an uneven distribution of byte values (e.g., text-based protocols such as HTTP). The appearance of an address-like string is uncommon for these protocols, as discovered in previous research [24, 39]. Furthermore, although binary protocols such as DNS could have an even distribution of byte values, the set T is usually small, occupying less than 0.1% of the virtual memory address space, and an exploit's parameters (except the injected code) are usually short, less than tens of bytes as we observed in our experiments. Therefore, it seems that the chance a byte sequence in T coincides with a necessary exploit parameter is small. In our research, we carefully studied 26 exploits, including attacks through binary protocols, and found none of their parameters were tampered with by our approach. In addition, those parameters are mostly dependent on the underlying vulnerability, which could leave an attacker little room to vary them.

Our randomization strategy also helps preserve exploit semantics: instead of scrambling the whole byte sequence, we only modify *one* byte—the most significant byte. We could extend the idea, for example, by generating three vaccines, each of which scrambles one of the three most significant bytes of the sequence. These vaccines can then used to probe an application in parallel. As a result, even if an exploit does use an address-like two-byte parameter (such as 0xbfff), we can still detect the exploit. Another approach involves a simple network anomaly detector (NAD) which narrows the search for address-like substrings to only part of an anomalous packet's payload. For example, a NAD monitoring the length of packets' application fields may identify an overlong CGI parameter; this allows a vaccine generator to scan only that field, avoiding randomizing other parameters even if they look like addresses. We can also whitelist well-known exploit tokens such as %n, and tokens present in normal traffic such as .ida?. All of these will then be kept intact during vaccine generation.

2.2 Exploit Detection and Vulnerability Diagnosis

Exploit attempts from vaccine packets are detected from the exceptions they cause in a vulnerable program, such as SEGV and ILL. Such exceptions happen with high probability if exploits' jump addresses have been scrambled.

The objective of vulnerability diagnosis is to reliably correlate an exception with one of the byte sequences being randomized, which identifies the location of the jump address on an exploit packet. This correlation is established by matching these byte sequences to the forensic data gathered from an exception, in which the corrupted pointer is of particular importance. On x86 systems, the corrupted pointer which causes a SEGV exception can be found in register CR2. It may also appear in EIP. Our approach logs the contents of these registers once an exception happens.

Formally, vulnerability diagnosis works as follows. Let $\tau_1, \tau_2, \ldots, \tau_n$ be the byte sequences (tokens) of a vaccine packet that have been scrambled (i.e., the high-order byte randomized) by the vaccine generator. Let p be the forensic string—the corrupted pointer collected from registers. If $p = \tau_i$ for $1 \leq i \leq n$, we correlate τ_i with the exception. This correlation can be validated using the following test: we randomize *all* bytes of τ_i to produce a new token τ and use it to generate a new vaccine; sending this vaccine to the vulnerable program, we check whether the exception happens again and the corrupted pointer also changes to τ. The validation test can be repeated to increase the confidence in the correlation.

2.3 Signature Generation

After vulnerability diagnosis, we have identified the jump address and its location in an exploit packet. The address alone, however, could be too general to be a signature, especially for binary protocols such as DNS. More information is required to form a high-quality signature. Here, we describe a *signature generation engine* that uses a known exploit as a template to generate vaccines and injects them into a vulnerable program to acquire key attributes of the underlying vulnerability. We call this technique *vaccine-injection* (VI). Our approach can generate signatures with or without application-specific information, as we elaborate below.

Application-independent Signature Generation. We can generate a signature without any knowledge about an application's protocol. Such a signature is in the form of a *token sequence*, which consists of an ordered sequence of byte strings (tokens) [22]. These tokens' locations in the exploit packet's payload could also be included as a part of the signature for a binary application protocol such as DNS. Our idea is to determine the roles played by individual bytes in an exploit by scrambling them to create vaccines and testing them in the vulnerable application, in an effort to identify the inputs necessary for the exploit to occur.

Let L be the byte length of an application-level exploit dataflow, and $B[i]$ be the ith byte on that dataflow, where $1 \leq i \leq L$. Suppose the scrambled jump address τ with a byte length l starts from the rth byte. The signature generation engine generates $L-l$ vaccines, $\{v_1, v_2, \ldots, v_{r-1}, v_{r+l}, \ldots, v_L\}$, such that v_i $(1 \leq i \leq L)$ randomizes the ith byte of the exploit payload and also keeps the token τ. Then, it injects all these vaccines into a vulnerable program. If v_i does *not* cause any exception, we record $B[i]$ (and also i for a binary protocol) as a signature token. A signature is formed using these tokens and the target address set T. A dataflow is deemed to match such a signature if it contains all these tokens and at least one byte sequence in T. We refer to this approach as *byte-based vaccine injection* (BVI).

Some servers process requests using multiple processes, such that crashing one does not affect the others. This property allows us to test many vaccines in *parallel*. Many exploits have exploit payload of a modest size, usually below 1kB. Therefore, we believe BVI can offer good performance. We also adopted a 'block-searching' technique to reduce the number of vaccines for generating a signature. We first test a vaccine which randomizes a *block* of contiguous bytes on an exploit packet. If the vaccine still causes the exception, we move on to randomize another byte block; otherwise, we test every byte inside that block to identify signature tokens. However, BVI could still be slow if the payload is large.

An attacker might duplicate an exploit token to several places. For example, the Code Red II worm (Figure 2) has multiple %u tokens, any of which is sufficient for the exploit to occur. This prevents the BVI algorithm from detecting that token, as randomization of one of its replicas does not make the exception disappear. We can solve this problem using an improved BVI algorithm described as follows. A vaccine v'_i scrambles the first i bytes on the exploit dataflow except all the signature tokens identified so far. If the vaccine does not cause any exception to the vulnerable program, the signature engine records the ith byte as a new signature token. Otherwise, our approach scrambles that byte before generating the next vaccine v'_{i+1}. This approach can capture one of the duplicated tokens. However, it is not parallelizable. Fortunately, such a duplication trick cannot be played on most tokens (e.g., .ida and GET) and thus the original BVI algorithm works in many cases.

Using Protocol Information. If an application's protocol specifications are available, in some cases we can generate a very accurate signature, close to a vulnerability-based signature. Such a signature makes use of the characteristics of buffer-overflow exploits and format-string exploits to describe a vulnerability. The algorithm for generating these signatures is also built upon the VI technique, and so we call the approach *application-based vaccine injection* (AVI).

Buffer-overflow exploits usually employ anomalously long fields [14]. Thus, a signature of the form (*application, command, field.name, max.field.size*) offers a good description of the vulnerability being exploited. Our signature generation engine first identifies the application field that includes the jump address, and then makes a quick estimate of that field's length using the number of the bytes prior to the address. This gives a coarse signature. To refine that signature, our approach iteratively alters the field size to generate new vaccines, and injects them into the vulnerable program. If a vaccine makes the exception disappear, we infer that the field is too short and then increase it. Otherwise, we shrink that field. Using a binary search, we can quickly determine the minimal length for the exploit to happen. The signature generated in this way can be pretty close to the size of a vulnerable buffer: for example, our experiment over ATP httpd (see Section 3.3) produced a signature only 23 bytes longer than the real size of the program's vulnerable buffer.

Format-string exploits usually contain the special symbol %n. In addition, the address token usually appears prior to this symbol. Therefore, a simple representation of the signature could be as follows: (*application, command, field.name*, %n). The accuracy of this signature can be verified by removing the %n from a vaccine to test the vulnerable program.

3. EVALUATION

We evaluated packet vaccine using a proof-of-concept implementation. In this section, we first describe this implementation and then present our experimental results and analysis on vaccine effectiveness and signature quality.

Our experiments were carried out on two Linux workstations: one with Redhat 7.3 operating system, Intel Pentium 4 1.5GHz CPU and 256MB memory, and the other with Redhat 6.2, Pentium 3 1GHz CPU and 256MB memory. We used the Redhat 7.3 system for all experiments except those involving the Bind TSIG exploit, which requires Redhat 6.2.

We also used several network traces to evaluate the quality of the signatures generated by our approach. Our dataset includes a trace of one million HTTP flows and one million DNS flows in and out of Indiana university.

3.1 Prototype Implementation

We implemented packet vaccine on Linux. The target address set T is extracted from an application's process proc files, including maps and status, and sent to a vaccine generation module. This module scans the dataflow of a recorded session for the byte sequences inside T, scrambles their most significant bytes, creates a socket to convert the new dataflow into vaccine packets and transports them to the application. On the systems running the application, we installed a process monitor developed using ptrace, which serves as a collector to gather the contents of important registers should an exception happen to the process being monitored. Registers important to vulnerability diagnosis are CR2 and EIP. However, CR2 can be accessed only in kernel mode. In our research, we developed a kernel patch for Linux 2.4.18 to read its content.

The signature generation engine has two components, a *prober* and a *verifier*. The prober tests an application using vaccines to identify signature tokens. It can work remotely. The verifier monitors processes for exception signals, and restarts the application if necessary. In our implementation, the verifier was embedded in the ptrace-based monitor. On starting signature generation, the prober first makes a persistent connection with the verifier, and then sends a vaccine packet to the application. If the application's process crashes, the verifier intercepts the exception signal and notifies the prober through the connection. Otherwise, the verifier waits for a period of time (longer than the maximum crash time) before signaling that no exception has occurred. Our implementation supports both the BVI and AVI algorithms and can generate token-sequence and application-level signatures. We implemented only sequential vaccine injection in our prototype system, which unfortunately introduced performance penalties. In our experiments, we found that some applications could take tens of milliseconds to crash. The delay caused by awaiting the crashes of multiple processes could be greatly reduced by a parallel approach.

3.2 Vaccine Effectiveness

A paramount question for packet vaccine is a vaccine's ability to detect an exploit. We address this question through an empirical evaluation reported in this section. We carried out experiments on real exploits of seven vulnerable applications obtained from SecurityFocus.[2] They have also been widely used for evaluating other techniques (e.g., [14, 40, 7]). In our research, we made sure that all these exploits were successful in the vulnerable applications by spawning a remote shell before testing them with our technique.

Packet vaccine successfully detected these exploits, and additionally diagnosed the related vulnerabilities to generate precise signatures. The details of exploits and detection results are listed in Table 1. While we implemented our proof-of-concept system only on Linux, we also analyzed another 19 exploits which include Windows-based exploits such as Code Red II. We found none of their semantics would be damaged by our approach. This implies that packet vaccine should also detect them.

Detecting a heap-based overflow turned out to be a little trickier. In the experiment on openssl, the value of the byte sequence we got from CR2 was larger than that of the randomized token by 12. We explain this as follows. The exploit took advantage of the free() function to overwrite a function's return address. The location of that address was faked as the content of a linking pointer in a bogus idle memory segment's heap management data structure. On the exploit's payload, the address of that segment's header was provided. That address was supposed to be lower than the linking pointer's address by 12. The exception happened when the heap management system attempted to access that linking pointer using the header's address which was randomized by our approach.

---

[2]Technical details of these exploits can be found by searching their Bugtraq ID from http://www.securityfocus.com.

Exploits	Bugtraq ID	Vulnerability Type	Exploit Packet Length	Detected	Number of Address-like Tokens
BIND tsig	2402	stack-based buffer overflow	510	Yes	3
Light httpd	6162	stack-based buffer overflow	231	Yes	13
ATP httpd	8709	stack-based buffer overflow	820	Yes	90
Samba	7294	stack-based buffer overflow	3097	Yes	26
OpenSSL v2	5363	heap-based buffer overflow	474	Yes	4
wu-ftpd	1378	format string attack	435	Yes	1
rpc.statd	1480	format string attack	1076	Yes	8

Table 1: Exploit Detection.

Exploits	Application Signature	Time(s)	Byte Sequence Signature	Time(s)
BIND tsig	—	—	4-12 (00, 01, 00, 00, 00, 00, 00, 01, 3c), 73 (3c), 134 (0c), 147 (31), 197 (0c), 210 (3e), 273 (3e), 336 (1e), 367 (10), 384 (3e), 447 (34), 500 (00), 505-507 (00, 00, fa)	4.881
Light httpd	(., 'GET', filename, 178)	0.345	0-3 (47, 45, 54, 20), 229-230 (0a, 0a)	1.360
ATP httpd	(., 'GET', filename, 703)	0.274	0-4 (47, 45, 54, 20, 2f), 818 (0a)	2.708
Samba	(., 'TRANS2_OPEN2', filename, 2000)	0.622	0-2 (00, 04, 08), 4-8 (ff, 53, 4d, 42, 32), 28-29 (01, 00), 32-33 (64, 00), 37-40 (d0, 07, 0c, 00), 55-56 (d0, 07), 58-60 (00, 0c, 00), 63-66 (01, 00, 00, 00)	7.636
OpenSSL v2	(., 'Master Key', arguments, 298)	0.358	0-11 (81, d8, 02, 01, 00, 80, 00, 00, 00, 80, 01, 4e)	5.012
wu-ftpd	(., 'SITE', 'EXEC', %n)	0.130	0-9 (53, 49, 54, 45, 20, 45, 58, 45, 43, 20), 431-432 (25, 6e)	4.228
rpc.statd	(., 'STAT', name, %n)	0.116	4-31 (00, 00, 00, 00, 00, 00, 00, 02, 00, 01, 86, b8, 00, 00, 00, 01, 00, 00, 00, 01, 00, 00, 00, 01, 00, 00, 00, 20), 36-39 (00, 00, 00, 00, 09), 60-63 (00, 00, 00, 00), 68-74(00, 00, 00, 00, 00, 00, 03), 164-165 (25, 6e)	5.780

Table 2: Signatures Generated. A token in a byte sequence signature is represented as $i-j(B_i,\ldots,B_j)$ $(i \leq j)$, where i and j are the positions of the individual bytes on the token and B_i is a byte's hexadecimal value. For example, **229-230(0a,0a)** indicates that the token **0x0a0a** lies between the **229th** and the **230th** bytes in the payload. The position information is *optional* and not useful for text-based protocols such as HTTP.

3.3 Signature Quality and Performance

A summary of results of our experiments on signature generation can be found in Table 2. To evaluate the quality of our signatures, we compared them with signatures reported in recent literature [3]. A vulnerability-based signature can prevent all possible exploits on a vulnerability [7]. Recently, Brumley et al. have proposed a gray-box approach to generate such a signature on the basis of static analysis of a vulnerable program's binary code [3]. Their technique intensively utilizes application information.

Brumley et al. describe in their paper two *monomorphic-execution-path* (MEP) signatures, one for Bind TSIG and the other for ATP httpd. MEP signatures computed from a single exploit are usually not vulnerability-based. Nevertheless, with the information extracted from the vulnerable application, they are still very accurate. Here, we analyze our signatures using these signatures.

Quality of the Token-Sequence Signature: Bind-TSIG. Bind is a very popular DNS server. It supports a secret-key transaction authentication in which messages bear transaction signatures (TSIG). There is a buffer-overflow vulnerability in Bind 8.2.x which allows an attacker to gain control of a system running Bind. This vulnerability can be exploited through both UDP and TCP queries. Our experiments were on UDP-based exploits and Bind 8.2.2. Figure 3 presents the MEP signature (the first row) and our token-sequence signature (the second row) computed using the BVI algorithm.[3]

Both signatures include bytes 6 to 10 which are zero and bytes 505 to 507 which are **0x0000fa** (a zero-length Qname followed by the field type TSIG). From Bind's source code,

---
[3]Our signature may also include the target address set T, which we believe does not make the signature too specific for a control-flow hijacking attack. This is because that set includes all possible jump targets, not a specific address.

we found that these bytes are the most important tokens for a successful exploit. Besides these tokens, our signature also contains some other bytes. Bytes 4 to 5 are the number of queries inside the packet. Byte 4 must be zero for the UDP-base exploit due to the size limit of a UDP-based packet. However, byte 5's content is unnecessarily specific because an exploit using more than one query could also succeed. On the other hand, byte 5 must be nonzero, which has not been pointed out by the MEP signature. Bytes 10-11 are the 'ARcount' field, which indicates the number of resource records in the additional records part. It must be nonzero to accommodate the TSIG field, but our signature is unnecessarily specific in fixing its value. Byte 12 appears in both signatures, but ours specifies its content. Ten bytes in the interval 73 to 447 in our signature are also unnecessarily specific. These ten bytes serve as the length octets in the 'Qname' field of a query, which are important for the successful parsing of a DNS query. However, an attacker may change the structure of the exploit packet to avoid these bytes. This problem is hard to avoid with only a single instance of the exploit and no application information at all.

The MEP signature also has some problems. It misses bytes 4 and 11, and also contains unnecessarily specific tokens, such as bytes 268 and 500. Byte 500 is also present in our signature. Both bytes signal the end of a query in a particular exploit. However, the attacker can avoid them by changing an exploit packet's structure, such as the number of questions and their sizes. For example, byte 268 has a nonzero value in the exploit used in our research.

A more accurate signature could be generated by our technique given more than one exploit instance. In our research, we compared another exploit of the Bind-TSIG vulnerability with the above one. These two exploit packets share 19 bytes at the same locations of their application payloads. Based on these 19 bytes, the BVI algorithm generated another signature (the third row in Figure 3) with 10 bytes.

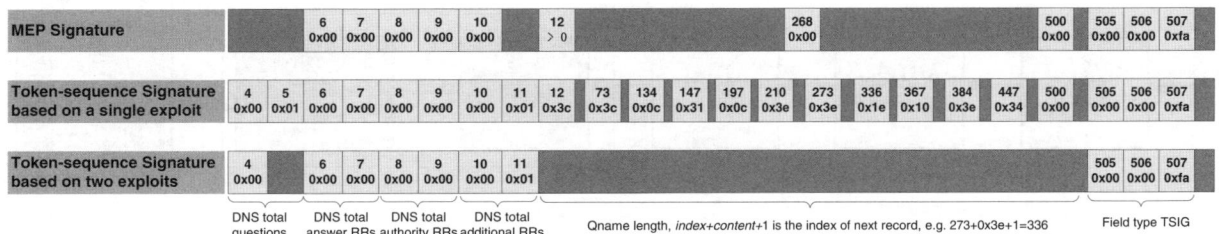

Figure 3: Signatures for Bind TSIG.

Only one of them, byte 11, is unnecessarily specific. This signature is comparable to the MEP signature in quality and capable of fending off many attacks on the vulnerability.

Using the block-searching technique, a sequential BVI algorithm took 4.881 seconds to generate the first token-sequence signature for Bind. We believe an optimized implementation and introduction of parallelization could improve that performance. The second signature was generated within 0.2 seconds.

Quality of the Application-level Signature: ATP-httpd. We also compared our application-level signature for ATP-httpd with the MEP signature in [3]. ATP-httpd contains a vulnerable buffer which will be overrun by a requested filename longer than 680 bytes. Built upon the analysis of the program's binary code, the MEP signature contains richer information than ours. It points out the HTTP command which leads to the vulnerability could be either 'GET' or 'HEAD', while our signature only identifies 'GET' from a single exploit instance. However, the MEP signature contains two specific tokens, '//' and '/', which actually are parts of the shell code. In addition, the total field length required by their signature is 812 bytes, which is not necessary for an exploit. Our signature offers a better estimate of the vulnerable buffer size. The AVI algorithm determined the maximal length of the field 'filename' as 703, 23 bytes longer than the vulnerable buffer. These 23 bytes turned out to be the local variables between the buffer and the pointer overwritten by the exploit. Our approach took 0.274 seconds to generate the signature. By comparison, the algorithm in [3] spent more than a second to complete a single step of signature generation which converts the results from static analysis into a signature.

In summary, it comes as little surprise that the MEP signatures are more accurate than our signatures in general. However, their quality advantages diminish somewhat with the availability of multiple exploit instances and application information. Furthermore, our black-box approach can perform significantly faster in some cases, and even works with obfuscated binaries which static analysis might not manage well.

Exploits	False + (Application Signature)	False + (Byte-Sequence Signature)
BIND tsig	—	w/ T, 0%, w/o T, 0%
Light httpd	0.602%	w/ T, 0%, w/o T, 0.0006%
ATP httpd	0.0077%	w/ T, 0%, w/o T, 0.142%

Table 3: False Positives. T refers to the target address set of the vulnerable application.

False Positives. We tested our signatures for Bind-TSIG, ATP-httpd and light-httpd using the aforementioned DNS and HTTP traces (Table 3). Surprisingly, most false positives come from application-level signatures, which are supposed to be very accurate! Further analysis offers the explanation: these signatures are application-dependent, only working for specific httpd servers, and supposed to be installed on the firewalls connecting to these servers. However, the HTTP traces were collected from edge routers, containing the traffic of other HTTP software that could accommodate a longer field.

4. EXAMPLE APPLICATION: PROTECTING INTERNET SERVERS

In the section, we present an architecture which applies packet vaccine to protect Internet servers from remote control-flow hijacking attacks. This architecture serves as an example to demonstrate the potential application of our technique. We also prototyped the architecture under Linux and empirically evaluated its performance.

4.1 Architecture

Figure 4 illustrates the architecture we propose. A service request is first intercepted and cached by a service proxy and parsed by a parser. The parser is optional here and only useful when we use application-level signatures. Then, the request is screened by a filter which identifies and drops known exploits using exploit signatures. Behind the filter, a *detector* examines the request and labels it as either *normal* or *suspicious*. The detector could simply be part of our packet vaccine mechanism, which classifies packets with regard to the appearance of address-like tokens in their payloads. Alternatively, we could employ other simple detection techniques, such as one which identifies packets with overlong fields. After classification, a *normal* request is forwarded to a server farm directly, while a *suspicious* request triggers the packet vaccine mechanism which acts as discussed in Section 2. If that request is determined to contain an exploit, packet vaccine generates a new signature and adds it to the filter. Otherwise, the proxy forwards the original request to the server farm.

The packet vaccine mechanism makes use of a small set of *test servers* in the server farm to test vaccine packets. A test server has a collector on it, which serves to glean information from registers' contents should an exception happen. In the case that the service being provided is stateful, the test server also needs a checkpoint/rollback (CR) mechanism to recover the state before each test. Such a rollback mechanism could be extremely lightweight (e.g., [8, 31]). Signature generation can also happen on a test server.

4.2 Performance Study

To implement a prototype system for HTTP service, we developed a service proxy and a filter (including an HTTP parser), and combined them with our implementation of

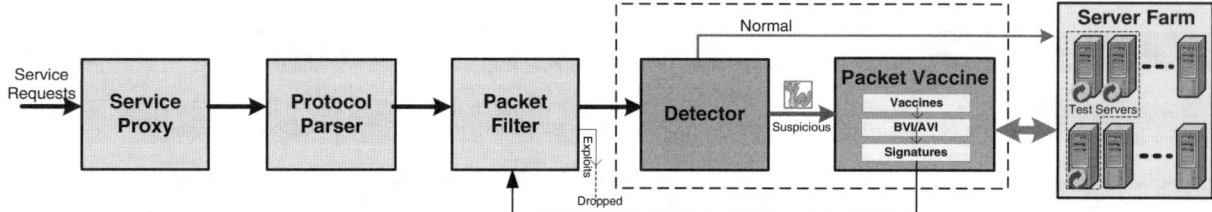

Figure 4: An architecture to protect Internet servers using packet vaccine.

packet vaccine (Section 3.1) which contains a detector. Since HTTP is a stateless service, we did not implement the process-level CR in this prototype.

Over the prototype system, we carried out a performance test. Two hosts were used in our experiment, one for both the proxy and the test server and the other for the web server. Both were equipped with 2.53GHz Intel Pentium 4 Processor and 1 GB RAM, and running Redhat Enterprise 2.6.9-22.0.1.EL. They were interconnected through a 100MB switch. We utilized an Apache 2.0.55 to provide web service. In our experiment, we evaluated the performance of our implementation from the following perspectives: (1) *Server overheads*, where we compared the workload capacity of our implementation with that of an unprotected Apache server; (2) *Client-side delay*, where we studied the average delay a client experiences under different test rates.

Server overheads. We tested the workload capacity using ApacheBench (ab) 2.0.41-dev, which comes bundled with the Apache source distribution. ApacheBench is a tool for benchmarking the Apache web server. In our experiment, we measured the workload capability in terms of requests processed per second (requests/second) under the following five server configurations: (**0**) 'Apache only', (**D0**) 'Apache and the proxy on different hosts', (**S0**) 'Apache and the proxy on the same host', (**D1**) 'Apache on one host, and the proxy and packet vaccine on another', (**S1**) 'Apache, proxy and packet-vaccine all on the same host'.

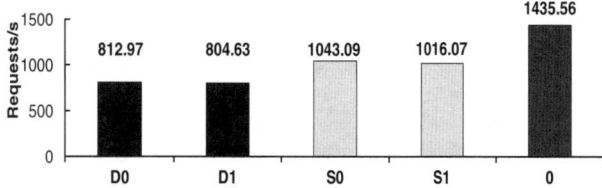

Figure 5: The workload capacities in five different server settings.

Figure 5 illustrates the experiment results. At a first glance, it seems that our implementation brought down the Apache's performance by about 44% in the setting (**D1**) and about 29% in the setting (**S1**), which is quite unpleasant. A close look at the results, however, reveals that the major performance penalty came from the service proxy. The homegrown proxy used in our proof-of-concept implementation could not keep up with the high-performance of Apache and therefore dragged down the performance of the whole system. Simply adding the proxy into the system introduced about 43% performance penalty in (**D0**) and 27% in (**S0**). On the other hand, the packet vaccine components worked pretty fast. They only affected the performance by 1% to 2%. Therefore, we tend to believe that a high-performance HTTP proxy could greatly improve the workload capability.

Client-side delay. Once the detector identifies a suspicious request, a round of exploit detection will be triggered to test that request. This introduces delay to a legitimate client if the request turns out to be innocent. Here, we call the ratio of service requests being tested (i.e., the fraction deemed suspicious) the *test rate*. If the test rate increases, the average delay experienced by a legitimate client will also increase. In our experiment, we studied the change of the client-side delay against different test rates. We carried out both a local experiment within IU's campus network and a cross-campus experiment between IU and NCSU. The experimental results are presented in Figure 6.

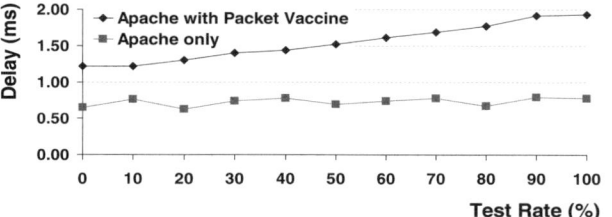

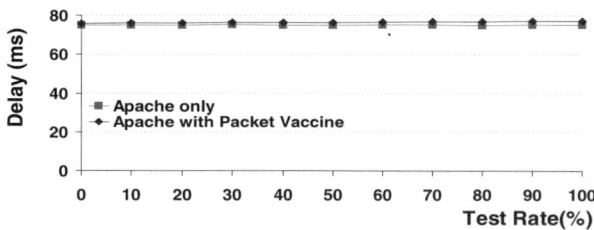

Figure 6: The average delay experienced by a local or remote client.

As we expected, the average delay for a local client increased almost linearly with the test rate. However, this result could be misleading, as the local client experienced much smaller round trip delay (RTD) than an average Internet user: the RTD in a campus we measured is around $300\mu s$, while the average RTD on the Internet is much larger. Therefore, an Internet client's perception of the presence of packet vaccine could be completely overshadowed by the RTD. This was confirmed in the cross-campus experiment: as presented in Figure 6, the $75ms$ RTD between the two campuses dominated the client-side delay, making the $1ms$ overhead of our protection mechanism negligible.

In summary, packet vaccine does introduce performance penalties to the server, but we believe this penalty is acceptable if weighed against the security enhancements it offers.

On the other hand, the client-side overhead is almost negligible, being dwarfed by the RTD an average Internet client experiences.

5. LIMITATIONS

Packet vaccine may have false negatives in exploit detection. For example, there is a possibility that the randomizations performed by our approach destroy the exploit's semantics. This seems more likely to occur for applications using binary protocols, though so far we have not found an example "in the wild". In general, our approach is more reliable in protecting applications using text-based protocols. Several ways to reduce the likelihood of this problem were discussed in Section 2.3. A simple approach is to generate multiple vaccines, each randomizing one byte of an address-like token. In this way, if the exploit semantics survives any of these randomizations, our approach will detect the exploit.

Our approach cannot work directly on packets with encrypted payload or checksums. In this case, we need an application-level proxy to decode these packets and construct new packets for vaccine generation.

Both types of signatures we use in our research are limited in their capabilities to represent necessary exploit conditions. For example, null-httpd contains a vulnerability that allows one to specify a smaller buffer while supplying a longer payload. An ideal signature is to check whether the real payload size matches the specified size. However, none of our signatures can describe this condition. We leave it to future work to examine how to use our black-box techniques to acquire information for more expressive signatures [38, 3].

6. RELATED WORK

Network anomaly detection (NAD) has been widely used to detect exploit attempts from network traffic [41, 39, 35, 12]. A typical network signature generator extracts common substrings from attack dataflow as an exploit signature. Examples include Earlybird [30], Honeycomb [11], Autograph [10], SweetBait [26], Polygraph [22], Hamsa [13] and PADS [32]. Signature generation solely relying on network information can be misled into generating an incorrect signature by carefully crafted attack packets, which helps a worm to evade detection [25] or causes legitimate packets to be dropped.

Host-based approaches make use of host information to detect anomalies and generate signatures. As exploits actually happen on a host, these approaches can be more accurate than network-based approaches. TaintCheck [23], VSEF [21], Minos [6], Vigilante [5] and DACODA [7] track dataflow through a process from the receipt of a network packet (or modification thereof [23]) to the point where an anomaly happens, e.g., jumping to an address offered by the input data. These approaches can slow the running process significantly, however, by an order of magnitude or more. In contrast, our vaccine mechanism tracks suspicious dataflow in a black-box fashion, which is significantly faster than these gray-box approaches and still preserves much of their accuracy in cases we have explored. Some host-based approaches apply static analysis [3] to identify a program's vulnerabilities. Such an approach no longer works over well-obfuscated binaries.

Liang et al. and Xu et al. proposed two approaches [40, 14] that use memory address-space randomization (ASR) to foil exploit attempts, and then automatically generate signatures through forensic analysis of the related exceptions. In particular, COVERS [14] was the first to propose a novel construction of application-level signature which uses field length to characterize a buffer overflow vulnerability. Although we also use this signature, our AVI technique augments their approach by making an accurate estimate of the field length. Our technique also offers a more reliable way to correlate exceptions with the exploit packets.

In an attempt to find a balance between performance and accuracy, several hybrid approaches combining network-based and host-based techniques have been developed [1, 15, 29]. However, many of them are based on instrumenting a vulnerable program's source code, and are therefore less suitable for protecting commodity software. HACQIT [27] invokes a test process after an exploit crashes a protected program, and replays suspicious packets to a sandbox running the same program to monitor whether the same exception happens again. However, this approach does not offer a reliable means to establish a correlation between the exception and the exploit inputs.

The vaccine technique can trace its root to software robustness testing, especially software-implemented fault injection (SWIFI) [18]. SWIFI is a software testing and evaluation method which involves inserting random faults into a system to determine its response to these faults. Some important SWIFI systems include the Crashme program [4], the Fuzz project [17], the FIAT system [2], the FERRARI system [9], the FTAPE system [36], and Ballista [33]. Our proposal differs fundamentally from these approaches in two respects. First, we rely on anomalous packets to guide vaccine generation, making our vaccines more likely to reveal a program's vulnerabilities than the random faults used in a typical SWIFI approach. Second, we aim at exploit prevention and will generate exploit signatures to shield the software vulnerabilities discovered.

7. CONCLUSIONS

In this paper, we presented packet vaccine, a fast, black-box technique for exploit detection, vulnerability diagnosis and signature generation. We described its design and examples for its application. We also implemented a proof-of-concept prototype, and evaluated our technique using it. Our experimental results demonstrate the effectiveness of our technique, which successfully captures real exploits and generates effective signatures, and its efficiency, which improves over gray-box approaches in many cases.

8. REFERENCES

[1] K. G. Anagnostakis, S. Siridoglou, P. Akritidis, K. Xinidis, E. Markatos, and A. Keromytis. Detecting targeted attacks using shadow honeypots. In *Proceedings of USENIX Security Symposium 2005*, August 2005.

[2] J. H. Barton, E. W. Czeck, Z. Z. Segall, and D. P. Siewiorek. Fault injection experiments using FIAT. *IEEE Trans. Comput.*, 39(4):575–582, 1990.

[3] David Brumley, James Newsome, Dawn Song, Hao Wang, and Somesh Jha. Towards automatic generation of vulnerability-based signatures. In *Proceedings of the 2006 IEEE Symposium on Security and Privacy*, 2006.

[4] George J. Carrette. CRASHME: Random input testing. http://people.delphiforums.com/gjc/crashme.html, as of March, 2006.

[5] Manuel Costa, Jon Crowcroft, Miguel Castro, Antony I. T. Rowstron, Lidong Zhou, Lintao Zhang, and Paul T. Barham. Vigilante: end-to-end containment of internet worms. In *SOSP*, pages 133–147, 2005.

[6] Jedidiah R. Crandall and Frederic T. Chong. Minos: Control data attack prevention orthogonal to memory model. In *MICRO*, pages 221–232, 2004.

[7] Jedidiah R. Crandall, Zhendong Su, and S. Felix Wu. On deriving unknown vulnerabilities from zero-day polymorphic and metamorphic worm exploits. In *CCS '05: Proceedings of the 12th ACM conference on Computer and communications security*, pages 235–248, 2005.

[8] George W. Dunlap, Samuel T. King, Sukru Cinar, Murtaza A. Basrai, and Peter M. Chen. Revirt: Enabling intrusion analysis through virtual-machine logging and replay. In *Proceedings of OSDI*, 2002.

[9] Ghani A. Kanawati, Nasser A. Kanawati, and Jacob A. Abraham. FERRARI: A flexible software-based fault and error injection system. *IEEE Trans. Comput.*, 44(2):248–260, 1995.

[10] Hyang-Ah Kim and Brad Karp. Autograph: Toward automated, distributed worm signature detection. In *Proceedings of 13th USENIX Security Symposium*, pages 271–286, San Diego, CA, USA, August 2004.

[11] Christian Kreibich and Jon Crowcroft. Honeycomb: creating intrusion detection signatures using honeypots. *SIGCOMM Computer Communication Review*, 34(1):51–56, 2004.

[12] C. Kruegel, E. Kirda, D. Mutz, W. Robertson, and G. Vigna. Polymorphic worm detection using structural information of executables. In *Proceedings of RAID'05*, pages 207–226, September 2005.

[13] Zhichun Li, Manan Sanghi, Yan Chen, Ming-Yang Kao, and Brian Chavez. Hamsa: Fast signature generation for zero-day polymorphicworms with provable attack resilience. In *SP '06: Proceedings of the 2006 IEEE Symposium on Security and Privacy (S&P'06)*, pages 32–47, 2006.

[14] Zhenkai Liang and R. Sekar. Fast and automated generation of attack signatures: a basis for building self-protecting servers. In *CCS '05: Proceedings of the 12th ACM conference on Computer and communications security*, pages 213–222, 2005.

[15] Michael E. Locasto, Ke Wang, Angelos D. Keromytis, and Salvatore J. Stolfo. Flips: Hybrid adaptive intrusion prevention. In *Proceedings of the 8th International Symposium on Recent Advances in Intrusion Detection (RAID)*, September 2005.

[16] MemView. http://www2.biglobe.ne.jp/~sota/memview-e.html, as of May, 2006.

[17] Barton Miller, David Koski, Cjin Pheow Lee, Vivekananda Maganty, Ravi Murthy, Ajitkumar Natarajan, and Jeff Steidl. Fuzz revisited: A re-examination of the reliability of UNIX utilities and services. Technical report, 1995.

[18] J.D. Musa, G. Fuoco, N. Irving, B. Juhlin, and D. Kropfl. *Handbook of Software Reliability Engineering*, chapter The Operational Profile, pages 167–216. McGraw-Hill, 1996.

[19] Gleb Naumovich and Nasir D. Memon. Preventing piracy, reverse engineering, and tampering. *IEEE Computer*, 36(7):64–71, 2003.

[20] Associate Press News. Microsoft warns against outside fixes. http://biz.yahoo.com/ap/060331/microsoft_s_security_snags.html?.v=4, March 31, 2006.

[21] James Newsome, David Brumley, and Dawn Song. Vulnerability-specific execution filtering for exploit prevention on commodity software. In *Proceedings of the 13th Annual Network and Distributed Systems Security Symposium*, 2005.

[22] James Newsome, Brad Karp, and Dawn Song. Polygraph: Automatically generating signatures for polymorphic worms. In *Proceedings of IEEE Symposium on Security and Privacy*, pages 226– 241, Oakland, CA, USA, May 2005.

[23] James Newsome and Dawn Song. Dynamic taint analysis for automatic detection, analysis, and signature generation of exploits on commodity software. In *Proceedings of the 12th Annual Network and Distributed System Security Symposium*, San Diego, CA, USA, Feburary 2005.

[24] A. Pasupulati, J. Coit, K. Levitt, S.F. Wu, S.H. Li, R.C. Kuo, and K.P. Fan. Buttercup: On network-based detection of polymorphic buffer overflow vulnerabilities. In *Proceedings of the9th IEEE/IFIP Network Operation and Management Symposium (NOMS'2004)*, May 2004.

[25] Roberto Perdisci, David Dagon, Wenke Lee, Prahlad Fogla, and Monirul Sharif. Misleading worm signature generators using deliberate noise injection. In *IEEE Symposium on Security and Privacy*, page to appear, May 2006.

[26] Georgios Portokalidis and Herbert Bos. SweetBait: Zero-hour worm detection and containment using honeypots. Technical Report IR-CS-015, Vrije Universiteit Amsterdam, May 2005.

[27] James C. Reynolds, James Just, Larry Clough, and Ryan Maglich. On-line intrusion detection and attack prevention using diversity, generate-and-test, and generalization. In *HICSS '03: Proceedings of the 36th Annual Hawaii International Conference on System Sciences (HICSS'03) - Track 9*, page 335.2, 2003.

[28] David W. Richardson, Steven D. Gribble, and Edward D. Lazowska. The limits of global scanning worm detectors in the presence of background noise. In *WORM '05: Proceedings of the 2005 ACM workshop on Rapid malcode*, pages 60–70. ACM Press, 2005.

[29] Stelios Sidiroglou, Michael E. Locasto, Stephen W. Boyd, and Angelos D. Keromytis. Building a reactive immune system for software services. In *USENIX Annual Technical Conference*, pages 149 – 161, April, 2005.

[30] Sumeet Singh, Cristian Estan, George Varghese, and Stefan Savage. Automated worm fingerprinting. In *Procedddings of OSDI*, pages 45–60, 2004.

[31] Sudarshan M. Srinivasan, Srikanth Kandula, Christopher R. Andrews, and Yuanyuan Zhou. Flashback: A lightweight extension for rollback and deterministic replay for software debugging. In *USENIX Annual Technical Conference, General Track*, pages 29–44, 2004.

[32] Yong Tang and Shigang Chen. Defending against internet worms: A signature-based approach. In *Proceedings of IEEE INFOCOM05*, Miami, Florida, USA, May 2005.

[33] The Ballista@ Project: COTS Software Robustness Testing. http://www.ece.cmu.edu/~koopman/ballista/, as of January, 2006.

[34] Microsoft Debugging Tools. http://www.microsoft.com/whdc/devtools/debugging/default.mspx, as of May, 2006.

[35] Thomas Toth and Christopher Krügel. Accurate buffer overflow detection via abstract payload execution. In *Proceedings of RAID*, pages 274–291, 2002.

[36] Timothy K. Tsai and Ravishankar K. Iyer. Measuring fault tolerance with the ftape fault injection tool. In *MMB '95: Proceedings of the 8th International Conference on Modelling Techniques and Tools for Computer Performance Evaluation*, pages 26–40. Springer-Verlag, 1995.

[37] Paul C. van Oorschot. Revisiting software protection. In *Proceedings of ISC*, pages 1–13, 2003.

[38] Helen J. Wang, Chuanxiong Guo, Daniel R. Simon, and Alf Zugenmaier. Shield: vulnerability-driven network filters for preventing known vulnerability exploits. In *SIGCOMM*, pages 193–204, 2004.

[39] Ke Wang and Salvatore J. Stolfo. Anomalous payload-based network intrusion detection. In *Proceedings of RAID Symposium 2004*, pages 203–222, 2004.

[40] Jun Xu, Peng Ning, Chongkyung Kil, Yan Zhai, and Chris Bookholt. Automatic diagnosis and response to memory corruption vulnerabilities. In *CCS '05: Proceedings of the 12th ACM conference on Computer and communications security*, pages 223–234, 2005.

[41] Vinod Yegneswaran, Jonathon T. Giffin, Paul Barford, and Somesh Jha. An architecture for generating semantics-aware signatures. In *Proceedings of USENIX Security Symposium 2005*, August 2005.

Protomatching Network Traffic for High Throughput Network Intrusion Detection

Shai Rubin
Computer Sciences Dept.
University of Wisconsin, Madison
shai@cs.wisc.edu

Somesh Jha
Computer Sciences Dept.
University of Wisconsin, Madison
jha@cs.wisc.edu

Barton P. Miller
Computer Sciences Dept.
University of Wisconsin, Madison
bart@cs.wisc.edu

ABSTRACT

Before performing pattern matching, a typical misuse-NIDS performs protocol analysis: it parses network traffic according to the attack protocol and normalizes the traffic into the form used by its signatures. For example, consider a NIDS that attempts to identify an HTTP-based attack. The NIDS must extract the URL from the raw traffic, convert HEX encoded characters into their equivalent ASCII form if necessary, and only then perform matching on the normalized URL. Protocol analysis is time consuming, especially in a NIDS that analyzes and normalizes all traffic just to discover that the majority of the traffic does not match any of its signatures.

We develop a technique called *protomatching* that combines protocol analysis, normalization, and pattern matching into a single phase. The goal of the protomatching signatures is to exclude non-attack traffic quickly before the NIDS performs any further time-consuming analysis. Protomatching is based on a novel signature with two properties. First, the signature ensures that the attack pattern appears in the context that enables successful attack. This saves the need for protocol analysis. Second, the signature matches both encoded and normalized forms of an attack and this saves the need for normalization.

We empirically show that a Snort implementation that uses protomatching is up to 49% faster than an unmodified Snort.

Categories and Subject Descriptors

K.6.5 [**Security and Protection**]: Metrics—*invasive software, unauthorized access*.

General Terms

Security, Management.

Keywords

Intrusion detection, signatures, protocol analysis.

1. INTRODUCTION

A misuse network intrusion detection system (NIDS) recognizes an attack via a signature, usually a string or a regular expression that matches a characteristic pattern of the attack. For example, to identify an HTTP-based attack called *DNS-tools* (CVE-2002-0613), which enables attackers to modify DNS entries, Snort [32] checks whether the network traffic contains the string "`dnstools.php`". However, not all traffic that contains this string is a real DNS-tools attack. To correctly identify the DNS-tools attack, Snort must ensure that the string "`dnstools.php`" is part of a URL in an HTTP request.

To achieve such level of accuracy, Snort performs *protocol analysis* before pattern matching. It parses HTTP traffic, finds the URL in any HTTP request, and looks for "`dnstools.php`" only in those places. To the best of our knowledge, commercial intrusion detection systems (e.g., [4, 5, 14, 44]) also perform protocol analysis for many commonly used protocols, such as FTP, HTTP, and SMTP.

During protocol analysis a NIDS also performs *traffic normalization*. Some protocols, such as FTP or HTTP, enable *multiple encodings* for the same payload. Most notoriously, HTTP allows URLs to be encoded using lower- or uppercase characters as well as hexadecimal ASCII values [9, 10]. A NIDS typically translates, or *normalizes*, the raw traffic into the form used by its signatures. For example, Snort translates hexadecimal encodings in a URL into lower-case characters. A NIDS that does not normalize network traffic is vulnerable to evasion attacks [12, 20, 23, 29, 46].

All common NIDS seem to use the methodology outlined above, which we call *analyze-normalize-match* (ANM). First, a NIDS encodes its signatures in a *normalized* form. Then, during runtime, the NIDS parses the traffic according to the protocol the attack uses and normalizes the traffic, if necessary. Last, the NIDS matches the normalized traffic against its normalized signatures.

Unfortunately, the ANM methodology incurs performance penalty. An ANM-based NIDS inspects the traffic twice: once during protocol analysis and once during matching. An ANM-based NIDS wastes time analyzing and normalizing benign traffic only to discover later that this traffic does not match any of its signatures. Indeed, our experiments reveal that Snort spends about 30% of its time analyzing and normalizing benign HTTP requests. Dreger *et al.* [8] noticed similar results for Bro [25] and recognized that HTTP analysis is a serious bottleneck for sites with a high volume of HTTP traffic.

We propose to replace ANM with a more efficient tech-

System	Size (MB)	% speedup (compared to default Snort configuration)	% speedup (compared to custom configuration)
Original Snort (ANM)	7	–	–
Snort+deterministic protomatcher (P–ANM)	30	45	25
Snort+hierarchical protomatcher (P–ANM)	13	49	27

Table 1: Summary of results on real HTTP traffic with all Snort's HTTP signatures (999 signatures).

nique: *protomatch—analyze-normalize-match* (P–ANM). As we explain below, the *protomatching* phase inspects the network traffic exactly once, but still performs protocol analysis, normalization, and pattern matching. The goal of the protomatching phase is to quickly identify traffic, either normalized or encoded, that can **never** match any signature in the NIDS database. In comparison, traffic that **might** match a signature, is further analyzed using a more computationally expensive technique, such as the ANM. A NIDS that is based on the P–ANM methodology is efficient because it inspects 99% of the traffic only once, instead of inspecting 100% of the traffic twice. Our experiments show that a P–ANM-based Snort is up to 49% faster than Snort that uses the ANM technique.

The first contribution of this paper is the concept of a *protomatching signature*, a regular expression with two properties. First, the expression ensures that the characteristics pattern of an attack appears in the context that is necessary for the attack to succeed. For example, based on the HTTP specification, we build a protomatching signature for the DNS-tools attack that identifies the string "dnstools.php" only if this string appears in a URL. Hence, a protomatching signature saves the need for protocol analysis.

Second, a protomatching signature matches both normalized and encoded versions of an attack. To do so, we represent alternate encodings as *substitutions* [13]: operations that map a character to a regular expression describing all possible encodings of that character. Then, we build a signature that matches all possible representations of an attack. For example, we change the signature to match encoded variants, such as "GET dnstools.%70h%70", in which the character 'p' is replaced with its hex encoding "%70".

Since a protomatching signature is a regular expression, we automatically compile it into a *protomatcher*: a deterministic finite state machine that matches every possible encoding of "dnstools.php" and ensures that this string appears only in a URL of a valid HTTP method.

Ideally, we would like to construct a *full-coverage protomatcher*, a protomatcher that contains a protomatching signature for every signature in the NIDS database. While we have built a full-coverage protomatcher for more than 50% of the 1022 HTTP signatures in Snort's database, we noticed that a full-coverage protomatcher cannot fit in 2GB of memory if it contains signatures that require complicated regular expressions. Therefore, we developed the *superset protomatcher* that requires a smaller memory footprint.

A *superset protomatcher* recognizes a superset of the traffic matched by a full-coverage protomatcher. For example, instead of recognizing the string "dnstools.php", the superset protomatcher recognizes the string "dnstools". This means three things. First, a superset protomatcher consumes less memory. Second, since it recognizes a superset of the traffic, it never misses traffic matched by the full-coverage protomatcher. However, it might produce *false matches*: traffic that matches the superset protomatcher but does not match any of the NIDS signatures. Third, traffic that does not match the superset protomatcher also does not match any signature in the NIDS database, so it can be immediately accepted as benign.

The superset protomatcher is the second contribution of this paper and the core of the P–ANM technique. Traffic, either normalized or encoded, that *does not* match the superset protomatcher is immediately ignored because it is benign. Traffic that *does* match the superset protomatcher is forwarded to the second phase: the traditional ANM technique.

This two-phase design is time-efficient because the superset protomatcher accepts 99% of the benign traffic and rarely utilizes the slower ANM-based path. This design fits into memory because the combination of superset protomatcher and the ANM-based matching consumes an order of magnitude less memory than a full-coverage protomatcher.

We discuss two possible protomatcher implementations. First, we implemented a protomatcher as a deterministic finite state machine. Second, to reduce the protomatcher memory footprint, we implement it as a *hierarchical protomatcher* that is based on two automata: a matcher and a normalizer. Unlike the ANM method that first fully analyzes and normalizes the traffic and then performs matching, the hierarchical protomatcher performs protocol analysis and matching until it encounters an encoded representation. Then, it passes control to the normalizer, which translates the encoding into a normalized form and returns it to the matcher. The hierarchical protomatcher consumes only 13MB and improves Snort's performance by 49%, with the default Snort configuration, and by 27% with our customized configuration (Table 1).

In summary, this paper makes the following contribution:

1. **The idea of a protomatching signature**. We developed a signature that matches both encoded and normalized variants of the attack and guarantees that the attack pattern appears in the right context.

2. **The idea of the superset protomatcher**. We show that when it is infeasible to build a protomatching-signature for every signature in the NIDS database, the idea of protomatching signatures can still be very beneficial. We use shorter protomatching signatures to build a superset protomatcher that filters 99% of the benign traffic and therefore omits the need for expensive protocol analysis.

3. **Feasibility study of the P–ANM methodology**. We implemented the P–ANM methodology in Snort. We show that a deterministic protomatcher improved Snort's performance by 45% when we use Snort's default configuration. Even after customizing Snort's configuration to maximize performance (Section 5), our implementation performed 25% faster. The hierarchical protomatcher consumed only 13MB and im-

proved Snort's performance by 49%, with the default configuration, and by 27% with our customized configuration (Table 1).

Since protomatcher-based Snort consumes more memory than the original Snort, it might suffer more cache misses. We investigated a *cache-poisoning* attack in which an attacker attempts to degrade the protomatcher's performance by forcing it to generate many cache misses. While this attack degrades the protomatcher by 2%, it degrades the performance of the original Snort by 4.5% (Section 5.3).

2. RELATED WORK

We review related work in the areas of protocol analysis and traffic normalization, efficient pattern matching, and intrusion detection for high-speed links.

Protocol analysis and traffic normalization. Typically, modern NIDS are based on the ANM methodology. This can be easily seen in Snort [32] and Bro [25] because their source code is available. Based on our discussions with NIDS developers, user manuals we read, and our practical experience using proprietary NIDS, we believe that the ANM methodology is also common in other commercial systems [4, 5, 44]. Since the ANM boosts task separation, it is attractive from the software engineering viewpoint. In this work, however, we challenge the ANM methodology on the point of efficiency: we combine analysis, normalization, and matching into a single protomatching phase.

Ptacek and Newsham [29] were the first to recognize that a NIDS that does not perform normalization is susceptible to evasion. To defend against evasion, Handley et al. [12] investigated normalization techniques for the TCP, IP, UDP, and ICMP protocols. They proposed a normalizer that reverses transformations before the NIDS analyzes the traffic. In this work, we focus on the normalization of higher level protocols that support alternate encodings, such as HTTP, TELNET, SMTP, or FTP.

The problem of alternate encodings is particularly painful for HTTP traffic. HTTP allows three encodings of URLs: alphabetic characters, HEX encoding, and UTF-8 [10]. While the Apache web server supports only these encodings, the IIS server supports five more encodings that are Microsoft-specific [31]. This assortment of HTTP encodings complicates HTTP normalization, a situation that has been extensively exploited to evade detection systems [9, 20, 22, 23, 46]. In addition, Dreger et al. [8] found that the overhead of HTTP traffic analysis can increase Bro [25] runtime by a factor of five. Given the high cost of HTTP analysis, and the fact that HTTP contributes more than 60% of the overall traffic to many organizations [8], we chose to evaluate our technology using a protomatcher for HTTP traffic.

Alternate encodings are not unique to HTTP. Attackers have injected TELNET control characters in the middle of FTP commands [30]. Furthermore, FTP [27] and SMTP [28] commands are case insensitive. Even when protocol specifications do not allow multiple encodings, a protocol implementation might allow it. For example, implementations of many protocols encode the carriage-return line-feed sequence either as the sequence "\r\n" or as a single "\n". While we focus on HTTP encodings, we believe that alternate encodings of data in other protocols can be handled using a protomatcher-based design.

Fast pattern matching for NIDS. There is a significant amount of work on efficient *string matching* for intrusion detection purposes. Researchers have proposed either software-based [1, 3, 6, 11, 47] or hardware-based [18, 39, 45] matching algorithms. This previous work does not address the problem of matching in the presence of alternate encodings and none of these algorithms mention protocol analysis as part of the matching algorithm.

Recent research has suggested that strings alone are not sufficient to accurately detect attacks. Instead, researchers have proposed using *regular expression* matching [24, 48]. To match regular expressions, Sommer and Paxson [36] used a DFA. However, unlike our work, they performed matching on already-normalized traffic. They also noticed that their DFA might not fit into memory, so they constructed the relevant portions of their DFA during matching. They acknowledge that this incremental approach degrades Bro's performance and might not be needed because their DFA did not grow beyond 20MB. We construct our protomatcher (our DFA) once and do not modify it further.

The pattern matching mechanism in Snort (since version 2.0 [32]) is based on a concept similar to protomatching. First, Snort uses a fast set-wise string matching algorithm (e.g., [1]) that identifies any rule that **may** match the traffic. Then, if such rules have been identified, Snort invokes a slower matcher that fully match the traffic against those rules. This is the same pre-filtering principle that is the core of the protomatching technique. However, unlike protomatching, Snort applies this two-stage process after it normalizes the traffic while protomatching saves the normalization phase. Indeed, this difference results in up to 49% improvement in Snort performance.

Dealing with high-speed links. Vendors of commercial NIDS advertise that their products can monitor high-speed links of at least 1 Gbps (e.g., [5, 38]). However, anecdotal evidence suggests that this is not always the case [42]. Schaelicke et al. [35] conducted an evaluation of several systems and emphasized pattern matching and protocol analysis as the factors limiting NIDS performance.

To deal with high-speed links, researchers have suggested a distributed NIDS that balances the network traffic such that each sensor monitors a different portion of the protected network [16, 37]. Our work focuses on the performance of a single sensor. Since protocol analysis and pattern matching are usually done by each sensor, sensors that use a protomatcher would further increase the throughput of such distributed designs.

3. THE BASICS OF PROTOMATCHING

We illustrate the concepts behind protomatching. We first formulate how the ANM methodology relates protocol analysis and matching. Then, we formulate the concept of a protomatching signature. Last, we develop the concept of the superset protomatcher, which is the basis of the P–ANM methodology. Throughout this section we use the DNS-tools attack as our running example.

The DNS-tools exemplary attack. The DNS-tools attack allows attackers to bypass access to a popular DNS administration tool and gain administrative privileges on a DNS server (CVE-2002-0613 [41]). An attacker that launches this attack typically uses an HTTP request as il-

```
input : A string w, L_protocol, N normalization
        function, S a signature.
output: Returns match if and only if
        (w ∈ L_protocol) ∧ (N(w) ∈ L(S)).
//Input conforms to the attack protocol?
1 if w ∉ L_protocol then
2     return no-match
//Compute normalized traffic
3 Compute w' = N(w) ;
//Normalized traffic matches signature?
4 if w' ∈ L(S) then
5     return match;
6 else
7     return no-match;
```

Algorithm 1: The analyze, normalize, match (ANM) method.

lustrated below, denoted R_1:

R_1: `GET dnstools.php?section=hosts&`
$\qquad$ `user_logged_in=true HTTP/1.1`

Consider an attacker who wants to use the DNS-tools attack and also wants to avoid detection by a NIDS. The attacker first encodes parts of the substring "`dnstools.php`" using hexadecimal encoding. For example, the attacker changes one 'o' and one 'p' into their hexadecimal values, obtaining the string "`dnsto%4fls.ph%50`". Then, the attacker further obfuscates the substring by mixing upper- and lower-case characters, for example, by converting 'h' into 'H'. The result of this process is the following request:

R_2: `GET dnsto%4fls.pH%50?section=hosts&`
$\qquad$ `user_logged_in=true HTTP/1.1`

The NIDS goal is to identify both R_1 and R_2 as instances of the DNS-tools attack.

3.1 The ANM Methodology

We use Snort's signature for the DNS-tools attack to illustrate how the ANM methodology attempts to identify both R_1 and R_2. Snort searches the URL of every incoming HTTP request for the following regular expression over the ASCII input $\Sigma = \{0 \ldots, 255\}$:

$S_{dns} = \Sigma^\star \cdot$ "`dnstools.php`" $\cdot \Sigma^+ \cdot$ "`user_logged_in=true`"

S_{dns} matches R_1, but does not match R_2. Therefore, to determine that R_2 is a DNS-tools attack, Snort performs the following steps. First, it parses R_2 and verifies that it is a valid HTTP request. Second, during this parsing, Snort normalizes the URL in R_2: Snort converts "`%4f`" into 'o', "`%50`" into 'p', and 'H' into 'h'. Last, it checks whether the normalized URL matches S_{dns}.

Formalizing the ANM methodology. Let $L_{protocol}$ be a language that defines valid messages of the attack protocol (e.g., the syntax of HTTP [10]). Let $N : \Sigma^\star \to \Sigma^\star$ be a *normalization* function, a function that translates a string into its normalized form. Let S be a signature (e.g., a regular expression), and denote the language that the signature defines as $L(S)$. Let w be the NIDS input (e.g., a TCP stream).

The ANM method first checks that w conforms to the syntax of the attack protocol, that is, whether $w \in L_{protocol}$.

Second, ANM computes the normalized version of the input, $N(w)$. Last, the ANM method checks whether the normalized input matches the signature language, that is, $N(w) \in L(S)$. Formally, the ANM method returns **match** if and only if $(w \in L_{protocol}) \wedge (N(w) \in L(S))$ (Algorithm 1).

It is easy to see why Algorithm 1 is inefficient. It performs a membership check (Line 1) and computes $N(w)$ (Line 3) on every input, for example a network packet, regardless of whether the input is benign or malicious. The P–ANM methodology addresses exactly this inefficiency. P–ANM attempts to determine whether a packet is benign using a single membership check.

3.2 A Protomatching Signature for the DNS-tools Attack

The basic idea behind protomatching is to convert protocol analysis and normalization into a single membership check. First, we expand S_{dns} such that it accounts for all possible encodings of the DNS-tools attack. Second, we expand S_{dns} such that it conforms to the HTTP specification.

Expanding S_{dns} to match all encodings of the DNS-tools attack. This step is based on the observation that alternate encodings are substitutions: operations that map characters to regular expressions [13]. We process S_{dns} and substitute each character that can be encoded in multiple ways with a regular expression that describes all possible variants of an attack, **with respect** to a given set of substitutions. These substitutions preserve language regularity, producing a regular expression that matches both normalized and encoded versions of an attack (see [33] for a proof of this claim).

We illustrate a substitution for the alternate encodings our attacker used to obfuscate R_1: upper-/lower-case and HEX encodings. Our implementation also handles encoding called `Uencode`, which is unique to Microsoft IIS server. We discuss other possible encodings in Section 4.1.

Consider the character 'd'. Our substitution, denoted N^{-1}, maps 'd' as follows:

$$N^{-1}(\text{d}) = [\text{d}|\text{D}|\text{\%44}|\text{\%64}]$$

We denote the substitution N^{-1} because it computes the inverse of the normalization function in the ANM technique (Section 3.1). Instead of mapping encodings to a normalized representation, N^{-1} maps the normalized representation to an expression describing every possible encoding.

We replace each character that appears in a URL in S_{dns} with its corresponding regular expression. We replace the character 'd' in the string "`dnstools`" with the regular expression $[\text{d}|\text{D}|\text{\%44}|\text{\%64}]$, then we replace the character 'n' with $[\text{n}|\text{N}|\text{\%4e}|\text{\%4E}|\text{\%6e}|\text{\%6E}]$, and so on. In the end, we obtain a signature similar to the expression PS^1_{dns} (for brevity, we omit the substitutions for characters beside 'd' and 'n'):

$$PS^1_{dns} = \Sigma^\star \cdot [\text{d}|\text{D}|\text{\%44}|\text{\%64}] \cdot [\text{n}|\text{N}|\text{\%4e}|\text{\%4E}|\text{\%6e}|\text{\%6E}] \cdot$$
$$\text{``stools.php''} \cdot \Sigma^+ \cdot \text{``user\_logged\_in=true''}$$

Formally speaking, $PS^1_{dns} = N^{-1}(S_{dns})$. As long as we can express alternate encodings as regular substitutions, the set of all possible variants of an attack is a regular language [33].

Expanding PS^1_{dns} to conform to HTTP specification. PS^1_{dns} integrates multiple encodings into a regular

Algorithm 2:
```
input  : A string w, L_protocol, N normalization
         function, S a signature.
output : Returns match if and only if
         w ∈ L_protocol ∩ L(N^{-1}(S)).
//1,2 are done once in a pre-computing phase
1  Pre-compute N^{-1}(S) ;
2  Pre-compute PS = L_protocol ∩ L(N^{-1}(S));
3  if w ∈ PS then
4      return match;
5  else
6      return no-match;
```
Algorithm 2: Protomatching method.

Algorithm 3:
```
input  : A string w, L_protocol, N normalization
         function, S a signature.
output : Returns match if and only if
         w ∈ L_protocol ∩ L(N^{-1}(S)).
1  Pre-compute Ŝ such that L(Ŝ) ⊇ L(S);   //Compute
   superset signature (pre computation phase).
2  Pre-compute N^{-1}(Ŝ);                 //Done once in a
   pre-computing phase
3  Pre-compute P̂S = L_protocol ∩ L(N^{-1}(Ŝ)) ; //Done once
   in a pre-computing phase
4  if w ∈ L(P̂S) then
5      return the result of Algorithm 1 on
         (w, L_protocol, N, S) ;
6  else
7      return no-match;
```
Algorithm 3: The protomatch, analyze, normalize, match (P–ANM) method.

expression, so it saves the need for a separate normalization phase. Now, our goal is to convert PS^1_{dns} to an expression that also conforms to the HTTP specification, so we can omit the protocol analysis phase.

The HTTP syntax is given in Backus-Naur form [26] in the HTTP specification [10]. To the best of our knowledge, whether the HTTP syntax can be expressed as a regular language has yet to be investigated. However, below we propose a regular expression that matches the HTTP *Request-Line*: the HTTP method followed by a URL followed by the HTTP version. We chose to model the Request-Line because it appears in more than 85% of Snort signatures. Currently, we construct the regular expressions for the HTTP protocol manually. In the future, we plan to use automated techniques [15] to construct a regular approximation directly from the BNF representation of the HTTP syntax.

We convert PS^1_{dns} into our final protomatching signature (Section 4.1 details the conversion process):

$$PS_{dns} = (\Sigma^* \cdot \text{``}\backslash n\backslash n\text{''})^* \cdot \text{``GET''} \cdot (SP)^+ (U)^* \cdot [d|D|\%44|\%64] \cdot$$
$$[n|N|\%4e|\%4E|\%6e|\%6E] \cdot \text{``stools.php''} \cdot$$
$$(U)^+ \cdot \text{``user\_logged\_in=true HTTP/1.1}\backslash n\text{''}$$

In the PS_{dns} expression, '\n' denotes a newline character, SP denotes white space characters, and U denotes characters that can appear in a URL according to the HTTP specification (e.g., '\n' cannot appear in a URL).

PS_{dns} ensures that the string "dnstools.php", is part of a URL of a valid HTTP method. It ensures that (i) the "GET" appears in the beginning of a line, (ii) only white spaces separate the "GET" from the URL, (iii) only valid characters appear in the URL, and (iv) the URL ends according to the HTTP protocol. Therefore, when the traffic we observe matches PS_{dns} we can be sure that the string "dnstools.php" (or any of its encoded versions) is part of a URL in a valid HTTP method. In other words, our conversion saves the need for protocol analysis.

Formalization of a protomatching approach. Algorithm 2 uses a protomatching signature for matching. First, the algorithm computes $N^{-1}(S)$ (Line 1). Second, the algorithm computes $L_{protocol} \cap N^{-1}(S)$ (Line 2), which is our protomatching signature, denoted by PS. In particular, in our example $PS_{dns} = L_{HTTP} \cap N^{-1}(S_{dns})$. Note that the expression for $L_{protocol} \cap L(N^{-1}(\hat{S}))$ can be pre-computed because it is independent of the input. Last, in Line 3 the algorithm checks whether the input matches our protomatching signature. It is possible to formally prove that protomatching is equivalent to the ANM methodology, that is, Algorithm 2 is equivalent to Algorithm 1 [33].

In Section 4.2 we discuss two techniques to convert a set of protomatching signatures into a protomatcher: a deterministic finite state machine that can be used at runtime to detect whether the network traffic matches any of the underlying signatures. Ultimately, our goal is to construct a *full-coverage protomatcher*, a protomatcher that contains a protomatching signature for every signature in the NIDS database. The advantage of a full-coverage protomatcher is its efficiency: it requires a single inspection of each network byte during the membership check (Line 3 in Algorithm 2).

Unfortunately, a full-coverage protomatcher is difficult to achieve. While we show that it is feasible to build a full-coverage protomatcher for more than 500 Snort signatures (Section 5.1), these signatures are short regular expressions. They usually require matching of two strings: the HTTP method and an additional string. However, researchers have shown that accurate signatures require sophisticated expressions [7, 24, 34, 36, 48]. For example, PS_{dns} requires matching three strings. Our experience shows that a protomatcher for all of the HTTP signatures in Snort's database requires more than 2GB of memory.

3.3 A Superset Protomatching Signature

The goal of the superset protomatcher is to reduce the memory footprint of the full-coverage protomatcher. It does so, by trading matching efficiency for memory consumption. Instead of using a signature like PS_{dns}, it uses a less-specific, or a *superset*, signature for the DNS-tools attack.

A superset signature omits some portions of the original signature in the NIDS database. Hence, its corresponding protomatcher consumes less memory. At the same time, a superset signature causes *false matches*, cases in which the traffic matches the signature but does not match the original signature in the NIDS database. Hence, in cases where the traffic matches a superset signature, we need to verify whether or not the traffic also matches the original signature. In our implementation, this slower matching process is based on the traditional ANM methodology.

This two-phase approach is the core of the P–ANM methodology. P–ANM is time-efficient because, under normal conditions, most of the traffic is benign. It is memory efficient because it splits the detection into two processes that together consume less memory than a full-coverage protomatcher.

Consider again the DNS-tools attack. Its superset signature can be:

$$\hat{S}_{dns} = \text{``GET''} \cdot \Sigma^\star \cdot \text{``dnstools.php''}$$

We follow the steps from Section 3.2 and construct the following superset protomatching signature (for brevity, we omit encodings of characters beside 'd' and 'n'):

$$\hat{PS}_{dns} = (\Sigma^\star \cdot \text{``\textbackslash n\textbackslash n''})^\star \cdot \text{``GET''} \cdot (SP)^+ (\text{U})^\star \cdot [\text{d}|\text{D}|\text{\%44}|\text{\%64}] \cdot [\text{n}|\text{N}|\text{\%4e}|\text{\%4E}|\text{\%6e}|\text{\%6E}] \cdot \text{``stools.php''}$$

There are several reasons to construct $\hat{S}_{dns}$ as we did above.

1. **$\hat{S}_{dns}$ preserves false negative correctness of S_{dns}.** The language of $\hat{S}_{dns}$ is a superset of the language of S_{dns}: that is, $L(\hat{S}_{dns}) \supset L(S_{dns})$. This property remains true even if we apply our substitutions, that is $\mathcal{N}^{-1}(\hat{S}_{dns}) \supset \mathcal{N}^{-1}(S_{dns})$. Therefore, a superset protomatcher recognizes any attack instance recognized by the full-coverage protomatcher and some additional traffic. By constructing $\hat{S}_{dns}$ as a superset of S_{dns} we ensure lack of false negatives.

2. **A protomatcher based on $\mathcal{N}^{-1}(\hat{S}_{dns})$ consumes exponentially less memory than a protomatcher based on $\mathcal{N}^{-1}(S_{dns})$.** The reason is that $\mathcal{N}^{-1}(\hat{S}_{dns})$ contains only two explicit substrings separated with the general term $\Sigma^\star$. Essentially, every $\Sigma^\star$ followed by a string doubles the memory that a DFA consumes. For example, to match user_logged_in=true in $\mathcal{N}^{-1}(S_{dns})$, a DFA must identify that the input stream already contains GET and dnstools.php, in that order. In a DFA, such knowledge is expressed by adding states. When we consider a set of signatures, each signature doubles the size of the protomatcher, resulting in an exponential growth of states.

3. **The majority of benign traffic does not match the superset protomatcher.** Like other researchers [2, 19], we have noticed that 99% of the HTTP requests in traffic we monitor does not match both S_{dns} and $\hat{S}_{dns}$. This seems intuitive because most HTTP traffic does not target the organization's DNS server.

4. PROTOMATCHING IMPLEMENTATION

We implemented the P–ANM methodology in Snort. We chose Snort because it is widely used and its source code as well as its signatures are publicly available. We describe how to convert Snort HTTP signatures into protomatching signatures and how to construct a protomatcher that can match those signatures during runtime. We focus on HTTP signatures because they account for 46% of all Snort signatures, and because Snort performs HTTP analysis and normalization that we would like to save.

While we use Snort, our protomatching signatures and runtime protomatcher can be used in the context of other NIDS. After all, our signatures are regular expressions that conform to the HTTP specification and our protomatcher is based on a finite state machine.

4.1 Automatically Converting Snort Signatures into Protomatching Signatures

Snort enables two types of patterns in an HTTP signature: one that must appear in the attack URL, denoted uri-content, and one that can appear anywhere in the HTTP request, denoted content. We split Snort signatures into four types, based on the combinations of these two patterns (Table 2).

We constructed two signatures from each Snort signature: A *full-coverage* signature, used by our full coverage protomatcher, and a *superset* signature, used by our superset protomatcher. For Type 3 signatures, we used the uri-content pattern as our superset signature. For Type 4 signatures, we used the longest pattern, under the assumption that longer patterns would cause fewer false matches. We leave other strategies to build superset signatures as future work.

In our current implementation, we do not translate other fields of a Snort's signature into a regular expression. For example, we ignore the *depth* and *offset* fields that specify portions of the packet in which the pattern should be found. Therefore, when a signature uses such a field (less than 10% of the signatures use these fields), we always treat its corresponding protomatching signature as a superset signature.

Expanding Snort signatures according to the HTTP syntax. To convert a Snort signature into a protomatching signature, we first expand it according to the syntax of HTTP. Before each pattern (e.g., a uri-content pattern), we add a regular expression that matches a valid HTTP method. In the case that we know the method necessary for the attack to succuss, for example "GET", we add only this method. In all other cases, we add a regular expression that matches either "GET", "HEAD", or "POST". We believe that these are the most common methods used in HTTP attacks; clearly other methods can be added as needed.

We ensure that an HTTP method always appears in the beginning of a line, as required by the HTTP specification. Between the method and the pattern, we allow only characters that are permitted in a URL; for example, we disallow white space characters. These expressions can be automatically added as defined by the regular expressions denoted by M, L, and P, in Table 2.

Alternate Encodings as substitutions. We implemented three types of HTTP alternate encodings: upper-/lower-case switching, HEX encoding, and the Microsoft-specific U-encoding [9]. We denoted these transformations as UL, HEX, and Uencode, respectively. The uri-content pattern requires normalization of the Uencode, HEX, and UL transformations, while the content pattern only requires normalization of the UL transformation.

In general, like the HEX encodings, the Uencode maps a character to a string containing the character's hexadecimal ASCII value, but does so using 4 instead of 2 bytes. We chose these transformations because they have been widely used for evasion [20, 23, 46] and represent transformations used by two popular web servers: Apache and IIS. To embed these multiple encodings into our protomatching signatures, we just replaced each character in the uri-content pattern with all of its possible encodings. Table 3 illustrates the three encodings defined as substitutions, and a simple DFA that identifies both encoded and normalized versions of the character 'o'.

UTF-8 is another transformation allowed by the HTTP specifications [10], and the only other transformation supported by Apache. We do not foresee any problems in supporting UTF-8. In the case of a protomatcher for Apache, UTF-8 would replace the Uencode. In the case of a proto-

Type	Patterns (re denotes a regular expression).	# of sig. in Snort database	Convert into a regular expression M: HTTP method (e.g., "GET", "HEAD"). S: White space. U: A valid URL character [10]. L: match beginning of a line i.e., $\Sigma^\star \cdot$ '\n' $\quad\mathcal{I}^{-1}$: UL substitution (Table 3). $\mathcal{N}^{-1}$: UL+HEX+Uencode substitutions (Table 3). $\quad \Sigma: \{0\ldots 255\}$	
1	$uri = re_1$	749	full-coverage:	$\text{L}\cdot\text{M}\cdot\text{S}^+\cdot\text{U}^\star\cdot\mathcal{N}^{-1}(re_1)$
			superset:	$\text{L}\cdot\text{M}\cdot\text{S}^+\cdot\text{U}^\star\cdot\mathcal{N}^{-1}(re_1)$
2	$content = re_1$	130	full-coverage:	$\text{L}\cdot\text{M}\cdot\text{S}^+\cdot\Sigma^\star\cdot\mathcal{I}^{-1}(re_1)$
			superset:	$\Sigma^\star\cdot\mathcal{I}^{-1}(re_1)$
3	$uri = re_1,$ $content = re_2$	122	full-coverage:	$\text{L}\cdot\text{M}\cdot\text{S}^+\cdot\text{U}^\star\cdot\mathcal{N}^{-1}(re_1)\cdot\text{U}^\star\cdot\text{S}\cdot\text{"HTTP"})\cap(\Sigma^\star\cdot\mathcal{I}^{-1}(re_2)\cdot\Sigma^\star)$
			superset:	$\text{L}\cdot\text{M}\cdot\text{S}^+\cdot\text{U}^\star\cdot\mathcal{N}^{-1}(re_1)$
4	$content_1 = re_1,$ $\ldots,$ $content_n = re_n$	21	full-coverage:	$\text{L}\cdot\text{M}\cdot\bigcap_i \Sigma^\star\cdot\mathcal{I}^{-1}(re_i)\cdot\Sigma^\star$
			superset:	$\text{L}\cdot\text{M}\cdot\Sigma^\star\cdot\mathcal{I}^{-1}(re_1)$

Table 2: Converting Snort signatures into protomatching signatures.

matcher for IIS, UTF-8 would increase the size of a protomatcher, and it might be that only a hierarchical protomatcher would be feasible.

There are five other esoteric transformations only supported by the IIS server [31]. Since they have a distinctive pattern, are rarely used, and their usage is highly suspicious, we believe that the best way to handle those is through signatures rather than normalization. For example, the *double HEX encoding* encodes the character '%' using the HEX encoding. That is, the substring "%2520" is first decoded into "%20" and then into the space character. In this case, we used the signature $\Sigma^\star\cdot$%25, as also done by Snort. For other Microsoft-specific encodings the reader is referred to [31].

4.2 Converting Protomatching Signatures into a Protomatcher

We describe two possible implementations for a protomatcher that can be used in practice to match our protomatching signatures, based on a *deterministic* finite state machine and based on an *hierarchical* state machine. We used both techniques to implement our protomatchers in our experiments (Section 5). To handle possible false matches of our superset protomatcher, we invoked Snort's analyzer, normalizer, and matcher. While this might not be the most time-efficient method, it was sufficient to illustrate the benefits of the P-ANM methodology.

Implementing a protomatcher using a deterministic FSM. Recall that a protomatching signature is a regular expression. Since regular expressions are closed under the union operation, there exists a single regular expression that recognizes all signatures in parallel. We used publicly available tools [21, 49] to construct this expression and then to automatically convert it into a deterministic finite state machine. We implemented the machine as an $M \times 256$ table where M is the number of states and 256 is the size of our alphabet. The table represents a function $f(i,j) = k$, that is, from state i with the input j the automaton moves to state k. We incorporate this table into Snort during compilation time.

Note that an implementation based on a deterministic FSM is highly efficient because it inspects each network byte exactly once. There is no need for any protocol analysis or normalization. Ultimately, our goal is a full-coverage protomatcher that only uses full-coverage signatures. Unfortunately, this was not always feasible because a deterministic protomatcher based only on full-coverage signatures consumed more than 2GB of memory when we added full-coverage versions of signatures of Type 2, 3 and 4. Hence, when a full-coverage protomatcher was infeasible, we constructed a superset protomatcher using the superset signatures (Section 5.1).

The *hierarchical protomatcher*, implementing a protomatcher using an hierarchical FSM. As we show in Section 5.1, a deterministic superset protomatcher is feasible but can consume more than 20MB of memory. Because of its large size, the protomatcher might cause many cache misses. Therefore, we investigated an implementation that reduces the memory footprint of our protomatcher.

The hierarchical protomatcher is a memory-efficient implementation of a protomatcher. It splits the protomatcher into two machines: a *matcher* and a *normalizer*. The matcher is responsible for protocol analysis and pattern matching. The normalizer is responsible for handling multiple encodings. Unlike the ANM method that first normalizes the whole HTTP request, the hierarchical protomatcher consults with the normalizer only when necessary. For example, when a matcher encounters the character '%' in "dnsto%4fls" (R_2 in Section 3), it calls the normalizer that interprets the string "%4f" and returns the character 'o' to the matcher, which continues with the matching process. Formally, the normalizer is implemented as a transducer [40], a finite state machine that outputs a symbol on each transition.

To implement a hierarchical protomatcher, we build a protomatching signature according to the rules in Table 2, but we do not apply the $\mathcal{N}^{-1}$ substitution. Hence, in the hierarchical protomatcher case, our protomatching signature accounts only for the UL encoding, while the normalizer implements the HEX and Uencode transformations.

The main advantage of a hierarchical protomatcher is its memory efficiency, it consumes an order of magnitude less memory that a deterministic protomatcher. The reason is that a deterministic implementation inlines the normalization of the HEX and Uencode encodings in every pattern, while in the hierarchical implementation we abstract this normalization as a function call.

We can use some of the memory we save to improve the accuracy of the underlying superset protomatcher. We pinpoint superset signatures that cause a significant number of false matches (Section 3.3) and convert them back to full coverage signatures. Since we do so only for a few signatures, we only increase the size of the hierarchical protomatcher to 6MB (Section 5.2.2), which is equivalent to the size of Snort's matcher. More importantly, this process significantly reduces the number of false matches of the under-

Name	$\mathcal{N}^{-1}(o)$	A DFA that identifies the HEX, Uencode, and UL substitutions for the character 'o'
HEX	([%6f|%6F|%4f|%4F])	
Uencode	([%u006f|%u006F|%u004f|%u004F|%U006f|%U006F|%U004f|%U004F])	
UL	([o|O])	

Table 3: Representing encodings as substitutions. The $\mathcal{I}^{-1}$ substitution is analogous to $\mathcal{N}^{-1}$ but without the Uencode substitution.

lying superset protomatcher. We believe that such a signature refinement process can be automated. For example, a NIDS can use an administrator responses to discover the signatures that cause the most false matches. However, such study is beyond the scope of this paper.

5. FEASIBILITY STUDY

We investigated the ability of the P–ANM methodology to improve the performance of Snort (version 2.3.3). We were interested in the following questions.

1. **Is a protomatcher-based Snort feasible?** Especially, we were interested in understanding the memory consumption of various protomatchers and the time it takes to construct them.

   In Section 5.1, we show that a protomatcher-based Snort is feasible. We show that, based on 999 protomatching signatures, our protomatchers consume up to 22.09MB or 6.47MB, when implemented using the deterministic or hierarchical implementations, respectively.

2. **How does a protomatcher affect Snort performance?** In Section 5.2, we compared the performance of a protomatcher-based Snort to a vanilla Snort: the original Snort that uses the ANM methodology.

   In Section 5.2.1, we show that a protomatcher-based Snort is 45% faster that a vanilla Snort. After we tuned Snort to use a Wu-Manber pattern-matching algorithm [47], our protomatcher-based Snort was still 25% faster. We show that, on real HTTP traffic, our protomatcher classified more than 99% of the traffic as benign. In other words, our protomatcher determined that more than 99% of the traffic was benign without utilizing the expensive analysis-normalizing-matching phases.

   In Section 5.2.2, we reduced the number of false matches using an hierarchical protomatcher. We identified superset signatures that cause the majority of false matches and replaced them back with their full-coverage versions. We reduced the number of false matches by more than 50% and improved our performance by an additional 2% (or 4% without the Wu-Manber pattern matching), beyond the 25% obtained with the deterministic protomatcher.

3. **How does the memory size of a protomatcher affect its performance?** Since a protomatcher consumes more memory than a vanilla Snort, a protomatcher-based Snort might suffer more cache misses. We investigated a *cache-poisoning* attack in which an attacker attempts to degrade the protomatcher's performance by forcing it to generate many cache misses. While this attack degrades the a protomatcher by 2%, it degrades the performance of the vanilla Snort by 4.5%.

Experimental methodology. Our Snort distribution contained 1022 HTTP signatures. We removed 23 (2.25%) of them from the experiments because they caused thousands of false alarms in our environment.

To compare the performance of our protomatcher-based Snort to a vanilla Snort, we used traces of live HTTP traffic that we collected from the gateway router of our department web server. Each trace contained between five and seven million packets with their full HTTP payload. Two traces, T_1 and T_2, were collected during the day, starting at 9:00am and 2:00pm, respectively. Trace T_3 was collected during the night, starting at 1:00am. All traces where collected during September 2005. On average, 4% of the TCP packets in each trace are requests.

Our performance metric is Snort's *Average per-Packet Processing Time* (ApPPT). We measured this time in CPU cycles. We started counting cycles when Snort gets a packet from libpcap [43] and stopped counting when Snort finishes handling the packet and is ready to accept the next one. We ran all our experiments on a Pentium 4, 3GHz, 1MB L2 cache, 2GB of memory, running Tao Linux 1.0.

5.1 Feasibility and Memory Consumption

We experimented with protomatchers containing different numbers of signatures. First, we built full-coverage protomatchers, using full-coverage Type 1 signatures (Table 2). Since we have 747 Type 1 signatures, we built protomatchers with 100, 300, 500, and 747 signatures. Second, we tried to add another 120 full-coverage Type 2 signatures. However, this attempt failed because building a full-coverage protomatcher consumed more than 2GB of memory. Hence, we switched to a superset protomatcher, using 120 Type 2 superset signatures and 747 full-coverage signatures. Finally, we added 132 Type 3 and Type 4 superset signatures, constructing a superset protomatcher with 999 signatures.

We build the protomatchers once using the deterministic FSM and once using an hierarchical FSM. Building times for the deterministic protomatchers span from 4 minutes for the smallest to 92 minutes for largest, when constructed with the UL, HEX, and Uencode transformations (Figure 1a). Building times for an hierarchical implementation are considerably lower. Recall that we implement a protomatcher as a separate table (Section 4.2) and we incorporate the table into Snort during the compilation process. Therefore, from the administrator viewpoint, the only differences between a protomatcher-based Snort and a vanilla Snort, is their compilation time. According to our measurements, given a protomatcher as a header file, the compilation time of a protomatcher-based Snort is longer by at most 10 minutes.

Number of signatures	100	300	500	747	867	999
Type of signatures (Table 2):	1	1	1	1	1,2	1,2,3,4
Protomatcher type:	full-coverage	full-coverage	full-coverage	full-coverage	superset	superset
Building time (min):	4	9	18	60	80	92
Memory size D-FSM (MB):	1.79	3.46	11.30	19.06	22.09	18.92
Memory size H-FSM (MB):	0.29	0.71	1.34	2.03	3.03	6.47

(a) Summary of feasibility study: building times and memory sizes are for protomatchers built using the UL, HEX, and Uencode transformations (Section 4.1).

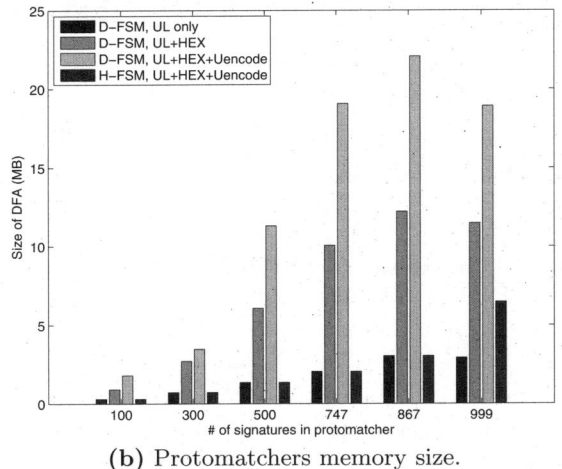

(b) Protomatchers memory size.

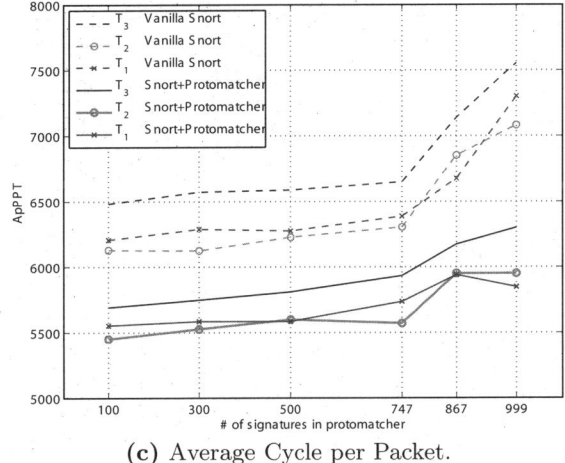

(c) Average Cycle per Packet.

Figure 1: Performance comparison of a protomatcher-based Snort with a vanilla Snort. (D-FSM and H-FSM denote a deterministic and hierarchical protomatcher, respectively.)

Figure 1a also presents the memory consumption of protomatchers that support normalization for the UL, HEX, and Uencode transformations. Protomatchers based on a deterministic implementation consume up to 22.09MB, or less than 1% of our workstation memory. Protomatchers based on the hierarchical implementation consume an order of magnitude less memory.

Figure 1b presents the memory consumption of protomatchers that support different types of normalization: the UL+HEX+Uencode that we used in our ApPPT experiment (Section 5.2.1), a UL+HEX protomatcher that is suitable to use with the Apache web server, and a UL protomatcher for comparison purposes. The HEX+UL protomatcher is about half the size of its HEX+UL+Uencode counterpart, indicating the heavy memory consumption imposed by the Uencode normalization. In comparison, the UL protomatchers consume less than 3MB, or half the size of Snort's matcher.

Note that the protomatcher with 867 signatures consumes more memory than a protomatcher with 999 signatures. The reason is that sometimes adding a signature reduces the number of states in an automaton. Consider, for example, an automaton with a signature $\Sigma^* \cdot$ cgi_bin/perl to which we add the signature $\Sigma^* \cdot$ perl. When we add the latter signature we reduce the number of states because the new automaton does not need to ensure the existence of substring "cgi_bin/".

In summary, our results show that a protomatcher-based NIDS is feasible. The memory consumption of our protomatchers is reasonable. Since construction of a protomatcher is done in a separate process, one can build a protomatcher without halting Snort. Then, incorporating a protomatcher into Snort just increases the compilation time by 10 minutes.

5.2 Performance Improvements

We first compared the ApPPT of our protomatcher-based Snort to a vanilla Snort. Then we investigated the capability of a protomatcher to quickly filter benign traffic.

5.2.1 Improved ApPPT

We ran our protomatcher-based and vanilla Snorts on our three traces and measured the ApPPT. Our protomatcher normalizes the UL, HEX and Uencode transformations. We report the results against a vanilla Snort configured with the Wu-Manber pattern matching algorithm. By default, Snort uses the Aho-Corasick algorithm [1]. When compared to the Aho-Corasick algorithm, our protomatcher-based Snort further improved the performance by additional 10%-15% **above** the numbers reported below.

Up to 25% improvement of Snort's ApPPT. The ApPPT of Snort with a superset protomatcher containing 999 signatures is lower by up to 25%, 19%, and 18% on T_1, T_2, and T_3, respectively (Figure 1c). Note the steep increase in Snort ApPPT as we add Type 3 and 4 signatures (the 867 and 999 marks). Inherently, matching complex regular expressions is a time consuming task. This illustrates the benefits of the P-ANM approach: the majority of the traffic does not match any signature and a protomatcher saves expensive analysis, normalization, and matching times.

Recall that an hierarchical protomatcher is less efficient than a deterministic protomatcher, because it must implement communication between the matcher and the normalizer. However, our hierarchical protomatcher further improved the ApPPT by 2%, on average (compare the "Hierarchical" and "Deterministic" lines in Figure 3). With 999 signatures the hierarchical protomatcher improved Snort's

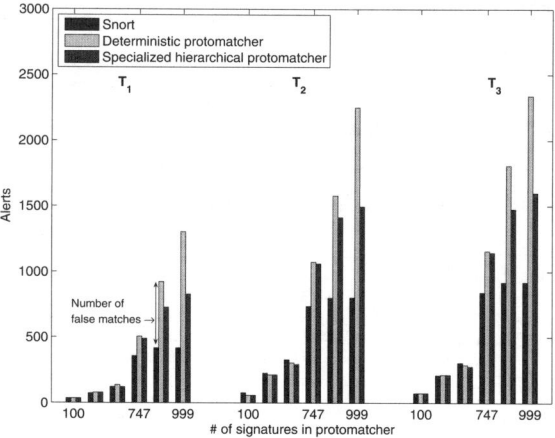

Figure 2: Snort alerts vs. protomatcher alerts. We define a false match as a case when traffic matches a protomatcher but does not produce an alert.

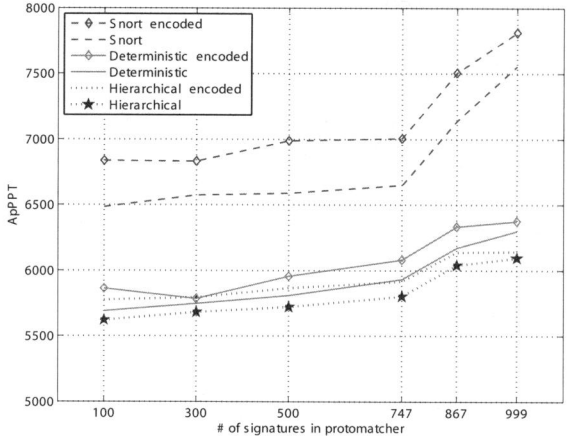

Figure 3: Effects of a cache-poisoning attack on the T_3 trace. The hierarchical protomatcher exhibits the best ApPPT across all traces.

ApPPT by 27%, 21%, and 23% on T_1, T_2, and T_3, respectively. The reason for this improvement is the smaller size of the hierarchical protomatcher, which reduced the number of cache misses. Therefore, Snort with a hierarchical protomatcher can handle up to 27% more traffic than a vanilla Snort.

Separate normalization is expensive. Although only 4% of the packets contain HTTP requests, our superset protomatcher improved Snort's ApPPT by up to 25%. This illustrates the large toll of normalization on the performance of a NIDS. The benefit from the superset protomatcher increases as we add complex signatures, that is, Type 3 and 4 (Table 3) signatures, to the system. This is evident in Figure 1c and Figure 3.

5.2.2 Utilization of the P–ANM Methodology

On average, a protomatcher classified 99% of the HTTP requests as benign. The superset protomatcher with 999 signatures classified as benign 99%, 98.8%, and 99% of the HTTP requests in T_1, T_2, and T_3, respectively. The ability of a protomatcher to quickly accept benign traffic, either normalized or encoded, is the core of its efficiency.

However, our superset protomatchers still produce false matches (Section 3.3). For example, our superset protomatcher with 999 signatures matches 885 HTTP requests in T_1 while Snort produces only 417 alerts (Compare "Snort" and "Deterministic protomatcher" bars in Figure 2). Since the hierarchical protomatcher is a compact implementation of a protomatcher, it provides plenty of opportunity to further reduce the number of false matches.

We discovered three Type 4 signatures that caused more than 50% of the false matches across all the three traces. We changed these signatures from superset signatures back to their full-coverage signatures. This increased the size of the hierarchical protomatcher from 3MB to 6MB (Figure 1b) but also decreased the number of false matches by more than 50% in the case of a hierarchical protomatcher with 999 signatures (compare "Deterministic protomatcher" and "Specialized hierarchical protomatcher" bars in Figure 2).

5.3 Sensitivity to Cache Poisoning Attack

A protomatcher is oblivious to the content of the input: the time it spends on a packet with n bytes is always equal to the time it takes to perform n table lookups. One way to degrade the performance of a protomatcher is to increase the table lookup time. To achieve this, we investigated a cache poisoning attack in which we increase the table lookup time by causing cache misses. We study one type of such attack, using URL HEX encoding. Of course, this initial study does not imply that our protomatcher is robust against all cache poisoning attacks; such a study is beyond the scope this paper. However, our study does show that our current protomatcher implementation can effectively sustain our chosen attack.

Our cache poisoning attack attempts to cause cache misses by forcing a protomatcher to visit more states. To do so, we encode all URLs in our traces using only hexadecimal representation. For example, we change a URL like `www.example.com` into `%77%77%77%2e...%2e%63%67%6d`. Note that the hexadecimal representation visits three times more states than the normalized request. This technique generates traffic that accesses real web pages, so it is less likely to be noticed. Note that our attack changes the character distribution inside URLs, so it could be detected using anomaly detection techniques [17]. However, currently Snort does not use such techniques.

We assumed that the attack would have a larger effect on a protomatcher-based Snort than on vanilla Snort. Recall that Snort normalizes all URLs before matching. Hence, in the vanilla Snort case, we assumed the same cache behavior during matching. Furthermore, Snort's normalizer consumes less that 2KB, so it easily fits into L1 cache. Hence, we assumed no significant difference in the cache behavior between the encoded and original traces during normalization. In the protomatcher case, however, we assumed that the cache misses would increase by a factor of three. Since we did not take any effort to increase the spatial locality of the protomatcher states, we believe that this is a reasonable assumption.

Figure 3 presents the effects of our cache poisoning attack.

Two observations should be noted:

1. The attack increases the ApPPT of all three systems (in Figure 3, compare "Snort-encoded" and "Snort", "Deterministic-encoded" and "Deterministic", and "Hierarchical-encoded" and "Hierarchical"). This increase is expected because the encoded URLs are three times longer than the original ones.

2. Snort is more affected by the attack than our protomatcher-based implementation. Snort's ApPPT increased by 4.5% (compare the "Snort-encoded" and "Snort" lines). In comparison, the ApPPT of Snort with a deterministic protomatcher only increased by 2% (compare the "Deterministic-encoded" and "Deterministic" lines).

This result is surprising since it contradicts our assumptions above. There might be two reasons for this result. First, the attack was ineffective in increasing the number of cache misses. It means that a more sophisticated cache poisoning attack is needed. Second, the attack was effective, but cache performance is only a minor component of the ApPPT. Further investigation of these issues is left for future work.

6. CONCLUSION AND FUTURE WORK

We formulated the concept of a protomatcher: a deterministic finite automaton that performs protocol analysis, normalization, and matching. We studied the performance of a protomatcher-based Snort and showed that it is both feasible and beneficial. We envision two main research directions that extend our work.

To become widely accepted, protomatching should be further automated. Currently, we manually convert a full-coverage signature into a superset one. We also manually refine a protomatcher to reduce the number of its false matches. We plan to investigate ways to automate these tasks in the future.

We also plan to further study the resiliency of protomatching and the P–ANM methodology against resource-consumption attack. Attackers may attempt to induce a protomatcher to generate many false matches. Such an attempt would excessively utilize the four phases in the P–ANM methodology, which would become more expensive than the three phases in the ANM methodology. Therefore, it is important to develop a mechanism that chooses the least expensive methodology based on the current network load.

We presented only an initial study on a protomatcher resiliency against cache poisoning attack. This issue should be study further. For example, it will be interesting to develop a cache-aware protomatcher whose states are organized in a way that increases their locality in memory.

Acknowledgments. We deeply thank Vinod Ganapathy and the anonymous referees for their useful comments that have helped us refine the concepts and experiments presented in this paper.

7. REFERENCES

[1] A. V. Aho and M. J. Corasick. Efficient string matching: an aid to bibliographic search. *Comm. of the ACM*, **18**(6), June 1975.

[2] S. Antonatos, M. Polychronakis, P. Akritidis, K. G. Anagnostakis, and E. P. Markatos. Piranha: Fast and memory-efficient pattern matching for intrusion detection. In *IFIP International Information Security Conference*, Chiba, Japan, May 2005.

[3] R. S. Boyer and J. S. Moore. A fast string searching algorithm. *Comm. of the ACM*, **20**(10), Oct. 1977.

[4] CheckPoint Software Technologies. InterSpec Internal Security. Available at www.checkpoint.com.

[5] Cisco Systems. Cisco IPS 4200 Series Sensors. Available at www.cisco.com.

[6] C. J. Coit, S. Staniford, and J. McAlemey. Towards faster string matching for intrusion detection or exceeding the speed of snort. In *DARPA Information Survivability Conference and Exposition (DISCEX II'01)*, Anaheim, CA, June 2001.

[7] J. R. Crandall, Z. Su, S. F. Wu, and F. T. Chong. On deriving unknown vulnerabilities from zero-day polymorphic and metamorphic worm exploits. In *ACM Conference on Computer and Communications Security*, Alexandria, VA, Nov. 2005.

[8] H. Dreger, A. Feldmann, V. Paxson, and R. Sommer. Operational experiences with high-volume network intrusion detection. In *ACM Conference on Computer and Communications Security*, Washington, DC, 2004.

[9] eEye Digital Security. %u encoding IDS bypass vulnerability, 2001. Available at www.eeye.com/html/Research/Advisories/AD20010705.html.

[10] R. Fielding, J. Gettys, J. Mogul, H. Frystyk, L. Masinter, P. Leach, and T. Berners-Lee. *RFC 2616 - Hypertext Transfer Protocol*. The Internet Engineering Task Force, June 1999.

[11] M. Fisk and G. Varghese. Applying fast string matching to intrusion detection. Technical Report CS2001-0670, University of California San Diego, May 2001. Updated version available at http://woozle.org/\~mfisk/.

[12] M. Handley and V. Paxson. Network intrusion detection: Evasion, traffic normalization, and end-to-end protocol semantics. In *USENIX Security Symposium*, Washington, DC, Aug. 2001.

[13] J. Hopcroft, R. Motwani, and J. Ullman. **Introduction to Automata Theory, Languages, and Computation**. Addison-Wesley, 2 edition, 2001.

[14] Internet Security Systems. RealSecure Network 10/100. Available at www.iss.net.

[15] J. C. Junqua and G. van Noord, editors. **Robustness in Language and Speech Technology**. Springer, 2001.

[16] C. Kruegel, F. Valeur, G. Vigna, and R. A. Kemmerer. Stateful intrusion detection for high-speed networks. In *IEEE Symposium on Security and Privacy*, Oakland, CA, May 2002.

[17] C. Krügel, T. Toth, and E. Kirda. Service specific anomaly detection for network intrusion detection. In *ACM Symposium on Applied Computing*, Madrid, Spain, March 2002.

[18] R.-T. Liu, N.-F. Huang, C.-N. Kao, and C.-H. Chen. A fast pattern matching algorithm for network processor-based intrusion detection system. In *IEEE International Conference on Performance, Computing, and Communications*, Phoenix, AZ, Apr. 2004.

[19] E. Markatos, S. Antonatos, M. Polychronakis, and

K. Anagnostakis. Exclusion-based signature matching for intrusion detection. In *IASTED International Conference on Communications and Computer Networks*, Cambridge, MA, Nov. 2002.

[20] R. Marti. THOR: A tool to test intrusion detection systems by variations of attacks. Master's thesis, Swiss Federal Institute of Technology, Mar. 2002.

[21] M. Mohri, F. C. N. Pereira, and M. D. Riley. AT&T Finite-State Machine Library. Available at www.research.att.com/sw/tools/fsm.

[22] D. Mutz, C. Krügel, W. Robertson, G. Vigna, and R. R. Kemmerer. Reverse engineering of network signatures. In *The AusCERT Asia Pacific Information Technology Security Conference*, Gold Coast, Australia, May 2005.

[23] D. Mutz, G. Vigna, and R. A. Kemmerer. An experience developing an IDS stimulator for the black-box testing of network intrusion detection systems. In *Annual Computer Security Applications Conference*, Las Vegas, NV, Dec. 2003.

[24] J. Newsome, B. Karp, and D. Song. Polygraph: Automatic signature generation for polymorphic worms. In *IEEE Symposium on Security and Privacy*, Oakland, CA, May 2005.

[25] V. Paxson. Bro: a system for detecting network intruders in real-time. *Computer Networks*, **31**(23/24), Dec. 1999.

[26] N. Peter. Revised report on the algorithmic language ALGOL 60. *Comm. of the ACM*, **3**(5), 1960.

[27] J. Postel and J. Reynolds. *RFC 959 - File Transfer Protocol*. The Internet Engineering Task Force, 1985.

[28] J. B. Postel. *RFC 821 - Simple Mail Transfer Protocol*. The Internet Engineering Task Force, 1982.

[29] T. H. Ptacek and T. N. Newsham. Insertion, evasion, and denial of service: Eluding network intrusion detection. Technical Report T2R-0Y6, Secure Networks, Inc., Calgary, AB, Canada, 1998.

[30] Robert Grahm. SideStep: IDS evasion tool, Jan. 2000.

[31] D. J. Roelker. HTTP IDS evasions revisited, Jan. 2003. Available at www.idsresearch.org.

[32] M. Roesch. Snort: the Open Source Network Intrusion Detection System. Available at www.snort.org.

[33] S. Rubin. *Formal Models and Tools to Improve NIDS Accuracy*. PhD thesis, University of Wisconsin-Madison, 2006.

[34] S. Rubin, S. Jha, and B. P. Miller. Language-based generation and evaluation of NIDS signatures. In *IEEE Symposium on Security and Privacy*, Oakland, CA, May 2005.

[35] L. Schaelicke, T. Slabach, B. Moore, and C. Freeland. Characterizing the performance of network intrusion detection sensors. In *International Symposium on Recent Advances in Intrusion Detection*, Pittsburgh, PA, Sep. 2003.

[36] R. Sommer and V. Paxson. Enhancing byte-level network intrusion detection signatures with context. In *ACM Conference on Computer and Communications Security*, Washington, DC, Oct. 2003.

[37] R. Sommer and V. Paxson. Exploiting independent state for network intrusion detection. In *Annual Computer Security Applications Conference*, Tucson, AZ, Dec. 2006.

[38] SourceFire Inc. SourceFire IS3000 Series. Available at www.sourcefire.com.

[39] L. Tan and T. Sherwood. A high throughput string matching architecture for intrusion detection and prevention. In *32nd International Symposium on Computer Architecture*, Madison, WI, June 2005.

[40] R. Teitelbaum. *Minimal Distance Analysis of Syntax Errors in Computer Programs*. PhD thesis, Computer Science Department, Carnegie-Mellon University, Sep. 1975.

[41] The National Institute of Standards and Technology (NIST). National vulnerability database. Available at nvd.nist.gov.

[42] The NSS Group. Intrusion prevention systems (IPS) group test (Edition 3), Aug. 2005. Available at www.nss.co.uk.

[43] The Tcpdump Group. TCPDUMP/LIBPCAP. Available at www.tcpdump.org.

[44] TippingPoint, a Division of 3Com. UnityOne, Intrusion Prevention Systems. Available at www.tippingpoint.com.

[45] G. Tripp. A finite-state-machine based string matching system for intrusion detection on high-speed networks. In *European Institute for Anti-Virus Research (EICAR) Annual Conference*, Malta, May 2005.

[46] G. Vigna, W. Robertson, and D. Balzarotti. Testing network-based intrusion detection signatures using mutant exploits. In *ACM Conference on Computer and Communications Security*, Washington, DC, Oct. 2004.

[47] S. Wu and U. Manber. A fast algorithm for multi-pattern searching. Technical Report TR94-17, Department of Computer Science at the University of Arizona, May 1994.

[48] V. Yegneswaran, J. Giffin, P. Barford, and S. Jha. An architecture for generating semantic-aware signatures. In *USENIX Security Symposium*, Washington, DC, Aug. 2005.

[49] S. Yu. Grail+: A symbolic computation environment for finite-state machines, regular expressions, and finite languages. Available at www.csd.uwo.ca/research/grail/grail.html.

Evading Network Anomaly Detection Systems: Formal Reasoning and Practical Techniques

Prahlad Fogla and Wenke Lee
College of Computing, Georgia Institute of Technology
Atlanta, Georgia, USA
prahlad@cc.gatech.edu, wenke@cc.gatech.edu

ABSTRACT

Attackers often try to evade an intrusion detection system (IDS) when launching their attacks. There have been several published studies in evasion attacks, some with available tools, in the research community as well as the "hackers" community. Our recent empirical case study showed that some payload-based network anomaly detection systems can be evaded by a polymorphic blending attack (PBA). The main idea of a PBA is to create each polymorphic instance in such a way that the statistics of attack packet(s) match the normal traffic profile. In this paper, we present a formal framework for the open problem: given an anomaly detection system and an attack, can one *automatically* generate its PBA instances? We show that in general, generating a PBA that optimally matches the normal traffic profile is a hard problem (NP-complete). However, the problem of finding a PBA can be reduced to the SAT or ILP problems so that solvers available in those domains can be used to find a near-optimal solution. We also present a heuristic (*hill-climbing*) to find an approximate solution. Our framework can not only expose how the IDS can be exploited by a PBA but also suggest how the IDS can be improved to prevent the PBA. We have experimented with our framework using the PAYL 1-gram and 2-gram anomaly detection system, and the results have validated our framework.

Categories and Subject Descriptors

C.2.0 [**Computer-Communication Networks**]: General—*Security and protection*

General Terms

Security

Keywords

anomaly detection, polymorphic blending attack, mimicry attack

1. INTRODUCTION

As defense techniques such as intrusion detection systems (IDSs) become widely deployed, attackers now have to consider how to defeat these mechanisms when launching their attacks. In particular, polymorphic attacks are designed to evade the detection of misuse detection systems. In such an attack, each polymorphic instance of an attack can look very different. Thus, there may not be an accurate or reliable pattern that an IDS can use as an attack signature. Anomaly detection systems offer a defense against polymorphic attacks because typically each attack instance still looks different from normal data. For example, the payload of an attack packet may contain some non-printable characters or unusual byte structure; whereas the payload of a normal packet predominantly contains ASCII characters with predefined structure, as required by the application protocol. Indeed, several network-based anomaly detection systems, e.g., PAYL [31, 32], which model the byte (or n-gram) frequency characteristics of the normal packets, have been shown to be effective against polymorphic attacks.

It is inevitable that attackers will attempt to evade anomaly detection systems. Our previous work [10] showed that a polymorphic blending attack (PBA) can evade PAYL. A PBA is similar to a mimicry attack [30], but is applied to network IDS rather than host-based IDS. The main idea of a PBA is that, after learning the normal profile using some normal packets, the attacker can mutate a given attack instance so that the byte characteristics of the final attack packet(s) match the normal profile. We showed that such mutations can be achieved using a simple byte substitution scheme followed by byte padding. However, these techniques are not general in that they are based on heuristics that work well for PAYL but not necessarily other anomaly detection systems.

Therefore, an important open problem with polymorphic blending attacks is: given an anomaly detection system and an attack, can one *automatically* generate the PBA instances? The motivation for solving this problem is to provide a defender the means to evaluate an IDS and even improve it.

In this paper, we present a formal framework for polymorphic blending attacks. We first show that a wide range of IDS can be represented using either a regular grammar or a stochastic regular grammar. The problem of generating a PBA then becomes finding a mutated attack instance that will be accepted by the IDS regular grammar. We show that in general, generating a PBA is a hard problem (NP-complete), but it can be translated to the SAT (satisfiability) or ILP (integer linear programming) problem. Thus, solvers in these problem domains can be used to find a near optimal solution for generating a PBA. In addition, we present heuristics to find approximate solutions very efficiently. We also show that our framework can not only expose how the IDS can be exploited by a PBA but also suggest how the IDS can be modified to prevent the PBA while maintaining the desired detection performance (e.g., low false positive rate). We have experimented with our framework using the PAYL 1-gram and 2-gram anomaly detection system, and the results have validated our framework.

The rest of the paper is organized as follows. Section 2 discusses the related work. In section 3, we present our framework that includes models for the class of IDS considered in this research, and the basic steps to generate a polymorphic blending attack. In section 4, we analyze the complexity of generating a polymorphic blending attack and present some methods to find approximate solutions for the problem. We show the results of our experiments in section 5. In Section 6, we describe how to use our framework to improve an IDS. Section 7 concludes the paper.

2. RELATED WORK

Various attack mutation techniques have been used by attackers to evade misuse detection systems. [29] presents some mutation techniques commonly used at the different layers of protocols. In [23], Rubin et al. modeled different attack transformations as inference rules. These rules can be used repeatedly to generate different attack instances from a given initial attack instance.

Several attack specification languages have been developed by researchers to represent the attack signatures efficiently and accurately. NETSTAT [6] represents the attack events using a state machine called STATL. Snort [21] represents a signature using regular expressions consisting of network bytes. It also uses other packet attributes. Liang et al. [17] presented extended FSA based attack signatures. GARD [22] is a tool for regular language based generation of attack instances. It is observed that many of the proposed signatures are based on a regular grammar or state machine.

Code polymorphism and metamorphism techniques [27] are used to mutate and obfuscate the shellcodes present in the attack. Tools like tPE, AdMutate, and CLET [5, 16, 33] perform advanced code polymorphism, and are available on the Web. Several approaches have been proposed to detect polymorphic attacks. [3] uses instruction semantics information to detect different instances of malware. Kruegel et al. [13] presented graph coloring based detection of malicious code. In [28], Toth et al. observed that the maximum binary executable section in a normal http packet is very small and may be used to detect the presence of shellcode in a packet. STRIDE [1] checks for the presence of NOP sleds in a packet to find buffer overflow attacks. Polygraph [20] generates signatures for worms by finding a set of longest substrings shared between different instances of the worm.

Payload-based anomaly IDS are used to detect application-layer attacks, including polymorphic attacks. PAYL [31, 32] records the byte (or n-gram) frequency characteristics of the normal packets. If the frequency characteristics of a packet differs significantly from the normal, the packet is deemed suspicious. NETAD and LERAD [14, 18] models the first few bytes of the application layer headers of a packet using a set of rules. In [15], Kruegel et al. proposed six different models, namely, length, character distribution, probabilistic regular grammar, token set, attribute presence, and attribute order for detection of http attacks. They modeled different http attributes like URL path, SQL query string, etc. Sekar et al. [25] presented an anomaly detection system based on network protocol specifications. Specification is defined using extended finite state automaton. Statistical features based on the state transitions are monitored to detect anomalies.

In host based IDS, system call sequence is commonly used to detect intrusions. Sekar et al. [24] proposed to use a program counter along with system calls to model the normal system call sequence as a DFA. [8, 7] proposed to use call stack information to reliably detect carefully crafted intrusions. Mimicry attack was first introduced by Wager et al. [30]. It modifies the system call sequence generated by the attack code so that it matches the normal system call sequence. An advanced mimicry attack was described in [12] that can evade the IDSs that uses call stack information. Polymorphic blending attack is similar to mimicry attacks because both try to evade anomaly IDS by matching the attack characteristics to the normal profile. Mimicry attacks target host-based IDSs, whereas polymorphic blending attacks target network IDSs.

[2] raises doubts on the security of machine learning based IDSs in the face of a determined and resourceful attacker. It discussed the possible manipulation of the IDS training process so that the normal space of an IDS is gradually moved to include attack packets. A polymorphic blending attack takes a different approach and modifies the attack instance to move it closer to the normal space. CLET [5] is a publicly available polymorphism tool which tries to perform preliminary blending. CLET adds extra padding bytes in a given attack packet in an effort to match the byte frequency distribution of the attack to the normal profile. Our recent work [10] explores polymorphic blending attacks and presents basic techniques for generating such attacks. It shows that polymorphic blending attacks are feasible, and presents a case study for PAYL 1-gram and 2-gram. In this paper, we present a methodical approach to polymorphic blending attacks. Our research includes formal models and theoretical results, general algorithms, and efficient heuristics.

3. A FORMAL FRAMEWORK

The basic observation behind a PBA is that a network IDS monitoring high-speed and high-volume traffic typically uses simple statistical measures instead of complex structural or semantic information to model the normal traffic. An attacker can exploit this simplicity or limitation to devise attacks capable of evading the IDS.

In a polymorphic blending attack, it is assumed that the attacker knows the features and the algorithms used in the IDS [10]. Given some normal data packets, the attacker can generate an artificial normal profile that is close to the normal profile being used by the IDS. The attacker can roughly guess the error threshold of the IDS using an estimation of the desired false positive and detection rate of the IDS. After the attacker estimates the normal profile used by the IDS, the attacker decides the various parameters (e.g., attack vector, decryptor, encryption scheme, etc.) and structure (e.g., placement of each attack section in the payload) of the attack data. Then the attacker carefully chooses an encryption key so that the encrypted attack body closely matches the desired normal profile. Lastly the attacker pads the attack data with some normal data to match the attack packet even closer to the normal profile. A detailed description of each of the above steps can be found in [10].

In this work, we study the problem: given an anomaly detection system and an attack, can one automatically generate the PBA instances? Our approach is to develop a formal framework that starts with the models for the IDS. Based on these models, we can then reason about the complexity of the problem of generating a PBA, and develop the general algorithms for solving the problem.

3.1 Modeling Intrusion Detection Systems

In our recent work [10], we considered a class of anomaly detection systems that use only simple byte statistics of the normal traffic. We would like to generalize the concept of polymorphic blending attack to include a wide range of anomaly detection systems that use other structural information of the normal traffic.

3.1.1 Anomaly Detection Systems

Since a polymorphic attack typically mutates only the packet payload, we limit our scope to payload-based anomaly IDS. These systems record the statistics and structure of the bytes present in the normal network traffic packets. Such anomaly IDS proposed by researchers include NETAD[18], LERAD [19], service-specific

IDS [14], PAYL [31], and structure based detection of Web attacks by Kruegel et al. [15]. We observe that these IDSs can be represented as stochastic Finite State Automaton (sFSA) or equivalently stochastic Regular Grammar (sRG). sFSA is similar to FSA and has a probability assigned to all the transitions in the FSA.

3.1.1.1 PAYL.

PAYL records the average frequency of different unique n-grams that appear in normal traffic packets. An n-gram model can be described using an sFSA where each state represents the unique $(n-1)$-gram corresponding to the last $(n-1)$ bytes in the packet. A transition from state $A(a_0 a_1 \cdots a_{n-2})$ to state $A'(a_1 a_2 \cdots a_{n-1})$ exists if and only if n-gram $(a_0 a_1 \ldots a_{n-2} a_{n-1})$ is present in the normal traffic. The probability of a transition is equal to the probability of the corresponding n-gram in the normal traffic. Every state is a start state and every state is an accept/end state. For example, 1-gram model can be represented using a single state FSA: for every unique byte in the normal traffic, there exists a transition from the state to itself; and the probability of the transition is the same as the frequency of the byte in normal traffic.

3.1.1.2 NETAD and LERAD.

Mahoney et al. presented a series of anomaly IDSs that use some network level data along with some payload data to detect intrusions. These systems use attributes such as bytes or words present at specific positions in the payload. A LERAD rule is of the form (if $word_1 = x1, \cdots, word_{m-1} = x_{m-1}$ then $word_m \in X = \{x_{1,m}, \cdots, x_{n,m}\}$). Such a rule can be seen as a regular grammar of the form $(x_1 x_2 \cdots x_{m-1} \{x_{1,m} | \cdots | x_{n,m}\})$. Multiple rules can be combined using the '|' term to obtain a single regular grammar.

3.1.1.3 Structure-Based Systems.

Kruegel et al. presented an IDS for Web services. A Web traffic packet is divided into attributes, and different attributes are recorded using different byte characteristics, including: attribute length, byte frequency, byte structure using sRG, and token set. As in PAYL, byte frequency can be represented using a sFSA with one state. Token set ($T = \{t_1, \cdots, t_n\}$) can be seen as a regular grammar of the form $(t_1 | \cdots | t_n)$. Models of the different attributes can be combined to form a single sFSA. The length constraint on an attribute is handled separately during the blending attack generation.

In summary, the above anomaly detection systems can be represented using a sFSA. An anomaly detection system normally allows some errors. Typically, it uses some distance metric to define the distance of an observed packet from the normal profile. If the distance is smaller than a threshold, the packet is considered normal. Otherwise, the packet is considered anomalous. The distance metric is orthogonal to the IDS models and cannot be directly incorporated in the corresponding sFSA. In later sections, we will show how the distance metric is accounted for when generating a polymorphic blending attack.

3.1.2 An Example

We have shown that a wide range of anomaly IDSs represent the byte statistics and/or structure of a normal packet using either FSA or sFSA. One main reason for using FSA (or equivalently, regular expression) is that determining whether a string is generated by a FSA is very fast, and can thus be used over high speed networks.

In this work, we assume that the attacker is trying to evade an IDS that can be represented using either an FSA or sFSA. Figure 1 shows a simple example of a sFSA IDS. The IDS accepts the strings containing only the following tuples: ab, ba, and bb. We use this simple IDS as a running example throughout the paper.

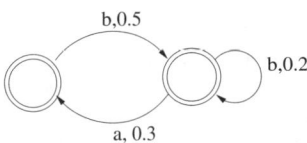

Figure 1: Simple sFSA IDS containing 3 tuples

$aaba$	cbb	$k_1 k_2$	$00h$ $01h$ $00h$ $00h$
AV	Dec	Key	Attack Code

$aaba$	cbb	$k_1 k_2$	$00h \oplus k_1$ $01h \oplus k_2$ $00h \oplus k_1$ $00h \oplus k_2$
AV	Dec	Key	Encrypted Attack Code

Figure 2: Simple attack example. Attack code is 4 byte string with `NUL` and `SOH` `ASCII` characters.

3.2 Polymorphic Blending Attack

As in [10], we focus on the attacks that allow arbitrary code execution. Thus, the attack packet contains a shellcode that is run on the victim host. A polymorphic attack contains five sections:

- Attack vector: exploits the vulnerability present on the victim host. Polymorphism of attack vector is achieved using different attack mutation techniques as discussed in [10].
- Attack code: the shellcode that the attacker wants to execute at the victim host. The attack code is stored encrypted in the attack packet.
- Polymorphic decryptor: decrypts the encrypted attack code and transfers control to attack code. Various code obfuscation techniques can be used to generate different instances of polymorphic decryptor.
- Decryption table: used to decrypt the attack code.
- Padding: extra (junk) data appended in order to closely match the normal profile.

In [10], we used a simple byte substitution scheme for encryption. During encryption, every attack character in the attack body is substituted by a normal character. To store the reverse substitution (or decoding) table of the simple byte substitution scheme, we use the same technique as in [10]: the index of the decoding table determines the attack character, and the entry at an index is the normal character used to substitute the corresponding attack character.

The polymorphic blending techniques studied in this paper include both *XOR* encryption scheme and byte substitution scheme. It is important to consider *XOR* because there are several existing polymorphism tools that use *XOR* based encryption already, and these tools may be extended to generate PBA. Unlike substitution, the decryption key for *XOR* is the same as the encryption key, and can be stored in a straight forward manner.

Both substitution and *XOR* are very simple schemes and are used in more complex encryption schemes. By studying PBA with these simple schemes, we hope to develop a understanding as well as solutions applicable to more complex schemes.

3.2.1 Generating An Attack Instance

Assume that the attacker has learned a (s)FSA corresponding to the (artificial) normal profile of the targeted IDS, the next step is to design an attack packet that can be accepted by the (s)FSA.

First, the attacker decides on the encryption scheme used for encrypting the attack code. Then a mutated instance of the attack vector and a polymorphic decryptor is generated, and their positions in the attack packet are determined.

The attack packet sections of the attack vector and polymorphic decryptor should preferably be accepted by the (s)FSA already. In some cases, there does not exist a path in the (s)FSA that corresponds to these attack packet sections (e.g., the mutated attack vector still contains characters not seen in normal packets), resulting in errors when the packet is matched with the IDS normal profile. If such error is greater than the IDS threshold, it means that it is not possible to generate a successful polymorphic blending attack.

The next step is to determine the encryption key. The main requirement is that the packet sections of the encrypted attack code and the decryption key should be accepted by the (s)FSA. For a sFSA, the additional requirement is for the whole attack packet to also match the transition probabilities. One approach is to first adjust the sFSA for what have been already matched by the attack packet sections of the attack vector and the polymorphic decryptor, then find a encryption key so that the sections of the encrypted attack code and the decryption key match the remaining parts of the sFSA. More specifically, the path taken by the attack vector and the decryptor in the sFSA is first identified. If there does not exist such a complete path (e.g., some transitions are not in the sFSA), there will be an error matching the attack packet with the normal profile already. If there exists multiple paths, the path with high probability transitions is chosen. Then the probabilities of the transitions present in the path are reduced according to the number of times a transition appears in the path. An encryption key is then chosen so that the sections of the encrypted attack code and the decryption key can match the adjusted sFSA closely.

The final, and sometimes optional, step is to pad the attack packet to have a desired packet length. For a sFSA, padding can be used to make the final attack packet match even more closely with the normal profile. The process works as follows. First, given the original sFSA representing the IDS, adjust the probabilities of the transitions in sFSA for the attack vector, the decryptor, the key table, and the encrypted attack code. That is, similar to the step of determining the encryption key discussed above, the sFSA is adjusted according to what have already been matched by these existing attack packet sections. Then more bytes are padded to the attack packet to match the remaining transitions and probabilities of the sFSA.

We use a simple blending attack (shown in Figure 2) to demonstrate the different concepts presented in the paper. We use an *XOR* encryption scheme with key length 2 (obviously, the encryption key and the decryption key is the same in an *XOR* scheme). The attack vector, decryptor, key, and attack code are concatenated in a given order to produce an attack packet payload. Although we assume an attack structure as shown in Figure 2, the techniques presented in this paper should work for other attack structures.

For convenience, we denote the string corresponding to the decryption key concatenated with the encrypted attack code as S_{key_ac}.

4. FORMAL ANALYSIS

As discussed in Section 3.2.1, one of the steps in generating a PBA is to find an encryption key so that the attack packet sections of the decryption key and the encrypted attack code, or S_{key_ac}, can be accepted by the FSA or the adjusted sFSA (for convenience, in this section we simply call the adjusted sFSA a sFSA). We will show that this is a hard problem even when using very simple encryption schemes, namely, substitution and *XOR*. As a corollary, the problem is hard when using more complex encryption schemes. The most direct and important corollary, however, is that the problem of generating a PBA is a hard problem.

A brute force approach to find the encryption key requires generating every possible key and checking the distance (as defined by the IDS) of the S_{key_ac} to the (s)FSA. For a simple substitution-based (encryption) scheme, this will take at least $^nP_m(n-m)^{n-m}$ iterations, where n is the number of unique normal characters and m is the number of unique attack characters. For *XOR* encryption with key of length l, finding an optimal polymorphic blending attack will take at least n^l iterations. These numbers can be very large, and thus a brute force approach is often impractical.

In this section, we first analyze the hardness of finding an encryption key that ensures S_{key_ac} is a valid string accepted by the FSA (without any transition probabilities) corresponding to the IDS. We show that this problem is NP-complete. Thus, it may not be solvable deterministically in polynomial time of the key length l or m. We show this result for both byte substitution based encryption and *XOR* based encryption. This result can be extended to show that even if we allow a solution to contain $\epsilon \|S_{key\_ac}\|$ number of invalid transitions, the problem is still NP-complete. We extend the above results and argue that the problem of finding an encryption key that optimally matches the S_{key_ac} to sFSA or finding a solution within an ϵ range of the optimal solution is also NP-complete.

4.1 NP-Completeness of the Blending Attack Problem

4.1.1 Substitution Based Encryption Scheme

We formally define the problem of finding a substitution key that ensures S_{key_ac} is accepted by the FSA of an IDS.

DEFINITION 4.1. *Given an attack code and the FSA of an IDS, the problem PBA_{sub}^{FSA} is to find a one-to-one mapping from attack characters to normal characters such that S_{key_ac} is accepted by the given FSA.*

THEOREM 4.1. *Problem PBA_{sub}^{FSA} is NP-complete.*

PROOF. For a problem to be NP-complete, the problem should be in NP and should be NP-hard.

A problem is in NP if a given solution can be verified for its correctness in polynomial time. Given a one-to-one mapping, we can easily generate the decryption key (a table) and the encrypted attack code. Since FSA is a decidable language, we can verify in polynomial time whether or not S_{key_ac} string will be accepted by the FSA. Thus, we can efficiently verify if the one-to-one mapping is correct or not.

To prove that the problem is NP-hard, we reduce the well known 3-SAT problem to PBA_{sub}^{FSA}. Consider a 3-SAT problem with q, $q \leq 128$, variables and r clauses. If q is smaller than 128, we add dummy unused variables to make the total number of variables to be 128. Suppose the 3-SAT problem is,

$$SAT = (x_{10} \vee x_{11} \vee x_{12}) \wedge (x_{20} \vee x_{21} \vee x_{22}) \wedge \cdots \wedge (x_{r0} \vee x_{r1} \vee x_{r2}),$$

where $x_{10}, x_{11}, \cdots, x_{r2} \in \{x_0, \overline{x_0}, x_1, \overline{x_1}, \cdots x_{127}, \overline{x_{127}}\}$.

Given the above 3-SAT, we design the PBA_{sub}^{FSA} problem as follows. For every variable x_i in the 3-SAT, we have an attack character att_i, two normal characters $norm_i$ and $norm_{i+128}$, and a corresponding entry e_{att_i} in the decryption table. e_{att_i}, $128 \leq i \leq 255$, is a dummy decryption table entry. The value of the variable x_i is related to e_{att_i} as follows.

$$x_i = 1, \text{ if and only if } e_{att_i} = norm_i \text{ and } e_{att_{i+128}} = norm_{i+128}$$
$$= 0, \text{ if and only if } e_{att_i} = norm_{i+128} \text{ and } e_{att_{i+128}} = norm_i \quad (1)$$

For every clause $clause_\alpha$, $1 \leq \alpha \leq r$ in the 3-SAT, we construct a section of FSA (FSA_α) as shown in Figure 3. First, we construct the truth table of the clause (see Table 1). For each entry in the truth table, we have a path containing three edges where each edge corresponds to the value of a variable in the truth table entry. In addition to FSA_α, $1 \leq \alpha \leq r$, we have a section of FSA (FSA_{KT}) of length 256 corresponding to the key.

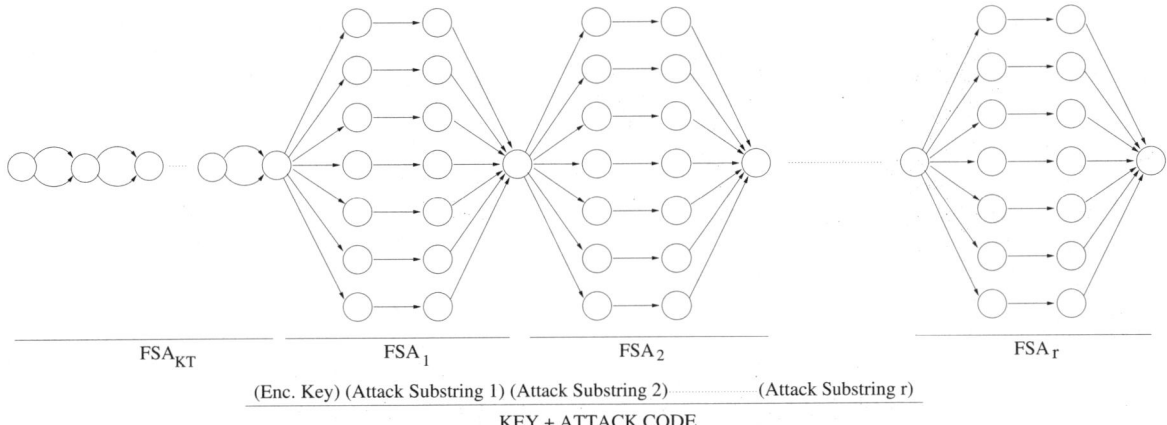

Figure 4: FSA and S_{key_ac} corresponding to the SAT problem

x_1	x_3	x_8	e_{att_1}	e_{att_3}	e_{att_8}
0	0	0	$norm_{129}$	$norm_{131}$	$norm_{136}$
0	0	1	$norm_{129}$	$norm_{131}$	$norm_8$
0	1	1	$norm_{129}$	$norm_3$	$norm_8$
1	0	0	$norm_1$	$norm_{131}$	$norm_{136}$
1	0	1	$norm_1$	$norm_{131}$	$norm_8$
1	1	0	$norm_1$	$norm_3$	$norm_{136}$
1	1	1	$norm_1$	$norm_3$	$norm_8$

Table 1: Truth table and corresponding key table for clause $x_1 \vee \overline{x_3} \vee x_8$

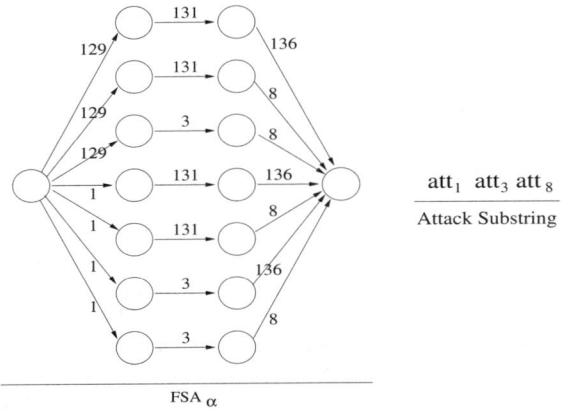

Figure 3: FSA_α **and attack substring for clause** $x_1 \vee \overline{x_3} \vee x_8$. **For convenience, we represent** $norm_i$ **by just** i.

Also, for every variable x_i or $\overline{x_i}$ in a $clause_\alpha$, we have an attack character att_i in the attack code. Thus for every clause $clause_\alpha$, we have a substring (str_α) of length 3 in the attack code. Figure 3 shows an example attack code substring for a hypothetical clause. The encoded attack substring will be $e_{att_1} e_{att_3} e_{att_8}$.

In Figure 3, we can observe that the encrypted str_α is accepted by the FSA_α if and only if the encoding of attack characters are chosen from one of the entries in the given encoding table shown in Table 1. Since every entry in the encoding table corresponds to an entry in the truth table of the $clause_\alpha$, the encrypted str_α is accepted by the FSA_α if and only if $clause_\alpha$ is *true*.

The final FSA, FSA_{SAT}, and the attack code corresponding to the 3-SAT problem are shown in Figure 4. The construction of the above FSA_{SAT} takes polynomial time. Please note that there exists a solution to the given PBA_{sub}^{FSA} problem if and only if the encrypted str_α is accepted by FSA_α for all $1 \leq \alpha \leq r$.

If the above PBA_{sub}^{FSA} problem has a solution mapping e_{att_i}, $0 \leq i \leq m-1$, then one can find assignments for variables x_i, $0 \leq i \leq 127$ using Equation 1. Since S_{key_ac} is accepted by FSA_{SAT} for mapping $e_{att_i}, 0 \leq i \leq m-1$, the encrypted str_α is accepted by FSA_α for all $1 \leq \alpha \leq r$. However, the encrypted str_α is accepted by FSA_α only if $clause_\alpha$ is *true*. Thus, all clauses of the 3-SAT problem is *true* and the 3-SAT is satisfied.

Also, if there exists an assignment of variables x_i such that the 3-SAT problem is satisfied, then we can compute e_{att_i} using Equation 1. Since 3-SAT is satisfied, all $clause_\alpha$, $1 \leq \alpha \leq r$, are *true*. But $clause_\alpha$ is *true* only if the encrypted str_α is accepted by FSA_α. Thus, all encrypted str_α, $1 \leq \alpha \leq r$, are accepted by FSA_α, and S_{key_ac} is accepted by FSA_{SAT}.

From above, we can conclude that PBA_{sub}^{FSA} is at least as hard as 3-SAT. Since 3-SAT is NP-hard, PBA_{sub}^{FSA} is also NP-hard. Since PBA_{sub}^{FSA} is also in NP, PBA_{sub}^{FSA} is an NP-complete problem. □

4.1.2 XOR Encryption Scheme

We formally define the problem statement of finding a XOR encryption key that ensures S_{key_ac} is accepted by the FSA of an IDS.

DEFINITION 4.2. *Given an attack code and the FSA of an IDS, the problem PBA_{xor}^{FSA} is to find an encryption key, of length l, so that S_{key_ac} is accepted by the given FSA.*

THEOREM 4.2. *Problem PBA_{xor}^{FSA} is NP-complete.*

PROOF. The proof of NP-completeness of PBA_{xor}^{FSA} is similar to the proof of PBA_{sub}^{FSA}. The proof is not provided in the paper due to the space restrictions. □

4.1.3 Corollaries

We have now proved that finding an encryption key that ensures S_{key_ac} is accepted by the FSA of an IDS is NP-complete. Suppose we allow the solution to have $\epsilon \| S_{key\_ac} \|$ number of invalid transitions, the problem still remains NP-hard because of the fact that $(1-\epsilon)$-SAT (or ϵ-UNSAT, $\epsilon < 1$) is an NP-hard problem.

Now consider the problem of finding an encryption key that optimally matches the S_{key_ac} to sFSA. This problem is considered harder than $PBA_{sub/xor}^{FSA}$ because in addition to the requirement of using only valid edges of the sFSA, we need to match the probability of each transition in the sFSA. Thus, finding the suitable encryption key for sFSA should be NP-hard. Following logic similar to above, we can also conclude that finding an encryption key that matches the S_{key_ac} to sFSA within ϵ bound of the optimal solution is also NP-hard.

Substitution and *XOR* are very basic encryption schemes. In fact, the more complex encryption schemes such AES and DES [11] use substitution and *XOR* as basic operations. Since we have shown that the problem is hard when the simpler schemes are in use, we can conclude that the problem is still hard when using the other more complex encryption schemes.

To conclude, we have shown that finding an encryption key that ensures S_{key_ac} is accepted by the (s)FSA of an IDS is a hard problem (NP-complete). The most direct and important corollary is that the problem of generating a polymorphic blending attack is hard. In fact, the problem of determining whether or not a polymorphic blending attack exists is also a hard problem. This follows from the fact that it is NP-hard to verify if a given SAT or ϵ-UNSAT problem has a solution or not.

4.2 Reduction to SAT and ILP

In Section 4.1, we showed that the problem of finding an appropriate encryption key for a polymorphic blending attack is very hard. That is, it may take time exponential to the key length or character size. Although an attacker may not have any time restrictions, a polynomial time solution is clearly more desirable. There are good heuristic solvers available for the SAT problems or Integer Linear Programming (ILP) problems. These solvers provide approximate solutions in very reasonable amount of time. If we have a non-stochastic, or FSA based, IDS, we can reduce the polymorphic blending attack problem to a SAT problem. For a sFSA based IDS, we can map the problem to ILP. Then a heuristic solver can be used to obtain solution for the reduced problem.

Before we show the SAT or ILP reduction, we would reduce the problem of finding an appropriate encryption key to the problem of finding a path from the start state to accept states in a Directed Acyclic Graph (DAG). An edge in the DAG may have constraints of the form $k_i = j$, $j \in U$, where k_i is the ith key and U is the set of all characters. The chosen path should contain a minimal number of conflicting constraints. In addition, we may have some restrictions on the frequency of occurrences of edges corresponding to the frequency matching requirement for sFSA. In the following sections, we use the example shown in Figures 1 and 2 to illustrate the concept.

4.2.1 Construction of DAG

Given a (s)FSA, a key length (l_k), and an attack code ac of length l_{ac}, we construct a DAG of depth $l_k + l_{ac}$ as follows. Suppose s_0 is the end state of the path in (s)FSA as traced by the attack vector and the decryptor. v_0 is the root vertex of the DAG corresponding to state s_0. At every depth d of the DAG, we have a set of vertices $V_d = \{v_{d_i}\}$ such that the state v_i is reachable from state s_0 in exactly d transitions in (s)FSA. The accept vertices of the DAG are the leaf vertices at depth $l_k + l_{ac}$, which correspond to accept states in the (s)FSA. There exists an edge $e_{d_{ij}}$ from v_{d_i} to $v_{(d+1)_j}$ if and only if there exists a transition t_{ij} from state v_i to state v_j in (s)FSA. The weight of the edge is proportional to the probability of transition t_{ij}. The constraint $constr_{d_{ij}}$ associated with an edge $e_{d_{ij}}$ at depth d is shown below.

$$\begin{aligned}constr_{d_{ij}} &= (k_d == char_{ij}), & \text{if } d < l_k, \\ &= (k_{ac[d-l_k]} == char_{ij}), & \text{if } d \geq l_k \text{ and substitution}, \\ &= (k_{(d \bmod l_k)} \oplus ac[d-l_k] == char_{ij}), & \text{if } d \geq l_k \text{ and XOR} \end{aligned}$$

where $char_{ij}$ is the normal character corresponding to transition t_{ij} and $ac[i]$ is the attack character at the i position of attack code. An example construction of DAG is shown in Figure 5. The DAG corresponds to the example FSA and example attack shown in Figure 1 and 2, respectively.

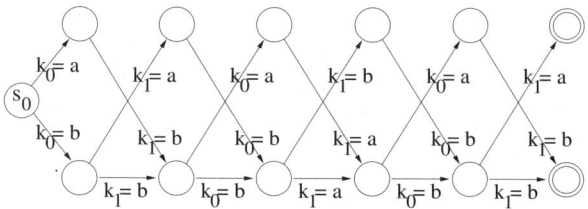

Figure 5: DAG corresponding to example FSA

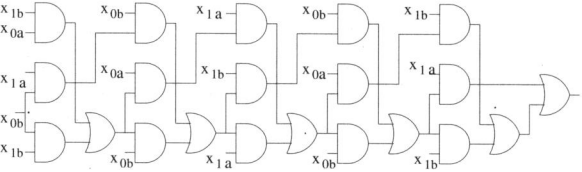

Figure 6: SAT representation of example DAG

The problem of finding an appropriate encryption key for a given attack code and FSA is equivalent to finding a path from the root vertex to an accept vertex in the DAG. Given a path P_{dag} in DAG consisting of edges with no (or minimal) conflicting constraints, we can find the encryption key by setting the constraints of the edges on the path to *true*. The path P_{fsa} followed by S_{key_ac} in the (s)FSA is similar to the path P_{dag}. If $e_{d_{ij}}$ is an edge at depth d in P_{dag} then transition t_{ij} is in P_{fsa} at depth d.

4.2.2 Translation to SAT

If the given IDS is an FSA with no probabilities on edges, the problem of finding a appropriate path in DAG can be translated to SAT. First, we translate the DAG problem to CIRCUIT-SAT [4]. Then we can efficiently translate CIRCUIT-SAT to SAT. For each constraint of the form $k_i = j$ in the DAG, we have a variable x_{ij} that is *true* if and only if the constraint is satisfied, and false otherwise. Now we can directly translate the DAG to CIRCUIT-SAT. A vertex v with input degree deg_{in} in DAG has a corresponding OR gate (OR_v) in CIRCUIT-SAT with deg_{in} inputs. For every outgoing edge (with some constraint $k_i = j$) of a vertex v, we have an AND gate whose input is x_{ij} and OR_v. The final output is OR of all the accept states. Figure 6 shows the conversion of our example DAG to CIRCUIT-SAT. We can then efficiently translate the CIRCUIT-SAT into a SAT problem. In the given SAT problem, we need to add additional requirement that a given key k_i is assigned to only one normal character. This means if x_{ij} is $true$ then $x_{ij'}$ is $false$ for all $j' \neq j$. Also, for substitution based encryption scheme, we need to add additional clauses in SAT to ensure that a normal character is assigned to a single attack character. These cardinality constraints can be efficiently represented in SAT [26]. Furthermore, for a substitution scheme, there exists empty entries in the decryption key table corresponding to the characters not present in the attack code. These characters can be mapped to any unassigned normal character. This requirement can be written as clause $\bigwedge_{j \in N}((\bigvee_{i \in M} x_{ij}) \vee (\bigvee_{i \in \overline{M}} x_{ij}))$, where M and N are the set of attack and normal characters, respectively. We can solve the final SAT problem using one of the several available SAT solvers, which are capable of solving huge SAT (or c-UNSAT) problems in reasonable amount of time.

4.2.3 Translation to ILP

For a sFSA IDS, we propose translating the problem of finding a good path in a DAG into an Integer Linear Programming problem. ILP is known to be NP-hard but there exists multiple good heuristics to solve big ILP problems. An ILP tries to find the minimum of a linear function over a set of variables defined by a finite number of linear constraints. All the variables in the solution should be integers. An ILP is called 0-1 ILP if all the variables are required to be either 0 or 1. Now we will show the reduction of finding the optimal path in a DAG to ILP problem.

For every edge $e_{d_{ij}}$ at level d in the DAG, we have a variable $h_{e_{d_{ij}}}$.

$$h_{e_{d_{ij}}} = 1 \text{ if edge } e_{d_{ij}} \text{ is in the } solution\ path \text{ of DAG,}$$
$$= 0 \text{ otherwise}$$

For every constraint $x_i = j$ in the DAG, we have a variable $constr_{ij}$.

$$constr_{ij} = 1 \text{ if constraint } x_i = j \text{ is } true \text{ in the solution of DAG,}$$
$$= 0 \text{ otherwise}$$

There is a valid path from the start vertex to an accept vertex if and only if the number of *selected* outgoing edges at the start state is one, the number of *selected* incoming edges is equal to the number of *selected* outgoing edges for all the intermediate vertices, and the number of *selected* incoming edges is equal to one for one of the end vertices. An edge is *selected* if it is in the solution path. These three conditions can be represented in terms of linear equations as follows:

$$\sum_{e \in OUT(v_0)} h_e + err_{v_0} = 1$$

$$\sum_{e \in IN(V_{accept})} h_e + err_{accept} = 1$$

$$\sum_{e \in IN(v_{di})} h_e + err_{di} = \sum_{e \in OUT(v_{di})} h_e, \forall\ v_{di} \text{ at depth } d, \forall 1 \leq d \leq l_k + l_{ac} - 1$$

where $IN(v)$ and $OUT(v)$ are the sets of in and out edges of vertex v, respectively. v_{di} is the ith vertex at depth d. The err terms account for invalid paths in the solution and are 0 if the path conditions are satisfied. In case the condition is not satisfied, err_{di} can be either 1 or -1 depending on the difference of the number of *selected* incoming and outgoing edges at the given node.

At any depth $d, 0 \leq d \leq l_k + l_{ac} - 1$, the number of edges from vertices at depth d to vertices at depth $d+1$ should be one. That is,

$$\sum_{e \in OUT(V_d)} h_e + err_{V_d} = 1, \forall 0 \leq d \leq l_k + l_{ac} - 1 \quad (2)$$

where V_d is the set of vertices at depth d. Again, the err_{V_d} term accounts for the errors. err_{V_d} can take values 0 or 1 depending on the number of outgoing edges at a given depth.

If an edge in the DAG is chosen in the path, the corresponding constraint should be satisfied. Suppose $constr_e$ represents the constraint associated with edge e. Then the requirement can be satisfied using following equation:

$$constr_e \geq h_e, \forall \text{ edge } e \in \text{DAG} \quad (3)$$

Further, we can ensure that a given key is assigned to only one character by using following equation:

$$\sum_{j \in U} constr_{ij} = 1, \forall 0 \leq i \leq l_k \quad (4)$$

where U is the set of all possible characters and l_k is the key length.

For a one-to-one byte substitution scheme, a normal character should not be assigned to multiple attack characters. That is,

$$\sum_{i \in M} constr_{ij} \leq 1, \forall j \in N \quad (5)$$

where M and N are the set of attack and normal characters, respectively. The following set of equations ensure that the characters not present in the attack character set are mapped only to normal characters not assigned to attack characters.

$$NAC_j \times \|\overline{M}\| \geq \sum_{i \in \overline{M}} x_{ij}, \forall j \in N,$$
$$NAC_j + \sum_{i \in M} x_{ij} \leq 1, \forall j \in N$$

The first equation makes sure that NAC_j is 1 if any non-attack character is mapped to a normal character j. The second equation ensures that if NAC_j is 1, then the normal character j is not assigned to any attack character, and vice-versa.

The above set of equations guarantee that there exists a path from the start vertex to the end vertex with some errors. Now we present the minimization criteria to reduce the errors and the distance of S_{key_ac} from the sFSA. Assume a distance metric of the form:

$$dist = \sum_{transition\ t \in sFSA} \kappa_t \times |p_t - \frac{num_t}{l_k + l_{ac}}| \quad (6)$$

where κ_t is some constant associated with transition t, p_t is the probability of the transition to be taken in sFSA, and num_t is the number of times the transition t is taken by the S_{key_ac}.

The minimization criteria for the ILP problem can be written as:

$$\sum_{transition\ t \in sFSA} const_t \times |p_t - \frac{\sum_d h_{d_t}}{l_k + l_{ac}}| + \sigma \times (\sum_{v \in V_{dag}} |err_v| + \sum_d err_{V_d}) \quad (7)$$

where V_{dag} is the set of vertices in the DAG and σ is the weight of the errors caused by taking a invalid edge in the sFSA. Note that the $|\alpha - \beta|$ term in minimization can be rewritten as abs_{diff} where, $\alpha - \beta \geq -abs_{diff}$ and $\alpha - \beta \leq abs_{diff}$.

Solving the above ILP for the given minimization criteria will provide the encryption key. Using this we can generate the encrypted attack code and prepare the polymorphic blending attack packet by concatenating attack vector, decryptor, decryption key, and the encrypted attack code. We can then perform padding to match the final attack packet even closer to the normal, as discussed in Section 3.2.1.

4.3 Heuristic Solutions

Rather than finding the optimal solution, an attacker may simply apply some heuristics that produce an approximate (or good enough) solution using much less resources (time and memory). Here we present a simple heuristic that finds a good approximate solution very efficiently.

The heuristic is based on the *hill climbing* algorithm, which is used widely in artificial intelligence and constraint solving. *Hill climbing* starts with an initial solution and iteratively improves it. At each step, the algorithm looks at the neighboring solutions and chooses one that is better than the current solution. The definition of neighbors depends on the problem domain.

We now present our heuristic. Given an IDS and an attack instance, we choose a random encryption key and calculate the distance between S_{key_ac} and (s)FSA. Now, we randomly choose a k_i to modify in the key. For all the possible character (c) values, we first temporarily assign $k_i = c$. For a substitution scheme, if c is already assigned to some attack character k_j, we temporarily swap the normal characters assigned to k_i and k_j. We then find the new distance to (s)FSA using the temporary key. We choose the character that reduces the distance by the maximum value and assign it to k_i. At this point, a new key position (k_j) is chosen to modify and the process is repeated. This above process is iterated for the desired number of iterations or until a satisfactory solution is produced.

It is possible in the above approach to reach a local maximum that is not very close to the optimal solution. We reach a local maximum if modifying any key increases the distance to (s)FSA. To overcome this problem, whenever we reach local maximum, we choose a small set of key positions and set them to some random values, and restart the above iterative process of finding solution. The idea is by randomly picking another starting point in the solution space, the new solution point may belong to a locale that has better local maximum.

The above heuristic can give us very good solution with sufficient number of iterations.

5. EXPERIMENTS AND RESULTS

The motivation of our research was to address the open problem: given an anomaly detection system and an attack, can one *automatically* generate the PBA instances? Thus, in our experiments, we wanted to directly compare our formal framework with the more ad-hoc approaches developed in our previous work [10]. The key elements of our experimental set-up were the same as in [10]. That is, we used the same anomaly detection systems, namely, PAYL 1-gram and 2-gram, as well as the same attack and same traffic datasets, as in [10]. The results showed that, although our framework is based on an abstract model of IDS and uses general algorithms, it automatically generated PBA instances that were more evasive (i.e. better matched the IDS normal profiles) than or at least as good as the PBA instances from the more PAYL specific algorithms in [10].

Briefly, the experimental dataset contains 15 days of Web traffic with 4.7 million packets. 14 days of traffic were used for training the IDS profile. A part of the remaining 1 day of traffic was used to generate/train an artificial profile used by the attacker. We generated PAYL 1-gram and 2-gram models (using the 14 days of traffic) for three different packet lengths, namely, 418, 730, and 1460. The attack vector is based on the implementation of `firew0rker` [9]. More information of the dataset can be found in [10].

For all the experiments, we divided the attack flow into multiple packets. The attack vector was placed at the start of the first packet. The decryptor was divided into several sections and allotted to different attack packets. The attack body was also divided into multiple chunks. The sFSA corresponding to the artificial profile was adjusted for attack vector and polymorphic decryptor. A separate encryption key for each attack body fragment was generated using our framework to match the adjusted artificial profile. Each attack body fraction was encrypted using the corresponding key and appended to the corresponding attack packet. Then each attack packet was padded to the desired packet length. The final attack packets were then used together to launch an attack.

5.1 PAYL 1-gram Evasion

We applied our framework to generate polymorphic blending attacks to evade 1-gram PAYL, using substitution-based encryption and XOR encryption, respectively. For the XOR scheme, we used a 64 byte key. For each encryption scheme, we translated the problem of finding the optimal encryption key for 1-gram evasion to an ILP problem. We used *ILOG CPLEX* to solve the ILP problems. *CPLEX* is a commercial optimization tool for solving Mixed Integer Programs (MIP). We obtained a near-optimal solution (i.e., encryption key) for the ILP problems. The attack code was then encrypted using this key, and padding was performed. For comparison, we also generated polymorphic blending attacks using the one-to-one local substitution scheme presented in [10].

It took 6.5 seconds on average to solve an ILP problem on a Pentium-M 2GHz machine. The solution provided was within 0.2% of the optimal solution. Figure 7 shows the distance of the attack flow from the artificial profile and the IDS profile. The results for both substitution and XOR encryption schemes, as well as the scheme from [10], are shown in the figure. The x-axis shows the number of packets attack flow was divided into and the y-axis shows distance of the attack flow from the artificial profile and IDS normal profile. This distance is the maximum of the distances of individual attack packets in a flow. A horizontal line corresponding to anomaly error threshold for the 1% IDS false positive is also shown.

The error distance of attacks generated using substitution based encryption with ILP is almost identical to the previous approach from [10]. Thus, the 1-gram blending approach in [10] also provides a near optimal substitution table.

The error distance for attacks generated using the XOR encryption scheme is much higher. This is expected. For substitution, by replacing attack characters with normal characters, we can ensure that only normal characters are present in the mutated attack packet. For *XOR*, it is harder to find an appropriate key such that it contains only normal characters and *XOR*ing it with attack characters also results in only normal characters. For packet length 418 and 730, the error distance of PBA generated using XOR based scheme is twice or more than the substitution-based scheme. For packet length 1460, the error distance for XOR based scheme is comparatively smaller. Also, the difference decreases as the number of attack packets in the attack flow increases. The large amount of padding space available masks the error produced by the attack code in XOR based scheme.

In the plots, all the attack points below the horizontal error threshold line will not be detected by the IDS with a 1% false positive rate. If the false positive rate is decreased, typically the anomaly error threshold is increased. That is, if the horizontal line is moved up, more attack points will be missed by the IDS.

For packet length 730 and 1460, we need only two packets to evade PAYL 1-gram when using substitution. The IDS can be evaded using attack flow of size as low as 1460. For packet length 418, we need 8 packets to evade the IDS. The XOR based scheme can also evade the IDS for packet lengths 730 and 1460, although with bigger attack size. For packet length 418, XOR needs to divide attack packets into many more packets in order to evade the IDS.

5.2 PAYL 2-gram Evasion

We also generated polymorphic blending attacks to evade PAYL 2-gram. We used the heuristic presented in Section 4.3 to generate such attacks. We started with a random solution and iterated the hill climbing steps 25000 times. The best encryption key seen during the process was recorded. The attack code was then encrypted using this key and the attack packet was padded to the desired length. The distance of the attack flow from the normal profiles was recorded. For comparison, we also generated the polymorphic blending attacks using the 2-gram blending algorithm presented in [10].

For substitution based encryption, it took $10min$ on average to perform 25000 iterations on a given problem. For XOR encryption, performing 25000 iterations took little more than an hour on average. The time of each iteration is dependent on the range of keys and the number of terms in the distance calculation. For substitution, the range of keys is all the normal characters; whereas

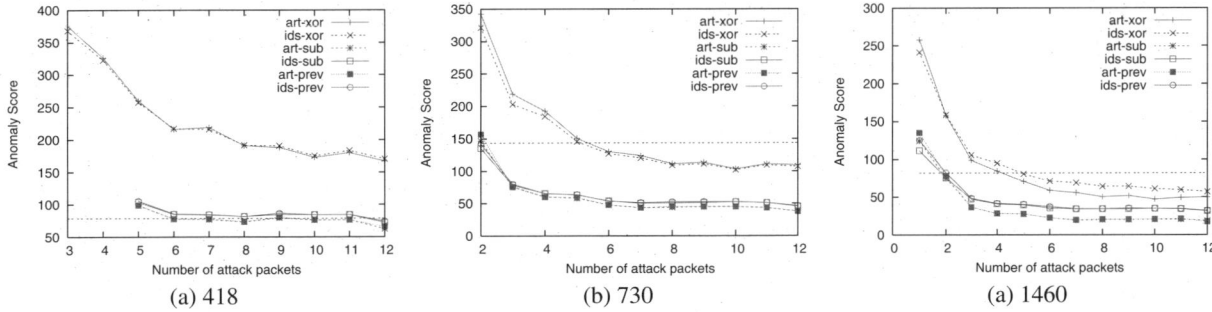

(a) 418 (b) 730 (a) 1460

Figure 7: Anomaly score or error distance of 1-gram blending attack. The plots with prefix *art* and *ids* corresponds to distance from the artificial profile and the IDS profile, respectively. *xor* and *sub* corresponds to the PBA generated for XOR and substitution based schemes using our framework. *prev* denotes the algorithm from previous paper.

in XOR, the range of keys can be the set of all the possible characters. Also, since XOR is not able to match the normal profiles closely, there are several new 2-grams in the attack packet and thus the number of terms in the distance calculation may be large. These two reasons account for the long running time for XOR.

The distances of the attack packets from the normal profiles are shown in Figure 8. The results for substitution-based encryption and XOR encryption, as well as the scheme from [10], are shown, along with a horizontal line corresponding to error threshold of the 1% IDS false positive rate.

Using substitution, our heuristic-based approach is able to better match the normal profile than the previous approach [10] for most attack instances. For packet length 418, when the number of attack packets is 8 or 9, the previous approach matches the normal profile better than the heuristic from our framework. Checking the distance of individual attack packets, we observed that for some packets, the heuristic got stuck in a bad local maximum for considerable number of iterations. Thus, the heuristic was not able to find a good solution in the given number of iterations. In such cases, one can restart the heuristic using another random solution or run the heuristic for more number of iterations.

Compared with PAYL 1-gram, we needed a larger number of attack packets to evade PAYL 2-gram. The minimum attack size required to evade the PAYL 2-gram is 2190, or 3 packets of length 730.

6. HARDENING THE IDS AGAINST POLYMORPHIC BLENDING ATTACKS

Our framework can be used to improve an stochastic IDS to make it harder for an attacker to evade the IDS. The basic idea is to find transitions (or edges) in the sFSA that are traversed frequently by multiple polymorphic blending attacks. If an edge contributes only a small error term in computing the distance of PBAs to normal profile but contributes a large error term for normal packets, we can exclude the edge when calculating the distance of a monitored packet from the normal file. The result will be increased detection rate against attacks, including the PBAs, with maybe only slightly increased false alarm rate.

We now explain in more details how to improve the IDS against PBA. First, we collect multiple attack instances for different known attacks on the system. For each of these instances, we generate multiple polymorphic blending attacks using techniques described in Section 4.2. For all of these attacks pba_i, we find paths $path_i$ taken by pba_i in the sFSA. Suppose the number of times an edge e_j is present in the $path_i$ is denoted by n_{ji} and their contributing term in distance calculation is $dist_{ji}$. Then the relative contribution of edge e_j in the distance between polymorphic blending attacks and sFSA is calculated as $pba\_rel\_dist_{e_j} = \frac{\sum_{pba_i} dist_{ji}}{\sum_{e_k \in E}(\sum_{pba_i} dist_{ki})}$, where E is the set of all edges in the sFSA. Similarly, we calculate the relative contribution of edge e_j in the distance between normal packets and sFSA as $nor\_rel\_dist_{e_j} = \frac{\sum_{nor_i \in T} dist_{ji}}{\sum_{e_k \in E}(\sum_{nor_i \in T} dist_{ki})}$, where T is the set of training data packets.

We would like to exclude the edges that have a high value of $\frac{nor\_rel\_dist_{e_j}}{pba\_rel\_dist_{e_j}}$ and that occur with high frequency in PBAs. Let $deg_{e_j} = \frac{nor\_rel\_dist_{e_j}}{pba\_rel\_dist_{e_j}} \times pba\_p_{e_j}$, where $pba\_p_{e_j}$ is the probability of an edge e_j being present in a PBA. deg_{e_j} denotes the degree with which an edge e_j is assisting the attacker in evading the IDS. We exclude the edges that have the higher values of deg_{e_j}. Note that we are not actually removing the edge from the sFSA. Rather, by excluding the distance term corresponding to edge e_j, we are marginally reducing the distance of blending attack packets from the IDS but significantly reducing the distance of normal packets from the IDS. Thus, the error distance threshold of the IDS can be reduced while keeping the false positive rate almost same and increasing the detection rate for a PBA. If an attacker launches a new PBA against the IDS using a new exploit, the IDS will have a high probability of detecting the new attack if it takes the similar path (in sFSA) as the attacks pba_i. We plan to perform more detailed analysis of above proposed technique.

7. CONCLUSION

In this paper, we presented a formal framework for polymorphic blending attacks. We modeled a variety of IDSs as either FSA or sFSA. We then reduced the problem of finding an optimal PBA that matches the IDS normal profile to the problem of finding an encryption key that optimally matches the string representing *the decryption key concatenated with the encrypted attack code* to the (s)FSA of the IDS. We showed that finding such an encryption key is an NP-complete problem. We presented some techniques to reduce this problem to a satisfiability problem and an nteger linear programming problem. Thus, optimization algorithms available for these problem domains can be used to generate a near-optimal encryption key and hence, a near-optimal PBA. We also proposed a heuristic that can be used to find good approximate encryption keys very efficiently. We validated our framework using PAYL 1-gram and 2-gram. We also proposed a technique to improve the performance of an IDS against PBAs.

The results of our experiments showed that our framework can automatically generate PBAs that can evade an IDS. The PBAs

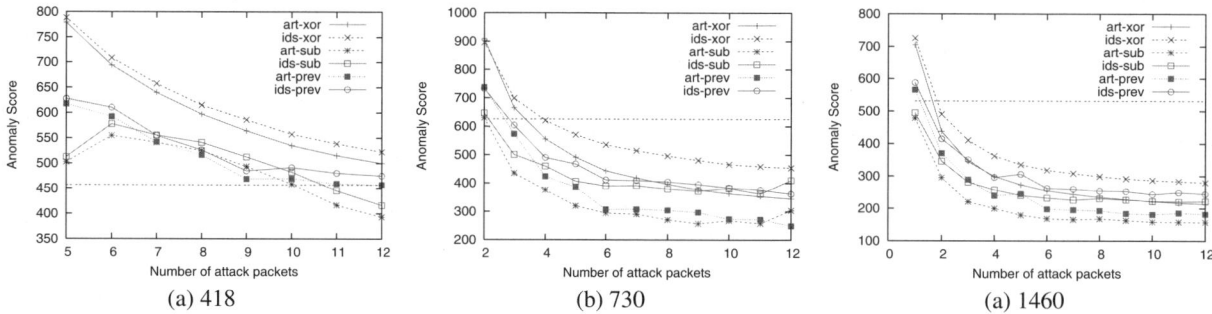

Figure 8: Anomaly scores of 2-gram blending attacks.

generated by our framework were able to match the normal profile more closely than the PBAs produced by the previously proposed IDS specific algorithms. The time required to solve the ILP problem to generate a PBA to evade PAYL 1-gram was only a few seconds. Generating attack packets to evade PAYL 2-gram took several minutes. A substitution-based encryption scheme was shown to be more effective than *XOR* encryption for evading an IDS.

A polymorphic blending attack is also achievable using other shellcode transformation techniques, for e.g., equivalent instruction substitution and garbage insertions. We plan to further study polymorphic blending attacks by incorporating different attack mutation techniques, metamorphism and code obfuscation. Our current framework focuses on algorithms for generating the (optimal) encryption key. We plan to extend it to automatically determine the best mutation techniques as well the optimal padding bytes.

Acknowledgements

This work is supported in part by NSF grant CCR-0133629 and Office of Naval Research grant N000140410735. The contents of this work are solely the responsibility of the authors and do not necessarily represent the official views of NSF and the U.S. Navy. The authors would like to thank Dr. Emilie Danna for her valuable suggestions.

8. REFERENCES

[1] P. Akritidis, E. P. Markatos, M. Polychronakis, and K. Anagnostakis. Stride: Polymorphic sled detection through instruction sequence analysis. *In 20th IFIP International Information Security Conference*, 2005.

[2] M. Barreno, B. Nelson, R. Sears, A. D. Joseph, and J. D. Tygar. Can machine learning be secure? *In Proceedings of the ACM Symposium on Information, Computer, and Communication Security (ASIACCS)*, 2006.

[3] M. Christodorescu, S. Jha, S. Seshia, D. Song, and R. Bryant. Semantics-aware malware detection. *In Proceedings of the IEEE Symposium on Security and Privacy*, 2005.

[4] T. H. Cormen, C. E. Leiserson, and R. L. Rivest. Introduction to algorithms. *The MIT Press/McGraw-Hill*, 1990.

[5] T. Detristan, T. Ulenspiegel, Y. Malcom, and M. Underduk. Polymorphic shellcode engine using spectrum analysis. *Phrack Issue 0x3d*, 2003.

[6] S. T. Eckmann, G. Vigna, and R. A. Kemmerer. Statl: An attack language for state-based intrusion detection. *JOURNAL OF COMPUTER SECURITY*, 10:71–104, 2002.

[7] H. Feng, J. Giffin, Y. Huang, S. Jha, W. Lee, and B. Miller. Formalizing sensitivity in static analysis for intrusion detection. *In Proceedings of the IEEE Symposium on Security and Privacy*, 2004.

[8] H. Feng, O. Kolesnikov, P. Fogla, W. Lee, and W. Gong. Anomaly detection using call stack information. *In Proceedings of the IEEE Symposium on Security and Privacy*, 2003.

[9] Firew0rker. Windows media services remote command execution exploit. *http://www.k-otik.com/exploits/07.01.nsiilog-titbit.cpp.php*, 2003.

[10] P. Fogla, M. Sharif, R. Perdisci, O. M. Kolesnikov, and W. Lee. Polymorphic blending attacks. *In 15th USENIX Security Symposium*, 2006.

[11] C. Kaufman, R. Perlman, and M. Speciner. Network security: Private communication in a public world. *Prentice Hall*, 2002.

[12] C. Kruegel, E. Kirda, D. Mutz, W. Robertson, and G. Vigna. Automating mimicry attacks using static binary analysis. *In 14th Usenix Security Symposium*, 2005.

[13] C. Kruegel, E. Kirda, D. Mutz, W. Robertson, and G. Vigna. Polymorphic worm detection using structural information of executables. *In Recent Advances in Intrusion Detection (RAID)*, 2005.

[14] C. Kruegel, T. Toth, and E. Kirda. Service specific anomaly detection for network intrusion detection. *In Proceedings of the ACM SIGSAC*, 2002.

[15] C. Kruegel and G. Vigna. Anomaly detection of web-based attacks. *In Proceedings of the ACM Conference on Computer and Communication Security (ACM CCS)*, pages 251–261, 2003.

[16] Ktwo. Admmutate: Shellcode mutation engine. *http://www.ktwo.ca/ADMmutate-0.8.4.tar.gz*, 2001.

[17] Z. Liang and R. Sekar. Fast and automated generation of attack signatures: a basis for building self-protecting servers. *Proceedings of the 12th ACM Conference on Computer and Communications Security (ACM CCS)*, pages 213 – 222, 2005.

[18] M. Mahoney. Network traffic anomaly detection based on packet bytes. *In Proceedings of the ACM SIGSAC*, 2003.

[19] M. Mahoney and P.K. Chan. Learning nonstationary models of normal network traffic for detecting novel attacks. *In Proceedings of the SIGKDD*, 2002.

[20] J. Newsome, B. Karp, and D. Song. Polygraph: Automatically generating signatures for polymorphic worms. *In Proceedings of the IEEE Symposium on Security and Privacy*, 2005.

[21] Martin Roesch. Snort-lightweight intrusion detection for networks. *In Proceedings of the 13th USENIX conference on System administration*, pages 229 – 238, 1999.

[22] S. Rubin, S. Jha, and B. P. Miller. Language-based generation and evaluation of nids signatures. *In Proceedings of the IEEE Symposium on Security and Privacy*, 2005.

[23] S. Rubin, S. Jha, and B.P. Miller. Automatic generation and analysis of nids attacks. *In Annual Computer Security Applications Conference (ACSAC)*, 2004.

[24] R. Sekar, M. Bendre, D. Dhurjati, and P. Bollineni. A fast automaton-based method for detecting anomalous program behaviors. *In Proceedings of the IEEE Symposium on Security and Privacy*, 2001.

[25] R. Sekar, A. Gupta, J. Frullo, T. Shanbhag, A. Tiwari, H. Yang, and S. Zhou. Specification-based anomaly detection: A new approach for detecting network intrusions. *In Proceedings of the ACM conference on Computer and communications security (CCS)*, 2002.

[26] C. Sinz. Towards an optimal cnf encoding of boolean cardinality constraints. *In Principles and Practice of Constraint Programming*, pages 827–831, 2005.

[27] P. Szor. Advanced code evolution techniques and computer virus generator kits. *The Art of Computer Virus Research and Defense*, 2005.

[28] T. Toth and C. Kruegel. Accurate buffer overflow detection via abstract payload execution. *In Recent Advances in Intrusion Detection (RAID)*, 2002.

[29] G. Vigna, W. Robertson, and D. Balzarotti. Testing network-based intrusion detection signatures using mutant exploits. *In Proceedings of the ACM Conference on Computer and Communication Security (ACM CCS)*, pages 21–30, 2004.

[30] D. Wagner and P. Soto. Mimicry attacks on host-based intrusion detection systems. *In Proceedings of the ACM Conference on Computer and Communication Security (ACM CCS)*, 2002.

[31] K. Wang and S. Stolfo. Anomalous payload-based network intrusion detection. *In Recent Advances in Intrusion Detection (RAID)*, 2004.

[32] K. Wang and S. Stolfo. Anomalous payload-based worm detection and signature generation. *In Recent Advances in Intrusion Detection (RAID)*, 2005.

[33] T. Yetiser. Polymorphic viruses: Implementation, detection, and protection. *Technical Report, VDS Advanced Research Group*, 1993.

Data Collection With Self-Enforcing Privacy

Philippe Golle
Palo Alto Research Center
pgolle@parc.com

Frank McSherry
Microsoft Research
mcsherry@microsoft.com

Ilya Mironov
Microsoft Research
mironov@microsoft.com

ABSTRACT

Consider a pollster who wishes to collect private, sensitive data from a number of distrustful individuals. How might the pollster convince the respondents that it is trustworthy? Alternately, what mechanism could the respondents insist upon to ensure that mismanagement of their data is detectable and publicly demonstrable?

We detail this problem, and provide simple data submission protocols with the properties that a) leakage of private data by the pollster results in evidence of the transgression and b) the evidence cannot be fabricated without breaking cryptographic assumptions. With such guarantees, a responsible pollster could post a "privacy-bond", forfeited to anyone who can provide evidence of leakage. The respondents are assured that appropriate penalties are applied to a leaky pollster, while the protection from spurious indictment ensures that any honest pollster has no disincentive to participate in such a scheme.

Categories and Subject Descriptors

E.3 [**Data**]: Data Encryption; K.4.1 [**Computers and Society**]: Public Policy Issues—*Privacy*

General Terms

Security

Keywords

Privacy, Data collection

1. INTRODUCTION

We study the problem of a pollster who wishes to collect private information from individuals of a population. Such information can have substantial value to the pollster, but the pollster is faced with the problem that participation levels and accuracy of responses drop as the subject matter becomes increasingly sensitive. Individuals are, understandably, unwilling to provide accurate sensitive data to an untrustworthy pollster who is unable to make concrete privacy assurances.

The same problem affects individuals who are compelled to provide sensitive data to an untrusted party. Examples such as the census and medical data highlight cases where individuals are compelled to accuracy, either through law or the threat of poor treatment, but the absence of "privacy oversight" leaves many uncomfortable. What mechanisms can be used to assure individuals that poor privacy discipline can be caught and publicly demonstrated?

We stress that this problem is different from the question of how the pollster or data collector can manage data to preserve privacy. Privacy preserving data mining research has blossomed of late and gives many satisfying answers to this question [1]. Instead, the problem we consider is that individuals may not trust the pollster to apply quality privacy protection, either because the pollster has poor privacy discipline, poor security, or simply because it is selling data on the side. Published research on privacy preserving data mining demonstrates techniques for use by a benevolent pollster, but gives no assurances to individuals who are not convinced of the benevolence of the pollster.

The focus of this paper is a mechanism for submitting data to an untrustworthy pollster, such that a) leakage of private data can be caught and publicly demonstrated, and b) if private data are not leaked, the probability of presenting evidence of a leak is arbitrarily small. We stress that both of these properties are critical; the individuals must be protected from a bad pollster as much as the pollster must be protected from fraudulent accusations.

We make a distinction between private, personally identifying data, and non-identifying data (such as, for example, a noisy average computed over aggregated respondents' data). Our schemes ensure that leakage of private data by the pollster is detected (and punished). But some of our schemes allow the pollster to publish aggregated, non-identifying data. This is not a limitation of our schemes, but rather a useful feature, since publication of non-identifying aggregated data is typically permitted and useful. Formal definitions of private and non-identifying data are given in section 2.

1.1 Overview of Existing Solutions

Much research has gone into the design of data analysis mechanisms that attempt to minimize the amount of sensitive information leaked. However compelling these solutions may be, their value is greatly diminished in the absence of any guarantee that they are being applied properly. They do give substantial value when the pollster is trusted, e.g., when the pollster and individuals from whom data are collected belong to the same organization, or when the pollster has legal rights to the data of the individuals.

There are several techniques to address the problem of an untrustworthy pollster, with varying features and drawbacks. Randomized response [23, 3] is a method in which respondents pre-sanitize their own data by randomly altering it before submission. For example, when asked to reveal their gender, an individual could flip a coin with bias $p < 1/2$ and alter her response if the coin comes up heads. So long as the parameters of the pre-sanitization (p, and any additional details of the pre-sanitization process) are understood, many analyses accommodate this sort of perturbation. However, the noise levels introduced can be quite substantial. In some contexts, e.g., medical histories, the introduction of noise is simply a non-option; one does not want to accidentally misreport the absence of a peanut allergy.

Another approach is the use of trusted third parties, and their emulation through secure function evaluation [24, 13]. In this case, the data are collected by a trusted third party, and the untrusted pollster is only permitted to ask the trusted party certain questions of the data. The drawback of this approach is that the existence of a trusted third party is a substantial assumption, and the computational overhead involved in removing this assumption through secure function evaluation can be significant.

A third approach is to anonymize the data before submission, so that one cannot correlate sensitive features with individual identities. Mix networks [7, 19] allow respondents to submit data to the pollster anonymously. Unfortunately, anonymity is not feasible in many practical contexts. Mix networks can only be used to submit data that do not contain personally identifiable information (PII), so that the data themselves do not disclose information about the identity of the submitter. Whether a particular datum serves as PII depends entirely on the context, and it is rarely safe to assume that a particular parcel of data will not be disclosive when presented publicly.

In addition, or as an alternative to the deployment of privacy-preserving techniques, one may consider methods of detecting or discouraging leaks of sensitive information. This self-enforcement approach has been explored in the literature, mostly in the context of digital rights management [4, 6, 10, 15, 17]. The cryptographic schemes proposed in these papers deter a user, or a coalition of users, from sharing access to digital content by making such behavior traceable or by conditioning shared access to content on sharing some sensitive data, such as credit card numbers.

Finally, one might draw a comparison between our work and the process of tainting data, wherein submitters introduce an identifiable tracer into their submissions. One primitive example would be to encode a nonce into the least significant bits of a submission. Should the submitter see this tracer attached to their data again, they are assured that the information must have originated from the pollster. However convincing such a scheme might be to the individual, who may now severe communications with the pollster, it does little to convince the public that the pollster has done anything wrong. A public demonstration of the tainted data only confirms that either the pollster or the individual leaked the data, and does not preclude the possibility that the individual is setting up the pollster.

Existing schemes to watermark or fingerprint data [5, 2], including a publicly verifiable scheme [21], are designed for a setting where one data-holder manages access to its information, which is typically some large relational database or a digital movie. These techniques are not applicable in a distributed scenario, where the data are contributed by many individual participants.

1.2 Overview of Our Techniques

At the heart of our approach is the assumed presence of opportunistic third parties that we will call the *bounty hunters*, who listen for leaks of private information and assemble a case against the pollster. The bounty hunters participate in the data collection, pretending to be simple respondents (in fact, they may be). However, rather than following the cryptographic protocol for data submission, they submit "baits", whose decrypted contents *provably* cannot be determined without access to a secret held by the pollster. A bounty hunter herself does not know the contents of the data she submits. Since the pollster is the only individual capable of decrypting and examining the submitted bit, any report of the actual data in this message must come from the pollster, and thereby incriminates the pollster of leaking private data. Collaboration between bounty hunters is allowed, but not necessary. A single bounty hunter can produce evidence that incriminates a dishonest pollster who leaks private data.

The technical details we must discuss are the data submission process that allows respondents to submit data to the pollster, and the indictment process, in which a case is made by one or several bounty hunters against a pollster who leaked private data. There are several desirable properties of the indictment process, foremost that leakage of private data, even probabilistically, results in a viable case and that non-leakage cannot result in a viable case with high probability. These details are examined in Section 5.

1.3 Paper Outline

We begin in Section 2 with a discussion of the model, and several preliminary definitions and assumptions that will form the basis of our approach. Moreover, we detail several cryptographic primitives and the properties we take advantage of. In Section 3 we describe a simple approach for the case where the pollster uses the data collected from respondents only for internal consumption, and need not be able to publish any information about it (not even sanitized non-identifying information). Section 4 outlines a submission protocol based on randomized response, which adds the property that every submission serves as bait but introduces some uncertainty into the submitted data. Section 5 describes an approach that allows submission of precise data, but introduces the need for an interactive indictment process. Schemes from Sections 4 and 5 permit limited public disclosure of analysis of the pollster's data if the pollster follows specific sanitization policies. Finally, in Section 6 we conclude with a summary of the results, as well as promising directions for future investigation.

2. MODEL

We start by introducing some terminology and describing the players in our data collection processes. First, there is a *pollster*, who is interested in collecting bits from a large collection of *respondents*. The pollster may also publish aggregated poll results, as long as doing so does not compromise the privacy of any respondent.

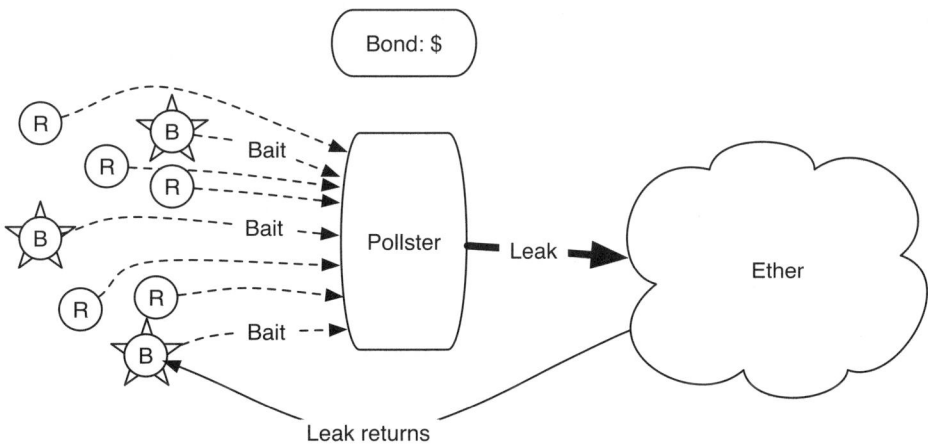

Figure 1: The pollster collects data from respondents. A number of bounty hunters, hidden among the respondents, submit baits. The pollster cannot distinguish baits from the data submitted by respondents. A privacy breach, or leak, occurs if the pollster releases private data. Baits allow bounty hunters to offer publicly verifiable evidence of a privacy breach.

The respondents have a vested interest in the privacy of their bits, and are assumed interested in participating in a protocol that enforces privacy. To this end, the pollster offers some form of *bounty*, which it must forfeit if a privacy violation is uncovered. The bounty could be explicit in the form of a bond, or implicit in the form of penalties imposed if privacy is violated.

Lurking among the respondents are some number of *bounty hunters*, who masquerade as one or more respondents and attempt to ensnare the pollster in a privacy violation. The bounty hunters submit *baits*, which the pollster cannot distinguish from legitimate data, and hope to learn from the pollster specific information about their baits that will constitute evidence of a privacy violation. If a bounty hunter uncovers a privacy violation, this evidence can be presented to claim the bounty. We say that our scheme offers *self-enforcing* privacy, since it is in the best interest of the pollster to preserve the privacy of the data collected from respondents.

Defining privacy. A self-enforcing data collection scheme ensures that a pollster who publishes sensitive data must forfeit a bounty. At the same time, the data collection scheme would ideally allow the pollster to publish aggregated poll results, as long as these results do not compromise the privacy of any respondent. Unfortunately, we do not how to define and enforce these properties in a strictly complementary way, i.e., in such a way that any publication of the pollster is classified as either safe or helping the bounty hunter. Instead, we introduce two notions of privacy:

- **Privacy breach.** A *privacy breach*, formally defined below along the lines of a *classical compromise* [16], is a clear violation of privacy. It amounts to the pollster releasing information that makes it possible to guess the sensitive bits confided by the respondents with a success probability non-negligibly greater than 1/2 *without* using any auxiliary information. A privacy breach can be thought of as a lower-bound on the privacy that the pollster must offer the respondents. We will prove that our schemes ensure that a privacy breach with certain parameters allows a bounty hunter to claim the bounty.

- **Differential privacy [12].** Differential privacy is a quantifiable definition of privacy-preserving functionality. We will show that our schemes ensure that a pollster who preserves differential privacy for some range of parameters cannot lose his bounty.

As noted above, there exists a "gap" between a privacy breach and differential privacy. In other words, the pollster may release data that violate differential privacy, but do not lead to a privacy breach.

2.1 Privacy Breach

It is important to formally describe what we mean by a breach of privacy, so that we can argue that we protect against such breaches. One appealing definition is that the pollster should not release specific information about respondents to other entities. However, such breaches will not generally be detectable, as the pollster could easily release the sensitive information to parties that will not themselves pass on the information, and the breach will not be detectable without their help. Instead, we will focus on breaches that are *detectable*, i.e. breaches for which the information released by the pollster finds its way back to the individuals who submitted that information, or agents acting on their behalf. Our focus on detectable breaches is justified by our assumption that the bounty hunters play an active role in monitoring the pollster and looking for data leaks. Naturally, the pollster should not be able to tell the bounty hunters from other agents interested in obtaining the pollster's data.

It will be critical that the respondents are able to identify a privacy breach as such. One example would be seeing one's private data made available, though less direct observations, such as for example being contacted on one's cell phone by a solicitor, can lead to similar conclusions.

Formally, we consider a model where the data received by the pollster are encoded as an n-bit vector $v = \{v_1, \ldots, v_n\}$. Note that the values received by the pollster may not be known by the respondents.

DEFINITION 2.1. *A (ℓ, ϵ)-privacy breach exists when ℓ indices $i_1, \ldots, i_\ell$ are identified such that any assignment of $v_{i_1}, \ldots, v_{i_\ell}$ consistent with the information published by the pollster agrees with v on at least $1/2 + \epsilon$-fraction of the entries.*

2.2 Differential Privacy

A natural question that arises in the presence of a posted privacy bond is whether the pollster can analyze and release any properties of the private data collected from respondents. Might it be that all useful functions reveal too much about the structure of baits so that the bond must be forfeited as soon as the pollster publishes any information at all about the data collected?

In this section, we introduce ϵ-differential privacy, a natural definition of privacy proposed in [11, 12]. We argue that the pollster can publish the results of any analysis that preserves differential privacy, without incurring a substantial risk of having to forfeit the bounty. Indeed, the chance of producing evidence of a privacy breach against a pollster is exponentially small if the pollster releases only information that preserves differential privacy.

DEFINITION 2.2. *A randomized function f over data sets gives ϵ-differential privacy if for any two data sets X_1 and X_2, which differ in at most one point, and $S \subseteq Range(f)$,*

$$\Pr[f(X_1) \in S] \leq \exp(\epsilon) \times \Pr[f(X_2) \in S].$$

In our application, the output of the function f is the information that the pollster releases about the data. The definition of differential privacy ensures that this information is not substantially affected by a respondent's presence in (or absence from) the data. Intuitively, if all the information published by the pollster preserves differential privacy, a bounty hunter cannot learn the data of any respondent and thus can also not learn information about any bait. We will use this property to show that publication of the results of analyses that preserve ϵ-differential privacy do not give bounty hunters enough information about baits to successfully claim the bounty with non-negligible probability.

Differential privacy is discussed in more detail in [11, 12], in which methods are presented for performing several common data analyses in a way that preserves differential privacy. Examples include histogram computations such as OLAP, as well as more algorithmic analyses such as Principal Components Analysis, k-means clustering, perceptron classification, and ID3 decision trees construction.

We stress that our data collection schemes are not bound to ϵ-differential privacy. This definition of privacy was chosen only to demonstrate that the privacy of our mechanisms can coexist with non-trivial data analyses. Differential privacy is among the stronger definitions of privacy, and is therefore easier to accommodate. Differential privacy is applied in Section 5.

2.3 Cryptographic Building Blocks

The approaches we present make use of cryptography to ensure that certain information is concealed from respondents, and that other information can be presented irrefutably. We now detail some of the cryptographic primitives that we use and their properties.

Secure channels. In our schemes, respondents will submit to the pollster data encrypted with homomorphic public-key encryption schemes, such as ElGamal or RSA. These encryption schemes naturally do not provide chosen-ciphertext security. It is thus imperative that these public-key ciphertexts be submitted over a secure channel, such as TLS. In fact, it is easy to demonstrate that the security of respondents' submissions is compromised if our schemes were used without an additional layer of (symmetric-key) encryption.

ElGamal cryptosystem. ElGamal is a randomized public-key encryption scheme. Let G be a group, and let $g \in G$ be a generator of a multiplicative subgroup G_q of order q where the Decisional Diffie-Hellman problem is hard. The secret key is an element x chosen at random from Z_q. The corresponding public key is the value $y = g^x$. The encryption of a plaintext $m \in G_q$ is a pair (g^r, my^r) for a value r chosen at random in Z_q. To decrypt a ciphertext (A, M), the value $m = M/A^x$ is computed. We will use two important properties of ElGamal:

- **Multiplicative homomorphism:** Consider two ElGamal ciphertexts $C_1 = (g^r, m_1 y^r)$ and $C_2 = (g^s, m_2 y^s)$ for plaintexts m_1 and m_2. The component-wise product $C_1.C_2 = (g^{r+s}, m_1 m_2 y^{r+s})$ is an ElGamal ciphertext for $m_1 m_2$.

- **Re-encryption.** Let (g^r, my^r) denote an encryption of a plaintext m. Let s be a random value in $\mathbb{Z}_q$. The pair (g^{r+s}, my^{r+s}) is also an encryption of m. The new pair is called a re-encryption of the first ciphertext. Note that a ciphertext can be re-encrypted without knowledge of m or of the secret key x.

Proof of plaintext knowledge (KPT). Let $E(m) = (g^r, my^r)$ be an encryption generated by a prover. The prover can prove to a verifier that she knows the plaintext m by proving that she knows $\log_g(g^r)$. This can be done with a protocol by Schnorr [22]. The protocol can be made non-interactive with the Fiat-Shamir heuristic. We denote an instance of this protocol for an ElGamal ciphertext C as $KPT(C)$.

Proof of correct decryption (PCD) [8]. A prover proves to an honest verifier that an ElGamal ciphertext (C, M) decrypts to a plaintext m. The proof consists of showing that $\log_g(y) = \log_C(M/m) = x$ without leaking any information about the secret key x. We denote an instance of this protocol to prove correct decryption of an ElGamal ciphertext C as $PCD(C)$.

Proof of correct re-encryption (PCR) [8]. A prover proves to an honest verifier that an ElGamal ciphertext (g^s, my^s) is a re-encryption of a ciphertext (g^r, my^r) without leaking any other information. The proof consists of showing that $\log_g(g^s/g^r) = \log_y((my^s)/(my^r)) = s - r$, without leaking any information about the value $s - r$. The computational cost of this protocol is 2 modular exponentiations for the prover and 4 modular exponentiations for the verifier. We denote an instance of this protocol to prove that an ElGamal ciphertext C_2 is a re-encryption of C_1 as $PCR(C_1 \leadsto C_2)$.

Discrete logarithm proof systems [9]. An efficient zero-knowledge proof can be constructed for any monotone boolean formula whose atoms consist of the protocols to prove plaintext knowledge (KPT), correct decryption (PCD) or correct re-encryption (PCR).

Verifiable mixing [14, 18]. Let $L = \{(A_i, M_i)\}$ and $L' = \{(A'_i, M'_i)\}$ be two lists of ElGamal ciphertexts. A verifiable mixing protocol allows a prover to prove to an honest verifier the existence of a permutation π and a sequence of exponents γ_j such that $(A'_j, M'_j) = (A_{\pi(j)} g^{\gamma_j}, M_{\pi(j)} y^{\gamma_j})$, without leaking any information about π or the values γ_j. Given n input ciphertexts, the computational cost of the most efficient verifiable mixing protocol [14] is $6n$ modular exponentiations for the prover and $6n$ modular exponentiations for the verifier.

3. SELF-ENFORCING PRIVACY WITH NO RELEASE OF DATA

In this section, we present a scheme that allows the pollster to collect data from respondents, but not to release any information about the data collected.

The scheme is structured as follows. The pollster commits to a secret binary string by publishing encryptions of the bits of the secret string under a randomized public-key encryption scheme, such as ElGamal, which is homomorphic and allows for re-encryption of ciphertexts. Each time a respondent submits a bit, she has a choice of either submitting an encrypted bit of her own data or preparing a bait by re-encrypting any of the pollster's secret bits. The pollster decrypts all the ciphertexts received and thus recovers the data submitted by respondents. Since the pollster cannot distinguish baits from regular submissions, some baits will unavoidably be decrypted if the pollster leaks a substantial fraction of the data. Decrypted baits reveal some of the bits of the pollster's secret string. Once enough of the secret bits are known to the injured parties, they can claim the bounty by proving knowledge of the secret string.

In this section and throughout the paper, we assume that respondents are labelled with unique identifiers $P_1, \ldots, P_n$.

Setup. The pollster outputs public parameters for a public-key encryption scheme E that is semantically secure under re-encryption and has a multiplicative homomorphism. In what follows, we use ElGamal. The public parameters are a group G and a generator $g \in G$ of a multiplicative subgroup G_q of order q in which the Decisional Diffie-Hellman problem is hard.

Commitment to the bounty. Let k be a security parameter (e.g., $k = 160$). The pollster chooses a k-bit secret value $\beta = b_1 \ldots b_k$. The pollster outputs $E(g^{b_i})$ for $i = 1, \ldots, k$ and proves that these ciphertexts are well-formed by showing that each ciphertext decrypts either to g^0 or g^1. This is done with a (disjunctive) discrete logarithm proof system consisting of two proofs of correct decryption (see section 2.3). Using the multiplicative homomorphism of E, the pollster computes $\prod_{i=1}^{k} E(g^{b_i})^{2^i} = E(g^\beta)$. The pollster then decrypts this value, proves correct decryption with the protocol $PCD(E(g^\beta))$ described in section 2.3, and outputs the commitment g^β. A bounty is then offered to anyone who recovers the secret value β.

Data submission. In the data submission step, a respondent sends to the pollster either one true bit of data, or a bait.

- **Sending one true bit of data.** To send a bit $b \in \{0, 1\}$ to the pollster, a respondent P_i computes the randomized ciphertext $E(g^b)$ and sends the resulting value to the pollster over a secure channel (e.g. using TLS). Recall that the encryption scheme E is semantically secure, so that it is computationally impossible to learn any information about the bit b from the (randomized) ciphertext $E(g^b)$.

- **Sending a bait.** To send a bait to the pollster, the respondent chooses a random index $r \in \{1, \ldots, k\}$, re-encrypts the ciphertext $E(g^{b_r})$ and sends the re-encrypted ciphertext to the pollster over a secure channel.

Data collection. The pollster receives ElGamal ciphertexts from respondents. Since ElGamal is semantically secure under re-encryption, the pollster cannot distinguish true bits from baits. The pollster then decrypts all ciphertexts $C = E(g^{b_i})$ and recovers the corresponding plaintexts. Only well-formed plaintexts (i.e. those that decrypt to g^0 or g^1) are tallied. Malformed plaintexts are discarded.

3.1 Claiming the Bounty

Honest pollster. This scheme does not allow the pollster to publish anything about the data collected. We show first that corrupt respondents cannot fraudulently claim the bounty of an innocent pollster. If the pollster leaks no information about data collected from respondents, claiming the bounty is equivalent to recovering the value β from the commitment g^β. Since the discrete logarithm problem is assumed hard in the group G generated by g, this problem is computationally intractable. Thus corrupt respondents cannot wrongly claim the bounty of an innocent pollster.

Dishonest pollster. We consider next a dishonest pollster, and show that the bounty can be recovered if the pollster publishes data that result in a privacy breach. Let us start with a simple example. If the pollster leaks $\ell < k$ baits, respondents can recover the secret β in time $2^{(k-\ell)/2}$ using the technique of [20] and present β as evidence of the pollster's misbehavior to claim the bounty. Note that the verification process is non-interactive: the correctness of β is verified against the commitment g^β, without communicating with the pollster. The correctness of the bounty is also publicly verifiable without the involvement of the pollster.

More generally, let us consider a pollster who publishes data that result in a privacy breach. For example, the pollster may leak the data collected from respondents with noise added. The following proposition shows that a privacy breach allows bounty hunters to recover all the bits of the pollster's secret with high probability.

PROPOSITION 3.1. *Consider a pollster who commits a (ℓ, ϵ)-privacy breach. Recall that k denotes the size of the pollster's secret. Let $0 < \alpha < 1$ denote the fraction of baits among the bits submitted by respondents and bounty hunters. If $\ell > k/(\alpha \epsilon^2)$, the bounty hunters can (with high probability) reconstruct the secret β with no computational effort.*

PROOF. Let us denote the data received by the pollster as an n-bit vector $v = \{v_1, \ldots, v_n\}$. By definition, an (ℓ, ϵ)-privacy breach means that a set of ℓ indices $i_1, \ldots, i_\ell$ is identified such that any assignment of $v_{i_1}, \ldots, v_{i_\ell}$ consistent with the information published by the pollster agrees with v on at least $1/2 + \epsilon$-fraction of the entries.

Among the values $v_{i_1}, \ldots, v_{i_\ell}$, the number of baits is $\alpha\ell$. Now let us consider a bit b_i of the pollster's secret β. The number of baits in $v_{i_1}, \ldots, v_{i_\ell}$ that are re-encryptions of the bit b_i is $\alpha\ell/k$. By definition of a privacy breach, each of these baits is correct with probability greater than $1/2 + \epsilon$. If a majority of these $\alpha\ell/k$ values are 0, we conclude that $b_i = 0$ (and otherwise $b_i = 1$).

Let X be a random variable defined by the sum of the $\alpha\ell/k$ baits that are re-encryptions of the bit b_i. According to the Chernoff bound,

$$\Pr[X < \alpha\ell/(2k)] < e^{-(\alpha\ell/k)(1/2+\epsilon)(1-1/(1+2\epsilon))^2/2}.$$

The probability of error is thus small if $\ell = O(k/(\alpha\epsilon^2))$. This concludes the proof. □

Let us consider a numerical example. If the pollster commits to a 160-bit secret ($k = 160$) and leaks correct bits with probability $1/2 + \epsilon$, where $\epsilon = 1/4$, and if the respondents submit a bait with probability $\alpha = 10\%$, then 12,800 bits are required to recover β with modest computational effort (2^{40} modular exponentiations).

We stress that this scheme is secure for the pollster only if it releases no information whatsoever about the data collected. The following example illustrates the danger for the pollster of releasing even seemingly innocuous data.

Consider a pollster who intends to publish the noisy gender majority for each ZIP code in the survey. For appropriately chosen parameters of the noise, this information can be disclosed without a privacy breach. Still, the scheme described in this section does not allow for the safe release of this information.

Indeed, an unscrupulous bounty hunter may create sufficiently many false identities in a given ZIP code area, and let all these identities submit as baits re-encryptions of the same secret bit of the pollster's secret. The bounty hunter may succeed in biasing the results of the poll so that the noisy majority will be equal to the value of this secret bit with high probability. Repeating this attack will eventually allow the bounty hunter to learn all the bits of the pollster's secret and claim the bounty.

In the rest of this paper, we propose improved schemes that will allow the pollster to safely release sanitized non-identifying information about the data collected.

4. SELF-ENFORCING PRIVACY AND RANDOMIZED RESPONSE.

The scheme described in the previous section requires different processes for submitting true answers and baits. It calls for proactive bounty hunters, who may have an incentive to create multiple fake identities that crowd out real contributors and compromise the poll's validity.

In this section, we propose a different scheme based on the concept of randomized response, where each response is a bait and the role of bounty hunters in the survey is strictly passive.

4.1 Basic Scheme With Randomized Response

Setup. The pollster outputs public parameters for an ElGamal encryption scheme denoted E. As in section 3, we denote $g \in G$ the generator of a multiplicative subgroup G_q of order q in which the Decisional Diffie-Hellman problem is hard.

Commitment. The pollster chooses k bits $b_1, \ldots, b_k$ independently at random from a biased distribution that assigns weight p to '0' and weight $1-p$ to '1', for some $1/2 < p < 1$. Let β denote the integer whose binary representation is $b_1, \ldots, b_k$. The pollster outputs $E(g^{b_i})$ for $i = 1, \ldots, k$. The pollster proves that the values b_i's are well-formed and are drawn from the correct distribution by applying to the set $E(g^{b_i})$ a permutation that reorders the bits in increasing order (all the bits 0 come before the bits 1). The pollster proves correct application of the permutation with one of the verifiable mixing protocols reviewed in section 2.3. Finally, the pollster provably decrypts the outputs of the mix, which allows everyone to verify that the key bits contain the correct proportion of 0's and 1's. The value p should be $p > 1/2$ for randomized response to make sense, but also $p < 1$ to protect the bounty. For example, the pollster can choose $p = 2/3$. Next, using the multiplicative homomorphism of E, the pollster computes $E(g^\beta) = \prod_{i=1}^{k} E(g^{b_i})^{2^i}$. The pollster then provably decrypts this value and outputs g^β. The bounty is placed on the value β.

Data submission. Let b denote the bit to be submitted by a respondent. The respondent chooses a random index $i \in \{1, \ldots, k\}$. If $b = 0$, the respondent sends to the pollster a re-encryption of the ciphertext $E(g^{b_i})$. If $b = 1$, the respondent uses the multiplicative homomorphism of ElGamal to compute the ciphertext $E(g^{1-b_i}) = E(g)/E(g^{b_i})$ and sends this ciphertext to the pollster over a secure channel. Let C denote the ElGamal ciphertext sent to the pollster.

The respondent must also submit a proof of correct operation. The respondent gives a proof to the pollster of the following discrete-log system (see section 2.3):

$$\Big(\bigvee_{i=1}^{k} PCR\Big(E(g^{b_i}) \rightsquigarrow C\Big)\Big) \vee \Big(\bigvee_{i=1}^{k} PCR\Big(E(g)/E(g^{b_i}) \rightsquigarrow C\Big)\Big).$$

According to [9], the cost of this proof is $6k - 1$ modular exponentiations for the prover (the respondent) and $6k$ modular exponentiations for the verifier (the pollster). The purpose of this proof it to prevent respondents from cheating by submitting non-randomized replies that would carry more weight than randomized ones.

Data collection and Claiming the bounty. These steps are exactly as in section 3.

It is natural to compare this approach with the simpler randomized response schemes described in the introduction, in which the respondents pre-randomize their own data. The values submitted in our scheme have no greater statistical fidelity than in the simpler scheme. The important distinction is that, in our scheme, the interests of the pollster are aligned with the privacy concerns of the participants: a value p close to one gives very accurate answers but puts the bounty at risk. The privacy of the individuals is not a result of choosing p close to $1/2$, as in classical randomized response, but inherent for all values of p.

4.2 Variant that Allows Release of Some Data

The data collected from respondents are most useful when the pollster is able to analyze it and can act on the analyses (or even publish the results of the analyses) without fear of forfeiting the bounty (as long as the results of the analyses do not compromise the privacy of respondents). While the scheme of section 4.1 ensures that privacy breaches are punished, the pollster would also like assurances about what sort of behavior (or publication) is allowed, based on the data collected. If the publication of certain data is allowed, because it poses no threat to the privacy of respondents, the publication of that data should not allow a bounty hunter to successfully claim the bounty.

When the pollster performs queries over the bits submitted by the respondents, it is in fact performing queries over bits of its own secret. Publishing the results of such queries raises the concern that the pollster may accidentally reveal information about its secret bits. The pollster would like to restrict itself to queries that guarantee the "privacy" of its own secret, so that it runs no risk of having to forfeit the bounty. The property desired by the pollster is the same as ϵ-differential privacy for respondent data: the distribution over results should not be substantially affected by the modification of one of the pollster's secret bits.

To achieve this property, we propose a simple variant of the data submission protocol of section 4.1. Recall from section 4.1 that the pollster outputs k ciphertexts $E(g^{b_i})$ for $i = 1, \ldots, k$ in the commitment step. Intuitively, the goal of the variant presented here is to prevent one respondent (or a set of colluding respondents) from all submitting the *same* ciphertext $E(g^{b_i})$. We achieve this with the following data submission protocol:

1. The pollster re-encrypts the ciphertexts $E(g^{b_i})$ for $i = 1, \ldots, k$ and permutes them according to a permutation π chosen uniformly at random and known only to the pollster. The pollster outputs the permuted set $E(g^{b_{\pi(i)}})$ for $i = 1, \ldots, k$.

2. Let b denote the bit to be submitted by a respondent. The respondent chooses a random index $j \in \{1, \ldots, k\}$. Let i denote the value (not known to the respondent) such that $j = \pi(i)$. If $b = 0$, the respondent computes a re-encryption of the ciphertext $E(g^{b_{\pi(i)}})$. If $b = 1$, the respondent uses the multiplicative homomorphism of ElGamal to compute the ciphertext $E(g^{1-b_{\pi(i)}}) = E(g)/E(g^{b_{\pi(i)}})$. Either way, let C denote the ciphertext computed by the respondent. The respondent sends the pollster a commitment to C.

3. The pollster reveals the permutation π and proves correct mixing in step 1 (see section 2.3 for details on how that is done). If the verification fails, the respondent aborts the data submission process.

4. The respondent outputs C, together with a proof of a discrete-log system that shows that C is either a re-encryption of $E(g^{b_{\pi(i)}})$ or of $E(g)/E(g^{b_{\pi(i)}})$, as in section 4.1.

5. The pollster checks C against the commitment received in step 2, and checks the discrete-log proof system. If both are correct, the bit from the respondent is accepted.

A malicious respondent may attempt to skew the distribution of the indices $\pi(i)$ by not completing step 4. To ensure a near-uniform distribution (with statistical distance from the uniform less than $1/k$), the pollster should use a random index if the submission protocol is aborted after the permutation π is revealed.

Now consider ϵ-differential privacy as applied to the respondent data. If the information released by the pollster preserves ϵ-differential privacy for the respondents, then the distribution over its outputs does not change substantially (as a function of ϵ) if any respondent changes its submitted value. Let s_i denote the number of respondents from a query set S whose submission is a re-encryption of bit b_i. Since a change in the value of the secret bit b_i results in a change of at most s_i values, any computation that preserves ϵ-differential privacy for the respondents' data also preserves (ϵn_i)-differential privacy for bit b_i of the pollster's secret.

THEOREM 4.1. *An ϵ-differential privacy query over the set S increases the probability of the bounty being claimed by at most $\exp(\epsilon 2(1-p)k \max_i s_i)$.*

PROOF. Consider the probability that the bounty hunter succeeds in identifying the $(1-p)k$ secret locations i for which $b_i = 1$, taken first over the randomness in the selection of the locations, and then over the randomness given by ϵ-differential privacy. Take $c = 2(1-p)k \max_i s_i$ as the largest number of respondents whose received data would change as a result of an arbitrary change in the $(1-p)k$ locations of non-zero bits. The bounty hunter's distribution over guesses is conditioned on the locations chosen, but differential privacy guarantees that no guess increases in probability by more than a factor of $\exp(\epsilon c)$. We can therefore remove the bounty hunter's dependence on the actual location at the cost of a factor of $\exp(\epsilon c)$.

We have

$$\Pr_{\text{location}} \Pr_{\text{guess}} [\text{guess} = \text{location} \mid \text{location}]$$
$$\leq \Pr_{\text{location}} \Pr_{\text{guess}} [\text{guess} = \text{location}] \exp(\epsilon c)$$
$$= \exp(\epsilon c) / \binom{k}{(1-p)k}.$$

The final step follows from the observation that no matter the distribution over the guess, the uniform distribution over the actual location makes the probability one over the number of possible locations. $\square$

While this may seem like a substantial increase in the probability that the bounty is claimed, one must keep in mind that the probability of forfeiting the bounty before any information is released is $1/\binom{k}{(1-p)k}$. If the query is independent of the distribution of respondents, $\sum_{i:b_i=1} s_i$ is unlikely to greatly exceed $(1-p)\|S\|$.

If the query is permitted to depend on the distribution, perhaps because the respondents themselves pose the questions in an attempt to trap the pollster, then $\sum_{i:b_i=1} s_i$ could be as large as $\|S\|$, but even in this case the pollster can still choose ϵ and k to yield meaningful results.

4.3 A Stronger Bound for Sum Queries

In the case where the query is independent of the assignment of respondents to bits, we can occasionally prove a stronger bound for the scheme of section 4.2. Consider the query that counts the number of respondents from S whose bit is set. If the pollster were to change the location of one of its non-zero bits, the total sum would change by at most the difference in the sums for the two locations. If the distribution of respondents is uniform, this difference can be substantially smaller than the sums themselves, improving substantially on the bound above. The following lemma is a standard balls-and-bins argument of the number of balls in a bin tightly concentrated around its expected value. As a corollary, the lemma implies that the difference between the number of balls in two bins is likely to be small compared to the total number of the balls in both bins, which corresponds to the change in a sum-query's answer if the location of a non-zero bit changes.

LEMMA 4.2. *Letting s_i be the random variable denoting the number of respondents in bin i, with probability at least $1 - \delta$, for all i we have $(s_i - \mu)^2 \leq 4(s/k)\ln(k/\delta)$, provided that $\delta > \exp(-s/k)$.*

Letting d be the change in the value of the sum above, an identical change can be attained by changing the values of d respondents. If the pollster maintains ϵ-differential privacy for the respondents, the pollster is assured of ϵd-differential privacy for the location of each of its non-zero bits, even though substantially more than d respondents may live at each location.

THEOREM 4.3. *For any counting query that is independent of the distribution of respondents to bins that maintains ϵ-differential privacy of the respondents data with probability at least $1 - \delta$ the probability of a bounty being claimed is at most $\exp(4\epsilon(1-p)k\sqrt{(s/k)\ln(k/\delta)})/\binom{k}{(1-p)k}$, which vanishes for large enough k.*

PROOF. We start with the observation that for each of the k possible locations for the set bits, the number of positive and negative respondents are within $c' = 2\sqrt{(s/k)\ln(k/\epsilon)}$ of their mean, with probability at least $1 - \epsilon$. Conditioned on this event holding, changing the location of the $(1-p)k$ bits results in a change of at most $c = 2(1-p)kc'$ to the sum. A change of c to the sum could be caused by the alteration of as many respondents data, but ϵ-differential privacy ensures that the probability of no event should increase by more than a factor of $\exp(\epsilon c)$ due to such a change. The proof follows in a form identical to that of Theorem 4.1. □

5. A DIFFERENT SCHEME THAT ALLOWS SOME RELEASE OF DATA

In this section, we propose another scheme that allows the pollster to release information about the data collected as long as it does not violate the privacy of any non-trivial fraction of respondents. In a nutshell, our scheme works as follows. The bounty hunter prepares encryptions of unknown bits and submit them as baits. Should the pollster leak information about these bits, the bounty hunter indicts the pollster by presenting the bits and a proof of the baits' validity in order to claim the bounty. After the indictment, the onus is on the pollster to refute the accusation, which can be achieved by proving that sufficiently many bits decrypt to different values than alleged by the bounty hunter.

Setup. Let E denote a semantically secure public-key encryption scheme (e.g., RSA in what follows) and let D denote the corresponding decryption function. The pollster outputs public parameters for E. Let h be a hash function and let f be another hash function whose image is the set of ciphertexts of E. In our proof of security, we model h and f as random oracles. In the real world the functions are instantiated based on cryptographically strong hashes, such as SHA-256.

Sending a bit to the pollster. To send a bit $b \in \{0, 1\}$ to the pollster, a respondent P_i chooses a value r such that the least significant bit of $h(P_i\|r)$ is b. The respondent sends P_i and $E(r)$ to the pollster.

Decryption. Given an identifier P and a ciphertext C, the pollster decrypts C to recover the plaintext r, then computes the least significant bit b of $h(P\|r)$.

Sending a bait to the pollster. To send a bait to the pollster, the respondent chooses a random value s, computes $f(s)$ and sends to the pollster P_i and $f(s)$. Notice that neither the decryption of $f(s)$ nor the bit recovered by the pollster is known to the bounty hunter.

Accusing the pollster. If the pollster releases uniquely identifiable bits, some of which can be linked to the baits, the bounty hunter can indict the pollster. The indictment consists of $n > n_0$ *distinct* triples of the form

$$\langle P_i,\ s_i,\ b_i \rangle,$$

which we call *exhibits*. An exhibit is *valid* if and only if the bit decrypted by the pollster, i.e. the least significant bit of $h(P_i\|D(f(s_i)))$ is equal to b_i.

The pollster can contest the indictment by demonstrating that at least $(1/2 - w_n)n$ of the alleged exhibits are invalid. The minimum number of exhibits n_0 and the exact form of w_n, which lies between 0 and $1/2$ and serves to protect the pollster, will be discussed later. The pollster proves that an exhibit is invalid by outputting $r_i = D(f(s_i))$, with a proof of correct decryption, and demonstrating that the least significant bit of $h(P_i\|r_i)$ is not b_i.

If the pollster cannot defend herself or refuses to do so, the bounty must be forfeited. Note that this solution requires the pollster to be online for the indictment process, but it does not rely on a trusted third party.

Security. We note first that the reason for using RSA in this scheme, instead of ElGamal as in previous schemes, is to give respondents the ability to select a random valid ciphertext for which they do not know the corresponding plaintext. We note also that properly constructed baits are indistinguishable from other submissions, and encode bits that are uncorrelated and provably unknown to the bounty hunter. Next, we show that a pollster whose data disclosure policy preserves ϵ-differential privacy cannot be convicted by an over-zealous bounty hunter.

PROPOSITION 5.1. *If the data queries answered by the pollster preserve ϵ-differential privacy, the probability that any bounty hunter can claim the bounty is less than*

$$\max_{n \geq n_0} \exp(n\epsilon - nw_n^2/2).$$

PROOF. We will argue that for any possible set of attack locations, a bounty hunter's probability of successful attack is at most $\exp(n\epsilon - nw_n^2/2)$. Even though the bounty hunter may choose which locations to attack based on the pollster's output, the bound ensures that any choice will be unlikely to succeed.

Fix any set of n locations that a bounty hunter may attempt to attack, submitting evidence of leakage from these locations. There is a distribution over the values the bounty hunter proposes as evidence, induced by the randomness present in the ϵ-differential privacy mechanism. We consider two worlds, one in which none of these bits were used in any queries, and one in which all bits are used, the latter being the world we are concerned with. In the first world, the attacked bits were not used in any queries, and the probability of success for the bounty hunter is at most $\exp(-nw_n^2/2)$ by a Chernoff bound. Using a hybrid argument, we introduce the bits into the query, increasing the probability of successful attack by at most $\exp(\epsilon)$ with each step. After n steps, we arrive at the world where no bits are excluded from the query, in which the probability of successful attack is at most $\exp(n\epsilon - nw_n^2/2)$. $\square$

The pollster must determine a value of ϵ that permits sufficient utility without compromising the security of the bounty. Safe values of ϵ in turn depend on the values n_0 and w_n that govern the indictment rules. These values must be chosen to permit a sufficient level of safe disclosure. At the same time, respondents should also insist on realistic settings of n_0 and w_n to ensure that bounty hunters are able to catch privacy leaks.

It is also worth noting that the probability that a bounty hunter succeeds in a fraudulent claim depends only on the number of exhibits n, and not on the total number of baits submitted. Were this not the case, there would be a strong incentive for bounty hunters to flood the system with baits, corrupting the integrity of the poll.

6. CONCLUSION

We have studied three data submission protocols that provide the ability to offer publicly verifiable evidence of data leaks. This evidence is convincing both in that actual leakage can be demonstrated, and in that a fraudulent indictment is highly unlikely to succeed in the absence of leakage. All three protocols assume the presence of proactive "bounty hunters" who submit "baits" to the data collector. Baits are indistinguishable from regular data but offer irrefutable evidence of a data leak when one occurs.

Our three protocols differ in the properties they offer. The first protocol allows for non-interactive bounty verification and relatively exact data collection. The second protocol uses a form of randomized response to collect data, and allows every input to serve as a bait. The third protocol permits the pollster to publicly disclose a limited amount of non-identifying information about the data collected, but it requires an interactive indictment process.

These three protocols demonstrate several desirable properties of a data collection mechanism with self-enforcing privacy. We leave open the problem of designing a protocol that offers all these properties simultaneously. Understanding which features are compatible with others, and which (if any) are mutually exclusive, is an interesting direction for future research.

7. REFERENCES

[1] Rakesh Agrawal and Ramakrishnan Srikant. Privacy-preserving data mining. In *Proceedings of the 2000 ACM SIGMOD International Conference on Management of Data*, pages 439–450, 2000.

[2] Rakesh Agrawal, Peter J. Haas, and Jerry Kiernan. Watermarking relational data: framework, algorithms and analysis. *The VLDB Journal*, 12(2):157–169, 2003.

[3] Andris Ambainis, Markus Jakobsson, and Helger Lipmaa. Cryptographic randomized response techniques. In *Public Key Cryptography—PKC 2004*, volume 2947 of *Lecture Notes in Computer Science*, pages 425–438, 2004.

[4] Alexandra Boldyreva and Markus Jakobsson. Theft-protected proprietary certificates. In *Digital Rights Management Workshop—DRM 2002*, volume 2696 of *Lecture Notes in Computer Science*, pages 208–220, 2003.

[5] Dan Boneh and James Shaw. Collusion-secure fingerprinting for digital data. *IEEE Transactions on Information Theory*, 44(5):1897–1905, 1998.

[6] Benny Chor, Amos Fiat, Moni Naor, and Benny Pinkas. Tracing traitors. *IEEE Transactions on Information Theory*, 46(3):893–910, 2000.

[7] David Chaum. Untraceable electronic mail, return addresses, and digital pseudonyms. In *Communications of the ACM*, 24(2):84-88, 1981.

[8] David Chaum and Torben P. Pedersen. Wallet databases with observers. In *Advances in Cryptology—CRYPTO '92*, volume 740 of *Lecture Notes in Computer Science*, pages 89–105, 1993.

[9] Jan Camenisch and Markus Stadler. Proof systems for general statements about discrete logarithms. Technical Report 260, Dept. of Computer Science, ETH Zurich, March 1997.

[10] Cynthia Dwork, Jeffrey B. Lotspiech, and Moni Naor. Digital signets: Self-enforcing protection of digital information. In *Proceedings of the Twenty-Eighth Annual ACM Symposium on Theory of Computing, 1996*, pages 489–498. ACM.

[11] Cynthia Dwork, Frank McSherry, Kobbi Nissim, and Adam Smith. Calibrating noise to sensitivity in private data analysis. In *TCC 2006*, volume 3876 of *Lecture Notes in Computer Science*, pages 265–284, 2006.

[12] Cynthia Dwork. Differential privacy. Invited talk. In *ICALP 2006, Part II*, volume 4052 of *Lecture Notes in Computer Science*, pages 1–12, 2006.

[13] Oded Goldreich, Silvio Micali, and Avi Wigderson. How to play any mental game, or a completeness theorem for protocols with honest majority. In *Proceedings of the 19th Annual ACM Symposium on Theory of Computing*, pages 218–229. ACM.

[14] Jens Groth. A verifiable secret shuffle of homomorphic encryptions. In *Public Key Cryptography—PKC 2003*, volume 2567 of *Lecture Notes in Computer Science*, pages 145–160, 2002.

[15] Markus Jakobsson, Ari Juels, and Phong Q. Nguyen. Proprietary certificates. In *Topics in Cryptology—CT-RSA 2002*, volume 2271 of *Lecture Notes in Computer Science*, pages 164–181. 2002.

[16] Krishnaram Kenthapadi, Nina Mishra, and Kobbi Nissim. Simulatable auditing. In *PODS 2005*, pages 118–127. ACM, 2005.

[17] N. Boris Margolin, Matthew Wright, Brian Neil Levine. Analysis of an incentives-based protection system. In *Proc. ACM Digital Rights Management Workshop 2004*, pages 22–30. ACM, 2004.

[18] C. Andrew Neff. A verifiable secret shuffle and its application to e-voting. In *ACM Conference on Computer and Communications Security*, pages 116–125, 2001.

[19] Wakaha Ogata, Kaoru Kurosawa, Kazue Sako, and Kazunori Takatani. Fault tolerant anonymous channel. In *ICICS '97*, volume 1334 of *Lecture Notes in Computer Science*, pages 440–444, 1997.

[20] John M. Pollard. Monte Carlo methods for index computation (mod p). *Mathematics of Computation*, 32:918–924, 1978.

[21] Birgit Pfitzmann and Matthias Schunter. Asymmetric fingerprinting (extended abstract). In *Advances in Cryptology—EUROCRYPT '96*, volume 1070 of *Lecture Notes in Computer Science*, pages 84–95, 1996.

[22] Claus-Peter Schnorr. Efficient signature generation by smart cards. *J. Cryptology*, 4(3):161–174, 1991.

[23] Stanley L. Warner. Randomized response: A survey technique for eliminating evasive answer bias. *The American Statistical Association*, 60(309):63–69, March 1965.

[24] Andrew Chi-Chih Yao. Protocols for secure computations (extended abstract). In *23rd Annual Symposium on Foundations of Computer Science*, pages 160–164, Chicago, Illinois, 3–5 November 1982. IEEE.

Searchable Symmetric Encryption: Improved Definitions and Efficient Constructions [*]

Reza Curtmola
Department of Computer Science
Johns Hopkins University
Baltimore, MD
crix@cs.jhu.edu

Juan Garay
Bell Labs – Lucent Technologies
Murray Hill, NJ
garay@research.bell-labs.com

Seny Kamara
Department of Computer Science
Johns Hopkins University
Baltimore, MD
seny@cs.jhu.edu

Rafail Ostrovsky
Department of Computer Science and
Department of Mathematics
UCLA
Los Angeles, CA
rafail@cs.ucla.edu

ABSTRACT

Searchable symmetric encryption (SSE) allows a party to outsource the storage of its data to another party (a server) in a private manner, while maintaining the ability to selectively search over it. This problem has been the focus of active research in recent years. In this paper we show two solutions to SSE that simultaneously enjoy the following properties:

1. Both solutions are more efficient than all previous constant-round schemes. In particular, the work performed by the server per returned document is *constant* as opposed to linear in the size of the data.

2. Both solutions enjoy stronger security guarantees than previous constant-round schemes. In fact, we point out subtle but serious problems with previous notions of security for SSE, and show how to design constructions which avoid these pitfalls. Further, our second solution also achieves what we call *adaptive* SSE security, where queries to the server can be chosen adaptively (by the adversary) during the execution of the search; this notion is both important in practice and has not been previously considered.

Surprisingly, despite being more secure and more efficient, our SSE schemes are remarkably simple. We consider the simplicity of both solutions as an important step towards the deployment of SSE technologies.

As an additional contribution, we also consider *multi-user* SSE. All prior work on SSE studied the setting where only the owner of the data is capable of submitting search queries. We consider the natural extension where an arbitrary group of parties other than the owner can submit search queries. We formally define SSE in the multi-user setting, and present an efficient construction that achieves better performance than simply using access control mechanisms.

Categories and Subject Descriptors

E.3 [**Data Encryption**]; H.3.3 [**Information Storage and Retrieval**]: Information Search and Retrieval.

General Terms

Algorithms, Security, Theory.

Keywords

Searchable symmetric encryption, multi-user searchable encryption, security definitions.

1. INTRODUCTION

Private-key storage outsourcing allows clients with either limited resources or limited expertise to store and distribute large amounts of symmetrically encrypted data at low cost. Since regular private-key encryption prevents one from searching over encrypted data, clients also lose the ability to selectively retrieve segments of their data. To address this, several techniques have been proposed for provisioning symmetric encryption with search capabilities [27, 16, 7, 10, 12]; the resulting construct is typically called *searchable encryption*. The area of searchable encryption has been identified by DARPA as one of the technical advances that can be used to balance the need for both privacy and national security in information aggregation systems [2]. In addition, it can allow services such as Google Desktop [1] to offer valuable features (e.g., the ability of searching a client's data across several computers) without sacrificing the client's privacy.

Searchable encryption can be achieved securely in its full generality using the work of Ostrovsky and Goldreich on software protection based on *oblivious RAMs* [24, 18]. While oblivious RAMs hide all information about the RAM use

---

[*] An extended version of the paper is available as IACR ePrint report 2006/210 [13].

from a remote and potentially malicious server with a polylogarithmic overhead in all parameters (including computation and communication), this comes at the cost of a logarithmic number of rounds of interaction for each read and write. In the same paper, they show a 2-round solution, but with considerably larger square-root overhead. Therefore, the previously mentioned work on searchable encryption achieves more efficient solutions (typically in one or two rounds) by weakening the privacy guarantees (e.g., revealing the access pattern).

We start by examining the definition of what it means to reveal the user's access and search patterns (precise definitions below) while "hiding everything else," and show that the existing security definitions have several important limitations. Additionally, we show that the current definitions only achieve what we call *non-adaptive* SSE security, while the more natural usage of searchable encryption calls for *adaptive* security (a notion that we make precise in Section 3). We propose new security definitions for both the non-adaptive and adaptive cases, and present efficient constructions for both based on any one-way function.

Our first construction is the most efficient non-adaptive SSE scheme to date in terms of computation on the server, and incurs a minimal (i.e., constant) cost for the user. Our second construction achieves adaptive security, which was not previously achieved by any constant-round solution. (Later on we perform a detailed comparison between our constructions and previous work—see Table 1.)

We also extend the problem of SSE to the multi-user setting, where a client wishes to allow an authorized group of users to search through its document collection.

Before providing a detailed comparison to existing work, we put our work in context by providing a classification of the various models for privacy-preserving searches.

On different models for private search. In recent years, there has been some confusion regarding three distinct models for searching with privacy: searching on *private-key* encrypted data (which is the subject of this work); searching on *public-key* encrypted data; and (single-database) *private information retrieval* (PIR).

Common to all three models is a server (sometimes called the "database") that stores data, and a user that wishes to access, search, or modify the data while revealing as little as possible to the server. There are, however, important differences between these three settings.

In the setting of **searching on private-key**-encrypted data, the user himself encrypts the data, so he can organize it in an arbitrary way (before encryption) and include additional data structures to allow for efficient access of relevant data. The data and the additional data structures can then be encrypted and stored on the server so that only someone with the private key can access it. In this setting, the initial work for the user (i.e., for preprocessing the data) is at least as large as the data, but subsequent work (i.e., for accessing the data) is very small relative to the size of the data for both the user and the server. Furthermore, everything about the user's access pattern can be hidden [24, 18].

In the setting of **searching on public-key**-encrypted data, users who encrypt the data (and send it to the server) can be different from the owner of the decryption key. In a typical application, a user publishes a public key while multiple senders send e-mails to the mail server [10, 3]. Anyone with access to the public key can add words to the index, but only the owner of the private key can generate "trapdoors" to test for the occurrence of a keyword. Although the original work on public-key encryption with keyword search (PEKS) by Boneh, di Crescenzo, Ostrosvky and Persiano [10] reveals the user's access pattern, recently Boneh, Kushilevitz, Ostvrosky and Skeith [11] have shown how to build a public-key encryption scheme that hides even the access pattern. This construction, however, has an overhead in search time that is proportional to the square root of the database size, which is far less efficient then the best private-key solutions.

Recently, Bellare, Boldyreva and O'Neill [6] introduced the notion of asymmetric *efficiently searchable encryption* (ESE) and proposed three constructions in the random oracle model. Unlike PEKS, asymmetric ESE schemes allow anyone with access to a user's public key to add words to the index *and* to generate trapdoors to search. While ESE schemes achieve optimal search time (same as our constructions – see below), they are inherently deterministic and therefore provide security guarantees that are weaker than the ones considered in this work.

In single-database **private information retrieval**, (or PIR) introduced by Kushilevitz and Ostrovsky [23], they show how a user can retrieve data from a server containing *unencrypted* data without revealing the access pattern and with total communication less then the data size. This was extended to keyword searching, including searching on streaming data [25]. We note, however, that since the data in PIR is always unencrypted, any scheme that tries to hide the access pattern must touch all data items. Otherwise, the server learns information: namely, that the untouched item was not of interest to the user. Thus, PIR schemes require work which is linear in the database size. Of course, one can amortize this work for multiple queries and multiple users in order to save work of the database per query, as shown in [21, 22], but the key feature of all PIR schemes is that the data is always unencrypted, unlike the previous two settings on searching on *encrypted* data.

Related work. We already mentioned the work on software protection and oblivious RAMs by Goldreich and Ostrovsky [18]. In an effort to reduce the round complexity associated with oblivious RAMs, Song, Wagner and Perrig [27] showed that a solution for searchable encryption was possible for a weaker security model. Specifically, they achieve searchable encryption by crafting, for each word, a special two-layered encryption construct. Given a trapdoor, the server can strip the outer layer and assert whether the inner layer is of the correct form. This construction, however, has some limitations: while the construction is proven to be a secure encryption scheme, it is not proven to be a secure *searchable* encryption scheme; the distribution of the underlying plaintexts is vulnerable to statistical attacks; and searching is linear in the length of the document collection.

The above limitations are addressed by the works of Goh [16] and of Chang and Mitzenmacher [12], who propose constructions that associate an "index" to each document in a collection. As a result, the server has to search each of these indexes, and the amount of work required for a query is proportional to the number of documents in the collection. Goh introduces a notion of security for indexes (IND-CKA, for "chosen-keyword attack," and the slightly stronger

IND2-CKA), and puts forth a construction based on Bloom filters [8] and pseudo-random functions. Chang and Mitzenmacher achieve a notion of security similar to IND2-CKA, except that it also tries to guarantee that the trapdoors do not leak any information about the words being queried. We discuss these security definitions and their shortcomings in more detail in Section 3.

As mentioned above, encryption with keyword search has also been considered in the public-key setting [10, 3], where anyone with access to a user's public-key can add words to an index, but only the owner of the private-key can generate trapdoors to test for the occurrence of a keyword. While related, the public-key solutions are suitable for different applications and are not as efficient as private-key solutions, which is the main subject of this work. Asymmetric ESE [6] achieves comparable efficiency, but at the price of providing weaker security guarantees. Further, we also note that the notion of multi-user SSE—which we introduce in this work—combined with a classical public-key encryption scheme, achieves a functionality similar to that of asymmetric ESE, with the added benefit of allowing the owner to revoke search privileges.

Our results. We now summarize our contributions.

1. We review existing security definitions for searchable encryption, including IND2-CKA [16] and the simulation-based definition in [12], and highlight their shortcomings. Specifically, we point out that IND2-CKA is not an adequate notion of security for SSE and then highlight (and fix) technical issues with Chang and Mitzenmacher's simulation-based definition. We address both of these issues by proposing new indistinguishability and simulation-based definitions that provide security for both indexes and trapdoors, and show their equivalence.

2. We introduce new adversarial models for SSE. The first, which we refer to as *non-adaptive*, only considers adversaries that make their search queries without taking into account the trapdoors and search outcomes of previous searches. The second—*adaptive*—considers adversaries that can choose their queries as a function of previously obtained trapdoors and search outcomes. All previous work on SSE (with the exception of oblivious RAMs) falls within the non-adaptive setting. The implication is that, contrary to the natural use of searchable encryption described in [27, 16, 12], these definitions only guarantee security for users that perform all their searches *at once*. We address this by introducing indistinguishability and simulation-based definitions in the adaptive setting, and show that they are equivalent.

3. We present two constructions which we prove secure under the new definitions. Our first scheme is only secure in the non-adaptive setting, but is the most efficient SSE construction to date. In fact, it achieves searches in one communication round, requires an amount of work on the server that is proportional to the actual number of documents that contain the queried word, requires constant storage on the client, and linear (in the size of the document collection) storage on the server. While the construction in [16] also performs searches in one round, it can induce false positives, which is not the case for our construction. Additionally, all the constructions in [16, 12] require the server to perform an amount of work proportional to the total number of documents in the collection.

Our second construction is secure against an adaptive adversary, but at the price of requiring a higher communication overhead per query and more storage at the server (comparable with the storage required by Goh's construction). While our adaptive scheme is conceptually simple, we note that constructing efficient and provably secure adaptive SSE schemes is a non-trivial task. The main challenge lies in proving such constructions secure in the simulation paradigm, since the simulator requires the ability to "commit" to a correct index before the adversary has even chosen its search queries—in other words, the simulator needs to commit to an index and then be able to perform some form of equivocation.

Table 1 compares our constructions (**SSE-1** and **SSE-2**) with the previous SSE schemes. To make the comparison easier, we assume that each document in the collection has the same (constant) size (otherwise, some of the costs have to scaled by the document size). The server computation row shows the costs per returned document for a query. Note that all previous work requires an amount of server computation at least linear with the number of documents in the collection, even if only one document matches a query. In contrast, in our constructions the server computation is constant per each document that matches a query, and the overall computation per query is proportional to the number of documents that match the query. In all the considered schemes, the computation and storage at the user is $O(1)$. We remark that, as an additional benefit, our constructions can also handle updates to the document collection in the sense of [12]. We point out an optimization which lowers the communication size and the server's computation per query from linear to logarithmic in the number of updates (see full version [13]).

4. Previous work on searchable encryption only considered the single-user setting. We also consider a natural extension of this setting, namely, the *multi-user* setting, where a user owns the data, but an arbitrary group of users can submit queries to search his document collection. The owner can control the search access by granting and revoking searching privileges to other users. We formally define searchable encryption in the multi-user setting, and present an efficient construction that does not require authentication, thus achieving better performance than simply using access control mechanisms.

Finally, we note that in most of the works mentioned above the server is assumed to be honest-but-curious. However, using techniques for memory checking [9] and universal arguments [5] one can make those solutions robust against malicious servers at the price of additional overhead. We restrict our attention to honest-but-curious servers as well, and postpone this extension to the full version.

Due to space limitations, full-fledged security definitions, security proofs and extensions are presented in the full version of the paper [13].

2. PRELIMINARIES

Let $\Delta = \{w_1, \ldots, w_d\}$ be a dictionary of d words, and 2^Δ be the set of all possible documents. Further, let $\mathcal{D} \subseteq 2^\Delta$ be a collection of n documents $\mathcal{D} = (D_1, \ldots, D_n)$ and 2^{2^Δ} be the set of all possible document collections. Let $\mathsf{id}(D)$ be the identifier of document D, where the identifier can be any string that uniquely identifies a document, such as

Properties	[24, 18]	[24, 18]-light	[27]	[16]	[12]	SSE-1	SSE-2
hides access pattern	yes	yes	no	no	no	no	no
server computation	$O(\log^3 n)$	$O(\sqrt{n})$	$O(n)$	$O(n)$	$O(n)$	$O(1)$	$O(1)$
server storage	$O(n \cdot \log n)$	$O(n)$	$O(n)$	$O(n)$	$O(n)$	$O(n)$	$O(n)$
number of rounds	$\log n$	2	1	1	1	1	1
communication	$O(\log^3 n)$	$O(\sqrt{n})$	$O(1)$	$O(1)$	$O(1)$	$O(1)$	$O(1)$
adaptive adversaries	yes	yes	no	no	no	no	yes

Table 1: Properties and performance (per query) of various SSE schemes. n denotes the number of documents in the document collection. For communication costs, we consider only the overhead and omit the size of the retrieved documents, which is the same for all schemes. For server computation, we show the costs per returned document. For simplicity, the security parameter is not included as a factor for the relevant costs.

a memory location. We denote by $\mathcal{D}(w)$ the lexicographically ordered list consisting of the identifiers of all documents in $\mathcal{D}$ that contain the word w. We sometimes refer to $\mathcal{D}(w)$ as the *outcome of a search* for w and to the sequence $(\mathcal{D}(w_1), \ldots, \mathcal{D}(w_n))$ as the *access pattern* of a client. We also define the *search pattern* of a client as any information that can be derived from knowing whether two arbitrary searches were performed for the same word or not.

We write $x \leftarrow \mathcal{X}$ to represent an element x being sampled from a distribution $\mathcal{X}$ and $x \xleftarrow{R} X$ to represent an element x being sampled uniformly from a set X. The output x of an algorithm $\mathcal{A}$ is denoted by $x \leftarrow \mathcal{A}$. We write $||$ to mean string concatenation. We call a function $\nu : \mathbb{N} \to \mathbb{N}$ negligible if for every polynomial $p(\cdot)$ and all sufficiently large k, $\nu(k) < \frac{1}{p(k)}$.

Model. The participants in a single-user searchable encryption scheme include a user that wishes to store an encrypted document collection $\mathcal{D} = (D_1, \ldots, D_n)$ on an honest-but-curious server S, while preserving the ability to search through them. We note that while we choose, for ease of exposition, to limit searches to be over documents, any SSE scheme can be trivially extended to search over lists of arbitrary keywords associated with the documents.

The participants in a multi-user searchable encryption scheme include a trusted owner O, an honest-but-curious server S, and a set of users N. O owns the document collection $\mathcal{D}$ and wants to grant and revoke searching privileges to a subset of users in N. We let $\mathsf{G} \subseteq \mathsf{N}$ be the set of users allowed to search. We assume that currently non-revoked users behave honestly. The honest-but-curious server S is a party that follows the protocol specification correctly, but may try to analyze the messages received during the protocol in order to learn additional information.

Basic primitives. A symmetric encryption scheme is a set of three polynomial-time algorithms $(\mathcal{G}, \mathcal{E}, \mathcal{D})$ such that $\mathcal{G}$ takes a security parameter k in unary and returns a secret key K; $\mathcal{E}$ takes a key K and an n-bit message m and returns a ciphertext c; $\mathcal{D}$ takes a key K and a ciphertext c and returns m if K was the key under which c was produced. Informally, a symmetric encryption scheme is considered secure if the ciphertexts it outputs do not leak any partial information about the plaintext even to an adversary that can adaptively query an encryption and a decryption oracle.

In addition to encryption schemes, we also make use of pseudo-random functions (PRF) and permutations (PRP), which are polynomial-time computable functions that cannot be distinguished from random functions by any probabilistic polynomial-time adversary.

3. REVISITING SSE DEFINITIONS

We begin by reviewing the definition of a SSE scheme.

DEFINITION 3.1. (SEARCHABLE SYMMETRIC ENCRYPTION SCHEME (SSE)) *A SSE scheme is a collection of four polynomial-time algorithms* (Keygen, BuildIndex, Trapdoor, Search) *such that:*

Keygen(1^k) *is a probabilistic key generation algorithm that is run by the user to setup the scheme. It takes a security parameter k, and returns a secret key K such that the length of K is polynomially bounded in k.*

BuildIndex$(K, \mathcal{D})$ *is a (possibly probabilistic) algorithm run by the user to generate indexes. It takes a secret key K and a polynomially bounded in k document collection $\mathcal{D}$ as inputs, and returns an index $\mathcal{I}$ such that the length of $\mathcal{I}$ is polynomially bounded in k.*

Trapdoor(K, w) *is run by the user to generate a trapdoor for a given word. It takes a secret key K and a word w as inputs, and returns a trapdoor T_w.*

Search$(\mathcal{I}, T_w)$ *is run by the server S in order to search for the documents in $\mathcal{D}$ that contain word w. It takes an index $\mathcal{I}$ for a collection $\mathcal{D}$ and a trapdoor T_w for word w as inputs, and returns $\mathcal{D}(w)$, the set of identifiers of documents containing w.*

A correct intuition. So far, establishing correct security definitions for searchable encryption has been elusive. Clearly, as we have discussed, one could use the general definitions from oblivious RAMs, but subsequent work (including ours) examines if more efficient schemes can be achieved by revealing *some* information. The first difficulty seems to be in correctly capturing this intuition as a formal security definition. In the literature, security for searchable encryption is typically characterized as the requirement that nothing be leaked beyond the outcome of a search (i.e., the identifiers of the documents returned from a search), however we are not aware of any previous work on SSE that satisfies this intuition. In fact, with the exception of oblivious RAMs, all previous constructions leak, in addition to the search outcomes, the user's search pattern. This is clearly the case for the schemes presented in [27, 16, 12] since their trapdoors are deterministic. Therefore, a more accurate characterization of the security notion achieved (or rather, sought) for SSE is that nothing should be leaked beyond the outcome and the *pattern* of a sequence of searches, where the pattern of a search is any information that can be derived from knowing whether two searches were performed for the same word or not.

Limitations of previous SSE definitions. The second issue seems to be in appropriately capturing the adversary's power. While Song, Wagner and Perrig proved their construction secure, the definition implicitly used in their work is that of a classical encryption scheme, where the adversary is not allowed to perform searches. This was partly rectified by Goh who proposed the notion of indistinguishability against chosen-keyword attacks (IND2-CKA) in [16][1]. Intuitively, the notion of security that IND2-CKA tries to achieve can be described as follows: given access to a set of indexes, the adversary (i.e., the server) is not able to learn any partial information about the underlying document that he cannot learn from using a trapdoor that was given to him by the client, and this holds even against adversaries that can trick the client into generating indexes and trapdoors for documents and keywords of its choice (i.e., chosen-keyword attacks). A formal specification of IND2-CKA is presented in [13] [2].

We remark that Goh's work addresses a larger problem than searchable encryption, namely that of secure indexes, which are secure data structures that have many uses, only one of which is searchable encryption. And though much work on searchable encryption uses IND2-CKA as a security definition [20, 26, 4], we note that it was never intended as such. This is simply because, as Goh remarks (*cf.* Note 1, p. 5 of [16]), IND2-CKA does not explicitly require that trapdoors be secure since this is not a requirement for all applications of secure indexes.

To remedy this, one might be tempted to introduce a second definition that exclusively guarantees the semantic security of trapdoors. One would then prove a construction secure under both IND2-CKA, and the new definition. While this might seem like a reasonable (though cumbersome) idea, the straightforward approach of requiring trapdoors to be indistinguishable does not work. In fact, as we show in [13], SSE schemes can be built with trapdoors that, taken independently, leak no partial information about the word being queried, but when combined with an index allow an adversary to recover the entire word. This illustrates that the security of indexes and the security of trapdoors are intrinsically linked.

Chang and Mitzenmacher propose a simulation-based definition that aims to guarantee privacy for indexes *and* trapdoors [12]. Similarly to the classical definition of semantic security for encryption [19], they require that anything that can be computed from the index and the trapdoors for various queries, can be computed from the search outcome of those queries. However, while the intuition seems correct, in the case of searchable encryption one must also take care in describing *how* the search queries are generated. In particular, whether they can be made adaptively (i.e., after seeing the outcome of previous queries) or non-adaptively (i.e., without seeing the outcome of any queries). This distinction is important because it leads to security definitions that achieve drastically different privacy guarantees. Indeed, while non-adaptive definitions only guarantee security to clients who generate all their queries at once, adaptive definitions guarantee privacy even to clients who generate queries as a function of previous search outcomes. Unfortunately, as we show in [13], the definition presented in [12] is not only *non-adaptive*, but can be trivially satisfied by any SSE scheme, even one that is insecure.

Our security definitions. We now address the above issues. For ease of readability, in this section we present our approach at a somewhat informal level, but a more rigorous treatment can be found in the full version of the paper [13].

Above, we mentioned our (and previous work's) willingness to let the outcome and the pattern of a sequence of searches be known to the adversary (i.e., the server) in order to achieve greater efficiency. This can be more formally specified as follows. First, we note that an interaction between the client and the server will be determined by a document collection and a set of words that the client wishes to search for (and that we wish to hide from the adversary); we call an instantiation of such an interaction a *history*. Given a history, we refer to what the adversary actually gets to "see" during an interaction as the history's *view*. In particular, the view will consist of the index (of the document collection) and the trapdoors (of the queried words). It will also contain some additional common information, such as the number of documents in the collection and their ciphertexts). However (if done properly) the view (i.e., the index and the trapdoors) should not reveal any information about the history (i.e., the documents and the queried words) besides the outcome and the pattern of the searches (i.e., the information we are willing to leak). This leads to the notion of the *trace* of an interaction/history, which consists of exactly the information we are willing to leak about the history and nothing else. More precisely, this should include the identifiers of the documents that contain each query word in the history (i.e., the outcome of each search), and information that describes which trapdoors in the view correspond to the same underlying words in the history (i.e., the pattern of the searches).

We are now ready to state our first security definition for SSE. First, we assume that the adversary generates the histories in the definition at once. In other words, it is not allowed to see the index of the document collection or the trapdoors of any query words it chooses before it has finished generating the history. We call such an adversary *non-adaptive*.

DEFINITION 3.2. (NON-ADAPTIVE INDISTINGUISHABILITY SECURITY FOR SSE—INFORMAL VERSION) *A SSE scheme is secure in the sense of non-adaptive indistinguishability if for any two adversarially constructed histories with equal length and trace, no (probabilistic polynomial-time) adversary can distinguish the view of one from the view of the other with probability non-negligibly better than $\frac{1}{2}$.*

Second, for each history the adversary generates, we give him the ability to choose the word queries as a function of the index and the trapdoors corresponding to the document collection and the previous queries it chose. More precisely,

---

[1] Goh also defines a weaker notion, IND-CKA, that allows an index to leak the number of words in the document.
[2] We note that, unlike the latter and our own definitions (see below), IND2-CKA applies to indexes that are built for individual documents, as opposed to indexes built from entire document collections.

for each history, the adversary must choose a document collection and multiple query words. So in this version of our definition, after he chooses a document collection, he will receive its corresponding index *before* he chooses his first query word. And he will then receive that query word's trapdoor *before* he chooses his next query word, and so on. What this implies is that for the two histories he constructs, he can choose query words as a function of the index and his previous query words' trapdoors. Intuitively, this could enable the adversary to perform more sophisticated attacks than in the previous case. We call such histories "(adversarially) adaptively constructed" (a formal specification of this process is described in [13]).

DEFINITION 3.3. (ADAPTIVE INDISTINGUISHABILITY SECURITY FOR SSE—INFORMAL VERSION) *A SSE scheme is secure in the sense of* adaptive indistinguishability *if for any two adaptively-constructed histories with equal length and trace, no (probabilistic polynomial-time) adversary can distinguish the view of one history from the view of the other with probability non-negligibly better than $\frac{1}{2}$.*

An alternative approach to security definitions is the so-called semantic security or "simulation-based" approach [19, 17]. At a high level, in such an approach the security guarantee is provided by the existence, for all adversaries, of a polynomial-time algorithm (the *simulator*) which, being given very little information (in our case, a history's trace), is able to compute whatever the adversary is able to compute from the given information (in our case, the history's view). In [13], we also present simulation-based definitions for SSE (overcoming the shortcomings of the simulation-based definition in [12]), both for the non-adaptive and adaptive settings, and, moreover, we are able to prove:

THEOREM 3.4. *Non-adaptive (respectively, adaptive) indistinguishability security of SSE is equivalent to non-adaptive (resp., adaptive) semantic security of SSE.*

We remark that the existence of such an equivalence proof typically vouches for the soundness of the definitions presented herein. Further, it allows us to state that an SSE scheme is simply non-adaptively (resp., adaptively) secure, without any reference to the proof methodology.

4. EFFICIENT AND SECURE SSE

In this section we present our efficient SSE constructions, and state their security in terms of the definitions presented in Section 3 (the security proofs are presented in the full version [13]). We start by introducing some additional notation and the data structures used by the constructions. Let $\Delta', \Delta \subseteq \Delta$, be the set of distinct words that exist in the document collection $\mathcal{D}$. We assume that words in Δ can be represented using at most p bits. Also, recall that $\mathcal{D}(w)$ is the set of identifiers of documents in $\mathcal{D}$ that contain word w ordered in lexicographic order.

We use several data structures, including arrays, linked lists and look-up tables. Given an array A, we refer to the element at address i in A as $\text{A}[i]$, and to the address of element x relative to A as $\text{addr}(\text{A}(x))$. So if $\text{A}[i] = x$, then $\text{addr}(\text{A}(x)) = i$. In addition, a linked list L, stored in an array A, is a set of nodes $\text{N}_i = \langle v_i; \text{addr}(\text{A}(\text{N}_{i+1})) \rangle$, where $1 \leq i \leq |\text{L}|$, v_i is an arbitrary string and $\text{addr}(\text{A}(\text{N}_{i+1}))$ is the memory address of the next node in the list.

4.1 An efficient SSE construction

We first give an overview of our one-round non-adaptive SSE construction. We associate a single index $\mathcal{I}$ with a document collection $\mathcal{D}$. The index $\mathcal{I}$ consists of two data structures:

– An array A, in which we store in encrypted form the set $\mathcal{D}(w)$, for each word $w \in \Delta'$, and
– a look-up table T, which contains information that enables one to locate and decrypt the appropriate elements from A, for each word $w \in \Delta'$.

We start with a collection of linked lists L_i, $w_i \in \Delta'$, where the nodes of each L_i are the identifiers of documents in $\mathcal{D}(w_i)$. We then write in the array A the nodes of all lists L_i, "scrambled" in a random order and encrypted with randomly generated keys. Before encryption, the j-th node of L_i is augmented with information about the index in A of the $(j+1)$-th node of L_i, together with the key used to encrypt it. In this way, given the position (index) in A and the decryption key for the first node of a list L_i, the server will be able to locate and decrypt all the nodes in L_i. Note that by storing in A the nodes of all lists L_i in a random order, the size of each L_i is hidden.

We now build a look-up table T that allows one to locate and decrypt the first element of each list L_i. Each entry in T corresponds to a word $w_i \in \Delta$ and consists of a pair <address,value>. The field value contains the index in A and the decryption key for the first element of L_i. value is itself encrypted using the output of a pseudo-random function. The other field, address, is simply used to locate an entry in T. The look-up table T is managed using *indirect addressing* (described below).

The user computes both A and T based on the un-encrypted $\mathcal{D}$, and stores them on the server together with the encrypted $\mathcal{D}$. When the user wants to retrieve the documents that contain word w_i, it computes the decryption key and the address for the corresponding entry in T and sends them to the server. The server locates and decrypts the given entry of T, and gets the index in A and the decryption key for the first node of L_i. Since each element of L_i contains information about the next element of L_i, the server can locate and decrypt all the nodes of L_i, which gives the identifiers in $\mathcal{D}(w_i)$.

Efficient storage and access of sparse tables. We describe the indirect addressing method that we use to efficiently manage look-up tables. The entries of a look-up table T are tuples <address,value>, in which the address field is used is used as a *virtual address* to locate the entry in T that contains some value field. Given a parameter p, a virtual address is from a domain of exponential size (i.e., from $\{0,1\}^p$). However, the maximum number of entries in a look-up table will be polynomial in p, so the number of virtual addresses that are used can be approximated as $\text{poly}(p)$. If, for a lookup-up table T, the address field is from $\{0,1\}^p$, the value field is from $\{0,1\}^v$ and there are at most m entries in T, then we say T is a $(\{0,1\}^p \times \{0,1\}^v \times m)$ look-up table.

Let Addr be the set of virtual addresses that are used for entries in a look-up table T. We can efficiently store T such that, when given a virtual address, it returns the associated value field. We achieve this by organizing the Addr set in a so-called *FKS dictionary* [15], an efficient data structure for

---

Keygen($1^k, 1^\ell$): Generate random keys $s, y, z \xleftarrow{R} \{0,1\}^k$ and output $K = (s, y, z, 1^\ell)$.

BuildIndex($K, \mathcal{D}$):

1. Initialization:
   a) scan $\mathcal{D}$ and build Δ', the set of distinct words in $\mathcal{D}$. For each word $w \in \Delta'$, build $\mathcal{D}(w)$;
   b) initialize a global counter ctr = 1.

2. Build array A:
   a) for each $w_i \in \Delta'$: // (build a linked list L_i with nodes $N_{i,j}$ and store it in array A)
      - generate $\kappa_{i,0} \xleftarrow{R} \{0,1\}^\ell$
      - for $1 \leq j \leq |\mathcal{D}(w_i)|$:
        - generate $\kappa_{i,j} \xleftarrow{R} \{0,1\}^\ell$ and set node $N_{i,j} = \langle \text{id}(D_{i,j}) \| \kappa_{i,j} \| \psi_s(\text{ctr}+1) \rangle$, where $\text{id}(D_{i,j})$ is the jth identifier in $\mathcal{D}(w_i)$;
        - compute $\mathcal{E}_{\kappa_{i,j-1}}(N_{i,j})$, and store it in $A[\psi_s(\text{ctr})]$;
        - ctr = ctr + 1
      - for the last node of L_i (i.e., $N_{i,|\mathcal{D}(w_i)|}$), before encryption, set the address of the next node to NULL;
   b) let $m' = \sum_{w_i \in \Delta'} |\mathcal{D}(w_i)|$. If $m' < m$, then set remaining $(m - m')$ entries of A to random values of the same size as the existing m' entries of A.

3. Build look-up table T:
   a) for each $w_i \in \Delta'$:
      - value = $\langle \text{addr}(A(N_{i,1})) \| \kappa_{i,0} \rangle \oplus f_y(w_i)$;
      - set $T[\pi_z(w_i)] = $ value.
   b) if $|\Delta'| < |\Delta|$, then set the remaining $(|\Delta| - |\Delta'|)$ entries of T to random values.

4. Output $\mathcal{I} = (A, T)$.

Trapdoor(w): Output $T_w = (\pi_z(w), f_y(w))$.

Search($\mathcal{I}, T_w$):

1. Let $(\gamma, \eta) = T_w$. Retrieve $\theta = T[\gamma]$. Let $\langle \alpha \| \kappa \rangle = \theta \oplus \eta$.
2. Decrypt the list L starting with the node at address α encrypted under key κ.
3. Output the list of document identifiers contained in L.

Figure 1: Efficient SSE construction (SSE-1)

storage of sparse tables that requires $O(|\text{Addr}|) \ (+o(|\text{Addr}|))$ storage and $O(1)$ look-up time. In other words, given some virtual address A, we are able to tell if $A \in \text{Addr}$ and if so, return the associated value in constant look-up time. Addresses that are not in Addr are considered undefined.

SSE-1 in detail. We are now ready to proceed to the details of the construction. Let k, ℓ be security parameters and let $(\mathcal{G}, \mathcal{E}, \mathcal{D})$ be a semantically secure symmetric encryption scheme with $\mathcal{E} : \{0,1\}^\ell \times \{0,1\}^r \to \{0,1\}^r$. In addition, we make use of one pseudo-random function f and two pseudo-random permutations π and ψ with the following parameters:

- $f : \{0,1\}^k \times \{0,1\}^p \to \{0,1\}^{\ell + \log_2(m)}$;
- $\pi : \{0,1\}^k \times \{0,1\}^p \to \{0,1\}^p$; and
- $\psi : \{0,1\}^k \times \{0,1\}^{\log_2(m)} \to \{0,1\}^{\log_2(m)}$.

Let m be the total size of the plaintext document collection, expressed in *units*. A *unit* is the smallest possible size for a word (e.g. one byte).[3] Let A be an array of size m. Let T be a $(\{0,1\}^p \times \{0,1\}^{\ell + \log_2(m)} \times |\Delta|)$ look-up table, managed using indirect addressing as described previously. Our construction SSE-1 = (Keygen, BuildIndex, Trapdoor, Search) is described in Fig. 1.

Consistent with our security definitions, SSE-1 reveals only the outcome and the pattern of a search, the total size of the encrypted document collection and the number of documents in $\mathcal{D}$. Recall that the array A can be seen as a collection of linked lists L_i, where each L_i contains the identifiers of documents containing word w_i. Let $m' = \sum_{w_i \in \Delta'} |L_i|$. If, for all $D_j \in \mathcal{D}$, a word does not appear more than once in document D_j, it is clear that $m = m'$. If the size of A is smaller than m, then the array A reveals that at least one document in $\mathcal{D}$ contains a word more than once. To avoid such leakage, we set the size of A equal to m and fill the $(m - m')$ remaining entries with random values. We follow the same line of reasoning for the look-up table T, which has at least one entry for each distinct word in $\mathcal{D}$. To avoid revealing the number of distinct words in $\mathcal{D}$, we add additional $(|\Delta| - |\Delta'|)$ entries in T, filled with random values, such that the number of entries in T is always equal to $|\Delta|$.

In the full version of the paper [13], we show:

THEOREM 4.1. *SSE-1 is a non-adaptively secure SSE scheme.*

Regarding efficiency, we remark that each query takes only one round, and $O(1)$ message size. In terms of storage, the demands are $O(1)$ on the user and $O(m)$ on the server; more specifically, in addition to the encrypted $\mathcal{D}$, the server stores

---
[3] If the documents are not encrypted with a length preserving encryption scheme or if they are compressed before encryption, then m is the maximum between the total size of the plaintext $\mathcal{D}$ and the total size of the encrypted $\mathcal{D}$.

> Keygen(1^k): **Generate random key** $s \xleftarrow{R} \{0,1\}^k$ **and output** $K = s$.
>
> BuildIndex($K, \mathcal{D}$):
> 1. **Initialization:**
>   - scan $\mathcal{D}$ and build Δ', the set of distinct words in $\mathcal{D}$. For each word $w \in \Delta'$, build $\mathcal{D}(w)$.
> 2. **Build look-up table** T:
>   a) for each $w_i \in \Delta'$:
>   - for $1 \leq j \leq |\mathcal{D}(w_i)|$:
>   - value = $\mathsf{id}(D_{i,j})$, where $\mathsf{id}(D_{i,j})$ is the jth identifier in $\mathcal{D}(w_i)$;
>   - set $\mathtt{T}[\pi_s(w_i||j)] = \mathtt{value}$.
>   b) let $m' = \sum_{w_i \in \Delta'} |\mathcal{D}(w_i)|$. If $m' < m$, then set values for the remaining $(m - m')$ entries such that for all $D \in \mathcal{D}$, it holds that value = $\mathsf{id}(D)$ for exactly max entries. Also, set the address field of these remaining entries to random values.
> 3. **Output** $\mathcal{I} = \mathtt{T}$.
>
> Trapdoor(w): **Output** $T_w = (T_{w_1}, \ldots, T_{w_{\max}}) = (\pi_s(w||1), \ldots, \pi_s(w||\mathtt{max}))$.
>
> Search($\mathcal{I}, T_w$): **For** $1 \leq i \leq \mathtt{max}$: **retrieve** $\mathsf{id} = \mathtt{T}[T_{w_i}]$ **and output** id.

Figure 2: Adaptively secure SSE construction (SSE-2)

the index $\mathcal{I}$, which contains the array A of size $O(m)$ and the look-up table T of size $O(|\Delta|)$. Since the size of encrypted $\mathcal{D}$ is $O(m)$, accommodating the auxiliary data structures used for searching does not change (asymptotically) the storage requirements for the server. The user spends $O(1)$ time to compute a trapdoor, while for a query for word w, the server spends time proportional to $|\mathcal{D}(w)|$.

4.2 Adaptive SSE security

While our SSE-1 construction is efficient, it was only proven secure against non-adaptive adversaries. We now show a second construction, SSE-2, which achieves semantic security against adaptive adversaries, at the price of requiring higher communication size per query and more storage on the server. (Asymptotically, however, costs are the same—see Table 1.)

The difficulty of proving our SSE-1 construction secure against an adaptive adversary stems from the difficulty of creating in advance a view for the adversary that would be consistent with future (unknown) queries. Given the intricate structure of the SSE-1 construction, with each word having a corresponding linked list whose nodes are stored encrypted and in a random order, building an appropriate index is quite challenging. We circumvent this problem as follows.

For a given word w and a given integer j, we derive a label for w by concatenating w with j (j is first converted to a string of characters). For example, if w is "coin" and j is 1, then $w||j$ is "coin1". We define the *family* of a word $w \in \Delta'$ to be the set of labels $\mathsf{F}_w = \{w||j : 1 \leq j \leq |\mathcal{D}(w)|\}$. For example, if the word "coin" appears in three documents, then $\mathsf{F}_w = \{$"coin1", "coin2", "coin3"$\}$. Now, for each word $w \in \Delta'$, we choose not to keep a list of nodes with the identifiers in $\mathcal{D}(w)$, but instead to simply derive the family F_w of w, and insert the elements of F_w into the index. Searching for w becomes equivalent with searching for all the labels in w's family. Since each label in w's family will appear in only one document, a search for it "reveals" only one entry in the index. Translated to the proof, this will allow the simulator to easily construct a view for the adversary that is indistinguishable from a real view.

We now give an overview of the SSE-2 construction. We associate with the document collection $\mathcal{D}$ an index $\mathcal{I}$, that consists of a look-up table T. For each label in a word w's family, we add an entry in T, whose value field is the identifier of the document that contains an instance of w. In order to hide the number of distinct words in each document, we have to "pad" the look-up table T such that the identifier of each document appears in the same number of entries. The search for a word w is slightly different than for the SSE-1 construction: a user needs to search for all the labels in w's family.

Let k be a security parameter. We use a pseudo-random permutation $\pi : \{0,1\}^k \times \{0,1\}^p \rightarrow \{0,1\}^p$. Recall that a *unit* is the smallest possible size for a word (e.g. one byte). Also, recall that Δ' is the set of distinct words that exist in $\mathcal{D}$. Let max be the size of the largest plaintext document in $\mathcal{D}$, expressed in units. Let $m = \mathtt{max} \cdot n$, where n is the number of documents in $\mathcal{D}$. Let T be a $(\{0,1\}^p \times \{0,1\}^{\log_2(n)} \times m)$ look-up table, managed using indirect addressing. The construction SSE-2 is described in Fig. 2. In [13] we prove:

THEOREM 4.2. *SSE-2 is an adaptively secure SSE scheme.*

Just like SSE-1, SSE-2 requires for each query one round of communication and an amount of computation on the server proportional with the number of documents that match the query (i.e., $O(|\mathcal{D}(w)|)$). Similarly, the storage and computational demands on the user are $O(1)$. The communication is equal to max and the storage on the server is increased by a factor of max when compared to SSE-1. We note that the communication cost can be reduced if in each entry of T corresponding to an element in some word w's family, we also store $|\mathcal{D}(w)|$ in encrypted form. In this way, after searching for a label in w's family, the user will know $|\mathcal{D}(w)|$ and can derive F_w. The user can then send in a single round all the trapdoors corresponding to the remaining labels in w's family.

5. MULTI-USER SSE

In this section we consider the natural extension to the SSE setting where a user owns a document collection, but an arbitrary group of users can submit queries to search his collection. A familiar question arises in this new setting,

> MKeygen($1^k, 1^\ell$): let $K^s \leftarrow$ Keygen($1^k, 1^\ell$) and $r \xleftarrow{R} \{0,1\}^k$. Output $K_\mathcal{O} = (K^s, r)$.
>
> MBuildIndex($K_\mathcal{O}, \mathcal{D}$): run $\mathcal{I}^s \leftarrow$ BuildIndex($K^s, \mathcal{D}$). Initialize the BE scheme. Set $\mathsf{R} = \{\varnothing\}$. Send r and $\mathcal{E}^{\mathsf{BE}}_{\mathsf{N}}(r)$ to the server. Output $\mathcal{I}^m = \mathcal{I}^s$.
>
> AddUser($K_\mathcal{O}, U$): send $K_U = (K^s, r)$ to user U, where r is the current key used for ϕ. Also send to U the long-lived secrets needed for the BE scheme.
>
> RevokeUser($K_\mathcal{O}, U$): $\mathsf{R} = \mathsf{R} \cup \{U\}$. Pick a new key $r' \xleftarrow{R} \{0,1\}^k$ and send r' and $\mathcal{E}^{\mathsf{BE}}_{\mathsf{N}\backslash\mathsf{R}}(r')$ to $\mathcal{S}$. $\mathcal{S}$ overwrites the old values of r and $\mathcal{E}^{\mathsf{BE}}_{\mathsf{N}\backslash\mathsf{R}\cup\{U\}}(r)$ with r' and $\mathcal{E}^{\mathsf{BE}}_{\mathsf{N}\backslash\mathsf{R}}(r')$, respectively.
>
> MTrapdoor(K_U, w): let $T^s_w \leftarrow$ Trapdoor(K^s, w). Retrieve $\mathcal{E}^{\mathsf{BE}}_{\mathsf{N}\backslash\mathsf{R}}(r)$ from $\mathcal{S}$ and use the long-lived BE secrets to recover r. Output $T^m_{U,w} = \phi_r(T^s_w)$.
>
> MSearch($\mathcal{I}^m, T^m_{U,w}$): recover $T^s_w = \phi_r^{-1}(T^m_{U,w})$; let $T^s_w = (\gamma, \eta)$. If γ is a valid virtual address, then run Search($\mathcal{I}^m, T^s_w$) and return its output. Otherwise, return $\perp$.

Figure 3: Multi-user SSE construction (M-SSE)

that of managing access privileges, but while preserving privacy with respect to the server. We first present a definition of a multi-user searchable encryption scheme (*MSSE*) and some of its desirable security properties, followed by an efficient construction which, in essence, combines a single-user SSE scheme with a broadcast encryption (BE) scheme [14]. Let N denote the set of all possible users, and G ⊆ N the set of users that are currently authorized to search.

DEFINITION 5.1. (MULTI-USER SEARCHABLE SYMMETRIC ENCRYPTION SCHEME) *A multi-user SSE scheme is a collection of six polynomial-time algorithms* M-SSE = (MKeygen, MBuildIndex, AddUser, RevokeUser, MTrapdoor, MSearch) *such that:*

MKeygen(1^k) *is a probabilistic key generation algorithm that is run by the owner O to setup the scheme. It takes a security parameter k, and returns an owner secret key, $K_\mathcal{O}$.*

MBuildIndex($K_\mathcal{O}, \mathcal{D}$) *is run by O to construct indexes. It takes the owner's secret key $K_\mathcal{O}$ and a document collection $\mathcal{D}$ as inputs, and returns an index $\mathcal{I}$.*

AddUser($K_\mathcal{O}, U$) *is run by O whenever it wishes to add a user to the group* G*. It takes the owner's secret key $K_\mathcal{O}$ and a user U as inputs, and returns U's secret key, K_U.*

RevokeUser($K_\mathcal{O}, U$) *is run by O whenever it wishes to revoke a user from* G*. It takes the owner's secret key $K_\mathcal{O}$ and a user U as inputs, and revokes the user's searching privileges.*

MTrapdoor(K_U, w) *is run by a user (including O) in order to generate a trapdoor for a given word. It takes a user U's secret key K_U and a word w as inputs, and returns a trapdoor $T_{U,w}$.*

MSearch($\mathcal{I}_\mathcal{D}, T_{U,w}$) *is run by the server $\mathcal{S}$ in order to search for the documents in $\mathcal{D}$ that contain word w. It takes the index $\mathcal{I}_\mathcal{D}$ for collection $\mathcal{D}$ and the trapdoor $T_{U,w}$ for word w as inputs, and returns $\mathcal{D}(w)$ if user $U \in$ G and $\perp$ if user $U \notin$ G.*

We briefly discuss notions of security that a multi-user SSE scheme should achieve. It should be clear that the (semantic) security of a multi-user scheme can be reduced to the semantic security of the underlying single-user scheme. The reason is that in the multi-user case, just like in the single-user case, we are only concerned with providing security against the server. One distinct property in this new setting is that of *revocation*, which essentially states that a revoked user no longer be able to perform searches on the owner's documents. A formal specification of this property is given in the full version of the paper [13].

DEFINITION 5.2 (CORRECTNESS). *Let $\mathcal{D}$ be a document collection and $\mathcal{I}_\mathcal{D}$ be its corresponding index. We say that a multi-user SSE scheme,* M-SSE = (MKeygen, MBuildIndex, AddUser, RevokeUser, MTrapdoor, MSearch)*, is correct if*

$$\Pr\left[\mathsf{MSearch}(\mathcal{I}_\mathcal{D}, T_{U,w}) = \mathcal{D}(w) : U \in \mathsf{G}\right] = 1.$$

Our construction makes use of a single-user SSE scheme and a broadcast encryption (BE) scheme. Recall that in BE, a center encrypts a message m to a group G of privileged users who are allowed to access the message. The group G can be dynamically changing, as users can be added to or removed from G. Although the encrypted message can be received by a larger set N of receivers, only the users in G can recover the message. When a user joins the system, it receives a set of secrets, referred to as *long-lived* secrets. The long-lived secrets are distinct for each user. Given an encrypted message, the long-lived secrets allow a user to decrypt it only if the user was non-revoked at the time the message was encrypted. We use off-the-shelf BE as a building block in our multi-user secure index construction in order to efficiently manage user revocation.

We now provide an overview of the construction. In order to retrieve the documents that contain the word w, an authorized user U computes a regular single-user trapdoor $T_{U,w}$, but applies on it a pseudo-random permutation ϕ keyed with a secret key r before sending it to the server. The server, upon receiving $\phi_r(T_{U,w})$, recovers the trapdoor by computing $T_{U,w} = \phi_r^{-1}(\phi_r(T_{U,w}))$. The key r currently used for ϕ is only known by the owner, by the set of currently authorized users and by the server. Each time a user is revoked, the owner picks a new r and stores it on the server encrypted such that only non-revoked users can decrypt it. Broadcast encryption provides an efficient method to distribute r to the set of non-revoked users. The server will use the new r to compute ϕ_r^{-1} for subsequent queries. Revoked users cannot recover the current r and, with overwhelming probability, their queries will not yield a valid trapdoor after the server applies ϕ_r^{-1}.

When the owner O of a document collection $\mathcal{D}$ gives a user U permission to search through $\mathcal{D}$, it sends to U all the secret information needed to perform searches in a single-user context [4]. The extra layer given by the pseudo-random permutation ϕ, together with the guarantees offered by the BE scheme and the assumption that the server is honest-but-curious, is what prevents users from performing successful searches once they are revoked.

Next we describe the multi-user SSE construction in detail. Let SSE = (Keygen, BuildIndex, Trapdoor, Search) be a single-user SSE scheme and BE = ($\mathcal{G}^{BE}, \mathcal{E}_G^{BE}, \mathcal{D}_G^{BE}$) be a broadcast encryption scheme. Though our MSSE construction is general and can be instantiated with any single-user SSE scheme, for ease of exposition, we describe it using our SSE-1 construction. We require a standard security notion for the BE scheme, namely that it provide revocation-scheme security against a coalition of all revoked users and that its key assignment algorithm satisfies key indistinguishability. Recall that we let N denote the set of all users, and G $\subseteq$ N the set of users (currently) authorized to search; let R denote the set of revoked users. Let ϕ be a pseudo-random permutation such that $\phi : \{0,1\}^k \times \{0,1\}^{p+\log_2(m)+\ell} \to \{0,1\}^{p+\log_2(m)+\ell}$. Our multi-user construction, M-SSE, is described in Fig. 3.

Our multi-user construction is very efficient on the server side: when given a trapdoor, the server only needs to evaluate a pseudo-random permutation in order to determine if the user is revoked. If access control mechanisms were used instead for this step, a "heavier" authentication protocol would be required. Refer to [13] for further details.

Acknowledgements. The authors thank Fabian Monrose for helpful discussions at early stages of this work. The first author was at Bell Labs during part of this work. The third author is supported by a Bell Labs Graduate Research Fellowship. The fourth author is supported in part by an IBM Faculty Award, a Xerox Innovation Group Award, a gift from Teradata, an Intel equipment grant, a UC-MICRO grant, and NSF Cybertrust grant No. 0430254.

6. REFERENCES

[1] Google Desktop. http://desktop.google.com.

[2] Privacy with Security. DARPA Information Science and Technology (ISAT) Study Group, December 2002. http://www.cs.berkeley.edu/~tygar/papers/ISAT-final-briefing.pdf.

[3] M. Abdalla, M. Bellare, D. Catalano, E. Kiltz, T. Kohno, T. Lange, J. M. Lee, G. Neven, P. Paillier, and H. Shi. Searchable encryption revisited: Consistency properties, relation to anonymous IBE, and extensions. In *CRYPTO 2005*, volume 3621 of *LNCS*, pages 205–222. Springer, 2005.

[4] L. Ballard, S. Kamara, and F. Monrose. Achieving efficient conjunctive keyword searches over encrypted data. In *Proceedings of the Seventh International Conference on Information and Communication Security (ICICS 2005)*, pages 414–426, 2005.

[5] B. Barak and O. Goldreich. Universal arguments and their applications. In *IEEE Conference on Computational Complexity*, pages 194–203, 2002.

[6] M. Bellare, A. Boldyreva, and A. O'Neill. Efficiently-searchable and deterministic asymmetric encryption. Cryptology ePrint archive, June 2006. report 2006/186, http://eprint.iacr.org/2006/186.

[7] S. Bellovin and W. Cheswick. Privacy-enhanced searches using encrypted Bloom filters. Technical Report 2004/022, IACR ePrint Cryptography Archive, 2004.

[8] B. Bloom. Space/time trade-offs in hash coding with allowable errors. *Communications of the ACM*, 13(7):422–426, 1970.

[9] M. Blum, W. S. Evans, P. Gemmell, S. Kannan, and M. Naor. Checking the correctness of memories. In *IEEE Symposium on Foundations of Computer Science*, pages 90–99, 1991.

[10] D. Boneh, G. di Crescenzo, R. Ostrovsky, and G. Persiano. Public key encryption with keyword search. In *Proc. EUROCRYPT 04*, pages 506–522, 2004.

[11] D. Boneh, E. Kushilevitz, R. Ostrovsky, and W. Skeith. Public-key encryption that allows PIR queries. Unpublished Manuscript, August 2006.

[12] Y. C. Chang and M. Mitzenmacher. Privacy preserving keyword searches on remote encrypted data. In *Applied Cryptography and Network Security Conference*, 2005.

[13] R. Curtmola, J. Garay, S. Kamara, and R. Ostrovsky. Searchable symmetric encryption: Improved definitions and efficient constructions. Cryptology ePrint archive, June 2006. report 2006/210, http://eprint.iacr.org/2006/210.

[14] A. Fiat and M. Naor. Broadcast encryption. In D. R. Stinson, editor, *Proc. CRYPTO 93*, volume 773 of *Lecture Notes in Computer Science*, pages 480–491. Springer-Verlag, 1994.

[15] M. Fredman, J. Komlós, and E. Szemerédi. Storing a sparse table with 0(1) worst case access time. *J. ACM*, 31(3):538–544, 1984.

[16] E.-J. Goh. Secure indexes. Technical Report 2003/216, IACR ePrint Cryptography Archive, 2003. See http://eprint.iacr.org/2003/216.

[17] O. Goldreich. *Foundations of Cryptography*. Cambridge University Press, 2001.

[18] O. Goldreich and R. Ostrovsky. Software protection and simulation on oblivious RAMs. *Journal of the ACM*, 43(3):431–473, 1996.

[19] S. Goldwasser and S. Micali. Probabilistic encryption. *JCSS*, 28(2):270–299, Apr. 1984.

[20] P. Golle, J. Staddon, and B. Waters. Secure conjunctive keyword search over encrypted data. In M. Jakobsson, M. Yung, and J. Zhou, editors, *Applied Cryptography and Network Security Conference (ACNS)*, volume 3089 of *LNCS*, pages 31–45. Springer-Verlag, 2004.

[21] Y. Ishai, E. Kushilevitz, R. Ostrovsky, and A. Sahai. Batch codes and their applications. In *36th Annual ACM Symposium on Theory of Computing (STOC '04)*, pages 262–271. ACM, 2004.

[22] Y. Ishai, E. Kushilevitz, R. Ostrovsky, and A. Sahai. Cryptography from anonymity. In *47th Annual IEEE Symposium on Foundations of Computer Science (FOCS '06)*. IEEE, 2006.

[23] E. Kushilevitz and R. Ostrovsky. Replication is NOT needed: SINGLE database, computationally-private information retrieval. In *IEEE Symposium on Foundations of Computer Science*, pages 364–373, 1997.

[24] R. Ostrovsky. Software protection and simulations on oblivious RAMs. In *Proceedings of 22nd Annual ACM Symposium on Theory of Computing*, 1990. MIT Ph.D. Thesis, 1992.

[25] R. Ostrovsky and W. Skeith. Private searching on streaming data. In *Advances in Cryptology - CRYPTO '05*, volume 3621 of *Lecture Notes in Computer Science*, pages 223–240. Springer, 2005.

[26] D. Park, K. Kim, and P. Lee. Public key encryption with conjunctive field keyword search. In *5th International Workshop WISA 2004*, volume 3325 of *LNCS*, pages 73–86. Springer, 2004.

[27] D. Song, D. Wagner, and A. Perrig. Practical techniques for searching on encrypted data. In *IEEE Symposium on Security and Privacy*, pages 44–55, May 2000.

---

[4] Note that O should possess an additional secret that will not be shared with U and that allows him to perform authentication with the server when he wants to update $\mathcal{D}$. This guarantees that only O can perform updates to $\mathcal{D}$.

Attribute-Based Encryption for Fine-Grained Access Control of Encrypted Data

Vipul Goyal*
UCLA
vipul@cs.ucla.edu

Omkant Pandey†
UCLA
omkant@cs.ucla.edu

Amit Sahai‡
UCLA
sahai@cs.ucla.edu

Brent Waters§
SRI International
bwaters@csl.sri.com

ABSTRACT

As more sensitive data is shared and stored by third-party sites on the Internet, there will be a need to encrypt data stored at these sites. One drawback of encrypting data, is that it can be selectively shared only at a coarse-grained level (i.e., giving another party your private key). We develop a new cryptosystem for fine-grained sharing of encrypted data that we call Key-Policy Attribute-Based Encryption (KP-ABE). In our cryptosystem, ciphertexts are labeled with sets of attributes and private keys are associated with access structures that control which ciphertexts a user is able to decrypt. We demonstrate the applicability of our construction to sharing of audit-log information and broadcast encryption. Our construction supports delegation of private keys which subsumes Hierarchical Identity-Based Encryption (HIBE).

Categories and Subject Descriptors: E.3 [Data Encryption]: Public key cryptosystems.

General Terms: Security.

---

*This research was supported in part by NSF ITR/Cybertrust grants 0456717 and 0627781.
†This research was supported in part by NSF ITR/Cybertrust grants 0456717 and 0627781.
‡This research was supported in part by an Alfred P. Sloan Foundation Research Fellowship, an Intel equipment grant, and NSF ITR/Cybertrust grants 0205594, 0456717 and 0627781.
§This research was supported in part by NSF, the US Army Research Office Grant No. W911NF-06-1-0316, and the Department of Homeland Security (DHS) and the Department of Interior (DOI) under Contract No. NBCHF040146. Any opinions, finding and conclusions or recommendations expressed in this material are those of the author(s) and do not necessarily reflect the views of the funding agencies.

Permission to make digital or hard copies of all or part of this work for personal or classroom use is granted without fee provided that copies are not made or distributed for profit or commercial advantage and that copies bear this notice and the full citation on the first page. To copy otherwise, to republish, to post on servers or to redistribute to lists, requires prior specific permission and/or a fee.
CCS'06, October 30–November 3, 2006, Alexandria, Virginia, USA.
Copyright 2006 ACM 1-59593-518-5/06/0010 ...$5.00.

Keywords: Attribute-based encryption, access control, audit logs, broadcast encryption, delegation, hierarchical identity-based encryption.

1. INTRODUCTION

There is a trend for sensitive user data to be stored by third parties on the Internet. For example, personal email, data, and personal preferences are stored on web portal sites such as Google and Yahoo. The attack correlation center, dshield.org, presents aggregated views of attacks on the Internet, but stores intrusion reports individually submitted by users. Given the variety, amount, and importance of information stored at these sites, there is cause for concern that personal data will be compromised. This worry is escalated by the surge in recent attacks and legal pressure faced by such services.

One method for alleviating some of these problems is to store data in encrypted form. Thus, if the storage is compromised the amount of information loss will be limited. One disadvantage of encrypting data is that it severely limits the ability of users to selectively share their encrypted data at a fine-grained level. Suppose a particular user wants to grant decryption access to a party to all of its Internet traffic logs for all entries on a particular range of dates that had a source IP address from a particular subnet. The user either needs to act as an intermediary and decrypt all relevant entries for the party or must give the party its private decryption key, and thus let it have access to *all* entries. Neither one of these options is particularly appealing. An important setting where these issues give rise to serious problems is *audit logs* (discussed in more detail in Section 7).

Sahai and Waters [32] made some initial steps to solving this problem by introducing the concept of Attributed-Based Encryption (ABE). In an ABE system, a user's keys and ciphertexts are labeled with sets of descriptive attributes and a particular key can decrypt a particular ciphertext only if there is a match between the attributes of the ciphertext and the user's key. The cryptosystem of Sahai and Waters allowed for decryption when at least k attributes overlapped between a ciphertext and a private key. While this primitive was shown to be useful for error-tolerant encryption with biometrics, the lack of expressibility seems to limit its applicability to larger systems.

Our Contribution. We develop a much richer type of attribute-based encryption cryptosystem and demonstrate its applications. In our system each ciphertext is labeled by the encryptor with a set of descriptive attributes. Each private key is associated with an access structure that specifies which type of ciphertexts the key can decrypt. We call such a scheme a Key-Policy Attribute-Based Encryption (KP-ABE), since the access structure is specified in the private key, while the ciphertexts are simply labeled with a set of descriptive attributes. [1]

We note that this setting is reminiscent of secret sharing schemes (see, e.g., [3]). Using known techniques one can build a secret-sharing scheme that specifies that a set of parties must cooperate in order to reconstruct a secret. For example, one can specify a tree access structure where the interior nodes consist of **AND** and **OR** gates and the leaves consist of different parties. Any set of parties that satisfy the tree can reconstruct the secret.

In our construction each user's key is associated with a tree-access structure where the leaves are associated with attributes. [2] A user is able to decrypt a ciphertext if the attributes associated with a ciphertext satisfy the key's access structure. The primary difference between our setting and secret-sharing schemes is that *while secret-sharing schemes allow for cooperation between different parties, in our setting, this is expressly forbidden*. For instance, if Alice has the key associated with the access structure "X **AND** Y", and Bob has the key associated with the access structure "Y **AND** Z", we would not want them to be able to decrypt a ciphertext whose only attribute is Y by colluding. To do this, we adapt and generalize the techniques introduced by [32] to deal with more complex settings. We will show that this cryptosystem gives us a powerful tool for encryption with fine-grained access control for applications such as sharing audit log information.

In addition, we provide a delegation mechanism for our construction. Roughly, this allows any user that has a key for access structure X to derive a key for access structure Y, if and only if Y is more restrictive than X. Somewhat surprisingly, we observe that our construction with the delegation property subsumes Hierarchical Identity-Based Encryption [24, 20] and its derivatives [1].

1.1 Organization

We begin with a discussion of related work in Section 2. Next, we give necessary background information and our definitions of security in Section 3. We then present our first construction and a proof of security in Section 4. We give a construction for the large universe case in Section 5. We then show how to add the delegation property in Section 6. We follow with a discussion of how our system applies to audit logs in Section 7. We discuss the application of our construction to broadcast encryption in Section 8. Finally, we discuss some interesting extensions and open problems in Section 9.

2. RELATED WORK

Fine-grained Access Control. Fine-grained access control systems facilitate granting differential access rights to a set of users and allow flexibility in specifying the access rights of individual users. Several techniques are known for implementing fine grained access control.

Common to the existing techniques (see, e.g., [26, 19, 36, 27, 23, 28] and the references therein) is the fact that they employ a trusted server that stores the data in clear. Access control relies on software checks to ensure that a user can access a piece of data only if he is authorized to do so. This situation is not particularly appealing from a security standpoint. In the event of server compromise, for example, as a result of a software vulnerability exploit, the potential for information theft is immense. Furthermore, there is always a danger of "insider attacks" wherein a person having access to the server steals and leaks the information, for example, for economic gains. Some techniques (see, e.g., [2]) create user hierarchies and require the users to share a common secret key if they are in a common set in the hierarchy. The data is then classified according to the hierarchy and encrypted under the public key of the set it is meant for. Clearly, such methods have several limitations. If a third party must access the data for a set, a user of that set either needs to act as an intermediary and decrypt all relevant entries for the party or must give the party its private decryption key, and thus let it have access to all entries. In many cases, by using the user hierarchies it is not even possible to realize an access control equivalent to monotone access trees.

In this paper, we introduce new techniques to implement fine grained access control. In our techniques, the data is stored on the server in an encrypted form while different users are still allowed to decrypt different pieces of data per the security policy. This effectively eliminates the need to rely on the storage server for preventing unauthorized data access.

Secret-Sharing Schemes. Secret-sharing schemes (SSS) are used to divide a secret among a number of parties. The information given to a party is called the share (of the secret) for that party. Every SSS realizes some access structure that defines the sets of parties who should be able to reconstruct the secret by using their shares.

Shamir [33] and Blakley [6] were the first to propose a construction for secret-sharing schemes where the access structure is a threshold gate. That is, if any t or more parties come together, they can reconstruct the secret by using their shares; however, any lesser number of parties do not get any information about the secret. Benaloh [5] extended Shamir's idea to realize any access structure that can be represented as a tree consisting of threshold gates. Other notable secret-sharing schemes are [25, 14].

In SSS, one can specify a tree-access structure where the interior nodes consist of AND and OR gates and the leaves consist of different parties. Any set of parties that satisfy the tree can come together and reconstruct the secret. Therefore in SSS, collusion among different users (or parties) is not only allowed but required.

In our construction each user's key is associated with a tree-access structure where the leaves are associated with attributes. A user is able to decrypt a ciphertext if the attributes associated with a ciphertext satisfy the key's access

---

[1] This contrasts with what we call Ciphertext-Policy Attribute-Based Encryption (CP-ABE), where an access structure (i.e. policy) would be associated to each ciphertext, while a user's private key would be associated with a set of attributes. KP-ABE and CP-ABE systems are useful in different contexts.

[2] In fact, we can extend our scheme to work for any access structure for which a Linear Secret Sharing Scheme exists (see full version of this paper for details [21]).

structure. In our scheme, contrary to SSS, users should be *unable* to collude in any meaningful way.

Identity-Based Encryption and Extensions. The concept of Attribute-Based Encryption was introduced by Sahai and Waters [32], who also presented a particular scheme that they called Fuzzy Identity-Based Encryption (FIBE). The Fuzzy-IBE scheme builds upon several ideas from Identity-Based Encryption [9, 34, 17]. In FIBE, an identity is viewed as a set of attributes. FIBE allows for a private key for an identity, ω, to decrypt to a ciphertext encrypted with an identity, ω', if and only if the identities ω and ω' are close to each other as measured by the "set overlap" distance metric. In other words, if the message is encrypted with a set of attributes ω', a private key for a set of attributes ω enables decrypting that message, if and only if $|\omega \cap \omega'| \geq d$, where d is fixed during the setup time. Thus, FIBE achieves error tolerance making it suitable for use with biometric identities. However, it has limited applicability to access control of data, our primary motivation for this work. Since the main goal in FIBE is error tolerance, the only access structure supported is a threshold gate whose threshold is fixed at the setup time.

We develop a much richer type of attribute-based encryption. The private keys of different users might be associated with different access structures. Our constructions support a wide variety of access structures (indeed, in its most general form, every LSSS realizable access structure), including a tree of threshold gates.

Yao et. al. [18] show how an IBE system that encrypts to multiple hierarchical identities in a collusion-resistant manner implies a forward secure Hierarchical IBE scheme. They also note how their techniques for resisting collusion attacks are useful in attribute-based encryption. However, the cost of their scheme in terms of computation, private key size, and ciphertext size increases exponentially with the number of attributes. We also note that there has been other work that applied IBE techniques to access control, but did not address our central concern of resisting attacks from colluding users [35, 13].

3. BACKGROUND

We first give formal definitions for the security of Key-Policy Attribute Based Encryption (KP-ABE). Then we give background information on bilinear maps and our cryptographic assumption.

3.1 Definitions

DEFINITION 1 (ACCESS STRUCTURE [3]). *Let following be a set of parties:* $\{P_1, \ldots, P_n\}$. *A collection* $\mathbb{A} \subseteq 2^{\{P_1, \ldots, P_n\}}$ *is monotone if* $\forall B, C$: *if* $B \in \mathbb{A}$ *and* $B \subseteq C$ *then* $C \in \mathbb{A}$. *An* access structure *(resp., monotone access structure) is a collection (resp., monotone collection)* $\mathbb{A}$ *of non-empty subsets of* $\{P_1, \ldots, P_n\}$, *i.e.,* $\mathbb{A} \subseteq 2^{\{P_1, \ldots, P_n\}} \setminus \{\emptyset\}$. *The sets in* $\mathbb{A}$ *are called the* authorized sets, *and the sets not in* $\mathbb{A}$ *are called the* unauthorized sets.

In our context, the role of the parties is taken by the attributes. Thus, the access structure $\mathbb{A}$ will contain the authorized sets of attributes. We restrict our attention to monotone access structures. However, it is also possible to (inefficiently) realize general access structures using our techniques by having the not of an attribute as a separate attribute altogether. Thus, the number of attributes in the system will be doubled. From now on, unless stated otherwise, by an access structure we mean a monotone access structure.

An (Key-Policy) Attribute Based Encryption scheme consists of four algorithms.

Setup This is a randomized algorithm that takes no input other than the implicit security parameter. It outputs the public parameters PK and a master key MK.
Encryption This is a randomized algorithm that takes as input a message m, a set of attributes γ, and the public parameters PK. It outputs the ciphertext E.
Key Generation This is a randomized algorithm that takes as input – an access structure $\mathbb{A}$, the master key MK and the public parameters PK. It outputs a decryption key D.
Decryption This algorithm takes as input – the ciphertext E that was encrypted under the set γ of attributes, the decryption key D for access control structure $\mathbb{A}$ and the public parameters PK. It outputs the message M if $\gamma \in \mathbb{A}$.

We now discuss the security of an ABE scheme. We define a selective-set model for proving the security of the attribute based under chosen plaintext attack. This model can be seen as analogous to the selective-ID model [15, 16, 7] used in identity-based encryption (IBE) schemes [34, 9, 17].

Selective-Set Model for ABE

Init The adversary declares the set of attributes, γ, that he wishes to be challenged upon.
Setup The challenger runs the Setup algorithm of ABE and gives the public parameters to the adversary.
Phase 1 The adversary is allowed to issue queries for private keys for many access structures $\mathbb{A}_j$, where $\gamma \notin \mathbb{A}_j$ for all j.
Challenge The adversary submits two equal length messages M_0 and M_1. The challenger flips a random coin b, and encrypts M_b with γ. The ciphertext is passed to the adversary.
Phase 2 Phase 1 is repeated.
Guess The adversary outputs a guess b' of b.

The advantage of an adversary $\mathcal{A}$ in this game is defined as $\Pr[b' = b] - \frac{1}{2}$.

We note that the model can easily be extended to handle chosen-ciphertext attacks by allowing for decryption queries in Phase 1 and Phase 2.

DEFINITION 2. *An attribute-based encryption scheme is secure in the Selective-Set model of security if all polynomial time adversaries have at most a negligible advantage in the Selective-Set game.*

3.2 Bilinear Maps

We present a few facts related to groups with efficiently computable bilinear maps.

Let $\mathbb{G}_1$ and $\mathbb{G}_2$ be two multiplicative cyclic groups of prime order p. Let g be a generator of $\mathbb{G}_1$ and e be a bilinear map, $e : \mathbb{G}_1 \times \mathbb{G}_1 \to \mathbb{G}_2$. The bilinear map e has the following properties:

1. Bilinearity: for all $u, v \in \mathbb{G}_1$ and $a, b \in \mathbb{Z}_p$, we have $$e(u^a, v^b) = e(u, v)^{ab}$$
2. Non-degeneracy: $e(g, g) \neq 1$.

We say that $\mathbb{G}_1$ is a bilinear group if the group operation in $\mathbb{G}_1$ and the bilinear map $e : \mathbb{G}_1 \times \mathbb{G}_1 \rightarrow \mathbb{G}_2$ are both efficiently computable. Notice that the map e is symmetric since $e(g^a, g^b) = e(g,g)^{ab} = e(g^b, g^a)$.

3.3 The Decisional Bilinear Diffie-Hellman (BDH) Assumption

Let $a, b, c, z \in \mathbb{Z}_p$ be chosen at random and g be a generator of $\mathbb{G}_1$. The decisional BDH assumption [7, 32] is that no probabilistic polynomial-time algorithm $\mathcal{B}$ can distinguish the tuple $(A = g^a, B = g^b, C = g^c, e(g,g)^{abc})$ from the tuple $(A = g^a, B = g^b, C = g^c, e(g,g)^z)$ with more than a negligible advantage. The advantage of $\mathcal{B}$ is

$$\left| \Pr[\mathcal{B}(A, B, C, e(g,g)^{abc}) = 0] - \Pr[\mathcal{B}(A, B, C, e(g,g)^z) = 0] \right|$$

where the probability is taken over the random choice of the generator g, the random choice of a, b, c, z in $\mathbb{Z}_p$, and the random bits consumed by $\mathcal{B}$.

4. CONSTRUCTION FOR ACCESS TREES

In the access-tree construction, ciphertexts are labeled with a set of descriptive attributes. Private keys are identified by a tree-access structure in which each interior node of the tree is a threshold gate and the leaves are associated with attributes. (We note that this setting is very expressive. For example, we can represent a tree with "AND" and "OR" gates by using respectively 2 of 2 and 1 of 2 threshold gates.) A user will be able to decrypt a ciphertext with a given key if and only if there is an assignment of attributes from the ciphertexts to nodes of the tree such that the tree is satisfied.

4.1 Access Trees

Access tree $\mathcal{T}$. Let $\mathcal{T}$ be a tree representing an access structure. Each non-leaf node of the tree represents a threshold gate, described by its children and a threshold value. If num_x is the number of children of a node x and k_x is its threshold value, then $0 < k_x \leq num_x$. When $k_x = 1$, the threshold gate is an OR gate and when $k_x = num_x$, it is an AND gate. Each leaf node x of the tree is described by an attribute and a threshold value $k_x = 1$.

To facilitate working with the access trees, we define a few functions. We denote the parent of the node x in the tree by $\text{parent}(x)$. The function $\text{att}(x)$ is defined only if x is a leaf node and denotes the attribute associated with the leaf node x in the tree. The access tree $\mathcal{T}$ also defines an ordering between the children of every node, that is, the children of a node are numbered from 1 to num. The function $\text{index}(x)$ returns such a number associated with the node x. Where the index values are uniquely assigned to nodes in the access structure for a given key in an arbitrary manner.

Satisfying an access tree. Let $\mathcal{T}$ be an access tree with root r. Denote by $\mathcal{T}_x$ the subtree of $\mathcal{T}$ rooted at the node x. Hence $\mathcal{T}$ is the same as $\mathcal{T}_r$. If a set of attributes γ satisfies the access tree $\mathcal{T}_x$, we denote it as $\mathcal{T}_x(\gamma) = 1$. We compute $\mathcal{T}_x(\gamma)$ recursively as follows. If x is a non-leaf node, evaluate $\mathcal{T}_{x'}(\gamma)$ for all children x' of node x. $\mathcal{T}_x(\gamma)$ returns 1 if and only if at least k_x children return 1. If x is a leaf node, then $\mathcal{T}_x(\gamma)$ returns 1 if and only if $\text{att}(x) \in \gamma$.

4.2 Our Construction

Let $\mathbb{G}_1$ be a bilinear group of prime order p, and let g be a generator of $\mathbb{G}_1$. In addition, let $e : \mathbb{G}_1 \times \mathbb{G}_1 \rightarrow \mathbb{G}_2$ denote the bilinear map. A security parameter, κ, will determine the size of the groups. We also define the Lagrange coefficient $\Delta_{i,S}$ for $i \in \mathbb{Z}_p$ and a set, S, of elements in $\mathbb{Z}_p$: $\Delta_{i,S}(x) = \prod_{j \in S, j \neq i} \frac{x-j}{i-j}$. We will associate each attribute with a unique element in $\mathbb{Z}_p^*$. Our construction follows.

Setup Define the universe of attributes $\mathcal{U} = \{1, 2, \ldots, n\}$. Now, for each attribute $i \in \mathcal{U}$, choose a number t_i uniformly at random from $\mathbb{Z}_p$. Finally, choose y uniformly at random in $\mathbb{Z}_p$. The published public parameters PK are

$$T_1 = g^{t_1}, \ldots, T_{|\mathcal{U}|} = g^{t_{|\mathcal{U}|}}, Y = e(g,g)^y \; .$$

The master key MK is:

$$t_1, \ldots, t_{|\mathcal{U}|}, y \; .$$

Encryption (M, γ, PK) To encrypt a message $M \in \mathbb{G}_2$ under a set of attributes γ, choose a random value $s \in \mathbb{Z}_p$ and publish the ciphertext as:

$$E = (\gamma, E' = MY^s, \{E_i = T_i^s\}_{i \in \gamma}) \; .$$

Key Generation $(\mathcal{T}, \text{MK})$ The algorithm outputs a key that enables the user to decrypt a message encrypted under a set of attributes γ if and only if $\mathcal{T}(\gamma) = 1$. The algorithm proceeds as follows. First choose a polynomial q_x for each node x (including the leaves) in the tree $\mathcal{T}$. These polynomials are chosen in the following way in a top-down manner, starting from the root node r.

For each node x in the tree, set the degree d_x of the polynomial q_x to be one less than the threshold value k_x of that node, that is, $d_x = k_x - 1$. Now, for the root node r, set $q_r(0) = y$ and d_r other points of the polynomial q_r randomly to define it completely. For any other node x, set $q_x(0) = q_{\text{parent}(x)}(\text{index}(x))$ and choose d_x other points randomly to completely define q_x.

Once the polynomials have been decided, for each leaf node x, we give the following secret value to the user:

$$D_x = g^{\frac{q_x(0)}{t_i}} \text{ where } i = \text{att}(x) \; .$$

The set of above secret values is the decryption key D.

Decryption (E, D) We specify our decryption procedure as a recursive algorithm. For ease of exposition we present the simplest form of the decryption algorithm and discuss potential performance improvements in the next subsection.

We first define a recursive algorithm $\text{DecryptNode}(E, D, x)$ that takes as input the ciphertext $E = (\gamma, E', \{E_i\}_{i \in \gamma})$, the private key D (we assume the access tree $\mathcal{T}$ is embedded in the private key), and a node x in the tree. It outputs a group element of $\mathbb{G}_2$ or $\perp$.

Let $i = \text{att}(x)$. If the node x is a leaf node then:

$$\text{DecryptNode}(E, D, x) = \begin{cases} e(D_x, E_i) = e(g^{\frac{q_x(0)}{t_i}}, g^{s \cdot t_i}) \\ = e(g,g)^{s \cdot q_x(0)} & \text{if } i \in \gamma \\ \\ \perp & \text{otherwise} \end{cases}$$

We now consider the recursive case when x is a non-leaf node. The algorithm $\text{DecryptNode}(E, D, x)$ then proceeds as follows: For all nodes z that are children of x, it calls DecryptNode(E, D, z) and stores the output as F_z. Let S_x be an arbitrary k_x-sized set of child nodes z such that $F_z \neq \perp$. If no such set exists then the node was not satisfied and the function returns $\perp$.

Otherwise, we compute:

$$\begin{aligned}F_x &= \prod_{z \in S_x} F_z^{\Delta_{i,S'_x}(0)}, \quad \text{where } \begin{matrix}i = \text{index}(z)\\ S'_x = \{\text{index}(z) : z \in S_x\}\end{matrix}\\ &= \prod_{z \in S_x} (e(g,g)^{s \cdot q_z(0)})^{\Delta_{i,S'_x}(0)}\\ &= \prod_{z \in S_x} (e(g,g)^{s \cdot q_{\text{parent}(z)}(\text{index}(z))})^{\Delta_{i,S'_x}(0)} \quad \text{(by constr.)}\\ &= \prod_{z \in S_x} e(g,g)^{s \cdot q_x(i) \cdot \Delta_{i,S'_x}(0)}\\ &= e(g,g)^{s \cdot q_x(0)} \quad \text{(using polynomial interpolation)}\end{aligned}$$

and return the result.

Now that we have defined our function DecryptNode, the decryption algorithm simply calls the function on the root of the tree. We observe that DecryptNode(E, D, r) = $e(g,g)^{ys}$ = Y^s if and only if the ciphertext satisfies the tree. Since, $E' = MY^s$ the decryption algorithm simply divides out Y^s and recovers the message M.

We discuss how to optimize the decryption procedure in the full version of this paper [21].

4.3 Proof of Security

We prove that the security of our scheme in the attribute-based Selective-Set model reduces to the hardness of the Decisional BDH assumption.

THEOREM 1. *If an adversary can break our scheme in the Attribute-based Selective-Set model, then a simulator can be constructed to play the Decisional BDH game with a non-negligible advantage.*

PROOF: Suppose there exists a polynomial-time adversary $\mathcal{A}$, that can attack our scheme in the Selective-Set model with advantage ϵ. We build a simulator $\mathcal{B}$ that can play the Decisional BDH game with advantage $\epsilon/2$. The simulation proceeds as follows:

We first let the challenger set the groups $\mathbb{G}_1$ and $\mathbb{G}_2$ with an efficient bilinear map, e and generator g. The challenger flips a fair binary coin μ, outside of $\mathcal{B}$'s view. If $\mu = 0$, the challenger sets $(A, B, C, Z) = (g^a, g^b, g^c, e(g,g)^{abc})$; otherwise it sets $(A, B, C, Z) = (g^a, g^b, g^c, e(g,g)^z)$ for random a, b, c, z. We assume the universe, $\mathcal{U}$ is defined.

Init The simulator $\mathcal{B}$ runs $\mathcal{A}$. $\mathcal{A}$ chooses the set of attributes γ it wishes to be challenged upon.

Setup The simulator sets the parameter $Y = e(A, B) = e(g,g)^{ab}$. For all $i \in \mathcal{U}$, it sets T_i as follows: if $i \in \gamma$, it chooses a random $r_i \in \mathbb{Z}_p$ and sets $T_i = g^{r_i}$ (thus, $t_i = r_i$); otherwise it chooses a random $\beta_i \in \mathbb{Z}_p$ and sets $T_i = g^{b\beta_i} = B^{\beta_i}$ (thus, $t_i = b\beta_i$). It then gives the public parameters to $\mathcal{A}$.

Phase 1 $\mathcal{A}$ adaptively makes requests for the keys corresponding to any access structures $\mathcal{T}$ such that the challenge set γ does not satisfy $\mathcal{T}$. Suppose $\mathcal{A}$ makes a request for the secret key for an access structure $\mathcal{T}$ where $\mathcal{T}(\gamma) = 0$. To generate the secret key, $\mathcal{B}$ needs to assign a polynomial Q_x of degree d_x for every node in the access tree $\mathcal{T}$.

We first define the following two procedures: PolySat and PolyUnsat.

PolySat($\mathcal{T}_x, \gamma, \lambda_x$) This procedure sets up the polynomials for the nodes of an access sub-tree with satisfied root node, that is, $\mathcal{T}_x(\gamma) = 1$. The procedure takes an access tree $\mathcal{T}_x$ (with root node x) as input along with a set of attributes γ and an integer $\lambda_x \in \mathbb{Z}_p$.

It first sets up a polynomial q_x of degree d_x for the root node x. It sets $q_x(0) = \lambda_x$ and then sets rest of the points randomly to completely fix q_x. Now it sets polynomials for each child node x' of x by calling the procedure PolySat($\mathcal{T}_{x'}, \gamma, q_x(\text{index}(x'))$). Notice that in this way, $q_{x'}(0) = q_x(\text{index}(x'))$ for each child node x' of x.

PolyUnsat($\mathcal{T}_x, \gamma, g^{\lambda_x}$) This procedure sets up the polynomials for the nodes of an access tree with unsatisfied root node, that is, $\mathcal{T}_x(\gamma) = 0$. The procedure takes an access tree $\mathcal{T}_x$ (with root node x) as input along with a set of attributes γ and an element $g^{\lambda_x} \in \mathbb{G}_1$ (where $\lambda_x \in \mathbb{Z}_p$).

It first defines a polynomial q_x of degree d_x for the root node x such that $q_x(0) = \lambda_x$. Because $\mathcal{T}_x(\gamma) = 0$, no more than d_x children of x are satisfied. Let $h_x \leq d_x$ be the number of satisfied children of x. For each satisfied child x' of x, the procedure chooses a random point $\lambda_{x'} \in \mathbb{Z}_p$ and sets $q_x(\text{index}(x')) = \lambda_{x'}$. It then fixes the remaining $d_x - h_x$ points of q_x randomly to completely define q_x. Now the algorithm recursively defines polynomials for the rest of the nodes in the tree as follows. For each child node x' of x, the algorithm calls:

- PolySat($\mathcal{T}_{x'}, \gamma, q_x(\text{index}(x'))$), if x' is a satisfied node. Notice that $q_x(\text{index}(x'))$ is known in this case.
- PolyUnsat($\mathcal{T}_{x'}, \gamma, g^{q_x(\text{index}(x'))}$), if x' is not a satisfied node. Notice that only $g^{q_x(\text{index}(x'))}$ can be obtained by interpolation as only $g^{q_x(0)}$ is known in this case.

Notice that in this case also, $q_{x'}(0) = q_x(\text{index}(x'))$ for each child node x' of x.

To give keys for access structure $\mathcal{T}$, simulator first runs PolyUnsat($\mathcal{T}, \gamma, A$) to define a polynomial q_x for each node x of $\mathcal{T}$. Notice that for each leaf node x of $\mathcal{T}$, we know q_x completely if x is satisfied; if x is not satisfied, then at least $g^{q_x(0)}$ is known (in some cases q_x might be known completely). Furthermore, $q_r(0) = a$.

Simulator now defines the final polynomial $Q_x(\cdot) = bq_x(\cdot)$ for each node x of $\mathcal{T}$. Notice that this sets $y = Q_r(0) = ab$. The key corresponding to each leaf node is given using its polynomial as follows. Let $i = \text{att}(x)$.

$$D_x = \begin{cases} g^{\frac{Q_x(0)}{t_i}} = g^{\frac{bq_x(0)}{r_i}} = B^{\frac{q_x(0)}{r_i}} & \text{if } i \in \gamma \\ g^{\frac{Q_x(0)}{t_i}} = g^{\frac{bq_x(0)}{b\beta_i}} = g^{\frac{q_x(0)}{\beta_i}} & \text{otherwise} \end{cases}$$

Therefore, the simulator is able to construct a private key for the access structure $\mathcal{T}$. Furthermore, the distribution of the private key for $\mathcal{T}$ is identical to that in the original scheme.

Challenge The adversary $\mathcal{A}$, will submit two challenge messages m_0 and m_1 to the simulator. The simulator flips a fair binary coin ν, and returns an encryption of m_ν. The ciphertext is output as:

$$E = (\gamma, E' = m_\nu Z, \{E_i = C^{r_i}\}_{i \in \gamma})$$

If $\mu = 0$ then $Z = e(g,g)^{abc}$. If we let $s = c$, then we have $Y^s = (e(g,g)^{ab})^c = e(g,g)^{abc}$, and $E_i = (g^{r_i})^c = C^{r_i}$. Therefore, the ciphertext is a valid random encryption of message m_ν.

Otherwise, if $\mu = 1$, then $Z = e(g,g)^z$. We then have $E' = m_\nu e(g,g)^z$. Since z is random, E' will be a random element of $\mathbb{G}_2$ from the adversaries view and the message contains no information about m_ν.

Phase 2 The simulator acts exactly as it did in Phase 1.

Guess $\mathcal{A}$ will submit a guess ν' of ν. If $\nu' = \nu$ the simulator will output $\mu' = 0$ to indicate that it was given a valid BDH-tuple otherwise it will output $\mu' = 1$ to indicate it was given a random 4-tuple.

As shown in the construction the simulator's generation of public parameters and private keys is identical to that of the actual scheme.

In the case where $\mu = 1$ the adversary gains no information about ν. Therefore, we have $\Pr[\nu \neq \nu'|\mu = 1] = \frac{1}{2}$. Since the simulator guesses $\mu' = 1$ when $\nu \neq \nu'$, we have $\Pr[\mu' = \mu|\mu = 1] = \frac{1}{2}$.

If $\mu = 0$ then the adversary sees an encryption of m_ν. The adversary's advantage in this situation is ϵ by definition. Therefore, we have $\Pr[\nu = \nu'|\mu = 0] = \frac{1}{2} + \epsilon$. Since the simulator guesses $\mu' = 0$ when $\nu = \nu'$, we have $\Pr[\mu' = \mu|\mu = 0] = \frac{1}{2} + \epsilon$.

The overall advantage of the simulator in the Decisional BDH game is $\frac{1}{2}\Pr[\mu' = \mu|\mu = 0] + \frac{1}{2}\Pr[\mu' = \mu|\mu = 1] - \frac{1}{2} = \frac{1}{2}(\frac{1}{2} + \epsilon) + \frac{1}{2}\frac{1}{2} - \frac{1}{2} = \frac{1}{2}\epsilon$.

Chosen-Ciphertext Security. Our security definitions and proofs have been in the chosen-plaintext model. Similar to [32], we notice that our construction can be extended to the chosen-ciphertext model by applying the technique of using simulation-sound NIZK proofs to achieve chosen-ciphertext security [31]. However, in Section 9 we describe how our delegation mechanism can be used with the techniques of Cannetti, Halevi, and Katz [16] to achieve a much more efficient CCA-2 system.

5. LARGE UNIVERSE CONSTRUCTION

In our previous constructions, the size of public parameters grows linearly with the number of possible attributes in the universe. Combining the tricks presented in section 4 with those in the large universe construction of Sahai and Waters [32], we construct another scheme which uses all elements in $\mathbb{Z}_p^*$ as the universe. Yet the size of public parameters only grow linearly in a parameter n. The parameter n is the maximum size of the set γ we can encrypt under. [3]

---
[3]If we are willing to accept random oracles [4], it is possible to overcome the size-limitation on γ by replacing the function $T(X)$ in our construction (see Setup) by a hash function

As noted in [32], having large universe allows us to apply a collision resistant hash function $H : \{0,1\}^* \to \mathbb{Z}_p^*$ and use arbitrary strings, that were not necessarily considered during public key setup, as attributes. For example we can add any verifiable attribute, such as "Lives in Beverly Hills", to a user's private key.

5.1 Description

Let $\mathbb{G}_1$ be a bilinear group of prime order p, and let g be a generator of $\mathbb{G}_1$. Additionally, let $e : \mathbb{G}_1 \times \mathbb{G}_1 \to \mathbb{G}_2$ denote the bilinear map. A security parameter, κ, will determine the size of the groups. Also define the Lagrange coefficient $\Delta_{i,S}$ for $i \in \mathbb{Z}_p$ and a set, S, of elements in $\mathbb{Z}_p$, exactly as before. The data will be encrypted under a set γ of n elements[4] of $\mathbb{Z}_p^*$. Our construction follows.

Setup (n) Choose a random value $y \in \mathbb{Z}_p$ and let $g_1 = g^y$. Now choose a random element g_2 of $\mathbb{G}_1$.

Next, choose $t_1, \ldots, t_{n+1}$ uniformly at random from $\mathbb{G}_1$. Let N be the set $\{1, 2, \ldots, n+1\}$. Define a function T, as:

$$T(X) = g_2^{X^n} \prod_{i=1}^{n+1} t_i^{\Delta_{i,N}(X)}.$$

Function T can be viewed as the function $g_2^{X^n} g^{h(X)}$ for some n degree polynomial h. The public parameters PK are: $g_1, g_2, t_1, \ldots, t_{n+1}$ and the master key MK is: y.

Encryption (m, γ, PK) To encrypt a message $m \in \mathbb{G}_2$ under a set of attributes γ, choose a random value $s \in \mathbb{Z}_p$ and publish the ciphertext as:

$$E = (\gamma, E' = me(g_1, g_2)^s, E'' = g^s, \{E_i = T(i)^s\}_{i \in \gamma}).$$

Key Generation $(\mathcal{T}, \text{MK}, \text{PK})$ The algorithm outputs a key which enables the user to decrypt a message encrypted under a set of attributes γ, if and only if $\mathcal{T}(\gamma) = 1$. The algorithm proceeds as follows. First choose a polynomial q_x for each non-leaf node x in the tree $\mathcal{T}$. These polynomials are chosen in the following way in a top down manner, starting from the root node r.

For each node x in the tree, set the degree d_x of the polynomial q_x to be one less than the threshold value k_x of that node, that is, $d_x = k_x - 1$. Now for the root node r, set $q_r(0) = y$ and d_r other points of the polynomial q_r randomly to define it completely. For any other node x, set $q_x(0) = q_{\text{parent}(x)}(\text{index}(x))$ and choose d_x other points randomly to completely define q_x.

Once the polynomials have been decided, for each leaf node x, we give the following secret values to the user:

$$\begin{aligned} D_x &= g_2^{q_x(0)} \cdot T(i)^{r_x} \quad \text{where } i = \text{att}(x) \\ R_x &= g^{r_x} \end{aligned}$$

where r_x is chosen uniformly at random from $\mathbb{Z}_p$ for each node x. The set of above secret pairs is the decryption key D.

(see [30] for details). This also improves the efficiency of the system.
[4]With some minor modifications, which we omit for simplicity, we can encrypt to all sets of size $\leq n$.

Decryption (E, D) As for the case of small universe, we first define a recursive algorithm DecryptNode(E, D, x) that takes as input the ciphertext $E = (\gamma, E', E'', \{E_i\}_{i \in \gamma})$, the private key D (we assume the access tree $\mathcal{T}$ is embedded in the private key), and a node x in the tree. It outputs a group element of $\mathbb{G}_2$ or $\perp$ as follows.

Let $i = \text{att}(x)$. If the node x is a leaf node then:

$$\text{DecryptNode}(E, D, x) = \begin{cases} \frac{e(D_x, E'')}{e(R_x, E_i)} = \frac{e(g_2^{q_x(0)} \cdot T(i)^{r_x}, g^s)}{e(g^{r_x}, T(i)^s)} \\ = \frac{e(g_2^{q_x(0)}, g^s) \cdot e(T(i)^{r_x}, g^s)}{e(g^{r_x}, T(i)^s)} \\ = e(g, g_2)^{s \cdot q_x(0)} \quad \text{if } i \in \gamma \\ \perp \text{ otherwise} \end{cases}$$

We now consider the recursive case when x is a non-leaf node. The algorithm DecryptNode(E, D, x) then proceeds as follows: For all nodes z that are children of x, it calls DecryptNode(E, D, z) and stores the output as F_z. Let S_x be an arbitrary k_x-sized set of child nodes z such that $F_z \neq \perp$. If no such set exists then the node was not satisfied and the function returns $\perp$.

Otherwise, we compute:

$$\begin{aligned} F_x &= \prod_{z \in S_x} F_z^{\Delta_{i, S'_x}(0)}, \quad \text{where } \begin{array}{l} i = \text{index}(z) \\ S'_x = \{\text{index}(z) : z \in S_x\} \end{array} \\ &= \prod_{z \in S_x} (e(g, g_2)^{s \cdot q_z(0)})^{\Delta_{i, S'_x}(0)} \\ &= \prod_{z \in S_x} (e(g, g_2)^{s \cdot q_{\text{parent}(z)}(\text{index}(z))})^{\Delta_{i, S'_x}(0)} \quad \text{(by constr.)} \\ &= \prod_{z \in S_x} e(g, g_2)^{s \cdot q_x(0) \cdot \Delta_{i, S'_x}(0)} \\ &= e(g, g_2)^{s \cdot q_x(0)} \quad \text{(using polynomial interpolation)} \end{aligned}$$

and return the result.

Now that we have defined our function DecryptNode, the decryption algorithm simply calls the function on the root of the tree. We observe that DecryptNode$(E, D, r) = e(g, g_2)^{ys} = e(g_1, g_2)^s$ if and only if the ciphertext satisfies the tree. Since, $E' = me(g_1, g_2)^s$ the decryption algorithm simply divides out $e(g_1, g_2)^s$ and recovers the message m.

For a security proof of this construction, see [21].

6. DELEGATION OF PRIVATE KEYS

In our large universe construction, individual users can generate new private keys using their private keys, which can then be delegated to other users. A user which has a private key corresponding to an access tree $\mathcal{T}$ can compute a new private key corresponding to ANY access tree $\mathcal{T}'$ which is *more restrictive* than $\mathcal{T}$ (i.e., $\mathcal{T}' \subseteq \mathcal{T}$). Thus, the users are capable to acting as a local key authority which can generate and distribute private keys to other users.

Computation of a new private key from an existing private key is done by applying a set of basic operations on the existing key. These operations are aimed at step by step conversion of the given private key for an access tree $\mathcal{T}$ to a private key for the targeted access tree $\mathcal{T}'$ (given that $\mathcal{T}' \subseteq \mathcal{T}$). In the following, a (t, n)-gate denotes a gate with threshold t and number of children n. The operations are as follows.

1) Adding a new trivial gate to $\mathcal{T}$

This operation involves adding a new node y *above* an existing node x. The new node y represents a $(1, 1)$ threshold gate which after adding becomes the parent of x. The former parent of x (if x is not the root node), say z, becomes the parent of y.

Since the threshold of y is 1, we are required to associate a 0 degree polynomial q_y with it such that $q_x(0) = q_y(x))$ and $q_y(0) = q_z(y))$. The second condition essentially fixes q_y and the first one is automatically satisfied since z was the parent of x earlier. Hence, no changes to the private key are required for this operation.

2) Manipulating an existing (t, n)-gate in $\mathcal{T}$

This operation involves manipulating a threshold gate so as to make the access structure more restrictive. The operation could be of the following three types.

2.1) Converting a (t, n)-gate to a $(t+1, n)$-gate with $(t+1) \leq n$

Consider a node x representing a (t, n)-gate. Clearly, the polynomial q_x has the degree $(t - 1)$ which has to be increased to t. Define a new polynomial q'_x as follows.

$$q'_x(X) = (X + 1) q_x(X)$$

Now, we change the key such that q'_x becomes the new polynomial of the node x. This is done as follows. For every child y of x, compute the constant $C_x = y) + 1$. For every leaf node z in the subtree[5] $\mathcal{T}_y$, compute the new decryption key as

$$D'_z = (D_z)^{C_x}, R'_z = (R_z)^{C_x}$$

The above results in the multiplication of all the polynomials in the subtree $\mathcal{T}_y$ with the constant C_x. Hence, $q'_y(0) = (y) + 1) q_y(0)$ which is indeed a point on the new polynomial q'_x. Note that since $q'_x(0) = q_x(0)$, no changes outside the subtree $\mathcal{T}_x$ are required.

The above procedure effectively changes x from a (t, n)-gate to a $(t + 1, n)$-gate and yields the corresponding new private key.

2.2) Converting a (t, n)-gate to a $(t + 1, n + 1)$-gate

This procedure involves adding a new subtree (with root say z) as a child of a node x while increasing the degree of x by 1 at the same time. Let z be the vth child of x so that $z) = v$. We shall change the polynomial q_x to the following.

$$q'_x(X) = (aX + 1) q_x(X) \text{ where } a = \frac{-1}{v}$$

As in the previous operation, for every (existing) child y of x, the polynomials in the subtree $\mathcal{T}_y$ are multiplied with the appropriate constant $C_x = a.y) + 1$. This ensures that $q'_y(0)$ is indeed a point on q'_x. Further, set $q_z(0) = 0 \ (= q'_x(v))$. Given $q_z(0)$, keys can be created for the subtree $\mathcal{T}_z$ as in the original key generation algorithm. Hence, the keys of the subtrees of all the children (old as well as new) of the node x have been made consistent with the new polynomial q'_x, thus achieving our goal.

2.3) Converting a (t, n)-gate to a $(t, n - 1)$-gate with $t \leq (n - 1)$

This operation involves deleting a child y of a node x. This can be easily achieved just by deleting decryption keys

---
[5]Recall that $\mathcal{T}_y$ denotes the subtree defined by y as its root.

corresponding to all the leaves of $\mathcal{T}_y$ from the original decryption key.

3) Re-randomizing the obtained key

Once we obtain a key for the desired access structure (by applying a set of operations of type 1 and 2), we apply a final re-randomization step to make it independent of the original key from which it was computed. Re-randomization of a node x with a (known) constant C_x is done as follows. Choose a random polynomial p_x of degree[6] d_x such that $p_x(0) = C_x$. Define the new polynomial q'_x as $q'_x(X) = q_x(X) + p_x(X)$. We have to change the key such that q'_x becomes the new polynomial of node x. This is done by recursively re-randomizing every child y of x with the constant $C_y = p_x(y)$. If y is a leaf node, the new decryption key corresponding to y is computed as follows.

$$D'_y = D_y.g_2^{C_y}.T(i)^{r_y},\ R'_y = R_y.g^{r_y}$$

Where $i = \text{att}(y)$ and r_y is chosen randomly.

Now, re-randomization of the private key is done just by re-randomizing the root node r with the constant $C_r = 0$. The key obtained is the final key ready to be distributed to other users.

In the following theorem, we prove that the above set of operations is complete. That is, give a key for an access tree $\mathcal{T}$, this set of operations is sufficient to compute a key for ANY access tree $\mathcal{T}'$ which is more restrictive than $\mathcal{T}$.

THEOREM 2 (COMPLETENESS THEOREM). *The given set of operations is complete.*

PROOF: We can obtain a key for an access tree $\mathcal{T}'$ from a key for $\mathcal{T}$ using the following general technique. Add a new trivial gate x (using operation 1) above the root node of $\mathcal{T}$ so that the new gate becomes the root node. Now we apply operation 2.2 to x to convert it from a $(1,1)$-gate to a $(2,2)$-gate. The new subtree added as a child of x corresponds to the access tree $\mathcal{T}'$. Finally, we re-randomize the key obtained (operation 3). This gives us a key for the access structure ($\mathcal{T}$ AND $\mathcal{T}'$). However, since $\mathcal{T}'$ is more restrictive than $\mathcal{T}$, this access structure is equivalent to $\mathcal{T}'$ itself. Hence, we have obtained a key for the access structure $\mathcal{T}'$. We point out that this is a general method for obtaining a more restrictive access structure; in practice users will likely use the delegation tools described above in a more refined manner to achieve shorter private key sizes and faster decryption times.

In the above setting, we may imagine an entity having multiple private keys (procured from different entities). We note that it is possible to use these multiple keys (for different access structures) to compute a key for the targeted access structure. Given n keys for the access trees $\mathcal{T}_1, \mathcal{T}_2, \ldots, \mathcal{T}_n$, using an operation similar to operation 1, we can connect them to obtain a single tree with an OR gate being the root node and $\mathcal{T}_1, \mathcal{T}_2, \ldots, \mathcal{T}_n$ each being a child subtree of that OR gate. Thus, we obtain a single key for the access structure $\mathcal{T} = (\mathcal{T}_1 \text{ OR } \mathcal{T}_2 \text{ OR } \ldots \text{ OR } \mathcal{T}_n)$. This key can then be used to generate new private keys.

---

[6]Recall that d_x is the degree of the polynomial q_x associated with the node x

7. AUDIT LOG APPLICATION

An important application of KP-ABE deals with secure forensic analysis: One of the most important needs for electronic forensic analysis is an "audit log" containing a detailed account of all activity on the system or network to be protected. Such audit logs, however, raise significant security concerns: a comprehensive audit log would become a prized target for enemy capture. Merely encrypting the audit log is not sufficient, since then any party who needs to legitimately access the audit log contents (for instance a forensic analyst) would require the secret key – thereby giving this single analyst access to essentially all secret information on the network. Such problematic security issues arise in nearly every secure system, and particularly in large-scale networked systems such as the Global Information Grid, where diverse secret, top secret, and highly classified information will need to appear intermingled in distributed audit logs.

Our KP-ABE system provides an attractive solution to the audit log problem. Audit log entries could be annotated with attributes such as, for instance, the name of the user, the date and time of the user action, and the type of data modified or accessed by the user action. Then, a forensic analyst charged with some investigation would be issued a secret key associated with a particular "access structure" – which would correspond to the key allowing for a particular kind of encrypted search; such a key, for example, would only open audit log records whose attributes satisfied the condition that "the user name is Bob, OR (the date is between October 4, 2005 and October 7, 2005 AND the data accessed pertained to naval operations off the coast of North Korea)". Our system would provide the guarantee that even if multiple rogue analysts collude to try to extract unauthorized information from the audit log, they will fail.

A more concrete example audit-log application of our ABE system would be to the ArmyCERT program, which uses netflow logs [29]. Basically, an entry is created for every flow (e.g. TCP connection), indexed by seven attributes: source IP address, destination IP address, L3 protocol type, source port, destination port, ToS byte (DSCP), and input logical interface (ifIndex). These aspects of every flow are in the clear, and the payload can be encrypted using our ABE system with these fields as attributes.

Note that in our scheme, we would need to assume that the attributes associated with audit log entries would be available to all analysts.[7] This may present a problem in highly secret environments where even attributes themselves would need to be kept hidden from analysts. We leave the problem of constructing KP-ABE systems where attributes associated with ciphertexts remain secret as an important open problem.

8. APPLICATION TO BROADCAST ENCRYPTION: TARGETED BROADCAST

We describe a new broadcast scenario that we call *targeted broadcast*. Consider the following setting.

- A broadcaster broadcasts a sequence of different items, each one labeled with a set of attributes describing

---

[7]We observe that this does not mean that the attributes need be "public." Our KP-ABE system's ciphertexts could be re-encrypted, with a key that corresponds to the general clearance level of all analysts.

the item. For instance, a television broadcaster might broadcast an episode of the show "24", and label this item with attributes such as the name of the program ("24"), the genre ("drama"), the season, the episode number, the year, month, and date of original broadcast, the current year, month, and date, the name of the director, and the name of the producing company.

- Each user is subscribed to a different "package". The user package describes an access policy, which along with the set of attributes describing any particular item being broadcast, determine whether or not the user should be able to access the item. For example, a television user may want to subscribe to a package that allows him view episodes of "24" from either the current season or Season 3. This could be encoded as policy as ("24" **AND** ("Season:5" **OR** "Season:3")).

The essential idea of Targeted Broadcast is to enjoy the economies-of-scale offered by a broadcast channel, while still being able to deliver programming targeted at the needs or wishes of individual users. The growing acceptability of such a model can be seen by the rising popularity of DVR systems such as TiVo, which allow users to easily record only the programming they want in order to watch it later. In the case of television, taking the approach that we envision here would allow for much more flexibility than just allowing users to select what channels they like.

Our KP-ABE system naturally offers a targeted broadcast system. A new symmetric key would be chosen and used to encrypt each item being broadcast, and then the KP-ABE system would be used to encrypt the symmetric key with the attributes associated with the item being broadcast. The KP-ABE system would precisely allow the flexibility we envision in issuing private keys for the unique needs of each user.

It is worth mentioning that handling such a situation with the best known broadcast encryption schemes [10, 22] (which allow encrypting to an arbitrary subset of users) is quite inefficient in comparison. The efficiency of such systems is dependent on the size of the authorized user set or the number of users in the system [10, 22], and would also require the broadcaster to refer to its database of user authorizations each time a different item is to be encrypted for broadcast. In our scheme, the encryption of an item would depend only on the properties of that item. The broadcaster could in principle even forget about the levels of access granted to each user after preparing a private key for the user.

9. DISCUSSION AND EXTENSIONS

We discuss various extensions to our scheme and open problems.

Construction for any LSSS-realizable access structure. In our previous constructions, the access structure is a tree consisting of threshold gates. In an attempt to accommodate more general and complex access structures, we construct a scheme that can support all LSSS-realizable access structures. Since for every LSSS-realizable access structure, there exists a monotone span program that computes the corresponding boolean function and vice versa [3], our new construction supports all MSP-based access structures. Details are given in the full version of this paper [21].

Achieving CCA-Security and HIBE from Delegation. We briefly outline how we can achieve efficient CCA-2 security and realize the Hierarchical Identity-Based Encryption by applying delegation techniques to the large universe construction.

To achieve CCA-2 security an encyrptor will chooses a set γ of attributes to encrypt the message under and then generate a public/private key pair for a one time signature scheme. We let VK denote the bitstring represenation of the public key and let γ' be the set $\gamma \cup$ VK. The encryptor encrypts the ciphertext under the attributes γ' and then signs the ciphertext with the private key and attaches the signature and the public key description. Suppose a user has a key for access structure X wishes to decrypt. The user first checks that the ciphertext is signed under VK and rejects the ciphertext otherwise. Then it creates an new key for the access structure of "X AND CCA : VK". By similar arguments to those in Canetti, Halevi, and Katz [16] this gives chosen-ciphertex security. We can also use other methods [11, 12] to achieve greater efficiency.

We can realize a HIBE by simply managing the the assignment of attributes in a careful manner. For example, to encrypt to the hierarchical identity "edu:ucla" one can encrypt to the set of attributes { "1-edu", "2-ucla" }. Someone who has the top-level key for edu will have a policy that requires the attribute "1-edu" to be present. To delegate a key for "edu:ucla" it simply creates a policy for "1-edu" AND "2-ucla" using our delegation techniques. We view the fact that a primitive HIBE follows so simply from our scheme as an attestation to the power of these techniques.

Ciphertext-Policy Attribute-Based Encryption. In this work, we considered the setting where ciphertexts are associated with sets of attributes, whereas user secret keys are associated with policies. As we have discussed, this setting has a number of natural applications. Another possibility is to have the reverse situation: user keys are associated with sets of attributes, whereas ciphertexts are associated with policies. We call such systems Ciphertext-Policy Attribute-Based Encryption (CP-ABE) systems. We note that the construction of Sahai and Waters [32] was most naturally considered in this framework. CP-ABE systems that allow for complex policies (like those considered here) would have a number of applications. An important example is a kind of sophisticated Broadcast Encryption, where users are described by (and therefore associated with) various attributes. Then, one could create a ciphertext that can be opened only if the attributes of a user match a policy. For instance, in a military setting, one could broadcast a message that is meant to be read only by users who have a rank of Lieutenant or higher, and who were deployed in South Korea in the year 2005. We leave constructing such a system as an important open problem.

Searching on Encrypted Data. Our current constructions do not hide the set of attributes under which the data is encrypted. However, if it were possible to hide the attributes, then viewing attributes as keywords in such a system would lead to the first general keyword-based search on encrypted data [8]. A search query could potentially be any monotone boolean formula of any number of keywords. We leave the problem of hiding the set of attributes as open.

10. REFERENCES

[1] Michel Abdalla, Dario Catalano, Alexander W. Dent, John Malone-Lee, Gregory Neven, and Nigel P. Smart. Identity-based encryption gone wild. In Michele Bugliesi, Bart Preneel, Vladimiro Sassone, and Ingo Wegener, editors, *ICALP (2)*, volume 4052 of *Lecture Notes in Computer Science*, pages 300–311. Springer, 2006.

[2] S.G. Akl and P.D. Taylor. Cryptographic Solution to a Multi Level Security Problem. In *Advances in Cryptology – CRYPTO*, 1982.

[3] A. Beimel. *Secure Schemes for Secret Sharing and Key Distribution*. PhD thesis, Israel Institute of Technology, Technion, Haifa, Israel, 1996.

[4] M. Bellare and P. Rogaway. Random oracles are practical: A paradigm for designing efficient protocols. In *ACM conference on Computer and Communications Security (ACM CCS)*, pages 62–73, 1993.

[5] J. Benaloh and Leichter J. Generalized Secret Sharing and Monotone Functions. In *Advances in Cryptology – CRYPTO*, volume 403 of *LNCS*, pages 27–36. Springer, 1988.

[6] G. R. Blakley. Safeguarding cryptographic keys. In *National Computer Conference*, pages 313–317. American Federation of Information Processing Societies Proceedings, 1979.

[7] D. Boneh and X. Boyen. Efficient Selective-ID Secure Identity Based Encryption Without Random Oracles. In *Advances in Cryptology – Eurocrypt*, volume 3027 of *LNCS*, pages 223–238. Springer, 2004.

[8] D. Boneh, G.D. Crescenzo, R. Ostrovsky, and G. Persiano. Public-Key Encryption with Keyword Search. In *Advances in Cryptology – Eurocrypt*, volume 3027 of *LNCS*, pages 506–522. Springer, 2004.

[9] D. Boneh and M. Franklin. Identity Based Encryption from the Weil Pairing. In *Advances in Cryptology – CRYPTO*, volume 2139 of *LNCS*, pages 213–229. Springer, 2001.

[10] D. Boneh, C. Gentry, and B. Waters. Collusion Resistant Broadcast Encryption with Short Ciphertexts and Private Keys. In *Advances in Cryptology – CRYPTO*, volume 3621 of *LNCS*, pages 258–275. Springer, 2005.

[11] Dan Boneh and Jonathan Katz. Improved efficiency for cca-secure cryptosystems built using identity-based encryption. In *CT-RSA*, pages 87–103, 2005.

[12] Xavier Boyen, Qixiang Mei, and Brent Waters. Direct chosen ciphertext security from identity-based techniques. In *ACM Conference on Computer and Communications Security*, pages 320–329, 2005.

[13] Robert W. Bradshaw, Jason E. Holt, and Kent E. Seamons. Concealing complex policies with hidden credentials. In *ACM Conference on Computer and Communications Security*, pages 146–157, 2004.

[14] E. F. Brickell. Some ideal secret sharing schemes. *Journal of Combinatorial Mathematics and Combinatorial Computing*, 6:105–113, 1989.

[15] R. Canetti, S. Halevi, and J. Katz. A Forward-Secure Public-Key Encryption Scheme. In *Advances in Cryptology – Eurocrypt*, volume 2656 of *LNCS*. Springer, 2003.

[16] R. Canetti, S. Halevi, and J. Katz. Chosen Ciphertext Security from Identity Based Encryption. In *Advances in Cryptology – Eurocrypt*, volume 3027 of *LNCS*, pages 207–222. Springer, 2004.

[17] Clifford Cocks. An identity based encryption scheme based on quadratic residues. In *IMA Int. Conf.*, pages 360–363, 2001.

[18] Y. Dodis, N. Fazio, A. Lysyanskaya, and D.F. Yao. ID-Based Encryption for Complex Hierarchies with Applications to Forward Security and Broadcast Encryption. In *ACM conference on Computer and Communications Security (ACM CCS)*, pages 354–363, 2004.

[19] Rita Gavriloaie, Wolfgang Nejdl, Daniel Olmedilla, Kent E. Seamons, and Marianne Winslett. No registration needed: How to use declarative policies and negotiation to access sensitive resources on the semantic web. In *ESWS*, pages 342–356, 2004.

[20] Craig Gentry and Alice Silverberg. Hierarchical id-based cryptography. In *ASIACRYPT*, pages 548–566, 2002.

[21] V. Goyal, O. Pandey, A. Sahai, and B. Waters. Attribute Based Encryption for Fine-Grained Access Conrol of Encrypted Data. Avaialble at: http://eprint.iacr.org/2006/.

[22] D. Halevy and A. Shamir. The LSD Broadcast Encryption Scheme. In *Advances in Cryptology – CRYPTO*, volume 2442 of *LNCS*, pages 47–60. Springer, 2002.

[23] Hugh Harney, Andrea Colgrove, and Patrick Drew McDaniel. Principles of policy in secure groups. In *NDSS*, 2001.

[24] Jeremy Horwitz and Ben Lynn. Toward hierarchical identity-based encryption. In Lars R. Knudsen, editor, *EUROCRYPT*, volume 2332 of *Lecture Notes in Computer Science*, pages 466–481. Springer, 2002.

[25] M. Ito, A. Saito, and T. Nishizeki. Secret Sharing Scheme Realizing General Access Structure. In *IEEE Globecom*. IEEE, 1987.

[26] Myong H. Kang, Joon S. Park, and Judith N. Froscher. Access control mechanisms for inter-organizational workflow. In *SACMAT '01: Proceedings of the sixth ACM symposium on Access control models and technologies*, pages 66–74, New York, NY, USA, 2001. ACM Press.

[27] Jiangtao Li, Ninghui Li, and William H. Winsborough. Automated trust negotiation using cryptographic credentials. In *ACM Conference on Computer and Communications Security*, pages 46–57, 2005.

[28] Patrick Drew McDaniel and Atul Prakash. Methods and limitations of security policy reconciliation. In *IEEE Symposium on Security and Privacy*, pages 73–87, 2002.

[29] Cisco Networks. http://netflow.cesnet.cz/n_netflow.php.

[30] M. Pirretti, P. Traynor, P. McDaniel, and B. Waters. Secure Atrribute-Based Systems. In *ACM conference on Computer and Communications Security (ACM CCS)*, 2006. To appear.

[31] A. Sahai. Non-malleable non-interactive zero knowledge and adaptive chosen-ciphertext security. In *IEEE Symposium on Foundations of Computer Science*, 1999.

[32] A. Sahai and B. Waters. Fuzzy Identity Based Encryption. In *Advances in Cryptology – Eurocrypt*, volume 3494 of *LNCS*, pages 457–473. Springer, 2005.

[33] A. Shamir. How to share a secret. *Commun. ACM*, 22(11):612–613, 1979.

[34] A. Shamir. Identity Based Cryptosystems and Signature Schemes. In *Advances in Cryptology – CRYPTO*, volume 196 of *LNCS*, pages 37–53. Springer, 1984.

[35] Nigel P. Smart. Access control using pairing based cryptography. In *CT-RSA*, pages 111–121, 2003.

[36] Ting Yu and Marianne Winslett. A unified scheme for resource protection in automated trust negotiation. In *IEEE Symposium on Security and Privacy*, pages 110–122, 2003.

Secure Attribute-Based Systems

Matthew Pirretti[*], Patrick Traynor, and Patrick McDaniel[†]
SIIS Laboratory, CSE, Pennsylvania State University
University Park, PA, USA
pirretti, traynor, mcdaniel@cse.psu.edu

Brent Waters[‡]
SRI International
Menlo Park, CA, USA
bwaters@csl.sri.com

ABSTRACT

Attributes define, classify, or annotate the datum to which they are assigned. However, traditional attribute architectures and cryptosystems are ill-equipped to provide security in the face of diverse access requirements and environments. In this paper, we introduce a novel secure information management architecture based on emerging attribute-based encryption (ABE) primitives. A policy system that meets the needs of complex policies is defined and illustrated. Based on the needs of those policies, we propose cryptographic optimizations that vastly improve enforcement efficiency. We further explore the use of such policies in two example applications: a HIPAA compliant distributed file system and a social network. A performance analysis of our ABE system and example applications demonstrates the ability to reduce cryptographic costs by as much as 98% over previously proposed constructions. Through this, we demonstrate that our attribute system is an efficient solution for securely managing information in large, loosely-coupled, distributed systems.

Categories and Subject Descriptors: K.6 Management of Computing and Information Systems: Security and Protection

General Terms: Security, Performance

Keywords: Attribute-based encryption, secure systems, applied cryptography

1. INTRODUCTION

Attributes define, classify, or annotate the datum to which they are assigned. The semantics of an attribute indicate some purpose or characteristic and, when used within larger collections, enable efficient identification and classification of like objects. For example, individuals in enterprise systems are often segregated into groups of common interest or duty based on a given set of attributes [27], e.g., function, department, university. These attributes are then used to associate sets of permissions and tasks to the specified individuals. Existing systems principally rely on the assignment and subsequent enforcement of policies by trusted and often centralized servers. However, these servers are acutely ill-equipped to deal with disconnected and asynchronous clients. Reliance upon centralized servers further limits scalability and mandates a single point of trust.

Attribute-based encryption (ABE) [26], a generalization of identity-based cryptosystems, incorporates attributes as inputs to its cryptographic primitives. Objects are encrypted using a set of attributes describing the intended receiver. A principal possessing this subset as part of their pool of attributes can recover the original plaintext. More flexible requirements are achievable through the use of a thresholding primitive, for which only k-of-n attributes are necessary to perform decryption. Furthermore, decryption under both the standard and threshold approaches is collusion-resistant as multiple parties are unable to meaningfully pool attributes. Such cryptographic mechanisms allow encryption to inextricably bind expressive, enforceable access policy to objects.

Attribute-based systems have enormous potential for providing data security in distributed environments. Peer-to-peer systems are an example of one such beneficiary: individuals may publish documents that implicitly target those users who are assigned the appropriate attributes. Moreover, such publishing can be completely transparent to the peer-to-peer system. For example, a user Bob looking for employment in the field of secure systems engineering could place a copy of his résumé in publicly accessible web space encrypted with the attributes "secure systems engineering" and "human resources manager". Only potential employers satisfying these attributes would be able to decrypt this information and contact Bob.

In this paper, we develop and evaluate a secure attribute system built on attribute-based encryption (ABE). A descriptive policy system is defined that predicates access on logical expressions over attributes. We show how these policies can be realized through applications of novel ABE

---

[*]Matthew Pirretti was supported by NSF 0093085 and NSF 0202007
[†]Patrick Traynor and Patrick McDaniel were supported by NSF grant CCF-0524132, Cisco and Motorola
[‡]Brent Waters was supported by NSF

constructions. We also demonstrate their semantic depth through their use in two applications: a HIPAA compliant distributed file system and a social network.

We have developed an extensive ABE implementation tailored for the rapid creation of attribute systems- the first known implementation and characterization of such cryptographic constructions. In an effort to aid development and subsequent system use, we perform an in-depth empirical analysis of the input parameter space. Our implementation includes several novel optimizations to the original ABE cryptosystem described by Sahai and Waters [26]. The major operations of the system, including system initialization, key generation, and the encryption and decryption of objects are benchmarked. We then measure the cost of implementing complex attribute policies. Whereas past work has suggested that these constructions were too expensive for use in real systems [25], this analysis shows that such policies are not only feasible, but can also be highly efficient. For instance, we demonstrate that the cost of key generation and encryption can be reduced by more than 80% and 98%, respectively, by using constructions secure in the random oracle model.

The remainder of this paper is organized as follows: Section 2 presents an overview of the cryptographic mechanisms supporting ABE; Section 3 compares ABE systems to PKI systems; Section 4 introduces a descriptive policy system for use in ABE-based systems; Section 5 offers sample policies for two example applications; Section 6 gives the results of our performance analysis; Section 7 explores relevant related work; Section 8 offers concluding remarks.

2. ATTRIBUTE-BASED ENCRYPTION

We now give an overview of Attribute-Based Encryption (ABE) algorithms. The Sahai-Waters [26] (ABE) cryptosystem as implemented in this paper is specifically detailed. We focus our efforts on providing the description of the scheme and intuition for its construction. For the proof of security see Sahai and Waters [26].

Attribute-Based Encryption can be viewed as a generalization of Identity-Based Encryption (IBE) [5,9,30]. In IBE a user's identity is a string such as "bobsmith@yahoo.com". A party in the system can encrypt a message to this particular user with only the knowledge of the recipient's identity and the system's public parameters. In particular the encryption algorithm does not need to have access to a separate public key certificate of the recipient.

In Attribute-Based Encryption a user's identity is composed of a *set*, S, of strings which serve as descriptive attributes of the user. For example, a user's identity could consist of attributes describing their university, department, and job function. A party in the system can then specify another set of attributes S' such that a receiver can only decrypt a message if his identity S has at least k attributes in common with the set S', where k is a parameter set by the system. Like traditional Identity-Based Encryption, a party in an Attribute-Based Encryption system only needs to know the receiver's description in order to determine their public key. However, the expressiveness of an ABE system is potentially much more powerful. For example, there could be several different recipients that are able to decrypt a message encrypted for a set S'.

ABE-based systems can also leverage "threshold constructions" where a user with identity S will be able to decrypt a message if it has at least k attributes that overlap with a set S' chosen by the encryptor. Although there could in theory exist even more expressive ABE systems, the threshold constructions described and illustrated in the following sections are sufficiently semantically deep that we can define complex and precise encryption policies.

2.1 ABE Algorithms

We now informally specify a threshold Attribute-Based Encryption system as a collection of four algorithms:

- **Setup**(k): The Setup algorithm is run by an authority in order to create a new ABE system. Setup takes as input a threshold value, k and outputs a master key MK and a set of public parameters PK.

- **Key-Gen**(S, MK): The authority executes the Key-Gen algorithm for the purpose of generating a new secret key SK. The algorithm takes as input the user's identity, S, as a set of strings representing a user's attributes and the master-key MK and outputs S's secret key SK.

- **Encrypt**(M, S', PK): The Encrypt algorithm is run by a user to encrypt a message M, with a target set S', and the public parameters. It outputs a ciphertext, C.

- **Decrypt**(C, S', S, SK): The Decrypt algorithm is run by a user with identity S and secret key SK to attempt to decrypt a ciphertext C that has been encrypted with S'. If the set overlap $|S \cap S'|$ is greater than or equal to k the algorithm will output the decrypted message M.

2.2 ABE Constructions

We have investigated the use of two separate ABE constructions: the Sahai-Waters construction and a variant of the Sahai-Waters construction that we refer to as the random oracle construction. The Sahai-Waters construction is from the Sahai-Waters Large Universe system as described in Section 6 of [26]. A formal explanation of both constructions can be found in Appendix B.

Both constructions use elliptic curves to perform pairing-based cryptography. Bilinear maps (pairings) $e : \mathbb{G}_1 \times \mathbb{G}_2 \to \mathbb{G}_T$ upon elements of an elliptic curve are the basis for pairing-based cryptosystems. In the Sahai-Waters construction, decryption is possible by performing pairings between k components of a ciphertext and k private key components. We refer the reader to the IBE paper by Boneh-Franklin [5] for more details on pairing-based cryptosystems.

2.2.1 Sahai-Waters Construction

We note two observations about the Sahai-Waters construction, which have led us to design the random oracle construction. Both observations pertain to the use of following function used in both the **Key-Gen** and **Encrypt** algorithms:

$$T(i) = g^{x^i} \prod_{j=1}^{n+1} t_j^{\Delta_{j,N}(i)}, \qquad (1)$$

where N is the set $\{1, \ldots, n+1\}$.

Our first observation is that because of T, the **Setup** algorithm must take as input a ciphertext size n in addition to the threshold value k. Without the techniques proposed in

Section 4.3, this construction mandates that each ciphertext must contain exactly n attributes and that the threshold must be a fixed value k for all ciphertexts.

Our second observation is based upon our experience in building an implementation of the Sahai-Waters construction. We have found T requires a great deal of computational effort. It is easily seen that the number of exponentiations required to solve T is equal to $n + 1$.

Because of these two observations we have proposed the following modification to the Sahai-Waters construction.

2.2.2 Random Oracle Construction

We drastically reduce computational overhead in the key generation and encryption algorithms by replacing T with a hash function used as a random oracle [4]. A simple argument shows that the random oracle can be "programmed" such that the security proof of Sahai and Waters still holds. We refer the reader to the literature [4,8] for further discussion on the random oracle model.

Implementing T as a random oracle has the following characteristics. First, ciphertexts can contain a variable number of attributes, rather than be required to contain n. Second, the $n + 1$ exponentiations needed to solve T in the Sahai-Waters construction have been replaced with a single cryptographic hash.

However, using model that requires the random oracle heuristic results in a slightly weaker security model; the use of random oracles makes the security of the cryptosystem dependent upon the security of the hash function used to compute T. In Section 6, we experimentally compare implementations of the original Sahai and Waters construction with our variant.

2.2.3 Decryption Optimization

Under both constructions, the dominant operations are pairings followed by exponentiations. Decryption as described by Sahai and Waters has the following form [26]:

$$M = E' \prod_{i \in S} \left(\frac{e(d_i, E_i)}{e(D_i, E'')} \right)^{\Delta_{i,S}(0)} \quad (2)$$

where e denotes a pairing operation. In the equation above, there are $2k$ pairings and k exponentiations. Decryption can be optimized to reduce the number of bilinear map operations by bringing the Lagrange coefficients in:

$$M = \frac{E' \prod_{i \in S} e(d_i^{\Delta_{i,S(0)}}, E_i)}{e(\prod_{i \in S} D_i^{\Delta_{i,S(0)}}, E'')} \quad (3)$$

This optimization reduces the number of bilinear map operations from $2k$ to $k + 1$ at the expense of increasing the number of exponentiations from k to $2k$. Because bilinear map operations are more computationally intensive than exponentiations, this optimization increases the overall speed of decryption.

3. ATTRIBUTE KEY INFRASTRUCTURE

ABE systems do not vouch for identity in the traditional sense, as users are represented by the summation of their attributes. Accordingly, as identities are no longer necessarily unique, there is no need to validate bindings between keys and users. This alleviates many of the managerial problems found in traditional PKI systems [12] (e.g., user name collisions, per-user revocation). However, new challenges arise.

We briefly consider these issues, as well as their similarity to traditional PKI systems. Note that this work is not intended to solve the problems of PKIs, but to apply available approaches where possible and invent others where needed.

The process by which users are certified in an ABE system is analogous to certification in a PKI. Similar to a traditional PKI, a user presents the authority with a set of credentials that prove their right to fulfill an attribute. Instead of mapping a user to an identity, certification establishes that the user fulfills the semantic of the attribute. Such semantics are specific to the supported community (e.g. job function in a business system, clubs belonged to in a social network). This process is repeated for all attributes appropriate to each user. Key distribution is significantly simplified in such a system, as public keys are simply the combination of the cryptosystem's public parameters and attribute names.

The revocation process is significantly different in a ABE system as attributes, not users or keys, are revoked. In fact, there is no way to revoke a user, save revoking all of his attributes. Like traditional PKI systems, revocation can impact all users who either have or use an attribute. Unlike traditional PKI systems, however, the compromise of a particular attribute may not mandate its revocation. As Section 4 details, it is the specific application of multiple attributes that defines policy. The compromise of any single attribute may therefore be a necessary but not sufficient condition for its revocation. Consequently, it may be desirable to revoke all, a subset or none of the compromised user attributes. Explored in depth in Section 6.3, we consider both online and offline revocation approaches.

A superficial reading of the above issues may lead one to falsely conclude that ABE systems must be online. The creation of keys, certification of users, and adding attributes are largely isomorphic to certification issuance operations present in current PKI. Revocation can also be handled offline (however, online approaches such as OCSP [24] are likely to be desirable in some environments). Hence, ABE systems can operate entirely offline in largely the same that current PKI systems do.

4. ATTRIBUTE POLICY

We now informally define an expository system for describing encryption policies in attribute-based systems. An *attribute policy* (or just policy throughout) is a specification of cryptographic operations carried out on a plaintext in the attribute-based system. Hence, through encryption, a party is able to embed expressive policies into objects themselves, allowing for the decentralized enforcement of such policies. Note that the following policy description is not particular to the specific constructions of our implementation, and is appropriate for defining policy in any ABE system that supports threshold constructions.

4.1 Definition

There are two components central to the definition of policies: attributes and objects. An *attribute* consists of a uniquely identifying string, $Name$, and its hash, $H(Name)$. The semantics of the $Name$ identifier are irrelevant to the policy itself and will be driven by the application it supports (see Section 5 for examples). The hash is necessary for the ABE construction (see Appendix B), but plays no role in the formulation of policy. We broadly refer to all encrypted or recovered data as *objects*. For example, objects in a distributed file system would be the files that it stores. Conversely, objects in a social network may include a mix of personal communiques (e.g. emails, instant messages, etc.), profile information, and pictures. We refer to the universe of all attributes as the set $A = \{a_1, a_2, \ldots a_x\}$, the set of objects $O = \{o_1, o_2, \ldots, o_y\}$. Where meaning is obvious or differentiation unnecessary, subscripts may be omitted.

The attribute policy is a specification of the attributes and threshold used to encrypt an object. For example, consider a policy P that mandates encryption using a single attribute a under a threshold of 1. We denote this policy:

$$P = T_1(a) \qquad (4)$$

Of course, policies are of most interest when they are applied to objects. The application of the example attribute policy on an object o is denoted:

$$E(o, P) \quad \text{or equivalently} \quad E(o, T_1(a)) \qquad (5)$$

which states that o has been encrypted under attribute a using a 1-out-of-1 threshold encryption function. This object can therefore only be decrypted by a user possessing this attribute. Now consider the application of a similar policy P' on o that encrypts using 2-out-of-3 threshold using attributes a_1, a_2, and a_3:

$$E(o, P') \quad \text{or equivalently} \quad E(o, T_2(a_1, a_2, a_3)) \qquad (6)$$

Now consider the most general case. An arbitrary policy P'' is defined:

$$P'' = T_k(S) \quad | \quad S \subseteq A, S \neq \emptyset, 1 \leq k \leq |S| \qquad (7)$$

which states that the following must be true for any legal policy, a) the set of attributes must be a nonempty subset of A, and b) the threshold must at least 1 and no more than the total number of attributes.

Note that policies can be arbitrarily nested. Users can build complex expressions of attributes, thresholds, and logical operators. For example one may wish to combine P and P' over o above to achieve,

$$E(E(o, P'), P) \text{ or equivalently } E(E(o, T_2(a_1, a_2, a_3)), T_1(a) \qquad (8)$$

which states that one must decrypt under policy P first, then P' to recover o. Explored more fully in the following section, logical conjunction and disjunction are expressible through attribute policy. For example, if P_i and P_j are policies, then:

$$P_k = (P_i) \wedge (P_j) \qquad P_l = (P_i) \vee (P_j) \qquad (9)$$

The semantics of these policies are straightforward. A logical 'and' policy states that one must be able to decrypt both under both policies to extract the plaintext. The logical 'or' policy requires that one must be able decrypt under either or both of the policies to obtain the plaintext. Such constructions are examined in greater detail in Section 5.

The remainder of this paper explores how we build constructions meeting the semantics of these policies and how they can be applied to build novel and interesting applications.

4.2 Implementing Policy

Implementing singular attribute or threshold policies is straightforward using ABE constructions. An example of a threshold policy is depicted in Figure 1. This policy states

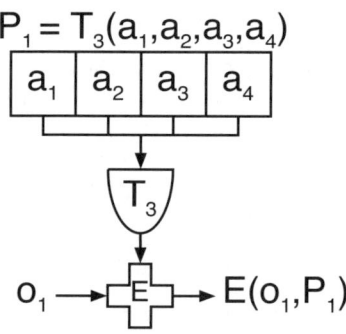

Figure 1: Encryption using threshold policy P_1. Object o_1 can only be decrypted by a principal in possession of at least three of the requisite attributes.

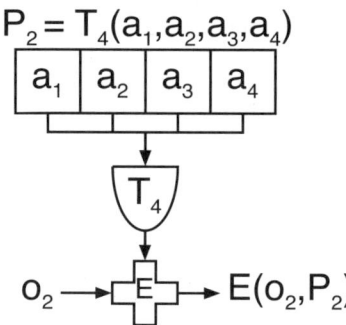

Figure 2: Encryption using and policy P_2. Object o_2 can be decrypted by principals who possess all four requisite attributes.

that decryption is possible if the party performing decryption possess at least three of the following attributes: a_1, a_2, a_3, a_4. This is illustrated by the requisite attributes being fed to the threshold operator T_3. The output of the threshold primitive is the desired policy P_1, which can then be used to define an encryption operation.

We refer to a policy where *k out of k* attributes are required to decrypt an object as an **and logic** policy, and 1 *in k* attributes as **or logic** policy. These policies can be easily implemented using the thresholding primitive, where the threshold is k in the case of **and logic** and 1 in **or logic**. Figure 2 illustrates an **and** policy. The policy P_2 requires that the party performing decryption must possess all four of the following attributes: a_1, a_2, a_3, a_4. P_2 is implemented by giving the threshold operator T_4 the four required attributes. An **or** policy is trivially similar, and is thus not illustrated.

Expressing policy becomes somewhat more complex when the input policies are not subsets of S, i.e., not expressions over atomic policies. Consider the case of an **or** policy spanning three (possibly complex) policies P_1, P_2, and P_3. In this case, one need only encrypt each of the input objects under each policy and concatenate them together; anyone able to decrypt at least one of these objects should be able to recover the underlying object. Denoting concatenation as ".", the ciphertext of an object o_i encrypted under a policy $P_1 \vee P_2 \vee P_3$ would be $E(o_i, P_1) \cdot E(o_i, P_2) \cdot E(o_i, P_3)$.

Now consider the case of an **and** policy spanning three

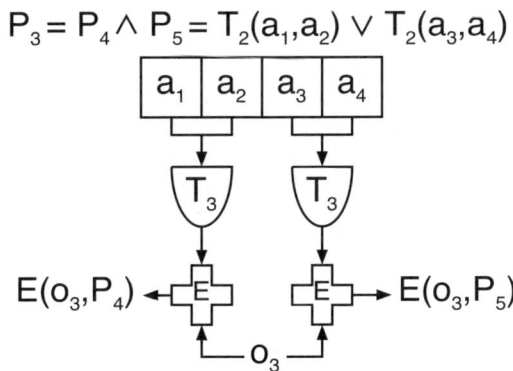

Figure 3: Encryption operation using and-or policy P_3. Principals who possess either attributes a_1 and a_2 or a_3 and a_4 are capable of decrypting object o_3.

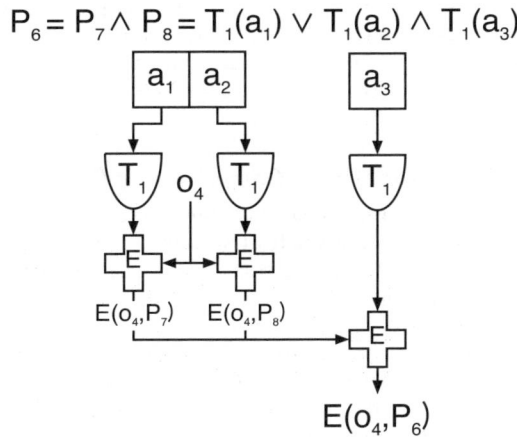

Figure 4: Encryption operation using or-and policy P_4. A principal possessing attribute a_3 and either a_1 or a_2 may decrypt object o_4.

(possibly complex) policies P_1, P_2, and P_3. One cannot simply use a threshold as above because the input policies do not reflect a threshold over atomic attributes. Hence, another construction must be used. Observe that we can achieve **and** semantics by sequentially encrypting the object with each policy. Thus the policy $P_1 \wedge P_2 \wedge P_3$ would be $E(E(E(o_i, P_1), P_2), P_3)$. This satisfies the policy semantic because only principals which possess the underlying attributes satisfying all policies can recover the plaintext.

Conjunction and disjunction constructions can be nested arbitrarily. In Figures 3 and 4 we illustrate policies that use both **and logic** and **or logic**. Specifically, in the **and-or** policy, any party performing decryption must possess either the attributes a_1 and a_2 or the attributes a_3 and a_4. The **or-and** policy requires the decrypting party possess a_1 or a_2 in addition to attribute a_3.

Observe that the conjunction constructor has a weaker security model than the original ABE constructions, where the base objects are encrypted under attributes. Whereas ABE encryption prevents any collusion attack, it is possible for adversaries to collude to recover the plaintext in this construction. To illustrate, in the **or-and** example[1] in Figure 4, two colluding parties satisfying P_7 and P_8 independently can recover the plaintext. The first adversary need decrypt the outer encryption using its a_3 assignment, then pass the inner $E(o_4, T_1(A_1))$ ciphertext to the second adversary who can then decrypt using a_1 to recover the original plaintext o_4. Work to improve this aspect of the associated cryptographic constructions is currently under way. Until then, our constructions are no weaker than standard cryptographic methods e.g. public key cryptosystems.

4.3 Extending the Flexibility of ABE

ABE natively supports a k-of-n threshold primitive. However the cryptographic constructions discussed in Appendix B mandate that k be a fixed constant across all ciphertext objects created by a given attribute system. Further, for the implementation without random oracles, the number of attributes in each ciphertext n, must also be fixed. These requirements greatly limit the liberty at which principals can draft policies; each policy must be created with a single type of threshold primitive. Thus, if $k = 4, n = 4$, then all policies would have to be written using T_4 threshold operators and each ciphertext would have to contain exactly 4 attributes. As a result the policy $P_{10} = T_2(a_1, a_2, a_3)$ could not be implemented.

Because we are interested in enabling the creation of highly expressive policies, we discuss three separate approaches which circumvent the fixed n, k requirement for constructions without random oracles[2]. The first two approaches were initially introduced by Sahai and Waters [26]. To better understand the difference between these three solutions it is helpful to introduce the following notation. Let (k_i, n_i) denote a valid pairing of k, n for a particular system.

The first solution is to provide all principals in a system with "default attributes", which act as placeholders and are devoid of semantic meaning (i.e. they are given to all users regardless of their attributes). The purpose of these attributes is to enable objects to contain any threshold operator, $T_{k'}$ such that $1 \leq k' \leq k$. To attain this end, each object must contain the maximum number of possible attributes n. The default attributes can then pad for any of the required n attributes. This scheme extends a system where only (k, n) is valid to a system where any of the following are valid: $(1, n - (k - 1)), (2, n - (k - 2)), \ldots, (k, n)$. Given $k, n = 10$, this method allows a single cryptosystem to express the ten **and** policies between (1,1) and (10,10); however, this example cryptosystem could not express any system in which k, n were not equal.

The second solution entails creating n separate cryptosystems, each with a different value of k, enabling policy to use any of the following: $(1, n), (2, n), \ldots, (n, n)$. Similar to the previous approach, this scheme extends policy expressive-

---

[1] This policy expression could be optimized to reduce the required number of encryptions. Specifically, $(a_1 \vee a_2) \wedge a_3 \equiv (a_1 \wedge a_3) \vee (a_2 \wedge a_3)$. This optimization would require two encryptions with T_2, as opposed to four encryptions with T_1. Such logic expression reductions have been thoroughly studied by other works [17] and are therefore not the focus of this work.

[2] While this discussion focuses on circumventing the fixed n, k in the constructions without random oracles, our approaches can be extended to constructions with random oracles, which is only limited by having a fixed k.

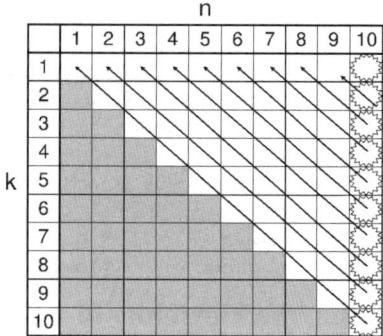

Figure 5: Example of our method for extending flexibility of ABE. ABE constructions without random oracles can only directly implement a *k out of n* policy where k and n are fixed. In this example we show how any possible pairing of $k \leq n$ and n, given that no more than 10 attributes will be present in any ciphertext. To attain this end, 10 separate cryptosystems (denoted by stars) are implemented. Arrows indicate the use of semantically void default attributes. These attributes can be used on each of the 10 cryptosystems to attain any k, n pairing.

ness, but requires a large number of systems to express a diverse set of policies.

Our approach is a hybrid of the above two techniques. Specifically, this approach enables any policy that is expressed with at most n attributes. More exactly, $\forall n_j \leq n$ and $\forall k_i \leq n_j, (k_i, n_j)$ are all valid pairings of k and n. This scheme is implemented by creating n separate cryptosystems as described in the second solution. From these cryptosystems, default attributes can be used to attain all the desired policies. This is illustrated in Figure 5, where $n = 10$. Each of these 10 cryptosystems are denoted with a "star," giving the following (k, n) pairs: $(1, 10), (2, 10), \ldots, (10, 10)$. The diagonal lines indicate the other pairings of (k, n) that are obtainable by using default attributes from within the 10 cryptosystems. Consider the $(9, 10)$ cryptosystem, in which default attributes allow for the expression of the policies $(1, 2), (2, 3), \ldots, (9, 10)$. Similar expressiveness is possible for the remaining cryptosystems.

This scheme can easily be extended to meet the needs of the target application. For instance, a system may opt to only create a subset of the possible cryptosystems (e.g. the values of k for powers of 2 less than or equal to n). Section 6 explores the performance trade-offs associated with using such a sampling. Trade-offs between expressibility, performance and overhead must be carefully considered.

5. APPLICATION OF POLICY

The threshold, conjunction and disjunction constructions discussed in Section 4 result in an expressive policy system. In this section we illustrate the use of policy in two separate applications: HIPAA compliant distributed storage systems and social networks.

5.1 Distributed File Systems

A *content-addressable* file system enables users to locate files based on attributes or keywords describing their contents. Accordingly, data becomes searchable in a more meaningful fashion than the traditional approach of specifying file paths. To date, most work on content-addressable file systems has focused on automatically generating descriptions of a file's contents [6, 14, 16]. The use of ABE strengthens the security properties of such systems. Because the access control policy of every object is embedded within it, *the enforcement of policy becomes an inseparable characteristic of the data itself.* This is in direct contrast to most currently available systems, which rely directly upon a trusted host to mediate access and administer policy. As file systems become more distributed in nature and rely upon domains of varying trust to control access, traditional approaches no longer provide adequate guarantees.

Example systems include large multisite research efforts such as the Human Genome Project, which was formed in order to map the sequence of chemical building blocks composing the human genome. While this multinational research effort requires a total of only 3 gigabytes of space to store the genome itself, the space estimated for additional annotations will sufficiently dwarf the initial sequence data in the system [2]. As this and other research projects begin to require petabytes of storage, the ability to securely store such information across multiple sites becomes increasingly critical.

5.1.1 Policy for HIPAA Compliant Medical Systems

We use a Health Insurance Portability and Accountability Act (HIPAA) compliant medical system as an example of a loosely-coupled, content-addressable file system with strict security requirements. HIPAA was designed to clearly enumerate the security requirements and provide information flow control for electronically stored medical information such that patient privacy is maintained [32]. While the system below is by far not a comprehensive example of HIPAA requirements, it demonstrates the ease with which a fully compliant system could be constructed using ABE.

In this example, a patient i's medical information is composed of several fields. Patients define the following privacy policies to best protect each component of their medical information: currently used medications ($P_{i,Med}$), medical history ($P_{i,Hist}$), contact information ($P_{i,CI}$), and insurance information ($P_{i,Ins}$). A patient's policy describes the attributes that must be possessed by medical personnel in order to access their medical information. As illustrated in Figure 6, these attributes describe various job functions of medical personnel as well as the health insurance plans they accept.

A patient, Robert Oppenheimer, supplements his limited insurance coverage through the ACME Corporation with a "Medicare D" prescription plan. His policy therefore stipulates that only doctors (Dr) and nurses (Rn) supporting his insurance plan can view his full medical history, contact information, and a listing of his current medications. Oppenheimer's policy also allows a pharmacist (Rx) in his plan to view the medications he is currently taking, so that he/she can ensure no conflicts between prescriptions exist. Further, a pharmacist is allowed access to contact information to notify him when prescriptions have been filled. Oppenheimer's policy also restricts the access of billing personnel ($Bill$) to his insurance and contact information such that charges can be filed with his insurance providers without danger of exposing private information. Lastly, without revealing his contact information, Oppenheimer allows the list

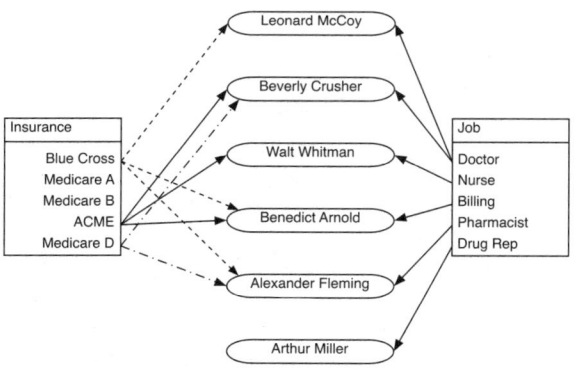

Figure 6: Mapping of attributes to principals in HIPAA compliant medical system.

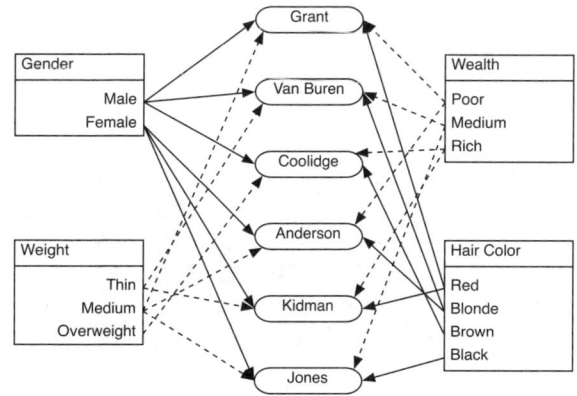

Figure 7: Mapping of attributes to principals in social network application.

of medications he is currently using to be made available to pharmaceutical representatives (*Rep*) analyzing the combination of drugs with which their products are prescribed in concert. As described above, Oppenheimer's policies are represented as follows:

$$\begin{aligned}
P_{O,Hist} &= T_1(Dr, Rn) \wedge T_1(ACME, MedicareD) \\
P_{O,CI} &= T_1(Dr, Rn, Bill, Rx) \\
&\quad \wedge T_1(ACME, MedicareD) \\
P_{O,Ins} &= T_1(Bill) \wedge T_1(ACME, MedicareD) \\
P_{O,Med} &= (T_1(Dr, Rn, Rx) \wedge T_1(ACME, MedicareD)) \\
&\quad \vee T1(Rep)
\end{aligned}$$

From the above policies, only Dr. Crusher and Nurse Whitman can access his medical history. Dr. Crusher, Nurse Whitman, Billing Secretary Arnold, and Pharmacist Fleming can access his contact information. Billing Specialist Arnold can also access Oppenheimer's insurance information. Dr. Crusher, Nurse Whitman, Pharmacist Fleming and Pharmaceutical Representative Miller are able to determine Oppenheimer's current regime of medication.

5.2 Social Networks and Online Communities

Social networks, such as orkut, Facebook and Friendster [1], are an online application which enable users to find other users with similar interests. To use these applications, users must reveal large quantities of personal information (e.g. name, age, address, personal interests, sexuality, etc.) into the public domain. Groups of people sharing similar attributes and friends are then automatically linked to each other. Currently, such systems provide only weak privacy guarantees; network membership allows access to the wealth of user information. Accordingly, user data can readily be mined and abused by undesirable parties.

ABE-based systems are well suited to provide user controlled-privacy, as users in these communities are already characterized by their attributes. In Friendster, for example, a user with the attribute "Anon U. Alumnus" is automatically enrolled in a group of the same name. Accordingly, the creation of "white-lists" for communication immediately becomes possible without requiring enumeration of all user identities. Constructing a social network using ABE also provides scalability. Current social networks require a trusted central server to store all profile information and enforce policy. Because ABE-based systems do not require a trusted storage system, profile information could be stored on untrusted servers, significantly decreasing the traffic and storage requirements incurred by a system. Further, in an ABE-based system, objects are embedded with policy, enabling distributed enforcement.

5.2.1 Policy in a Social Network

We now demonstrate policy in a social network through an application where the principals are users of an online dating service. Each user dictates their own policy in order to restrict access to their personal information.

Figure 7 illustrates a sample network. A principal's policy can be viewed as a description of attributes they find desirable in other principals. Possession of the attributes described in the policy is therefore a prerequisite to being able to access another principal's personal information. We begin with a relatively simple policy. Van Buren is only interested in meeting women with black hair, medium wealth, and medium weight. His policy is represented as:

$$\begin{aligned}
P_V &= T_4(Female \wedge BlackHair \wedge MedWealth \\
&\quad \wedge MedWeight).
\end{aligned}$$

Of the above principals, only Jones can access Van Buren's profile. This policy, which is equivalent to policy P_2 as depicted in Figure 2, can be expressed using a single threshold primitive T_4. It is therefore possible to directly implement P_V with a single ABE encryption. Accordingly, data encrypted under this policy will be resistant to collusion.

Grant's policy, whereby only blonde or red haired women can access his profile information, is represented as:

$$P_G = T_1(Female) \wedge (T_1(Blonde) \vee T_1(Red)).$$

As such, only Anderson or Kidman can access his information. Notice that Grant's policy is equivalent to policy P_6 in Figure 4 and therefore cannot be implemented using a single threshold operator. Accordingly, P_G is less resistant to collusion than P_V.

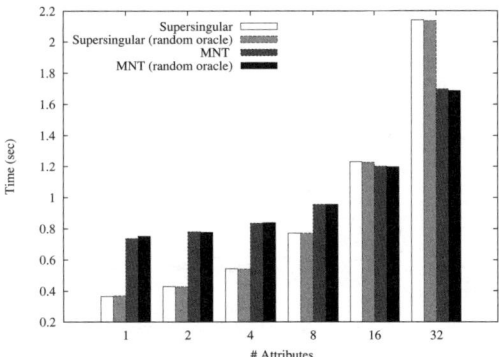

Figure 8: System_Setup: As the number of attributes grows, MNT curves become more efficient.

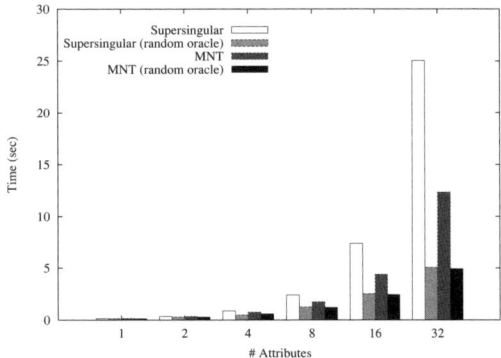

Figure 9: Key_Generation: Performance becomes nearly identical for systems using either type of curve with random oracles.

Lastly, Anderson is interested in hearing from men who possess at least two of the following attributes: red hair, medium weight, overweight, or medium wealth. Her policy can be represented as:

$$P_A = T_1(Male) \wedge T_2(Red, MedWeight, Overweight, MedWealth).$$

Given Anderson's policy Coolidge and Grant can access her information. Notice, however, that a principal's policy is not necessarily symmetric. For instance, of these two, only Grant has a policy that would allow Anderson to contact him.

6. SYSTEM EVALUATION

The policies discussed in the previous section illustrate the potential expressibility of ABE-based systems. In this section, we characterize the performance of systems providing such functionality. We begin by exploring the cost of the base cryptographic constructions. We then determine the cost of implementing a selection of the previously defined policies. We finish by comparing the performance of an ABE-based system to a comparable system implemented with RSA cryptographic primitives.

As demonstrated by numerous others (e.g. [10]), the selection of cryptographic parameters can have a drastic impact on system performance. In this section, we characterize the

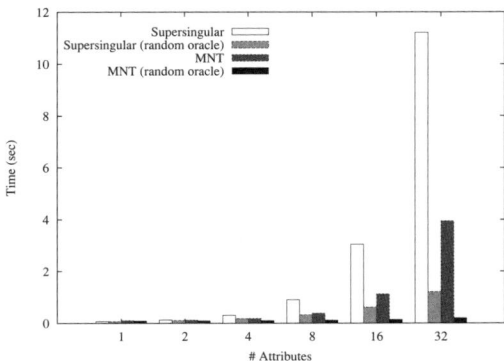

Figure 10: Encryption: SS constructions are significantly slower than MNT constructions.

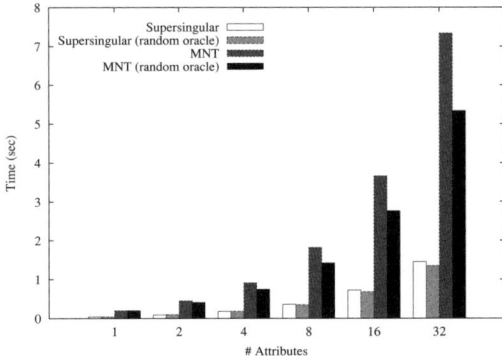

Figure 11: Decryption: SS constructions are significantly faster than MNT constructions.

parameter space by profiling the performance of attribute systems under different input parameters. Such analysis is necessary to optimize the system for a particular application or environment. All experiments were carried out on a 2.0 GHz Apple Xserve G5 with 4GB memory running Mac OS X Server 10.3.9. All disk operations were performed on a 1.82TB RAID 5 disk array. All results are calculated from an average of 500 iterations of the measured operation.

We have implemented an ABE library upon which secure attribute systems can be constructed. This C library contains approximately 5,200 lines of code and has been tested on Solaris, OS X and Linux platforms. To our knowledge, we are the first to implement, measure, and characterize the theoretical mechanisms of attribute-based encryption. Accordingly, we explore a wide range of potential inputs and settings system architects should consider when designing new secure attribute-based environments. For instance, systems using our API can choose between the two most studied elliptic curve groups providing bilinear maps: *supersingular elliptic curves* (SS), which enable fast cryptographic pairing operations [22], and *MNT elliptic curves*, which are used to obtain small ciphertext sizes [23]. We use the Pairing-Based Crypto library [19] for the underlying implementation of these groups and OpenSSL [3] for providing a supporting Key Encapsulation Mechanism (KEM) [31].

The following analysis measures the four central functions of the attribute system as defined in Section 2: Setup_System, Key_Generation, Encryption, and Decrypt-

Table 1: Base cryptographic operations for the major attribute functions.

operation	Random Oracle			No Random Oracle		
	hashes	expon.	pairings	hashes	expon.	pairings
System_Setup		1	1		1	1
Key_Generation (x attributes)	x	$3x$			$3x + (n*x)$	
Encryption (y attributes)	y	$2 + y$			$2 + y + (n*y)$	
Decryption (threshold k)		$2k$	$k + 1$		$2k$	$k + 1$

Table 2: Table of R-Square values (No Random Oracles)

	# Attributes	Data Length	Curve Type	Rand Init
Initialize_Randomness	$2.083E{-}5$	$1.043E{-}5$	$1.440E{-}6$	0.9721
System_Setup	0.8052	$1.321E{-}4$	0.0138	$2.355E{-}6$
New_Attribute	0.0442	$1.499E{-}4$	$6.959E{-}4$	$2.282E{-}5$
Key_Generation	0.8297	$2.480E{-}4$	0.0369	$1.158E{-}9$
Encryption	0.7134	$2.120E{-}4$	0.0692	$6.470E{-}9$
Decryption	0.5355	$1.733E{-}4$	0.2222	$5.466E{-}9$

Table 3: Table of R-Square values (Random Oracles)

	# Attributes	Data Length	Curve Type	Rand Init
Initialize_Randomness	$1.167E{-}5$	$1.254E{-}5$	$3.363E{-}7$	0.9721
System_Setup	0.7908	$8.915E{-}7$	0.0176	$3.827E{-}7$
New_Attribute	0.0394	$1.014E{-}4$	$8.343E{-}5$	$3.719E{-}6$
Key_Generation	0.9997	$9.792E{-}8$	$1.551E{-}4$	$3.916E{-}8$
Encryption	0.4781	$3.167E{-}7$	0.1993	$5.456E{-}8$
Decryption	0.5608	$4.543E{-}9$	0.2041	$4.078E{-}9$

ion. For reference, Table 1 provides an outline of the base cryptographic operations for each of the base operations. See Appendix A for greater detail on the use of these functions and the design of our attribute system API. All source code and documentation are available at:

http://siis.cse.psu.edu/attribute.html

These experiments indicate several important properties of the parameter space. Firstly, MNT is faster than SS for encryption whereas the opposite is true for decryption. Secondly, encryption costs are significantly improved by the use of random oracles. Hence, the curve selected should be a reflection of the relative number of encryptions and decryptions performed in the system, as well as the capabilities of the encryptor and the intended recipients. Lastly, the ability to express complex policies with ABE allows for practical use of attribute-based systems.

6.1 Experimental Results

The first set of experiments measure the degree to which different system parameters affect performance: we vary the number of attributes, length of data, elliptic curve and initialization of randomness parameters. We then perform an R-squared or coefficient of determination analysis over the measured results. This technique identifies the portion of observed variance in one variable that is directly attributable to a second. On a scale from zero to one, numbers closer to one represent a significant correlation between variables. For precision, we also include measurements for two additional subfunctions: Initialize_Randomness preloads random bytes from the local entropy pool and New_Attribute allocates a new attribute to a principal.[3] The results of these tests reveal that the number of attributes followed by the elliptic curve used are the dominant factors, as shown in Tables 2 and 3. Lastly, to characterize the growth of execution time against the number of attributes, we run a regression analysis for the worst case and present our findings in the standard linear form, i.e., $y = mx + b$.

Figure 8 shows the cost of System_Setup as a function of the number of attributes. Systems using a SS curve without random oracles average 0.366 seconds ($\sigma = 0.049$) and 2.141 seconds ($\sigma = 0.133$) for 1 and 32 attributes, respectively. A system using the MNT elliptic curve without random oracles averages between 0.737 seconds ($\sigma = 0.202$) and 1.699 seconds ($\sigma = 0.284$) for the same range. Execution time for both curves scales linearly in the number of attributes (SS w/o random oracles: $y = 0.572x + 0.3126; r^2 = 0.9999$). Random oracles have no role in system setup, and hence have no bearing on performance. System setup therefore poses no significant computational burden in real systems.

Figure 9 illustrates the cost of key generation, which is consistently cheaper for MNT curves. For a system using 32 attributes without random oracles, MNT curves require an average of 12.355 seconds ($\sigma = 0.035$) to generate a user key, compared to 25.05 seconds ($\sigma = 0.052$) for SS curves. Random oracle constructions are significantly faster - systems built on SS and MNT curves perform similarly at all numbers of attributes, e.g., at 5.051 ($\sigma = 0.017$) and 4.927 ($\sigma = 0.017$) seconds, respectively for 32 attributes. Execution time scales linearly for both curves with and without the use of random oracles (SS w/o random oracles: $y = 0.8003x - 2.37; r^2 = 0.9584$). Note that key generation for each user is performed infrequently (likely once). If the user community is fairly static, such costs will be amortized by operations on data. Conversely, in environments where users may join frequently, it behooves the administrator to select parameter choices that minimize these costs, e.g., MNT elliptic curves using random oracles.

As shown in Figure 10 for both SS and MNT elliptic curves, the construction without random oracles requires an average of 11.213 ($\sigma = 0.031$) and 3.946 ($\sigma = 0.017$)

---
[3] In all tests in this section, Initialize_Randomness is included in measurements of System_Setup, and New_Attribute is included by Key_Generation.

seconds to encrypt data using 32 attributes. Systems implementing the construction with random oracles experience dramatically improved encryption performance, i.e., 1.207 ($\sigma = 0.009$) and 0.204 ($\sigma = 0.006$) seconds, respectively. Here, MNT elliptic curves are approximately 65% to 85% faster than their SS counter-parts (in constructions with and without random oracles, respectively). Systems using MNT curves with random oracles are in fact 98% faster those using SS curves without random oracles. Both systems scale linearly in the number of attributes with and without random oracles (SS w/o random oracles: $y = 0.3590x - 1.148; r^2 = 0.9487$). Conversely, as illustrated in Figure 11, a system of 32 attributes with and without random oracles exhibits a decryption time of 1.452 ($\sigma = 0.009$) and 1.348 ($\sigma = 0.044$) for an SS construction and 7.341 ($\sigma = 0.029$) and 5.342 ($\sigma = 0.841$) seconds in MNT, respectively. Execution time for both systems scales linearly in the number of attributes with and without random oracles (MNT w/o random oracles: $y = 0.2298x - 0.103; r^2 = 0.9999$). Note the systems using SS curves experience approximately 80% faster performance than their MNT counterparts.

Lastly, we compare the performance of ABE against traditional cryptographic techniques. From OpenSSL's benchmarking tool [3], the platform used for ABE benchmarking is capable of performing RSA public key encryption in 0.0003 and 0.00097 seconds for 1024 and 2048-bit keys, respectively. To offer similar semantic expressiveness and prevent the need for 2^N-1 keys (there are 2^N-1 nonempty subsets in a set of size N), we assume that each attribute in an ABE system has a corresponding RSA key pair. For simple policies, encryption under a single attribute/key is 300 and 93 times faster under RSA (0.0003 and 0.00097 vs 0.09 seconds). ABE's thresholding primitive, however, allows much more efficient execution. For example, a policy requiring a threshold of 2 of 32 attributes has nearly identical execution times (0.1488 vs 0.2043 seconds) for both RSA-1024 and ABE with MNT curves and random oracles. RSA-2048 requires approximately 0.5 seconds to achieve the same ends. A system requiring 16 of 32 attributes would also require 0.2043 seconds for an ABE system; however, the equivalent RSA systems would require approximately 33.4 and 107.97 *days*, over 46.6 million times slower, to achieve the same. ABE's inherent expressibility makes it a practical means of constructing real attribute systems.

6.1.1 HIPAA System Policy Analysis

We now determine the cost incurred for implementing expressive policies. Encrypting with the policy $P_{O,CI}$ from Section 5.1.1 requires an initial encryption of the principal Oppenheimer's contact information using T_1(ACME, Medicare D), a process requiring E_2 time to complete. This ciphertext object is then independently re-encrypted with each of the following attributes: Dr, Rn, Billing, Rx. Each of these encryptions requires 1 attribute, and thus takes E_1 time to complete. This policy could alternatively be implemented using two encryptions if the **or** construction is replaced with T_1(Dr, Rn,Bill, Rx). Table 4 shows the timing values for this optimized policy, noted as $P_{O,CI} = E(E(CI, T_1(\text{ACME,Medicare D})), T_1(\text{Dr,Rn,Bill,Rx}))$.

Decrypting data encrypted under $P_{O,CI}$ requires two operations. The first decryption occurs with any of the following attributes: Dr, Rn, Billing, Rx. The second decryption, which enables recovery of the plaintext, requires decryption

Table 4: Average performance (sec) for $P_{O,CI}$

	No Rand Oracles		Rand Oracles	
	SS	MNT	SS	MNT
E_2	0.13	0.12	0.10	0.10
E_4	0.31	0.18	0.18	0.10
D_2	0.09	0.46	0.10	0.42
D_4	0.18	0.91	0.18	0.75
$E(P_{O,CI})$	0.44	0.30	0.28	0.20
$D(P_{O,CI})$	0.27	1.37	0.28	1.17

Table 5: Average performance (sec) for P_G.

	No Rand Oracles		Rand Oracles	
	SS	MNT	SS	MNT
E_1	0.07	0.10	0.07	0.09
D_1	0.04	0.20	0.04	0.20
$E(P_G)$	0.28	0.40	0.28	0.36
$D(P_G)$	0.08	0.40	0.08	0.40

of T_1(ACME, Medicare D). Table 4 shows average execution time.

6.1.2 Social Network Analysis

We now examine the cost of expressing policy as described in Section 5.2.1 for a social network application. Consider the time required to encrypt a message under Grant's policy, P_G. Grant's information I_G is first independently encrypted under T_1(Red Hair) and T_1(Blonde). Both values, noted as $I'_G = E(I_G, T_1(\text{Red Hair}))$ and $I''_G = E(I_G, T_1(\text{Blonde}))$ respectively, are then encrypted under T_1(Female), yielding $E(I'_G, T_1(\text{Female}))$ and $E(I''_G, T_1(\text{Female}))$. Note that the total number of encryptions can be halved if the **or** semantic is equivalently implemented as T_1(Red Hair, Blonde). The total time to encrypt P_G is given by $E(P_G)$.

Table 5 details the time required to perform the unoptimized operations required to formulate P_G. These values represent the encryption and decryption operations for SS and MNT elliptic curves with and without random oracles.

In the case of decryption for P_G, two decryptions are required. The decrypting party initially performs two decryptions with a_3. From this, only a_1 or a_2 must be decrypted in order to recover the original plaintext. The total time to decrypt P_G is given by $D(P_G)$.

6.2 Ciphertext Size and User Key Length

Ciphertext size and key length are important to some classes of applications, e.g., in high traffic volume or low bandwidth networks or on resource poor devices. Here, we briefly detail the size of ciphertexts and the size of a user's private key. Specifically, we quantify ciphertext length and user key length as described in Appendix B. Because the focus of this paper is on attribute systems and ABE, we do not include structured data framing or data encrypted with symmetric cryptography in our treatment of ciphertext size.

We shall first discuss a discrepancy between MNT curves and SS curves that is necessary to understand our analysis. Recall that the Sahai-Waters construction makes use of a bilinear group $\mathbb{G}$ to perform bilinear map operations: $e : \mathbb{G} \times \mathbb{G} \to \mathbb{G}_T$. This type of bilinear map is said have symmetric groups. A bilinear map that is asymmetric has the following form: $e : \mathbb{G}_1 \times \mathbb{G}_2 \to \mathbb{G}_T, \mathbb{G}_1 \neq \mathbb{G}_2$. SS curves are characterized by having symmetric bilinear groups. Both $\mathbb{G}$ and $\mathbb{G}_T$ require 512 bits to be represented. MNT curves are characterized by having asymmetric bilinear groups. $\mathbb{G}_1$

is represented with 170 bits while $\mathbb{G}_2, \mathbb{G}_T$ are represented with 510 bits.

Each attribute i possessed by a principal corresponds to two private key components d_i and D_i. For SS curves both of these components are members of $\mathbb{G}$. For MNT curves $D_i \in \mathbb{G}_1, d_i \in \mathbb{G}_2$. This yields (for a private key with n attributes):

$$Supersingular\ KeySize(n) = 2 \cdot n \cdot 512 \text{bits} \quad (10)$$
$$MNT\ KeySize(n) = (170 + 510) \cdot n \text{bits} \quad (11)$$

A ciphertext C scales with the number of attributes it contains as follows. A ciphertext with n attributes is composed of C', C'', and n elements C_i. For SS curves $C'', C_i \in \mathbb{G}$ and $C' \in \mathbb{G}_T$. For MNT curves $C'' \in \mathbb{G}_2$, $C' \in \mathbb{G}_T$, and $C_i \in \mathbb{G}_1$. This yields (for a n attribute ciphertext):

$$Supersingular\ CTLen(n) = (n+2) \cdot 512 \text{ bits} \quad (12)$$
$$MNT\ CTLen(n) = 2 \cdot 510 + 170n \text{ bits} \quad (13)$$

6.3 Attribute Revocation Issues

We now address some of the practical issues relevant to constructing ABE-based systems. An in-depth discussion of the implementation and the associated parameters is provided in Appendix A.

Revocation of users and keys in systems is a well studied but nontrivial problem [21]. Revocation is even more difficult in attribute systems, given that each attribute is conceivably possessed by multiple different users, whereas public/private key pairs are uniquely associated with a single principal. While an in-depth discussion of revocation is out of the scope of this paper, we give a brief overview of one method by which revocation could be implemented.

One revocation technique would require each attribute to contain a time frame within which it is valid. For instance, the attribute "*Staff Member-December 31st 2006*" denotes that the usefulness of the current attribute expires at the end of 2006. Affixing temporal information to each attribute necessitates the system administrator periodically releasing the latest version of attributes and periodically reissue user keying information. Removal of an attribute from this system would be accomplished by the administrator not releasing the latest version of the attribute. Similarly, revoking an attribute from an individual requires the administrator to withdraw the updated attribute in the user's private key. There are significant trade-offs between the load placed upon the administrator and the amount of time that can elapse before an attribute/user can be purged. We therefore leave more efficient solutions to future work.

7. RELATED WORK

Securing the sharing of information between groups is a fundamental problem that arises in numerous applications. Such applications include multilevel security, secure multicast, collaborative online communities, and distributed file systems. The fundamental importance of the secure exchange of information has resulted in a wide range of solutions.

Traditional access control mechanisms can be categorized into three groups: mandatory access control (MAC) [11], discretionary access control (DAC) [18, 28], and role-based access control (RBAC) [13, 27]. In MAC, an administrative mechanism enforces centralized access control on every object. Systems implementing DAC require the owner of an object to dictate policy. Under RBAC, a user's role in an organization inherently dictates their ability to access and manipulate data. Each role in an RBAC system is associated with a set of permissions required to carry out that role. While these mechanisms are highly effective at controlling access for systems under a single administrative authority, they have been largely unsuccessful at providing the same for unconnected and distributed environments.

ABE can enforce access control policy in such environments because it cryptographically binds objects to their policies. Only users possessing the requisite set of attributes are able to view and/or manipulate data. The ability to make policy portable through cryptography is not new. Several works have attempted to use a public key infrastructure (PKI) [15] or secure group communications mechanisms [7, 20] to provide similar access control mechanisms. The difficulty with applying standard cryptographic techniques is they are designed to control access to **single** groups. In real systems, however, users are often members of **multiple** groups. Unique keys must therefore be assigned or negotiated for each of the subgroups for which a user is a member. Such solutions do not scale for complex organizations with significant communication across groups. In contrast, users in ABE-based systems automatically belong to every possible attribute subset group without the need for additional keying.

By using cryptographic mechanisms that are in and of themselves able to express complex policies, ABE-based systems become a highly practical means of ensuring the efficient and secure exchange of information between groups.

8. CONCLUSION

This paper has presented a novel secure information management architecture and implementation. We extended existing constructions for attribute-based encryption (ABE) and promoted them as a practical systems building block. The needs of complex attribute applications were met via the introduction of a policy system and an associated implementation for its enforcement. We illustrated the infrastructure through the creation and performance evaluation of two applications: a HIPAA compliant distributed file system and a social network. A further empirical study shows that a careful selection of parameters and use of construction optimizations can lead to significant cost savings. These analyses demonstrate that our attribute approach is an attractive solution for securely managing information in large, loosely-coupled, distributed systems.

9. REFERENCES

[1] Friendster. http://www.friendster.com, 2006.
[2] The human genome project. http://www.ornl.gov/sci/techresources/Human_Genome/home.shtml, 2006.
[3] The OpenSSL project. http://www.openssl.org, 2006.
[4] M. Bellare and P. Rogaway. Random oracles are practical: A paradigm for designing efficient protocols. In *ACM Conference on Computer and Communications Security*, pages 62–73, 1993.
[5] D. Boneh and M. K. Franklin. Identity-based encryption from the weil pairing. In *Proceedings of the 21st Annual International Cryptology Conference on Advances in Cryptology*, pages 213–229. Springer-Verlag, 2001.
[6] M. Bowman, C. Dharap, M. Baruah, B. Camargo, and S. Potti. A file system for information management. In *Proceedings of the ISMM International Conference on Intelligent Information Management Systems*, March 1994.

[7] R. Canetti, J. Garay, G. Itkis, D. Micciancio, M. Naor, and B. Pinkas. Multicast security: A taxonomy and some efficient constructions. In *Proceedings of IEEE INFOCOM'99*, 1999.

[8] R. Canetti, O. Goldreich, and S. Halevi. The random oracle methodology, revisited (preliminary version). In *STOC*, pages 209–218, 1998.

[9] C. Cocks. An identity based encryption scheme based on quadratic residues. In *IMA Int. Conf.*, pages 360–363, 2001.

[10] E. Cronin, S. Jamin, T. Malkin, and P. McDaniel. On the Performance, Feasibility, and Use of Forward Secure Signatures. In *Proceedings of 10th ACM Conference on Computer and Communications Security (CCS)*, pages 131–144. ACM, October 2003. Washington, DC.

[11] D. E. Denning. A lattice model of secure information flow. *Commun. ACM*, 19(5):236–243, 1976.

[12] C. Ellison and B. Schneier. Ten Risks of PKI: What You're Not Being Told About Public i Key Infrastructure. *Computer Security Journal*, 16(1):1–7, 2000.

[13] D. F. Ferraiolo, R. Sandhu, S. Gavrila, D. R. Kuhn, and R. Chandramouli. Proposed NIST standard for role-based access control. *ACM Trans. Inf. Syst. Secur.*, 4(3):224–274, 2001.

[14] B. Gopal and U. Manber. Integrating content-based access mechanisms with hierarchical file systems. In *OSDI '99: Proceedings of the third symposium on Operating systems design and implementation*, pages 265–278, Berkeley, CA, 1999. USENIX Association.

[15] T. Hardjono and B. Weis. The Multicast Group Security Architecture. RFC 3740 (Informational), Mar. 2004.

[16] D. R. Hardy and M. F. Schwartz. Essence: A resource discovery system based on semantic file indexing. In *Proceedings of the USENIX Winter Conference*, pages 361–374, Berkeley, CA, January 1993. USENIX Association.

[17] F. J. Hill and G. R. Peterson. *Computer aided logical design with emphasis on VLSI*. Wiley, 4 edition, 1993.

[18] B. Lampson. Protection. In *Proceedings of the 5th Annual Princeton Conference on Information Sciences and Systems*, pages 437–443, Princeton University, 1971.

[19] B. Lynn. PBC library. http://rooster.stanford.edu/~ben/pbc/, 2006.

[20] P. McDaniel, A. Prakash, and P. Honeyman. A flexible framework for secure group communication. In *USENIX Security Symposium*, pages 99–114, 1999.

[21] P. McDaniel and A. D. Rubin. A response to "can we eliminate certificate revocation lists?". In *FC '00: Proceedings of the 4th International Conference on Financial Cryptography*, pages 245–258, London, UK, 2001. Springer-Verlag.

[22] A. J. Menezes, T. Okamoto, and S. A. Vanstone. Reducing elliptic curve logarithms to logarithms in a finite field. *IEEE Transactions On Information Theory*, 39(5):1639–1646, September 1993.

[23] A. Miyaji, M. Nakabayashi, and S. Takano. New explicit conditions of elliptic curve traces for FR-reduction. *IEICE Transactions on Fundamentals*, E84-A(5):1234–1243, 2001.

[24] M. Myers, R. Ankney, A. Malpani, S. Galperin, and C. Adams. X.509 Internet Public Key Infrastructure: Online Certificate Status Protocol - OCSP. http://www.ietf.org/rfc/rfc2560.txt, 1999.

[25] D. Nali, C. Adams, and A. Miri. Using threshold attribute-based encryption for practical biometric-based access control. 1(3):173–182, November 2005.

[26] A. Sahai and B. Waters. Fuzzy identity based encryption. In *Eurocrypt 2005*, 2005.

[27] R. S. Sandhu, E. J. Coyne, H. L. Feinstein, and C. E. Youman. Role-based access control models. *Computer*, 29(2):38–47, 1996.

[28] R. S. Sandhu and P. Samarati. Access control: Principles and practice. *IEEE Communications Magazine*, 32(9):40–48, 1994.

[29] A. Shamir. How to share a secret. *Commun. ACM*, 22(11):612–613, 1979.

[30] A. Shamir. Identity-based cryptosystems and signature schemes. In *Proceedings of CRYPTO 84 on Advances in cryptology*, pages 47–53. Springer-Verlag New York, Inc., 1985.

[31] V. Shoup. Using hash functions as a hedge against chosen ciphertext attack. In *EUROCRYPT*, pages 275–288, 2000.

[32] United States Department of Health and Human Services. Health Insurance Portability and Accountability Act. http://aspe.hhs.gov/admnsimp/pl104191.htm, 1996.

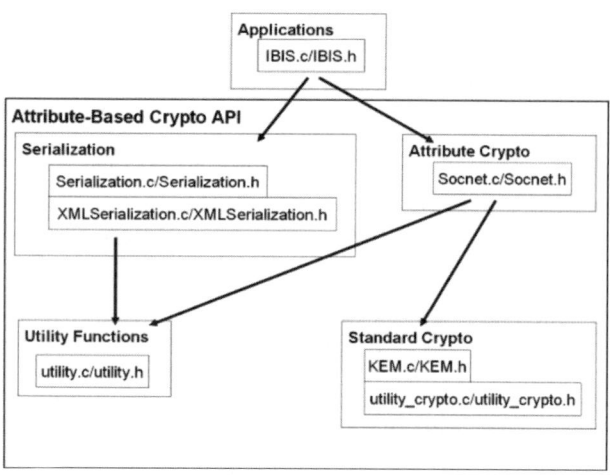

Figure 12: Components of attribute-based cryptosystem API.

APPENDIX
A. ABE SYSTEMS DESIGN & ISSUES

Figure 12 illustrates the architecture of ABE systems built using our ABE API. Specifically there are two main components: the ABE API and the application code.

ABE API – We have created the ABE API to enable rapid development of systems and applications which use attribute-based cryptography. Our API uses the PBC library [19] to implement our attribute-based cryptography. This C language API has been specifically designed to enable a programmer with no knowledge of ABE to quickly write applications; the complex cryptography inherent to ABE is entirely handled by the API.

For didactic purposes we present the API as four distinct modules: attribute-based cryptography, standard cryptography, serialization, and utility functions.

Attribute-based Cryptography – The majority of application level code interacts with the API through the attribute-based cryptography component. This module was specifically designed for ease of use, consisting of seven simple functions: Setup_System, Create_User, New_Attribute, Give_Attribute, Key_Generation, Encryption, and Decryption.

The Setup_System function creates and initializes a new attribute-based cryptosystem. Specifically, this instantiates two key structures: global_params and authority_priv. global_params contains global parameters required to perform encryption and decryption operations. authority_priv contains the master secret, from which all attribute keys are defined. authority_priv must be kept secret in order to ensure the security of the system.

Setup_System must be given pbc_param_file_name, the name of an XML file defining an elliptic curve from which all of the API's ABE cryptography is formulated. Included with the API are two such parameter files, a_param.xml (Supersingular curve) and c159_param.xml (MNT curve). Supersingular curves are optimized for fast cryptographic pairings, and MNT curves are optimized to result in small cryptographic group elements.

The nature of ABE cryptography is such that every ciphertext in a given cryptosystem is of a fixed length n. The user can specify what this length is by providing the API with `ct_len`.

To increase the flexibility of the API, `Setup_System` creates several "default" attributes. The default attributes can be included in a ciphertext to take the place of non-default attributes, enabling the user to create ciphertexts with less than `ct_len` attributes.

The `New_Attribute` function is used to add a new attribute, whose name is specified by `att_name`, to the universe of attributes in the system. Upon completion of this function the new attribute's name and the hash of its name can be made publicly available. At this point the new attribute can be used for encryption operations.

The `Create_User` function adds a user named `user_id` to the system. This function instantiates `user_id`, a structure which stores the user's name, the user's default and non-default attribute information, and a polynomial. Each user is given a unique polynomial. Tieing each user's per-attribute keying information to their polynomial prevents users from colluding in order to attain more attributes.

The `Give_Attribute` function is used to give a user a new attribute. Specifically this function is used to update the user's attribute data structures and does **not** generate any keying information. The `Key_Generation` function is used to create a user's keying information based on the attributes that they possess. Keeping key generation separate enables the `Give_Attribute` function to be executed with fewer trust assumptions than is needed to perform the `Key_Generation` function.

The `Encryption` function is used by a user to create a new ciphertext, `ciphertext`. The user specifies, `message`, a string they would like to encrypt and, `uid`, a list of attributes that they would like to encrypt to. The user can encrypt with at most `ct_len` attributes. The API will pad the ciphertext with as many default attributes as is necessary to make the ciphertext contain a total of `ct_len` attributes. A list of the attributes used to perform encryption are included in each ciphertext in order for the party performing decryption to know which attributes are required to decrypt the message.

Encryption is significantly more complicated than the API's function calls would seem to indicate. Specifically, the ABE constructions mandate that a ciphertext's payload must be a group element. To enable ABE to carry non-group element payloads we use the Key Encapsulation Mechanism(KEM). In our API, KEM takes a group element payload and uses SHA-1 to convert it into a HMAC key and an AES key. The AES key is then used to encrypt the user's message.

`Decryption` decrypts a ciphertext encrypted by the `Encryption`. This process begins with the decrypting party verifying that they have the required attributes. The party performing decryption will then use their attributes to decrypt the decrypt the ciphertext in order to obtain the AES and HMAC key. The party will then use the HMAC key to verify the ciphertext. If the ciphertext can be verified, then the AES key will be used to decrypt the actual payload.

Standard Cryptography – In addition to the attribute-based cryptography we have also used standard cryptographic tools. The implementation of these tools are contained in the `crypto_utility` and `KEM` code. The `crypto_utility` code implements some of the low level cryptographic operations required by ABE. `KEM` implements all of the operations required to enable ABE to encrypt non-group member payloads.

Serialization – The serialization routines enable the API data structures to be written out to disk for long term storage. There are two different implementations of this functionality. `Serialization` stores API data structures into byte-encoded files. `XMLSerialization` stores API data structures into XML files. `XMLSerialization` is human readable, has better platform independence, and is more fault tolerant. `Serialization` results in slightly less disk space.

Utility Functions – The `utility` functions are a group of functions that increase the ease of programming with the attribute-based cryptography API. Included in the `utility` routines are functions that print API data structures and conversion routines.

B. ATTRIBUTE-BASED ENCRYPTION

For our system we use a variant of the Sahai-Waters Large Universe system [26](Section 6) which we now describe.

In this construction we will make use of a bilinear group $\mathbb{G}$ of prime order p. The group will have an efficiently computable bilinear map $e : \mathbb{G} \times \mathbb{G} \to \mathbb{G}_T$ that maps two elements from the bilinear group into an element of the "target group". The salient feature of these groups is that if g is a generator of $\mathbb{G}$ then for all $a, b \in \mathbb{Z}_p$ we have that $e(g^a, g^b) = e(g, g)^{ab}$. We refer the reader to the IBE paper of Boneh-Franklin [5] for more details on bilinear groups.

The Sahai-Waters construction works by computing a bilinear map between k components of the ciphertext with corresponding pieces of the private key. The result of these are interpolated using the secret sharing method of Shamir (in the exponent). We first define the following Lagrangian coefficients, which we will use in our construction, as the following function over $\mathbb{Z}_p$:

$$\Delta_{i,S}(X) = \prod_{j \in S, j \neq i} \frac{x-j}{i-j}.$$

Additionally, we will assume all systems will work in some predetermined bilinear group $\mathbb{G}$ of appropriate size.

The cryptosystem follows:

Setup(k): The setup algorithms first chooses a random exponent $y \in \mathbb{Z}_p$ and lets the public parameter be $Y = e(g,g)^y$ and the threshold value k. It keeps the public key and the secret exponent y as the master key.

Key-Gen(S, MK): Let $H : \{0,1\}^* \to \mathbb{Z}_p$ be a collision-resistant hash function and let $T : \mathbb{Z}_p \to \mathbb{G}$ be a function that we will model as a random oracle [4].

First let Γ be the set defined as $\Gamma = \bigcup_{s \in S} H(s)$. The set Γ is essentially the set of the hash of all attributes. (Note that since H is collision-resistant Γ should contain $|S|$ unique elements of $\mathbb{Z}_p$.) Then the authority will choose a new random degree $k-1$ polynomial $q(x)$ over $\mathbb{Z}_p$ such that $q(0) = y$ and for all $i \in \Gamma$ the authority chooses a random r_i. Then for all $i \in \Gamma$ the private keys components are:

$$D_i = g^{q(i)} T(i)^{r_i}, d_i = g^{r_i}$$

Encrypt(M, S', PK): The encryption algorithm first computes the set $\Gamma' = \bigcup_{s \in S'} H(s)$. Next, it chooses a random exponent $t \in \mathbb{Z}_p$. The ciphertext is output as:

$$C = \left(C' = MY^t, C'' = g^t, \{C_i = T(i)^t : i \in \Gamma'\}\right).$$

Notice that both the size of the ciphertext and the encryption time grows linearly with the size of the set S.

Decrypt(C, S', S, SK): The decryption algorithm first computes the sets Γ and Γ' as before. If the size of the intersection $|\Gamma \cap \Gamma'| < k$ the algorithm aborts, this will occur if the overlap between the private key attribute set S and the ciphertext set S' is below the threshold k. Otherwise it chooses an arbitrary set U such that $|U| = k$ and $U \subseteq \Gamma \cap \Gamma'$. For each $i \in U$ the decryptor computes a temporary value

$$A_i = \frac{e(D_i, C'')}{e(d_i, C_i)} = \frac{e(g^{q(i)}T(i)^{r_i}, g^t)}{e(g^{r_i}, T(i)^t)} = e(g,g)^{tq(i)}.$$

This computation gives k shares of the polynomial $tq(i)$ in the exponent. Using polynomial interpolation the algorithm recovers the blinding value $e(g,g)^{yt}$ and divides it out by computing:

$$M = C' / \left(A_i^{\Delta_{i,U}(0)} \right) = C'/e(g,g)^{tq(0)} = C'/e(g,g)^{ty} = M.$$

The decryption algorithm interpolates a polynomial in the exponent using Shamir's [29] secret sharing method. However, since a new random polynomial is chosen for each private key, the system is secure against collusion attacks such that different users are unable to combine their separate attributes.

The difference between the construction given here and that of Sahai and Waters is in the computation of the function $T(i)$. In their construction there is a upper bound, n, on the number of attributes that can label a ciphertext which is set at setup. The setup function publishes values $t_1, \ldots, t_n$. The function $T(i)$ is computed as:

$$T(i) = g^{x^i} \prod_{j=1}^{n+1} t_j^{\Delta_{j,N}(i)}$$

where N is the set $\{1, \ldots, n+1\}$.

It is easily seen that the number of exponentiations required to compute $T(i)$ is equal to $n+1$ in the original Sahai and Waters construction. We drastically reduce the computation overhead replacing the computation of T with a hash function as a random oracle. A simple argument shows that the random oracle can be "programmed" such that the simulation in the security proof of Sahai and Waters goes through. We refer the reader to the literature [4,8] for further discussion on the random oracle model. In Section 6 we experimentally compare implementations of the original Sahai and Waters construction with our variant.

Resiliency Policies in Access Control

Ninghui Li
Dept. of Computer Science
Purdue University
ninghui@cs.purdue.edu

Mahesh V. Tripunitara
Motorola Labs
tripunit@motorola.com

Qihua Wang
Dept. of Computer Science
Purdue University
wangq@cs.purdue.edu

ABSTRACT

We introduce the notion of resiliency policies in the context of access control systems. Such policies require an access control system to be resilient to the absence of users. An example resiliency policy requires that, upon removal of any s users, there should still exist d disjoint sets of users such that the users in each set together possess certain permissions of interest. Such a policy ensures that even when emergency situations cause some users to be absent, there still exist independent teams of users that have the permissions necessary for carrying out critical tasks. The Resiliency Checking Problem determines whether an access control state satisfies a given resiliency policy. We show that the general case of the problem and several subcases are intractable (**NP**-hard), and identify two subcases that are solvable in linear time. For the intractable cases, we also identify the complexity class in the polynomial hierarchy to which these problems belong. We discuss the design and evaluation of an algorithm that can efficiently solve instances of nontrivial sizes that belong to the intractable cases of the problem. Finally, we study the consistency problem between resiliency policies and static separation of duty policies.

Categories and Subject Descriptors

D.4.6 [**Operating Systems**]: Security and Protection—*Access controls*; K.6.5 [**Management of Computing and Information Systems**]: Security and Protection; F.2.2 [**Analysis of Algorithms and Problem Complexity**]: Nonnumerical Algorithms and Problems—*Complexity of proof procedures*

General Terms

Security, Theory

Keywords

Access Control, Fault-tolerant, Policy Design

1. INTRODUCTION

While policy analysis has been a main research area in access control for several decades, almost all existing work focuses on properties which ensure that users who should not have access do not get access. For example, safety analysis [9, 17, 22] studies whether an access right can be leaked to unauthorized users. Separation of duty (SoD) policies [2, 20] ensure that no single user (or a set of users of size less than some threshold) is able to perform a sensitive task. Such focus on safety properties probably stems from the fact that access control has been mostly viewed as a tool for restricting access. However, an equally important aspect of access control is to enable access (selectively).

We introduce the notion of resiliency policies which state properties about enabling access in access control. Resiliency policies require that the access control state is resilient to absent users. For example, the access control system of an institution has three separate permissions regarding release of funds: one permission is an endorsement that the request for funds is legitimate, the second permission is the issuance of a check, and the third one is for logging the transaction. The institution's financial office, which takes charge of funding, is composed of a senior treasurer and a number of junior treasurers. In compliance of the separation of duty principal, the senior treasurer has all permissions except the one for logging, while each of the junior treasurers has only one of the three permissions. Since issuing funds is a critical task, the institution would like to ensure that even if a few (e.g., two) treasurers (that may include the senior treasurer) are absent (e.g., due to sickness), the remaining personnel in the financial office still have enough privileges to release funds.

Another example resiliency policy requirement is as follows: There must exist three mutually disjoint sets of users such that each set has no more than four users and the users in each set together have all permissions to carry out a critical task. Such a policy would be needed when one needs to be able to send up to three teams of users to different sites to perform a certain task, perhaps in response to some events. One needs to ensure that each team has enough permissions to perform the task, and each team consists of no more than four users (e.g., due to the limit of transportation means).

Such policies are particularly useful when evaluating whether the access control configuration of a system is ready for emergency response. These policies ensure that even when emergency situations cause some users to be absent, there still exist independent teams of users that have the necessary permissions for carrying out critical tasks. In other words, these policies mandate that there is a certain level of redundancy in assigning permissions to users so that the system can tolerate some users being absent.

Our contributions in this paper are as follows:

1. We introduce the notion of Resiliency Policies which express requirements about enabling access rather than restricting access. We give a concrete formulation for a resiliency policy which captures the intuition discussed above.

2. We study computational complexities of the Resiliency Checking Problem, which determines whether an access control state satisfies a given resiliency policy. We show that this problem is **NP**-hard in the general case and is in $\mathbf{coNP^{NP}}$, a complexity class in the Polynomial Hierarchy. We show that several subcases are **NP**-complete. We identify two subcases that are solvable in linear time.

3. We show that, notwithstanding the intractability results, many instances of the Resiliency Checking Problem of nontrivial sizes may still be efficiently solvable. We present an algorithm for the Resiliency Checking Problem. Our algorithm uses a pruning technique that reduces the number of combinations that need to be considered. The experimental results show that this pruning technique can reduce the search space by several orders of magnitude. Our algorithm also takes advantage of the observation that the problem of checking whether the state can tolerate the removal of a particular absent set can be naturally formulated as the boolean satisfiability problem. This enables us to use existing SAT solvers in our implementation and benefit from several decades of research in designing SAT solvers. Our experimental results show that our algorithm can efficiently solve instances of nontrivial sizes.

4. Resiliency policies may conflict with safety-oriented policies such as static separation of duty (SSoD) policies [14]. We study the policy consistency problem between resiliency policies and SSoD policies. We demonstrate how to simplify the problem and present criteria for determining consistency for a number of special cases. Finally, we show that determining consistency is both **NP**-hard and **coNP**-hard, but is in $\mathbf{NP^{NP}}$.

The remainder of this paper is organized as follows. In Section 2, we define resiliency policies and the Resiliency Checking problem. We present computational complexities of the Resiliency Checking problem in Section 3, and an algorithm for the problem and an implementation of the algorithm in Section 4. In Section 5, we explore the policy consistency problem. We discuss related work in Section 6. Finally, we conclude and present open problems related to the concept of resiliency in Section 7.

2. RESILIENCY POLICIES AND THE RESILIENCY CHECKING PROBLEM

Definition 1 (Resiliency Policies). A resiliency policy takes the form
$$\mathsf{rp}\langle P, s, d, t\rangle$$
where rp is a keyword, $P = \{p_1, \ldots, p_n\}$ is a set of permissions, $s \geq 0$ and $d \geq 1$ are integers, and t is either a positive integer or the special symbol ∞.

We say that an access control state satisfies such a resiliency policy if and only if upon removal of any set of s users, there still exist d mutually disjoint sets of users such that each set contains no more than t users and the users in each set together are authorized for all permissions in P.

Example 1. Consider the access control state from Figure 1. It relates to the example we introduce in Section 1. To issue funds, all three permissions *Endorse*, *Issue* and *Log* must be possessed by a set of users. In our resiliency policy, we set $P = \{Endorse, Issue, Log\}$. If we set $s = 1$ in our policy, then we want the system to be resilient to the absence of any (one) user. If we set $d = 2$, this means that we require two sets of users such that users in each set together possess all permissions. If we set $t = \infty$, this means that the set of users that together possess all permissions can be of any size.

We observe that in our example, $\mathsf{rp}\langle P, 1, 2, \infty\rangle$ is satisfied. For instance, after removing *Alice*, the two users *Carl* and *Earl* together have all three permissions, as are *Bob* and *Doris*. The cases in which another user is removed can be verified similarly. However, $\mathsf{rp}\langle P, 2, 2, \infty\rangle$ is not satisfied because if *Alice* and *Bob* are absent, the only user that possesses *Endorse* is *Carl*, and one user cannot belong to two disjoint sets. Similarly, $\mathsf{rp}\langle P, 2, 1, \infty\rangle$ is satisfied, but $\mathsf{rp}\langle P, 3, 1, \infty\rangle$ is not satisfied because if *Alice*, *Bob* and *Carl* are absent, then no user possesses *Endorse*. And finally, we observe that $\mathsf{rp}\langle P, 1, 1, 2\rangle$ is satisfied, but not $\mathsf{rp}\langle P, 1, 1, 1\rangle$ because for the latter case, there exists no single user that has all three permissions.

Intuitively, a resiliency policy $\mathsf{rp}\langle P, s, d, t\rangle$ specifies a fault tolerance requirement with respect to a certain critical task. The set P includes all permissions that are needed to carry out the task. The faults that we would like to tolerate are absent users. The parameter s specifies the number of absent users that we want to be able to tolerate. The parameter d is motivated by the requirement that several teams may be needed to carry out multiple instances of the task. If only one team is needed, then d can be set to 1. The parameter t specifies the size limit of each team. This is motivated by limitations on the maximal number of users that can be involved in any instance of task. If no such limitation exists, then t can be set to ∞.

The two parameters s and d are related. If an access control state satisfies $\mathsf{rp}\langle P, s, d, t\rangle$, then it also satisfies $\mathsf{rp}\langle P, s+i, d-i, t\rangle$ for any i such that $0 < i < d$. For example, if, after removing any 2 users, there exist 3 mutually disjoint sets of users such that each set covers all permissions in P, then after removing any 3 users, there are at least 2 sets left. However, if a state satisfies $\mathsf{rp}\langle P, s+1, d-1, t\rangle$, it may not satisfy $\mathsf{rp}\langle P, s, d, t\rangle$. For our example shown in Figure 1, we observe that $\mathsf{rp}\langle P, 1, 2, \infty\rangle$ is satisfied. However $\mathsf{rp}\langle P, 0, 3, \infty\rangle$ is not satisfied because we need the 3 users *Alice*, *Bob* and *Carl* that possess *Endorse* to belong to distinct sets; this still leaves one permission that needs to covered by each set, and we have only two users that remain.

Resiliency policies can be defined in any access control system in which there are users and permissions. This includes almost all access control systems, including Discretionary Access Control systems [13, 8] and Role Based Access Control systems [25]. We assume that an access control state is given by a binary relation $UP \subseteq \mathcal{U} \times \mathcal{P}$, where $\mathcal{U}$ represents the set of all users, and $\mathcal{P}$ represents the set of all permissions. Note that by assuming that a state is given by a binary relation $UP \subseteq \mathcal{U} \times \mathcal{P}$, we are not assuming permissions are directly assigned to users; rather, we assume only that one can calculate the relation UP from the access control state.

Definition 2 (Resiliency Checking Problem (RCP)). Given a resiliency policy r and an access control state UP, determining whether UP satisfies r is called the Resiliency Checking Problem (RCP).

A resiliency policy has three parameters: s, d, and t. In some situations, one may need to consider only those policies with one or more of these parameters degenerated. The parameter s, which denotes the number of absent users that the system needs to tolerate, may be degenerated to always be 0. The parameter d, which denotes the number of sets of users required, may be degenerated to always be 1. Finally, the parameter t, which denotes the size bound on each set, may be degenerated to always be ∞. There are eight

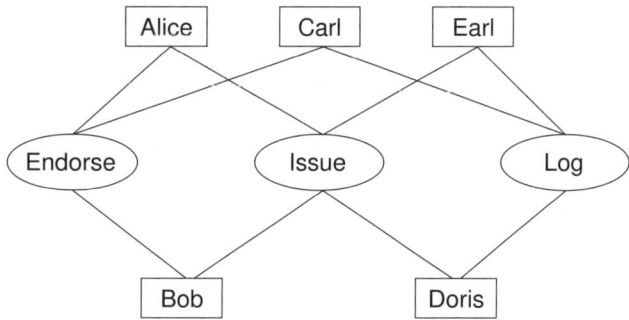

Figure 1: An example of an access control state with 5 users, *Alice, Bob, Carl, Doris* **and** *Earl*, **and 3 permissions,** *Endorse, Issue* **and** *Log*. **A line segment connects a user (e.g.,** *Alice*) **to a permission (e.g.,** *Endorse*) **to indicate that the user has the permission. This corresponds to the example from Section 1 on releasing funds; all three permissions must be possessed by a group of users that together want to release funds.**

cases where some of the three parameters are degenerated. For example, a resiliency policy in the subcase RCP$\langle s = 0, d = 1 \rangle$ has the form rp$(P, 0, 1, t)$, which asks whether there exists a set of users of size at most t that together have all permissions in P; while the subcase RCP$\langle t = \infty \rangle$ asks whether there exist several distinct sets of users (d sets) each of whose users together have all permissions in P, even after any set of s users is removed from the state. In particular, RCP$\langle \ \rangle$ is the general case of the problem.

3. COMPUTATIONAL COMPLEXITIES OF THE RESILIENCY CHECKING PROBLEM

The following theorem summarizes the computational complexity results for RCP and its various subcases. These results are also shown in Figure 2.

Theorem 1. *The computational complexities of the Resiliency Policy Checking problem are as follows.*

- RCP$\langle \ \rangle$, *the most general case, is* **NP**-*hard and is in* **coNP**$^{\textbf{NP}}$, *as are the two subcases* RCP$\langle d = 1 \rangle$ *and* RCP$\langle t = \infty \rangle$.

- RCP$\langle s = 0, d = 1 \rangle$, RCP$\langle s = 0, t = \infty \rangle$, *and* RCP$\langle s = 0 \rangle$ *are* **NP**-*complete.*

- RCP$\langle d = 1, t = \infty \rangle$ *and* RCP$\langle s = 0, d = 1, t = \infty \rangle$ *can be solved in linear time.*

Our complexity results show that RCP is in **coNP**$^{\textbf{NP}}$. This means that the complement of RCP can be solved by a nondeterministic Oracle Turing Machine that has oracle access to a machine that can answer any **NP** queries. (See Appendix A for a brief overview of Oracle Turing Machines.) Intuitively, given an access control state and a resiliency policy $r = $ rp(P, s, d, t), to decide nondeterministically that the state does not satisfy r, one can guess a set of s users to removed, and then query the **NP** oracle whether the remaining users contain d mutually disjoint sets of users such that each set is of size at most t and the users in each set together have all permissions in P.

Another way to understand the computational complexity of RCP is to observe that an RCP instance has the form $\forall$ size-s subset, $\exists d$ sets of users that satisfy some requirements that can be efficiently verified. Problems in **NP** have the form of $\exists$ an evidence that satisfies some polynomial-time verifiable requirements. Problems in **coNP** has the form $\forall$ choices, some polynomial-time ver-

ifiable requirements hold. RCP has one alternation of $\forall$ followed by $\exists$, which makes it in **coNP**$^{\textbf{NP}}$.

We have shown that RCP (and its two subcases RCP$\langle d = 1 \rangle$ and RCP$\langle t = \infty \rangle$) are **NP**-hard and are in **coNP**$^{\textbf{NP}}$. It remains open whether these three problems are **coNP**$^{\textbf{NP}}$-complete or not. Readers who are familiar with computational complexity theory will recognize that **coNP**$^{\textbf{NP}}$ is a complexity class in the Polynomial Hierarchy. (See Appendix A for a brief introduction to the Polynomial Hierarchy.) Because the Polynomial Hierarchy collapses when $\textbf{P} = \textbf{NP}$, showing that an **NP**-hard decision problem is in the Polynomial Hierarchy, although is not equivalent to showing that the problem is **NP**-complete, has the same consequence: the problem can be solved in polynomial time if and only if $\textbf{P} = \textbf{NP}$.

In the rest of this section, we prove the results in Theorem 1. The following lemmas prove that RCP$\langle s = 0 \rangle$ is in **NP**, RCP$\langle s = 0, d = 1 \rangle$ and RCP$\langle s = 0, t = \infty \rangle$ are **NP**-hard, RCP$\langle \ \rangle$ is in **coNP**$^{\textbf{NP}}$, and RCP$\langle d = 1, t = \infty \rangle$ is in **P**. The complexities of other subcases can be implied from these results.

Lemma 2. RCP$\langle s = 0 \rangle$ *is in* **NP**.

PROOF. An instance consists of an access control state UP and a policy rp$\langle P, 0, d, t \rangle$. UP satisfies rp$\langle P, 0, d, t \rangle$ if and only if there exist d mutually disjoint sets of users such that the users in each set together cover all permissions in P and each set has at most t users. If these d sets are given, they can be verified in polynomial time. Therefore, RCP$\langle s = 0 \rangle$ is in **NP**.

Lemma 3. RCP$\langle s = 0, d = 1 \rangle$ *is* **NP**-*hard*.

PROOF. We reduce the **NP**-complete SET COVERING problem [19] (also referred to as MINIMUM COVERING problem in [6]) to RCP$\langle s = 0, d = 1 \rangle$. In SET COVERING, we are given a set S, n subsets of S: $S_1, \ldots, S_n$, and a budget K, and need to determine whether the union of K subsets is the same as S. An instance of RCP$\langle s = 0, d = 1 \rangle$ asks whether an access control state UP satisfies a policy rp$\langle P, 0, 1, t \rangle$. In our reduction, each element in S is mapped to a permission in P and each subset S_i is mapped to a user u_i. In other words, if the subset S_i contains an element, then u_i is authorized for the permission corresponding to the element. We now argue that the mapping ensures that there exists a set of users of size at most K together have all the permissions in P if and only if K subsets cover S. Assume that a set of users of size at most K exists such that those users together have all the permissions in P. Then, we pick the subsets that are mapped to those users, and their union gives us S. For the other direction, assume

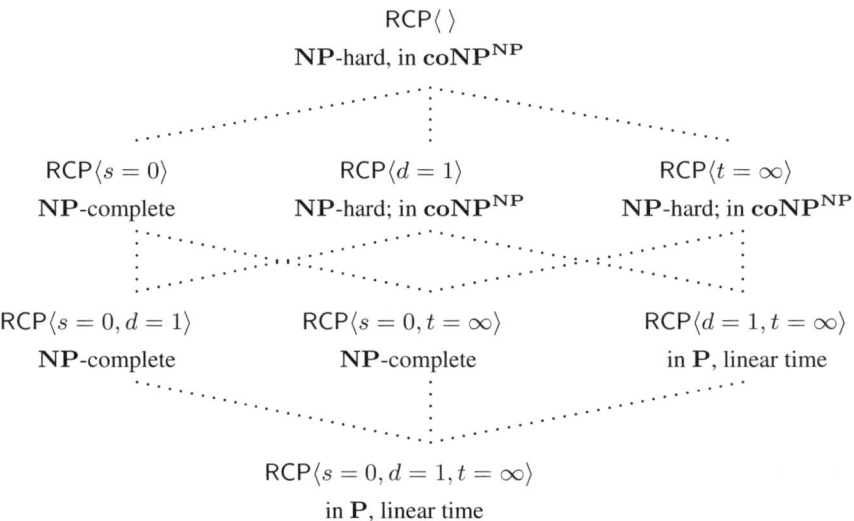

Figure 2: Time complexity of the Resiliency Checking Problem (RCP) and its various subcases.

that K subsets cover S. Then, the K users to which the subsets are mapped together have all permissions in P.

Lemma 4. $RCP\langle s = 0, t = \infty \rangle$ *is* **NP**-*hard*.

PROOF. We reduce the **NP**-complete DOMATIC NUMBER problem [6] to $RCP\langle s = 0, t = \infty \rangle$. Given a graph $G(V, E)$, the DOMATIC NUMBER problem asks whether V can be partitioned into k disjoint sets $V_1, V_2, \cdots, V_k$, such that each V_i is a dominating set for G. V' is a dominating set for $G = (V, E)$ if for every node u in $V - V'$, there is a node v in V' such that $(u, v) \in E$. An instance of $RCP\langle s = 0, t = \infty \rangle$ asks whether an access control state UP satisfies a policy $\mathsf{rp}\langle P, 0, d, \infty \rangle$. Given a graph $G = (V, E)$, we construct an access control state UP with n users $u_1, u_2, \cdots, u_n$ and n permissions $p_1, p_2, \cdots, p_n$, where n is the number of nodes in V. Each user corresponds to a node in G, and $v(u_i)$ denotes the node corresponding to user u_i. In UP, user u_i is authorized for the permission p_j if and only if either $i = j$ or $(v(u_i), v(u_j)) \in E$. Let P denote the set $\{p_1, p_2, \cdots, p_n\}$. A dominating set in G corresponds to a set of users that together have all permissions in P. UP satisfies $\mathsf{rp}\langle P, 0, k, \infty \rangle$ if and only if V contains k disjoint dominating sets.

Lemma 5. $RCP\langle \ \rangle$ *is in* $\mathbf{coNP^{NP}}$.

PROOF. We show that the complement of $RCP\langle \ \rangle$ is in $\mathbf{NP^{NP}}$. Assume that we have an oracle that decides the Resiliency Checking problem when $s = 0$, which, as we know, is **NP**-complete. We construct a nondeterministic oracle Turing machine M that accepts UP and $\mathsf{rp}\langle P, s, d, t \rangle$ when UP does not satisfy $\mathsf{rp}\langle P, s, d, t \rangle$. M nondeterministically removes s users, and then queries the oracle. If the oracle machine returns "yes", M rejects; otherwise, M accepts, because it has found a set of users, the removal of which violates the Resiliency policy. The construction of M shows that the complement of $RCP\langle \ \rangle$ is in $\mathbf{NP^{NP}}$. Therefore, $RCP\langle \ \rangle$ is in $\mathbf{coNP^{NP}}$.

Lemma 6. $RCP\langle d = 1, t = \infty \rangle$ *can be solved in linear time*.

An instance in $RCP\langle d = 1, t = \infty \rangle$ asks whether an access control state satisfies a policy $\mathsf{rp}\langle P, s, 1, \infty \rangle$. We observe that the answer is "no" if and only if some permission in P is assigned to no more than s users. In this case, removing the s users who have that permission would result in no user having that permission. On the other hand, if each permission is assigned to at least $s+1$ users, after removing any set of s users, each permission is still assigned to at least one user, which means that the set of all remaining users together have all permissions in P.

Definition 3 (The Tolerance Bound). Given an access control state UP and a set $\{p_1, \cdots, p_m\}$ of permissions, we define the *tolerance bound* of UP and $\{p_1, \cdots, p_m\}$, denoted by $tb(UP, \{p_1, \cdots, p_m\})$, to be $\min_{1 \leq i \leq m} \#(p_i)$, where $\#(p_i)$ denotes the number of users who are authorized for p_i in the state UP.

Given an $RCP\langle d = 1, t = \infty \rangle$ instance that asks whether UP satisfies $\mathsf{rp}\langle P, s, 1, \infty \rangle$, the answer is yes if and only if the tolerance bound is at least $s + 1$. More generally, given an RCP instance that asks whether UP satisfies $\mathsf{rp}\langle P, s, d, t \rangle$, if $s + d > tb(UP, P)$, then the answer is "no". On the other hand, when $d \geq 2$ and $s + d \leq tb(UP, P)$, we do not immediately know whether UP satisfies $\mathsf{rp}\langle P, s, d, t \rangle$ or not.

We now give a linear-time algorithm for calculating the tolerance bound. This, together with the above observations, suffices to prove Lemma 6. The algorithm maintains a counter for each permission. It first goes through all pairs in UP to count how many users each permission is assigned to. It then returns the minimal value among the counters.

4. AN ALGORITHM FOR RCP

The fact that RCP is intractable (**NP**-hard) means that there exist difficult problem instances that take exponential time in the worst case. Many instances that will be encountered in practice may still be efficiently solvable. We now describe an algorithm for RCP. We first describe an algorithm for the subcase $RCP\langle t = \infty \rangle$, i.e., there is no limit on the number of users in any of the d mutually disjoint sets. We then describe how to extend the algorithm to deal with the parameter t when it is not degenerated. In Section 4.2 we discuss our implementation of this algorithm and its effectiveness using experimental results.

4.1 Description of the algorithm

To determine whether UP satisfies $\mathsf{rp}\langle P, s, d, \infty \rangle$, a straightforward algorithm is to enumerate all sets of s users, and for each such set A (which we call an *absent set*), remove the users in A from UP and check whether among the remaining users there are d mutually disjoint sets of users such that each set covers the permissions. If the answer is "no" for any absent set, then we know that UP does not satisfy rp $\langle P, s, d, \infty \rangle$. If we have enumerated through all absent sets, and the answer is "yes" for each of them, then we know that UP satisfies rp $\langle P, s, d, \infty \rangle$. Our algorithm adds the following improvements, which greatly reduces the running time.

Preprocessing Given the state UP and the policy rp$\langle P, s, d, \infty \rangle$, we first remove (u, p) from UP if $p \notin P$, as we do not need to consider permissions not in the policy. Also, we only consider those users who are authorized for at least one permission in P. Finally, we calculate the tolerance bound $tb(UP, P)$, using the methods described in the end of Section 3. If $s + d > tb(UP, P)$, then we know the answer is "no".

Reduction to SAT A key step to solve RCP is to determine whether, after removing a certain set of users, there still exist d mutually disjoint sets of users such that each set covers all permissions in P. We observe that such a problem can be translated into a SAT instance. This enables us to benefit from the extensive research on SAT and to use existing SAT solvers. SAT has been studied extensively for several decades (see, for example, [5]), and many clever algorithms have been developed. Problems in many fields, including databases, planning, computer-aided design, machine vision and automated reasoning, have been reduced to SAT and solved using SAT solvers. Often times, this results in better performance than using existing domain-specific algorithms for those problems.

The translation works as follows. Let U be the set of users after removing users in an absent set. For each user u_i in U and each integer j from 1 to d, we have a propositional variable $v_{i,j}$. This variable is true if the i'th user is assigned to the j'th group. Then we have the following two kinds of clauses. The first kind of clauses ensure that all permissions are covered in each of the d groups: For each permission p in P, let $u_{i_1}, u_{i_2}, \cdots, u_{i_x}$ be users in U who are authorized for p. Then for each j from 1 to d, we add the clause $v_{i_1,j} \vee v_{i_2,j} \vee \cdots \vee v_{i_x,j}$. There are $|P| \cdot d$ of such clauses. The second kind of clauses ensure that no user is selected in two groups at the same time: For each user u_i, and for each pair k, ℓ such that $0 < k < \ell \leq d$, we add the clause $\neg v_{i,k} \vee \neg v_{i,\ell}$. There are $nd(d-1)/2$ such clauses, where n is the number of users. It is clear that the total number of clauses added is polynomial to the size of the RCP instance.

Static Pruning The number of size-s user sets among n users is close to n^s when s is small compared with n. For example, there are more than one billion such sets for $s = 6$ and $n = 100$. We observe that not all these sets need to considered. There is a partial order relation among these sets such that if A_1 *dominates* A_2, and the RCP instance can tolerate the removal of A_1, then it can also tolerate the removal of A_2. This means that we only need to consider A_1. We now explain this pruning technique.

Definition 4 (Absent Set Domination). Among all users in UP, we say a user u_1 *dominates* another user u_2 if u_1's set of permissions is a superset (not necessarily strict superset) of u_2's. We say a set of users, A_1, *dominates* another set A_2 if there is a bijection between users in A_2 and A_1 such that for every user u in A_2, the corresponding user in A_1 dominates the user u.

Lemma 7. *Assuming that A_1 dominates A_2, if an RCP instance can tolerate removing A_1, then it can also tolerate removing A_2.*

PROOF. We need to show that, if after removing A_1, there are d mutually disjoint sets of users such that each set covers all permissions in P, then after removing A_2, there are also d mutually disjoint sets each of which covers all permissions in P.

By definition, if A_1 dominates A_2, then there exists a bijection f between A_2 and A_1, such that $f(u) = v$ implies user $v \in A_1$ dominates user $u \in A_2$. Without loss of generality, we assume that f satisfies the property that if $u \in A_1 \cap A_2$, then $f(u) = u$. Observe that if f does not satisfy this property for some $u \in A_1 \cap A_2$, then there exist $u_1 \in A_1$ and $u_2 \in A_2$ such that $f(u) = u_1$ and $f(u_2) = u$. It follows that u_1 dominates u and u dominates u_2. Because the domination relation is transitive, we have u_1 dominates u_2. We can then assign $f(u) = u$ and $f(u_2) = u_1$. By repeating this process, we can arrive at a bijection f such that if $u \in A_1 \cap A_2$, then $f(u) = u$. This property implies that if $u \in A_2 \setminus A_1$, then $f(u) \in A_1 \setminus A_2$.

Let $S_1, \cdots, S_d$ be the disjoint sets of users after the removal of A_1, we now construct $S_1', \cdots, S_d'$ such that (1) these sets consists of only users not in A_2, (2) they are mutually disjoint, and (3) users in each set together have all permissions in P.

For each $k \in [1, d]$, S_k' is constructed as follows: for every user u in S_k, if $u \in A_2$, then u is replaced with $f(u)$. Observe that because $u \in S_k$, then $u \notin A_1$, and thus $u \in A_2 \setminus A_1$ and $f(u) \in A_1 \setminus A_2$. Therefore, each S_k' includes only users not in A_2. To show that they are mutually disjoint, we need to show, for each $w \in S_k'$, that w does not appear in S_j', where $j \neq k$. There are two cases. Case 1: w is the result of replacing $x \in A_2$, in which case $w = f(x)$ is a member of A_1, implying w does not appear in S_j. Hence, if w also appears in S_j', it must also be from replacement of x. This is impossible, because x cannot appear both in S_k and S_j. Case 2: w appears in S_k, in which case $w \notin S_j$. Furthermore, $w \notin A_1$, and therefore w cannot be used as replacement for any other user. Therefore, w does not appear in S_j'. Finally, by definition of dominance, user $f(u)$'s set of permissions is a superset of u's. Since S_k has all permissions in P, S_k' also has all permissions in P.

Enumerate all absent sets that need to be considered

We would like to systematically generate only size-s user sets that we need to consider. That is, we need to ensure that (1) any size-s user set is dominated by at least one generated user set, and (2) we do not generate two sets such that one of them dominates the other. The naïve way of finding all such sets is to generate all size-s user sets and, for each such set, check whether it is dominated by any other size-s set. However, this would be very inefficient. We now describe an algorithm that directly generates only the user sets that need to be considered.

The algorithm works as follows. First of all, we sort all users based on the number of permissions they have, in decreasing order, and assign each user an index, that is, users are listed as $u_0, \cdots, u_{n-1}$. If $0 \leq i < j \leq n - 1$, then u_i has at least as many permissions as u_j. By definition of dominance, if u_i dominates u_j, then either $i < j$ or u_i and u_j have exactly the same set of permissions. Secondly, we use an index e that initially has value $s - 1$. We generate the first size-s set $\{u_0, \cdots, u_e\}$, and then increase the index e by one each time and generate all user sets that include u_e and are not dominated by any other set generated before. A key observation is that we only need to generate user sets that have the *closure property*. We now explain this observation.

Definition 5 (Closure Property). Given a set of users $U = \{u_0, \cdots, u_{n-1}\}$, we say a set $A \subseteq U$ has the *closure property* if and only if for any $u_k \in A$, and any $u_i \in U$ such that $i < k$ and u_i dominates u_k, we have $u_i \in A$.

In other words, if a set A has the closure property, then any user that dominates a user in A and comes before that user must also be in A. The relationships between the closure property and the set dominance relation are established in the following two lemmas.

Lemma 8. *Let A be a size-s user set that satisfies the closure property and let e be the index of the user with largest index in A, then there is no size-s subset of $\{u_0, u_1, \cdots, u_{e-1}\}$ that dominates A.*

PROOF. Because A satisfies the closure property, then u_e and all users among $\{u_0, u_1, \cdots, u_{e-1}\}$ that dominate u_e are also in A. Let k be the number of users in $\{u_0, u_1, \cdots, u_{e-1}\}$ that dominate u_e, then A has $k+1$ users that dominate u_e (including u_e itself). By the definition of set domination, any set that dominates A must have at least $k+1$ users that dominate u_e. Whereas any subset of $\{u_0, u_1, \cdots, u_{e-1}\}$ has at most k users that dominate u_e. Therefore, no subset of $\{u_0, u_1, \cdots, u_{e-1}\}$ dominates A.

Lemma 8 shows that if A satisfies the closure property, then none of the sets that have been considered so far dominates A, so A needs to be considered.

Lemma 9. *Let A be a size-s user set that does not satisfy the closure property and let e be the index of the user with largest index in A, then there exists a size-s subset of $\{u_0, u_1, \cdots, u_{e-1}, u_e\}$ that dominates A and satisfies the closure property.*

PROOF. Since A does not have the closure property, there is a user $u_k \in A$ such that there exists u_i such that $i < k$, u_i dominates u_k, and $u_i \notin A$. We change A to A_1 by substituting u_k with u_i, that is, $A_1 = A \setminus \{u_k\} \cup \{u_i\}$. Clearly, A_1 dominates A. If A_1 still does not satisfy the closure property, we can repeat the substitution process until the resulting set has closure property.

Lemma 9 shows that if A does not satisfy the closure property, then there must exist a set that dominates A and either has been considered or will be generated and considered, so there is no need to consider A. The above two lemmas together show that we need to generate only the users sets that satisfy the closure property.

Dynamic Pruning When an absent set A is generated, we invoke a SAT solver to evaluate whether after users in A are removed, the remaining users still satisfy the requirements. If the answer is "yes", then we would get back a solution, which consists of d sets of users such that each set covers all permissions. Let E be the set of all users that appear in any of the d sets; we call E a solution set for A. Let U be the set of all users in UP. Clearly, $E \subseteq U - A$. If E contains fewer users than $U - A$, then it is possible that when another set A' is generated we have $E \cap A' = \emptyset$. When this happens, we know that we do not need to consider A', as E is also a solution set for A'. Based on this observation, one can store the solution sets returned by the SAT solver, and use them to check whether absent sets generated later need to be considered.

Handling the case that $t \neq \infty$ The reduction to SAT described above works only when $t = \infty$. To handle the case that $t \neq \infty$, we can use pseudo boolean constraints. In Pseudo-Boolean (PB) constraints, all variables take values of either 0 (false) or 1 (true). Constraints are linear inequalities with integer coefficients, for example, $2x + y + z \geq 2$ is a PB constraint. A disjunctive clause encountered in SAT is a special case of PB constraints; for example, $x \lor y \lor z$ is equivalent to $x + y + z \geq 1$. Many SAT solvers also support PB constraints. In particular, the SAT solver we use, SAT4J [4], supports PB constraints.

When $t \neq \infty$, we can translate the problem of determining whether d sets of size no more than t exist to the satisfiability problem with PB constraints. The translation works as follows. For each user u_i and each integer j from 1 to d, we have a propositional variable $v_{i,j}$. This variable is true if the i'th user is assigned to the j'th group. Then we have the following three kinds of constraints. The first kind ensures that all permissions are covered:

For each permission p in P, let $u_{i_1}, u_{i_2}, \cdots, u_{i_x}$ be the users who are authorized for the permission p. Then, for each j from 1 to d, we add the constraint $v_{i_1,j} + \cdots + v_{i_x,j} \geq 1$. There are $|P| \cdot d$ of such constraints. The second kind ensures that each set contains at most t users: for each j from 1 to d, we add the constraint $v_{0,j} + v_{1,j} + \cdots + v_{n-1,j} \leq t$. There are d such constraints. The third kind ensures that no user is selected in two groups: For each user i, add the constraint $v_{i,1} + \cdots + v_{i,d} \leq 1$. There are n such constraints, where n is the number of users.

4.2 Implementation and Evaluation

We have implemented the algorithm described in Section 4.1, and performed several experiments using randomly generated instances. Our goals of implementing the algorithm and performing these experiments are to understand the effectiveness of the pruning techniques developed in Section 4 and to understand how well the algorithm scales with different parameters.

The implementation of our algorithm was written in Java. We use SAT4J [4], an open source satisfiability library in Java. Experiments were carried out on a PC with an Intel Pentium 4 CPU running at 3.2 GHz with 1 GB of RAM running Microsoft Windows XP Professional 2002. Our time units are milliseconds. In this subsection, n, s and d denote the number of total users, the number of users that may be absent, and the number of disjoint sets of users we seek after the removal a set of users respectively. The methodology that we use in generating testing instances is explained in Appendix B.

Our experimental results show that our algorithm is able to solve nontrivial size of RCP instance in reasonable amount of time. For example, our implementation spent around 500ms on instances with 60 to 100 users, 10 permissions, $s = 3$ and $d = 6$; and around 2 seconds on instances with 80 to 100 users, 10 permissions, $s = 3$ and $d = 4$. We discuss our observations from the experiments in the rest of this section.

The algorithm scales reasonably well with n when d is small; however when d is over about 8, the algorithm stops scaling. The running time of the algorithm depends on the total number of absent sets that need to be examined and the time it takes for the SAT solver to solve each SAT instance. The time spent in the SAT solver is greatly influences by d, which is the number of distinct sets of users we seek after an absent set of users is removed. In Figure 3, we plot the running time of the algorithm for cases in which the instance is true, for increasing n (number of users) and d. We observe that up to a particular value for d (7 in this case), the algorithm scales well as n increases. For example, for $n = 100$ and $d = 6$, the algorithm takes only about 1.7 seconds. However, as d becomes larger, the algorithm stops scaling. A major reason is that, as d increases beyond a certain threshold (8 in our case), each SAT instance that is generated is time-consuming for the SAT solver to solve. Consequently, lots of time is spent in the SAT solver, which results in increase of running time of our algorithm. This threshold of around 8 seems to hold for many other experiments we have performed.

Static pruning is very effective Table 1 shows the effect of static pruning for increasing values of n (number of users) and s (size of absent sets). While static pruning always reduces the number of absent sets to be considered, its effect is especially pronounced for large values of n and s. For example, for $n = 100$ and $s = 8$, we see a reduction of 7 orders of magnitude in the number of absent sets that need to be considered. We point out also that the effect of static pruning is increasingly pronounced for larger values of n

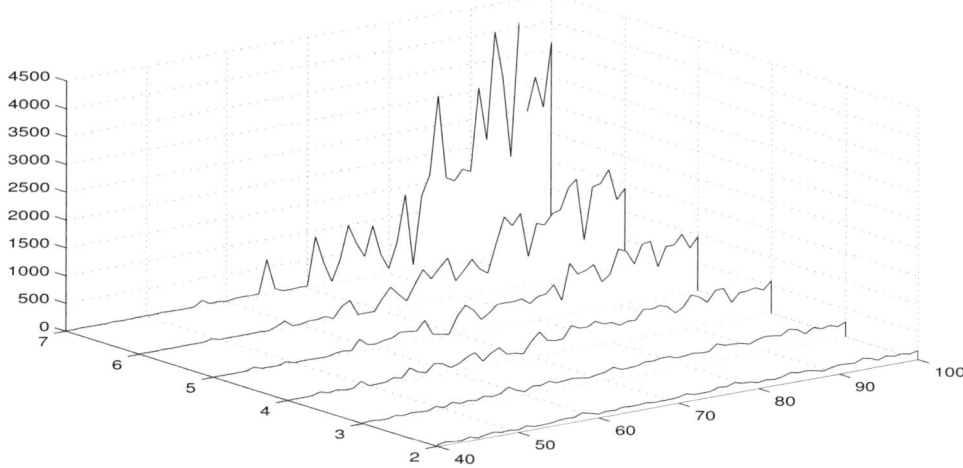

Figure 3: This graph shows the effect on running time (in milliseconds) as the number of users n and the number of disjoint sets d increase. The size of absent sets is 3 and there are 10 permissions. The value of n increases from 40 to 100 and the value of d increases from 2 to 7. For smaller values of n (say, $n = 40$), increasing d has almost no effect on the running time so long as d is no larger than 7. The reason is that relatively few absent sets need to be considered. However, for larger values of n (say, $n = 90$), increasing d has a pronounced effect on the running time.

when s is constant. For example, for $s = 6$ and increasing n from 40 to 100, the reduction in the number of absent sets that need to be considered improves from a difference of 3 orders of magnitude to 6. For a fixed number of permissions (10 in this case), occurrences of dominance may increase as n increases (because there are likely more users who have a lot of permissions that dominate other users). This explains why the number of absent sets after pruning is fewer, for example, for $s = 4, n = 100$ (640 absent sets) than for $s = 4, n = 40$ (1042 absent sets).

Dynamic pruning is not effective The basic idea of dynamic pruning is to store, for each absent set A, the set E of users that are used in the solution returned by the SAT solver. When encountering another absent A', we check whether $A' \cap E = \emptyset$; if so, then we can skip A'. Somewhat unexpected for us, it turns out that dynamic pruning is not effective. In fact, using dynamic pruning is often slower than without dynamic pruning. After analyzing this effect, the reason became clear. Dynamic pruning adds additional processing time for each absent set. It is cost effective only when invoking the SAT solver is expensive so that it is worthwhile to take more effort to further decrease the number of absent sets needed to be examined. However, when invoking the SAT solver is expensive, i.e., when it is difficult to find d mutually disjoint sets of users such that each set has all permissions, the solution returned by the SAT solver likely includes all users that are not in A, which means that that solution set will not be able to prune any other absent set.

5. ON THE CONSISTENCY OF RESILIENCY AND SEPARATION OF DUTY POLICIES

As we have discussed in the introduction, resiliency policies are a natural complement to traditional safety policies in access control. Consequently, a question arises regarding the consistency of resiliency policies with other policies. In this section, we explore the co-existence of resiliency policies with static separation of duty (SSoD) policies.

The intent of an SSoD policy is to preclude any group of users from possessing too many permissions. We adopt the concrete formulation of such policies from Li et al. [14]. An SSoD policy is of the form $\mathsf{ssod}\langle P, k \rangle$, where P is a set of permissions and $1 < k \le |P|$ is an integer. An access control state satisfies the policy if there exists no set of fewer than k users that together possess all permissions in P. In the policy $\mathsf{ssod}\langle P, k \rangle$, P denotes the set of permissions that are needed to perform a sensitive task, and k denotes the minimal number of users that are allowed to perform the task. If the policy is satisfied, then no set of $k - 1$ users can together perform the task, because they do not have all the permissions; thus at least k users need to be involved, achieving the goal of separation of duty. For example, the policy $\mathsf{ssod}\langle \{p_1, p_2\}, 2 \rangle$ means that no single user is allowed to have both p_1 and p_2.

In many cases, it is desirable for an access control system to have both resiliency and SSoD policies. If an access control system has only resiliency policies, then they can be satisfied by giving all permissions to all users, resulting in each single user can perform any task. Similarly, if an access control system has only SSoD policies, then they can be satisfied by not giving any permission to any user, resulting in no task can be performed. It is clear that neither kind of policies by itself is sufficient to capture the security requirements. When both kinds of policies coexist, safety and functionality requirements can all be specified.

Due to their opposite focus, resiliency policies and separation of duty policies can conflict with each other. For example, a separation of duty policy $\mathsf{ssod}\langle P, 2 \rangle$ requires that no user possess all permissions in P. A resiliency policy $\mathsf{rp}\langle P, s, d, 1 \rangle$ requires the existence of a user that has all permissions in P. Clearly, the two policies cannot be satisfied simultaneously. We formally define our notion of consistency amongst such policies in the following definition.

Definition 6. Given a set F of resiliency and separation of duty policies, the policies in F are *consistent* if and only if there exists an access control state UP such that UP satisfies every policy in F. Determining whether F is consistent is called the *Policy Consistency Checking Problem* (PCCP).

s \ n	40	60	80	100
2	45	28	40	36
	780	1770	3160	4950
4	1042	694	684	640
	9.1×10^4	4.9×10^5	1.6×10^6	3.9×10^6
6	9713	9248	5310	6653
	3.8×10^6	5.0×10^7	3.0×10^8	1.2×10^9
8	7.7×10^6	6.1×10^4	1.2×10^5	8.7×10^4
	7.7×10^7	2.6×10^9	2.9×10^{10}	1.9×10^{11}

Table 1: A table that shows that static pruning is effective. The columns are values for n (number of users) and rows are values for s (size of the absent set). The number of permissions is 10. For each cell in the table, the entry above the dotted line is the number of absent sets that need to be considered with static pruning in effect, and the number below the dotted line is the number of absent sets to be considered without pruning (i.e., $\binom{n}{s}$). We observe that the effect of static pruning is pronounced, especially for large values of n and s. There is always an improvement of at least 1 order of magnitude, and when $n = 100$ and $s = 8$, there is an improvement of 7 orders of magnitude.

The following lemma asserts that the actual value of s and d in a resiliency does not affect its compatibility with SSoD policies. This enables us to replace all resiliency policies in the form of $\mathsf{rp}\langle P_i, s_i, d_i, t_i\rangle$ in F with the special form $\mathsf{rp}\langle P_i, 0, 1, t_i\rangle$ when studying $\mathsf{PCCP}\langle F\rangle$. This greatly simplifies the problem.

Lemma 10. *F is a set of policies and $R = \mathsf{rp}\langle P, s, d, t\rangle \in F$. Let $R' = \mathsf{rp}\langle P, 0, 1, t\rangle$ and $F' = (F - \{R\}) \cup \{R'\}$. F is consistent if and only if F' is consistent.*

PROOF. It is clear that if F is consistent then F' is consistent. In the following, we prove that if F' is consistent then F is consistent. Assume that state UP' satisfies all policies in F'. UP' satisfying R' implies that there is a set U of no more than t users together have all permissions in P. We then construct a new state UP by adding $s + d - 1$ copies of all users in U to UP'. Note that adding copies of existing users in UP' will not lead to violation of SSoD policies in F'. In this case, UP satisfies R plus all policies in F'. In other words, UP satisfies all policies in F and F is consistent.

The following theorem gives the computational complexity results about general cases of PCCP. Observe that the case with one SSoD policy and an arbitrary number of resiliency policies is **coNP**-hard, and the case with one resiliency policy and an arbitrary number of SSoD policies is **NP**-hard. Therefore, it is unlikely that the general case is in **NP** or in **coNP**; however, we show that the problem is in $\mathbf{NP^{NP}}$.

Theorem 11. *The computational complexities for PCCP are as follows:*

1. *PCCP $\langle 1, n\rangle$ is **coNP**-hard, where PCCP $\langle 1, n\rangle$ denotes the subcase that there is a single SSoD policy, and an arbitrary number of resiliency policies.*

2. *PCCP $\langle m, 1\rangle$ is **NP**-hard, where PCCP $\langle m, 1\rangle$ denotes the subcase that there is an arbitrary number of SSoD policies, and a single resiliency policy.*

3. *PCCP $\langle m, n\rangle$, i.e., the most general case of PCCP, is in $\mathbf{NP^{NP}}$.*

The proof for Theorem 11 is in Appendix C. It is of course possible that there are special cases of PCCP that are efficiently solvable. Detailed analysis of the tractable subcases of PCCP is beyond the scope of this paper.

6. RELATED WORK

To our knowledge, there is no prior work in resiliency policies in the context of access control. Prior analysis work in access control deals mostly with safety and security analysis, and separation of duty.

Simple safety analysis, i.e., determining whether an access control system can reach a state in which an unsafe access is allowed, was first formalized by Harrison et al. [9] in the context of the well-known access matrix model [8, 13], and was shown to be undecidable in the HRU model [9]. Following that, there have been various efforts in designing access control systems in which simple safety analysis is decidable or efficiently decidable, e.g., the take-grant model [17], the schematic protection model [22], and the typed access matrix model [24]. Koch et al. [11] considered safety in RBAC with the RBAC state and state-change rules posed as a graph formalism [12]. Li et al. [15] proposed the notion of security analysis which generalizes safety analysis; it was considered in the context of a trust management framework. Security analysis has since been considered also in the context of RBAC [16].

Separation of duty (SoD) has long existed in the physical world, sometimes under the name "the two-man rule", for example, in the banking industry and the military. To our knowledge, in the information security literature the notion of SoD first appeared in Saltzer and Schroeder [20] under the name "separation of privilege." Clark and Wilson's commercial security policy for integrity [2] identified SoD along with well-formed transactions as two major mechanisms of fraud and error control. Separation of Duty policies were also studied in [1, 3, 7, 10, 14, 18, 21, 23, 26].

Another related concept is availability policies in [15, 16], which asks whether a user always possesses certain permissions across state changes. In that work, checking whether an availability policy is satisfied in a state is straightforward; the challenges arises from the fact that the access control state may be changed by administrative operations, and the possible state space may be infinite. Unlike availability policies, resiliency policies such as the ones we consider in this paper do not specify a permission requirement on any individual user; rather, they specify requirements about tolerating absent users and the overall ability of groups of users to perform critical tasks. Consequently, resiliency policies are more powerful and checking whether a state satisfies a resiliency policy is a challenging problem in itself.

7. CONCLUSION AND FUTURE WORK

We have introduced the notion of resiliency policies in the context of access control systems. Unlike most existing work on policy analysis in access control, resiliency policies are about enabling access rather than restricting access. Resiliency policies are particularly useful when evaluating whether the access control configuration of a system is ready for emergency response. To the best of our knowledge, such resiliency policies have not been previously studied in access control.

We have shown that the problem of checking whether an access control state satisfies a resiliency policy in the general case is intractable (**NP**-hard), and is in the Polynomial Hierarchy (in $\mathbf{coNP^{NP}}$). We have shown also that several subcases of the problem remain intractable. Notwithstanding these intractability results, many instances that will be encountered in practice may be efficiently solvable. In an effort to seek an efficient solution for practical instances of the problem and to understand what the hard instances are, we have designed and implemented an algorithm for RCP. Our algorithm takes advantages of an effective static pruning approach and the existence of fast SAT solvers. Our experimental results have shown that the algorithm is capable to solve RCP instances of nontrivial sizes in a reasonable amount of time. We have also explored the co-existence of resiliency policies with static separation of duty (SSoD) policies. In particular, we have presented several computational complexity results on checking whether a set of resiliency policies and SSoD policies are consistent.

Open problems To our knowledge, this is the first work in access control research to clearly formulate properties on enabling access, rather than restricting access. Because this work opens up a new area, even though we have presented a number of results in this paper, many more interesting problems remain open. One fruitful area of future research lies in the interaction between resiliency policies and other policies. In the study of the consistency problem with SSoD policies and resiliency policies, we do not consider the total number of available users as a limiting factor. In practice, the number of users in any organization is bounded. This makes it harder to satisfy both resiliency policies (which require users to possess more permissions) and SSoD policies (which require users to possess fewer permissions). Hence, it would be interesting to consider the consistency problem with an upper bound on the number of users in the access control state.

In addition to resiliency and separation of duty policies, other kinds policies may exist. For example, an *assignment range policy* states that a set of permissions can be possessed only by a certain set of users. This may be motivated by the fact that not all users are qualified to receive these permissions. For example, the permission to install software on campus-wide network servers may be assigned only to qualified and authorized staff, and should not be given to others. The interaction among resiliency policies, SSoD policies, and assignment range policies is an interesting and challenging problem for future work.

Another open area lies in designing techniques for enforcing resiliency policies: if a state does not satisfy an existing set of policies, how do we alter the state to make it satisfy these policies? This problem seems to be particularly interesting in the context of Role-Based Access Control systems, where one changes the role assignments of users to satisfy existing policies. Another approach for achieving resiliency is to use delegation; that is, when a user is absent, some of his permissions can be automatically and temporarily assigned to one or more other users. However, we may require such delegation to satisfy other coexisting policies such as separation of duty.

Acknowledgement

This work is supported by NSF CNS-0448204 (CAREER: Access Control Policy Verification Through Security Analysis And Insider Threat Assessment), by Purdue Research Foundation, and by sponsors of CERIAS. We thank the anonymous reviewers for their helpful comments.

8. REFERENCES

[1] G.-J. Ahn and R. S. Sandhu. Role-based authorization constraints specification. *ACM Transactions on Information and System Security*, 3(4):207–226, Nov. 2000.

[2] D. D. Clark and D. R. Wilson. A comparision of commercial and military computer security policies. In *Proceedings of the 1987 IEEE Symposium on Security and Privacy*, pages 184–194. IEEE Computer Society Press, May 1987.

[3] J. Crampton. Specifying and enforcing constraints in role-based access control. In *Proceedings of the Eighth ACM Symposium on Access Control Models and Technologies (SACMAT 2003)*, pages 43–50, Como, Italy, June 2003.

[4] Daniel Le Berre (project leader). SAT4J: A satisfiability library for Java. URL http://www.sat4j.org/, Jan. 2006.

[5] D. Du, J. Gu, and P. M. Pardalos, editors. *Satisfiability Problem: Theory and Applications*, volume 35 of *DIMACS Series in Discrete Mathematics and Theoretical Computer Science*. AMS Press, 1997.

[6] M. R. Garey and D. J. Johnson. *Computers And Intractability: A Guide to the Theory of NP-Completeness*. W.H. Freeman and Company, 1979.

[7] V. D. Gligor, S. I. Gavrila, and D. F. Ferraiolo. On the formal definition of separation-of-duty policies and their composition. In *Proceedings of IEEE Symposium on Research in Security and Privacy*, pages 172–183, May 1998.

[8] G. S. Graham and P. J. Denning. Protection — principles and practice. In *Proceedings of the AFIPS Spring Joint Computer Conference*, volume 40, pages 417–429. AFIPS Press, May 16–18 1972.

[9] M. A. Harrison, W. L. Ruzzo, and J. D. Ullman. Protection in operating systems. *Communications of the ACM*, 19(8):461–471, Aug. 1976.

[10] T. Jaeger and J. E. Tidswell. Practical safety in flexible access control models. *ACM Transactions on Information and System Security*, 4(2):158–190, May 2001.

[11] M. Koch, L. V. Mancini, and F. Parisi-Presicce. Decidability of safety in graph-based models for access control. In *Proceedings of the Seventh European Symposium on Research in Computer Security (ESORICS 2002)*, pages 229–243. Springer, Oct. 2002.

[12] M. Koch, L. V. Mancini, and F. Parisi-Presicce. A graph-based formalism for RBAC. *ACM Transactions on Information and System Security*, 5(3):332–365, Aug. 2002.

[13] B. W. Lampson. Protection. In *Proceedings of the 5th Princeton Conference on Information Sciences and Systems*, 1971. Reprinted in ACM Operating Systems Review, 8(1):18-24, Jan 1974.

[14] N. Li, Z. Bizri, and M. V. Tripunitara. On mutually-exclusive roles and separation of duty. In *Proceedings of the 11th ACM Conference on Computer and Communications Security (CCS-11)*, pages 42–51. ACM Press, Oct. 2004.

[15] N. Li, J. C. Mitchell, and W. H. Winsborough. Beyond proof-of-compliance: Security analysis in trust management. *Journal of the ACM*, 52(3):474–514, May 2005. Preliminary version appeared in *Proceedings of 2003 IEEE Symposium on Security and Privacy*.

[16] N. Li and M. V. Tripunitara. Security analysis in role-based access control. In *Proceedings of the Ninth ACM Symposium on Access Control Models and Technologies (SACMAT 2004)*, pages 126–135, June 2004.

[17] R. J. Lipton and L. Snyder. A linear time algorithm for deciding subject security. *Journal of the ACM*, 24(3):455–464, 1977.

[18] M. J. Nash and K. R. Poland. Some conundrums concerning separation of duty. In *Proceedings of IEEE Symposium on Research in Security and Privacy*, pages 201–209, May 1990.

[19] C. H. Papadimitriou. *Computational Complexity*. Addison Wesley Longman, 1994.

[20] J. H. Saltzer and M. D. Schroeder. The protection of information in computer systems. *Proceedings of the IEEE*, 63(9):1278–1308, September 1975.

[21] R. Sandhu. Separation of duties in computerized information systems. In *Proceedings of the IFIP WG11.3 Workshop on Database Security*, Sept. 1990.

[22] R. S. Sandhu. The schematic protection model: Its definition and analysis for acyclic attenuating systems. *Journal of the ACM*, 35(2):404–432, 1988.

[23] R. S. Sandhu. Transaction control expressions for separation of duties. In *Proceedings of the Fourth Annual Computer Security Applications Conference (ACSAC'88)*, Dec. 1988.

[24] R. S. Sandhu. The typed access matrix model. In *Proceedings of the 1992 IEEE Symposium on Security and Privacy*, pages 122–136. IEEE Computer Society Press, May 1992.

[25] R. S. Sandhu, E. J. Coyne, H. L. Feinstein, and C. E. Youman. Role-based access control models. *IEEE Computer*, 29(2):38–47, February 1996.

[26] T. T. Simon and M. E. Zurko. Separation of duty in role-based environments. In *Proceedings of The 10th Computer Security Foundations Workshop*, pages 183–194. IEEE Computer Society Press, June 1997.

APPENDIX

A. BACKGROUND ON ORACLE TURING MACHINES AND POLYNOMIAL HIERARCHY

Oracle Turing Machines An oracle Turing machine, with oracle L, is denoted as M^L. L is a language. M^L can use the oracle to determine whether a string is in L or not in one step. More precisely, M^L is a two-tape deterministic Turing machine. The extra tape is called the oracle tape. M^L has three additional states: $q_?$ (the query state), and q_{yes} and q_{no} (the answer states). The computation of M^L proceeds like in any ordinary Turing machine, except for transitions from $q_?$. When M^L enters $q_?$, it checks whether the contents of the oracle tape are in L. If so, M^L moves to q_{yes}. Otherwise, M^L moves to q_{no}. In other words, M^L is given the ability to "instantaneously" determine whether a particular string is in L or not.

Polynomial Hierarchy The polynomial hierarchy provides a more detailed way of classifying NP-hard decision problems. The complexity classes in this hierarchy are denoted by $\Sigma_k \mathbf{P}, \Pi_k \mathbf{P}, \Delta_k \mathbf{P}$, where k is a nonnegative integer. They are defined as follows:
$$\Sigma_0 \mathbf{P} = \Pi_0 \mathbf{P} = \Delta_0 \mathbf{P} = \mathbf{P},$$
and for all $k \geq 0$,
$$\Delta_{k+1}\mathbf{P} = \mathbf{P}^{\Sigma_k \mathbf{P}},$$
$$\Sigma_{k+1}\mathbf{P} = \mathbf{NP}^{\Sigma_k \mathbf{P}},$$
$$\Pi_{k+1}\mathbf{P} = \mathbf{co\text{-}}\Sigma_{k+1}\mathbf{P} = \mathbf{coNP}^{\Sigma_k \mathbf{P}}.$$

Some classes in the hierarchy are
$$\Delta_1 \mathbf{P} = \mathbf{P}, \Sigma_1 \mathbf{P} = \mathbf{NP}, \Pi_1 \mathbf{P} = \mathbf{coNP},$$
$$\Delta_2 \mathbf{P} = \mathbf{P}^{\mathbf{NP}}, \Sigma_2 \mathbf{P} = \mathbf{NP}^{\mathbf{NP}},$$
$$\Pi_2 \mathbf{P} = \mathbf{coNP}^{\mathbf{NP}}.$$

B. METHODOLOGY FOR GENERATING TESTING INSTANCES

Our goals of implementing the algorithm and performing experiments are to understand the effectiveness of the pruning techniques developed in Section 4 and to understand how well the algorithm scales with different parameters. To achieve such goals, We try to generate instances to approximate realistic instances. We generate instances for testing using combinations of the following approaches.

- *Purely Random*: For each permission p_i and user u_j, we assign p_i to u_j with a certain probability. The probability is an adjustable parameter which is called the *density* parameter.

- *With Constraints*: Often times, an access control system may include (explicit or implicit) constraints that restrict user-permission assignment. For example, there may be requirement that no user is authorized for permissions p_i and p_j at the same time. To model this aspect, mutual exclusion constraints among permissions are randomly generated. Two permissions are mutually exclusive if no user can be authorized for both permissions. The total number of pairs of permissions is $p(p-1)/2$. The number of constraints to be generated is determined by an adjustable parameter that specifies the ratio of the the constraints to $p(p-1)/2$. After the generation of constraints and user-permission assignment, if a user is assigned to two permissions that are mutually exclusive, we randomly remove one permission from the assignment.

- *Density Variation*: In situations where resiliency is an issue, it is likely that some permissions are assigned only to a small number of people. To model these situations, we assign different permissions with different densities. We have two parameters that specify the lower bound and the upper bound for the permission assignment densities respectively. The sequence of all permissions $p_1, \cdots, p_m$ will be assigned with nondecreasing density, with p_1 being assigned with the lower bound density and p_m with the upper bound density.

Finally, if a user is not assigned any permission, we randomly assign one permission to the user, so that we do not have a useless user in the generated instance.

C. PROOFS FOR THEOREM 11

Without loss of generality, we assume that for any static separation of duty policy $\mathsf{ssod}\langle P, k\rangle$, we have $k \leq |P|$. We also assume that in any resiliency policy $\mathsf{rp}\langle P, s, d, t\rangle$, we have either $t = \infty$ or $t \leq |P|$.

Lemma 12. PCCP $\langle 1, n\rangle$ *is* **coNP**-*hard, where* PCCP $\langle 1, n\rangle$ *denotes the subcase that there is a single SSoD policy, and an arbitrary number of resiliency policies.*

PROOF. We reduce the **NP**-complete SET COVERING problem [19] (also referred to as MINIMUM COVERING problem in [6]) to the complement of PCCP. In SET COVERING, we are given a set $X = \{e_1, \cdots, e_m\}$, n subsets of X: $X_1, \ldots, X_n$, and a budget b, and need to determine whether the union of b subsets is the same as X. Given an instance of the SET COVERING problem, we construct one SSoD policy $S = \mathsf{ssod}\langle P, b + 1\rangle$ and b rp policies $R_i = rp\langle P_i, 0, 1, 1\rangle$ ($1 \leq i \leq b$), where $P = \{p_1, \cdots, p_m\}$ corresponds to X and $P_i = \{p_j \mid e_j \in X_i\}$ corresponds to X_i. Let $F = \{S, R_1, \cdots, R_n\}$. In the following, we prove that F is inconsistent if and only if the answer to the SET COVERING problem is "yes".

On the one hand, if F is inconsistent, there does not exist any state that satisfies all polices in F. In other words, if a state satisfies all resiliency policies in F, there exists no more than b users in the state who together have all permission in P. Let UP be a state with n users $u_1, \cdots, u_n$ such that $(u_i, p_j) \in UP$ if and only if $p_j \in P_i$. It is clear that UP satisfies all resiliency policies in F, and hence there exist no more than b users together have all permissions in P. In other words, there exist no more than b elements in $\{P_1, \cdots, P_n\}$ whose union is P. Thus, the answer to the set covering problem is "yes".

On the other hand, if the answer to the set covering problem is "yes", then there exist no more than b elements in $\{P_1, \cdots, P_n\}$ whose union is P. For any state UP that satisfies all resiliency policies in F, let U be the set of users that satisfy at least one resiliency policy. $u \in U$ if and only if there exists P_i such that u has all permissions in P_i. In this case, there exist no more than b users in U who together have all permissions in P. Hence, UP does not satisfy S, which implies that no state satisfies all policies in F.

Lemma 13. PCCP $\langle m, 1\rangle$ *is* **NP**-*hard, where* PCCP $\langle m, 1\rangle$ *denotes the subcase that there is an arbitrary number of SSoD policies, and a single resiliency policy.*

PROOF. We reduce the **NP**-complete SET SPLITTING problem to PCCP$\langle F\rangle$. In the SET SPLITTING problem, we are given a set $X = \{e_1, \cdots, e_n\}$, m subsets of X: $X_1, \ldots, X_m$, and need to determine whether there exist Y_1 and Y_2 such that $Y_1 \cup Y_2 = X$ and there does not exist X_i ($1 \leq i \leq m$) such that $X_i \subseteq Y_1$ or $X_i \subseteq Y_2$. Given an instance of the SET SPLITTING problem, construct a resiliency policy $R = \mathsf{rp}\langle P, 0, 1, 2\rangle$ and m SSoD policies $S_i = \mathsf{ssod}\langle P_i, 2\rangle$ ($1 \leq i \leq m$), where $P = \{p_1, \cdots, p_n\}$ corresponds to X and $P_i = \{p_j \mid e_j \in X_i\}$ corresponds to X_i. Let $F = \{R, S_1, \cdots, S_m\}$. In the following, we prove that F is consistent if and only if the answer to the SET SPLITTING problem is "yes".

On the one hand, if F is consistent, then there exists a state UP that satisfies all policies in F. UP satisfying R implies that there exist two users u_1 and u_2 in UP such that u_1 and u_2 together have all permissions in P. Furthermore, UP satisfying S_i implies that neither u_1 nor u_2 has all permissions in P_i. Let $Y_1 = \{e_i \mid (u_1, p_i) \in UP\}$ and $Y_2 = \{e_i \mid (u_2, p_i) \in UP\}$. We have $Y_1 \cup Y_2 = X$ and neither Y_1 nor Y_2 is a superset of any X_i. The answer to the set splitting problem is "yes".

On the other hand, if the answer to the set splitting problem is "yes", then such Y_1 and Y_2 exist. We construct a state UP containing only two users u_1 and u_2 such that $(u_i, p_j) \in UP$ ($1 \leq i \leq 2$) if and only if $p_j \in Y_i$. Since $Y_1 \cup Y_2 = X$, u_1 and u_2 together have all permissions in P. Furthermore, since there does not exist X_i such that X_i is a subset of Y_1 or Y_2, neither u_1 nor u_2 has all permissions in P_i, which implies that UP satisfies S_i. Therefore, UP satisfies all policies in F.

Lemma 14. *Let* $F = \{S_1, S_2, \cdots S_m, R_1, \cdots, R_n\}$, *where* $S_i = \mathsf{ssod}\langle P_i, k_i\rangle$ ($1 \leq i \leq m$) *and* $R_j = \mathsf{rp}\langle Q_j, s_j, d_j, t_j\rangle$ ($1 \leq j \leq n$). PCCP$\langle F\rangle$ *is in* $\mathbf{NP^{NP}}$.

PROOF. We construct a set of policies F' by replacing every R_i ($1 \leq i \leq n$) in F with $\mathsf{rp}\langle P_i, 0, 1, t_i\rangle$. From Lemma 10, F is consistent if and only if F' is consistent.

We construct a nondeterministic Oracle Turing machine M that makes use of an **NP** oracle machine to determine whether F' is consistent. M first nondeterministically selects an integer a such that $max(k_1, \cdots, k_m) \leq a \leq \Sigma_{i=1}^n |Q_i|$ and then generates a users. Note that at least $max(k_1, \cdots, k_m)$ users are needed to satisfy all SSoD policies in F', and at most $\Sigma_{i=1}^n |Q_i|$ users are needed to satisfy all resiliency policies in F'. (The state can have more than $\Sigma_{i=1}^n |Q_i|$ users, but in order to show that all resiliency policies in F' are satisfied, at most $\Sigma_{i=1}^n |Q_i|$ users need to be involved.) Then M constructs a state UP by nondeterministically assigning a subset of Q to u, where $Q = \bigcup_{i=1}^n Q_i$ is the set of all permissions appear in the resiliency policies. Next, M nondeterministically construct n sets $U_1, \cdots, U_n$ of users in UP, and then, for every $i \in [1, n]$, checks whether users in U_i together have all permissions in P_i and $|U_i| \leq t_i$. If the answer is "no", then M returns False. Finally, M invokes the **NP** oracle to to check whether UP violates any SSoD policy. (In order to prove that a state violates a static separation of duty policy $\mathsf{ssod}\langle P, k\rangle$, we just need to present a set of no more than k users in the state who together have all permissions in P. Therefore, checking whether a state violates an SSoD policy is in **NP**.) If the oracle machine answers "yes", M returns False. Otherwise, M returns True, which means that UP satisfies all policies in F' and hence F' is consistent. It is clear that M terminates in polynomial time if the oracle machine returns an answer instantaneously. Therefore, PCCP is in $\mathbf{NP^{NP}}$ in general.

Safety and Consistency in Policy-Based Authorization Systems

Adam J. Lee
adamlee@cs.uiuc.edu

Marianne Winslett
winslett@cs.uiuc.edu

Department of Computer Science
University of Illinois at Urbana-Champaign
Urbana, IL 61801

ABSTRACT

In trust negotiation and other distributed proving systems, networked entities cooperate to form proofs that are justified by collections of certified attributes. These attributes may be obtained through interactions with any number of external entities and are collected and validated over an extended period of time. Though these collections of credentials in some ways resemble partial system snapshots, these systems currently lack the notion of a consistent global state in which the satisfaction of authorization policies should be checked. In this paper, we argue that unlike the notions of consistency studied in other areas of distributed computing, the level of consistency required during policy evaluation is predicated solely upon the security requirements of the policy evaluator. As such, there is little incentive for entities to participate in complicated consistency preservation schemes like those used in distributed computing, distributed databases, and distributed shared memory. We go on to show that the most intuitive notion of consistency fails to provide basic safety guarantees under certain circumstances and then propose several more refined notions of consistency which provide stronger safety guarantees. We provide algorithms that allow each of these refined notions of consistency to be attained in practice with minimal overheads.

Categories and Subject Descriptors: C.2.4 [Distributed Systems]: Distributed applications; D.4.6 [Operating Systems]: Security and Protection—*access controls, authentication*;K.6.5 [Management of Computing and Information Systems]: Security and Protection

General Terms: Security

Keywords: Consistency, credentials, distributed proving, trust negotiation

1. INTRODUCTION

It is difficult to design flexible and secure authorization systems for environments in which trust relationships cannot be determined a priori. Two proposed authorization techniques for these types of environments are trust negotiation [4, 5, 12, 15, 16, 22, 24, 26] and distributed proving [3, 19, 25]. In these types of systems, participants collect certified credentials that describe their attributes, environmental conditions, and other state information from any number of external entities. These credentials can then be used when attempting to satisfy the authorization policies protecting sensitive resources in the system.

To some extent, the collection of credentials used to satisfy a given authorization policy acts as a partial snapshot of the system within which the policy is evaluated. This is an abuse of terminology, however, as this snapshot is collected over a variable-length window of time and thus may not actually represent a system state that ever existed; to avoid confusion, in this paper we will refer to these collections of credentials as *views*. Clearly, the correctness of an authorization decision depends on the validity and stability of the view used during policy evaluation. If we assume that each credential is stable (i.e., that the assertion stated in the credential remains true until its pre-ordained expiration time) then policy evaluation can be reduced to the problem of stable predicate evaluation on distributed snapshots [8]. However, because it is possible for credentials to become invalidated prematurely, this somewhat naive model of policy evaluation can erode the safety guarantees of the underlying authorization system. This is especially worrisome in trust negotiation and distributed proving, as interactions typically involve multiple rounds of interaction and credential exchange. Consider the following two examples:

Example 1. Bob works in the Finance department of Acme Petroleum Corporation (APeC), though he also spends part of his time "on loan" to the Petroleum Operations group helping manage their operational budget. While consulting for the operations group, Bob is given a PetrolOps group credential to allow him basic access to the operations group's resources. To aid his research, Bob wishes to access an online geological database provided by GeoTech, a third-party vendor. GeoTech allows operations group members at Department of Energy certified Oil Companies trial access to the database, provided that their company authorizes them to make purchases of over $10,000 (the cost of a department subscription to the database). Bob submits his PetrolOps group credential and APeC's OilCorp credential to GeoTech along with a policy stating that it must provide proof of membership in the Better Business Bureau to see his pur-

---

Permission to make digital or hard copies of all or part of this work for personal or classroom use is granted without fee provided that copies are not made or distributed for profit or commercial advantage and that copies bear this notice and the full citation on the first page. To copy otherwise, to republish, to post on servers or to redistribute to lists, requires prior specific permission and/or a fee.
CCS'06, October 30–November 3, 2006, Alexandria, Virginia, USA.
Copyright 2006 ACM 1-59593-518-5/06/0010 ...$5.00.

chase authorization. GeoTech verifies Bob's **PetrolOps** credential and ΛPcC's **OilCorp** credential and then sends Bob its **BBB** credential. As a consultant to the operations group, Bob is not authorized to make purchases of more than $200, so he should not be able to satisfy this policy. However, Bob can make purchases of this size for the Finance group. Bob then activates his **Finance** group credential (which invalidates his **PetrolOps** credential) and obtains a certified **Purchase** attestation authorizing him to make purchases of up to $10,000 dollars, which he then submits to GeoTech. GeoTech verifies this credential and grants Bob access to the database. The inconsistent system view used by the database leads to the permission of an undesirable access.

Example 2. Alice is a PhD student studying infectious diseases at State University. As part of her research, Alice wishes to access an outbreak incident database hosted by the Center for Disease Control. The CDC requires that academic users of this data be US citizens and members of an NSF-sponsored epidemiology project. To this end, Alice discloses her **Student** credential issued by State University and her **ProjectSpread** credential issued by the NSF. Alice considers her citizenship private, however, and requires that she first receive a certified privacy policy that she manually reviews prior to releasing her citizenship credential. Alice submits a policy to this effect to the CDC. The CDC verifies Alice's **Student** and **ProjectSpread** credentials and then discloses its certified **PrivacyPolicy** to Alice. Just then, Alice's research adviser notifies her that effective immediately, she will no longer be supported by the Spread project; the NSF then revokes her **ProjectSpread** credential. Alice then reviews the **PrivacyPolicy** submitted by the CDC and decides that it is safe to disclose her **USCitizen** credential. The CDC verifies this credential and permits Alice to access the requested data, as it did not detect that her project membership had been revoked prior to policy satisfaction.

The types of safety problems that emerged in the above examples occur because credentials are collected over a non-instantaneous window of time. In general, credential and policy instabilities can arise from one or more of the following four causes. First, the *natural expiration* of a credential can cause problems if a previously-valid credential expires before other required credentials can be validated. Second, *inter-credential dependencies* can give rise to problems if, for example, the activation of a new role causes the revocation of a previously activated role (as in Example 1). Third, an *external event* might cause the invalidation of a certain credential after it is validated, but prior to the entire policy being satisfied. For example, the removal of Alice from the Spread project in Example 2 caused credential revocation. Lastly, an *unstable environment* could cause instability if the policy is predicated on some aspect of the environment, such as the time of day or occupancy status of a room.

To the best of our knowledge, the problem of enforcing view consistency in trust negotiation and distributed proving systems has not been discussed elsewhere in the literature. Though similar to the consistency problems studied in distributed systems [21], distributed databases [7], and distributed shared memory [1], it is in many ways their dual. In these previous works, ensuring a consistent global state has been the concern of both data providers and users, as many entities can update the values of data fields replicated at a number of sites; this provides all parties with the incentive to cooperate. However, since a credential revocation can be made only by the issuer of that credential (and thus consistent update sequences can be attained trivially), the problem studied in this paper becomes the concern only of data consumers. In fact, the degree to which each data consumer is concerned with this problem may even vary based on the criticality of the policy being evaluated. For instance, a hardware store offering a discount to students of a particular university will probably not be concerned if a student ID credential is revoked after it has been issued for the semester, much less if it is revoked during a policy evaluation; an electronic door lock protecting access to expensive laboratory equipment at the university would care, however. Heavyweight solutions that require the cooperation of groups of certificate authorities (CAs) and users are not suitable, as the consistency property required will vary from user to user and preserving the autonomy of entities in the open system is of the utmost importance.

In this paper, we make several contributions regarding the level of safety attainable when evaluating policies in authorization systems that employ trust negotiation or other forms of distributed proving. To the best of our knowledge, we present the first formalization of the view consistency problem for trust negotiation and distributed proving systems and show how naive approaches to policy evaluation can lead to the permission of undesirable accesses to system resources in the face of prematurely invalidated credentials (Section 2). We then define several levels of credential view consistency, each of which provides different guarantees on the types of inappropriate access conditions that can be prevented (Section 3). We provide algorithms that can be incorporated into existing trust negotiation and distributed proving systems to attain these levels of consistency and assert the correctness of each algorithm (Section 4). We also demonstrate other desirable characteristics of these algorithms, including the fact that they require only minimal cooperation between the users engaged in the the trust negotiation or distributed proving protocol and no cooperation between groups of CAs or other users (Section 5). Finally, we comment on previous related work (Section 6) and examine potential areas for future work (Section 7).

2. ASSUMPTIONS AND DEFINITIONS

In this section, we present our assumptions regarding the open systems in which trust negotiation and distributed proving protocols are used. We then formally describe the problem of determining the consistency level of a system view used to evaluate an authorization policy.

2.1 System Model

An open system consists of a possibly infinite set $\mathcal{E}$ of entities, each of which is a resource provider, client, or both. *Resource providers* are entities who wish to offer resources or services to other entities in the system, while *clients* are entities that access the functionality offered by resource providers. Resource providers may wish to enforce authorization checks on the resources or services that they provide. Since the lack of pre-existing trust relationships in open systems hinders the use of traditional identity-based solutions to this problem, trust negotiation or distributed proving approaches to authorization will be used.

We place no limitations on the temporal duration of a trust negotiation or distributed proving session other than those imposed by the underlying protocol. For example, many trust negotiation protocols halt if no measurable progress is made during a particular round of the negotiation [15, 26]; we do not prevent this, nor do we require any such constraints be in place. We assume that the credentials used by an entity during the execution of one of these protocols may be obtained dynamically at runtime. This assumption allows portions of a distributed proof to be "outsourced" to other entities (as in [3, 19, 25]) and permits entities to acquire new attribute certificates while a trust negotiation session is in progress. These assumptions indicate that the collection of credentials used as the view in which an authorization policy is satisfied may be composed of the observations of an arbitrary number of entities and be collected over a variable-width window of time.

We assume that the certified attribute and environmental state information used to satisfy trust negotiation policies or form distributed proofs will be issued by an arbitrary number of CAs that exist in the system. All credentials issued will have an expiration time but may also be revoked prematurely by the issuing CA (as was the case with Alice's ProjectSpread credential in Section 1). In the remainder of this paper, we will denote the set of all credentials by $\mathcal{C}$. Given a credential $c \in \mathcal{C}$, we denote by $\alpha(c)$ the earliest time at which the issuing CA would possibly consider c to be valid. In the case of X.509 certificates [10], $\alpha(c)$ would be the time indicated in the "Not Before" field of the certificate; if no such field exists, then $\alpha(c)$ indicates the issue time of the credential. Similarly, we denote the expiration time of a credential c by $\omega(c)$. We assume that once a credential is revoked, it will never again become valid. Since only the issuing CA may revoke a credential, each CA can ensure that an omniscient view of the credentials that it has issued remains consistent at all times. We assume that each CA offers an online method that allows any entity to check the current status of a particular credential issued by the CA. This functionality could be provided through the Online Certificate Status Protocol (OCSP) [20] or by an online CA such as COCA [27].

2.2 Problem Definition

Prior to accepting a given credential as evidence that can be used to satisfy some portion of an authorization policy, the policy evaluator must first verify that the credential is valid. In this paper, we are concerned with two types of credential validity: syntactic and semantic.

DEFINITION 1 (SYNTACTIC VALIDITY). *A credential c is syntactically valid if the following conditions hold: (i) it is formatted properly, (ii) it has a valid digital signature, (iii) the time $\alpha(c)$ has passed, and (iv) the time $\omega(c)$ has not yet passed.*

DEFINITION 2 (SEMANTIC VALIDITY). *A credential c issued at time t_i is semantically valid at time t if an online method of verifying c's status indicates that c was not revoked at time t' and $t_i \leq t \leq t'$.*

Informally, if a credential is syntactically valid, it is well-formed. The semantic validity of a credential at a given time implies that the credential has not been revoked prior to that time; that is, the credential issuer asserts that the meaning of the credential is still valid. To ground these definitions with a real-world example, in the case of credit card validation, verifying syntactic validity involves checking that the signature on the back of the card matches the signature on the charge slip, the card has an appropriate issuer logo on the front, and the expiration date has not passed. Semantic validation occurs when the credit card clearinghouse authorizes a transaction. Note that in the degenerate case that a credential is assumed to be a stable assertion, syntactic validity implies semantic validity. We now define the more general concept of validity and derive two propositions and a corollary that will be useful later in the paper.

DEFINITION 3 (VALIDITY). *A credential c is valid at time t if it is syntactically and semantically valid at time t.*

PROPOSITION 1. *If a credential c is found to be syntactically valid at a time t' such that $\alpha(c) \leq t' < \omega(c)$, then c is syntactically valid at all times t where $\alpha(c) \leq t < \omega(c)$.*

PROPOSITION 2. *If a credential c is semantically valid at a time $t' \geq \alpha(c)$, then c is semantically valid at all times t where $\alpha(c) \leq t \leq t'$.*

COROLLARY 1. *If a credential c is valid at a time t' such that $\alpha(c) \leq t' < \omega(c)$, then c is valid at all times t where $\alpha(t) \leq t \leq t'$.*

As was observed earlier, each credential collected by an entity during a trust negotiation or distributed proving protocol constitutes a piece of evidence attesting to a small portion of the global state of the network. During a trust negotiation or the construction of a distributed proof, these pieces of evidence are collected over time and used to incrementally satisfy a given authorization policy. We now more precisely define one entity's *view* of the system in terms of the credentials acquired during a particular trust negotiation or distributed proving session.

DEFINITION 4 (CREDENTIAL STATE). *Let the set T contain all possible timestamps and the null value $\bot$. The state of a credential c as observed by an entity e is defined as $s_c^e = \langle c, r, syn, sem_v, sem_i \rangle \in \mathcal{C} \times (T \setminus \{\bot\}) \times \mathbb{B} \times T \times T$. The value r indicates the local time at which c was received by e. The boolean value syn is true if c is syntactically valid, false otherwise. The values sem_v and sem_i denote the most recent time that c was verified to be semantically valid and the first time that c was found to be semantically invalid, respectively. If the semantic validity of c has not yet been checked, both sem_v and sem_i will be set to $\bot$, otherwise at least one of them will contain a valid timestamp from the set $T \setminus \{\bot\}$. We use $\mathcal{S}$ to denote the set of all possible credential state tuples. Throughout this paper, we will use dot notation to access fields of these state tuples (e.g., $s_c^e.r$ represents the receipt time of the credential whose state is stored in s_c^e).*

DEFINITION 5 (VIEW). *A set of credential states observed by an entity e is called one of e's views of the system. A view contains at most one credential state tuple for any particular credential c.*

Given the above definitions, we now have a precise vocabulary for describing each entity's knowledge about the state of the system. Since this state information is gathered over time, it cannot be considered to be a precise snapshot of the global state and thus the consistency of an entity's view of the system becomes important to consider.

DEFINITION 6 (RELEVANCE). *A credential c is considered* relevant *to a policy P by entity e at time t if e has received c and considers the satisfaction of P in some way dependent on c at time t. Given a view V_e, $V_e^{P,t}$ is the subset of V_e containing state information for credentials that e considers relevant to P at time t.*

DEFINITION 7 (VIEW CONSISTENCY). *A view $V_e^{P,t}$ is ϕ-consistent if $V_e^{P,t}$ satisfies a predicate ϕ that places temporal constraints on the times at which e observes the validity of each credential c whose state information is stored in $V_e^{P,t}$.*

Definition 6 is very subtle, as it can vary from user to user. For instance, a naive user might consider every credential that she has ever received to be relevant to a policy P, while another user might only consider credentials explicitly mentioned in P to be relevant. Further, the set of credentials considered relevant to a policy P by a single user might change over time. For example, if Alice is evaluating the policy $P = c_1 \wedge (c_2 \vee c_3)$, she may initially consider c_1 and c_2 relevant to P and determine whether a consistent view can be constructed using these credentials. If this fails, then she may decide that c_1 and c_3 are relevant to P and again attempt to construct a consistent view. Consistency is fundamentally tied to the concept of relevance by Definition 7 and can thus be undermined by a faulty interpretation of relevance (for instance, by assuming that nothing is relevant to P). At a minimum, entities should consider the set of credentials used to satisfy P to be relevant to P and may also include other credentials in this set (for instance, credentials used to satisfy the release policies protecting credentials disclosed during the authorization protocol invoked to satisfy P). Resource providers have the autonomy and local knowledge necessary to decide which credentials are relevant at each moment and should thus be subjected to consistency requirements.

2.3 Practical Considerations for Consistency Enforcement

In this paper, we focus on limiting the unexpected behaviors that trust negotiation and distributed proving systems may manifest as a result of inconsistent views. To this end, we define several enforceable notions of view consistency, discuss the guarantees provided by each, and provide algorithms to attain these levels of view consistency in practice. In proposing mechanisms for view consistency enforcement, we will keep several high-level requirements in mind.

Loose clock synchronization A minimal level of clock synchronization is necessary, as otherwise the expiration times stored in credentials could not be reliably interpreted. However, we cannot assume that clocks are closely synchronized (e.g., seconds).

Minimal cooperation View consistency is a concern *only* for the policy evaluator. We cannot assume that groups of CAs, groups of CAs and users, or large groups of users will be willing to cooperate, as there is no incentive for this.

Minimal impact to existing protocols Trust negotiation and distributed proving have been active areas of research over the course of the last several years. As such, view consistency enforcement should require minimal changes to existing trust negotiation and distributed proving protocols.

In the remainder of this paper, our discussion proceeds bearing these requirements in mind. In Section 5 we discuss how our solutions for enforcing view consistency satisfy these requirements.

3. LEVELS OF CONSISTENCY

In this section, we present four increasingly-powerful levels of view consistency. We show that the guarantees afforded by each of these consistency levels can be strengthened if assumptions can (safely) be made about which of the above reasons for credential invalidation can be expected to occur during the course of the authorization protocol. This indicates that like many other aspects of trust negotiation and distributed proving, the choice of consistency level is likely to be a strategic choice made independently by each protocol participant. We defer all discussion pertaining to unstable environments until Section 5.

3.1 Incremental Consistency

To most people, the idea of satisfying a policy is straight forward. This usually entails presenting the evidence required to justify each clause of the policy. For instance, if Alice wishes to cash a check and is asked for two forms of ID, she could, for example, produce a driver's license and a passport during her transaction with the bank teller. The teller can verify that both IDs show Alice's picture and list the same home address and thus be reasonably satisfied that Alice is indeed who she says that she is. The teller is convinced that her view of the "system" is consistent because Alice could produce valid instances of the required documents during the course of their interaction. We call this intuitive notion of consistency *incremental consistency*. To formally define incremental consistency, we first define the predicates $checked : \mathcal{S} \to \mathbb{B}$ and $\phi_{inc} : 2^{\mathcal{S}} \to \mathbb{B}$.

$$checked(s) \equiv (s.syn = true) \wedge (s.sem_v \neq \bot) \quad (1)$$
$$\phi_{inc}(V) \equiv \forall s \in V : checked(s) \wedge (\alpha(s.c) \leq s.r \leq s.sem_v) \quad (2)$$

The predicate $checked(s)$ is satisfied if and only if the syntactic validity of $s.c$ has been verified and $s.c$ was ever observed to be semantically valid. The predicate $\phi_{inc}(V_e)$ is satisfied if and only if each credential in the view V_e was valid at the point that it was received by e. Note that Corollary 1 is used when computing the endpoints of each credential's observed validity period. Thus, the formal definition of incremental consistency is as follows.

DEFINITION 8 (INCREMENTAL CONSISTENCY). *A view $V_e^{P,t}$ is incrementally consistent iff $\phi_{inc}(V_e^{P,t})$ is true.*

Incremental consistency works for Alice and the bank teller, as it is exceedingly unlikely that Alice's driver's license or passport will be revoked or become invalid during their transaction. In addition to being intuitively useful, incremental consistency is also widely used in practice. Current trust

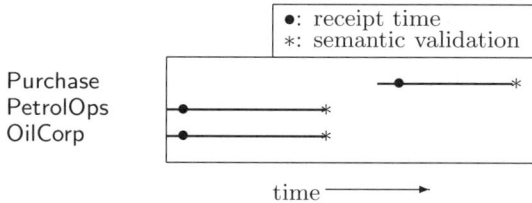

Figure 1: An incrementally consistent view.

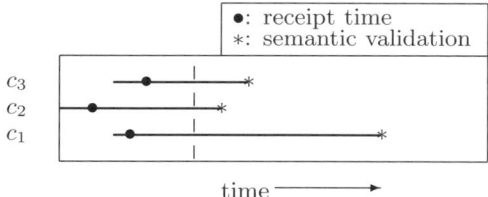

Figure 2: An internally consistent view.

negotiation prototypes (e.g., [4, 5, 12, 24]) implement incremental consistency by validating credentials as they are received. This approach to credential validation is also discussed in many papers that present protocols and strategies for trust negotiations and distributed proving that, to the best of our knowledge, have not yet been implemented (e.g., [6, 15, 22, 25], to name a few).

Incremental consistency works especially well when authorization policies are stable predicates. A stable predicate is a condition that remains true once it becomes true. Example stable predicates include "Alice has paid her 2005 income taxes" and "process X has terminated." If all relevant user attributes and environmental conditions are stable, then incremental consistency allows us to conclude that all credentials used to satisfy a given policy were simultaneously valid at the time of policy satisfaction. This, of course, assumes that we verify that no credential expired naturally before the final decision was made.

If policy predicates are not stable, however, incremental consistency cannot guarantee that all relevant credentials were ever valid simultaneously. For example, recall Example 1 presented in Section 1. Figure 1 shows GeoTech's view of Bob's credentials in this system, where the validity periods of each credential are indicated with horizontal lines. GeoTech never observed Bob's PetrolOps and Purchase credentials to be valid simultaneously. With inter-credential dependencies, such as that between Bob's PetrolOps and Finance credentials, incremental consistency is not always a good choice.

Although incremental consistency is the only form of view consistency supported by existing trust negotiation prototypes, we believe that this is only because until now, the issue of view consistency has not received any attention. The trust negotiation and distributed proving literature is full of examples motivating the use of these systems in grid computing, dynamic coalitions, and ubiquitous computing. These environments are all highly dynamic and, in some cases, could involve the use of mutually-exclusive roles and access rights; under these conditions incremental consistency is likely to be unsatisfactory. We now present three stronger notions of view consistency that are easily enforceable in practice and discuss the guarantees that each provides.

3.2 Internal Consistency

In this section, we define and discuss a stronger notion of view consistency that we will call *internal consistency*. Informally, if an authorization decision is made using an internally consistent view, then all credentials relevant to the authorization decision were valid *simultaneously* at some point in time during the authorization protocol. To formally define internal consistency, we first define the functions $start : 2^S \to T$ and $end : 2^S \to T$, and the predicate $\phi_{int} : 2^S \to \mathbb{B}$.

$$start(V) = min(\{s.r \mid s \in V\}) \quad (3)$$
$$end(V) = max(\{s.r \mid s \in V\}) \quad (4)$$
$$\begin{aligned}\phi_{int}(V) \equiv\ & (\forall s \in V : checked(s)) \\ & \wedge (max(\{\alpha(s) \mid s \in V\}) < \\ & \quad min(\{s.sem_i \mid s \in V\})) \\ & \wedge (max(\{\alpha(s) \mid s \in V\}) < end(V)) \\ & \wedge (min(\{\omega(s) \mid s \in V\}) > start(V))\end{aligned} \quad (5)$$

The function $start(V)$ is the earliest local time at which a credential in V was received; similarly, $end(V)$ is the latest local time at which a credential in V was received. For a given view, V, these functions effectively bound the duration of the interactive portion of the associated authorization protocol. The predicate ϕ_{int} holds true if and only if (i) each credential in the view was at one point observed to be valid, (ii) the last credential to become valid does so before the minimum known endpoint of any credential's validity period, (iii) the last credential to become valid does so before the end of the authorization protocol, and (iv) the minimum known endpoint of any credential's validity period occurs after the start of the authorization protocol.

DEFINITION 9 (INTERNAL CONSISTENCY). *A view $V_e^{P,t}$ is internally consistent iff $\phi_{int}(V_e^{P,t})$ is true.*

Internal consistency does not imply that all relevant credentials used to satisfy a policy are valid simultaneously at the moment the policy is decided to be satisfied. Rather, it implies that all relevant credentials are valid simultaneously at *some* point during the authorization protocol. Given a graphic representation of an internally consistent view, one should be able to draw at least one vertical line that intersects each credential's validity interval (see Figure 2). If external events cannot cause the revocation of a credential, then all credentials in an internally consistent view can be shown to be valid at the time of policy satisfaction. However, should an external revocation occur, this is not the case. Recall Example 2, in which all of Alice's credentials were valid at the start of the authorization protocol, but due to the NSF's revocation of her ProjectSpread credential, they were not all valid at the time that the decision was made.

3.3 Stronger Levels of Consistency

In some cases, it might be desirable not only to have the guarantee that each relevant credential in a given view was valid simultaneously at *some* point during the authorization protocol, but rather that they were all valid simultaneously at the endpoint of the authorization protocol. In others, perhaps it is required that each relevant credential is valid

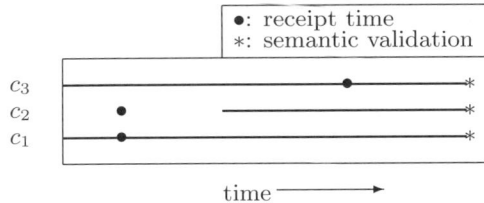

Figure 3: An endpoint consistent view.

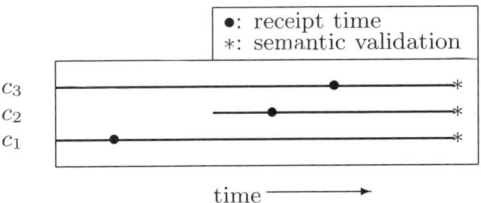

Figure 4: An interval consistent view.

from the time that it is received until the decision point of the authorization protocol. We will call these levels of consistency *endpoint consistency* and *interval consistency*, respectively (see Figures 3 and 4). These consistency levels are defined in terms of the $\phi_{end} : 2^S \to \mathbb{B}$ and $\phi_{interval} : 2^S \to \mathbb{B}$ predicates.

$$\phi_{end}(V) \equiv \forall s \in V : checked(s) \wedge \\ (\alpha(s.c) \leq end(V) \leq s.sem_v) \quad (6)$$

$$\phi_{interval}(V) \equiv \forall s \in V : checked(s) \wedge \\ (\alpha(s.c) \leq s.r \leq end(V) \leq s.sem_v) \quad (7)$$

DEFINITION 10 (ENDPOINT CONSISTENCY). *A view $V_e^{P,t}$ is endpoint consistent iff $\phi_{end}(V_e^{P,t})$ is true at the decision point t.*

DEFINITION 11 (INTERVAL CONSISTENCY). *A view $V_e^{P,t}$ is interval consistent iff $\phi_{interval}(V_e^{P,t})$ is true at the decision point t.*

Guaranteeing the interval consistency of the view used to evaluate an authorization policy clearly affords the policy evaluator a high level of confidence in the outcome of the authorization decision. In Sections 3.1 and 3.2, we showed that if certain assumptions could be made about the likelihood of inter-credential dependencies and external causes of revocation, that incrementally consistent and internally consistent views can actually be treated as endpoint consistent. Given the above definitions, it should be clear that the following proposition holds.

PROPOSITION 3. *An interval consistent view is also endpoint and incrementally consistent, and an endpoint consistent view is also internally consistent.*

One could imagine an extension of interval consistency requiring that all relevant credentials remain valid from the time that they are received until the end of the interaction between the two parties participating in the authorization protocol. That is, if Bob negotiates with GeoTech to gain access to their database (as in Example 1), GeoTech might want to guarantee that it could detect if any of Bob's credentials were revoked after the end of the authorization protocol and consequently prevent Bob from further accessing their database. In [19], the authors propose an access control system for pervasive computing environments that accomplishes this under the assumption that credential issuers will proactively push revocation information to endpoints in the system. As discussed in Section 2.1, there is no incentive for CAs to maintain the local state necessary to do this in a large open system. In fact, the soundness of algorithms requiring these types of assumptions depends on the reliability with which revocation information is propagated. Enforcement algorithms for the consistency levels discussed in this paper can be proven sound without making such assumptions.

4. ENFORCEMENT ALGORITHMS

In this section, we discuss the enforcement of the view consistency levels previously presented. We first enumerate the characteristics of an ideal algorithm for consistent view construction and argue that such an algorithm is likely to be impossible to construct in practice. We then discuss two practical algorithms for consistent view construction and use these algorithms to define two extreme points on a multidimensional spectrum of trade-offs affecting view consistency algorithms. We evaluate the costs associated with each of these algorithms and analyze the "distance" from these practical algorithms from the idealized case.

4.1 Comments on the Ideal Case

Each algorithm that we present in this paper, and in fact the entire notion of view consistency, is based on the conclusions that can be drawn from the observations of a single entity. As such, the soundness of an algorithm designed to create ϕ-consistent views is only one concern of interest to entities wishing to use that algorithm. Another important goal is quantifying the completeness of this algorithm when compared to an algorithm run by an omniscient entity with complete knowledge of the state of all credentials at all times; we will refer to this as *ideal completeness*. Since entities in any realistic system cannot know the global state of the system at any given time, ideal completeness provides an interesting best case to which the algorithms that we develop can be compared. As we develop the algorithms in this section, we will quantify the shortcomings of these algorithms with respect to ideal completeness. Since incremental consistency is easily implementable, we begin our discussion with an algorithm for constructing internally consistent views.

4.2 Internal Consistency

Algorithm 1 ensures that the views used for authorization policy evaluation are internally consistent. We make the following assumptions in Algorithm 1 (and later algorithms):

- The notation $\leftarrow_r$ denotes random assignment. For example, $s \leftarrow_r \{0,1\}^m$ assigns to s a random value chosen from the set of all length-m binary strings.

- Each entity $e \in \mathcal{E}$ has a set of credentials $C_e = \{c_1, \ldots, c_{n_e}\}$.

- There exists a globally agreed-upon cryptographic hash function $h : \{0,1\}^* \to \{0,1\}^l$ where l is the (fixed) output length of $h(\cdot)$.

Algorithm 1 Internal Consistency
1: // Initialize a connection with entity e'
2: **Function** INIT($e' \in \mathcal{E}$) = COMMIT(e')
3:
4: // Commit credentials to entity e'
5: **Function** COMMIT($e' \in \mathcal{E}$) =
6: $\quad s \leftarrow_r \{0,1\}^m$ // create a salt
7: $\quad k \leftarrow k_e - |C_e|$ // need k fake credentials
8: $\quad$ **for** $i = 1$ to k **do**
9: $\quad\quad r_i \leftarrow_r \{0,1\}^m$ // generate fake credentials
10: $\quad CC_e \leftarrow \{h(s \mid c_1), \ldots, h(s \mid c_n), h(r_1), \ldots, h(r_k)\}$
11: $\quad$ Shuffle CC_e randomly
12: $\quad$ Send $\langle e, s, CC_e \rangle$ to e'
13:
14: // Receive committed credentials from entity e'
15: **Function** RCV($e' \in \mathcal{E}, s' \in \{0,1\}^m, CC_{e'} \in 2^{\{0,1\}^l}$) =
16: **if** $EntityInfo.lookup(e') \neq \bot$ **then**
17: $\quad$ **for all** $\langle c, r, syn, sem_v, sem_i \rangle \in View$ **do**
18: $\quad\quad$ **if** c is semantically valid **then**
19: $\quad\quad\quad View.store(c, \langle c, r, true, NOW, sem_i \rangle)$
20: $\quad\quad$ **else**
21: $\quad\quad\quad View.delete(c)$
22: $EntityInfo.store(e', \langle NOW, s', CC_{e'} \rangle)$
23:
24: // Receive a credential c from entity e'
25: **Function** RCV($e' \in \mathcal{E}, c \in \mathcal{C}$) =
26: $\langle rcv, s', CC_{e'} \rangle = EntityInfo.lookup(e')$
27: **if** $h(s' \mid c) \notin CC_{e'}$ **then**
28: $\quad$ Reject c
29: **else if** ((c is syntactically valid) and ($\alpha(c) \leq rcv$) and (c is semantically valid)) **then**
30: $\quad View.store(c, \langle c, NOW, true, NOW, \bot \rangle)$
31: **else**
32: $\quad$ Reject c

Algorithm 2 Endpoint and Interval Consistency
1: // Receive a credential c from entity e'
2: **Function** RCV($e' \in \mathcal{E}, c \in \mathcal{C}$) =
3: **if** c is syntactically valid **then**
4: $\quad View.store(c, \langle c, NOW, true, \bot, \bot \rangle)$
5: **else**
6: $\quad$ Reject c
7:
8: // Invoked at the end of the access control protocol
9: **Function** VALIDATEALL($RelevantCreds \in 2^\mathcal{C}$) =
10: **for all** $\langle c, r, syn, sem_v, sem_i \rangle \in View$ **do**
11: $\quad$ **if** $c \in RelevantCreds$ **then**
12: $\quad\quad$ **if** ($\omega(c) > NOW$) and (c is semantically valid) **then**
13: $\quad\quad\quad View.store(c, \langle c, r, true, NOW, \bot \rangle)$
14: $\quad\quad$ **else**
15: $\quad\quad\quad$ Fail and report that c is invalid

- Each entity e chooses a parameter k_e used to hide the number of credentials that she possesses.
- Each entity maintains a hash table, $EntityInfo$, mapping entity names to state information. The function $EntityInfo.store : \mathcal{E} \times (T \setminus \{\bot\}) \times \{0,1\}^m \times 2^{\{0,1\}^l} \to \bot$ stores state information, while the function $EntityInfo.lookup : \mathcal{E} \to (T \setminus \{\bot\}) \times \{0,1\}^m \times 2^{\{0,1\}^l}$ retrieves state information.
- Each entity maintains a hash table, $View$, mapping credential identifiers to credential state information. The function $View.store : \mathcal{C} \times \mathcal{S} \to \bot$ stores credential state information, while $View.delete : \mathcal{C} \to \bot$ deletes state information.
- The current time as observed by some entity $e \in \mathcal{E}$ is accessible via the local variable NOW.

Algorithm 1 works as follows. At the start of the authorization protocol, each entity calls the INIT method to commit her credentials and a strategically chosen amount of random noise to the remote party. Each entity then stores her remote partner's set of committed credentials in the $EntityInfo$ hash table. As credentials are received from the remote party during the authorization protocol, the receiver checks to see if the credential was previously committed. If so, the credential state information for this credential is created and stored; if not, the credential is removed from $View$. Should one entity acquire new credentials at runtime, she can recommit her credential set to the remote party by directly using the COMMIT method. If this occurs, the remote party must immediately recheck the semantic validity of each credential stored in the current view and update its associated credential state information (lines 17–22).

This credential recommit process involves fairly high communication overheads for the recipient, as it must contact up to $|View|$ servers to revalidate all potentially relevant credentials. To mitigate denial of service attacks against implementations of this algorithm, entities should require that a recommit message be accompanied by a credential that (i) is relevant at the moment it is received, (ii) was not included in the previous credential set commitment, and (iii) was issued within some fixed window of the time of the last negotiation round. This will ensure that unless parties receive legitimate new credentials, they cannot force excess semantic validity checks. We now highlight several interesting properties of Algorithm 1.[1]

PROPOSITION 4. *Any view created using Algorithm 1 is incrementally consistent.*

PROPOSITION 5. *All credentials accepted by Algorithm 1 were held by their bearer at the time of the most recent credential recommit.*

THEOREM 1. *If e's execution of a trust negotiation or distributed proving protocol for target policy P succeeds at time t while using Algorithm 1 to enforce view consistency, then the view, $V_e^{P,t}$ is internally consistent.*

PROPOSITION 6. *Algorithm 1 does not disclose credential contents (e.g., credential types or attribute values) to the remote party. Further, if $h(\cdot)$ approximates a random oracle, then no entity can guess the number of credentials held by their communication partner during a given run of the algorithm, nor can they guess the number of new credentials committed during a recommit.*

Although Theorem 1 asserts the soundness of Algorithm 1, this algorithm is not ideally complete as discussed in Section 4.1. That is, it is possible for all credentials to be valid simultaneously at the time of the last recommit even if Algorithm 1 fails. Consider the case that Bob commits several credentials to Alice, all of which are valid at the moment the committed credential set is sent to Alice. However, before Alice can verify some credential c that was committed by Bob, c's issuing CA revokes the credential. Alice thus cannot tell that c was valid at the time that the credential set was committed, though an omniscient entity could. In the full version of this paper [14], we propose an online credential status protocol that would allow Algorithm 1 to more closely approximate ideal completeness.

---
[1] Full proofs of the claims made in this section can be found in the full version of this paper [14].

4.3 Endpoint and Interval Consistency

Algorithm 2 guarantees that all executions of an authorization protocol that succeed do so using interval consistent views. In general, the strategy adopted by this algorithm is similar to that taken in optimistic concurrency control algorithms for transaction management. That is, credentials are syntactically validated as they arrive, as this can be done without external interaction, but are assumed to be semantically valid. When a decision point is reached, the VALIDATEALL method is invoked to check the semantic validity of each relevant credential in the view and terminates the protocol if any credentials are found to be invalid. Because e has reached the decision point, it will have the clearest idea yet as to which submitted credentials are actually relevant. If one of these credentials is invalid, however, e can continue to search for another set that satisfies the policy; this new set can then be checked for validity, and so on. If only endpoint consistent views are required, then both the semantic and syntactic validity checks can be delayed until the VALIDATEALL method.

THEOREM 2. *If an execution of a trust negotiation or distributed proving protocol for a target policy P succeeds at time t while using Algorithm 2 to enforce view consistency, then the view $V_e^{P,t}$ is interval, endpoint, internally and incrementally consistent.*

Although Algorithm 2 is sound (by Theorem 2) it is not ideally complete. Since the VALIDATEALL method takes some finite, but non-instantaneous, amount of time to check the semantic validity of each c_i whose state is stored in V, it is entirely possible that each c_i was valid at $end(V)$, but one such credential was revoked before its semantic validity could be checked by the algorithm. An omniscient entity could detect this event, even though it would go undetected by Algorithm 2. The well known limitations of causal orderings and virtual clocks [13, 9] lead us to conjecture that any sound and complete algorithm for endpoint consistency will require synchronized clocks.

4.4 Trade-offs in Consistency Enforcement

In examining Algorithms 1 and 2, a clear trade-off emerges. By deferring semantic validation checks until the end of the protocol, Algorithm 2 reduces the work for the verifier by allowing her to semantically validate only the credentials that were ultimately determined to be relevant to the satisfaction of the policy. This reduction in work comes at a price. In the case that the policy being satisfied uses guard conditions to protect the disclosure of more sensitive portions of the policy (e.g., as in [6, 15]), optimistically assuming that credentials are semantically valid could leak sensitive policy information to unauthorized viewers. To correct this problem, each set of guard conditions must be viewed as a sub-negotiation in its own right, so that the semantic validity of the credentials satisfying the guard conditions is checked before access is granted to the remaining policy. Alternatively, Algorithm 1 can be modified to call the VALIDATEALL method at its conclusion. However, Algorithm 1 incurs much higher overheads for the verifier, as each credential received must be validated throughout the protocol, as its relevance cannot be fully determined until the end of the protocol.

These algorithms are two extreme points on the spectrum of possible consistency enforcement algorithms. In some cases, an entity may prefer to aggressively monitor the validity of some credentials received over the course of the authorization protocol, while deferring checks on other credentials. For instance, for a policy $P = c_1 \wedge (c_2 \vee c_3)$, it is clear that c_1 is relevant to the satisfaction of P. Thus c_1 could be monitored more aggressively (using a scheme like that in Algorithm 1), while checks on the validity of credentials c_2 and c_3 could be delayed until the end of the protocol. Designing consistency enforcement algorithms that balance this trade-off between relevance, work for the verifier, and information leakage will be an interesting challenge.

5. DISCUSSION

In this section, we discuss several interesting facets of view consistency. In particular, we show that the algorithms presented in this paper satisfy the requirements presented in Section 2.3, consider the effects of an unstable environment on view consistency, and introduce the notion of strategic algorithms for view consistency enforcement.

5.1 Requirements Revisited

In Section 2.3 we presented three requirements that view consistency algorithms should satisfy: loose clock synchronization, minimal cooperation, and minimal impact on existing protocols. Each algorithm presented in this paper relies only on its local perception of time and causal event orderings; no synchronization with external sources is necessary. Further, only a small amount of cooperation between entities is required for these algorithms to function correctly. Specifically, in Algorithm 1, only the two parties engaged in the authorization protocol need to cooperate to form a consistent view. The only way that the remote party can fail to cooperate in these algorithms is to incorrectly commit her credential values; this failure can only deny her access to the requested resource. Algorithm 2 requires no cooperation between entities in the system to succeed. Lastly, the algorithms presented in this paper have virtually no impact on existing trust negotiation and distributed proving protocols, as they were designed to wrap the functionality already provided by existing protocols and systems. By disabling credential verification in existing systems and using wrapper code that implements the consistency checking algorithms presented in this paper, existing systems can enforce stronger levels of view consistency.

5.2 Dynamic Environments

In context-rich environments like smart buildings and grid computing systems, it is entirely possible for authorization policies to be predicated on the state of the surrounding environment. For instance, authorization policies may consider the time of day or the occupancy status of a room. A malicious client can attempt to alter the state of their surrounding environment in unexpected ways to twist the outcome of an authorization protocol. The environmental inputs to an authorization protocol can consist of either certified environmental information collected by the client (or some agent acting on his behalf) or observations made by the resource provider. In the event that only certified environmental information is used, then the endpoint and interval consistency algorithms presented in this paper can ensure that all environmental assertions remain true throughout the duration of the authorization protocol. However, ensuring that observational data regarding system context does not become invalidated is a more difficult task. The resource provider must either continuously monitor the pertinent state information or register to be alerted should its value change. Periodically checking the state is insufficient, as the value could fluctuate between checks and not be detected. If the resource provider has the capability to register such alerts, then this mechanism combined with one of the algorithms presented in this paper can ensure that the consistency of their view can be protected from the effects of unstable environmental conditions that are either naturally occurring or maliciously induced.

5.3 Strategic Algorithm Design

Trust negotiation and distributed proving are dynamic processes, the properties of which depend on the strategies or tactics adopted by their participants [3, 23, 26]. Similarly, the level of view consistency required by a given entity is to some extent also a strategic decision (this is a further extension of the trade-off noted in Section 4.4). The levels of view consistency presented in this paper were designed to enforce various levels of safety, and thus the algorithms provided focused on satisfying only this criteria. However, safety may not always be the only concern for some resource providers. Rather, they may wish to enforce some level of safety but require algorithms with stronger guarantees regarding the availability of their services or privacy preservation than those provided by the algorithms in this paper.

For instance, recall that Algorithm 1 allows an entity Alice to hide her credentials in a set of credentials and fake commitments of size k_e. To do this, however, requires that she compute and disclose the results of k_e hashes; the overhead of this process quickly becomes burdensome as k_e increases. In the full version of this paper [14], we explore the consequences of replacing the commitment algorithm used in Algorithm 1 with an alternate commitment scheme based on Merkle trees [17]. We show that this scheme allows Alice to use a very large k_e to hide her real credentials with far less overhead than the commitment scheme presented in Algorithm 1. We also prove that this commitment scheme has the same security properties of that used in Algorithm 1, which implies that simply changing the commitment scheme used by Algorithm 1 allows us to tune both the performance and privacy guarantees of Algorithm 1 without effecting the consistency property that it enforces. This suggests that further analysis of these types of strategic trade-offs in view consistency algorithms may be an interesting area of future research.

5.4 A Note Regarding CA Clock Skew

The algorithms in this paper assume that the times $\alpha(c)$ and $\omega(c)$ are interpreted relative to the local clock, as is done in commodity software like web browsers. That is, if the local clock indicates that $\omega(c)$ not yet passed, then c is accepted as syntactically valid. While in many cases this is a safe assumption to make, especially if online semantic validity checks are made, it can in some cases lead to troubles if CA clocks are poorly synchronized. For example, consider the case in which an entity receives credentials c_1 (issued by CA 1 and expiring at time t_1) and c_2 (pre-issued by CA 2 and becoming valid at time $t_2 \leq t_1$) as part of an authorization protocol. Based on the local interpretation of t_1 and t_2, the validity period of these credentials overlaps. However, if the clock at CA 2 is slower than the clock at CA 1 by an amount of at least $t_1 - t_2$, then despite *appearing* to overlap, the validity intervals of c_1 and c_2 never *actually* overlap.

Fortunately, the widespread use of time synchronization protocols such as NTP [18] by service providers reduces the likelihood of this type of error randomly occurring between unrelated credentials. However, it is of the utmost importance that if two or more CAs issue mutually-exclusive certificates, their clocks be closely synchronized to ensure that no misleading apparent overlaps can occur. Such apparent overlaps are not introduced by the algorithms developed in this paper, but rather by the widespread notion of using a local interpretation of certificate expiration times. Fortunately, there is no way for an attacker to exploit this type of error without proactively altering at least one CA's clock prior to the issuance of some credential used in the negotiation that the attacker wishes to alter.

6. RELATED WORK

Safety in trust negotiation has been discussed in several previous works, though the definitions of safety used in these works differs from that considered in this paper. In [26] Yu and Winslett describe the notion of "safe disclosure sequences." Informally, they consider a trust negotiation safe if each resource disclosed during the negotiation was "unlocked" (i.e., its authorization policy was satisfied) at the time that it was disclosed. Winsborough and Li [23] note that under this notion of safety, private information that is not explicitly revealed during a trust negotiation can still be inferred based on the way that an entity carries out the negotiation. They propose several more refined notions of safety for trust negotiation protocols based on the concept of indistinguishability, each of which gives users stronger guarantees regarding the amount of private information leaked during the negotiation. Irwin and Yu [11] propose another definition of safety based on the idea of information gain. Our work is orthogonal to these previous works in that we are concerned with safety problems that emerge as a result of the consistency of the underlying state information used during policy evaluation rather than those that arise due to information leakage during a negotiation. It would be prudent for system designers to consider both types of safety.

Another area of closely related work is that of concurrency control and consistency enforcement in distributed systems, distributed databases, and distributed shared memory. Each of these areas has a rich body of literature, surveys of which can be found in [21], [7], and [1], respectively. In general, these problem domains assume that multiple entities will be updating values stored at multiple locations within the system and as such, maintaining data consistency is of concern to everyone. Therefore, solutions to transaction management in these domains typically involve the cooperation of multiple entities, as every entity has incentive to cooperate. However, as was mentioned in Sections 1 and 2.1, groups of entities have no incentive to cooperate in solving the view consistency problem for trust negotiation and distributed proving since this problem is of concern only to a particular resource provider evaluating a particular policy. Therefore, the solutions developed in the distributed systems, distributed databases, and distributed shared memory literature are unsuitable for our problem domain; the solutions that we develop in this paper require only the cooperation of, at most, the two parties participating in the authorization protocol.

A final area of related work is the collection of system state snapshots in distributed systems. Collecting consistent snapshots that can be used to evaluate stable predicates over the system state is a well-known problem, to which an elegant solution is presented in [8]. This algorithm is not directly applicable to the problem addressed in this paper, however, due to the unstable nature of credential statuses. There exist algorithms for collecting distributed state snapshots that can be used to evaluate unstable predicates (for a survey, see [2]), though these algorithms have very high overheads and make unreasonable assumptions about process cooperation for our problem domain.

7. CONCLUSIONS AND FUTURE WORK

In this paper, we presented the notion of view consistency in policy-based authorization systems. We showed that failing to consider the consistency of the system views used during executions of these protocols can cause a marked decrease in the safety of the decisions made by the underlying authorization system. We then defined the incremental, internal, endpoint, and interval consistency levels and demonstrated algorithms to attain these consistency levels in practice. We asserted the soundness of these algorithms and commented on their completeness when compared to an ideal algorithm run by an omniscient entity. These algorithms require at most the cooperation of the two parties involved in the authorization process; should any entity not cooperate, the algorithms will fail rather than violate the consistency conditions that they were designed to enforce.

There are several areas of interesting future work relating to view consistency. As alluded to in Section 5.3, the design of consistency enforcement algorithms that make a variety of trade-offs regarding safety, availability, and privacy-preservation properties could prove to be a fruitful area of investigation. Given the autonomous nature of the entities participating in trust negotiation and distributed proving authorization protocols, it would be beneficial to explore the notion of interoperable families of algorithms for consistency enforcement (as was done in [26] for trust negotiation strategies). This would allow each entity to acquire the consistency level she requires without placing unnecessary constraints on her communication partners. Another area of future work involves the development of consistent views shared by several entities in the system. Given the falling costs associated with fine-grained clock synchronization via technologies such as GPS and an increased interest in distributed authorization, interesting notions of view consistency are likely to emerge from a study of this topic.

Acknowledgments. This research was supported by the NSF under grants IIS-0331707, CNS-0325951, and CNS-0524695 and by Sandia National Laboratories under grant number DOE SNL 541065. Lee was also supported by a Motorola Center for Communications graduate fellowship. The authors wish to thank an anonymous reviewer who suggested the Merkle commitment scheme in Section 5.3.

8. REFERENCES

[1] S. V. Adve and K. Gharachorloo. Shared memory consistency models: A tutorial. *IEEE Computer*, pages 66–76, Dec. 1996.

[2] O. Babaoğlu and K. Marzullo. Consistent global states of distributed systems: Fundamental concepts and mechanisms. In S. J. Mullender, editor, *Distributed Systems*, pages 55–96. Addison-Wesley, 1993.

[3] L. Bauer, S. Garriss, and M. K. Reiter. Distributed proving in access-control systems. In *IEEE Symposium on Security and Privacy*, May 2005.

[4] M. Y. Becker and P. Sewell. Cassandra: Distributed access control policies with tunable expressiveness. In *IEEE International Workshop on Policies for Distributed Systems and Networks*, 2004.

[5] E. Bertino, E. Ferrari, and A. C. Squicciarini. Trust-X: A peer-to-peer framework for trust establishment. *IEEE Transactions on Knowledge and Data Engineering*, 16(7):827–842, Jul. 2004.

[6] P. Bonatti and P. Samarati. Regulating service access and information release on the web. In *ACM Conference on Computer and Communications Security*, 2000.

[7] W. Cellary, E. Gelenbe, and T. Morzy. *Concurrency Control in Distributed Database Systems*. Elsevier Science Publishing Company, Inc., 1988.

[8] K. M. Chandy and L. Lamport. Distributed snapshots: Determining global states of distributed systems. *ACM Transactions on Computer Systems*, 3(1):63–75, Feb. 1985.

[9] D. R. Cheriton and D. Skeen. Understanding the limitations of causally and totally ordered communication. In *ACM Symposium on Operating Systems Priniciples*, 1993.

[10] R. Housely, W. Ford, W. Polk, and D. Solo. Internet X.509 Public Key Infrastructure Certificate and CRL Profile. IETF RFC 2459, Jan. 1999.

[11] K. Irwin and T. Yu. Preventing attribute information leakage in automated trust negotiation. In *ACM Conference on Computer and Communications Security*, Nov. 2005.

[12] H. Koshutanski and F. Massacci. Interactive credential negotiation for stateful business processes. In *International Conference on Trust Management*, May 2005.

[13] L. Lamport. Time, clocks, and the ordering of events in a distributed system. *Communications of the ACM*, 21(7):558–565, Jul. 1978.

[14] A. J. Lee and M. Winslett. Safety and consistency in policy-based authorization systems (extended version). Technical Report UIUCDCS-R-2006-2761, University of Illinois at Urbana-Champaign, Aug. 2006.

[15] J. Li, N. Li, and W. H. Winsborough. Automated trust negotiation using cryptographic credentials. In *ACM Conference on Computer and Communications Security*, Nov. 2005.

[16] N. Li and J. Mitchell. RT: A role-based trust-management framework. In *DARPA Information Survivability Conference and Exposition*, Apr. 2003.

[17] R. C. Merkle. *Secrecy, authentication, and public key systems*. PhD thesis, Stanford University, 1979.

[18] D. L. Mills. Network Time Protocol (Version 3) Specification, Implementation and Analysis. IETF RFC 1305, Mar. 1992.

[19] K. Minami and D. Kotz. Scalability in a secure distributed proof system. In *International Conference on Pervasive Computing*, May 2006.

[20] M. Myers, R. Ankney, A. Malpani, S. Glaperin, and C. Adams. X.509 Internet public key infrastructure online certificate status protocol - OCSP. IETF RFC 2560, Jun. 1999.

[21] A. S. Tanenbaum and M. van Steen. *Distributed Systems: Principles and Paradigms*. Prentice Hall, 2002.

[22] W. H. Winsborough and N. Li. Towards practical automated trust negotiation. In *IEEE International Workshop on Policies for Distributed Systems and Networks*, Jun. 2002.

[23] W. H. Winsborough and N. Li. Safety in automated trust negotiation. In *IEEE Symposium on Security and Privacy*, May 2004.

[24] M. Winslett, T. Yu, K. E. Seamons, A. Hess, J. Jacobson, R. Jarvis, B. Smith, and L. Yu. The TrustBuilder architecture for trust negotiation. *IEEE Internet Computing*, 6(6):30–37, Nov./Dec. 2002.

[25] M. Winslett, C. Zhang, and P. A. Bonatti. PeerAccess: A logic for distributed authorization. In *ACM Conference on Computer and Communications Security*, Nov. 2005.

[26] T. Yu, M. Winslett, and K. E. Seamons. Supporting structured credentials and sensitive policies through interoperable strategies for automated trust negotiation. *ACM Transactions on Information and System Security*, 6(1), Feb. 2003.

[27] L. Zhou, F. B. Schneider, and R. van Renesse. COCA: A secure distributed online certification authority. *ACM Transactions on Computer Systems*, 20(4):329–368, Nov. 2002.

On the Modeling and Analysis of Obligations

Keith Irwin
North Carolina State University
kirwin@ncsu.edu

Ting Yu
North Carolina State University
yu@csc.ncsu.edu

William H. Winsborough
University of Texas at San Antonio
wwinsborough@acm.org

ABSTRACT

Traditional security policies largely focus on access control requirements, which specify who can access what under what circumstances. Besides access control requirements, the availability of services in many applications often further imposes obligation requirements, which specify what actions have to be taken by a subject in the future as a condition of getting certain privileges at present. However, it is not clear yet what the implications of obligation policies are concerning the security goals of a system.

In this paper, we propose a formal metamodel that captures the key aspects of a system that are relevant to obligation management. We formally investigate the interpretation of security policies from the perspective of obligations, and define secure system states based on the concept of *accountability*. We also study the complexity of checking a state's accountability under different assumptions about a system.

Categories and Subject Descriptors: K.6.5 [Management of Computing and Information Systems]: Security and Protection

General Terms: Security, Theory

Keywords: Policy, Obligations

1. INTRODUCTION

Security policies are widely used in the management of sensitive information and valuable resources in a variety of applications and systems. Traditional security policies largely focus on the specification and management of access control requirements, i.e., what principals are allowed to access what objects and when. Besides controlling principals' privileges, the availability of services as well as the correct operation of a system often imposes obligation requirements which specify what actions a subject is *obliged* to perform in the future in order to allow certain actions to be taken at present. For example, if a customer is allowed to subscribe to a certain service, then she is obliged to pay the subscription fee by the end of each month. In some situations, one's obligations may also result from the action performed by others instead of by oneself. For example, in a conference reviewing system, once a paper is assigned by the program chair to a reviewer, the reviewer is obliged to submit her review before a certain deadline.

Obligation requirements have traditionally been hard-coded into applications. But recently, obligations are increasingly being expressed explicitly as part of security policies. As is the case for many policy-based systems, this approach offers advantages in terms of flexible management and easy maintenance of obligations. It allows a system to change obligation requirements quickly when flaws in existing policies are found, to react to new circumstances promptly, and to accurately enforce complex obligation requirements. In particular, supporting obligations in policies can be an important part of translating high level security goals into low level policies.

As traditional access control policies are only concerned with permitting or denying subjects the ability to take certain actions, they cannot be used directly to express obligation requirements. Recently, several security policy languages have been proposed to support the specification of obligation requirements [9, 12, 15, 19]. Works have also been done on monitoring the fulfillment of obligations [4, 5].

The introduction of obligations inevitably complicates the management of security policies. Traditionally, a reference monitor of a system only needs to determine whether an access request should be allowed, purely from an access control perspective. When obligations are added, it is not immediately clear how they should affect the reference monitor's decision. In other words, the relationship between access control and obligations is not yet well studied. Further, existing work on obligations focuses on the responsibility of the subjects who receive obligations. However, to ensure the correct operation of a system, a reference monitor should not blindly assign obligations to subjects and eventually check whether they have been fulfilled. Instead, a system should only allow obligations to be assigned when the receiving subject will have sufficient privileges as well as other resources in the system to successfully fulfill the obligation. That is, a diligent user should always be able to fulfill her obligations. Otherwise, the obligation should be not assigned in the first place.

Conceptually, a security policy defines what the secure states are for a system, and the job of a reference monitor is to ensure the system stays in a secure state and prevent it from transitioning into insecure states. With obligations introduced in security policies, the questions of how to define secure states and how to ensure the security of a system, to the best of our knowledge, have not yet been adequately investigated. These are the focus of this paper. More specifically, the contributions of this paper are as follows.

1. We abstract common system components that are relevant to the management of obligations, and propose a formal metamodel to capture a system and its possible states from the perspective of obligations.

2. Built on the above metamodel, we propose a formal definition of secure states for obligation management. Our definition is based on the concept of "accountability". Intuitively, considering subjects as autonomous entities, a system cannot prevent a subject from failing to fulfill its obligations. Violation of obligations may not be avoidable. However, once an obligation is violated, the system should be able to clearly identify who is responsible. In this paper, we view an obligation as a contract between a system and a subject. A system is accountable if and only if all the obligations will be fulfilled supposing all the subjects are diligent. In other words, a secure state implies that any obligation violation can only be due to the lack of diligence of subjects.

3. We further study the problem of checking whether a state is accountable. We identify a set of conditions, which, when satisfied, allow the efficient checking of a state's accountability. We also show that when some of the conditions do not hold, the problem becomes intractable in the worst case, when only considering the abstract construct of the metamodel.

4. We study the accountability problem in the context of a simple yet expressive authorization system with obligations, and show the tractability of the accountability checking problem even though the system does not satisfy the above identified conditions.

The rest of the paper is organized as follows. In section 2, we propose a set of criteria for the specification and management of obligations, and further elaborate the problem addressed in the paper. Section 3 presents a formal metamodel for systems with obligations as well as obligation policies. In section 4, we formally define the concept of accountability in terms of state transitions. In section 5, we study the problem of determining a state's accountability, based on the proposed model. In section 6, we show that the accountability problem can be efficiently solved in an authorization system with obligations. We report closely related work in section 7, and conclude this paper in section 8.

2. PROPERTIES OF OBLIGATIONS

An obligation is a requirement for a subject to take some action at some time in the future. Such obligations are sometimes also referred to as positive obligations, in contrast to "negative obligations" (or *refrainments*), where a subject is required not to take some action at some time in the future. For example, in privacy-enhanced systems, one common negative obligation is that a company should not share users' private information without their consent. In this paper, we focus on positive obligations. Although we believe that our model could handle negative obligations, we do not include them because in centralized systems negative obligations can be easily transformed into access control requirements, and thus can be enforced directly by the reference monitor. Positive obligations, on the other hand, cannot be enforced by a system in a direct way. After all, if a system could cause an action to happen, then it would no longer be a user action, but instead a system action. Negative obligations are more necessary in distributed systems, a topic we will explore in later papers.

Unenforcable. As mentioned above, an action required by an obligation cannot be forced to happen by a system. For example, a system cannot force a subscriber to pay his fee on time. Similarly, a program chair cannot ensure that every reviewer will submit their reviews before the deadline. This is an essential difference from access control requirements, which must be enforcable by the systems.

Monitorable. Though a system cannot force an obligation to be fulfilled, it should be able to monitor the status of an obligation, i.e., whether it has been fulfilled. Although it is conceivable that a system could give a user an obligation which it cannot monitor, such obligations are clearly not relevant to any decisions or analysis which the system undertakes.

If an obligation is to be feasibly monitored, there are two conditions which must be met. The first is that it must be clear when an obligation has been fulfilled. And equally importantly, it must be clear when a user has failed to fulfill an obligation with which he has been charged. One implication of this requirement is that an obligation must have some sort of deadline before which it must be accomplished. Otherwise there is no point at which it can be said that a user has violated his obligation. As such, there is no real motivation for users to fulfill their obligations. Therefore, a time window should be an intrinsic property of an obligation.

Obligations arise in a system when a user is allowed to take an action with the condition that they also accept an obligation or when a user's action causes someone else to receive an obligation. For instance, a policy may specify that users are allowed to run certain tests, but only if they agree to submit a report about the results of those tests within a week afterwards. Or a policy might say that employees are allowed to submit vacation requests, but when they do, this gives the human resources department an obligation to review those requests and take action on them within a week.

3. A METAMODEL FOR SYSTEMS WITH OBLIGATIONS

In this section, we present a formal metamodel that encompasses the basic constructs of a system from the perspective of obligations. The model serves as the foundation for our later discussion on the accountability of states and checking of accountability.

First let us discuss how we model obligations themselves. An obligation is an action which some subject must carry out during some timeframe. Thus, we model an obligation as a tuple $obl(s, a, O, [t_s, t_e])$, where s is a subject, a is an action, $[t_s, t_e]$ is a time window during which s is obliged to take action a, and O is a finite sequence of zero or more objects on which the action must be performed.

An obligation system consists of the following components:

- $\mathcal{T}$: a countable set of time values. For simplicity we take $\mathcal{T}$ to be the non-negative integers, with 0 indicating the system start time and each value indicating a point in time after that by a multiple of some appropriate, unspecified time interval.

- $\mathcal{S}$: a set of subjects that could be added to the state of the system.
- $\mathcal{O}$: a set of objects with $\mathcal{S} \subseteq \mathcal{O}$.
- $\mathcal{A}$: a finite set of actions that can be initiated by subjects. Each action's behavior is given by a function that takes as input the current system state (defined just below), the subject performing the action, and a finite sequence of zero or more objects. It outputs a new system state (except for the new time which is managed separately).
- $\mathcal{B} = \mathcal{S} \times \mathcal{A} \times \mathcal{O}^* \times \mathcal{T} \times \mathcal{T}$: a set of obligations that subjects can incur. Given an obligation $b \in \mathcal{B}$, we use $b.s$ to refer to the subject that is obligated, $b.a$ for the action the subject is obligated to perform, $b.O$ for the finite sequence of zero or more objects that are parameters to the action, and $b.t_s$ and $b.t_e$ to refer to the start and end, respectively, of the period in which the subject is obligated to perform the action.
- $\mathcal{ST} = \mathcal{T} \times \mathcal{FP}(\mathcal{S}) \times \mathcal{FP}(\mathcal{O}) \times \Sigma \times \mathcal{FP}(\mathcal{B})$: the set of system states. Here, we use $\mathcal{FP}(\mathcal{X}) = \{X \subset \mathcal{X} | X \text{ is finite}\}$ to denote the set of finite subsets of the given set. We use $st = \langle t, S, O, \sigma, B \rangle$ to denote systems states, where t is the time in the system, B is the set of pending obligations, and σ is a fully abstract representation of all other features of the system state. Σ is possibly infinite. We use $st_{cur} = \langle t_{cur}, S_{cur}, O_{cur}, \sigma_{cur}, B_{cur} \rangle$ to denote the current state of the system.
- $\mathcal{P}$: a set of policy rules. Each policy rule specifies an action that can be taken, under what circumstances it may be taken, and what obligations (if any) results from that action. We denote policy rules of this type by using the notation $a(st, s, O) \leftarrow cond : F_{obl}$, where $a \in \mathcal{A}$ and $cond$ is a predicate in $\mathcal{S} \times \mathcal{T} \times \Sigma \times \mathcal{O} \to \{true, false\}$, indicating that subject s is authorized to perform action a on objects O at time t with the system in state σ if $cond(s, t, \sigma, O)$ is true. F_{obl} is an *obligation function*, which takes the current state of the system σ, the current time, the subject s, and the arguments O as its input and outputs a finite set $B \subset \mathcal{B}$ of obligations caused by the action. Note that obligations in B may not necessarily be incurred by the same subject. If the action a is taken, then all the obligations in B should be fulfilled.

We allow multiple action rules to have the same action as their heads. As long as the condition of one rule is true, the action is allowed, causing the obligations resulting from that rule to be incurred.

As such, action rules can be used to express obligation requirements where a subject has a choice to take one of several actions to fulfill its obligation. We can simply have several rules with the same head and condition but with different obligations.

The conditions in action rules essentially model access control policies of a system. Here we adopt a closed world assumption. If none of the conditions of the action rules for an action is true, then a subject is not allowed to take that action.

An obligation $obl(s, a, O, [t_s, t_e])$ may be in one of four states: *invalid*, *pending*, *fulfilled*, or *violated*. If it is the case that the t_e is already passed when it is assigned, then the obligation is *invalid*, as it, on its face, can clearly not be carried out. If an obligation has been assigned and its action has been carried out during the time window $[t_s, t_e]$, then it has been *fulfilled*. If it has been assigned, has not been *fulfilled*, and is not *invalid*, but t_e has passed, then it is *violated*. If an obligation is not *invalid* but has not yet become *fulfilled* or *violated*, then it is *pending*.

3.1 State Transition

As indicated above, we assume system time is discrete, and model it as a non-negative integer representing the number of clock ticks from some predetermined start time. For simplicity, we assume that each action can be finished in a single clock tick, and its effect will be reflected in the state of the next clock tick. Suppose the state of a system at time t_0 is st_0, and Alice takes an action at t_0. This action will not change st_0. Instead the state st_1 at time $t_0 + 1$ will be affected by Alice's action. When multiple actions are attempted at the same time, we assume the system uses a function f_{trans} to get the state of the next clock tick. More specifically, f_{trans} takes the current state of the system (which also includes the current time) and a set of actions attempted by subjects, and returns for the next clock tick the state obtained by applying the permitted actions in some unspecified, fixed order which has the property that at the point each action is taken the state satisfies a policy rule which allows that action. In other words, we adopt a deterministic model of state transition; from the current state and the actions taken at present, we can uniquely determine the next state.

In fact, since an action may have multiple action rules associated with it, when a subject takes an action, it may choose different rules to authorize the action. This may result in different obligations assigned. Thus, when describing state transitions, besides specifying what actions subjects take, it is also necessary to indicate the policy rules applied for these actions. Let AP be a set of tuples (s, a, r), where s is a subject, a is an action and r is a policy rule for a. (AP stands for "action plan".) Let st be a state at time t. We use $st \vdash_{AP} st'$ to denote that st has transitioned to state st' at time $t + 1$ after actions in AP have been taken at time t. Legal state transitions must have the property that for each $(s, a, r) \in AP$, the condition of r is true at the state that is current when (s, a, r) is reached in the unspecified, fixed order in which members of AP are performed by f_{trans}. If $AP = \emptyset$, then st' corresponds to the system state when the time is incremented by one clock tick, but no other component of the system state is changed.

DEFINITION 1. *We say* $st \vdash_{AP} st'$ *is an* obligation-abiding transition, *if (1) there are no two tuples* (s_1, a_1, r_1) *and* (s_2, a_2, r_2) *in AP such that* $s_1 = s_2$ *and* $a_1 = a_2$; *and (2) for any* $(s, a, r) \in AP$, *there exists a pending obligation* $(s, a, [t_s, t_e])$ *in* $st.B$. *An obligation-abiding transition is* valid *if no pending obligations in st become violated in st'.*

An obligation-abiding transition corresponds to the system evolution where subjects take actions only to fulfill their obligations. A sequence of valid obligation-abiding transitions corresponds to the situation where subjects are diligent and always fulfill their obligations. Note that although a transition in which no users take any actions will be guaranteed to be obligation-abiding, it will not be guaranteed to be valid.

3.2 An Example Obligation System

We use a simple conference reviewing system as an example to show how a system with obligations can be represented in the above metamodel.

In this system, after collecting submitted papers, the program chair of a conference assigns papers to reviewers. Once the assignment is done, each reviewer is obliged to submit their reviews by a certain deadline. A reviewer can also submit reviews for papers not assigned to them. If a reviewer submits a review for a paper, she is obliged to attend the discussion of the paper, which decides whether the paper should be accepted.

This system can be modeled as follows.

- Subjects s are the registered users in the system.
- Objects o are submitted papers (and the subjects).
- The actions allowed in the system include assigning papers to reviewers, submitting a review and joining discussion of a paper.
- The σ-portion of the system state is no longer fully abstract, but instead represents attributes of subjects and objects. For instance, the set of roles of subject s is given by $\sigma.roles(s)$.

The policy of the system may be the following:

- $assign\_reviewer(st, s_1, \{s_2, o\}) \leftarrow$
  $prog\_chair \in st.\sigma.roles(s_1) \land$
  $reviewer \in st.\sigma.roles(s_2) \land$
  $st.\sigma.name(s_2) \notin st.\sigma.author\_list(o):$
  $\{obl(s_2, submit\_review(s_2, o),$
  $[06/01/06, 07/15/06])\}$.

- $submit\_review(st, s, \{o\}) \leftarrow$
  $reviewer \in st.\sigma.roles(s) \land$
  $st.\sigma.name(s) \notin st.\sigma.author\_list(o):$
  $\{obl(s, discuss(s, o), [07/22/06, 07/22/06]),$
  $obl(s, vote(s, o), [07/23/06, 07/23/06])\}$.

  (Once a reviewer submits a review of a paper, he or she is added into the reviewer list of the paper.)

- $discuss(st, s, \{o\}) \leftarrow reviewer \in st.\sigma.roles(s) \land$
  $s.name \notin st.\sigma.author\_list(o) : \emptyset$.

- $vote(st, s, \{o\}) \leftarrow$
  $s.name \in st.\sigma.reviewer\_list(o) : \emptyset$.

We do not show the semantics of each action since most of them are straightforward. Suppose on 06/01/06 the program chair assigns Alice to review papers p_1, p_2 and p_3. Then three pending obligations are added into the system: $obl_1 = obl(\text{Alice}, submit\_review(\text{Alice}, p_1), [06/01/06, 07/15/06])$, $obl_2 = obl(\text{Alice}, submit\_review(\text{Alice}, p_2), [06/01/06, 07/15/06])$ and $obl_3 = obl(\text{Alice}, submit\_review(\text{Alice}, p_3), [06/01/06, 07/15/06])$. On 07/10/06, Alice submits her review for paper p_1. Then in the state of the system on 07/11/06, the status of obl_1 becomes *fulfilled*, and a new pending obligation $obl_4 = obl(\text{Alice}, discuss(\text{Alice}, p_1), [07/23/06, 07/23/06])$ is created. Note that the transition from the state of the system on 07/10/06 to that on 07/11/06 is obligation-abiding, as the action Alice takes is required by one of her obligations. Also, since on 07/11/06 no pending obligations become *violated*, the transition is a valid obligation-abiding one.

On the other hand, suppose *Bob*, who is not assigned as one of the reviewers for p_1, is interested in the paper, and also submits a review for p_1 on 07/10/06. Then the transition is not obligation-abiding, since Bob's action is not required by any obligation.

Further, suppose Alice fails to submit a review for p_3 before 07/15/06. Then the transition from the state on 07/15/06 to that on 07/16/06 is still obligation-abiding according to our definition. However, it is not a valid one since obl_3 becomes *violated* on 07/16/06.

4. SECURITY GOALS IN SYSTEMS WITH OBLIGATIONS

Conceptually, a system's security policy divides system states into two disjoint sets: secure states and insecure states. The goal of security is to ensure that a system always stays in secure states and never transits into insecure states. Under the same principle, we study the question of how we should interpret a security policy which includes obligations, i.e., what states are considered secure under policies with obligations.

One straightforward approach is to define secure states as those that have no obligations being violated. Such states are certainly desirable. However, due to the unenforceable nature of obligations, a system can never guarantee that an obligation will be fulfilled. Instead, it seems more appropriate for a system to ensure that all obligations *can* be fulfilled, in the sense that the obligated user has the necessary authorizations to perform the obligatory action. However, as we will see, this alone does not provide sufficiently clear guidance. Certainly if the system determines that it is impossible for a user to fulfill the obligation which would be incurred by his performing some requested action, then the system should deny that action. Likewise, if the system is certain that a user will have sufficient privileges to fulfill an obligation before its deadline, then it should allow the requested action. But what is the appropriate thing for the system to do if the ability of the user to perform the obligation depends on whether or not actions are taken to change his privileges?

Suppose there exists one sequence of actions or events which would cause the user to be unable to fulfill his obligations and another which would cause the user to be able to. One highly conservative response would be always to deny requests that would incur obligations that the obligated user might not be able to fulfill. However, in any system with a superuser, this would result in virtually every request being denied. In fact, so long as there existed any remedy which could take rights away from the user in any circumstance, all requests from that user for actions which carry obligations would be denied. One can imagine a scenario in which a CEO is denied the right to edit a file because it is theoretically possible that the board of directors could oust her from her position in the next five minutes.

Similarly, if one took the completely optimistic approach, the opposite scenario rears its head. So long as there exists some possible way that the user might be able to fulfill the incurred obligation, the requested action would be permitted, even if the events necessary for the user to have the required authorizations are highly unlikely. For example, Bob, a mailroom employee, could be allowed an action even though he could only fulfill the associated obligation if the board fired the CEO and hired Bob in his place within the next five minutes. As such, it is clear that neither strategy is acceptable.

In this paper, we offer the concept of *accountability*. Rather than requiring that it be impossible for obligations to be violated, instead we assume that it is possible that obligations go unfulfilled, but when they do, we would like to clearly identify whose fault it is. Obviously, an obligation can go unfulfilled because a subject simply fails to take the required action before the deadline, even if he has sufficient privileges and resources. It is desirable that this is the only reason that an obligation will go unfulfilled.

Intuitively, if it is the case that all users have sufficient privileges and resource to carry out their obligations so long as every other user carries out his or her obligation, then a system is said to be in an *accountable state*, because we can know that whoever first fails to carry out an obligation is responsible for the violation and anything which results from it. In this paper, we describe systems which attempt to maintain accountability, but we reserve for future work issues pertaining to assigning blame in more complex scenarios such as failures after the system has left an accountable state.

Note that when determining whether a state is accountable, we only consider those actions that are required to be taken by obligations. Although a user may actively take some actions which may interfere with another user's obligations, such actions can be controlled by a system. Once the system determines whether the resulting state will be accountable or not, it can take appropriate actions. For example, it may either prevent the user from taking the action, or it may discharge or change the interfered obligations so that the resulting state is still accountable. In this paper, we focus on the definition and checking of accountable states. We will briefly discuss the handling of actions that may lead a system into an unaccountable state.

Before we give a formal definition of accountable states, it is necessary to discuss in more detail what situation it is in which a user is deemed as failing to fulfill her obligation. Intuitively, when an obligation $(s, a, [t_s, t_e])$ is assigned, we can view it as a contract between a subject s and a system. From the subject's perspective, it has promised to take action a during the given time window. On the other hand, from the system's perspective, it also implicitly promises that if everybody else fulfills their obligations, then s should have the needed privileges and resources to take action a. Depending on the interpretation of obligations, we may have different definitions of accountability.

Here we consider two types of interpretations. In the first type, if everybody else fulfills their obligations, then a system guarantees that Alice can take action a at any time point during $[t_s, t_e]$. This is a very strong promise from the system, which means that the condition of the rule for action a is always true for Alice during the obligation's time window.

In the second type, a system only promises that, if everybody else fulfills their obligations, then Alice can at least take action a at the end point t_e of the obligation's time window. Note that it does not mean that Alice has to take the action at t_e; it only means that in the worst case Alice will still be able to fulfill the obligation at t_e. Clearly, this type of promise is weaker than the first one, since it only requires that the condition of the rule for action a is true at t_e instead of during the whole time window of the obligation.

Generally speaking, the first type of accountability is suitable for systems in which users may have additional restrictions of which the system is not aware. If there are additional constraints on when a user is available to fulfill her obligations, then the system should ensure that the full time is available to the user so that whenever the user has the opportunity to fulfill one of her obligations, she has the necessary authorizations to do so. For instance, a system at a company with flexible work hours would probably prefer strong accountability since it would not know when employees would be available to fulfill their obligations. By contrast, in a more all-encompassing system, the weaker accountability is sufficient: since the system is aware of all scheduling constraints, whenever the user chooses to attempt to fulfill the obligation, either she will be able to fulfill it at that time, or the system will have ensured that she will again be available at a later time prior to the deadline. Hence, a system at a military base, where all users can be expected to be available precisely when the policies say that they should, might be able to use weak accountability, and derive benefit from the fact that it is a weaker requirement and therefore easier to ensure.

There is a third possible type of accountability, in which the system ensures only that there exists some time within the frame when the user will be able to fulfill his obligation. This type, however, is not likely to be generally suitable for ordinary users since it would require that the user discover that time before it passes. It may be suitable for systems with automated agents which could regularly poll the system to see if the obligation can be fulfilled. However, for reasons of space, we do not formalize or otherwise discuss the implications of that third type of accountability in this paper.

Next, we formally define accountable states for each of the first two interpretations. Recall from Definition 1 that a transition is valid and obligation-abiding if all the actions in its action plan are from existing obligations and no obligations are violated.

DEFINITION 2. *Let st be a system state with time t and pending obligations $obl_1, \ldots, obl_n$. We say st is a* type-1 undesirable state *if there exists an obligation $obl_i = obl(s_i, a_i, [t_{si}, t_{ei}])$, $1 \leq i \leq n$, such that (1) the condition cond of each action rule for a_i is false for s_i, i.e., $cond(s_i, t, st.\sigma) = false$; and (2) $t_{si} \leq t \leq t_{ei}$.*

A state is type-1 undesirable if a subject cannot fulfill an obligation although the current time is within the time window of the obligation.

DEFINITION 3. *A state st is* strongly accountable *if there exists no sequence of valid obligation-abiding transitions that lead st to a type-1 undesirable state.*

This definition of accountability corresponds to the first interpretation of obligations. For the second interpretation, we have the following definition.

DEFINITION 4. *Let st be a system state with time t and pending obligations $obl_1, \ldots, obl_n$. We say st is a* type-2 undesirable state *if there exists an obligation $obl_i = obl(s_i, a_i, [t_{si}, t_{ei}])$, $1 \leq i \leq n$, such that (1) the condition cond of each action rule for a_i is false for s_i, i.e., $cond(s_i, t, st.\sigma) = false$; and (2) $t = t_{ei}$.*

A state is type-2 undesirable if the current time is the deadline of a pending obligation, which however cannot be fulfilled at present.

DEFINITION 5. *A state st is* weakly accountable *if there exists no sequence of valid obligation-abiding transitions that lead st to a type-2 undesirable state.*

Obviously, if a state is strongly accountable, it is also weakly accountable. Let us consider the following scenarios. Suppose in state st at time 0 Alice does not have the privilege to read file f, but she has an obligation to read f between time 10 and time 20. Meanwhile, Bob, whose is the owner of f, has an obligation to grant the read privilege of f to Alice between time 5 and time 15. According to our definition, st is not strongly accountable, as it is possible that Alice decides to fulfill her obligation at time 12, but she cannot do so because she lacks the read privilege. In this case Bob is not to blame, since Bob can decide to fulfill his obligation at time 14 for example. In other words, st may possibly transit into a future state where a subject cannot fulfill its obligation, which however is not due to any subject's negligence.

On the other hand, st is weakly accountable. This is because Alice can always fulfill her obligation at time 20. If she still does not have the privilege to read f at time 20, it must be due to Bob's negligence.

5. THE ACCOUNTABILITY PROBLEM

In this section, we study the *accountability problem*, i.e., given a state in a system, determining whether it is accountable. This problem is naturally faced by a reference monitor when a subject makes a request to take an action which, if allowed, would result in the assignment of obligations.

Note that since the constructs of our metamodel are very abstract, it can accommodate systems with arbitrarily complex internal structures. It is not hard to see that, purely based on the metamodel without any constraints on a system's properties, the accountability problem can easily be undecidable. In the following, we show a reduction of the halting problem to the accountability problem.

Given a Turing machine T, we can construct a system which emulates that Turing machine by specifying that the system state, σ, includes a potentially infinite tape, a position on that tape, a current machine state, and a boolean variable which describes whether or not the system has halted. Then we define a single action "Advance", which changes the state and the tape in accordance with the state transition rules of T and, if T halts, sets the halt variable to true. For purposes of simplicity, we define a single subject, s. To make the Turing machine operate, we define an initial obligation of $(s, Advance, [1, 1])$. Then we define a policy rule for Advance so that the state can be advanced so long as the machine has not halted and causes an obligation to advance the state again. Explicitly, the policy rule is $Advance(s) \leftarrow halt = false : f(\sigma, t, s) = (s, Advance, curr+1, curr+1)$, where $curr$ denotes the current time of a system. As such, if the machine ever halts, there will be an obligation which cannot be fulfilled. But if it never halts, there will not be one. Therefore, the question of whether or not the state is unaccountable is equivalent to the question of whether or not the Turing machine will halt. As such, in the worst case, the accountability problem is undecidable, when only considering the constructs of the metamodel without any constraints.

To describe such a reduction is, of course, by no means to say that the accountability problem is undecidable in all obligations systems. In specific systems, determining accountability may often be quite easy. In particular, we are interested in identifying the properties of obligation systems which allow us to efficiently solve the accountability problem.

Let us consider obligation systems that satisfy the following conditions:

- No cascading obligations. In the metamodel, the action to fulfill an obligation may also incur further obligations. If a system does not have such cascading obligations, then each obligation only involves actions whose policies do not carry obligations.

- Monotonicity. From a given state, if the condition on a policy is true for a subject, it will remain true in all future states. In other words, the set of rules that a subject can satisfy does not decrease during state transitions. As such, the set of rules is monotonic relative to time.

- Commutative actions. For any two actions, a_1 and a_2 if the conditions of the policy rules of both a_1 and a_2 are met, then taking action a_1 followed by a_2 has the same effect on the system state as taking action a_2 followed by a_1.

These properties might seem a little draconian. But as we will show later, if we remove any one of them, without considering other specifics of a system, the accountability problem is intractable. Again, this does not mean that particular systems which do not have these properties cannot be efficiently solved, since any particular system would have additional properties which we do not assume here. In fact, in section 6 we will present a class of systems which do not conform to all of these assumptions, but do have an efficient algorithm for the accountability problem. With that said, any time we have a system where the above three properties hold, we can efficiently solve the accountability problem, without looking at other specifics of that system.

THEOREM 1. *Given a system that satisfies the above three properties, the problem to check whether a given state is weakly accountable is tractable.*

PROOF. Sketch. In a monotonic system, once an obligatory action becomes enabled, it remains so in later states. Thus, weak accountability in this context is equivalent to requiring that each obligatory action is enabled at the end of its time window. The scenario that will be most challenging with respect to enabling a given obligatory action is the one in which all other obligatory actions are taken at the latest possible time, right before their deadlines. Therefore, we can evaluate a particular state by examining what happens when each obligatory action in that state is attempted at the obligation's deadline. If by so doing, every obligation can be fulfilled (presuming that all other obligations are fulfilled), then the state is weakly accountable. If there exists an obligation in the state for whose action there exists no policy rule whose condition is satisfied when the obligation's deadline has arrived, then the state is unaccountable.

Assuming it takes a constant time to check whether a condition is satisfied in a state, the complexity of the above algorithm is $O(nm)$, where n is the number of pending obligations in a state, and m is the number of action rules in the policy. □

We have a similar result for the checking of strong accountability.

THEOREM 2. *Given a system that satisfies the above three properties, the problem to check whether a given state is strongly accountable is tractable.*

PROOF. Sketch. In strong accountability, it must be the case that an obligation can be fulfilled at any point during its time frame. In a monotonic system, this is equivalent to being able to be fulfilled at the start of its time period. However, if we were to simply schedule all obligations for the start points of their time windows, we would not have a worst case scenario, since all the conditions would be fulfilled as soon as they possibly could.

Instead, we iterate through each obligation individually and consider whether or not it will be able to be fulfilled at the beginning of its time period assuming that the other obligations all happen as late as possible. We can do this by assuming that all obligations whose end time is before the current obligation's start time have occurred and seeing if the state will allow the current obligation to be fulfilled. If all obligations can be fulfilled under these conditions, then the state is accountable, elsewise it is unaccountable. The complexity of this algorithm is $O(n^2 m)$, where n is the number of pending obligations in a state, and m is the number of action rules in the policy. □

If we remove one of the above three properties, then the accountability problem becomes intractable, when only given the constructs of the metamodel.

THEOREM 3. *Given a system which only satisfies two of above three properties, i.e., the no cascading obligation, the monotonicity and commutative action properties, the problem of determining whether a state of the system is strongly/weakly accountable is intractable.*

Instead of attempting to reduce NP-complete problems to the accountability problem, it is in fact much easier to reduce them to the problem of checking of unaccountability, and as such we prove that under the above conditions, the accountability problem is Co-NP Hard. Due to space limits, we present one of these three reductions in appendix A and the other two reductions only in the technical report version of this paper [16].

6. A CONCRETE MODEL

As we saw in the previous section, in the abstract model, determining the accountability of a system can be done efficiently provided several restrictions are placed on the system. In particular, one requirement was that actions perform state transitions that cause the set of enabled actions in the system to either increase (with respect to $\subset$) or stay the same. In the abstract model, this could not be relaxed without losing tractability. However in practice the restriction is unlikely to be satisfactory, as it means that once a subject is able to perform a given action, they will always be able to perform that action.

By contrast with our meta-model, in practice, permission states are structured objects. It turns out that entirely realistic assumptions about that structure enable us to remove the assumption that actions must increase while preserving tractability.

In this section we present a concrete model in which Σ, the set of abstract states, is instantiated to be $\mathcal{M} = 2^{\mathcal{S} \times \mathcal{O} \times \mathcal{R}}$, the set of permission sets, in which $\mathcal{R}$ is a set of access rights subjects can have on objects. We denote permission sets by M and individual permissions by $m = (s, o, r)$. Each permission is a triple consisting of a subject, an object, and an access right, and signifies that the subject has the right on the object.

Actions are also modified so as to operate on permission sets. Each action $a \in \mathcal{A}$ is now assumed to perform a finite sequence of operations that each either add or remove a single permission from the permission set ($grant(m)$ and $revoke(m)$). Clearly, a subject or an object with no associated permissions has no effect on the system, so we assume that in every state st, an object or a subject exists in $st.O$ and/or $st.S$ if and only if it occurs in some permission in the permission set $st.M$.

As we show below, the following restrictions on the model make the problem of determining whether a state is accountable tractable.

1. Policy rule conditions consist of a Boolean combination of permission tests ($m \in M_{cur}$ or $m \notin M_{cur}$) expressed in conjunctive normal form.

2. Actions are partitioned into two sets—the first consists of actions whose policy rules can impose obligations and the second consists of actions that can occur in those obligations. This means that one cannot become obligated to take actions in the first set, but performing such actions voluntarily can incur obligations to perform actions in the second set. Performing these latter actions does not incur any obligations.

6.1 Basic algorithm for dealing with the simplified concrete model

Given a current state $st_{cur} = \langle t_{cur}, M_{cur}, B_{cur} \rangle$ that is known to be accountable, we can use the procedure below to determine whether or not adding a new obligation b leaves the system in an accountable state. Given a set of obligations B that need to be added, we can use the procedure by considering the elements of the set one at a time, adding each obligation in turn to B_{cur} unless it would leave the system in an unaccountable state, in which case we stop and return the result that B cannot be added while preserving accountability.

We are given an obligation $b = \langle b.s, b.a, b.t_s, b.t_e \rangle$. Let the policy rule that governs $a = b.a$ be $a \leftarrow cond(s, t, M)$. Under the restrictions identified above, $cond(s, t, M)$ is a Boolean combinations of permission membership tests expressed in conjunctive normal form (i.e., a disjunction of conjunctions). The following steps are used to check each permission membership test in turn to see whether it is guaranteed. If all tests in one of the disjuncts (i.e., one of the conjunctions) are guaranteed, then the obligation can be added to obtain an accountable system. Essentially, all we are doing is checking existing obligations to ensure that the last obligation which touches the same permission has left it in the state we want. If it has revoked the right we need or if it is not clear which obligation will be done last due to overlapping, conflicting obligations, then we reject.

1. Check rights. We do the following for each test which is part of the precondition. We assume the test is positive (*i.e.*, let it take the form $((s, o, r) \in M)$. If in fact the test is negative $((s, o, r) \notin M)$, then the procedure is obtained from the following by reversing the roles of "grant" and "revoke" and negating permission membership tests.

   (a) If there is an overlapping revoke action, *i.e.*, some $br \in B_{cur}$ has $br.a = revoke(b.s, o, r)$ and the intervals $[b.t_s, b.t_e]$ and $[br.t_s, br.t_e]$ intersect, then the test cannot be guaranteed.

   (b) Otherwise, if the privilege exists in the current state, *i.e.*, $(b.s, o, r) \in M_{cur}$ then

   i. If there is a prior revoke action, *i.e.*, some $br \in B_{cur}$ with $br.a = revoke(b.s, o, r)$ has $br.t_e < \mathbf{b.t_s}$, then pick such a $br \in B_{cur}$ so as to maximize $br.t_e$ (subject to $br.t_e < \mathbf{b.t_s}$). The test can be guaranteed only if there exists an obligation that someone grant the permission (again) after br but before b, *i.e.*, only if some $bg \in B_{cur}$ has $bg.a = grant(b.s, o, r)$, $br.t_e < bg.t_s$ and $bg.t_e < \mathbf{b.t_s}$.

   ii. Otherwise, the test can be guaranteed.

   (c) Otherwise, if the privilege does not exist in the current state then

   i. If there is some grant obligation for the tested permission $bg \in B_{cur}$ that ends before b **starts**, then pick some bg so as to maximize $bg.t_s$ while satisfying $bg.t_e < \mathbf{b.t_s}$. The test can be guaranteed only if no revoke obligation for the tested permission $br \in B_{cur}$ that overlaps with the interval $[bg.t_s, b.t_e]$.

   ii. Otherwise, the test cannot be guaranteed.

2. Check effect of b on obligations it overlaps. If b revokes or grants a right which could cause the condition of an obligation it overlaps to be false, then the state is not accountable.

3. Check effect of b on later obligations. The obligation b either grants or revokes some right. Obligations which depend on the presence or absence of this right need to be considered. To check them, we repeat step 1 of this algorithm for each of them.

The equivalent algorithm for weak accountability is related, but more complex. Obviously the relevant deadlines differ, but there is also a matter of dependency analysis which must be performed. Due to space constraints, we do not include it here. It can be found in the tech report [16].

THEOREM 4. *Under the restrictions identified above, the problems of determining whether a concrete system is strongly accountable and weakly accountable are in* ***P***.

7. RELATED WORK

Several policy languages have been proposed recently that support the specification of obligations in security policies. XACML [28] and KAoS [29] both have a limited model of obligations. Specifically, they model obligations assigned to a system and cannot describe user obligations, *i.e.*, obligations assigned to ordinary users who are not always trusted to fulfill obligations. Ponder [9] and Rei [19] both support the specification of user obligations. However, in the basic constructs of both languages, time constraints of obligations, *e.g.*, deadlines, cannot be directly expressed.

Heimdall [12] is a prototype obligation monitoring platform which keeps track of pending obligations. It detects when obligations are fulfilled or violated. This requires the modeling of time constraints in obligations, which are explicitly supported in its policy language xSPL. Sailer and Morciniec [24] propose a means of using a third party to monitor obligation compliance in contracts in web services settings.

Bettini et al. [4] studied the problem of choosing appropriate policy rules to minimize the provisions and obligations that a user receives in order to take certain actions. In their policy model, each privilege inference rule is associated with a set of obligations and provisions, *i.e.*, actions that have to be taken *before* a request can be granted. Different from the policy model used in this paper, they assume actions in provisions and obligations are disjoint from those requiring privileges. In other words, they can always be fulfilled. Clearly, with such policies, a system state is always accountable. Bettini et al. [5,6] further extended their policy model to express the handling of obligation violations.

While the above works focus on the specification and monitoring of obligations, this paper formally defines secure states of a system with the presence of obligations, and studies the complexity of checking whether a state is secure. Therefore, this research is complementary to the above mentioned works.

There have been other attempts to analyze systems with obligations to determine whether or not parties have sufficient rights to carry out their obligations. Firozabadi et al. [11] describe a system for reasoning about obligations and policies in virtual organizations. However, their policies and obligations are both just static allotments of resources at specified time periods, making the comparison quite simple at the cost of being very specific to their model. Kamoda et al. [20] attempt something a little more comprehensive in their model of policies in a web services setting. They model obligations and privileges which combine subjects, actions, and roles including a model of role hierarchies. However, their model is still limited because they model privileges and obligations as being triggered only by events which are independent of the actions in the system. As such they are unable to model any situation in which user actions can change the state of the system.

A large amount of work has been done on access control policies. A variety of policy languages and models have been proposed. Some of them are generic (e.g., [10, 17, 18, 25, 30]) while others are for specific applications (e.g., [1, 8, 23, 27]) or data models (e.g., [2, 3, 13, 21]). A common problem in access control is compliance checking, i.e., whether an access should be allowed according to an access control policy. Depending on access control models, compliance checking may be very simple (e.g., checking an access control list), or quite complex (e.g., in distributed trust management [7]). The problem of determining whether a state is accountable is analogous to compliance checking in access control, in that it must be performed to determine whether to allow a requested operation. However, since it needs to consider the fulfillment of obligations in future states, the determining accountability is inherently more complicated.

Another important class of problems in access control is static policy analysis, e.g., safety analysis [14, 26] and availability analysis [22]. It is interesting to investigate what types of policy analysis can be performed in obligation policies, but existing work does not address obligations. The analysis presented in this paper is dynamic, but accountability can be used in static analysis of systems.

8. CONCLUSION

Obligations can be important to the correct operation and availability of systems and applications. The specification of obligations thus has increasingly been integrated into security policies. In this paper, we have proposed a useful means of modeling this combination of security policies with obligation policies. This has lead us to the concept of accountability, which we have formally defined. Beyond that, we proven several results concerning the complexity of determining whether or not a system state is accountable. We have also described a reasonable, more concrete system in our meta-model and outlined an algorithm for checking the accountability of states in that system. In short, we have investigated, formally, the relationship between obligations and security policies.

However, many interesting open problems still exist in the management and analysis of obligations. In particular, we would like to investigate the following problems in the future.

- When a reference monitor determines that an action will cause a system to become unaccountable, there are several possible actions it could take. Simple denial may have a negative impact on the availability of services. Another possible approach is to allow the action, but meanwhile adjust the obligations in the system to make sure the resulting state is still accountable. For example, if Bob would like to delete Alice from a system, then the system may either transfer Alice's obligations to Bob or discharge them. What decisions the reference monitor makes should also be part of obligation management policies.
- In some systems it may be complex to check a state's accountability dynamically. However, there may be static accountability analysis which can be done. In particular, we would like to investigate those properties of obligation policies which ensure that when a system enforces its policies it will never become unaccountable. For example, if the conditions of the rules of those actions involved in obligations are always true, then all the system states are trivially accountable. We would like to identify more such properties in the future.
- We present a concrete model for authorization systems with obligations in section 6. It would be interesting to make it more specific to support commonly available features in today's access control systems such as roles and cascading delegations.
- Obligations may also be assigned due to the occurence of unexpected events instead of explicit action requests [9]. For example, a policy may specify that a system administrator is obliged to restore the file server within 24 hours after a system crash. We plan to extend our metamodel to support event triggered obligations, and investigate how the concept of accountability may be affected after introducing them.

Acknowledgments

This research was sponsored by the NSF through IIS CyberTrust grant 0430166 (NCSU), NSF ITR award CCR-0325951 (through a sub-award from Brigham Young University), and NSF award CCF-0524010. We also thank our anonymous reviewers for their helpful comments.

9. REFERENCES

[1] R. J. Anderson. A security policy model for clinical information systems. In *Proc. IEEE Symposium on Security and Privacy*, pages 30–43, 1996.

[2] E. Bertino, F. Buccafurri, E. Ferrari, and P. Rullo. A logical framework for reasoning on data access control policies. In *Proc. 12th IEEE Computer Security Foundations Workshop*, pages 175–189, 1999.

[3] E. Bertino, S. Castano, and E. Ferrari. On specifying security policies for web documents with an XML-based language. In *Proc. 6th ACM Symposium on Access Control Models and Technologies*, Chantilly, VA, May 2001.

[4] C. Bettini, S. Jajodia, X. S. Wang, and D. Wijesekera. Provisions and obligations in policy management and security applications. In *VLDB*, Hong Kong, China, Aug. 2002.

[5] C. Bettini, S. Jajodia, X. S. Wang, and D. Wijesekera. Obligation monitoring in policy management. In *IEEE International Workshop on Policies for Distributed Systems and Networks (POLICY 2003)*, Lake Como, Italy, June 2003.

[6] C. Bettini, S. Jajodia, X. S. Wang, and D. Wijesekera. Provisions and obligations in policy rule management. *J. Network Syst. Manage.*, 11(3), 2003.

[7] M. Blaze, J. Feigenbaum, and M. Strauss. Compliance Checking in the PolicyMaker Trust Management System. In *Financial Cryptography*, British West Indies, Feb. 1998.

[8] C. Bussler and S. Jablonski. Policy resolution for workflow management systems. In *Proc. Hawaii International Conference on System Science*, Maui, Hawaii, January 1995.

[9] D. Damianou, N. Dulay, E. Lupu, and M. Sloman. The Ponder Policy Specification Language. In *2nd International Workshop on Policies for Distributed Systems and Networks*, Bristol, UK, Jan. 2001.

[10] N. Damianou, N. Dulay, E. Lupu, and M. Sloman. The ponder policy specification language. In *Proc. International Workshop on Policies for Distributed Systems and Networks*, pages 18–38, 2001.

[11] B. S. Firozabadi, M. Sergot, A. Squicciarini, and E. Bertino. A framework for contractual resource sharing in coalitions. In *5th IEEE International Workshop on Policies for Distributed Systems and Networks (POLICY 2004*, Yorktown Heights, New York, June 2004.

[12] P. Gama and P. Ferreira. Obligation policies: An enforcement platform. In *6th IEEE International Workshop on Policies for Distributed Systems and Networks (POLICY 2005)*, Stockholm, Sweden, June 2005.

[13] P. Griffiths and B. Wade. An authorization mechanism for a relational database systems. *ACM Transactions on Database Systems*, 1(3), 1976.

[14] M. A. Harrison, W. L. Ruzzo, and J. D. Ullman. Protection in operating systems. *Communications of the ACM*, 19(8):461–471, Aug. 1976.

[15] IBM. *Enterprise Privacy Authorization Language (EPAL 1.1) Specification.* http://www.zurich.ibm.com/security/enterprise-privacy/epal/.

[16] K. Irwin, T. Yu, and W. Winsborough. On the modeling and analysis of obligations. Technical Report NCSU CS TR 2006-26, North Carolina State University, 2006. ftp://ftp.ncsu.edu/pub/unity/lockers/ftp/csc_anon/tech/2006/TR-2006-26.%pdf.

[17] S. Jajodia, P. Samarati, and V. S. Subrahmanian. A logical language for expressing authorizations. In *Proc. 1997 IEEE Symposium on Security and Privacy*, pages 31–42, 1997.

[18] S. Jajodia, P. Samarati, V. S. Subrahmanian, and E. Bertino. A unified framework for enforcing multiple access control policies. In *Proc. ACM SIGMOD International Conference on Management of Data*, pages 474–485, 1997.

[19] L. Kagal, T. W. Finin, and A. Joshi. A policy language for a pervasive computing environment. In *IEEE International Workshop on Policies for Distributed Systems and Networks (POLICY 2003)*, Lake Como, Italy, June 2003.

[20] H. Kamoda, M. Yamaoka, S. Matsuda, K. Broda, and M. Sloman. Policy conflict analysis using free variable tableaux for access control in web services environments. In *Policy Management for the Web Workshop*, Chiba, Japan, May 2005.

[21] M. Kudo and S. Hada. XML document security based on provisional authorization. In *Proc. ACM Conference on Computer and Communication Security*, Athens, Greece, November 2000.

[22] N. Li, W. H. Winsborough, and J. C. Mitchell. Beyond proof-of-compliance: Safety and availability analysis in trust management. In *Proceedings of IEEE Symposium on Security and Privacy*, pages 123–139. IEEE Computer Society Press, May 2003.

[23] T. Ryutov and C. Neuman. Representation and evaluation of security policies for distributed system services. In *Proc. DARPA Information Survivability Conference and Exposition*, January 2000.

[24] M. Sailer and M. Morciniec. Monitoring and execution for contract compliance. Technical Report TR 2001-261, HP Labs, 2001.

[25] R. Sandhu, V. Bhamidipati, and Q. Munawer. The ARBAC97 model for role-based aministration of roles. *ACM Transactions on Information and Systems Security*, 2(1):105–135, Feb. 1999.

[26] R. S. Sandhu. The Schematic Protection Model: Its definition and analysis for acyclic attenuating systems. *Journal of ACM*, 35(2):404–432, 1988.

[27] E. Sirer and K. Wang. An access control language for web services. In *Proc. 7th ACM Symposium on Access Control Models and Technologies*, Monterey, CA, June 2002.

[28] X. TC. Oasis extensible access control markup language (xacml). http://www.oasis-open.org/committees/xacml/.

[29] A. Uszok, J. M. Bradshaw, R. Jeffers, N. Suri, P. J. Hayes, M. R. Breedy, L. Bunch, M. Johnson, S. Kulkarni, and J. Lott. Kaos policy and domain services: Toward a description-logic approach to policy representation, deconfliction, and enforcement. In *IEEE International Workshop on Policies for Distributed Systems and Networks (POLICY 2003)*, Lake Como, Italy, June 2003.

[30] OASIS eXtensible Access Control Markup Language (XACML). http://www.oasis-open.org/committees/xacml/, 2005.

APPENDIX
A. REDUCTION

In section 5, we describe three system properties under which we can solve the problem of accountability efficiently. Here we show that if we remove one of these properties, the problem of unaccountability become NP-Hard, and hence the problem of accountability is Co-NP-Hard. Specifically, in [16] we show reductions for systems where we relax each of the three assumptions. Due to space constraints, we only present one of them here, specifically, we show that if we do not assume monotonic systems then set covering problem can be reduced to the unaccountability problem.

In set covering problem, there is a set, S, and a set of n subsets of S, $\{S_1, \ldots, S_n\}$. There is also a parameter, k, and the question is whether or not there is a set of k of the subsets such that the union of those sets is equal to S. That is, is every member of S "covered" by one of the k subsets.

In this case we are not going to have cascading obligations. As such, we cannot have a system in which the obligations force us to make choices between different options. Instead we will structure things such that the different options are represented by different schedules of the same actions.

As such, our state, σ is represented by a set S' and an integer m. We have a single subject, s. We have $n+1$ actions. For $i = 1, \ldots, n$ we define an action $remove_i$ which removes S_i from S' and increments the value of m by one. We also define an action $check$ which has no effect.

Next we define the policies for the actions. For $i = 1, \ldots, n$ we define policies $remove_i(s) \leftarrow true : f(\sigma, t, s) = \emptyset$ and $check(s) \leftarrow \neg(m = k \wedge S' = \emptyset) : f(\sigma, t, s) = \emptyset$.

Our initial state is the state such that $m = 0$, $S' = S$ and we have a set of obligations $\{(s, remove_i, 0, 2n) | i = 0, \ldots, n\} \bigcup \{s, check, 0, 2n\}$. All of the subset actions can be fulfilled. But only those which are fulfilled before the check obligation matter to the check of accountability. So, the question of accountability is the question of whether or not it's possible to schedule k of the subset obligations before the check obligation which together cover S. If it is possible, then the initial state is unaccountable. If it is impossible, then the initial state is accountable, since that means that there is no schedule which will result in an obligation failure. As such, we have reduced Set Cover to unaccountability.

We finish by verifying that our other two assumptions hold in our system. Clearly, there are no cascading obligations as no policy contains any new obligations. And it is the case that all actions are commutative since for any two actions, $remove_i$ and $remove_j$, the result of apply them is that S' takes on the value of $S' \backslash (S_i \bigcup S_j)$ and m takes on the value $m + 2$. Therefore, we have successfully demonstrated that accountability is Co-NP Hard when only two of our three conditions are assumed.

RoleMiner: Mining Roles using Subset Enumeration

Jaideep Vaidya
MSIS Department and CIMIC
Rutgers University
Newark, NJ USA
jsvaidya@cimic.rutgers.edu

Vijayalakshmi Atluri
MSIS Department and CIMIC
Rutgers University
Newark, NJ USA
atluri@rutgers.edu

Janice Warner
MSIS Department and CIMIC
Rutgers University
Newark, NJ USA
janice@cimic.rutgers.edu

ABSTRACT

Role engineering, the task of defining roles and associating permissions to them, is essential to realize the full benefits of the role-based access control paradigm. Essentially, there are two basic approaches to accomplish this: the *top-down* and the *bottom-up*. The top-down approach relies on a careful analysis of the business processes to define job functions and then specify appropriate roles from them. While this approach can aid in defining roles more accurately, it is tedious and time consuming since it requires that the semantics of the business processes be well understood. Moreover, it ignores existing permissions within an organization and does not utilize them. On the other hand, the bottom-up approach starts with existing permissions and attempts to derive roles from them, thus helping to automate role definition. In this paper, we present an unsupervised approach called *RoleMiner* that mines roles from existing user-permission assignments. Since a role is nothing but a set of permissions, when no semantics are available, the task of role mining is essentially that of clustering users that have same (or similar) permissions. However, unlike the traditional applications of data mining that ideally require identification of non-overlapping clusters, roles will have overlapping permission needs and thus permission sets that define roles should be allowed to overlap. It is this distinction from traditional clustering that makes the problem of role mining non-trivial. Our experiments with real and simulated data sets indicate that our role mining process is quite accurate and efficient.

Categories and Subject Descriptors

K.6 [**Management of Computing and Information Systems**]: Security and Protection

*This work is supported in part by the National Science Foundation under grant IIS-0306838.

Permission to make digital or hard copies of all or part of this work for personal or classroom use is granted without fee provided that copies are not made or distributed for profit or commercial advantage and that copies bear this notice and the full citation on the first page. To copy otherwise, to republish, to post on servers or to redistribute to lists, requires prior specific permission and/or a fee.
CCS'06, October 30–November 3, 2006, Alexandria, Virginia, USA.
Copyright 2006 ACM 1-59593-518-5/06/0010 ...$5.00.

Keywords

RBAC, role engineering, role mining

1. INTRODUCTION

Role-based access control (RBAC) has now become a well-accepted model in many organizations for enforcing security. Roles represent organizational agents that perform certain job functions within the organization. Users in turn are assigned appropriate roles based on their qualifications. According to the RBAC reference model [4], roles describe the relationship between users and permissions. Users are human beings and permissions are a set of many-to-many relations between objects and operations. Roles can be hierarchically structured, where senior roles generally inherit the permissions assigned to junior roles. Additionally, constraints such as separation of duties may be associated with the roles.

Employing RBAC is not only convenient but reduces the complexity of access control because the number of roles in an organization is significantly smaller than that of users. Moreover, the use of roles as authorization subjects, instead of users, avoids having to revoke and re-grant authorizations whenever users change their positions and/or duties within the organization. RBAC essentially decouples users and permissions, thereby reducing the burden of security administration. As a result, RBAC has been implemented successfully in a variety of commercial systems, and has become the norm in many applications.

It has been identified by Edward Coyne [2] that the goal of *role engineering* is to define a set of roles that is complete, correct and efficient. He has pointed out that this task is essential before all the benefits of RBAC can be realized. In particular, role engineering requires defining roles and assigning permissions to them. Often, this is a cooperative process where various authorities from different disciplines should understand the semantics of operations of one another. According to a study by NIST [5], role engineering is the costliest component of RBAC implementation. Since role engineering is a time-consuming and a costly process, organizations are often reluctant to move to RBAC. Therefore, a systematic and efficient means of accomplishing definition of roles is essential to minimizing the expenses of RBAC implementation and maintenance.

There are essentially two ways of accomplishing the task of role engineering: *top-down* and *bottom-up*. In the top-down approach, roles are defined by carefully analyzing and decomposing business processes into smaller units in a functionally independent manner. These functional units are then associated with permissions on information systems. In other words, this approach begins with defining a particular job function and then creating a role for this job function by associating needed permissions. With dozens of business processes, tens of thousands of users and millions of authorizations, this is seemingly a difficult task. Therefore, relying on a top-down approach in most cases is not viable, although, as shown in case studies such as that presented at SACMAT 2001 by Schaad et al. [13], is being done successfully by some organizations. In contrast, the bottom-up approach starts from the existing permissions before RBAC is implemented and aggregates them into roles. One may also use a mixture of these two approaches to conduct role engineering. While the top-down model is likely to ignore the existing permissions, a bottom-up model may not consider business functions of an organization [8]. However, the bottom-up approach is advantageous in that, much of the role engineering process can be automated. Role mining can be used as a tool, in conjunction with a top-down approach, to identify potential or candidate roles which can then be examined to determine if they are appropriate given existing functions and business models.

In this paper, we propose a bottom-up approach to *mine* roles from the existing permissions to aid in migrating to role based access control. Our approach not only eases the task of role engineering, it also helps in providing the security administrators an insight into the user-role assignment. It serves as a tool to assist the security administrator in discovering potential roles. It does, however, require an expert review of the results to choose which discovered roles are most advantageous to implement.

In order to mine roles, one must first start with a definition of a role. We begin with the fact that a role is simply a set of permissions. With respect to mining roles from permissions, we make the following observations.

- Roles can be assigned overlapping permissions.

- The above also implies that a particular permission might be held by members of different roles. That is, permissions are not exclusive to roles nor are they exclusive to a hierarchy.

- A user may play several different roles, and the user may have a certain permission due to more than one of those roles (since multiple roles may include the same permission).

In our approach, we take into account all the above observations. However, it is important to note that ignoring certain observations would make the job of role mining easier, but may result in more inaccurate set of roles. Unlike the traditional applications of data mining that ideally require identification of clusters that do not overlap, role mining indeed requires identification of overlapping clusters. This is because it is often the case that a user may belong to more than one role. It is this distinction that makes the problem of role mining non-trivial. Therefore, aggregation of permissions to create roles by employing a simple clustering algorithm will likely not result in the discovery of actual roles. This is where we differ from earlier research. In particular, our proposed technique of mining roles is accomplished via subset enumeration. Our algorithm computes all possible interesting roles. The roles are then prioritized according to certain metrics, currently based on counts of users who have the permission set of the discovered roles, and presented as candidate roles.

When compared to the earlier role mining proposals, our approach enjoys several advantages. First, as mentioned earlier, it is capable of detecting roles with overlapping permission sets. Second, given the set of user-permission assignments, the subset enumeration employed in our RoleMiner allows us to identify the set of *all possible roles* within an organization, ordered such that the most useful roles are found at the top of the set. Third, our approach does not force every permission to be part of a role, but instead identifies the permissions that distinguish the role. Other permissions held by users could be identified as anomalous. Fourth, unlike some of the previously proposed approaches that are sensitive to the order in which initial role sets are considered, our approach is order-independent. Finally, it is incremental in that when new users and permissions are added, our RoleMiner updates the roles in an incremental fashion, rather than starting the discovery process from scratch.

This paper is organized as follows. Section 2 discusses some preliminaries to our work, including RBAC and Data Mining. Section 3 presents our algorithm for mining roles. Section 4 presents our experimental evaluation and summarizes the results found. Section 5 discusses some of the consequences of the algorithm. Related work is presented in Section 6. Finally, Section 7 concludes the paper and presents avenues for future research.

2. PRELIMINARIES

In this section, we briefly review the RBAC and clustering concepts.

2.1 Role-based Access Control

We adopt the NIST standard of the Role Based Access Control (RBAC) model [4]. For the sake of simplicity, we do not consider sessions or separation of duties constraints.

DEFINITION 1. [RBAC]

- $U, ROLES, OPS,$ and RES are the set of users, roles, operations, and resources.

- $UA \subseteq U \times ROLES$, a many-to-many mapping user-to-role assignment relation.

- PRMS (the set of permissions) $\subseteq \{(op, res)|op \in OPS \bigwedge res \in RES\}$

- $PA \subseteq PRMS \times ROLES$, a many-to-many mapping of permission-to-role assignments.

- $assigned\_users(R) = \{u \in U|(u,R) \in UA\}$, the mapping of role R onto a set of users.

- $assigned\_permissions(R) = \{p \in PRMS|(p,R) \in PA\}$, the mapping of role R onto a set of permissions.

2.2 Clustering

Clustering divides a data set into different groups. The goal of clustering is to find groups that are very different from one another, and whose members are very similar to one another. In clustering, at start, one does not know what the clusters will be, or even which attributes will be used to cluster the data. Consequently, domain experts are needed to interpret the results. Once reasonable clusters have been found, these can then be used to classify new data. Some of the common algorithms used to perform clustering include Kohonen feature maps [9] and K-means. Berkhin [1] provides an excellent survey of clustering.

Clustering seems to be a ready solution to our problem, since our end goal is to cluster the permissions into roles. One problem is that clustering typically applies to numeric data, while user/permissions data is categorical (in fact, boolean). There has been recent work in the data mining field on clustering categorical data [6, 7]. However, even this suffers from the problem that typically there is single assignment of points to clusters. Thus, in our case, a permission would be assigned to a single role – which is clearly an artificial constraint. This is precisely the drawback of Schlegelmilch and Steffens's role mining approach [14] which does in fact use an agglomerative clustering algorithm to find the roles. Some sort of alternative approach is required to handle the drawbacks.

3. ROLE MINING THROUGH SUBSET ENUMERATION

Our approach to role mining is to discover roles from the data set of permissions possessed by all users. We begin with the assumption that roles are nothing but groups of permissions. In this sense, if the number of total different permissions is k, the number of all possible roles is 2^k. However, the meaningful and existing roles are a subset of these. Clearly, it is infeasible to enumerate all roles. A data driven technique is necessary to efficiently discover roles. First, one must consider the interesting properties of the data that are relevant to role discovery.

We now look at some guiding principles that are relevant to automated mining of roles.

Principle 1: Defacto role definitions are embedded in existing permissions. This basic assumption related to the presence of a role is as follows: If none of the users of a corporation have a particular permission set such as $\{p_a, p_b\}$, then this permission set probably does not form a role. The reason for using the word "probably" is that it is possible that such a role exists but no user has been assigned to that role. However, in such a case, role mining is not going to provide a meaningful answer without supervision. It is impossible for an automated program to distinguish between whether no role exists or no user belonging to such a role is present. Thus, we limit our attention to those subset of permissions that are owned by at least one user.

A lazy technique might work where we enumerate a role only if it is present. Thus, for a user, one could enumerate all subsets of the permissions the user possesses – however, this is also flawed, since very often we have users having hundreds of permissions or even more. It is impossible to enumerate all possible roles even in this limited case.

Principle 2: Identify roles with as many permissions as possible. This principle states that if two roles *always* co-occur in the dataset (i.e., any user will have either both of the roles or neither of them), then these two roles will be together detected as one role consisting of all the permissions. For example, if all users either have both the permissions p_a and p_b or neither of the permissions p_a and p_b, then the permission set $\{p_a, p_b\}$ will be detected as one role. This happens, even if in reality, there are two roles, one consisting of the permission $\{p_a\}$ while the other consists of the permission $\{p_b\}$. However, this is unavoidable in a bottom-up algorithm that utilizes no semantics. Without semantics, it is impossible to differentiate between a single role with all the permissions and two roles each having a subset of the permissions.

This also implies that only those subsets of permissions that are maximally common between at least two users should be enumerated as a role. For example, consider a dataset of only 2 users. If the first user has the permissions $\{p_a, p_b, p_c, p_d\}$, while the other has the permissions $\{p_a, p_b, p_e\}$, $\{p_a, p_b\}$ should be enumerated as a role. Only $\{p_a\}$ or only $\{p_b\}$ should not be considered as roles, since there is no way without the use of semantics to justify the decision. However, if there is a third user with perhaps the permissions $\{p_b, p_f\}$, then $\{p_b\}$ should also be identified as a potential role.

Ideally, for role engineering, we must identify potential roles as well as express some confidence in them. These are two different steps – identifying all possible roles, and then prioritizing / selecting the final set of roles from among them. Our proposed approach to discovering roles, called *RoleMiner* takes into consideration the above guiding principles. The RoleMiner algorithm consists of two parts: role identification and role prioritization. We present the details below.

3.1 Role Identification

We give two algorithms to perform role identification. The first algorithm, called CompleteMiner, is complete in the respect that all potential roles are identified. However, the runtime complexity is exponential – thus it is infeasible, except for fairly small data sets. This is presented here for clarity. The second algorithm, called FastMiner, is fast (the complexity is only n^2). The drawback is that it identifies only a subset of the potential roles. However, as will be argued later, the roles discovered by this algorithm are sufficient for practical purposes.

3.1.1 CompleteMiner

The CompleteMiner consists of three phases, described below:

1. **Identification of Initial Set of Roles:** In this phase, we group all users who have the exact same set of permissions. This can be done in a single pass over the data by maintaining a hash table of the sets of permissions seen. This significantly reduces the size of the data set since all users who have the same roles have the exact same permissions. These form the initial set of roles, say *InitRoles*.

2. **Subset Enumeration**: In this phase, we determine all of the potentially interesting roles by computing all possible intersection sets of all the roles created in the initial phase. Let this set be *GenRoles*. Each unique intersection set is added to the set of generated roles (thus only the unique set of intersections is maintained). Though the number of intersections is equal to the size of the power set of the initial roles ($2^{InitRoles}$), the actual roles enumerated are much smaller (since many intersections result in the empty set, or in the same intersection set). Note that the generated roles are the maximal set of interesting roles. Basically any set of permissions unique to a group of users is detected as a role. However, there is no need-

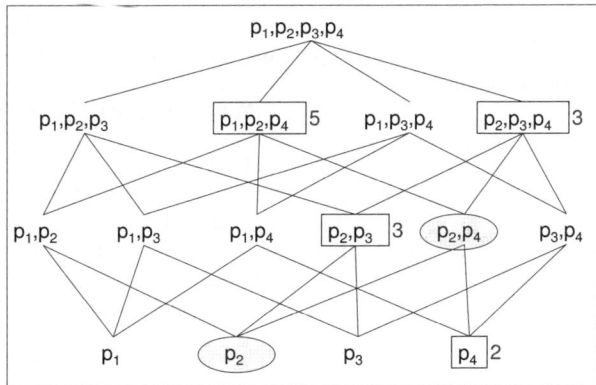

Figure 1: Example

(a) Sample Dataset for an example organization

User	p_1	p_2	p_3	p_4
u_1	0	0	0	0
u_2	1	1	0	1
u_3	0	1	1	0
u_4	1	1	0	1
u_5	1	1	0	1
u_6	0	1	1	1
u_7	0	1	1	1
u_8	0	1	1	0
u_9	0	1	1	0
u_{10}	0	0	0	1
u_{11}	0	0	0	1
u_{12}	0	0	0	0
u_{13}	1	1	0	1
u_{14}	1	1	0	1
u_{15}	0	1	1	1

(b) Results for the sample dataset

	Original Count	Generated Count
p_1, p_2, p_4	5	0
p_2, p_3, p_4	3	0
p_2, p_3	3	3
p_2, p_4	0	8
p_2	0	11
p_4	2	8

Table 1: Example data and results

less bifurcation of permissions. Thus if $\{p_1, p_2, p_3\}$ are owned by several users, but no users have any subset of those permissions, then only $\{p_1, p_2, p_3\}$ is reported as a role.

3. **User Count Computation:** In this phase, for each generated role in *GenRoles*, we count the number of users who have the permissions associated with that role. We actually maintain two sets of counts: (i) the *orig_count(i)*, the original number of users who have exactly the set of permissions corresponding to role i and nothing else, and (ii) *count(i)*, an updated count of users whose permissions are a superset of the permissions associated with this role i. It should be obvious that each generated role will have the original count set to 0, whereas each initial role will have an updated count greater than its original count if and only if it is a subset of one of the other initial roles.

Algorithm 1 gives the detailed steps. The first phase consists of lines 4-11. The *for* loop iterates over all users while the *if* statement at line 5 either increments the count of the role (if present) or adds it to the initial set. Phase 2 consists of lines 12-21. The *for* loop in line 13 iterates over all of the roles initially created. The *for* loop in lines 15-17 intersect this set with all of the remaining roles initially created and adds all the unique intersections formed to GenRoles. The *for* loop in lines 18-20 ensures the intersection with all of the roles formed in GenRoles as well. This is necessary to ensure that all possible intersections take place. Phase 3 consists of lines 22-29. The *for* loop in line 23 iterates over all the generated roles. The *for* loop in line 24 iterates over all the initial roles. Line 25 checks for existence of the subset relationship. If a generated role is a subset of an initial role, the count of users of the initial role gets added to its count (in line 26).

EXAMPLE 1. The following toy example demonstrates the working of the algorithm. Assume a hypothetical organization with 15 users and 4 permissions. Table 1(a) shows one sample database with the assignment of permissions to users. Since there are 4 permissions, and a role is defined as a collection of permissions, the number of possible different roles is $2^4 = 16$ (i.e., the size of the powerset). Figure 1

Algorithm 1 CompleteMiner

Require: Dataset $D \equiv (U, P)$
Require: $P(u)$ gives the set of permissions assigned to user u
Require: $R(x)$ represents a role consisting of the set of permissions x
Require: $Count(i)$ gives the count of users associated with role i
1: {Cluster users into initial roles based on exact match of the set of permissions}
2:   $InitRoles \leftarrow \{\}$
3:   $GenRoles \leftarrow \{\}$
4:   **for** each user $u \in U$ **do**
5:     **if** $R(P(u)) \notin InitRoles$ **then**
6:       Set orig_count of $R(P(u))$ to 1
7:       $InitRoles \leftarrow InitRoles \cup R(P(u))$
8:     **else**
9:       Increment $orig\_count$ of $R(P(u))$
10:    **end if**
11: **end for**
12: {Enumerate all possible interesting roles}
13: **for** each Role $i \in InitRoles$ **do**
14:   $InitRoles \leftarrow InitRoles - i$
15:   **for** each Role $j \in InitRoles$ **do**
16:     $GenRoles \leftarrow GenRoles \cup (i \cap j)$
17:   **end for**
18:   **for** each Role $j \in GenRoles$ **do**
19:     $GenRoles \leftarrow GenRoles \cup (i \cap j)$
20:   **end for**
21: **end for**
22: {Count number of users belonging to each candidate role}
23: **for** each Role $i \in GenRoles$ **do**
24:   **for** each Role $j \in InitRoles$ **do**
25:     **if** $i \subset j$ **then**
26:       $Count(i) \leftarrow Count(i) + orig\_count(j)$
27:     **end if**
28:   **end for**
29: **end for**

depicts 15 of those roles (all excluding the empty set - i.e., the role with no permissions).

Now, in the first phase of our algorithm, the set *InitRoles* gets initialized to $\{\{p_1, p_2, p_4\}, \{p_2, p_3, p_4\},$ $\{p_2, p_3\}, \{p_4\}, \{\}\}$. These roles along with their corresponding counts are rectangled in Figure 1. The empty set, along with its count of 2 is not shown in the figure since it does not add to the computation in phase 2 or 3. In phase 2, we enumerate all possible unique intersection sets of the initial roles found in phase 1. These result in two additional roles, $\{\{p_2, p_4\}, \{p_2\}\}$, which are ovaled in the figure. Since $\{p_2, p_3\}, \{p_4\}$ and $\{\}$ are also the result of some intersections, at the end of phase 2, *GenRoles* gets set to the roles $\{\{p_2, p_3\}, \{p_2, p_4\}, \{p_2\}, \{p_4\}, \{\}\}$. In phase 3, the generated roles are matched to the corresponding counts which are 6,8,11,10, and 2. These are shown in table 1(b).

3.1.2 FastMiner

The key problem with the CompleteMiner algorithm is its computational complexity. Since we compute *all* possible intersections, the running time is exponential. This is quite infeasible, except for very small data sets. Also, phase 3 of Algorithm 1 is unnecessary. An efficient implementation should be able to count number of users in phase 2 itself. This brings us to the efficient algorithm for role mining: The FastMiner algorithm has two main improvements. First, the only intersections performed are between pairs of initial roles. Also, the user counts are computed while intersecting the initial roles. Thus, the computational complexity is $O(n^2)$, instead of exponential. The first phase (lines 4-11) is identical to the CompleteMiner. In phase 2 (lines 13-33), all intersecting roles between pairs of users are found. Lines 17-31 update the counts for the generated role. The key is to maintain a list of contributing initial roles for each generated role (and only add counts if an initial role has not already contributed). This ensures that there is no double counting of users. Algorithm 2 provides the details.

Algorithm 2 FastMiner

Require: Dataset $D \equiv (U, P)$
Require: $P(u)$ gives the set of permissions assigned to user u
Require: $R(x)$ represents a role consisting of the set of permissions x
Require: $Count(i)$ gives the count of users associated with role i
1: {Cluster users into initial roles based on exact match of the set of permissions}
2:   $InitRoles \leftarrow \{\}$
3:   $GenRoles \leftarrow \{\}$
4:   **for** each user $u \in U$ **do**
5:     **if** $R(P(u)) \notin InitRoles$ **then**
6:       Set orig_count of $R(P(u))$ to 1
7:       $InitRoles \leftarrow InitRoles \cup R(P(u))$
8:     **else**
9:       Increment $orig\_count$ of $R(P(u))$
10:    **end if**
11: **end for**
12: {Enumerate all intersecting roles between pairs of users}
13: **for** each Role $i \in InitRoles$ **do**
14:   $InitRoles \leftarrow InitRoles - i$
15:   **for** each Role $j \in InitRoles$ **do**
16:     $NewRole \leftarrow i \cap j$
17:     **if** $NewRole \notin GenRoles$ **then**
18:       $Count(NewRole) \leftarrow Count(NewRole) + orig\_count(i)$
19:       $Count(NewRole) \leftarrow Count(NewRole) + orig\_count(j)$
20:       Add i, j to the list of contributors for NewRole
21:       $GenRoles \leftarrow GenRoles \cup NewRole$
22:     **else**
23:       **if** i has not contributed before to NewRole **then**
24:         $Count(NewRole) \leftarrow Count(NewRole) + orig\_count(i)$
25:         Add i to the list of contributors for NewRole
26:       **end if**
27:       **if** i has not contributed before to NewRole **then**
28:         $Count(NewRole) \leftarrow Count(NewRole) + orig\_count(j)$
29:         Add j to the list of contributors for NewRole
30:       **end if**
31:     **end if**
32:   **end for**
33: **end for**

Key features

While the FastMiner algorithm is quite practical, the trade off is that only a subset of the roles are identified. Specifically, only those roles that are maximally common to any 2 users are found. Thus, any role that is common to at least 3 users, but not maximally common to any 2 of them, will not be identified by the algorithm. Figure 2 highlights this problem. Figure 2(a), 2(b) and 2(c) show collections of permissions that will be detected as roles by FastMiner. However, the collection of permissions shown in Figure 2(d) will not be detected by FastMiner. These will be detected as a role only if we consider intersections between triples of initial roles (rather than pairs). This would result in an $O(n^3)$ algorithm. However, even in this case intersections between quadruples would not be detected. In the most general case, this degenerates back to the CompleteMiner algorithm. One might think that as we increase the number of roles considered in the intersections, we would also increase the accuracy of the results. While increasing complexity would improve accuracy in most data mining problems, it is not necessarily the case in role mining. Essentially, performing more intersections would result in a large number of candidate roles, making identification of useful roles more difficult. Indeed, it is not clear how one could formally quantify the correctness of the algorithm. This would require that we *know* the real set of roles. However, different system administrators might choose to have different roles from the same set of candidates due to differing situations and personalities. For example, one system administrator might choose to merge two roles and create a third super-role for the sake of convenience even though having the original two roles is sufficient to describe the organization. There are no right or wrong ways of specifying roles – the final set of roles is unique to each situation. A formal metric is needed to help understand what set of discovered candidate roles are "good" or useful roles. By quantifying this, one could determine if increasing the number of intersections (and thus the complexity) is worth the additional results. We plan on doing this as part of future work. For now, we argue that intersections between pairs are typically sufficient for detecting most roles, as long as most users have few roles. The experimental results further elucidate this point. Interestingly, the FastMiner algorithm gives the exact same results as the CompleteMiner algorithm for the toy example given earlier.

3.2 Role Prioritization

Since the FastMiner algorithm typically identifies lots of potential roles, these need to be prioritized/ordered in some way. Without the use of semantics, there is very little that we can do. One possibility is to simply order the roles according to the number of users having that role. However, instead of simply using the final Count, we use the following predicate to sort: $(orig\_count * priority + Count)$. Here, priority is simply the multiplication factor used to bias the results towards the roles found in the initial phase (i.e., do not report a generated set as a role unless it is really interesting). Thus, this ensures that a well supported original role does not get lost in the chaff of generated roles (i.e., roles that are not part of *InitRoles*). A priority factor of 0 simply sorts the roles according to the user count. We experimented with several different values for priority.

One problem with this procedure is that rare roles will not be found easily (since they will be significantly lower in the sorted list). As can be seen, the postprocessing in this step is currently quite simple. There are many other possibilities, and this is a significant task for further research. The use of semantics will also make this step significantly better. In the ideal, we wish to develop a confidence metric for a candidate role that would enable us to determine how certain we are about the suitability of the role.

Algorithm 3 CreateTestData(NRoles, NUsers, NPermissions, MRolesUsr, MPermissionsRole)

Require: NRoles, the number of roles to be generated
Require: NUsers, the number of users to be generated
Require: NPermissions, the total number of permissions
Require: MRolesUsr, the maximum number of roles a user can have
Require: MPermissionsRole, the maximum number of permissions a role can have
1: {Create the Roles}
2: **for** $i = 1 \ldots NRoles$ **do**
3:    Set nrt to a random number between $1, \ldots, MPermissionsRole$
4:    Set Roles[i] to nrt randomly chosen permissions
5: **end for**
6: {Create the Users}
7: **for** $i = 1 \ldots NUsers$ **do**
8:    Set nrl to a random number between $0, \ldots, MRolesUsr$
9:    Randomly select nrl roles from Roles
10:   Set Users[i] to the permissions in the union of the selected roles
11: **end for**
12: Return Users

---

4. RESULTS SUMMARY

In order to validate our algorithm, we created a test data generator. The test data generator performs as follows: First a set of roles are created. For each role, a random number of permissions up to a certain maximum are chosen to form the role. The maximum number of permissions to be associated with a role is set as a parameter to the algorithm. Next, the users are created. For each user, a random number of roles are chosen. Again, the maximum number of concurrent roles a user can have is set as a parameter to the algorithm. Finally, the user permissions are set according to the roles the user has. In some cases, the number of roles randomly chosen is 0 – indicating that the user has no roles – and therefore no permissions. Algorithm 3 gives the details.

Since the test data creator algorithm is randomized, we actually ran it 5 times on each particular set of parameters to generate the datasets. Our role mining algorithm was run on each of the created data sets. All results reported for a specific parameter set are averaged over the 5 runs.

For each set of experiments, we report the speed as well as the accuracy of the fast miner algorithm. To measure the accuracy, we find the percentage of true roles found in 1x the number of real roles, as well as in 2x the number of roles in the results. Thus, if the real number of roles is 200, we count the percentage of real roles found in the first 200, as well as the first 400 roles reported in the results of the algorithm.

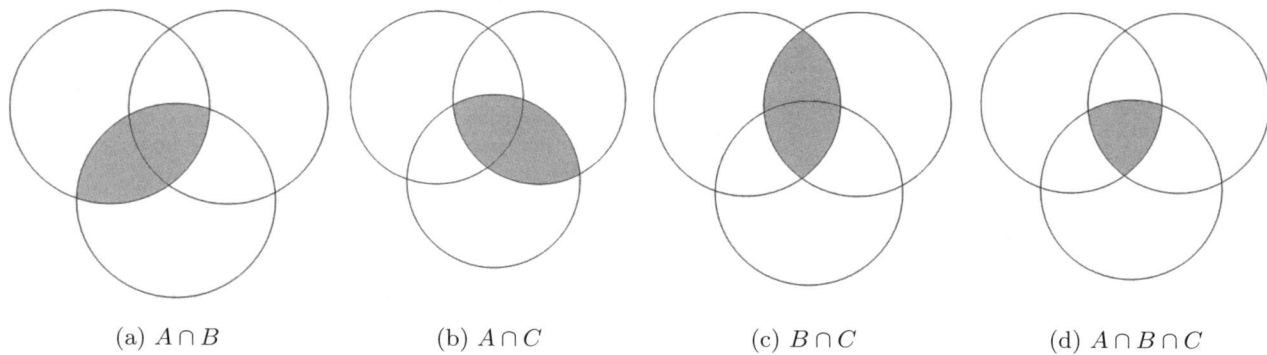

(a) $A \cap B$ (b) $A \cap C$ (c) $B \cap C$ (d) $A \cap B \cap C$

Figure 2: Limitations of FastMiner

Dataset	Parameters				
	NRoles	NUsers	NPermissions	MRolesUsr	MPermissionsRole
data1	10	2000	100	3	10
data2	100	2000	500	3	50
data3	100	2000	1000	3	100
data4	100	2000	2000	3	200

Table 2: constant number of users/roles, varying permissions

In the first set of experiments, we kept the number of users and roles constant, while changing the number of permissions (and correspondingly, the number of permissions per role). Table 2 describes the test parameters. Figure 3(b) shows the time taken by the algorithm, while Figure 3(a) shows the accuracy of the algorithm. It can be clearly seen that the accuracy is close to 100% when we look at twice the number of real roles, and close to 85% when we look at exactly the number of real roles in the results. This is quite good. The algorithm also runs quite fast, taking around 5 minutes on the largest datasets.

In the second set of experiments, we kept the number of roles and permissions constant while varying the number of users. Table 3 describes the test parameters. Figure 4(a) shows the time taken by the algorithm, while Figure 4(b) shows the accuracy of the algorithm. Here, we can see that accuracy actually increases with the number of users. The algorithm performs quite well as long as the number of roles is not more than 1/10th the number of users. This is quite realistic, and therefore the algorithm should perform well in real situations. Again, the speed of the algorithm is quite good, with the largest datasets (of 5000 users) taking approximately 23 minutes.

In the third set of experiments we keep the permissions constant, and vary the number of users as well as the number of roles. Table 4 describes the test parameters. Figure 5(a) shows the time taken by the algorithm, while Figure 5(b) shows the accuracy of the algorithm. This actually gives a surprising result – the accuracy of the algorithm significantly drops off at the largest data point (with 5000 users/500 roles). Similarly, the running time drastically increases as well. However, this is not due to the number of users (as the earlier experiment showed), but rather due to the number of roles vis-a-vis the number of permissions. When the ratio of permissions to roles is small (in this case, $1500/500 = 3$), the algorithm performs quite poorly. This is because many spurious intersections are formed which do not contribute to real roles. As long as the number of permissions is at least 5 times the number of roles, the algorithm performs quite well.

We also ran experiments on the data set provided by Ulrike Steffens [14], consisting of over 6000 users and 1671 permissions. This dataset was processed by the FastMiner algorithm in 17 minutes, which is quite fast. Interestingly, around 1400 users had no permissions at all. Out of the 679 roles discovered by OR-CA, our algorithm discovered 118 matching roles. The rest of the roles discovered by our algorithm were different from the roles in OR-CA. This is at least partially due to the fact that many of the roles we discovered have overlapping rights while this only occurs in a single tree path along a role hierarchy in ORCA. These results have been communicated to Steffens for verification (since the data was anonymized, we had no way of checking the results). We are now awaiting further communication with them.

5. DISCUSSION OF OUR APPROACH

In this section, we discuss some of the key properties of our approach. As part of identifying roles, the algorithm can also make suggestions on candidate roles that should be more heavily considered. For example, the top 100 roles reported by the algorithm are good candidates to be selected as roles based on the data set. This can some times show fallacies in pre-existing role sets.

Another useful property of the algorithm is that it is order independent. Irrespective of the order of users or permissions, the same subsets are generated. Therefore the same set of roles is generated. This is not true of many clustering algorithms and therefore is a very valuable property to have.

The role prioritization phase should actually be much more complex. For example, one could find the minimal set of roles such that all users are covered by those roles.

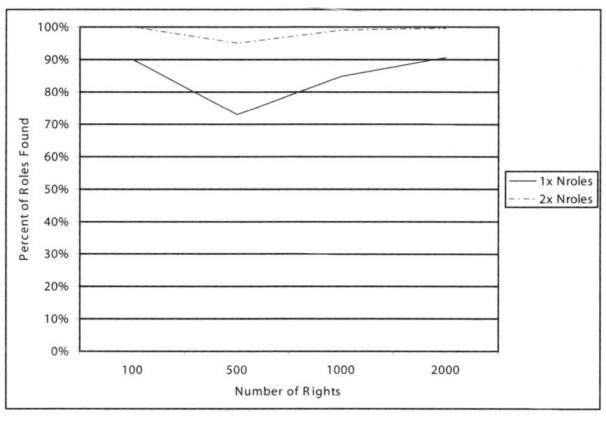
(a) Accuracy

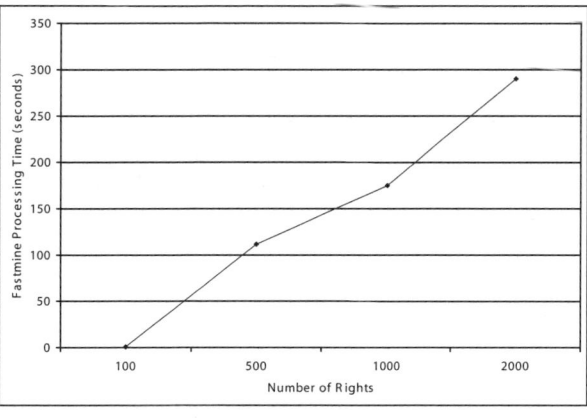
(b) Speed

Figure 3: Constant number of users

Dataset	Parameters				
	NRoles	NUsers	NPermissions	MRolesUsr	MPermissionsRole
data1	200	500	1500	3	150
data2	200	1000	1500	3	150
data3	200	3000	1500	3	150
data4	200	5000	1500	3	150

Table 3: Constant number of permissions/roles, varying users

Organization structure could be used to arrange roles in a hierarchy. Other analysis is also possible. We are experimenting with several metrics currently.

Furthermore, the algorithm is incremental. Only the final list of roles along with their associated information needs to be maintained. When a new user is added to the system, the users set of permissions can be intersected with the existing roles to determine if any additional roles are created. Updating the existing roles can also be done at the same time. This especially makes a lot of sense when a new batch of users is to be added. All of the prior work on intersection no longer needs to be repeated.

6. RELATED WORK

A number of approaches have been proposed in the literature to accomplish the task of role engineering, which can be categorized into three approaches: top-down, bottom-up, and hybrid. While the top-down approach defines roles by examining the business processes, the bottom-up approach typically aggregates existing permissions to come up with roles.

Coyne [2] is the first to describe the role engineering problem, and to present the concepts of the top-down approach. Later, Roeckle et al. [12] present a process-oriented approach, which first analyzes business processes to deduce roles and access permissions on systems are assigned to the roles. Shin et al. [15] present a system-centric approach that examines backward and forward information flows and employs UML to conduct a top-down engineering of roles. Thomsen et al. [16] propose a bottom-up approach, which derives permissions from objects and their methods and then derive roles derived from these permissions. Neumann and Strembeck [11] consider usage scenarios as a semantic unit for deriving permissions, which are then aggregated into roles. Epstein and Sandhu [3] propose to use UML for facilitating role engineering, where roles can be defined in either a top-down or a bottom-up manner. Kern et al. [8], propose a life-cycle approach, an iterative-incremental process, that considers different stages of the role life-cycle including role analysis, role design, role management, and role maintenance.

Kuhlmann, Shohat, and Schmipf [10] present another bottom-up approach, which employs a clustering technique similar to the k-means clustering. As such, it is required to first pre-define the number of clusters.

Comparison with ORCA: More recently, Schlegelmilch and Steffens [14] have proposed a role mining approach (known as ORCA). Since our proposed approach is close to this work, we provide more details on ORCA.

The fundamental difference between our RoleMiner and ORCA lies in the way clusters are formed. ORCA clusters access permissions (which they call rights) based on the users having those permissions. These initial clusters are then merged on the basis of maximal overlap between users assigned to the permissions to form a role hierarchy. Specifically, the ORCA algorithm begins with one cluster for every permission (c_r) which initially make up the set Clusters. ORCA defines permissions(c_r) = the set of permissions associated with the cluster and members(c_r) = the set of people that have permission r. It starts with a partially ordered set of clusters ($\prec$), the cluster hierarchy. Initially ($\prec$) is an empty set. Note that, just because many people have that set of permissions, that cluster is not necessarily a role.

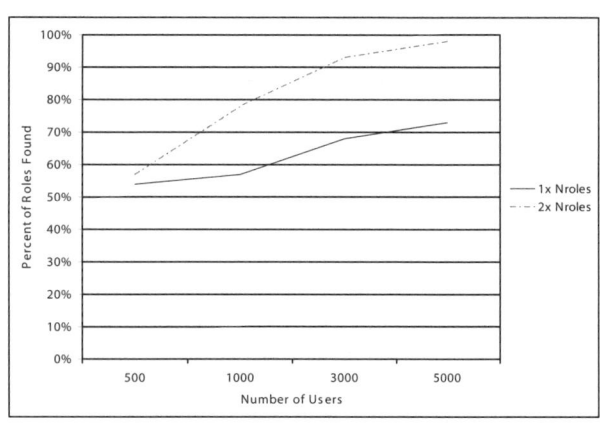
(a) Accuracy

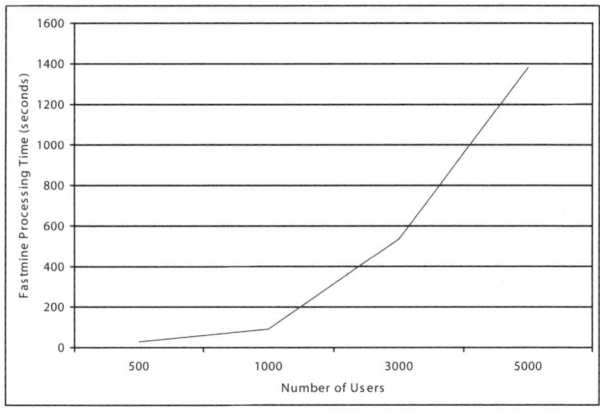
(b) Speed

Figure 4: Constant number of permissions/roles, varying users

Dataset	Parameters				
	NRoles	NUsers	NPermissions	MRolesUsr	MPermissionsRole
data1	10	100	1500	3	150
data2	50	500	1500	3	150
data3	100	1000	1500	3	150
data4	500	5000	1500	3	150

Table 4: Constant number of permissions, varying users/roles

Therefore, ORCA may not result in the actual set of roles.

Pairs of clusters with maximal overlap among their members are first merged and placed in the $\prec$. That is, for every pair of clusters $\langle c_i, c_j \rangle$, members($\langle c_i, c_j \rangle$) = members(c_i) $\cap$ members(c_i) and permissions($\langle c_i, c_j \rangle$) = permissions(c_i) $\cup$ permissions(c_j). The pairs of clusters having the most number of members in common are considered, and the pair with the largest set of permissions selected. If there is more than one pair with the largest set of permissions, one cluster is picked randomly. Let this selected cluster be c_s merged from c_a and c_b. c_s is added as a super-cluster of c_a and c_b, and c_a and c_b are no longer considered. The steps of finding a pair of clusters with maximal overlap and maximum permissions is then repeated until no more clusters can be found. It could be the case that the overlapping users may be members of two roles, however, ORCA assumes that the merged one is always a super-role of the two.

One of the main limitations of ORCA is that permission cannot be associated with more than one role unless the roles are along the same path of the role hierarchy tree. The authors indeed point out that this is a major issue stating that "permissions are seldom used by one role only and may be necessary for incomparable roles". Another issue is that their algorithm assumes a cluster is a good candidate for a role if it has the maximal number of people in common. In reality, this is not necessarily true, and a good role might as well have very few members. A third issue is that when clusters are merged, it is assumed that the merged cluster is a super-role of the two from which it is formed. While this may sometimes be the case, it may also be the case that the merged roles are unrelated except that several people are members of both roles. In addition, the algorithm proposed in ORCA is more computationally intensive because it begins by forming a cluster for every permission.

Since roles are nothing but groups of permissions, all users playing a role will have similar (or the same set of) permissions. Therefore, we cluster users having similar access permissions and assume this as our initial set of roles. We begin by forming a cluster for every group of users having the exact same permissions. Unless every user has a completely different set of permissions, the number of initial clusters must be smaller than the initial set obtained by ORCA.

7. CONCLUSIONS AND FUTURE RESEARCH

Role mining is a critical process for migrating existing systems with many access permissions and users to RBAC. It takes an agglomerative approach of finding inherent roles given assigned permissions. In this paper, we presented an unsupervised role mining process, *RoleMiner* which has three major advantages as compared to earlier role mining proposals. First, it is capable of detecting roles with overlapping permission sets. Second, it employs subset enumeration on permissions to allow us to identify the set of *all possible roles* in the organization. Third, we do not assume that every permission must be part of a role. Permissions that distinguish a role are found. Other permissions held by users could be identified as anomalous and be added upon further analysis.

While our process was shown to be very accurate in finding candidate roles, there is much additional work that we plan to accomplish. We have only a rudimentary approach to deciding which candidate roles are most useful based on

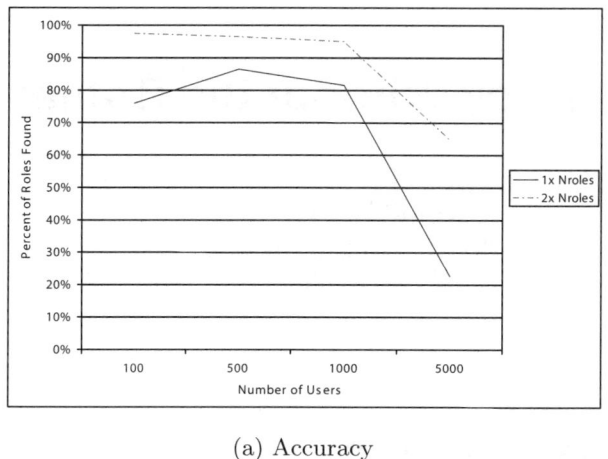
(a) Accuracy

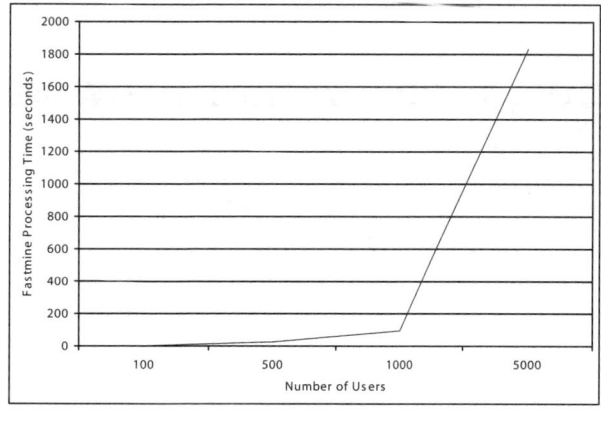
(b) Speed

Figure 5: Constant number of permissions, varying users/roles

user counts. We are investigating other metrics that could additionally or alternatively be used to identify the best roles to consider.

Other work includes using the semantics associated with permissions. The only data set available to us had no semantics and so our process found roles purely on the basis of whether a user had a permission or not. If information on the type of permissions or resources were available, it could be used to further refine the roles. Permissions that are semantically related to different functions could be separated when post-processing the results, thus uncovering whether a potential role (cluster of permissions) was actually a role or a blend of two or more roles. Semantics would also be helpful in pre-processing data. By knowing the semantics of the resources, permissions associated with both organizational as well as functional roles might be distinguishable, making further refinement of roles found more accurate.

8. REFERENCES

[1] Pavel Berkhin. Survey of clustering data mining techniques. Technical report, Accrue Software, San Jose, CA, 2002.

[2] E.J.Coyne. Role-engineering. In *1st ACM Workshop on Role-Based Access Control*, 1995.

[3] P. Epstein and R. Sandhu. Engineering of role/permission assignment. In *17th Annual Computer Security Application Conference*, December 2001.

[4] D. Ferraiolo, R. Sandhu, S. Gavrila, D. Kuhn, and R. Chandramouli. Proposed NIST standard for role-based access control. *TISSEC*, 2001.

[5] M. P. Gallagher, A.C. O'Connor, and B. Kropp. The economic impact of role-based access control. *Planning report 02-1, National Institute of Standards and Technology*, March 2002.

[6] Venkatesh Ganti, Johannes Gehrke, and Raghu Ramakrishnan. Cactus: Clustering categorical data using summaries. In *KDD '99: Proceedings of the fifth ACM SIGKDD international conference on Knowledge discovery and data mining*, pages 73–83, New York, NY, USA, 1999. ACM Press.

[7] Sudipto Guha, Rajeev Rastogi, and Kyuseok Shim. ROCK: A robust clustering algorithm for categorical attributes. *Information Systems*, 25(5):345–366, 2000.

[8] A. Kern, M. Kuhlmann, A. Schaad, and J. Moffett. Observations on the role life-cycle in the context of enterprise security management. In *7th ACM Symposium on Access Control Models and Technologies*, June 2002.

[9] Teuvo Kohonen. The self organizing map. *IEEE Transactions on Computers*, 78(9):1464–1480, 1990.

[10] Martin Kuhlmann, Dalia Shohat, and Gerhard Schimpf. Role mining - revealing business roles for security administration using data mining technology. In *Symposium on Access Control Models and Technologies (SACMAT)*. ACM, June 2003.

[11] G. Neumann and M. Strembeck. A scenario-driven role engineering process for functional rbac roles. In *7th ACM Symposium on Access Control Models and Technologies*, June 2002.

[12] Haio Roeckle, Gerhard Schimpf, and Rupert Weidinger. Process-oriented approach for role-finding to implement role-based security administration in a large industrial organization. In *ACM RBAC*, 2000.

[13] A. Schaad, J. Moffett, and J. Jacob. The role-based access control system of a European bank: A case study and discussion. *In Proceedings of ACM Symposium on Access Control Models and Technologies*, pages 3–9, May 2001.

[14] Jürgen Schlegelmilch and Ulrike Steffens. Role mining with ORCA. In *Symposium on Access Control Models and Technologies (SACMAT)*. ACM, June 2005.

[15] Dongwan Shin, Gail-Joon Ahn, Sangrae Cho, and Seunghun Jin. On modeling system-centric information for role engineering. In *8th ACM Symposium on Access Control Models and Technologies*, June 2003.

[16] D. Thomsen, D. O'Brien, and J. Bogle. Role based access control framework for network enterprises. In *14th Annual Computer Security Application Conference*, pages 50–58, December 1998.

Doppelganger: Better Browser Privacy Without the Bother

Umesh Shankar
ushankar@cs.berkeley.edu
UC Berkeley

Chris Karlof
ckarlof@cs.berkeley.edu
UC Berkeley

ABSTRACT

We introduce Doppelganger, a novel system for creating and enforcing fine-grained, privacy preserving browser cookie policies with low manual effort. Browser cookies pose privacy risks, since they can be used to track users' actions in detail, but some cookies also enable useful functionality, like personalization features. Web browsers currently lack an effective cookie management mechanism. Users must choose between two unpalatable options: a permissive, privacy-compromising policy for every site they visit, or a seemingly endless series of questions to which they must supply underinformed opinions. Doppelganger takes a big step forward: it makes automated determinations of cookies' value to enable a cost-benefit analysis, and offers an automated recovery system when that mechanism—or the user—makes an incorrect judgment. Doppelganger leverages client-side parallelism to automatically and simultaneously explore multiple cookie policies, enabling each user to create her ideal cookie policy. We tackle important and difficult subproblems along the way: mechanisms for recording and replaying web sessions; improved handling of third-party cookies; and enforcing fine-grained, per-site cookie mediation. We implemented Doppelganger as a Firefox extension; we discuss experimental results comparing it to various browser settings, as well as lessons learned from the real-world engineering challenges we faced in our implementation.

Categories and Subject Descriptors

H.4.3 [**Information Systems Applications**]: Communication Applications – Information Browsers

General Terms

Security, Human Factors

Keywords

Cookies, Web Privacy, Usable Security

1. INTRODUCTION

1.1 Background

An *HTTP cookie* (from here on, simply "cookie") is a small data item sent by a web site to a web browser, then sent back to the originating site on subsequent requests. While the original intent of cookies was to provide a session state mechanism for the stateless HTTP protocol, cookies have since been used not just for things like shopping carts and authentication, but also for tracking users' web surfing habits and building targeted advertising profiles. The result is that site operators or third parties can gain undesirable insight into users' habits and browsing history. Cookies can identify a user at sites where she believes herself to be anonymous and track her actions across sites and browsing sessions. The problem is exacerbated if web sites can correlate the collected data to users' real-world identities.

The difficulty, therefore, is in deciding which cookies are worth accepting and which are not. Ideally, a user should be able to compare the privacy cost of a cookie with the functionality benefit the cookie enables. Most users are not equipped to make these decisions manually and accept the global defaults in their browsers, which often apply a single policy to all sites. These defaults tend to err on the side of functionality rather than privacy. Since web site features such as shopping carts and logins often require cookies and users may become confused or annoyed if these features don't work, the default policies liberally accept cookies. Although this approach will minimize users' frustration, it will also accept many useless tracking cookies which unnecessarily violate users' privacy. Our goal is to get the best of all worlds: a cookie policy that protects users' privacy while simultaneously retaining the desired functionality and, perhaps most importantly, not pestering users so much that they disable the system.

1.2 A solution: Doppelganger

We introduce Doppelganger, a web browser privacy tool to help each user formulate her ideal cookie policy. Doppelganger is a system that, in effect, simulates a world in which the user has accepted cookies and compares it to the (default) world in which the user has not. If there is no change in the user's experience between the two worlds, then we can fairly say that the cookies are not useful. Thus, Doppelganger essentially creates a hidden twin of the user who is constantly exploring the value of cookies on the sites the user browses and who informs the user when accepting cookies may be a good tradeoff; useless cookies are rejected by default, to preserve privacy. Another key component of Doppelganger is an automated error recovery module, which users may invoke with a single click. Error recovery attempts not only to correct the cookie

Goal	Mechanism
Automatically determine useful cookies	Mirror user session in hidden browser window (the *fork* window) that accepts additional cookies; look for differences in output (see Section 3.2)
Detect differences in pages	Compare page titles; look for user's name/ID in mirrored page; see if a click cannot be mirrored
Determine privacy implications of cookies	Parse and interpret site's P3P policy
Recover from errors	Enable additional cookies and replay user session, using information from the log (see Section 3.3)
Record user session, to enable error recovery	Central log of user's mouse clicks, form field values, and browser state changes (START and STOP events for each page load)

Figure 1: Summary of Doppelganger's cookie management mechanisms.

policy, but also takes action to restore the user's session to a good state, as though cookies had been accepted from the start.

We followed two main principles in designing Doppelganger. The first principle is that users don't like to be constantly interrupted with questions or alerts, and when this happens, they will tend to disable or ignore the offending mechanism [14, 26]. In particular, users should not be asked to do anything manually that can be done automatically. Doppelganger uses *client-side parallelism* to explore alternate policies in the background, trying to find those which result in a positive cost-benefit analysis between privacy loss and functionality gain. In a perfect world, Doppelganger would be able to automatically deduce the ideal cookie policy with no user interaction. In reality, some interaction is needed because users have different privacy preferences and valuations of web site features. Therefore, we assume it is reasonable to expect users to make a small number of high-level privacy decisions.

The second principle we followed in designing Doppelganger is that users don't care about cookies so much as *privacy* and *functionality*. Previous work suggests that: (1) the lack of privacy information in an easily-digestible form may be a significant obstacle to achieving good outcomes for users [1]; (2) users are sensitive to privacy protections, and are more willing to accept a privacy risk if data about them is protected [11]; and (3) users are willing to compromise some amount of privacy if they are offered meaningful incentives to do so [15]. Doppelganger seeks to address all these issues. Instead of requiring users to make uninformed low-level decisions about cookies directly, Doppelganger reformulates cookie decisions as cost-benefit analyses between privacy loss and functionality gains, which are presented to the user. This enables users to make informed decisions regarding their privacy and accept privacy loss when there is commensurate compensation.

1.3 Contributions:

- We introduce Doppelganger, a system for creating and enforcing fine-grained, privacy-preserving cookie policies with low manual effort.

- We show that Doppelganger improves the handling of third-party cookies in Firefox, especially with respect to redirection and inline frames.

- We show how to use client-side parallelism to explore multiple cookie policies simultaneously and find the right balance of privacy and functionality for each user.

- We leverage concepts from Recovery Oriented Computing [5] to implement an automated single-click recovery mechanism.

- We present empirically tuned algorithms for recording and replicating user actions.

- We evaluate the effectiveness of Doppelganger in establishing functional and privacy-preserving cookie policies for typical web browsing habits and compare the results against those obtained with available browser settings.

We summarize Doppelganger's cookie management mechanisms in Figure 1.

2. HTTP COOKIES

HTTP cookies are a general mechanism for web servers to store and retrieve persistent state on web clients [21]. Since HTTP is a stateless protocol, cookies enable web applications to store persistent state over multiple HTTP requests. For example, web shopping applications can use cookies to track which items a user adds to her shopping cart.

When a client makes an HTTP request to a server, the server has the option of including one or more Set-Cookie headers in its response. Clients will return these cookies in subsequent HTTP requests using the Cookie header. The Set-Cookie header has one required field, a name/value pair of the form $NAME = VALUE$. A web server uses this field to encode the state information it wishes to store on the client. There are also four optional fields: expires=$DATE$, domain=$DOMAIN$, path=$PATH\ PREFIX$, and secure.

The expires field indicates how long the cookie is valid. After that date, the client's web browser should delete the cookie. If the expires field is omitted, then the cookie is called a *session cookie* and should be deleted when user closes the web browser. Cookies with an expires field are called *persistent cookies*.

The domain and path fields indicate for which HTTP requests clients should send back cookies. To determine which cookies to include with an HTTP request, the client searches its cookie jar for cookies for domains which suffix-match the domain of the request and paths which prefix-match the path of the request. For example, if the user requests the URL http://online.foobar.com/store/index.html, then a cookie with domain=.foobar.com and path=/store would be included with this request, but a cookie with domain=pics.foobar.com would not. The same-origin policy in web browsers prohibits one domain from setting cookies for another. The final optional field, secure, indicates whether the cookie should be only sent over encrypted HTTPS connections.

Cookies are also characterized by the context in which they are sent or received. Suppose a user clicks on a link for a particular document, and then the web browser issues a request for that document. After the browser receives the HTML page from the web server, it parses the page for references to elements needed to render the page, and issues additional HTTP request for these elements. Examples of additional elements include images, Javascript files, stylesheets, Flash objects, and sub-documents. Some of these requests may be to the same domain of the requested document, but some requests may be to different domains. The latter is often the case with advertisements. Content whose URL matches the domain of the main page (i.e., the one in the URL bar) is considered to be *first-party*. All other elements are in *third-party* context. For example, if a user is visiting www.x.com, then content served from *.x.com is first-party, whereas content on the page from www.y.com, such as an ad, is third-party.

2.1 Uses of cookies

Cookies have many purposes: session state, personalization, authentication, and tracking. Web sites use cookies for personalization to remember users' preferences and settings. For example, Google allows users to customize the format of their search results and uses cookies to remember these preferences. Web sites with user accounts also use cookies to authenticate users' sessions [12]: after a user logs in, a web site can set an session cookie on the user's machine to authenticate her subsequent requests. Web sites can set persistent cookies to remember users and not require a login on subsequent visits. Lastly, web sites can use cookies to track users and their actions. For example, e-commerce sites can track customers' browsing history to make purchase suggestions, and advertising sites can track users to conduct targeted advertising. However, tracking cookies have troubling privacy implications. By tracking the pages a web surfer visits, the web searches she makes, and the items she browses and purchases, web site operators and Internet advertisers can construct sophisticated profiles of users for targeted advertising, data mining, and information sharing with other companies.

Tracking cookies also make cookie management difficult. Many users might prefer not to accept tracking cookies due to the privacy risks; recent studies [19] have found that about 58% of users have deleted their cookies at some point. To prevent her web surfing habits from being tracked, a privacy-conscious user might decide not to accept or send any cookies, but blocking all cookies causes a significant loss in functionality on the web. Most web mail services, e-commerce, and banking sites require users to accept and send cookies for authentication, and blocking cookies also denies users personalization features. Blocking all cookies is consequently impractical for most users.

2.2 Web browser cookie management

Rather than blocking all cookies, the average privacy-conscious user would probably be willing to accept some cookies from the web services she derives some benefit from, but would like to block cookies that compromise her privacy "too much" or provide her no value. Sadly, web browsers provide few useful options to users who wish to customize their cookie settings to this end. Users can configure their browsers to accept only first-party cookies, accept only session cookies, prompt for a decision, and combinations of the above policies.

These options are inadequate. Accepting only first-party cookies is a good start; most web sites do not require clients to accept third-party cookies to operate correctly and advertising companies such as DoubleClick use advertisements and web bugs [3] in conjunction with third-party tracking cookies to correlate users' web browsing across multiple sites. However, current web browsers' implementations of a "first-party only" policy fall short of expectations. For example, Firefox misclassifies IFRAME content as first-party, so advertisers embed ads in IFRAMEs [23] to trick browsers into accepting and sending their otherwise third-party cookies. Also, click-tracking services and advertisers use HTTP redirection [22] to evade third-party cookie blockers. Suppose www.xyz.com hires a click-tracking service www.trackyou.com to record statistics about its site usage. As a user navigates www.xyz.com, say by clicking on a link that seems to point to news articles on www.xyz.com, the target of the link may actually be something like www.trackyou.com/redirect?target=www.xyz.com/news.html. The user's request first visits www.trackyou.com, enabling www.trackyou.com to record the request and then redirect the browser to the real target, www.xyz.com/news.html. However, since the first request is for www.trackyou.com, a browser with a first-party only cookie policy will allow www.trackyou.com to set a cookie on the user's machine. The danger here is that if a third site www.abc.com and www.xyz.com both use the same click-tracker www.trackyou.com, this enables www.abc.com and www.xyz.com to collude with www.trackyou.com to determine their common users and track their browsing habits. Furthermore, if a user has an account on either www.xyz.com or www.abc.com that reveals her real name, this enables both sites to associate her browsing history with her real identity.

Accepting only session cookies also seems like a good idea, since it limits the ability of web sites to track users across browsing sessions. However, blocking all persistent cookies denies users the option of web site personalization and authentication without logging in or another more heavyweight solution. In addition, broadband connections and more effective computer power management make it convenient for users to leave their computers on and browsers open for longer time periods. We anticipate these factors will increase the length of users' average browsing session. A session cookie used over the course of a long browsing session (say, a week) could violate a user's privacy as much as a persistent cookie.

The only existing option for users who want a fine-grained cookie policy is for the web browser to prompt the user for every decision. With this policy, when the browser receives a cookie from a web site foo.com, it opens a dialog notifying the user it has received a cookie from foo.com, and asks the user whether it should accept the cookie, accept the cookie for each session only, or block it. The dialog also offers the option to apply the decision to every cookie from the same domain. Although in theory this mechanism enables the user to tailor her cookie policy at a fine level of granularity, the usability costs are severe [18]. First, despite the option for the browser to remember her decisions for each domain, a user will often receive a barrage of these interruptive dialogs in a browsing session. Second, although the dialog informs the user that a web site is trying to set a cookie, the user is given no information on how the cookie will be used by the web site and must often make policy decisions before she has even viewed the site's home page. If a user makes a mistake in her policy (e.g., deciding to block cookies at a site she later needs authenticator cookies to login), she must navigate several confusing browser menus (up to three levels deep) to correct her decision. Also, choosing which cookies to accept is non-obvious. She may know she needs to enable cookies for a particular domain to make it "work", but should she enable session cookies or persistent cookies? A user may discover she must enable cookies after she has already taken a series of actions on the web site. In the worst case, she must repeat all these actions after she makes the necessary changes in the browser settings to correct her cookie policy.

2.3 "This site requires cookies"

Web sites do little to help with the cookie management problem. A web site can easily detect whether a particular user's browser will accept or deny cookies by using Javascript or a series of redirects. Many sites require cookies. If such a site detects the user is blocking cookies, it will inform the user that she must enable cookies to use the site and give the user instructions on how to enable cookies. The directions given by many web sites, however, instruct the user to enable cookies for all web sites, including third-party cookies. This sort of directive is easy for sites to issue, but can have big consequences for the hapless user's privacy not only at that site but every one the user visits. Naturally none of these negative effects are suffered by the site giving the instructions. Furthermore, few web sites give users information on how the site makes use of cookies. Without this information, users cannot easily decide whether they should accept cookies from the site.

2.4 Cookie management: The state of the art

Previous work does little to help users make informed decisions about cookie policies. Several Firefox extensions try to make the user interface for managing cookies less cumbersome. Cookie Button [8], Cookie Toggle [10], and Permit Cookies [20] add toolbars and enable keyboard shortcuts to help users quickly change cookie policies for the current domain. Add'n'Edit Cookies [2], Cookie Culler [9], and View Cookies [25] add shortcuts to easily view and delete cookies stored for a particular domain. Although these tools help alleviate the difficulty and annoyance of navigating the browser menus to change cookie policies and view previously set cookies, their focus is still on the low-level mechanism of cookie management, which few users understand and fewer still know how to manipulate. They do not help users decide the correct policy for a domain, nor do they cast the problem in more intuitive terms. A much more promising system is Acumen [13], which works on social recommendations for accepting cookies; users are notified how many other users accept the cookies in question. This system does not protect users' privacy itself, though, as it does central data collection of users' choices. It also does not take into account users' inability to make good choices without information. Such a system, with appropriate anonymization, is complementary to ours and could serve as another line of defense before users are burdened.

The Platform for Privacy Preferences (P3P) Project [24] is a protocol developed by the World Wide Web Consortium to help inform users of the privacy guarantees of the web sites they visit. P3P envisions users configuring their web browsers with specifications of their privacy requirements while surfing the web. Then, when a user visits a web site, that site will send a compact P3P policy specifying how it uses personal information, and the browser will determine whether the user's and site's policies are compatible. If not, the browser would inform the user of the incompatibility. P3P seems useful for helping users make informed decisions about their cookies policies, but in practice P3P has many problems [7]. Companies have been reluctant to adopt its complicated protocol structure, policy configuration is cumbersome for users, and the barrage of privacy warnings and notifications while web browsing becomes burdensome and confusing. Recently, though, there are more tools for writing and understanding P3P policies [4, 6, 16], and we hope that either P3P or some other privacy standard emerges to help us accurately gauge privacy risks.

Felten et al. have explored techniques to increase users' peripheral awareness of cookies and improve their ability to make informed decisions about cookie policies [18]. Their Cookie Watcher tool notifies users of cookie events and gives some limited information on the risks of accepting cookies. For example, it notifies users that a third-party persistent cookie could be used to track users across sites and web browsing sessions. Although Cookie Watcher may help users understand the risks of accepting cookies from a web site, it does little to help users evaluate the benefits of accepting a cookie. Likewise, Bugnosis [3] alerts users to the presence of "web bugs"—invisible images used for tracking, sometimes via cookies—but does nothing to mitigate their effect.

3. HOW DOPPELGANGER WORKS

If a user wants to decide whether or not a particular cookie is beneficial, she must determine whether the benefit she receives from accepting the cookie outweighs the attendant privacy loss she suffers. Thus, her ideal cookie policy is one that accepts only those cookies for which the cost-benefit analysis yields a positive result. Although each user values privacy risks and functionality gains differently, we want to avoid interruption when the answer is clear.

To this end, we developed Doppelganger, a web browser privacy tool to help each user perform this cost/benefit analysis and formulate her ideal cookie policy. Doppelganger's main goal is to identify useful cookies and their privacy implications automatically. Doppelganger relies on the following principle to identify useful cookies: if a cookie from a domain confers some benefit, it should be evident in the user's experience. If no such benefit is found, then we may assume that cookies from that site may be blocked.

Doppelganger uses two main techniques to identify cookies beneficial to the browsing experience: mirroring and user initiated error recovery. Network bandwidth and CPU power have been increasing rapidly, and web browsing clients often have excess bandwidth and CPU available. We leverage that spare bandwidth and computing power to take a "partial derivative" with respect to the cookies whose benefit we are trying to measure. When Doppelganger encounters a domain in the user's browsing session for which it hasn't determined a cookie policy, it mirrors the user's web session in a hidden parallel session whose only difference is the cookies accepted and sent. We refer to this hidden parallel session as the *fork window* since it represents a forking of the browser state. Correspondingly, we refer to the cookies speculatively used by the fork window as *fork cookies*. We show an overview of Doppelganger's architecture in Figure 2.

When Doppelganger detects a difference between the main window and fork window, it reveals the fork window and asks the user to compare the two. The benefit of the fork cookies is any advantageous service present in the fork window which is not in the user's main browsing window. To evaluate the cost of these fork cookies, Doppelganger provides the user a condensation of the domain's P3P policy (if available) and a description of the kind of tracking enabled by the cookie. Doppelganger records the result and automatically uses it for future cookie policy decisions for that domain.

The second technique Doppelganger uses to identify beneficial cookies is user initiated error recovery. The user interface for this error recovery is a single button labeled Fix Me on the browser status bar. Fix Me is a rewind-and-playback mechanism. Doppelganger maintains a log of a user's actions and browser state changes, and invokes the Fix Me mechanism when the user indicates to the system that something is wrong, perhaps due to an error message or missing functionality which the mirroring system missed. The idea is that if a lack of cookies was the problem, then

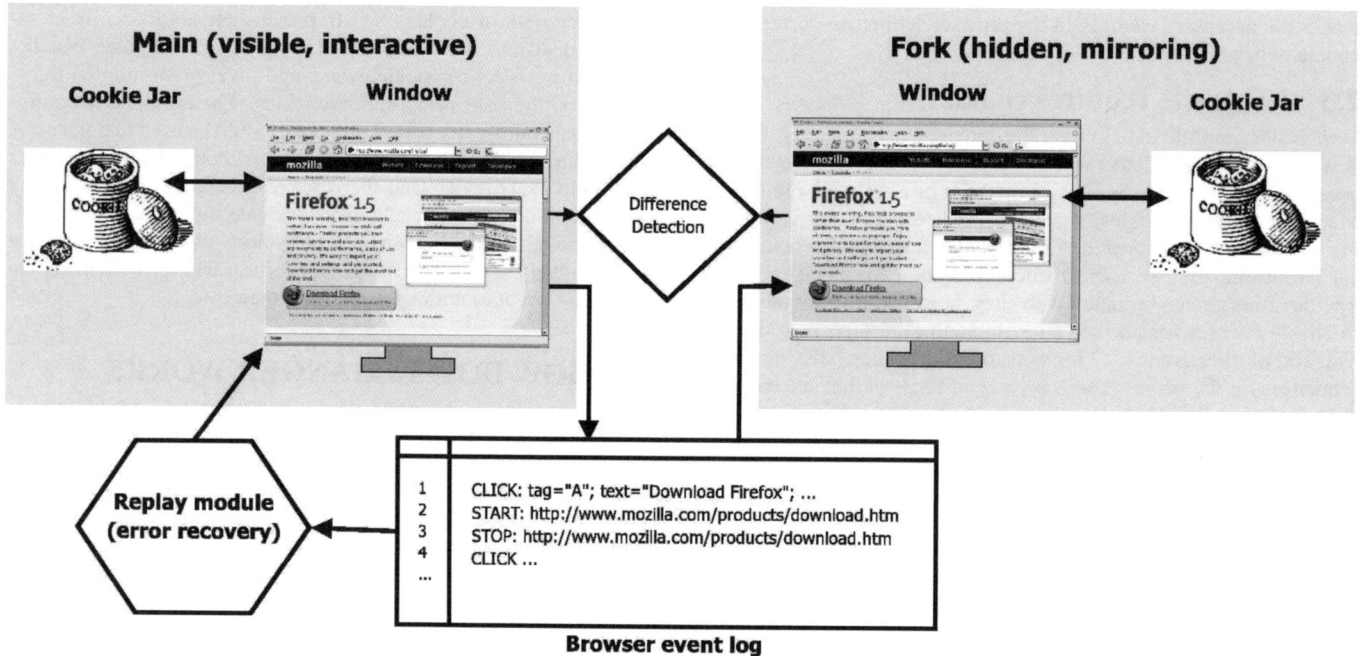

Figure 2: An overview of Doppelganger. Doppelganger mirrors the user's web session in a hidden fork session whose only configuration difference is the cookies accepted and sent. When Doppelganger detects a difference between the contents of the main window and fork window, it reveals the fork window and asks the user to compare the two (see Figure 5). Doppelganger also maintains a log of the user's actions for error recovery.

we may enable cookies and replay the user's actions, simulating what the user's session *would* have been if cookies had been enabled in the first place enabled.

Doppelganger can operate in three different configuration modes: *high paranoia*, *medium paranoia*, and *low paranoia*. These modes differ primarily in how Doppelganger handles session cookies. The privacy loss of most session cookies is relatively small, but some uses of session cookies pose higher privacy risks (Section 2.2). In low paranoia mode, Doppelganger accepts all first-party session cookies for all domains, and in medium and high paranoia modes, Doppelganger determines a per-domain policy for session cookies. In all modes, Doppelganger determines a per-domain policy for persistent cookies. We discuss these modes and their usability trade-offs in more detail in Section 3.4.

We implemented Doppelganger primarily as a Firefox extension in about 6000 lines of Javascript code. We also made a small (30 line) change to the main C++ source code, which we have submitted for inclusion into the mainline. Since Doppelganger is implemented in Javascript, it is portable to any operating system on which Firefox runs. The primary user interface is limited to the Fix Me button on the browser taskbar. For debugging purposes, we appended a tab to the LiveHTTPHeaders extension [17], used for watching HTTP request and response traffic, that enabled us to monitor and configure our system.

3.1 An example

Before discussing Doppelganger in detail, we first present a more elaborate example of Doppelganger in operation where a user interacts with a fictitious web site www.xyz.com. To illustrate all of Doppelganger's features, we assume Doppelganger has been configured in high paranoia mode, the most conservative configuration. At the end of the example, Doppelganger will have determined a complete cookie policy for www.xyz.com.

Suppose a user visits an e-commerce site, www.xyz.com, for the first time. The default policy for the main user window in Doppelganger is to block all cookies. At the same time, the hidden fork window will also visit www.xyz.com, but will accept (and send back) first-party cookies from the site, with the aim of deciding whether first-party session cookies from www.xyz.com are beneficial. For the next few page loads on www.xyz.com (details, Section 3.2), Doppelganger mirrors each user action and form field value in the fork window (Section 3.5); after each page load, Doppelganger compares the resulting main and fork windows for differences, and alerts the user if it judges them significant.

Suppose the user adds an item to her shopping cart, an action which requires cookies to be enabled. The fork window will contain the shopping cart page, but the main window will remain on the item page (a common failure mode). Doppelganger will detect the two pages as different, triggering a user-choice dialog.[1] Doppelganger displays the primary window and fork window side-by-side. A dialog box will give an estimation of the privacy risk from switching to the fork window (the cost) and the user can see what additional features are offered on the fork side (the benefit; in this case a functioning shopping cart). The user can then choose one of three options: switch to the fork side and accept the cookies, stay with the main browser and reject the cookies, or defer judgment and continue mirroring. Let us assume the user chooses to switch to the fork window, accepting the cookies. Since the user indicated that first party session cookies provided some benefit, Dop-

---

[1] Low and medium paranoia modes eliminate this dialog box.

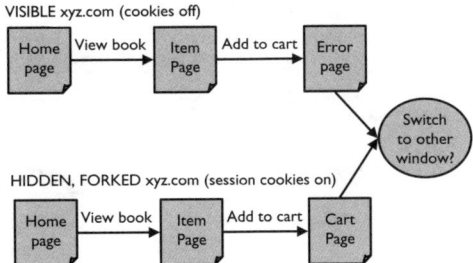
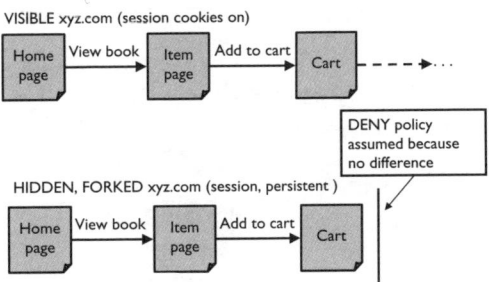

(a) In the first session, the user must accept session cookies to add an item to her cart, but Doppelganger disables them by default. The mirroring process automatically detects this and offers the user the choice to switch to the fork browser, which shows the desired shopping cart.

(b) In the second session, Doppelganger accepts session cookies in the main window per the earlier decision. The fork window also sends back persistent cookies to test their benefit. Doppelganger finds no difference between the fork and main windows, so it assumes persistent cookies are unnecessary and appends a rule to block persistent cookies to the policy for www.xyz.com.

Figure 3: Graphical representation of mirroring for the example in Section 3.1.

pelganger records the decision to accept first-party session cookies at www.xyz.com for future sessions.

While the user has indicated that first-party session cookies from www.xyz.com have benefits, Doppelganger still hasn't determined whether first-party persistent cookies from www.xyz.com offer any benefits. However, Doppelganger will store first-party persistent cookies from www.xyz.com from this session in a separate fork cookie space for additional investigation during the next session. Now, suppose the user ultimately decides not to purchase the item yet, and closes her browser.

During her next session, she navigates to www.xyz.com again. The browser has deleted the www.xyz.com session cookies from the previous session, and Doppelganger is keeping the www.xyz.com persistent cookies aside in the fork window's cookie space. To determine if those persistent cookies have value, Doppelganger repeats the mirroring process again. This time, the main window accepts session cookies as per the earlier decision, but the fork window not only accepts new session cookies but also sends the persistent cookies it received before. For our example, let us say that the persistent cookies do not enable any additional features. After the user has visited a few pages on the site, if Doppelganger does not detect significant differences between the fork window and the main browsing window, it will decide that persistent cookies are not necessary. Doppelganger will record the decision automatically and stop the mirroring without any user interaction.

The persistent cookies in the fork window will, however, be retained for future error recovery if the user later finds that some desired feature does not work. Suppose that the user had entered some personalization features in her first browsing session at www.xyz.com which affect the browsing experience in relatively subtle ways that Doppelganger missed. The user may notice this problem after Doppelganger already made an automatic policy decision to reject persistent cookies for www.xyz.com. Doppelganger provides the Fix Me button on the status bar of the main browsing window to recover from these errors. When the user presses Fix Me while browsing www.xyz.com, Doppelganger rewinds the user's browsing session on www.xyz.com, enables the next most permissive cookie acceptance policy (in this case, accepting first-party persistent cookies), and automatically replays the user's session at

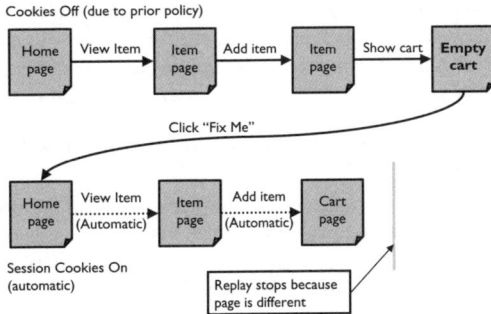

Figure 4: An example use of Doppelganger's error recovery mechanism. Suppose either Doppelganger or the user made a decision to not accept cookies at a particular site, but it turns out cookies are needed to maintain a shopping cart. If needed, the user can indicate that cookies may be needed by clicking the Fix Me button; Doppelganger rewinds to the start of the session at the site, enables cookies, and replays all the user's actions without any further user intervention.

www.xyz.com. The user gets the same final page as when she pressed Fix Me, but now with any additional benefits of sending the www.xyz.com persistent cookies received in the first browsing session. We discuss Doppelganger's error recovery mechanism further in Section 3.3.

3.2 Mirroring

In this section we discuss Doppelganger's mirroring system in more detail. During a user's browsing session, Doppelganger observes all page loads in the main window. When it encounters a page load for a domain for which it has doesn't have a complete policy, it begins to mirror the session in the fork window. Doppelganger mirrors the session by replicating the user's main window actions in the fork window and then looking for differences between the two. Mirroring user events is non-trivial; we discuss it in depth in Section 3.5. In the rest of this section, we will describe

 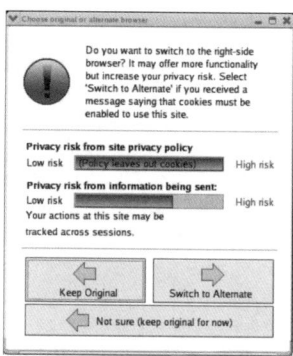 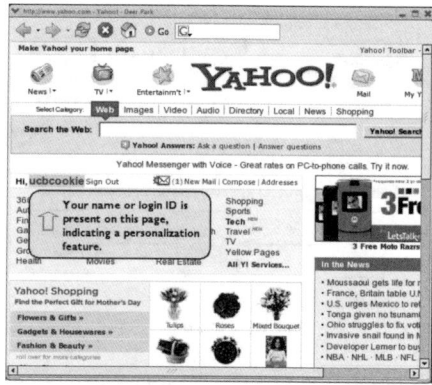

(a) Main window (b) Comparison dialog (c) Fork (mirroring) window

Figure 5: A screenshot of the Doppelganger's comparison dialog. When Doppelganger detects a significant difference between the main and fork windows, it prompts the user for a decision. Doppelganger provides some indication of the difference and a measure of the privacy risk from accepting cookies. In this case, Doppelganger detects the presence of a personalization feature and alerts the user to it.

how Doppelganger formulates cookies policies and how it maintains two separate cookies spaces for the fork and main windows. We then show how Doppelganger uses the fork window to make automatic decisions affecting the cookie policy and expose additional functionality enabled by cookies to the user.

3.2.1 Fork window cookie policies and cookie name spaces

Doppelganger formulates cookie policies based on tail domains. Tail domains are the last two components in the host name of URLs (e.g., yahoo.com).[2] Doppelganger applies the same cookie policy to all cookies and pages matching the tail domain.

Doppelganger enforces one of five possible first-party cookie policies for each tail domain D:

Policy	Session Cookies	Persistent Cookies
$P_{?,?}$	?	?
$P_{S,?}$	accept	?
$P_{S,P}$	accept	accept
$P_{S,X}$	accept	downgrade
$P_{X,X}$	deny	deny

In the table, "?" means that Doppelganger does not yet know the correct policy for D, and "downgrade" means that persistent cookies are converted to session cookies.

Doppelganger currently blocks all third-party cookies because we have not encountered any sites where they provide any benefit, although it is capable of enforcing third-party cookie policies on a per-site basis. If we discover some sites for which third-party cookies prove beneficial, we can easily enable that feature.[3] Note that since Doppelganger enforces per-site policies, enabling third-party cookies for one site would only allow tracking with other sites

---

[2]There is much debate over the right way to decide how many trailing domain name components to use; presently we use a simple heuristic, as do most browsers, to use an additional component for international TLDs (two letter suffixes).

[3]Since a site's context can be out of its control, such as being included in a frame by another site, an intended first-party cookie can become a third-party cookie. It is uncommon for external documents that are not advertisements to be put in frames, and many sites do not work properly when framed in any case.

that also had third-party cookies enabled. This is in sharp contrast to browsers' settings, which have only global policies for third-party cookies.

There is another problem related to third-party cookies: the mechanism Firefox uses to identify third-party cookies is vulnerable to IFRAME [23] and redirection [22] tricks. IFRAMEs are entire documents embedded in HTML pages, and for various reasons, Firefox incorrectly determines the context of HTTP requests generated by IFRAMEs. For the purposes of cookie management, Firefox classifies an IFRAME request as an independent request for a top-level HTML page rather than as a request for an element of a larger page. Therefore, Firefox classifies cookies for IFRAME requests as first-party instead of third-party with respect to the enclosing page. Doppelganger addresses the IFRAME problem by more reliably determining the context of an HTTP request by matching its tail domain against that of the topmost page's URL. We discuss the countermeasure to redirection tricks in Section 3.2.5.

In order to send different sets of cookies to the main window and the fork window, Doppelganger partitions the cookie space to allow multiple copies of a cookie with a specific NAME=VALUE pair and impose different access controls on them. Doppelganger achieves this by implementing `nsICookieConsent`, an optional Firefox interface designed for deferring cookie policy decisions to an external module. Its original intention was for use with P3P, but that functionality has since been disabled. `nsICookieConsent` was designed to be applied to classes of cookies at once, since the browser's controls do not have provisions for individual cookies. However, Doppelganger must impose differential access controls according to the particular cookie being set and which window (fork or main) is using it. Addressing this problem required a small change to the interface and a small (30 line) change to the main C++ source code to use this new interface.

3.2.2 Mirroring in the fork window

When the user visits a domain D, Doppelganger checks its cookie policy for D. If it has a complete policy (i.e., $P_{S,P}$, $P_{S,X}$, or $P_{X,X}$), Doppelganger does no mirroring and simply applies the policy to the main window. Suppose Doppelganger has no policy for D (e.g., it is the user's first visit to the domain). Then the fork window starts sending and receiving first-party cookies for

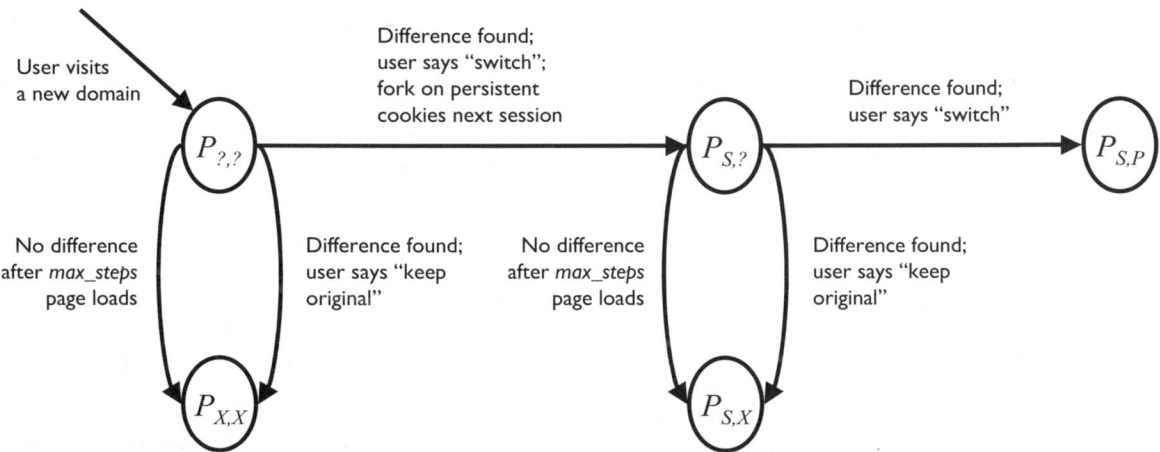

Figure 6: How Doppelganger determines a cookie policy for a domain during the mirroring process. When Doppelganger detects a difference between the main and fork windows, it prompts the user to decide whether the additional features are worth the potential privacy risk. Doppelganger makes an automatic policy decision if it does not detect any differences after *max_steps* page loads. We omit some additional transitions present in low and medium paranoia modes (see Section 3.4).

that domain. Doppelganger mirrors the user's actions for a constant number (*max_steps*) of page loads on D and monitors the fork window for differences. If it detects no difference after *max_steps* page loads, Doppelganger concludes that cookies at D provide no benefit, stops mirroring, and sets the cookie policy for D to deny all cookies ($P_{X,X}$).

Alternatively, if Doppelganger detects a difference, it prompts the user to decide whether the additional features are worth the privacy risk by attempting to highlight benefits and display privacy risks. For an example comparison screenshot, see Figure 5. If the user answers "keep original", Doppelganger stops mirroring and sets the cookie policy for D to $P_{X,X}$. If "switch to alternate", it stops mirroring and sets the cookie policy for D to $P_{S,?}$. Recall that $P_{S,?}$ accepts session cookies, but has an undetermined policy for persistent cookies. Doppelganger then transfers the state of the fork window to the main window to automatically provide the user the benefit of the cookies. For the remainder of the session, Doppelganger accepts all first-party cookies from D.

Now, suppose the user closes her browser, restarts it the next day, and revisits D. The policy for D is now $P_{S,?}$ and the browser may have persistent cookies from D from the previous session. Since Doppelganger has not yet determined whether persistent cookies from D are beneficial, it begins to "fork" on these cookies. Doppelganger loads persistent cookies for D from the previous session into the fork cookie space and clears all of D's cookies from the main cookie space. Doppelganger then proceeds as it was when forking on session cookies, except now, both windows accept session cookies instead of just the fork window.

The difference is the fork window may have persistent state from the previous session which positively affects the user's experience. Doppelganger tries to detect this. Again, Doppelganger mirrors the user's actions for a constant (*max_steps*) number of page loads on D and monitors the fork window for differences. If it detects no difference after *max_steps* page loads, Doppelganger concludes persistent cookies at D provide no benefit, stops mirroring, and sets the cookie policy for D to block persistent cookies ($P_{S,X}$). Otherwise, if it detects a difference, it prompts the user for a decision whether the difference is beneficial. If the user answers "no", Doppelganger stops mirroring and sets the cookie policy for D to $P_{S,X}$. If "yes", it stops mirroring, sets the cookie policy for D to accept persistent cookies ($P_{S,P}$), and transfers the state of the fork window to the main window. We summarize the mirroring process in Figure 6.

Presently, *max_steps* is a constant. We want it to be small enough that we do not end up effectively accepting more cookies via the fork window, but large enough to see differences due to cookies. Large-scale trials are needed to determine a good value; in testing we set *max_steps* to 5.

3.2.3 Difference detection

Doppelganger must be able to detect when the fork and main windows significantly differ in function or personalization enough to warrant interrupting the user for a decision. Doppelganger should ignore things like advertisements, randomized placement of news items, or other sources of natural nondeterminism. In our difference detection algorithm, we must address a tradeoff: if Doppelganger reports too many page pairs are different, the user will be asked to make too many decisions, whereas if the system fails to detect meaningful differences, cookies will be rejected too aggressively and the user must detect problems manually and initiate error recovery (Section 3.3). In both cases the user is needlessly inconvenienced. At present we use a coarse mechanism: we compare page titles (to detect obvious errors) and we look for the presence of the user's name or login ID in the fork window (and its absence in the main window) to detect personalization. In addition, if a user action cannot be replicated in the fork window, we assume the pages are different. A better heuristic is the source of ongoing work. We do not consider an error in loading a page as a significant difference; instead the mirroring process re-starts at the next page after re-syncing the fork window to the main one.

3.2.4 Exposing the cost of cookies

Even beneficial cookies carry privacy risks. When Doppelganger detects a potential benefit of accepting cookies at a domain D, it tries to measure and expose the privacy risks when it prompts the user to compare the fork and main windows. One measure of the risk is the type of cookies Doppelganger must enable for the user to benefit (i.e, session or persistent). We also assess risk by interpreting the domain's P3P policy, if one exists; we borrowed some P3P parsing code from [4] for this purpose. Doppelganger represents the privacy risk with two bars, one derived from the site's privacy policy, and one representing the risk from the type of cookie allowed. For an example of Doppelganger's risk assessment during a comparison, see Figure 5.

3.2.5 Addressing ephemeral site visits

Doppelganger uses a slightly different strategy to address domains which may be visited often but never for very long. This situation arises in several situations: click-tracking and advertisement redirect tricks [22] (discussed in Section 2.2), certain web portals, and search engines. Web portals and search engines contain links to other domains that are the user's ultimate goal; in the meantime, though, the portals use cookies to track the user's actions. Also, shopping search portals use redirects through advertising trackers (e.g., DoubleClick and Dealtime) which set persistent cookies to track the offsite links that users follow. All these cookies appear to the browser as, technically, first-party cookies, but we want to block most of them since they confer no benefit.

The risk is that Doppelganger will perpetually mirror visits to these sites. Since users will likely never have *max_steps* consecutive page loads on these domains, Doppelganger will never arrive at a policy decision for them. Doppelganger would therefore invoke the mirroring process on every visit to these domains to try to determine their cookies' value, in effect enabling cookies forever. To address this problem, we maintain a lifetime hit count for domains with an undetermined cookie policy and set the domain's cookie policy to $P_{X,X}$ when the hit count exceeds a constant, *max_visits*. We are still determining an optimal value for this constant; in testing, we set *max_visits* to 8. The end result is that a policy decision is made for every site after a finite amount of time.

3.2.6 Logins

Doppelganger optimizes cookie management for sites where a user logs in. When Doppelganger detects a user logging into a domain, it automatically enables session cookies for that domain. The rationale for this policy is that if a user has a relationship with a site which requires a login, then accepting session cookies is unlikely to cause additional privacy loss, and we want to avoid unnecessary user interruptions. Doppelganger detects logins by looking for form submissions containing username and password fields.

3.3 User initiated error recovery

The second major component of Doppelganger is user-initiated automated error recovery. The first line of defense is the mirroring mechanism described above, but Doppelganger's comparison function may be imprecise, mirroring may end prematurely, or the user may change her mind regarding the cost/benefit of cookies from a domain. Doppelganger invokes the error recovery mechanism when the user notices some feature is not working properly, or when she sees a cookie-related error message Doppelganger did not automatically detect. The user interface is simple: Doppelganger installs a single button labeled Fix Me on the browser status bar that the user can click when necessary. Our techniques for error recovery borrow ideas from Recovery-Oriented computing; in particular, we use the "Three R" model of recovery introduced by Brown et. al [5]: Rewind, Repair, Replay.

Doppelganger handles recovery differently depending whether it is mirroring a session or not. If Doppelganger is mirroring a session, it simply uses the mirroring comparison dialog to show the user what recovery would look like. If Doppelganger is not currently mirroring a session, it must achieve the same effect. To do this, Doppelganger enables the next most permissive cookie policy setting (as the fork window would have) and replays the user's session at the current site from the beginning by replaying all user-initiated UI events (e.g., clicks, form submissions). We do not replay across site boundaries.

Of course, strict replaying is not the goal: we want the result to be different (and better). Doppelganger manages the replay with a state machine which watches page loads and sends user events. If Doppelganger cannot replay a user event, an expected page does not load, or an unexpected page loads, Doppelganger stops the replay. Since one of these events is evidence of a page not present in the original sequence, Doppelganger optimistically assumes the problem is fixed; since the desired outcome is one that we have not yet seen, there is no way to know if it is the correct one automatically. If the problem has in fact not been fixed, the user may click the button again, and Doppelganger will enable the next most permissive cookie setting (if possible) and replay again. If this, too, fails, then likely a lack of cookies was not the source of the problem.

There are two problematic cases for replaying. The first is the nonlinearity of many sessions: what if the user had hit the Back or Forward buttons during the original session? Our current approach is to replay those buttons (but not Reloads) during the replay as well; this seems to work in practice. Another case is that of HTTP `POST` requests. According to the specification, `GET` requests, the most common kind, are to be used for idempotent requests, and `POST` requests for non-idempotent ones like transactions. Although we believe the danger is low—after all, if the transaction completed, why would the user be invoking the replay?—we do not replay through `POST`s. Some sites abuse the `POST` request for idempotent actions, which would block the replay. This misuse is bad policy, since it makes it difficult for users to go back and forward, reduces the effectiveness of proxy servers, and reduces the effectiveness of our replay system while making their users do more work to accept cookies. Others misuse `GET` requests for non-idempotent actions, which is very dangerous since the back and forward buttons can easily and inadvertently trigger the action again; proxy caching could also break. In short, there are many existing reasons for sites to use `POST` and `GET` requests appropriately, and if a site does misuse `POST` or `GET`, problems will probably surface regardless of Doppelganger.

3.4 Higher-privacy modes

Always-on Internet connections and more effective power management have conspired to make very long sessions not only possible but easy. Accordingly, we have implemented multiple modes of operation for Doppelganger, characterized by "paranoia level", which have different session cookie policies. *Low paranoia* mode always accepts session cookies, and is thus the least intrusive, least private mode. *High paranoia* never accepts session cookies by default, using the same mirroring-and-recovery tandem on session cookies as on persistent cookies. It is the most privacy-preserving, but most intrusive mode; remember, though, that comparisons only must be made when the mirroring process detects a difference.

As a compromise between the two, *medium paranoia* mode uses mirroring, but when a difference is detected, automatically enables session cookies without asking the user. Since the privacy risks of session cookies are generally low, the net benefit of accepting them is likely positive at a domain where the mirroring process detects a benefit, and we can avoid interrupting the user to make a decision. In addition, medium paranoia mode enables session cookies when

Mode	Session Cookie Policy	Persistent Cookie Policy	Notes
Low paranoia	Accept all	Per-domain	Requires least user interaction
Medium paranoia	Per-domain (never ask user)	Per-domain	Automatically enables session cookies on POST or when a difference is detected during mirroring
High paranoia	Per-domain	Per-domain	Highest privacy; requires the most user interaction

Figure 7: Summary of Doppelganger's different privacy modes.

a POST request is seen. A main benefit of medium is that it automatically denies cookies from tracking sites which are visited using redirection, but never requires users to make left-or-right comparisons for session cookies. However, if the mirroring process fails to detect a useful difference, the user may need to use the Fix Me button. We summarize Doppelganger's different privacy modes in Figure 7.

3.5 Replicating individual user actions

In this section we show how Doppelganger replicates the two dominant types of user interactions in web browsers: mouse clicks and form submissions. Doppelganger replicates user actions both during mirroring and during error recovery. In the former case, we replicate user actions in the fork window immediately as they occur; in the latter case, we replay a series of actions from the log into the main window. In both cases, the proximate mechanism of replication is the same, and we describe the algorithms in this section. By replaying at the level of user actions rather than page loads, relevant Javascript code on the page is triggered automatically.

3.5.1 Mouse clicks

Replicating clicks turns out to be difficult in practice for two main reasons: document elements do not have unique IDs, and there is a significant amount of nondeterminism in what results are returned for a given URL. This latter problem can arise from naturally changing pages, e.g., news sites and search engines, but this problem also surfaces in advertisements and stochastic link rewriting for click tracking. The basic click replication mechanism involves three steps: (1) record information about the click; (2) try to find the matching target in the fork browser; and (3) send the appropriate click event to the target.

3.5.1.1 Recording the click.

Our goal in recording clicks is to capture enough information about the click that we can replicate it, and to record it in a way that tolerates small changes to the document in which it is being replayed. Our initial algorithm constructed a path in the DOM tree from the document root to the clicked element and tried to reconstruct this path in the DOM tree of fork window document. This approach failed because it was too precise; it could not adapt to small perturbations in the document. In the end, we used a heuristic refined through experimentation.

First, when the user clicks an element, we record some identifying information about the event:

- The URL of the page in which the click occurred
- If the click was in a frame, the topmost document's URL
- The HTML tag name of the target of the click (e.g., "DIV")
- The mouse coordinates of the click (for imagemaps)
- The element's attributes (e.g, "href", "id", "name")
- The text content within the element

- If it is a form element, information about the enclosing form
- Which mouse button the user pressed

To be more precise, we do not always record the immediate target of the element; there may be many HTML fragments of the form

`<B>click here</B>`

for example, and the `<B>` tag is not interesting for a click perspective. Instead, we start backtracking to the root of the document to find the nearest ancestor of the target element which initiates some action. For example, if the HTML code reads

`<A href='/x'>To get $$, <B>click here</B></A>`

we would record the click on the `<A>` tag, not the `<B>` tag. In general, we stop at elements which have an `href` attribute, an `onclick` attribute, or are input elements of a form. This reduces ambiguity considerably when trying to replicate the click.

3.5.1.2 Finding a match in the target window.

When Doppelganger replicates a click on an element in the source window, a primary challenge is locating the analogous element in the target window. If the web were static, each request for a URL would yield the same response and the task would be straightforward. Instead, there is a fair bit of nondeterminism in the responses. For example, on news sites, a new article may change the locations of the previous articles. Some search engines rewrite their search results links to track which ones are clicked most often. We therefore implemented a "best-match" algorithm, which compares candidate elements against the information recorded about the original click.

First, we narrow candidates to elements with the same tag name, e.g., "A" or "INPUT". We then build a match record for each one, comparing on several characteristics:

- Exact match (index 0)
- The "id" (1), "href" (2), "name" (3), "type" (4), and "value" (5) attributes
- The text content of the element (6)
- Information about element's parent form (if any) (7)

Say the candidate element is E; we denote a characteristic of E by $E.c$ where c is, e.g., the value of the `id` attribute. Denote the log-recorded value of c by $O.c$.

Then we make a match record R as follows: for each characteristic c with index i,

$$R[i] = \begin{cases} 0 : E.c \neq O.c \\ 1 : E.c = null \land O.c = null \\ 2 : E.c = O.c \end{cases}$$

Then the best match record is selected by comparing the record values in order, i.e., $R_i[0]$ vs. $R'_j[0]$, then $R_i[1]$ vs. $R'_j[1]$, et seq.

Finally, it may be that the best possible match is not in fact a very good match. We empirically determined a cutoff in match scores and only consider the element a match if it passes this cutoff.

3.5.1.3 Mirroring the click.

After we have correctly identified the element in the target document, mirroring the click is relatively straightforward. We construct a click event object and deploy it to the target element. We also precede each click event with a focus event, as it would be if the user in fact clicked the mouse on it. This step is important because many web pages use "onfocus" Javascript handlers to change the page dynamically when an element is focused.

3.5.2 Forms

In addition to clicks, we must fill in web forms in the target window with the same data as in the original. Some of the challenges here are similar to the click problem above, in that forms do not always have unique identifying information such as a "name" attribute. If the "name" attribute is missing, we use the form's array index in the `document.forms` array; unlike clickable elements, it is unusual for forms to be added and removed by chance. The form elements virtually always have "name" properties because that is how their values are identified when the form is submitted, so identifying them within a form is easy.

There is a more subtle issue here, though, because forms often have hidden fields which are meant to be unique for each instance. Common fields of this sort are session IDs and nonces for password protocols. We do not modify the values of hidden form fields; in practice this has not been a problem since the user is only expected to fill in visible fields anyhow.

We fill in forms in the target document when replicating each click, not just when the form is submitted. Correspondingly, we store all form values at the time we record the click information. This is because Javascript on the page may modify the page based on the form values prior to submission, sometimes even reloading the page in response.

3.5.3 Other issues with replicating user actions

One complication we encountered in replicating user actions is the presence of frames; in the interest of space, a full discussion is omitted. The main idea for handling frames is that many of the above techniques can be implemented recursively, effectively treating all the frames as one giant page.

Another question is how precise to be in recording and replicating user actions. Omitting certain frequent events yields an efficiency benefit. So far we have not found it necessary to replicate each keystroke or mouse movement the user makes, although in principle these are events that the page can handle and respond to. Most often keyboard events are used either for scrolling or text entry, neither of which must be replicated at that granularity.[4] Mouse movement events generally result in, at most, superficial changes to a page, e.g., a dropdown menu when hovering over an element. So long as the menu links are accessible through the DOM for the page, it does not matter if they are visible or not during replay. If it becomes necessary, we can easily log and replay these events as well.

---

[4]Keyboard shortcuts are becoming more common, and we plan to record non-navigation keystrokes in the future.

Site(s)	Purpose / actions
Yahoo!	Check Yahoo! mail, news, TV listings
Netflix	Research movie reviews
GMail	Check email
CNN	Read news
Verizon Wireless	Research cell phone plans
Google, etc.	Research MP3 player purchase

Figure 8: Summary of browsing session for evaluation.

4. EVALUATION

We evaluated the effectiveness of Doppelganger versus various built-in browser settings by performing a script of common browsing tasks, summarized in Figure 8. In testing, we simulated a user who is willing to accept a certain amount of privacy loss for convenience at sites with whom he has a relationship (in this case, Yahoo!, Netflix, and GMail) but is more cautious at sites with whom he has no relationship (CNN, PC Magazine, Vanns.com, ComputerHQ.com, BeachCamera.com).

For each setting, we measured (1) the number of sites whose cookies were accepted, categorized by persistence and context, and (2) the inconveniences suffered by the user, including dialog boxes and lost functionality. An ideal scheme would incur a low number of each. The five settings we tested were four global settings (same for all sites): (1) All cookies enabled, (2) First-party cookies only, (3) First-party session cookies only; (4) Ask the user what to do for each cookie that is sent; and finally (5) using Doppelganger. We measured the number of sites rather than the number cookies because multiple cookies of the same type from the same site are equivalent from a privacy perspective. We executed each script by hand three times consecutively for each setting, retaining any state between runs. The idea was to capture the effects of session cookies, persistent cookies, and, for Doppelganger, the change in policy and behavior over time. We cleared all cookie-related state before changing settings.

The accepted-cookie results are shown in Figure 9, and the details of each setting and its corresponding user experience are described below. Two values are particularly interesting. The number of sites setting persistent cookies is significant because they allow users to be tracked over many sessions, and the number of sites setting third-party cookies is perhaps more so because they let the user be tracked across multiple sites. Third-party persistent cookies combine the worst of both.

There are three ways in which the user can be "inconvenienced" during our script: he can be asked to answer a browser's yes-or-no cookie dialog (see picture); he can be asked a left-or-right browser decision by Doppelganger (see Figure 5); or he can be forced to login upon each visit to a site where he has an account. This latter case occurs when his browser could have accepted a persistent cookie that would serve as an authenticator, but did not for some reason.

In the following discussion, we use figures from the last session of each setting, as that most closely represents a steady-state figure.

4.1 All cookies

This is the default setting in Firefox, and, perhaps predictably due to its permissiveness, led to the acceptance of the most cook-

Run	Number of sites setting:				Total persistent (FP-P + TP-P)
	FP-S cookies	FP-P cookies	TP-S cookies	TP-P cookies	
All cookies on (Run 1)	9	8	3	13	21
All cookies on (Run 2)	8	8	3	16	24
All cookies on (Run 3)	8	8	3	16	24
FP only (Run 1)	9	8	1	4	12
FP only (Run 2)	8	8	2	5	13
FP only (Run 3)	8	8	2	7	15
FP session only (Run 1)	9	0	9	0	0
FP session only (Run 2)	9	0	6	0	0
FP session only (Run 3)	9	0	8	0	0
Ask user (Run 1)	4	3	0	0	3
Ask user (Run 2)	3	3	0	0	3
Ask user (Run 3)	3	3	0	0	4
Doppelganger (Run 1)	4	0	0	0	0
Doppelganger (Run 2)	3	3	0	0	3
Doppelganger (Run 3)	3	3	0	0	3

FP = "First party" TP = "Third party" S = "Session" P = "Persistent"

Figure 9: Number of sites setting cookies while performing some common tasks. A script of common browsing tasks was run three times in succession for a variety of cookie management policies, and we measured the number of sites setting various kinds of cookies. "First party" here refers to sites that the user intended to visit, and "third-party" refers to all other sites.

ies of any setting: 24 sites set persistent cookies, including 16 in third-party context. Virtually every domain we visited set a persistent cookie (none sent only session cookies). This suggests that any future browsing sessions would be extensively tracked. The advantage of this permissive policy was that we were never asked any questions during the session.

4.2 First-party only

Since the danger of third-party cookies was recognized years ago, many users disabled third-party cookies in their browsers; the size of this group has forced sites to avoid any dependence on these cookies. Thus a first-party only setting in the browser is, in practice, a big win over the "allow all" setting. Indeed, we accepted persistent cookies from a little more than half as many sites as in the default setttig. However, the browser still accepted many cookies that either do not confer any benefit or were not worthwhile. These include persistent cookies from 7 sites we did not mean to interact with; these were accepted because of redirection and framing tricks. Furthermore, there were unneeded cookies from first-party sites—8 vs. 3 for Doppelganger—since we did not need cookies from, e.g., cnn.com. The main advantage of first-party only vs. more restrictive settings is that we were not asked any questions.

4.3 First-party session only

This setting downgrades all persistent cookies to session cookies with the aim of eliminating long term tracking. It was still vulnerable to tricks which forced us to accept session cookies from 6 to 9 sites (it varied across runs, because advertisements change) that we did not mean to interact with. The session-only restriction came with a significant downside: we were forced to log in to each site with which we had a relationship during every session, and would have had to do so indefinitely. All personalization features that do not require a login would also be lost.

4.4 Ask the user

The final built-in browser setting we tested was one in which the browser asks the user whether to accept each cookie as it was offered. This dialog box allowed a decision to apply to all cookies from the same site, and we checked this box each time to reduce the number of dialogs.

An example of the dialog box is shown here:

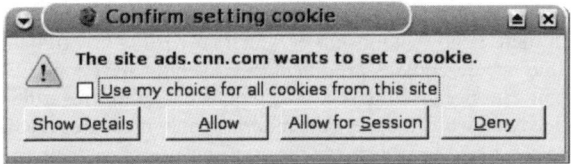

This setting poses something of a dilemma: in principle, we do not know which cookies to accept and which to deny, especially since most of the dialogs appear before a site's home page even finishes loading. Our solution was to use Doppelganger to determine which cookies were useful and, thus, always answer questions correctly, including when to accept persistent cookies and when to downgrade them to session cookies. While this assumes a somewhat oracular user, it helps put a lower bound on the difference between Doppelganger and the best-case scenario for existing browser settings.

Unsurprisingly, the Ask setting resulted in a dramatic reduction in the number of accepted cookies compared with other browser settings: there were only 4 sites which set persistent cookies, and no third-party cookies were accepted.

There was no loss of functionality with this setting, as there was with the "First-party session only" setting, but there was a significant problem: we were shown 26 dialog boxes during our session. Of these, 13 were due to first-party sites, and 13 due to third-party sites using various tricks. These dialogs presented almost no information to help the user decide whether to accept the cookie, except for the domain name.

4.5 Doppelganger

Our last test used Doppelganger for cookie management. Unsurprisingly, the cookie results were virtually identical to the Ask policy, since we used information gained by Doppelganger to answer the browser's questions during that trial. The total number of persistent cookies rose from Run 1 to Run 2 because Doppelganger reserves judgment on persistent cookies until it can test their usefulness in a subsequent session; ultimately, 3 sites' persistent cookies were accepted vs. 4 for the Ask setting.

We did not have to answer nearly as many questions using Doppelganger as we did with Ask. During the first run, there was a comparison dialog on the netflix.com homepage, because session cookies were required to use the site. This dialog does not appear in low or medium paranoia mode. An error message at verizonwireless.com prompted a click of the Fix Me button; this solved the problem without further effort; this is not necessary in low paranoia mode. During the second run, yahoo.com, netflix.com, and mail.google.com (Gmail) all had automatic login features if persistent cookies were enabled. This resulted in side-by-side comparisons, and in each case we chose to accept the cookie in exchange for the convenience. At verizonwireless.com, a persistent cookie remembered our zip code, prompting a comparison; we chose not to accept a persistent cookie, because we did not plan to visit the site often. During the third run, there were no dialogs at all. Indeed, by that time, Doppelganger had already silently decided not to accept cookies from cnn.com, pcmag.com, and dealtime.com (a site through which we were redirected each time we clicked on vendors from pcmag.com).

It is important to note that none of the dialogs we saw would ever recur; once a decision has been made, it is remembered for future sessions. While the same is true for the Ask policy, the test script is heavily weighted towards sites the simulated user has a relationship with. Thus, new sites the user encounters are likely to be ones with which there is no relationship or which are visited less frequently, and thus where cookies are much less likely to have value. This is significant because the Ask policy pops up dialogs regardless of cookie value, while Doppelganger does so only if cookies are likely to be useful, and in fact shows the user *how* useful, as well as relevant privacy information. Turning on the optimization above would make the number of Doppelganger dialogs smaller still.

5. WEB SITE COUNTERMEASURES

If Doppelganger becomes widely deployed, web sites might try to circumvent or fool its mechanisms to store persistent data on users' machines. In this section, we discuss various approaches they might take and how those approaches affect Doppelganger.

5.1 Always require cookies

A web site might require users to always accept cookies by redirecting users to an error page if it detects cookies are being blocked. Many sites currently do this. However, addressing this problem only requires Doppelganger to accept session cookies from the site, which have limited privacy risks; and if the user is using low or medium paranoia mode, Doppelganger will do so automatically. For very privacy-conscious users in high paranoia mode, Doppelganger will expose the sites' cookies requirements for inspection. In addition, web sites might try to require users to accept persistent cookies by requiring an extensive "sign-up" procedure if it does not detect a persistent cookie on the user's machine. This might encourage users to accept persistent cookies to avoid repeating this procedure on subsequent visits. However, this approach would likely alienate users; privacy conscious users may not want to accept persistent cookies or might routinely delete all their cookies, and for privacy reasons, public kiosks and library terminals must delete all cookies after each user's session.

5.2 Cause spurious differences

A web site might try to always create subtle or inconsequential differences between pages requested with cookies and those requested without cookies to frequently trigger Doppelganger's comparison dialog. A user who receives excessive comparison dialogs for a site might become annoyed and decide to accept the cookies to prevent future interruptions, or worse, disable Doppelganger entirely. Doppelganger's current difference detection algorithm is simple: we compare the page titles and look to see if the user's name or ID is only in the fork browser. Developing sophisticated and robust difference detection is important future research for Doppelganger, and we see two promising directions. One approach is compare pages structurally by examining their DOM trees. This is based on the assumption that substantive changes often result in the addition or removal of whole page elements. Another complementary approach is a visual comparison, comparing screen captures of the fork and main windows. Both of these approaches require filtering advertisements and other sources of randomness before comparison.

5.3 Other persistent objects

If a web site detects its persistent cookies are being blocked, it might resort to storing other persistent objects on the user's machine (e.g., flash objects, images, Javascript) and retrieving these objects in subsequent sessions. To fully address privacy with respect to persistent web objects, Doppelganger must apply its cookie policy for a domain to manage other cached objects from that domain as well. However, there may be well-intentioned sites that don't use cookies at all, but cache web objects on users' machines to improve performance. Fully understanding the effect of this approach on the user's web browsing experience requires further study.

6. FUTURE WORK

Currently, Doppelganger only has mechanisms to incrementally relax cookie policies, but not make them more restrictive. Users may later change their minds about the cost/benefit tradeoff for certain domains and want to make their cookie policies for those domains more restrictive. Exploring usable mechanisms for "tightening" the cookie policy is an area for future exploration.

Comprehensive user studies are needed to understand the usability of Doppelganger's mirroring and automated recovery mechanisms. We have recently conducted a controlled user study to examine the usability and performance of Doppelganger and plan to publish the results in the near future.

7. CONCLUSION

We introduced Doppelganger, a novel system for creating and enforcing fine-grained, privacy preserving cookie policies in web browsers with low manual effort. We showed how Doppelganger automatically identifies cookies which provide users additional functionality and exposes the costs and benefits of accepting those cookies. As with most markets, more complete information has the potential to lead to more efficient outcomes. In this case, that means that users will be able to select those sites that offer benefits commensurate with the users' privacy loss over sites with less favorable exchanges. Indeed, since subjective feelings of trust in users often induce users to accept more privacy costs, steps by sites to increase transparency—such as publishing a useful privacy policy—may actually increase the amount of usable personal information they obtain. In short, we believe that systems like the one we describe here can lead to better incentives for both parties. Thus, while our work here certainly does not expose all costs and benefits, and only deals with one aspect of online privacy (viz. tracking), we believe that it represents a meaningful step forward down the right path.

Acknowledgements

We would like to thank David Wagner, Marti Hearst, and Doug Tygar for their very valuable insights and suggestions. AJ Shankar, Naveen Sastry, and Marco Barreno were patient testers and supplied valuable usability feedback. We would also like to thank the anonymous referees for comments that have helped improve both the content and form of the paper.

8. REFERENCES

[1] Alessandro Acquisti and Jens Grosslags. Privacy and rationality in individual decision making. *IEEE Security and Privacy*, 3(1):26–33, 2005.

[2] Add & Edit Cookies. http://addneditcookies.mozdev.org/.

[3] Adil Alsaid and David Martin. Detecting web bugs with bugnosis: Privacy advocacy through education. In *Proceedings of the 2002 Workshop on Privacy Enhancing Technologies*, 2002.

[4] Fahd Arshad. Privacy Fox - A JavaScript-based P3P Agent for Mozilla Firefox. http://privacyfox.mozdev.org/PaperFinal.pdf, 2004.

[5] Aaron Brown and David Patterson. Undo for operators: Building an undoable e-mail store. In *USENIX 2003 Annual Technical Conference*, June 2003.

[6] Simon Byers, Lorrie Faith Cranor, and David Kormann. Automated analysis of P3P-enabled web sites. In *ICEC '03: Proceedings of the 5th International Conference on Electronic Commerce*, pages 326–338, New York, NY, USA, 2003. ACM Press.

[7] Electronic Privacy Information Center and Junkbusters. Pretty poor privacy: An assessment of P3P and Internet privacy. http://www.epic.org/reports/prettypoorprivacy.html, June 2000.

[8] Cookie Button. http://basic.mozdev.org/cookiebutton/.

[9] Cookie Culler. http://cookieculler.mozdev.org/index.html.

[10] Cookie Toggle. http://cookietoggle.mozdev.org/.

[11] Mary J. Culnan and Pamela K. Armstrong. Information privacy concerns, procedural fairness, and impersonal trust: An empirical investigation. *Organization Science*, 10(1):104–115, 1999.

[12] Kevin Fu, Emil Sit, Kendra Smith, and Nick Feamster. Dos and Don'ts of client authentication on the web. In *10th USENIX Security Symposium*, pages 251–268, August 2001.

[13] Jeremy Goecks and Elizabeth D. Mynatt. Social approaches to end-user privacy management. In Lorrie Faith Cranor and Simson Garfinkel, editors, *Security and Usability: Designing Secure Systems That People Can Use*, chapter 25, pages 523–545. O'Reilly, 2005.

[14] Nathan Good, Rachna Dhamija, Jens Grosslags, David Thaw, Steven Aronowitz, Deirdre Mulligan, and Jospeh Konstan. Stopping spyware at the gate: A user study of notice, privacy and spyware. In *Symposium on Usable Privacy and Security (SOUPS) 2005*, July 2005.

[15] Il-Horn Hann, Kai-Lung Hui, Tom S. Lee, and I. P. L. Png. Online information privacy: Measuring the cost-benefit trade-off. In *Proceedings of the Twenty-Third International Conference on Information Systems*, pages 1–8, 2002.

[16] Stephen E. Levy and Carl Gutwin. Improving understanding of website privacy policies with fine-grained policy anchors. In *WWW '05: Proceedings of the 14th International Conference on World Wide Web*, pages 480–488, New York, NY, USA, 2005. ACM Press.

[17] LiveHTTPHeaders Firefox extenstion. http://livehttpheaders.mozdev.org/.

[18] Lynette Millett, Batya Friedman, and Edward Felten. Cookies and web browser design: Toward realizing informed consent online. In *Proceedings of the CHI 2001 Conference on Human Factors in Computing Systems*, pages 46–52, April 2001.

[19] Gavin O'Malley. Jupiter analyst: Nielsen research confirms users delete cookies. http://publications.mediapost.com/index.cfm?fuseaction=Articles.san&s=2%8883&Nid=12855&p=297686.

[20] Permit Cookies. http://gorgias.de/mfe/.

[21] Persistent client state: HTTP cookies, Preliminary specification. http://wp.netscape.com/newsref/std/cookie_spec.html.

[22] Kevin Poulsen. Microsoft cookies jump domains. http://www.securityfocus.com/news/83, September 2000.

[23] Sites use IFrames to bypass cookie prefs. https://bugzilla.mozilla.org/show_bug.cgi?id=158463.

[24] The Platform for Privacy Preferences Project (P3P). http://www.w3.org/TR/P3P/.

[25] View Cookies. http://www.bitstorm.org/extensions/view-cookies/.

[26] Ka-Ping Yee. Guidelines and strategies for secure interaction design. In Lorrie Faith Cranor and Simson Garfinkel, editors, *Security and Usability: Designing Secure Systems That People Can Use*, chapter 13, pages 247–273. O'Reilly, 2005.

Fourth-Factor Authentication: Somebody You Know

John Brainard
RSA Laboratories
Bedford, MA, USA
jbrainard@rsasecurity.com

Ari Juels
RSA Laboratories
Bedford, MA, USA
ajuels@rsasecurity.com

Ronald L. Rivest
MIT CSAIL
Cambridge, MA, USA
rivest@mit.edu

Michael Szydlo
RSA Laboratories
Bedford, MA, USA
mszydlo@rsasecurity.com

Moti Yung
RSA Laboratories
Bedford, MA, USA
myung@rsasecurity.com

ABSTRACT

User authentication in computing systems traditionally depends on three factors: something you have (e.g., a hardware token), something you are (e.g., a fingerprint), and something you know (e.g., a password). In this paper, we explore a fourth factor, the social network of the user, that is, *somebody you know*.

Human authentication through mutual acquaintance is an age-old practice. In the arena of computer security, it plays roles in privilege delegation, peer-level certification, help-desk assistance, and reputation networks. As a direct means of logical authentication, though, the reliance of human being on another has little supporting scientific literature or practice.

In this paper, we explore the notion of *vouching*, that is, peer-level, human-intermediated authentication for access control. We explore its use in emergency authentication, when primary authenticators like passwords or hardware tokens become unavailable. We describe a practical, prototype vouching system based on SecurID, a popular hardware authentication token. We address traditional, cryptographic security requirements, but also consider questions of social engineering and user behavior.

Categories and Subject Descriptors

H.m [**Information Systems**]: [Miscellaneous]

General Terms

Security, Human Factors

Keywords

authentication, hardware tokens, vouchers

Permission to make digital or hard copies of all or part of this work for personal or classroom use is granted without fee provided that copies are not made or distributed for profit or commercial advantage and that copies bear this notice and the full citation on the first page. To copy otherwise, to republish, to post on servers or to redistribute to lists, requires prior specific permission and/or a fee.
CCS'06, October 30–November 3, 2006, Alexandria, Virginia, USA.
Copyright 2006 ACM 1-59593-518-5/06/0010 ...$5.00.

1. INTRODUCTION

Passwords remain the most common mechanism for user authentication in computer-security systems. Their various drawbacks, like poor selection by users and vulnerability to capture, are prompting a rapidly mounting adoption of hardware authentication tokens. Despite stronger security guarantees, though, hardware tokens share a limitation with passwords: inconsistent availability. Users frequently forget passwords. Similarly, they often lose, forget, and break their hardware tokens.

As a result, a workable authentication system requires at least *two* modes of authentication. There is the primary mode of authentication, the password or token employed by the user in the normal course of events. Then there is the form of *emergency* authentication for cases when the primary authenticator is unavailable to a user.

On the Internet, the most common form of emergency authenticator is e-mail. When a user forgets her password for a particular site, she often has the option of having the password itself or password-reset instructions sent to a pre-registered e-mail account. Another common emergency authentication mechanism is "life questions." The user is prompted to authenticate herself by furnishing answers to previously registered personal questions, e.g., "What is the name of your first pet?" In corporate environments, the preferred emergency authentication mechanism is the help desk: employees telephone support help-desk personnel for assistance in re-establishing their access privileges. Some consumer Web sites offer this option as well.

An authentication system is, of course, only as secure as its weakest component. It is desirable, therefore, that an emergency authenticator provide security as least as strong as a good primary authenticator. Life questions, as we explain later, often fall short in this regard, as their answers are vulnerable to guessing by attackers and sometimes subject to attacks involving mining of public databases. Similarly, since e-mail is rarely encrypted, and e-mail accounts are often password-protected, e-mail is generally an inadequate emergency authentication mechanism in systems where the primary authenticator is a hardware token. Help desks can provide emergency authentication with any of a range of security assurances, depending on the manner in which staff identify callers. Help-desk staff can ask life questions, identify the voices of callers they know personally, verify caller-

ID, and so forth. But corporations commonly dislike help desks as a emergency-authentication mechanism because of high labor costs. Additionally, human intermediation at help desks introduces a vulnerability to social-engineering attacks.

Passwords and life questions are often categorized in the abstract as "something you know," while hardware tokens are "something you have." A third category of authenticator is "something you are," that is, a biometric. Systems that authenticate users based on physical characteristics—particularly voice and fingerprints—are enjoying ever-rising popularity. The general consensus of the security community, however, is that biometrics are not suitable as primary authenticators. Biometrics are often not secret. People publicly expose their voices and fingers in various ways on a regular basis, creating the possibility of biometric spoofing. (Various countermeasures can alleviate but probably not eliminate this problem.) Users generally don't forget or lose their biometrics permanently—but chapped fingers and laryngitis can lead to temporary loss. Finally, some biometrics, like fingerprints, require special-purpose reader hardware. Thus, despite their attractions, biometrics do not provide a comprehensive answer to the problem of emergency authentication.

1.1 Our work

In this paper, we explore a fourth category of authenticator: "some*body* you know." The use of human relationships for authentication is by no means new. In social interactions, introducing one person to another is the most common way of identifying (and implicitly authenticating) acquaintances. This is true in the physical world and also to a large extent in cyberspace: e-mail is a popular informal channel of authentication. In the realm of computer security, however, the natural mechanism of authentication through social networks has seen little in the way of direct formal use or exploration.

Our particular focus here is a process that we call *vouching*. Vouching is peer-level authentication in which one user, the *helper*, leverages her primary authenticator in order to assist a second user, the *asker*, to perform emergency authentication. To lend clarity to our discussion, we sometimes refer to the asker as Alice and the helper as Harry. Based on simple security principles and social-engineering and usability considerations, we design and describe a prototype vouching system for SecurID, a popular hardware authentication token typically used in conjunction with a user-specific PIN. In our system, a helper can use her SecurID token to help grant temporary access privileges to an asker who has lost the ability to use her own SecurID token, but who remembers her PIN. In a nutshell, the helper obtains a temporary passcode that we call a *vouchcode*. The helper furnishes this vouchcode to the asker as a substitute for the asker's SecurID token, i.e., as a kind of replacement one-time passcode. Using the vouchcode, the asker can authenticate without her token.

While vouching is straightforward at a loose conceptual level, a number of subtle design issues arise on closer scrutiny that do not apply to traditional authenticators. A (randomly assigned) password, for instance, carries a certain measurable level of entropy; some rough characterization is possible even when a password is user-selected. Similarly, the bit-length of the cryptographic key in a hardware authentication token, the underlying cryptographic primitives, and the output format permit rigorous characterization of the token's security properties. Vouching systems rely on traditional primary authenticators like PINs and passcodes—but in a more complicated context.

Usability and social-engineering considerations are particularly tricky facets of voucher-system design. A voucher system that is difficult to use will merely drive users to call help desks; a system that is too easy to use could evolve into a primary authenticator, eroding user vigilance around the vouching process. It is also important to ensure that askers properly authenticate the people that they are vouching for. If helpers respond to vouching requests via e-mail or to vouching requests from strangers over the telephone, then a vouching system will offer little real security. For such reasons, a vouching system must rely on trustworthy interactions within a tight social network. The problem of creating a vouching system that places helpers in a strong position to attest to the identities of askers is an important aspect of our work.

1.2 Organization

In section 2, we survey the literature on concepts related to our vouching proposal. We treat the problem of modeling in section 3. We propose and analyze a vouching system for hardware tokens in 4 and describe a prototype implementation for SecurID tokens in section 5. In section 6, we discuss social engineering and related issues. We conclude in section 7 with thoughts on further avenues of research.

2. RELATED WORK

2.1 Backup authenticators

The limitations of passwords as authenticators are well known in the security community—from the general problem of poor selection by users [12] to the disquieting inclination of users to reveal their passwords to strangers [13]. Phishing, the fraudulent use of e-mail to capture user passwords (and other information), has exacerbated the problem of password capture. Researchers have proposed a variety of countermeasures to these problems, such as browser extensions that hash passwords with domain names [17].

Life questions are another "something-you-know" authenticator, but one that serves typically in emergency authentication. Life questions have received considerably less study than passwords. Some of the security vulnerabilities of password systems, like pharming, are also applicable to life questions. Life questions, though, have their own particular features. The fact that they serve most often as a emergency authenticator—and thus receive infrequent use—can render them difficult for users to answer consistently. Moreover, as shown recently by Griffith and Jakobsson, the answers to certain popular life questions are vulnerable to attacks that involve mining of public databases [9]. The answers to many commonly used life questions, like "What was the make of your first car?" have little underlying entropy.[1] Finally, users show a surprising willingness to divulge personal information to strangers in ways that can undermine the security of life-question systems. In a recent "live phishing" experiment, in which passersby in New York City's Central

---

[1]General Motors, for example, had about a 43% market share in the United States in 1983 [11].

Park were offered T-shirts in exchange for filling out surveys, more than 70% divulged their mother's maiden name, while more than 90% revealed their place and date of birth [6].

Consumers' mobile phones offer a platform for emergency authentication that is increasingly favored by financial institutions for high-risk online transactions (but less often for password recovery). Mobile-phone based authentication can operate in several ways. A financial institution can initiate a call to the phone of a customer to request transaction confirmation via an automated voice recognition or keypad-based entry system—or even by means of (biometric) speaker recognition. Alternatively an institution can transmit an authentication code to a phone via SMS messaging, and request that the user enter the code into a web form [16]. These techniques are excellent adjuncts to more traditional forms of authentication, and caller ID is a simple and attractive form of authentication (although somewhat vulnerable to spoofing). They can also be useful in emergency authentication, although such use is limited at present—particularly within corporations, where mobile phones are generally not under administrator control.

Password-reset via help-desk calls is a very common practice, but it's expensive. Vendors of password-reset products claim that each password-reset request involving live staff costs some $15-30 [1, 3]. This cost alone is a strong impetus for creating alternatives. Moreover, because help-desk calls often involve interactions between strangers, they are vulnerable to social-engineering attacks, such as those described by Kevin Mitnick in his *Art of Deception* [15].

2.2 Social relationships and authentication

Reliance on "somebody you know" is an age-old vehicle for authentication in everyday life. When you introduce one friend to another, you are effectively performing an authentication protocol based on a social relationship. Social-network-based authentication also has a pervasive but largely informal role in the security infrastructures of organizations. When an employee holds open an access-controlled door for a familiar colleague, when a system administrator resets a password for a colleague whose voice she recognizes on the phone, or when a manager brings a new employee to a corporate badging center, a form of vouching is taking place.

Familiar social interactions of this kind have given rise to analogous systems in cyberspace. People regularly send e-mail to colleagues and friends in order to effect introductions and request grants of privileges (to set up accounts, make payments, etc.). *Reputation systems* are an extension of the vouching principle to large on-line communities; eBay provides a familiar example in which account holders provide mutual ratings of commercial integrity. Given their loose authentication of participants, though, reputation systems are unsuitable mechanisms for authentication of individual identities.

Peer-level public-key infrastructures (PKIs) such as PGP and SPKI/SDSI [2, 8] provide a more rigorous basis for authentication through social networks. In such systems, principals make local decisions about whether to endorse the identities (certificates) of other principals. In the case of PGP, a social network called a "Web of Trust" helps authenticate e-mail addresses. Human intermediation can serve as a component in peer-level (and hierarchical) PKIs. It is a straightforward matter to define a credential asserting attestation of identity based on human contact ("I have met so-and-so face-to-face") and to create supporting policies and software interfaces. Trust-management systems like PolicyMaker [5] and its successor, KeyNote [4], permit general policy decisions around access to resources on the basis of digitally signed credentials, and can in principle take human relationships into account.

PKIs and trust-management systems, however, are abstractions for the creation and management of digital credentials. In contrast, in this paper we consider "somebody you know" as a starting point for authentication; our investigation is predicated on fundamental usability and human-interaction issues, rather than reference to a particular cryptographic mechanism. In this view, Carl Ellison's concept of "ceremonies" is particularly important prior work [7]; Ellison proposes a model to capture the human behavior surrounding cryptographic authentication protocols. Also pertinent are recent authentication systems designed with human factors as first principles. A recent example is "Seeing is Believing," a system that exploits visual contact (2D barcodes) as a physical mechanism for trust [14].

3. MODELING

In this section we model the security properties that we would like a voucher system to achieve. We begin by describing the parties and communication channels involved in the system. Then we discuss our assumptions concerning the two pieces of authentication data used. These items are the PIN and the tokencode. We describe the types of adversaries we intend to protect against, and present security requirements for such adversaries.

Parties and Channels: The principal parties involved in the primary authentication mechanism are the *User* and the *Server*. For the vouching protocol two types of users play distinct roles. These are the *Asker* and the *Helper*. We also consider a malicious external party *Adversary* with various capabilities.

Users or parties are denoted by a capital letter (e.g., X), and to each party we associate an identifier, which will be denoted with the corresponding lowercase letter. For example, the identifier of a user X will be denoted by x. Within an invocation of a protocol among users, we use the notation $X(y)$ to indicate that party X has represented itself as having claimed identifier y. With this notation, an honest party X will always be represented as $X(x)$. Our notation suggests that parties use consistent identifiers throughout the protocol, and we make a remark whenever this may not be the case.

Users have certain initial conditions imposed on them that can be expressed as relations. Namely, some users are enrolled as helpers, and users are organized in a (helper, asker) relation **H** where each helper is assigned a subset of users as askers; we say $(Y, X) \in \mathbf{H}$ if a party X is an asker for party Y acting as a helper.

In our model, the communication channels to and from the server are considered to be secure. In particular, this means that no adversary can obtain any information by eavesdropping over such channels. In practice, channels between an entity and the server may be achieved with SSL. The channel between an asker and a helper may be less secure, and we need to carefully design our protocol setting to mitigate eavesdropping risks on this channel.

Authentication Data Items: The basic authentication protocol as well as the vouching protocol involve two primary factors that the user presents to the system in order to be accepted as authenticated. The basic two factors are the PIN, PN, which is a user memorizable string, and the tokencode, TK, which is generated from the hardware token. Other factors are possible. In particular, a third piece of authentication data is the vouchcode, an ephemeral item used within the vouching protocol.

To model the information made available in a protocol message by parties, we introduce the notation $(X(u) : VA)$ to indicate the event wherein a party X sends a message claiming identity u and correctly presents the value item of VA. The following assumptions concern the probabilities of an adversary being able to present the user's PN and TK at any point within the protocol.

1. For any party X and any identifier y, the probabilities $prob\{X(y) : PN\}$ and $prob\{X(y) : TK\}$ are independent.

2. The legitimate user U who chose the PIN value and possesses the token and enrolled in the system as u can always present the PIN and tokencode. That is, $prob\{U(u) : PN\} = prob\{U(u) : TK\} = 1$. These assumptions amount to a user being able to recall their PIN value and also being in posession of their token.

3. An adversary, A, who is not U, and has not obtained U's PIN value PN, can only present it with small probability. That is $prob\{A(u) : PN\} < \epsilon$.

4. An adversary, A, (where $A \neq U$), can only present a tokencode with small probability. That is $prob\{A(u) : PN\} < \delta$.

Note that the adversary's probability of successfully presenting the required values (PN or TK) is the probability bound on the union of two events: (1) the adversary guessing the value (which is typically assumed long enough and to contain certain amount of entropy), and (2) the adversary stealing or otherwise obtaining the value.

Authentication Ceremony: An authentication ceremony AC is a sequence of interactions between a number of parties. It was introduced to describe not just a protocol between the user software agent and the system, but to designate that a party may act via his software agent and also act personally. Once defined, a ceremony is invoked by various parties (with certain relationships among them) who follow a sequence of actions. We use notation AC_T to refer to a ceremony of type T or an instance of this type of ceremony. To specify the parties P_i and the identifiers p_i' presented by these parties we use the notation $AC_T(P_1(p_1'), P_2(p_2'), \ldots)$.

In an authentication ceremony parties exchange messages via secure channels, and one of the parties will be the server S. The server is assumed to be trusted and its role is to accept the legal executions and reject the adversarial ones. The correctness and security definitions of an authentication ceremony are defined in terms of the server's final state, which is always either "accept" or "reject."

Interactive Logging and Detection: Given a ceremony, we may consider an extended ceremony where in addition to message channels there exist also "logging channels" where parties post information "about" the ceremony: e.g., posting the time of the invocation, the parties, and the state of the execution). These logging channels are not erasable by the parties, and can be viewed as a bulletin board maintained by the trusted server, and can be implemented by getting e-mail upon demand from the server.

By considering the extended ceremony that includes the original messages that are designed to stop impersonation (i.e., to reduce the probability of an attack), logging and log evaluation further allow for the detection of (what we hope are rare) successful attacks. For example, a party evaluating the log may detect unexpected behavior such an authentication under its identifier in which it did not take part. We denote the logging channels available to party U as $LG(U)$. The party reads and evaluates the log and decides whether there as been an attack or not. If an attack is detected in this way appropriate countermeasures can be taken. In actual implementation, the logging functionality may be implemented as a combination of e-mail notifications and server logs.

Next we discuss correctness and security definitions.

3.0.1 Correctness

The correctness requirement of the system is that when the parties act honestly, the authentication attempt will succeed, i.e., for a given authentication ceremony $AC = AC(P_i(p_i))$ (where in all cases party P_i and its true identifier p_i are involved and where the parties satisfy the initial required relationships) then the server "accepts."

3.0.2 Security Properties

To describe the security properties of a ceremony, we define adversaries (outsiders and insiders) who attack other parties. In an authentication ceremony, the adversarial goal is to impersonate a party. There are two types of security properties: (1) Prevention: where the server accepts an invocation with a party claiming a wrong identifier only with small probability. (2) Detection: where successful attacks are realized by parties.

4. VOUCHER SYSTEM

In this section we describe a voucher system for hardware authentication tokens such as RSA Security Inc.'s SecurID. A SecurID token typically takes the form of a key fob or card that displays a fresh numerical value, called a *tokencode* or *passcode*, every sixty seconds. To authenticate to a computer application, the user must type the current tokencode together with a user-specific PIN or password.

The tokencodes in a SecurID installation are validated by a specially designated authentication server. This server shares a secret seed with every token and also the PIN of the user, allowing it to validate tokencode/PIN pairs. The technical details of tokencode computation are largely unpublished. Briefly, though, a tokencode is computed as a cryptographic function of the current time and a secret key shared between the token and authentication server [10]. Our system can be applied equally well to other token-types, e.g., Verisign authentication tokens.

We will describe the regular system AC_R and the vouching system AC_V. In both cases we will describe the messaging and the extended ceremony with logging.

4.1 Regular Authentication

We first review the regular two-factor authentication protocol, AC_R, which does not include vouching. This illustrates the basic system upon which a vouching system will be constructed.

Enrollment: The process gives each user a PIN value and a hardware token that at each time unit produces tokencodes. At the server side the user identifier is associated with its chosen PIN and with the specific tokencode. Thus the server is able to produce the authentication values and compare them to the ones presented by the user.

Authentication Session: Once enrolled, a user U authenticates herself by presenting her identifier u, her PIN PN and the current TK to the server via a secure channel. Once the server gets the values it checks them against computed/stored values. If the values match the server "accepts" and otherwise it "rejects."

If logging is implemented, the server logs to $LG(U)$ that a session with u has taken place at its given time unit. This logging event is available to user U (and is not manipulated by it), and U can evaluate the log.

Properties of the regular protocol:

- Correctness: for the legitimate user U claiming to be itself $prob\{U(u) : PN\} = prob\{U(u) : TK\} = 1$; thus the server will always accept.

- Security: (prevention) The adversary A performs the authentication session with identity u which is not his own. Since $prob\{A(u) : PN\} < \epsilon$ and independently, $prob\{A(u) : PN\} < \delta$, the server accepts with probability at most $\epsilon\delta$, which is very small, thus impersonation is prevented.

- Security: (detection) In case the server maintain logs that are accessible by the user at all times, in the unlikely event that A is successful, U will learn about a session initiated under her identity at a time when she herself did not participate. Thus she will detect the break.

4.2 Concrete protocol steps

The vouching system we describe is designed to deal with the case in which a user does not have her token available, but does recall her PIN. In other words, it is designed to deal with unavailability of one of the two authentication factors. Expressed succinctly, the vouching process involves this user (asker) contacting a pre-registered helper for assistance in authenticating herself. The helper obtains a temporary tokencode from the server, called a *vouchcode*. The helper communicates the vouchcode to the asker. The vouchcode aids the asker in the emergency authentication process. The vouching system involves the server and the user (or asker) and helper. The vouching ceremony is denoted $AC_V = AC(S, U, H)$. We now explain in detail the steps involved in the vouching process.

Enrollment: In this stage formally the relation $\mathbf{H}(X, Y)$ is created. The server records this relation and each user learns its helpers and its askers.

It is a matter of system policy which users may act as helpers for a given asker—and how those helpers are designated. One possibility is for an asker to be enrolled in the voucher system concurrently with the original provisioning of the hardware token. At this time an appropriate administrator (or perhaps the asker's supervisor) is notified by e-mail to take action. This notification directs the administrator to a management interface through which the administrator can explicitly specify which helpers are be allowed to

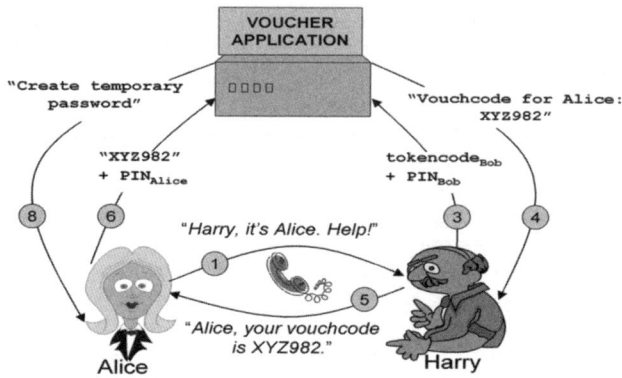

Figure 1: A schematic of the basic vouching process: Harry the helper aids Alice to obtain a temporary password. Step numbers correspond to those in text description (and some are omitted).

vouch for this asker. The same management interface allows for subsequent modifications to the association of helpers to askers. Each helper is automatically sent e-mail notifying him of this responsibility, and must acknowledge and agree to the corporate policy concerning vouching. This may be implemented in a very convenient way, for example embedding an "accept" button in the e-mail notification. We believe that the explicit nature of assuming responsibility as a helper is central to clarifying the accountability of security risks.

The process of assigning helpers to askers can also be partially automated by populating the management interface with existing relationship data, such as that contained in corporate organizational charts. A variant of the enrollment procedure might allow askers to choose their own helpers, although this could increase risks if a supervisor is not involved.

Once enrolled, an asker who has lost the ability to use her token may cooperate with a helper to achieve emergency authentication via the following steps:

1. Asker Contacts Helper: The vouching process begins with the asker U contacting the helper H via an out-of-band channel and claiming to be u. The channel can be the telephone or even face-to-face contact. As we explain later, e-mail contact should be deprecated or prohibited.

2. Helper Authenticates Asker: The helper verifies the identity of the asker. Formally, the process has to assure H with probability at least $1 - \mu$ (for small μ) that the claimed identifier u is the asker's true identifier. In implementing this step over a telephone communication, the helper makes sure that he recognizes the asker's voice (and her telephone, if possible, using caller-ID); the helper may, if in doubt, ask questions that help confirm the asker's identity. In the case of face-to-face contact, the helper authenticates the asker as a matter of course. It is also assumed that the asker identifies the helper and that they both recognize that they are in a helper-asker relationship.

3. Helper Authenticates to Server: Using his own client machine, the helper accesses a vouching-specific web page. Through this interface, the helper strongly authenticates himself to the server using two factor authentication—

i.e., using his token TK and PIN value PN as in the regular authentication session described above. The helper asserts identity h and declares that he is helping user u.

4. Helper Obtains Vouchcode: In response to successful authentication, the helper is prompted by the server S for the name of the asker. The server verifies that H and U are in the helper-asker relationship. If the verification succeeds, the helper then receives an asker-specific *vouchcode* VC (we assume that VC is guessable with probability at most δ). The server further marks the fact that there is an ongoing vouching session involving H and U. The vouchcode assumes an alphanumeric form amenable to verbal communication. For easy communication, it should also be relatively short; it contains 20 bits of entropy in our prototype.

5. Helper Gives Vouchcode to Asker: Next, H informs U of the value VC —orally in the case of telephone communication, and either orally or as a written value in the case of face-to-face contact.

6. Asker Enters Vouchcode: U asserts identity u and presents his PN and VC to S. In our implementation, using a special-purpose web interface on her own client machine, the asker authenticates by entering her username, the vouchcode, and her PIN.

7. Server Authenticates Asker: Upon receipt of this information, the server identifies in its database an active vouching session for this asker. A vouching session is considered active if it has been initiated by a valid helper within a short time period as specified by the system policy (e.g., 3 minutes). (This timeout helps ensure tight synchronization of the vouching process, and that vouchcodes are not saved for later use.) The server verifies the asker's PN and VC. The server either "accepts" the authentication attempt or "rejects", i.e., aborts the vouching session; either way, the asker is informed of the result.

8. Temporary Password: If the asker has successfully authenticated, she is prompted to choose a temporary password TP and granted access for the current session. The asker can use this temporary password together with PN for subsequent sessions, while the vouchcode becomes invalided. The asker's temporary password expires after a time period specified by system policy (e.g., one or two days).

One important reason to have the user convert her vouchcode into a temporary password is that a user-selected password is likely to be easier to remember than a system-generated vouchcode. We touch on security-related reasons in our discussion below.

9. Logging: An important component of a voucher system is its support for detection mechanisms based on user notification and administrator audit. We propose that confirmations of successful/failed asker authentication be sent to the helper and to the asker (and, as a system option, to an administrator). Additionally, an audit log should record all transactions in the vouching system.

Briefly, then, in order to obtain a vouchcode for use by the asker, the helper makes use of the registered helper-asker association and his ability to perform strong authentication. The asker employs this vouchcode and her PIN to obtain a temporary password. This password serves effectively as a token substitute: the asker may use it in lieu of a tokencode wherever system policy permits.

4.3 Defining and Claiming the Voucher System Properties

4.3.1 Correctness

The **correctness** property requires that a legitimate user U (claiming to be u) when asking H (claiming to be h) for help will will always be successful in getting the server S to accept provided $\mathbf{H}(H,U)$. To see that correctness holds, we simply examine the steps of the protocol.

1. In steps 1 and 2, H and U will recognize each other and U has a way to assure H of her identity with probability 1.

2. In steps 3 and 4, H will cause S to accept its authentication with probability 1. This is based on the correctness claim of the regular authentication protocol.

3. Successful authentication will result in a vouchcode VC that will get to U in step 5. The server S is expecting this value within a vouching session. Since $\mathbf{H}(H,U)$, S will produce a VC associated with H and U with probability 1.

4. In steps 6, 7, and 8, U will be able to present her identifier u, PN, and VC correctly to the server, which will accept with probability 1. Furthermore, U will be able to enter his chosen temporary password TP.

5. Finally, the log messages will be always be received by U and H, who expect them as a positive feedback.

4.3.2 Security

We now discuss the **security** of our proposed vouching system by enumerating several formal properties and discussing specific types of attacks. Practical security concerns such as social engineering will be treated in the next section.

We assume in our discussion here that an adversary does not benefit from collusive attacks—as should be the case in well-designed systems enforcing independence among user PINs and authenticators. In fact, when an adversary is attacking a user in the system, we may assume it controls all other users who are not involved in the session and their secrets and we merely concentrate on a specific invocation. We also assume the trustworthiness of the enrollment process, and the system administration and server S. If these assumptions did not hold, the primary authentication mechanism itself would be vulnerable.

To describe the desired security properties we need to first clarify the capabilities of the Adversary. By an *Outside attacker* we mean an attacker who is not enrolled in the systems as either a user, or a helper. By an *Inside attacker* we mean an attacker who is either an asker, helper, or another inside party who acts under a false identifier or without the preconditioned relationship holding.

Prevention Requirements: In an authentication ceremony, the most important aspect of security is to prevent unauthorized access. We define the prevention aspect of security in terms of the following adversarial invocations. Any adversary taking part in a protocol in any of these instances should lead to S rejecting with high probability.

- (User impersonation by an outsider): An instance where $AC_V(S, A(u), H(x))$ where $A \neq U$ should not be accepted. The actual helper H may or may not be present.

- (Helper impersonation by an outsider): An instance where $AC_V(S, U(x), A(h))$, where $A \neq H$, should not be accepted. Such an instance should fail regardless of the user.

- (Helper not registered:) When $(U', H') \notin \mathbf{H}$, an instance of the form $AC_V(S, U'(u'), H'(h'))$, should not be accepted. Such an instance should fail even if U' and H' are colluding.

- (H attacking U:) An instance $AC_V(S, H(u), H(h))$, where H is an adversary trying to impersonate U, should not be accepted.

- (U attacking H:) An instance $AC_V(S, U(u), U(h))$, where U is an adversary trying to impersonate H, should not be accepted.

Detection Requirements: The following detection properties deal with the case in which an attacker has managed to get the server to accept the authentication attempt. In such a case, detection is a second line of defense.

- (Detecting user impersonation by an outsider): In case of a successful instance where $AC_V(S, A(u), H(x))$ with $A \neq U$ (user impersonation, with or without the actual helper H being present), U will detect it in the log.

- (Detecting helper impersonation by an outsider): In case of a successful instance $AC_V(S, U(x), A(h))$, where $A \neq H$, i.e., helper impersonation by a third party, regardless of the user, H will detect it in the log.

- (Illegal help request:) In case of a successful instance where $AC_V(S, U'(u'), H'(h'))$ where (U', H') not in $\mathbf{H}$ (U' and H' are colluding in this attack), U and H will detect it in the log.

- (Detecting H attacking U:) In case of a successful instance where $AC_V(S, H(u), H(h))$, where H is an adversary trying to impersonate U, U will detect it in the log.

- (Detecting U attacking H:) In case of a successful instance where $AC_V(S, U(u), U(h))$, where U is an adversary trying to impersonate H, H will detect it in the log.

Next we discuss how the above definition is satisfied by the protocol.

Outside Attacker Impersonates User: We first consider the case of an outside attacker who targets user U enrolled in the voucher system as an asker. Of course, the attacker can—irrespective of the presence of the vouching system—impersonate U by obtaining or guessing her current TK and PN and using the regular authentication session, but this was shown to be an event of small ($\epsilon\delta$) probability.

The adversary A may start an instance as user U, with $A \neq U$, but he will fail with overwhelming probability to convince H that he is U in step 2.

The adversary might also try to obtain U's vouchcode and PIN. The first way for the attacker to obtain U's vouchcode is by guessing it (knowing that U has asked for help and a VC exists). But guessing PN and VC is not easier than guessing TK and PN.

In the unlikely event that the user has been impersonated, she will be able to detect it in the log.

Outside Attacker Impersonates Helper: Next the attacker may attempt to impersonate the helper. This is an instance of the protocol $AC(S, U(u), A(h))$ where $A \neq H$. In this case step 3 will fail with probability $1 - \epsilon\delta$. Again, in the unlikely event of success the helper will detect the attempt in the log.

Unregistered Helper: Another attack on the system can be an instance where the user U' and the helper H' do not have the precondition relation $(H', U') \in \mathbf{H}$. While a helper can obtain a vouchcode for a user who is a registered asker for H', the system will not produce a vouchcode in step 4 for user U' who is unregistered. In the unlikely event that a legal helper H is impersonated by an adversary, H will recognize an interactive logging event that he has not actually taken part in.

Helper attacks Asker: Once an asker is assigned a helper in the enrollment process there is a risk that the helper attacks the asker. H can obtain a vouchcode for U without even involving her. By guessing U's PN (with probability ϵ in our model), H can then impersonate the asker in the voucher system.

Briefly, then, the main new risk that the vouching protocol introduces is that security against malicious helpers is reduced to that of obtaining the asker's PIN. In case of a successful attack, U finds a session in the log in which she did not participate.

We remark that in our definition of an authentication ceremony we assume that a party presents itself with a consistent identity throughout the protocol instance. In reality, in the vouching ceremony, an adversary acting as a helper may present itself to a user under one identity and to the system under another identity. The analysis of such attacks with divergent identities in very similar to our analyses with consistent identities.

Asker attacks Helper: When an asker U attempts to attack the helper in our model, it can only ask for help, authenticate itself, and obtain a vouchcode. No helper-specific information is leaked or made available to the asker. Thus, an asker can only attack a helper as an outsider, and no better.

4.4 Pragmatic consideration of Attacks Beyond the Model

Remarks on Outsider Attacks: We have formally modelled attacks on vouchcodes via a guessing or misbehaving adversarial helper. We note that in practice an avenue for an outside attacker is to obtain a vouchcode in transit, for example if it is communicated over insecure e-mail from H to U. This is one reason why our system discourages or prohibits e-mail on the asker-helper channel. As we discuss below, an attacker can alternatively attempt to extract U's vouchcode from H via social engineering. To ensure that such refined attacks do not weaken the vouchcode system, it is desirable that their collective probability be at most a small value δ. (Quantifying this probability rigorously in real life systems is a challenge, of course.) One reason to have the asker convert a vouchcode into a passcode (i.e., step 8 of our vouching protocol) is to minimize the risk of vouchcode compromise by an outside attacker. If, for example, the helper provides the vouchcode to the asker on a slip of paper, then it is particularly desirable that the vouchcode expire quickly.

Remarks on Illegal Helping: The system should avoid situations where it is easy for a helper to get the PIN of the asker, e.g., when the asker is tempted to use the helper machine (for example in the helper's office) in order to get a temporary password, while the helper has a malicious software that logs and snatches the asker's PIN.

Remarks on Asker Attacking Helper: Playing the role of the helper in an actual real world setting carries some risk. U could ask H to vouch for her—i.e., request a vouchcode on her behalf—on her own machine. U can make a compelling case for this, since it may be convenient or make H look impolite if he refuses. U, however, if malicious, could have keystroke-logging software on her computer that captures H's PIN.

Such attacks are even more of a concern when two parties are enrolled to vouch for one another. As discussed above, it is enough for a helper to know a user's PIN in order to impersonate her.

Of course, if H authenticates on U's machine—or any other untrusted machine—for any purpose, he exposes himself to the possibility that hidden malicious software will capture his tokencode and PIN in real time and impersonate him completely. This problem is not specific to vouching: an organization that allows any sharing of machines exposes itself to such vulnerabilities whether or not a vouching system is in place. It is possible, though, that the presence of a vouching system may encourage sharing of machines. Some countermeasures to asker-helper attacks are possible. For example, the vouching system can enforce a policy whereby an asker is removed from the helper's list of helpers until the helper changes his PIN.

5. AN IMPLEMENTATION

RSA Laboratories developed a prototype implementation of Voucher-based authentication. The system consisted of three components: an asker web application, a helper web application, and an administrative program. Both web applications consisted of sequences of Common Gateway Interface (CGI) applications developed in C++ and running on the Microsoft® Internet Information Services (IIS) web server. The administrative application was a Windows® application developed in C++ using the Microsoft Foundation Classes (MFC). The underlying database, common to all the applications, was built in Microsoft Access.

The asker application simulates a generic web application protected by RSA SecurID®. The subsequent pages walk the asker through the steps required to authenticate without possession of a token. These web pages are as follows:

- Asker Page 1: The normal user login page for SecurID, with fields for user name and PASSCODE. The special, added feature for vouching is a button labeled "Forgot/lost my Token."

- Asker Page 2: Instructions for asker to contact a helper, with a form to enter the helper's login name.

- Asker Page 3: A form for entering the asker's PIN and the vouchcode supplied to the asker by the helper.

- Asker Page 4: If the PIN and vouchcode are both determined to be correct, this page displays another form that allows the asker to enter and confirm a temporary password. (The system may enforce security restrictions on the form of the password. For example, in our prototype, passwords must be at least eight characters in length.)

- Asker Page 5: If the password is confirmed, the asker is presented with another login page where the PIN and temporary password are used to log in.

- Asker Page 6: If the PIN and temporary password are both verified successfully, the asker is presented with a page confirming successful authentication.

The helper application allows the helper to retrieve vouchcodes for askers. It consists of only two pages:

- Helper Page 1: The helper is presented with a login form with fields for his user name and passcode, along with the name of the asker being assisted.

- Helper Page 2: If the helper's passcode is successfully verified, a vouchcode is generated for the specified asker. The vouchcode is displayed to the helper who is then responsible for conveying the value to the asker.

The administrative application controls overall system configuration and the vouching capabilities of individual askers. The application uses two dialogs, a general "Voucher Management" dialog and an asker-specific "Helper Management" dialog. The Voucher Management dialog controls several system parameters, such as the length and validity periods of both vouchcodes and temporary passwords. The Helper Management dialog specifies rules for the set of permissible asker-helper relationships. In our prototype, these relationships are confined within groups that are pre-defined by an administrator database. In a real application, this database might be imported from an external source, such as an HR database, in order to take advantage of previously established groups. In our prototype, a given user may be designated as permitted to help anyone in the same group, permitted to help selected askers in the same group, or prohibited from helping any askers.

5.1 Implementation issues

Two significant questions arose in the implementation of our prototype voucher system. The first is this: should the application provide the asker with a list of her helpers? The difficulty, of course, is that since the asker has not yet authenticated to the system, such a list would be accessible to anyone. Helper lists could open the system up to social engineering attacks, since an attacker could learn the list of helpers for an asker he wishes to impersonate. We could, of course, require an asker to enter her PIN in order to see her list of helpers. This approach, however, would expose the PIN to a new vector of guessing attack. Omitting helper lists entirely could prove problematic because it may not be reasonable to expect askers to remember all of their helpers, especially if vouching is used infrequently. In our prototype, we have chosen not to list helpers, but this should in general be a matter of policy.

The other question involves the seeding of suitable groups in the Helper Management application. In an organization of any size, having individual helpers authorized to vouch for all askers is problematic, since there may be many askers whom a given helper does not know. A logical alternative is to have supervisors act as helpers for their subordinates, but this may prove to be an unacceptable burden for the supervisors—and an embarrassment for their subordinates. Our conclusion is that the most suitable arrangement is for groups to consist of organizational peers. This should work well: we can generally assume the askers within a group are familiar with each other and willing to act as helpers. Some form of peer-level group structure exists in most organizations—and, indeed, most HR databases—and need not be created just for the purpose of using vouchers.

We are presently in the process of conducting a small pilot study to understand user interaction with and refine our prototype system.

6. SOCIAL ENGINEERING

The security of a voucher system depends critically, of course, on how users interact within its framework. In addition to ordinary data-security considerations, as reflected in our security model, it is vital to consider the various potential vulnerabilities to social engineering. Simple refinements to a vouching system—some of them already incorporated into our prototype—can considerably strengthen the security of the system as a whole. In this section we explore the threat of social engineering and related problems and some possible countermeasures.

6.1 Tailgating

Politeness is often at odds with security. For example, the practice of "tailgating," in which employees allow people to follow them when they unlock doors, undermines the security of physical access-control systems in office buildings. A desire for politeness often causes employees to grant access even to strangers, rather than to challenge them. A naïvely deployed voucher system can be similarly vulnerable to tailgating. If Harry receives a telephone call from a colleague Alice asking for his help, Harry may well feel uncomfortable refusing—even if he doesn't know Alice well (or at all).

In contrast to a physical-access system, a voucher system can impose restrictions on asker-helper relationships and activity. In our prototype, the administrator defines the relationships permissible in the vouching system, allowing vouching, for example, only between peers of well-defined corporate groups. The aim of such features is, of course, to ensure against situations in which Harry feels pressure to offer help to an inappropriate asker Alice.

One can imagine a broad range of alternative mechanisms for constraining helper-asker relationships. The possibilities are particularly rich when an interface exists between a vouching system and other communication systems. For instance, a vouching system integrated with a telephone system might automatically enroll Alice as an asker for Harry if the two speak on the telephone on a regular basis. (Such solutions require sensitivity to privacy concerns, of course.) Additionally, Harry might only be permitted to vouch for Alice when she calls from a telephone number that Harry himself has recently called. For deployment of voucher systems in consumer environments, where administrative oversight and user awareness of security may limited, such automated approaches could be particularly attractive. They are probably not practical in today's environments, but could become so in future, when VoIP and similar technologies result in tighter integration of communication systems.

6.2 Weakly authenticated contact

Similar in spirit to the problem of tailgating is that of weakly authenticated contact. Suppose that Alice is registered as an asker for Harry, and Harry receives a request via e-mail from someone purporting to be Alice. Should he offer his help? Such e-mail could, of course, originate with an entirely unauthenticated source, e.g., a public e-mail account. Without proper safeguards, Harry could easily end up helping an impostor.

It is important that vouching policies prohibit poorly authenticated contact by an asker. Helping should be forbidden to askers who communicate by e-mail—unless, perhaps, the e-mail is internal to a company, followed up with a phone call, or otherwise appropriately authenticated. The safest policy is to mandate telephone or face-to-face contact.

Policy enforcement, however, is a tricky matter. Ideally, as suggested in our discussion of tailgating, an integrated platform might automate policy enforcement. In existing systems, with their loose affiliation among modes of communication, such automation would be difficult to achieve. The most practical approach, therefore, is to require helpers to indicate how they have been contacted by askers, and to authorize or deny vouching transactions accordingly. We are in the process of implementing a pull-down menu (in Helper Page 1) for this purpose in our prototype; the descriptions "E-mail," "Telephone," "In person," and "Other" specify the form of asker contact. If the helper chooses "E-mail" or "Other," the system warns the helper that the mode of contact is prohibited by system policy and denies the vouching transaction.

We emphasize the importance of allowing the helper to specify the form of asker contact accurately, even if the form of contact violates system policy. Otherwise, the helper may be tempted simply to bypass the protective mechanism. For example, if a menu of contact options lists only the choices "Telephone" and "In person," users contacted by e-mail may be tempted just to select "Telephone" because of the lack of a more accurate option (in spite of any warnings). Delicate design decisions can have a big impact. Even the order of menu choices is important: the pull-down menu in our prototype gives "E-mail" as a first choice, so that careless menu selection results in rejection of the helping request.

Another mechanism that can help prevent inappropriate use of vouching requests by e-mail is reduction or elimination of the helper's ability to cut and paste voucher codes. For example, a vouchcode might be displayed as an image, rather than text. (Some people display their e-mail addresses on web pages this way to discourage bots from harvesting them.)

Of course, a user who wishes to bypass the contact policy of the vouching system can do so. Our belief, however, is that most users will willingly comply with system policy provided that such compliance is not onerous. Of course, administrator-level system auditing can be a powerful additional mechanism for policy enforcement. We believe that the e-mail notices in our prototype—a kind of peer-level audit mechanism—will also help curb abuses.

6.3 Spidering

Any systemic weakness in the second factor used alongside voucher codes (the PIN in our prototype system) poses a special hazard. Given such weakness, an attacker that has compromised one helper can readily compromise other accounts. Consider, for example, an organization in which PINs are user-selected and many users choose the PIN '1234' for convenience. An attacker who has compromised Harry's account can simulate a voucher request from Alice, perhaps guess Alice's PIN, compromise Alice's account, proceed to attack Alice's enrolled askers, and so forth. An organization with a loose social network and thus high exposure to social engineering would be similarly at risk.

The helper-asker relationships in a voucher system may be represented very simply as a digraph D in which each user is represented by a node and each enrolled helper-asker relationship as a directed edge. A natural strategy for an attacker that has compromised one node is to exploit weak adjacent edges in order to compromise as much of the system as possible. We refer to such an attack—and its exploitation of the web of helper-asker relationships—as *spidering*.

We emphasize that traversal of any edge in D by an attacker requires compromise of the second factor in the vouching system (users' PINs). Thus the best defense against spidering is a sound vouching process combined with a good second factor. Other defensive mechanisms are possible, however. Among these are:

1. *Dynamic digraphs:* The digraph D might evolve as a function not just of user relationships, but also recent voucher assignments. For example, if Alice has authenticated using a voucher assigned by Bob, Alice herself might lose her permission to assign a voucher to another employee for a certain period of time or until she re-authenticates in a stronger manner. Other throttling rules might also apply. For example, suppose that an edge in D is colored if vouching has taken place recently between the corresponding helper and asker. The vouching system might prohibit vouching or alert administrators if the diameter of a colored subgraph grows beyond some pre-specified threshold.

2. *Multiple vouchers:* To extend the digraph-based view of a voucher system, a user may be able to authenticate only as the result of a joint operation among several connected users. For example, Alice might only be able to authenticate in a voucher system on receiving vouchers from two authorized helpers, rather than one.

6.4 Lazy vouching

Of course, a voucher system should be easy to use. If we make it too easy, though, users may be tempted to rely on the voucher mechanism as a primary authenticator. We refer to such undue reliance on vouching as *lazy vouching*.

For example, Alice might decide never to bother obtaining a SecurID token from system administrators, but instead rely on Harry whenever she needs to authenticate. In some environments, this may not be a problem. After all, lazy vouching does not introduce a new technical vector of attack. Lazy vouching can undermine system security in two ways, however: (1) In a voucher system with administrative audit, lazy vouching can make vouching seem unexceptional, and thereby obscure system breaches; (2) By desensitizing users to the process of vouching and the surrounding checks, lazy vouching can increase risks of social engineering.

For these reasons, we believe that vouching systems should be designed and configured to discourage lazy vouching. In practice, we believe that the light burden of having to fill out browser forms in our prototype is a good deterrent.

7. CONCLUSIONS

This paper introduces the concept of vouching as a tool for on-line authentication. Vouching directly leverages human relationships, and this work can be seen as part of a broad exploration of the interplay between social networks and user authentication. Breaking the common conceptual identification of a user with her client machine, and common modality of single-user involvement in authentication, we have extended the classic authentication triad of "what you have," "what you know," and "what you are." In addition to introducing the concept of vouching as a tool for on-line authentication, this paper provides a rough security model for analyzing vouching, and discusses the special social-engineering concerns of a multi-user protocol. We present a specific protocol for vouching with hardware tokens, our implementation with the well known SecurID product, and thereby demonstrate both feasibility and usability.

The basic vouching system we have outlined has been designed to be as simple and secure as possible, but several variants and extensions are worth further exploration:

Alternate Factors: Vouching need not be restricted to hardware tokens and accompanying PINs. The basic vouching procedure could work with a different primary authentication factor. As an example, in the protocol we describe for SecurID, a voucher system could permit swapping of the roles of the tokencode and PIN to deal with the case where an asker has forgotten her PIN but still has her token.

Restricted privileges: To enhance the security of a vouching protocol and discourage lazy vouching, a voucher system might grant restricted privileges to askers. For example, the system might permit viewing of e-mail, but not sending, or might limit access to company-sensitive documents.

Asker Selection of Helpers: In some deployments the involvement of an administrator or supervisor in enrollment might be cumbersome. It may in some cases be desirable for askers to designate their own helpers. The security risks of this type of approach deserve further scrutiny.

8. REFERENCES

[1] v-GO SSPR 5.0 product description. Referenced 2006 at www.passlogix.com.

[2] Simple Distributed Security Infrastructure (SDSI) web page, 2001. Referenced 2006 at http://theory.lcs.mit.edu/~cis/sdsi.html.

[3] PeopleSoft and Courion deliver integrated password management solution, 27 August 2001. Press release. Referenced 2006 at www.courion.com.

[4] M. Blaze, J. Feigenbaum, and A. D. Keromytis. The KeyNote trust management system. In *Security Protocols International Workshop*, pages 59–63. Springer-Verlag, 1998. LNCS no. 1550.

[5] M. Blaze, J. Feigenbaum, and M. Strauss. Compliance-checking in the PolicyMaker trust-management system. In *Financial Cryptography*, pages 251–265. Springer-Verlag, 1998. LNCS no. 1465.

[6] W. Eazel. 'Live phishing experiment nets consumers hook, line, and sinker. *SC Magazine*, 8 November 2005. Referenced 2006 at www.scmagazine.com.

[7] C. Ellison. UPnP security ceremonies design document: For UPnP device architecture 1.0, 3 October 2003. Referenced 2006 at http://www.upnp.org.

[8] C. Ellison. IETF RFC 2692: SPKI requirements, September 1999.

[9] V. Griffith and M. Jakobsson. Messin' with Texas: Deriving mothers maiden names using public records. In J. Ioannidis, A. D. Keromytis, and M. Yung, editors, *Applied Cryptography and Network Security (ACNS)*, pages 91–103. Springer-Verlag, 2005. LNCS no. 3531.

[10] RSA Security Inc. RSA SecurID authenticators, 2006. Product Specification. Referenced 2006 at www.rsasecurity.com.

[11] J. Jubak. Globalization isn't what's killing GM. *MSN Money*, 29 November 2005. Referenced 2006 at moneycentral.msn.com.

[12] D. V. Klein. Foiling the cracker: A survey of and improvements to, password security. In *UNIX Security II: USENIX Workshop Proceedings*, pages 5–14, Berkeley, CA, 1990.

[13] J. Leyden. Office workers give away passwords for a cheap pen. *The Register*, 18 April 2003. Referenced 2006 at www.theregister.co.uk.

[14] J. M. McCune, A. Perrig, and M. K. Reiter. Seeing-is-believing: Using camera phones for human-verifiable authentication. In *IEEE Symposium on Security and Privacy*, pages 110–124, 2005.

[15] K. D. Mitnick and W. L. Simon. *The Art of Deception: Controlling the Human Element of Security*. Wiley, 2002.

[16] T. Pullar-Strecker. NZ bank adds security online. *Sidney Morning Herald*, 8 November 2004. Referenced 2006 at www.smh.com.au.

[17] B. Ross, C. Jackson, N. Miyake, D. Boneh, and J. Mitchell. Stronger password authentication using browser extensions. In P. McDaniel, editor, *USENIX Security*, pages 17–32, 2005.

An Effective Defense Against Email Spam Laundering

Mengjun Xie, Heng Yin, Haining Wang
Department of Computer Science
The College of William and Mary, Williamsburg, VA 23187
{mjxie, hyin, hnw}@cs.wm.edu

ABSTRACT

Laundering email spam through open-proxies or compromised PCs is a widely-used trick to conceal real spam sources and reduce spamming cost in underground email spam industry. Spammers have been plaguing the Internet by exploiting a large number of spam proxies. The facility of breaking spam laundering and deterring spamming activities close to their sources, which would greatly benefit not only email users but also victim ISPs, is in great demand but still missing. In this paper, we reveal one salient characteristic of proxy-based spamming activities, namely packet symmetry, by analyzing protocol semantics and timing causality. Based on the packet symmetry exhibited in spam laundering, we propose a simple and effective technique, DBSpam, to on-line detect and break spam laundering activities inside a customer network. Monitoring the bi-directional traffic passing through a network gateway, DBSpam utilizes a simple statistical method, Sequential Probability Ratio Test, to detect the occurrence of spam laundering in a timely manner. To balance the goals of promptness and accuracy, we introduce a noise-reduction technique in DBSpam, after which the laundering path can be identified more accurately. Then, DBSpam activates its spam suppressing mechanism to break the spam laundering. We implement a prototype of DBSpam based on *libpcap*, and validate its efficacy through both theoretical analyses and trace-based experiments.

Categories and Subject Descriptors: C.2.0 [Computer Communication Networks]: Security and protection

General Terms: Security.

Keywords: Spam, Proxy, SPRT.

1. INTRODUCTION

As a side-product of free email services, spam has become a serious problem that afflicts every Internet user in recent years. According to MessageLabs [1], currently over 60% email traffic is spam. Although a number of anti-spam mechanisms have been proposed and deployed to foil spammers, spam messages continue swarming into Internet users' mailboxes. A more effective spam detection and suppression mechanism close to spam sources is critical to dampen the dramatically-grown spam volume.

At present, proxies such as off-the-shelf SOCKS and HTTP proxies play an important role in the spam epidemic. Spammers launder email spam through these proxies to conceal their real identities and reduce spamming cost. The popularity of proxy-based spamming is mainly due to the anonymous characteristic of a proxy and the availability of a large number of spam proxies. The IP address of a spammer is obfuscated by a spam proxy during the protocol transformation, which hinders the tracking of real spam origins. According to Composite Blocking List [7] which is a highly-trusted DNSBL (DNS-based Blackhole List), the number of available spam proxies and bots in August 2006 is more than 3,200,000. Such numerous spam proxies facilitate the formation of email spam laundering, by which a spammer has great flexibility to change spam paths and bypass anti-spam barriers.

To break this spam laundering, we propose a simple and effective mechanism, called *DBSpam*, which detects and blocks spam proxies' activities inside a customer network and further traces the corresponding spam sources outside the network. DBSpam is designed to be placed at a network vantage point such as the edge router or gateway that connects a customer network to the Internet. The customer network could be a regional broadband (cable or DSL) customer network, a regional dialup network, or a campus network. It detects ongoing proxy-based spamming by monitoring bi-directional traffic. Due to the protocol semantics of SMTP and timing causality, the behavior of proxy-based spamming demonstrates the unique characteristics of connection correlation and packet symmetry. Utilizing this distinctive spam laundering behavior, we can easily identify the suspicious TCP connections involved in spam laundering. Then, we can single out the spam proxies, trace the spam sources behind them, and block the spam traffic. Based on *libpcap*, we implement a prototype of DBSpam and evaluate its effectiveness against email spam laundering through theoretical analyses and trace-based experiments.

In general, DBSpam is distinctive from all previous anti-spam approaches in the following two aspects.

- First, DBSpam pushes the defense line towards spam sources. DBSpam enables an ISP (Internet Service Provider) to on-line detect spam laundering activities and spam proxies inside its customer networks. The quick responsiveness of DBSpam offers the ISP an opportunity to suppress laundering activities and quarantine the identified spam proxies.

- Second, DBSpam has no need to scan message contents, and has very few assumptions about the connections between a spammer and its proxies. DBSpam works even if (1) these connections are encrypted and the message contents are compressed; and (2) a spammer uses proxy chains inside the monitored network.

One additional benefit of DBSpam is that once spam laundering is detected, fingerprinting spam messages at the sender side is viable and spam signatures may be distributed to accelerate spam detection at other places. In addition to all these advantages, DBSpam is complementary to existing anti-spam techniques and can be incrementally deployed over the Internet.

The remainder of the paper is organized as follows. Section 2 briefly presents spam laundering mechanisms. Section 3 surveys commonly-used anti-spam techniques. Section 4 describes the unique behavior of proxy-based spamming. Section 5 details the working mechanism of DBSpam. Section 6 evaluates the effectiveness of DBSpam through the trace-based experiments. Section 7 discusses the robustness of DBSpam against potential evasions. Finally, we conclude the paper with Section 8.

2. SPAM LAUNDERING MECHANISMS

Spam laundering studied in this paper refers to the spamming process, in which only proxies are involved in origin disguise. The proxy refers to the application such as SOCKS that simply performs "protocol translation" (i.e., rewrite IP addresses and port numbers) and tunnels packets through. Different from an email relay, which first receives the whole message and then forwards it to the next mail server, an email proxy requires that the connections on both sides of the proxy synchronize during the message transferring. More importantly, unlike an email relay which inserts the information—"Received From" that records the IP address of sender and the timestamp when the message is received—in front of the message header before relaying the message, an email proxy does not record such trace information during protocol transformation. Thus, from a recipient's perspective, the email proxy, instead of the original sender, becomes the source of the message. It is this identity replacement that makes email proxy a favorite choice of spammers.

Initially, spammers just seek open proxies on the Internet, which usually are mis-configured proxies allowing anyone to access their services. There are many Web sites and free software providing open proxy search function. However, once such mis-configurations are corrected by system administrators, spammers have to find other available "open" proxies.

It is ideal for a spammer to own many "private" and stable proxies. Unsecured home PCs with broadband connections are the best candidates for this purpose. To achieve this, malicious software including specially-designed worms and viruses, such as SoBig and Bagle, is used to hijack home PCs. Equipped with Trojan horse or backdoor programs, these compromised machines are available zombies. After proxy programs such as SOCKS or Wingate are installed, these zombies are ready to be used as proxies to pump out email spam. Without serious performance degradation, most non-professional Windows users are not aware of the ongoing spamming. Recent research on the network-level behavior of spammers [28] also confirms that most sinked spam is originated from compromised Windows hosts.

To counter the soaring growth of spam volume, many ISPs have adopted the policy of blocking port 25 (SMTP port), in which outbound email from a subscriber must be relayed by the ISP-designated email server. In other words, the ISP's edge routers only forward the SMTP traffic from some designated IP addresses to the outside. However, spammers have easily evaded such simple SMTP port blocking mechanisms. The spam laundry is simple: having zombies send spam messages to their ISP email servers first. In February 2005, Spamhaus [2] reported that over the past few months a number of major ISPs had witnessed far more spam messages coming directly from the email servers of other ISPs. This change in proxy-based spamming activity is mainly caused by the use of new stealth spamware, which instructs the hijacked proxy (e.g., zombie) to send spam messages via the legitimate email server of the proxy's ISP.

3. ANTI-SPAM TECHNIQUES

Many anti-spam techniques have been proposed and deployed to counter email spam from different perspectives. Based on the placement of anti-spam mechanisms, these techniques can be divided into two categories: recipient-oriented and sender-oriented. In terms of fighting spam at the source, HoneySpam [13] might be the closest work to ours. In the following, we briefly describe recipient-oriented and sender-oriented techniques, respectively, and then compare our work with HoneySpam.

3.1 Recipient-oriented Techniques

This class of techniques either (1) block/delay email spam from reaching the recipient's mailbox or (2) remove/mark spam in the recipient's mailbox. Due to the flourish of techniques in this category, we further divide them into content-based and non-content-based sub-categories.

3.1.1 Content-based Techniques

The techniques in this sub-category detect and filter spam by analyzing the content of received messages, including both message header and message body.

Email address filters: Email address filters are simply whitelists or blacklists. Whitelists consist of all acceptable email addresses and blacklists are the opposite. Blacklists can be easily broken when spammers forge new email addresses, but using whitelists alone makes the world enclosed. Garriss et al. [18] developed a new whitelisting system, which can automatically populate whitelists by exploiting friend-of-friend relationships among email correspondents.

Heuristic filters: The features that are rare in normal messages but appear frequently in spam, such as non-existing domain names and spam-related keywords, can be used to distinguish spam from normal email. SpamAssassin [3] is such an example. Each received message is verified against the heuristic filtering rules. Compared with a pre-defined threshold, the verification result decides whether the message is spam or not.

Machine learning based filters: Since spam detection can be converted into the problem of text classification, many content-based filters utilize machine-learning algorithms for filtering spam. Among them, Bayesian-based approaches [15, 19, 24, 35] have achieved outstanding accuracy and have been widely used. As these filters can adapt their classification engines with the change of message content, they outperform heuristic filters.

3.1.2 Non-content-based Techniques

The techniques in this sub-category use non-content spam characteristics, such as source IP address, message sending rate, and violation of SMTP standards, to detect email spam.

DNSBLs: DNSBLs are distributed blacklists, which record IP addresses of spam sources and are accessed via DNS queries. When an SMTP connection is being established, the receiving MTA (Mail Transfer Agent) can verify the sending machine's IP address by querying the subscribed DNSBL. Even DNSBLs have been widely used, their effectiveness [22, 28] and responsiveness [27] are still under study.

MARID: MARID (MTA Authorization Records In DNS) is a class of techniques to counter forged email addresses by enforcing sender authentication. MARID is also based on DNS and can be seen as a distributed whitelist of authorized MTAs. Multiple MARID drafts [10] have been proposed, in which [8, 12] have been deployed in some places.

Challenge-Response (C-R): C-R is used to keep the merit of whitelist without losing important messages. Incoming messages, whose sender email addresses are not in the recipient's whitelist, are bounced back with a challenge that needs to be solved by a human being. After a proper response is received, the sender's address can be added into the whitelist.

Tempfailing: Tempfailing [30] is based on the fact that legitimate SMTP servers have implemented the retry mechanism as required by SMTP, but a spammer seldom retries if sending fails. It usually works with a greylist that records the failed messages and the MTAs failed on their first tries.

Delaying: As a variation of rate limiting, delaying is triggered by an unusually high sending rate. Most delaying mechanisms, such as [17, 20, 33, 34] are applied at receiving MTAs.

Sender Behavior Analysis: This technique distinguishes spam from normal email by examining behavior of incoming SMTP connections. Messages from the machine exhibiting characteristics of malicious behavior such as directory harvest are blocked before reaching mailbox [4].

3.2 Sender-oriented Techniques

Usage Regulation: To effectively throttle spam at the source, ISPs and ESPs (Email Service Providers) have taken various measures such as blocking port 25, SMTP authentication, to regulate the usage of email services. Message submission protocol [9] has been proposed to replace SMTP, when a message is submitted from an MUA (Mail User Agent) to its MTA.

Cost-based approaches: Borrowing the idea of postage from regular mail systems, many cost-based anti-spam techniques [5, 11, 14, 23, 32] attempt to shift the cost of thwarting spam from receiver side to sender side. All these techniques assume that the average email cost for a normal user is trivial and negligible, but the accumulative charge for a spammer will be high enough to drive them out of business. Cost concept may have different forms in different proposals. Bonded Sender [5] advocates associating email with real money, while SHRED [23] proposes affixing electronic stamps to messages. Both centralized [11, 23] and distributed [32] cost enforcement mechanisms have been proposed.

3.3 HoneySpam

HoneySpam [13] is a specialized honeypot framework based on honeyd [25] to deter email address harvesters, poison spam address databases, and intercept or block spam traffic that goes through the open relay/proxy decoys set by HoneySpam. With the network virtualization offered by honeyd, HoneySpam can set up multiple fake web servers, open proxies, and open relays. Fake web servers provide specially crafted webpages to trap email address harvesting bots. Fake open proxies or open relays are used to track spammers exploiting them and block spam going through them.

HoneySpam shares the same motivation of countering spam at the source as DBSpam, and both deal with spam proxies. However, the role of proxy and anti-spam approaches in HoneySpam are quite different from those in DBSpam. The proxies of HoneySpam are intentionally set on end hosts, and spam sources are logged by HoneySpam. Thus, spam tracking is very easy. In contrast, detecting spam proxies is the major task of DBSpam, and proxy identification and spam tracking can only be accomplished through traffic analysis. On the other hand, these two tracing and blocking systems are complementary to each other. Moreover, both of them can be used for spam signature generation, spam forensic and law enforcement.

4. PROXY-BASED SPAM BEHAVIOR

In this section, we delineate the distinct behavior of proxy-based spamming, which directly inspire the design of our detecting algorithm. Figure 1 depicts a typical scenario of proxy-based spamming in a customer network such as a Cox regional residential network. Although spammers can conceal their real identities from destination MTAs by exploiting spam proxies, they cannot make the connection between a spam source and its proxy *invisible* to the edge router or gateway that sits in between. Here we assume that there is a network vantage point where we can monitor all the bi-directional traffic passing through the customer network, and the location of the gateway (or firewall) of the customer network (e.g. edge router R in Figure 1) that connects to the Internet is such a point.

4.1 Laundry Path of Proxy Spamming

As shown in Figure 1, there is a customer network N, in which spam proxies reside. Both spammer S and receiving MTA M are connected to customer network N via edge router R. S may be the original spam source or just another spam proxy (but it must be closer to the real spam source). M is the outside MTA.

Note that for the customer network that has its own mail server(s) such as campus (or enterprise) networks, the monitored network N may not be the whole network, but one of its protected sub-networks. Usually such campus/enterprise networks are divided into multiple sub-networks for security and management concerns. Their mail servers are placed in DMZ (DeMilitarized Zone) or a special sub-network that is separated from other sub-networks such as wireless, dormitory, or employee sub-networks. It is one of these loosely-

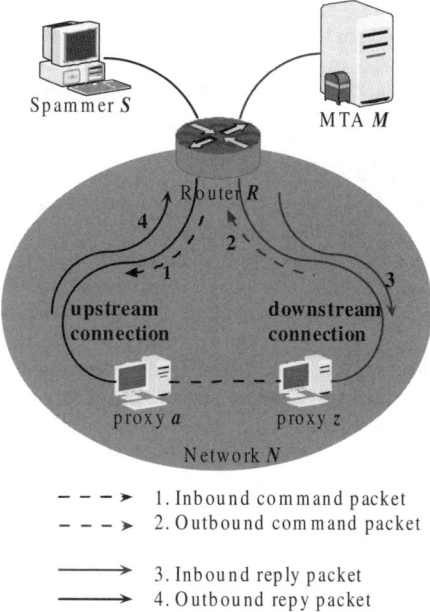

Figure 1: Scenario of Proxy-based Spamming

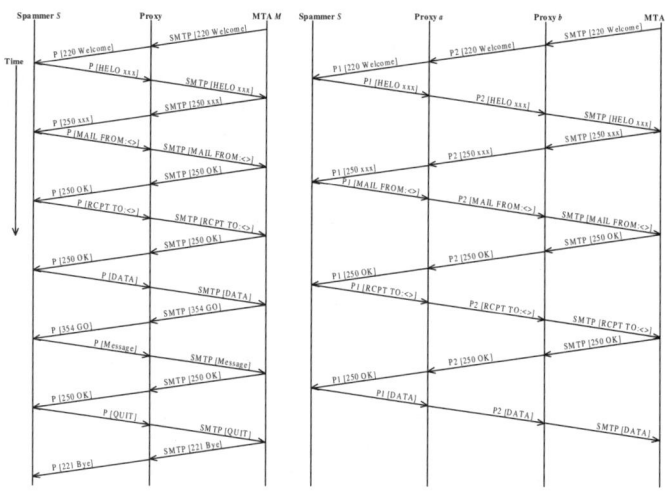

Figure 2: Time-line of Spamming Processes for Single Proxy (left) and Proxy Chain (right)

managed sub-networks that becomes the monitored network N and the router/gateway connecting the sub-networks becomes the vantage point R. Thus, the assumption of exterior MTA M is valid even when the MTA is under the the same administration domain as network N.

Inside monitored network N, S may use a single or multiple spam proxies. If multiple proxies are employed, they may either launder spam messages individually or be organized into one or multiple proxy chains, depending on the spammer's strategy. Without loss of generality, only one chain is shown in Figure 1. Spammer S usually communicates with spam proxies through SOCKS or HTTP. The spam message sent from S to a may even be encrypted. If it is a proxy chain, the spam message can be conveyed by different proxy protocols at different hops. For instance, SOCKS 4 is used between S and a, while HTTP is employed between a and z. However, all these protocol variations and message content encryptions cannot change the fact: it is last-hop proxy z [1] that does the protocol transformation and forwards the spam message to the MTA via SMTP.

We define the connection between spammer S and first-hop proxy a as the *upstream* connection, and define the connection between last-hop proxy z and MTA M as the *downstream* connection. The upstream and downstream connections plus the proxy chain form the spam laundry path, which is shown in Figure 1.

4.2 Connection Correlation

There is a one-to-one mapping between the upstream and downstream connections along the spam laundry path. While this kind of connection mapping is common for proxy-based spamming, it is very unusual for normal email transmission. In normal email delivery, there is only one connection, i.e., the connection between sender and receiving MTA. The existence of such connection correlation is a strong indication of spam laundering and provides valuable clue for spammer

---

[1] proxy z and proxy a are the same in the single proxy scenario.

tracking. Here we assume that the downstream connection is an SMTP connection. For the upstream connection we have no restriction except that it should be a TCP connection. The packets in the upstream connection may be encrypted and even compressed.

The detection of such spam-proxy-related connection correlation is challenging due to the following three reasons. First, content-based approaches could be ineffective as spammers may use encryption to evade content examination. Second, because such a detection mechanism is usually deployed at network vantage points, the induced overhead should be affordable, which is critical to the success of its deployment. Third, since spam traffic is machine-driven and could be delayed by proxy at will, those timing-based correlation detection algorithms such as [36] may not work well in this environment.

4.3 Packet Symmetry

Figure 2 illustrates the detailed communication processes of spam laundering for both single proxy and proxy chain cases at the application layer, in which the message format is "PROTOCOL [content]". For simplicity, P/P1/P2 stands for different application protocols, including SOCKS (v4 or v5), HTTP, etc. For SMTP, its packet content is in plain-text. But for application protocols P/P1/P2, their packet contents may be encrypted. For ease of presentation, the small delays introduced by message processing at end hosts and intermediate proxies are ignored. The initial proxy handshaking process is also omitted as it has no effect on email transactions. Without losing any generality, here we only show the shortest SMTP transaction process for the single-proxy case and parts of SMTP transaction process for the proxy-chain case.

Due to protocol semantics, the process of proxy-based spamming is similar to that of an interactive communication. The appearance of one inbound SOCKS-encapsulated (or HTTP-encapsulated) [2] SMTP command message on the upstream connection will trigger the occurrence of one outbound SMTP command message on the downstream connection later. Similarly, for each inbound SMTP reply message

---

[2] For the ease of presentation, we only use SOCKS in the rest of paper, although HTTP can be used as well.

on the downstream connection, later on there will be one corresponding outbound SOCKS-encapsulated reply message carried by TCP on the upstream connection. We term this communication pattern as *message symmetry*.

This message symmetry leads to the *packet symmetry* at the network layer with a few exceptions, in which the one-to-one packet[3] mapping between the upstream and downstream connections may be violated. The exceptions can be caused by (1) packet fragmentation, (2) packet compression, (3) packet retransmission occurring along the laundry path. However, due to the fact that SMTP reply messages are very short (usually less than 300 bytes including packet header) and Path MTUs for most customer networks are above 500 bytes, the occurrence of (1) and (2) is very rare. Moreover, the packet retransmission problem can be easily resolved by checking TCP sequence numbers. In general, the packet symmetry between the inbound and outbound reply packets holds most of time.

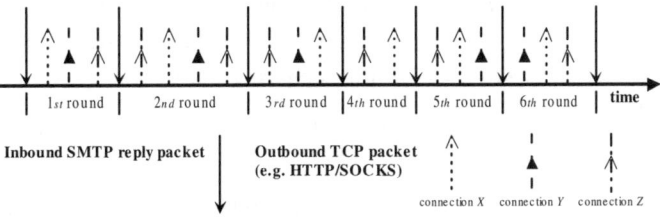

Figure 3: Example of Reply Round and TCP Correlation

Such packet symmetry is exemplified in Figure 3, where the arrow with long solid line stands for the arrival of an inbound SMTP reply packet of the suspicious SMTP connection. In addition to the inbound SMTP connection, there are three outbound TCP connections X, Y, and Z, as shown in Figure 3. Three kinds of arrows with different dotted lines stand for the arrivals of outbound TCP packets belonging to these outbound TCP connections, respectively. The upward arrow indicates that the packet is leaving the monitored network, while the downward arrow indicates the packet is entering the network.

All of the inbound SMTP reply packets shown in Figure 3 belong to the same suspicious SMTP connection. We define a *reply round* as the time interval between the arrivals of two consecutive reply packets on an SMTP connection. Thus, the n_{th} reply round is the time interval between the arrival of the n_{th} reply packet and that of the $(n+1)_{th}$ reply packet. Even for the simplified SMTP transaction, it has six reply rounds as shown in Figure 3. Within one reply around, the number of arrows with a specific dotted line indicates the number of outbound TCP packets of the corresponding TCP connection.

According to the one-to-one mapping of packet symmetry, each SMTP reply packet observed on the downstream SMTP connection should cause *one and only one* TCP packet appeared on the upstream connection. As Figure 3 shows, if one connection among X, Y, and Z is the suspicious upstream connection, one and only one outbound TCP packet must be observed from that connection in every reply round. Based on this rule, only TCP connection X meets this "one and only one" requirement and can be classified as the suspicious upstream connection with high probability. In the second reply round, more than one packets appear on connection Z; and in the fourth round, no packet occurs on connection Y. Thus, we can easily filter out TCP connections Y and Z as normal background traffic. Note that the order of packet arrivals in a reply round does not affect the checking result of packet symmetry.

This *packet symmetry* is the key to distinguish the suspicious upstream and downstream connections along the spam laundry path from normal background traffic. It simply captures the fundamental feature of chained interactive communications, and does not assume any specific time distribution of packet arrivals. We use this simple rule to detect the laundry path of proxy-based spamming, and the detection scheme is robust against any possible time perturbation induced by spammers. Note that the *one and only one* mapping of packet symmetry can be relaxed, which we will discuss in Section 7.

5. WORKING MECHANISM OF DBSPAM

DBSpam consists of two major components: spam detection module and spam suppression module, in which the detection module is the core of DBSpam. To the best of our knowledge, so far there is no effective technique which can on-line detect both spam proxies and the corresponding spammers behind them. We envisage that DBSpam may achieve the following goals: (1) fast detection of spam laundering with high accuracy; (2) breaking spam laundering via throttling or blocking after detection; (3) support for spammer tracking and law enforcement; (4) support for spam message fingerprinting; and (5) support for global forensic analysis.

In essence, the detection module of DBSpam is a simple and efficient *connection correlation detection* algorithm to identify the laundry path of spam messages (i.e., the suspicious downstream and upstream connections) and the spam source[4] that drives spamming behind the proxies.

5.1 Deployment of DBSpam

Like other network intrusion detection systems, DBSpam needs to be placed at a network vantage point that connects a customer network to the Internet, where it can monitor the bi-directional traffic of the customer network. For a single-homed network, it is easy to locate such a network vantage point (an edge router or a firewall) and deploy DBSpam on it. For a multi-homed network, it may not be possible to locate a single network vantage point that can monitor all the bi-directional traffic passing through the customer network.

However, on one hand, many customer networks use multi-homing not for load-balance, but for reliability and fault-tolerance. Therefore, in case of the backup multi-homing, DBSpam works well if deployed at the primary ISP edge router. On the other hand, even in the load-balance multi-homing scenario, as long as the packets that belong to the same proxy chain go through the same ISP edge router or firewall, DBSpam still can work at different ISP edge routers or firewalls without coordination. Moreover, there are special network devices (e.g., [6]) which can passively aggregate traffic from multiple network segments. By hooking up to

---

[3]TCP control packets such as SYN, ACK are not counted here.

[4]Or just another spam proxy that is outside the customer network but at least one more step closer to the real source.

such devices, DBSpam can still have the complete view of network traffic.

5.2 Design Choices and Overview

Our goal is to detect the spam laundry path promptly and accurately, once a proxy-based spamming activity occurs on the monitored network. We show in the previous section that packet symmetry is the inherent characteristic of proxy-based spamming behavior. Since legitimate messages are rarely delivered along the path illustrated in Figure 1, the possibility of a normal SMTP connection being consistently correlated with an unrelated TCP connection is very small in terms of packet symmetry. Hence, frequent observations of connection correlation is a strong indication of occurrence of spam laundering.

According to the packet symmetry rule, for the upstream TCP connection along a spam laundry path, its outbound packet[5] number in each reply round of the downstream SMTP connection is always one. For a normal TCP connection, however, this rule can only be satisfied with a very small probability. Thus, a simple and intuitive correlation detection method is to count the number of outbound packets observed on suspicious TCP connections in sequential reply rounds of an SMTP connection. Given the characteristic of successive arrival of observations, this correlation detection problem is well suited for the statistical method of *Sequential Probability Ratio Test* (SPRT) developed by Wald [31].

As a simple and powerful mathematical tool, SPRT has been used in many areas such as portscan detection [21] and wireless MAC protocol misbehavior detection [26]. Basically, an SPRT can be viewed as an one-dimensional random walk. The walk starts from a point between two boundaries and can go either upward or downward with different probabilities. With each arrival of observation, the walk makes one step in the direction determined by the result of observation. Once the walk firstly hits or crosses either the upper boundary or the lower boundary, it terminates and the corresponding hypothesis is selected. For SPRT, its actual false positive probability and false negative probability are bounded by predefined values. It has been proved that SPRT minimizes the average number of required observations to reach a decision among all sequential and non-sequential tests, which do not have larger error probabilities than SPRT.

We utilize the packet symmetry of SMTP reply packets to detect proxy-based spamming activity. Basically, we monitor the inbound SMTP traffic first, then apply the rule of packet symmetry for detecting the spam laundry path inside the customer network. In other words, DBSpam focuses on the clock-wise reply packet flow as shown in Figure 1, instead of the counter-clock-wise command packet flow, for connection correlation detection. The arrivals of inbound SMTP reply packets, which delimit the reply rounds and drive the progress of connection correlation detection, become a self-setting clock of the detection algorithm. SPRT terminates by either selecting the hypothesis that C_tcp is correlated with C_smtp or choosing the opposite hypothesis.

There are two benefits of using SMTP reply messages to drive SPRT. First, as mentioned earlier, SMTP reply messages are very small, which minimizes the occurrence of packet fragmentation; and we can significantly increase the processing capacity of DBSpam by monitoring small packets only. Second, being either the spam target or the relay, the remote SMTP servers are usually very reliable; and the implementation and listening port of these servers strictly follow the SMTP protocol semantics. Thus, the packet symmetry rule always holds, and SMTP packets can be easily identified based on the port number of TCP header.

In the rest part of the section, we first briefly describe the basic concept of SPRT, then present the detection module of DBSpam, which include two phases: SPRT detection and noise reduction.

5.3 Sequential Probability Ratio Testing

Let X_i, $i = 1, 2, \ldots$, be random variables representing the events observed sequentially. The SPRT for a simple hypothesis H_0 against a simple alternative H_1 has the following form:

$$\begin{aligned}\Lambda_n \geq B &\Longrightarrow \text{accept } H_1 \text{ and terminate test,}\\ \Lambda_n \leq A &\Longrightarrow \text{accept } H_0 \text{ and terminate test,} \quad (1)\\ A < \Lambda_n < B &\Longrightarrow \text{conduct another observation.}\end{aligned}$$

where two constants or boundaries A and B satisfy $0 < A < B < \infty$, and Λ_n is the log-likelihood ratio defined as follows:

$$\Lambda_n = \lambda(X_1, \ldots, X_n) = \ln \frac{\Pr(X_1, \ldots, X_n | H_1)}{\Pr(X_1, \ldots, X_n | H_0)}. \quad (2)$$

Assume $X_1, \ldots, X_i$ are independent and identically distributed (i.i.d) Bernoulli random variables with

$$\Pr(X_i = 1|\theta) = 1 - \Pr(X_i = 0|\theta) = \theta, \quad (3)$$

then

$$\Lambda_n = \ln \frac{\prod_1^n \Pr(X_i|H_1)}{\prod_1^n \Pr(X_i|H_0)} = \sum_1^n \ln \frac{\Pr(X_i|H_1)}{\Pr(X_i|H_0)} = \sum_1^n Z_i, \quad (4)$$

where $Z_i = \ln \frac{\Pr(X_i|H_1)}{\Pr(X_i|H_0)}$. Λ_n can be viewed as a random walk (or more properly a family of random walks[6]) with steps Z_i which proceeds until it first crosses boundary A or B. Suppose the distributions for H_1 and H_0 are θ_1 and θ_0, respectively. Λ_n moves up with step length $\ln \frac{\theta_1}{\theta_0}$ when $X_i = 1$, and goes down with step length $\ln \frac{1-\theta_1}{1-\theta_0}$ when $X_i = 0$.

In SPRT, we define two types of error

$$\alpha = \Pr(S_1|H_0), \qquad \beta = \Pr(S_0|H_1),$$

where $\Pr(S_i|H_j)$ denotes the probability of selecting H_i but in fact H_j is true. If we call the selection of H_1 detection and the selection of H_0 normality, the event of $S_1|H_0$ can be viewed as a false positive. So, α represents the false positive probability. Likewise, the event of $S_0|H_1$ can be termed a false negative and β represents false negative probability.

Let α^* and β^* be user-desired false positive and false negative probabilities, respectively. According to (1), we can derive[7] the *Wald boundaries* as follows:

$$A = \ln \frac{\beta^*}{1-\alpha^*}, \qquad B = \ln \frac{1-\beta^*}{\alpha^*}, \quad (5)$$

---

[5] Here packets refer to non-retransmitted, non-zero-payload TCP packets.

[6] It is a family of random walks, since the distribution of the steps depends on which hypothesis is true.

[7] Due to the space limitation, the derivations of (5), (6), and (7) are omitted here. See [21, 31] for details.

and the derived relationships between actual error probabilities and user-desired error probabilities are:

$$\alpha \leq \frac{\alpha^*}{1-\beta^*}, \qquad \beta \leq \frac{\beta^*}{1-\alpha^*}, \qquad (6)$$

$$\alpha + \beta \leq \alpha^* + \beta^*. \qquad (7)$$

Inequality (6) suggests that the actual error probabilities α and β can only be slightly larger than their expected values α^* and β^*. For example, if the desired α^* and β^* are both 0.01, then their actual values α and β will be no greater than 0.0101. Inequality (7) can be interpreted as that the sum of actual error probabilities is bounded by the sum of their desired values.

According to Wald's theory, $E[N] = E[\Lambda_N]/E[Z_i]$. Suppose hypothesis H_1 is true and Bernoulli variable X_i has distribution θ_1 which implies that Λ_n steps up with probability θ_1 or goes down with probability $1-\theta_1$, we have

$$E[Z_i|H_1] = \theta_1 \ln \frac{\theta_1}{\theta_0} + (1-\theta_1) \ln \frac{1-\theta_1}{1-\theta_0}. \qquad (8)$$

If the user-desired false negative probability of the test is β^*, then the true positive probability is $1-\beta^*$ and

$$\begin{aligned} E[\Lambda_N|H_1] &= \beta^* A + (1-\beta^*)B \\ &= \beta^* \ln \frac{\beta^*}{1-\alpha^*} + (1-\beta^*) \ln \frac{1-\beta^*}{\alpha^*}. \end{aligned} \qquad (9)$$

With (8) and (9), we have

$$E[N|H_1] = \frac{\beta^* \ln \frac{\beta^*}{1-\alpha^*} + (1-\beta^*) \ln \frac{1-\beta^*}{\alpha^*}}{\theta_1 \ln \frac{\theta_1}{\theta_0} + (1-\theta_1) \ln \frac{1-\theta_1}{1-\theta_0}}. \qquad (10)$$

Likewise, we can derive

$$E[N|H_0] = \frac{(1-\alpha^*) \ln \frac{\beta^*}{1-\alpha^*} + \alpha^* \ln \frac{1-\beta^*}{\alpha^*}}{\theta_0 \ln \frac{\theta_1}{\theta_0} + (1-\theta_0) \ln \frac{1-\theta_1}{1-\theta_0}}. \qquad (11)$$

Apparently the average observation number $E[N]$ of SPRT is determined by four parameters: predefined error probabilities α^*, β^* and distribution parameters θ_0 and θ_1. The determination of these values and their effect on $E[N]$ will be discussed with our correlation detection algorithm in the following.

5.4 SPRT Detection Algorithm

According to the principle of packet symmetry, within each reply round, there must be one and only one outbound TCP packet appearing on the corresponding upstream connection. By contrast, those connections that have none or more than one TCP packet can be classified as innocent connections. Within the framework of SPRT, this correlation detection problem can be easily transformed into an SPRT, in which we test the hypothesis H_1 that C_{tcp} is correlated with C_{smtp} against the hypothesis H_0 that two connections are uncorrelated by counting the number of TCP packets appearing on C_{tcp} in each reply round of C_{smtp}.

If we use a Bernoulli random variable X_i to represent the observation result on C_{tcp} in i-th reply round of C_{smtp} and assume that these variables in different rounds are i.i.d, we have the following distribution:

$$X_i|H_1 = \begin{cases} \theta_1 & \text{if one outbound TCP packet observed} \\ 1-\theta_1 & \text{otherwise} \end{cases}$$

Algorithm 1 Detect-Correlation

1: Input: C_{tcp}, C_{smtp}
2: Para: A, B
3: Output: C_{tcp} is correlated with C_{smtp} or not
4: **repeat**
5:   **for** each reply round of C_{smtp} **do**
6:     **if** # of outbound packets on C_{tcp} is 1 **then**
7:       $\Lambda_n \leftarrow \Lambda_n + \ln \frac{\theta_1}{\theta_0}$
8:     **else**
9:       $\Lambda_n \leftarrow \Lambda_n + \ln \frac{1-\theta_1}{1-\theta_0}$
10:     **end if**
11:     **if** $\Lambda_n \geq B$ **then**
12:       C_{tcp} is correlated with C_{smtp} and the test stops
13:     **else if** $\Lambda_n \leq A$ **then**
14:       C_{tcp} is not correlated with C_{smtp} and the test stops
15:     **else**
16:       wait for observation in next reply round
17:     **end if**
18:   **end for**
19: **until** C_{smtp} is closed

$$X_i|H_0 = \begin{cases} \theta_0 & \text{if one outbound TCP packets observed} \\ 1-\theta_0 & \text{otherwise} \end{cases}$$

The algorithm of detecting connection correlation can be expressed in Algorithm 1.

For proxy-based spamming, given that packet symmetry holds most of time, the major reason that correlation cannot be detected without is mainly attributed to the packet misses by the monitoring system. For example, when the traffic volume exceeds the capacity that the monitoring system can handle, packets may be dropped by the monitoring system. If the packet conveying SMTP reply message is dropped on either the downstream connection or the upstream connection, the correlation detection will fail in this reply round. So we can use packet miss rate to estimate the probability of a proxy connection being correlated when spamming occurs, i.e. θ_1. From the conservative perspective, we take 0.01 as the packet miss rate which in fact is fairly high[8] considering only small packets (say less than 300 bytes) need attention and only packet header information is required for detection algorithm. So θ_1 is 0.99 in this case.

To estimate θ_0, we employ the mathematical model given in [16]. We assume that the uni-directional packet arrivals of a normal TCP connection can be modeled as a non-homogeneous Poisson process, which can be approximated by a sequence of Poisson processes with varying rates, and over varying time periods that could be arbitrarily small. For example, let $M(t)$ denote the number of packets sent in an outbound TCP connection during time interval t. Process $\{M(t), t \geq 0\}$ can be represented by a sequence of Poisson processes $(\lambda_1, \Delta t_1)$, $(\lambda_2, \Delta t_2)$, $\cdots$, where $t = \Delta t_1 + \Delta t_2 + \cdots$. The advantage of this model is to approximate almost any distribution. More importantly, the number of packets observed during any given time interval T, can be represented by a Poisson process M with a single rate $\hat{\lambda}_T$. Here $\hat{\lambda}_T$ is the weighted mean of the rates of all the Poisson processes during T.

With this model, we can easily compute the probability of

---

[8]In practice, the miss rate is usually below 0.005 in our campus network.

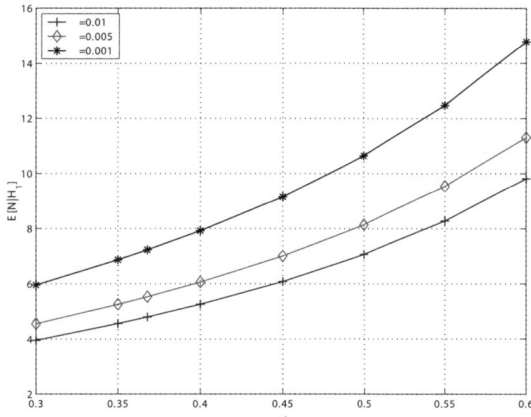

Figure 4: $E[N|H_1]$ **vs.** θ_0 **and** α^* ($\theta_1 = 0.99, \beta^* = 0.01$)

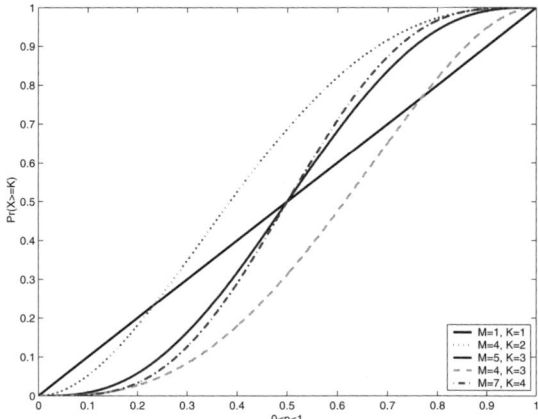

Figure 5: $\Pr(X \geq K)$ **vs.** p **and** (M, K)

one and only one packet sent in a reply round if T denotes the duration of a reply round. From

$$\Pr(M = i) = e^{-(\hat{\lambda}_T T)} \frac{(\hat{\lambda}_T T)^i}{i!}, \qquad (12)$$

we have

$$\Pr(M = 1) = e^{-(\hat{\lambda}_T T)} (\hat{\lambda}_T T) \leq e^{-1}. \qquad (13)$$

In (13) $\Pr(M = 1)$ reaches its maximum value e^{-1} when $\hat{\lambda}_T T = 1$. Although this is a theoretical derivative, we find that it is valid on almost all of the evaluated traces. Thus, we set $\theta_0 = e^{-1}$.

If we choose 0.005 for false positive probability α^* and 0.01 for false negative probability β^*, with $\theta_0 = e^{-1}$ and $\theta_1 = 0.99$, $E[N|H_1]$ is 5.5 and $E[N|H_0]$ is 2.02, respectively. Figure 4 shows how $E[N|H_1]$ varies with the changes of α^* and θ_0, when β^* and θ_1 are fixed. In general, $E[N|H_1]$ increases when θ_0 gets bigger or α^* gets smaller. Intuitively, this prolonged random walk is a natural result of smaller step length $\ln \frac{\theta_1}{\theta_0}$ or enlarged distance $\ln \frac{1-\beta^*}{\alpha^*}$ for the walk towards upper threshold.

From the perspective of anomaly detection, it is desirable that error probabilities, especially the false positive probability, can be as low as possible. In the framework of SPRT, this implies that $E[N|H_1]$ goes up, i.e., the average detection time is prolonged. However, given that not all SMTP transactions (the shortest one has only 6 reply rounds) can be longer enough to make the SPRT reach a decision when α is too small, a tradeoff between lowering false positive and false negative has to be made. In DBSpam, we set $\alpha^* = 0.005$ so that even the shortest spam transactions can be captured.

5.5 Noise Reduction

To further lower the false positives of SPRT, we introduce a simple and effective noise reduction technique in DBSpam. In a series of correlation tests, we define the active spam sources and proxies that are prone to be identified many times as signals, and define those innocent IP addresses that may be accidentally captured as noises. We utilize the dichotomy between signal and noise to distinguish spam sources and proxies from innocent end-hosts. We call this procedure *noise reduction*. The noise reduction are executed in two steps: first, we maintain a set S_i of external IP addresses that appear in the correlation results for each time window Δ; second, in the consecutive M time windows, we single out the external IP addresses, which appear no fewer than K times, as the spam sources and the corresponding proxy addresses as the spam proxies.

The time window Δ is determined by the lower-bound of spamming rate v (in replies/s) and the number of reply rounds N:

$$\Delta \geq N/v. \qquad (14)$$

Hence, a spammer sending spam faster than v must appear in S_i at least once in each time window Δ. Assume that the appearance of an IP address in S_i is independent, with a constant probability p. Then, the number of occurrences of the IP address among M time windows follows the *binomial* distribution.

$$\Pr(X = i) = \binom{M}{i} p^i (1-p)^{M-i}. \qquad (15)$$

The probability of having no fewer than K occurrences in the binomial distribution is:

$$\Pr(X \geq K) = \sum_{i=K}^{M} \binom{M}{i} p^i (1-p)^{M-i}. \qquad (16)$$

Figure 5 illustrates the dynamics of $\Pr(X \geq K)$ with the variation of probability p for several pre-determined tuples of (M, K). The diagonal line shows the case of tuple $(M = 1, K = 1)$, in which $\Pr(X \geq K)$ is equal to p. Clearly, if p is smaller than 0.2, all other curves are below this diagonal line, indicating that their values of $\Pr(X \geq K)$ are smaller than that of tuple $(M = 1, K = 1)$. In contrast, if p is larger than 0.8, these curves are above the diagonal line, indicating that their values of $\Pr(X \geq K)$ are larger than that of tuple $(M = 1, K = 1)$.

The value of p for an innocent address depends on the false positive rate of the correlation detection, which should be closer to zero than one. The left part of Figure 5 illustrates the noise reduction can further lower the chance of an innocent address being mis-classified as a spam source. On the other hand, the value of p for a spam source is related to the complementary of the false negative rate of the correlation detection, which should be closer to one than zero as shown in the right part of Figure 5. This indicates that noise reduction increases the probability of a spam source being identified as well. Therefore, both false positives and false negatives are reduced after noise reduction. Figure 5

Table 1: Trace Information

Attribute	S-1-A	S-1-B	S-1-C	S-2-A	S-2-B	S-2-C	N-1	N-2
duration (sec)	770	674	756	654	1,385	1,398	5,116	14,944
# of packets	3,872,550	4,178,567	4,509,336	12,036,413	26,422,563	26,172,898	24,434,518	297,733,228
avg packet/sec	5,029	6,200	5,965	18,404	19,078	18,722	4,776	19,923
trace size	295MB	318MB	343MB	931MB	2,044MB	2,018MB	1,851MB	22.4GB
packet miss rate	< 0.001	< 0.001	< 0.001	0.008	0.005	0.005	<0.001	0.006
# of threads/spammer	1	3	1	1	3	1	-	-

shows that when M is fixed, the probability $\Pr(X \geq K)$ goes smaller with bigger K. For example, $\Pr(X \geq 3|M = 4)$ is much smaller than $\Pr(X \geq 2|M = 4)$. Moreover, the noise reduction algorithm works very well even with very small M and K. For example, with $(M = 4, K = 3)$, pre-noise-reduction false positive rate, which is 0.1, can be significantly lowered to 0.0037 after noise reduction. These two rules of thumb may guide the selection of (M, K) in practice. We will further discuss the parameter setup of Δ, M and K, and demonstrate the effectiveness of the noise reduction technique in Section 6.3.2.

6. SYSTEM EVALUATION

We implemented a prototype of DBSpam using *libpcap* on Linux. Due to access limitation, we cannot deploy our prototype in an ISP network environment to evaluate its on-line performance. Alternatively, we collected traces from a middle-sized campus network and conducted a series of trace-based experiments to validate the efficacy of DBSpam.

By replaying the collected traces with our prototype, we attempt to answer the following questions: (1) how fast DBSpam can detect spam laundering; (2) how accurate the detection result of DBSpam is; (3) how many system resources DBSpam consumes.

6.1 Data Collection

The campus network is connected to the Internet via an OC-3 data link. A Snort-based NIDS [29] is deployed on the edge router of the campus network to block any suspicious proxy traffic (e.g. SOCKS and HTTP) via signature checking. All outgoing email messages must go through the main email server and secure authentication is enforced.

This well-protected campus network provides an ideal platform to assess the false positive ratio of DBSpam on normal network traffic. According to the IT department, proxy-based spamming activities on this campus network are very rare. To evaluate the detection time and accuracy of DBSpam on spam laundering, we generate "spam" traffic, including both plain-text and encrypted proxy traffic, with the cooperation of the IT department. Although the monitoring systems of IT can detect plain-text proxy traffic by checking content, our encrypted proxy traffic successfully evades their detection.

The generated spamming scenario is similar to the one shown in Figure 1. The campus network plays the role of network N. We use two home PCs outside the campus network, which are located in two different ISP broadband networks, to emulate two spam sources. The spam sink (MTA M in Figure 1) is located in the dark net of the campus network. The dark net is a special subnet that directly links to the edge router and is used to dump all malicious traffic. Two SOCKS and HTTP proxies run in two different subnets of the campus network to form a proxy chain. We use a common spamware and *sockschain* [9] to emulate proxy-chain spamming. The spam messages are sent from the two home PCs, through the proxy chain and destined to the spam sink. The data collection point is just before the edge router and can see all the traffic passing through the edge router. We use *tcpdump* to capture all small bi-directional TCP packets with the `snaplen` set to 75 bytes.

We collected multiple traces of normal and spam traffic in two different months. The detailed information of the traces is listed in Table 1, and additional explanations are given below. First, we only captured small TCP packets with packet length less than 300 bytes as DBSpam only utilizes the SMTP reply messages for detection, which are usually conveyed by TCP packets with length less than 300 bytes. Second, We collected two kinds of traces to evaluate the performance of DBSpam, one with generated spam traffic and the other without generated spam traffic. All traces include the normal background SMTP traffic passing through the campus network. The name of a trace follows the format "{S|N}-{1|2}-{A|B|C}". S (N) indicates that the trace has Spam (No spam) traffic. 1 (2) refers to the different month of trace collection. A (B, C) is only for spam traces and stands for different spam scenario. Third, in order to validate DBSpam for detecting both plain-text and encrypted spam traffic, we injected encrypted and compressed spam traffic through SSH tunneling into traces S-*-C (* is either 1 or 2), and injected plain-text spam traffic into S-*-A and S-*-B. Fourth, a multi-threaded spamming technique was used in S-*-B to validate the efficacy of DBSpam in a multi-threaded spamming scenario. The N-threaded spamming means up to N upstream connections may be issued simultaneously from the spam source to a proxy for spam laundering.

6.2 Detection Time

The overall detection time of DBSpam is determined by SPRT detection time, the noise-reduction time window Δ, and the number of consecutive windows M. Among these three factors, SPRT detection time is the fundamental one, which bounds the value of time window Δ. In the following, we focus on the estimation of SPRT detection time.

6.2.1 SPRT Detection Time

We evaluate SPRT detection time from two perspectives: the number of observations needed to reach a decision and the actual time spent by SPRT.

Number of Observations N: The theoretical average number of observations under spam hypothesis ($E[N|H_1]$) and non-spam hypothesis ($E[N|H_0]$) can be easily computed based on Equations (10) and (11). In our evaluation, they are rounded to 6 and 3, respectively, with $\alpha^* = 0.005$, $\beta^* = 0.01$, $\theta_0 = e^{-1}$, and $\theta_1 = 0.99$. Table 2 shows the distribution of $N|H_1$ in six spam traces. The results clearly

---
[9] Both are binary Windows programs so that we cannot modify any code.

Table 2: Distribution of $N|H_1$

Trace	$N = 6$	$N = 11$	$N >= 16$
S-1-A	970 (100%)	0	0
S-1-B	5019 (96.9%)	139 (2.7%)	21 (0.4%)
S-1-C	2245 (92.8%)	169 (7.0%)	6 (0.2%)
S-2-A	433 (99.1%)	3 (0.7%)	1 (0.2%)
S-2-B	4298 (94.7%)	198 (4.4%)	40 (0.9%)
S-2-C	1758 (98.9%)	16 (1.0%)	3 (0.1%)

demonstrate the dominance of ($N = 6$) in all traces. The comparatively low percentage of ($N = 6$) in trace S-1-C is mainly caused by the abnormally high packet-miss-rate of the spam traffic but not the whole traffic. Note that due to the characteristics of SPRT, the detection of connection correlation (H_1) can only be reached after certain number of observations, such as 6 and 11.

Figure 6 shows the distribution of $N|H_0$ for non-spam traces N-1 and N-2. The curves indicate that SPRT can filter out at least 95% of normal connections within four observations. The distributions of $N|H_0$ for spam traces are similar to those for non-spam traces.

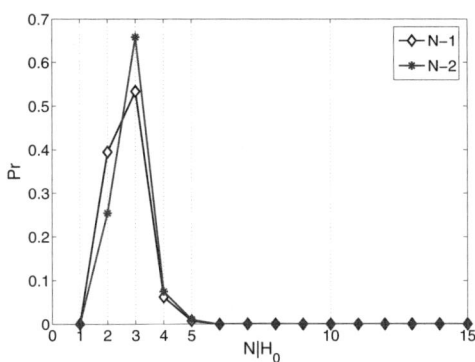

Figure 6: Distribution of $N|H_0$

Actual Detection Time of SPRT: After recording the start and end points for each SPRT on six spam traces, we derive all the detection time in these traces and draw their CDF (cumulative distribution function) in Figure 7. The detection time is approximated by ceiling for CDF drawing, e.g., 1.2s is ceiled to 2s. We classify the results from six traces into two groups: "S-1" and "S-2", since the results in each group are very similar. As shown in Figure 7, 95% detections are made within 5 seconds. Note that the actual detection time is roughly the duration of 6 reply rounds of SMTP connection, since the computation overhead of SPRT is negligible. The curve difference between "S-1" and "S-2" is due to the inferior link quality in "S-2" experiments.

6.3 Detection Accuracy

Since the detection module of DBSpam has two phases—SPRT detection and noise reduction, we first evaluate the false positive and false negative of SPRT detection, and then present the overall detection accuracy of DBSpam after noise reduction.

6.3.1 Accuracy of SPRT

False Positives: The upper part of Table 3 shows the false positives of SPRT in different traces. The "detection" row is the total number of correlations reported by SPRT, and "True Positives (TP)" and "False Positives (FP)" rows list the outcome of detections. The "True Negatives (TN)"

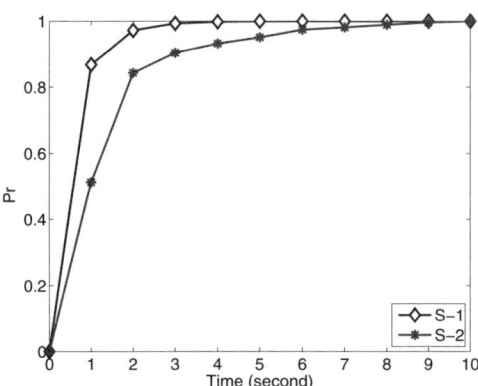

Figure 7: CDF of Detection Time for SPRT

row lists the number of tests on normal connections that are correctly identified. According to the definition of false positive probability $\alpha = \frac{FP}{FP+TN}$, the probabilities in all traces are well below 0.0002, indicating that the false positive probability of SPRT is fairly small in practice.

False Negatives: We estimate the false negatives by counting the number of proxy connections that are missed by SPRT, and compute the ratio of missed spam connections, which are shown in the lower part of Table 3. The false negatives of SPRT are attributed to the missed packets in the spam traces. The three spam traces S-2-A/B/C contain both long SMTP connections (more than 10 reply rounds) and short SMTP connections (six reply rounds). More than 70% of the total connections are short SMTP connections. For those short spam connections with only six reply rounds, if any packet on either the upstream connection or the downstream connection is missed in the trace, SPRT cannot reach a decision, leading to a false negative. A simple estimation shows the feasibility of the missing ratio of spam connections. For simplicity, we assume that the packet miss rate p is constant through the trace. Then, the probability of one packet missing in six reply rounds is $\binom{12}{1}p(1-p)^{11}$. If $p = 0.005$ (the packet miss rate of traces S-2-B/C), the probability is around 0.057, which is more than the miss ratio as shown in Table 3.

6.3.2 DBSpam Accuracy after Noise Reduction

To investigate the efficacy of noise-reduction, we first need to determine the value of time window Δ. Figure 7 shows that over 80% of all SPRTs on spam traces terminate within 2 seconds. So, we set the time window Δ to 2 seconds. For (M, K), we test several combinations and the final detection results are shown in Table 4, where the data format is "number of FP/number of overall detections". From the table, we can see that noise reduction eliminates the majority of false positives of SPRT, due to the fact that most of wrongly-classified correlations only occur sporadically. The false positive of DBSpam approaches zero, when (1) M and K are relatively large and (2) the gap between M and K is small. Such dynamics of false positive reduction fits well with the analysis in Section 5.5. For our traces, any combination with 4/5 for M and 3/4 for K can achieve fairly high accuracy. Of course, the high detection accuracy is achieved at the cost of lowering detection sensitivity. It always exists a tradeoff between accuracy and sensitivity in network anomaly detection. However, even when the time window

Table 3: False Positives and False Negatives of SPRT

Attribute	S-1-A	S-1-B	S-1-C	S-2-A	S-2-B	S-2-C	N-1	N-2
Detection	970	5,179	2,420	437	4,536	1,777	66	2,368
True Positives	966	5,108	2,369	320	3,510	1,558	-	-
False Positives	4	71	51	117	1,026	219	66	2,368
True Negatives	290,889	1,156,085	596,979	1,634,307	8,895,993	4,266,100	687,390	15,941,150
FP/(FP+TN)	1.4e-5	6.1e-5	8.5e-5	7.2e-5	1.2e-4	5.1e-5	9.6e-5	1.5e-4
Spam Connections	958	570	324	329	1,351	969	-	-
Missed Connections	8	2	0	6	27	13	-	-
Missed Conn Ratio	0.008	0.004	0	0.018	0.020	0.013	-	-

Δ is set to 2 seconds and M is set to 5, the overall delay of DBSpam detection is just 10 seconds but with much higher accuracy.

Currently most false positives of DBSpam are induced by P2P applications. The capacity of spawning thousands of connections in a second and the behavior of periodic PING/PONG communications make P2P applications have a much higher probability of being correlated than any other applications. Due to its hog overwhelming proportion in bandwidth consumption, many ISPs and university networks in US have restricted the maximal connections that P2P applications can establish, which helps reduce the false positives of DBSpam.

Table 4: Overall False Positives of DBSpam ($\Delta = 2s$)

Trace	(M, K)			
	(3, 2)	(4, 3)	(5, 3)	(5, 4)
S-1-A	0/188	0/138	0/124	0/110
S-1-B	0/162	0/126	0/103	0/103
S-1-C	0/194	0/150	0/124	0/123
S-2-A	0/65	0/36	0/52	0/27
S-2-B	13/335	3/243	4/216	0/186
S-2-C	0/193	0/124	0/135	0/94
N-1	0/0	0/0	0/0	0/0
N-2	7/7	1/1	2/2	0/0

*Data Format: # of false positives / # of total detections

6.4 Resource Consumption

According to Table 1, the arrival rate of small TCP packets at the edge router can reach around 20,000 packets per second (pps), at which DBSpam must be able to handle. Current high-end PCs can meet this requirement without much difficulty. Using a Dell Precision 360 machine with Pentium-4 3GHz CPU and 512MB memory, we run the prototype of DBSpam on each trace multiple times. We use *time* and *ps* to measure the CPU and memory usage. The results are listed in Table 5. The average packet processing rate of DBSpam is computed by dividing the total packet number of the trace over the processing time ("CPU Time"). The processing rates clearly demonstrate the capability of DBSpam working at high-speed networks. Even in the worst case, DBSpam still can handle 241,965 pps, which is over 10 times more than the required processing speed.

Table 5: Resource Consumption

Trace	CPU Util	CPU Time	pps	Peak Mem
S-1-A	36.3%	9.0s	430,283	2.2MB
S-1-B	37.7%	9.8s	426,384	1.6MB
S-1-C	24.0%	9.3s	484,875	1.2MB
S-2-A	58.0%	36.8s	327,076	11.9MB
S-2-B	84.3%	109.2s	241,965	10.5MB
S-2-C	57.1%	78.6s	332,989	2.8MB
N-1	21.7%	51.1s	478,171	5.6MB
N-2	32.1%	789.9s	376,925	8.4MB

Memory consumption of DBSpam is mainly determined by two factors: the number of active SMTP connections and the number of outbound TCP connections during each SMTP reply rounds. So, the peak memory consumption is not necessarily determined by the network traffic volume. As DBSpam only needs to maintain very few states, and only a very small portion (false positive probability) of connections need to maintain states for relatively long time (lifespan of SMTP connections), the overall memory consumption should not be a problem. Also note that the memory management of our prototype is quite naive since our focus is mainly on the correctness, not on the performance.

6.5 Suppressing Spam Activities

Once the spam laundering activities are identified, DBSpam can effectively stifle them by activating the suppression module. Since spam suppression mechanisms such as blocking and throttling are straightforward to implement, the evaluation results of the suppression module are not included due to space limit.

7. POTENTIAL EVASIONS

In such an ongoing arms race between spammers and anti-spammers, we envision that sufficiently aggressive spammers will seek sophisticated techniques to evade DBSpam. This is especially true for a spammer who is able to fully control remote spam proxy machines and deploy arbitrarily customized software. It may use non-off-the-shelf proxy programs, which can manipulate the traffic between the spam source and the first-hop proxy, to break packet symmetry. One possible way is to split a single reply packet from SMTP server into n fragmented packets on the first-hop proxy and then to transfer them back to the spam source.

However, as long as enough observations are collected, DBSpam can still capture such potential evasions. Recall that the effect of this packet splitting on SPRT model is just the change of the value of θ_0, which measures the probability of 1 to n outbound TCP packets observed in a reply round. So, instead of $\theta_0 = \Pr(M = 1)$, now $\theta_0 = \Pr(M = 1) + \ldots + \Pr(M = n)$. According to Equation (10), without changing other parameters, the augmented value of θ_0 renders more average number of observations needed to detect a spam proxy. On the other hand, not all SMTP transactions have enough reply rounds for detection. Due to enlonged observations, short-living spamming activities may not be detected.

To demonstrate the capability of DBSpam in capturing such evasions, we relax the definition of packet symmetry, in which one or two data packets may appear in one reply round, and adjust θ_0 to 0.5.[10] Then, we estimate the overall false positives of DBSpam, which are listed in Table 6 under the parameter setting of $M = 5$, $K = 4$, and $\Delta = 2s$. For comparison, the results without relaxation are listed in the first row, while the results with relaxation are listed in

---
[10]Note that θ_0 never exceeds 0.5 in all our traces with various packet lengths from 150 to 300 bytes.

Table 6: False Positive Comparisons ($M = 5$, $K = 4$, $\Delta = 2s$)

θ_0	α^*	$E[N|H_1]$	S-1-A	S-1-B	S-1-C	S-2-A	S-2-B	S-2-C	N-1	N-2
e^{-1}	0.005	5.5	0/110	0/103	0/123	0/27	0/186	0/94	0/0	0/0
0.5	0.005	8.1	0/0	0/103	0/120	0/0	0/97	0/32	0/0	8/8
0.5	0.02	6.0	0/110	2/105	0/121	0/27	7/194	1/94	0/0	21/21

the second row. Clearly, the short-living spamming activities are missed by DBSpam, with zero detection for S-*-A traces and much fewer detections for S-2-B and S-2-C traces. However, those spamming activities with more reply rounds can still be accurately detected. Since parameter α^* is tunable, we vary its value, from 0.005 to 0.02, to accommodate the shortest SMTP transactions for the examination of DBSpam. The third row in Table 6 lists the results after this adjustment, showing that DBSpam can capture almost all spamming activities as same as before but at the cost of slightly more false positives, which is the necessary tradeoff in capturing evasive spam proxy traffic.

Moreover, instead of employing off-the-shelf proxy software, any advanced evasion technique will inevitably induce the modifications on the current spam methods and degrade the spam laundering efficiency. The customized proxy software also increases the cost of spamming. Overall, DBSpam indeed significantly raises the protection bar against email spam, breaking the laundering and tracing out the real spam sources, in the anti-spam-vs-spam arms race.

8. CONCLUSION

In this paper, we present a simple and effective system, DBSpam, to detect and break proxy-based email spam laundering activities inside a customer network and to trace out the corresponding spam sources outside the network. Instead of content checking, DBSpam leverages the protocol semantics and timing causality of proxy-based spamming to identify spam proxies and real spam sources behind them. Based on connection correlation and packet symmetry principles, DBSpam monitors the bi-directional traffic passing through a network gateway, and utilizes a simple statistical method, Sequential Probability Ratio Test, to quickly filter out innocent connections and identify the spam laundry path with high probability. To further reduce false positives and false negatives, we propose a noise reduction technique to make spammer-tracking more accurate after gathering consecutive correlation detection results. We implement a prototype of DBSpam using *libpcap* on Linux, and conduct trace-based experiments to evaluate its effectiveness. Our experimental results reveal that DBSpam can be tuned to detect spam proxies and sources with low false positives and false negatives in seconds. After detecting spam proxies and related spam sources, DBSpam can effectively throttle or block spam traffic.

9. REFERENCES

[1] http://www.messagelabs.com/Threat_Watch/.
[2] http://www.spamhaus.org/news.lasso?article=156.
[3] http://spamassassin.apache.org/.
[4] http://www.postini.com.
[5] http://www.senderscorecertified.com.
[6] http://www.toplayer.com.
[7] Composite blocking list. http://cbl.abuseat.org.
[8] Domainkeys: Proving and protecting email sender identity. http://antispam.yahoo.com/domainkeys.
[9] http://www.rfc-editor.org/rfc/rfc2476.txt.
[10] MTA Authorization Records in DNS. http://www.ietf.org/html.charters/OLD/marid-charter.html.
[11] The penny black project. http://www.research.microsoft.com/research/sv/PennyBlack/.
[12] SPF. http://www.openspf.org.
[13] M. Andreolini, A. Bulgarelli, M. Colajanni, and F. Mazzoni. Honeyspam: Honeypots fighting spam at the source. In *Proc. USENIX SRUTI 2005*, Cambridge, MA, July 2005.
[14] A. Back. Hashcash. http://www.hashcash.org/.
[15] J. Blosser and D. Josephsen. Scalable centralized bayesian spam mitigation with bogofilter. In *Proc. USENIX LISA 2004*, Atlanta, GA, November 2004.
[16] A. Blum, D. X. Song, and S. Venkataraman. Detection of interactive stepping stones: Algorithms and confidence bounds. In *Proc. RAID 2004*, Sophia Antipolis, France, September 2004.
[17] L. Donnerhacke. Teergrubing faq. http://www.iks-jena.de/mitarb/lutz/usenet/teergrube.en.html.
[18] S. Garriss, M. Kaminsky, M. J. Freedman, B. Karp, D. Mazieres, and H. Yu. Re: Reliable email. In *Proc. USENIX NSDI 2006*, San Jose, CA, May 2006.
[19] P. Graham. A plan for spam. http://www.paulgraham.com/spam.html.
[20] T. Hunter, P. Terry, and A. Judge. Distributed tarpitting: Impeding spam across multiple servers. In *Porc. USENIX LISA 2003*, San Diego, CA, October 2003.
[21] J. Jung, V. Paxson, A. W. Berger, and H. Balakrishnan. Fast portscan detection using sequential hypothesis testing. In *Proc. IEEE Symposium on Security and Privacy 2004*, Oakland, CA, May 2004.
[22] J. Jung and E. Sit. An empirical study of spam traffic and the use of dns black lists. In *Proc. ACM SIGCOMM Internet Measurement Conference*, Taormina, Italy, October 2004.
[23] B. Krishnamurthy and E. Blackmond. SHRED: Spam harassment reduction via economic disincentives. http://www.research.att.com/~bala/papers/shred-ext.pdf.
[24] K. Li and Z. Zhong. Fast statistical spam filter by approximate classifications. In *Proc. ACM SIGMETRICS 2006*, St. Malo, France, June 2006.
[25] N. Provos. A virtual honeypot framework. In *Proc. USENIX Security 2004*, San Diego, CA, August 2004.
[26] S. Radosavac, J. S. Baras, and I. Koutsopoulos. A framework for mac protocol misbehavior detection in wireless networks. In *Proc. 4th ACM workshop on Wireless security*, Cologne, Germany, September 2005.
[27] A. Ramachandran, D. Dagon, and N. Feamster. Can DNS-based blacklists keep up with bots? In *CEAS 2006*, Mountain View, CA, July 2006.
[28] A. Ramachandran and N. Feamster. Understanding the network-level behavior of spammers. In *Proc. ACM SIGCOMM 2006*, Pisa, Italy, September 2006.
[29] M. Roesch. Snort - lightweight intrusion detection for networks. In *Proc. USENIX LISA 1999*, Seattle, WA, November 1999.
[30] R. D. Twining, M. M. Williamson, M. Mowbray, and M. Rahmouni. Email prioritization: Reducing delays on legitimate mail caused by junk mail. In *Proc. USENIX Annual Technical Conference 2004*, Boston, MA, June 2004.
[31] A. Wald. *Sequential Analysis*. Dover Publications, 2004.
[32] M. Walfish, H. Zamfirescu, H. Balakrishnan, D. Karger, and S. Shenker. Distributed quota enforcement for spam control. In *Proc. USENIX NSDI 2006*, San Jose, CA, May 2006.
[33] M. M. Williamson. Design, implementation and test of an email virus throttle. In *Proc. 19th Annual Computer Security Applications Conference*, Las Vegas, Nevada, December 2003.
[34] D. Woolridge, J. Law, and M. Kawasaki. The qmail spam throttle mechanism. http://spamthrottle.qmail.ca/man/qmail-spamthrottle.5.html.
[35] B. Yerazunis. Crm114 - the controllable regex mutilator. http://crm114.sourceforge.net.
[36] Y. Zhang and V. Paxson. Detecting stepping stones. In *Proc. USENIX Security 2000*, Denver, CO, August 2000.

Forward-Secure Signatures with Untrusted Update

Xavier Boyen *   Hovav Shacham †   Emily Shen ‡   Brent Waters §¶

ABSTRACT

In most forward-secure signature constructions, a program that updates a user's private signing key must have full access to the private key. Unfortunately, these schemes are incompatible with several security architectures including Gnu Privacy Guard (GPG) and S/MIME, where the private key is encrypted under a user password as a "second factor" of security, in case the private key storage is corrupted, but the password is not.

We introduce the concept of forward-secure signatures with untrusted update, where the key update can be performed on an encrypted version of the key. Forward secure signatures with untrusted update allow us to add forward security to signatures, while still keeping passwords as a second factor of security. We provide a construction that has performance characteristics comparable with the best existing forward-secure signatures. In addition, we describe how to modify the Bellare-Miner forward secure signature scheme to achieve untrusted update.

Categories and Subject Descriptors

E.3 [**Data**]: Data Encryption—*Public key cryptosystems*

General Terms

Design, Security, Performance.

Keywords

Digital Signatures, Forward Security, Untrusted Storage, Two-factor Authentication, Software Implementation.

---

*Voltage Security Inc. — xb@boyen.org
†Weizmann Inst. Sci. — hovav.shacham@weizmann.ac.il
  Supported by a Koshland Scholars Program fellowship.
‡Stanford University — emily@cs.stanford.edu
§SRI International — bwaters@csl.sri.com
¶Supported by NFS and DHS. This material is based upon work supported by the Department of Homeland Security (DHS) and the Department of Interior (DOI) under Contract No. NBCHF040146. Any opinions, finding and conclusions or recommendations expressed in this material are those of the author(s) and do not necessarily reflect the views of DHS and DOI.

Permission to make digital or hard copies of all or part of this work for personal or classroom use is granted without fee provided that copies are not made or distributed for profit or commercial advantage and that copies bear this notice and the full citation on the first page. To copy otherwise, to republish, to post on servers or to redistribute to lists, requires prior specific permission and/or a fee.
CCS'06, October 30–November 3, 2006, Alexandria, Virginia, USA.
Copyright 2006 ACM 1-59593-518-5/06/0010 ...$5.00.

1. INTRODUCTION

One problem commonly faced in security research is how to limit damage when an attacker compromises a system and private secrets are exposed. Ross Anderson, in an invited talk, originally proposed a signature scheme known as Forward-Secure Signatures [3] that was meant to mitigate the damage when private signature keys were exposed. In his proposal, each signature would be associated with the current time period in addition to the signed message. After each time interval, a user's private signing key is updated such that it can no longer be used to sign for past time periods. In this manner, if a user's private key is compromised at a given time period, the attacker is unable to forge signatures that appear to come from any earlier time period.

Anderson's original solution was quite simple. At setup time a user would issue himself a separate certificate and private key for every time period. As each time period passed, the private key for that period would be deleted. While the solution was quite elegant, it had the drawback that the private key size of the user grew linearly with the number of time periods, making it rather inefficient and infeasible for systems requiring a small storage space (such as smartcards). Bellare and Miner [4] formalized the notion of Forward-Secure Signatures and provided two schemes with improved private storage requirements. Several constructions followed [25, 22, 19, 21, 1, 2] that provided different tradeoffs in the metrics of private key storage, signature size, and the computational time required for setup, key update, signing, and verifying.

While there has been considerable (and successful) effort put into achieving constructions with sufficient efficiency for practical use, in order for a new primitive to be adopted it must integrate into existing software architectures. In addition, it should be as transparent as possible—the added burden on the user should be minimal. One feature of several cryptographic software suites, including the popular Pretty Good Privacy suite [34], is that a user's private key is encrypted under some additional secret, typically a password. The added value of encrypting the private key is that it provides a "second factor" for security in case the storage is compromised.

Unfortunately, most existing forward-secure signature constructions are difficult to integrate into this model. The primary difficulty is that the "update" algorithms need to access the private key unencrypted in order to move it forward. Given this limitation, the software will either need to require that the user intervene each time it updates the key or it will need to forgo the second factor altogether. The former option is undesirable for several reasons: first, it places a burden on the users to take an action at regular intervals to update their keys, something they will most likely come to resent; second, it makes some update schedules (say, every hour) infeasible; third, since updates are not automatic, keys are likely to be often not fully updated, and, if an non-updated key is compromised, the attacker can produce signatures for past time periods—precisely the threat that forward-secure signatures are supposed to prevent. The second choice is also unfavorable since it is unclear that achieving forward security is worth the tradeoff of abandoning the extra security provided by keeping the keys encrypted under the second factor; developers of security software will be unlikely to adopt a new feature in favor of dropping an old one.

1.1 Our Contribution

We introduce *forward-secure signatures with untrusted update*, a signature primitive that allows a program to update, i.e., move forward in time, an "encrypted" version of the private key. To sign a message an algorithm must have an additional secret key, the second factor, which in practice is a password provided by the user. The required security properties are both that an attacker with access to just the encrypted key cannot forge signatures and that the scheme will maintain the traditional forward-secure properties.

We observe that the original proposal of Anderson can actually be easily modified to achieve these properties—simply encrypt each of the private keys under the secret key. However, as discussed above, this solution requires an unreasonable amount of private key storage. Our goal is to design a forward secure scheme with comparable or better performance and security properties than existing schemes.

We create a (non-trivial) forward-secure signature scheme with untrusted updates. Our construction is "tree-based" and the underlying structure is similar to recent hierarchical identity-based encryption schemes [8, 6] where the private keys are composed of elements from bilinear groups. Using this structure, we are able to encrypt private keys in such a way that a third party can perform a homomorphic key update operation on them.

Our scheme is quite efficient with signatures consisting of three group elements, private keys of $O(\log(n)^2)$ group elements, and constant encryption, verification, update, and setup times. In addition, our scheme is provably secure without random oracles. These features are actually interesting in their own right, and we regard the forward-secure construction as being of interest even if we disregard the untrusted update property.

While our primary contribution is the introduction of a new forward-secure signature scheme with untrusted update, one might desire to realize untrusted update in a different signature scheme that has different security or performance tradeoffs. We show that the concept of untrusted update can be applied in other schemes by sketching how to modify the Bellare-Miner [4] forward-secure signature scheme to achieve untrusted update.

Finally, to demonstrate the practical applicability of our scheme, we provide an implementation of our construction. We create an API for our construction.

1.2 Organization

We begin by discussing related work in Section 2. Next, we give a formal description of a forward-secure signature scheme with untrusted update and definitons of security in Section 3. We present some background information on bilinear groups and our assumptions in Section 4. Then we present our construction and a proof of its security in Section 5. Additionally, we describe in Section 6 how to equip the Bellare-Miner [4] forward-secure signature with an untrusted update. We follow by describing a software implementation of our scheme in Section 7. Finally, we conclude in Section 8.

2. RELATED WORK

Originally, forward security was introduced for key exchange protocols [16]. Anderson's original suggestion was for the user to store a separate private key for each time period. Bellare and Miner [4] later formalized the notion of forward-secure signatures.

Following work can roughly be divided into two classes. The first comprises generic constructions that need not necessarily require random oracles. The first of these is the tree construction of Bellare and Miner [4]. In this construction a generic signature scheme is used to build a binary tree from chains of certificates where leaves correspond to time periods. The private key storage, signature time, and verification time will all be a multiplicative factor of $O(\log(T))$ longer than the original signature scheme, where T is the number of time periods. Malkin, Micciancio, and Miner [25] apply Merkle trees [26] so that signing and verifying requires $O(\log(n))$ hashes (instead of signing or verifying operations). This comes at some additional expense during setup and key generation, though. They additionally show a method for combining various tree-based schemes in order to make various tradeoffs. Cronin et al. [9] provide an evaluation of the practical performance of these schemes and create an open-source forward-secure signature library. Krawczyk [22] presents a generic method for keeping a short private secret key on trusted storage; however, one must still maintain some (possibly untrusted) storage linear in T for the signer.

The other class of forward-secure signatures comprises specific random oracle-based schemes. The first of these was due to Bellare and Miner [4] in which they achieve short signatures with fast key update by applying Ong-Schnorr signatures [28]; however, the verification procedure is linear in T. Abdalla and Reyzin [2] later show tradeoffs in the computational time with signature and public key size. Itkis and Reyzin [19] propose a scheme with highly efficient signature and verification times based on Guillou-Quisquater signatures [15]. Although their basic technique requires an expensive update, they show how to apply certain pebbling techniques to achieve constant update time while storing just $O(\log(T))$ elements. Finally, Kozlov and Reyzin [21] give a scheme with fast key update.

In addition, there has been work on related subjects such as key-insulated update and intrusion resilient signature schemes [10, 11, 20, 18] and applications of forward security to group signatures [32] and threshold cryptography [1].

Canetti, Halevi and Katz [8] show how Hierarchical Identity-Based Encryption [14, 17] (HIBE) can be used to achieve forward-secure encryption. Boneh, Boyen, and Goh [6] later show how certain HIBE ciphertexts can be compressed to a constant size. In our work we apply Naor's observation (stated in [7]) that private keys from an Identity-Based Encryption [31] system can be viewed as signatures on a given identity. We use a particular version of hierarchical signatures to achieve forward security with short signatures without random oracles.

3. DEFINITIONS

We describe forward-secure signature schemes and give a formal definition for their security. A forward-secure signature scheme is made up of four algorithms:

KeyGen(T): The setup algorithm takes in an integer, T, the number of time periods and outputs a public verification key VK, the encrypted signing key EncSK, and another second factor secret decryption key DecK. The current time period identifier, ID, is initially set to 1. The time period is embedded within the encrypted signing key.

Update(EncSK, ID'): The update algorithm takes in the encrypted signing key at some time period ID and outputs a new encrypted signing key EncSK' for time period ID' > ID. After this the previous signing key is erased. If ID' $\geq T$, then the old key is just erased and there is no new key. This algorithm does not require the decryption key.

Sign(EncSK, DecK, M): The signing algorithm takes as input the encrypted signing key EncSK, the second factor decryption key DecK, and a message M. It outputs a signature S for the time period ID that is embedded in the signing key. The time period may be included as part of the signature.

Verify(S, M, VK): The verification algorithm takes as input a signer's verification key VK, a message M, and a signature S. It outputs either valid or invalid.

3.1 Security Model

We now define the security of forward-secure signatures with untrusted update in terms of two games.

3.1.1 Forward Security

The first security game captures the "traditional" notions of existential unforgeability and forward security. The game is played between an adversary $\mathcal{A}$ and a challenger $\mathcal{B}$, and proceeds in three phases.

Key Generation. The challenger runs the Setup algorithm and gives the adversary the verification key VK, and the second factor decryption key DecK. The time period t is set to 1.

Interactive Queries. In the query phase the adversary can issue three types of requests in an adaptive, interactive manner:

Sign: The adversary can query the challenger to sign a message M on the current time period ID; the challenger will then return a signature S.

Update: The adversary can request that the challenger execute the update algorithm, in which case the time period will be increased to a new value ID' chosen by the adversary.

Corrupt: The adversary can request that the challenger hand out all its keys at the current time period. The challenger returns the encrypted signing key EncSK for the current time period.

The adversary can repeatedly make Sign and Update queries; however, once he makes a Corrupt query the game moves to the next phase.

Final Forgery. Let ID' be the time period at which the Corrupt query was issued. The adversary produces a forgery, consisting of a time, message, signature tuple (ID*, M^*, S^*). The adversary is successful if ID* < ID', the signature verifies for time ID*, and the adversary had not queried for a signature on M^* at the exact time period ID*.

We let $\mathsf{AdvFS}_\mathcal{A}$ denote the advantage of an algorithm $\mathcal{A}$ in the forward-security game.

3.1.2 Update Security

This game captures the notion of security against an adversary that controls the storage of the encrypted signing key, but not the second factor decryption key.

Key Generation. The challenger runs the Setup algorithm and gives the adversary the verification key, VK, and the initial encrypted signing key, EncSK. The time period ID is set to 1.

Query Phase. The adversary can issue two types of interactive requests:

Sign: The adversary can query the challenger to sign a message M under a key EncSK' specified by the adversary, for the current time period ID. The challenger *must* output a signature S if the given key EncSK' appears well formed (for the current time period ID). It *may* return an error symbol $\perp$ whenever it can demonstrate that EncSK' is not well formed.

Update: The adversary can request that the challenger update the clock to a new time ID' > ID, of the adversary's choice.

The adversary can repeatedly make Sign and Update queries, and at some point will choose to move to the next phase.

Final Forgery. At last, the adversary produces a time, message, and signature tuple (ID*, M^*, S^*). The adversary is successful if the signature verifies for time ID*, and the adversary had not queried for a signature on M^* at time period ID*.

We let $\mathsf{AdvUS}_\mathcal{A}$ denote the advantage of an algorithm $\mathcal{A}$ in the update security game.

The "encrypted" signing key EncSK need not be encrypted in the traditional semantically secure sense. The only requirement is that the key be blinded or rendered inoperative in such a way that no adversary can gain a non-negligible advantage in the update security game. In other words, the encrypted signing key EncSK should be useless by itself to produce signatures.

The decryption key DecK is viewed in this model as being output by the key generation algorithm. In Section 7 we will address the issue of letting DecK be derived from a user password, as will often be done in practice.

4. BACKGROUND

Before describing our scheme, we briefly review a few notions.

4.1 Bilinear Groups and Pairings

We review the usual notions of bilinear groups and bilinear maps defined over them [13, 29]. We use a multiplicative notation for the group operations. For simplicity, we restrict our attention to "symmetric" bilinear maps, while noting that our constructions can be advantageously generalized to use asymmetric pairings.

Let $\mathbb{G}$ and $\mathbb{G}_t$ be two cyclic groups of prime order p, and let g be a generator of $\mathbb{G}$. A symmetric bilinear map over $\mathbb{G}$ is a non-constant function $e : \mathbb{G} \times \mathbb{G} \to \mathbb{G}_t$ such that $e(u^a, v^b) = e(u, v)^{ab}$ for all $u, v \in \mathbb{G}$ and all $a, b \in \mathbb{Z}$. We say that an (infinite) family of groups $\mathbb{G}$ with these properties forms a bilinear group family if the group operation and the bilinear map admit $O(\text{poly} \log |p|)$-time algorithms. It is common to refer to $\log |p|$, or an appropriately rounded multiple thereof, as the security parameter.

4.2 Computational Complexity Assumptions

Many complexity assumptions have been proposed in the context of bilinear pairings. In this paper, we make use of the Computational Diffie-Hellman assumption in bilinear groups (CDH), and the Bilinear Diffie-Hellman Inversion assumption (BDHI).

The CDH assumption in a bilinear group $\mathbb{G}$ is very similar to the familiar CDH assumption: given group elements $g, g^a, g^b \in \mathbb{G}$, it assumes that it is infeasible to compute $g^{ab} \in \mathbb{G}$. One important distinction, however, is that in a bilinear group the corresponding DDH problem is easy: given $g, g^a, g^b, Z \in \mathbb{G}$ we can tell whether $Z = g^{ab}$ by testing whether the equality $e(g^a, g^b) = e(g, Z)$ holds in $\mathbb{G}_t$. The CDH assumption in a bilinear group thus makes a stronger statement than it does classically.

The BDHI assumption in a bilinear group $\mathbb{G}$ originates from [27, 5]. Given a parameter $\ell \geq 1$, the ℓ-BDHI assumption in $\mathbb{G}$ states the following: given $g, g^\alpha, g^{\alpha^2}, \ldots, g^{\alpha^\ell} \in \mathbb{G}$, it is infeasible to compute $e(g, g)^{1/\alpha} \in \mathbb{G}_t$.

DEFINITION 4.1. *We say that the ℓ-BDHI assumption holds in $\mathbb{G}$ if no efficient algorithm can solve a random instance with non-negligible probability. For $i = 1, \ldots, \ell$, let $g_i = g^{(\alpha^i)} \in \mathbb{G}$. An algorithm $\mathcal{A}$ has an advantage ϵ in solving the ℓ-BDHI problem in $\mathbb{G}$ if*

$$\Pr\left[\mathcal{A}(g, g_1, \ldots, g_\ell) = e(g, g)^{(\alpha^{-1})}\right] \geq \epsilon \ .$$

The probability is over the random choice of $g \in \mathbb{G}$ and α in $\mathbb{Z}_p$, and the random bits used by $\mathcal{A}$. We then say that the computational (t, ϵ, ℓ)-BDHI assumption holds in $\mathbb{G}$ if no t-time algorithm has advantage at least ϵ in solving a random instance of the ℓ-BDHI problem in $\mathbb{G}$.

Clearly, for all ℓ, the $(\ell+1)$-BDHI assumption is at least as strong as the ℓ-BDHI assumption. In addition, the 1-BDHI assumption is itself at least as strong as the CDH assumption in $\mathbb{G}$. We shall still mention and use the CDH assumption in $\mathbb{G}$ as it makes certain proofs clearer.

5. CONSTRUCTION

It is instructive to first understand the intuition behind the forward-secure signature scheme without the untrusted update property. The scheme is roughly based on a hierarchical identity-based encryption (HIBE) structure [17, 14]. We use a similar approach to Canetti, Halevi, and Katz [8] in that we use a binary tree hierarchy to represent a discrete notion of time: we map the leaves of the tree to the corresponding time periods. A secret key holder will store the private keys material for a set of at most ℓ nodes, from which the private keys for the current and all future time periods can be derived, where ℓ is the height or depth of the binary tree.

To sign a message, the signer will use the private key of the current time period. By applying the compression techniques of Boneh, Boyen, and Goh [6], this private key can be expressed in two group elements, even though it is for an ℓ-level identity. The signature is obtained by appending to the hierarchy a final "identity" that is dependent on the message; we use the Waters hash [33] at this level to obtain existential unforgeability.

By combining all of these ideas we can obtain a forward-secure scheme with constant size signatures that is provably secure without random oracles. To obtain the untrusted update property, at key creation time we simply multiply the initial private keys by a second factor "decryption key" DecK, which we will assume to be a secret group element. (In practice, one can hash a secret bitstring to obtain the group element.) Since the private keys are blinded by DecK, an attacker with access to the private storage will not be able to sign messages. However, the user can divide out DecK and recover the true private keys of the original scheme outlined above in order to sign messages. Finally, we observe that the update procedure of our particular HIBE-based signature scheme will produce private key nodes that are still a factor of DecK away from a "real" private key. The fact that this remains consistent during an HIBE *Derive* (i.e., key delegation) procedure is what allows an untrusted entity perform an update.

We now give a detailed construction of our scheme and then state our formal theorems of security.

5.1 Scheme

Let $\mathbb{G}$ be a bilinear group of prime order p, and let $e : \mathbb{G} \times \mathbb{G} \to \mathbb{G}_t$ be a symmetric bilinear map.

Messages to be signed are taken as fixed-length binary strings of m bits, for simplicity. Furthermore, for certain proofs (such as of the Update Security property), it is convenient to forbid that signatures on the same message be ever issued at two different time periods. One simple way to enforce this restriction is to embed the time period within the m bits to be signed, for example, as $M = \text{Time} \| \text{Msg}$ (with the drawback of reducing the available space for the message itself). Another way is to hash the message together with the time using a collision-resistant hash function $H : \{0, 1\}^* \to \{0, 1\}^m$, and sign the result $M = H(\text{Time} \| \text{Msg})$. Yet another way involves a universal one-way hash family $H : \{0, 1\}^{m'} \times \{0, 1\}^* \to \{0, 1\}^{m''}$ such that $m' + m'' = m$; the signer would pick a random index $R \in \{0, 1\}^{m'}$ into the UOWHF family, and then sign the concatenation $M = R\|H(R, \text{Time}\|\text{Msg})$.

Using a hash function such as these also has the advantage of letting us sign arbitrarily long messages. In the description that follows, we shall keep all this in mind though we treat M as an m-bit fixed-length binary string.

Time is modeled discretely as a sequence of 2^ℓ atomic time periods, arranged as the leaves of a binary tree of depth ℓ, in chronological order. A time period is identified by an ℓ-bit integer $\mathsf{ID} = (I_1, \ldots, I_\ell) \in \{0,1\}^\ell$ (though for convenience we exclude the zero-th period 0^ℓ and let the first period be represented by $0^{\ell-1}1$). The bits I_1 to I_ℓ are ordered from the top to the bottom of the tree, while a 0 and a 1 respectively indicate the first and second branch in the order of traversal. It follows that the traversal of the leaves, the chronology of the time periods, and the numerical values of their identifiers, all obey the same ordering.

For $j = 1, \ldots, \ell+1$, we define a time period's "second sibling at depth j", as

$$\mathsf{sibling}(j, \mathsf{ID}) = \begin{cases} (I_1, \ldots, I_{j-1}, 1) & \text{if } j < \ell+1 \text{ and } I_j = 0, \\ \bot & \text{if } j < \ell+1 \text{ and } I_j = 1, \\ \mathsf{ID} & \text{if } j = \ell+1. \end{cases}$$

A second sibling at depth j is either $\bot$, or a j-bit string that is never a prefix of ID — except that for notational convenience we pose $\mathsf{sibling}(\ell+1, \mathsf{ID}) = \mathsf{ID}$ at the fictitious depth $\ell+1$.

Last, we let $\mathsf{bit}(i, S)$ denote the i-th bit from the string $S \in \{0,1\}^*$, that is, for $1 \leq i \leq n$ and $S = (s_1, s_2 \ldots, s_n) \in \{0,1\}^n$, we have $\mathsf{bit}(i, S) = s_i \in \{0,1\}$.

Our forward-secure signature scheme with untrusted updates works as follows:

KeyGen(ℓ, m): Let 2^ℓ be the number of time periods, and $\{0,1\}^m$ the message space. The generation of a random initial set of keys proceeds as follows. Select two random integers $\nu, \omega \in \mathbb{Z}_p$, and a few random group elements $g, h_0, h_1, \ldots, h_\ell, f_0, f_1, \ldots, f_m \in \mathbb{G}$. We fix $\mathsf{ID} = 0^{\ell-1}1$.

For each $j = 1, \ldots, \ell+1$, we let $k_j = \mathsf{sibling}(j, \mathsf{ID})$. Specifically, for the initial value of ID, we have

$$(k_1, k_2, \ldots, k_\ell, k_{\ell+1}) = (1, 01, \ldots, 0^{\ell-2}1, \bot, 0^{\ell-1}1).$$

For each k_j, a private key component K_j is computed as follows. If $k_j = \bot$, then $K_j = \bot$. Otherwise, pick a random integer $r_j \in \mathbb{Z}_p$, and let

$$K_j = \left(g^{\nu+\omega} \cdot \left(h_0 \cdot \underbrace{h_1^{\mathsf{bit}(1,k_j)} \cdot \ldots \cdot h_j^{\mathsf{bit}(j,k_j)}}_{|k_j| \text{ factors}}\right)^{r_j}, g^{r_j},\right.$$

$$\left.\underbrace{h_{j+1}^{r_j}, \ldots, h_\ell^{r_j}}_{\ell - |k_j| \text{ components}}\right).$$

Since for $j < \ell$ we have $k_j = 0^{j-1}1$, we obtain $K_j = \left(g^{\nu+\omega} \cdot (h_0 \cdot h_j)^{r_j}, g^{r_j}, h_{j+1}^{r_j}, \ldots, h_\ell^{r_j}\right)$. For $j = \ell$, we get $K_\ell = \bot$. For $j = \ell+1$, we end up with $K_{\ell+1} = \left(g^{\nu+\omega} \cdot (h_0 \cdot h_\ell)^{r_{\ell+1}}, g^{r_{\ell+1}}\right)$ by letting $h_{\ell+1}^0 = 1$ in the general expression above even though $h_{\ell+1}$ is not defined.

The encrypted signing key for period $\mathsf{ID} = 0^{\ell-1}1$ and the "second factor" decryption key are

$$\mathsf{EncSK}_{0^{\ell-1}1} = \left(\mathsf{ID}, K_1, K_2, \ldots, K_{\ell-1}, \bot, K_{\ell+1}\right),$$
$$\mathsf{DecK} = g^{-\omega}.$$

Also compute $V = e(g,g)^\nu$ and $W = e(g,g)^\omega$. The public verification key is given by

$$\mathsf{VK} = \left(g, V, W, h_0, h_1, \ldots, h_\ell, f_0, f_1, \ldots, f_m\right).$$

CheckKey$(\mathsf{EncSK}_\mathsf{ID}, \mathsf{VK})$: To verify that an encrypted key $\mathsf{EncSK}_\mathsf{ID}$ is valid for a public key VK, proceed as follows.

Parse $\mathsf{EncSK}_\mathsf{ID}$ as the tuple $(\mathsf{ID}, K_1, \ldots, K_\ell, K_{\ell+1})$. For $j = 1, \ldots, \ell+1$, let $k_j = \mathsf{sibling}(j, \mathsf{ID})$, and check that $k_j = \bot$ if and only if $K_j = \bot$. Then, for each j such that $k_j \neq \bot$, parse K_j as $(a_0, a_1, b_{j+1}, \ldots, b_\ell)$, verify that $1 = V \cdot W \cdot e(a_0, g^{-1}) \cdot e(a_1, h_0 \prod_{i=1}^j h_i^{\mathsf{bit}(i,k_j)})$ in $\mathbb{G}_t$ and that $\forall i = j+1, \ldots, \ell : e(a_1, h_i) = e(b_i, g)$ in $\mathbb{G}_t$. If all equalities are verified, output $\mathtt{valid}$, otherwise output $\mathtt{invalid}$.

Update$(\mathsf{EncSK}_\mathsf{ID}, \mathsf{ID}', \mathsf{VK})$: To update an encrypted signing key $\mathsf{EncSK}_\mathsf{ID}$ from time period ID to time period ID', given the verification key VK (but not the decryption key), proceed as follows. To start, parse $\mathsf{EncSK}_\mathsf{ID}$ as $(\mathsf{ID}, K_1, \ldots, K_\ell, K_{\ell+1})$, and ascertain that $0^\ell < \mathsf{ID} \leq \mathsf{ID}' \in \{0,1\}^\ell$.

Let j, j' denote indices in the range $1, \ldots, \ell+1$. Let $k_j = \mathsf{sibling}(j, \mathsf{ID})$ and $k'_{j'} = \mathsf{sibling}(j', \mathsf{ID}')$. By construction, each non-$\bot$ string $k'_{j'}$ contains exactly one of the strings k_j as a prefix. Formally: $\forall j' \in \{1, \ldots, \ell+1\}$, either $k'_{j'} = \bot$, or $\exists! j \in \{1, \ldots, j'\}$ s.t. $k'_{j'} = k_j \| s$ for some string $s \in \{0,1\}^{|k'_{j'}|-|k_j|}$.

For all $j' = 1, \ldots, \ell+1$, we construct the j'-th component $K'_{j'}$ of the updated key as follows. If $k'_{j'} = \bot$, then set $K'_{j'} = \bot$. If $k'_{j'} = k_j$, then set $K'_{j'} = K_j$. Otherwise, determine the index $j < j'$ such that $k'_{j'} = k_j \| s$ for some suffix string s, parse the j-th component of $\mathsf{EncSK}_\mathsf{ID}$ as $K_j = (a_0, a_1, b_{j+1}, \ldots, b_\ell)$, pick a fresh random $r_{j'} \in \mathbb{Z}_p$, and set

$$K'_{j'} = (a'_0, a'_1, b'_{j'+1}, \ldots, b'_\ell)$$
$$= \left(a_0 \cdot \left(\underbrace{b_{j+1}^{\mathsf{bit}(j+1, k'_{j'})} \cdot \ldots \cdot b_{j'}^{\mathsf{bit}(j', k'_{j'})}}_{|k'_{j'}| - |k_j| \text{ factors}}\right)\right.$$
$$\left. \cdot \left(h_0 \cdot \underbrace{h_1^{\mathsf{bit}(1, k'_{j'})} \cdot \ldots \cdot h_{j'}^{\mathsf{bit}(j', k'_{j'})}}_{|k'_{j'}| \text{ factors}}\right)^{r_{j'}},\right.$$
$$\left. a_1 \cdot g^{r_{j'}}, \; b_{j'+1} \cdot h_{j'+1}^{r_{j'}}, \; \ldots, \; b_\ell \cdot h_\ell^{r_{j'}}\right).$$

Once all the $K'_{j'}$ have been computed, output the new encrypted key for period ID' as

$$\mathsf{EncSK}_{\mathsf{ID}'} = \left(\mathsf{ID}', K'_1, \ldots, K'_{\ell+1}\right).$$

Sign$(\mathsf{EncSK}_\mathsf{ID}, \mathsf{DecK}, \mathsf{VK})$: To sign a message $M \in \{0,1\}^m$ using the encrypted signing key $\mathsf{EncSK}_\mathsf{ID}$ and the decryption key DecK, proceed as follows.

Parse $\mathsf{EncSK_{ID}} = (\mathsf{ID}, K_1, \ldots, K_\ell, K_{\ell+1})$, and then parse $K_{\ell+1} = (a_0, a_1) \neq \bot$. The elements a_0 and a_1 and the time period ID are all that we need from $\mathsf{EncSK_{ID}}$. At this point, if the key is not trusted (e.g., being the result of an untrusted update), the signer needs to ensure that $1 = V \cdot W \cdot e(a_0, g^{-1}) \cdot e(a_1, h_0 \prod_{i=1}^{\ell} h_i^{\mathsf{bit}(i,\mathsf{ID})})$ in $\mathbb{G}_t$. If this test fails, output $\bot$ and halt.

Otherwise, to produce a signature, pick two random integers $r, s \in \mathbb{Z}_p$, and output

$$S_{\mathsf{ID}}(M) = \Bigg(\mathsf{ID},$$
$$\mathsf{DecK} \cdot a_0 \cdot \Big(h_0 \cdot \prod_{i=1}^{\ell} h_i^{\mathsf{bit}(i,\mathsf{ID})}\Big)^r \cdot \Big(f_0 \cdot \prod_{j=1}^{m} f_j^{\mathsf{bit}(j,M)}\Big)^s,$$
$$a_1 \cdot g^r, \quad g^s \Bigg).$$

Verify(S, M, VK): To verify a signature $S = (\mathsf{ID}, s_0, s_1, s_2)$ on a message $M \in \{0,1\}^m$ with respect to a verification key VK, it suffices to check the following equality in $\mathbb{G}_t$:

$$1 = V \cdot e(s_0, g^{-1}) \cdot e\Big(s_1, h_0 \prod_{i=1}^{\ell} h_i^{\mathsf{bit}(i,\mathsf{ID})}\Big)$$
$$\cdot e\Big(s_2, f_0 \prod_{j=1}^{m} f_j^{\mathsf{bit}(j,M)}\Big).$$

Output `valid` if the equality holds, and `invalid` if it does not.

Observe that for any time period, the signature contains only three group elements in addition to the time period identifier ID. Signature verification is also fairly fast as it requires only three pairings.

Validating Untrusted Keys

The test procedure *CheckKey* serves to completely validate an encrypted signing key $\mathsf{EncSK_{ID}}$. Without this check, certain corruptions of $\mathsf{EncSK_{ID}}$ may not be immediately apparent, but will creep up in the update process, and eventually surface as we try to sign at some future time ID'. The basic check in the *Sign* algorithm is sufficient to prevent a signature from being created incorrectly, and thus *CheckKey* is superfluous in the formal proof of security. In practice, however, *CheckKey* is a useful discretionary test that should be performed before overwriting an old key with an new key from an untrusted source.

We also note that, as written, *CheckKey* implicitly assumes that the checking algorithm has an uncorrupted version of the public key VK when validating EncSK, in addition to DecK. This is potentially problematic if an attacker controlling the key storage can corrupt VK as well as EncSK. However, in practice we can protect against this attack by having the key DecK include a MAC key in addition to the second factor decryption key. The MAC key will be used by *KeyGen* to append an authentication code to the public key VK, and later by *CheckKey* to check the integrity of VK against that code. For simplicity we stick to the original model in the formal proofs.

Re-randomization Issues

For performance reasons, the *Update* procedure does not fully re-randomize the encrypted key $\mathsf{EncSK_{ID}}$ upon all invocations. In particular, running a "zero-step" update of the form *Update*($\mathsf{EncSK_{ID}}, \mathsf{ID}', \mathsf{VK}$) with $\mathsf{ID}' = \mathsf{ID}$ simply outputs the given key, unaltered. We could have designed *Update* to recompute each component $K'_{j'}$, even in the case where $k'_{j'} = k_j \neq \bot$, causing the key to be fully re-randomized no matter how small the update. However, the selective re-randomization strategy gives us amortized constant time for single-step updates, which the indiscriminate strategy does not.

In the *Sign* procedure, we re-randomize the two group elements from $\mathsf{EncSK_{ID}}$ that intervene in the signing process, before each signature. This ensures that the signatures are jointly uniformly distributed (over the space of valid signatures for given times and messages), which helps us keep the security proofs reasonably simple.

Very Fine Time Granularities

Our update algorithm is general in the sense that it lets us jump to any time period in the future in a single operation, as opposed to the next period only. The possibility of making large jumps greatly simplifies the updating task in the case where many time periods have elapsed since the last update. In turn, this makes it possible to have an extremely fine-grained discretization of time (such as $1\mu s$ periods over a 10-year span), without significant performance degradation.

Achieving Constant Update Time

Another desirable feature to have is a true constant-time update when the time increment is 1, rather than the *amortized* constant-time we currently have. Canetti, Halevi, and Katz [8] show a tree-based method for achieving constant-time updates by associating interior nodes with time periods and doing an in-order traversal through the tree. We could apply the same method; however, we choose to use a simpler structure to better expose the novel features of our scheme. The CHK method also loses its advantage over the simpler approach under fine-grained discretizations of time—where updates are to cover many *micro-periods* at once—, which we expect to be more important in practice.

5.2 Security

We now state our two main security theorems. We refer the reader to the full version of the paper for proofs.

THEOREM 5.1. *Let $\mathcal{A}$ be an adversary that produces an existential forgery, in the forward security attack model, against the signature scheme instantiated for m-bit messages and T time periods. Assume that $\mathcal{A}$ makes no more than q queries, and succeeds with probability ϵ in time τ. Then there exists an algorithm $\mathcal{B}$ that solves the ℓ-BDHI problem in $\mathbb{G}$ in time $\tilde{\tau} \approx \tau$ with success probability $\tilde{\epsilon} \geq \epsilon/(4mqT)$.*

THEOREM 5.2. *Let $\mathcal{A}$ be an adversary that produces an existential forgery, in the update security attack model, against the forward secure signature scheme instantiated to accept m-bit messages. Assume that $\mathcal{A}$ makes no more than q queries, and succeeds with probability ϵ in time τ. Then there exists an algorithm $\mathcal{B}$ that solves the CDH problem in $\mathbb{G}$ in time $\tilde{\tau} \approx \tau$ with success probability $\tilde{\epsilon} \geq \epsilon/(4mq)$.*

6. ADDING UNTRUSTED UPDATE TO THE BELLARE-MINER SCHEME

In the previous section we achieved untrusted update from a new forward-secure signature scheme derived from HIBE. Our primary leverage for achieving untrusted update was the particular algebraic structure of the underlying scheme. This leads to the natural question of whether other existing number-theoretic forward-secure signature schemes have similar properties and can be modified to achieve untrusted update.

In this section we describe how to modify the Bellare-Miner [4] scheme to achieve untrusted update. We give short, intuitive descriptions of the schemes and proofs, and omit the details.

We briefly describe the number theoretic scheme of Bellare and Miner [4]. The key generation algorithm chooses a Blum-Williams integer $N = pq$ and publishes the public key as N and ℓ random points $U_1, \ldots, U_\ell \in \mathbb{Z}_N$, where ℓ depends on the security parameter of the scheme. If the scheme has T total time periods, the initial private keys for time period 1 will be $S_{1,1}, \ldots, S_{\ell,1} \in \mathbb{Z}_N$, where $S_{i,1}$ is the 2^{T+1}-th root of U_i. The factorization is discarded after key generation.

Key update is done simply by squaring each value of the secret key. The private keys for time period $j+1$ are computed as $S_{i,j+1} = S_{i,j}^2$ for all $1 \le i \le \ell$. To sign a message at a time period t, the signer proves non-interactively that he has the 2^{T+2-t}-th roots for all U_i using Fiat-Shamir [12] heuristic techniques in the random oracle model.

Adding untrusted update is rather straightforward. Let the second factor DecK be ℓ elements $\text{DecK}_1, \ldots, \text{DecK}_\ell$ of $\mathbb{Z}_N$. (In practice these can be generated by applying a hash function modeled as a random oracle to a shorter secret.) The new initial secret keys are constructed by multiplying in the blinding factors as $S'_{1,i} = S_{1,i}\text{DecK}_i$ for all $1 \le i \le \ell$, where $S_{1,1}, \ldots, S_{\ell,1}$ are the initial secret keys from the original scheme. The key update algorithm is the same; it simply squares each component of the secret key. At time period t, the secret key components will be $S'_{t,i} = S_{t,i}\text{DecK}^{2^{t-1}}$. To recover the private key component of the original scheme $S_{t,i}$, the signer simply needs to compute $\text{DecK}^{2^{t-1}}$ and divide it out from $S'_{t,i}$; after this step the signer can sign the message as before. One drawback is that the signing algorithm will need to perform a number of squarings that is linear in ℓT. We can informally, argue that the scheme is secure against untrusted storage since the stored private keys will be just random group elements in $\mathbb{Z}_n$.

It would be interesting to see untrusted update added to other signature schemes. In particular the Itkis-Reyzin [19] scheme is of particular interest due to its desirable performance parameters. However, the straightforward methods to add untrusted update remove the performance benefits that made the scheme interesting to begin with. We leave the addition of untrusted update to other existing forward secure signature schemes (without significantly degrading performance) as future work.

7. IMPLEMENTATION & APPLICATIONS

We present the implementation of our signature scheme. We first describe the core functionality and the interface to our implementation. Then we discuss our performance measurements. Finally, we describe some issues that arise in integrating our code with security application programs.

7.1 Software Implementation

We describe the API to our forward-secure signature functionality and report timing numbers. For the elliptic curve operations underlying our crypto code, we used Ben Lynn's PBC library [24]. PBC uses the GMP library [30] for its bignums and other math code. We expect to make the source for our library available under a GPL-compatible license.

7.1.1 Interface

We describe our API and explain some of the choices we made. In addition to the core functions we describe here, there are, of course, routines for such mundane operations as reading keys from and writing them to disk.

```
void    fs_gen_sys_param( fs_sys_param_t  param,
                          pairing_t       pairing );

void    fs_gen( fs_public_key_t   pk,
                fs_private_key_t  sk,
                fs_dec_key_t      dk,
                fs_sys_param_t    param,
                unsigned int      msg_len_bits,
                unsigned int      num_intervals );
```

Key generation first requires system parameter generation. It is expected that applications will ship with preselected system parameters, so that users need not generate their own. The actual key generation algorithm outputs a public key pk, an encrypted secret key sk, and a second-factor dk that can be used to decrypt sk for the signing operation. We expect that applications will use a password obtained from the user to encrypt the second-factor dk. The algorithm takes several arguments: the system parameters generated earlier; the length of messages to be signed, typically 160 and the output of a collision-resistant hash function like SHA-1; and the number of intervals T over which the key can evolve.

```
int    get_time_from_sk( fs_private_key_t  sk );

int    fs_check_key( fs_private_key_t  sk,
                     fs_public_key_t   pk );

void   fs_update( fs_private_key_t  sk,
                  fs_public_key_t   pk );
```

Three functions are called for managing time periods. None of these requires the second factor dk to operate, so they can all be called without the user's involvement. The first, get_time_from_sk, returns the time period to which the (encrypted) private key has been updated. The second, fs_check_key, checks that the private key is indeed valid for signing and has not been tampered with. The idea is that the application uses this function as a self-test before requiring the user to enter a password so that she can sign a message. The third, fs_update, updates the key by a single time period. It can, obviously be called repeatedly to update the key to an arbitrary period.

```
void    fs_sign( unsigned char     *sig,
                 unsigned char     *msg,
                 unsigned int      msg_len_bits,
                 fs_public_key_t   pk,
```

Table 1: Times, in seconds, of forward-secure signing operations, for various total numbers of time intervals.

Operation	$T = 16$	$T = 64$	$T = 256$
Key generation	37.365	38.636	41.026
Signature generation	0.134	0.135	0.137
Signature verification	0.328	0.333	0.341
Key update (1 step)	0.109	0.168	0.202
Key validity check	0.205	0.208	0.214

Table 2: Times, in seconds, of forward-secure signing operations, for large total numbers of intervals. Times are averaged over the first 1024 time intervals.

Operation	$T = 2^{10}$	$T = 2^{20}$	$T = 2^{30}$
Key generation	41.99	51.23	66.98
Signature generation	0.14	0.14	0.14
Signature verification	0.34	0.33	0.33
Key update (1 step)	0.22	0.22	0.23
Key validity check	0.21	0.21	0.21

```
                 fs_private_key_t   sk,
                 fs_dec_key_t       dk );

int   fs_verify( unsigned char     *sig,
                 unsigned char     *msg,
                 unsigned int      msg_len_bits,
                 int               time,
                 fs_public_key_t   pk );
```

Finally, two functions are provided for signing and verifying messages. The signing operation, `fs_sign`, transforms a message in `msg` to a signature in `sig`; the verification operation `fs_verify` checks that `sig` is a correct signature on `msg`. The signing operation requires the second-factor key `dk` along with the (encrypted) secret key `sk`; the verification algorithm obviously, requires only the public key `pk`.

The signing function generates a signature for the present time period, so it does not take as input the time period to use. An application should make sure, before signing, that it has updated the key to the correct time period. This check could be performed within `fs_sign` itself, but it could then be the case that the application and the library have different ideas of what the present time period should be.

Our verification function, on the other hand, takes a time period j. It is the application's job to determine what time period it expects the signature to be from. As with the signing algorithm, the intention is that all clock-to-time-period conversion be handled by a single location in the code to avoid what is (more literally here than usually) a time-of-check–time-of-use error. The algorithm required for this conversion is quite simple for most applications, doing simple arithmetic on the start and end times for the key's validity and the number T of time periods through which it evolves, along with the time at which the signature was generated. Some applications might impose additional constraints, however, so the code should be implemented in the application to ensure a consistent answer.

7.1.2 Timing

In Table 1, we present timing numbers for the basic operations we expose. We test our code for several choices of total time intervals: $T \in \{16, 64, 256\}$. The setup uses an MNT curve chosen for 1024-bit security. We sign 160-bit messages. To smooth out the timing, we ran the following procedure five times: generate key, then, for each of the T time periods, verify that the key is valid, sign and verify a message and update to the next period. Measurements were taken on a 2.8 GHz Pentium IV machine with 512 MB RAM, running OpenBSD 3.8.

In additon, in Table 2, we present timing numbers for our system when used with a large number of intervals: $T \in \{2^{10}, 2^{20}, 2^{30}\}$. To generate this table we followed the same procedure as above, but we ran the timing procedure only once and averaged the single-step key update time only through the first 1024 intervals.

All operations except key generation take less than one second to complete. We note that efficient pairing computation is an active research area, and new algorithms are likely to decrease the times we see.

7.2 Practical Considerations

We describe some details that must be considered when integrating our forward-secure signature code with an application.

7.2.1 Storing the Second Factor

In our implementation, the second factor key `dk` is generated uniformly at random in the course of the `fs_gen` function. It is expected that the application encrypt this second factor on disk by means of a user password. In our analysis of update security, we expected that `dk` is kept secret from the adversary. Thus if a password is used to encrypt `dk`, it is important that this password contain sufficient entropy to deter offline password guessing. The same caveat, of course, applies to any private key stored encrypted with a password. If a better source of entropy is available, the application can make use of it.

An alternative design strategy would have allowed the application to supply the second-factor key ω on its own, rather than having ω randomly chosen by `fs_gen`. The application could then generate ω using a user-supplied password. The resulting scheme would be somewhat more efficient. However, it would require tighter integration between our code's representation of bignums and the application's routines for extracting entropy from passwords, a requirement we deemed inadvisable.

7.2.2 Key Storage and Unattended Updates

The natural and correct choice for implementing automatic updates is for the application to set up a `cron` job on the user's behalf that updates the key at the appropriate interval. The cron job can check if a key update has been skipped (for example, because the system was powered down) and apply the updates necessary to bring the key to the current time period. An application making use of our library can also check whether updates have been missed when it is run by the user.

The security guarantees of our scheme mean that it is also possible for the encrypted private key to be stored on a remote server whose job it is to update it at every interval.

The danger of this approach is that if the server retains old versions of the secret key then a compromise of the second factor will cause these old versions to be revealed, not just the version corresponding to the present time period.

7.2.3 Mapping Times to Time Periods

The security of a forward-secure signature scheme relies crucially on a proper mapping of signatures to the time period in which they were generated. For example, consider a backdated signature that claims, "I was made in time period j," but was in fact made in time period $j' > j$, and verifies only with the public key for period j'. An application that accepts this signature will violate the forward-security semantics, though no cryptographic flaw exists. As with much about signatures, the semantics of signature verification are what is important, but also what is hard to pin down; cf. Laurie and Bohm [23].

It is also important, but less so, that the signer correctly calculate the the appropriate time period with which to generate a signature. In the worst case, the resulting signature will not verify, since it will be deemed to have been backdated. The calculations are also easier for the signer, since it relies on the current time as reported by the OS (or a clock server), not the signing time listed in a maliciously generated signature. The secret signing key should be kept updated on disk, of course.

An upshot of this is that an application that verifies signatures from untrusted sources must be sure to make the user aware of the time in which it understands the signature to have been generated and, ideally, also the start and end of that time period. This is an extension of the backdating attack mentioned earlier. Even if a signature that claims to have been issued in time period j and was in fact issued then, the contents of the message might lead the user astray. For example, suppose document signing functionality with forward security is added to a word processing application. If the displays a checkmark next to the document to indicate that the signature was valid, an attacker who compromises Alice's signing key today can create a signature (correctly using the current time period) on a document that, in its body, prominently lists a date in the past. Now Bob, opening the document, will be tricked into believing that Alice generated it earlier, and might not discount it even if he hears from Alice that her key was recently compromised.

8. CONCLUSION

We introduced the notion of forward-secure signatures with untrusted update. With these signatures, private keys can be updated forward in time as they are kept in encrypted form, without first requiring decryption. This allows practical applications such as GPG to adopt the benefits of forward security, without forgoing the practice of encrypting the private keys under a second factor such as a user password.

We presented and proved secure a very efficient construction of forward-secure signatures with untrusted update, based on pairings. We also showed how to retrofit untrusted update into some existing forward-secure schemes, based on factoring. To validate the concept, we implemented our main (pairing-based) construction using the open-source GMP and PBC libraries, and obtained performance measurements.

9. REFERENCES

[1] Michel Abdalla, Sara K. Miner, and Chanathip Namprempre. Forward-secure threshold signature schemes. In *CT-RSA*, pages 441–456, 2001.

[2] Michel Abdalla and Leonid Reyzin. A new forward-secure digital signature scheme. In *ASIACRYPT*, pages 116–129, London, UK, 2000. Springer-Verlag.

[3] Ross Anderson. Invited Lecture. 4th ACM Computer and Communications Security, 1997.

[4] Mihir Bellare and Sara K. Miner. A forward-secure digital signature scheme. In *CRYPTO*, pages 431–448, 1999.

[5] Dan Boneh and Xavier Boyen. Efficient selective-id secure identity-based encryption without random oracles. In *EUROCRYPT*, pages 223–238, 2004.

[6] Dan Boneh, Xavier Boyen, and Eu-Jin Goh. Hierarchical identity based encryption with constant size ciphertext. In *EUROCRYPT*, volume 3494 of *Lecture Notes in Computer Science*, pages 440–456. Berlin: Springer-Verlag, 2005.

[7] Dan Boneh and Matthew K. Franklin. Identity-based encryption from the Weil pairing. *SIAM J. Computing*, 32(3):586–615, 2003.

[8] Ran Canetti, Shai Halevi, and Jonathan Katz. A forward-secure public-key encryption scheme. In *EUROCRYPT*, pages 255–271, 2003.

[9] Eric Cronin, Sugih Jamin, Tal Malkin, and Patrick McDaniel. On the performance, feasibility, and use of forward-secure signatures. In *CCS*, pages 131–144, New York, NY, USA, 2003. ACM Press.

[10] Yevgeniy Dodis, Jonathan Katz, Shouhuai Xu, and Moti Yung. Key-insulated public key cryptosystems. In *EUROCRYPT*, pages 65–82, 2002.

[11] Yevgeniy Dodis, Jonathan Katz, Shouhuai Xu, and Moti Yung. Strong key-insulated signature schemes. In *Public Key Cryptography*, pages 130–144, 2003.

[12] Amos Fiat and Adi Shamir. How to prove yourself: Practical solutions to identification and signature problems. In *CRYPTO*, pages 186–194, 1986.

[13] Steven Galbraith. Pairings. In Ian F. Blake, Gadiel Seroussi, and Nigel Smart, editors, *Advances in Elliptic Curve Cryptography*, volume 317 of *London Mathematical Society Lecture Notes*, chapter IX, pages 183–213. Cambridge University Press, 2005.

[14] Craig Gentry and Alice Silverberg. Hierarchical id-based cryptography. In *ASIACRYPT*, pages 548–566, 2002.

[15] Louis C. Guillou and Jean-Jacques Quisquater. A "paradoxical" identity-based signature scheme resulting from zero-knowledge. In *CRYPTO*, pages 216–231, 1988.

[16] Christoph G. Günther. An identity-based key-exchange protocol. In *EUROCRYPT*, pages 29–37, 1989.

[17] Jeremy Horwitz and Ben Lynn. Toward hierarchical identity-based encryption. In *EUROCRYPT*, pages 466–481, 2002.

[18] Gene Itkis, Robert McNerney, and Scott Russell. Intrusion-resilient secure channels. In *ACNS*, pages 238–253, 2005.

[19] Gene Itkis and Leonid Reyzin. Forward-secure signatures with optimal signing and verifying. In *CRYPTO*, pages 332–354, London, UK, 2001. Springer-Verlag.

[20] Gene Itkis and Leonid Reyzin. Sibir: Signer-base intrusion-resilient signatures. In *CRYPTO*, pages 499–514, 2002.

[21] Anton Kozlov and Leonid Reyzin. Forward-secure signatures with fast key update. In *SCN*, pages 241–256, 2002.

[22] Hugo Krawczyk. Simple forward-secure signatures from any signature scheme. In *CCS*, pages 108–115, New York, NY, USA, 2000. ACM Press.

[23] Ben Laurie and Nicholas Bohm. Signatures: An interface between law and technology, January 2003. Online: http://www.apache-ssl.org/tech-legal.pdf.

[24] Ben Lynn. PBC library. Online: http://rooster.stanford.edu/~ben/pbc/.

[25] Tal Malkin, Daniele Micciancio, and Sara K. Miner. Efficient generic forward-secure signatures with an unbounded number of time periods. In *EUROCRYPT*, pages 400–417, 2002.

[26] Ralph C. Merkle. A digital signature based on a conventional encryption function. In *CRYPTO*, pages 369–378, 1987.

[27] Shigeo Mitsunari, Ryuichi Sakai, and Masao Kasahara. A new traitor tracing. *IEICE Transactions on Fundamentals*, E850A(2):481–484, 2002.

[28] H. Ong and Claus-Peter Schnorr. Fast signature generation with a fiat shamir-like scheme. In *EUROCRYPT*, pages 432–440, 1990.

[29] Kenneth Paterson. Cryptography from pairings. In Ian F. Blake, Gadiel Seroussi, and Nigel Smart, editors, *Advances in Elliptic Curve Cryptography*, volume 317 of *London Mathematical Society Lecture Notes*, chapter X, pages 215–51. Cambridge University Press, 2005.

[30] GMP Project. The Gnu multiprecision arithmetic library. Online: http://www.swox.com/gmp/.

[31] Adi Shamir. Identity-based cryptosystems and signature schemes. In *CRYPTO*, pages 47–53, 1984.

[32] Dawn Xiaodong Song. Practical forward secure group signature schemes. In *CCS*, pages 225–234, New York, NY, USA, 2001. ACM Press.

[33] Brent Waters. Efficient identity-based encryption without random oracles. In *EUROCRYPT*, pages 114–127, 2005.

[34] Philip R. Zimmermann. *The Official PGP User's Guide*. The MIT Press, 1995. ISBN 0-262-74017-6.

How to Win the Clone Wars: Efficient Periodic n-Times Anonymous Authentication

Jan Camenisch
Zurich Research Lab
IBM Research
jca@zurich.ibm.com

Susan Hohenberger
Zurich Research Lab
IBM Research
sus@zurich.ibm.com

Markulf Kohlweiss
Dept. of Electrical Engineering
Katholieke Universiteit Leuven
mkohlwei@esat.kuleuven.be

Anna Lysyanskaya
Computer Science Dept.
Brown University
anna@cs.brown.edu

Mira Meyerovich
Computer Science Dept.
Brown University
mira@cs.brown.edu

ABSTRACT

We create a credential system that lets a user anonymously authenticate at most n times in a single time period. A user withdraws a dispenser of n e-tokens. She shows an e-token to a verifier to authenticate herself; each e-token can be used only once, however, the dispenser automatically refreshes every time period. The only prior solution to this problem, due to Damgård et al. [29], uses protocols that are a factor of k slower for the user and verifier, where k is the security parameter. Damgård et al. also only support one authentication per time period, while we support n. Because our construction is based on e-cash, we can use existing techniques to identify a cheating user, trace all of her e-tokens, and revoke her dispensers. We also offer a new anonymity service: glitch protection for basically honest users who (occasionally) reuse e-tokens. The verifier can always recognize a reused e-token; however, we preserve the anonymity of users who do not reuse e-tokens too often.

Categories and Subject Descriptors: K.6.5 [**Security and Protection**]:Authentication.

General Terms: Security, Algorithms.

Keywords: n-anonymous authentication, clone detection, credentials.

1. INTRODUCTION

As computer devices get smaller and less intrusive, it becomes possible to place them everywhere and use them to collect information about their environment. For example, with today's technology, sensors mounted on vehicles may report to a central traffic service which parts of the roads are treacherous, thus assisting people in planning their commutes. Some have proposed mounting sensors in refrigerators to report the consumption statistics of a household, thus aiding in public health studies, or even mounting them in people's bodies in an attempt to aid medical science. In all these areas, better information may ultimately lead to a better quality of life.

Yet this vision appears to be incompatible with privacy. A sensor installed in a particular car will divulge that car's location, while one installed in a fridge will report the eating and drinking habits of its owner.

A naive solution would be to supply only the relevant information and nothing else.[1] A report about the road conditions should not say which sensor made the measurement. However, then nothing would stop a malicious party from supplying lots of false and misleading data. We need to authenticate the information reported by a sensor without divulging the sensor's identity. We also need a way to deal with rogue sensors, i.e., formerly honest sensors with valid cryptographic keys that are captured by a malicious adversary and used to send lots of misleading data.

The same problem arises in other scenarios. Consider an interactive computer game. Each player must have a license to participate, and prove this fact to an on-line authority every time she wishes to play. For privacy reasons, the player does not wish to reveal anything other than the fact that she has a license. How can we prevent a million users from playing the game for the price of just one license?

A suite of cryptographic primitives such as group signatures [27, 21, 1, 6] and anonymous credentials [25, 30, 39, 14, 16, 17] has been developed to let us prove that a piece of data comes from an authorized source without revealing the identity of that particular source. However, none of the results cited above provide a way to ensure anonymity and unlinkability of honest participants while at the same time guaranteeing that a rogue cannot undetectably provide misleading data in bulk. Indeed, it seems that the ability to provide false data is a consequence of anonymity.

---

[1] Divulging the relevant information alone may already constitute a breach of privacy. This has to do with statistical properties of the data itself. See Sweeney [46] and Chawla et al. [28] on the challenges of determining which data is and is not safe to reveal.

Recently Damgård, Dupont and Pedersen [29] presented a scheme that overcomes this seeming paradox. The goal is to allow an honest participant to anonymously and unlinkably submit data at a small rate (for example, reporting on road conditions once every fifteen minutes, or joining one game session every half an hour), and at the same time to have a way to identify participants that submit data more frequently. This limits the amount of false information a rogue sensor can provide or the number of times that a given software license can be used per time period.

While the work of Damgård et al. is the first step in the right direction, their approach yields a prohibitively expensive solution. To authenticate itself, a sensor acts as a prover in a zero-knowledge (ZK) proof of knowledge of a relevant certificate. In their construction, the zero-knowledge property crucially depends on the fact that the prover must make some random choices; should the prover ever re-use the random choices he made, the prover's secrets can be efficiently computed from the two transcripts. The sensor's random choices are a pseudorandom function of the current time period (which must be proven in an additional ZK proof protocol). If a rogue sensor tries to submit more data in the same time period, he will have to use the same randomness in the proof, thus exposing his identity. It is very challenging to instantiate this solution with efficient building blocks. Damgård et al. use the most efficient building blocks available, and also introduce some of their own; their scheme requires that the user perform $57+68k$ exponentiations to authenticate, where k is the security parameter (a sensor can cheat with probability 2^{-k}).

We provide a completely different approach that yields a practical, efficient, and provably secure solution. We relate the problem to electronic cash (e-cash) [23, 24] and in particular, to compact e-cash [12]. In our approach, each participant obtains a set of e-tokens from the central server. Similar to the withdrawal protocol of e-cash, the protocol through which a participant obtains these e-tokens does not reveal any information to the server about what these e-tokens actually look like. Our protocol lets a participant obtain all the e-tokens it will ever need in its lifetime in one efficient transaction. The user performs only 3 multi-base exponentiations to obtain e-tokens, and 35 multi-base exponentiations to show a single e-token. If the user is limited to one e-token per time period (as in the Damgård et al.'s scheme), the scheme can be further simplified and the user will need to do only 13 multi-base exponentiations to show an e-token. We provide more details on efficiency in §4.3.

Distributed sensors can use an e-token to anonymously authenticate the data they send to the central server. In the on-line game scenario, each e-token can be used to establish a new connection to the game. Unlike e-cash, where it is crucial to limit the amount of money withdrawn in each transaction, the number of e-tokens obtained by a participant is unlimited, and a participant can go on sending data or connecting to the game for as long as it needs. The e-tokens are anonymous and unlinkable to each other and to the protocol where they were obtained. However, the number of e-tokens that are valid during a particular time period *is* limited. Similarly to what happens in compact e-cash, reusing e-tokens leads to the identification of the rogue participant. We also show how to reveal all of its past and future transactions.

Thus, in the sensor scenario, a sensor cannot send more than a small number of data items per time period, so there is a limit to the amount of misleading data that a rogue sensor can submit. Should a rogue sensor attempt to do more, it will have to reuse some of its e-tokens, which will lead to the identification of itself and possibly all of its past and future transactions. Similarly, in the on-line game scenario, a license cannot be used more than a small number of times per day, and so it is impossible to share it widely.

OUR CONTRIBUTION Our main contribution is the new approach to the problem, described above, that is an order of magnitude more efficient than the solution of Damgård et al. In Section 4, we present our basic construction, which is based on previously-proposed complexity theoretic assumptions (SRSA and y-DDHI) and is secure in the plain model.

Our construction builds on prior work on anonymous credentials [14, 38], so that it is easy to see which parts need to be slightly modified, using standard techniques, to add additional features such as an anonymity revoking trustee, identity attributes, etc. The computational cost of these additional features is a few additional modular exponentiations per transaction.

In Section 5, we extend our basic solution to make it tolerate occasional glitches without disastrous consequences to the anonymity of a participant. Suppose that a sensor gets reset and does not realize that it has already sent in a measurement. This should not necessarily invalidate all of the sensor's data. It is sufficient for the data collection center to notice that it received two measurements from the same sensor, and act accordingly. It is, of course, desirable, that a sensor that has too many such glitches be discovered and replaced. Our solution allows us to be flexible in this respect, and tolerates m such glitches (where m is specified ahead of time as a system-wide parameter) at the additive cost of $O(km)$ in both efficiency and storage, where k is the security parameter. This does not add any extra computational or set-up assumptions to our basic scheme.

In Section 6, we consider more variations of our basic scheme. We show, also in the plain model, how to enable the issuer and verifiers to prove to third parties that a particular user has (excessively) reused e-tokens (this is called *weak exculpability*); and enable the issuer and verifiers to trace all e-tokens from the same dispenser as the one that was excessively reused (this is called *tracing*). We also show, in the common-parameters and random-oracle models, how to achieve *strong exculpability*, where the honest verifiers can prove to third parties that a user reused a particular e-token. Finally, we explain how e-token dispensers can be revoked; this requires a model where the revocation authority can continuously update the issuer's public key.

A NOTE ON TERMINOLOGY Damgård et al. call the problem at hand "unclonable group identification," meaning that, should a user make a copy of his sensor, the existence of such a clone will manifest itself when both sensors try to submit a piece of data in the same time period. We extend the problem, and call the extended version "periodic n-times anonymous authentication," because it is a technique that allows one to provide anonymous authentication up to n times during a given time period. For $n = 1$ (when there is only one e-token per user per time period) our scheme solves the same problem as the Damgård et al. scheme.

RELATED WORK Anonymity, conditional anonymity, and revocable anonymity, are heavily researched fields; due to space constraints, we compare ourselves only to the most relevant and the most recent work. Anonymous credentials allow one to prove that one has a set of credentials without revealing anything other than this fact. Revocable anonymity [27, 10, 18, 37] allows a trusted third party to discover the identity of all otherwise anonymous participants; it is not directly relevant to our efforts since we do not assume any such TTP, nor do we want anyone to discover the identity of honest users. Conditional anonymity requires that a user's transactions remain anonymous until some conditions are violated; our results fall within that category. With the exception of Damgård et al.'s work [29], no prior literature on conditional anonymity considered conditions of the form "at most n anonymous transactions per time period are allowed." Most prior work on conditional anonymity focused on e-cash [26, 9, 12], where the identity of double-spenders could be discovered. A recent variation on the theme is Jarecki and Shmatikov's [36] work on anonymous, but *linkable*, authentication where one's identity can be discovered after one carries out too many transactions. Another set of recent papers [47, 42] addressed a related problem of allowing a user to show a credential anonymously and unlinkably up to k times. Showing an anonymous credential more than k times allows a verifier to link the $k+1$ st showing to a previous transaction, but, in contrast to our scheme, does not in any way lead to the identification of the misbehaving user.

2. DEFINITION OF SECURITY

Our definitions for periodic n-times anonymous authentication are based on the e-cash definitions of [12] and [13]. We define a scheme where users $\mathcal{U}$ obtain e-token dispensers from the issuer $\mathcal{I}$, and each dispenser can dispense up to n anonymous and unlinkable e-tokens per time period, but no more; these e-tokens are then given to verifiers $\mathcal{V}$ that guard access to a resource that requires authentication (e.g., an on-line game). $\mathcal{U}$, $\mathcal{V}$, and $\mathcal{I}$ interact using the following algorithms:

- IKeygen(1^k, *params*) is the key generation algorithm of the e-token issuer $\mathcal{I}$. It takes as input 1^k and, if the scheme is in the common parameters model, these parameters *params*. It outputs a key pair $(pk_\mathcal{I}, sk_\mathcal{I})$. Assume that *params* are appended as part of $pk_\mathcal{I}$ and $sk_\mathcal{I}$.
- UKeygen($1^k, pk_\mathcal{I}$) creates the user's key pair $(pk_\mathcal{U}, sk_\mathcal{U})$ analogously.
- Obtain($\mathcal{U}(pk_\mathcal{I}, sk_\mathcal{U}, n), \mathcal{I}(pk_\mathcal{U}, sk_\mathcal{I}, n)$) At the end of this protocol, the user obtains an e-token dispenser D, usable n times per time period and (optionally) the issuer obtains tracing information t_D and revocation information r_D. $\mathcal{I}$ adds t_D and r_D to a record $R_\mathcal{U}$ which is stored together with $pk_\mathcal{U}$.
- Show($\mathcal{U}(D, pk_\mathcal{I}, t, n), \mathcal{V}(pk_\mathcal{I}, t, n)$). Shows an e-token from dispenser D in time period t. The verifier outputs a token serial number (TSN) S and a transcript τ. The user's output is an updated e-token dispenser D'.
- Identify($pk_\mathcal{I}, S, \tau, \tau'$). Given two records (S, τ) and (S, τ') output by honest verifiers in the Show protocol, where $\tau \neq \tau'$, computes a value $s_\mathcal{U}$ that can identify the owner of the dispenser D that generated TSN S.

The value $s_\mathcal{U}$ may also contain additional information specific to the owner of D that (a) will convince third parties that $\mathcal{U}$ is a violator (weak exculpability), that (b) will convince third parties that $\mathcal{U}$ double-showed this e-token (strong exculpability), or that (c) can be used to extract all token serial numbers of $\mathcal{U}$ (traceability).

A periodic n-times anonymous authentication scheme needs to fulfill the following three properties:

Soundness. Given an honest issuer, a set of honest verifiers are guaranteed that, collectively, they will not have to accept more than n e-tokens from a single e-token dispenser in a single time period. There is a knowledge extractor $\mathcal{E}$ that executes u Obtain protocols with all adversarial users and produces functions, $f_1, \ldots, f_u$, with $f_i : \mathbb{T} \times \mathbb{I} \to \mathbb{S}$. $\mathbb{I}$ is the index set $[0..n-1]$, $\mathbb{T}$ is the domain of the time period identifiers, and $\mathbb{S}$ is the domain of TSN's. Running though all $j \in \mathbb{I}$, $f_i(t, j)$ produces all n TSNs for dispenser i at time $t \in \mathbb{T}$. We require that for every adversary, the probability that an honest verifier will accept S as a TSN of a Show protocol executed in time period t, where $S \neq f_i(j, t), \forall 1 \leq i \leq u$ and $\forall 0 \leq j < n$ is negligible.

Identification. There exists an efficient function ϕ with the following property. Suppose the issuer and verifiers $\mathcal{V}_1, \mathcal{V}_2$ are honest. If $\mathcal{V}_1$ outputs (S, τ) and $\mathcal{V}_2$ outputs (S, τ') as the result of Show protocols, then Identify($pk_\mathcal{I}, S, \tau, \tau'$) outputs a value $s_\mathcal{U}$, such that $\phi(s_\mathcal{U}) = pk_\mathcal{U}$, the violator's public key. In the sequel, when we say that a user has *reused* an e-token, we mean that there exist (S, τ) (S, τ') that are both output by honest verifiers.

Anonymity. This property is captured as follows: the adversary, acting as the issuer, may run many Obtain protocols with many honest users. Then this adversary may invoke Show protocols with users of his choice, up to n times per time period with the same user. The adversary should not be able to distinguish whether he is indeed interacting with real users or with a simulator $\mathcal{S}$ that pretends to be real users *without* knowing anything about them, including which users it is supposed to be at any point in time, and without access to any secret or public key, or the user's e-token dispenser D. A formal definition for *anonymity* appears in the full version of this paper.

Additional Extensions In §5, we provide definitions of security in the context of *glitches*, i.e., re-use of e-tokens that do not occur "too often."

In §6, we discuss natural extensions to our basic construction that build on prior work on anonymous credentials and e-cash, namely the concepts of weak and strong exculpability, tracing, and revocation. We now define the corresponding algorithms and security guarantees for these extensions:

- VerifyViolator($pk_\mathcal{I}, pk_\mathcal{U}, s_\mathcal{U}$) publicly verifies that the user with public key $pk_\mathcal{U}$ has double-spent at least one e-token.
- VerifyViolation($pk_\mathcal{I}, S, pk_\mathcal{U}, s_\mathcal{U}$) publicly verifies that the user with public key $pk_\mathcal{U}$ is guilty of double-spending the e-token with TSN S.
- Trace($pk_\mathcal{I}, pk_\mathcal{U}, s_\mathcal{U}, R_\mathcal{U}, n$), given a valid proof $s_\mathcal{U}$ and the user's tracing record $R_\mathcal{U}$, computes all TSNs corresponding to this user. Suppose the user has obtained u e-token dispensers, Trace outputs functions $f_1, \ldots, f_u$ such that by running though all $j \in [0..n-1]$, $f_i(t, j)$ produces all n TSNs for e-token dispenser D_i at time t. If $s_\mathcal{U}$ is invalid, i.e. VerifyViolator($pk_\mathcal{I}, pk_\mathcal{U}, s_\mathcal{U}$) rejects, Trace does nothing.

- Revoke($pk_\mathcal{I}, r_D, RD$) takes as input a revocation database RD (initially empty) and revocation information r_D that corresponds to a particular user (see **Obtain**). It outputs the updated revocation database RD. In the sequel, we assume that RD is part of $pk_\mathcal{I}$.

These algorithms should fulfill the following properties:

Weak exculpability. An adversary cannot successfully blame an honest user $\mathcal{U}$ for reusing an e-token. More specifically, suppose an adversary can adaptively direct a user $\mathcal{U}$ to obtain any number of dispensers and show up to n e-tokens per dispenser per time period. The probability that the adversary produces $s_\mathcal{U}$ such that VerifyViolator($pk_\mathcal{I}, pk_\mathcal{U}, s_\mathcal{U}$) accepts is negligible.

Strong exculpability. An adversary cannot successfully blame a user $\mathcal{U}$ of reusing an e-token with token serial number S, even if $\mathcal{U}$ double-showed some other e-tokens. More specifically, suppose an adversary can adaptively direct a user to obtain any number of dispensers and show any number of e-tokens per dispenser per time period (i.e. he can reset the dispenser's state so that the dispenser reuses some of its e-tokens). The probability that the adversary outputs a token serial number S that was *not* reused and a proof $s_\mathcal{U}$ such that VerifyViolation($pk_\mathcal{I}, S, pk_\mathcal{U}, s_\mathcal{U}$) accepts is negligible.

Tracing of violators. The token serial numbers of violators can be efficiently computed. More specifically, given a value $s_\mathcal{U}$ such that VerifyViolator($pk_\mathcal{I}, pk_\mathcal{U}, s_\mathcal{U}, n$) accepts, and supposing $\mathcal{U}$ has obtained u e-token dispensers, Trace($pk_\mathcal{I}, pk_\mathcal{U}, s_\mathcal{U}, R_\mathcal{U}, n$) produces functions $f_1, \ldots, f_u$ such that by running though all $j \in [0..n-1]$, $f_i(t, j)$ produces all n TSNs for e-token dispenser i at time t.

Dynamic revocation. The **Show** protocol will only succeed for dispensers D that have not been revoked with **Revoke**. (Recall that **Show** takes as input the value $pk_\mathcal{I}$ that contains the database DB of revoked users.)

3. PRELIMINARIES

Our e-token system can be shown secure under several different complexity assumptions. **Notation:** we write $\mathbb{G} = \langle g \rangle$ to denote that g generates the group $\mathbb{G}$.

BILINEAR MAPS. Let Bilinear_Setup be an algorithm that, on input the security parameter 1^k, outputs the parameters for a bilinear map as $\gamma = (q, g_1, h_1, \mathbb{G}_1, g_2, h_2, \mathbb{G}_2, \mathbb{G}_T, \mathbf{e})$. Each group $\mathbb{G}_1 = \langle g_1 \rangle = \langle h_1 \rangle$, $\mathbb{G}_2 = \langle g_2 \rangle = \langle h_2 \rangle$, and $\mathbb{G}_T$ are of prime order $q \in \Theta(2^k)$. The efficient mapping $\mathbf{e} : \mathbb{G}_1 \times \mathbb{G}_2 \to \mathbb{G}_T$ is both: (*Bilinear*) for all $g_1 \in \mathbb{G}_1, g_2 \in \mathbb{G}_2$, and $a, b \in \mathbb{Z}_q^2$, $\mathbf{e}(g_1^a, g_2^b) = \mathbf{e}(g_1, g_2)^{ab}$; and (*Non-degenerate*) if g_1 is a generator of $\mathbb{G}_1$ and g_2 is a generator of $\mathbb{G}_2$, then $\mathbf{e}(g_1, g_2)$ generates $\mathbb{G}_T$.

COMPLEXITY ASSUMPTIONS. The security of our scheme relies on the following assumptions:

Strong RSA Assumption [4, 34]: Given an RSA modulus n and a random element $g \in \mathbb{Z}_n^*$, it is hard to compute $h \in \mathbb{Z}_n^*$ and integer $e > 1$ such that $h^e \equiv g \bmod n$. The modulus n is of a special form pq, where $p = 2p' + 1$ and $q = 2q' + 1$ are safe primes.

Additionally, our constructions require *one* of y-DDHI or SDDHI, depending on the size of the system parameters. Alternatively, we can substitute DDH for either of these assumptions, where the cost is an increase in our time and space complexity by a factor roughly the security parameter.

y-Decisional Diffie-Hellman Inversion (y-DDHI) [5, 32]: Suppose that $g \in \mathbb{G}$ is a random generator of order $q \in \Theta(2^k)$. Then, for all probabilistic polynomial time adversaries $\mathcal{A}$,

$$\Pr[a \leftarrow \mathbb{Z}_q^*;\ x_0 = g^{1/a};\ x_1 \leftarrow \mathbb{G};\ b \leftarrow \{0,1\};$$
$$b' \leftarrow \mathcal{A}(g, g^a, g^{a^2}, \ldots, g^{a^y}, x_b) : b = b'] < 1/2 + 1/\text{poly}(k).$$

In the full version of this paper, we show that the SDDHI assumption holds in generic groups.

Strong DDH Inversion (SDDHI): Suppose that $g \in \mathbb{G}$ is a random generator of order $q \in \Theta(2^k)$. Let $\mathcal{O}_a(\cdot)$ be an oracle that, on input $z \in \mathbb{Z}_q^*$, outputs $g^{1/(a+z)}$. Then, for all probabilistic polynomial time adversaries $\mathcal{A}^{(\cdot)}$ that do not query the oracle on x,

$$\Pr[a \leftarrow \mathbb{Z}_q^*;\ (x, \alpha) \leftarrow \mathcal{A}^{\mathcal{O}_a}(g, g^a);\ y_0 = g^{1/(a+x)};\ y_1 \leftarrow \mathbb{G};$$
$$b \leftarrow \{0,1\};\ b' \leftarrow \mathcal{A}^{\mathcal{O}_a}(y_b, \alpha) : b = b'] < 1/2 + 1/\text{poly}(k).$$

Additionally, our constructions require *one* of the following assumptions. XDH requires non-supersingular curves, whereas SF-DDH may reasonably be conjectured to hold in any bilinear group.

External Diffie-Hellman Assumption (XDH) [35, 44, 40, 6, 3]: Suppose Bilinear_Setup(1^k) produces the parameters for a bilinear mapping $\mathbf{e} : \mathbb{G}_1 \times \mathbb{G}_2 \to \mathbb{G}_T$. The XDH assumption states that the Decisional Diffie-Hellman (DDH) problem is hard in $\mathbb{G}_1$.

Sum-Free Decisional Diffie-Hellman Assumption (SF-DDH) [31]: Suppose that $g \in \mathbb{G}$ is a random generator of order $q \in \Theta(2^k)$. Let L be any polynomial function of k. Let $\mathcal{O}_{\vec{a}}(\cdot)$ be an oracle that, on input a subset $I \subseteq \{1, \ldots, L\}$, outputs the value $g_1^{\beta_I}$ where $\beta_I = \prod_{i \in I} a_i$ for some $\vec{a} = (a_1, \ldots, a_L) \in \mathbb{Z}_q^L$. Further, let R be a predicate such that $R(J, I_1, \ldots, I_t) = 1$ if and only if $J \subseteq \{1, \ldots, L\}$ is DDH-independent from the I_i's; that is, when $v(I_i)$ is the L-length vector with a one in position j if and only if $j \in I_i$ and zero otherwise, then there are no three sets I_a, I_b, I_c such that $v(J) + v(I_a) = v(I_b) + v(I_c)$ (where addition is bitwise over the integers). Then, for all probabilistic polynomial time adversaries $\mathcal{A}^{(\cdot)}$,

$$\Pr[\vec{a} = (a_1, \ldots, a_L) \leftarrow \mathbb{Z}_q^L; (J, \alpha) \leftarrow \mathcal{A}^{\mathcal{O}_{\vec{a}}}(1^k); y_0 = g^{\prod_{i \in J} a_i};$$
$$y_1 \leftarrow \mathbb{G}; b \leftarrow \{0,1\}; b' \leftarrow \mathcal{A}^{\mathcal{O}_{\vec{a}}}(y_b, \alpha)\ :\ b = b' \wedge$$
$$R(J, Q) = 1] < 1/2 + 1/\text{poly}(k),$$

where Q is the set of queries that $\mathcal{A}$ made to $\mathcal{O}_{\vec{a}}(\cdot)$.

KEY BUILDING BLOCKS. We summarize the necessary information about our system components.

DY Pseudorandom Function (PRF). Let $\mathbb{G} = \langle g \rangle$ be a group of prime order $q \in \Theta(2^k)$. Let a be a random element of $\mathbb{Z}_q^*$. Dodis and Yampolskiy [32] showed that $f_{g,a}^{DY}(x) = g^{1/(a+x)}$ is a pseudorandom function, under the y-DDHI assumption, when either: (1) the inputs are drawn from the restricted domain $\{0,1\}^{O(\log k)}$ only, or (2) the adversary specifies a polynomial-sized set of inputs from $\mathbb{Z}_q^*$ *before* a function is selected from the PRF family (i.e., before the value a is selected). For our purposes, we require something stronger: that the DY construction work for inputs drawn arbitrarily and adaptively from $\mathbb{Z}_q^*$.

THEOREM 3.1. *In the generic group model, the Dodis-Yampolskiy PRF is adaptively secure for inputs in $\mathbb{Z}_q^*$.*

The proof is included in the full version of this paper.

Pedersen and Fujisaki-Okamoto Commitments. Recall the Pedersen commitment scheme [43], in which the public parameters are a group $\mathbb{G}$ of prime order q, and generators $(g_0, \ldots, g_m)$. In order to commit to the values $(v_1, \ldots, v_m) \in \mathbb{Z}_q^m$, pick a random $r \in \mathbb{Z}_q$ and set $C = \text{PedCom}(v_1, \ldots, v_m; r) = g_0^r \prod_{i=1}^m g_i^{v_i}$.

Fujisaki and Okamoto [34] showed how to expand this scheme to composite order groups.

CL Signatures. Camenisch and Lysyanskaya [16] came up with a secure signature scheme with two protocols: (1) An efficient protocol for a user to obtain a signature on the value in a Pedersen (or Fujisaki-Okamoto) commitment [43, 34] without the signer learning anything about the message. (2) An efficient proof of knowledge of a signature protocol. Security is based on the Strong RSA assumption. Using bilinear maps, we can use other signature schemes [17, 6] for shorter signatures.

Verifiable Encryption. For our purposes, in a verifiable encryption scheme, the encrypter/prover convinces a verifier that the plaintext of an encryption under a known public key is equivalent to the value hidden in a Pedersen commitment. Camenisch and Damgård [11] developed a technique for turning any semantically-secure encryption scheme into a verifiable encryption scheme.

Bilinear El Gamal Encryption. We require a cryptosystem where g^x is sufficient for decryption and the public key is $\phi(g^x)$ for some function ϕ. One example is the bilinear El Gamal cryptosystem [7, 2], which is semantically secure under the DBDH assumption; that is, given (g, g^a, g^b, g^c, Q), it is difficult to decide if $Q = \mathbf{e}(g, g)^{abc}$. DBDH is implied by y-DDHI or Sum-Free DDH.

AGREEING ON THE TIME. Something as natural as time becomes a complex issue when it is part of a security system. First, it is necessary that the time period t be the same for all users that show e-tokens in that period. Secondly, it should be used only for a single period, i.e., it must be unique. Our construction in §4 allows for the use of arbitrary time period identifiers, such as those negotiated using the hash tree protocol in [29]. For Glitch protection, §5, we assume a totally ordered set of time period identifiers.

4. A PERIODIC N-TIMES ANONYMOUS AUTHENTICATION SCHEME

4.1 Intuition Behind our Construction

In a nutshell, the issuer and the user both have key pairs. Let the user's keypair be $(pk_\mathcal{U}, sk_\mathcal{U})$, where $pk_\mathcal{U} = g^{sk_\mathcal{U}}$ and g is a generator of some group G of known order. Let f_s be a pseudorandom function whose range is the group G. During the **Obtain** protocol, the user obtains an e-token dispenser D that allows her to show up to n tokens per time period. The dispenser D is comprised of seed s for PRF f_s, the user's secret key $sk_\mathcal{U}$, and the issuer's signature on $(s, sk_\mathcal{U})$. We use CL signatures to prevent the issuer from learning anything about s or $sk_\mathcal{U}$. In the **Show** protocol, the user shows her i^{th} token in time period t: she releases TSN $S = f_s(0, t, i)$, a double-show tag $E = pk_\mathcal{U} \cdot f_s(1, t, i)^R$ (for a random R supplied by the verifier), and runs a ZK proof protocol that (S, E) correspond to a valid dispenser for time period t and $0 \leq i < n$ (the user proves that S and E were properly formed from values $(s, sk_\mathcal{U})$ signed by the issuer). Since f_s is a PRF, and all the proof protocols are zero-knowledge, it is computationally infeasible to link the resulting e-token to the user, the dispenser D, or any other e-tokens corresponding to D. If a user shows $n+1$ e-tokens during the same time interval, then two of the e-tokens *must* use the same TSN. The issuer can easily detect the violation and compute $pk_\mathcal{U}$ from the two double-show tags, $E = pk_\mathcal{U} \cdot f_s(1, t, i)^R$ and $E' = pk_\mathcal{U} \cdot f_s(1, t, i)^{R'}$. From the equations above, $f_s(1, t, i) = (E/E')^{(R-R')^{-1}}$ and $pk_\mathcal{U} = E/f_s(1, t, i)^R$.

4.2 Our Basic Construction

Let k be a security parameter and $l_q \in O(k)$, l_x, l_{time}, and l_{cnt} be system parameters such that $l_q \geq l_x > l_{\text{time}} + l_{\text{cnt}} + 3$ and $2^{l_{\text{cnt}}} - 1 > n$, where n is the number of tokens we allow per time period.

In the following, we assume implicit conversion between binary strings and integers, e.g., between $\{0,1\}^l$ and $[0, 2^l - 1]$. Let $F_{(g,s)}(x) := f_{g,s}^{DY}(x) := g^{1/(s+x)}$ for $x, s \in \mathbb{Z}_q^*$ and $\langle g \rangle = \mathbb{G}$ being of prime order q. For suitably defined l_{time}, l_{cnt}, and l_x define the function $c : \{0,1\}^{l_x - l_{\text{time}} - l_{\text{cnt}}} \times \{0,1\}^{l_{\text{time}}} \times \{0,1\}^{l_{\text{cnt}}} \to \{0,1\}^{l_x}$ as:

$$c(u, v, z) := \left(u 2^{l_{\text{time}}} + v\right) 2^{l_{\text{cnt}}} + z \ .$$

Issuer Key Generation: In $\mathsf{IKeygen}(1^k, \textit{params})$, the issuer $\mathcal{I}$ generates two cyclic groups:

1. A group $\langle \mathbf{g} \rangle = \langle \mathbf{h} \rangle = \mathbf{G}$ of composite order $\mathbf{p'q'}$ that can be realized by the multiplicative group of quadratic residues modulo a special RSA modulus $N = (2\mathbf{p'} + 1)(2\mathbf{q'} + 1)$. In addition to CL signatures, this group will be needed for zero-knowledge proofs of knowledge used in the sequel. Note that soundness of these proof systems is computational only and assumes that the prover does not know the order of the group.
2. A group $\langle g \rangle = \langle \tilde{g} \rangle = \langle h \rangle = \mathbb{G}$ of prime order q with $2^{l_q - 1} < q < 2^{l_q}$.

The issuer must also prove in zero-knowledge that N is a special RSA modulus, and $\langle \mathbf{g} \rangle = \langle \mathbf{h} \rangle$ are quadratic residues modulo N. In the random oracle model, one non-interactive proof may be provided. In the plain model, the issuer must agree to interactively prove this to anyone upon request.

Furthermore, the issuer generates a CL signature key pair (pk, sk) set in group $\mathbf{G}$. The issuer's public-key will contain $(\mathbf{g}, \mathbf{h}, \mathbf{G}, g, \tilde{g}, h, \mathbb{G}, pk)$, while the secret-key will contain all of the information.

User Key Generation: In $\mathsf{UKeygen}(1^k, pk_\mathcal{I})$, the user chooses a random $sk_\mathcal{U} \in \mathbb{Z}_q$ and sets $pk_\mathcal{U} = g^{sk_\mathcal{U}} \in \mathbb{G}$.

Get e-Token Dispenser: $\mathsf{Obtain}(\mathcal{U}(pk_\mathcal{I}, sk_\mathcal{U}, n), \mathcal{I}(pk_\mathcal{U}, sk_\mathcal{I}, n))$. Assume that $\mathcal{U}$ and $\mathcal{I}$ have mutually authenticated. A user $\mathcal{U}$ obtains an e-token dispenser from an issuer $\mathcal{I}$ as follows:

1. $\mathcal{U}$ and $\mathcal{I}$ agree on a commitment C to a random value $s \in \mathbb{Z}_q$ as follows:

(a) $\mathcal{U}$ selects s' at random from $\mathbb{Z}_q$ and computes $C' = \mathrm{PedCom}(sk_\mathcal{U}, s'; r) = g^{sk_\mathcal{U}} \tilde{g}^{s'} h^r$.

(b) $\mathcal{U}$ sends C' to $\mathcal{I}$ and proves that it is constructed correctly.

(c) $\mathcal{I}$ sends a random r' from $\mathbb{Z}_q$ back to $\mathcal{U}$.

(d) Both $\mathcal{U}$ and $\mathcal{I}$ compute $C = C'\tilde{g}^{r'} = \mathrm{PedCom}(sk_\mathcal{U}, s' + r'; r)$. $\mathcal{U}$ computes $s = s' + r' \bmod q$.

2. $\mathcal{I}$ and $\mathcal{U}$ execute the CL signing protocol on commitment C. Upon success, $\mathcal{U}$ obtains σ, the issuer's signature on $(sk_\mathcal{U}, s)$. This step can be efficiently realized using the CL protocols [16, 17] in such a way that $\mathcal{I}$ learns nothing about $sk_\mathcal{U}$ or s.

3. $\mathcal{U}$ initializes counters $T := 1$ (to track the current period) and $J := 0$ (to count the e-tokens shown in the current time period). $\mathcal{U}$ stores the e-token dispenser $D = (sk_\mathcal{U}, s, \sigma, T, J)$.

Use an e-Token: $\mathrm{Show}(\mathcal{U}(E, pk_\mathcal{I}, t, n), \mathcal{V}(pk_\mathcal{I}, t, n))$. Let t be the current time period identifier with $0 < t < 2^{l_\mathrm{time}}$. (We discuss how two parties might agree on t in Section 3.) A user $\mathcal{U}$ reveals a single e-token from a dispenser $D = (sk_\mathcal{U}, s, \sigma, T, J)$ to a verifier $\mathcal{V}$ as follows:

1. $\mathcal{U}$ compares t with T. If $t \neq T$, then $\mathcal{U}$ sets $T := t$ and $J := 0$. If $J \geq n$, abort!

2. $\mathcal{V}$ sends to $\mathcal{U}$ a random $R \in \mathbb{Z}_q^*$.

3. $\mathcal{U}$ sends to $\mathcal{V}$ a token serial number S and a double spending tag E computed as follows:
$$S = F_{(g,s)}(c(0, T, J)), \qquad E = pk_\mathcal{U} \cdot F_{(g,s)}(c(1, T, J))^R$$

4. $\mathcal{U}$ and $\mathcal{V}$ engage in a zero-knowledge proof of knowledge of values $sk_\mathcal{U}, s, \sigma$, and J such that:

   (a) $0 \leq J < n$,
   (b) $S = F_{(g,s)}(c(0, t, J))$,
   (c) $E = g^{sk_\mathcal{U}} \cdot F_{(g,s)}(c(1, t, J))^R$,
   (d) $\mathrm{VerifySig}(pk_\mathcal{I}, (sk_\mathcal{U}, s), \sigma) = \mathrm{true}$.

5. If the proof verifies, $\mathcal{V}$ stores (S, τ), with $\tau = (E, R)$, in his database. If he is not the only verifier, he also submits this tuple to the database of previously shown e-tokens.

6. $\mathcal{U}$ increases counter J by one. If $J \geq n$, the dispenser is empty. It will be refilled in the next time period.

Technical Details. The proof in Step 4 is done as follows:

1. $\mathcal{U}$ generates the commitments $C_J = g^J h^{r_1}$, $C_u = g^{sk_\mathcal{U}} h^{r_2}$, $C_s = g^s h^{r_3}$, and sends them to $\mathcal{V}$.

2. $\mathcal{U}$ proves that C_J is a commitment to a value in the interval $[0, n-1]$ using standard techniques [22, 19, 8].

3. $\mathcal{U}$ proves knowledge of a CL signature from $\mathcal{I}$ for the values committed to by C_u and C_s in that order. This step can be efficiently realized using the CL protocols [16, 17].

4. $\mathcal{U}$ as prover and $\mathcal{V}$ as verifier engage in the following proof of knowledge, using the notation by Camenisch and Stadler [21]:

$$PK\{(\alpha, \beta, \delta, \gamma_1, \gamma_2, \gamma_3) : g = (C_s g^{c(0,t,0)} C_J)^\alpha h^{\gamma_1} \wedge$$
$$S = g^\alpha \wedge g = (C_s g^{c(1,t,0)} C_J)^\beta h^{\gamma_2} \wedge$$
$$C_u = g^\delta h^{\gamma_3} \wedge E = g^\delta (g^R)^\beta\}.$$

$\mathcal{U}$ proves she knows the values of the Greek letters; all other values are known to both parties.

Let us explain the last proof protocol. From the first step we know that C_J encodes some value $\hat{J}$ with $0 \leq \hat{J} < n$, i.e., $C_J = g^{\hat{J}} h^{\hat{r}_J}$ for some $\hat{r}_J$. From the second step we know that C_s and C_u encoded some value $\hat{u}$ and $\hat{s}$ on which the prover $\mathcal{U}$ knows a CL signature by the issuer. Therefore, $C_s = g^{\hat{s}} h^{\hat{r}_s}$ and $C_u = g^{\hat{u}} h^{\hat{r}_u}$ for some $\hat{r}_s$ and $\hat{r}_u$. Next, recall that by definition of $c(\cdot, \cdot, \cdot)$ the term $g^{c(0,t,0)}$ corresponds to $g^{t 2^{l_\mathrm{cnt}}}$. Now consider the first term $g = (C_s g^{c(0,t,0)} C_J)^\alpha h^{\gamma_1}$ in the proof protocol. We can now conclude the prover $\mathcal{U}$ knows values $\hat{a}$ and $\hat{r}$ such that $g = g^{(\hat{s} + t 2^{l_\mathrm{cnt}} + \hat{J})\hat{a}} h^{\hat{r}}$ and $S = g^{\hat{a}}$. From the first equation it follows that $\hat{a} = (\hat{s} + (t 2^{l_\mathrm{cnt}} + \hat{J}))^{-1} \pmod{q}$ must hold provided that $\mathcal{U}$ is not privy to $\log_g h$ (as we show via a reduction in the proof of security) and thus we have established that $S = F_{(g,\hat{s})}(c(0, t, \hat{J}))$ is a valid serial number for the time period t. Similarly one can derive that $E = g^{\hat{u}} \cdot F_{(g,\hat{s})}(c(1, t, \hat{J}))^R$, i.e., that E is a valid double-spending tag for time period t.

Identify Cheaters: $\mathrm{Identify}(pk_\mathcal{I}, S, (E, R), (E', R'))$. If the verifiers who accepted these tokens were honest, then $R \neq R'$ with high probability, and proof of validity ensures that $E = pk_\mathcal{U} \cdot f_s(1, T, J)^R$ and $E' = pk_\mathcal{U} \cdot f_s(1, T, J)^{R'}$. The violator's public key can now be computed by first solving for $f_s(1, T, J) = (E/E')^{(R-R')^{-1}}$ and then computing $pk_\mathcal{U} = E/f_s(1, T, J)^R$.

THEOREM 4.1. *Protocols* IKeygen, UKeygen, Obtain, Show, *and* Identify *described above achieve* soundness, identification, *and* anonymity *properties in the plain model assuming* Strong RSA, *and* y-DDHI *if* $l_x \in O(\log k)$ *or* SDDHI *otherwise.*

In the full version of this paper, we prove the theorem. Recall that l_x dictates the number of time periods and the number of allowed shows per time period; when these values are small, security is based only on Strong RSA and y-DDHI.

4.3 Efficiency Discussion

To analyze the efficiency of our scheme, it is sufficient to consider the number of (multi-base) exponentiations the parties have to do in $\mathbb{G}$ and $\mathbf{G}$. In a decent implementation, a multi-base exponentiation takes about the same time as a single-base exponentiation, provided that the number of bases is small. For the analysis we assume that the Strong RSA based CL-signature scheme is used.

Obtain: both the user and issuer perform 3 exponentiations in $\mathbf{G}$. Show: the user performs 12 multi-base exponentiation in $\mathbb{G}$ and 23 multi-base exponentiations in $\mathbf{G}$, while the verifier performs 7 multi-base exponentiation in $\mathbb{G}$ and 13 multi-base exponentiations in $\mathbf{G}$. If n is odd, the user only needs to do 12 exponentiations in $\mathbf{G}$, while the verifier needs to do 7. To compare ourselves to the Damgård et al. [29] scheme, we set $n = 1$. In this case, Show requires that the user perform 12 multi-base exponentiation in $\mathbb{G}$ and 1 multi-base exponentiations in $\mathbf{G}$ and the verifier perform 7 multi-base exponentiation in $\mathbb{G}$ and 1 multi-base exponentiations in $\mathbf{G}$. Damgård et al. requires $57 + 68r$ exponentiations in $\mathbb{G}$, where r is the security parameter (i.e., 2^{-r} is the probability that the user can cheat). Depending on the application, r should be at least 20 or even 60. Thus, our scheme is an order of magnitude more efficient than Damgård et al.

5. GLITCH PROTECTION EXTENSION

In our periodic n-times anonymous authentication scheme, a user who shows two tokens with the same TSN becomes identifiable. (Recall that only n unique TSN values are available to a user per time period.) A user might accidentally use the same TSN twice because of hardware breakdowns, clock desychronization, etc. We want to protect the anonymity of users who occasionally cause a *glitch* (repeat a TSN in two different tokens), while still identifying users who cause an excessive amount of glitches. A user might be permitted up to m glitches per *monitoring interval* (e.g., year). Any TSN repetition will be detected, but the user's anonymity will not be compromised until the $(m+1)$st glitch. A token that causes a glitch is called a *clone*.

Suppose a user has u glitches in one monitoring interval. Our goal is to design a scheme where:

- if $u = 0$, all shows are anonymous and unlinkable;
- if $1 \leq u \leq m$, all shows remain anonymous, but a link-id L is revealed, making all clones linkable;
- if $u > m$, the user's public key is revealed.

One can think of link-id L as a pseudonym (per monitoring interval) that is hidden in each token released by the same user (much in the same way that the user's public key was hidden in each token released by a user in the basic scheme). If tokens (S, τ) and (S, τ') caused a glitch, then we call (S, τ, τ') a *glitch tuple*, where by definition $\tau \neq \tau'$. We introduce a new function GetLinkId that takes as input a glitch tuple and returns the link-id L. Once $m + 1$ clones are linked to the same pseudoym L, there is enough information from these collective original and cloned transcripts to compute the public key of the user.

We continue to use identifier $t \in \mathbb{T}$ for (indivisible) time periods. Identifier $v \in \mathbb{V}$ refers to a monitoring interval. We give two glitch protection schemes: §5.1 considers disjoint monitoring intervals, while §5.2 works on overlapping monitoring intervals. For the first scheme, we assume the existence of an efficient function $M_\mathbb{V}$ that maps every time period t to its unique monitoring interval $v \in \mathbb{V}$.

5.1 Basic Glitch Protection

Our basic glitch protection scheme tolerates up to m clones per monitoring interval v; monitoring intervals are disjoint.

We informally define the protocols and security properties of a periodic authentication scheme with glitch protection:

- ShowGP($\mathcal{U}(D, pk_\mathcal{I}, t, n, m), \mathcal{V}(pk_\mathcal{I}, t, n, m)$). Shows an e-token from dispenser D in time period t and monitoring interval $v = M_\mathbb{V}(t)$. The verifier obtains a token serial number S and a transcript τ.
- GetLinkId($pk_\mathcal{I}, S, \tau, \tau'$). Given e-tokens (S, τ, τ'), where $\tau \neq \tau'$ by definition, computes a link-id value L.
- IdentifyGP($pk_\mathcal{I}, (S_1, \tau_1, \tau'_1), \ldots, (S_{m+1}, \tau_{m+1}, \tau'_{m+\ell+1})$). Given $m+1$ glitch tuples where for each i, GetLinkId(S_i, τ_i, τ'_i) produces the same link-id L, computes a value $s_\mathcal{U}$ that can be used to compute the public key of the owner of the dispenser D from which the TSNs came.

We give two new security properties *GP Anonymity* and *GP Identification* that supercede the *Anonymity* and *Identification* properties of §2.

GP Anonymity. An adversarial issuer, even when cooperating with verifiers and other dishonest users, and adaptively directing honest users to show e-tokens and up to m clones of his choice for every monitoring interval $v \in \mathbb{V}$, cannot learn anything about a user's e-token usage behavior except what is available from side information from the environment. This property is captured by a simulator $\mathcal{S}$ which can interact with the adversary as if he were the user. $\mathcal{S}$ doesn't have access to the user's secret or public key, or her e-token dispenser D. However, when the adversary asks $\mathcal{S}$ to make a clone, the environment passes $\mathcal{S}$ a link-id. In the full version of this paper, we provide a formal definition.

GP Identification. Suppose the issuer and verifiers are honest and they receive $m + 1$ glitch tuples Input $= (S_1, \tau_1, \tau'_1)$, ..., $(S_{m+1}, \tau_{m+1}, \tau'_{m+1})$ with the same $L = $ GetLinkId($pk_\mathcal{I}, S_i, \tau_i, \tau'_i$) for all $1 \leq i \leq m + 1$. Then with high probability algorithm IdentifyGP($pk_\mathcal{I}$, Input) outputs a value $s_\mathcal{U}$ for which there exists an efficient function ϕ such that $\phi(s_\mathcal{U}) = pk_\mathcal{U}$, identifying the violator.

Intuition behind construction. Recall that in our basic scheme, an e-token has three logical parts: a serial number $S = F_{(s,g)}(c(0, T, J))$, a tag $E = pk_\mathcal{U} \cdot F_{(s,g)}(c(1, T, J))^R$, and a proof of validity. If the user shows a token with TSN S again, then he must reveal $E' = pk_\mathcal{U} \cdot F_{(s,g)}(c(1, T, J))^{R'}$, where $R \neq R'$, and the verifier can solve for $pk_\mathcal{U}$ from (E, E', R, R').

Now, in our glitch protection scheme, an e-token has four logical parts: a serial number $S = F_{(s,g)}(c(0, T, J))$, a tag K that exposes the link-id L if a glitch occurs, a tag E that exposes $pk_\mathcal{U}$ if more than m glitches occur, and a proof of validity.

We instantiate $K = L \cdot F_{(g,s)}(c(2, T, J))^R$. Now a double-show reveals L just as it revealed $pk_\mathcal{U}$ in the original scheme. The link-id for monitoring interval v is $L = F_{(s,g)}(c(1, v, 0))$.

Once the verifiers get $m+1$ clones with the same link-id L, they need to recover $pk_\mathcal{U}$. To allow this, the user includes tag $E = pk_U \cdot \prod_{i=1}^{m} F_{(g,s)}(c(3, v, i))^{\rho_i} \cdot F_{(s,g)}(c(4, T, J))^R$. (Here, it will be critical for anonymity that the user and the verifier *jointly* choose the random values $R, \rho_1, \ldots, \rho_m$.)

Now, suppose a user causes $m + 1$ glitches involving ℓ distinct TSNs. Given $(E, R, \rho_1, \ldots, \rho_m)$ from each of these $(m+\ell+1)$ tokens, the public key of the user can be computed by repeatedly using the elimination technique that allowed the discovery of L from (K, K', R, R'). We have $(m+\ell+1)$ equations E and $(m+\ell+1)$ unknown *bases* including $pk_\mathcal{U}$ and the $F_{(s,g)}(.)$ values. Thus, solving for $pk_\mathcal{U}$ simply requires solving a system of linear equations.

Construction. ShowGP and IdentifyGP replace the corresponding Show and Identify algorithms of the basic construction in §4.

ShowGP($\mathcal{U}(D, pk_\mathcal{I}, t, n, m), \mathcal{V}(pk_\mathcal{I}, t, n, m)$). Let $v = M_\mathbb{V}(t)$. A user $\mathcal{U}$ shows a single e-token from a dispenser $D = (sk_\mathcal{U}, s, \sigma, T, J)$ to a verifier $\mathcal{V}$ as follows:

1. $\mathcal{U}$ compares t with T. If $t > T$, then $\mathcal{U}$ sets $T := t$ and $J := 0$. If $J \geq n$, abort!
2. $\mathcal{V}$ and $\mathcal{U}$ jointly choose $R, \rho_1, \ldots, \rho_m$ uniformly at random from $\mathbb{Z}_q^*$. The user and verifier can use coin-flipping to generate a seed x and then use x to generate the other values (either via a PRF like $F_{g,x}(.)$, or by treating some hash function $H(.)$ as a random oracle).

3. $\mathcal{U}$ sends $\mathcal{V}$ an interval serial number S, a double spending tag K encoding the link-id L, and a special $(m+1)$-cloning tag E:

$$S = F_{(g,s)}(c(0,T,J)),$$
$$K = F_{(g,s)}(c(1,v,0)) \cdot F_{(g,s)}(c(2,T,J))^R,$$
$$E = pk_\mathcal{U} \cdot F_{(g,s)}(c(3,v,1))^{\rho_1} \cdots$$
$$F_{(g,s)}(c(3,v,m))^{\rho_m} \cdot F_{(g,s)}(c(4,T,J))^R$$

4. $\mathcal{U}$ performs a zero-knowledge proof that the values above were correctly computed.

5. If the proof verifies, $\mathcal{V}$ stores (S, τ), where $\tau = (K, E, R, \rho_1, \ldots, \rho_m)$, in his database.

6. $\mathcal{U}$ increments counter J by one. If $J \geq n$ the dispenser is empty. It will be refilled in the next time period.

GetLinkId$(pk_\mathcal{I}, S, (K, E, R, \vec{\rho}), (K', E', R', \vec{\rho}'))$. Returns
$$L = \frac{K}{(K/K')^{(R-R')^{-1}R}}.$$

IdentifyGP$(pk_\mathcal{I}, (S_1, \tau_1, \tau_1'), \ldots, (S_{m+1}, \tau_{m+1}, \tau_{m+1}'))$. Let the $m+1$ glitch tuples include ℓ distinct TSN values. We extract the values $(E_i, R, \rho_1, \ldots, \rho_m)$ (or $(E_i', R', \rho_1', \ldots, \rho_m')$) from all $m + \ell + 1$ unique transcripts. Now, we use the intuition provided above to solve for $pk_\mathcal{U}$.

THEOREM 5.1. *The scheme described above is a secure periodic n-times anonymous authentication scheme with basic glitch protection. It fulfills the soundness, GP anonymity and GP identification properties.*

The proof can be found in the full version of this paper.

5.2 Window Glitch Protection

The basic glitch protection scheme prevents users from creating more than m clones in a single monitoring interval. If two neighboring time periods fall in different monitoring intervals, then a malicious user can create m clones in each of them. We want to catch users who make more than m clones within any W consecutive time periods.

We define an interval of consecutive time-periods to be a window. For convenience, we will consider each time period identifier t to be an integer, and time periods t and $t+1$ to be neighbors. Each time period is in W different windows of size W. If we let a time period define the *end* of a window, then time period t would be in windows $t, t+1, \ldots, t+W-1$.

(m, W)-Window glitch protection allows a user to clone at most m e-tokens during any window of W consecutive time periods. We describe the new protocols associated with a window glitch protection scheme:

- **ShowWGP**$(\mathcal{U}(D, pk_\mathcal{I}, t, n, m, W), \mathcal{V}(pk_\mathcal{I}, t, n, m, W))$. Shows an e-token from dispenser D for time period t. The verifier obtains a serial number S and a transcript τ.
- **GetLinkIds**$(pk_\mathcal{I}, S, \tau, \tau')$. Given two e-tokens (S, τ) and (S, τ'), outputs a *list* of W link-ids $L_1, \ldots, L_W$.
- **IdentifyWGP**$(pk_\mathcal{I}, (S_1, \tau_1, \tau_1'), \ldots, (S_{m+1}, \tau_{m+1}, \tau_{m+1}'))$. Given $m+1$ glitch tuples where for each i, the same link-id L is in the list of link-ids produced by GetLinkId(S_i, τ_i, τ_i'), computes a value $s_\mathcal{U}$ that can be used to compute the public key of the owner of the dispenser D from which the TSNs came.

We modify the *GP Anonymity* and *GP Identification* properties to apply to window glitch protection.

WGP Anonymity. This property is the same as for basic glitch protection, except that now the adversary cannot ask a user to create more than m clones within any window of W consecutive time periods. Since each time period is part of W different windows, the environment will pass the simulator a *set* of link-ids.

WGP Identification. Suppose the issuer and verifiers are honest. Should they receive a list of $m+1$ glitch tuples Input $= (S_1, \tau_1, \tau_2'), \ldots, (S_{m+1}, \tau_{m+1}, \tau_{m+1}')$, such that $\exists L : \forall i : L \in$ GetLinkIds$(pk_\mathcal{I}, S_i, \tau_i, \tau_i')$, then with high probability IdentifyWGP$(pk_\mathcal{I}, \text{Input})$ outputs a value $s_\mathcal{U}$ for which there exists an efficient function ϕ such that $\phi(s_\mathcal{U}) = pk_\mathcal{U}$, identifying the violator.

Construction. Intuitively, we replicate our basic glitch solution W times for overlapping windows of W time periods.

ShowWGP$(\mathcal{U}(D, pk_\mathcal{I}, t, n, m, W))$. We modify the ShowGP protocol as follows. In step 3, the user and verifier jointly choose random numbers $R_1, \ldots, R_W$ and $\rho_{1,1}, \ldots, \rho_{W,m}$. In step 4, the user calculates essentially the same values S, K, E, except that now she calculates separate K_i and E_i tags for every window in which time period T falls:

$$S = F_{(s,g)}(c(0,T,J))$$
$$K_i = F_{(s,g)}(c(1,T+i,0)) \cdot F_{(s,g)}(c(2,T,J))^{R_i}$$
$$E_i = pk_\mathcal{U} \cdot F_{(s,g)}(c(3,T+i,1))^{\rho_{i,1}} \cdots$$
$$F_{(s,g)}(c(3,T+i,m)))^{\rho_{i,m}} \cdot F_{(s,g)}(c(4,T,J))^{R_i}$$

Finally, in step 5, the user proves to the verifier that the values $S, K_1, \ldots, K_W, E_1, \ldots, E_W$ are formed correctly. That, along with the random numbers generated in step 3, forms the transcript stored in steps 6. Step 7 is unchanged.

GetLinkIds$(pk_\mathcal{I}, S, \tau, \tau')$. Returns the link-ids:
$$L_i = \frac{K_i}{(K_i/K_i')^{(R_i-R_i')^{-1}R_i}}, \quad 1 \leq i \leq m+1.$$

IdentifyWGP$(pk_\mathcal{I}, (S_1, \tau_1, \tau_2'), \ldots, (S_{m+1}, \tau_{m+1}, \tau_{m+1}'))$. For all i, let $L \in$ GetLinkIds$(pk_\mathcal{I}, S_i, \tau_i, \tau_i')$, that is, let L be the link-id each glitch tuple has in common. Let these $m+1$ glitch tuples include ℓ distinct TSN values. We extract the values $(E_{i,j}, R_i, \rho_{i,1}, \ldots, \rho_{i,m})$ (or $(E_{i,j}', R_i', \rho_{i,1}', \ldots, \rho_{i,m}')$) from all $m + \ell + 1$ unique transcripts, where j depends on where L falls in the list GetLinkIds$(pk_\mathcal{I}, S_i, \tau_i, \tau_i')$. Now, we use the same techniques as before to solve for $pk_\mathcal{U}$.

THEOREM 5.2. *The scheme described above is a secure periodic n-times anonymous authentication scheme with window glitch protection. It fulfills the soundness, WGP anonymity and WGP identification properties.*

The proof can be found in the full version of the paper.

6. ADDITIONAL EXTENSIONS

One advantage of our approach to periodic anonymous authentication is that its modular construction fits nicely with previous work [12, 15]. Thus, it is clear which parts of our system can be modified to enable additional features.

6.1 Weak Exculpability

Recall that *weak exculpability* allows an honest verifier (or group of verifiers) to prove in a sound fashion that the user with public key $pk_\mathcal{U}$ reused *some* token. This convinces everyone in the system that the user with $pk_\mathcal{U}$ is untrustworthy.

To implement weak exculpability, we need to define algorithm VerifyViolator and to slightly adapt the IKeygen, UKeygen, Show, and Identify algorithms. IKeygen' now also runs Bilinear_Setup, and the parameters for the bilinear map $\mathbf{e}: \mathbb{G}_1 \times \mathbb{G}_2 \to \mathbb{G}_T$ are added to $pk_\mathcal{I}$. UKeygen' selects a random $sk_\mathcal{U} \in \mathbb{Z}_q^*$ and outputs $pk_\mathcal{U} = \mathbf{e}(g_1, g_2)^{sk_\mathcal{U}}$. In the Show' protocol, the double-spending tag is calculated as $E = g_1^{sk_\mathcal{U}} \cdot F_{(g_1,s)}(c(1,T,J))^R$. Consequently the value s_U, returned by Identify', is $g_1^{sk_\mathcal{U}}$ – which is secret information! Thus, the VerifyViolator algorithm is defined as follows: VerifyViolator$(pk_\mathcal{I}, pk_\mathcal{U}, s_\mathcal{U})$ accepts only if $\mathbf{e}(s_\mathcal{U}, g_2) = \mathbf{e}(g_1^{sk_\mathcal{U}}, g_2) = pk_\mathcal{U}$. Intuitively, because $g_1^{sk_\mathcal{U}}$ is *secret* information, its release signals that this user misbehaved.

A subtle technical problem with this approach is that tag E is now set in a bilinear group $\mathbb{G}_1$, where DDH may be easy, and we need to ensure that the DY PRF is still secure in this group. Indeed, in groups where DDH is easy, the DY PRF is *not* secure. There are two solutions [12]: (1) make the XDH assumption, i.e., DDH is hard in $\mathbb{G}_1$, and continue to use the DY PRF, or (2) make the more general Sum-Free DDH assumption and use the CHL PRF [12], which works in groups where (regular) DDH is easy.

THEOREM 6.1. *The above scheme provides* weak exculpability *under the Strong RSA, y-DDHI if $l_x \in O(\log k)$ or SDDHI, and either XDH or Sum-Free DDH assumptions.*

6.2 Strong Exculpability

Recall that *strong exculpability* allows an honest verifier (or group of verifiers) to prove in a sound fashion that the user with public key $pk_\mathcal{U}$ reused an e-token with TSN S.

For strong exculpability, we need to define VerifyViolation and to adapt the Show and the Identify algorithms. In Show'', the ZK proof of validity is transformed into a non-interactive proof, denoted Π, using the Fiat-Shamir heuristic [33]. The proof Π is added to the coin transcript, denoted τ. And Identify''$(pk_\mathcal{I}, S, \tau_1, \tau_2)$ adds both transcripts τ_1, and τ_2 to its output $s_\mathcal{U}$. (The function $\phi(s_\mathcal{U}) = pk_\mathcal{U}$ ignores the extra information.)

Thus, the VerifyViolation algorithm is defined as follows: VerifyViolation$(pk_\mathcal{I}, S, pk_\mathcal{U}, s_\mathcal{U})$ parses $\tau_1 = (E_1, R_1, \Pi_1)$ and $\tau_2 = (E_2, R_2, \Pi_2)$ from $s_\mathcal{U}$. Then, it checks that $\phi(s_\mathcal{U}) = pk_\mathcal{U}$ and that Identify''$(pk_\mathcal{I}, S, \tau_0, \tau_2) = s_\mathcal{U}$. Next, it verifies both non-interactive proofs Π_i with respect to (S, R_i, T_i). If all checks pass, it accepts; else, it rejects.

A subtlety here is that, for these proofs to be sound even when the issuer is malicious, the group $\mathbf{G}'$ that is needed as a parameter for zero-knowledge proofs here must be a system parameter generated by a trusted third party, such that no one, including the issuer, knows the order of this group. So in particular, $\mathbf{G}'$ cannot be the same as $\mathbf{G}$ [20].

THEOREM 6.2. *The above scheme provides* strong exculpability *under the Strong RSA, and y-DDHI if $l_x \in O(\log k)$ or SDDHI assumptions in the random oracle model with trusted setup for the group $\mathbf{G}'$.*

6.3 Tracing

We can extend our periodic n-times authentication scheme so that if a user reuses even one e-token, *all* possible TSN values she could compute using *any* of her dispensers are now publicly computable. We use the same IKeygen', UKeygen', Show', and Identify' algorithms as weak exculpability, slightly modify the Obtain protocol, and define a new Trace algorithm.

In UKeygen', the user's keypair $(\mathbf{e}(g_1, g_2)^{sk_\mathcal{U}}, sk_\mathcal{U})$ is of the correct form for the bilinear ElGamal cryptosystem, where the value $g_1^{sk_\mathcal{U}}$ is sufficient for decryption. Now, in our modified Obtain', the user will provide the issuer with a verifiable encryption [11] of PRF seed s under *her own* public key $pk_\mathcal{U}$. The issuer stores this tracing information in $R_\mathcal{U}$. When Identify' exposes $g_1^{sk_\mathcal{U}}$, the issuer may run the following trace algorithm:

Trace$(pk_\mathcal{I}, pk_\mathcal{U}, s_\mathcal{U}, R_\mathcal{U}, n)$. The issuer extracts $g_1^{sk_\mathcal{U}}$ from $s_\mathcal{U}$, and verifies this value against $pk_\mathcal{U}$; it aborts on failure. The issuer uses $g_1^{sk_\mathcal{U}}$ to decrypt all values in $R_\mathcal{U}$ belonging to that user, and recovers the PRF seeds for *all* of the user's dispensers. For seed s and time t, all TSNs can be computed as $f_s(t,j) = F_{(\mathbf{e}(g_1,g_2),s)}(c(0,t,j))$, for all $0 \le j < n$.

THEOREM 6.3. *The above scheme provides* tracing of violators *under the Strong RSA, y-DDHI if $l_x \in O(\log k)$ or SDDHI, and either XDH or Sum-Free DDH assumptions.*

6.4 Dynamic Revocation

Implementing dynamic revocation requires modifying the Obtain and Show protocols in the basic scheme, and defining a new Revoke algorithm.

The mechanisms introduced in [15] can be used for revoking CL signatures. In an adjusted CL protocol for obtaining a signature on a committed value, the user obtains an additional witness $\mathbf{w} = \mathbf{v}^{e^{-1}}$, where $\mathbf{v}$ is the revocation public key and e is a unique prime which is part of the CL signature σ. In the CL protocol for proving knowledge of a signature, the user also proves knowledge of this witness. Violators with prime $\tilde{e}$ can be excluded by updating the revocation public key $\mathbf{v}$, such that $\mathbf{v}' = \mathbf{v}^{\tilde{e}^{-1}}$, and publishing $\tilde{e}$. While all non-excluded users can update their witness by computing function $f(e, \tilde{e}, \mathbf{v}', \mathbf{w}) = \mathbf{w}'$, without knowing the order of $\mathbf{G}$, this update does *not* work when $e = \tilde{e}$.

Thus, our e-token dispensers can be revoked by revoking their CL signature σ. Obtain''' is adapted to provide users with a witness $\mathbf{w}$ and to store the corresponding e as r_D. Show''' is adapted to update and prove knowledge of the witness. The Revoke$(pk_\mathcal{I}, r_D)$ algorithm is defined as follows: Compute $\mathbf{v}' = \mathbf{v}^{r_D^{-1}}$ and publish it together with update information r_D. Additional details are in [15].

THEOREM 6.4. *The above scheme provides* dynamic revocation *under the Strong RSA, and y-DDHI if $l_x \in O(\log k)$ or SDDHI assumptions.*

7. ACKNOWLEDGMENTS

Part of Jan Camenisch's work reported in this paper is supported by the European Commission through the IST Programme under Contracts IST-2002-507932 ECRYPT and IST-2002-507591 PRIME. The PRIME projects receives research funding from the European Community's Sixth Framework Programme and the Swiss Federal Office for Education and Science. Part of Susan Hohenberger's work is supported by an NDSEG Fellowship. Markulf Kohlweiss is supported by the European Commission through the IST Programme under Contract IST-2002-507591 PRIME. Anna Lysyanskaya is supported by NSF Grant CNS-0347661. Mira Meyerovich is supported by a U.S. Department of Homeland Security Fellowship and NSF Grant CNS-0347661. All opinions expressed in this paper are the authors' and do not necessarily reflect the policies and views of EC, DHS, and NSF.

8. REFERENCES

[1] G. Ateniese, J. Camenisch, M. Joye, and G. Tsudik. A practical and provably secure coalition-resistant group signature scheme. In *CRYPTO*, vol. 1880, p. 255–270, 2000.

[2] G. Ateniese, K. Fu, M. Green, and S. Hohenberger. Improved Proxy Re-encryption Schemes with Applications to Secure Distributed Storage. In *NDSS*, p. 29–43, 2005.

[3] L. Ballard, M. Green, B. de Medeiros, and F. Monrose. Correlation-Resistant Storage. Johns Hopkins University, Technical Report # TR-SP-BGMM-050705, 2005.

[4] N. Barić and B. Pfitzmann. Collision-free accumulators and fail-stop signature schemes without trees. In *EUROCRYPT '97*, volume 1233, p. 480–494, 1997.

[5] D. Boneh and X. Boyen. Short signatures without random oracles. In *EUROCRYPT*, v.3027 of LNCS, p. 56–73, 2004.

[6] D. Boneh, X. Boyen, and H. Shacham. Short group signatures using strong Diffie-Hellman. In *CRYPTO*, volume 3152 of LNCS, p. 41–55, 2004.

[7] D. Boneh and M. Franklin. Identity-based encryption from the Weil pairing. In *CRYPTO*, v.2139, p. 213–229, 2001.

[8] F. Boudot. Efficient proofs that a committed number lies in an interval. In *EUROCRYPT*, vol. 1807, p. 431–444, 2000.

[9] S. Brands. *Rethinking Public Key Infrastructure and Digital Certificates— Building in Privacy*. PhD thesis, Eindhoven Inst. of Tech. The Netherlands, 1999.

[10] E. Brickell, P. Gemmel, and D. Kravitz. Trustee-based tracing extensions to anonymous cash and the making of anonymous change. In *SIAM*, p. 457–466, 1995.

[11] J. Camenisch and I. Damgård. Verifiable encryption, group encryption, and their applications to group signatures and signature sharing schemes. In *ASIACRYPT*, volume 1976 of *LNCS*, p. 331–345, 2000.

[12] J. Camenisch, S. Hohenberger, and A. Lysyanskaya. Compact E-Cash. In *EUROCRYPT*, volume 3494 of LNCS, p. 302–321, 2005.

[13] J. Camenisch, S. Hohenberger, and A. Lysyanskaya. Balancing accountability and privacy using e-cash. In *SCN (to appear)*, 2006.

[14] J. Camenisch and A. Lysyanskaya. Efficient non-transferable anonymous multi-show credential system with optional anonymity revocation. In *EUROCRYPT*, volume 2045 of LNCS, p. 93–118, 2001.

[15] J. Camenisch and A. Lysyanskaya. Dynamic accumulators and application to efficient revocation of anonymous credentials. In *CRYPTO*, 2442 of LNCS, p. 61-76, 2002.

[16] J. Camenisch and A. Lysyanskaya. A signature scheme with efficient protocols. In *SCN 2002*, volume 2576 of *LNCS*, p. 268–289, 2003.

[17] J. Camenisch and A. Lysyanskaya. Signature schemes and anonymous credentials from bilinear maps. In *CRYPTO 2004*, volume 3152 of LNCS, p. 56–72, 2004.

[18] J. Camenisch, U. Maurer, and M. Stadler. Digital payment systems with passive anonymity-revoking trustees. In *ESORICS 96*, volume 1146 of LNCS, p. 33–43, 1996.

[19] J. Camenisch and M. Michels. Proving in zero-knowledge that a number n is the product of two safe primes. In *EUROCRYPT '99*, volume 1592, p. 107–122, 1999.

[20] J. Camenisch and M. Michels. Separability and efficiency for generic group signature schemes. In *CRYPTO '99*, volume 1666 of *LNCS*, p. 413–430, 1999.

[21] J. Camenisch and M. Stadler. Efficient group signature schemes for large groups. In *CRYPTO '97*, volume 1296 of *LNCS*, p. 410–424, 1997.

[22] A. Chan, Y. Frankel, and Y. Tsiounis. Easy come – easy go divisible cash. In *EUROCRYPT*, v. 1403, p. 561–575, 1998.

[23] D. Chaum. Blind signatures for untraceable payments. In *CRYPTO '82*, p. 199–203. Plenum Press, 1982.

[24] D. Chaum. Blind signature systems. In *CRYPTO '83*, p. 153–156. Plenum, 1983.

[25] D. Chaum. Security without identification: Transaction systems to make big brother obsolete. *Communications of the ACM*, 28(10):1030–1044, Oct. 1985.

[26] D. Chaum, A. Fiat, and M. Naor. Untraceable electronic cash. In *CRYPTO*, volume 403 of *LNCS*, p. 319–327, 1990.

[27] D. Chaum and E. van Heyst. Group signatures. In *EUROCRYPT '91*, volume 547 of *LNCS*, p. 257–265, 1991.

[28] S. Chawla, C. Dwork, F. McSherry, A. Smith, and H. Wee. Toward privacy in public databases. In *TCC*, volume 3378 of *LNCS*, p. 363–385, 2005.

[29] I. Damgard, K. Dupont, and M. O. Pedersen. Unclonable group identification. In *EUROCRYPT*, volume 4004 of *LNCS*, p. 555–572, 2006.

[30] I. B. Damgård. Payment systems and credential mechanism with provable security against abuse by individuals. In *CRYPTO*, volume 403 of *LNCS*, p. 328–335, 1990.

[31] Y. Dodis. Efficient construction of (distributed) verifiable random functions. In *PKC*, volume 2567, p. 1–17, 2003.

[32] Y. Dodis and A. Yampolskiy. A Verifiable Random Function with Short Proofs an Keys. In *PKC*, volume 3386 of LNCS, p. 416–431, 2005.

[33] A. Fiat and A. Shamir. How to prove yourself: Practical solutions to identification and signature problems. In *CRYPTO*, volume 263 of LNCS, p. 186–194, 1986.

[34] E. Fujisaki and T. Okamoto. Statistical zero knowledge protocols to prove modular polynomial relations. In *CRYPTO '97*, volume 1294 of *LNCS*, p. 16–30, 1997.

[35] S. D. Galbraith. Supersingular curves in cryptography. In *ASIACRYPT*, volume 2248 of LNCS, p. 495–513, 2001.

[36] S. Jarecki and V. Shmatikov. Handcuffing big brother: an abuse-resilient transaction escrow scheme. In *EUROCRYPT*, volume 3027 of *LNCS*, p. 590–608, 2004.

[37] A. Kiayias, M. Yung, and Y. Tsiounis. Traceable signatures. In *EUROCRYPT*, vol. 3027, p. 571–589, 2004.

[38] A. Lysyanskaya. *Signature Schemes and Applications to Cryptographic Protocol Design*. PhD thesis, Massachusetts Institute of Technology, Sept. 2002.

[39] A. Lysyanskaya, R. Rivest, A. Sahai, and S. Wolf. Pseudonym systems. In *SAC*, vol. 1758, p. 184-199, 1999.

[40] N. McCullagh and P. S. L. M. Barreto. A new two-party identity-based authenticated key agreement. In *CT-RSA*, volume 3376 of LNCS, p. 262–274, 2004.

[41] V. I. Nechaev. Complexity of a determinate algorithm for the discrete log. *Mathematical Notes*, 55:165–172, 1994.

[42] L. Nguyen and R. Safavi-Naini. Dynamic k-times anonymous authentication. In *ACNS*, volume 3531 in LNCS, p. 318–333, 2005.

[43] T. P. Pedersen. Non-interactive and information-theoretic secure verifiable secret sharing. In *CRYPTO*, volume 576 of *LNCS*, p. 129–140, 1992.

[44] M. Scott. Authenticated ID-based key exchange and remote log-in with simple token and PIN number, 2002. http://eprint.iacr.org/2002/164.

[45] V. Shoup. Lower bounds for discrete logarithms and related problems. In *EUROCRYPT*, LNCS, p. 256–266, 1997. Update: http://www.shoup.net/papers/.

[46] L. Sweeney. k-anonymity: a model for protecting privacy. *International Journal on Uncertainty, Fuzziness and Knowledge-based Systems*, 10(5):557–570, 2002.

[47] I. Teranishi, J. Furukawa, and K. Sako. k-times anonymous authentication (extended abstract). In *Asiacrypt*, volume 3329 of LNCS, p. 308–322, 2004.

A Fully Collusion Resistant Broadcast, Trace, and Revoke System

Dan Boneh*
Stanford University
Stanford, CA
dabo@cs.stanford.edu

Brent Waters†
SRI International
Menlo Park, CA
bwaters@csl.sri.com

ABSTRACT

We introduce a simple primitive called *Augmented Broadcast Encryption* (ABE) that is sufficient for constructing broadcast encryption, traitor-tracing, and trace-and-revoke systems. These ABE-based constructions are resistant to an arbitrary number of colluders and are secure against *adaptive adversaries*. Furthermore, traitor tracing requires no secrets and can be done by anyone. These broadcast systems are designed for broadcasting to arbitrary sets of users. We then construct a secure ABE system for which the resulting concrete trace-and-revoke system has ciphertexts and private keys of size $\sqrt{N}$ where N is the total number of users in the system. In particular, this is the first example of a fully collusion resistant broadcast system with sub-linear size ciphertexts and private keys that is secure against adaptive adversaries. The system is publicly traceable.

Categories and Subject Descriptors

E.3 [**Data**]: Data Encryption

General Terms

Security

1. INTRODUCTION

A **broadcast encryption** system [15] enables a broadcaster to encrypt a message for an arbitrary subset $S \subseteq \{1, \ldots, N\}$ of users who are listening on a broadcast channel. Any user in S can decrypt the broadcast using his private key. Moreover, even if *all* users outside of S collude they obtain no information about the contents of the broadcast. Such systems are said to be collusion resistant. **Traitor tracing** [10] is an orthogonal problem. Here a broadcaster encrypts messages so that *all* N users can decrypt the resulting ciphertexts. Suppose a coalition of users $T \subseteq \{1, \ldots, N\}$

*Supported by NSF and the Packard Foundation
†Supported by NSF

get together and build a pirate decoder $\mathcal{D}$. Then there is a tracing algorithm *Trace* that takes the public key PK as input and interacts with $\mathcal{D}$ as a black-box oracle. The algorithm outputs the identity of at least one of the users who created $\mathcal{D}$. That is, $\emptyset \ne Trace^\mathcal{D}(\text{PK}) \subseteq T$. Note, however, that there is no way to revoke the traitor — broadcasts can always be decrypted by all users. The tracing algorithm, as described above, needs no secrets and can be run by anyone. Such systems are said to be publicly traceable.

Trace and Revoke [25, 24] systems provide both broadcast encryption and traitor tracing. They are motivated by content protection on various platforms such as PCs, DVD players, and general content viewers. When the system is first rolled out, broadcasts are encrypted for some subset of users $S \subseteq \{1, \ldots, N\}$ authorized to receive them. The goal is to then revoke users when their keys are compromised. Suppose a pirate builds a pirate decoder $\mathcal{D}$ using the private keys of users $T \subseteq \{1, \ldots, N\}$. The tracing algorithm then interacts with $\mathcal{D}$ and identifies one of the active keys in the pirate's possession, namely a key of user $t \in T \cap S$. We write $\emptyset \ne Trace^\mathcal{D}(\text{PK}, S) \subseteq T \cap S$. The broadcaster revokes user t by encrypting future broadcasts to the set $S' \leftarrow S \setminus \{t\}$. If the pirate decoder $\mathcal{D}$ can still decrypt these broadcasts, we run the tracing algorithm $Trace^\mathcal{D}(\text{PK}, S')$ again and obtain another pirate key $t' \in T \cap S'$. Again, t' is revoked by setting $S'' \leftarrow S' \setminus \{t'\}$ and so on. Roughly speaking, the trace and revoke system is secure if this process eventually disables $\mathcal{D}$ without revoking any innocent party. We give precise definitions later in the paper. Note that the broadcaster can add or remove recipients from S at will.

Our Contribution. In this paper we focus on constructing public-key trace and revoke systems that are fully collusion resistant and have short ciphertexts and private keys. The system is publicly traceable in the sense that anyone can run the tracing algorithm — no additional secrets are needed. Since a party performing the tracing needs no secrets, the overall system remains secure even if this party is compromised. For message privacy we only consider chosen plaintext attacks. Rather than directly build a trace and revoke system, we instead construct a simpler primitive we call **Augmented Broadcast Encryption** or ABE for short. We then show that ABE implies a trace and revoke system.

An ABE contains the same algorithms as a public-key broadcast encryption system, namely

$$(Setup_{\text{ABE}}, \quad Encrypt_{\text{ABE}}, \quad Decrypt_{\text{ABE}})$$

The encryption algorithm $Encrypt_{\text{ABE}}(S, \text{PK}, i, M)$, however, takes one additional parameter i. Here PK is the public key, M is a message, S is a subset of $\{1, \ldots, N\}$, and i is an additional special input $1 \le i \le N+1$. The encryption algorithm outputs a ciphertext that can be decrypted by any user in $S \cap \{i, \ldots, N\}$. We require that

- The output of $Encrypt_{\text{ABE}}(S, \text{PK}, N+1, M)$ contains no information about M, and
- For $i \in S$ the distribution generated by algorithm $Encrypt_{\text{ABE}}(S, \text{PK}, i, M)$ is indistinguishable from the distribution generated by $Encrypt_{\text{ABE}}(S, \text{PK}, i+1, M)$ for any attacker that does not possess the secret key of user i. When $i \notin S$ the two distributions are indistinguishable to anyone.

We give precise definitions in the next section. We show that an ABE system directly gives a secure and fully collusion resistant broadcast encryption system. To encrypt message M to set S we run $Encrypt_{\text{ABE}}(S, \text{PK}, 1, M)$, namely setting $i = 1$. Values of i greater than 1 are only used in the proof of security and for tracing. The resulting broadcast encryption system is secure against *adaptive adversaries* — adversaries that choose adaptively the subset of users to attack. We then show that this broadcast system is publicly traceable (and hence is a trace and revoke system). The tracing algorithm is based on a standard tracing technique that was previously used in [3, 24, 21] and was recently made explicit in [6]. The tracing system of [6], however, requires a secret tracing key held by a trusted party. Here, tracing requires no secrets so that there is no need for a trusted party.

In summary, the two simple ABE security properties are sufficient for obtaining a trace and revoke system that is fully collusion resistant, is secure against adaptive adversaries, and is publicly traceable. We view this as the preamble leading to our main results. The main part of the paper builds a secure ABE system where the size of private keys and ciphertexts is $\sqrt{N}$. We thus obtain a trace and revoke system with $\sqrt{N}$ size ciphertext and private keys that is fully collusion resistant and is secure against adaptive adversaries. Some previous fully collusion resistant broadcast systems [4, 7] for arbitrary sets were only secure against static adversaries and were not traceable. Building a broadcast system secure against adaptive adversaries was left as an open problem in [4].

1.1 Related work

Broadcast encryption systems are often designed for the case when the pirate has fewer than some t private keys [15, 35, 36, 1, 37, 25, 13, 18]. Several elegant constructions [24, 12, 20, 19], primarily designed for broadcasting to sets where a small number of users are revoked, resist arbitrary collusion, but the size of the ciphertext grows linearly with the number of revoked users. A recent system based on pairings [4] resists arbitrary collusion and has constant size ciphertext and private keys, but does not support traitor tracing. The system is only proven secure for static adversaries, namely adversaries that commit to the set they wish to attack before seeing the public key. Broadcast encryption secure against adaptive attacks was defined in [13], but the resulting system had linear size ciphertexts when broadcasting to an arbitrary set S. The system in this paper provides adaptive security with sub-linear ciphertexts and private keys.

Similarly, traitor tracing systems are often designed for the case when the pirate has fewer than t private keys [10, 34, 32, 33, 23, 26, 3, 16, 11, 29, 2, 31, 30, 37, 21, 22]. A recent system based on pairings [6] resists arbitrary collusion and has constant size private keys and $\sqrt{N}$ size ciphertexts. That system is the basis of our tracing mechanism. Many tracing traitors systems, including [6], assume the tracer is a trusted party and require a secret tracing key. The system in this paper is publicly traceable meaning that tracing requires no secrets. Other publicly traceable systems are provided in [27, 28, 38, 22, 9].

Several trace and revoke systems are available [25, 17, 24, 37, 13, 14, 20, 19] that are designed for broadcasting to large sets. Table 1 summarizes the existing sub-linear size fully collusion resistant systems currently available. Here N is the total number of users in the system. As usual, all the expressions in the table should be multiplied by the security parameter.

2. AUGMENTED BROADCAST ENC.

Our goal is to build a fully collusion resistant trace and revoke system secure against adaptive adversaries. In particular, this gives a broadcast encryption system secure against adaptive adversaries. However, instead of directly building a trace and revoke system we build a simpler primitive we call *Augmented Broadcast Encryption* or ABE for short. We begin by defining what ABE is and then explain how it gives a trace and revoke system. Then in the next section we build an efficient ABE.

2.1 Augmented Broadcast Encryption: Definitions

An ABE is a public-key broadcast system comprising of the following algorithms:

Setup$_{\text{ABE}}(N, \lambda)$ A probabilistic algorithm that takes as input N, the number of users in the system, and a security parameter λ. The algorithm runs in polynomial time in λ and outputs a public key PK and private keys $\text{SK}_1, \ldots, \text{SK}_N$, where SK_u is given to user u.

Encrypt$_{\text{ABE}}(S, \text{PK}, i, M)$ Takes as input a subset of users $S \subseteq \{1, \ldots, N\}$, a public key PK, an integer i satisfying $1 \le i \le N+1$, and a message M. It outputs a ciphertext C. This algorithm encrypts a message to a set $S \cap \{i, \ldots, N\}$.

Decrypt$_{\text{ABE}}(S, j, \text{SK}_j, C, \text{PK})$ Takes as input a subset $S \subseteq \{1, \ldots, N\}$, the private key SK_j for user j, a ciphertext C, and the public key PK. The algorithm outputs a message M or $\bot$.

Correctness property. The system must satisfy the following correctness property:

for all subsets $S \subseteq \{1, \ldots, N\}$, all $i, j \in \{1, \ldots, N+1\}$ (where $j \le N$), and all messages M:

Let $(\text{PK}, (\text{SK}_1, \ldots, \text{SK}_N)) \xleftarrow{R} Setup_{\text{ABE}}(N, \lambda)$

and $C \xleftarrow{R} Encrypt_{\text{ABE}}(S, \text{PK}, i, M)$.

If $\mathbf{j \in S}$ and $\mathbf{j \ge i}$ then

$$Decrypt_{\text{ABE}}(S, j, \text{SK}_j, C, \text{PK}) = M.$$

	system type	ciphertext size	private key size	public key size	comment
[4]	Broadcast Encryption	$O(1)$	$O(1)$	$O(N)$	Static attacker
[6]	Traitor Tracing	$O(\sqrt{N})$	$O(1)$	$O(\sqrt{N})$	Private tracing
This paper	Trace and Revoke	$O(\sqrt{N})$	$O(\sqrt{N})$	$O(\sqrt{N})$	adaptive attacker and public tracing

Table 1: Sub-linear size fully collusion resistant systems

Security. We define security of an ABE system using two games. The first game is a **message hiding game** and says that a ciphertext created using index $i = N+1$ is unreadable by anyone. The second game is an **index hiding game** and captures the intuition that a broadcast ciphertext created using index i reveals no non-trivial information about i. We will consider all these games for a fixed number of users, N.

For simplicity, we define our games for security against an adversary that mounts a chosen-plaintext attack (CPA). We can easily extend them to handle chosen-ciphertext attacks (CCA) by giving the adversary access to a decryption oracle for each user in the system.

Game 1. The first game, called **Message Hiding**, says that an adversary cannot break semantic security when encrypting using index $i = N + 1$. The game proceeds as follows:

- **Setup** The challenger runs $Setup_{ABE}(N, \lambda)$ and gives the adversary PK and all secret keys $\{SK_1, \ldots, SK_N\}$.

- **Challenge** The adversary outputs a set $S \subseteq \{1, \ldots, N\}$ and two equal length messages M_0, M_1. The challenger flips a coin $\beta \in \{0, 1\}$ and sends
$$C \xleftarrow{R} Encrypt_{ABE}(S, PK, N + 1, M_\beta)$$
to the adversary.

- **Guess** The adversary returns a guess $\beta' \in \{0, 1\}$ of β.

We define the advantage of adversary $\mathcal{A}$ in winning the game as $\mathsf{MH\,Adv}_\mathcal{A} = |\Pr[\beta' = \beta] - 1/2|$.

Game 2. The second game, called **Index Hiding**, says that an adversary cannot distinguish between an encryption to index i and one to index $i + 1$ without the key SK_i. Additionally, it says that an adversary cannot distinguish between an encryption to index i and one to index $i + 1$ when i is not in the target set S even with the key SK_i. The game takes as input a parameter $i \in \{1, \ldots, N\}$ which is given to both the challenger and the adversary. The game proceeds as follows:

- **Setup** The challenger runs $Setup_{ABE}(N, \lambda)$ and gives the adversary PK and the set of private keys
$$\{SK_j \text{ s.t. } j \neq i\}$$

- **Query** The adversary outputs a bit $\tilde{s} \in \{0, 1\}$. If $\tilde{s} = 1$ the challenger sends SK_i to the adversary. Otherwise the challenger does nothing.

- **Challenge** The adversary gives the challenger a set $S \subseteq \{1, \ldots, N\}$ and a message M. The only restriction is that if $\tilde{s} = 1$ then $i \notin S$. The challenger flips a coin $\beta \in \{0, 1\}$ and sends $C \xleftarrow{R} Encrypt_{ABE}(S, PK, i+\beta, M)$ to the adversary.

- **Guess** The adversary returns a guess $\beta' \in \{0, 1\}$ of β.

We define the advantage of adversary $\mathcal{A}$ as the quantity $\mathsf{IH\,Adv}_\mathcal{A}[i] = |\Pr[\beta' = \beta] - 1/2|$. In words, the game captures two properties. The case $\tilde{s} = 0$ captures the fact that even if all users other than i collude they cannot distinguish whether i or $i + 1$ was used to create a ciphertext C. The case $\tilde{s} = 1$ captures the fact that when $i \notin S$ then even if everyone colludes they cannot distinguish whether i or $i + 1$ was used to create C. Indeed, when $i \notin S$ the key SK_i gives little additional information.

Now that the games are established we are ready to define secure ABE.

DEFINITION 2.1. *We say that an N-user Augmented Broadcast System (ABE) is secure if for all polynomial time adversaries $\mathcal{A}$ we have that $\mathsf{MH\,Adv}_\mathcal{A}$ and $\mathsf{IH\,Adv}_\mathcal{A}[i]$ for $i = 1, \ldots, N$, are negligible functions of λ.*

2.2 Using Augmented Broadcast Encryption

We first show that a secure ABE is a broadcast encryption system secure against adaptive attackers. We then show that this system is traceable, thus obtaining a trace and revoke system. From here on, whenever we refer to an adversary we mean an adversary whose running time is polynomial in the security parameter λ.

2.2.1 Broadcast encryption secure against adaptive attacks

Let $\mathcal{E} = (Setup_{ABE}, Encrypt_{ABE}, Decrypt_{ABE})$ be a secure ABE system. Define
$$Encrypt(S, PK, M) = Encrypt_{ABE}(S, PK, 1, M)$$

We show that $\mathcal{E}_{BE} = (Setup_{ABE}, Encrypt, Decrypt_{ABE})$ is a fully collusion resistant broadcast encryption system secure against adaptive attackers.

First we need a slightly more elaborate message hiding game. In addition to N, λ, the extended message hiding game takes as input a parameter $i \in \{1, \ldots, N+1\}$ which is only given to the challenger. The game proceeds as follows:

- **Setup** The challenger runs $Setup_{ABE}(N, \lambda)$ and gives the adversary PK.

- **Query** The adversary issues *adaptive* private key queries: it repeatedly sends values $j \in \{1, \ldots, N\}$ to the challenger and the challenger responds with SK_j. Let $S_0 \subseteq \{1, \ldots, N\}$ denote the entire set of private keys requested by the adversary during the query phase. Let $\overline{S_0} = \{1, \ldots, N\} \setminus S_0$.

- **Challenge** The adversary outputs a set $S \subseteq \overline{S_0}$ and two equal length messages M_0, M_1. The challenger flips a coin $\beta \in \{0, 1\}$ and sends
$$C \xleftarrow{R} Encrypt_{\text{ABE}}(S, \text{PK}, i, M_\beta)$$
to the adversary. This is the only place where i is used in this game.

- **Guess** The adversary returns a guess $\beta' \in \{0, 1\}$ of β.

We define the advantage of adversary $\mathcal{A}$ in winning the game as $\text{MH Adv}_\mathcal{A}[i] = \big| \Pr[\beta' = \beta] - 1/2 \big|$.

The main point is that $\text{MH Adv}_\mathcal{A}[1]$ is the same quantity used to define broadcast encryption security against adaptive attackers [13, 4] for $\mathcal{E}_{BE}$. Hence, if we prove that $\text{MH Adv}_\mathcal{A}[1]$ is negligible then $\mathcal{E}_{BE}$ is a broadcast system that is fully collusion resistant and secure against adaptive adversaries.

THEOREM 2.2. *If $\mathcal{E}$ is a secure ABE then $\text{MH Adv}_\mathcal{A}[1]$ is a negligible function of λ for any polynomial time adversary $\mathcal{A}$.*

PROOF SKETCH. Suppose $\text{MH Adv}_\mathcal{A}[1] > \epsilon$ for some adversary $\mathcal{A}$ and non-negligible ϵ. Since $\mathcal{E}$ is a secure ABE we know that $\text{MH Adv}_\mathcal{A}$ (defined in Game 1) is negligible. It follows that $\text{MH Adv}_\mathcal{A}[N+1]$ is negligible. For simplicity, say $\text{MH Adv}_\mathcal{A}[N+1] = 0$. Then, by the standard hybrid argument there exists a $j \in \{1, \ldots, N\}$ such that
$$\big| \text{MH Adv}_\mathcal{A}[j] - \text{MH Adv}_\mathcal{A}[j+1] \big| > \epsilon/N$$
In other words, this $\mathcal{A}$ is somehow able to distinguish
$$Encrypt_{\text{ABE}}(S, \text{PK}, j, M) \text{ from } Encrypt_{\text{ABE}}(S, \text{PK}, j+1, M)$$
for some M and S. But then $\mathcal{A}$ can be directly used to win the ABE index hiding game.

More precisely, we show in the full version that for all adversaries $\mathcal{A}$ there exists an adversary $\mathcal{B}$ such that for all $i = 1, \ldots, N$ we have
$$\big| \text{MH Adv}_\mathcal{A}[i] - \text{MH Adv}_\mathcal{A}[i+1] \big| \leq 2 \cdot \text{IH Adv}_\mathcal{B}[i] \quad (1)$$
Then
$$\big| \text{MH Adv}_\mathcal{A}[1] - \text{MH Adv}_\mathcal{A}[N+1] \big| \leq$$
$$\sum_{i=1}^{n} \big| \text{MH Adv}_\mathcal{A}[i] - \text{MH Adv}_\mathcal{A}[i+1] \big| \leq$$
$$2 \sum_{i=1}^{n} \text{IH Adv}_\mathcal{B}[i]$$

But since $\mathcal{E}$ is a secure ABE we know that $\text{MH Adv}_\mathcal{A}[N+1]$ and $\text{IH Adv}_\mathcal{B}[i]$ for $i = 1, \ldots, N$ are negligible for any polynomial time $\mathcal{A}$. Therefore, $\text{MH Adv}_\mathcal{A}[1]$ is negligible, as required. □

2.2.2 Trace and Revoke

A trace and revoke system is a broadcast system with a tracing algorithm. We formally define trace and revoke systems in Appendix A along with the games used to define security. We show here that a secure ABE directly gives a trace and revoke system. In particular, we show a tracing algorithm for the broadcast system $\mathcal{E}_{BE}$ above. The tracing algorithm uses a general tracing method, previously used in [3, 24, 21, 6]. We use the notation from Appendix A. For a given $\epsilon > 0$ and a set $S_\mathcal{D}$ the tracing algorithm $Trace^\mathcal{D}(S_\mathcal{D}, \text{PK}, \epsilon)$ works as follows.

1. Initialize set T to the empty set.
2. For $i = 1$ to N, do the following:
   (a) The algorithm repeats the following steps $8\lambda(N/\epsilon)^2$ times:
      i. Sample M from the finite message space at random.
      ii. Let $C \xleftarrow{R} Encrypt_{\text{ABE}}(S_\mathcal{D}, \text{PK}, i, M)$.
      iii. Call oracle $\mathcal{D}$ on input C, and compare the output of $\mathcal{D}$ to M.
   (b) Let $\hat{p}_i$ be the fraction of times that $\mathcal{D}$ decrypted the ciphertexts correctly.
   (c) If $\hat{p}_i - \hat{p}_{i+1} \geq \epsilon/(4N)$, then add i to set T.
3. Output the set T.

Note that the running time of $Trace$ is cubic in N. It can be made (almost) quadratic using binary search instead of a linear scan.

Let $\mathcal{E}_{TR} = (Setup_{\text{ABE}}, Encrypt, Decrypt_{\text{ABE}}, Trace)$ be the resulting trace and revoke system. Note $\mathcal{E}_{TR}$ is just the broadcast system $\mathcal{E}_{BE}$ with the tracing algorithm $Trace$. We show that $\mathcal{E}_{TR}$ is secure in the sense of Definition A.1, namely fully collusion resistant against an adaptive adversary.

THEOREM 2.3. *If $\mathcal{E}$ is a secure ABE then for $\mathcal{E}_{TR}$ the quantity $\text{TR Adv}_\mathcal{A}$ defined in Appendix A.1 is negligible.*

PROOF SKETCH. Let $(\mathcal{D}, S_\mathcal{D})$ be the pirate decoder and the recipient set output by the adversary. Define
$$p_i = \Pr[\mathcal{D}(Encrypt_{\text{ABE}}(S_\mathcal{D}, \text{PK}, i, M)) = M]$$
We know that $p_1 \geq \epsilon$ and p_{N+1} is negligible. The former follows from the fact that $\mathcal{D}$ is a useful decoder. The later follows directly from the ABE message hiding game. Then there must exist some $j \in \{1, \ldots, N\}$ such that $p_j - p_{j+1} \geq \epsilon/(2N)$. By the Chernoff bound it follows that with overwhelming probability, $\hat{p}_j - \hat{p}_{j+1} \geq \epsilon/(4N)$. Hence, the set T output by $Trace^\mathcal{D}(S_\mathcal{D}, \text{PK}, \epsilon)$ is non-empty.

It remains to show that whenever $\hat{p}_j - \hat{p}_{j+1} > \epsilon/(4N)$ we have that $j \in S_\mathcal{D} \cap U$. For such j we know, by Chernoff, that with overwhelming probability $p_j - p_{j+1} \geq \epsilon/(8N)$. We can now show that the ABE index hiding game implies $j \in U$ and $j \in S_\mathcal{D}$. Clearly $j \in S_\mathcal{D}$ since otherwise, even given all the secret keys, there is no hope of distinguishing p_j from p_{j+1}. But if $j \in S_\mathcal{D}$ and $j \notin U$ then $\mathcal{D}$ must distinguish p_j from p_{j+1} without the key SK_j. Again, such a $\mathcal{D}$ can be directly used to win the ABE index hiding game. Hence, $j \in S_\mathcal{D} \cap U$. We give the proof details in the full version of the paper. □

3. BACKGROUND AND COMPLEXITY ASSUMPTIONS

Our traitor tracing system uses bilinear groups of composite order. We review the definition of such groups and then state our complexity assumptions. We follow [5] in which composite order bilinear groups were first introduced.

The assumptions we make are a little stronger than the ones in [6]. These stronger assumptions are needed to make the system publicly traceable, namely not requiring a secret tracing key. Without public traceability a direct variant of the system in this paper can be proven secure under the exact same assumptions used in [6].

Bilinear groups of composite order. Let $\mathcal{G}$ be an algorithm called a *group generator* that takes as input a security parameter $\lambda \in \mathbb{Z}^{>0}$ and outputs a tuple $(p, q, \mathbb{G}, \mathbb{G}_T, e)$ where p, q are two distinct primes, $\mathbb{G}$ and $\mathbb{G}_T$ are two cyclic groups of order $n = pq$, and e is a function $e : \mathbb{G}^2 \to \mathbb{G}_T$ satisfying the following properties:

- (Bilinear) $\forall u, v \in \mathbb{G}, \forall a, b \in \mathbb{Z}, e(u^a, v^b) = e(u, v)^{ab}$.

- (Non-degenerate) exists $g \in G$ such that $e(g, g)$ has order n in $\mathbb{G}_T$.

We assume that the group action in $\mathbb{G}$ and $\mathbb{G}_T$ as well as the bilinear map e are all computable in polynomial time in λ. Furthermore, we assume that the description of $\mathbb{G}$ and $\mathbb{G}_T$ includes a generator of $\mathbb{G}$ and $\mathbb{G}_T$ respectively.

To summarize, $\mathcal{G}$ outputs the description of a group $\mathbb{G}$ of order $n = pq$ with an efficiently computable bilinear map. We will use the notation $\mathbb{G}_p, \mathbb{G}_q$ to denote the respective subgroups of order p and order q of $\mathbb{G}$.

3.1 Complexity assumptions

Next we review three complexity assumptions needed for proving security of our system. The first assumption is in a prime order subgroup $\mathbb{G}_p$ and the last two are over the composite order group $\mathbb{G}$.

Decision (Modified) 3-party Diffie-Hellman Assumption. For a given group generator $\mathcal{G}$ define the following distribution $P(\lambda)$:

$$(p, q, \mathbb{G}, \mathbb{G}_T, e) \xleftarrow{R} \mathcal{G}(\lambda), \quad n \leftarrow pq, \quad g_p \xleftarrow{R} \mathbb{G}_p$$
$$a, b, c \xleftarrow{R} \mathbb{Z}_p$$
$$\bar{Z} \leftarrow \left((n, \mathbb{G}, \mathbb{G}_T, e), \ g_p, \ g_p^a, \ g_p^b, \ g_p^c, \ g_p^{(b^2)} \right)$$
$$T \leftarrow g_p^{abc}$$
$$\text{Output } (\bar{Z}, T)$$

For an algorithm $\mathcal{A}$, define $\mathcal{A}$'s advantage in solving the decision 3-party Diffie-Hellman problem for $\mathcal{G}$ as:

$$\mathsf{D3DH\,Adv}_{\mathcal{G},\mathcal{A}}(\lambda) := \left| \Pr[\mathcal{A}(\bar{Z}, T) = 1] - \Pr[\mathcal{A}(\bar{Z}, R) = 1] \right|$$

where $(\bar{Z}, T) \xleftarrow{R} P(\lambda)$ and $R \xleftarrow{R} \mathbb{G}_p$.

DEFINITION 3.1. *We say that $\mathcal{G}$ satisfies the decision (modified) 3-party Diffie-Hellman assumption (D3DH) if for any polynomial time algorithm $\mathcal{A}$ we have that $\mathsf{D3DH\,Adv}_{\mathcal{G},\mathcal{A}}(\lambda)$ is a negligible function of λ.*

The assumption is a little stronger than the corresponding assumption in [6] since we also give $g_p^{(b^2)}$ to the adversary.

Diffie-Hellman Subgroup Decision Assumption. The Diffie-Hellman subgroup decision assumption states that a random element in $\mathbb{G}_q$ is indistinguishable from a random element in $\mathbb{G}$, even when an ElGamal encryption of a $\mathbb{G}_p$ element is provided. More precisely, for a given group generator $\mathcal{G}$ define the following distribution $P(\lambda)$:

$$(p, q, \mathbb{G}, \mathbb{G}_T, e) \xleftarrow{R} \mathcal{G}(\lambda), \quad n \leftarrow pq$$
$$g, h \xleftarrow{R} \mathbb{G}, \quad v_p \xleftarrow{R} \mathbb{G}_p$$
$$a \xleftarrow{R} \mathbb{Z}_q, \quad b \xleftarrow{R} \mathbb{Z}_n,$$
$$\bar{Z} \leftarrow \left((n, \mathbb{G}, \mathbb{G}_T, e), \ g, \ h, \ g^b v_p, \ h^b, \ g^{pa}, \ h^{pa} \right)$$
$$\text{Output } \bar{Z}$$

For an algorithm $\mathcal{A}$, define $\mathcal{A}$'s advantage in solving the Diffie-Hellman Subgroup Decision problem for $\mathcal{G}$ as:

$$\mathsf{DHSD\,Adv}_{\mathcal{G},\mathcal{A}}(\lambda) := \left| \Pr[\mathcal{A}(\bar{Z}, T) = 1] - \Pr[\mathcal{A}(\bar{Z}, R) = 1] \right|$$

where $\bar{Z} \xleftarrow{R} P(\lambda)$, $T \xleftarrow{R} \mathbb{G}_q$, and $R \xleftarrow{R} \mathbb{G}$.

DEFINITION 3.2. *We say that $\mathcal{G}$ satisfies the Diffie-Hellman subgroup decision assumption (DHSD) if for any polynomial time algorithm $\mathcal{A}$ we have that $\mathsf{DHSD\,Adv}_{\mathcal{G},\mathcal{A}}(\lambda)$ is a negligible function of λ.*

The Diffie-Hellman subgroup decision assumption is a little stronger than the subgroup decision assumption introduced in [5] and also used in [6]. In our definition the adversary is also given an ElGamal encryption of an element v_p that is known to be in $\mathbb{G}_p$. Furthermore, the adversary is given $g^{pa}, h^{pa} \in \mathbb{G}_q$.

Bilinear Subgroup Decision Assumption. The Bilinear Subgroup Decision (BSD) assumption states that a random order p element in $\mathbb{G}_T$ is indistinguishable from a random element in $\mathbb{G}_T$ when $g_p, g_q \in \mathbb{G}$ are given. More precisely, for a given group generator $\mathcal{G}$ define the following distribution $P(\lambda)$:

$$(p, q, \mathbb{G}, \mathbb{G}_T, e) \xleftarrow{R} \mathcal{G}(\lambda), \quad n \leftarrow pq, \quad g_p \xleftarrow{R} \mathbb{G}_p, \quad g_q \xleftarrow{R} \mathbb{G}_q,$$
$$\bar{Z} \leftarrow \left((n, \mathbb{G}, \mathbb{G}_T, e), \ g_p, \ g_q \right)$$
$$\text{Output } \bar{Z}$$

For an algorithm $\mathcal{A}$, define $\mathcal{A}$'s advantage in solving the bilinear subgroup decision problem for $\mathcal{G}$ as:

$$\mathsf{BSD\,Adv}_{\mathcal{G},\mathcal{A}}(\lambda) :=$$
$$\left| \Pr[\mathcal{A}(\bar{Z}, e(T, g)) = 1] - \Pr[\mathcal{A}(\bar{Z}, e(R, g)) = 1] \right|$$

where $\bar{Z} \xleftarrow{R} P(\lambda)$, $T \xleftarrow{R} \mathbb{G}_p$, and $R \xleftarrow{R} \mathbb{G}$. Here g is an arbitrary generator of $\mathbb{G}$.

DEFINITION 3.3. *We say that $\mathcal{G}$ satisfies the bilinear subgroup decision assumption (BSD) if for any polynomial time algorithm $\mathcal{A}$ we have that $\mathsf{BSD\,Adv}_{\mathcal{G},\mathcal{A}}(\lambda)$ is a negligible function of λ.*

4. AN EFFICIENT AUGMENTED BROADCAST ENCRYPTION SYSTEM

We construct an Augmented Broadcast Encryption (ABE) system that has ciphertexts and private keys of size $O(\sqrt{N})$. We begin by offering some intuition into the design and technical novelty of our scheme. An Augmented Broadcast Encryption system must have both broadcast and tracing properties. To achieve this we will use some techniques from the broadcast encryption system of [4] and the traitor tracing system of [6].

4.1 Difficulty of Achieving Trace and Revoke

A trace and revoke system cannot be constructed by naively combining a broadcast encryption system and a tracing system. Consider the following (misguided) approach. Suppose we created both a broadcast encryption and traitor tracing system each for N users, where each user has the same index in both systems. To encrypt a message M, an algorithm splits the message randomly into two pieces M_b, M_t such that $M_b \cdot M_t = M$ and then encrypts M_b under the broadcast system and M_t under the tracing system. In order to decrypt a message a single user will need to be able to decrypt under both systems. However, if two users, Alice and Bob collude to make a pirate decoder they can break this construction. They will simply use Alice's key to decrypt the ciphertext from the broadcast system and Bob's key to decrypt the ciphertext from the tracing system. The tracing algorithm will identify Bob as a traitor. However, after Bob is revoked (from the broadcast system) the decoder will still be functional and moreover will continue to identify Bob as the traitor even though he was already revoked!

4.2 Our Approach

The principle behind resisting this type of attack is construct user's private keys in such a way that they must be simultaneously used for both the broadcast and tracing portions of a trace and revoke system. We are able to construct a secure ABE system by preventing colluding users from decomposing the two systems — the two sub-systems are essentially intertwined. In order to achieve this we multiply keys of the two portions together. Additionally, unlike [4] and [6] private keys in our system are randomized for each user to prevent such attacks. The more straightforward combination of [4] and [6] results in a scheme that is insecure since a revoked user can still break the "privacy" property and prevent other colluders from being traced, although we do not show this here. A consequence of using randomized keys is that each user must have $O(\sqrt{N})$ size private key storage.

The other primary contribution of our scheme is that it allows for public traceability. This comes from the fact that there exists a public key algorithm for encrypting to arbitrary index. In [6] the public encryption algorithm could only be used to broadcast a message to everyone and a secret tracing key was required for encrypting to arbitrary indices. The reason behind this was that "column" ciphertexts needed to be randomized in the $\mathbb{G}_p$ subgroup, while kept well-formed in the $\mathbb{G}_q$ subgroup. The most natural way to do this is to give an element of $\mathbb{G}_p$ as part of the public parameters. However, giving out such an element allows an attacker to break the scheme. In our construction we construct the parameters in such a way that allows for this type of public encryption without giving out an element of $\mathbb{G}_p$.

4.3 Notation

We will express our ABE system using the same two index notation as the tracing traitors system of [6]. We assume that the number of users, N in the system equals m^2 for some m. If the number of users is not a square we can add "dummy" users to pad out to the next square. We arrange the users in an $m \times m$ matrix. Each user is assigned and identified by an unique tuple (x, y) where $1 \leq x, y \leq m$.

We must have a linear ordering of the users that we can traverse. The first user in the system will be the user at matrix position $(1, 1)$ and from there we will order the users by traversing one row at a time. More precisely, the user at matrix position (x, y) will have the index $u = (x-1)m+y$ in our ordering. Additionally, an encryption to position (i, j) means that a user at position (x, y) will be able to decrypt the message if either $x > i$ or both $x = i$ and $y \geq j$. With this notation, the Index Hiding property states that:

- For $j < m$ it is difficult to distinguish between an encryption of a message to (i, j) from $(i, j+1)$ without the key of user $(x = i, y = j)$.

- For $j = m$ it is difficult to distinguish an encryption of a message to position $(i, j = m)$ to that of one to $(i+1, j = 1)$ without the key of user $(i, j = m)$.

We emphasize that the use of pairwise notation (i, j) is purely a notational convenience for describing our system.

4.4 ABE Construction

We will assume there are $N = m^2$ users in the system and we will address each user will be assigned a unique pair of indexes (x, y) where $1 \leq x, y \leq m$. A detailed description of the algorithms follows:

$Setup_{\text{ABE}}(N = m^2, \lambda)$.
The setup algorithm takes as input the number of users N and a security parameter λ. It first generates an integer $n = pq$ where p, q are random primes (whose size is determined by the security parameter). The algorithm creates a bilinear group $\mathbb{G}$ of composite order n. It next creates random generators $g_p, h_p \in \mathbb{G}_p$ and $g_q, h_q \in \mathbb{G}_q$ and sets $g = g_p g_q, h = h_p h_q \in \mathbb{G}$. Additionally it chooses random elements $u_{p,1}, \ldots, u_{p,m} \in \mathbb{G}_p$, $u_{q,1}, \ldots, u_{q,m} \in \mathbb{G}_q$, and defines $u_i = u_{p,i} u_{q,i}$ for $i = 1, \ldots, m$. Next it chooses random exponents

$$\delta, r_1, \ldots, r_m, \quad c_1, \ldots, c_m, \quad \alpha_1, \ldots, \alpha_m \in \mathbb{Z}_n$$
$$\beta \in \mathbb{Z}_q, \quad \gamma \in \mathbb{Z}_p$$

The public key PK includes the description of the group and the following elements:

$$\begin{bmatrix}
g, \ h, \ \tilde{V} = g^\delta g_p^\gamma, \ V = h^\delta \\
E_q = g_q^\beta, \ E_1 = g^{\beta r_1}, \ \ldots, \ E_m = g^{r_m}, \\
E_{q,1} = g_q^{\beta r_1}, \ \ldots, \ E_{q,m} = g_q^{\beta r_m}, \\
F_1 = h^{r_1}, \ \ldots, \ F_m = h^{r_m}, \\
F_{q,1} = h_q^{\beta r_1}, \ \ldots, \ F_{q,m} = h_q^{\beta r_m} \\
G_1 = e(g,g)^{\alpha_1}, \ \ldots, \ G_m = e(g,g)^{\alpha_m}, \\
G_{q,1} = e(g_q, g_q)^{\beta \alpha_1}, \ \ldots, \ G_{q,m} = e(g_q, g_q)^{\beta \alpha_m} \\
H_1 = g^{c_1}, \ \ldots, \ H_m = g^{c_m}, \\
U_1 = u_1, \ \ldots, \ U_m = u_m, \\
U_{q,1} = u_{q,1}^\beta, \ \ldots, \ U_{q,m} = u_{q,m}^\beta
\end{bmatrix}$$

The authority creates the private key for user (x, y) by first choosing a random exponent $\sigma_{x,y} \in \mathbb{Z}_n$ then generates it as:

$$\text{SK}_{x,y} = \left(d'_{x,y},\ d''_{x,y},\ d_1, \ldots, d_{y-1},, d_{y+1}, \ldots, d_m\right) =$$

$$= \left(g^{\alpha_x} g^{r_x c_y} \cdot u_y^{\sigma_{x,y}},\qquad g^{\sigma_{x,y}},\right.$$

$$\left. u_1^{\sigma_{x,y}}, \ldots, u_{y-1}^{\sigma_{x,y}},\, u_{y+1}^{\sigma_{x,y}}, \ldots, u_m^{\sigma_{x,y}}\right)$$

The public parameters $u_{q,1}^{\beta}, \ldots, u_{q,m}^{\beta}$ are related to the broadcast portion of the system, while the other parameters are related to the traitor tracing portion of the system. The secret key component $d'_{x,y}$ contains the secret key g^{α_x} blinded by $g^{r_x c_y}$, which is related to traitor tracing and $u_y^{\sigma_{x,y}}$, which is related the broadcast encryption system. An important technical point is that since $d'_{x,y}$ contains both pieces multiplied together, an attacker will be unable to separate these pieces out and decrypt the tracing and broadcast portions of the system separately. Thus, for a key to be useful for decrypting a ciphertext it must be both in the broadcast set of the ciphertext and have an index greater than or equal to the encrypted index.

$Encrypt_{\text{ABE}}(S, \text{PK}, (i, j), M)$.
The $Encrypt_{\text{ABE}}$ algorithm is primarily used for tracing. It encrypts a message M to the subset of receivers that are in S and that have row values greater than i or both row value equal to i and column values greater than j. The algorithm encrypts messages $M \in \mathbb{G}_T$. It first chooses random

$$t, \kappa, w_1, \ldots, w_m, s_1, \ldots, s_m \in \mathbb{Z}_n$$
$$b_1, \ldots, b_{j-1} \in \mathbb{Z}_n$$
$$(\nu_{1,1}, \nu_{1,2}, \nu_{1,3}), \ldots, (\nu_{i-1,1}, \nu_{i-1,2}, \nu_{i-1,3}) \in \mathbb{Z}_n^{(3)}$$

Let S_x denote the set of all values y such that the user (x, y) is in the set S. For each row x we create five ciphertext components $(R_x, \tilde{R}_x, T_x, A_x, B_x)$ as follows:

if $x > i$:
$R_x = E_{q,x}^{s_x} \qquad \tilde{R}_x = F_{q,x}^{\kappa s_x} \qquad A_x = E_q^{s_x t}$
$T_x = (\prod_{k \in S_x} U_{q,k})^{s_x t} \qquad B_x = M G_{q,x}^{s_x t}$

if $x = i$:
$R_x = E_x^{s_x} \qquad \tilde{R}_x = F_x^{\kappa s_x} \qquad A_x = g^{s_x t}$
$T_x = (\prod_{k \in S_x} U_k)^{s_x t} \qquad B_x = M G_x^{s_x t}$

if $x < i$:
$R_x = g^{\nu_{x,1}} \qquad \tilde{R}_x = h^{\kappa \nu_{x,1}} \qquad A_x = g^{\nu_{x,2}}$
$T_x = (\prod_{k \in S_x} U_k)^{\nu_{x,2}} \qquad B_x = e(g,g)^{\nu_{x,3}}$

For each column y the algorithm creates values $(C_y, \tilde{C}_y)$ as:

if $y \geq j$: $C_y = H_y^t h^{\kappa w_y} \qquad \tilde{C}_y = g^{w_y}$
if $y < j$: $C_y = H_y^t h^{\kappa w_y} V^{\kappa b_y} \qquad \tilde{C}_y = g^{w_y} \tilde{V}^{b_y}$

We note that for $y < j$ the $\mathbb{G}_p$ subgroup will be completely random in C_y.

The final ciphertext, containing $O(\sqrt{N} = m)$ group elements, consists of

$$\left((R_x, \tilde{R}_x, T_x, A_x, B_x)_{x=1}^m,\quad (C_y, \tilde{C}_y)_{y=1}^m\right)$$

The T_x values can be viewed as a broadcast encryption to all members of the row x that are in the sub-target set S_x. We can also see how the parameters allow for public encryption (which in turn gives public traceability) to an arbitrary index (i, j). The public parameters that are from the $\mathbb{G}_q$ subgroup are used for the encryption to rows greater than i. The public parameters values $V, \tilde{V}$ are used to make column components that are well formed in the $\mathbb{G}_q$ subgroup and random in the $\mathbb{G}_p$ subgroup. By forming the parameters in this way we can accomplish this without giving out a group element from $\mathbb{G}_p$, which would break the difficulty of subgroup hiding.

$Decrypt_{\text{ABE}}(S, (x, y), \text{SK}_{x,y}, C, \text{PK})$.
If user $(x, y) \in S$ it can attempt to decrypt by first computing a temporary key

$$K'_{x,y} = d'_{x,y} \prod_{\substack{k \in S_x \\ k \neq y}} d_{x,y,k}$$

Then it computes:

$$B_x / \left(e(K'_{x,y}, A_x) e(\tilde{R}_x, \tilde{C}_y) / \left(e(R_x, C_y) e(T_x, d''_{x,y})\right)\right).$$

Suppose that the ciphertext was encrypted to index (i, j) and that $x > i$ then in decryption, the pairing $e(K'_{x,y}, A_x)$ gives the value

$$e(g, g_q)^{\alpha_x s_x t} e(g, \prod_{k \in S_x} u_{q,k})^{s_x t \theta_{x,y}} e(g, g_q)^{s_x t r_x c_y}$$

The other pairings are used to divide out

$$e(g, \prod_{k \in S_x} u_{q,k})^{s_x t \theta_{x,y}} e(g, g_q)^{s_x t r_x c_y}$$

and get the blinding factor $e(g, g_q)^{\alpha_x s_x t}$. If $x = i$ and $y \geq j$ then decryption can be explained in a similar way except the target groups are in $\mathbb{G}_T$ instead of the subgroup $\mathbb{G}_{T,q}$,

5. SECURITY

We prove security of our Augmented Broadcast Encryption system by showing that it is secure under both games defined in Section 2. The proof structure is similar to that of [6]. There are two cases where there are important differences between what we prove here and what was given in [6]. The first is to show that it is difficult to distinguish between an encryption to indices (i, j) and indices $(i, j + 1)$ even when the attacker has key $K_{i,j}$ if user (i, j) is revoked. This is shown in the proof of Lemma 5.2.

Secondly, we need to prove that the public parameters given out to allow for public encryption to arbitrary indices do not break the security of our scheme. This is reflected in our proof of Claim 5.7, which shows that deciding subgroups is still hard even if the adversary has access to our public parameters. The other portions of the proofs are conceptually similar to [6], however, we include them for completeness.

5.1 Proof of Security for Game 1 (Message Hiding)

The argument for security of the Message Hiding game is very straightforward since an encryption to index $(m + 1, 1)$ contains no information about the ciphertext. The simulator simply runs the actual Setup algorithm and encrypts message M_β to set S and index $N+1$. Since, all B_x values contain no information about the ciphertext the bit β is perfectly hidden and the adversary's advantage is 0.

5.2 Proof of Security for Game 2 (Index-Hiding)

For clarity we present our Index-Hiding proofs in a structure similar to that of [6]. We state our main theorem and prove its security from a series of claims and lemmas whose proofs are given in the full version of this paper.

THEOREM 5.1. *Suppose that the (Modified) 3-party Diffie-Hellman, Bilinear Subgroup Decision, and Diffie-Hellman Subgroup Decision assumptions hold. Then no polynomial time adversary $\mathcal{A}$ can win the Index-Hiding game with non-negligible advantage.*

First we consider the case where the adversary $\mathcal{A}$ chooses to distinguish between an encryption to indices (i, j) and $(i, j + 1)$ where $j < m$. We state the following lemma whose proof is given in the full version.

LEMMA 5.2. *Suppose that the Decision (Modified) 3-party Diffie-Hellman, assumption holds. Then no polynomial time adversary $\mathcal{A}$ can distinguish between an encryption to (i, j) and an encryption to $(i, j + 1)$ in the Index Hiding game with non-negligible advantage.*

In this game we build a simulator that will guess the bit $\tilde{s}$. If $\tilde{s} = 0$ and $(i, j) \in S$ then the proof is very similar to that from the traitor tracing system of [6]. However, if $\tilde{s} = 1$ and $(i, j) \notin S$ then our simulator will need to generate the key $K_{i,j}$ for the adversary and still simulate the challenge ciphertext. The proof of this case captures the security that is gained by our particular method of composing a broadcast and traitor tracing system to make an Augmented Broadcast Encryption system.

We now consider the case when the adversary $\mathcal{A}$ attempts to distinguish between an encryption to (i, m) and one to $(i + 1, 1)$ for some $1 \leq i < m$. We refer to the rows with ciphertexts in the $\mathbb{G}_q$ subgroup as "greater than" rows and the the row with well formed ciphertexts in $\mathbb{G}$ as a "target" row. Additionally, when we say we "encrypt to column j" this means that we create ciphertexts for which C_y is well formed in the $\mathbb{G}_p$ subgroup for all $y \geq j$. We state our lemma and then discuss its proof.

LEMMA 5.3. *Suppose that the Decision (Modified) 3-party Diffie-Hellman, Bilinear Subgroup Decision, and Diffie Hellman Subgroup Decision assumptions hold. Then no polynomial time adversary $\mathcal{A}$ can distinguish between an encryption to (i, m) and an encryption to $(i+1, 1)$ in the Index Hiding game with non-negligible advantage.*

To prove the lemma we define a sequence of hybrid experiments:

- H_1: Encrypt to column m, row i is target row, i+1 is a "greater than" row.
- H_2: Encrypt to column $m+1$, row i is target row, i+1 is a "greater than" row.
- H_3: Encrypt to column $m+1$, row i is less than row, i+1 is a "greater than" row (no target row exists).
- H_4: Encrypt to column 1, row i is less than row, i+1 is "greater than" row (no target row exists).
- H_5: Encrypt to column 1, row i is less than row, i+1 is target row.

The following claims, whose proof is given in the full version, show that these hybrid games are indistinguishable.

CLAIM 5.4. *Suppose that the Decision (Modified) 3-party Diffie-Hellman assumption holds. Then no polynomial time adversary can distinguish between experiments H_1 and H_2 with non-negligible advantage.*

CLAIM 5.5. *Suppose that the Decision (Modified) 3-party Diffie-Hellman and the Bilinear Subgroup Decision assumptions hold. Then no polynomial time adversary can distinguish between experiments H_2 and H_3 with non-negligible advantage.*

CLAIM 5.6. *Suppose that the Decision (Modified) 3-party Diffie Hellman assumption holds. Then no polynomial time adversary can distinguish between experiments H_3 and H_4 with non-negligible advantage.*

CLAIM 5.7. *Suppose that the Diffie-Hellman Subgroup Decision assumption holds. Then no polynomial time adversary can distinguish between experiments H_4 and H_5 with non-negligible advantage.*

Lemma 5.3 follows by summing the maximum adversarial advantages across the hybrid experiments. Theorem 5.1 follows from Lemma 5.2 and Lemma 5.3. □

6. CONCLUSION

We constructed a fully collusion resistant trace and revoke system for arbitrary sets S where the size of ciphertexts and private keys is $O(\sqrt{N})$. The system is publicly traceable and secure against adaptive adversaries which is unusual for algebraic constructions. Instead of directly building a trace and revoke system we constructed a simpler primitive called Augmented Broadcast Encryption (ABE) with $O(\sqrt{N})$-size ciphertexts and private keys. We showed that ABE is sufficient for both broadcast encryption and tracing traitors. While we proved our broadcast secure under plaintext attacks, it is not difficult to modify it slightly and apply the methods of Canetti, Halevi, and Katz [8] for security against CCA attacks. We hope that future research will lead to ABEs with even shorter ciphertexts and private keys.

REFERENCES

[1] J. Anzai, N. Matsuzaki, and T. Matsumoto. A quick key distribution scheme with entity revocation. In *Proc. of Asiacrypt '99*, pages 333–347. Springer-Verlag, 1999.

[2] O. Berkman, M. Parnas, and J. Sgall. Efficient dynamic traitor tracing. In *Proceedings of SODA '00*, 2000.

[3] Dan Boneh and Matthew K. Franklin. An efficient public key traitor tracing scheme. In *CRYPTO '99: Proceedings of the 19th Annual International Cryptology Conference on Advances in Cryptology*, pages 338–353, London, UK, 1999. Springer-Verlag.

[4] Dan Boneh, Craig Gentry, and Brent Waters. Collusion resistant broadcast encryption with short ciphertexts and private keys. In *CRYPTO '05*, pages 258–275, 2005.

[5] Dan Boneh, Eu-Jin Goh, and Kobbi Nissim. Evaluating 2-dnf formulas on ciphertexts. In Joe Kilian, editor, *Proceedings of Theory of Cryptography Conference 2005*, volume 3378 of *LNCS*, pages 325–342. Springer, 2005.

[6] Dan Boneh, Amit Sahai, and Brent Waters. Fully collusion resistant traitor tracing with short ciphertexts and private keys. In *Eurocrypt '06*, 2006.

[7] Dan Boneh and Alice Silverberg. Applications of multilinear forms to cryptography. *Contemporary Mathematics*, 324:71–90, 2003.

[8] Ran Canetti, Shai Halevi, and Jonathan Katz. Chosen-ciphertext security from identity-based encryption. In *Proceedings of Eurocrypt 2004*, LNCS, pages 207–222, 2004.

[9] Hervé Chabanne, Duong Hieu Phan, and David Pointcheval. Public traceability in traitor tracing schemes. In *EUROCRYPT '05*, pages 542–558, 2005.

[10] B. Chor, A. Fiat, and M. Naor. Tracing traitors. In *Proceedings of Crypto '94*, volume 839 of *LNCS*, pages 257–270, 1994.

[11] Benny Chor, Amos Fiat, Moni Naor, and Benny Pinkas. Tracing traitors. *IEEE Transactions on Information Theory*, 46(3):893–910, 2000.

[12] Yevgeniy Dodis and Nelly Fazio. Public key broadcast encryption for stateless receivers. In *Proceedings of the Digital Rights Management Workshop 2002*, volume 2696 of *LNCS*, pages 61–80. Springer, 2002.

[13] Yevgeniy Dodis and Nelly Fazio. Public key trace and revoke scheme secure against adaptive chosen ciphertext attack. In *Public Key Cryptography - PKC 2003*, volume 2567 of *LNCS*, pages 100–115, 2003.

[14] Yevgeniy Dodis, Nelly Fazio, Aggelos Kiayias, and Moti Yung. Scalable public-key tracing and revoking. *Distributed Computing*, 17(4):323–347, 2005. Extended abstract in PODC '03.

[15] A. Fiat and M. Naor. Broadcast encryption. In *Proceedings of Crypto '93*, volume 773 of *LNCS*, pages 480–491. Springer-Verlag, 1993.

[16] Amos Fiat and T. Tassa. Dynamic traitor tracing. In *Proceedings of Crypto '99*, volume 1666 of *LNCS*, pages 354–371, 1999.

[17] Eli Gafni, Jessica Staddon, and Yiqun Lisa Yin. Efficient methods for integrating traceability and broadcast encryption. In *CRYPTO '99: Proceedings of the 19th Annual International Cryptology Conference on Advances in Cryptology*, pages 372–387, London, UK, 1999. Springer-Verlag.

[18] Juan A. Garay, Jessica Staddon, and Avishai Wool. Long-lived broadcast encryption. In *CRYPTO '00: Proceedings of the 20th Annual International Cryptology Conference on Advances in Cryptology*, pages 333–352. Springer-Verlag, 2000.

[19] M. T. Goodrich, J. Z. Sun, , and R. Tamassia. Efficient tree-based revocation in groups of low-state devices. In *Proceedings of Crypto '04*, volume 2204 of *LNCS*, 2004.

[20] D. Halevy and A. Shamir. The lsd broadcast encryption scheme. In *Proceedings of Crypto '02*, volume 2442 of *LNCS*, pages 47–60, 2002.

[21] Aggelos Kiayias and Moti Yung. On crafty pirates and foxy tracers. In *ACM Workshop in Digital Rights Management – DRM 2001*, pages 22–39, London, UK, 2001. Springer-Verlag.

[22] Aggelos Kiayias and Moti Yung. Breaking and repairing asymmetric public-key traitor tracing. In Joan Feigenbaum, editor, *ACM Workshop in Digital Rights Management – DRM 2002*, volume 2696 of *Lecture Notes in Computer Science*, pages pp. 32–50. Springer, 2002.

[23] K. Kurosawa and Y. Desmedt. Optimum traitor tracing and asymmetric schemes. In *Proceedings of Eurocrypt '98*, pages 145–157, 1998.

[24] D. Naor, M. Naor, and J. Lotspiech. Revocation and tracing schemes for stateless receivers. In *Proceedings of Crypto '01*, volume 2139 of *LNCS*, pages 41–62, 2001.

[25] M. Naor and B. Pinkas. Efficient trace and revoke schemes. In *Financial cryptography 2000*, volume 1962 of *LNCS*, pages 1–20. Springer, 2000.

[26] Moni Naor and Benny Pinkas. Threshold traitor tracing. In *CRYPTO '98: Proceedings of the 18th Annual International Cryptology Conference on Advances in Cryptology*, pages 502–517, London, UK, 1998. Springer-Verlag.

[27] B. Pfitzmann. Trials of traced traitors. In *Proceedings of Information Hiding Workshop*, pages 49–64, 1996.

[28] B. Pfitzmann and M. Waidner. Asymmetric fingerprinting for larger collusions. In *Proceedings of the ACM Conference on Computer and Communication Security*, pages 151–160, 1997.

[29] Reihaneh Safavi-Naini and Yejing Wang. Sequential traitor tracing. In *Proceedings of Crypto '00*, volume 1880 of *LNCS*, pages 316–332, 2000.

[30] Alice Silverberg, Jessica Staddon, and Judy L. Walker. Efficient traitor tracing algorithms using list decoding. In *Proceedings of ASIACRYPT '01*, volume 2248 of *LNCS*, pages 175–192, 2001.

[31] Jessica N. Staddon, Douglas R. Stinson, and Ruizhong Wei. Combinatorial properties of frameproof and traceability codes. Cryptology ePrint 2000/004, 2000.

[32] D. Stinson and R. Wei. Combinatorial properties and constructions of traceability schemes and frameproof codes. *SIAM Journal on Discrete Math*, 11(1):41–53, 1998.

[33] D. Stinson and R. Wei. Key preassigned traceability schemes for broadcast encryption. In *Proceedings of SAC '98*, volume 1556 of *LNCS*, 1998.

[34] D. R. Stinson and R. Wei. Combinatorial properties and constructions of traceability schemes and frameproof codes. *SIAM J. Discret. Math.*, 11(1):41–53, 1998.

[35] Doug R. Stinson. On some methods for unconditionally secure key distribution and broadcast encryption. *Des. Codes Cryptography*, 12(3):215–243, 1997.

[36] Doug R. Stinson and Tran Van Trung. Some new results on key distribution patterns and broadcast

encryption. *Des. Codes Cryptography*, 14(3):261–279, 1998.

[37] W. Tzeng and Z. Tzeng. A public-key traitor tracing scheme with revocation using dynamic shares. In *Proc. of PKC 2001*, pages 207–224. Springer-Verlag, 2001.

[38] Yuji Watanabe, Goichiro Hanaoka, and Hideki Imai. Efficient asymmetric public-key traitor tracing without trusted agents. In *Proceedings CT-RSA '01*, volume 2020 of *LNCS*, pages 392–407, 2001.

APPENDIX
A. TRACE AND REVOKE SYSTEMS

A Trace and Revoke (TR) system is a broadcast encryption system with an additional tracing algorithm. We describe TR systems where encryption is public-key and tracing requires no secrets. We refer to [6] for more information on the definition of traitor tracing systems. A TR system consists of four algorithms:

***Setup**(N, λ)* A probabilistic algorithm that takes as input N, the number of users in the system, and a security parameter λ. The algorithm runs in polynomial time in λ and outputs a public key PK and private keys $SK_1, \ldots, SK_N$, where SK_u is for user u.

***Encrypt**($S, \mathbf{PK}, M$)* Takes as input a subset $S \subseteq \{1, \ldots, N\}$, a public key PK, and a message M. It outputs a ciphertext C intended for a specific recipient set S.

***Decrypt**($S, j, \mathbf{SK}_j, C, \mathbf{PK}$)* Takes as input a subset $S \subseteq \{1, \ldots, N\}$, the private key SK_j for user j, a ciphertext C, and the public key PK. The algorithm outputs a message M or $\bot$.

***Trace**$^\mathcal{D}(S, \mathbf{PK}, \epsilon)$* The tracing algorithm is an oracle algorithm that interacts with a pirate decoder $\mathcal{D}$. The algorithm is given as input a set $S \subseteq \{1, \ldots, N\}$, the public key PK and a parameter ϵ, and runs in time polynomial in the security parameter λ and $1/\epsilon$. Only values of ϵ that are polynomially related to λ are considered valid inputs to *Trace*. The tracing algorithm outputs a set $T \subseteq \{1, \ldots, N\}$.

Correctness. The system must satisfy the same **correctness property** as for broadcast encryption, namely: for all subsets $S \subseteq \{1, \ldots, N\}$, all $j \in \{1, \ldots, N\}$, and all messages M:

Let $(PK, (SK_1, \ldots, SK_N)) \xleftarrow{R} Setup_{\text{ABE}}(N, \Lambda)$

and $c \xleftarrow{R} Encrypt(S, PK, M)$.

If $j \in S$ then $Decrypt(S, j, SK_j, c, PK) = M$.

A.1 Security

We define security of a trace and revoke system using two natural games. The **Message Hiding Game** is the same as for a broadcast encryption system secure against an adaptive attacker [13, 4]. We described the game in Section 2.2.1. We let $\mathsf{MH\,Adv}_\mathcal{A}[1]$ denote the advantage of $\mathcal{A}$ in winning the game.

The **Tracing Game** ensures that the tracing algorithm successfully traces any pirate decoder, no matter how many secret keys were used to create the decoder. The adversary's goal is to build a pirate decoder $\mathcal{D}$ that will decrypt all ciphertexts encrypted for a certain set $S_\mathcal{D}$. The tracing algorithm's goal is extract from $\mathcal{D}$ at least one of the keys $u \in S_\mathcal{D}$ that were used to construct $\mathcal{D}$. The broadcaster will encrypt all future messages to the set $S' = S_\mathcal{D} \setminus \{u\}$. If the decoder $\mathcal{D}$ can decode ciphertexts encrypted for S' then we run the tracing algorithm again, this time giving it the set S', to extract another of the pirate's keys in $S_\mathcal{D}$. This process continues, iteratively shrinking S', until $\mathcal{D}$ stops functioning.

The following game ensures that this process will eventually disable $\mathcal{D}$ without disabling any innocent parties. The game is defined between a challenger and an adversary $\mathcal{A}$ (both are given N, λ and ϵ as input):

1. The challenger runs $Setup(N, \lambda)$ to obtain PK and private keys $SK_1, \ldots, SK_N$. It provides PK to $\mathcal{A}$.

2. The adversary issues *adaptive* private key queries. It repeatedly sends values $j \in \{1, \ldots, N\}$ to the challenger and the challenger responds with SK_j. Let $U \subseteq \{1, \ldots, N\}$ be the total set of keys obtained by the adversary.

3. Finally, the adversary $\mathcal{A}$ outputs a set $S_\mathcal{D} \subseteq \{1, \ldots, N\}$ and a pirate decoder $\mathcal{D}$ which is a probabilistic circuit that takes as input ciphertexts in C and outputs some message M.

4. The challenger now runs $Trace^\mathcal{D}(S_\mathcal{D}, PK, \epsilon)$ to obtain a set $T \subseteq \{1, \ldots, N\}$.

We say that the adversary $\mathcal{A}$ wins the game if the following two conditions hold:

- For a randomly chosen M in the finite message space, we have that
$$\Pr[\mathcal{D}(Encrypt(S_\mathcal{D}, PK, M)) = M] \geq \epsilon$$
A pirate decoder satisfying this condition is said to be a *useful decoder*.

- The set T is either empty, or is not a subset of $U \cap S_\mathcal{D}$.

We denote by $\mathsf{TR\,Adv}_\mathcal{A}$ the probability that adversary $\mathcal{A}$ wins this game.

The game above places no limit on the size of the coalition under the control of the adversary. Furthermore, the pirate decoder need not be perfect. It need only decrypt valid content with probability ϵ. Finally, note that we are modeling a stateless (resettable) pirate decoder — the decoder is just an oracle and maintains no state between activations. Non stateless decoders were studied in [21].

DEFINITION A.1. *We say that an N-user Trace and Revoke system is secure if for all polynomial time adversaries $\mathcal{A}$ we have that $\mathsf{MH\,Adv}_\mathcal{A}[1]$ and $\mathsf{TR\,Adv}_\mathcal{A}$ are negligible functions of λ.*

Puppetnets: Misusing Web Browsers as a Distributed Attack Infrastructure

V. T. Lam, S. Antonatos[*], P. Akritidis[†], K. G. Anagnostakis
Systems and Security Department, Institute for Infocomm Research
21 Heng Mui Keng Terrace, Singapore
{vtlam,antonat,akritid,kostas}@s3g.i2r.a-star.edu.sg

ABSTRACT

Most of the recent work on Web security focuses on preventing attacks that *directly* harm the browser's host machine and user. In this paper we attempt to quantify the threat of browsers being *indirectly* misused for attacking third parties. Specifically, we look at how the existing Web infrastructure (e.g., the languages, protocols, and security policies) can be exploited by malicious Web sites to remotely instruct browsers to orchestrate actions including denial of service attacks, worm propagation and reconnaissance scans. We show that, depending mostly on the popularity of a malicious Web site and user browsing patterns, attackers are able to create powerful botnet-like infrastructures that can cause significant damage. We explore the effectiveness of countermeasures including anomaly detection and more fine-grained browser security policies.

Categories and Subject Descriptors

D.4.6 [**Operating Systems**]: Security and Protection—*Invasive software*

General Terms

Security, Measurement, Experimentation

Keywords

Web security, malicious software, distributed attacks

1. INTRODUCTION

In the last few years researchers have observed two significant changes in malicious activity on the Internet [48, 60, 52]. The first is the shift from amateur proof-of-concept attacks to professional profit-driven criminal activity. The second is the increasing sophistication of the attacks. Although significant efforts are made towards addressing the underlying vulnerabilities, it is very likely that attackers will try to adapt to *any* security response, by discovering new ways of exploiting systems to their advantage [39]. In this arms race, it is important for security researchers to proactively explore and mitigate new threats before they materialize.

This paper discusses one such threat, for which we have coined the term *puppetnets*. Puppetnets rely on Web sites that *coerce* Web browsers to (unknowingly) participate in malicious activities. Such activities include distributed denial-of-service, worm propagation and reconnaissance probing, and can be engineered to be carried out in stealth, without any observable impact on an otherwise innocent-looking Web site. Puppetnets exploit the high degree of flexibility granted to the mechanisms comprising the Web architecture, such as HTML and Javascript. In particular, these mechanisms impose few restrictions on how remote hosts are accessed. A malicious Web site can thereby transform a collection of Web browsers into an *impromptu* distributed system that is effectively controlled by the attacker. Puppetnets expose a deeper problem in the design of the Web. The problem is that the security model is focused almost exclusively on protecting browsers and their host environment from malicious Web servers, as well as servers from malicious browsers. As a result, the model ignores the potential of attacks against third parties.

Web sites controlling puppetnets could be either legitimate sites that have been subverted by attackers, malicious "underground" Web sites that can lure unsuspected users by providing interesting services (such as free Web storage, illegal downloads, etc.), or Web sites that openly invite users to participate in vigilante campaigns. We must note however that puppetnet attacks are different from previous vigilante campaigns against spam and phishing sites that we are aware of. For instance, the Lycos "Make Love Not Spam" campaign[53] required users to install a screensaver in order to attack known spam sites. Although similar campaigns can be orchestrated using puppetnets, in puppetnets users may not be aware of their participation, or may be coerced to do so; the attack can be launched stealthily from an innocent-looking Web page, without requiring any extra software to be installed, or any other kind of user action.

Puppetnets differ from botnets in three fundamental ways. First, puppetnets are not heavily dependent on the exploitation of specific implementation flaws, or on social engineer-

---

[*]S. Antonatos is with FORTH-ICS and the University of Crete, Greece. This work was done while visiting I²R.
[†]P. Akritidis is with FORTH-ICS, Greece. This work was done while visiting I²R.

ing tactics that trick users into installing malicious software on their computer. They exploit *architectural* features that serve purposes such as enabling dynamic content, load distribution and cooperation between content providers. At the same time, they rely on the *amplification* of vulnerabilities that seem insignificant from the perspective of a single browser, but can cause significant damage when abused by a popular Web site. Thus, it seems harder to eliminate such a threat in similar terms to common implementation flaws, especially if this would require sacrificing functionality that is of great value to Web designers. Additionally, even if we optimistically assume that major security problems such as code injection and traditional botnets are successfully countered, some puppetnet attacks will still be possible. Furthermore, the nature of the problem implies that the attack vector is pervasive: puppetnets can instruct *any* Web browser to engage in malicious activities.

Second, the attacker does not have complete control over the actions of the participating nodes. Instead, actions have to be composed using the primitives offered from within the browser sandbox – hence the analogy to puppets. Although the flexibility of puppetnets seems limited when compared to botnets, we will show that they are surprisingly powerful.

Finally, participation in puppetnets is dynamic, making them a moving target, since users join and participate unknowingly while surfing the net. Thus, it seems easy for the attackers to maintain a reasonable population, without the burden of having to look for new victims. At the same time, it is harder for the defenders to track and filter out attacks, as puppets are likely to be relatively short-lived.

A fundamental property of puppetnet attacks, in contrast to most Web attacks that directly harm the browser's host machine, is that they only *indirectly* misuse browsers to attack third parties. As such, users are less likely to be vigilant, less likely to notice the attacks, and have lesser incentive to address the problem. Similar problems arise at the server side: if puppetnet code is installed on a Web site, the site may continue to operate without any adverse consequences or signs of compromise (in contrast to defacement and other similar attacks), making it less likely that administrators will react in a timely fashion.

In this paper we experimentally assess the threat from puppetnets. We discuss the building blocks for engineering denial-of-service attacks, worm propagation and other puppetnet attacks, and attempt to quantify how puppetnets would perform. Finally, we examine various options for guarding against such attacks.

2. PUPPETNETS: DESIGN AND ANALYSIS

We attempt to map out the attackers' opportunity space for misusing Web browsers. In lieu of the necessary formal tools for analyzing potential vulnerabilities, neither the types of attacks nor their specific realizations are exhaustive enough to provide us with a solid worst-case scenario. Nevertheless, we have tried to enhance the attacks as much as possible, in an attempt to approximately determine in what ways and to what effect the attacker could capitalize on the underlying vulnerabilities. In the rest of this section, we explore in more detail a number of ways of using puppetnets, and attempt to quantify their effectiveness.

2.1 Distributed Denial of Service

The flexibility of Web architecture provides many ways

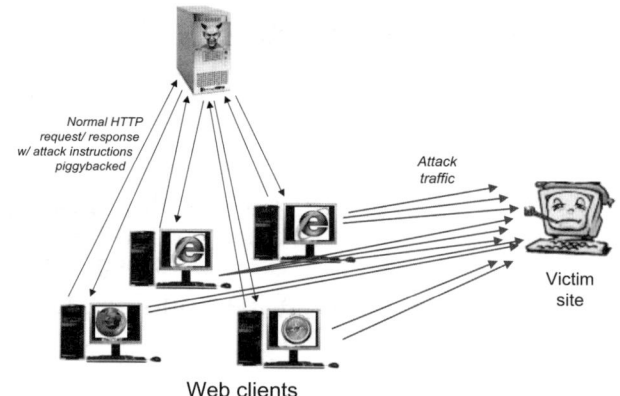

Figure 1: DDoS using puppetnets

for launching DoS attacks using puppetnets. The common component of the attack in all of its forms is an instruction that asks the remote browser to access some object from the victim. There are several ways of embedding such instructions in an otherwise legitimate Web page. The simplest way is to add an image reference, as commonly used in the vast majority of Web pages. Other ways include opening up pop-up windows, creating new frames that load a remote object, and loading image objects through Javascript. We are not aware of any browser that imposes restrictions on the location or type of the target referenced through these mechanisms.

We assume that the intention of the attacker is to maximize the effectiveness of the DDoS attack, at the lowest possible cost, and as stealthily as possible. An attack may have different objectives: maximize the amount of ingress traffic to the victim, the egress traffic from the victim, connection state, *etc*. Here we focus on raw bandwidth attacks in both directions, but emphasize on ingress traffic as it seems harder to defend against: the host has full control over egress traffic, but usually limited control over ingress traffic.

To create a large number of requests to the target site, the attacker can embed a sequence of image references in the malicious Web page. This can be done using either a sequence of `IMG SRC` instructions, or a Javascript loop that instructs the browser to load objects from the target server. In the latter case, the attack seems to be much more efficient in terms of *attack gain*, e.g., the effort (in terms of bandwidth) that the attacker has to spend for generating a given amount of attack traffic. This assumes that the attacker either targets the same URL in all requests, or is able to construct valid target URLs through Javascript without wasting space for every URL. To prevent client-side caching of requests, the attacker can also attach an invariant modifier string to the attack URL that is ignored by the server but considered by the client in the context of deciding whether the object is already cached [1].

Another constraint is that most browsers impose a limit on the number of simultaneous connections to the same

---

[1] The URL specification [10] states that URLs have the form `http://host:port/path?searchpart`. The `searchpart` is ignored by Web servers such as Apache if included in a normal file URL.

server. For IE and Firefox the limit is two connections. However, we can circumvent this limit using aliases of the same server, such as using the DNS name instead of the IP address, stripping the "www" part from or adding a trailing dot to the host name, etc. Most browsers generally treat these as different servers. Servers with "virtual hosting" are especially vulnerable to this form of amplification.

To make the attack stealthy in terms of not getting noticed by the user, the attacker can employ hidden (*e.g.*, zero-size) frames to launch the attack-bearing page in the background. To maximize effectiveness the requests should not be rendered within a frame and should not interfere with normal page loading. To achieve this, the attacker can employ the same technique used by Web designers for pre-loading images for future display on a Web page. The process of requesting target URLs can then be repeated through a loop or in an event-driven fashion. The loop is likely to be more expensive and may interfere with normal browser activity as it may not relinquish control frequently enough for the browser to be used for other purposes. The event-driven approach, using Javascript timeouts, appears more attractive.

2.1.1 Analysis of DDoS attacks

We explore the effectiveness of puppetnets as a DDoS infrastructure. The "firepower" of a DDoS attack will be equal to the number of users concurrently viewing the malicious page on their Web browser (henceforth referred to as *site viewers*) multiplied by the amount of bandwidth each of these users can generate towards the target server. Considering that some Web servers are visited by millions of users every day, the scale of the potential threat becomes evident. We consider the *size* of a puppetnet to be equal to the site viewers for the set of Web servers controlled by the same attacker. Although it is tempting to use puppetnet size for a direct comparison to botnets, a threat analysis based on such a comparison alone may be misleading. Firstly, a bot is generally more powerful than a puppet, as it has full control over the host, in contrast to a puppet that is somewhat constrained within the browser sandbox. Secondly, a recent study [17] observes a trend towards smaller botnets, suggesting that such botnets may be more attractive, as they are easier to manage and keep undetected, yet powerful enough for the attacker to pursue his objectives. Finally, an attacker may construct a hybrid, two-level system, with a small botnet consisting of a number of Web servers, each controlling a number of puppets.

To estimate the firepower of puppetnets we could rely on direct measurements of site viewers for a large fraction of the Web sites in the Internet. Although this would be ideal in terms of accuracy, to the best of our knowledge there is no published study that provides such data. Furthermore, carrying out such a large-scale study seems like a daunting task. We therefore obtain a rough estimate using "second-hand" information from Web site reports and Web statistics organizations. There are several sources providing data on the number of daily or monthly visitors:

- Many sites use tools such as Webalizer [8] and Web-Trends [24] to generate usage statistics in a standard format. This makes them easy to locate through search engines and to automatically post-process. We have obtained WebTrends reports for 249 sites and Webalizer reports for 738 sites, covering a one-month period in December 2005. Although these sites may not be entirely random, as the sampling may be influenced by the search engine, we found that most of them are non-commercial sites with little content and very few visits.

- Some Web audit companies such as ABC Electronic provide public databases for a fairly large number of their customers [3]. These sites include well-known and relatively popular sites. We obtained 138 samples from this source.

- Alexa [5] tracks the access patterns of several million users through a browser-side toolbar and provides, among other things, statistics on the top-500 most popular sites. Although Alexa statistics have been criticized as inaccurate because of the relatively small sample size [7], this problem applies mostly to less popular sites and not the top-500. Very few of these sites are tracked by ABC Electronic.[2]

We have combined these numbers to get an estimate on the number of visitors per day for different Web sites. Relating the number of visitors per day to the number of site viewers is relatively straightforward. Recall that Little's law [35] states that if λ is the arrival rate of clients in a queuing system and T the time spent in the system, then the number of customers active in the system N is $N = \lambda T$.

To obtain the number of site viewers we also need, for each visit to a Web site, the time spent by users viewing pages on that site. None of the sources of Web site popularity statistics mentioned above provide such statistics. We therefore have to obtain a separate estimate of Web browsing times, making the assumption that site popularity and browsing times are not correlated in a way that could significantly distort our rough estimates.

Note that although we base our analysis on estimates of "typical" Web site viewing patterns, the attacker may also employ methods for increasing viewing times, such as incentivizing users (*e.g.*, asking users to keep a pop-up window open for the download to proceed), slowing down the download of legitimate pages, and providing more interesting content in the case of a malicious Website.

Web session time measurements based entirely on server log files may not accurately reflect the time spent by users viewing pages on a particular Web site. These measurements compute the session time as the difference between the last and first request to the Web site by a particular user, and often include a timeout threshold between requests to distinguish between different users. The remote server usually cannot tell whether a user is *actively* viewing a page or whether he has closed the browser window or moved to a different site. As we have informally observed, many users leave several browser instances open for long periods of time, we were concerned that Web session measurements may not be reliable enough by themselves for the purposes of this study. We thus considered the following three data sources for our estimates:

---

[2] Alexa only provides relative measures of daily visitors as a fraction of users that have the Alexa toolbar installed, and not absolute numbers of daily visitors. To obtain the absolute number of daily visitors we compare the numbers from Alexa to those from ABC Electronic, for those sites that appear on both datasets. This gives us a (crude) estimate of the Internet population, which we then use to translate visit counts from relative to absolute.

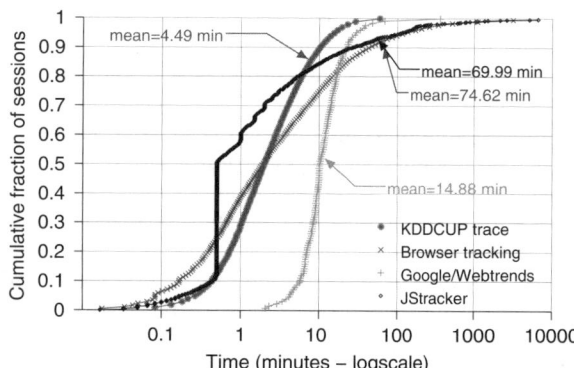

Figure 2: Web site viewing times

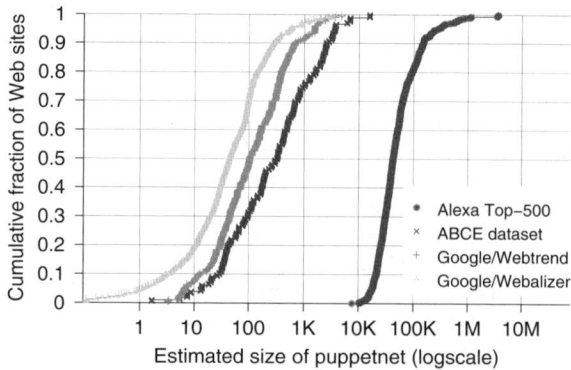

Figure 3: Estimated size of puppetnets

- We obtain *real* browsing times through a small-scale experiment: we developed browser extensions for both IE and Firefox that keep track of Web page viewing times and regularly post anonymized usage reports back to our server. The experiment involved roughly 20 users and resulted in a dataset of roughly 9,000 page viewing reports.
- We instrumented all pages on the server of our institution to include Javascript code that makes a small request back to the server every 30 seconds. This allows us to infer how long the browser points to one of the instrumented pages. We obtained data on more than 3,000 sessions over a period of two months starting January 2006. These results are likely to be optimistic, as the instrumented Web site is not particularly deep or content-heavy.
- We analyzed the KDD Cup 2000 dataset [29] which contains clickstream and purchase data from a defunct commercial Web site. The use of cookies, the size of the dataset, and the commercial nature of the measured Website suggest that the data are reasonably representative for many Web sites.
- We obtained, through a search engine, WebTrends reports on Web session times from 249 sites, similar to the popularity measurements, which provide us with mean session time estimates per site.

The distributions of estimated session times, as well as the means of the distributions, are shown in Figure 2. As suspected, the high-end tail of the distribution for the more reliable browser-tracking measurements is substantially larger than that for other measurement methods. This confirms our informal observation that users tend to leave browser windows open for long periods of time, and our concern that logfile-based session time measurements may underestimate viewing times. The Javascript tracker numbers also appear to confirm this observation. As in the case of DDoS, we are interested in the mean number of active viewers. Our results show that because of the high-end tails, the mean time that users keep pages on their browser is around 74 minutes, 6-13 times more than the session time as predicted using logfiles.[3]

---

[3]Note that the WebTrends distribution seems to have much lower variance and a much higher median than the other two sources. This is an artifact, as for WebTrends we have a distribution of *means* for different sites, rather than the distribution of session times.

From the statistics on daily visits and typical page viewing times we estimate the size of a puppetnet. The results for the four groups of Web site popularity measurements are shown in Figure 3. The main observation here is that puppetnets appear to be comparable in size to botnets. Most top-500 sites appear highly attractive as targets for setting up puppetnets, with the top-100 sites able to form puppetnets controlling more than 100,000 browsers at any time. The sizes of the largest potential puppetnets (for the top-5 sites) seem comparable to the largest botnets seen [27], at 1-2M puppets. Although one could argue that top sites are more likely to be secure, the figures for sites other than the top-500 are also worrying: More than 20% of typical commercial sites can be used for puppetnets of 10,000 nodes, while 4-10% of randomly selected sites can be popular enough for hosting puppetnets of more than 1,000 nodes.

As discussed previously, however, the key question is not how big a puppetnet is but whether the firepower is sufficient enough for typical DDoS scenarios. To estimate the DDoS firepower of puppetnets we first need to determine how much traffic a browser can typically generate under the attacker's command.

We experimentally measure the bandwidth generated by puppetized browsers, focusing initially only on ingress bandwidth, since it is harder to control. Early experiments with servers and browsers in different locations (not presented here in the interest of space) show that the main factor affecting DoS strength is the RTT between client and server. We therefore focus on precisely quantifying DoS strength in a controlled lab setting, with different line speeds and network delays emulated using *dummynet* [43], and an Apache Web server running on the victim host. We consider two types of attacks: a simple attack aiming to maximize SYN packets (maxSYN), and one aiming to maximize the ingress bandwidth consumed (maxURL). For the maxSYN attack, the sources of ten Javascript image objects are set to be non-existent URLs repeatedly every 50 milliseconds. Upon renewal of the image source, old connections are stalled and new connections are established. For the maxURL attack we load a page with several thousand requests for non-existent URLs of 2048 bytes each (as IE can handle URLs of up to 2048 characters). The link between puppet and server was set to 10 Mbit/s in all experiments.

In Figure 4, the ingress bandwidth of the server is plotted against the RTT between the puppet and the server, for the case of 3 aliases. The effectiveness of the attack decreases for high RTTs, as requests spend more time "in-flight" and the

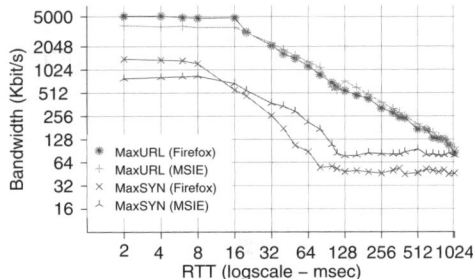

Figure 4: **Ingress bandwidth consumed by one puppet vs. RTT between browser and server**

	Firefox	Explorer
maxSYN 2 aliases	83.97 Mbit/s	106.30 Mbit/s
maxSYN 3 aliases	137.26 Mbit/s	173.28 Mbit/s
maxURL 2 aliases	664.74 Mbit/s	502.06 Mbit/s
maxURL 3 aliases	1053.79 Mbit/s	648.33 Mbit/s

Figure 5: **Estimated bandwidth of ingress DDoS from 1000 puppets**

connection limit to the same server is capped by the browser. For the maxSYN experiment, a puppet can generate up to 300 Kbit/s to 2 Mbit/s when close to the server, while for high RTTs around 250 msec the puppet can generate only around 60 Kbit/s. For the maxURL attack, these numbers become 3-5 Mbit/s and 200-500 Kbit/s respectively. The results seem to differ for both browsers: IE is more effective for maxSYN, while Firefox is more effective for maxURL. We have not been able to determine the cause of the difference, mostly due to the lack of source code for IE. The same figures apply for slower connections, with RTTs remaining the dominant factor determining puppet DoS performance.

Using the measurements of Figure 4, the distribution of RTTs measured in [49] and the capacity distribution from [47], we estimate the firepower of a 1000-node puppetnet, for different aliasing factors, as shown in Table 5. From these estimates we also see that around 1000 puppets are sufficient for consuming a full 155 Mbit/s link using SYN packets alone, and only around 150 puppets are needed for a maxURL attack on the same link. These estimates suggest that puppetnets can launch powerful DDoS attacks and should therefore be considered as a serious threat.

Considering the analysis above, we expect the following puppetnet scenarios to be more likely. An attacker owning a popular Web page can readily launch puppetnet attacks; many of the top-500 sites are highly suspect offering "warez" and other illegal downloads. Furthermore, we have found that some well-known underground sites, not listed in the top-500, can create puppetnets of 10,000-70,000 puppets (see [33]). Finally, the authors of reference [58] report that by scanning the most popular one million Web pages according to a popular search engine, they found 470 malicious sites, many of which serve popular content related to celebrities, song lyrics, wallpapers, video game cheats, and wrestling. These malicious sites were found to be luring unsuspected users with the purpose of installing malware on their machines by exploiting client-side vulnerabilities. The compromised machines are often used to form a botnet, but visits to these popular sites could be used for staging a puppetnet attack instead.

Another way to stage a puppetnet attack is by compromising and injecting puppetnet code to a popular Web site. Although popular sites are more likely to be secure, checking the top-500 sites from Alexa against the defacement statistics from zone-h[61] reveals that in the first four months of 2006 alone, 7 pages having the same domain as popular sites were defaced. For the entire year 2005 this number reaches 18. We must note, however, that the defaced pages were usually not front pages, and therefore their hits are likely to be less than those of the front pages. We also found many of them running old versions of Apache and IIS, although we did not go as far as running penetration tests on them to determine whether they were patched or not.

2.2 Worm propagation

Puppetnets can be used to spread worms that target vulnerable Web sites through URL-encoded exploits. Vulnerabilities in Web applications are an attractive target for puppetnets as these attacks can usually be encoded in a URL and embedded in a Web page. Web applications such as blogs, wikis, and bulletin boards are now among the most common targets of malicious activity captured by honeynets. The most commonly targeted applications according to recent statistics [41] are Awstats, XMLRPC, PHPBB, Mambo, WebCalendar, and PostNuke.

A Web-based worm can enhance its propagation with puppetnets as follows. When a Web server becomes infected, the worm adds puppetnet code to some or all of the Web pages on the server. The appearance of the pages could remain intact, just like in our DDoS attack, and each unsuspected visitor accessing the infected site would automatically run the worm propagation code. In an analogy to real-world diseases, Web servers are *hosts* of the worm while browsers are *carriers* which participate in the propagation process although they are not vulnerable themselves. Besides using browsers to increase the aggregate scanning rate, a worm could spread entirely through browsers. This could be particularly useful if behavioral blockers prevent servers from initiating outbound connections. Furthermore, puppetnets could help worms penetrate NAT and firewall boundaries, thereby extending the reach of the infection to networks that would otherwise be immune to the attack. For example, the infected Web server could instruct puppets to try propagating on private addresses such as 192.168.x.y. The scenarios for worm propagation are shown in Figure 6.

2.2.1 Analysis of worm propagation

To understand the factors affecting puppetnet worm propagation we first utilize an analytical model, and then proceed to measure key parameters of puppetnet worms and use simulation to validate the model and explore the resulting parameter space.

Analytical model: We have developed an epidemiological model for worm propagating using puppetnets. The details of our model are described elsewhere [33]. Briefly, we have extended classical homogeneous models to account for how clients and servers contribute to worm propagation in a puppetnet scenario. The key parameters of the model are the browser scanning rate, the puppetnet size, and the time spent by puppets on the worm-spreading Web page.

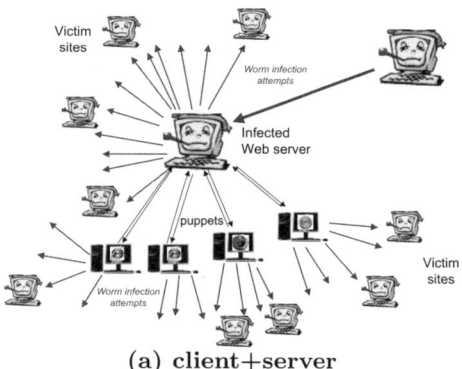
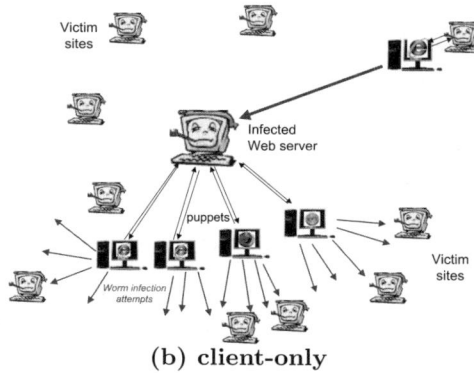

(a) client+server   (b) client-only

Figure 6: Two different ways that puppetnets could be used for worm propagation: (a) illustrates an infected server that uses puppets to propagate the worm, and (b) a server that propagates only through the puppet browsers.

Scanning performance: If the attacker relies on simple Web requests, the scanning rate is constrained by the default browser connection timeout and limits imposed by the OS and the browser on the maximum number of outstanding connections. In our proof-of-concept attack, we have embedded a hidden HTML frame with image elements into a normal Web page, with each image element pointing to a random IP address with a request for the attack URL. Note that the timeout for each round of infection attempts can be much lower than the time needed to infect all possible targets (based on RTTs). We assume that the redundancy of the worm will ensure that any potential miss from one source is likely to be within reach from another source.

Experimentally, we have found that both IE and Firefox on an XP SP2 platform can achieve maximum worm scanning rates of roughly 60 scans/min, mostly due to OS connection limiting. On other platforms, such as Linux, we found that a browser can perform roughly 600 scans/min without noticeable impact on regular activities of the user. These measurements were consistent across different hardware platforms and network connections.

Impact on worm propagation: We simulate a puppetnet worm outbreak and compare results with the prediction of our analytical model. We consider CodeRed [12] as an example of a worm targeting Web servers and use its parameters for our experiments. To simplify the analysis, we ignore possible human intervention such as patching, quarantine, and the potential effect of congestion resulting from worm activity.

We examine three different scenarios: (a) a normal worm where only compromised servers can scan and infect other servers, (b) a puppetnet-enhanced worm where both the compromised servers and their browsers propagate the infection, and (c) a puppetnet-only worm where servers only push the worm solely through puppets to achieve stealth or bypass defenses.

We have extended a publicly available CodeRed simulator [62] to simulate puppetnet worms. We adopt the parameters of CodeRed as used in [62]: a vulnerable population of 360,000 and a server scanning rate of 358 scans/min. In the simulation, we directly use these parameters, while in our analytical model, we map these parameters to analytical model parameters and numerically solve the differential equations. Note that our model is a special case of the Kephart-White Susceptible-Infected-Susceptible (SIS) model [28] with no virus curing. The compromise rate is $K = \beta \times \langle k \rangle$ where β is the virus birth rate defined on every directed edge from an infected node to its neighbors, and $\langle k \rangle$ is the average node out-degree. Assuming the Internet is a fully connected network, $\langle k \rangle_{CodeRed} = 360,000$ and $\beta_{CodeRed,server} = 358/2^{32}$, we have $K_s = 0.03$. Our simulation and analytical model also include the delay in accessing a Web page as users have to click or reload a newly-infected Web page to start participating in worm propagation.

We obtain our simulation results by taking the mean over five independent runs. For this experiment, we use typical parameters measured experimentally: browsers performing 36 scans/min (i.e., an order of magnitude slower than servers), and Web servers with about 13 concurrent users, and an average page holding time of 15 minutes. To study the effect of these parameters, we vary their values and estimate worm infection time as shown in Figures 8 and 9.

Figure 7 illustrates the progress of the infection over time for the three scenarios. In all cases the propagation process obeys the standard S-shaped logistic growth model. The simulated virus propagation matches reasonably well with the analytical model. Both agree on a worm propagation time of 50 minutes for holding times in the order of $t_h = 15min$ (that is, compared to the case of zero holding time). A client-only worm can perform as well as a normal worm, suggesting that puppetnets are quite effective at propagating worm epidemics.

Figure 8 illustrates the time needed to infect 90% of the vulnerable population for different browser scanning rates. When browsers scan at 45 scans/min, the client-only scenario is roughly equivalent to the server-only scenario. At the maximum scan rate of this experiment (which is far more than the scan rate for IE, but only a third of the scan rate for Linux), a puppetnet can infect 90% of the vulnerable population within 19 minutes. This is in line with Warhol worms and an order of magnitude faster than CodeRed.

Figure 9 confirms that the popularity of compromised servers plays an important role in worm performance. The break-even point between the server-only and client-only cases is when Web servers have 16 concurrent clients on average. For large browser scanning rate or highly popular compromised servers, the client-only scenario converges to the client-server scenario. That means that infection attempts launched from browsers are so powerful that they dominate the infection process.

Finally, in a separate experiment we found that if the worm uses a small initial hitlist to specifically target busy

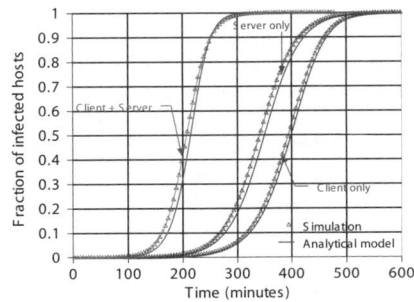 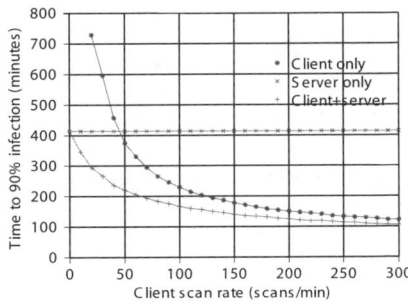 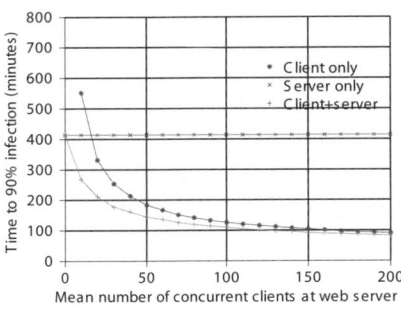

Figure 7: Worm propagation with puppetnet

Figure 8: Worm infection for different browser scan rates

Figure 9: Worm infection versus popularity of Web servers

Web servers with more than 150 concurrent visitors, the infection time is reduced to less than two minutes, similar to flash worms [50].

2.3 Reconnaissance probes

We discuss how malicious Web sites can orchestrate distributed reconnaissance probes. Such probes can be useful for the attacker to locate potential targets before launching an actual attack. The attacker can thereby operate in stealth, rather than risk triggering detectors that look for aggressive opportunistic attacks. Furthermore, as in worm propagation, puppets can be used to scan behind firewalls, NATs and detection systems. Finally, probes may also enable attackers to build *hitlists* that have been shown to result in extremely fast-spreading worms[50].

As with DDoS, the attacker installs a Web page on the malicious Web site that contains a hidden HTML frame that performs all the attack-related activities. The security model of modern browsers imposes restrictions on how the attacker can set up probing. For instance, it is not possible to ask the browser to request an object from a remote server and then forward the response back to the malicious Website. This is because of the so-called "same domain" (or "same origin") policy [46], which is designed to prevent actions such as stealing passwords and monitoring user activity. For the same reason, browsers refuse access to the contents of an inline frame, unless the source of the frame is in the same domain with the parent page.

Unfortunately, there are workarounds for the attacker to indirectly infer whether a connection to a remote host is successful. The basic idea is similar to the timing attack of [19]. We "sandwich" the probe request between two requests to the malicious Web site.

We can infer whether the target is responding to a puppet by measuring the time difference between the first and third request. If the target does not respond, the difference will be either very small (*e.g.*, because of an ICMP UNREACHABLE message) or very close to the browser request timeout. If the target is responsive, then the difference will vary but is unlikely to coincide with the timeout.

Because browsers can launch multiple connections in parallel, the attacker needs to serialize the three requests. This can be done with additional requests to the malicious Web site in order to consume all but one connection slots. However, this would require both discovering and also keeping track of the available connection slots on each browser, making the technique complex and error-prone. A more attractive solution is to employ Javascript, as modern browsers provide hooks for a default action after a page is loaded (the *onLoad* handler) and when a request has failed (the *onError* handler). Using these controls, the attacker can easily chain requests to achieve serialization without the complexity of the previous technique. We therefore view the basic sandwich attack as a backup strategy in case Javascript is disabled.

We have tested this attack scenario as shown in Figure 10. In a hidden frame, we load a page containing several image elements. The script points the source of each image to the reconnaissance target. Setting the source of an image element is an asynchronous operation. That is, after we set the source of an image element, the browser issues the request in a separate thread. Therefore, the requests to the various scan targets start at roughly the same time. After the source of each image is set, we wait for a timeout to be processed through the *onLoad* and *onError* handlers for every image. We identify the three cases (*e.g.*, unreachable, live, or non-responsive) similar to the sandwich attack but instead of issuing a request back to the malicious Web site to record the second timestamp we collect the results through the *onLoad/onError* handlers.

After the timeout expires, the results can be reported to the attacker, by means such as embedding timing data and IP addresses in a URL. The script can then proceed to another round of scanning. Each round takes time roughly equal to the timeout, which is normally controlled by the OS. It is possible for the attacker to use a smaller timeout through the *setTimeout()* primitive, which speeds up scanning at the expense of false negatives. We discuss this trade-off in Section 2.3.1.

There are both OS and browser restrictions on the number of parallel scans. On XP/SP2, the OS enforces a limit of no more than ten "outstanding"[4] connection requests at any given time [6]. Some browsers also impose limits on the number of simultaneous established connections. IE and Opera on Windows (without SP2), and browsers such as Konqueror on Linux, impose no limits, while Firefox does not allow more than 24. The attacker can choose between using a safe common-denominator value or employing Javascript to identify the OS and browser before deciding on the number of parallel scans.

The same process can be used to identify services other than Web servers. When connecting to such a service, the browser will issue an HTTP request as usual. If the remote server responds with an error message and closes the connection, then the resulting behavior is the same as probing

---

[4]A connection is characterized outstanding when the SYN packet has been sent but no SYN+ACK has been received.

Web servers. This is the case for many services, including SSH: the SSH daemon will respond with an error message that is non-HTTP-compliant and cannot be understood by the browser, the browser will subsequently display an error page, but the timing information is still relevant for reconnaissance purposes. This approach, however, does not work for all services, as some browsers block certain ports: IE blocks ports FTP, SMTP, POP3, NNTP and IMAP to prevent spamming through Web pages (we return to this problem in Section 2.4); Firefox blocks a larger number of ports[1]; interestingly, Apple's Safari does not impose any restrictions. The attacker can rely on the "User-agent" string to trigger browser-specific code.

It is important to note that puppetnets are limited to determining only the *liveness* of a remote target. As the same-domain policy restricts the attack to timing information only, the attack script cannot relay back to the attacker information on server software, OS, protocol versions, *etc.*, which are often desirable. Although this is a major limitation, distributed liveness scans can be highly valuable to an attacker. An attacker could use a puppetnet to evade detectors that are on the lookout for excessive numbers of failed connections, and then use a smaller set of sources to obtain more detailed information about each live target.

2.3.1 Analysis of reconnaissance probing

There is a subtle difference between worm scanning and reconnaissance scanning. In worm scanning, the attacker can opportunistically launch probes and does not need to observe the result of each probe. In contrast, reconnaissance requires data collection and reporting.

There are two parameters in the reconnaissance attack that we need to explore experimentally: the timeout for considering a host non-responsive, and the threshold for considering a host unreachable. The attacker can tune the timeout and trade off accuracy for speed of data collection. The unreachable threshold does not affect scanning speed, but if it is too large it may affect accuracy, as it would be difficult to distinguish between unreachable hosts and live hosts. Both parameters depend on the properties of the network through which the browser is performing the scans.

In our first experiment we examine how the choice of timeout affects reconnaissance speed and accuracy and whether the unreachable threshold may interfere with reconnaissance accuracy. As most browsers under the attacker's control are expected to stay on the malicious page only for a few minutes, the attacker may want to maximize the scanning rate. If the timeout is set too low, the attacker will not be able to discover hosts that would not respond within the timeout.

Note that in the case of XP/SP2, the timeout must be higher than the default connection timeout of the OS, which is 25 seconds. The reason is that the scanning process has to wait until outstanding connections of the previous round are cleared before issuing new ones. The analysis below is therefore more relevant to non-XP/SP2 browsers.

We measure the time needed to download the main index file for roughly 50,000 unique Websites, obtained through random queries to a search engine. We perform our measurements from four hosts in locations with different network characteristics. The distributions of the download times are presented in Figure 11. We see that in all cases, a threshold of 200-300 msec would result in a loss of around 5% of the live targets, presumably those within very short RTT distance from the scanning browser. We consider this loss to be acceptable.

Recall that the goal of the attacker may be speed rather than efficiency. That is, the attacker may not be interested in finding all servers, but finding a subset of them very quickly. We use simulation, driven by our measured distributions, to determine the discovery rate for a puppet using different timeout values, assuming 200 msec as the unreachable threshold. The results are summarized in Figure 12. For the four locations in our study, the peak discovery rate differs in absolute value, and is maximized at different points, suggesting that obtaining optimal results would require effort to calibrate the timeout on each puppet. However, all sources seem to perform reasonably well for small timeouts of 1-2 seconds.

In our second experiment we look at a 2-day packet trace from our institution and try to understand the behavior of ICMP unreachable notifications. We identify roughly 23,000 distinct unreachable events. The RTTs for these notifications were between 5 msec and 18 seconds. Nearly 50% responded within 347 msec, which, considering the response times of Figure 11, would result in less than 5% false negatives if used as a threshold. The remaining 50% of unreachables that exceed the threshold will be falsely identified as live targets, but as reported in [9], only 6.36% of TCP connections on port 80 receive an ICMP unreachable as response. As a result, we expect around 3% of the scan results to be false positives.

2.4 Protocols other than HTTP

One limitation of puppetnets is that they are bound to the use of the HTTP protocol. This raises the question of whether any other protocols can be somehow "tunneled" on top of HTTP. This can be done, in some cases, using the approach of [55, 13]. Briefly, it is possible to craft HTML forms that embed messages to servers understanding other protocols. The browser is instructed to issue an HTTP POST request to the remote server. Although the request contains the standard HTTP POST preamble, the actual post data can be fully specified by the HTML form. Thus, if the server fails gracefully when processing the HTTP part of the request (*e.g.,* ignoring them, perhaps with an error message, but without terminating the session), all subsequent messages will be properly processed. Two additional constraints for the attack to work is that the protocol in question must be text-based (since the crafted request can only contain text) and asynchronous (since all messages have to be delivered in one pass).

In this scenario, SMTP tunneling is achieved by wrapping the SMTP dialogue in a HTTP POST request that is automatically triggered through a hidden frame on the malicious Web page. For IRC servers that do not require early handshaking with the user (*e.g.,* the *identd* response), a browser can be instructed to login, join IRC channels and even send customized messages to the channel or private messages to pre-selected list of users sitting in that channel. This feature enables the attacker to use puppetnet for certain attacks such as triggering botnets, flooding and social engineering. The method is pretty similar to SMTP. An example of how a Web server could instruct puppets to send spam is provided in [33].

Although this vulnerability has been discussed previously, its potential impact in light of a puppetnet-like attack in-

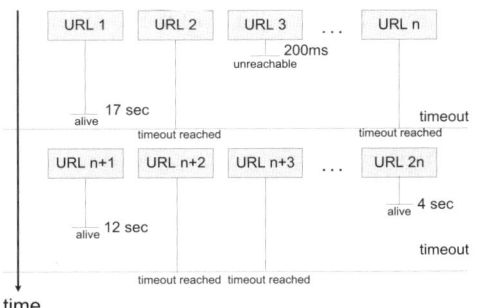
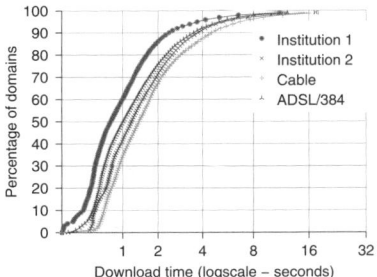
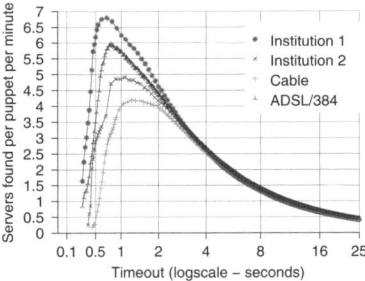

Figure 10: Illustration of reconnaissance probing method.

Figure 11: CDF of time to get main index from different sites.

Figure 12: Discovery rate, per puppet.

frastructure has not been considered, and vendors may not be aware of the implications of the underlying vulnerability. We have found that although IE refuses outgoing requests to a small set of ports (including standard ports for SMTP, NNTP, *etc.*) and Firefox blocks a more extensive list of ports, Apple's Safari browser as well as IE5.2 on Mac OSX do not impose *any* similar port restrictions[5]. Thus, although the extent of the threat may not be as significant as DDoS and worm propagation, popular Web sites with a large Apple/Safari user base can be easily turned into powerful spam conduits.

2.5 Exploiting cookie-authenticated services

A large number of Web-based services rely on cookies for maintaining authentication state. A typical example is Web-based mail services that offer a "remember me" option to allow return visits without re-authentication. Such services could be manipulated by a malicious site that coerces visitors to post forms created by the attacker with the visitors' credentials. There are three constraints for this attack. First, the inline frame needs to be able to post cookies; this works on Firefox, but not IE. Second, the attacker needs to have knowledge about the structure and content of the form to be posted, as well as the target URL; this depends on the site design. Finally, the attacker needs to be able to instruct browsers to automatically post such forms; this is possible in all browsers we tested.

We have identified sites that are vulnerable to this attack.[6] As proof-of-concept, we have successfully launched an attack to one of our own accounts on such a site. Although this seems like a wider problem (*e.g.*, it allows the attacker to forward the victim's email to his site, *etc.*), in the context of puppetnets, the attacker could be on the lookout for visitors that happen to be pre-authenticated to one of the vulnerable Web sites, and could use them for purposes such as sending spam or performing mailbomb-type DoS attacks.

Given the restriction to Firefox and the need to identify visitors that are pre-authenticated to particular sites, it seems that this attack would only have significant impact on highly popular sites, or moderately popular sites with unusually high session times, or sites that happen to have an unusually large fraction of Firefox visitors. Considering these constraints, the attack may seem weak compared to the ubiquitous applicability of DoS, scanning, and worm

propagation. Nevertheless, none of these three scenarios can be safely dismissed as unlikely.

2.6 Distributed malicious computations

So far we have described scenarios of puppetnets involved in network-centric attacks. However, besides network-centric attacks, it is easy to imagine browsers unwillingly participating in malicious computations. This is a form of Web-based computing which, to the best of our knowledge, has not been considered as a platform for malicious activity. Projects such as RC5 cracking [21], use the Web as a platform for distributed computation but this is done with the users' consent. Most large-scale distributed computing projects rely on stand-alone clients, similar to SETI@home [30].

It is easy to instruct a browser to perform local computations and send the results back to the attacker. Computation can be done through Javascript, Active-X or Java applets. By default, Active-X does not appear attractive as it requires user confirmation. Javascript offers more stealth as it is lightweight and can be made invisible. Sneaking Java applets into hidden frames on malicious Web sites seems easy, and although the resources needed for instantiating the Java VM might be noticeable (and an "Applet loaded" message may be displayed on the status bar), it is unlikely to be considered suspect by a normal user.

To illustrate the extent of the problem we measured the performance of Javascript and Java applets for MD5 computations. On a low-end desktop, the Javascript implementation can perform around 380 checksums/sec, while the Java applet within the browser can compute roughly 434K checksums/sec – three orders of magnitude faster than Javascript. Standalone Java can achieve up to 640K checks/sec. In comparison, an optimized C implementation computes around 3.3M checks/sec. Hence, a 1,000-node puppetnet can crack an MD5 hash as fast as a 128-node cluster.

3. DEFENSES

In this section we examine potential defenses against puppetnets. The goal is to determine whether it is feasible to address the threat by tackling the source of the problem, rather than relying on techniques that attempt to mitigate the resulting attacks, such as DDoS, which may be hard to implement right at a global scale.

We discuss various defense strategies and the tradeoffs they offer. We concentrate on defenses against DDoS, scanning and worm propagation. Detecting malicious computations seems hard, and well beyond the scope of this paper. Cookie-authenticated services seem trivial to protect

---

[5] We have informed Apple about this vulnerability.
[6] These sites include a very popular Web-based mail service, the name of which we would prefer to disclose only upon request.

by adding non-cookie session state that is communicated to the browser when the user wishes to re-authenticate.

Disabling Javascript. The usual culprit, when it comes to Web security problems, is Javascript, and it is often suggested that many problems would go away if users disable Javascript and/or Web sites refrain from using it. However, the trade-off between quality content and security seems unfavorable: the majority of Web sites employ Javascript, there is growing demand for feature-rich content, especially in conjunction with technologies such as Ajax[20], and most browsers are shipped with Javascript enabled. It is interesting to note, however, that a recently-published Firefox extension that selectively enables Javascript only for "trusted" sites [36] has been downloaded 7 million times roughly one month after its release on April 9th, 2006.

In the case of puppetnets, disabling Javascript could indeed alter the threat landscape, but it would only reduce rather than eliminate the threat. The development of our attacks suggests that even without Javascript, it would still be feasible to launch DDoS, perform reconnaissance probes and propagate worms, although the effectiveness of the attacks would be at least one order of magnitude less than with Javascript enabled. Considering these observations, disabling Javascript does not seem like an attractive proposition towards completely eliminating the puppetnet threat.

Careful implementation of existing defenses. We observe that in most cases the attacks we developed were quite sensitive to minor tweaks. That is, although simple versions of the attack were quite easy to construct, maximizing their effectiveness required a lot more effort. Particularly the connection rate limiter implemented in XP/SP2 had a profound effect on the performance of worm propagation and reconnaissance. Unfortunately, we were able to demonstrate that the rate limiter can be partially bypassed. In particular, by reloading the core attack frame we were able to clear the TCP connection cache, presumably because a frame reload results in the underlying socket being closed and the TCP connection state entry being removed.

In this particular case it appears easy to address the problem by means of separating the actual connection state table from the state of the connection rate limiter. The rate limiter could either mirror the regular connection state table but choose to retain entries for closed sockets up to a time-out, or keep track of aggregate connection state statistics. This could reduce the effectiveness of worm propagation of up to an order of magnitude.

Another case that suggests that existing defenses are not always properly implemented is the Spam distribution attack described in Section 2.4. Although both IE and Firefox have mitigated this problem, at least in part, through blocking certain ports, Apple's Safari and the OSX version of IE5.2 did not properly address this known vulnerability.

However, careful implementation of existing defenses is insufficient for addressing the whole range of threats posed by puppetnets.

Filtering using attack signatures. We consider whether it is practical to develop IDS/IPS signatures for puppetnet attacks. In some cases it seems fairly easy to construct such a signature. For example, in the case of puppetnet-delivered spam it is easy to scan traffic for messages to the SMTP port that contain evidence of both a HTTP POST request *and* legitimate SMTP commands. This should cover most other protocols tunneled through POST requests.

Can we develop signatures for puppetnet DoS attacks? We could consider signatures of malicious Web pages that contain unusually high numbers of requests to third-party sites. However, our example attack suggests that there are many possible variations to the attack, making it hard to obtain a complete set of signatures. Additionally, because the attacks are implemented in HTML and Javascript, it appears unlikely that simple string matching or even regular expressions would be sufficient for expressing the attack signatures. Instead, more expensive analyzers, such as HTML parsers, would be needed.

Furthermore, obfuscation of HTML and Javascript seems to be both feasible and effective [51, 44], allowing the attacker to compose obfuscated malicious Web page on-the-fly. For example, one could use the *document.write()* method to write the malicious page into an array in completely random order before execution. This makes the attack difficult to detect using static analysis alone, a problem that is also found in shellcode polymorphism [14, 31, 42]. Although we must leave room for the possibility that such a unusual use of *document.write()* or similar approaches may be amenable to detection, such analysis seems complex and is likely to be expensive and error-prone.

Client-side behavioral controls. Another possible defense strategy is to further restrict browser policies for accessing remote sites. It seems relatively easy to have a client-side solution deployed with a browser update, as browser developers seem to be concerned about security, and the large majority of users rely on one among 2-3 browsers.

One way to restrict DoS, scanning and worm propagation is to establish bounds on how a Web page can instruct the browser to access "foreign" objects, *e.g.*, objects that do not belong to the same domain. These resource bounds should be persistent, to prevent attackers from resetting the counters using page reloads. For similar reasons, the bounds should be tied to the requesting server, and not to a page or frame instance, to prevent attackers from evading the restriction through multiple frames, chained requests, *etc*.

We consider whether it makes sense to impose controls on foreign requests from a Web page. We attempt to quantify whether such a policy would break existing Web sites, and what impact it would have on DDoS and other attacks. We first look at whether we can limit the total number of different objects (*e.g.*, images, embedded elements and frames) that the attacker can load from foreign servers, considering all servers to be foreign except for the one offering the page. This restriction should be enforced across multiple automatic refreshes or frame updates, to prevent the attacker from resetting the counters upon reaching the limit. (Counters would only be reset only when a user clicks on a link.) Of course, this is likely to "break" sites such as those that use automatic refresh to update banner ads. Given that ads are loaded periodically, *e.g.*, one refresh every few minutes, it seems reasonable to further refine the basic policy with a timeout (or leaky bucket mechanism) that occasionally resets or tops-up the counters.

To evaluate the effectiveness of this policy, we have obtained data on over 70,000 Web pages by crawling through a search engine. For each Web page we obtain the number

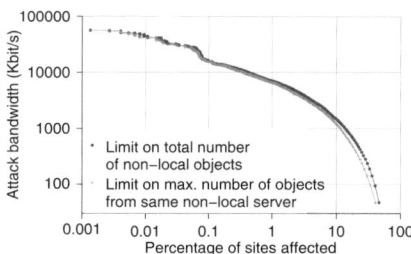
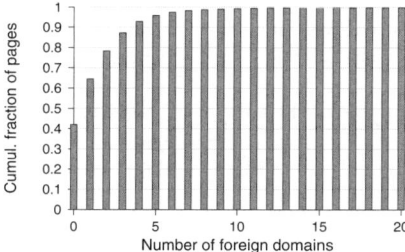

Distribution	Firepower
KDDCUP	47.03 Kbit/sec
Google/Webtrends	14.39 Kbit/sec
JSTracker	3.05 Kbit/sec
Web tracking	2.87 Kbit/sec

Figure 13: Effectiveness of remote request limits

Figure 14: Cumul. histogram of foreign domains referenced by Web sites

Figure 15: Impact of ACT defense on 1000-puppet DDoS attack

of non-local embedded object references. We then compute for each upper bound N of non-local references, the fraction of sites that would be corrupted should such a policy be implemented, against the effective DoS bandwidth of a 1000-node puppetnet under the same policy. A variation of the above policy involves a cap on the maximum number of non-local references to the same non-local server.

The results are shown in Figure 13. We observe that this form of restriction is somewhat effective when compared to the DDoS firepower of Figure 5, providing a 3-fold reduction in DDoS strength when trying to minimize disruption to Web sites. The strength of the attack, however, remains significant, at 50 Mbit/s for 1000 puppets. Obtaining a 10x reduction in DDoS strength would disrupt around 0.1% of all Web sites, with DDoS reduced to 10 Mbit/s. Obtaining a further 10x reduction seems impractical, as the necessary request cap would negatively affect more than 10% of Web pages. The variation limiting the max. number of requests to the same non-local server also does not offer any notable improvement. Given the need to defend against not only 1000-node but even larger puppetnets, we conclude that although the improvement offered is significant, this defense is not good enough to address the DDoS threat.

The above policy only targets DDoS. To defend against worms and reconnaissance probes, we look at the feasibility of imposing limits on the number of distinct remote servers to which embedded object references are made. The cumulative histogram is shown in Figure 14. We see that most Web sites access very few foreign domains: around 99% of Web sites access 11 or less foreign domains; around 99.94% of Web sites access less than 20 foreign domains. Not visible on the graph is a handful of Web sites, typically "container" pages, that access up to 33 different domains. Based on this profile, it seems reasonable to implement a restriction of around 10 foreign domains, keeping in mind that the limit should be set as low as possible, given that a large fraction of puppets have a very short lifetime in the system. Note that sites that are out of profile could easily avoid violating the proposed policy, by proxying requests to the remote servers. We repeated the worm simulation of Section 2.2.1 to determine the impact of such a limit on worm propagation. As expected, this policy almost completely eliminates the speed-up for the client-server worm compared to server-only, as puppets can perform only a small fraction of the scans they could perform without this policy. Similar effects apply to scanning as well.

Unfortunately, the above heuristic can be circumvented if the attacker has access to a DNS server. The attacker could map all foreign target hosts to identifiers that appear to be in the same domain but are translated by the attacker's DNS server to IP addresses in the foreign domain. Attacks aiming at consuming egress bandwidth from servers that rely on the "Host:" tag in the HTTP request would be less effective, but all other attacks are not affected.

Server-side controls and puppetnet tracing. Considering the DoS problem and the difficulty in coming up with universally acceptable thresholds for browser-side connection limiting, one could argue that it is the Web developers who should specify how their sites are accessed by third parties.

One way for doing that is for servers to use the "Referer" tag of HTTP requests to determine whether a particular request is legitimate and compliant, similar to [57]. The server could consult the appropriate access policy and decide whether to honor a request. This approach would protect servers against wasting their egress bandwidth, but does not allow the server to exercise any control over incoming traffic.

Another use of referrer information can be to trace the source of the puppetnet attack, and take action to shutdown the control Web site. That is, puppetnets have a single point of failure. However, this process is relatively slow as it involves human coordination. Thus, attackers may already have succeeded in disrupting service. Moreover, even when the controlling site has been taken down, existing puppets will continue to perform an attack – the attack will only subside once all puppet browsers have been pointed elsewhere, which is likely to be in the order of 10-60 minutes, based on the viewing time estimates of Section 2.1.1.

However, as shown in [33], we have been able to circumvent the default behavior of browsers that set referrer information, making puppetnet attacks more difficult to filter and trace. It is unclear at this point if minor modifications could address the loss of referrer-based defenses. Thus, referrer-based filtering does not currently offer much protection and may not be sufficient, even in the longer-term, for adequately protecting against puppetnet attacks.

Server-directed client-side controls. To protect against unauthorized incoming traffic from puppets, we examine the following approach. If we assume that the attacker cannot tamper with the browser software, a server can communicate site access policies to a browser during the first request. In our implementation, we embed Access Control Tokens (ACTs) in the server response through an extension header ("X-ACT:") that is either a blanket "permit/deny" or a Javascript function, similar to proxy autoconfiguration[38]. This script is executed on the browser side for each request to the server to determine whether a request is legitimate or not. The use of Javascript offers flexibility for site developers to design their own policies, without having to standardize specific behaviors or a new policy language.

Perhaps the simplest policy would be to ask browsers to completely stop issuing requests if the server is under attack. More fine-grained policies might restrict the total number or rate of requests in each session, or may impose custom restrictions based on target URL, referrer information, navigation path, etc. One could envision a tool for site owners to extract behavioral profiles from existing logs, and then turn these profiles into ACT policies. For a given policy, the owners can also compute the exposure in terms of potential puppetnet DDoS firepower, using the same methodology used in this paper. The specifics of profiling and exposure estimation are beyond the scope of this paper.

ACTs require at least one request-response pair for the defense to kick in, given that the browser may not have communicated with the server in the past. After the first request, any further unauthorized requests can be blocked on the browser side. Thus, ACTs can reduce the DoS attack strength to one request per puppet, which makes them quite attractive. On the other hand, this approach requires modifications to both servers and clients.

To illustrate the effectiveness of this approach, we estimate, using simulation, the firepower of a 1000-puppet DDoS attack where all users support ACTs on their browsers. The puppet viewing time on the malicious site is taken from the distributions shown in Figure 2. The victim site follows the most conservative policy: if a request comes from a non-trusted referrer then user is not allowed to make any further requests. The results are summarized in Table 15. As the attack is restricted to one request per user, the firepower is limited to only a few Kbit/sec.

In theory, it is possible to prevent the first unauthorized request to the target if policies are communicated to the browser out-of-band. One could directly embed ACTs in URL references, through means such as overloading the URL. Given that an ACT needs to be processed by the browser, it must fully specify the policy "by value". To prevent the attacker from tampering with the ACT, it must be cryptographically signed by the server. Besides being cumbersome, this also requires the browser to have a public key to verify the ACTs, which makes this proposal less attractive.

4. RELATED WORK

Web security has attracted a lot of attention in recent years, considering the popularity of the Web and the observed increase in malicious activity. Rubin *et al.* [45] and Claessens *et al.* [16] provide comprehensive surveys of problems and potential solutions in Web security, but do not discuss any third-party attacks like puppetnets. Similarly, most of the work on making the Web more secure focuses on protecting the browser and its user against attacks by malicious Web sites (c.f., [32, 25, 18, 15, 26]).

The most well-known form of HTML tag misuse is known as cross-site scripting (or XSS) and is discussed in a CERT advisory in 2000 [11]. The advisory focuses primarily on the threat of attackers injecting scripts into sites such as message boards, and the implications that such scripts could have on users browsing those sites, including potential privacy loss. Although XSS and puppetnet attacks both exploit weaknesses of the Web security architecture, there are two fundamental differences. First, puppetnet attacks require the attacker to have more control over a Web server, in order to maximize exposure of users to the attack code. Injecting puppetnet code on message boards in a XSS fashion is also an option, but is less likely to be effective. The second important difference is that puppetnets exploit browsers for attacking third parties, rather than attacking the browser executing the malicious script.

During the course of our investigation we became aware of a report [56] describing a DDoS attack that appears to be very similar to the one described in this paper. The report, published in early December 2005, states that a well-known hacker site was attacked using a so-called "xflash" attack which involves a "secret banner" encoded on Web sites with large numbers of visitors redirecting users to the target. According to the same report, the attack generated 16,000 SYN packets per second towards the target. As we have not been able to obtain a sample of the attack code, we cannot directly compare it to the one described here. However, from the limited available technical information, it seems likely that attackers are already considering puppetnet-style techniques as part of their arsenal.

Another example of a puppetnet-like attack observed in the wild is "referer spamming" [23], where a malicious Web site floods some other site's logs to make its way into top referer lists. The purpose of the attack is to trick search engines that rank sites based on link counts, since the victims will include the malicious sites in their top referer lists.

The work that is most closely related to ours is a short paper by Alcorn[4] discussing "XSS viruses", developed independently and concurrently[2] to our investigation. The author of this work imagines attacks similar to ours, focusing on puppetnet-style worm propagation and also mentions the possibility of DDoS and spam distribution. The main difference is that our work offers a more in-depth analysis of each attack as well as concrete experimental assessment of the severity of the threat. For instance, a proof-of-concept implementation of an XSS virus that is similar to our puppetnet worm is provided albeit without analyzing its propagation characteristics. Similarly, DDoS and spam are mentioned as *potential* attacks but without any further investigation. The author discusses referer-based filtering as a potential defense, which, as we have shown, can be currently circumvented and is also unlikely to be sufficient in the long term. One major difference in the attack model is that we consider popular malicious or subverted Web sites as the primary vector for controlling puppetnets, while [4] focuses on first infecting Web servers in order to launch other types of attacks. Similar ideas are also discussed in [37]. While the work of [4] and [37] are both interesting and important, we believe that raising awareness and convincing the relevant parties to mobilize resources towards addressing a threat requires not just a sketch or proof-of-concept artifact of a potential attack, but extensive analysis and experimental evidence. In this direction, we hope that our work provides valuable input.

The technique we used for sending spam was first described by Jochen [55], although we independently developed the same technique as part of our investigation on puppetnets. Our work goes one step further by exploring how such techniques can be misused by attackers that control a large number of browsers. A scanning approach that is somewhat similar to how puppets could propagate worms is imagined by Weaver et al. in [59], but only in the context of a malicious Web page directing a client to create a large number of requests to nonexistent servers with the purpose of abusing scan blockers. The misuse of Javascript for attacks such as scanning behind firewalls was independently invented by Grossman and Niedzialkowski[22] while our study was in review[2].

The reconnaissance technique relies on the same principle used for timing attacks against browser privacy [19]. Similar to our probing, this attack relies on timing accesses to a particular Web site. In our case, we use timing information to infer whether the target site exists or is unreachable. In the case of the Web privacy attack, the information is used to determine if the user recently accessed a page, in which case it can be served instantly from the browser cache.

Puppetnets are malicious distributed systems, much like reflectors and botnets. Reflectors have been analyzed extensively by Paxson [40]. Reflectors are regular servers that, if targeted by appropriately crafted packets, can be misused for DDoS attacks against third parties. The value of reflectors lies both in allowing the attacker to bounce attack packets through a large number of different sources, hereby making it harder for the defender to develop the necessary packet filters, as well as acting as amplifiers, given that a single packet to a reflector can trigger the transmission of multiple packets from the reflector to the victim.

There are several studies discussing botnets. Cooke et al. [17] have analyzed IRC-based botnets by inspecting live traffic for botnet commands as well as behavioral patterns. The authors also propose a system for detecting botnets with advanced command and control systems using correlation of alerts. Other studies of botnets include [54, 34]. From our analysis it becomes evident that botnets are much more powerful than puppetnets and therefore a much larger threat. However, they are currently attracting a lot of attention, and may thus become increasingly hard to setup and manage, as end-point and network-level security measures continue to focus on botnets.

5. CONCLUDING REMARKS

We have explored a new class of Web-based attacks that involve malicious Web sites manipulating their visitors towards attacking third parties. We have shown how attackers can set up powerful malicious distributed systems, called Puppetnets, that can be used for distributed DoS, reconnaissance probes, worm propagation and other attacks. We have attempted to quantify the effectiveness of these attacks, demonstrating that the threat of puppetnets is significant. We have also discussed several directions for developing defenses against puppetnet attacks. None of the strategies were completely satisfying, as most of them offered only partial solutions. Nevertheless, if implemented, they are likely to significantly reduce the effectiveness of puppetnets.

Acknowledgments

We thank S. Sidiroglou, S. Ioannidis, M. Polychronakis, E. Athanasopoulos, E. Markatos, M. Greenwald, the members of the Systems and Security Department at I^{2}R and the anonymous reviewers for very insightful comments and suggestions on earlier versions of this work. We also thank Blue Martini Software for the KDD Cup 2000 data.

6. REFERENCES

[1] Mozilla Port Blocking. http://www.mozilla.org/projects/netlib/PortBanning.html, December 2004.

[2] PuppetNet Project Web Site. http://s3g.i2r.a-star.edu.sg/proj/puppetnets, September 2005.

[3] ABC Electronic. ABCE Database. http://www.abce.org.uk/cgi-bin/gen5?runprog=abce/abce&noc=y, 2006.

[4] W. Alcorn. The cross-site scripting virus. http://www.bindshell.net/papers/xssv/xssv.html. Published: 27th September, 2005. Last Edited: 16th October 2005.

[5] Alexa Internet Inc. Global top 500. http://www.alexa.com/site/ds/top_500, 2006.

[6] S. Andersen and V. Abella. Changes to Functionality in Microsoft Windows XP Service Pack 2, Part 2: Network Protection Technologies. Microsoft TechNet, http://www.microsoft.com/technet/prodtechnol/winxppro/maintain/sp2netwk%.mspx, November 2004.

[7] Anonymous. About the Alexa Toolbar and traffic monitoring service: How accurate is Alexa? http://www.mediacollege.com/internet/utilities/alexa/, 2004.

[8] B. L. Barrett. Home of the webalizer. http://www.mrunix.net/webalizer, August 2005.

[9] V. Berk, G. Bakos, and R. Morris. Designing a framework for active worm detection on global networks. In *Proceedings of the IEEE International Workshop on Information Assurance*, March 2003.

[10] T. Berners-Lee, L. Masinter, and M. McCahill. Uniform Resource Locators (URL). *RFC 1738*, Dec. 1994.

[11] CERT. Advisory CA-2000-02: Malicious HTML Tags Embedded in Client Web Requests. http://www.cert.org/advisories/CA-2000-02.html, February 2000.

[12] CERT. Advisory CA-2001-19: 'Code Red' Worm Exploiting Buffer Overflow in IIS Indexing Service DLL. http://www.cert.org/advisories/CA-2001-19.html, July 2001.

[13] CERT. Vulnerability Note VU#476267: Standard HTML form implementation contains vulnerability allowing malicious user to access SMTP, NNTP, POP3, and other services via crafted HTML page. http://www.kb.cert.org/vuls/id/476267, August 2001.

[14] R. Chinchani and E. V. D. Berg. A fast static analysis approach to detect exploit code inside network flows. In *Proceedings of the International Symposium on Recent Advances in Intrusion Detection (RAID)*, Sept. 2005.

[15] N. Chou, R. Ledesma, Y. Teraguchi, and J. Mitchell. Client-side defense against web-based identity theft. In *Proceedings of the 11th Annual Network and Distributed System Security Symposium (NDSS '04)*, February 2004.

[16] J. Claessens, B. Preneel, and J. Vandewalle. A tangled world wide web of security issues. *First Monday*, 7(3), March 2002.

[17] E. Cooke, F. Jahanian, and D. McPherson. The Zombie Roundup: Understanding, Detecting, and Disrupting Botnets. In *Proceedings of the 1st USENIX Workshop on Steps to Reducing Unwanted Traffic on the Internet (SRUTI 2005)*, July 2005.

[18] E. W. Felten, D. Balfanz, D. Dean, and D. S. Wallach. Web Spoofing: An Internet Con Game. In *Proceedings of the 20th National Information Systems Security Conference*, pages 95–103, October 1997.

[19] E. W. Felten and M. A. Schneider. Timing attacks on Web privacy. In *Proceedings of the 7th ACM Conference on Computer and Communications Security (CCS'00)*, pages 25–32, New York, NY, USA, 2000. ACM Press.

[20] J. J. Garrett. Ajax: A New Approach to Web Applications. http://www.adaptivepath.com/publications/essays/archi-ves/000385.php, February 2005.

[21] P. Gladychev, A. Patel, and D. O'Mahony. Cracking RC5 with Java applets. *Concurrency: Practice and Experience*, 10(11-13):1165–1171, 1998.

[22] J. Grossman and T. Niedzialkowski. Hacking intranet websites from the outside - javascript malware just got a lot more dangerous. Blackhat USA, August 2006.

[23] M. Healan. Referer spam. http://www.spywareinfo.com/articles/referer_spam/, Sept. 2003.

[24] W. Inc. Webtrends web analytics and web statistics. http://www.webtrends.com, 2006.

[25] S. Ioannidis and S. M. Bellovin. Building a Secure Browser. In *Proceedings of the Annual USENIX Technical Conference, Freenix Track*, June 2001.

[26] C. Jackson, A. Bortz, D. Boneh, and J. C. Mitchell. Protecting browser state from Web privacy attacks. In *Proceedings of the WWW Conference*, 2006.

[27] G. Keizer. Dutch botnet bigger than expected. http://informationweek.com/story/showArticle.jhtml?articleID=172303265, October 2005.

[28] J. O. Kephart and S. R. White. Directed-graph epidemiological models of computer viruses. In *Proceedings of the 1991 IEEE Computer Society Symposium on Research in Security and Privacy*, May 1991.

[29] R. Kohavi, C. Brodley, B. Frasca, L. Mason, and Z. Zheng. KDD-Cup 2000 organizers' report: Peeling the onion. *SIGKDD Explorations*, 2(2):86–98, 2000.

[30] E. Korpela, D. Werthimer, D. Anderson, J. Cobb, and M. Lebofsky. SETI@home – Massively Distributed Computing for SETI. *Computing in Science & Enginering*, 3(1):78–83, 2001.

[31] C. Kruegel, E. Kirda, D. Mutz, W. Robertson, and G. Vigna. Polymorphic worm detection using structural information of executables. In *Proceedings of the International Symposium on Recent Advances in Intrusion Detection (RAID)*, Sept. 2005.

[32] C. Kruegel and G. Vigna. Anomaly detection of Web-based attacks. In *Proceedings of the 10th ACM Conference on Computer and Communications Security (CCS'03)*, pages 251–261, New York, NY, USA, 2003. ACM Press.

[33] V. T. Lam, S. Antonatos, P. Akritidis, and K. G. Anagnostakis. Puppetnets: Misusing web browsers as a distributed attack infrastructure (extended version). Technical Report, http://s3g.i2r.a-star.edu.sg/proj/puppetnets, August 2006.

[34] J. Li, T. Ehrenkranz, G. Kuenning, and P. Reiher. Simulation and analysis on the resiliency and efficiency of malnets. In *Proceedings of the 19th Workshop on Principles of Advanced and Distributed Simulation (PADS'05)*, pages 262–269, Washington, DC, USA, 2005. IEEE Computer Society.

[35] J. D. C. Little. A Proof of the Queueing Formula $L = \lambda W$. *Operations Research*, (9):383–387, 1961.

[36] G. Maone. Firefox add-ons: Noscript. https://addons.mozilla.org/firefox/722/, May 2006.

[37] D. Moniz and H. Moore. Six degrees of xssploitation. Blackhat USA, August 2006.

[38] Mozilla.org. End User Guide: Automatic Proxy Configuration (PAC). http://www.mozilla.org/catalog/end-user/customizing/enduserPAC.html, August 2004.

[39] C. Nachenberg. Computer virus-antivirus coevolution. *Commun. ACM*, 40(1):46–51, 1997.

[40] V. Paxson. An analysis of using reflectors for distributed denial-of-service attacks. *ACM Computer Communication Review*, 31(3):38–47, 2001.

[41] Philippine Honeynet Project. Philippine Internet Security Monitor - First Quarter of 2006. http://www.philippinehoneynet.org/docs/PISM20061Q.pdf

[42] M. Polychronakis, K. G. Anagnostakis, and E. P. Markatos. Network-level polymorphic shellcode detection using emulation. In *Proceedings of the GI/IEEE SIG SIDAR Conference on Detection of Intrusions and Malware and Vulnerability Assessment (DIMVA)*, July 2006.

[43] L. Rizzo. Dummynet: a simple approach to the evaluation of network protocols. *ACM Computer Communication Review*, 27(1):31–41, 1997.

[44] B. Ross, C. Jackson, N. Miyake, D. Boneh, and J. C. Mitchell. Stronger password authentication using browser extensions. In *Proceedings of the 14th Usenix Security Symposium*, 2005.

[45] A. D. Rubin and D. E. G. Jr. A Survey of Web Security. *IEEE Computer*, 31(9):34–41, 1998.

[46] J. Ruderman. The Same Origin Policy. http://www.mozilla.org/projects/security/components/same-origin.html, August 2001.

[47] S. Saroiu, P. Gummadi, and S. Gribble. A measurement study of peer-to-peer file sharing systems. In *Proceedings of Multimedia Computing and Networking (MMCN)*, 2002.

[48] B. Schneier. Attack trends 2004 and 2005. *ACM Queue*, 3(5), June 2005.

[49] F. Smith, J. Aikat, J. Kapur, and K. Jeffay. Variability in TCP round-trip times. In *Proceedings of the 3rd ACM SIGCOMM Conference on Internet measurement*, 2003.

[50] S. Staniford, D. Moore, V. Paxson, and N. Weaver. The top speed of flash worms. In *Proc. ACM WORM*, Oct. 2004.

[51] Stunnix. Stunnix javascript obfuscator - obfuscate javascript source code. http://www.stunnix.com/prod/jo/overview.shtml, 2006.

[52] Symantec. Internet Threat Report: Trends for January 05-June 05. Volume VIII. Available from www.symantec.com, September 2005.

[53] TechWeb.com. Lycos strikes back at spammers with dos screensaver. http://www.techweb.com/wire/security/54201269, 2004.

[54] The Honeynet Project. Know your enemy: Tracking botnets. http://www.honeynet.org/papers/bots/, March 2005.

[55] J. Topf. HTML Form Protocol Attack. http://www.remote.org/jochen/sec/hfpa/, August 2001.

[56] VNExpress Electronic Newspaper. Website of largest Vietnamese hacker group attacked by DDoS. http://vnexpress.net/Vietnam/Vi-tinh/2005/12/3B9E4A6D/, December 2005.

[57] D. Wang. HOWTO: ISAPI Filter which rejects requests from SF_NOTIFY_PREPROC_HEADERS based on HTTP Referer. http://blogs.msdn.com/david.wang, July 2005.

[58] Y.-M. Wang, D. Beck, X. Jiang, R. Roussev, C. Verbowski, S. Chen, and S. Kin. Automated Web Patrol with Strider HoneyMonkeys: Finding Web Sites That Exploit Browser Vulnerabilities. In *Proceedings of the 13th Annual Network and Distributed System Security Symposium (NDSS '06)*, February 2006.

[59] N. Weaver, S. Staniford, and V. Paxson. Very Fast Containment of Scanning Worms. In *Proceedings of the 13th USENIX Security Symposium*, pages 29–44, August 2004.

[60] A. T. Williams and J. Heiser. Protect your PCs and Servers From the Bothet Threat. Gartner Research, ID Number: G00124737, December 2004.

[61] zone-h. Digital attacks archive. http://www.zone-h.org/en/defacements/, 2006.

[62] C. C. Zou, W. Gong, and D. Towsley. Code Red Worm Propagation Modeling and Analysis. In *Proceedings of the 9th ACM Conference on Computer and Communications Security (CCS)*, pages 138–147, November 2002.

A Natural Language Approach to Automated Cryptanalysis of Two-time Pads

Joshua Mason
Johns Hopkins University
josh@cs.jhu.edu

Kathryn Watkins
Johns Hopkins University
kwatkins@jhu.edu

Jason Eisner
Johns Hopkins University
jason@cs.jhu.edu

Adam Stubblefield
Johns Hopkins University
astubble@cs.jhu.edu

ABSTRACT

While keystream reuse in stream ciphers and one-time pads has been a well known problem for several decades, the risk to real systems has been underappreciated. Previous techniques have relied on being able to accurately guess words and phrases that appear in one of the plaintext messages, making it far easier to claim that "an attacker would never be able to do *that*." In this paper, we show how an adversary can automatically recover messages encrypted under the same keystream if only the *type* of each message is known (e.g. an HTML page in English). Our method, which is related to HMMs, recovers the most probable plaintext of this type by using a statistical language model and a dynamic programming algorithm. It produces up to 99% accuracy on realistic data and can process ciphertexts at 200ms per byte on a $2,000 PC. To further demonstrate the practical effectiveness of the method, we show that our tool can recover documents encrypted by Microsoft Word 2002 [22].

Categories and Subject Descriptors

E.3 [**Data**]: Data Encryption

General Terms

Security

Keywords

Keystream reuse, one-time pad, stream cipher

1 Introduction

Since their discovery by Gilbert Vernam in 1917 [20], stream ciphers have been a popular method of encryption. In a stream cipher, the plaintext, p, is exclusive-ORed (XORed) with a keystream, k, to produce the ciphertext, $p \oplus k = c$. A special case arises when the keystream is truly random: the cipher is known as a one-time pad, proved unbreakable by Shannon [18].

It is well known that the security of stream ciphers rests on never reusing the keystream k [9]. For if k *is* reused to encrypt two different plaintexts, p and q, then the ciphertexts $p \oplus k$ and $q \oplus k$ can be XORed together to recover $p \oplus q$. The goal of this paper is to complete this attack by recovering p and q from $p \oplus q$. We call this the "two-time pad problem."

In this paper we present an automated method for recovering p and q given only the "type" of each file. More specifically, we assume that p and q are drawn from some known probability distributions. For example, p might be a Word document and q might be a HTML web page. The probability distributions can be built from a large corpus of examples of each type (e.g. by mining the Web for documents or web pages). Given the probability distributions, we then transform the problem of recovering p and q into a "decoding" problem that can be solved using some modified techniques from the natural language processing community. Our results show that the technique is extremely effective on realistic datasets (more than 99% accuracy on some file types) while remaining efficient (200ms per recovered byte).

Our attack on two-time pads has practical consequences. Proofs that keystream reuse leaks information hasn't stopped system designers from reusing keystreams. A small sampling of the systems so affected include Microsoft Office [22], 802.11 WEP [3], WinZip [11], PPTP [17], and Soviet diplomatic, military, and intelligence communications intercepted [2, 21]. We do not expect that this problem will disappear any time soon. Indeed, since NIST has endorsed the CTR mode for AES [7], effectively turning a block cipher into a stream cipher, future systems that might otherwise have used CBC with a constant IV may instead reuse keystreams. The WinZip vulnerability is already of this type.

To demonstrate this practicality more concretely, we show that our tool can be used to recover documents encrypted by Microsoft Word 2002. The vulnerability we focus on was known before this work [22], but could not be exploited effectively.

1.1 Prior Work

Perhaps the most famous attempt to recover plaintexts that have been encrypted with the same keystream is the National Security Agency's VENONA project [21]. The NSA's forerunner, the Army's Signal Intelligence Service, noticed that some encrypted Soviet telegraph traffic appeared to reuse keystream material. The program to reconstruct the messages' plaintext began in 1943 and did not end until 1980. Over 3,000 messages were at least partially recovered. The project was partially declassified in 1995, and many of the decryptions were released to the public [2]. However, the ciphertexts and cryptanalytic methods remain classified.

There is a "classical" method of recovering p and q from $p \oplus q$ when p and q are known to be English text. First guess a word likely to appear in the messages, say `the`. Then, attempt to XOR `the` with each length-3 substring of $p \oplus q$. Wherever the result is something that "looks like" English text, chances are that one of the messages has `the` in that position and the other message has the result of the XOR. By repeating this process many times, the cryptanalyst builds up portions of plaintext. This method was somewhat formalized by Rubin in 1978 [15].

In 1996, Dawson and Nielsen [5] created a program that uses a series of heuristic rules to automatically attempt this style of decryption. They simplified matters by assuming that the plaintexts used only 27 characters of the 256-character ASCII set: the 26 English uppercase letters and the space. Given $p \oplus q$, this assumption allowed them to unambiguously recover non-coinciding spaces in p and q, since in ASCII, an uppercase letter XORed with a space can not be equal to any two uppercase letters XORed together. They further assumed that two characters that XORed to 0 were both equal to space, the most common character. To decode the words between the recovered spaces, they employed lists of common words of various lengths (and a few "tricks"). They chose to test their system by running it on subsets of the same training data from which they had compiled their common-word lists (a preprocessed version of the first 600,000 characters of the English Bible). They continued adding new tricks and rules until they reached the results shown in Figure 1. It is important to note that the rules they added were *specifically designed to get good results on the examples they were using for testing* (hence are not guaranteed to work as well on other examples).

We were able to re-attempt Dawson and Nielsen's experiments on the King James Bible[1] using the new methodology described in this paper *without any special tuning or tricks*. Dawson and Nielsen even included portions of all three test passages they used, so our comparison is almost completely apples-to-apples. Our results are compared with theirs in Figure 1.

2 Our Method

Instead of layering on heuristic after heuristic to recover specific types of plaintext, we instead take a more principled and general approach. Let x be the known XOR of the two ciphertexts. A feasible solution to the two-time pad problem is a string pair (p, q) such that $p \oplus q = x$. We assume that p and q were independently drawn from known probability distributions $\Pr_1$ and $\Pr_2$, respectively. We then seek the most probable of the feasible solutions: the (p, q) that maximizes $\Pr_1(p) \cdot \Pr_2(q)$.

To define $\Pr_1$ and $\Pr_2$ in advance, we adopt a parametric model of distributions over plaintexts—known as a *language model*—and estimate its parameters from known plaintexts in each domain. For example, if p is known to be an English webpage, we use a distribution $\Pr_1$ that has previously been fit against a *corpus* (naturally occurring collection) of English webpages. The parametric form we adopt for $\Pr_1$ and $\Pr_2$ is such that an exact solution to our search problem is tractable.

This kind of approach is widely used in the speech and natural language processing community, where recovering the most probable plaintext p given a speech signal x is actually known as "decoding."[2] We borrow some well-known techniques from that community: smoothed n-gram language models, along with dynamic programming (the "Viterbi decoding" algorithm) to find the highest-probability path through a hidden Markov model [14].

2.1 Smoothed n-gram Language Models

If the plaintext string $p = (p_1, p_2, \ldots p_\ell)$ is known to have length ℓ, we wish $\Pr_1$ to specify a probability distribution over strings of length ℓ. In our experiments, we simply use an n-gram character language model (taking $n = 7$), which means defining

$$\Pr_1(p) = \prod_{i=1}^{\ell} \Pr_1(p_i \mid p_{i \dotminus n+1}, \ldots p_{i-2}, p_{i-1}) \qquad (1)$$

where $i \dotminus n$ denotes $\max(i - n, 0)$. In other words, the character p_i is assumed to have been chosen at random, where the random choice may be influenced arbitrarily by the previous $n - 1$ characters (or $i - 1$ characters if $i < n$), but is otherwise independent of previous history. This independence assumption is equivalent to saying that the string p is generated from an $(n-1)$st order Markov process.

Equation (1) is called an n-gram model because the numerical factors in the product are derived, as we will see, from statistics on substrings of length n. One obtains these statistics from a training corpus of relevant texts. Obviously, in practice (and in our experiments) one must select this corpus without knowing the plaintexts p and q.[3] However, one may have side information about the *type* of plaintext ("genre"). One can create a separate model for each type of plaintext that one wishes to recover (e.g. English corporate email, Russian military orders, Klingon poetry in Microsoft Word format). For example, our HTML language model was derived from a training corpus that we built by searching Google on common English words and crawling the search results.

---

[1] We used the Project Gutenberg edition which matches the excerpts from [5], available at www.gutenberg.org/dirs/etext90/kjv10.txt

[2] That problem also requires knowing the distribution $\Pr(x \mid p)$, which characterizes how text strings tend to be rendered as speech. Fortunately, in our situation, the comparable probability $\Pr(x \mid p, q)$ is simply 1, since the observed x is a deterministic function (namely XOR) of p, q. Our method could easily be generalized for imperfect (noisy) eavesdropping by modeling this probability differently.

[3] It would be quite possible in future work, however, to choose or build language models based on information about p and q that our methods themselves extract from x. A simple approach would try several choices of $(\Pr_1, \Pr_2)$ and use the pair that maximizes the probability of observing x. More sophisticated and rigorous approaches based on [1, 6] would use the imperfect decodings of p and q to *reestimate* the parameters of their respective language models, starting with a generic language model and optionally iterating until convergence. Informally, the insight here is that the initial decodings of p and q, particularly in portions of high confidence, carry useful information about (1) the genres of p and q (e.g., English email), (2) the particular topics covered in p and q (e.g., oil futures), and (3) the particular n-grams that tend to recur in p and q specifically. For example, for (2), one could use a search-engine query to retrieve a small corpus of documents that appear similar to the first-pass decodings of p and q, and use them to help build "story-specific" language models $\Pr_1$ and $\Pr_2$ [10] that better predict the n-grams of documents on these topics and hence can retrieve more accurate versions of p and q on a second pass.

(a)	Correct pair recovered		Incorrect pair recovered		Not decrypted	
	[5]	This work	[5]	This work	[5]	This work
$P_0 \oplus P_1$	62.7%	100.0%	17.8%	0%	20.5%	0%
$P_1 \oplus P_2$	61.5%	99.99%	17.6%	0.01%	20.9%	0%
$P_2 \oplus P_0$	62.6%	99.96%	17.9%	0.04%	19.5%	0%

(b)	Correct when keystream used three times		Incorrect when keystream used three times		Not decrypted	
	[5]	This work	[5]	This work	[5]	This work
P_0	75.2%	100.0%	12.3%	0%	12.5%	0%
P_1	76.3%	100.0%	11.4%	0%	12.3%	0%
P_2	75.4%	100.0%	11.8%	0%	12.8%	0%

Figure 1: These tables show a comparison between previous work [5] and this work. All results presented for previous work are directly from [5]. Both systems were trained on the exact same dataset (the first 600,000 characters of the King James Version of the Bible, specially formatted as in [5] — all punctuation other than spaces were removed and all letters converted to upper case) and were tested on the same three plaintexts (those used in [5], which were included in the training set). Unlike the prior work, our system was tuned *automatically* on the training set, and not tuned at all for the test set. (a) The first table shows the results of recovering the plaintexts from the listed xor combinations. The reported percentages show the recovery status for the pair of characters in each plaintext position, *without necessarily being in the correct plaintext*. For example, the recovered P_0 could contain parts of P_0 and parts of P_1. (b) The second table shows the results when the same keystream is used to encrypt all three files, and $P_0 \oplus P_1$ and $P_1 \oplus P_2$ are fed as inputs to the recovery program *simultaneously*. Here the percentages show whether a character was correctly recovered in the *correct* file.

It is tempting to define $\Pr_1(\mathtt{s} \mid \mathtt{h, o, b, n, o, b})$ as the fraction of occurrences of hobnob in the $\Pr_1$ training corpus that were followed by s: namely $c(\mathtt{hobnobs})/c(\mathtt{hobnob?})$, where $c(\ldots)$ denotes count in the training corpus and ? is a wildcard. Unfortunately, even for a large training corpus, such a fraction is often zero (an underestimate!) or undefined. Even positive fractions are unreliable if the denominator is small. One should use standard "smoothing" techniques from natural language processing to obtain more robust estimates from corpora of finite size.

Specifically, we chose parametric Witten-Bell backoff smoothing, which is about the state of the art for n-gram models [4]. This method estimates the 7-gram probability by interpolating between the naive count ratio above and a recursively smoothed estimate of the 6-gram probability $\Pr_1(\mathtt{s} \mid \mathtt{o, b, n, o, b})$. The latter, known as a "backed-off" estimate, is less vulnerable to low counts because shorter contexts such as obnob pick up more (albeit less relevant) instances. The interpolation coefficient favors the backed-off estimate if observed 7-grams of the form hobnob? have a low count on average, indicating that the longer context hobnob is insufficiently observed.

Notice that the first factor in equation (1) is simply $\Pr(p_1)$, which considers no contextual information at all. This is appropriate if p is an arbitrary packet that might come from the middle of a message. If we know that p starts at the beginning of a message, we prepend a special character BOM to it, so that $p_1 = \text{BOM}$. Since $p_2, \ldots p_n$ are all conditioned on p_1 (among other things), their choice will reflect this beginning-of-message context. Similarly, if we know that p ends at the end of a message, we append a special character EOM, which will help us correctly reconstruct the final characters of an unknown plaintext p. Of course, for these steps to be useful, the messages in the training corpus must also contain BOM and EOM characters. Our experiments only used the BOM character.

2.2 Finite-State Language Models

Having estimated our probabilities, we can regard the 7-gram language model $\Pr_1$ defined by equation (1) as a very large edge-labeled directed graph, G_1, which is illustrated in Figure 2d, Figure 2a. Each vertex or "state" of G_1 represents a context—not necessarily observed in training data—such as the 6-gram not_se.

Sampling a string of length ℓ from $\Pr_1$ corresponds to a random walk on G_1. When the random walk reaches some state, such as hobnob, it next randomly follows an outgoing edge; for instance, it chooses the edge labeled s with independent probability $\Pr_1(\mathtt{s} \mid \mathtt{h, o, b, n, o, b})$. Following this edge generates the character s and arrives at a new 6-gram context state obnobs. Note that the h has been safely forgotten since, by assumption, the choice of the *next* edge depends only on the 6 most recently generated characters. Our random walk is defined to start at the empty, 0-gram context, representing ignorance; it proceeds immediately through 1-gram, 2-gram, ... contexts until it enters the 6-gram contexts and continues to move among those.

The probability of sampling a particular string p by this process, $\Pr_1(p)$, is the probability of the (unique) path labeled with p. (A path's label is defined as the concatenation of its edges' labels, and its probability is defined as the product of its edges' probabilities.)

In effect, we have defined $\Pr_1$ using a probabilistic finite-state automaton.[4] In fact, our attack would work for any language models $\Pr_1, \Pr_2$ defined in this way, not just n-gram language models. In the general finite-state case, different states could remember different amounts of context—or non-local context such as a "region" in a document. For example, n-gram probabilities might be significantly differ-

---
[4]Except that G_1 does not have final states; we simply stop after generating ℓ characters, where ℓ is given. This is related to our treatment of BOM and EOM.

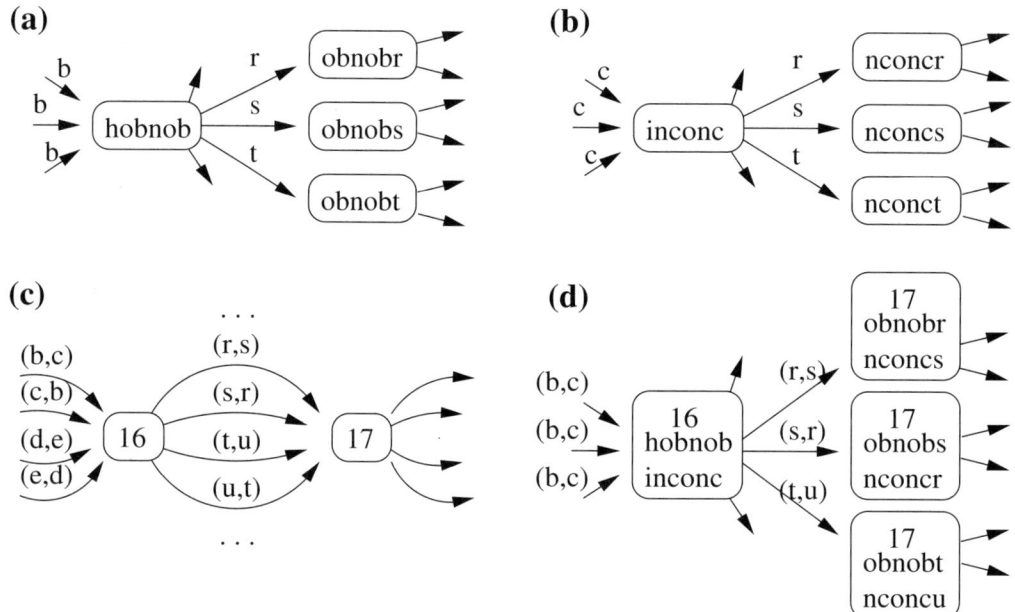

Figure 2: Fragments of the graphs built lazily by our algorithm. (a) shows G_1, which defines Pr_1. If we are ever in the state hobnob (a matter that is yet to be determined), then the next character is most likely to be b, s, space, or punctuation—as reflected in arc probabilities not shown—though it could be anything. (b) similarly shows G_2. inconc is most likely to be followed by e, i, l, o, r, or u. (c) shows X, a straight-line automaton that encodes the observed stream $x = p\,\text{xor}\,q$. The figure shows the unlikely case where $x = (\ldots, 1, 1, 1, 1, 1, 1, 1, \ldots)$: thus all arcs in X are labeled with (p_i, q_i) such that $p_i \oplus q_i = x_i = 1$. All paths have length $|x|$. (d) shows G_x. This produces exactly the same pair sequences of length $|x|$ as X does, but the arc probabilities now reflect the product of the two language models, requiring more and richer states. $(16, \text{hobnob}, \text{inconc})$ is a reachable state in our example since $\text{hobnob} \oplus \text{inconc} = 111111$. Of the 256 arcs (p_{17}, q_{17}) leaving this state, the only reasonably probable one is s,r, since both factors of its probability $\text{Pr}_1(\text{s} \mid \text{hobnob}) \cdot \text{Pr}_2(\text{r} \mid \text{inconc})$ are reasonably large. Note, however, that our algorithm might choose a less probable arc (from this state or from a competing state also at time 16) in order to find the *globally* best path of G_x that it seeks.

ent in a message header vs. the rest of the message, or an HTML table vs. the rest of the HTML document. Beyond remembering the previous $n-1$ characters of context, a state can remember whether the previous context includes a `<table>` tag that has not yet been closed with `</table>`. Useful non-local properties of the context can be manually hard-coded into the FSA, or learned automatically from a corpus [1].

2.3 Cross Product of Language Models

We now move closer to our goal by constructing the joint distribution $\text{Pr}(p, q)$. Recall our assumption that p and q are sampled *independently* from the genre-specific probability distributions Pr_1 and Pr_2. It follows that $\text{Pr}(p, q) = \text{Pr}_1(p) \cdot \text{Pr}_2(q)$. Replacing $\text{Pr}_1(p)$ by its definition (1) and $\text{Pr}_2(q)$ by its similar definition, and rearranging the factors, it follows that

$$\text{Pr}(p, q) = \prod_{i=1}^{\ell} \text{Pr}(p_i, q_i \mid p_{i\dot{-}n+1}, \ldots p_{i-2}, p_{i-1},$$
$$q_{i\dot{-}n+1}, \ldots q_{i-2}, q_{i-1}) \quad (2)$$

where

$$\text{Pr}(p_i, q_i \mid p_{i\dot{-}n+1}, \ldots, p_{i-1}, q_{i\dot{-}n+1}, \ldots, q_{i-1})$$
$$= \text{Pr}_1(p_i \mid p_{i\dot{-}n+1}, \ldots, p_{i-1})$$
$$\cdot \text{Pr}_2(q_i \mid q_{i\dot{-}n+1}, \ldots, q_{i-1}) \quad (3)$$

We can regard equation (2) as defining an even larger graph, G (similar to Figure 2d), which may be constructed as the cross product of $G_1 = (V_1, E_1)$ and $G_2 = (V_2, E_2)$. That is, $G = (V_1 \times V_2, E)$, where E contains the labeled edge $(u_1, u_2) \xrightarrow{(char_1, char_2)\,:\,prob_1 \cdot prob_2} (v_1, v_2)$ iff E_1 contains $u_1 \xrightarrow{char_1\,:\,prob_1} v_1$ and E_2 contains $u_2 \xrightarrow{char_2\,:\,prob_2} v_2$. The weight $prob_1 \cdot prob_2$ of this edge is justified by (3). Again, we never *explicitly* construct this enormous graph, which has more than 256^{14} edges (for our situation of $n = 7$ and a character set of size 256).

This construction of G is similar to the usual construction for intersecting finite-state automata [8], the difference being that we obtain a (weighted) automaton over character *pairs* would still apply even if, as suggested at the end of the previous section, we used finite-state language models other than n-gram models. It is known as the "same-length cross product construction."

2.4 Constructing and Searching the Space of Feasible Solutions

Given x of length ℓ, the feasible solutions (p, q) correspond to the paths through G that are compatible with x. A path $e_1 e_2 \ldots e_\ell$ is compatible with x if for each $1 \leq i \leq \ell$, the edge e_i is labeled with some (p_i, q_i) such that $p_i \oplus q_i = x_i$. As a special case, if p_i and/or q_i is known to be the special character BOM or EOM, then $p_i \oplus q_i$ is unconstrained (indeed undefined).

We now construct a new weighted graph, G_x, that represents just the feasible paths through G. All these paths have length ℓ, so G_x will be acyclic. We will then find the most probable path in G_x and read off its label (p, q).

The construction is simple. G_x, shown in Figure 2d, contains precisely all edges of the form

$$(i-1, \ (p_{i \dot- n+1} \ldots, p_{i-1}), \ (q_{i \dot- n+1} \ldots, q_{i-1}))$$
$$\xrightarrow{(p_i, q_i) : prob} (i, \ (p_{i \dot- n+2} \ldots, p_i), \ (q_{i \dot- n+2} \ldots, q_i)) \quad (4)$$

such that $p_j \oplus q_j = x_j$ for each $j \in [i \dot- n + 1, i]$ and $prob = \Pr_1(p_i \mid p_{i \dot- n+1}, \ldots p_{i-2}, p_{i-1}) \cdot \Pr_2(q_i \mid q_{i \dot- n+1}, \ldots q_{i-2}, q_{i-1})$.

G_x may also be obtained in finite-state terms as follows. We represent x as a graph X (Figure 2c) with vertices 0, 1, ... ℓ. From vertex $i - 1$ to vertex i, we draw 256 edges,[5] labeled with the 256 (p_i, q_i) pairs that are compatible with x_i, namely $(0, 0 \oplus x_i), \ldots (255, 255 \oplus x_i)$. We then compute $G_x = (V_x, E_x)$ by intersecting X with the language-pair model G as one would intersect finite-state automata. This is like the cross-product construction of the previous section, except that here, the edge set E_x contains $(i - 1, u) \xrightarrow{(char_1, char_2) : 1 \cdot prob} (i, v)$ iff the edge set of X contains $(i-1) \xrightarrow{(char_1, char_2) : 1} i$ and E contains $u \xrightarrow{(char_1, char_2) : prob} v$.

Using dynamic programming, it is now possible in $O(\ell)$ time to obtain our decoding by finding the best length-ℓ path of G_x from the initial state $(0, (), ())$. Simply run a single-source shortest-path algorithm to find the shortest path to any state of the form $(\ell, \ldots)$, taking the length of each edge to be the negative logarithm of its probability, so that minimizing the sum of lengths is equivalent to maximizing the product of probabilities.[6] It is not even necessary to use the full Dijkstra's algorithm with a priority queue, since G_x is acyclic. Simply iterate over the vertices of G_x in increasing order of i, and compute the shortest path to each vertex $(i, \ldots)$ by considering its incoming arcs from vertices $(i-1, \ldots)$ and the shortest paths to *those* vertices. This is known as the Viterbi algorithm; it is guaranteed to find the optimal path.

The trouble is the size of G_x. On the upside, because x_j constrains the pair (p_j, q_j) in equation (4), there are at most $\ell \cdot 256^6$ states and $\ell \cdot 256^7$ edges in G_x (not $\ell \cdot 256^{12}$ and $\ell \cdot 256^{14}$). Unfortunately, this is still an astronomical number. It can be reduced somewhat if $\Pr_1$ or $\Pr_2$ places hard restrictions on characters or character sequences in p and q, so that some edges have probability 0 and can be omitted. As a simple example, perhaps it is known that each (p_j, q_j) must be a pair of *printable* (or even alphanumeric) characters for which $p_j \oplus q_j = x_j$. However, additional techniques are usually needed.

Our principal technique at present is to prune G_x drastically, sacrificing the optimality guarantee of the Viterbi algorithm. In practice, as soon as we construct the states $(i, \ldots)$ at time i, we determine the shortest path from the initial state to each, just as above. But we then keep only the 100 best of these time-i states according to this metric (less greedy than keeping only the 1 best!), so that we need to construct at most $100 \cdot 256$ states at time $i + 1$. These are then evaluated and pruned again, and the decoding proceeds. More sophisticated multi-pass or A* techniques are also possible, although we have not implemented them.[7]

2.5 Multiple Reuse

If a keystream k is used more than twice, the method works even better. Assume we now have three plaintexts to recover, p, q, and r, and are given $p \oplus q$ and $p \oplus r$ (note that $q \oplus r$ adds no further information). A state of G or G_x now includes a triple of language model states, and an edge probability is a product of 3 language model probabilities.

The Viterbi algorithm can be used as before to find the best path through this graph given a pair of outputs (those corresponding to $p \oplus q$ and $p \oplus r$). Of course, this technique can be extended beyond three plaintexts in a similar fashion.

3 Implementation

Our implementation of the probabilistic plaintext recovery can be separated cleanly into two distinct phases. First, language models are built for each of the types of plaintext that will be recovered. This process only needs to occur once per type of plaintext since the resulting model can be reused whenever a new plaintext of that type needs to be recovered. The second phase is the actual plaintext recovery.

All our model building and cracking experiments were run on a commodity Dell server (Dual Xeon 3.0 GHz, 8GB RAM) that cost under $2,000. The server runs a Linux kernel that supports the Xeon's 64-bit extensions to the x86 instruction set. The JVM used is BEA's freely available JRockit since Sun's JVM does not currently support 64-bit memory spaces on x86.

3.1 Building the Language Models

To build the models, we used an open source natural language processing (NLP) package called LingPipe [4].[8] LingPipe is a Java package that provides an API for many common NLP tasks such as clustering, spelling correction, and part-of-speech tagging. We only used it to build a character based n-gram model based on a large corpus of documents (see section 4 for details of the corpora used in our experiments). Internally, LingPipe stores the model as a trie with greater length n-grams nearer the leaves. We had LingPipe "compile" the model down to a simple lookup table based representation of the trie. Each row of the table, which corresponds to a single n-gram, takes 18 bytes except for the rows which correspond to leaf nodes (maximal length n-grams) which take only 10 bytes. If an n-gram is never seen in the training data, it will not appear in the table; instead, the longest substring of the n-gram that does appear in the table will be used. The extra 8 bytes in these nodes specify how to compute the probability in this "backed-off" case. All probabilities in both LingPipe and our Viterbi implementation are computed and stored in log-space to avoid issues with integer underflow. All of the language models used in this paper have $n = 7$. The language models take several

---

[5] Each edge has weight 1 for purposes of weighted intersection or weighted cross-product. This is directly related to footnote 2.

[6] Using logarithms also prevents underflow.

[7] If we used our metric to prioritize exploration of G_x instead of pruning it, we would obtain A* algorithms (known in the speech and language processing community as "stack decoders"). In the same A* vein, the metric's accuracy can be improved by considering right context as well as left: one can add an estimate of the shortest path from $(i, \ldots)$ through the remaining ciphertext to the final state. Such estimates can be batch-computed quickly by decoding x from end-to-beginning using smaller, m-gram language models ($m < n$).

[8] Available at: http://www.alias-i.com/lingpipe

hours to build, and the non-compiled trie representation can use many gigabytes of memory for large corpora. For example, our biggest compiled model is 872 MB and took over 8 hours to generate. It contains the results of looking at more than 7 billion training characters from 300,000 HTML files.

Language modeling has a rich literature, and there are many options that could be useful in particular scenarios, especially for non-natural-language or unknown-genre plaintexts. For example, if we had used less training data, we would have been able to take far more context into account without exceeding the available RAM. LingPipe can build practical 32-gram character language models when the training data is limited to 10 million words [4]. An intermediate strategy would be to include long contexts (e.g., 31-grams) in the language model only when they are very frequent, otherwise backing off to shorter contexts (e.g., 6-grams). There also exist modeling techniques for considering discontiguous or long-distance contexts; for adapting to input properties (cf. Lempel-Ziv); and for training effectively on heterogeneous corpora that consist of several (labeled or unlabeled) genres.

3.2 Recovering the Plaintext

The optimized Viterbi search represents the bulk of our implementation. Ideally, we would first generate the full automaton from the LingPipe language model tables. Unfortunately, this creates a state explosion since each state in one model can be paired with every state in the other model. Instead, we generate states on the fly from the tables as they are needed. This is not quite as expensive as it seems.

Because of the way that our automaton is constructed, each state transition represents adding a single character to each of the n-grams in the current state. The number of single characters that can be added to an n-gram (given x) is usually relatively small. If we assume the underlying plaintext can be modeled using only non-binary characters, there are only 128 possible choices. In our experiments, the actual number of observed 1-grams when modeling HTML, e-mail, or other plain-text protocols hovers around the number of printable characters. This means each combination state only has approximately 96 transitions.

Unfortunately, the number of possible states still grows exponentially in the length of the plaintexts being recovered. To deal with this, we use the "beam search" heuristic optimization: low probability partial paths are pruned early so as to limit the number of states that we must keep in memory. This path pruning means that the search is no longer guaranteed to end up with the best path (it might have been pruned), but the technique seemed to work in practice in our experiments. The pruning is implemented as a binary tree of the current states sorted by the probability of the path; the tree is pruned after each output byte is considered. Processing each byte in our implementation takes approximately 200ms.

After the newly created tree is sorted and pruned, the surviving nodes' state numbers, along with their parents' state numbers, are written to disk. The program writes a file containing this information for each byte in the input file. At the end of the Viterbi search, the program reassembles the final Viterbi crack path by parsing these files. The result is the pair of plaintext messages that represent the best path through the graph.

4 Results

In this section, we present the results of a variety of experiments we performed using our implementation. We examined three different types of files: unstructured English text files (emails, with headers), English text files with text-based structure (HTML documents), and English text files with binary structure (Microsoft Word documents). We examine the effect of such factors as the amount of training data available, the number of times that a keystream was reused, and whether having the plaintexts be of different types affects the reconstruction. We always assumed that both plaintexts started with BOM, and we took ℓ to be the length of the XOR stream, determined by the shorter of the plaintexts. We did not use EOM.

4.1 Data Collection and Testing Methodology

All of the data that we used in our experiments is publicly available. Our HTML dataset was the easiest to gather. We searched Google for common English words and used wget to crawl the results. Our largest HTML training corpus consists of 300,000 different files (over 7 billion characters).

Emails were slightly harder to come by — we didn't find anyone willing to allow us to experiment on their inboxes. Fortunately, the Federal Energy Regulatory Commission (FERC) has made available the emails sent by the senior managers of Enron. These emails were collected during FERC's investigation into Enron's business practices.[9] The emails do not include attachments, and some of the emails have been redacted or removed due to requests from the employees involved. However, it is, to our knowledge, the best email corpus publicly available. Our largest email training corpus consists of 500,000 emails (more than 4 billion characters).

To collect Microsoft Word documents we again turned to Google, this time filtering so that only .doc files would be returned. The largest Word training corpus we use is 90,000 files (more than 450 million characters). We only train on the first 5,000 bytes of each file as Word files are typically larger than HTML documents or email messages; this is, however, more than sufficient to get past the Word header information.

In all cases, we randomly reserved some of the files we collected for experimentation or evaluation. All of the results in this section are reported on documents that were not used in the training of the model or design of the method.

4.2 Basic Results

We begin by examining how our reconstruction works on each type of plaintext we consider (i.e. HTML, email, Word). We randomly selected 100 files of each type and XORed pairs of the same type together to create 50 different XORed streams for each type. This corresponds to a likely real-world case: when a system or protocol that exhibits keystream reuse is generally used with a single type of file. We tried to recover the plaintexts from each of these streams under models built using training corpora of varying size. The results are shown in Figure 3. It is first worth noting that increasing the training corpus size has a relatively small effect on the results. It turns out that at the corpus sizes we consider, the most important factor is the "variety" of documents in the corpus. When we were initially experimenting, we failed to randomize our choices of which documents from our full corpus would be included in each training set. This led to

---
[9]The 400MB zip file is available from http://www.cs.cmu.edu/~enron/.

the excellent results on test files that happened to be related to those that were trained on and terrible results on all other files. Randomizing the selection of the training set fixed this problem in the HTML and Word training sets. The email corpus consists of messages from only around 150 users. This provides a low degree of diversity and so the email models seem to more easily become "unbalanced." The 200K email corpus appears to do much better than the other email corpora on many tests (leading to its high median recovery percentage). However, unlike the other email models, it can recover only 50% of characters in some files.

The HTML results are by far the best, with more than 99% of characters correctly being decoded. The Word results are the worst at 44%, likely due to Word files having less predictable structure and more possible byte values than the other two data sets. The email results fall in the middle, with 82% of the characters correctly recovered on average. We examine why the email reconstruction is worse than the HTML reconstruction in the following sections and show some techniques that can be used to improve it.

4.3 The Switching Streams

The email results seem to be far worse than those for HTML. This is only true because of the particular way we decided to measure success. If we instead said that a reconstruction succeeded on a particular byte when the two bytes that are returned are correct, regardless of whether they are in the correct file, the results would change dramatically as shown in Figure 4. This disparity is due to an artifact of the way we recover the plaintexts. Our n-gram models only look at the last few characters that are recovered. Usually this allows for the streams to be reconstructed sequentially — the next character recovered in each plaintext is likely to complete a word already begun in that plaintext. However, assume that the reconstruction fails to recover correct bytes for a short stretch. When the reconstruction gets back on track, it has no idea to which stream the following correct bytes belong. Therefore, it will sometimes choose incorrectly, and correct bytes will be added to the incorrect file until another such switch occurs. We now consider two methods by which this problem can be ameliorated.

4.4 Improving the Results

Our first attempt at improving the results simply assumes that the keystream was used more than twice. Here, the model should prevent switching as each plaintext is now matched with two other plaintexts (i.e. two of the plaintexts can no longer be switched because their XOR differences with the third will no longer be valid). This technique is quite effective as shown in Figure 5.

There are times when an attacker may be unwilling to wait for multiple keystream reuses; perhaps they are very rare events. However, if the two plaintexts are of two different types, as could be the case in encrypted protocols that transport multiple types of documents (e.g. WEP), a similar effect occurs. After a period of errors, the model can recover correctly since the distributions of the two plaintexts are different and thus the probabilities of the switched stream paths will be lower than the probabilities of the true path. The results of recovering HTML documents XORed with email messages is shown in Figure 6.

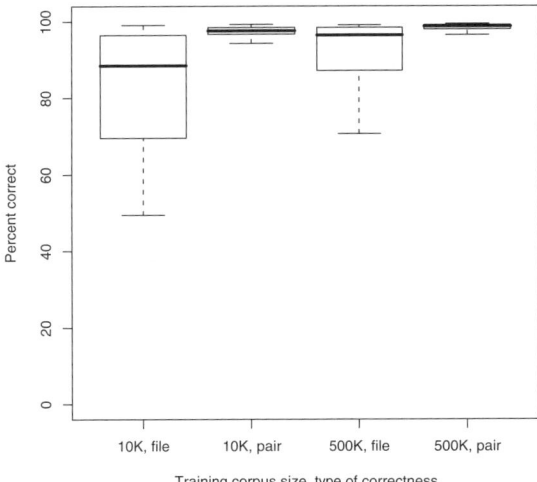

Figure 4: This graph illustrates the major problem that is encountered during the reconstruction. After a short string of errors occurs and the reconstruction recovers, it's no longer clear to which recovered plaintext the correct characters should be added. This creates reconstructions where parts of each original plaintext occur between short errors. Here, we show the difference between computing correctness based on whether a character was assigned the correct *file* vs. simply whether each *pair* of recovered characters was correct. All plots are of 100 randomly chosen email messages xored in pairs.

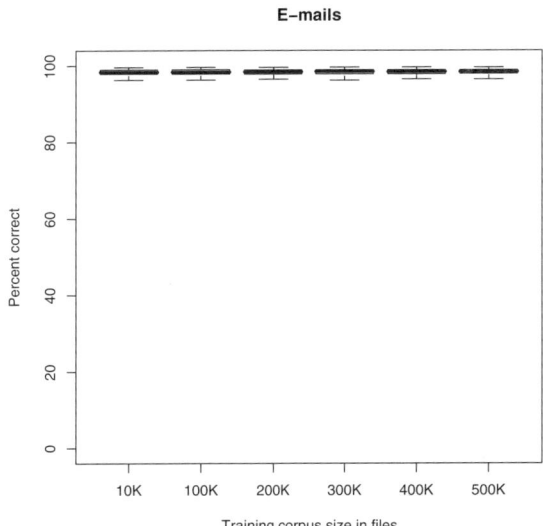

Figure 5: This graph shows the difference between having a keystream reused once, yielding $p \oplus q$, and having it reused twice, giving $p \oplus q$ and $p \oplus r$. Each scenario was evaluated using 50 randomly chosen instances from the email corpus.

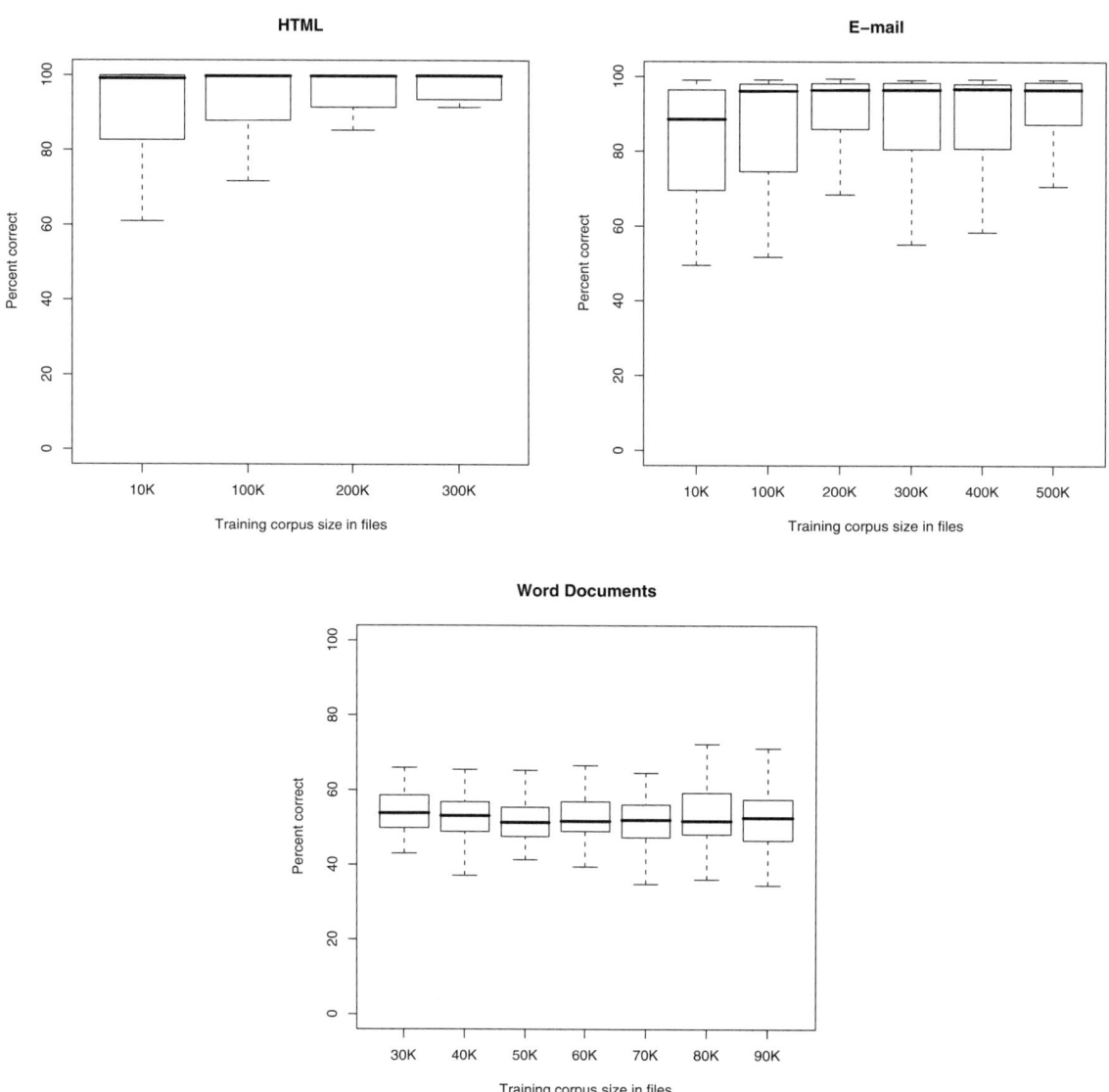

Figure 3: These graphs show the basic results when cracking two files of the same type xored together at different training corpus sizes. Fifty pairs of files were cracked at each training corpus size. The models should not be directly compared to one another: different types of files have different lengths and so influence the model differently. Plotting based on the number of characters read for each model would also be deceptive as it would not indicate the differing number of start bytes following bom (which we find to be important given our heavy pruning). See section 4.1 for information on the number of bytes in the models.

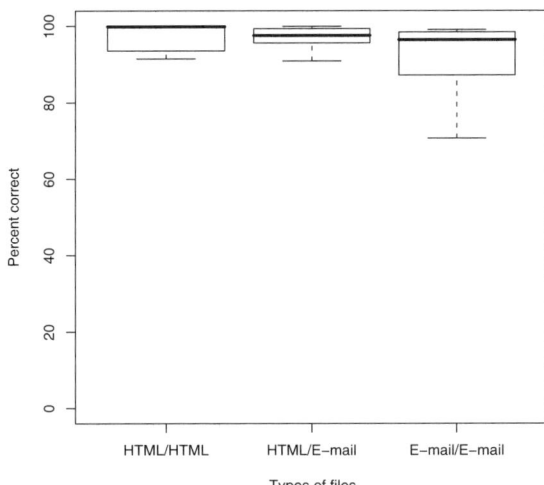

Figure 6: Here, the results show the effect of using reusing a keystream on two different types of documents. Fifty streams were recovered first for two xored HTML files, then for an HTML file xored with an email message, and finally with two email messages xored together.

5 Attacking Word 2002

This section shows how our technique can be used to attack a real system: Microsoft Word 2002 RC4 encryption as shown vulnerable in [22]. Microsoft Word 2002 allows users to encrypt their documents with a password. The user can select from one of many cipher suites and simply enter his password, encrypting the document using the chosen cipher. Among the choices for cipher suite is a popular stream cipher, RC4. RC4 [16] takes a key, k, and uses it to generate a keystream, $RC4(k)$, that is then xored with the plaintext.

When Microsoft Word encrypts a document with RC4, the document is assigned a randomly generated initialization vector, IV. The initialization vector is then concatenated with the user's encryption password and cryptographically hashed forming the key array for RC4:

$$k = H(IV || password)$$

The problem arises when the document is edited and saved. The initialization vector is not regenerated after editing. This means the original document and the edited document both use the same keying material. Depending which editing changes are made, an attacker possessing both the original encrypted document, $p \oplus RC4(k)$ and its edited version, $p' \oplus RC4(k)$, can use our method to gain portions of the original plaintexts.

Not all changes to the document allow recovery, however. For example, if only a single character is changed in p, $p \oplus p'$ will be almost completely zeros. While this is interesting and useful information to an attacker, it is not a full recovery. Fortunately, most edits do not affect only a few bytes of the file: inserting a single character near the beginning of the document is sufficient as all of the other characters will be offset.

Aside from adding characters, there are several features that a user can change that may or may not yield portions of the plaintext. For instance, making a character at the top of a Word document bold yields no results. Adding a footnote, though, follows a similar pattern as adding a character. If a footnote is added at the top of a document, a large portion of the original document can be obtained. However, a footnote at the end of a document is the same as appending a character to the end of a document. The track changes feature follows the same pattern, but yields slightly more information should the editor append a character to the end of a document. Deleting and re-adding a paragraph with the exact same formatting, as well as double spacing an entire document, yields no useful results.

In order to test our method, we used Google to search for a Word document with two available revisions. This models a real set of changes that could occur between two saved versions of a document. We were not able to find such a pair that was encrypted, so we used Word 2002 to encrypt the pair ourselves so that the IV was reused. We then applied our tool using the Word corpus build from 90,000 other Word documents. The recovery was 54% accurate (84% pairwise accurate), which agreed with the experiments from section 4.2. A portion of the recovered text is shown in figure 7.

```
November 13, 2002#ATA/ATAPI Host Adapters Standard
(ATF;#h Packet)#This is no internal working document of
T13, a Technical Committee of Accredited Standards Com-
mittee INCITS. The T13 Technical Committee may modify the
contents. This document is made available and has not been
approved. The contents may be modified by the T13 Technical
technical committees, and their associated task groups to re-
produce this document for the purposes of
```

```
November 13, 2002#ATA/ATAPI Host Adapters Standard
(ATA # Adapter)#This is an internal working document of
T13, a Technical Committee of Accredited Standards Com-
mittee INCITS. The T13 Technical Committee may modify
the contents. This document is made available for review and
comment only.#Permission is granted to members of INCITS,
its technical committees, and their associated task groups to
reproduce this document for the purposes of
```

Figure 7: At top, the cracked file resulting from adding a character to an encrypted Word document. Characters that are underlined were recovered incorrectly while characters that are wavy underlined were recovered into the wrong file (see section 4.3). Hash characters (#) represent unprintable ASCII values (i.e. formatting). Underneath, the original corresponding plaintext is shown.

6 Related Work

Markov models (hidden or otherwise) have been previously used for several purposes in security and cryptography such as improving dictionary attacks [13], mounting side channel attacks on protocols [19], recoving keystrokes based on the way they sound [23] and solving simple substitution ciphers [12], among others.

7 Conclusions

We have shown that keystream reuse is a real problem—allows a practical attack—when the data being encrypted comes from a known, non-uniform distribution. Our attack is general and can be easily applied to new types of files as the need arises. We have achieved over 99% accurate recovery in some instances and shown how to improve our results for other types of files under specific conditions such as a keystream being reused multiple times. Finally, footnotes 3 and 7 outlined opportunities for future improvements in accuracy and speed.

The technique does not directly apply for plaintexts that have a near uniform distribution, such as files that have been compressed. In theory, a language modeling attack could still be used in this case—one simply searches for p, q, such that $p \oplus q = x$ and $\Pr_1(decompress(p)) \cdot \Pr_2(decompress(q))$ is maximized. However, dynamic programming can no longer be used to render this brute-force attack tractable.

Acknowledgments

This material is based in part upon work supported by the National Science Foundation under Grant No. 0347822 to the third author. We thank Yoshi Kohno, David Molnar, Kevin Fu, Erika McCallister, Fabian Monrose, and Charles Wright for their helpful comments on this work.

7.1 References

[1] L. E. Baum. An inequality and associated maximization technique in statistical estimation of probabilistic functions of a Markov process. *Inequalities*, 3, 1972.

[2] R. L. Benson and M. Warner. *Venona: Soviet Espionage and the American Response 1939-1957*. Central Intelligence Agency, Washington, D.C., 1996.

[3] N. Borisov, I. Goldberg, and D. Wagner. Intercepting mobile communications: The insecurity of 802.11. In *MOBICOM 2001*, 2001.

[4] B. Carpenter. Scaling high-order character language models to gigabytes. In *Association for Computational Linguistics Workshop on Software*, Ann Arbor, MI, 2005.

[5] E. Dawson and L. Nielsen. Automated cryptanalysis of xor plaintext strings. *Cryptologia*, 20(2):165–181, April 1996.

[6] A. P. Dempster, N. M. Laird, and D. B. Rubin. Maximum likelihood from incomplete data via the EM algorithm. *J. Royal Statist. Soc. Ser. B*, 39(1):1–38, 1977. With discussion.

[7] M. Dworkin. Recommendation for block cipher modes of operation. NIST Special Publication 800-38A, 2001.

[8] J. E. Hopcroft and J. D. Ullman. *Introduction to Automata Theory, Languages and Computation*. Addison-Wesley, Reading, MA, 1979.

[9] D. Kahn. *The Codebreakers*. Scribner, New York, NY, 1996.

[10] S. Khudanpur and W. Kim. Contemporaneous text as side information in statistical language modeling. *Computer Speech and Language*, 18(2):143–162, 2004.

[11] T. Kohno. Attacking and repairing the winzip encryption scheme. In *11th ACM Conference on Computer and Communications Security*, pages 72–81, Oct 2004.

[12] D. Lee. Substitution deciphering based on hmms with applications to compressed document processing. *IEEE Transactions on Pattern Analysis and Machine Intelligence*, 24(12):1661–1666, Dec 2002.

[13] A. Narayanan and V. Shmatikov. Fast dictionary attacks on human-memorable passwords using time-space tradeoff. In *12th ACM Conference on Computer and Communications Security*, pages 364–372, Washington, D.C., Nov 2005.

[14] L. R. Rabiner. A tutorial on Hidden Markov Models and selected applications in speech recognition. *Proceedings of the IEEE*, 77(2):257–286, Feb 1989.

[15] F. Rubin. Computer methods for decrypting random stream ciphers. *Cryptologia*, 2(3):215–231, July 1978.

[16] B. Schneier. *Applied Cryptography: Protocols, Algorithms, and Source Code in C*. John Wiley & Sons, Inc., New York, NY, USA, 1993.

[17] B. Schneier, Mudge, and D. Wagner. Cryptanalysis of microsoft's pptp authentication extensions (ms-chapv2). In *CQRE '99*, 1999.

[18] C. E. Shannon. A mathematical theory of communication. *Bell System Technical Journal*, 27:379—423, July 1948.

[19] D. X. Song, D. Wagner, and X. Tian. Timing analysis of keystrokes and timing attacks on ssh. In *10th USENIX Security Symposium*, Aug 2001.

[20] G. Vernam. Secret signaling system. U.S. Patent 1310719, July 1919.

[21] P. Wright. *Spy Catcher*. Viking, New York, NY, 1987.

[22] H. Wu. The misuse of rc4 in microsoft word and excel. Cryptology ePrint Archive, Report 2005/007, 2005. http://eprint.iacr.org/.

[23] L. Zhuang, F. Zhou, and J. D. Tygar. Keyboard acoustic emanations revisited. In *12th ACM Conference on Computer and Communications Security*, pages 373–382, Washington, D.C., Nov 2005.

Dictionary Attacks Using Keyboard Acoustic Emanations

Yigael Berger
Dept. of Computer Science
Tel Aviv University
Ramat Aviv 69978, ISRAEL
yigael.berger@gmail.com

Avishai Wool
School of Electrical Engineering
Tel Aviv University
Ramat Aviv 69978, ISRAEL
yash@acm.org

Arie Yeredor
School of Electrical Engineering
Tel Aviv University
Ramat Aviv 69978, ISRAEL
arie@eng.tau.ac.il

ABSTRACT

We present a dictionary attack that is based on keyboard acoustic emanations. We combine signal processing and efficient data structures and algorithms, to successfully reconstruct single words of 7-13 characters from a recording of the clicks made when typing them on a keyboard. Our attack does not require any training, and works on an individual recording of the typed word (may be under 5 seconds of sound). The attack is very efficient, taking under 20 seconds per word on a standard PC. We demonstrate a 90% or better success rate of finding the correct word in the top 50 candidates identified by the attack, for words of 10 or more characters, and a success rate of 73% over all the words we tested. We show that the dominant factors affecting the attack's success are the word length, and more importantly, the number of repeated characters within the word. Our attack can be used as an effective acoustic-based password cracker. Our attack can also be used as part of an acoustic long-text reconstruction method, that is much more efficient and requires much less text than previous approaches.

Categories and Subject Descriptors

K.6.5 [**Security and Protection**]: Authentication; H.5.5 [**Sound**]: Signal analysis, synthesis, and processing

General Terms

Algorithms, Security

Keywords

Keyboard acoustics, Dictionary attacks, Password cracking

1. INTRODUCTION

1.1 Background

The study of signals emanating from electronic or mechanical devices goes back a long way in history. In the mid 1950's this subject gained immense public interest and awareness with the United States government coming out with its TEMPEST program [13], aimed at preventing the possibility of exploiting these compromising emanations, leaking from devices handling sensitive information. More recently, extracting information out of various types of emanations was demonstrated: electromagnetic emanations leaking video information [6], optical emanations such as in [7] and acoustic emanations from mechanical devices such as dot-matrix printers [3].

This paper deals with acoustic signals emanating from a PC keyboard in reaction to a person typing on it, and is a step further in the direction set by the recent works of [1] and [14]. The main observation, and the reason why keyboard acoustic emanations leak information, is that different keys on the keyboard make different click sounds.

1.2 Related Work

Attacks against emanations caused by human typing have attracted interest in recent years. In particular, the seminal works of [1] and [14] showed that keyboard acoustic emanations do leak information that can be exploited to reconstruct the typed text.

[1] used tagged recordings of a person typing on a keyboard, to train a neural network that would recognize subsequent keystrokes. They extracted *fft* coefficients out of the press segments of keystrokes and used them as features for comparison. Training a back-propagation neural net with 100 training samples per each of the 30 keys they were able to demonstrate an 80% recognition rate. They showed that performance degrades when applying the trained net on a different keyboard or person. They also showed that their method can be applied to other push-button devices such as telephone and ATM pads. However, the reliance on tagged samples significantly limits the scope of the attack.

Subsequently, [14] suggested a method to uncover the typed text without tagged training samples. In this work, Cepstrum [11] features were preferred over *fft* coefficients. Their attack is split into two phases, with the first being the *Unsupervised Training* phase, and the second being the *Recognition* phase that is based on the outcome of the first. Their method requires about 10 minutes of recording and roughly 30 minutes of computation to uncover up to 96% of the typed text.

The earlier work of [12] shows that human typing patterns leak information via timing information, and that this timing is noticeable even through SSH encryption. Inter-keystroke timing is also available from keyboard acoustic

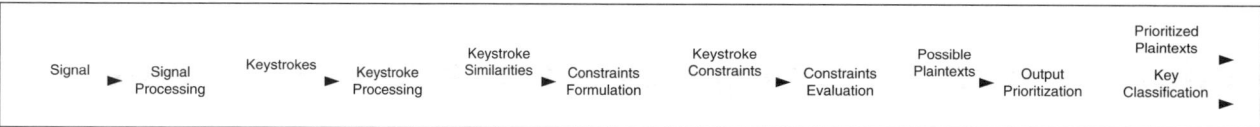

Figure 1: Attack stages.

emanations, so this type of attack can be combined with ours.

1.3 Contribution

Beyond the basic observation that different keystrokes produce different sounds, we make two new observations that are central to our attack: Firstly, the sounds that keystrokes make correlate to their physical positioning on the board. Specifically, keys such as Q, W and E, that are located close to one another, sound more alike than keys positioned far apart like Z and P. Secondly, we discovered that if we work at a granularity of *words*, rather than individual keys, we can exploit the statistical properties of the language in addition to the properties of the signal.

Based on these observations, we present a dictionary attack that is based on keyboard acoustic emanations. We combine signal processing and efficient data structures and algorithms, to successfully reconstruct single words of 7-13 characters from a recording of the clicks make when typing them on a keyboard. Our attack does not require any training, and works on an individual recording of the typed word (may be under 5 seconds of sound).

The attack is very efficient, taking under 20 seconds per word on a standard PC. We demonstrate a 90% or better success rate of finding the correct word in the top 50 candidates identified by the attack, for words of 10 or more characters, and a success rate of 73% over all the words we tested. We show that the dominant factors affecting the attack's success are the word length, and more importantly, the number of repeated characters within the word. Along the way, we tested various signal processing primitives, and discovered that the simple cross-correlation primitive is more effective in this attack than other known methods (like *fft*, and *Cepstrum*) used by previous authors.

Our attack can be used as an effective acoustic-based password cracker. The attack can also be used as part of an acoustic long-text reconstruction method, that is much more efficient and requires much less text than previous approaches.

Organization: The next section gives a general overview of the whole attack. Section 3 describes the signal processing we used in the attack. Section 4 describes the combinatorial constraint generation methods, and Section 5 describes our constraint satisfaction algorithms and their implementation. In Section 6 we analyze the attack's performance, and in Section 7 we present our conclusions and suggestion for further research.

2. ATTACK OVERVIEW

Our attack takes as input an audio signal containing a recording of a single word typed by a single person on a keyboard, and a dictionary of words. We assume that the typed word is present in the dictionary. The aim of the

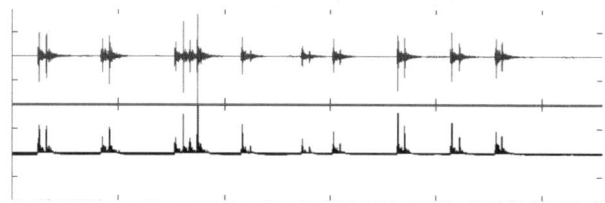

Figure 2: Example of a bare signal (top) and its corresponding energy bins (bottom) for the keystrokes of the word "difference".

attack is to reconstruct the original word from the signal. We concentrate on handling extremely short audio signals containing a single word, with seven or more characters long. This means that the signal is only a few seconds long. It is well known that such short words are often chosen as a password (cf. [5]). The attack does not require any training.

Our attack comprises of several stages: (see Figure 1) (i) Signal Processing and Feature Extraction; (ii) Keystroke Processing; (iii) Constraint Formulation; (iv) Constraint Evaluation; and (v) Outcome Prioritization. Below we provide a general overview of each stage. In Sections 3, 4, and 5 we provide the full details.

2.1 Signal Processing

We begin by processing the signal in order to separate and extract the keystrokes from it (see Figure 2). Assume that the signal contains an N characters long word. Note that each keystroke produces two separate sound segments, as noted in [1, 14], generated by the press of the key button and its release. Let $PRESS_i$ ($RELEASE_i$) denote the i'th key press (release) in the signal. The output of the signal processing stage consists of the two arrays of signal segments, PRESS and RELEASE.

2.2 Keystroke Processing

A basic capability we need is a method to calculate the similarity between each pair of keystrokes. What we demonstrate is that a good similarity metric not only tells us how similar two keystrokes sound, but also lets us deduce information about the keys' physical proximity on the keyboard. Specifically, for a metric sim, if $sim(K_i, K_j) > sim(K_i, K_k)$ then with a significant amount of confidence we can say that K_i and K_j are positioned more closely on the keyboard than K_i and K_k. Without this property, it would not have been possible to employ our method on such short signals.

There are several possible methods of calculating a similarity metric between two acoustic signals. As part of our study we tested three of these metrics (see Section 3.3). Somewhat surprisingly, we found that the best performance was obtained using the simple cross-correlation metric rather than *fft* or Cepstrum.

Type	Notation	Meaning
EQ	$=$	$K_i = K_j$ means that the i'th keystroke and the j'th keystroke stem from the same key on the keyboard.
ADJ	$\simeq$	$K_i \simeq K_j$ means that the j'th keystroke stems from a key that is adjacent to the key which the i'th keystroke stems from. For example, $Q \simeq W$ but not $Q \simeq E$ since E is located two positions away from Q on a QWERTY keyboard.
NEAR	$\sim$	$K_i \sim K_j$ means that K_i and K_j are at most two keys apart on the keyboard, e.g., keys NEAR G include R, D, N, J, etc.
DIST	$\nsim$	Distant keys are those that are not NEAR to each other.

Table 1: The four constraint types

Once we have a similarity metric, we measure the similarity of each PRESS$_i$ to every other press, and the same is done for every RELEASE$_i$. This produces two $N \times N$ matrices of key-to-key similarities. We then combine the two matrices into a single $N \times N$ similarity matrix M_{ij}. We evaluated 5 possible methods for combining the PRESS similarities with the RELEASE similarities. We found that using an unweighted average outperforms the other possibilities we examined, in being resilient and sustaining good performance across different keyboards.

2.3 Constraint Formulation

The similarity matrix M is used to formulate constraints on the recorded word. A constraint is a binary operator expressing a relation on a pair of keystrokes. For example, the constraint $K_i = K_j$ means that the i'th and j'th keystrokes stem from the same key on the keyboard. Note that the constraint does not state what the actual key is, but only that the examined text complies with the given condition. In this manner we define four types of constraints that are listed in Table 1.

A given word produces a specific set of constraints—but the opposite does not necessarily hold; a specific set of constraints may be true of several words. Consider the word "help". Under ideal conditions, this word produces the following constraints: $K_1 \nsim K_2$, $K_1 \nsim K_3$, $K_1 \nsim K_4$, $K_2 \nsim K_3$, $K_2 \nsim K_4$, $K_3 \simeq K_4$. However, the same constraints are also true for the words "iraq", "nose", "path" and more. The full specification of the key relations can be found in appendix A.

2.4 Constraint Evaluation

We use the similarity matrix M to infer as many correct constraints as possible, and use them to postulate on the value of the text. Assuming that we know the length of the text word, N, and that all the constraints that we inferred are correct, we can evaluate the constraints: Go over all possible dictionary words of length N and output all those that match the constraints. This search can be made very efficient using suitable data structures.

However, inferring constraints from the similarity matrix is inherently inaccurate. Any inference policy will either fail to infer all the possible constraints, or produce false ones, or both. It is important to understand that even a single false constraint is enough to cause us to discard the correct word.

Our first method of reducing errors is the policy we use to infer constraints. The success of such a policy is measured by two metrics (for each constraint type):

- *Precision* measures the fraction of constraints that hold true for the real word, relative to the number of constraints produced, for that specific constraint-type.

- *Recall* measures the fraction of true constraints with respect to the total number of possible constraints in that category.

There is always a tension between being these two metrics, forcing us to balance between coming up with as many of the true constraints as possible while being as precise as possible.

Our preferred policy of formulating constraints is called the ***BestFriendsPickPolicy***. We found that it performs well in both the *Precision* and *Recall* of the produced constraints, per constraint type (*EQ, ADJ, NEAR, DIST*). The *BestFriendsPickPolicy* is specified in Section 4.1.

2.5 Dealing with False Constraints

The next method we use to mitigate the false constraints that are inferred is by using constraint combinations. The main idea is to select many different subsets ("combinations") of constraints and evaluate each combination. If enough constraints are correct, then many combinations will be consistent with the correct word.

However, if implemented naively, this method can be extremely inefficient computationally. Let C denote the set of constraints extracted out of the similarity matrix, using the constraint inference policy. For this given C, there are $2^{|C|}$ possible constraint combinations. Our method may produce a few dozens of constraints for a 7-13 character word: we clearly face a combinatorial explosion in the number of possible combinations

To overcome the combinatorial explosion, we *randomly* choose a relatively small collection of the possible combinations. We have empirically found that about 1000 combinations usually suffice. Section 5.3 details how we choose the collection, and evaluate the effectiveness of our choice.

Having chosen the combinations of constraints, each combination c is evaluated against the possible words in the dictionary. This yields a list of possible words L_c that all conform with the constraints of combination c.

2.6 Outcome Prioritization

Our next goal is to produce a unified list U of candidate words, prioritized based on the L_c lists produced for the various constraint combinations. For each word w in the dictionary we count the number of combinations c for which $w \in L_c$, and sort the words in decreasing order.

Recall that any erroneous constraint will necessarily preclude the correct word from appearing in the L_c list of any

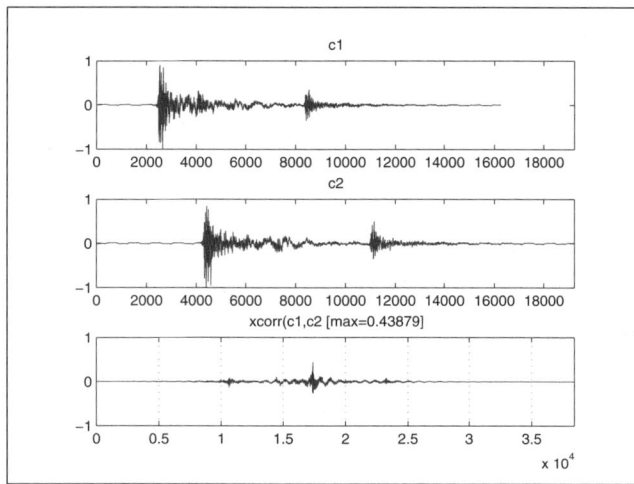

Figure 3: Two recordings of the key C, with their cross-correlation function.

combination c that includes this constraint. Therefore, such an L_c will include only false choices—that are randomly distributed. On the other hand, a correct combination of constraints c will include the correct word in its L_c. Therefore, after trying enough combinations, the correct word will appear near the top of the sorted unified list U.

3. SIGNAL PROCESSING

3.1 Recording

We recorded the keyboard acoustic emanations using a cheap omni-directional clip microphone, at a 44,100 Hz sampling rate. We used three different keyboards: an IBM KB-8923, a Genius K295 and an old VISION keyboard (model unknown).

3.2 Feature Extraction

The purpose of the feature extraction process is to be able to locate and compare the different keystrokes in the given input signal (recall Figure 2). This entails identifying the start and end of each keystroke, and then splitting it into the press and release segments.

We break the signal into windows of 2 milliseconds (100 samples), then sum the *fft* coefficients obtained on each window. This gives us an indication of the amount of energy contained in each window. We normalize the energy bins to values between 0 and 1, and using them, we compute the difference between each window and its predecessor, which gives us a "delta vector". Each element of the vector is an indication of a rise in energy in that time frame.

We now go through this delta vector and look for surges of energy that can account for a key press. Once we have homed on a specific peak we look for a fall followed by a resurgence of energy that fits the description of a key-release phase. The two peaks (press and release) are checked to appear within a boundary of 100 ms. If the energy in the whole press is found to be too weak, then this segment is ignored and we move on to the next peak. Figure 2 (bottom) shows the computed energy peaks.

3.3 Measuring Similarity between Keystrokes

Once we identify where each key press and release begins and ends, we can measure the similarity between each of them. There are several possible measures of similarity, so before proceeding let us first describe the properties of a good measure.

Let K_i denote the recording of some key α_i on the keyboard. A similarity measure $sim(K_i, K_j)$ is a function with real output between 0 and 1 with the following properties:

- **Adjacency.** We would like $sim(K_i, K_j) > sim(K_i, K_k)$ to be true if α_i and α_j are physically closer to one another on the keyboard than α_i and α_k are. This criterion is crucial to the success of our method.

- **Symmetry.** $sim(K_i, K_j) = sim(K_j, K_i)$.

- **Reflexivity.** A signal K_i should obviously come out most similar to itself. Ideally $sim(K_i, K_i) = 1$.

Note that **transitivity** is not a requirement, otherwise we would end up with the two farthest keys on the keyboard being similar to one another.

A "good" similarity measure should also have two more properties: **a) universality**, i.e., that it performs well across different keyboards, and **b) computational efficiency**.

There are several known methods for measuring the similarity between two signals, using two major approaches: **a)** methods that rely on the **shape** of the signals, viewing the signal in its time-domain representation and, **b)** methods that concentrate on the **spectral** content, looking at the frequency-domain representation of the signal. We evaluated the following candidates: (1) The simple cross-correlation function (time domain); (2) *fft* coefficients (frequency domain); and (3) Cepstrum Coefficients (quefrency[1] domain). Other possible measures which we did not try include: Wavelets Analysis, Zero-Crossing/Wave Fluctuation rates, EMD[2] [8], and more.

3.3.1 A Similarity Measure based on the Cross Correlation Function

One of the basic primitives in signal processing is the cross-correlation function [9]. The cross-correlation function operates on signals in their time-domain representation, and is computed as follows: given two digitized signals $x[\cdot]$ and $y[\cdot]$, let

$$CC[x, y, t] = \sum_k x[k] \cdot y[t+k].$$

This is, in effect, a sliding dot-product of the two signals, where signal y is shifted by t samples over x. The cross-correlation is maximized at the offset t where the two signals are most similar. We define $sim^{xcorr}(x, y) = max_t(CC(x, y, t))$ to be the similarity measure between two given signals. Figure 3 shows a plot of two signals and their cross-correlation function, clearly demonstrating the peak at the point where the shapes of the two signals match the most.

For convenience, the signals and the outcome of their cross-correlations are normalized. Thus, the similarity of a signal to itself (its auto-correlation) comes out exactly 1, as required.

---
[1]This terminology was chosen by the Cepstrum inventors [11] to denote a signal in a hybrid time-frequency domain.

[2]EMD is the Earth Movers Distance measure used in music research, by [8]

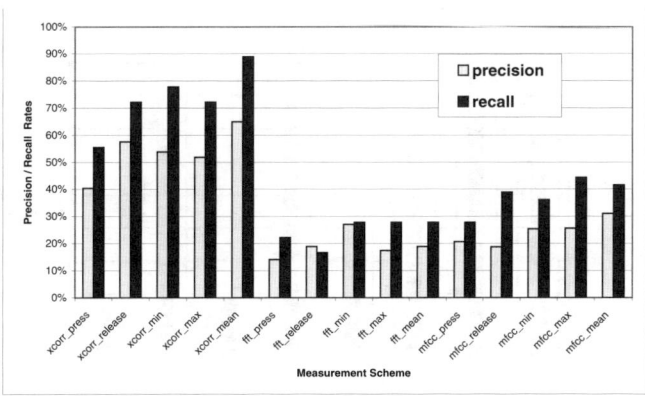

Figure 4: *Precision/Recall* **rates for the various measurement schemes.**

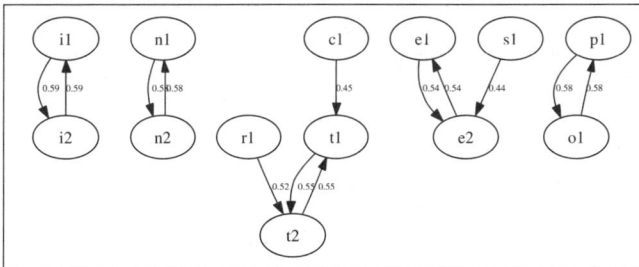

Figure 5: Top *xcorr* values for the word *interceptions*. Each key is represented by a node. An arrow from K_i to K_j indicates that K_j is the top similarity rank for K_i. The numbers on the arrows are the values of the similarity measure. Note that four out of the five best-friends loops are correct (letters *e*,*i*,*n*,*t*) but the fifth loop (for letters *o*,*p*) is erroneous.

If the two signals are M and N samples long, then computing the correlation requires $O(MN)$ operations of multiplication and addition. Since we are working on signals that are supposed to be more or less aligned in phase due to the procedure we employ in earlier stages (recall section 3.2), then there is no need to slide both signals all the way from start to end but rather only for short intervals around their start and end. In our prototype computation, we used the built-in Matlab function ***xcorr*** to compute the CC function.

3.3.2 A Similarity Measure based on FFT Coefficients

In essence, this method computes the Euclidean distance between the spectrum of the signals. Specifically:

1. Truncate each K_i to some fixed length L. In our case of dealing with key presses and releases, this L would normally be 2 milliseconds worth of sampling. If K_i is shorter than L, then pad it with 0.

2. Compute *fft* on each K_i. We ignore the phase information and use only the coefficients that represent the energy for each frequency.

3. Group the *fft* coefficients in equally spaced bands. Each band will be represented by the sum of the coefficients belonging to it. Let B_i denote the vector of bands for K_i.

4. Compute $M_{i,j}$ by calculating the Euclidean distance between B_i and B_j. Note that $M_{i,i} = 0$, so complement each value to 1 (0 becomes 1, 0.1 becomes 0.9, etc.)

To compute the *fft* coefficients we use Matlab's ***fft*** function.

3.3.3 A Similarity Measure based on Cepstrum Coefficients

This is done very much like in the section above describing the usage of the *fft* coefficients, only that here we use the Mel Frequency Cepstral Coefficients instead of the bands. This is the method used by [14]. We used the *mfcc* function provided by the Auditory Toolbox package [10].

3.3.4 Press vs. Release

So far we have listed several methods of extracting and comparing features of the given signal. Using these methods we are able to measure the similarity between all the press segments, and separately, between all the release segments. In our experiments we also considered how best to combine the press and release similarities into a single matrix M.

For each of the three measurement methods mentioned above (*xcorr*, *fft*, *mfcc*), we tested five different schemes to combine the press and the release similarities. Let sp_{ij} denote the press similarity and let sr_{ij} denote the release similarity, between K_i and K_j. The schemes we tested are: **1)** press only: $M_{ij} = sp_{ij}$; **2)** release only: $M_{ij} = sr_{ij}$; **3)** *min*: $M_{ij} = \min(sp_{ij}, sr_{ij})$; **4)** *max*: $M_{ij} = \max(sp_{ij}, sr_{ij})$; **5)** *mean*: $M_{ij} = [sp_{ij} + sr_{ij}]/2$;

min, *max* and *mean* are defined as follows: Let $sim^{press}(K_i, K_j)$ denote the similarity measure between the presses of α_i and α_j and $sim^{release}(K_i, K_j)$ is the similarity measure between the releases of α_i and α_j.

We define the similarity between K_i and K_j based on the *min* scheme to be:

$$min(sim^{press}(K_i, K_j), sim^{release}(K_i, K_j)).$$

The *max* scheme is

$$max(sim^{press}(K_i, K_j), sim^{release}(K_i, K_j))$$

and the *mean* scheme is

$$[(sim^{press}(K_i, K_j) + sim^{release}(K_i, K_j))]/2.$$

3.3.5 Selecting the Similarity Measure

We used the two quality criteria: *Precision* and *Recall* as defined in Section 2.4. Figure 4 shows a comparison of the *Precision* and *Recall* rates for all the different schemes and measurement methods we discussed. It is clear that the best scheme to use is the one using the ***xcorr*** function with the *mean* scheme, namely ***xcorr_mean***.

We have found that the *xcorr_mean* scheme performs well over different keyboards, word lengths and constraint types. We used *xcorr_mean* throughout the remainder of this paper.

	1	2	3	4	...	N−1	N
1	EQ	EQ	ADJ	NEAR			
2	EQ	ADJ	NEAR				
3	ADJ	NEAR					
4	NEAR						
⋮							
N−1							DIST
N						DIST	DIST

Table 2: Constraint formulation rules.

4. CONSTRAINTS AND COMBINATIONS

4.1 Constraint Formulation

We now describe the method used to extract constraints of the various types, out of a given key-similarities matrix $M_{i,j}$ obtained in previous stages of the attack. Naturally, there are many ways to do this. Some of the methods we examined produce constraints with a very high *Precision* rate but with a low recall value, and vice versa. After much experimentation we arrived at a method that achieves a balance between the two, and performs well on different keyboards. We call it the ***BestFriendsPickPolicy***. We omit the details of the less successful alternatives.

Let $rank(i,j)$ denote the position of keystroke j in row M_{i*} of the similarity matrix (excluding M_{ii}), sorted in decreasing order. In other words, if $rank(i,j) = 1$ then K_j is the keystroke most similar to K_i.

We now introduce the notion of keystroke friends. Assume that $rank(i,j) = 1$. As discussed in Section 3.3, this indicates that K_i and K_j may lie physically close to one another on the keyboard.

Let us now look at i's rank in j's row, $rank(j,i)$. We say that i and j are *best friends* if i is also j's top rank, i.e., $rank(j,i) = 1$. If K_i and K_j are best friends, then it is reasonable for us to assume that they are derived from the same key α and we can infer the *EQ* constraint $K_i = K_j$ (see Figure 5).

The *BestFriendsPickPolicy* uses a generalization of this notion of friends, to infer keystroke constraints. In general, for a given pair of keystrokes K_i and K_j, we use $rank(i,j)$ and $rank(j,i)$ as indexes to a cell in Table 2. For example, if $rank(i,j) = 3$ and $rank(j,i) = 2$ then we infer a *NEAR* constraint $K_i \sim K_j$. On the other hand, if $rank(i,j) = N$ and $rank(j,i) = N$ then they are very dis-similar and we infer a *DIST* constraint $K_i \nsim K_j$.

Figure 6 demonstrates the performance of the *BestFriendsPickPolicy*. As we can see, the policy produces *Precision* and *Recall* rates that are roughly consistent over different keyboards. The figure clearly shows that the *Precision/Recall* rates are at their best for the *EQ* constraint, while the inferred *DIST* constraints have high *Precision* but low *Recall*.

Note that we do not use a predetermined threshold to classify keystrokes. Our experience shows that each keyboard and every recording results in different similarity values, thus calibrating a threshold may be non-trivial and keyboard-dependent, and may possibly require significant training data.

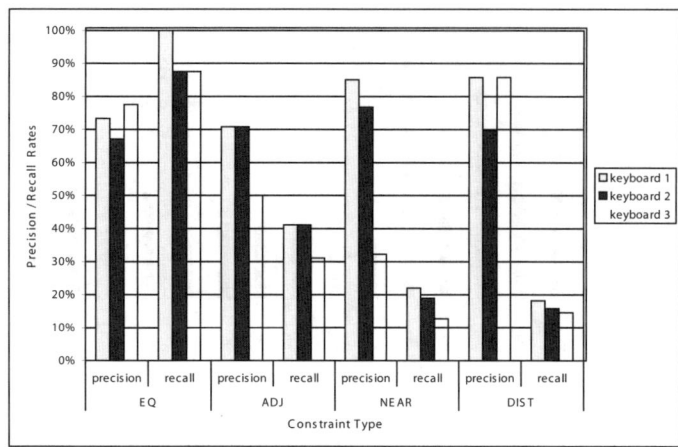

Figure 6: *BestFriendsPickPolicy* performance over 27 words.

5. THE KEY-CONSTRAINTS SATISFACTION ALGORITHM

Once we inferred some constraints we need to evaluate them against the dictionary to extract the words that are consistent with them.

We begin by describing the algorithm for evaluating a single constraint, and based on that, we specify the algorithm for evaluating multiple combinations of constraints.

5.1 Evaluating a Single Constraint

Let N denote the number of letters in the word we are trying to discover. Let Σ denote all the possible letters in the alphabet and let $\square$ denote a constraint type, $\square \in \{=, \simeq, \sim, \nsim\}$. Define $\Sigma_l^{\square} = \{\lambda | \lambda \in \Sigma, (l, \lambda) \in \square\}$ to be the set of letters that match letter l under the the given constraint type $\square$. For a full specification of the relations, see Appendix A.

For a given constraint $\xi_m = K_i \square K_j$, we run through all the possible values $l \in \Sigma$ for K_i and collect out of the dictionary words of length N whose i'th letter is l and j'th letter is in $\Sigma_l^{\square}$. By pre-processing the dictionary into a suitable search data structure, this step can be implemented very efficiently. The outcome of the evaluation of the constraint is the list of unique words collected above. Let $EVAL(\xi_i)$ denote the outcome of the evaluation of a single constraint ξ_i. The $EVAL$ algorithm is shown in Figure 7.

5.2 Evaluating a Combination of Constraints

Evaluating a single constraint is only our basic building block. The next step is evaluating a combination of constraints $c = \{\xi_m\}$. This can be done in the obvious way:

$$EVAL(c) = \bigcap_m EVAL(\xi_m),$$

using the $EVAL$ algorithm of 5.1.

However, our general attack requires us to produce multiple combinations of constraints, and to evaluate each combination. Doing so naively is extremely time consuming. Therefore, we have come up with an efficient way to simultaneously evaluate all these combinations of constraints, using Boolean matrix algebra. We now describe this method:

As before, let $c = \{\xi_m\}$ denote a combination of con-

Input:
1. The number of keys in the given word N.
2. Two indexes, i and j, and a constraint of the form $K_i \square K_j$ where $\square \in \{=, \simeq, \sim, \infty\}$.

Let $Words(N, i, l) =$
$\{w | w \in dictionary, ||w|| = N, w[i] = l\}$

For each possible letter $l \in \Sigma$:
  Let $\Sigma_l^\square \leftarrow \{l' | (l, l') \in \square\}$
  Let $Possible_l \leftarrow \bigcup_{l' \in \Sigma_l^\square} \left\{ Words(N, i, l) \bigcap Words(N, j, l') \right\}$
Let $Possible \leftarrow \bigcup_{l \in \Sigma} Possible_l$

Output: the set of words in $Possible$.

Figure 7: The *EVAL* algorithm for evaluating a single constraint on a single word. The set $Words(N, i, l)$, of all the words of length N that have letter l in position i, is implemented using the search data structure produced from the pre-processed dictionary.

straints and let w_k denote the k'th word in the dictionary. We construct a Boolean matrix W that encodes the separate evaluation of each of the constraints, i.e., $W_{m,k} = 1$ iff $w_k \in EVAL(\xi_m)$. Let $\vec{c}$ denote the characteristic vector of the combination c, i.e., $\vec{c}_m = 1$ iff constraint ξ_m is included in the combination c, and 0 otherwise.

Using this notation, the evaluation of the combination c is the Boolean multiplication of the $\vec{c}$ with the matrix W, i.e.,

$$EVAL(c) = \vec{c} \cdot W.$$

Note that $\cdot$ is computed as the Boolean product of the two.

In a computing environment such as Matlab, Boolean multiplication can be implemented by using the normal (integer) matrix multiplication operator, which is then divided by the sum of the elements of $\vec{c}$ and truncated so that a value of 1 is produced only if the sum of terms is exactly $|c| = \sum_t \vec{c}_t$. In other words,

$$\vec{c} \cdot W = \left\lfloor \frac{\vec{c} * W}{|c|} \right\rfloor.$$

The advantage of using such matrix notation is that it lets us evaluate *multiple* combinations simultaneously, while running the *EVAL* only once per constraint, as follows. Let $\mathcal{C} = \{\}_{\setminus}$ be a collection of combinations of constraints. Let Γ be the characteristic matrix of the combinations: $\Gamma_{n,m} = 1$ iff constraint ξ_m belongs to combination c_n. Then we can evaluate *all* the combinations with a single Boolean matrix operation:

$$R = \Gamma \cdot W,$$

where $R_{n,k} = 1$ iff the evaluation $EVAL(c_n)$ includes the word w_k.

To prioritize the output, for each word w_k we compute the number of combinations it appeared in: $r(w_k) = \sum_n R_{n,k}$. The final output U is the list of dictionary words sorted in decreasing order of $r(w_k)$.

5.3 Dealing with Errors

Neither the similarity measure, nor the constraint inference policy, is perfect. However, it is important to keep in mind that even a single false constraint of any kind, will preclude us from finding the right word. We deal with these errors by incorporating randomness into our attack.

Instead of relying on *all* the constraints we extracted in earlier stages, we look at multiple combinations of them. Since there may be a huge number of possible combinations, we do not generate all the possible combinations, but rather generate a random sample of combinations. Each combination is constructed by running through the list of constraints $\{\xi_m\}$ and including constraint ξ_m in the combination with probability p. A single combination is expected to include $|\xi| * p$ constraints.

When dealing with single words, 7 characters or more, we typically infer 13-17 constraints. We used $p = 0.2$ in all cases. Finally, in most cases 100-200 combinations were more than enough, with rare cases that required 1000 combinations.

Evaluating this collection of combinations is performed efficiently as described in Section 5.2, using the matrix Γ to represent the whole collection of random combinations.

Note that Γ may have a large number of rows, depending on the number of choose to use. W, on the other hand, has a column per dictionary word: in our case, around 60,000 columns. Thus, the multiplication of Γ with W may yield an enormous matrix. Luckily, the result of the Boolean multiplication is sparse (almost all the entries are 0), despite the fact that Γ itself is not sparse. There are well known advanced data structures that operate and maintain such matrices very efficiently. We used Matlab's `sparse` function for this purpose.

6. PERFORMANCE EVALUATION

6.1 Overall Success Rate

Our attack produces a sorted list U of word candidates. We measure the efficiency of our attack by marking the position of the correct word in U. The position of a word in U is called the **Rank** of the word, where a *Rank* of 1 is the most likely candidate. Based on the *Rank*, we calculate the frequency of the correct word appearing in the top 10, top 25, top 50, top 100 and top 500 places over all the tests we conducted.

We tested 27 words with lengths of 7-13 characters (see Table 3 in the appendix for the complete set of words). Each word was recorded on three keyboards (see Section 3.1). Each word recording was processed 7 times using different random choices.

The word corpus we used is an aggregation of the corncob wordlist [4] and the English words file from the SCOWL-6 package [2].

Figure 8 above summarizes the overall success rate of our attack, on the tested words. For example, we can see that in 73% of the tests, the correct word was located within the top 50 candidates.

6.2 Influential Factors

We tested the success of the attack as a function of two major factors:

1. Character repetitions in the word. As we saw in Sec-

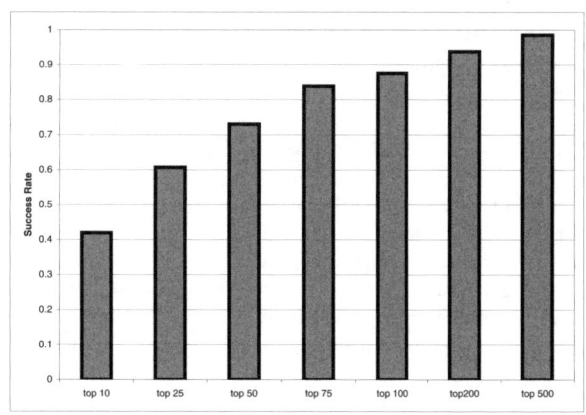

Figure 8: Overall effectiveness of the attack.

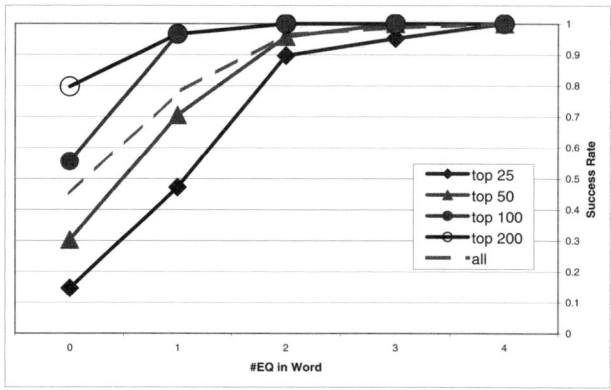

Figure 10: The success rate as a function of the number of repetitions.

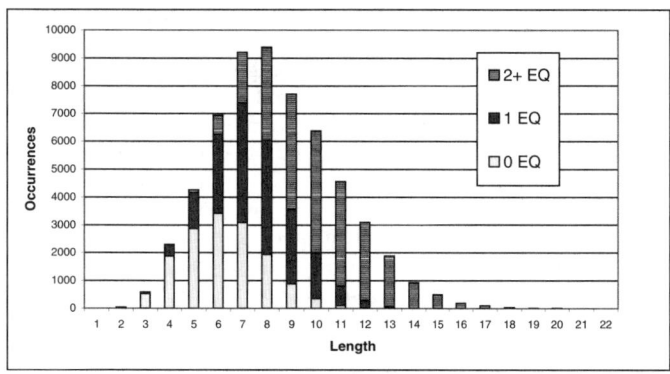

Figure 9: Distribution of word lengths in the word corpus used in the attack. The lower, middle and upper sections of each bar are proportional to amount of words with 0, 1, 2+ EQ in them, for that length category.

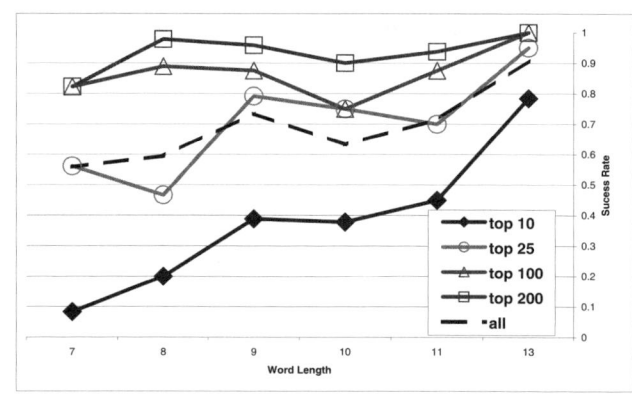

Figure 11: The success rate as a function of the word length.

tion 4.1, EQ constraints have the best *Precision* and *Recall* rates among all our constraint types. Naturally, we expect that the attack will work better against words that produce more EQ constraints.

2. Word length. The length of the word has two effects. First, longer words produce more constraints, so we expect better results. Second, the distribution of word lengths in the dictionary is non-uniform (see Figure 9), with words of length 7,8 being the most frequent. If there are many possible candidates of length N in the dictionary then we expect the attack success to degrade. However, for length $N \geq 7$, longer words are less frequent, so again, we expect the success of the attack to improve for longer words.

6.2.1 Character repetitions in the word

Figure 10 shows that the success rate is strongly influenced by the number of key repetitions in the word. We can see that the success rate grows dramatically with additional EQ constraints: the attack finds the correct word in the top 25 at a rate of 90% for words with 2 or more EQ constraints.

A closer inspection of the data (omitted) shows that for words of length 7-9 the major improvement occurs at 2 EQs. For words with length 10 and above, 1 EQ constraint is usually enough to place the word in the top 10 candidates.

6.2.2 Word Length

Figure 11 shows the dependence of the attack's success on the length of the word. The figure clearly shows that, as expected, we have better success against longer words. However, the dependence is less clear-cut than that of the number of repetitions.

7. CONCLUSIONS AND FUTURE WORK

We have presented a procedure that makes it possible to efficiently uncover a word out of audio recordings of keyboard click sounds. The effectiveness of our attack depends mainly on the properties of the hidden text, namely how long it is and how many repetitions are in it. Our methods require no training, work on signals as short as 5 seconds, and are resilient to different different keyboards.

We believe that our attack can be used as an effective

acoustic-based dictionary password cracker—and is another reason *not* to use dictionary words as passwords. However, counter-intuitively, choosing longer passwords *improves* the success of the attack: it seems that to withstand our attack, the ideal password length should be 7-8 characters.

We have some promising preliminary results which show that this attack can also be used as part of an acoustic long-text reconstruction method. The key observation is that the Space key has a very distinct sound and can be identified as a first step, allowing us to identify individual words. The current attack, in combination with inter-word statistics would then let us reconstruct whole sentences and paragraphs.

Clearly, we can integrate inter-keystroke timing information with our approach. We have looked at this direction briefly and it seems viable.

To improve the capabilities of our method for password cracking, we would need to use password dictionaries (rather than English word lists), and we would need to deal with Shift keys (for upper-case letters), punctuation marks, and digits. This seems possible since the Shift key would produce only a PRESS (or only a RELEASE) segment.

We also believe that our signal processing is quite unsophisticated, and developing more accurate models of the keyboard acoustic signals may be a fruitful direction of research for signal processing experts.

Acknowledgments

We would like to thank Miriam Furst-Yust, Zvi Gutterman, and Benny Ben-Ami for many useful discussions and tips. We also thank Roni Rosenfeld, Ron Hecht and Ami Navon for their help with Cepstrum Analysis.

8. REFERENCES

[1] D. Asonov and R. Agrawal. Keyboard acoustic emanations. In *IEEE Symposium on Security and Privacy*, pages 3–11, Oakland, CA, 2004.

[2] K. Atkinson. Scowl - spell checker oriented word lists, 2004. http://wordlist.sourceforge.net/.

[3] R. Briol. Emanation: How to keep your data confidential. *Symposium on Electromagnetic Security For Information Protection*, 1991.

[4] CornCob. The corncob list. http://www.mieliestronk.com/wordlist.html.

[5] D. Klein. Foiling the cracker: A survey of, and improvements to, password security. In *Proc. UNIX Security Workshop II*, Aug. 1990.

[6] M. G. Kuhn. Compromising emanations: Eavesdropping risks of computer displays. Technical Report UCAM-CL-TR-577, University of Cambridge, Computer Laboratory, Dec. 2003.

[7] J. Loughry and D. A. Umphress. Information leakage from optical emanations. *ACM Trans. Info. Sys. Security*, 5(3):262–289, 2002.

[8] Y. Rubner, C. Tomasi, and L. Guibas. The earth mover's distance as a metric for image retrieval. *International Journal of Computer Vision*, 40(2):99–122, 2000.

[9] Time domain processing: Correlation. http://www.bores.com/courses/intro/time/2_ave.htm.

[10] M. Slaney. Auditory toolbox, 1998. http://rvl4.ecn.purdue.edu/~malcolm/interval/1998-010/.

[11] S. W. Smith. *The Scientist and Engineers Guide to Digital Sound Processing*. California Technical Publishing, 1997.

[12] D. Song, D. Wagner, and X. Tian. Timing analysis of keystrokes and timing attacks on SSH. In *10th USENIX Security Symposium*, 2001.

[13] Tempest 101. http://www.tscm.com/TSCM101tempest.html.

[14] L. Zhuang, F. Zhou, and J. D. Tygar. Keyboard acoustic emanations revisited. In *CCS '05: Proceedings of the 12th ACM conference on Computer and communications security*, pages 373–382, New York, NY, USA, 2005. ACM Press.

APPENDIX
A. DATA TABLES

Word	length
paediatrician	13
interceptions	13
abbreviations	13
impersonating	13
soulsearching	13
hydromagnetic	13
inquisition	11
pomegranate	11
feasibility	11
polytechnic	11
obfuscating	11
difference	10
wristwatch	10
processing	10
unphysical	10
institute	9
extremely	9
sacrament	9
dangerous	9
identity	8
emirates	8
platinum	8
homeland	8
security	8
between	7
spanish	7
nuclear	7

Table 3: List of words used to test the attack.

key	ADJ	NEAR	DIST
Q	Q,W,S,A	Q,W,A,E,S,Z,D,X	B,C,F,G,H,I,J,K,L,M,N,O,P,R,T,U,V,Y
A	A,Q,W,S,Z	A,Q,Z,W,S,X,E,D	B,C,F,G,H,I,J,K,L,M,N,O,P,R,T,U,V,Y
Z	Z,A,S,X	Z,Q,A,W,S,X,E,D,C	B,F,G,H,I,J,K,L,M,N,O,P,R,T,U,V,Y
W	W,Q,A,S,D,E	W,Q,A,Z,S,X,E,D,C,R,F	B,G,H,I,J,K,L,M,N,O,P,T,U,V,Y
S	S,Q,A,Z,X,D,E,W	S,Q,A,Z,W,X,E,D,C,R,F	B,G,H,I,J,K,L,M,N,O,P,T,U,V,Y
X	X,Z,A,S,D,C	X,Q,A,Z,W,S,E,D,C,F,V	B,G,H,I,J,K,L,M,N,O,P,R,T,U,Y
E	E,W,S,D,F,R	E,Q,A,Z,W,S,X,D,C,R,F,V,T,G	B,H,I,J,K,L,M,N,O,P,U,Y
D	D,E,W,S,X,C,F,R	D,Q,A,Z,W,S,X,E,C,R,F,V,T,G	B,H,I,J,K,L,M,N,O,P,U,Y
C	C,X,D,F,V	C,W,S,Z,E,S,X,R,D,F,V,T,G,B	A,H,I,J,K,L,M,N,O,P,Q,U,Y
R	R,E,D,F,G,T	R,W,S,X,E,D,C,F,V,T,G,B,Y,H	A,I,J,K,L,M,N,O,P,Q,U,Z
F	F,R,E,D,C,V,G,T	F,W,S,X,E,D,C,R,V,T,G,Y,H,B	A,I,J,K,L,M,N,O,P,Q,U,Z
V	V,C,D,F,G,B	V,E,D,X,R,F,C,T,G,B,Y,H,N	A,I,J,K,L,M,O,P,Q,S,U,W,Z
T	T,R,F,G,H,Y	T,E,D,C,R,F,V,G,Y,H,B,U,J,N	A,I,K,L,M,O,P,Q,S,W,X,Z
G	G,T,R,F,V,B,H,Y	G,E,D,C,R,F,V,T,B,Y,H,N,U,J	A,I,K,L,M,O,P,Q,S,W,X,Z
B	B,V,G,H,N	R,D,C,T,F,V,G,Y,H,N,U,J,M	A,B,E,I,K,L,O,P,Q,S,W,X,Z
Y	Y,T,G,H,J,U	Y,R,F,V,T,G,B,U,H,I,J,N	A,C,D,E,K,L,M,O,P,Q,S,W,X,Z
H	H,Y,T,G,B,N,J,U	H,R,F,V,T,G,B,Y,U,J,N,I,K,M	A,C,D,E,L,O,P,Q,S,W,X,Z
N	N,B,H,J,M	N,T,F,V,Y,G,B,H,U,J,M,I,K	A,C,D,E,L,O,P,Q,R,S,W,X,Z
U	U,Y,H,J,K,I	U,T,G,B,Y,H,N,I,J,O,K,M,L	A,C,D,E,F,P,Q,R,S,V,W,X,Z
J	J,U,Y,H,N,M,K,I	J,T,G,B,Y,H,U,N,I,K,M,O,L	A,C,D,E,F,P,Q,R,S,V,W,X,Z
M	M,N,J,K	M,Y,H,B,U,J,N,I,K,O,L	A,C,D,E,F,G,P,Q,R,S,T,V,W,X,Z
I	I,U,J,K,L,O	I,Y,H,N,U,J,M,O,K,P,L	A,B,C,D,E,F,G,Q,R,S,T,V,W,X,Z
K	K,I,U,J,M,L,O	K,Y,H,N,U,J,M,I,O,L,P	A,B,C,D,E,F,G,Q,R,S,T,V,W,X,Z
O	O,I,K,L,P	O,U,J,M,I,K,P,L	A,B,C,D,E,F,G,H,N,Q,R,S,T,V,W,X,Y,Z
L	L,O,I,K,P	L,U,J,N,I,K,M,O,P	A,B,C,D,E,F,G,H,Q,R,S,T,V,W,X,Y,Z
P	P,O,L	P,I,J,M,I,K,O,L	A,B,C,D,E,F,G,H,N,Q,R,S,T,U,V,W,X,Y,Z

Table 4: *ADJ*, *NEAR*, *DIST* tables for each of the keys, as used in our algorithms.

Inferring the Source of Encrypted HTTP Connections

Marc Liberatore and Brian Neil Levine
Department of Computer Science
University of Massachusetts, Amherst
Amherst, MA 01003-9264
liberato@cs.umass.edu, brian@cs.umass.edu

ABSTRACT

We examine the effectiveness of two traffic analysis techniques for identifying encrypted HTTP streams. The techniques are based upon classification algorithms, identifying encrypted traffic on the basis of similarities to features in a library of known profiles. We show that these profiles need not be collected immediately before the encrypted stream; these methods can be used to identify traffic observed both well before and well after the library is created. We give evidence that these techniques will exhibit the scalability necessary to be effective on the Internet. We examine several methods of actively countering the techniques, and we find that such countermeasures are effective, but at a significant increase in the size of the traffic stream. Our claims are substantiated by experiments and simulation on over 400,000 traffic streams we collected from 2,000 distinct web sites during a two month period.

Categories and Subject Descriptors

C.2.0 [**Computer-Communications Networks**]: General—*Security and protection (e.g., firewalls)*; C.2.3 [**Computer-Communications Networks**]: Network Operations—*Network monitoring*; I.5.4 [**Pattern Recognition**]: Applications; K.4.1 [**Computers and Society**]: Public Policy Issues—*Privacy*

General Terms

Measurement, Security

Keywords

low-latency anonymity, network forensics, traffic analysis

This research was supported in part by National Science Foundation awards CNS-0133055 and ANI-0325868.

Permission to make digital or hard copies of all or part of this work for personal or classroom use is granted without fee provided that copies are not made or distributed for profit or commercial advantage and that copies bear this notice and the full citation on the first page. To copy otherwise, to republish, to post on servers or to redistribute to lists, requires prior specific permission and/or a fee.
CCS'06, October 30–November 3, 2006, Alexandria, Virginia, USA.
Copyright 2006 ACM 1-59593-518-5/06/0010 ...$5.00.

1. INTRODUCTION

Encrypted connections provide confidentiality at many different network layers. In combination with a proxy, this setup forms the basis of private communication over the Internet. Many such systems prevent observers from determining the true destination of IP traffic: multi-proxy tunnels using the Tor anonymous communication system [6]; simple SSH tunnels to a single proxy; and IPSec ESP mode tunnels to a remote VPN concentrator. WPA link layer connections also hide the destination IP address and content of a connection from an observer.

In this paper, we evaluate traffic analysis techniques that infer the source of a web page retrieved under the cover of an encrypted tunnel. These techniques identify sources by comparing observed traffic to profiles of known sites created from packet lengths, and are referred to as *profiling attacks*. A previous study [1] has shown the attack is feasible, with a method achieving about 25% accuracy.

This type of attack on web traffic is an area of concern for advocates of privacy enhancing technologies. This includes the developers of Tor, who have recognized that this type of web page fingerprinting is a significant problem [5]. Currently deployed low-latency anonymity systems do not significantly adjust traffic to prevent comparison to profiles. The advances we detail in this paper increase the importance of addressing the attack in implementations — our study includes results with an accuracy of up to 90% for realistic scenarios.

Privacy is the dual problem of digital forensics — accordingly, the attack is also an important method of investigation that advances the field of digital forensics. This is because commonly used forensic techniques for gathering and analyzing network data are limited to traffic with overt IP headers and data [2, 3]. Encrypted tunnels thwart the legitimate gathering of evidence by authorized law enforcement. Advances in the field of forensics require moving beyond this limitation. This paper provides guidance for such advances.

Contributions. Our results are based on traces we gathered of encrypted communications to 2,000 web sites, which we collected four times a day for two months. We have made these traces publicly available for validation, collaboration, and future work. To our knowledge, this is the largest public collection of such traffic for the study of profiling attacks. We built two systems that identify traffic, one based on the naive Bayes classifier and one on Jaccard's coefficient, a straightforward similarity metric. Both systems rely on packet lengths but discard timing information.

Despite this simplification, we found that under reasonable assumptions, traces were identifiable between 66–90% of the time. On the basis of our experiments, we expect performance to scale well, and we assume that more sophisticated methods could do better. We also examine in simulation the effectiveness of per-packet padding (that is, increasing the length of packets) in an attempt to defeat our profiling system. We find that this approach is reasonably effective, lowering predictive accuracy to less than 8% while increasing traffic volume by 145%.

While profiling requires a set of candidate sites, we believe that it would not be difficult for an observer to profile all publicly accessible web sites on the Internet in time for the attack to succed. In part, this is due to our finding that classification accuracy degrades very slowly over time, giving the observer at least four weeks to collect the profile after the encrypted traffic is observed in many cases. Moreover, using the technique we evaluated, an investigator could store profiles of the front pages of all web sites on the Internet with about 13GB of storage.

The remainder of this paper is organized as follows. Section 2 describes related work. In Sections 3 and 4, we describe our attack model and data collection methodology, and in Section 5, we describe our experimental methodology and results. We discuss the implication of these results in Section 6, and conclude in Section 7.

2. RELATED WORK

Traffic analysis is a large field; in this section, we survey work in that field that is related to the analysis of anonymity systems. In particular, we discuss prior work on passive logging, profiling (or fingerprinting) attacks, and analysis of countermeasures to these attacks, as these are the most relevant to our work. We focus on work that examines HTTP and secure HTTP and the vulnerabilities and exposures inherent in those protocols. For a broader overview of the field, consult Raymond [11] or the Free Haven anonymity bibliography[1].

Wright, et al. [15, 16] and others have shown that low-latency anonymity systems are vulnerable to passive logging and an intersection attack. These results are complementary to our own: The attack allows identification of the endpoints in an anonymous communication system with path changes, whereas our technique allows an observer to infer the content (and thus, the endpoint) from the traffic itself.

Hintz [9], Sun, et al. [12], and Bissias, et al. [1] present profiling attacks of encrypted connections, though our study differs from each of these in important ways. Hintz's work is a preliminary proof-of-concept, examining total data sent over SSL connections. While the technique is reasonably effective, SSL and its successor, TLS, are not designed to hide traffic patterns [4, 7]. One of our classification techniques is drawn from the work presented by Sun, et al., though the problem we study differs from theirs in a nontrivial way. In their work, they compare the size of web objects, rather than packets. This comparison is possible due to their strong simplifying assumption that objects can be differentiated by examining the timing of TCP connections. This assumption is not valid for WEP/WPA links, VPN connections, and SSH tunnels, and in the presence of widely-supported pipelined HTTP connections. We make the weaker assumption that packets, not objects, can be distinguished, and we base our identification of pages on the individual packets that compose these pages. Bissias, et al. present work similar in its assumptions to ours. They examine packet lengths and timings as the basis of identification, though they use only the crude metric of cross-correlation to determine similarity. The study also is preliminary in nature, attempting to distinguish among only 100 web sites. While appropriate for an initial study, such a small sample is hard to generalize from; in this study, we show results for a larger, more robust data set.

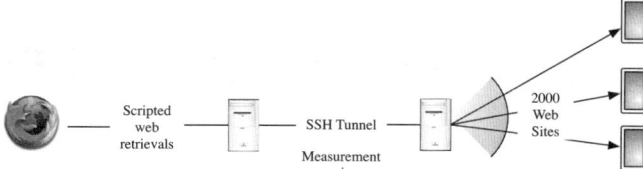

Figure 1: An illustration of the measurement setup. This figure also shows the position an observer may take to utilize profiling.

There are few careful studies of countermeasures to profiling attacks upon low-latency anonymity systems on the Internet. Fu, et al. [8], examine the technique of introducing dummy packets on links to defeat some types of traffic analysis. Their results are largely theoretical in nature, and designed to thwart general traffic analysis rather than specific profiling attacks. They find that variable intervals between dummy packets are more effective that constant intervals. Levine, et al. [10] examine dropping packets to foil statistical correlation. A generalization of partial-path cover traffic [13], this technique and their analysis is concerned mainly with foiling timing attacks. We believe that there is much work yet to be done in examining latency and bandwidth tradeoffs in cover traffic and delays for low-latency anonymity systems.

3. MODEL

In this section, we describe our model of an observer that can execute traffic analysis attacks to profile web sites and identify encrypted traffic on the basis of these profiles. We first describe our assumptions about how data can be collected by an observer. Then, we describe two specific methods to create these profiles, one based upon a similarity metric (Jaccard's coefficient) and one based upon a supervised learning technique (the naive Bayes classifier).

3.1 Observer Model

Figure 1 illustrates the network setup that we use throughout this paper. In it, the client connects to a remote proxy over an encrypted transport layer. The proxy makes requests on the client's behalf, and returns the results over the encrypted connection. The observer is limited to examining the encrypted traffic and creates a log of packet lengths (and interarrival times, if desired) corresponding to each distinct page load. Our observer has unlimited storage for these logs. We assume that the observer is not able to discriminate among individual objects as in Sun, et al. We assume the client uses a modern browser for retrieval, as described in Section 4.2, which prevents the observer from obtaining this information.

---

[1]http://freehaven.net/anonbib/topic.html

Because our techniques focus on packet lengths, it is not a requirement that the observer create profiles on the same link that she observes traffic. However, for simplicity, we assume this is the case.

We assume that the observer is able to determine where discrete communications begin and end (such as the loading of a web page and its associated objects). This is possible, for example, by observing sender think times that separate requests. In future, we will determine mechanisms for separating multiple requests and requests that appear with background traffic.

The proxy we evaluate in this paper is the OpenSSH implementation of a one-hop SOCKS proxy. However, we expect our results hold for VPN proxies and WPA base stations. Neither of these systems significantly alter packet lengths since they perform no buffering and packet aggregation or fragmentation. The Tor and JAP[2] low-latency anonymity systems provide only very limited aggregation and fragmentation. We believe that these systems will also be vulnerable to a form of this attack.

To launch the class of traffic analysis attacks that we evaluate in this paper, the observer requires a library of traffic traces between a client and a list of known web sites. We show in Section 5 that this library can be collected before or after the attack. Using the traces of connections to known destinations, the observer attempts to decide, in some fashion, which of the known traces most closely resembles the encrypted, unknown trace, as we explain below.

3.2 Profiling Methods

To determine similarity to known traces, the observer describes each trace in terms of *attributes* and lets each attribute range over many possible values. The problem then becomes an instance of supervised learning, as the observer has a set of labeled training instances (the traces gathered by the observer) and one or more unlabeled test instances (the observed, encrypted traces).

In the remainder of this paper, we denote each trace together with its attributes as an *instance*. Each instance has an attribute denoting the URL, or site, to which it corresponds. We also refer to this as the *class* of the instance. Each instance also has attributes that describe every packet in the trace. These attributes take the form of a tuple, *(direction, length)*. The *direction* denotes whether the packet went from the client to the server referenced in the URL, or vice versa; the length denotes the total length, in bytes, of the packet. The value assigned to each attribute is the number of packets observed in that trace with the corresponding *direction* and *length*.

Our first method for identifying unknown instances is to use a similarity metric; we use Jaccard's coefficient to measure similarity and thus determine the class of an instance. For two sets X and Y, Jaccard's coefficient S is defined as:

$$S(X,Y) = \frac{|X \cap Y|}{|X \cup Y|} \quad (1)$$

We build a model for Jaccard's coefficient based classification as follows. Each site in the model represented by a set. If there is only one training instance per site, then the model is built as follows: For each packet length and direction in a training instance, a *(direction, length)* tuple is inserted into the set corresponding to that site. If there is more than one instance in the training set per site, then such tuples are inserted into the set iff they are present in the majority of the instances in the training set. To convert the similarity metric S to an estimate of class membership probability, we normalize $S(X,Y)$ for a given X by dividing by $\sum_{Y \in U} S(X,Y)$, where U is the set of all training sets.

Our second method for identifying unknown instances is the naive Bayes classifier. We do not give a full explanation of the classifier here: a recent textbook such as Witten and Frank [14] will provide details. In short, the naive Bayes classifier assumes independence between all attributes, and estimates the probability of a set of value $\overline{A} = A_1, \ldots, A_n$ belonging to a particular class C_i as:

$$p(C_i|\overline{A}) \propto p(C_i) \prod_{j=1}^{n} (p(A_j|C_i)) \quad (2)$$

In our experiments, which are described below, we use Witten and Frank's Weka toolkit implementation, `weka.classifiers.bayes.NaiveBayes`, with normal kernel density estimation enabled.

4. DATA COLLECTION

To investigate the effectiveness of a profiling attack, we required a data set consisting of logs of encrypted traffic between a client and many servers.[3] We imposed several requirements on this data set. First, it had to be of sufficiently large size to make the problem non-trivial, while not so large to prevent multiple collections each day with our limited resources. Second, it had to reflect a real-world group of users. Finally, it had to be collected in a fashion analogous to the manner in which an actual observer would attempt. In this section, we present the details of these processes so that others can validate or recreate our collection process.

4.1 Initial Data Collection

We used our department's Internet traffic as a basis for choosing sites to profile. This network is used by an estimated population of over three hundred faculty, staff, and students. By monitoring DNS requests within the department, we gathered a list of remote hosts to which users connected. We heuristically refined this list into a set of HTTP URLs which we believe are reasonably representative of the web browsing habits of users in our department.

The initial step was to log all requests to the department's DNS server from 01 December 2005 through 04 January 2006. This month of logs yielded 44,305,203 requests from 828 hosts. We removed all requests that were not for address (`A`) records, as HTTP traffic would not have generated them. We removed all requests that were from outside the department, as we were studying users within the department. We removed all requests that were for names within our domain, as these tend to correlate with intra-department service accesses (such as secure shell access) and not HTTP requests.

We then removed requests we judged to be the result of automatic processes. Specifically, we removed all requests for a given name that were made from the same host with

---

[2] http://anon.inf.tu-dresden.de/

[3] The anonymized logs we collected and used are available at http://traces.cs.umass.edu/

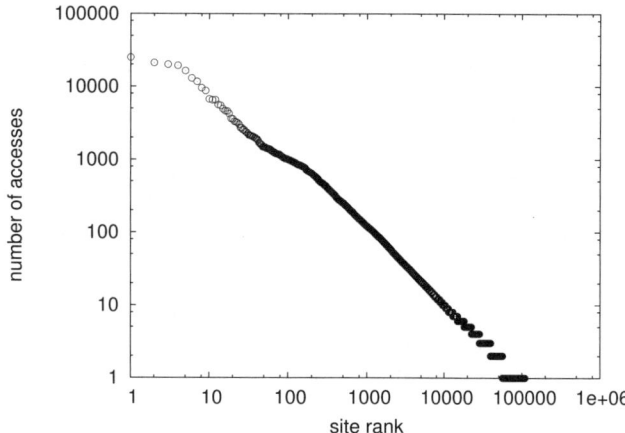

Figure 2: The relationship between site rank and number of accesses. Sites are ranked according to total number of accesses to each site observed.

an average frequency of greater than once per five minutes in any eight hour period. We then heuristically removed most requests for PlanetLab[4] machines.

From these requests, we constructed a list of URLs of the form `http://ipaddress/`, along with the associated count for each. On 01 February 2006, we attempted to contact each of these sites with an HTTP GET request on port 80. We removed from the list all sites that were unreachable, refused the connection, or returned an HTTP error that was not a redirection. We also replaced the `ipaddress` with the actual hostname the remote machine used, and we resolved all redirections. Finally, we summed the counts of all sites that redirected to the same URL. This final list of *(count, URL)* pairs was 109,479 pairs in length. For our experiments, presented in the next section, we focused on the 2,000 most accessed sites, which accounted for 64% of all web requests. The relationship between rank and number of accesses is show in Figure 2.

4.2 Page Retrieval

To gather realistic traces, we set up a client host with a recent GNU/Linux distribution. We used Mozilla Firefox 1.5[5] to retrieve each URL via a SOCKS proxy. OpenSSH 4.2p1[6] was set up to perform application level dynamic port forwarding (the `-D` option) and act as a proxy. This created an encrypted channel over which the HTTP requests and responses were forwarded, and it made distinguishing individual objects infeasible, as Firefox generally makes multiple simultaneous connections to load a web page that are multiplexed over the secure channel.

We configured Firefox to not cache data between retrievals, which allowed us to focus on the specific question of identifying encrypted streams by their profiles. We installed the latest Macromedia Flash plugin[7], as many of the URLs in our list contained content rendered by this plugin. We also configured Firefox to not attempt various extraneous connections, due to live bookmarks, automatic update checks,

---

[4] http://www.planet-lab.org/
[5] http://www.mozilla.org/projects/firefox/
[6] http://www.openssh.org/
[7] http://www.macromedia.com/

and the like. While disabling these features makes the resulting traffic somewhat less realistic, we believe it to be a reasonable trade-off to allow us to focus on the specific problem under investigation. As Firefox loaded each URL in our list, we used tcpdump 3.9.4[8], linked against libpcap 0.9.4, to log the first 68 bytes of each packet. This length is sufficient to capture IP and TCP headers of the packet and thus determine the total packet length. The information in these logs form the basis of the profiles that we detail in the next section.

5. EVALUATION

In this section, we describe the experimental methodology we used to evaluate our proposed classification methods as well as the results of that evaluation, based upon two months of data we collected. We give evidence that the two methods, one based upon Jaccard's coefficient and the other upon the naive Bayes classifier, have several properties of interest: We show that the Jaccard-based classifier's accuracy, under reasonable assumptions, is over 60%. We give evidence that it will scale reasonable well to large data sets. We show that many profiles, once constructed, remain valid for long periods of time. We show that training data can be gathered before or after test data, with a negligible effect upon accuracy. We show evidence that the identifiability of traces using these methods is a result of the distinctiveness in both the individual packet length as well as overall trace length. Finally, we examine the robustness of our method to several forms of static countermeasures, and find that users using these countermeasures must be willing to incur a bandwidth penalty of 145% in order to drive an observer's accuracy below 8%.

5.1 Experiment Setup

To evaluate the effectiveness of the profiling attack described above, we collected traces of encrypted traffic as described in Section 4. Specifically, each *sample* consists of the log of a retrieval of each of the 2,000 most-visited web sites. We created a new sample once every six hours for a period of two months, for a total of 480,000 samples.

From these samples, we performed individual experiments of several variables. Each experiment utilizes a set of training samples and a distinct set of testing samples. Each of the following three variables are relative to some sample i, the *initial sample*.

- t describes the number of sequential samples that form the training set.

- s describes the number of sequential samples that form the test set.

- Δ describes the number of sequential samples between the training and test sets. $\Delta = 0$ indicates the test set starts with the sample immediately following the last sample in the training set.

- N describes how many of the 2,000 sites we considered in a particular experiment. If it is less than 2,000, then we reduced the number of traces in each sample to this number by removing the traces corresponding to the same sites from all samples. We removed the least-popular sites first.

---

[8] http://www.tcpdump.org/

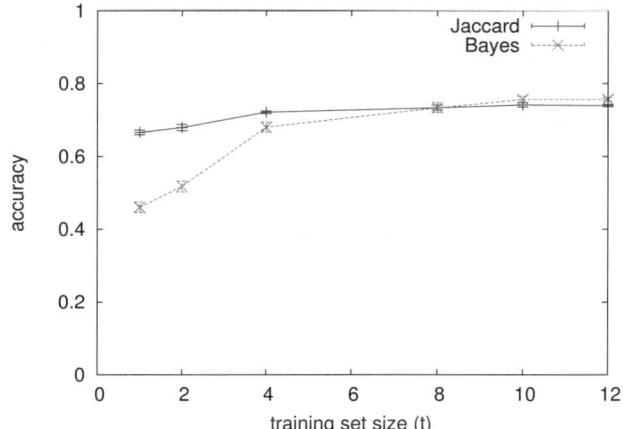

Figure 3: Effect upon accuracy of varying training set size.

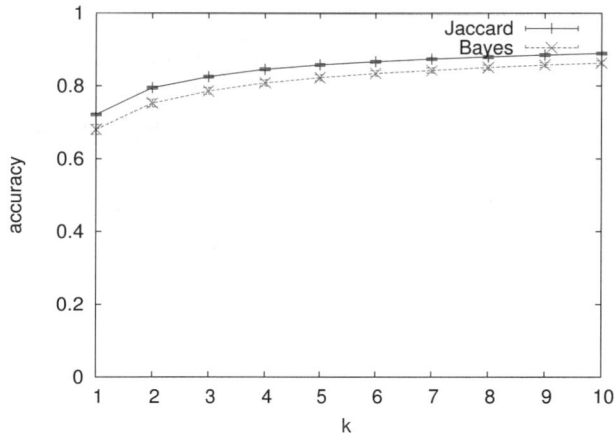

Figure 4: Effect upon accuracy of different values of k.

The result of each experiment is a table of probabilities of class membership for each of the instances in the test set. From this, we determine the *k-identifiability* of each instance. The k-identifiability for an instance is defined as 1 if the actual class of the instance is in the top k of the predicted class list, as ordered by probability estimate (predicted classes with the same estimate are ordered in an arbitrary but fixed manner) or 0 if the actual class is not in the top k. The *k-accuracy* for an experiment is the average of the k-identifiability value for each instance in the test set.

5.2 Classifier Performance

We evaluated the effect of changing each of the independent variables listed above. Unless the variable was the isolated and changing variable, each of the following graphs assumes $k = 1$, a training set size of $t = 4$ (one day of data), a test set size of $s = 4$, $\Delta = 3$ so that the training and test sets are one day apart, and $N = 1000$ sites. We chose random initial sample numbers such that 10 individual experiments were run with all other variables equal — these otherwise identical experiments are the source of the 95% confidence intervals in the graphs. Often the intervals are too small to be observed. Figures 3, 4, 5, and 7 show the results of these experiments.

In Figure 3, we show the effect of varying the training set size. While accuracy increases as more samples are added to the training sets, the rate of increase slows when $t = 4$. Thus, even small training sets give good accuracy, and additional training data give diminishing returns on accuracy. In Figure 5, we show the effect of varying Δ. Larger delays between the training and test sets result in lower accuracy, but the decrease appears to be linear in the delay and tolerably small (a drop from 73% to 63%), even after four weeks of delay. Thus, training data remains useful even after a significant amount of time has passed. This implies that it need not be collected too frequently, increasing the utility of large sets of such data. In Figure 4, the effect of a larger k is shown. Unsurprisingly, allowing the observer more chances to identify a trace allows for higher accuracy. At $k = 10$, the observer can expect to have correctly identified the trace 90% of the time.

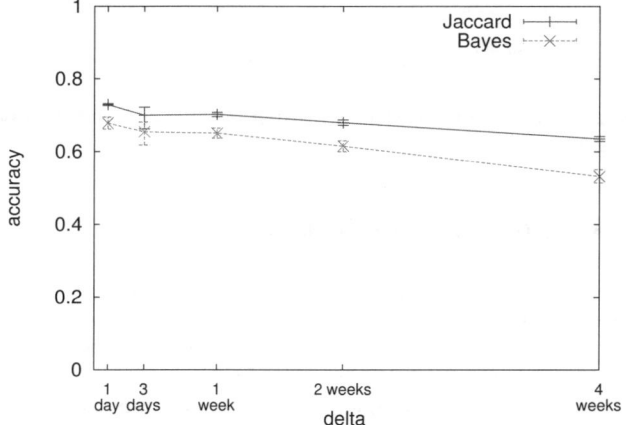

Figure 5: Effect upon accuracy of varying the delta (time between training and testing).

Figure 7 shows the effect of varying the number of sites in the training and test sets. The drop in accuracy appears to follow a relationship of the form:

$$acc = A \log_2 N + B \quad (3)$$

A linear regression analysis is presented in Table 1 which shows that a reasonably close log-linear relationship exists. This relationship implies that the profiling attack has good scalability properties, even as N grows to the size of the Internet.

5.3 Forensics Feasibility

This method of identifying encrypted traffic does not require gathering profile data prior to observing the traffic being analyzed. Such a situation occurs when the profiler is an investigator attempting to identify traces gathered in the past. Figure 6 shows the results of the experiments from Figure 5 with one important difference: the training set occurs after, rather than before, the test set. For otherwise identical parameters, the relative decrease in accuracy ($\frac{original - reversed}{original}$) is less than 3% ($p < 0.01$) across all of the experiments.

method	A	B	R-squared	squared error	F(1,18)	prob(F)
Jaccard	-0.03420	1.0679	0.9374	0.03398	269.6	0.0000
Bayes	-0.03901	1.0678	0.9358	0.03055	262.3	0.0000

Table 1: Regression analysis for $acc = A \log_2 N + B$

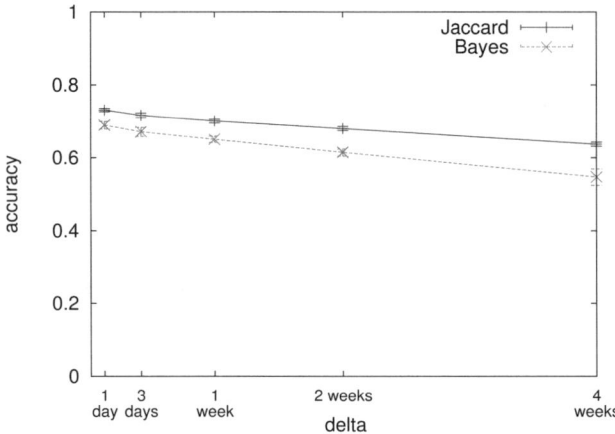

Figure 6: Results after swapping the test and training sets.

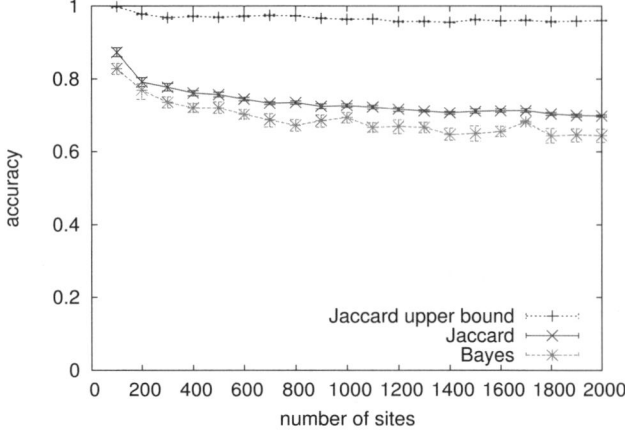

Figure 7: Effect upon accuracy of varying the number of total sites considered. Also shown in the theoretical maximum accuracy that the Jaccard-based classifier can achieve.

5.4 Explaining Performance

As the Jaccard-based classifier only sees occurrences of *(length, direction)* and not counts, some traces will be indistinguishable from others. If we assume no changes in the traces between the training and test sets, the accuracy of the Jaccard-based classifier is bounded by the number of unique traces within each sample. Since sites changes over time, and thus training and test sets can differ, this upper bound is well above the accuracy we observed in practice. Figure 7 shows the fraction of unique samples in the training set and the accuracy of the Jaccard-based classifier on these sets.

To model the source of this uniqueness, we examined the underlying distribution of (packet size, direction) tuples across all instances in the training sets for the 10 experiments corresponding to 2,000 sites. This distribution is shown in Figures 8 and 9. The distribution of per-trace occurrences has an entropy, as defined by:

$$H(x) = -\sum_{i=1}^{n} p(i) \log_2 p(i) \quad (4)$$

of 7.53 bits. We observed that traces have an average of 36.87 unique tuples, drawn without replacement, from such a distribution. If we assume that the appearance of each tuple in each log is independent, then the expected information yielded to the Jaccard-based classifier by a trace is bounded by $7.53 \cdot 36.87 - \log_2 36! \approx 137$ bits. We estimated, via a Monte Carlo procedure, the actual number of bits to be slightly lower (≈ 130), as each tuple can appear only once in a trace. This is far more information than is necessary to distinguish among the number of sites observed, yet as described above, not all sites are unique. We ascribe this discrepancy to a faulty assumption of independence between tuples. Future work could develop a better model to describe identifiability, but as we show below, privacy-conscious system designers will be likely to utilize padding in their designs and render this analysis unnecessary.

5.5 Countermeasure Effectiveness

If an initiator suspects the presence of an observer, he may attempt to obscure his traffic patterns through the use of padding. Here, we examine the effects of static per-packet padding (that is, the padding of each packet with dummy bytes to some predetermined size) on the identification method we propose. We examined four methods of padding in simulation:

- **linear:** Pad each packet to the nearest multiple of 50. This naive method adds a minimal number of bytes to each packet, and reduces the number of distinct sizes by up to a factor of 50.

- **exponential:** Pad each packet to the next largest power of two or the MTU, whichever is smaller. This method significantly reduces the number of distinct sites, to $\lceil \log_2 \text{MTU} \rceil$, while doubling the size of each packet, in the worst case. In practice, we found that this increased packet length by less than 9%, likely because most large packets are already of length equal to the MTU.

- **mice and elephants:** Pad each packet to either 100 or 1500. This method reduces the information available to the classifier even further than exponential padding. All small packets (mostly ACKs) are padded to one size, and all other packets to another. Use of this method comes at a cost of nearly 50% growth in data transmitted.

- **MTU:** Pad each packet to the MTU. This method dramatically increases overall data transmitted (by nearly

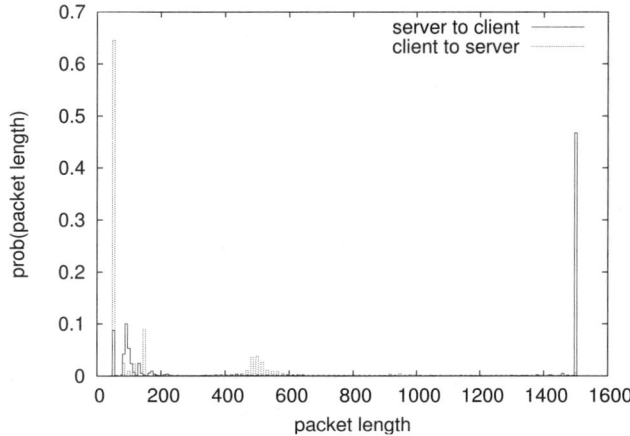

Figure 8: Overall packet length distribution. This distribution is based on every observed packet.

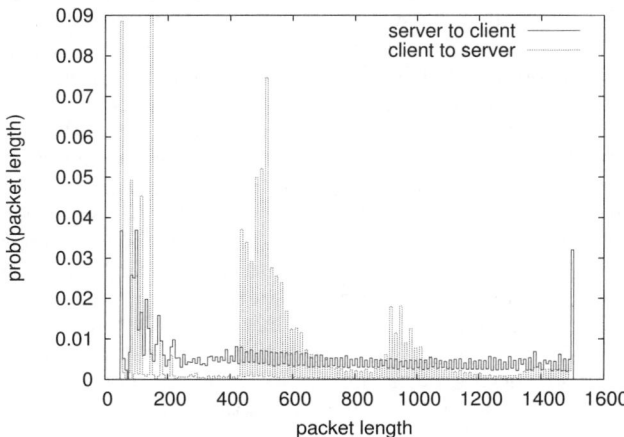

Figure 9: Per-trace occurrence distribution. Packet lengths are included in this distribution at most once per occurrence in each trace. This distribution more accurately reflects the Jaccard-based classifier's input than that of Figure 8.

150%) but renders all packets indistinguishable on the basis of packet length.

We assume the observer is able to determine the padding method being utilized, and can adjust his training sets by padding them in the same fashion. Thus, we evaluate accuracy on the basis of training and test sets that have been padded in identical fashions.

In Figure 10, we show each method's effect upon accuracy. (Recall that k-accuracy is defined in Section 5.1.) Table 2 lists each method's effect upon accuracy as well as the relative number of bytes transmitted. The Bayes-based classifier utilizes packet counts as well as packet size, and thus is better able to discriminate among instances with identical *(direction, length)* attributes but differing counts. The Bayes-based classifier retains enough accuracy, even under the MTU padding method, to be of concern to a privacy-conscious user. Based upon this result, we believe that designers of anonymous communications systems must utilize strong per-packet padding to preserve user privacy.

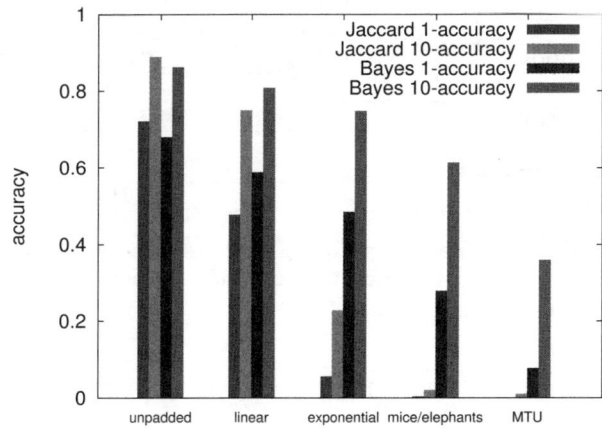

Figure 10: The effects of per-packet padding upon accuracy.

6. DISCUSSION

The main implication of this study is clear: encryption is not enough to protect user privacy. In most low-latency anonymity system designs, an encrypted connection to a proxy (or proxy network) is the basis for the privacy properties of the system. As our study shows, when an observer can utilize external knowledge, such as a library of trace profiles and knowledge of probable user behavior, the content of the data on the connection can be inferred.

Thus we offer the following advice to designers of low-latency systems: Pad packets entering the system to one of a small number of sizes. A system with this property will greatly reduce the amount of available information to an observer, and correspondingly reduce their ability to identify encrypted streams. For example, the current version of the Tor utilizes fixed size packets within its network of proxies. We conjecture this approach is insufficient to counter a traffic analysis attack. A preliminary examination of traces sent through the Tor system leads us to believe that we can discern underlying packet sizes at a finer level of granularity than this fixed size. We believe this discernment can be achieved by grouping packets based an interarrival time threshold, as Tor does not introduce deliberate delays to hide this information.

Building a library of profiles of encrypted HTTP traffic is a reasonable activity for both research and law enforcement purposes. Many forensic investigators will lack the time and tools to gather such data. We have shown the collecting such data is straightforward. Further, we have shown the such data can be used to build an identification system based solely upon packet size — such profiles can be collected from anywhere on the Internet and will not differ for most sites. It is certainly feasible to build software that will enable researchers or investigators to collaborate and create a profile of the entire Internet in a distributed fashion. This library would be similar to a distributed version of the National Software Reference Library[9] which maintains forensic hashes of many commercial applications and operating systems. Given the log-linear scaling of our clas-

---

[9]http://www.nsrl.nist.gov/

Padding	Jaccard 1-accuracy	Jaccard 10-accuracy	Bayes 1-accuracy	Bayes 10-accuracy	Data transmitted
none	0.721	0.889	0.680	0.862	1.000
linear	0.477	0.750	0.588	0.808	1.034
exponential	0.056	0.228	0.485	0.748	1.089
mice / elephants	0.003	0.020	0.279	0.614	1.478
MTU	0.001	0.010	0.077	0.359	2.453

Table 2: Per-packet padding and its effects upon accuracy and amount of data transmitted. The rightmost column shows the amount of data transmitted when using the specified padding method, relative to no padding.

sification method, we expect that such a library would have signification forensic value.

On average, our unoptimized measurement infrastructure retrieved one site every six seconds. Thus, we can collect traces of at least 600 sites an hour from just one computer, or over 100,000 sites in a week. Having more computers doing collection will increase the rate linearly, subject to bandwidth constraints.

The Netcraft survey[10] shows that are approximately 38 million active web sites on the Internet as of February 2006. If 400 volunteers profiled 600 sites an hour, then the entire Internet could be updated in a distributed library once a week. A greater number of volunteers would reduce the amount of work each has to do, or increase the accuracy of the profiles by updating the profiles more frequently. The mean size of an naive Bayes profile in our existing experiment is about 350 bytes; an archive of Internet top-level pages can thus be stored with less than 13GB. Of course, each Internet site consists of many pages. We conjecture that this problem can be addressed because many pages are based on common templates. For example, all Google web searches result in the same layout. The web site of the NY Times looks the same for our observer even though the content changes quite frequently. In general, we expect that sites with many, dynamically generate pages will follow templates out of a necessity for manageable administration; sites that with static content that is different on every page are likely to be updated only infrequently, and therefore require infrequent profiling.

7. CONCLUSION AND FUTURE WORK

We have shown that an observer can infer the contents of encrypted HTTP streams using a library of profiles collected before or after the encrypted stream. These profiles do not require much training data to construct, degrade slowly over time, and are compact. These properties make them useful to the forensic investigator and worrisome to privacy-conscious users. We believe that this study provides some guidance for protocol designers to prevent this attack in future systems, as well as a method for forensic investigators to gather evidence based upon current systems.

In the future, we plan to extend this work in several ways. First, we will examine the effect of introducing timing information into the profile. While this addition has the potential to significantly improve accuracy, it comes at a potentially high cost: the loss of location-neutrality in gathering the profiles. Second, we will examine the efficacy of more complex classification methods. Both of the methods we utilized

[10] http://www.netcraft.com

in this study assume independence among the attributes in the traces. We have given evidence that this independence assumption is flawed, and we expect that classifiers which model inter-attribute dependence will have improved accuracy. Third, we will evaluate this identification method on larger data sets and with a more realistic network substrate. While we expect performance to remain consistent, it is possible that real-world systems may have unanticipated effects upon our method's effectiveness. Fourth, we will examine other padding schemes. We suspect that non-deterministic padding of packets will have a lower packet size overhead than an equivalently strong deterministic scheme. Finally, we will examine more carefully the effects of padding, fragmentation, and packet delays upon our classifier. We suspect these techniques will interact with TCP/IP implementations is non-obvious ways. The designers of low-latency anonymity systems are reluctant to unnecessarily manipulate the traffic stream, for fear of introducing excess latency into their systems. We would like to verify that reasonable security properties can still be attained in the face of more advanced attacks, and we will model the performance penalty that must be incurred to mitigate such attacks.

8. REFERENCES

[1] George Dean Bissias, Marc Liberatore, and Brian Neil Levine. Privacy vulnerabilities in encrypted HTTP streams. In *Proceedings of the Privacy Enhancing Technologies workshop (PET 2005)*, May 2005.

[2] Eoghan Casey. *Digital Evidence and Computer Crime: Forensic Science, Computers and the Internet.* Elsevier, 2nd edition, 2004.

[3] Eoghan Casey. Network traffic as a source of evidence: tool strengths, weaknesses, and future needs. *Journal of Digital Investigation*, 1(1):28–43, 2004.

[4] T. Dierks and C. Allen. RFC 2246: The TLS protocol version 1, January 1999.

[5] Roger Dingledine, Nick Mathewson, and Paul Syverson. Challenges in deploying low-latency anonymity. http://tor.eff.org/cvs/tor/doc/design-paper/challenges.pdf.

[6] Roger Dingledine, Nick Mathewson, and Paul Syverson. Tor: The second-generation onion router. In *Proceedings of the 13th USENIX Security Symposium*, August 2004.

[7] Alan O. Freier, Philip Karlton, and Paul C. Kocher. *Secure Socket Layer*. IETF Draft, November 1996. http://home.netscape.com/eng/ssl3.

[8] Xinwen Fu, Bryan Graham, Riccardo Bettati, and Wei Zhao. Analytical and empirical analysis of

countermeasures to traffic analysis attacks. In *Proceedings of the 2003 International Conference on Parallel Processing*, pages 483–492, 2003.

[9] Andrew Hintz. Fingerprinting websites using traffic analysis. In *Proceedings of the Privacy Enhancing Technologies workshop (PET 2002)*. Springer-Verlag, LNCS 2482, April 2002.

[10] Brian N. Levine, Michael K. Reiter, Chenxi Wang, and Matthew K. Wright. Timing attacks in low-latency mix-based systems. In *Proceedings of Financial Cryptography (FC '04)*. Springer-Verlag, LNCS 3110, February 2004.

[11] Jean-François Raymond. Traffic Analysis: Protocols, Attacks, Design Issues, and Open Problems. In H. Federrath, editor, *Proceedings of Designing Privacy Enhancing Technologies: Workshop on Design Issues in Anonymity and Unobservability*, pages 10–29. Springer-Verlag, LNCS 2009, July 2000.

[12] Qixiang Sun, Daniel R. Simon, Yi-Min Wang, Wilf Russell, Venkata N. Padmanabhan, and Lili Qiu. Statistical identification of encrypted web browsing traffic. In *Proceedings of the 2002 IEEE Symposium on Security and Privacy*, Berkeley, California, May 2002.

[13] Paul Syverson, Gene Tsudik, Michael Reed, and Carl Landwehr. Towards an analysis of onion routing security. In *Proceedings of Designing Privacy Enhancing Technologies: Workshop on Design Issues in Anonymity and Unobservability*, pages 96–114. Springer-Verlag, LNCS 2009, July 2000.

[14] Ian H. Witten and Eibe Frank. *Data Mining: Practical machine learning tools and techniques*. Morgan Kaufmann, San Francisco, 2nd edition, 2005.

[15] Matthew Wright, Micah Adler, Brian Neil Levine, and Clay Shields. An analysis of the degradation of anonymous protocols. In *Proceedings of the Network and Distributed Security Symposium - NDSS '02*. IEEE, February 2002.

[16] Matthew Wright, Micah Adler, Brian Neil Levine, and Clay Shields. Defending anonymous communication against passive logging attacks. In *Proceedings of the 2003 IEEE Symposium on Security and Privacy*, May 2003.

TinySeRSync: Secure and Resilient Time Synchronization in Wireless Sensor Networks

Kun Sun, Peng Ning
Computer Science Dept.
NC State University
Raleigh, NC 27695
{ksun3,pning}@ncsu.edu

Cliff Wang
Army Research Office
RTP, NC 27709
cliff.wang@us.army.mil

An Liu, Yuzheng Zhou
Computer Science Dept.
NC State University
Raleigh, NC 27695
{aliu3,yzhou3}@ncsu.edu

ABSTRACT

Accurate and synchronized time is crucial in many sensor network applications due to the need for consistent distributed sensing and coordination. In hostile environments where an adversary may attack the networks and/or the applications through external or compromised nodes, time synchronization becomes an attractive target due to its importance. This paper describes the design, implementation, and evaluation of TinySeRSync, a secure and resilient time synchronization subsystem for wireless sensor networks running TinyOS. This paper makes three contributions: First, it develops a *secure single-hop pairwise time synchronization* technique using *hardware-assisted, authenticated medium access control (MAC) layer timestamping*. Unlike the previous attempts, this technique can handle high data rate such as those produced by MICAz motes (in contrast to those by MICA2 motes). Second, this paper develops a *secure and resilient global time synchronization* protocol based on a novel use of the μTESLA broadcast authentication protocol for *local authenticated broadcast*, resolving the conflict between the goal of achieving time synchronization with μTESLA-based broadcast authentication and the fact that μTESLA requires loose time synchronization. The resulting protocol is secure against external attacks and resilient against compromised nodes. The third contribution consists of an implementation of the proposed techniques on MICAz motes running TinyOS and a thorough evaluation through field experiments in a network of 60 MICAz motes.

Categories and Subject Descriptors

C.2.0 [**Computer-Communication Networks**]: General—*Security and protection*; C.2.1 [**Computer-Communication Networks**]: Network Architecture and Design—*Wireless communication*

---
*This work is supported by the National Science Foundation (NSF) under grant CAREER-0447761 and by the US Army Research Office (ARO) under grant W911NF-04-D-0003-0001.

Copyright 2006 Association for Computing Machinery. ACM acknowledges that this contribution was authored or co-authored by an employee, contractor or affiliate of the U.S. Government. As such, the Government retains a nonexclusive, royalty-free right to publish or reproduce this article, or to allow others to do so, for Government purposes only.
CCS'06, October 30–November 3, 2006, Alexandria, Virginia, USA.
Copyright 2006 ACM 1-59593-518-5/06/0010 ...$5.00.

General Terms

Security, Design, Algorithms

Keywords

Sensor Networks, Security, Time Synchronization

1. INTRODUCTION

Recent technological advances have made it possible to develop distributed sensor networks consisting of a large number of low-cost, low-power, and multi-functional sensor nodes that communicate over short distances through wireless links [6]. Such sensor networks are ideal candidates for a wide range of applications such as monitoring of critical infrastructures, data acquisition in hazardous environments, and military operations. The desirable features of distributed sensor networks have attracted many researchers to develop protocols and algorithms that can fulfill the requirements of these applications (e.g., [6,14,16,21,30,31,36]).

Accurate and synchronized time is crucial in many sensor network applications, particularly due to the need for consistent distributed sensing and coordination. However, due to the resource constraints on typical sensor nodes such as MICA motes [8], traditional time synchronization protocols (e.g., NTP [27]) cannot be directly applied in sensor networks.

A number of time synchronization protocols (e.g., [11,13, 17, 23, 26, 28, 33, 37]) have been proposed for sensor networks to achieve *pairwise* and/or *global* time synchronization. Pairwise time synchronization aims to establish relative clock offsets between pairs of sensor nodes, while global time synchronization aims to provide a network-wide time reference for all the sensor nodes in a network. Existing pairwise or global time synchronization techniques are all based on *single-hop* pairwise time synchronization, which discovers the clock difference between two neighbor nodes that can communicate with each other directly. Two approaches have been used for single-hop pairwise clock synchronization: *receiver-receiver synchronization* (e.g., RBS [11]), in which a reference node broadcasts a reference packet to help a pair of receivers identify their clock differences, or *sender-receiver synchronization* (e.g., TPSN [13]), where a sender communicates with a receiver to estimate their clock difference. Multi-hop pairwise clock synchronization protocols and most of the global clock synchronization protocols (e.g., [11, 13, 37]) establish multi-hop paths in a sensor network, so that all the nodes in the network can synchronize their clocks to the source based on the single-hop pairwise clock differences between adjacent nodes in these paths. Alternatively, diffusion based global synchronization protocols [23] have the nodes' clocks converge by spreading synchronization information locally.

1.1 Threats to Time Synchronization in Wireless Sensor Networks

Most of existing time synchronization techniques developed for sensor networks assume benign environments. However, in hostile environments, an adversary may certainly attack the time synchronization protocol due to its importance. Note that all time synchronization protocols rely on *time-sensitive* message exchanges. To mislead these protocols, the adversary may forge and modify time synchronization messages, jam the communication channel to launch Denial of Service (DoS) attacks, and launch pulse-delay attacks [12] by first jamming the receipt of time synchronization messages and then later replaying buffered copies of these messages. The adversary may also launch wormhole attacks [18] by creating low latency and high bandwidth communication channels between different locations in the network, and (selectively) delay or drop time synchronization messages transmitted through the wormholes. The adversary may use Sybil attacks [10, 29], where one node presents multiple identities, to defeat typical fault tolerant mechanisms. Though message authentication can be used to validate message sources and contents, it cannot validate the *timeliness* of messages, and thus is unable to defend against all of these attacks.

Moreover, the adversary may compromise some nodes, and exploit the compromised nodes in arbitrary ways to attack time synchronization. For example, the adversary may instruct the compromised nodes to (selectively) delay or drop time synchronization messages, and launch Sybil attacks [29] using the identities and keying materials of compromised nodes if message authentication is enabled. The adversary may also instruct the compromised nodes not to cooperate with others, and inject false time synchronization messages. The compromised nodes may collude with each other to cause the worst damage to the network.

1.2 Inadequacy of Current Solutions

It is natural to consider fault-tolerant time synchronization techniques, which have been studied extensively in the context of distributed systems (e.g., [7, 9, 22, 32, 39]). However, these techniques require either digital signatures (e.g., HSSD [9], CSM [22]), exponential copies of messages (e.g., COM [22]), or a completely connected network (e.g., CNV [22]) to prevent malicious nodes from modifying or destroying clock information sent by normal nodes. Thus, they are impractical for resource-constrained sensor nodes. A recent work provides an efficient fault-tolerant time synchronization protocol for a cluster of fully connected nodes by exploiting the broadcast nature of wireless communication [40]. However, it requires a trusted entity during network initialization to avoid heavy communication overhead, and does not allow incremental deployment of additional sensor nodes.

There have been several recent studies for secure and resilient time synchronization in sensor networks [12, 25, 38, 41]. Ganeriwal et al. proposed several techniques for secure pairwise synchronization (SPS), secure multi-hop synchronization, and group-wise synchronization [12]. The SPS technique provides authentication for medium access control (MAC) layer timestamping by adding timestamp and message integrity code (MIC) as the messages being transmitted. This approach works for low data rate sensor radios (e.g., CC1000 on MICA2 motes with 38.4Kbps data rate); however, it cannot keep up with recent IEEE 802.15.4 [20] compliant sensor radios such as CC2420 on MICAz and TelosB, whose data rate is 250Kbps. The group synchronization in [12] uses pairwise authentication to synchronize a group of nodes, thus introducing high computation and communication overheads. Moreover, the group synchronization assumes all nodes in a group can communicate with each other directly. Extensions to groups with multi-hops are speculated, but no specific solution is provided.

Manzo et al. discussed a few attacks against existing time synchronization protocols and several countermeasures to protect single-hop and multi-hop time synchronization [25]. However, there was no mechanism to authenticate the timeliness of synchronization messages, and thus no protection against, for example, pulse-delay attacks [12] and worm-hole attacks [19], in which the adversary may delay authenticated synchronization messages. Moreover, though μTESLA was suggested as a way to authenticate broadcast synchronization messages, no solution was given to resolve the conflict between the goal of achieving time synchronization and the fact that μTESLA requires loose time synchronization.

Sun et al. proposed a resilient time synchronization protocol that can deal with various attacks including compromised nodes [41]. However, similar to [25], the proposed techniques cannot authenticate the timeliness of synchronization messages, thus suffering from pulse-delay [12] and wormhole attacks [19]. In addition, the approaches in [41] use authenticated unicast communication to propagate global synchronization messages. This introduces substantial communication overhead as well as frequent message collisions in dense sensor networks [41].

Song et al. investigated countermeasures against attacks that mislead sensor network time synchronization by delaying synchronization messages [38]. They proposed two methods for detecting and tolerating delay attacks: one transforms attack detection into statistical outliers detection, and the other detects attacks by deriving the bound of the time difference between two nodes through message exchanges. Unfortunately, [38] only addresses synchronization of neighbor nodes, but does not support global time synchronization in multi-hop sensor networks.

1.3 Our Contributions

To address the aforementioned problems, we develop a secure and resilient time synchronization subsystem called *TinySeRSync* for wireless sensor networks, targeting common sensor platforms such as MICAz and TelosB running TinyOS [16]. Our solution offers a novel way to integrate (broadcast) authentication into time synchronization, which successfully provides authentication of the source, the content, and the timeliness of synchronization messages. Our solution not only addresses secure time synchronization between neighbor nodes, but also the global synchronization of an entire sensor network.

We make three contributions in this paper:

1. We develop a *secure single-hop pairwise time synchronization* technique using *hardware-assisted, authenticated medium access control (MAC) layer timestamping*. Unlike the previous attempts, this technique can

handle high data rate such as those produced by MICAz and TelosB motes (in contrast to those by MICA2 and MICA2DOT motes).

2. We develop a *secure and resilient global time synchronization* protocol based on a novel use of the μTESLA broadcast authentication protocol for *local authenticated broadcast*, resolving the conflict between the goal of achieving time synchronization with μTESLA-based broadcast authentication and the fact that μTESLA requires loose time synchronization. The resulting protocol is secure against external attacks and resilient against compromised nodes.

3. We provide an implementation of the proposed techniques on TinyOS and a thorough evaluation through field experiments in a network of 60 MICAz motes. The evaluation results indicate that TinySeRSync is a practical system for secure and resilient time synchronization in wireless sensor networks.

1.4 Organization of the Paper

The rest of the paper is organized as follows. The next section gives an overview of the TinySeRSync system. Section 3 and Section 4 describe the secure pairwise time synchronization and the secure and resilient global time synchronization in TinySeRSync, respectively. Section 5 provides the security and performance analysis of TinySeRSync. Section 6 discusses a few implementation issues. Section 7 presents the experimental evaluation of TinySeRSync in a network of 60 MICAz motes. Section 8 concludes this paper and points out future research directions.

2. OVERVIEW OF PROPOSED APPROACH

We assume that a sensor network consists of a large number of resource constrained motes such as MICA series of motes. We assume there is a *source node S* that is well synchronized to the external clock, for example, through a GPS receiver. We would like to synchronize the clocks of all the sensor nodes in the network to that of the source node. We assume the source node is trusted, and all the other nodes know the identity of the source node. The assumption of the single, trusted source node is to simplify the discussion in this paper. Our approach can be easily modified to accommodate multiple source nodes in order to enhance the performance, improve the availability of source nodes, and/or tolerate potentially compromised source nodes.

To deal with the ad hoc deployments of sensor networks and the lack of initial synchronization among sensor nodes, we propose to achieve global time synchronization in a sensor network in two *asynchronous* phases: *Phase I–secure single-hop pairwise synchronization*, and *Phase II–secure and resilient global synchronization*. In Phase I, pairs of neighbor nodes exchange messages with each other to obtain single-hop pairwise time synchronization. Phase I uses authenticated MAC layer timestamping and a two message exchange to ensure the authentication of the source, the content, and the timeliness of synchronization messages. Nodes run Phase I periodically to compensate (continuous) clock drifts and maintain certain pairwise synchronization precision, providing the foundation for global time synchronization as well as the μTESLA-based local broadcast authentication in Phase II.

Phase II uses *authenticated local (re)broadcast* to achieve global time synchronization, starting with a broadcast synchronization message from the source node. Phase II adapts μTESLA to ensure the timeliness and the authenticity of the local broadcast synchronization messages. To be resilient against potential compromised nodes, each node estimates multiple candidates of the global clock using synchronization messages received from multiple neighbor nodes, and chooses the median. Nodes that are synchronized to the source node further rebroadcast the synchronization messages locally. This process continues until all the nodes are synchronized. Phase II also runs periodically to maintain certain global time synchronization precision.

We would like to emphasize that the two phases are *asynchronous*. In other words, secure single-hop pairwise synchronization (Phase I) is executed by nodes individually and independently, while secure and resilient global synchronization (Phase II) is controlled by the source node and propagated throughout the network. The only requirement is that a node finishes Phase I before entering Phase II. Also note that both Phase I and Phase II are executed periodically. Though a node that has not performed Phase I synchronization with its neighbor nodes cannot participate in a global synchronization, it may join the next round of global synchronization once it finishes Phase I. Thus, our approach supports incremental deployment of sensor nodes, which is an important property required by many sensor network applications.

We present the two phases of TinySeRSync in detail in the next two sections.

3. PHASE I: SECURE SINGLE-HOP PAIRWISE TIME SYNCHRONIZATION

The goal of secure single-hop pairwise time synchronization is to ensure two neighbor nodes can obtain their clock difference through message exchanges in a secure fashion. This requires the authentication of the source, the content (i.e., the timing information), and the timeliness of each message used for such synchronization.

In the following, we first discuss how we provide authentication of the source and the timing information in synchronization messages, and then describe a secure two-way pairwise time synchronization protocol for a node to obtain the clock difference from a neighbor node.

3.1 Authenticated MAC Layer Timestamping

MAC layer timestamping has been widely accepted as an effective way to reduce the synchronization error during the message exchanges since it was proposed in [13]. By adding (on the sender's side) and retrieving (on the receiver's side) timestamps in the MAC layer, this approach avoids the uncertain delays introduced by application programs and medium access, and thus has more accurate synchronization precision.

To ensure the integrity of pairwise time synchronization, we may authenticate a synchronization message by adding a MIC once the MAC layer timestamp is added, assuming the two nodes performing pairwise synchronization share a secret pairwise key through, for example, TinyKeyMan [24]. This, however, introduces a potential problem due to the extra delay required by the MIC generation: It is necessary to have a MAC layer timestamp that marks the exact transmission time of a certain bit in the message at the sender's side, but the MIC generation and insertion require extra delay and have to be done after the timestamp is inserted into the message.

The delay introduced by MIC generation and insertion can be tolerated for sensor platforms with low data rate radio components, such as MICA2 motes. In an earlier study [12], Ganeriwal et al. attempted to provide authenticated MAC layer timestamping for MICA2 motes (38.4 kbps data rate) by generating MIC on the fly. Specifically, when the radio component of a sensor node begins to transmit the first byte of a synchronization message, it appends the current timestamp into the message, calculates the MIC, and appends the MIC into the message being transmitted. Due to the low data rate (38.4 kbps), the MAC layer timestamp and the MIC can be added into the packet before the corresponding bytes are transmitted [12]. However, with the increased data rate on recent sensor platforms with IEEE 802.15.4 compliant radio components (250 kbps data rate [20]), such as MICAz and TelosB motes, there is not enough time to generate and insert the MIC before the transmission of the MIC bytes due to the delay introduced by the MIC calculation [12].

We propose a prediction-based approach to address the above problem. In the following, we describe our approach, with a specific target of the IEEE 802.15.4 compliant radio component ChipCon CC2420 [3], which is commonly used in recent sensor platforms such as MICAz and TelosB motes. We also assume the sensor nodes use TinyOS [16], the open source operating system for networked sensor nodes.

3.1.1 Prediction-Based MAC Layer Timestamping and Hardware-Assisted Authentication

We observe that the code for generating a MIC is deterministic, and the time required for a MIC generation for messages with a given length (or, more precisely, a given number of blocks) is fixed. In addition, the process to transmit a packet (starting from observing the channel vacancy to the actual transmission of data payload) in CC2420 is also deterministic. Thus, when we put a timestamp into a synchronization message to be authenticated in the MAC layer, we may predict the time required by MIC generation and at the same time predict the delay between the start of transmission and the transmission a given bit in the packet.

Let us review how a sensor node (such as a MICAz mote) equipped with a CC2420 radio component handles packet transmission on TinyOS. Figure 1 shows the transmission and receiving process. When a node has a message to send, its micro-controller first transmits the message to the RAM (TXFIFO buffer) of the CC2420 radio component. After the buffering is done, CC2420 sends a signal to the micro-controller. At this time, if the radio channel is clear, the micro-controller signals CC2420 to send out the packet with a STXON strobe. Otherwise, it will back off randomly and then test the channel again. After receiving a STXON signal, CC2420 first sends 12 symbol periods, with 4 bits in each symbol, and then sends 4 byte preamble and 1 byte of Start of Frame Delimiter (SFD) field, followed by 1 byte length field and the MAC Protocol Data Unit (MPDU). The sequence of events follows strict timing, and the delays introduced by all of them are predictable.

We use the last bit of the SFD byte as the reference point for time synchronization. In other words, the sender takes the transmission (completion) time of the last bit of SFD as the MAC layer transmission timestamp, and the receiver marks the receiving time of the same bit as the receiving timestamp. To allow the sender to perform MAC layer timestamping and authentication, as mentioned earlier, we can predict the time when the last bit of SFD will be transmitted.

Sender Side: Now let us describe our proposed sending process in detail. Assume the sender has started sending a synchronization message to the RAM (TXFIFO buffer) of CC2420. At this time, the timestamp field in the message has not been filled. Upon completion of the transfer, CC2420 sends a signal to the micro-controller, which then starts handling the signal in the MAC layer. If the radio channel is clear, the micro-controller generates a timestamp by adding the current time with a constant offset Δ. This constant offset Δ is the time delay from checking the current time to the transmission of the last bit of SFD. The micro-controller then writes the timestamp directly to the corresponding bytes in CC2420's TXFIFO. Next, if the radio channel is still clear, it signals CC2420 to send out the message with a STXON strobe. Otherwise, it backs off for a random period of time and then repeats the above process. (Note that this back-off will force the micro-controller to write the MAC layer timestamp again when the same message is to be re-transmitted.) Upon receiving the STXON signal, as described earlier, CC2420 starts transmitting the symbol periods, the preamble, the SFD, and the MPDU. In the case when CC2420 can successfully transmit the packet, the execution and the data transmission are both deterministic, and the delay Δ is a constant. The delay Δ we obtained on MICAz motes is 399.28 μs. This includes the total transmission time for the 12 symbol periods, preamble and SFD ($(12 * 4 + (4 + 1) * 8)/250,000 = 0.000352$ s $= 352$ μs) and the execution time between checking the timestamp and starting the transmission (47.28 μs).

In our implementation, we have CC2420 start the inline authentication to generate the MIC of the message at the time when it begins to transmit the symbol periods. According to the manual of CC2420 [3], the inline authentication component in CC2420 can generate a 12-byte MIC on a 98-byte message in 99 μs. Thus, we can easily see the MIC generation can be completed before it is transmitted. Besides the MIC, CC2420 also generates a 2-byte Frame Check Sequence (FCS) using Cyclic Redundancy Check (CRC).

Receiver Side: After an approximately 2 μs propagation delay [3], the radio component CC2420 on the receiver node will receive the preamble of an incoming message. Once the SFD field is completely received by CC2420, the SFD pin will go high to signal the micro-controller, which then records the current time as the receiving timestamp. When the FIFOP pin goes low, the micro-controller will be signaled to read the data from CC2420's RXFIFO buffer, in which the first byte indicates the length of the message. During the receiving process, CC2420 performs inline verification of the MIC (and the CRC) in the message, using the pairwise key shared between the sender and the receiver. The micro-controller examines the verification result, and copies the whole packet if the packet is authenticated. All these operations are performed in the MAC layer, and transparent to the application layer.

Unlike the deterministic delay on the sender's side, the delay affecting the receiving timestamps on the receiver's side

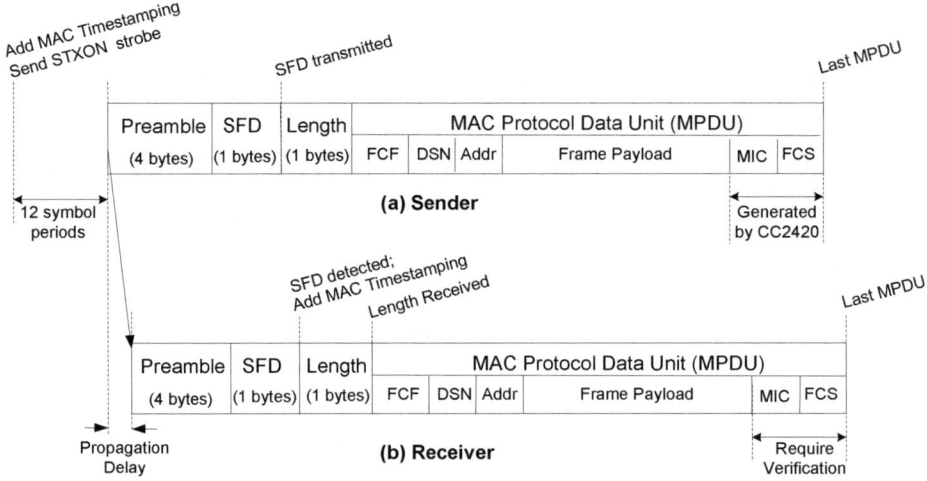

Figure 1: Packet Sending and Receiving Process

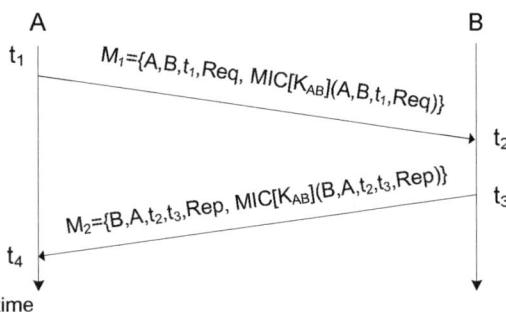

Figure 2: (Revised) Secure Pairwise Synchronization

is not entirely deterministic. When interrupt is disabled, the micro-controller will not be able to get the SFD signal immediately, and the resulting delay will be uncertain.

3.2 Secure Pairwise Synchronization

Given the authenticated MAC layer timestamping capability, we can now describe how two neighbor nodes can perform secure pairwise time synchronization.

Let us take a look at our options. RBS uses a receiver-receiver approach to synchronize nodes [11], in which a reference node broadcasts a reference packet to help pairs of receivers to identify their clock difference. However, an adversary can compromise it by simply launching a pulse-delay attack [12] or wormhole attack against one of the nodes to manipulate the packet transmission delay [19], so that the two nodes receive the reference packet at different times. In some protocols such as FTSP [26], one node passes its own time to the other by directly sending a MAC layer timestamped packet to the latter. This works well in benign environments, as demonstrated in [26]. However, in hostile environments, it suffers from the same problems mentioned above. TPSN uses a sender-receiver approach (through one request and one reply message) to help the sender obtain its clock difference from the receiver [13]. This approach was later improved with security in Secure Pairwise Synchronization (SPS) [12] to deal with pulse-delay and wormhole attacks. Specifically, it authenticates the messages be-

ing exchanged, and uses the timestamp information to estimate both the clock difference and the message transmission delay. Pulse-delay and wormhole attacks that manipulate packet transmission delay will introduce extra delay in message transmission, and will be detected.

We adopt the SPS approach [12] with a slight modification. SPS uses a random nonce to prevent replay of a previously transmitted reply message. In our case, we simply use the sender's timestamp in the reply message to prevent replay attack, so that the message size is further reduced.

Figure 2 shows the revised SPS protocol, in which all messages are timestamped and authenticated with the key K_{AB} shared by nodes A and B, as described in Section 3.1. Node A initiates the synchronization by sending message M_1. The message contains M_1's sending time t_1. Node B receives the message at t_2. After verifying the message, at time t_3, node B sends a message M_2 that includes t_2, t_3 to node A. When node A receives the message at t_4, it can calculate the clock difference $\Delta_{A,B} = \frac{(t_2-t_1)-(t_4-t_3)}{2}$, and the estimated one-way transmission delay $d_A = \frac{t_2-t_1+t_4-t_3}{2}$. Since all messages are authenticated, any modification to any message will be detected. To prevent the pulse-delay attacks [12] and wormhole attacks [19], node A verifies that the one-way transmission delay is less than the maximum expected delay. In fact, this approach can detect any attack that attempts to mislead single-hop pairwise time synchronization by introducing significant extra message delays. As a result, sender A can easily detect attempts to affect the timeliness of the synchronization messages. Thus, our revised SPS achieves the same degree of security as the original one with smaller messages.

Note that the revised SPS protocol only enables the sender (A) to obtain the clock difference with the receiver (B). If the receiver (B) also needs this information, it has to initiate this protocol with the sender (A) as well. An alternative is to perform a three-way message exchange so that both nodes will get the clock difference at the end of the protocol execution. However, in such a three-way protocol, both the sender and the receiver have to maintain their states at the intermediate protocol steps, and each node has to carefully maintain its states to avoid interference when it is involved in multiple concurrent synchronizations with different neigh-

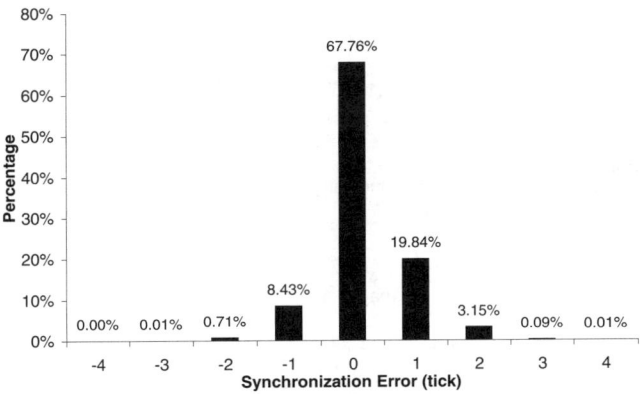

Figure 4: Synchronization Error Distribution in Secure Pairwise Synchronization (1 Tick = 8.68 μs)

bors. The additional space requirement and the increased software complexity do not strictly justify the possibly reduced communication overhead.

Despite the use of MAC layer timestamping, there is still possible uncertainty in pairwise time synchronization, which may affect the precision of time synchronization.

Figure 3 shows the sources of packet delays in the revised SPS protocol. On the target platform (i.e., MICAz motes), the time for the CPU to access current time is deterministic and less than 1 tick. The time for adding timestamp in the buffered packet is deterministic, too, which is less than 2 ticks. The encoding time is for the radio to encode and transform a part of the packet (4 bits a time in case of CC2420) to electromagnetic waves, and the decoding time is for the radio to transform and decode electromagnetic waves into binary data. These times are controlled by a chip sequence at 2 MChips/s, and thus are deterministic. The propagation time depends on the distance between the sender and the receiver, which is also deterministic. The uncertainty in the packet delay is mainly because of the jitter of the interrupt handling, which is caused by temporarily disabled interrupt handling on the receiving side. For example, the receiver may be executing a piece of code that disables interrupt handling when the SFD is received by the RF module, and thus cannot access the receiving timestamp promptly.

We tested 30 pairs of nodes in our lab to obtain the synchronization precision. For each pair of nodes, we ran 500 rounds of pairwise time synchronization. After two nodes finish a pairwise time synchronization, a third reference node broadcasts a query to them. Each of the node records the MAC layer receiving time of the broadcast message and sends the receiving time to the reference node. This allows the reference node to calculate the synchronization error. Figure 4 shows the distribution of the pairwise synchronization error.

4. PHASE II: SECURE AND RESILIENT GLOBAL TIME SYNCHRONIZATION

In this section, we present our method for secure and resilient global time synchronization, assuming all the sensor nodes perform secure single-hop time synchronization periodically.

4.1 Basic Approach

Given the secure pairwise synchronization protocol, the remaining threats to global time synchronization are twofold. First, an external attacker may fake or replay (local) broadcast messages used for global synchronization to mislead the regular nodes. To defend against this threat, we need to integrate authentication of both the *content* and the *timeliness* of broadcast synchronization messages during global time synchronization. Second, a compromised node may provide misleading synchronization information to disrupt the global time synchronization. Thus, our global time synchronization protocol must be resilient to compromised nodes.

We propose a distributed, resilient protocol integrated with (local) broadcast authentication to provide secure and resilient global time synchronization. Intuitively, the source node broadcasts synchronization messages periodically to adjust the clocks of all sensor nodes. The synchronization messages are flooded throughout the network to reach nodes that cannot communicate with the source node directly. Specifically, when receiving a synchronization message for the first time, each node rebroadcasts it (after a random delay to avoid collisions). To reduce the impact of processing delays at intermediate nodes on synchronization precision, our approach focuses on the clock differences without involving the delays directly in the computation. The timely transmission of all these messages are authenticated. Moreover, each node obtains synchronization information from multiple neighbor nodes, so that it can tolerate compromised nodes to a certain extent.

Specifically, each node i maintains a *local clock* C_i. The local clock C_S of the source node S is the desired global clock. For each neighbor node j, each node i can obtain a *single-hop pairwise clock difference* $\Delta_{i,j} = C_j - C_i$ using the secure pairwise time synchronization in Section 3. Each node i maintains a *source clock difference* $\delta_{i,S}$ between its local clock and the clock of the source node S. Node i can directly obtain it if it is a neighbor node of S. Otherwise, node i needs to estimate $\Delta_{i,S}$. When the source node S decides to send a synchronization message, it broadcasts such a message to its direct neighbors. Each of the neighbors can determine its source clock difference directly as mentioned earlier. These neighbor nodes then broadcast their source clock differences to help those that cannot receive the synchronization messages from S directly.

To tolerate up to t compromised neighbor nodes, each node i that receives source clock differences from non-source neighbor nodes computes at least $2t+1$ *candidate* source clock differences through different neighbor nodes. Specifically, for each source clock difference $\Delta_{j,S}$ that node i receives from node j, node i computes a candidate source clock difference $\Delta_{i,S}^j = C_S - C_i = (C_S - C_j) + (C_j - C_i) = \Delta_{j,S} + \Delta_{i,j}$. Given at least $2t+1$ such candidate clock differences, node i picks the median as the estimated source clock difference $C_{i,S}$. It is easy to see that node i can tolerate up to t compromised neighbor nodes that provide misleading synchronization information.

Each node i can estimate the *global clock* by using its local clock and its source clock difference, i.e., the *estimated global clock* $\hat{C}_S^i = C_i + \Delta_{i,S}$. After computing its own source clock difference, each node broadcasts this value to help the nodes that have not been synchronized estimate their source clock differences as well as the global clock.

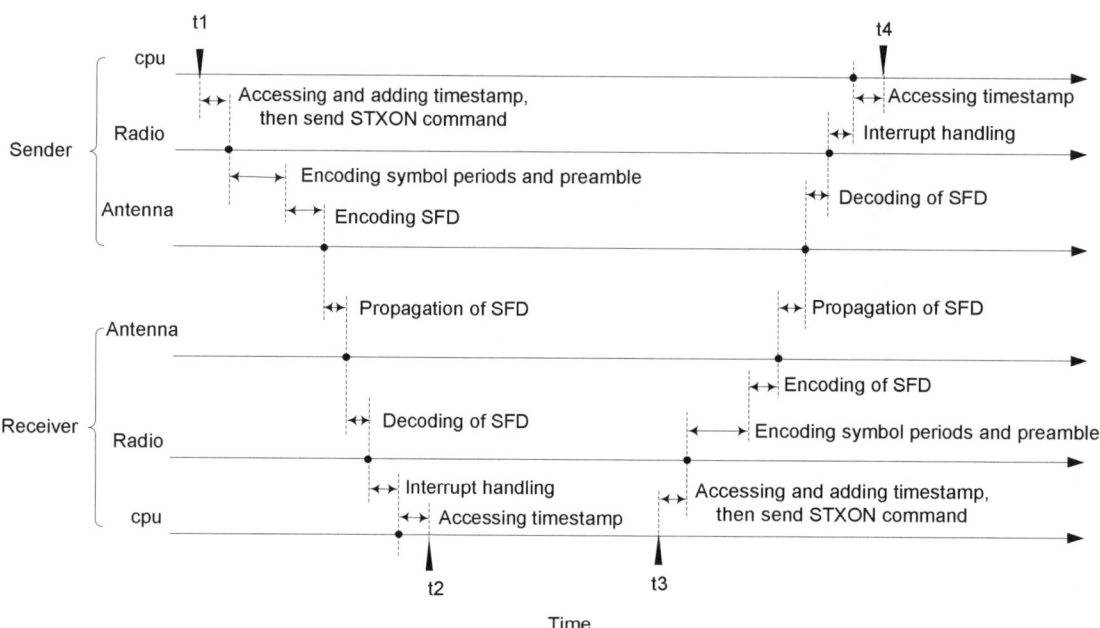

Figure 3: Delay Uncertainty (Certain Details are Omitted for Clarity)

Our approach is similar to the resilient clock estimation approach proposed in [41] in that both approaches estimate the clock difference between each node and the source node through multiple nodes that have already been synchronized. However, our approach has a critical difference from that approach: Our approach uses *authenticated local broadcast* to propagate synchronization messages, while the approach in [41] uses authenticated unicast that leads to substantial communication overhead as well as message collisions as shown in the performance study in [41]. This difference represents a key step that enables practical secure and resilient time synchronization in sensor networks.

The ability to authenticate local broadcast messages is the cornerstone of the proposed protocol. In the following, we describe in detail how this is done in TinySeRSync.

4.2 Authentication of Local Broadcast Synchronization Messages

As discussed earlier, the signaling messages for global time synchronization are broadcast in nature, and must be transmitted in a timely and authenticated way. There are two general solutions for authenticating broadcast messages in sensor networks: digital signatures and μTESLA [34, 36]. Though it is possible to verify digital signatures on sensor platforms, as shown in [15], signature operations are still multiple order of magnitude more expensive than secret key based solutions such as μTESLA. Using digital signatures for time synchronization may quickly exhaust the battery power of sensor nodes. Moreover, it is also an attractive target of Denial of Service (DoS) attacks: An attacker may broadcast synchronization messages with false digital signatures to force sensor nodes to perform expensive signature verifications.

μTESLA [36] relies on symmetric cryptography, and thus does not suffer from the above problems. However, μTESLA requires loose time synchronization between the broadcast sender and the receivers. Considering the goal of having the source node synchronize the clocks of all the sensor nodes, there seems to be a conflict in using μTESLA for authenticating broadcast time synchronization messages.

We can indeed avoid the above conflict. We observe that two neighbor nodes may securely perform single-hop pairwise time synchronization using the techniques in Section 3. Consider an arbitrary node A. Assume node A have synchronized with all its neighbor nodes so that node A and any of its neighbor nodes know the clock difference between them. As a result, if node A needs to broadcast a synchronization message to all its neighbor nodes, it may certainly use μTESLA for broadcast authentication, since the "loose synchronization" requirement needed by μTESLA is already satisfied. In other words, we only use μTESLA *locally* to avoid the above conflict.

Specifically, we adapt μTESLA for local broadcast authentication to protect the broadcast messages from a node to its neighbors, assuming the Phase I neighbor synchronization has completed. In the following, we first give a brief introduction to μTESLA, and then discuss the adaptation of μTESLA in TinySeRSync.

4.2.1 Overview of μTESLA

An asymmetric mechanism such as public key cryptography is generally required for broadcast authentication [34]. Otherwise, a malicious receiver can easily forge any message from the sender, as discussed earlier. μTESLA introduces asymmetry by delaying the disclosure of symmetric keys [36]. A sender broadcasts a message with a MIC generated with a secret key K, which is disclosed after a certain period of time. When a receiver gets this message, if it can ensure that the message was sent before the key was disclosed, the receiver buffers this message and authenticates the message when it later receives the disclosed key. To continuously authenticate broadcast messages, μTESLA divides the time period for broadcast into multiple intervals, assigning different keys to different time intervals. All mes-

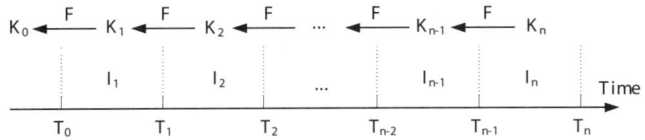

Figure 5: μTESLA

sages broadcast in a particular time interval are authenticated with the key assigned to that time interval.

To authenticate the broadcast messages, a receiver first authenticates the disclosed keys. μTESLA uses a one-way key chain for this purpose. The sender selects a random value K_n as the last key in the key chain and repeatedly performs a (cryptographic) hash function F to compute all the other keys: $K_i = F(K_{i+1}), 0 \leq i \leq n-1$, where the secret key K_i (except for K_0) is assigned to the i-th time interval. Because of the one-way property of the hash function, given K_j in the key chain, anybody can compute all the previous keys $K_i, 0 \leq i \leq j$, but nobody can compute any of the later ones $K_i, j+1 \leq i \leq n$. Thus, with the knowledge of the initial key K_0, which is called the *commitment* of the key chain, a receiver can authenticate any key in the key chain by merely performing hash function operations. When a broadcast message is available in the i-th time interval, the sender generates a MIC for this message with a key derived from K_i, broadcasts this message along with its MIC, and discloses the key K_{i-d} for time interval I_{i-d} in the broadcast message (where d is the disclosure lag of the authentication keys). Figure 5 illustrates the division of the time line and the assignment of authentication keys in μTESLA.

Each key in the key chain will be disclosed after some delay. As a result, the attacker can forge a broadcast message by using the disclosed key. μTESLA uses a security condition to prevent such situations. When a receiver receives an incoming broadcast message in time interval I_i, it checks the security condition $\lfloor (T_c + \Delta - T_1)/T_{int} \rfloor < i + d - 1$, where T_c is the local time when the message is received, T_1 is the start time of the time interval 1, T_{int} is the duration of each time interval, and Δ is the maximum clock difference between the sender and itself. If the security condition is satisfied, i.e., the sender has not disclosed K_i yet, the receiver accepts this message. Otherwise, the receiver simply drops it.

4.2.2 Short Delayed μTESLA: Adapting μTESLA for Global Synchronization

Distribution of μTESLA Parameters: In order to use μTESLA, the sender needs to transmit a number of parameters to all the receivers before the actual broadcast messages. These include the key chain ID, the key chain commitment, the duration of each time interval, and the starting time of the first time interval. We can fix the duration of time intervals and the length of each key chain as network wide parameters. However, the other parameters have to be communicated from each node to its neighbors. To reduce communication cost, we piggy-back the transmission of these μTESLA parameters with the single-hop pairwise synchronization between neighbors. In other words, each node sends the parameters of its own μTESLA key chain to a neighbor node during secure single-hop pairwise synchronization. When one key chain is about to expire, each node needs to communicate with each neighbor node again to transmit the parameters for the next key chain.

Balancing Key Chain Size and Authentication Delay: A direct application of μTESLA to authenticate the local broadcast synchronization messages faces a risk. μTESLA is subject to DoS attacks [35], in which an attacker overhearing a valid broadcast message may use the disclosed key in the message to forge broadcast synchronization messages. A receiver has to buffer all such (forged) messages claimed to be from some neighbor until it receives the disclosed key. As a result, the receiver may not have enough memory to buffer synchronization messages from other neighbor nodes. The immediate authentication mechanism proposed in [35] cannot be applied here, because it requires that the sender know the next message to be transmitted before sending the current message.

One possible way to mitigate the threat of DoS attacks in global synchronization is to exploit the tight time synchronization established during Phase I. Specifically, when using μTESLA for local broadcast authentication, we may use very short time intervals to limit the duration vulnerable to DoS attacks. Because the neighbor nodes have been tightly synchronized with each other during phase I, the broadcast sender can use very short time intervals and disclose an authentication key right after the corresponding interval is over. When the time interval is short enough, it does not give enough time to an attacker to forge broadcast messages using the disclosed key it just learns from the valid broadcast message. A short enough interval duration also offers authentication of the *timeliness* of the synchronization messages; it disallows a replayed message to be transmitted in the valid time interval, and thus enables receivers to detect and remove them.

However, this approach comes with a significant cost: To cover a certain period of time (e.g., 30 minutes), the sender needs to generate a fairly long key chain due to the short time intervals, and most of the keys will be wasted. Reducing the key chain length will force al the neighbor nodes to exchange the key chain commitments frequently, leading to heavy communication overhead.

We propose to adapt the μTESLA broadcast authentication protocol to address the above conflict. Specifically, we propose to use two different durations in one μTESLA instance, a short duration r and a long duration R. The short intervals and the long intervals are interleaved, as shown in Figure 6. As in the original μTESLA, each time interval is still associated with an authentication key, which is used to authenticate messages sent in this time interval. Each node broadcasts a message authenticated with μTESLA only during the short intervals, while broadcasting the disclosed key in the following long interval (possibly multiple times to tolerate message losses).

Upon receiving a broadcast message, a receiver first checks the security condition using the (MAC layer) message receipt time. Because each receiver and the sender have synchronized tightly with each other, the receiver can easily transform the receipt time into the time point in the sender's clock, and verify if the corresponding authentication key has been disclosed when the receiver receives the message.

Consider Figure 6. Suppose the receiver B receives a synchronization message M_i from the sender A at its local time t_i (taken in the MAC layer), and the start time of A's μTESLA instance is T_0 in A's clock. B may calcu-

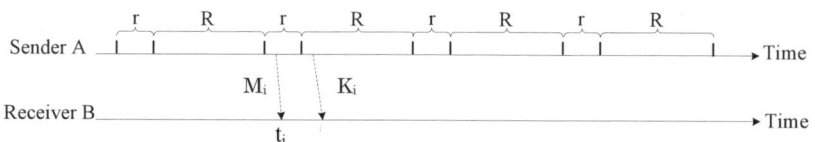

Figure 6: Short Delayed μTESLA

late $i = \lfloor \frac{t_i - T_0}{r+R} \rfloor$ and checks the following security condition: $t_i - T_0 + \Delta_{B,A} + \delta_{max} < i*(R+r) + r$, where $\Delta_{B,A}$ is the pairwise clock difference between A and B, and δ_{max} maximum synchronization error between two neighbor nodes. B stores the message and i only if this check is successful. Otherwise, B simply drops the message. After node B obtains the disclosed key K_i, it verifies $F^{i-j}(K_i) = K_j$ with a previously received key or commitment K_j where $j < i$. If the key is valid, B then uses K_i to verify the MIC included in the broadcast synchronization message M_i.

5. ANALYSIS

5.1 Security Analysis

Phase I. Phase I uses hardware-assisted inline authentication, providing authentication of the source and the content of synchronization messages. Moreover, Phase I uses a two-way message exchange to estimate both the clock difference between direct neighbors and the transmission delay, and can detect attacks that attempt to mislead time synchronization by introducing extra message delays. Thus, Phase I provides protection of the source, the content, and the timeliness of single-hop pairwise synchronization messages. Specifically, Phase I effectively defeats external attacks that attempt to mislead single-hop pairwise time synchronization, including forged and modified messages, pulse-delay attacks, and wormhole attacks that introduce extra delays. Phase I protocol cannot handle DoS attacks that completely jam the communication channel. Nevertheless, no existing protocol can survive such extreme DoS attacks.

Phase II. Phase II adapts μTESLA to provide local broadcast authentication. The security of this μTESLA variation follows directly from the original scheme [34]. Besides local broadcast authentication, another benefit of using μTESLA is the authentication of the timeliness of local broadcast synchronization messages, since a delayed message will be automatically discarded due to the violation of the security condition. Thus, similar to Phase I, by authenticating the source, the content, and the timeliness of local broadcast synchronization messages, Phase II can successfully defeat all the external attacks that are intended to mislead the time synchronization.

Since the source node is trusted, in Phase II, each direct neighbor node of the source node can directly estimate the global clock securely. However, the other nodes may receive false synchronization information from compromised nodes. The solution used by Phase II is to have each node use the source clock differences received from $2t+1$ neighbor nodes to estimate $2t+1$ candidate source clock differences, and select the median one as its own source clock difference.

We say a source clock difference obtained by a normal node is *correct* if it has no more error than one obtained using information *only* from normal nodes. The resilience property of the Phase II protocol can be seen by induction.

As discussed earlier, the source node is trusted. Consider a normal node i. Suppose the source clock difference is obtained through node j, that is, $\delta_{i,S} = \delta_{i,j} + \delta_{j,S}$. There are two cases. (1) If node j is a normal node, both $\delta_{j,S}$ and $\delta_{i,j}$ must be correct according to the induction assumption, and thus $\delta_{i,S} = \delta_{i,j} + \delta_{j,S}$ is correct by definition. (2) Suppose node j is malicious. Because there are at most t malicious nodes, $\delta_{i,S}$, which is the median of the $2t+1$ candidate source clock differences, must be between two candidate source clock differences obtained through two normal nodes. This implies that the synchronization error introduced by a compromised node is no more than the error introduced by a normal node. As a result, the source clock difference $\delta_{i,S}$ is still correct. Thus, if every node has no more than t compromised neighbor nodes, Phase II can successfully synchronize all the normal nodes as long as they have enough number of neighbor nodes. However, similar to Phase I, Phase II cannot handle DoS attacks that completely jam the communication channel.

In conclusion, TinySeRSync provides a comprehensive solution to providing secure and resilient time synchronization in wireless sensor networks. It can successfully defeat all non-DoS external attacks against time synchronization, and is resilient to compromised nodes.

5.2 Performance Analysis

Synchronization Precision and Coverage. TinySeRSync uses predication-based MAC layer timestamping in Phase I, avoiding many places that could introduce uncertainty during time synchronization. In Phase II, TinySeRSync tries to estimate the global clock through the estimation of source clock differences, and thus greatly reduces the impact generated by the propagation delays of synchronization messages. Thus, TinySeRSync can provide high precision time synchronization. Moreover, TinySeRSync employs flooding-based propagation of global synchronization messages; this allows all the nodes that have enough number of neighbor nodes to be synchronized.

Communication, Computation, and Storage Overheads. TinySeRSync uses message exchanges between direct neighbor nodes for Phase I synchronization. All these message exchanges are local, and do not introduce wide area interference. In Phase II, TinySeRSync adopts local broadcast for the propagation of global synchronization messages, effectively harnessing the broadcast nature of wireless communication. Thus, TinySeRSync is efficient in terms of communication.

TinySeRSync uses efficient symmetric cryptography for message authentication. In particular, it exploits the hardware cryptographic support provided by the CC2420 radio component. Thus, TinySeRSync introduces very light computation overhead for cryptographic operations.

TinySeRSync does increase the storage overhead on sensor nodes due to the need to maintain cryptographic keys, buffer the local broadcast messages, and store the source clock differences received from $2t+1$ neighbor nodes. A critical issue is the maintenance of the μTESLA key chain required for authenticating outgoing synchronization messages. Our adaptation of μTESLA greatly reduces the number of keys in each key chain. In addition, we use another approach to further reduce the memory requirement and the delay: After generating a key chain, each node only saves some select keys called *key anchors* (e.g., 1 of every 10 keys), and also caches the keys before the next key anchor to be used (e.g., the first 10 keys). When a μTESLA key is required for authentication, if the key is available in the cache, the node can directly use it. Otherwise, the node can regenerate and fill the key cache using the next key anchor.

Incremental Deployment. As discussed earlier, TinySeRSync uses two asynchronous phases, both of which are executed periodically. Thus, TinySeRSync works well with incremental deployment of sensor nodes. The newly deployed nodes first obtain the pairwise time differences and the commitments of the key chains from its neighbor nodes in Phase I, and then join the Phase II global time synchronization. Our experimental results in Section 7 will show the performance when there are incrementally deployed nodes.

6. IMPLEMENTATION DETAILS

Our implementation of TinySeRSync is targeted at MICAz motes [2]. (However, our implementation can be used with slight modification for other sensor platforms that also use CC2420 radio components, such as TelosB [4] and Tmote Sky [5].) A MICAz mote has an 8-bit micro-controller *ATMega128L* [1], which has 128 kB program memory and 4 kB SRAM. As discussed earlier, MICAz is equipped with the ChipCon CC2420 radio component [3], which works at 2.4GHz radio frequency and provides up to 250 kbps data rate. CC2420 is an IEEE 802.15.4 compliant RF transceiver that features hardware security support.

In the following, we give a few details that are critical for repeating our implementation.

6.1 Exploiting Hardware Security Support in CC2420

The hardware security support featured by CC2420 provides two types of security operations: *stand-alone encryption operation* and *in-line security operation*. The stand-alone encryption operation provides a plain AES encryption, with 128 bit plaintext and 128 bit keys. To encrypt a plaintext, a node first writes the plaintext to the stand-alone buffer *SABUF*, and then issues a SAES command to initiate the encryption operation. When the encryption is complete, the ciphertext is written back to the stand-alone buffer, overwriting the plaintext.

The in-line security operations can provide encryption, decryption, and authentication on frames within the receive buffer (RXFIFO) and the transmit buffer (TXFIFO) of CC2420 on a per frame basis. It supports three modes of security: *counter mode (CTR)*, *CBC-MIC*, and *CCM*. CTR mode performs encryption on the outgoing MAC frames in the TXFIFO buffer, and performs decryption on the incoming MAC frames in the RXFIFO buffer. CBC-MIC mode can generate and verify the message integrity code (MIC) of the messages. The length of MIC can be adjusted. CCM mode combines CTR mode encryption and CBC-MIC authentication in one operation. All the three security modes are based on AES encryption/decryption using 128 bit keys.

We use the CBC-MIC mode to authenticate both pairwise and global synchronization messages. A sender can use in-line CBC-MIC mode to generate the MIC for both pair-wise and global synchronization messages in the MAC layer after the message has been written to the TXFIFO buffer.

The receiver side, however, is slightly different. When a receiver receives a pair-wise synchronization message, since it already knows the secrete key shared with the sender, it can use the in-line CBC-MIC mode to verify the MIC before the message is read from the RXFIFO buffer. However, for the global synchronization messages, before receiving the disclosed key, the receiver cannot use in-line authentication to verify the MIC in the message. Because the receiver still needs the RXFIFO buffer to receive other messages, it cannot buffer the message in the RXFIFO buffer while waiting for the disclosed key. Thus, we have the receiver read the message from RXFIFO and buffer it in its local memory. When the key is received, the receive uses the stand-alone mode to authenticate the buffered global synchronization messages. Since the stand-alone mode only provides single-block encryption functionality, we implemented the CBC mode based on the hardware support.

6.2 Handling Timers

Using timers on MICAz is a tricky issue; improper uses usually lead to unexpected results. The micro-controller ATMega128 provides two 8-bit timers (Timer0, Timer2) and two 16-bit timers (Timer1, Timer3) [1]. In TinyOS, Timer 0 is mainly used as one-shot or repeat timers for applications. For MICAz, Timer 2 is used by CC2420 as a high precision timer (32 μs per tick) to back off the sending packets for a short period of time. Timer 1 is used by CC2420 for capturing radio packet transmit and receive events. In TinySeRSync, we use the remaining 16-bit Timer 3 to maintain the local clock and schedule the message transmission.

ATMega128L uses a 7.3728 MHz crystal oscillator as I/O clock source, whose accuracy is $\pm 40ppm$ [3]. In our implementation, we divide the I/O clock by 64 as the source of Timer 3, thereby achieving a 115.2 kHz Timer 3, with a 8.68 μs time resolution. Timer3 provides three compare match registers ($OCR3A/B/C$), each connected with an interrupt vector. If the compare match interrupt is enabled, whenever the value of Timer3 (TCNT3) equals to the value of one compare match register, it will trigger an interrupt to handle the event. Each node uses compare match register A to maintain a 48-bit logical clock. The value of Timer3 (TCNT3) is 16 bits, and it will overflow every 568.8 ms. We add another 32 bits to have a logical clock that will not overflow for over 77 years Each node sets compare match register B to launch pair-wise synchronization with its neighbors periodically. The source node will use compare match register B to initiate the global synchronization periodically. Each node uses compare match register C to send its global synchronization message in its nearest short μTESLA interval and disclose the key in the adjacent long μTESLA interval.

7. EXPERIMENT RESULTS

We performed a series of experiments in a network of 60 MICAz motes to evaluate the performance of TinySeRSync in real deployment. We focused on the performance metrics in normal situations, while relying on the analysis in Section 5 for the security properties.

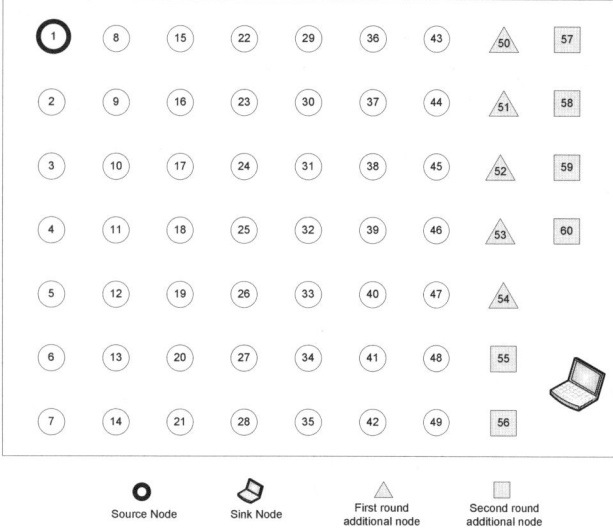

Figure 7: Network Topology

Table 1: Code Size

Memory	Size (bytes)
RAM	1,977
ROM	24,814

7.1 Configuration

Figure 7 shows the sensor network test-bed used in our experiments. (The different node shapes represent nodes deployed at different times during incremental deployment, which will be explained in Section 7.4.) The test-bed consists of 60 nodes, among which node 1 (with the solid circle) is configured as the source node.

We use a number of parameters in our evaluation. Each node performs a secure single-hop pairwise synchronization with its neighbor nodes for every $d_1 = 4$ seconds. During this synchronization, the node informs its neighbor nodes its μTESLA parameters. The source node starts a global synchronization every d_2 seconds. In our experiments, we use $d_2 = 5$ or 10 seconds. The degree of tolerance (against compromised neighbor nodes) is represented as t, as used throughout this paper. In our experiments, we use $t = 0, 1, 2, 3, 4$ to examine the various performance metrics.

We use a sink node to help collecting data from each sensor node. Periodically, the sink node broadcasts an anchor message with the highest power to all the nodes. Upon receiving this message, each node marks the receiving time and converts it to the global time using its source clock difference. The sink node then queries each node individually to get the receiving time (in the estimated global clock) along with other auxiliary information. This allows us to discover the synchronization error on each individual sensor node, the synchronization coverage, as well as the number of synchronization levels each node has to go through.

7.2 Code Size

Let us first look at the code size before presenting the performance results. The code size is related to the maximum number of compromised nodes we would like to tolerate. For each neighbor node, a node will spend 46 bytes to save the pairwise key, current key in key chain, and clock differences, etc. In our experiments, each node saves 10 keys for a key chain with 100 keys. Each node reserves a buffer to store at most 6 unauthenticated global synchronization messages, which increase the size of RAM.

7.3 Performance in Static Deployment

Let us first look at the performance of TinySeRSync in static deployments. In our experiments, we use the following metrics to evaluate the performance and the overhead of TinySeRSync: the average and the maximum synchronization errors, the synchronization rate (i.e., the percentage of nodes that can be synchronized), the synchronization level (i.e., the maximum number of hops that global synchronization messages have to go through before a sensor node can be synchronized.

Average and Maximum Synchronization Error: Figure 8(a) shows the maximum and the average synchronization error with different global synchronization intervals and different degrees of tolerance against compromised neighbor nodes. In all cases, the maximum synchronization error is below 14 ticks (121.52 μs), and the average synchronization error is below 6 ticks (52.08 μs). Figure 8(a) also shows that as the global synchronization interval increases, the maximum and the average synchronization errors both increase.

Synchronization Rate: Figure 8(b) shows the synchronization rate (i.e., the percentage of nodes that can be synchronized by TinySeRSync) after one, two, and three rounds of global synchronization. When the tolerance against compromised neighbor nodes increases, as we expected, the synchronization rate decreases. However, after three rounds of global synchronization, even in the worst case, about 95% of the nodes can be synchronized to the source node.

Synchronization Level: Figure 9(a) shows the maximum and the average number of hops the global synchronization messages have to traverse before all the nodes are synchronized. In our test-bed, in all cases, the average synchronization level is around 3. An interesting issue is that the maximum synchronization level initially decreases as the tolerance t increases, but then goes up as t is greater than 2. This is because when t is very small (i.e., $t = 0, 1$), a node can broadcast the synchronization message almost immediately after it is synchronized. The synchronization triggered by these fast nodes may be propagated to many nodes that have not been synchronized. However, when t is large enough, synchronizing a node with increased t requires receiving synchronization messages from more neighbor nodes, thus resulting in an increasing trend for maximum synchronization levels.

Communication Overhead: We measure the communication overhead by assessing the number of messages each node has to transmit per time unit. For each neighbor node, a node sends one message to obtain the pairwise time difference. In one round of global time synchronization, each node at most broadcasts one synchronization message and one key disclosure message. Suppose each node has n neighbor nodes, the pairwise synchronization interval is d_1, and the global synchronization interval is d_2. In a given long time interval T, each node sends at most $n \cdot \frac{T}{d_1} + \frac{2T}{d_2}$ messages. Figure 9(b) shows the communication overhead per hour for a configuration where each node has 10 neighbor nodes, the pairwise time synchronization interval is 4 seconds, and the global time synchronization interval is 10 seconds.

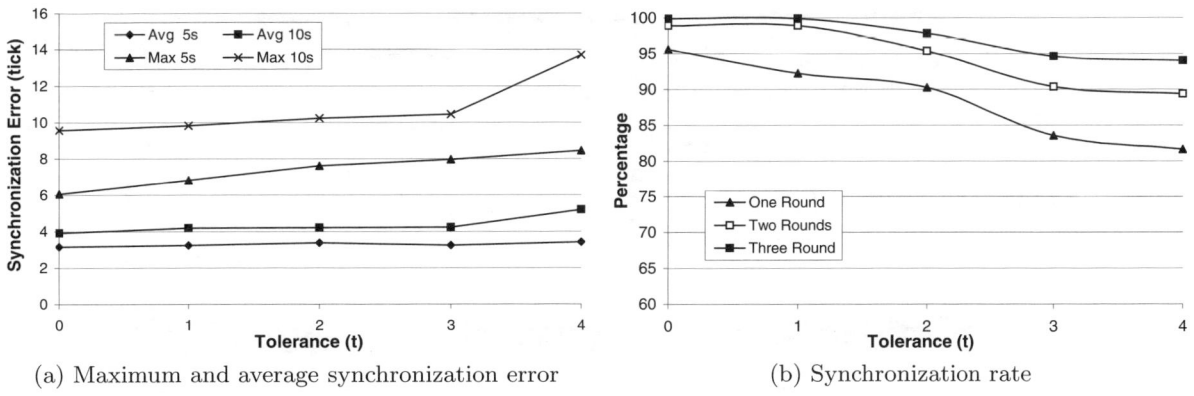

(a) Maximum and average synchronization error

(b) Synchronization rate

Figure 8: Synchronization Error and Synchronization Rate

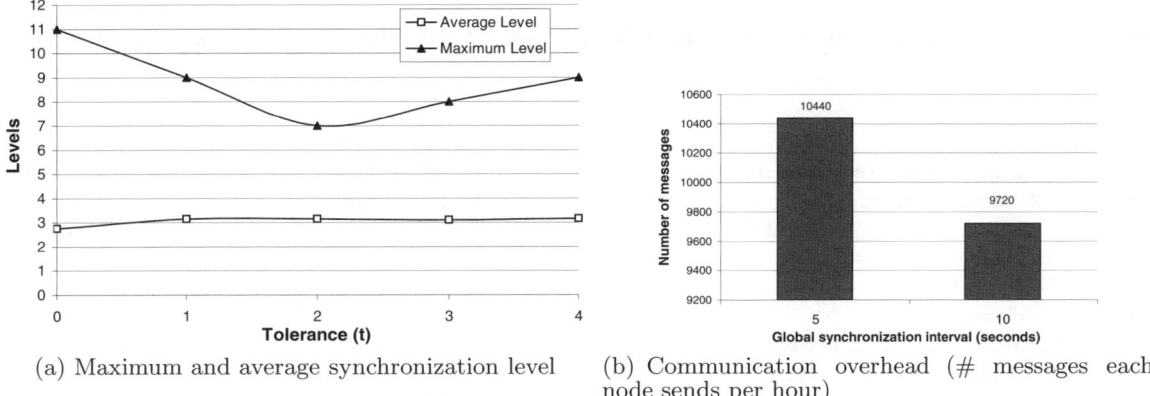

(a) Maximum and average synchronization level

(b) Communication overhead (# messages each node sends per hour)

Figure 9: Synchronization Level and Communication Overhead

7.4 Incremental Deployment

We evaluated the performance of TinySeRSync when there were incremental deployments. Consider Figure 7. At the beginning of the experiment, we deployed the 49 nodes marked as circles. We then added 5 new nodes into the network about 10 minutes later, and added another 6 new nodes about 1 minute later. In this experiment, we set $t = 2$, and the global synchronization interval is set to 10s. Figure 10 shows the history of the average synchronization error and the coverage in this experiment. As shown in the figure, when new nodes were just added into the network, they could not be synchronized immediately, and the average synchronization error was large and the synchronization rate dropped to around 90%. However, after a few rounds of global synchronization, all these new nodes were correctly synchronized, resulting in a low average synchronization error and 100% synchronization coverage.

8. CONCLUSION AND FUTURE WORK

In this paper, we presented the design, implementation, and evaluation of TinySeRSync, a secure and resilient time synchronization subsystem for wireless sensor networks running TinyOS. TinySeRSync includes a comprehensive suite of techniques, including a secure single-hop pairwise time synchronization protocol based on hardware-assisted, authenticated MAC layer timestamping, and a secure and resilient global time synchronization protocol based on a novel use of the μTESLA broadcast authentication protocol. These techniques exceed the capability of previous solutions. In particular, unlike the previous attempts, the secure single-hop pairwise synchronization technique can handle high data rate such as those produced by MICAz motes (in contrast to those by MICA2 motes). Moreover, our novel use of μTESLA in global time synchronization successfully resolved the conflict between the goal of achieving time synchronization and the fact that μTESLA requires loose time synchronization. The resulting protocol is secure against external attacks and resilient against compromised nodes.

Our future research is two-fold. First, we will investigate additional techniques that can improve the synchronization precision. One potential solution is to adapt the linear regression technique proposed in [26] to compensate the constant clock drifts. Second, we will look into the integration of TinySeRSync in sensor network applications.

9. ACKNOWLEDGMENTS

The authors would like to thank the anonymous reviewers for their valuable comments.

10. REFERENCES

[1] ATmega128(L) Complete Technical Documents. http://www.atmel.com/dyn/resources/prod_documents/doc2467.pdf.

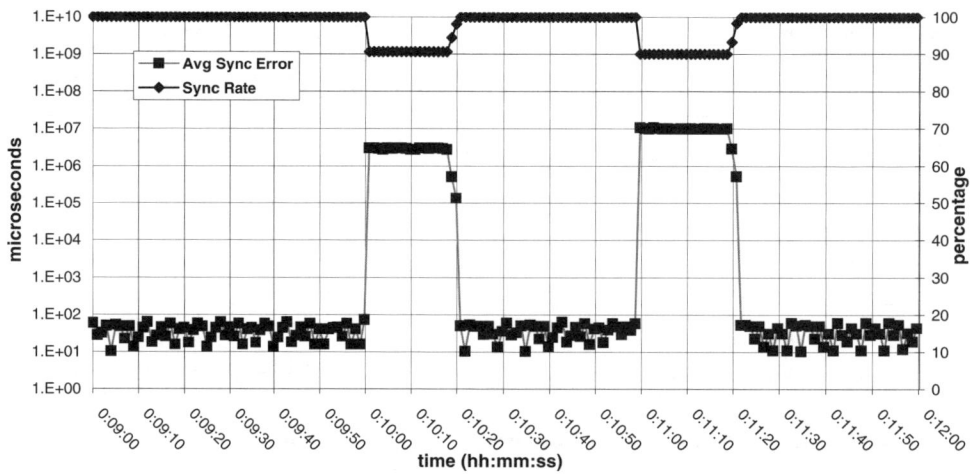

Figure 10: Average Synchronization Error (Left Y-axis) and Coverage (Right Y-axis) During Incremental Deployment ($t = 2$)

[2] MICAz: Wireless measurement system.
http://www.xbow.com/Products/Product_pdf_files/Wireless_pdf/MICAz_Datasheet.pdf.

[3] SmartRF CC2420 Datasheet (rev 1.3), 2005-10-03.
http://www.chipcon.com/files/CC2420_Data_Sheet_1_3.pdf.

[4] TelosB mote platform.
http://www.xbow.com/Products/Product_pdf_files/Wireless_pdf/TelosB_Datasheet.pdf.

[5] Tmote Sky: Reliable low-power wireless sensor networking eases development and deployment.
http://www.moteiv.com/products-tmotesky.php.

[6] I.F. Akyildiz, W. Su, Y. Sankarasubramaniam, and E. Cayirci. Wireless sensor networks: A survey. *Computer Networks*, 38(4):393–422, 2002.

[7] B. Barak, S. Halevi, A. Herzberg, and D. Naor. Clock synchronization with faults and recoveries. In *Proceedings of the 19th Annual ACM Symposium on Principles of Distributed Computing*, pages 133–142, 2000.

[8] Crossbow Technology Inc. Wireless sensor networks.
http://www.xbow.com/Products/Wireless_Sensor_Networks.htm.

[9] D. Dolev, J. Y. Halpern, B. Simons, and R. Strong. Dynamic fault-tolerant clock synchronization. *Journal of the ACM*, 42(1):143–185, 1995.

[10] J. R. Douceur. The sybil attack. In *First International Workshop on Peer-to-Peer Systems (IPTPS'02)*, Mar 2002.

[11] J. Elson, L. Girod, and D. Estrin. Fine-grained network time synchronization using reference broadcasts. *ACM SIGOPS Operating Systems Review*, 36:147–163, 2002.

[12] S. Ganeriwal, S. Capkun, C. Han, and M. B. Srivastava. Secure time synchronization service for sensor networks. In *Proceedings of 2005 ACM Workshop on Wireless Security (WiSe 2005)*, pages 97–106, September 2005.

[13] S. Ganeriwal, R. Kumar, and M. B. Srivastava. Timing-sync protocol for sensor networks. In *Proceedings of the First International Conference on Embedded Networked Sensor Systems (SenSys)*, pages 138–149, 2003.

[14] D. Gay, P. Levis, R. von Behren, M. Welsh, E. Brewer, and D. Culler. The nesC language: A holistic approach to networked embedded systems. In *Proceedings of Programming Language Design and Implementation (PLDI 2003)*, June 2003.

[15] N. Gura, A. Patel, A. Wander, H. Eberle, and S.C. Shantz. Comparing elliptic curve cryptography and RSA on 8-bit CPUs. In *Proceedings of Workshop on Cryptographic Hardware and Embedded Systems (CHES 2004)*, August 2004.

[16] J. Hill, R. Szewczyk, A. Woo, S. Hollar, D.E. Culler, and K. S. J. Pister. System architecture directions for networked sensors. In *Architectural Support for Programming Languages and Operating Systems*, pages 93–104, 2000.

[17] A. Hu and S. D. Servetto. Asymptotically optimal time synchronization in dense sensor networks. In *Proceedings of the Second ACM International Workshop on Wireless Sensor Networks and Applications (WSNA)*, pages 1–10, September 2003.

[18] Y. Hu, A. Perrig, and D. B. Johnson. Wormhole detection in wireless ad hoc networks. Technical Report TR01-384, Department of Computer Science, Rice University, Dec 2001.

[19] Y.C. Hu, A. Perrig, and D.B. Johnson. Packet leashes: A defense against wormhole attacks in wireless ad hoc networks. In *Proceedings of INFOCOM 2003*, April 2003.

[20] IEEE Computer Society. IEEE 802.15.4: Ieee standard for information technology – telecommunications and information exchange between systems local and metropolitan area networks – specific requirements part 15.4: Wireless medium access control (MAC) and physical layer (PHY) specifications for low-rate wireless personal area networks (LR-WPANs).
http://standards.ieee.org/getieee802/download/802.15.4-2003.pdf, October 2003.

[21] C. Intanagonwiwat, R. Govindan, and D. Estrin.

Directed diffusion: A scalable and robust communication paradigm for sensor networks. In *Proceedings of the sixth annual international conference on Mobile computing and networking (Mobicom '00)*, pages 56–67, August 2000.

[22] L. Lamport and P.M. Melliar-Smith. Synchronizing clocks in the presence of faults. *Journal of the ACM*, 32(1):52–78, 1985.

[23] Q. Li and D. Rus. Global clock synchronization in sensor networks. In *Proceedings of IEEE INFOCOM 2004*, pages 214–226, March 2004.

[24] D. Liu, P. Ning, and R. Li. TinyKeyMan: Key management for sensor networks. http://discovery.csc.ncsu.edu/software/TinyKeyMan/.

[25] M. Manzo, T. Roosta, and S. Sastry. Time synchronization attacks in sensor networks. In *Proceedings of the 3rd ACM workshop on Security of ad hoc and sensor networks*, pages 107–116, 2005.

[26] M. Maroti, B. Kusy, G. Simon, and A. Ledeczi. The flooding time synchronization protocol. In *Proceedings of the Second ACM Conference on Embedded Networked Sensor Systems (SenSys'04)*, pages 39–49, Nov 2004.

[27] D.L. Mills. Internet time synchronization: The network time protocol. *IEEE Transactions on Communications*, 39(10):1482–1493, 1991.

[28] M. Mock, R. Frings, E. Nett, and S. Trikaliotis. Clock synchronization for wireless local area networks. In *Proceedings of the 12th Euromicro Conference on Real-Time Systems (Euromicro-RTS 2000)*, June 2000.

[29] J. Newsome, R. Shi, D. Song, and A. Perrig. The sybil attack in sensor networks: Analysis and defenses. In *Proceedings of IEEE International Conference on Information Processing in Sensor Networks (IPSN 2004)*, April 2004.

[30] J. Newsome and D. Song. GEM: graph embedding for routing and data-centric storage in sensor networks without geographic information. In *Proceedings of the First ACM Conference on Embedded Networked Sensor Systems (SenSys '03)*, pages 76–88, Nov 2003.

[31] D. Niculescu and B. Nath. Ad hoc positioning system (APS). In *Proceedings of IEEE GLOBECOM '01*, 2001.

[32] A. Olson and K.G. Shin. Fault-tolerant clock synchronization in large multicomputer systems. *IEEE Transactions on Parallel and Distributed Systems*, 5(9):912–923, 1994.

[33] S. PalChaudhuri, A.K. Saha, and D.B. Johnson. Adaptive clock synchronization in sensor networks. In *Information Processing in Sensor Networks (IPSN)*, pages 340–348, April 2004.

[34] A. Perrig, R. Canetti, D. Song, and D. Tygar. Efficient authentication and signing of multicast streams over lossy channels. In *Proceedings of the 2000 IEEE Symposium on Security and Privacy*, May 2000.

[35] A. Perrig, R. Canetti, D. Song, and D. Tygar. Efficient and secure source authentication for multicast. In *Proceedings of Network and Distributed System Security Symposium*, February 2001.

[36] A. Perrig, R. Szewczyk, V. Wen, D. Culler, and D. Tygar. SPINS: Security protocols for sensor networks. In *Proceedings of Seventh Annual International Conference on Mobile Computing and Networks*, pages 521–534, July 2001.

[37] M.L. Sichitiu and C. Veerarittiphan. Simple, accurate time synchronization for wireless sensor networks. In *IEEE Wireless Communications and Networking Conference WCNC03*, 2003.

[38] H. Song, S. Zhu, and G. Cao. Attack-resilient time synchronization for wireless sensor networks. In *Proceedings of IEEE International Conference on Mobile Ad-hoc and Sensor Systems (MASS'05)*, 2005.

[39] T. K. Srikanth and S. Toueg. Optimal clock synchronization. *Journal of the ACM*, 34(3):626–645, 1987.

[40] K. Sun, P. Ning, and C. Wang. Fault-tolerant cluster-wise clock synchronization for wireless sensor networks. *IEEE Transactions on Dependable and Secure Computing (TDSC)*, 2(3):177–189, July–September 2005.

[41] K. Sun, P. Ning, and C. Wang. Secure and resilient clock synchronization in wireless sensor networks. *IEEE Journal on Selected Areas in Communications*, 24(2), February 2006.

Secure Hierarchical In-Network Aggregation in Sensor Networks

Haowen Chan
Carnegie Mellon University
haowenchan@cmu.edu

Adrian Perrig
Carnegie Mellon University
perrig@cmu.edu

Dawn Song
Carnegie Mellon University
dawnsong@cmu.edu

ABSTRACT

In-network aggregation is an essential primitive for performing queries on sensor network data. However, most aggregation algorithms assume that all intermediate nodes are trusted. In contrast, the standard threat model in sensor network security assumes that an attacker may control a fraction of the nodes, which may misbehave in an arbitrary (Byzantine) manner.

We present the first algorithm for provably secure hierarchical in-network data aggregation. Our algorithm is guaranteed to detect any manipulation of the aggregate by the adversary beyond what is achievable through direct injection of data values at compromised nodes. In other words, the adversary can never gain any advantage from misrepresenting intermediate aggregation computations. Our algorithm incurs only $O(\Delta \log^2 n)$ node congestion, supports arbitrary tree-based aggregator topologies and retains its resistance against aggregation manipulation in the presence of arbitrary numbers of malicious nodes. The main algorithm is based on performing the SUM aggregation securely by first forcing the adversary to commit to its choice of intermediate aggregation results, and then having the sensor nodes independently verify that their contributions to the aggregate are correctly incorporated. We show how to reduce secure MEDIAN, COUNT, and AVERAGE to this primitive.

Categories and Subject Descriptors

C.2.0 [**Computer-Communication Networks**]: General—*Security and Protection*

General Terms

Security, Algorithms

---

*This research was supported in part by CyLab at Carnegie Mellon under grant DAAD19-02-1-0389 from the Army Research Office, and grant CNS-0347807 from the National Science Foundation, and by a gift from Bosch. The views and conclusions contained here are those of the authors and should not be interpreted as necessarily representing the official policies or endorsements, either express or implied, of ARO, Bosch, Carnegie Mellon University, NSF, or the U.S. Government or any of its agencies.

Permission to make digital or hard copies of all or part of this work for personal or classroom use is granted without fee provided that copies are not made or distributed for profit or commercial advantage and that copies bear this notice and the full citation on the first page. To copy otherwise, to republish, to post on servers or to redistribute to lists, requires prior specific permission and/or a fee.
CCS'06, October 30–November 3, 2006, Alexandria, Virginia, USA.
Copyright 2006 ACM 1-59593-518-5/06/0010 ...$5.00.

Keywords

Secure aggregation, Sensor Networks, Data aggregation

1. INTRODUCTION

Wireless sensor networks are increasingly deployed in security-critical applications such as factory monitoring, environmental monitoring, burglar alarms and fire alarms. The sensor nodes for these applications are typically deployed in unsecured locations and are not made tamper-proof due to cost considerations. Hence, an adversary could undetectably take control of one or more sensor nodes and launch active attacks to subvert correct network operations. Such environments pose a particularly challenging set of constraints for the protocol designer: sensor network protocols must be highly energy efficient while being able to function securely in the presence of possible malicious nodes within the network.

In this paper we focus on the particular problem of securely and efficiently performing aggregate queries (such as MEDIAN, SUM and AVERAGE) on sensor networks. In-network data aggregation is an efficient primitive for reducing the total message complexity of aggregate sensor queries. For example, in-network aggregation of the SUM function is performed by having each intermediate node forward a single message containing the sum of the sensor readings of all the nodes downstream from it, rather than forwarding each downstream message one-by-one to the base station. The energy savings of performing in-network aggregation have been shown to be significant and are crucial for energy-constrained sensor networks [9, 11, 20].

Unfortunately, most in-network aggregation schemes assume that all sensor nodes are trusted [12, 20]. An adversary controlling just a few aggregator nodes could potentially cause the sensor network to return arbitrary results, thus completely subverting the function of the network to the adversary's own purposes.

Despite the importance of the problem and a significant amount of work on the area, the known approaches to secure aggregation either require strong assumptions about network topology or adversary capabilities, or are only able to provide limited probabilistic security properties. For example, Hu and Evans [8] propose a secure aggregation scheme under the assumption that at most a single node is malicious. Przydatek et al. [17] propose Secure Information Aggregation (SIA), which provides a statistical security property under the assumption of a single-aggregator model. In the single-aggregator model, sensor nodes send their data to a single aggregator node, which computes the aggregate and sends it to the base station. This form of aggregation reduces communications only on the link between the aggregator and the base station, and is not scalable to large multihop sensor deployments. Most of the algorithms in SIA (in particular, MEDIAN, SUM and AVERAGE) cannot be directly adapted to a hierarchical aggregation model since they involve sorting all of the input values; the final aggregator in the hierarchy thus needs to access all the data values of the sensor nodes.

In this paper, we present the first provably secure sensor network data aggregation protocol for general networks and multiple adversarial nodes. The algorithm limits the adversary's ability to manipulate the aggregation result with the tightest bound possible for general algorithms with no knowledge of the distribution of sensor data values. Specifically, an adversary can gain no additional influence over the final result by manipulating the results of the in-network aggregate computation as opposed to simply reporting false data readings for the compromised nodes under its control. Furthermore, unlike prior schemes, our algorithm is designed for general hierarchical aggregator topologies and multiple malicious sensor nodes. Our metric for communication cost is *congestion*, which is the maximum communication load on any node in the network. Let n be the number of nodes in the network, and Δ be the maximum degree of any node in the aggregation tree. Our algorithm induces only $O(\Delta \log^2 n)$ node congestion in the aggregation tree.

2. RELATED WORK

Researchers have investigated resilient aggregation algorithms to provide increased likelihood of accurate results in environments prone to message loss or node failures. This class of algorithms includes work by Gupta et al. [7], Nath et al. [15], Chen et al. [3] and Manjhi et al. [14].

A number of aggregation algorithms have been proposed to ensure *secrecy* of the data against intermediate aggregators. Such algorithms have been proposed by Girao et al. [5], Castelluccia et al. [2], and Cam et al. [1].

Hu and Evans [8] propose securing in-network aggregation against a single Byzantine adversary by requiring aggregator nodes to forward their inputs to their parent nodes in the aggregation tree. Jadia and Mathuria [10] extend the Hu and Evans approach by incorporating privacy, but also considered only a single malicious node.

Several secure aggregation algorithms have been proposed for the single-aggregator model. Przydatek et al. [17] proposed Secure Information Aggregation (SIA) for this topology. Also for the single-aggregator case, Du et al. [4] propose using multiple *witness* nodes as additional aggregators to verify the integrity of the aggregator's result. Mahimkar and Rappaport [13] also propose an aggregation-verification scheme for the single-aggregator model using a threshold signature scheme to ensure that at least t of the nodes agree with the aggregation result. Yang et al. [19] describe a probabilistic aggregation algorithm which subdivides an aggregation tree into subtrees, each of which reports their aggregates directly to the base station. Outliers among the subtrees are then probed for inconsistencies.

Wagner [18] addressed the issue of measuring and bounding malicious nodes' contribution to the final aggregation result. The paper measures how much damage an attacker can inflict by taking control of a number of nodes and using them solely to inject erroneous data values.

3. PROBLEM MODEL

In general, the goal of secure aggregation is to compute aggregate functions (such as SUM, COUNT or AVERAGE) of the sensed data values residing on sensor nodes, while assuming that a portion of the sensor nodes are controlled by an adversary which is attempting to skew the final result. In this section, we present the formal parameters of the problem.

3.1 Network Assumptions

We assume a general multihop network with a set $S = \{s_1, \ldots, s_n\}$ of n sensor nodes and a single (untrusted) base station R, which is able to communicate with the querier which resides outside of the network. The querier knows the total number of sensor nodes n, and that all n nodes are alive and reachable.

We assume the aggregation is performed over an *aggregation tree* which is the directed tree formed by the union of all the paths from the sensor nodes to the base station (one such tree is shown in Figure 1(a)). These paths may be arbitrarily chosen and are not necessarily shortest paths. The optimisation of the aggregation tree structure is out of the scope of this paper—our algorithm takes the structure of the aggregation tree as given. One method for constructing an aggregation tree is described in TaG [11].

3.2 Security Infrastructure

We assume that each sensor node has a unique identifier s and shares a unique secret symmetric key K_s with the querier. We further assume the existence of a broadcast authentication primitive where any node can authenticate a message from the querier. This broadcast authentication could, for example, be performed using μTESLA [16]. We assume the sensor nodes have the ability to perform symmetric-key encryption and decryption as well as computations of a collision-resistant cryptographic hash function H.

3.3 Attacker Model

We assume that the attacker is in complete control of an *arbitrary number* of sensor nodes, including knowledge of all their secret keys. The attacker has a network-wide presence and can record and inject messages at will. The sole goal of the attacker is to launch what Przydatek et al. [17] call a *stealthy attack*, i.e., to cause the querier to accept a false aggregate that is higher or lower than the true aggregate value.

We do not consider denial-of-service (DoS) attacks where the goal of the adversary is to prevent the querier from getting any aggregation result at all. While such attacks can disrupt the normal operation of the sensor network, they are not as potentially hazardous in security-critical applications as the ability to cause the operator of the network to accept arbitrary data. Furthermore, any maliciously induced extended loss of service is a detectable anomaly which will (eventually) expose the adversary's presence if subsequent protocols or manual intervention do not succeed in resolving the problem.

3.4 Problem Definition and Metrics

Each sensor node s_i has a data value a_i. We assume that the data value is a *non-negative* bounded real value $a_i \in [0, r]$ for some maximum allowed data value r. The objective of the aggregation process is to compute some function f over all the data values, i.e., $f(a_1, \ldots, a_n)$. Note that for the SUM aggregate, the case where data values are in a range $[r_1, r_2]$ (where r_1, r_2 can be negative) is reducible to this case by setting $r = r_2 - r_1$ and add nr_1 to the aggregation result.

Definition 1 *A direct data injection attack occurs when an attacker modifies the data readings reported by the nodes under its direct control, under the constraint that only legal readings in $[0, r]$ are reported.*

Wagner [18] performed a quantitative study measuring the effect of direct data injection on various aggregates, and concludes that the aggregates addressed in this paper (truncated SUM and AVERAGE, COUNT and Φ-QUANTILE) can be resilient under such attacks.

Without domain knowledge about what constitutes an anomalous sensor reading, it is impossible to detect a direct data injection attack, since they are indistinguishable from legitimate sensor readings [17, 19]. Hence, if a secure aggregation scheme does not make assumptions on the distribution of data values, it cannot limit the adversary's capability to perform direct data injection. We can thus define an optimal level of aggregation security as follows.

Definition 2 *An aggregation algorithm is **optimally secure** if, by tampering with the aggregation process, an adversary is unable to induce the querier to accept any aggregation result which is not already achievable by direct data injection.*

As a metric for communication overhead, we consider node *congestion*, which is the worst case communication load on any single sensor node during the algorithm. Congestion is a commonly used metric in ad-hoc networks since it measures how quickly the heaviest-loaded nodes will exhaust their batteries [6, 12]. Since the heaviest-loaded nodes are typically the nodes which are most essential to the connectivity of the network (e.g., the nodes closest to the base station), their failure may cause the network to partition even though other sensor nodes in the network may still have high battery levels. A lower communication load on the heaviest-loaded nodes is thus desirable even if the trade-off is a larger amount of communication in the network as a whole.

For a lower bound on congestion, consider an unsecured aggregation protocol where each node sends just a single message to its parent in the aggregation tree. This is the minimum number of messages that ensures that each sensor node contributes to the aggregation result. There is $\Omega(1)$ congestion on each edge on the aggregation tree, thus resulting in $\Omega(d)$ congestion on the node(s) with highest degree d in the aggregation tree. The parameter d is dependent on the shape of the given aggregation tree and can be as large as $\Theta(n)$ for a single-aggregator topology or as small as $\Theta(1)$ for a balanced aggregation tree. Since we are taking the aggregation tree topology as an input, we have no control over d. Hence, it is often more informative to consider per-edge congestion, which can be independent of the structure of the aggregation tree.

Consider the simplest solution where we omit aggregation altogether and simply send all data values (encrypted and authenticated) directly to the base station, which then forwards it to the querier. This provides perfect data integrity, but induces $O(n)$ congestion at the nodes and edges nearest the base station. For an algorithm to be practical, it must cause only sublinear edge congestion.

Our goal is to design an **optimally secure** aggregation algorithm with only **sublinear edge congestion**.

4. THE SUM ALGORITHM

In this section we describe our algorithm for the SUM aggregate, where the aggregation function f is addition. Specifically, we wish to compute $a_1 + \cdots + a_n$, where a_i is the data value at node i. We defer analysis of the algorithm properties to Section 5, and discuss the application of the algorithm to other aggregates such as COUNT, AVERAGE and MEDIAN in Section 6.

We build on the aggregate-commit-prove framework described by Przydatek et al. [17] but extend their single aggregator model to a fully distributed setting. Our algorithm involves computing a cryptographic commitment structure (similar to a hash tree) over the data values of the sensor nodes as well as the aggregation process. This forces the adversary to choose a fixed aggregation topology and set of aggregation results. The individual sensor nodes then independently audit the commitment structure to verify that their respective contributions have been added to the aggregate. If the adversary attempts to discard or reduce the contribution of a legitimate sensor node, this necessarily induces an inconsistency in the commitment structure which can be detected by the affected node. This basic approach provides us with a lower bound for the SUM aggregate. To provide an upper-bound for SUM, we can re-use the same lower-bounding approach, but on a complementary aggregate called the COMPLEMENT aggregate. Where SUM is defined as $\sum a_i$, COMPLEMENT is defined as $\sum(r - a_i)$ where r is the upper bound on allowable data values. When the final aggregates are computed, the querier enforces the constraint that SUM + COMPLEMENT = nr. Hence any adversary that wishes to increase SUM must also decrease COMPLEMENT, and vice-versa, otherwise the discrepancy will be detected. Hence, by enforcing a lower-bound on COMPLEMENT, we are also enforcing an upper-bound on SUM.

The overall algorithm has three main phases: query dissemination, aggregation-commit, and result-checking.

Query dissemination. The base station broadcasts the query to the network. An *aggregation tree*, or a directed spanning tree over the network topology with the base station at the root, is formed as the query is sent to all the nodes, if one is not already present in the network.

Aggregation commit. In this phase, the sensor nodes iteratively construct a commitment structure resembling a hash tree. First, the leaf nodes in the aggregation tree send their data values to their parents in the aggregation tree. Each internal sensor node in the aggregation tree performs an aggregation operation whenever it has heard from all its child sensor nodes. Whenever a sensor node s performs an aggregation operation, s creates a commitment to the set of inputs used to compute the aggregate by computing a hash over all the inputs (including the commitments that were computed by the children of s). Both the aggregation result and the commitment are then passed on to the parent of s. After the final commitment values are reported to the base station (and thus also to the querier), the adversary cannot subsequently claim a different aggregation structure or result. We describe an optimisation to ensure that the constructed commitment trees are perfectly balanced, thus requiring low congestion overhead in the next phase.

Result-checking. The result-checking phase is a novel distributed verification process. In prior work, algorithms have relied on the querier to issue probes into the commitment structure to verify its integrity [17, 19]. This induces congestion nearest the base station, and moreover, such algorithms yield at best probabilistic security properties. We show that if the verification step is instead fully distributed, it is possible to achieve provably *optimal* security while maintaining sublinear edge congestion.

The result-checking phase proceeds as follows. Once the querier has received the final commitment values, it disseminates them to the rest of the network in an authenticated broadcast. At the same time, sensor nodes disseminate information that will allow their peers to verify that their respective data values have been incorporated into the aggregate. Each sensor node is responsible for checking that its own contribution was added into the aggregate. If a sensor node determines that its data value was indeed added towards the final sum, it sends an authentication code up the aggregation tree towards to the base station. Authentication codes are aggregated along the way with the XOR function for communication efficiency. When the querier has received the XOR of all the authentication codes, it can then verify that all the sensor nodes have confirmed that the aggregation structure is consistent with their data values. If so, then it accepts the aggregation result.

We now describe the details of each of the three phases in turn.

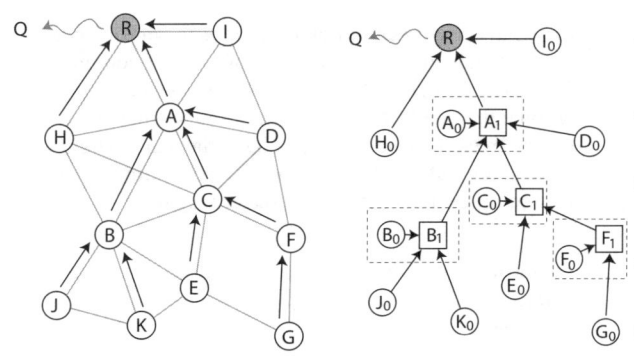

(a) Example network graph. Arrows: Aggregation tree. R: Base station. Q: Querier.

(b) Naive commitment tree, showing derivations of some of the vertices. For each sensor node X, X_0 is its leaf vertex, while X_1 is the internal vertex representing the aggregate computation at X (if any). On the right we list the labels of the vertices on the path of node G to the root.

Figure 1: Aggregation and naive commitment tree in network context

4.1 Query Dissemination

First, an aggregation tree is established if one is not already present. Various algorithms for selecting the structure of an aggregation tree may be used. For completeness, we describe one such process, while noting that our algorithm is directly applicable to any aggregation tree structure. The Tiny Aggregation Service (TaG) [11] uses a broadcast from the base station where each node chooses as its parent in the aggregation tree, the node from which it first heard the tree-formation message.

To initiate a query in the aggregation tree, the base station originates a query request message which is distributed following the aggregation tree. The query request message contains an attached nonce N to prevent replay of messages belonging to a prior query, and the entire request message is sent using an authenticated broadcast.

4.2 Aggregation-Commit Phase

The goal of the aggregation-commit phase is to iteratively construct a series of cryptographic commitments to data values and to intermediate in-network aggregation operations. This commitment is then passed on to the querier. The querier then rebroadcasts the commitment to the sensor network using an authenticated broadcast so that the rest of the sensor network is able to verify that their respective data values have been incorporated into the aggregate.

4.2.1 Aggregation-Commit: Naive Approach

We first describe a naive approach that yields the desired security properties but has suboptimal congestion overhead when sensor nodes perform their respective verifications. In the naive approach, when each sensor node performs an aggregation operation, it computes a cryptographic hash of all its inputs (including its own data value). The hash value is then passed on to the parent in the aggregation tree along with the aggregation result. Figure 1(b) shows a *commitment tree* which consists of a series of hashes of data values and intermediate results, culminating in a set of final commitment values which is passed on by the base station to the querier along with the aggregation results. Conceptually, a commitment tree is a hash tree with some additional aggregate accounting information attached to the nodes. A definition follows. Recall that N is the query nonce that is disseminated with each query.

Definition 3 *A **commitment tree** is a tree where each vertex has an associated label representing the data that is passed on to its parent. The labels have the following format:*

⟨count, value, complement, commitment⟩

Where count *is the number of leaf vertices in the subtree rooted at this vertex;* value *is the* SUM *aggregate computed over all the leaves in the subtree;* complement *is the aggregate over the* COMPLEMENT *of the data values; and* commitment *is a cryptographic commitment. The labels are defined inductively as follows:*

*There is one leaf vertex u_s for each sensor node s, which we call the **leaf vertex of** s. The label of u_s consists of* count=1, value=a_s *where a_s is the data value of s,* complement=$r-a_s$ *where r is the upper bound on allowable data values, and* commitment *is the node's unique ID.*

Internal vertices represent aggregation operations, and have labels that are defined based on their children. Suppose an internal vertex has child vertices with the following labels: $u_1, u_2, \ldots, u_q$, where $u_i = \langle c_i, v_i, \bar{v}_i, h_i \rangle$. Then the vertex has label $\langle c, v, \bar{v}, h \rangle$, with $c = \sum c_i$, $v = \sum v_i$, $\bar{v} = \sum \bar{v}_i$ and $h = H[N||c||v||\bar{v}||u_1||u_2||\cdots||u_q]$.

For brevity, in the remainder of the paper we will often omit references to labels and instead refer directly to the count, value, complement or commitment of a vertex.

While there exists a natural mapping between vertices in a commitment tree and sensor nodes in the aggregation tree, a vertex is a logical element in a graph while a sensor node is a physical device. To prevent confusion, we will always refer to the *vertices* in the commitment tree; the term *nodes* always refers to the physical sensor node device.

Since we assume that our hash function provides collision resistance, it is computationally infeasible for an adversary to change any of the contents of the commitment tree once the final commitment values have reached the root.

With knowledge of the root commitment value, a node s may verify the aggregation steps between its leaf vertex u_s and the root of the commitment tree. To do so, s needs the labels of all its *off-path* vertices.

Definition 4 *The set of **off-path** vertices for a vertex u in a tree is the set of all the siblings of each of the vertices on the path from u to the root of the tree that u is in (the path is inclusive of u).*

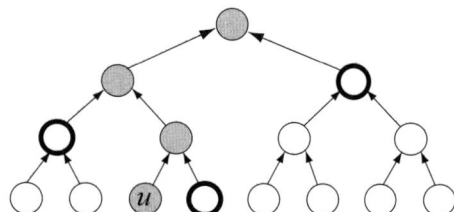

Figure 2: Off-path vertices for u are highlighted in bold. The path from u to the root of its tree is shaded grey.

Figure 2 shows a pictorial depiction of the off-path vertices for a vertex u in a tree. For a more concrete example, the set of off-path commitment tree vertices for G_0 in Figure 1 is $\{F_0, E_0, C_0, B_1, A_0, D_0, H_0, I_0\}$. To allow sensor node G to verify its contribution to the aggregate, the sensor network delivers labels of each off-path vertex to G_0. Sensor node G then recomputes the sequence of computations and hashes and verifies that they lead to the correct root commitment value.

Consider the congestion on the naive scheme. Let h be the height of the aggregation tree and Δ be the maximum degree of any node inside the tree. Each leaf vertex has $O(h\Delta)$ off-path vertices, and it needs to receive all their labels to verify its contribution to the aggregate, thus leading to $O(h\Delta)$ congestion at the leaves of the commitment tree. For an aggregation tree constructed with TaG, the height h of the aggregation tree depends on the diameter (in number of hops) of the network, which in turn depends on the node density and total number of nodes n in the network. In a 2-dimensional deployment area with a constant node density, the best bound on the diameter of the network is $O(\sqrt{n})$ if the network is regularly shaped. In irregular topologies the diameter of the network may be $\Omega(n)$.

4.2.2 Aggregation-Commit: Improved Approach

We present an optimization to improve the congestion cost. The main observation is that, since the aggregation trees are a subgraph of the network topology, they may be arbitrarily unbalanced. Hence, if we decouple the structure of the commitment tree from the structure of the aggregation tree, then the commitment tree could be perfectly balanced.

In the naive commitment tree, each sensor node always computes the aggregate sum of *all* its inputs. This can be considered a strategy of *greedy aggregation*. Consider instead the benefit of *delayed aggregation* at node C_1 in Figure 1(b). Suppose that C, instead of greedily computing the aggregate sum over its own reading (C_0) and both its child nodes E_0 and F_1, instead computes the sum *only* over C_0 and E_0, and passes F_1 directly to A along with $C_1 = C_0 + E_0$. In such a commitment tree, F_1 becomes a child of A_1 (instead of C_1), thus reducing the depth of the commitment tree by 1. Delayed aggregation thus trades off increased communication during the aggregation phase in return for a more balanced commitment tree, which results in lower verification overhead in the result-checking phase. Greenwald and Khanna [6] used a form of delayed aggregation in their quantile summary algorithm.

Our strategy for delayed aggregation is as follows: we perform an aggregation operation (along with the associated commit operation) if and only if it results in a *complete, binary* commitment tree.

We now describe our delayed aggregation algorithm for producing balanced commitment trees. In the naive commitment tree, each sensor node passes to its parent a single message containing the label of the root vertex of its commitment subtree T_s. In the delayed aggregation algorithm, each sensor node now passes on the labels of the root vertices of a *set* of commitment subtrees $F = \{T_1, \ldots, T_q\}$. We call this set a *commitment forest*, and we enforce the condition that the trees in the forest must be complete binary trees, and no two trees have the same height. These constraints are enforced by continually combining equal-height trees into complete binary trees of greater height.

Definition 5 *A commitment forest is a set of complete binary commitment trees such that there is at most one commitment tree of any given height.*

A commitment forest has at most n leaf vertices (one for each sensor node included in the forest, up to a maximum of n). Since all the trees are complete binary trees, the tallest tree in any commitment forest has height at most $\log n$. Since there are no two trees of the same height, any commitment forest has at most $\log n$ trees.

In the following discussion, we will for brevity make reference to "communicating a vertex" to another sensor node, or "communicating a commitment forest" to another sensor node. The actual data communicated is the *label* of the vertex and the *labels of the roots* of the trees in the commitment forest, respectively.

The commitment forest is built as follows. Leaf sensor nodes in the aggregation tree originate a single-vertex commitment forest, which they then communicate to their parent sensor nodes. Each internal sensor node s originates a similar single-vertex commitment forest. In addition, s also receives commitment forests from each of its children. Sensor node s keeps track of which root vertices were received from which of its children. It then combines all the forests to form a new forest as follows.

Suppose s wishes to combine q commitment forests $F_1, \ldots, F_q$. Note that since all commitment trees are complete binary trees, tree heights can be determined by inspecting the count field of the root vertex. We let the intermediate result be $F = F_1 \cup \cdots \cup F_q$, and repeat the following until no two trees are the same height in F: Let h be the smallest height such that more than one tree in F has height h. Find two commitment trees T_1 and T_2 of height h in F, and merge them into a tree of height $h+1$ by creating a new vertex that is the parent of both the roots of T_1 and T_2 according to the inductive rule in Definition 3. Figure 3 shows an example of the process for node A based on the topology in Figure 1.

The algorithm terminates in $O(q \log n)$ steps since each step reduces the number of trees in the forest by one, and there are at most $q \log n + 1$ trees in the forest. Hence, each sensor node creates at most $q \log n + 1 = O(\Delta \log n)$ vertices in the commitment forest.

When F is a valid commitment forest, s sends the root vertices of each tree in F to its parent sensor node in the aggregation tree. The sensor node s also keeps track of every vertex that it created, as well as all the inputs that it received (i.e., the labels of the root vertices of the commitment forests that were sent to s by its children). This takes $O(d \log n)$ memory per sensor node.

Consider the communication costs of the entire process of creating the final commitment forest. Since there are at most $\log n$ commitment trees in each of the forests presented by any sensor node to its parent, the per-node communication cost for constructing the final forest is $O(\log n)$. This is greater than the $O(1)$ congestion cost of constructing the naive commitment tree. However, no path in the forest is longer than $\log n$ hops. This will eventually enable us to prove a bound of $O(\log^2 n)$ edge congestion for the result-checking phase in Section 5.2.

Once the querier has received the final commitment forest from the base station, it checks that none of the SUM or COMPLEMENT aggregates of the roots of the trees in the forest are negative. If

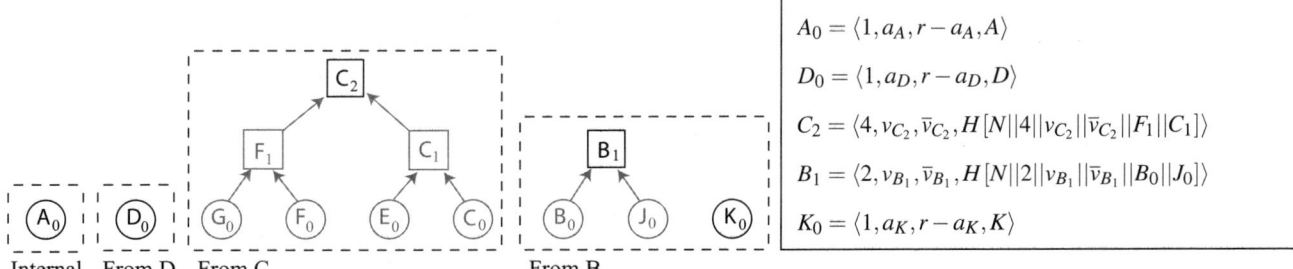

(a) Inputs: A generates A_0, and receives D_0 from D, C_2 from C, and (B_1, K_0) from B. Each dashed-line box shows the commitment forest received from a given sensor node. The solid-line box shows the vertex labels, each solid-line box below shows the labels of the new vertices.

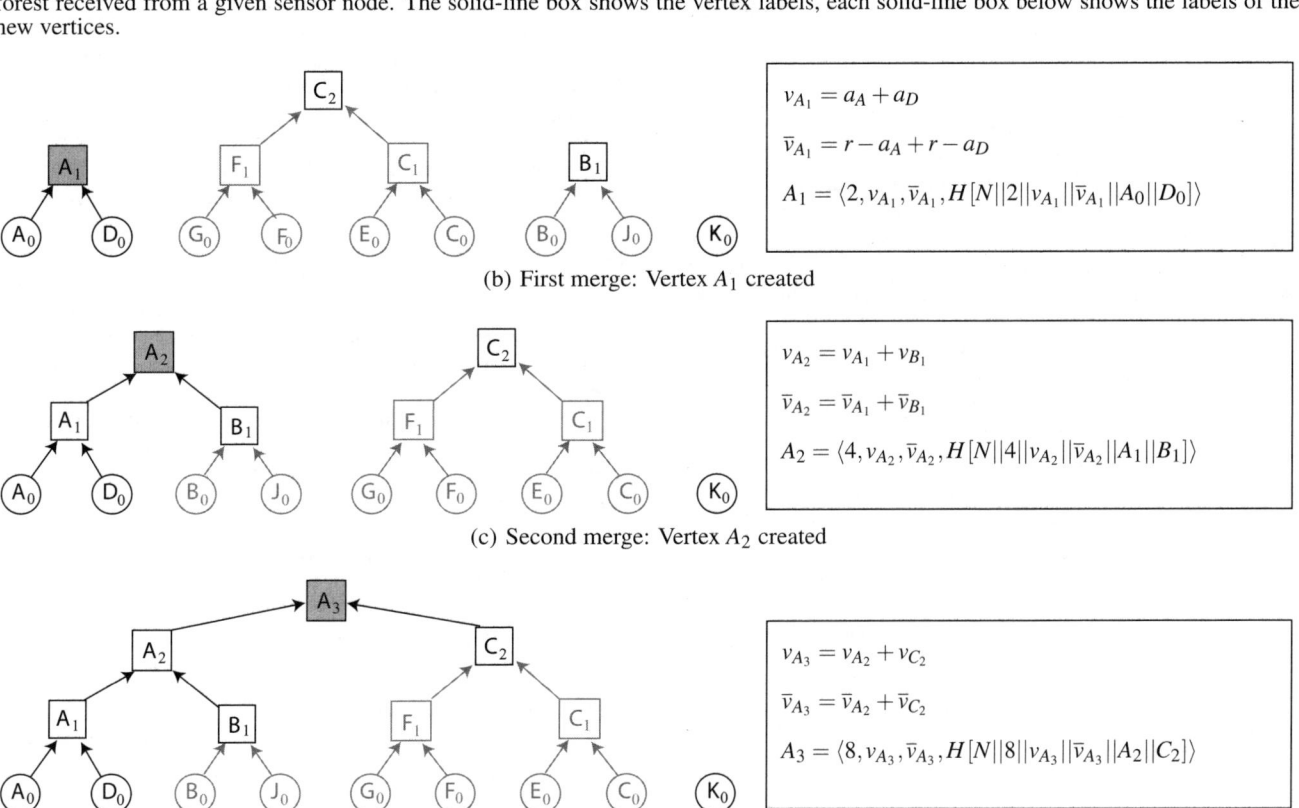

(b) First merge: Vertex A_1 created

(c) Second merge: Vertex A_2 created

(d) Final merge: Vertex A_3 created. A_3 and K_0 are sent to the parent of A in the aggregation tree.

Figure 3: Process of node A (from Figure 1) deriving its commitment forest from the commitment forests received from its children.

any aggregates are negative, the querier rejects the result and raises an alarm: a negative aggregate is a sure sign of tampering since all the data values (and their complements) are non-negative. Otherwise, the querier then computes the final pair of aggregates SUM and COMPLEMENT. The querier verifies that SUM + COMPLEMENT $= nr$ where r is the upper bound on the range of allowable data values on each node. If this verifies correctly, the querier then initiates the *result-checking* phase.

4.3 Result-checking phase

The purpose of the result-checking phase is to enable each sensor node s to independently verify that its data value a_s was added into the SUM aggregate, and the complement $(r - a_s)$ of its data value was added into the COMPLEMENT aggregate. The verification is performed by inspecting the inputs and aggregation operations in the commitment forest on the path from the leaf vertex of s to the root of its tree; if all the operations are consistent, then the root aggregate value must have increased by a_s due to the incorporation of the data value. If each legitimate node performs this verification, then it ensures that the SUM aggregate is at least the sum of all the data values of the legitimate nodes. Similarly, the COMPLEMENT aggregate is at least the sum of all the complements of the data values of the legitimate nodes. Since the querier enforces SUM + COMPLEMENT $= nr$, these two inequalities form lower and upper bounds on an adversary's ability to manipulate the final result. In Section 5 we shall show that they are in fact the tightest bounds possible.

A high level overview of the process is as follows. First, the aggregation results from the aggregation-commit phase are sent using authenticated broadcast to every sensor node in the network. Each sensor node then individually verifies that its contributions to the respective SUM and COMPLEMENT aggregates were indeed counted. If so, it sends an authentication code to the base station. The authentication code is also aggregated for communication effi-

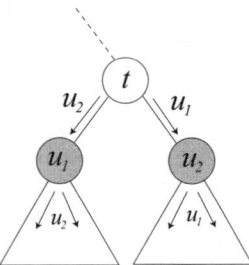

Figure 4: Dissemination of off-path values: t sends the label of u_1 to u_2 and vice-versa; each node then forwards it to all the vertices in their subtrees.

ciency. When the querier has received all the authentication codes, it is then able to verify that all sensor nodes have checked that their contribution to the aggregate has been correctly counted.

For simplicity, we describe each step of the process with reference to the commitment tree visualised as an *overlay network* over the actual aggregation tree. Hence, we will refer to *vertices* in the commitment tree sending information to each other; in the physical world, it is the sensor node that created the vertex is the physical entity that is responsible for performing communications and computations on behalf of the vertex. Each edge in the commitment tree may involve multiple hops in the aggregation tree; the routing on the aggregation tree is straightforward.

Dissemination of final commitment values. After the querier has received the labels of the roots of the final commitment forest, the querier sends each of these labels to the entire sensor network using authenticated broadcast.

Dissemination of off-path values. To enable verification, each leaf vertex must receive all its off-path values. Each internal vertex t in the commitment forest has two children u_1 and u_2. To disseminate off-path values, t sends the label of u_1 to u_2, and vice-versa (t also attaches relevant information tagging u_1 as the right child and u_2 as the left child). Vertex t also sends any labels (and left/right tags) received from its parent to both its children. See Figure 4 for an illustration of the process. The correctness of this algorithm in delivering all the necessary off-path vertex labels to each vertex is proven in Theorem 14 in Section 5.2. Once a vertex has received all the labels of its off-path vertices, it can proceed to the verification step.

Verification of inclusion. When the leaf vertex u_s of a sensor node s has received all the labels of its off-path vertices, it may then verify that no aggregation result-tampering has occurred on the path between u_s and the root of its commitment tree. For each vertex t on the path from u_s to the root of its commitment tree, u_s derives the label of t (via the computations in Definition 3). It is able to do so since the off-path labels provide all the necessary data to perform the label computation. During the computation, u_s inspects the off-path labels: for each node t on the path from u_s to the root, u_s checks that the input values fed into the aggregation operation at t are never negative. Negative values should never occur since the data and complement values are non-negative; hence if a negative input is encountered, the verification fails. Once u_s has derived the label of the root of its commitment tree, it compares the derived label against the label with the same count that was disseminated by the querier. If the labels are identical, then u_s proceeds to the next step. Otherwise, the verification fails and u_s may either immediately raise an alarm (for example, using broadcast), or it may simply do nothing and allow the aggregate algorithm to fail due to the absence of its confirmation message in the subsequent steps.

Collection of confirmations. After each sensor node s has successfully performed the verification step for its leaf vertex u_s, it sends an authentication code to the querier. The authentication code for sensor node s is $\text{MAC}_{K_s}(N||OK)$ where OK is a unique message identifier and K_s is the key that s shares with the querier. The collation of the authentication codes proceeds as follows (note that we are referring to the *aggregation* tree at this point, not the commitment tree). Leaf sensor nodes in the aggregation tree first send their authentication codes to their parents in the aggregation tree. Once an internal sensor node has received authentication codes from all its children, it computes the XOR of its own authentication code with all the received codes, and forwards it to its parent. At the end of the process, the querier will receive a single authentication code from the base station that consists of the XOR of all the authentication codes received in the network.

Verification of confirmations. Since the querier knows the key K_s for each sensor node s, it verifies that every sensor node has released its authentication code by computing the XOR of the authentication codes for all the sensor nodes in the network, i.e., $\text{MAC}_{K_1}(N||OK) \oplus \cdots \oplus \text{MAC}_{K_n}(N||OK)$. The querier then compares the computed code with the received code. If the two codes match, then the querier accepts the aggregation result. Otherwise, the querier rejects the result. A rejection may indicate the presence of the adversary in some unknown nodes in the network, or it may be due to natural factors such as node death or message loss. The querier may either retry the query or attempt to determine the cause of the rejection. For example, it could directly request the leaf values of every sensor node: if rejections due to natural causes are sufficiently rare, the high cost of this direct query is incurred infrequently and can be amortised over the other successful queries.

5. ANALYSIS OF SUM

In this section we prove the properties of the SUM algorithm. In Section 5.1 we prove the security properties of the algorithm, and in Section 5.2 we prove bounds on the congestion of the algorithm.

5.1 Security Properties

We assume that the adversary is able to freely choose **any** arbitrary topology and set of labels for the final commitment forest. We then show that any such forest which passes all the verification tests must report an aggregate result that is (optimally) close to the actual result. First, we define the notion of an *inconsistency*, or evidence of tampering, at a given node in the commitment forest.

Definition 6 *Let* $t = \langle c_t, v_t, \bar{v}_t, H_t \rangle$ *be an internal vertex in a commitment forest. Let its two children be* $u_1 = \langle c_1, v_1, \bar{v}_1, H_1 \rangle$ *and* $u_2 = \langle c_2, v_2, \bar{v}_2, H_2 \rangle$. *There is an* **inconsistency** *at vertex* t *in a commitment tree if either (1)* $v_t \neq v_1 + v_2$ *or* $\bar{v}_t \neq \bar{v}_1 + \bar{v}_2$ *or (2) any of* $\{v_1, v_2, \bar{v}_1, \bar{v}_2\}$ *is negative.*

Informally, an inconsistency occurs at t if the sums don't add up at t, or if any of the inputs to t are negative. Intuitively, if there are no inconsistencies on a path from a vertex to the root of the commitment tree, then the aggregate value along the path should be non-decreasing towards the root.

Definition 7 *Call a leaf-vertex* u **accounted-for** *if there is no inconsistency at any vertex on the path from the leaf-vertex* u *to the root of its commitment tree, including at the root vertex.*

Lemma 8 *Suppose there is a set of accounted-for leaf-vertices with distinct labels* $u_1, \ldots, u_m$ *and committed data values* $v_1, \ldots, v_m$ *in* the commitment forest. Then the total of the aggregation values at the roots of the commitment trees in the forest is at least $\sum_{i=1}^{m} v_i$.

Lemma 8 can be rigorously proven using induction on the height of the subtrees in the forest (see Appendix A). Here we present a more intuitive argument.

PROOF. (Sketch) We show the result for $m = 2$; a similar reasoning applies for arbitrary m. Case 1: Suppose u_1 and u_2 are in different trees. Then, since there is no inconsistency on any vertex on the path from u_1 to the root of its tree, the root of the tree containing u_1 must have an aggregation value of at least v_1. By a similar reasoning, the root of the tree containing u_2 must have an aggregation value of at least v_2. Hence the total aggregation value of the two trees containing u_1 and u_2 is at least $v_1 + v_2$.

Case 2: Now suppose u_1 and u_2 are in the same tree. Since they have distinct labels, they must be distinct vertices, and they must have a lowest common ancestor t in the commitment tree. The vertices between u_1 and t (including u_1) must have aggregation value at least v_1 since there are no inconsistencies on the path from u_1 to t, so the aggregation value could not have decreased. Similarly, the vertices between u_2 and t (including u_2) must have aggregation value at least v_2. Hence, one of the children of t has aggregation value at least v_1 and the other has aggregation value at least v_2. Since there was no inconsistency at t, vertex t must have aggregation value at least $v_1 + v_2$. Since there are no inconsistencies on the path from t to the root of the commitment tree, the root also must have aggregation value at least $v_1 + v_2$.

Negative root aggregate values are detected by the querier at the end of the aggregate-commit phase, so the total sum of the aggregate values of the roots of all the trees is thus at least $v_1 + v_2$. □

The following is a restatement of Lemma 8 for the COMPLEMENTARY SUM aggregate; its proof follows an identical structure and is thus omitted.

Lemma 9 *Suppose there is a set of accounted-for leaf vertices with distinct labels $u_1, \ldots, u_m$ with committed complement values $\bar{v}_1, \ldots, \bar{v}_m$ in the commitment forest. Then the total COMPLEMENT aggregation value of the roots of the commitment trees in the forest is at least $\sum_{i=1}^{m} \bar{v}_i$.*

Lemma 10 *A legitimate sensor node will only release its confirmation MAC if it is accounted-for.*

PROOF. By construction, each sensor node s only releases its confirmation MAC if (1) s receives an authenticated message from the querier containing the query nonce N and the root labels of all the trees in the final commitment forest and (2) s receives all labels of its off-path vertices (the sibling vertices to the vertices on the path from the leaf vertex corresponding to s to the root of the commitment tree containing the leaf vertex in the commitment forest), and (3) s is able to recompute the root commitment value that it received from the base station and correctly authenticated, and (4) s verified that all the computations on the path from its leaf vertex u_s to the root of its commitment tree are correct, i.e., there are no inconsistencies on the path from u_s to the root of the commitment tree containing u_s. Since the hash function is collision-resistant, it is computationally infeasible for an adversary to provide s with false labels that also happen to compute to the correct root commitment value. Hence, it must be that s was accounted-for in the commitment forest. □

Lemma 11 *The querier can only receive the correct final XOR check value if all the legitimate sensor nodes replied with their confirmation MACs.*

PROOF. To compute the correct final XOR check value, the adversary needs to know the XOR of all the legitimate sensor nodes that did not release their MAC. Since we assume that each of the distinct MACs are unforgeable (and not correlated with each other), the adversary has no information about this XOR value. Hence, the only way to produce the correct XOR check value is for all the legitimate sensor nodes to have released their relevant MACs. □

Theorem 12 *Let the final SUM aggregate received by the querier be S. If the querier accepts S, then $S_L \leq S \leq (S_L + \mu r)$ where S_L is the sum of the data values of all the legitimate nodes, μ is the total number of malicious nodes, and r is the upper bound on the range of allowable values on each node.*

PROOF. Suppose the querier accepts the SUM result S. Let the COMPLEMENT SUM received by the querier be $\bar{S}$. The querier accepts S if and only if it receives the correct final XOR check value in the result-checking phase, and $S + \bar{S} = nr$. Since the querier received the correct XOR check value, we know that each legitimate sensor node must have released its confirmation MAC (Lemma 11), and so the leaf vertices of each legitimate sensor node must be accounted-for (Lemma 10). The set of labels of the leaf vertices of the legitimate nodes is distinct since the labels contain the (unique) node ID of each legitimate node. Since all the leaf vertices of the legitimate sensor nodes are distinct and accounted-for, by Theorem 8, $S \geq S_L$ where S_L is the sum of the data values of all the legitimate nodes. Furthermore, by Theorem 9, $\bar{S} \geq \bar{S_L}$, where $\bar{S_L}$ is the sum of the complements of the data values of all the legitimate nodes. Let L be the set of legitimate sensor nodes, with $|L| = l$. Observe that $\bar{S_L} = \sum_{i \in L} r - a_i = lr - S_L = (n - \mu)r - S_L = nr - (S_L + \mu r)$. We have that $S + \bar{S} = nr$ and $\bar{S} \geq nr - (S_L + \mu r)$. Substituting, $S = nr - \bar{S} \leq S_L + \mu r$. Hence, $S_L \leq S \leq (S_L + \mu r)$. □

Note that nowhere was it assumed that the malicious nodes were constrained to reporting data values between $[0, r]$: in fact it is possible to have malicious nodes with data values above r or below 0 without risking detection if $S_L \leq S \leq (S_L + \mu r)$.

Theorem 13 *The SUM algorithm is optimally secure.*

PROOF. Let the sum of the data values of all the legitimate nodes be S_L. Consider an adversary with μ malicious nodes which only performs direct data injection attacks. Recall that in a direct data injection attack, an adversary only causes the nodes under its control to each report a data value within the legal range $[0, r]$. The lowest result the adversary can induce is by setting all its malicious nodes to have data value 0; in this case the computed aggregate is S_L. The highest result the adversary can induce is by setting all μ nodes under its control to yield the highest value r. In this case the computed aggregate is $S_L + \mu r$. Clearly any aggregation value between these two extremes is also achievable by direct data injection. The bound proven in Theorem 12 falls exactly on the range of possible results achievable by direct data injection, hence the algorithm is optimal by Definition 2. □

The optimal security property holds regardless of the number or fraction of malicious nodes; this is significant since the security property holds in general, and not just for a subclass of attacker multiplicities. For example, we do not assume that the attacker is limited to some ρ fraction of the nodes in the network.

5.2 Congestion Complexity

We now consider the congestion induced by the secure SUM algorithm. Recall that node congestion is defined as the communication load on the most heavily loaded sensor node in the network, and edge congestion is the heaviest communication load on a given link in the network. We only need to consider the case where the adversary is not performing an attack. If the adversary attempts to send more messages than the proven congestion bound, legitimate nodes can easily detect this locally and either raise an alarm or refuse to respond with their confirmation values, thus exposing the presence of the adversary. Recall that when we refer to a *vertex* sending and receiving information, we are referring to the commitment tree overlay network that lies over the actual physical aggregation tree.

Theorem 14 *Each vertex u receives the labels of its off-path vertices and no others.*

PROOF. Since, when the vertices are disseminating their labels in the result-checking phase, every vertex always forwards any labels received from its parents to both its children, it is clear that when a label is forwarded to a vertex u', it is eventually forwarded to the entire subtree rooted at u'.

By definition, every off-path vertex u_1 of u has a parent p which is a node on the path between u and the root of its commitment tree. By construction, p sends the label of u_1 to its sibling u_2 which is on the path to u (i.e., either u_2 is an ancestor of u, or $u_2 = u$). Hence, the label u_1 is eventually forwarded to u. Every vertex u'_1 that is not an off-path vertex has a sibling u'_2 which is not on the path between u and the root of its commitment tree. Hence, u is not in the subtree rooted at u'_2. Since the label of u'_1 is only forwarded to the subtree rooted at its sibling and nowhere else, the label of u'_1 never reaches u. □

Theorem 15 *The SUM algorithm induces $O(\log^2 n)$ edge congestion (and hence $O(\Delta \log^2 n)$ node congestion) in the aggregation tree.*

PROOF. Every step in the algorithm except the label dissemination step involves either broadcast or convergecast of messages that are at most $O(\log n)$ size. The label-dissemination step is the dominating factor.

Consider an arbitrary edge in the commitment-tree between parent vertex x and child vertex y. In the label dissemination step, messages are only sent from parent to child in the commitment tree. Hence the edge xy carries exactly the labels that y receives. From Theorem 14, y receives $O(\log n)$ labels, hence the total number of labels passing through xy is $O(\log n)$. Hence, the edge congestion in the *commitment* tree is $O(\log n)$. Now consider an arbitrary aggregation tree edge with parent node u and child node v. The child node v presents (i.e., sends) at most $\log n$ commitment-tree vertices to its parent u, and hence the edge uv is responsible for carrying traffic on behalf of at most $\log n$ commitment-tree edges — these are the edges incident on the commitment tree vertices that v presented to u. Note that v may not be responsible for creating all the vertices that it presents to u, but v is nonetheless responsible for forwarding the messages down to the sensor nodes which created those vertices. Since each edge in the commitment tree has $O(\log n)$ congestion, and each edge in the aggregation tree carries traffic for at most $\log n$ commitment-tree edges, the edge congestion in the aggregation tree is $O(\log^2 n)$. The node-congestion bound of $O(\Delta \log^2 n)$ follows from the $O(\log^2 n)$ edge congestion and the definition of Δ as the greatest degree in the aggregation tree. □

6. OTHER AGGREGATION FUNCTIONS

In this section we briefly discuss how to use the SUM algorithm as a primitive for the COUNT, AVERAGE and Φ-QUANTILE aggregates.

The COUNT Aggregate. The query COUNT is generally used to determine the total number of nodes in the network with some property; without loss of generality it can be considered a SUM aggregation where all the nodes have value either 1 (the node has the property) or 0 (otherwise). More formally, each sensor node s has a data value $a_s \in \{0, 1\}$, and we wish to compute $f(a_1, \ldots, a_n) = a_1 + a_2 + \cdots + a_n$. Since count is a special case of SUM, we can use the basic algorithm for SUM without modification.

The AVERAGE Aggregate. The AVERAGE aggregate can be computed by first computing the SUM of data values over the nodes of interest, and then the COUNT of the number of nodes of interest, and then dividing the SUM by the COUNT.

The Φ-QUANTILE Aggregate. In the Φ-QUANTILE aggregate, we wish to find the value that is in the Φn-th position in the sorted list of data values. For example, the median is a special case where $\Phi = 0.5$. Without loss of generality we can assume that all the data values are distinct; ties can be broken using unique node IDs.

If we wished to verify the correctness of a proposed Φ-quantile q, we can perform a COUNT computation where each node s presents a value $a'_s = 1$ if its data value $a_s \leq q$ and presents $a'_s = 0$ otherwise. If q is the Φ-quantile, then the computed sum should be equal to Φn. Hence, we can use any insecure approximate Φ-quantile aggregation scheme to compute a proposed Φ-quantile, and then securely test to see if the result truly is within the approximation bounds of the Φ-quantile algorithm.

7. CONCLUSION

In-network data aggregation is an important primitive for sensor network operation. The strong standard threat model of multiple Byzantine nodes in sensor networks requires the use of aggregation techniques that are robust against malicious result-tampering by covert adversaries.

We present the first optimally secure aggregation scheme for arbitrary aggregator topologies and multiple malicious nodes. This contribution significantly improves on prior work which requires strict limitations on aggregator topology or malicious node multiplicity, or which only yields a probabilistic security bound. Our algorithm is based on a novel method of distributing the verification of aggregation results onto the sensor nodes, and combining this with a unique technique for balancing commitment trees to achieve sublinear congestion bounds. The algorithm induces $O(\Delta \log^2 n)$ node congestion (where Δ is the maximum degree in the aggregation tree) and provides the strongest security bound that can be proven for any secure aggregation scheme without making assumptions about the distribution of data values.

8. REFERENCES

[1] H. Cam, S. Ozdemir, P. Nair, D. Muthuavinashiappan, and H. O. Sanli. Energy-efficient secure pattern based data aggregation for wireless sensor networks. *Computer Communications*, 29:446–455, 2006.

[2] C. Castelluccia, E. Mykletun, and G. Tsudik. Efficient aggregation of encrypted data in wireless sensor networks. In *Proceedings of The Second Annual International Conference on Mobile and Ubiquitous Systems*, 2005.

[3] J.-Y. Chen, G. Pandurangan, and D. Xu. Robust computation of aggregates in wireless sensor networks: distributed randomized algorithms and analysis. In *Proceedings of the Fourth International Symposium on Information Processing in Sensor Networks*, 2005.

[4] W. Du, J. Deng, Y. Han, and P. K. Varshney. A witness-based approach for data fusion assurance in wireless sensor

networks. In *Proceedings of the IEEE Global Telecommunications Conference*, 2003.

[5] J. Girao, M. Schneider, and D. Westhoff. CDA: Concealed data aggregation in wireless sensor networks. In *Proceedings of the ACM Workshop on Wireless Security*, 2004.

[6] M. B. Greenwald and S. Khanna. Power-conserving computation of order-statistics over sensor networks. In *Proceedings of the twenty-third ACM SIGMOD-SIGACT-SIGART symposium on Principles of database systems*, 2004.

[7] I. Gupta, R. van Renesse, and K. P. Birman. Scalable fault-tolerant aggregation in large process groups. In *Proceedings of the International Conference on Dependable Systems and Networks*, 2001.

[8] L. Hu and D. Evans. Secure aggregation for wireless networks. In *Workshop on Security and Assurance in Ad hoc Networks*, 2003.

[9] C. Intanagonwiwat, D. Estrin, R. Govindan, and J. Heidemann. Impact of network density on data aggregation in wireless sensor networks. In *Proceedings of the 22nd International Conference on Distributed Computing Systems*, 2002.

[10] P. Jadia and A. Mathuria. Efficient secure aggregation in sensor networks. In *Proceedings of the 11th International Conference on High Performance Computing*, 2004.

[11] S. Madden, M. J. Franklin, J. M. Hellerstein, and W. Hong. TAG: a tiny aggregation service for ad-hoc sensor networks. *SIGOPS Oper. Syst. Rev.*, 36(SI):131–146, 2002.

[12] S. Madden, M. J. Franklin, J. M. Hellerstein, and W. Hong. The design of an acquisitional query processor for sensor networks. In *Proceedings of the 2003 ACM International Conference on Management of Data*, 2003.

[13] A. Mahimkar and T. Rappaport. SecureDAV: A secure data aggregation and verification protocol for sensor networks. In *Proceedings of the IEEE Global Telecommunications Conference*, 2004.

[14] A. Manjhi, S. Nath, and P. B. Gibbons. Tributaries and deltas: efficient and robust aggregation in sensor network streams. In *Proceedings of the ACM International Conference on Management of Data*, 2005.

[15] S. Nath, P. B. Gibbons, S. Seshan, and Z. R. Anderson. Synopsis diffusion for robust aggregation in sensor networks. In *Proceedings of the 2nd International Conference on Embedded Networked Sensor Systems*, 2004.

[16] A. Perrig, R. Szewczyk, J. D. Tygar, V. Wen, and D. E. Culler. SPINS: Security protocols for sensor networks. *Wirel. Netw.*, 8(5):521–534, 2002.

[17] B. Przydatek, D. Song, and A. Perrig. SIA: Secure information aggregation in sensor networks. In *Proceedings of the 1st International Conference on Embedded Networked Sensor Systems*, 2003.

[18] D. Wagner. Resilient aggregation in sensor networks. In *Proceedings of the 2nd ACM Workshop on Security of Ad-hoc and Sensor Networks*, 2004.

[19] Y. Yang, X. Wang, S. Zhu, and G. Cao. SDAP: A secure hop-by-hop data aggregation protocol for sensor networks. In *Proceedings of the ACM International Symposium on Mobile Ad Hoc Networking and Computing*, 2006.

[20] Y. Yao and J. Gehrke. The COUGAR approach to in-network query processing in sensor networks. *SIGMOD Rec.*, 31(3):9–18, 2002.

APPENDIX
A. PROOF OF LEMMA 8

We first prove the following:

Lemma 16 *Let F be a collection of commitment trees of height at most h. Suppose there is a set U of accounted-for leaf-vertices with distinct labels $u_1, \ldots, u_m$ and committed values $v_1, \ldots, v_m$ in F. Let the set of trees that contain at least one member of U be T_F. Define $val(X)$ for any forest X to be the total of the aggregation values at the roots of the trees in X. Then $val(T_F) \geq \sum_{i=1}^{m} v_i$.*

PROOF. Proof: By induction on h.

Base case: $h = 0$. Then all the trees are singleton-trees. The total aggregation value of all the singleton-trees that contain at least one member of U is exactly $\sum_{i=1}^{m} v_i$.

Induction step: Assume the theorem holds for h, and consider an arbitrary collection F of commitment trees with at most height $h + 1$ where the premise holds. If there are no trees of height $h + 1$ then we are done. Otherwise, let the set R be all the root vertices of the trees of height $h + 1$. Consider $F' = F \backslash R$, i.e., remove all the vertices in R from F. The result is a collection of trees with height at most h. Let $T_{F'}$ be the set of trees in F' containing at least one member of U. The induction hypothesis holds for F', so $val(T_{F'}) \geq \sum_{i=1}^{m} v_i$. We now show that replacing the vertices from R cannot produce an T_F such that $val(T_F) < val(T_{F'})$. Each vertex r from R is the root of two subtrees of height h in F. We have three cases:

Case 1: Neither subtree contains any members of U. Then the new tree contains no members of U, and so is not a member of T_F.

Case 2: One subtree t_1 contains members of U. Since all the members of U are accounted-for, this implies that there is no inconsistency at r. Hence, the subtree without a member of U must have a non-negative aggregate value. We know that r performs the aggregate sum correctly over its inputs, so it must have aggregate value at least equal to the aggregate value of t_1.

Case 3: Both subtrees contain members of U. Since all the members of U are accounted-for, this implies that there is no inconsistency at r. The aggregate result of r is exactly the sum of the aggregate values of the two subtrees.

In case 2 and 3, the aggregate values of the roots of the trees of height $h + 1$ that were in T_F, was no less than the sum of the aggregate values of their constituent subtrees in $T_{F'}$. Hence, $val(T_F) \geq val(T_{F'}) \geq \sum_{i=1}^{m} v_i$. □

Let the commitment forest in Lemma 8 be F. Let the set of trees in F that contain at least one of the accounted-for leaf-vertices be T. By the above lemma, $val(T) \geq \sum_{i=1}^{m} v_i$. We know that there are no root labels with negative aggregation values in the commitment forest, otherwise the querier would have rejected the result. Hence, $val(F) \geq val(T) \geq \sum_{i=1}^{m} v_i$. ∎

Provably-Secure Time-Bound Hierarchical Key Assignment Schemes

Giuseppe Ateniese[1,*], Alfredo De Santis[2], Anna Lisa Ferrara[2], and Barbara Masucci[2]

[1]Department of Computer Science, The Johns Hopkins University, Baltimore, MD 21218, USA
ateniese@cs.jhu.edu

[2]Dipartimento di Informatica ed Applicazioni, Università di Salerno, 84084 Fisciano (SA), Italy
{ads, ferrara, masucci}@dia.unisa.it

ABSTRACT

A *time-bound hierarchical key assignment scheme* is a method to assign time-dependent encryption keys to a set of classes in a partially ordered hierarchy, in such a way that the key of a higher class can be used to derive the keys of all classes lower down in the hierarchy, according to temporal constraints.

In this paper we design and analyze time-bound hierarchical key assignment schemes which are provably-secure and efficient. We first consider the *unconditionally secure setting* and we show a tight lower bound on the size of the private information distributed to each class. Then, we consider the *computationally secure setting* and obtain several results: We first prove that a recently proposed scheme is insecure against collusion attacks. Hence, motivated by the need for *provably-secure* schemes, we propose two different constructions for time-bound hierarchical key assignment schemes. The first one is based on symmetric encryption schemes, whereas, the second one makes use of bilinear maps. These appear to be the first constructions of time-bound hierarchical key assignment schemes which are simultaneously *practical* and *provably-secure*.

Categories and Subject Descriptors: K.6.5 [Management of Computing and Information Systems]: Security and Protection

General Terms: Security.

Keywords: Access control, key assignment, provable security.

1. INTRODUCTION

Users of a computer system could be organized in a hierarchy of disjoint classes. These classes, called *security classes*, are positioned and ordered within the hierarchy based on

*Supported in part by NSF.

the fact that some users have more access rights than others. For instance, supervisors may have the right to access data stored by subordinates while subordinates cannot access any of the supervisors' data. Such cases abound in several areas, particularly in the government and military.

A *hierarchical key assignment scheme* is a method to assign an encryption key and some private information to each class in the hierarchy. The encryption key will be used by each class to protect its data by means of a symmetric cryptosystem, whereas, the private information will be used by each class to compute the keys assigned to all classes lower down in the hierarchy. This assignment is carried out by a central authority, the Trusted Authority (TA), which is active only at the distribution phase. Akl and Taylor [1] first proposed an elegant hierarchical key assignment scheme. Subsequently, many researchers have proposed schemes that either have better performance or allow insertion and deletion of classes in the hierarchy (e.g., [2, 22, 24, 27, 28, 31]). Despite the large number of proposed schemes, many of them lack a formal security proof and have been shown to be insecure against collusive attacks [11, 33, 40]. Atallah et al. [2] first addressed the problem of formalizing security requirements for hierarchical key assignment schemes. A scheme is *provably-secure* under a complexity assumption if the existence of an adversary A breaking the scheme is equivalent to the existence of an adversary B breaking the computational assumption. The usual method of construction of B uses the adversary A as a black-box. Atallah et al. [2] proposed a first provably-secure construction based on pseudorandom functions and a second one requiring the use of a symmetric encryption scheme secure against chosen-ciphertext attacks. Atallah et al. [2, 3] also considered the problem of reducing the number of steps required to perform key derivation in hierarchical key assignment schemes. In [17] constructions for efficient provably-secure key assignment schemes are presented. In particular, one construction provides constant private information and public information linear in the number of the classes.

All the above schemes would assign keys that never expire and new keys are generated only after inserting or deleting classes in the hierarchy. However, in practice, it is likely that a user may be assigned to a certain class for only a certain period of time. In such cases, users need a different key for each time period which implies that the key derivation procedure should also depend on the time period other than the hierarchy of the classes. Once a time period expires, users in a class should not be able to access any subsequent keys if they are not authorized to do so. As pointed out by Tzeng [35], there are several applications requiring a time-based access control. For example, a web-based electronic newspaper company could offer several types of subscription packages, covering different topics. Each user may decide to subscribe to one package for a certain period of time (e.g., a week, a month, or a year). Subscription packages could be structured to form a partially ordered hierarchy where leaf nodes represent different topics. For each time period, an encryption key is then assigned to each leaf node in the hierarchy. This key is then computed by each user that subscribes to that package and for that period of time. A similar solution was employed by Bertino et al. [6], who showed how to control access to an XML document according to temporal constraints.

A basic and straightforward way to achieve a time-based access control is to require each user to memorize encryption keys assigned to all classes lower down in the hierarchy for each time period in which the user is allowed to access their data. Tzeng [35] first addressed the problem of reducing the inherent complexity of such a solution and proposed a *time-bound* hierarchical key assignment scheme that requires each user to store information whose size does not depend on the number of keys that the user has access to or on the number of time periods. However, his scheme is very costly since each user must perform expensive computations in order to compute a legitimate key. Most importantly, Tzeng's scheme has been shown to be insecure against collusive attacks, whereby two or more users, assigned to some classes in distinct time periods, collude to compute a key to which they are not entitled [38]. Subsequently, Chien [12] proposed an efficient time-bound hierarchical key assignment scheme based on tamper-resistant devices. However, it was shown that malicious users can collusively misuse their devices to gain unauthorized accesses in [15, 37], where countermeasures were also proposed. Another time-bound hierarchical key assignment scheme was proposed by Huang and Chang [23] and later shown to be insecure against collusive attacks [34]. An RSA-based time-bound hierarchical key assignment scheme was proposed by Yeh [36], who claimed his scheme to be secure against collusive attacks. Recently, a modification of the Akl-Taylor scheme was used by Wang and Laih [39] to construct a time-bound hierarchical key assignment scheme.

1.1 Our Results

In this paper we design and analyze time-bound hierarchical key assignment schemes which are efficient and provably-secure with respect to *key indistinguishability*. Security with respect to key indistinguishability formalizes the requirement that the adversary is not able *to learn any information* about a key that it should not have access to, i.e., it is not able to distinguish it from a random string having the same length.

- We first consider an *information-theoretic* approach to time-bound hierarchical key assignment schemes. We prove a tight lower bound on the size of the private information distributed to each class.

- Afterwards, we prove that a recently proposed scheme [36] is insecure against collusive attacks.

- Finally, we propose two different constructions for time-bound key assignment schemes. The first one is based on symmetric encryption schemes, whereas, the second one makes use of bilinear maps. These appear to be the first constructions of time-bound hierarchical key assignment schemes which are simultaneously *practical* and *provably-secure*.

2. TIME-BOUND HIERARCHICAL KEY ASSIGNMENT SCHEMES

Consider a set of users divided into a number of disjoint classes, called *security classes*. A binary relation $\preceq$ that partially orders the set of classes V is defined in accordance with authority, position, or power of each class in V. The poset $(V, \preceq)$ is called a *partially ordered hierarchy*. For any two classes u and v, the notation $u \preceq v$ is used to indicate that the users in v can access u's data. We denote by A_v the set $\{u \in V : u \preceq v\}$, for any $v \in V$. The partially ordered hierarchy $(V, \preceq)$ can be represented by a directed graph, where each class corresponds to a vertex in the graph and there is an edge from class v to class u if and only if $u \preceq v$. Further, this graph could be simplified by eliminating all self-loops and edges which can be implied by the property of the transitive closure. We denote by $G = (V, E)$ the resulting directed acyclic graph.

In this paper we consider the case where a user may be in a class for only a period of time. We consider a sequence $T = (t_1, \ldots, t_{|T|})$ composed of distinct time periods. In the following we denote by $t \in T$ the fact that the time period t belongs to the sequence T. Each user may belong to a class for a certain non-empty contiguous subsequence λ of T. Let $\mathcal{P}$ be the set of all nonempty contiguous subsequences of T. Such a set is called the *interval-set* over T. A *time-bound hierarchical key assignment scheme* is a method to assign a private information $s_{v,\lambda}$ to each class $v \in V$ for each time sequence $\lambda \in \mathcal{P}$ and an encryption key $k_{u,t}$ to each class $u \in V$ for each time period $t \in T$. The generation and distribution of the private information and keys is carried out by a trusted third party, the TA, which is connected to each class by means of a secure channel. The encryption key $k_{u,t}$ can be used by users belonging to class u in time period t to protect their sensitive data by means of a symmetric cryptosystem, whereas, the private information $s_{v,\lambda}$ can be used by users belonging to class v for the time sequence λ to compute the key $k_{u,t}$ for any class $u \in A_v$ and each time period $t \in \lambda$.

An ideal time-bound hierarchical key assignment scheme should have low storage requirements and provide for efficient key derivation procedures. In addition, unauthorized users should not be able to compute keys to which they have no access right. More precisely, for each class $u \in V$ and each time period $t \in T$, the key $k_{u,t}$ should be protected against a coalition of users belonging to each class v such that $u \notin A_v$ in all time periods, and users belonging to each class w such that $u \in A_w$ in all time periods but t. We denote by $F_{u,t}$ the set $\{(v, \lambda) \in V \times \mathcal{P} : u \notin A_v \text{ or } t \notin \lambda\}$, corresponding to all users which are not allowed to compute the key $k_{u,t}$.

We refer to an *unconditionally secure* time-bound hierarchical key assignment scheme if its security relies on the theoretical impossibility of breaking it, despite the computational power of the coalition, whereas, we refer to a *computationally secure* time-bound hierarchical key assignment scheme if its security relies on the computational infeasibility of breaking it, according to some specific computational assumptions.

3. THE UNCONDITIONALLY SECURE SETTING

In this section we formally define unconditionally secure time-bound hierarchical key assignment schemes by using the entropy function (we refer the reader to [13] for a complete treatment of Information Theory). The same approach has been used in [16] to analyze key assignment schemes without temporal constraints.

For any class $u \in V$ and any time sequence $\lambda \in \mathcal{P}$, we denote by $S_{u,\lambda}$ and $K_{u,t}$ the sets of all possible values that $s_{u,\lambda}$ and $k_{u,t}$ can assume, respectively. Given a set of pairs $X = \{(u_1, \lambda_1), \cdots, (u_\ell, \lambda_\ell)\} \subseteq V \times \mathcal{P}$, we denote by S_X the set $S_{u_1,\lambda_1} \times \cdots \times S_{u_\ell,\lambda_\ell}$. In the following, with a boldface capital letter, say $\mathbf{Y}$, we denote a random variable taking values on a set, denoted by the corresponding capital letter Y, according to some probability distribution $\{Pr_{\mathbf{Y}}(y)\}_{y \in Y}$. The values such a random variable can take are denoted by the corresponding lower case letter. Given a random variable $\mathbf{Y}$, we denote by $H(\mathbf{Y})$ the Shannon entropy of $\{Pr_{\mathbf{Y}}(y)\}_{y \in Y}$. An unconditionally secure time-bound hierarchical key assignment scheme is defined as follows.

Definition 1. Let $G = (V, E)$ be the directed acyclic graph corresponding to a partially ordered hierarchy, let T be a sequence of distinct time periods, and let $\mathcal{P}$ be the interval-set over T. An *unconditionally secure time-bound hierarchical key assignment scheme* for G and $\mathcal{P}$ is a method to assign a private information $s_{u,\lambda}$ to each class $u \in V$, for each time sequence $\lambda \in \mathcal{P}$, and an encryption key $k_{u,t}$ to each class $u \in V$, for each time period $t \in T$, in such a way that the following two properties are satisfied:

Correctness. *Each user can compute the key held by any class lower down in the hierarchy for each time period in which it belongs to its class.*

Formally, for each class $v \in V$, each class $u \in A_v$, each time sequence $\lambda \in \mathcal{P}$, and each time period $t \in \lambda$, it holds that $H(\mathbf{K}_{u,t}|\mathbf{S}_{v,\lambda}) = 0$.

Security. *Any coalition of users have absolutely no information about any key the coalition is not entitled to obtain.*

Formally, for each class $u \in V$, each time period $t \in T$, and each coalition of users $X \subseteq F_{u,t}$, it holds that $H(\mathbf{K}_{u,t}|\mathbf{S}_X) = H(\mathbf{K}_{u,t})$.

The next theorem, whose proof can be found in the full version of this paper [4], states a lower bound on the size of the private information distributed to each user in any unconditionally secure time-bound hierarchical key assignment scheme. Such a result applies to the general case of arbitrary entropies of keys, but, for the sake of simplicity, we consider the case when all entropies of keys are equal. We denote this common entropy by $H(\mathbf{K})$.

THEOREM 1. *Let $G = (V, E)$ be the directed acyclic graph corresponding to a partially ordered hierarchy, let T be a sequence of distinct time periods, and let $\mathcal{P}$ be the interval-set over T. In any unconditionally secure time-bound hierarchical key assignment scheme for G and $\mathcal{P}$, for any pair* $(u, \lambda) \in V \times \mathcal{P}$, *it holds that*

$$H(\mathbf{S}_{u,\lambda}) \geq |A_u| \cdot |\lambda| \cdot H(\mathbf{K}),$$

where $|\lambda|$ denotes the number of time periods in the time sequence λ.

The above bound is tight. Indeed the straightforward unconditionally secure time-bound key assignment scheme, in which each user memorizes the encryption keys assigned to all classes lower down in the hierarchy, for all time periods in which it is entitled to access their data, meets the above bound with equality.

4. THE COMPUTATIONALLY SECURE SETTING

In this section we consider time-bound key assignment schemes based on specific computational assumptions. In this setting, we obtain several results of interest. We first provide a notion of security with respect to *key indistinguishability* against attacks carried out by *static* adversaries. Afterwards, we prove that Yeh's scheme [36] is not secure against collusive attacks. Finally, we propose two different constructions of time-bound hierarchical key assignment schemes. The first one is based on symmetric encryption schemes, whereas, the second one makes use of bilinear maps. Both constructions are provably-secure and efficient.

We use the standard notation to describe probabilistic algorithm and experiments following [21]. If $A(\cdot, \cdot, \ldots)$ is any probabilistic algorithm then $a \leftarrow A(x, y, \ldots)$ denotes the experiment of running A on inputs $x, y, \ldots$ and letting a be the outcome, the probability being over the coins of A. Similarly, if X is a set then $x \leftarrow X$ denotes the experiment of selecting an element uniformly from X and assigning x this value. If w is neither an algorithm nor a set then $x \leftarrow w$ is a simple assignment statement. A function $\epsilon : N \rightarrow R$ is *negligible* if for every constant $c > 0$ there exists an integer n_c such that $\epsilon(n) < n^{-c}$ for all $n \geq n_c$.

We define a time-bound hierarchical key assignment scheme as follows.

Definition 2. A *time-bound hierarchical key assignment scheme* is a pair (Gen, Der) of algorithms satisfying the following conditions:

1. The *information generation algorithm Gen* is probabilistic polynomial-time. It takes as inputs the security parameter 1^τ, the directed acyclic graph $G = (V, E)$ corresponding to a partially ordered hierarchy, and the interval-set $\mathcal{P}$ over a sequence of distinct time periods T, and produces as outputs

    (a) a private information $s_{u,\lambda}$, for any class $u \in V$ and any time sequence $\lambda \in \mathcal{P}$;

    (b) a key $k_{u,t}$, for any class $u \in V$ and any time period $t \in T$;

    (c) a public information pub.

We denote by (s, k, pub) the output of the algorithm Gen on inputs 1^τ, G, and $\mathcal{P}$, where s and k denote the sequences of private information and of keys, respectively.

2. The *key derivation algorithm Der* is deterministic polynomial-time. It takes as inputs the security parameter 1^τ, the directed acyclic graph $G = (V, E)$ corresponding to a partially ordered hierarchy, the interval-set $\mathcal{P}$ over a sequence of distinct time periods T, two classes u and v such that $u \in A_v$, a time sequence $\lambda \in \mathcal{P}$, the private information $s_{v,\lambda}$ assigned to class v for the time sequence λ, a time period $t \in \lambda$, and the public information pub, and produces as output the key $k_{u,t}$ assigned to class u at time period t.

We require that for each class $v \in V$, each class $u \in A_v$, each time sequence $\lambda \in \mathcal{P}$, each time period $t \in \lambda$, each private information $s_{v,\lambda}$, each key $k_{u,t}$, each public information pub which can be computed by Gen on inputs 1^τ, G, and $\mathcal{P}$, it holds that

$$Der(1^\tau, G, \mathcal{P}, u, v, \lambda, s_{v,\lambda}, t, pub) = k_{u,t}.$$

We consider security with respect to *key indistinguishability*. Such a requirement, first introduced by Atallah et al. [2], formalizes the fact that the adversarial coalition is not able to *distinguish* a key, that should not be accessible by any user of the coalition, from a random string of the same length. We consider a *static adversary* $\mathtt{STAT}_u$ which wants to attack a class $u \in V$ and which is able to corrupt *all* users in F_u. We define an algorithm $Corrupt_{u,t}$ which, on input the private information s generated by the algorithm Gen, extracts the secret values $s_{v,\lambda}$ associated to all pairs $(v, \lambda) \in F_{u,t}$. We denote by $corr$ the sequence output by $Corrupt_{u,t}(s)$. The computations performed by the adversary involve all public information generated by the algorithm Gen, as well as the private information $corr$ held by the corrupted users. Two experiments are considered. In the first one, the adversary is given the key $k_{u,t}$, whereas, in the second one, it is given a random string ρ having the same length as $k_{u,t}$. It is the adversary's job to determine whether the received challenge corresponds to $k_{u,t}$ or to a random string. We require that the adversary will succeed with probability only negligibly different from $1/2$.

Definition 3. [IND-ST] Let $G = (V, E)$ be the directed acyclic graph corresponding to a partially ordered hierarchy, let T be a sequence of distinct time periods, let $\mathcal{P}$ be the interval-set over T, and let (Gen, Der) be a time-bound hierarchical key assignment scheme for G and $\mathcal{P}$. Let $\mathtt{STAT}_{u,t}$ be a static adversary which attacks a class $u \in V$ in a time period $t \in T$. Consider the following two experiments:

$$\begin{aligned}
&Experiment\ \mathbf{Exp}^{\text{IND-1}}_{\mathtt{STAT}_{u,t}}(1^\tau) \\
&\quad (s, k, pub) \leftarrow Gen(1^\tau, G, \mathcal{P}) \\
&\quad corr \leftarrow Corrupt_{u,t}(s) \\
&\quad d \leftarrow \mathtt{STAT}_{u,t}(1^\tau, G, \mathcal{P}, pub, corr, k_{u,t}) \\
&\quad \mathbf{return}\ d
\end{aligned}$$

$$\begin{aligned}
&Experiment\ \mathbf{Exp}^{\text{IND-0}}_{\mathtt{STAT}_{u,t}}(1^\tau) \\
&\quad (s, k, pub) \leftarrow Gen(1^\tau, G, \mathcal{P}) \\
&\quad corr \leftarrow Corrupt_{u,t}(s) \\
&\quad \rho \leftarrow \{0,1\}^{length(k_{u,t})} \\
&\quad d \leftarrow \mathtt{STAT}_{u,t}(1^\tau, G, \mathcal{P}, pub, corr, \rho) \\
&\quad \mathbf{return}\ d
\end{aligned}$$

The advantage of $\mathtt{STAT}_{u,t}$ is defined as

$\mathbf{Adv}^{\text{IND}}_{\mathtt{STAT}_{u,t}}(1^\tau)$
$= |Pr[\mathbf{Exp}^{\text{IND-1}}_{\mathtt{STAT}_{u,t}}(1^\tau) = 1] - Pr[\mathbf{Exp}^{\text{IND-0}}_{\mathtt{STAT}_{u,t}}(1^\tau) = 1]|.$

The scheme is said to be *secure in the sense of* IND-ST if, for each class $u \in V$ and each time period $t \in T$, the function $\mathbf{Adv}^{\text{IND}}_{\mathtt{STAT}_{u,t}}(1^\tau)$ is negligible, for each static adversary $\mathtt{STAT}_{u,t}$ whose time complexity is polynomial in τ.

In Definition 3 we have considered a static adversary attacking a class. A different kind of adversary, the *adaptive* one, could also be considered. Such an adversary is first allowed to access all public information as well as all private information of a number of users of its choice; afterwards, it chooses the class u it wants to attack. In the full version of this paper [4], it is shown that security against adaptive adversaries is (polynomially) equivalent to the security against static adversaries. Hence, in this paper we will only consider static adversaries.

4.1 A Collusion Attack to Yeh's Scheme

In this section we describe a security weakness of Yeh's scheme [36], in particular we show that in some cases users can collude and compute encryption keys that they should not know. Yeh's scheme is described in the following.

Algorithm $Gen(1^\tau, G, \mathcal{P})$
The TA performs the following steps:

1. Randomly chooses two distinct large primes p and q and computes $n = p \cdot q$, having bitlength τ, and $\phi(n) = (p-1)(q-1)$;
2. For each class $u \in V$, randomly chooses a public integer e_u such that $gcd(e_u, \phi(n)) = 1$;
3. For each time period $t \in T$, randomly chooses a public integer g_t such that $gcd(g_t, \phi(n)) = 1$;
4. Let pub be the sequence of public information computed in the previous two steps;
5. For each class $u \in V$, computes the secret integer d_u, such that $e_u \cdot d_u = 1 \bmod \phi(n)$;
6. For each time period $t \in T$, computes the secret integer h_t such that $g_t \cdot h_t = 1 \bmod \phi(n)$;
7. Chooses a random integer k_0, where $1 < k_0 < n$, and for each class $u \in V$ computes a class key $k_u = k_0^{\prod_{v \in A_u} d_v} \bmod n$;
8. For each class $u \in V$ and each time sequence $\lambda \in \mathcal{P}$, computes the private information $s_{u,\lambda} = k_u^{\prod_{r \in \lambda} h_r} \bmod n$;
9. For each class $u \in V$ and each time period $t \in T$, computes the key $k_{u,t} = k_u^{h_t} \bmod n$.

Algorithm $Der(1^\tau, G, \mathcal{P}, u, v, \lambda, s_{u,\lambda}, t, pub)$
A user belonging to a class $u \in V$ for a time sequence $\lambda \in \mathcal{P}$ can use its private information $s_{u,\lambda}$ along with the public information pub to compute the key $k_{v,t}$, for each class v such that $v \in A_u$ and each time period $t \in \lambda$, as follows:

$$(s_{u,\lambda})^{\prod_{w \in A_u \setminus A_v} e_w \prod_{r \in \lambda\ \&\ r \neq t} g_r} = k_v^{h_t} \bmod n = k_{v,t}.$$

In order to show our attack we need the next lemma, which is a simple generalization of a result due to Shamir [32].

LEMMA 1. *Let n be the product of two distinct large primes. Given four integers $\alpha, \beta \in Z_n^*$ and $x, y \in Z$, such that $\beta^x = \alpha^y \bmod n$, it is easy to compute $\gamma \in Z_n^*$ such that $\gamma^x \bmod n = \alpha^{gcd(x,y)} \bmod n$.*

Consider the hierarchy of Figure 1 and let A and B be two users assigned to classes u and v in time sequences (t_1, t_2) and (t_2, t_3), respectively. Moreover, let $gcd(e_u, g_{t_3}) = 1$. In the following we show how users A and B can collude to compute the key k_{u,t_3}, that they should not be able to obtain. Let $k_{u,t_2} = k_0^{d_u d_w h_{t_2}} \bmod n$ and $s_{v,(t_2,t_3)} = k_0^{d_v d_w h_{t_2} h_{t_3}} \bmod n$. Let $\beta = k_{u,t_2}$ and $\alpha = (s_{v,(t_2,t_3)})^{e_v} \bmod n$. Since $\beta^{e_u} = \alpha^{g_{t_3}} \bmod n$, from Lemma 1 users A and B can efficiently compute the value γ such that $\gamma^{e_u} \bmod n = \alpha$, that is, $\gamma = k_0^{d_u d_w h_{t_2} h_{t_3}} \bmod n$. Thus, they can compute the key $k_{u,t_3} = k_0^{d_u d_w h_{t_3}} \bmod n = \gamma^{g_{t_2}} \bmod n$.

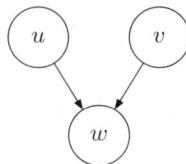

Figure 1: A partially ordered hierarchy.

More generally, let u and v be two distinct classes such that $v \notin A_u$ and $A_u \setminus \{u\} \subseteq A_v$. Let A and B be two users assigned to classes u and v in time sequences $(t_x, \ldots, t_y)$ and $(t_i, \ldots, t_j)$, respectively, where $t_1 \leq t_x < t_i \leq t_y < t_j \leq t_{|T|}$. Let $t_i \leq t \leq t_y$ and $g_t = \delta \cdot \rho$, where $\rho \geq 1$ and $\delta = gcd(e_u, g_{t'})$, for some $t_y < t' \leq t_j$. In the following we show how users A and B can collude to compute the key $k_{u,t'}$, that they should not be able to obtain. Let $\beta = k_{u,t}$ and

$$\alpha = (s_{v,(t_i,\ldots,t_j)})^{\prod_{r \in (t_i,\ldots,t_j) \; \& \; r \neq t,t'} g_r \prod_{w \in A_v \setminus A_u} e_w} \bmod n.$$

It is easy to see that

$$\beta^{e_u} = k_0^{h_t \prod_{w \in A_u \setminus \{u\}} d_w} \bmod n = \alpha^{g_{t'}} \bmod n.$$

Hence, from Lemma 1 users A and B can efficiently compute the value γ such that $\gamma^{e_u} \bmod n = \alpha^\delta \bmod n$, that is, $\gamma = k_0^{h_t \cdot h_{t'} \cdot \delta \prod_{w \in A_u} d_w} \bmod n$. Afterwards, they can compute the value

$$\gamma^\rho \bmod n = k_0^{h_{t'} \prod_{w \in A_u} d_w} \bmod n = k_{u,t'}.$$

4.2 A Scheme based on Symmetric Encryption Schemes

In this section we first show how to construct a time-bound key assignment scheme using as a building block a symmetric encryption scheme. Afterwards, we prove that the security property of the resulting time-bound key assignment scheme depends on the security property of the underlying encryption scheme. We first recall the definition of a symmetric encryption scheme.

Definition 4. A *symmetric encryption scheme* is a triple $\Pi = (\mathcal{K}, \mathcal{E}, \mathcal{D})$ of algorithms satisfying the following conditions:

1. The *key-generation algorithm* $\mathcal{K}$ is probabilistic polynomial-time. It takes as input the security parameter 1^τ and produces as output a string *key*.

2. The *encryption algorithm* $\mathcal{E}$ is probabilistic polynomial-time. It takes as inputs 1^τ, a string *key* produced by $\mathcal{K}(1^\tau)$, and a message $m \in \{0,1\}^*$, and produces as output the ciphertext y.

3. The *decryption algorithm* $\mathcal{D}$ is deterministic polynomial-time. It takes as inputs 1^τ, a string *key* produced by $\mathcal{K}(1^\tau)$, and a ciphertext y, and produces as output a message m. We require that for any string *key* which can be output by $\mathcal{K}(1^\tau)$, for any message $m \in \{0,1\}^*$, and for all y that can be output by $\mathcal{E}(1^\tau, key, m)$, we have that $\mathcal{D}(1^\tau, key, y) = m$.

In the following we describe a time-bound hierarchical key assignment scheme using $\Pi = (\mathcal{K}, \mathcal{E}, \mathcal{D})$ as a building block. The first step of the algorithm *Gen* performs a graph transformation, starting from the graph $G = (V, E)$ and $\mathcal{P}$. The output of such a transformation is a graph $G_{\mathcal{P}T} = (V_{\mathcal{P}T}, E_{\mathcal{P}T})$, where $V_{\mathcal{P}T} = V_{\mathcal{P}} \cup V_T$ and $V_{\mathcal{P}} \cap V_T = \emptyset$, constructed as follows:

- for each class $u \in V$ and each time sequence $\lambda \in \mathcal{P}$, we place a class u_λ in $V_{\mathcal{P}}$;
- for each class $u \in V$ and each time period $t \in T$, we place a class u_t in V_T;
- for each class $u \in V$, each time sequence $\lambda \in \mathcal{P}$, and each time period $t \in \lambda$, we place an edge between u_λ and u_t in $G_{\mathcal{P}T}$, i.e., $(u_\lambda, u_t) \in E_{\mathcal{P}T}$;
- for each pair of classes u and v connected by a path in G, each time sequence $\lambda \in \mathcal{P}$, and each time period $t \in \lambda$, we place an edge between u_λ and v_t in $G_{\mathcal{P}T}$, i.e., $(u_\lambda, v_t) \in E_{\mathcal{P}T}$.

Figure 2 shows an example of the graph transformation described above, where $\mathcal{P} = \{\lambda_1, \lambda_2, \lambda_3\}$, $\lambda_1 = (t_1)$, $\lambda_2 = (t_1, t_2)$, and $\lambda_3 = (t_2)$.

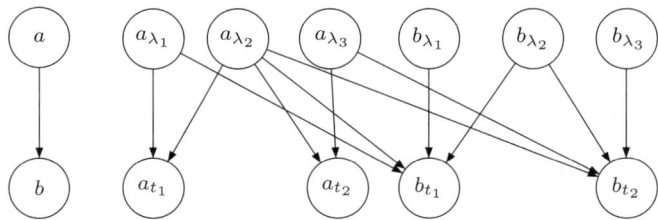

Figure 2: The graph transformation used in our construction.

Notice that in the two-level partially ordered hierarchy obtained by the above transformation the classes at the first level do not need to be assigned encryption keys, since they have no data to be protected. On the other hand, the classes at the second level do not need to perform key derivations, since there are no classes that can be accessed by them.

Algorithm $Gen(1^\tau, G, \mathcal{P})$
The TA performs the following steps:

1. Performs a graph transformation in order to obtain the two-level partially ordered hierarchy $G_{\mathcal{PT}} = (V_{\mathcal{PT}}, E_{\mathcal{PT}})$, where $V_{\mathcal{PT}} = V_{\mathcal{P}} \cup V_T$;
2. For each class u_λ in $V_{\mathcal{P}}$, let $s_{u,\lambda} \leftarrow \mathcal{K}(1^\tau)$;
3. For each class u_t in V_T, randomly chooses a secret value $k_{u,t} \in \{0,1\}^\tau$;
4. For any pair of classes $(u_\lambda, v_t) \in V_{\mathcal{P}} \times V_T$ such that $(u_\lambda, v_t) \in E_{\mathcal{PT}}$, computes the public information
$$p_{(u,\lambda),(v,t)} = \mathcal{E}_{s_{u,\lambda}}(k_{v,t});$$
5. Let pub be the sequence of public information computed in the previous step.

Algorithm $Der(1^\tau, G, \mathcal{P}, u, v, \lambda, s_{u,\lambda}, t, pub)$
A user belonging to a class $u_\lambda \in V_{\mathcal{P}}$ can use the private information $s_{u,\lambda}$ along with the public value $p_{(u,\lambda),(v,t)}$ to compute the key $k_{v,t}$ for any class $v_t \in V_T$ such that $(u_\lambda, v_t) \in E_{\mathcal{PT}}$, as follows:
$$\mathcal{D}_{s_{u,\lambda}}(p_{(u,\lambda),(v,t)}) = k_{v,t}.$$

4.2.1 Analysis of the Scheme

In order to analyze the security of our time-bound key assignment scheme we first need to define what we mean by a *secure* symmetric encryption scheme. We consider security with respect to plaintext indistinguishability, which is an adaption of the notion of *polynomial security* as given in [20]. We imagine an adversary $A = (A_1, A_2)$ running in two stages. In advance of the adversary's execution, a random key *key* is chosen and kept hidden from the adversary. During the first stage, the adversary A_1 outputs a triple $(x_0, x_1, state)$, where x_0 and x_1 are two messages of the same length, and *state* is some state information which could be useful later. One message between x_0 and x_1 is chosen at random and encrypted to give the challenge ciphertext y. In the second stage, the adversary A_2 is given y and *state* and has to determine whether y is the encryption of x_0 or x_1. Informally, the encryption scheme is said to be secure with respect to a non-adaptive chosen plaintext attack, denoted by IND-P1-C0 in [26], if every polynomial-time adversary A, which has access to the encryption oracle only during the first stage of the attack and has never access to the decryption oracle, succeeds in determining whether y is the encryption of x_0 or x_1 with probability only negligibly different from $1/2$.

THEOREM 2. *If the encryption scheme $\Pi = (\mathcal{K}, \mathcal{D}, \mathcal{E})$ is secure in the sense of IND-P1-C0, then our time-bound hierarchical key assignment scheme is secure in the sense of IND-ST.*

PROOF. The proof uses a standard hybrid argument. Let $u_{t^*} \in V_T$ be a class and assume there exist m classes in $V_{\mathcal{P}}$ which are able to access u_{t^*}. W.l.o.g., let $u_{1_{\lambda_1}}, \ldots, u_{m_{\lambda_m}}$ be such classes. Let STAT_{u,t^*} be a static adversary attacking class u_{t^*}. We construct a sequence of $m+1$ experiments $\mathbf{Exp}^1_{u,t^*}, \ldots, \mathbf{Exp}^{m+1}_{u,t^*}$, all defined over the same probability space. In each experiment we modify the way the view of STAT_{u,t^*} is computed, while maintaining the view's distributions indistinguishable among any two consecutive experiments. For any $q = 1, \ldots, m+1$, experiment $\mathbf{Exp}^q_{u,t^*}$ is defined as follows:

Experiment $\mathbf{Exp}^q_{u,t^*}(1^\tau)$
$(s, \alpha, pub) \leftarrow \widetilde{Gen}^q(1^\tau, G, \mathcal{P})$
$corr \leftarrow Corrupt_{u,t^*}(s)$
$d \leftarrow \text{STAT}_{u,t^*}(1^\tau, G, \mathcal{P}, pub, corr, \alpha_{u,t^*})$
return d

The algorithm $\widetilde{Gen}^q$ used in $\mathbf{Exp}^q_{u,t^*}$ is the same algorithm Gen used in our scheme with the following modification: for any $h = 1, \ldots, q-1$, the public value $p_{(v_h, \lambda_h),(u,t^*)}$ is computed as the encryption, with the key s_{v_h, λ_h}, of a random value $\beta_q \in \{0,1\}^\tau$, instead of the encryption of the key assigned to u_{t^*}, which is denoted by α_{u,t^*}. Notice that experiment $\mathbf{Exp}^1_{u,t^*}$ is the same as $\mathbf{Exp}^{\text{IND}-1}_{\text{STAT}_{u,t^*}}$. Indeed, the adversary STAT_{u,t^*} is given the value α_{u,t^*} and for each $h = 1, \ldots, m$, the public value $p_{(v_h, \lambda_h),(u,t^*)}$ computed by $\widetilde{Gen}^1$ corresponds to the encryption of α_{u,t^*}. On the other hand, experiment $\mathbf{Exp}^{m+1}_{u,t^*}$ is the same as $\mathbf{Exp}^{\text{IND}-0}_{\text{STAT}_{u,t^*}}$. Indeed, the adversary STAT_{u,t^*} is given the value α_{u,t^*} and, for each $h = 1, \ldots, m$, the public value $p_{(v_h, \lambda_h),(u,t^*)}$ computed by $\widetilde{Gen}^{m+1}$ corresponds to the encryption of the value β_{m+1}.

In the following we show that, for any $q = 2, \ldots, m+1$, the adversary's view in the $(q-1)$-th experiment is indistinguishable from the adversary's view in the q-th one, i.e., the random variables associated to such views are exactly the same. Hence, it follows that also the adversary's views in experiments $\mathbf{Exp}^{\text{IND}-1}_{\text{STAT}_{u,t^*}}$ and $\mathbf{Exp}^{\text{IND}-0}_{\text{STAT}_{u,t^*}}$ are indistinguishable.

Assume by contradiction that there exists a polynomial-time distinguisher B_q which is able to distinguish between the adversary STAT_{u,t^*}'s views in experiments $\mathbf{Exp}^{q-1}_{u,t^*}$ and $\mathbf{Exp}^q_{u,t^*}$ with non-negligible advantage. We show how to construct a polynomial-time adversary $A = (A_1, A_2)$, using B_q, which breaks the security of the encryption scheme $\Pi = (\mathcal{K}, \mathcal{E}, \mathcal{D})$ in the sense of IND-P1-C0. The algorithm A_1, on input 1^τ, makes queries to the encryption oracle $\mathcal{E}_{key}(\cdot)$ and outputs a triple $(x_0, x_1, state)$, where $x_0, x_1 \in \{0,1\}^\tau$, and *state* is some state information.

Algorithm $A_1^{\mathcal{E}_{key}(\cdot)}(1^\tau)$
$x_0 \leftarrow \{0,1\}^\tau$
$x_1 \leftarrow \{0,1\}^\tau$
//*construction of secret values*
for any $w_\lambda \in V_{\mathcal{P}} \setminus \{v_{q_{\lambda_q}}\}$
$\quad s_{w,\lambda} \leftarrow \mathcal{K}(1^\tau)$
for any $w_t \in V_T \setminus \{u_{t^*}\}$
$\quad k_{w,t} \leftarrow \{0,1\}^\tau$
//*construction of public values*
for $h = 1, \ldots, q-1$
$\quad p_{(v_h, \lambda_h),(u,t^*)} \leftarrow \mathcal{E}_{s_{v_h, \lambda_h}}(x_1)$
for $h = q+1, \ldots, m$
$\quad p_{(v_h, \lambda_h),(u,t^*)} \leftarrow \mathcal{E}_{s_{v_h, \lambda_h}}(x_0)$
for any $w_t \in V_T \setminus \{u_{t^*}\}$ s.t. $(v_{q_{\lambda_q}}, w_t) \in E_{\mathcal{PT}}$
$\quad p_{(v_q, \lambda_q),(w,t)} \leftarrow \mathcal{E}_{key}(k_{w,t})$
for any $z_\lambda \in V_{\mathcal{P}} \setminus \{v_{q_{\lambda_q}}\}, w_t \in V_T \setminus \{u_{t^*}\}$ s.t. $(z_\lambda, w_t) \in E_{\mathcal{PT}}$
$\quad p_{(z,\lambda),(w,t)} \leftarrow \mathcal{E}_{s_{z,\lambda}}(k_{w,t})$
//*construction of the view*
$pub' \leftarrow$ all public values constructed as above
$corr \leftarrow$ secret values held by classes in
$\quad \{w_\lambda \in V_{\mathcal{P}} : (w_\lambda, u_{t^*}) \notin E_{\mathcal{PT}}\}$
$state \leftarrow (pub', corr, x_0, x_1)$
return $(x_0, x_1, state)$

Let y be the challenge for the algorithm A, corresponding to the encryption of either x_0 or x_1 with the unknown key

key. The algorithm A_2 constructs the view for the distinguisher B_q, adding the value $p_{(v_q,\lambda_q),(u,t^*)} = y$ to the public information pub' constructed by A_1, and outputs the same output as B_q on inputs such a view, the class u, the time period t^*, and x_0. More formally, the algorithm A_2 is defined as follows:

Algorithm $A_2(1^\tau, y, state)$
   let $state = (pub', corr, x_0, x_1)$
   $pub \leftarrow pub'$ with $p_{(v_q,\lambda_q),(u,t^*)}$ set equal to y
   $d \leftarrow B_q(1^\tau, G, \mathcal{P}, pub, corr, x_0)$
   return d

Notice that if y corresponds to the encryption of x_1, then the random variable associated to the adversary's view is exactly the same as the one associated to the adversary view in experiment $\mathbf{Exp}^{q-1}_{u,t^*}$, whereas, if y corresponds to the encryption of x_0, it has the same distribution as the one associated to the adversary's view in experiment $\mathbf{Exp}^{q}_{u,t^*}$.

Hence, if the algorithm B_q is able to distinguish between such views with non negligible advantage, it follows that algorithm A is able to break the security of the encryption scheme $\Pi = (\mathcal{K}, \mathcal{E}, \mathcal{D})$ in the sense of IND-P1-C0. Contradiction.

Hence, for any $q = 2, \ldots, m+1$, the adversary's view in the $(q-1)$-th experiment is indistinguishable from the adversary's view in the q-th one. Therefore, the adversary's view in experiment $\mathbf{Exp}^{\text{IND}-1}_{\text{STAT}_{u,t^*}}$ is indistinguishable from the adversary's view in experiment $\mathbf{Exp}^{\text{IND}-0}_{\text{STAT}_{u,t^*}}$. This concludes the proof. □

4.2.2 Performance Evaluation

In this section we evaluate our time-bound hierarchical key assignment scheme taking into account several parameters. Regarding space requirements, the scheme requires a public value for each edge in the graph $G_{\mathcal{PT}}$ used in the construction. It is easy to see that $|E_{\mathcal{PT}}| = O(|V|^2) \cdot \sum_{i=1}^{|T|} i \cdot (|T| - i + 1) = O(|V|^2 \cdot |T|^3)$. On the other hand, each user belonging to a certain class for a time sequence has to store a single secret value. Moreover, users are required to perform a single decryption in order to derive a key.

To obtain a scheme secure in the sense of IND-ST we need to construct an encryption scheme secure in the sense of IND-P1-C0. To this aim, we could use a *pseudorandom function family*, an important cryptographic primitive originally defined by Goldreich, Goldwasser, and Micali [19]. Loosely speaking, a distribution of functions is pseudorandom if it satisfies the following requirements: 1) It is easy to sample a function according to the distribution and to evaluate it at a given point; 2) It is hard to tell apart a function sampled according to the distribution from a uniformly distributed function, given access to the function as a block-box. The two more efficient constructions of pseudorandom functions were proposed by Naor and Reingold [30]. In their constructions, the cost of evaluating such functions is comparable to two modular exponentiations.

Consider the following construction, called the *XOR construction* [5], of a symmetric encryption scheme $\Pi_{XOR,\mathcal{F}} = (\mathcal{K}_{XOR}, \mathcal{E}_{XOR}, \mathcal{D}_{XOR})$ which is based on a pseudorandom function family $\mathcal{F} : \{0,1\}^\tau \times \{0,1\}^\tau \rightarrow \{0,1\}^\tau$: The key generation algorithm $\mathcal{K}_{XOR}$ outputs a random τ-bit key ρ for the pseudorandom function family $\mathcal{F}$, thus specifying a function F_ρ of the family. The encryption algorithm $\mathcal{E}_{XOR}$ considers the message x to be encrypted as a sequence of τ-bits blocks $x = x_1 \cdots x_n$ (padding is done on the last block, if necessary), chooses a random string r of τ bits and computes, for $i = 1, \ldots, n$ the value $y_i = F_\rho(r+i) \oplus x_i$. The ciphertext is $r\|y_1 \cdots y_n$, where $\|$ denotes string concatenation. The decryption algorithm $\mathcal{D}_{XOR}$, on input a ciphertext z, parses it as $r\|y_1 \cdots y_n$ and computes, for $i = 1, \ldots, n$ the value $x_i = F_\rho(r+i) \oplus y_i$. The corresponding plaintext is $x = x_1 \cdots x_n$. The encryption scheme $\Pi_{XOR,\mathcal{F}}$ has been shown to be secure in the sense of IND-P1-C0 (see [5, 26]), assuming that $\mathcal{F}$ is a pseudorandom function family. Therefore, $\Pi_{XOR,\mathcal{F}}$ could be used to obtain a time-bound hierarchical key assignment scheme secure in the sense of IND-ST.

Notice that if the message x to be encrypted has length τ, the XOR construction reduces to compute the ciphertext as $r\|y$, where $y = F_\rho(r) \oplus x$ and r is a random string of τ bits. Such a construction has been used by Atallah et al. [2] to design a hierarchical key assignment scheme without temporal constraints. In their scheme, for each edge $(u, v) \in E$ there is a public value $y_{u,v} = F_{k_u}(\ell_v) \oplus k_v$, corresponding to the encryption of the key k_v assigned to class v, where the key k_u specifies a function F_{k_u} of the pseudorandom function family $\mathcal{F}$, and ℓ_v is a public label associated to v.

4.3 A Scheme based on Bilinear Maps

In this section we design a time-bound hierarchical key assignment scheme where the amount of public information does not depend on the number of time periods. Our scheme uses as a building block a bilinear map between groups. Bilinear maps have been used in cryptography to construct key exchange schemes [25], public-key cryptosystems [7, 8, 10], signature schemes [9], etc. We first recall the definition of a bilinear map.

A function $e : G_1 \times \hat{G}_1 \rightarrow G_2$ is said to be a *bilinear map* if the following properties are satisfied: 1) G_1 and $\hat{G}_1$ are two groups of the same prime order q; 2) For each $\alpha, \beta \in Z_q$, each $g \in G_1$, and each $h \in \hat{G}_1$, the value $e(g^\alpha, h^\beta) = e(g, h)^{\alpha\beta}$ is efficiently computable; and 3) The map is non-degenerate (i.e., if g generates G_1 and h generates $\hat{G}_1$, then $e(g, h)$ generates G_2). Typically, the group G_1 is a subgroup of the additive group of points of an elliptic curve $E(F_p)$, where p denotes the size of the field where the elliptic curve is defined. The group $\hat{G}_1$ is a subgroup of $E(F_{p^\eta})$, where $\eta > 0$ is the *embedding degree* of the map, whereas, the group G_2 is a subgroup of the multiplicative group of the finite field $F^*_{p^\eta}$.

In the following we describe a time-bound hierarchical key assignment scheme based on a bilinear map. For simplicity, we focus on symmetric bilinear maps (i.e., such that $G_1 = \hat{G}_1$), but our scheme works in the more general asymmetric setting (in particular, this implies that we could use the highly efficient MNT curves [29]).

We consider a two-level partially ordered hierarchy, where each level contains the same number of classes and there are no edges between classes at the same level. We remark that this is not a restriction, since any directed graph representing an access control policy can be transformed in a two-level partially ordered hierarchy having the above features, using a technique proposed in [14]. For the reader's convenience, we first explain how such a graph transformation works. Let $G = (V, E)$ be the graph corresponding to a partially ordered hierarchy. We can construct a two-level partially ordered hierarchy $G' = (V', E')$, where $V' = V_\ell \cup V_r$ and $V_\ell \cap V_r = \emptyset$, as follows:

- for each class $u \in V$, we place two classes u^ℓ and u^r in V', where $u^\ell \in V_\ell$ and $u^r \in V_r$;

- for each class $u \in V$, we place the edge (u^ℓ, u^r) in E';

- for each pair of classes v and u connected by a path in G, we place the edge (v^ℓ, u^r) in E'.

It is easy to see that the graphs G and G' define exactly the same access control policy. Figure 3 shows an example of the graph transformation described above.

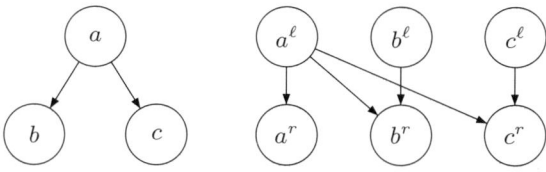

Figure 3: The graph transformation used in our construction.

Given a security parameter τ, let $\mathcal{G}$ be a randomized algorithm, called a *BDH parameter generator*, which, on input 1^τ, outputs a prime number q of τ bits, the description of two groups G_1 and G_2 of order q, and the description of a bilinear map $e : G_1 \times G_1 \to G_2$. The running time of $\mathcal{G}$ is polynomial in τ. We denote the output of $\mathcal{G}$ by $\mathcal{G}(1^\tau) = <q, G_1, G_2, e>$. Our scheme is as follows:

Algorithm $Gen(1^\tau, G', \mathcal{P})$
The TA performs the following steps:

1. Runs $\mathcal{G}(1^\tau)$ to generate a prime q, two groups G_1 and G_2 of order q and a bilinear map $e : G_1 \times G_1 \to G_2$;

2. Chooses a generator $g \in G_1^*$;

3. For each class $u^\ell \in V_\ell$, randomly chooses a secret value $\pi_u^\ell \in Z_q$;

4. For each class $u^r \in V_r$, randomly chooses a secret value $\pi_u^r \in Z_q$;

5. For each pair of classes $u^r \in V_r$ and $v^\ell \in V_\ell$ connected by an edge, i.e., such that $(v^\ell, u^r) \in E'$, computes the public information $p_{v,u} = g^{\pi_u^r / \pi_v^\ell}$;

6. Let pub be the sequence of public information computed in the previous step, along with the bilinear map e and the generator g;

7. For each time period $t \in T$, randomly chooses a secret value $\delta_t \in Z_q$;

8. For each class $v^\ell \in V_\ell$ and each time period $t \in T$, computes the private information $s_{v,t} = g^{\pi_v^\ell \cdot \delta_t}$;

9. For each class $v^\ell \in V_\ell$ and each time sequence $\lambda \in \mathcal{P}$, where $\lambda = (t_x, \ldots, t_y)$, computes the private information $s_{v,\lambda} = (s_{v,t_x}, \ldots, s_{v,t_y})$;

10. For each class $u^r \in V_r$ and each time period $t \in T$, computes the key $k_{u,t} = e(g,g)^{\pi_u^r \cdot \delta_t}$.

Algorithm $Der(1^\tau, G', \mathcal{P}, v^\ell, u^r, \lambda, s_{v,\lambda}, t, pub)$
A user belonging to a class v^ℓ in a time period $t \in T$ can use the private information $s_{v,t} = g^{\pi_v^\ell \cdot \delta_t}$ along with the public value $p_{v,u} = g^{\pi_u^r / \pi_v^\ell}$ to compute the key $k_{u,t}$ for any class u^r such that $(v^\ell, u^r) \in E'$, as follows:

$$\begin{aligned} e(s_{v,t}, p_{v,u}) &= e(g^{\pi_v^\ell \cdot \delta_t}, g^{\pi_u^r / \pi_v^\ell}) \\ &= e(g,g)^{\pi_u^r \cdot \delta_t} \\ &= k_{u,t}. \end{aligned}$$

4.3.1 Analysis of the Scheme

The security of our scheme is based on the assumption that the Bilinear Decisional Diffie-Hellman (formally introduced in [8]) is computationally hard. The *Bilinear Decisional Diffie-Hellman Problem (BDDH)* in $<G_1, G_2, e>$ is as follows: given the tuple $(g, g^\alpha, g^\beta, g^\gamma, x)$, for randomly chosen $\alpha, \beta, \gamma \in Z_q^*$, $x \in G_2$, and a random generator g of G_1, decide whether $x = e(g,g)^{\alpha \cdot \beta \cdot \gamma}$.

Definition 5. Let $\mathcal{G}$ be a BDH parameter generator. The advantage of an algorithm A in solving the BDDH Problem for $\mathcal{G}$ is defined as $\mathbf{Adv}_{\mathcal{G},A}^{\mathsf{BDDH}}(1^\tau) = |Pr[A(g, g^\alpha, g^\beta, g^\gamma, x) = 1] - Pr[A(g, g^\alpha, g^\beta, g^\gamma, e(g,g)^{\alpha \cdot \beta \cdot \gamma}) = 1]|$, where the probability is over the random choices of $\mathcal{G}(1^\tau)$, the random choice of g in G_1^*, the random choice of α, β, γ in Z_q^*, the random choice of x in G_2, and the random bits of A.

The BDDH problem is said to be hard in groups generated by $\mathcal{G}$ if the function $\mathbf{Adv}_{\mathcal{G},A}^{\mathsf{BDDH}}(1^\tau)$ is negligible, for each randomized algorithm A whose time complexity is polynomial in τ.

Now we are ready to prove that if the BDDH problem is hard in groups generated by $\mathcal{G}$, then our time-bound hierarchical key assignment scheme is secure in the sense of IND-ST.

THEOREM 3. *Our time-bound hierarchical key assignment scheme is secure in the sense of* IND-ST, *assuming the BDDH problem is hard in groups generated by $\mathcal{G}$.*

PROOF. We show that any polynomial-time adversary breaking the security of the scheme in the sense of IND-ST can be turned into a polynomial-time adversary solving the BDDH problem. Assume there exists a static adversary $\mathtt{STAT}_{u,t^*}$ whose advantage $\mathbf{Adv}_{\mathtt{STAT}_{u,t^*}}^{\mathsf{IND}}(1^\tau)$ is non negligible. In the following we show how to construct a polynomial-time adversary A that, given an instance $(g, g^\alpha, g^\beta, g^\gamma, x)$ of the BDDH problem, uses the adversary $\mathtt{STAT}_{u,t^*}$ to decide whether $x = e(g,g)^{\alpha \cdot \beta \cdot \gamma}$. The adversary A, on input the instance $(g, g^\alpha, g^\beta, g^\gamma, x)$, constructs the inputs for the adversary $\mathtt{STAT}_{u,t^*}$ by means of a simulation of the scheme, as shown in the following. In order to construct the public information pub to be given as input to $\mathtt{STAT}_{u,t^*}$, the adversary A performs the following steps:

1. For each class $v^\ell \in V_\ell$, randomly chooses a value $\sigma_v^\ell \in Z_q$;

2. For each class $v^r \in V_r$, randomly chooses a value $\sigma_v^r \in Z_q$;

3. For each pair of classes connected by an edge, computes the public information according to the three following distinct cases:

   (a) For each class $v^\ell \in V_\ell$ such that $(v^\ell, u^r) \in E'$, computes the value $p_{v,u} = (g^\beta)^{\sigma_u^r / \sigma_v^\ell}$. Note that

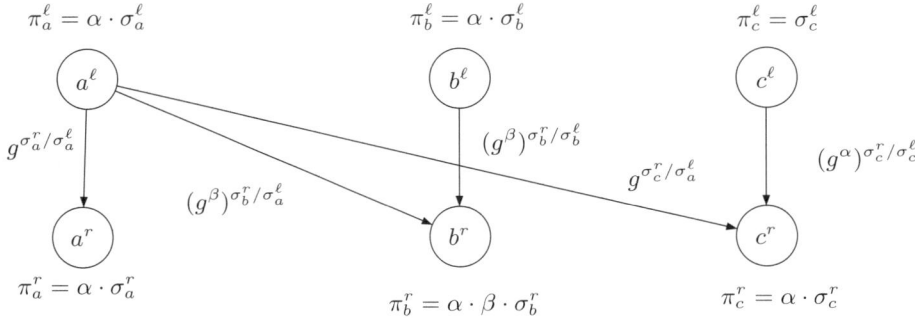

Figure 4: The two-level hierarchy of Figure 3 with the public information constructed by A and the secret values corresponding to the classes.

this means that the secret values π_u^r and π_v^ℓ associated to the classes v^ℓ and u^r during the initialization phase of the simulated scheme correspond to the values $\alpha \cdot \beta \cdot \sigma_h^r$ and $\alpha \cdot \sigma_v^\ell$, respectively;

(b) For each pair of classes $(v^\ell, w^r) \in V_\ell \times V_r \setminus \{u^r\}$ such that $(v^\ell, w^r) \in E'$ and $(v^\ell, u^r) \in E'$, computes the public information $p_{v,w} = g^{\sigma_w^r/\sigma_v^\ell}$. Note that this means that the secret values π_w^r and π_v^ℓ associated to the classes w^r and v^ℓ during the initialization phase of the simulated scheme correspond to the values $\alpha \cdot \sigma_w^r$ and $\alpha \cdot \sigma_v^\ell$, respectively;

(c) For each pair of classes $(v^\ell, w^r) \in V_\ell \times V_r \setminus \{u^r\}$ such that $(v^\ell, w^r) \in E'$ and $(v^\ell, u^r) \notin E'$, computes the public information $p_{v,w} = (g^\alpha)^{\sigma_w^r/\sigma_v^\ell}$. Note that this means that the secret values π_w^r and π_v^ℓ associated to the classes w^r and v^ℓ during the initialization phase of the simulated scheme correspond to the values $\alpha \cdot \sigma_w^r$ and σ_v^ℓ, respectively.

Observe that each pair of classes connected by an edge in E' is involved in exactly one of the above three cases. On the other hand, each single class may be involved in more than one case. However, it is easy to see that the secret value corresponding to each class is consistent with the others. Clearly, such secret values cannot be computed by the adversary A, but we have outlined the correspondence between each class and its secret value in order to fuel intuition over the reader. Figure 4 shows the two-level hierarchy of Figure 3 with the public information constructed by A and the secret values corresponding to the classes, assuming b^r is the attacked class.

In order to construct the private information *corr* held by corrupted classes, to be given as input to STAT_{u,t^*}, the adversary A performs the following steps:

1. For each time period $t \neq t^*$, randomly chooses a value $\delta_t \in Z_q$ and for each class $v^\ell \in V_\ell$, computes the private information $s_{v,t} = g^{\pi_v^\ell \cdot \delta_t}$, where the value π_v^ℓ corresponds either to $\alpha \cdot \sigma_v^\ell$ or to σ_v^ℓ according to the above construction. More precisely, we distinguish the following two cases:

   (a) For each class $v^\ell \in V_\ell$ such that $(v^\ell, u^r) \in E'$, A computes the value $s_{v,t} = (g^\alpha)^{\sigma_v^\ell \cdot \delta_t}$;

   (b) For each class $v^\ell \in V_\ell$ such that $(v^\ell, u^r) \notin E'$, A computes the value $s_{v,t} = g^{\sigma_v^\ell \cdot \delta_t}$.

2. For the time period t^*, randomly chooses a value $\varphi \in Z_q$ and for each class $v^\ell \in V_\ell$ such that $(v^\ell, u^r) \notin E'$, computes the private information $s_{v,t^*} = (g^\gamma)^{\sigma_v^\ell \cdot \varphi}$. Note that this means that the secret value δ_{t^*} associated to the time period t^* during the initialization phase of the simulated scheme corresponds to the value $\gamma \cdot \varphi$.

The last input for STAT_{u,t^*}, corresponding either to the key k_{u,t^*} or to a random value having the same length as k_{u,t^*}, is computed as $x^{\sigma_u^r \cdot \varphi}$.

It is easy to see that the adversary STAT_{u,t^*}'s view in the above simulation cannot be distinguished from the one obtained in a real execution of the scheme, since the random variables associated to such views are exactly the same. Moreover, all the computations needed to construct STAT_{u,t^*}'s view can be performed in polynomial-time.

Clearly, since STAT_{u,t^*} distinguishes the key k_{u,t^*} from a random string having the same length, with non negligible advantage, it follows that the adversary A decides whether x is equal to $e(g,g)^{\alpha \cdot \beta \cdot \gamma}$ with non negligible advantage. Hence, the theorem holds. □

4.3.2 Performance Evaluation

With respect to storage requirements, notice that the scheme requires a public value for each edge in the graph $G' = (V', E')$ used in the construction, thus the total number of public values is $|E'| = O(|V|^2)$, which does not depend on the number $|T|$ of time periods. This means that the number of time periods for which the scheme must be active does not need to be known in advance. Moreover, we stress that each public value is typically 171 bits long. On the other hand, each user belonging to a certain class for a time sequence has to store as many secret values as the number of time periods in the sequence. Hence, the number of private values for each user is $O(|T|)$. Moreover, users are required to evaluate the bilinear map at two given points, in order to perform key derivations.

Finally, notice that BDH parameter generators believed to satify the BDDH assumption can be efficiently constructed from the (modified) Weil [7] and Tate pairings [18] defined within elliptic or hyperelliptic curves over finite fields.

5. CONCLUSIONS

In this paper we have designed and analyzed time-bound hierarchical key assignment schemes which are provably-secure and efficient. We have considered both the *unconditionally secure* and *computationally secure* setting. We have proposed two different constructions for time-bound hierarchical key assignment schemes in the computationally secure setting. Both constructions are *provably-secure* and *efficient*. Moreover, they can be also used to implement more general access control policies (i.e., which cannot be represented by a partially ordered hierarchy).

6. ACKNOWLEDGMENTS

We would like to thank the anonymous referees for their careful reading and useful comments.

7. REFERENCES

[1] S. G. Akl and P. D. Taylor, *Cryptographic Solution to a Problem of Access Control in a Hierarchy*, ACM Trans. on Computer Systems, 1(3), 239–248, 1983.

[2] M. J. Atallah, M. Blanton, N. Fazio, and K. B. Frikken, *Dynamic and Efficient Key Management for Access Hierarchies*, CERIAS Tech. Rep. TR 2006-09, Purdue University. Prelim. version in Proc. of the 2005 ACM Conf. on Comput. and Commun. Security, 190–201, 2005.

[3] M. J. Atallah, M. Blanton, and K. B. Frikken, *Key Management for Non-Tree Access Hierarchies*, in Proc. of the 2006 ACM Symp. on Access Control Models and Technologies, 11–18, 2006.

[4] G. Ateniese, A. De Santis, A. L. Ferrara, and B. Masucci, *Provably-Secure Time-Bound Hierarchical Key Assignment Schemes*, IACR ePrint Archive, Report 2006/225.

[5] M. Bellare, A. Desai, E. Jokipii, and P. Rogaway, *A Concrete Security Treatment of Symmetric Encryption*, in Proc. of the 38th IEEE Symp. on Found. of Computer Sci., 394–403, 1997.

[6] E. Bertino, B. Carminati, and E. Ferrari, *A Temporal Key Management Scheme for Secure Broadcasting of XML Documents*, in Proc. of the 2002 ACM Conf. on Comput. and Commun. Security, 31–40, 2002.

[7] D. Boneh and X. Boyen, *Efficient Selective-ID Secure Identity-based Encryption without Random Oracles*, in Proc. of Eurocrypt 2004, LNCS, 3027, 223–238, 2004.

[8] D. Boneh and M. Franklin, *Identity-based Encryption from the Weil Pairing*, SIAM Journal Comput., 32(3), 586–615, 2003.

[9] D. Boneh, B. Lynn, and H. Shacham, *Short Signatures from the Weil Pairing*, Journal of Cryptology, 17(4), 297–319, 2004.

[10] R. Canetti, S. Halevi, and J. Katz, *A Forward-Secure Public-Key Encryption Scheme*, in Proc. of Eurocrypt 2003, LNCS, 2656, 255-271, 2003.

[11] T. Chen and Y. Chung, *Hierarchical Access Control based on Chinese remainder Theorem and Symmetric Algorithm*, Comput. & Security, 21(6), 565–570, 2002.

[12] H. Y. Chien, *Efficient Time-Bound Hierarchical Key Assignment Scheme*, IEEE Trans. on Know. and Data Eng., 16(10), 1301–1034, 2004.

[13] T. M. Cover, J. A. Thomas, *Elements of Information Theory*, John Wiley & Sons, 1991.

[14] A. De Santis, A. L. Ferrara, and B. Masucci, *Cryptographic Key Assignment Schemes for any Access Control Policy*, Inf. Proc. Lett., 92(4), 199–205, 2004.

[15] A. De Santis, A. L. Ferrara, and B. Masucci, *Enforcing the Security of a Time-Bound Hierarchical Key Assignment Scheme*, Inf. Sci., 176(12), 1684–1694, 2006.

[16] A. De Santis, A. L. Ferrara, and B. Masucci, *Unconditionally Secure Key Assignment Schemes*, Discrete Applied Math., 154(2), 234–252, 2006.

[17] A. De Santis, A. L. Ferrara, and B. Masucci, *Efficient Provably-Secure Key Assignment Schemes*, manuscript.

[18] S. D. Galbraith, K. Harrison, and D. Soldera, *Implementing the Tate Pairing*, in Proc. of the Algorithmic Number Theory Symp., LNCS, 1838, 385–394, 2000.

[19] O. Goldreich, S. Goldwasser, and S. Micali, *How to Construct Random Functions*, Journal of the ACM, 33(4), 792–807, 1986.

[20] S. Goldwasser and S. Micali, *Probabilistic Encryption*, Journal of Comput. and Syst. Sci., 28, 270–299, 1984.

[21] S. Goldwasser, S. Micali, and R. Rivest, *A Digital Signature Scheme Secure against Adaptive Chosen Message Attacks*, SIAM Journal Comput., 17(2), 281–308, 1988.

[22] L. Harn and H. Y. Lin, *A Cryptographic Key Generation Scheme for Multilevel Data Security*, Comput. and Security, 9(6), 539–546, 1990.

[23] H. F. Huang and C. C. Chang, *A New Cryptographic Key Assignment Scheme with Time-Constraint Access Control in a Hierarchy*, Comput. Stand. & Int., 26, 159–166, 2004.

[24] M. S. Hwang, *A Cryptographic Key Assignment Scheme in a Hierarchy for Access Control*, Math. and Comput. Modeling, 26(1), 27–31, 1997.

[25] A. Joux, *A One-round Protocol for Tripartite Diffie-Hellman*, in Proc. of the Algorithmic Number Theory Symp., LNCS, 1838, 385–394, 2000.

[26] J. Katz and M. Yung, *Characterization of Security Notions for Probabilistic Private-Key Encryption*, Journal of Cryptology, 19, 67–95, 2006.

[27] H. T. Liaw, S. J. Wang, and C. L. Lei, *A Dynamic Cryptographic Key Assignment Scheme in a Tree Structure*, Comput. and Math. with Appl., 25(6), 109–114, 1993.

[28] C. H. Lin, *Dynamic Key Management Schemes for Access Control in a Hierarchy*, Comput. Commun., 20, 1381–1385, 1997.

[29] A. Miyaji, M. Nakabayashi, and S. Takano, *New Explicit Conditions for Elliptic Curve Traces for FR-Reduction*, IEICE Trans. Fund., E-84(5), 1234–1243, 2001.

[30] M. Naor and O. Reingold, *Number-Theoretic Constructions of Efficient Pseudo-Random Functions*, Journal of the ACM, 51(2), 231–262, 2004.

[31] R. S. Sandhu, *Cryptographic Implementation of a Tree Hierarchy for Access Control*, Inf. Proc. Lett., 27, 95–98, 1988.

[32] A. Shamir, *On the Generation of Cryptographically Strong Pseudorandom Sequences*, ACM Trans. on Comput. Sys., 1, 38–44, 1983.

[33] V. Shen and T. Chen, *A Novel Key Management Scheme based on Discrete Logarithms and Polynomial Interpolations*, Comput. & Security, 21(2), 164–171, 2002.

[34] Q. Tang and C. J. Mitchell, *Comments on a Cryptographic Key Assignment Scheme*, Comput. Standards & Interfaces, 27, 323–326, 2005.

[35] W.-G. Tzeng, *A Time-Bound Cryptographic Key Assignment Scheme for Access Control in a Hierarchy*, IEEE Trans. on Knowl. and Data Eng., 14(1), 182–188, 2002.

[36] J. Yeh, *An RSA-Based Time-Bound Hierarchical Key Assignment Scheme for Electronic Article Subscription*, in Proc. of the 2005 ACM CIKM Conf. on Information and Knowledge Management, 285–286, 2005.

[37] X. Yi, *Security of Chien's Efficient Time-Bound Hierarchical Key Assignment Scheme*, IEEE Trans. on Knowl. and Data Eng., 17(9), 1298–1299, 2005.

[38] X. Yi and Y. Ye, *Security of Tzeng's Time-Bound Key Assignment Scheme for Access Control in a Hierarchy*, IEEE Trans. on Knowl. and Data Eng., 15(4), 1054–1055, 2003.

[39] S.-Y. Wang and C.-Laih, *Merging: An Efficient Solution for a Time-Bound Hierarchical Key Assignment Scheme*, IEEE Trans. on Dependable and Secure Comput., 3(1), 91–100, 2006.

[40] T. Wu and C. Chang, *Cryptographic Key Assignment Scheme for Hierarchical Access Control*, Int. Journal of Comput. Syst. Sci. and Eng., 1(1), 25–28, 2001.

Optimizing BGP Security by Exploiting Path Stability

Kevin Butler   Patrick McDaniel
SIIS Laboratory
Computer Science and Engineering
The Pennsylvania State University
University Park, PA USA
{butler,mcdaniel}@cse.psu.edu

William Aiello
Department of Computer Science
University of British Columbia
Vancouver, BC Canada
aiello@cs.ubc.ca

ABSTRACT

The Border Gateway Protocol (BGP) is the de facto interdomain routing protocol on the Internet. While the serious vulnerabilities of BGP are well known, no security solution has been widely deployed. The lack of adoption is largely caused by a failure to find a balance between deployability, cost, and security. In this paper, we consider the design and performance of BGP path authentication constructions that limit resource costs by exploiting route stability. Based on a year-long study of BGP traffic and indirectly supported by findings within the networking community, we observe that routing paths are highly stable. This observation leads to comprehensive and efficient constructions for path authentication. We empirically analyze the resource consumption of the proposed constructions via trace-based simulations. This latter study indicates that our constructions can reduce validation costs by as much as 97.3% over existing proposals while requiring nominal storage resources. We conclude by considering operational issues related to incremental deployment of our solution.

Categories and Subject Descriptors

D.4.6 [**Operating Systems**]: Security and Protection

General Terms

Security

Keywords

routing, security, path stability, BGP

1. INTRODUCTION

The Border Gateway Protocol (BGP) [38, 37] is the dominant interdomain routing protocol on the Internet. BGP establishes and maintains associations between IP address *prefixes* [34] (addresses) and source specific paths to the autonomous systems (networks) in which they reside. Each AS selects the best paths based on the advertised paths and routing policy. However, the BGP protocol is largely devoid of any security [33, 42, 3, 26, 31, 6]. One critical vulnerability resulting from this lack of security allows an adversary to manipulate *paths*: a malicious network can force IP traffic destined for a victim to be routed through themselves, prevent the network from being reachable, or simply destabilize the routes toward the victim network.

While many approaches have been proposed to address BGP security [42, 18, 29, 46, 7, 9, 19, 44, 43, 45, 49], none have been widely deployed. The lack of adoption is largely caused by a failure by the community to find an acceptable balance between cost and security. For example, the S-BGP protocol [18] offers comprehensive security by authenticating routing artifacts (e.g., prefix and path advertisements, withdrawals, etc.) using asymmetric cryptography. However, the computational and storage costs of performing strong S-BGP style authentication are viewed to be prohibitive in many environments [7, 30, 10, 45]. Recent works in BGP have sought optimizations that reduce these costs. For example, among others, these works have used advanced cryptography [9, 10], out-of-band security [7], relaxed guarantees [29], or address usage patterns [2] to reduce security costs. In recent work, Zhao et al. exploited the structure of the BGP protocol to implement an AS-local optimization for path validation [49].

In this work, we mitigate the cost of path validation by exploiting BGP's natural *path stability*. We posit and confirm that ASes offer few distinct paths for a prefix, and that those paths are largely static. Our study of a year's worth of BGP traffic at 40 globally distributed ASes shows that in the average case, less than 2% of prefixes were advertised using more than 10 paths, and less than 0.06% were advertised with more than 20 paths during a single month. The observed stability of BGP paths led to the design of efficient cryptographic structures for path authentication. ASes using these construction create long-lifetime cryptographic proof systems [23, 27] that validate all paths that they are likely to advertise. Efficiently validatable tokens reflecting current best paths are derived from these proof systems and distributed throughout the Internet. In this way, the costs of heavyweight cryptographic operations are amortized over many validations.

We further compare the computational cost of our solutions against other BGP security solutions via trace-based simulation. Our simulations demonstrate that our techniques reduce the costs through reducing signature validations by up to 97.3% over proposed solutions, and the storage costs at validating ASes are nominal. Note that schemes such as Nicol's optimize BGP in ways orthogonal to our solutions, and incorporating them could lead to even greater reductions in computational costs. However, we defer the analysis of the joint advantage of these solutions to future work.

We begin in the following section by outlining the operation and security requirements of BGP.

Permission to make digital or hard copies of all or part of this work for personal or classroom use is granted without fee provided that copies are not made or distributed for profit or commercial advantage and that copies bear this notice and the full citation on the first page. To copy otherwise, to republish, to post on servers or to redistribute to lists, requires prior specific permission and/or a fee.
CCS'06, October 30–November 3, 2006, Alexandria, Virginia, USA.
Copyright 2006 ACM 1-59593-518-5/06/0010 ...$5.00.

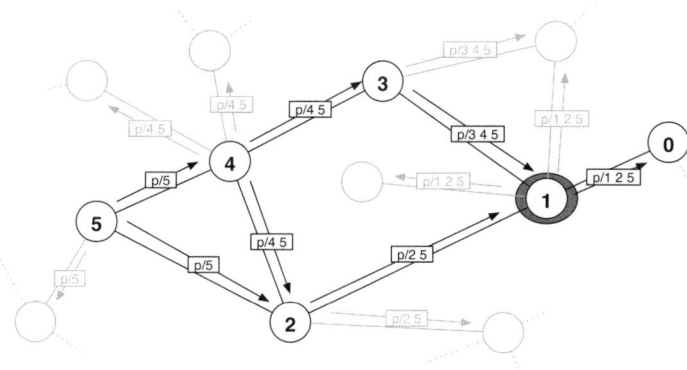

Figure 1: BGP Path discovery - AS5 originates the prefix p by announcing it to its neighbors (e.g., AS4). AS4 further propagates the prefix to its neighbors AS2 and AS3 after prepending its AS number to the prefix. AS1 (highlighted) receives routes from AS2 and AS3, and selects the best route (arbitrarily {2 4 5}), which is then propagated further (to AS0 and others).

2. INTERDOMAIN ROUTING

BGP provides two essential services[1]: the mapping of address prefixes (e.g., 192.168.0.0/16) onto the ASes that own them, and the construction of source specific paths to each reachable prefix. The interdomain routing topology is defined by physical links between adjacent ASes. Each AS *originates* the prefixes associated with a network by identifying and enumerating them in an UPDATE message sent to its neighbors (adjacent ASes). Received announcements are recursively concatenated with local AS numbers [12] and propagated, AS by AS, to form a routing path. This path (also called a *route*) is used to forward network traffic to the origin. Note that an AS may receive many paths for a single prefix. The AS identifies the "best" path using the *path selection algorithm*. The selection algorithm determines the best route by evaluating path length, policy, and other factors. Only the selected best path is propagated. IP traffic is routed, hop-by-hop, based on the best path known by the AS. Figure 1 illustrates route advertisement and path selection.

Which route represents the best path is re-evaluated each time a new route for a prefix is received. Suppression of non-best routes prevents undesirable routes from polluting the larger Internet, and is a key ingredient to the scalability of BGP. Recursive propagation of best routes ensures that every AS on the Internet acquires a route for every reachable prefix. A route is *withdrawn* when the AS discovers that the prefix is no longer reachable.

The ubiquity of BGP is also one of its greatest weaknesses. The number of ASes and complexity of their interaction affords an adversary opportunities to monitor, disrupt, or manipulate the routing process. The Routing Protocol Security (rpsec) working group of the IETF postulate a universe of possible effects of routing vulnerabilities [4]. Traffic congestion, black-holing, routing loops, slowed or prevented convergence, instability, traffic eavesdropping, network partitioning, and increased delay were deemed the most damaging consequences. The group's analysis led to a statement of general routing security requirements [36], and more specifically, to requirements for BGP security [26]. We consider the vulnerabilities germane to current work below and review broader classes of vulnerabilities and solutions in Section 7.

BGP security concerns are often classified by the three broad categories of data exchanged [31, 6]: signaling, prefix origins, and paths. Attacks on BGP signaling frustrate the session by incorrectly reporting errors, masquerading as other entities, or by consuming the victim's resources [26]. Authenticating an AS's right to advertise (originate) a prefix is essential to securing BGP. Failure to perform this authentication invites prefix hijacking: an adversary can steal address space simply by advertising it [18, 41, 2].

This paper investigates *path authentication*. Hu et al. identified the following classes of path attacks [10]: (*a*) *path forgery* - the adversary may attempt to forge paths in order to influence packet routing, (*b*) *path modification* - an adversary may add, remove, or alter data in the path or policy, (*c*) *denial of service* - an adversary consumes a victim's resources by sending spurious routes, and (*d*) *worm-holing* - in which colluding adversaries create false AS to AS links. Note that the first two classes are attacks, whereas the second two could be more accurately classified as consequences. Moreover, *worm-holing* is less of an attack on paths, but more of an attack on the topology. The false topology generated can be used to introduce incorrect paths, even if a path validation approach is perfectly implemented and deployed. With the exception of soBGP [29] (see Section 7), few security proposals address worm-holing, as it requires validation of BGP peering.

If an adversary can forge or modify routes, then it can *black-hole* traffic routed to it. To accomplish this, the adversary announces a highly desirable route that is incident to the path, e.g., by advertising a very short path. Traffic flowing to that prefix will be routed to the adversary and filtered. If the adversary wants to destabilize the network while remaining relatively clandestine, it can randomly drop a percentage of the traffic (called *grey-holing*). Note that it takes few drops to vastly reduce the throughput between the victim and the destination: each drop causes the congestion control algorithm to aggressively throttle traffic [35]. Connection recovery is slow, and the attacker gains advantage with little effort [48]. Paths may also be manipulated to route traffic through malicious ASes for monitoring [6]. That is, if an adversary can redirect traffic (as above), then it can monitor, record, or even modify that traffic as it transits its network. Furthermore, an AS's ability to filter or rapidly advertise and withdraw advertisements leads to a range of DoS attacks [26, 47] that may easily render targeted networks unreachable.

3. PATH VALIDATION CONSTRUCTIONS

In this section, we derive constructions for path authentication that will be subsequently examined and evaluated throughout the rest of the paper. As indicated in the previous section, any solution that secures the path must provide at least the following simplified guarantees: an AS receiving a route must be able to *a*) authenticate the source of an advertisement, *b*) authenticate that the ASes in the path advertised the sub-paths in the order which they are listed (i.e., no ASes were added or removed), and *c*) validate the times at which each of the (sub)advertisements occurred. Note that in reality the security guarantees are somewhat more subtle, but these definitions are sufficient to motivate the following discussion. Interested readers are directed to Appendix A for a formal definition of BGP path security requirements and the following constructions.

Consider S-BGP attestations [18]. As shown in figure 1, a BGP speaker sending a route announcement to its peer signs each an-

---
[1]Throughout we refer to the AS to AS communication protocol *eBGP* generically as BGP. The intra-AS *iBGP* protocol governs the way in which eBGP speaking edge-routers within an AS exchange routing information. iBGP is explictly outside the scope of this work.

nouncement as it propagates across the network. If the path to a given network prefix changes, a new announcement is signed and sent to the peer. For example, assume that prefix P, originated by AS 5, is being advertised by AS 1, which knows three paths to the destination: $\{2\ 5\}$, $\{2\ 4\ 5\}$ and $\{3\ 4\ 5\}$[2]. If the advertised path changes across three time periods t_1 through t_3, the attestations issued by AS 1 will include:

$$[P, \{2\ 5\}, t_1]_{S_2}$$
$$[P, \{2\ 4\ 5\}, t_2]_{S_2}$$
$$[P, \{3\ 4\ 5\}, t_3]_{S_3}$$

where S_n represents a digital signature issued for the route attestation by AS n. The signature authenticates AS 1 as being the verifiable source of the announcement. S-BGP announcements are recursively signed: signed attestation proves not only that the peer vouches for the path, but that each hop in the path also vouches for the included sub-path. For example, assume at time t_n that AS 1 receives the path $\{2\ 5\}$; in reality, the received attestation will have the logical form

$$[[[P, \{5\}, t_{n-2}]_{S_5}]P, \{2\ 5\}, t_{n-1}]_{S_2}$$

as originating AS 5 initially signs the path and AS 2 signs that original attestation and itself as part of the path vector for prefix P. When AS 1 advertises this route, it will cumulatively sign over the other attestations and the new path vector, as follows:

$$[[[[[P, \{5\}, t_{n-2}]_{S_5}]P, \{2\ 5\}, t_{n-1}]_{S_2}]P, \{1\ 2\ 5\}, t_n]_{S_1}$$

In this manner, the path can be recursively verified by validating each AS path signature back to the route origin. Finally, because the timestamp of each announcement is included, replay attacks are avoided. Thus, S-BGP attestations meet the requirements for path authentication.

It is obvious to see that these attestations can be costly in practice: there are currently over 200,000 prefixes being advertised by 22,000 ASes in the Internet [11]. This can lead to huge numbers of signatures and validations at each AS. We now introduce several novel approaches that attempt to mitigate these costs.

One opportunity to optimize cost is through signature aggregation. For example, we can exploit the fact that paths are stable: only a few paths are likely to be advertised for most prefixes. We propose that a *hash chain* [20] be initially generated for each distinct path associated with a particular prefix. The first value generated for each path is sent to the peer, with the entire message signed. An *authentication token* consisting of the next value in the hash chain for the new path is sent to the peer whenever a different route is advertised. The peer hashes the token forward to verify that it arrives at the anchor value of the hash chain. Hashing is approximately three orders of magnitude faster than a signature validation in software [10]. Thus, validation costs are greatly reduced. When the hash chain has been exhausted, a new announcement containing all paths and the signed tokens is sent to the BGP peer. Returning to the previous example, AS 1 sends all its paths in a single list, along with the tokens representing the hash chain anchors as follows:

$$\begin{bmatrix} P, \{2\ 5\}, h^{365}(x_1) \\ P, \{2\ 4\ 5\}, h^{365}(x_2) \\ P, \{3\ 4\ 5\}, h^{365}(x_3) \\ P, \{3\ 4\ 2\ 5\}, h^{365}(x_4) \end{bmatrix}_{S_1}$$

---

[2]We apply the convention that paths grow from right to left, with the originating AS occupying the rightmost value in the path vector.

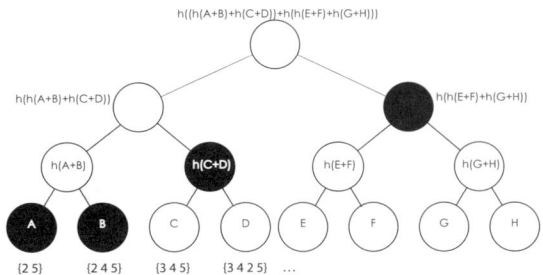

Figure 2: Tree construction for path aggregation. As in the list construction, we assume that A has been selected as representing the optimal path. Only the black nodes are hashed, and only the root is signed. There are $\lceil \log_2 n \rceil$ hashes that need to be computed for n leaves of the tree.

where x_1, x_2, and x_3 are the randomly-generated seed values for the hash chains for the paths $\{2\ 5\}$, $\{2\ 4\ 5\}$, $\{3\ 4\ 5\}$ and $\{3\ 4\ 2\ 5\}$, respectively, $h^n(x)$ represents a hash chain of length n with seed x, and the hash chain length of 365 is an example construction parameter, e.g., representing a chain that generates a token once a day for a year [3]. At time t_n, the authentication token associated with that time period is sent that represents the route advertised at that time:

$$t_1 \rightarrow h^{365-1}(x_1)$$
$$t_2 \rightarrow h^{365-2}(x_2)$$
$$t_3 \rightarrow h^{365-3}(x_3)$$

The token provides replay protection due to the infeasibility of generating a token representing a later time value. Note that these three authentication tokens fulfill the same security guarantees as their equivalent S-BGP attestations, i.e., $t_1 \rightarrow h^{365-1}(x_1)$ has the equivalent security guarantees to $[P\{2\ 5\}, t_1]_{S_2}$, etc. There is a minor security loss that is contingent on the size of the construction parameter. Because a signature is only generated when the hash chain is exhausted, a malicious peer can advertise any of the paths sent in the aggregate signature with the appropriate authentication token, regardless of whether it is optimal. There is no validation of routes until the next signature is generated. For example, a peer advertising path a and subsequently advertising path b can again advertise a even if it has been withdrawn by an upstream peer. However, a peer can always suppress an advertisement with any variant of BGP; the additional threat posed by an attacker advertising a pre-existing, validated path is minimal. The window for these threats can be reduced by making the construction parameter smaller, at the cost of having to generate signatures more frequently.

An advertising AS forwards not only the authentication tokens for its advertisement, but also tokens it received for the included sub-paths. This provides both verification of the peer announcement and recursive validation of all encompassed announcements back to the origin AS. In this way, the approach achieves similar security guarantees to that of S-BGP attestation: paths cannot be forged, sub-routes can be validated, and the timing of the announcements can be validated.

While the preceding construction mitigates the computational costs of recursively signed advertisements, it introduces other resource costs. Because each signed list contains all paths associ-

---

[3]AS 1 also transmits lists received from each previous peer and onion-signs those attestations as in the S-BGP example. We omit the full details for clarity.

ated with a given prefix, the bandwidth and storage costs associated with processing these lists may be prohibitive. For example, current routers have exceedingly small amounts of available main memory [25], and hard-disks induce considerably higher access latencies and often fall victim to more frequent failures.

As has been demonstrated in many domains, transmission and storage costs associated with authenticated material can be mitigated by using cryptographic proof systems, e.g., hash trees [23] and authenticated dictionaries [27, 8]. Our tree path authentication construction is based on the Merkle hash tree. In this construction, a *succinct* set-membership proof is generated by the announcing peer; as shown in Figure 2, each advertised path forms a leaf in the Merkle tree. When a path is announced, only a hash of the leaf's sibling, the parent's sibling, etc, up to the root node, are required. The root of the tree is signed. The computational costs are slightly greater than with the list construction, as a number of hashes proportional to the height of the tree must be computed by the peer receiving the path announcement; however, because of the very low cost of hashing, the extra effort is minimal. Hence, the tree construction provides an attractive balance between computational, storage, and bandwidth costs. Generation of hash chains for paths follows the same process as with the list construction; the leaves of the hash tree are associated with the generated hash tree value, such that only the authentication token is necessary to be sent for each route announcement.

We now consider a number of alternate constructions based on *how* paths are aggregated. In each case, hash chains are generated for each path as described above. However, we construct a different trees whose structure relates to the how aggregation is performed, to further amortize costs and exploit different computational and storage trade-offs. We refer to the approach described above as the *prefix* scheme: the AS constructs a tree where the leaves represent all of the distinct paths it advertises for that prefix. By contrast, in the *origin* scheme, the AS constructs a tree where the leaves represent all of the distinct (prefix,paths) pairs it advertises with the given origin AS. The *all* AS construction creates a single tree where the leaves represent all the paths the AS historically advertises. One can view these approaches as simply as different partionings of the paths an AS may advertise: the *prefix* scheme creates a tree with all the paths for each unique prefix, the *origin* scheme creates a tree for every unique AS that originates a prefix, and the *all* creates a single tree for all the advertisements that the AS emits.

Note that in both the list and tree structures, if a new announcement is received that contains a previously unseen new path, representing a new path to a given prefix, the announcement is sent to peers as an S-BGP type route attestation. When the aggregate constructions are resent to peers, this new path will be part of the aggregation. Hence, in the degenerate case where all advertised paths are new, the scheme reverts to S-BGP style advertisements.

We consider the stability of path advertisements in the following section. The degree to which paths are stable will determine how well the optimizations perform in practice. We explore the computational overheads of our proposed approaches via trace-based simulation in section 5. While our scheme does not explicitly address optimizations such as enumerating peer routes with bit vectors as shown in [49], as they are orthogonal to our goals, such methods can be employed to further reduce validation costs.

4. PATH STABILITY

This section analyzes the central hypothesis upon which our cryptographic constructions are based: the set of paths for a prefix or emitted from a AS are small and stable over time, i.e., ASes exhibit path reference locality. The following experiments evaluate path

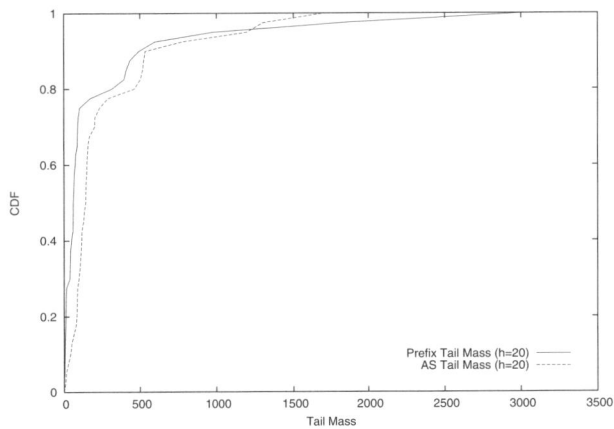

Figure 3: Tail mass - CDF of tail mass for 40 Route Views listening points during February 2004.

Tail Mass Test	Min (LP)	Median (LP)	Max (LP)
Prefix (h=10)	67 (#23)	1,178 (#8)	17,784 (#40)
Prefix (h=20)	0 (#23)	63 (#17)	3,027 (#40)
AS (h=10)	163 (#23)	1,135 (#8)	4,967 (#40)
AS (h=20)	10 (#23)	142 (#8)	1,701 (#40)

Table 1: Listening point tail mass

density (number of distinct paths observed from peers and other points across the Internet) and *stability* (rate of discovery of new paths). In these experiments, we examine data from the 40 listening points of the Route Views [24] BGP repository. Each listening point data-set represents a transcript of all UPDATE messages received by a monitored AS (called a listening point).

We are not the first to characterize path stability. Other studies use the available BGP data to investigate the number of unique paths to a prefix assuming connectivity to two listening points over a single day [10], to estimate the number of cryptographic operations required for prefix validation [45], to establish a delegation hierarchy [2], and to examine address allocation and routing table growth [5], scalability of router memories [28] and table fragmentation [21], or to ascertain the stability of popular routes [39]. We found these past analysis instructive but incomplete for our purposes. These analyses focused on instantaneous table size or growth over time, or considered only a small subset of prefixes. The current work required a characterization of total unique paths an observer sees per-AS and per-prefix on a continuing basis. Hence, while past studies largely focus on growth trends, our analysis required a finer characterization of path *churn*. Detailed below, these requirements prompted the study of AS/prefix *tail mass* and path *rates of discovery*.

We begin our analysis by using *tail mass* to measure path stability. Tail mass $T_h(k)$ is the number of unique values above a threshold h encountered by observer k. This study is concerned with number of unique paths, so we calculate tail mass as the number of prefixes or ASes that have more than h unique path vectors associated with them. Intuitively, tail mass shows how many prefixes or ASes have a "large" number of paths associated with them (as defined by a threshold h). The following is based on the analysis of the 217,707,968 updates observed by the 40 listening points during February 2004.

Figure 3 shows a cumulative distribution function of the prefix and AS tail masses of each listening point when the threshold is 20

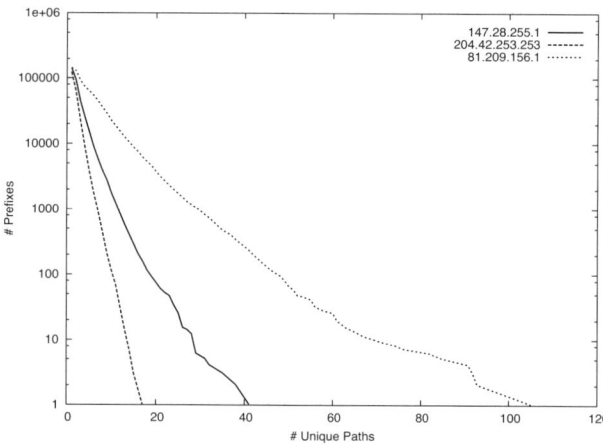

Figure 4: CCDFs of unique paths per prefix measured from multiple Route Views listening points, for February 2004.

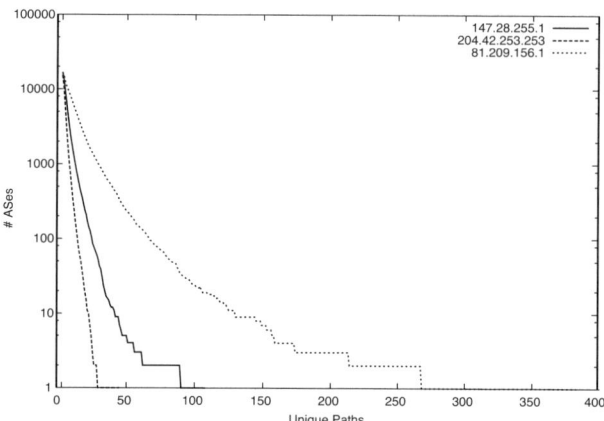

Figure 5: CCDFs of unique paths per AS measured from multiple Route Views listening points, for February 2004.

(h=20). A striking aspect of this data is its density, where 80% of the listening points have a tail mass less than 500, and 67% have masses less than 200. This indicates significant stability at the listening points.

Table 1 summarizes the most, least, and median-stable listening points as represented by tail mass, measured across several experiments. The data suggests candidate *representative* listening points as models for minimum, maximum, and typical stability. As such, we select listening point 23 (204.42.253.253) as maximally stable (i.e., has the smallest tail mass), point 40 (81.209.156.1) as minimally stable, and point 8 (147.28.255.1) as typical in the following experiments.

We now use the representative listening points to more closely scrutinize path stability. Figure 4 shows a CCDF for the unique number of paths observed by the listening point associated with various prefixes. In the average case, less than 2% of prefixes have more than 10 paths associated with them, and less that 0.06% more than 20. In the worst case, 15.3% of prefixes have more than 10 unique paths, 2.57% have more than 20, and 1.17% have more than 25.

Figure 5 shows a CCDF for the observation of unique paths by AS. Because the number of ASes a listener sees is little more than 10% of the total number of prefixes seen, we would expect that the number of unique paths per AS would be correspondingly larger than in the per-prefix case. However, the difference is not as pronounced because many prefixes originating from the same AS will have the same path. This vector will be counted n times for n different prefixes, but only one if they all originate from the same AS. For the average case, we found that 6.90% of ASes have more than 10 unique paths for at last one prefix in the AS, and only 1.00% have more than 20 unique paths. In our worst case, 33.2% of ASes have more than 10 unique paths, 11.1% have more than 20, and 5.17% have more than 30.

The path lengths for the minimally stable listener (81.209.156.1) are considerably longer than for other listeners. A WHOIS lookup and traceroutes to the destination show this router belongs to LambdaNet Communications Deutschland AG in Ashburn, VA. The reasons for its distinctly different global view of paths relates to route filtering policies and other policy or connectivity issues unrelated to this study (see full length technical report for details).

A final series of tests assess the stability of the set of observed paths. Centrally, these tests attempted to estimate listening point *rates of discovery*. The experiments compute the frequency with which new paths are observed. We classify newness with respect to the AS (new when the AS has never advertised the particular path before) and prefix (the prefix has never been advertised with the path). Using the previously defined listening points, we examine the period between January 2003 and March 2004; the rates of discovery are shown in figures 6 and 7.

Two trends emerge from this study. First, there is nearly an order of magnitude difference between the number of new paths discovered per AS versus per prefix. An AS can have many different prefixes, each advertising the same AS path. Hence, when classification is done by origin AS, the path is only counted once, versus n times for n different origin prefixes. Although difficult to observe in the figures, a second trend shows strong discovery periodicity. We found that regular periods of little discovery corresponded to weekends. The network is at its most stable on the weekend, and hence little activity was observable in the BGP feeds.

The preceding results support our intuition that the set of known paths are not only stable over time, but the amount of churn between known paths is relatively small. Hence, there is an opportunity to exploit the reference locality. We explore how our constructions use this fact to implement efficient security in the next section.

5. EVALUATION

In this section, we evaluate the efficiency of the constructions defined in the proceeding sections via trace-based simulation. We compare our solutions against S-BGP and its variants, and draw general conclusions about the effectiveness of the proposed optimizations.

5.1 Experimental Setup

Developed specifically for this work, the *pasim* simulator models a single AS on the Internet and measures the computational and bandwidth costs associated with the validation of received paths. Computation is measured by the number of signature validations, which dominates all other computational costs (e.g., buffer handling, etc.) making it a good cost approximation. The simulations measure the amount of bandwidth consumed by the received proofs, but do not consider bandwidth consumed by other non-security related bandwidth costs (e.g., control traffic). We do not simulate the costs associated with the generation of proofs. Because structures are signed with low frequency (days), these costs

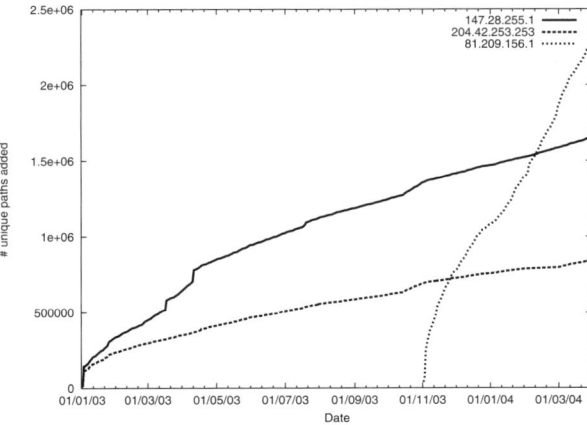

Figure 6: Rate of discovery CDFs for new paths per prefix as seen by multiple Route Views listening points, Jan 2003 to Mar 2004.

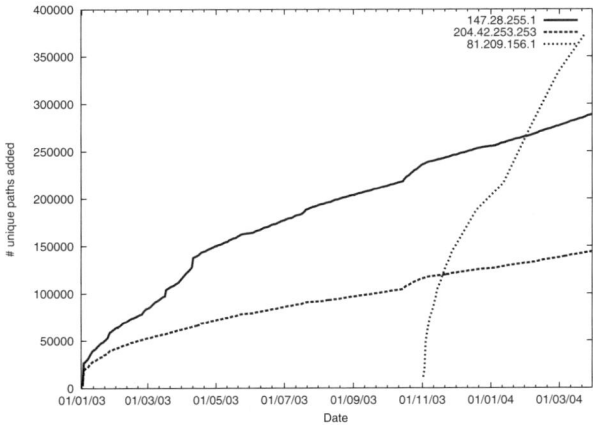

Figure 7: Rate of discovery CDFs for new paths per AS as seen by multiple Route Views listening points, Jan 2003 to Mar 2004.

will be dominated by validation. The simulations reported in this section use BGP update data collected during January 2004. Based on the results from the previous section, we ran simulations for the "typical" listening point (147.28.255.1).[4]

We simulate S-BGP route attestations and the signature amortization scheme proposed by Nicol et al. [30], which groups route updates into intervals and sends when the 30-second BGP timer is triggered; these updates are signed over a Merkle hash tree [5]. We contrast these schemes with simulations of our constructions: the prefix scheme, origin AS scheme, and the all AS paths scheme as defined in the preceding sections.[6] Each timed UPDATE in the trace data is played back to the simulated BGP router and processed according to the simulation solution. Unless described otherwise, all tests in this section assume that received signatures are hashed and kept in a 16 MB cache (described in further detail below), with simulated tree-based proof systems regenerated every 24 hours and authentication proofs issued every hour.

The simulation of our tree-based proposed schemes requires knowledge of all the paths advertised by an AS, which cannot be determined from a single listening point. One observation we make is that we are likely to see more unique paths from those ASes we are closest to. We approximate this by assuming unique paths comprise 7/8 of the paths observed from those ASes one hop away, 6/8 from ASes two hops away, etc., and adjust the tree size appropriately.[7] More precisely, if u unique paths associated with a proof system for an AS h hops away are seen, the proof system size is approximated to be $u(2 - h/8)$, e.g., $h = 3, u = 16 \rightarrow s = 16(13/8) = 26$. Note that an over or under estimate will affect the simulated size of the proofs, but not impact the amount of computational resources needed to validate them.

---

[4]We repeated the tests in the most and least stable listening points. In all cases, the costs scaled with the number of unique paths and rates of discovery as discussed in the preceding section.

[5]We do not model the aggregate signatures introduced in [49] as these optimizations are orthogonal to our main goal in comparing constructions; such optimizations are considered for future work.

[6]We simulated operation of the final variant of our scheme described in section 3, where expiration time of the attestation could be different from expiration time of the set-tag signature. The results differed from our origin AS scheme by a small factor. Hence, for clarity we omit these results from the graphs.

[7]We conservatively chose 8, as we observed that paths of four or more hops from the core were typically originated by stub ASes.

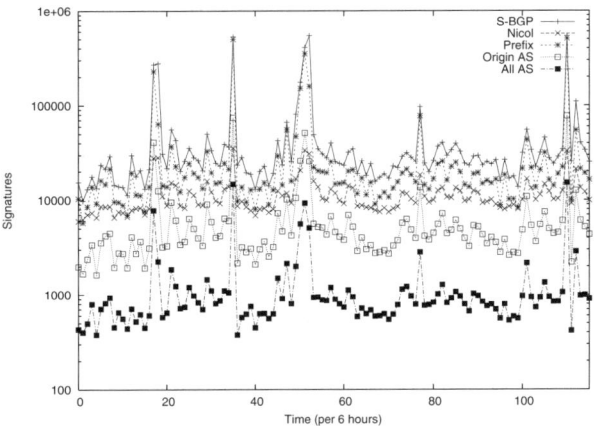

Figure 8: Validation cost - signatures validated per hour for *S-BGP*, the *Nicol et al.* scheme, and our constructions: *prefix path*, *origin paths*, and *all path* validation.

5.2 Simulation Results

Our initial simulations compare computation and bandwidth usage. Figure 8 shows the number of signatures used by each scheme. S-BGP consumes the most computational resources validating signatures. The Nicol optimization effectively reduces these costs by half. This drop is due to the amortization of signatures across the 30-second time period. Interestingly, this indicates that, on average, only a few paths propagate through an AS in a given time period. Because of the sustained load, the data lets us posit that optimizations over short periods (such as Nicol et al.) are likely to be less effective than longer periods, even if the latter may require more resources. The tree-based solutions require fewer validations than S-BGP. The prefix solution reduces the load by about 1/3. This is the effect of amortization over prefixes. Prefixes are largely stable and offer few paths, particularly over short time scales. Announcements for most prefixes will only be observed one or a few times per day. Hence, there is little opportunity to optimize. Note, however, that schemes such as SPV amortize costs in a fashion orthogonal to ours. Using our constructions in conjunction with those schemes could potentially reduce computational costs even further. The remaining AS path optimization schemes dominate all others:

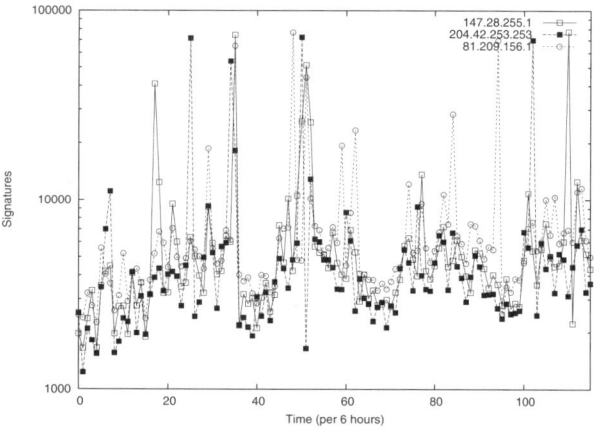

Figure 9: Signature validations by listening point - validations for the origin paths scheme per 6 hour period.

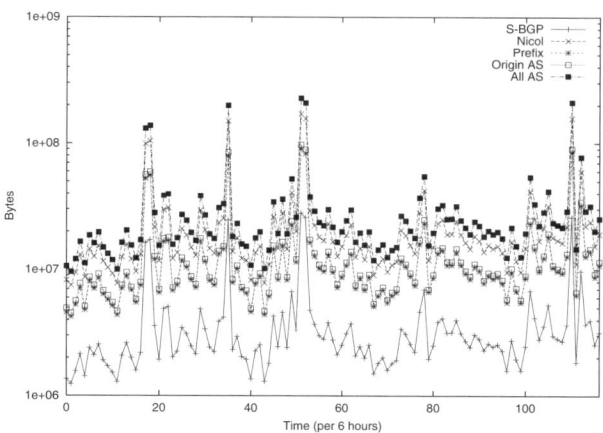

Figure 10: Bandwidth Cost - the number of bytes consumed by the transmission of the simulated path validation approaches.

the origin paths scheme represents an 86.3% reduction, and the all paths a 97.3% reduction in signature validations over S-BGP. In a given 24 hour period, the maximum number of signatures encountered will be two times the number of active ASes (assuming that all path proofs expire at some point during the day, and are recreated). The origin paths are somewhat more costly because they fail to fully exploit the opportunity to amortize cost.

Hashing typically consumes vanishingly small amounts of computational resources compared to signature validation; it is approximately 1,000 times faster than RSA signature validation [10]. However, in some schemes, hashing can be performed frequently enough that it potentially impacts performance. For instance, we found in the *all-path* construction, because the tree was so large, the computational cost was equivalent to one and a half signature validations. However, in all other cases, hashing was dominated by the signature validation costs.

Not shown for space considerations, the *instantaneous rate* of signature validations per router indicates the number of signatures per time quantum (in this case, 1 minute). We found many bursts where many validations are necessary per minute, particularly in the prefix scheme (where on average a burst would require less than 30 signature validations, but rare peaks would require a hundred or more). The origin scheme, which strikes the best compromise between validations and bandwidth, generally requires under 10 validations per minute, or one every six seconds on average.

Figure 9 shows the number of validations required for the origin scheme at the three listening points. The listening point demonstrating worst-case behavior has a number of bursty points with significant numbers of validations required; however, this burstiness is evident in all schemes and is constant across listening points.

Demonstrated in Figure 10, the bandwidth costs are largely the inverse of signature costs. S-BGP consumed far less bandwidth than the other approaches, because it generates small proofs. The prefix and Origin AS approaches were significantly more costly, consuming 3.35 and 3.57 times more resources than S-BGP, respectively. Interestingly, Nicol was second only to the all path scheme in consuming resources. The Nicol scheme creates a tree for every 30-second quantum, and subsequently sends a potentially large set of succinct proofs every period. The all path scheme was by far the most costly approach, consuming about 6 times as much bandwidth as S-BGP. In this case, the average bandwidth consumed per 6 hour period is 77 kilobytes. However, this approach may be prohibitive due to short bursts, which required as much as 139 megabytes in a single minute.

Any path authentication scheme must allocate storage resources for security relevant state (e.g., cryptographic proofs). In S-BGP, the additional space requirements to hold route attestations is estimated to be between 30 and 35 MB per BGP peer, though it is suggested that memory requirements in asymmetric peering relationships, such as between a large ISP router and a number of smaller peers, would be lower [15]. The storage requirements of the schemes proposed in this paper are unique to their design. Recall that the prefix approach requires every prefix to have a proof structure, while the all path approach requires a proof per AS; these two schemes form maximal and minimal requirements, respectively. Our simulations show that the total cost of storing *all proofs* across all peers ranges from approximately 55-60 MB for the prefix scheme to under 10 MB for the all path scheme. In the origin AS scheme, the total cost is approximately 25 MB.

The simulations illustrated in Figure 8 assume a proof cache of 16 MB. In our simulation model, this cache is separate from the storage space for the full set of proofs. We make this design decision so that the cache could be accessed more rapidly by the router as part of its fast path packet processing, but retain access to the proofs in stable storage (as needed for announcement creation). The additional stable storage costs are not onerous, and could likely be stored in memory itself on larger routers. Alternately, even smaller routers (e.g., Cisco 3600 series) include slots for flash memory, and are capable of accepting cards with 256 MB or greater, well above the requirements of our scheme. We assume that in real systems, to keep the cache size at a minimum, a hash of a received signature is stored in cache, rather than the signature itself. The router hashes the signature of an incoming update and checks whether it appears in the cache. If it is, a signature validation is not necessary. Hashed signatures are expired from the cache on a LRU basis. When sending an update, the full proofs to be sent are retrieved from stable storage.

6. DISCUSSION

A major difficulty of retrofitting security is the need for *incremental deployment*. Simply put, there are large portions of the Internet that will adopt solutions slowly or not at all. Any feasible solution must be designed such that communities of interested parties can work collaboratively to provide a working, secure system. Moreover, functionality can not come at the expense of poorly equipped enterprises. Such approaches would disenfranchise people and networks, and reduce universality of the Internet. However, those who do not participate need not receive benefit from deployment.

Past systems such as IRV [7] addressed incremental deployment by performing security *out-of-band*. They allow parties to exchange data without any change to BGP. Those who wish to exchange security relevant data do so freely over any mechanism that is available and convenient. However, this approach only works when the network is otherwise healthy or alternate channels are available. psBGP takes another tack in which the parties police each other's activities [45]. The incremental deployment approach in psBGP is one of a mutual embrace: like soBGP, communities of peers must work in concert to achieve a larger security posture.

We adopt this latter scheme, where communities of like-minded organizations will organically form *unions* of ASes. These unions will mutually authenticate credentials to be used in the issuance of proofs of authentication (a formal analysis of our scheme that includes a discussion of authentication proofs may be found in the Appendix). At the protocol layer, we adopt a similar strategy to S-BGP of signing transitions to and from non-adopting ASes. Of course, knowing which ASes are participating in the protocol is essential for ascertaining the validity of received routes. In a sense, our approach is similar to the S-BGP protocol, and as such can make use of its procedures and structures for incremental deployment.

Preventing *Worm-holing* is enormously difficult. There is nothing preventing an AS from achieving an arbitrary connectivity, and as such there is little one can do within a security protocol. Protocols such as soBGP do an approximate job of prevention by authenticating the network structure in the topology database. This prevents transient AS compromise from affecting the system as a whole, but does nothing against the truly adversarial AS. We argue that The real solutions to worm-hole prevention lie in good network management. For example, a large ISP should and often does filter multi-hop advertisements from stub ASes (ASes with no other connectivity other than provided by the ISP). Taken more generally, experience and formal relationships between networks are accurate sources of information for what constitutes good and bad connectivity.

Kent et al. [16] have suggested a path validation optimization aimed at reducing the load on validating S-BGP speaking routers. This optimization dictates that paths are validated only when they are selected as the *best paths*. However, it is not clear the degree to which this optimization will mitigate the computational costs of S-BGP. Consider an AS A with k neighbors. Any prefix p will be reachable through j neighbors, where $0 \leq j \leq k$, and j routes will be held by the AS. The fractional computational savings f for a given prefix on a given router over a period of time Δ is just the ratio of updates sent for that prefix during Δ divided by the total number of updates received for that prefix during Δ. Of course, f will vary from router to router and prefix to prefix, but f is likely to be on the order of $1/j$ for j defined above. For the data collected in our study, the median number of unique paths per prefix was 2.5 and the mean value was 2.8. A careful study of f remains for future work. But we note here that the same optimization can be used for our authentication proofs based on set-membership proofs. We will also achieve a factor f computational speedup. That is, when the optimization is applied to both schemes, the ratio of the computational overheads will remain the same.

7. RELATED WORK

Interdomain routing security has been studied for some time [33, 42], but comprehensive and efficient solutions remain elusive. The following considers how several of these efforts address path security.

Possibly the most comprehensive solution advanced to date, the Secure Border Gateway Protocol (S-BGP) [18, 17, 41] uses a public key infrastructure to support the authentication of routing artifacts. The S-BGP PKI maintains certificates for each AS and S-BGP-speaking router. Every router includes a *route attestation* with each advertisement. The route attestation is a signed statement of the AS identity, the paths, the prefix and the AS to which the announcement is directed. The S-BGP speaker also includes the route attestation of the route on which the advertisement is based. This prevents an adversary from adding or removing ASes from the path. While the authors of S-BGP have introduced a number of optimizations that reduce resource consumption [16], the costs associated with it are viewed as limiting factor in many environments [7, 10, 45]. For example, Nicol et al. showed that, under a set of timing and cost assumptions, such costs can double the path convergence time [30]. However, Nicol et al. did not model optimizations reported in [16]. It is not clear if and how the optimization would affect convergence times. While some argue that co-processors and protocol optimizations may make computation feasible, storage remains a major problem. Kent estimates that S-BGP will require an additional 30-35 megabytes of storage per peer [15]. Such costs are manageable in routers with a few peers, but are problematic in large ISPs or exchanges. However, Kent further argues that there are asymmetric configurations where only a few routes are accepted (as in customer/ISP peering), and hence these situations would require fewer resources.

Partially in deference to the costs associated with more comprehensive solutions, the soBGP and IRV projects sought other means of addressing BGP security. The soBGP [29] protocol uses a topology database to validate that advertised paths are consistent with the signed statements of connectivity between ASes. While this approach provides a limited security guarantee, it is effective in preventing a wide array of path hijacking and worm-holing attacks. However, soBGP does *not* provide path authentication, but simply implements a mechanism for detecting routes that are inconsistent with the authenticated topology. Philosophically similar to the earlier routing registry projects [22], the Interdomain Routing Validation (IRV) [7] project was motivated by the observation that any solution requiring a change to BGP was likely to be adopted slowly, if at all. IRV servers use an out-of-band (e.g., external to BGP sessions) protocol to exchange validation information. IRV is reliant on the routing infrastructure to extract and exchange routing data. Hence, unless some other infrastructure is put in place (e.g., static routes), the system is unable to function when connectivity is not available.

Validation of prefix ownership is essential to secure BGP. If not provided, an adversary can *hijack* entire networks by simply advertising the prefixes associated with them. Originally studied by Kent et al. [18, 41], an origin authentication (OA) service validates that an AS has the right to be the origin of a prefix. In a later work, Aiello et al. extended the study of OA by considering the semantics and efficient cryptographic constructions of origin authentication [2]. Principally, they explored formal semantics of the use and delegation of the IP address space. The set of all delegations between ICANN [13], registries, and organizations is modeled as a delegation hierarchy. Recently, Tan et al. suggested a alternative low cost, but weak form of origin authentication in which all BGP neighbors police and attest to the validity of the prefixes that an

AS originates [45]. However, this is limited, as colluding ASes can forge origin information.

Several proposals have sought efficient constructions for BGP security. Hu et al. introduced the concept of cumulative authentication for securing route advertisements in path vector protocols [9]. They use the TESLA timed key release authentication to validate announcements using low cost symmetric key cryptography. TESLA is limited in that it requires tight time bounds on message transmission, which is in conflict with protocols built on asynchronous propagation protocols such as BGP. More recently, Hu et al. introduced the Secure Path Vector Protocol (SPV) [10], which also seeks to implement BGP path security using low cost cryptography. SPV creates cascading authenticators over many (low cost) one time signature structures.

The Whisper protocol [43] uses a mechanism that detects inconsistencies in received routes using RSA-style [40] cryptographic operations. To simplify, any conflicts between routes received from multiple peers emanating from the same original advertisement is detectable. In the same work, Subramanian et al. introduce the Listen protocol, which does not provide comprehensive path authentication, but simply detects a class of attack.

Another approach that does not rely on a PKI or any form of cryptography is Pretty Good BGP [14], which relies on the stability of pre-existing routes as an indicator of their veracity. Longer-lived, more stable routes are preferred over newly appearing routes, which may require a secondary verification to determine if they are valid. Because of the lack of provable security, this solution is considered a stopgap measure to provide a modicum of protection until a cryptographic solution is implemented.

8. CONCLUSIONS

In this paper we have explored a range of cryptographic optimizations for securing BGP paths. Centrally, we exploit the stability of path advertisements to amortize cryptographic operations over many validations. This stability is confirmed via empirical analysis: the number of paths used by a particular AS for a given prefix is both small and largely constant over time. Through trace-based simulation, we show that our constructions reduce the computational costs of path authentication by as much as 97% over existing approaches, and show that other storage and bandwidth costs are nominal.

The problems of BGP security are sufficiently important to warrant discussion in the United States National Strategy to Secure Cyberspace [32]. This work studies tradeoffs between computational, bandwidth and storage costs for a range of BGP security path authentication mechanisms and is a step in a larger communal effort to design and deploy BGP security. The ultimate goal is to develop a comprehensive understanding of the security, cost, and manageability tradeoffs for BGP, to inform sound engineering decisions for future deployments. To this end, we plan to extend our evaluations to a range of realistic network environments, and to study the integration of optimizations suggested by others.

Acknowledgements

We are grateful to Jennifer Rexford for her detailed comments and patient answers to the many questions we posed to her. In addition, we would like to thank Nick Feamster, Tim Griffin, and Zhouqing Morley Mao for their feedback, and members of the SIIS lab at Penn State for commenting on multiple iterations of the paper.

9. REFERENCES

[1] W. Aiello, K. Butler, and P. McDaniel. Implications of Path Stability for Efficient Authentication in Interdomain Routing. Technical Report NAS-TR-0002-2004, Networking and Security Research Center, Department of Computer Science and Engineering, Pennsylvania State University, University Park, PA, Oct. 2004. Revised October 2005.

[2] W. Aiello, J. Ioannidis, and P. McDaniel. Origin Authentication in Interdomain Routing. In *Proceedings of ACM CCS '03*, October 2003.

[3] M. Baltatu, A. Lioy, F. Maino, and D. Mazzocchi. Security issues in control, management and routing protocols. *Computer Networks (Amsterdam, Netherlands: 1999)*, 34(6):881–894, 2000. Elsevier Editions, Amsterdam.

[4] A. Barbir, S. Murphy, and Y. Yang. Generic Threats to Routing Protocols (*Draft*). *IETF*, April 2004.

[5] S. Bellovin, R. Bush, T. Griffin, and J. Rexford. Slowing routing table growth by filtering based on address allocation policies. http://www.research.att.com/jrex/, June 2001.

[6] K. Butler, T. Farley, P. McDaniel, and J. Rexford. A Survey of BGP Security Issues and Solutions. Technical Report TD-5UGJ33, AT&T Labs - Research, Florham Park, NJ, Feb. 2004. (*revised June 2004*).

[7] G. Goodell, W. Aiello, T. Griffin, J. Ioannidis, P. McDaniel, and A. Rubin. Working around BGP: An incremental approach to improving security and accuracy of interdomain routing. In *Proceedings of NDSS '03*, Feb. 2003.

[8] M. Goodrich, R. Tamassia, and A. Schwerin. Implementation of an authenticated dictionary with skip lists and commutative hashing. In *Proceedings of DARPA Information Survivability Conference and Exposition II (DISCEX)*. IEEE Computer Society Press, June 2001. Los Angeles, CA.

[9] Y. Hu, A. Perrig, and D. Johnson. Efficient security mechanisms for routing protocols. In *Proceedings of NDSS '03*, Feb. 2003.

[10] Y.-C. Hu, A. Perrig, and M. Sirbu. SPV: Secure Path Vector Routing for Securing BGP. In *ACM SIGCOMM*. ACM, August 2004.

[11] G. Huston. BGP Reports, May 2005. http://bgp.potaroo.net/.

[12] IANA. Autonomous System Numbers, March 2003.

[13] ICANN. The Internet Corporation for Assigned Names and Numbers, July 2004. http://www.icann.org/.

[14] J. Karlin, S. Forrest, and J. Rexford. Pretty Good BGP: Protecting BGP by Cautiously Selecting Routes. Technical Report TR-CS-2005-37, University of New Mexico, Albuquerque, NM, USA, Oct. 2005.

[15] S. Kent. Securing the Border Gateway Protocol. *The Internet Protocol Journal*, 6(3), Sep. 2003.

[16] S. Kent. Securing the Border Gateway Protocol: A status update. In *Seventh IFIP TC-6 TC-11 Conference on Communications and Multimedia Security*, Oct. 2003.

[17] S. Kent, C. Lynn, J. Mikkelson, and K. Seo. Secure Border Gateway Protocol (S-BGP) Real World Performance and Deployment Issues. In *Proceedings of NDSS '00*, Feb. 2000.

[18] S. Kent, C. Lynn, and K. Seo. Secure Border Gateway Protocol (S-BGP). *IEEE Journal on Selected Areas in Communications*, 18(4), Apr. 2000.

[19] C. Kruegel, D. Mutz, W. Robertson, and F. Valeur. Topology-based detection of anomalous BGP messages. In *Proceedings of RAID '03*, Sept. 2003.

[20] L. Lamport. Password Authentication with Insecure Communication. *Commun. ACM*, 24(11):770–772, Nov. 1981.

[21] X. Meng, Z. Xu, L. Zhang, and S. Lu. An analysis of BGP routing table evolution. Technical Report TR030046, Computer Science Department, UCLA, Jan. 2003.

[22] Merit Network. The Internet Routing Registry, July 2004. http://www.irr.net/.

[23] R. Merkle. Protocols for public key cryptosystems. Oakland, CA, Apr. 1980. IEEE Symposium on Research in Security and Privacy.

[24] D. Meyer. The Route Views Project, Nov. 2006. http://www.routeviews.org/.

[25] D. Meyer and A. Partan. BGP Security, Availability,and Operator Needs. NANOG 28, June 2003.

[26] S. Murphy. BGP Security Vulnerabilities Analysis. RFC 4272, Jan. 2006.

[27] M. Naor and K. Nissim. Certificate revocation and certificate update. In *Proceedings of the 7th USENIX Security Symposium*, Jan. 1998.

[28] H. Narayan, R. Govindan, and G. Varghese. The impact of address allocation and routing on the structure and implementation of routing tables. In *Proceedings of ACM SIGCOMM '03*, Karlsruhe, Germany, Aug. 2003. ACM.

[29] J. Ng. Extensions to BGP to support secure origin BGP (soBGP). Internet Draft, Oct. 2002.

[30] D. Nicol, S. Smith, and M. Zhao. Evaluation of efficient security for BGP route announcements using parallel simulation. *Simulation Modelling Practice and Theory*, 12(3–4):187–216, July 2004.

[31] O. Nordström and C. Dovrolis. Beware of BGP attacks. *Computer Communications Review*, 34(2):1–8, Apr. 2004.

[32] Office of the President of the United States. Priority II: A National Cyberspace Security Threat and Vulnerability Reduction Program. National Strategy to Secure Cyberspace, Nov. 2004.

[33] R. Perlman. *Network layer Protocols with Byzantine Robustness*. PhD thesis, Massachusetts Institute of Technology, Cambridge, MA, Oct. 1988. MIT/LCS/TR-429.

[34] J. Postel. Internet Protocol. RFC 791, Sept. 1981.

[35] J. Postel. Transmission Control Protocol - DARPA Internet Protocol Program Specification. *IETF*, Sep. 1981. RFC 793.

[36] J. Puig, M. Achemlal, E. Jones, and D. McPherson. Generic Security Requirements for Routing Protocols (*Draft*). *IETF*, July 2004.

[37] Y. Rekhter and P. Gross. Application of the Border Gateway Protocol in the Internet. RFC 1772, Mar. 1995.

[38] Y. Rekhter and T. Li. A Border Gateway Protocol 4 (BGP-4). RFC 4271, Jan. 2006.

[39] J. Rexford, J. Wang, Z. Xiao, and Y. Zhang. BGP routing stability of popular destinations. In *IMW '02: Proceedings of the 2nd ACM SIGCOMM Workshop on Internet measurment*, pages 197–202, New York, NY, USA, 2002. ACM Press.

[40] R. Rivest, A. Shamir, and L. Adleman. A Method for Obtaining Digital Signatures and Public-Key Cryptosystems. *Communications of the ACM*, 21(2):120–126, Feb. 1978.

[41] K. Seo, C. Lynn, and S. Kent. Public-Key Infrastructure for the Secure Border Gateway Protocol (S-BGP). In *IEEE DARPA Information Survivability Conference and Exposition II*, June 2001.

[42] B. Smith and J. Garcia-Luna-Aceves. Securing the Border Gateway Routing Protocol. In *Proceedings of IEEE Global Internet 1996*, London, UK, Nov. 1996.

[43] L. Subramanian, V. Roth, I. Stoica, S. Shenker, and R. Katz. Listen and Whisper: Security mechanisms for BGP. In *Proceedings of NSDI'04*, Mar. 2004.

[44] S. Teoh, K. Ma, S. Wu, D. Pei, L. Wang, L. Zhang, D. Massey, and R. Bush. Visual-Based Anomaly Detection for BGP Origin AS Change (OASC) Events. In *Proceedings of IEEE/IFIP DSOM '03*, October 2003.

[45] T. Wan, E. Kranakis, and P. C. van Oorschot. Pretty Secure BGP (psBGP). In *Proc. of NDSS '05*. Internet Society (ISOC), Feb. 2005.

[46] L. Wang, X. Zhao, D. Pei, R. Bush, D. Massey, A. Mankin, S. F. Wu, and L. Zhang. Protecting BGP Routes to Top Level DNS Servers. In *Proceedings of the 23rd International Conference on Distributed Computing Systems (ICDCS)*, May 2003.

[47] K. Zhang, S.-T. Teoh, S.-M. Tseng, C.-N. Chuah, K.-L. Ma, and F. Wu. Performing BGP experiments on a semi-realistic internet environment. North American Network Operators Group (NANOG), October 2004.

[48] X. Zhang, S. Wu, Z. Fu, and T.-L. Wu. Malicious Packet Dropping: How It Might Impact the TCP Performance and How We Can Detect It. In *Proceedings of ICNP 2000*, Nov. 2000.

[49] M. Zhao, S. W. Smith, and D. M. Nicol. Aggregated path authentication for efficient BGP security. In *Proceedings of the 12th ACM Conference on Computer and Communications Security (CCS'05)*, Nov. 2005. Alexandria, VA, USA.

APPENDIX
A. PATH VALIDATION CONSTRUCTIONS

In this section, we define what we mean by attestations and route attestation tags and formally state their security properties. Route attestation tags are very similar to route attestations as defined in [18, 17, 16] with several minor differences highlighted in this section. We assume that the reader has a general familiarity with cryptographic primitives such as hash chains, hash tables and digital signatures. These constructions are explored in greater detail later in the section. We begin in the following subsection with a brief overview of our approach to path authentication, and continue with a formal description of its semantics, operation, and security.

A.1 Attestations and Route Attestation Tags

In previous works, route attestations were defined as a sequence of statements signed by routers using public key signatures. Our route attestation tags are also a sequence of attestations by routers, but here we allow the attestations to be more general public key authentication methods. In particular, an attestation may be either a signature or a *set-membership proof*. Set-membership proofs are essentially signatures of Merkle hash trees [23].

We first state several definitions that will be used below. We then state formally the definition of a set-membership proof, its definition of security, and several examples. Next, we define a *route attestation tag* or RAT as a sequence of attestations. Finally, we describe our scheme, as well as the schemes of [18] and [30] as instantiations of the general set-up. These descriptions are used in the subsequent sections, where the performance tradeoffs of these schemes are that empirically analyzed. For space considerations, we direct readers interested in a formal definition of security for a

RAT to our technical report [1], where we reduce the security of a RAT to the security of the attestations used in the RAT. While this does not appear to be surprising, the adversary used as the basis of the definition of security is quite powerful. The technical report also contains details on incremental deployability using these path authentication schemes.

The formalization below is general enough to capture not only our proposed schemes but the S-BGP scheme and the scheme of Nicol et al. as well. At the same time, it is specific enough to allow for a precise definition of existential forgery of a route announcement and a reduction to the security of standard cryptographic primitives.

A.2 Notation

Let $\mathcal{ASN} = \{1, \ldots, 2^{16} - 1\}$ be the set of all unique identifiers for an Autonomous System. These are the so-called Autonomous System Numbers. An *AS path* is a sequence, possibly empty, of AS numbers. Given a path $p \in \mathcal{ASN}^*$, let p_i, $i \geq 1$, denote the ith element in the sequence. Furthermore, let $p_i^\leq$, $i \geq 1$, denote the subsequence of the first i elements of p. For example, if $p = (23, 1708, 229)$, then $p_2 = 1708$ and $p_2^\leq = (23, 1708)$.

In BGP, AS padding is allowed. That is, a legitimate AS path can have a sequence of consecutive values that are identical. This is equivalent to saying that BGP allows paths with self loops (but not other kinds of loops). We call a path *almost simple* if it has no loops except for self loops.

Let $G = (\mathcal{ASN}, \mathcal{E})$ denote the AS graph. A pair of AS numbers (a_1, a_2) is in $\mathcal{E}$ if AS a_1 and AS a_2 have a service level agreement (SLA) to be eBGP neighbors. Note that $\mathcal{E}$ does not capture which pairs of ASes have active eBGP sessions between routers at the current time. That is, if (a_1, a_2) is in $\mathcal{E}$, there may be no current eBGP session between a router in a_1 and a router in a_2. Nonetheless, the edge in the graph G is maintained as long as the neighbors have an eBGP SLA. A path p is denoted *topology respecting* if every edge in the path is also an edge in G.

A route is a pair consisting of an address block and an AS path. Given an address block b and a path $p = (a_1, a_2, \ldots, a_k)$, the route r for b and p is written as $r = (b, p)$ or as $r = (b; a_1, a_2, \cdots, a_k)$. Considering the latter as a sequence, r_i, $i \geq 0$, denotes the ith element of the sequence and $r_i^\leq$ denotes the subsequence of r from r_0 to r_i, inclusive. Note that $r_0 = b$, and that for $i \geq 1$, $r_i = p_i$, and $r_i^\leq = (b, p_i^\leq)$. These definitions will be useful when defining the cryptographic mechanisms for protecting entire routes.

A.3 Signatures and Set-Membership Proofs

For completeness, recall the definition of a signature scheme. A signature scheme consists of three functions:

1. a randomized generation algorithm G takes as input a security parameter (e.g., the desired length of the output) and generates a public/private key pair (pk, sk);
2. a signing algorithm S that takes as input a secret key sk and a value a and computes a signature σ; and,
3. a verification algorithm V which takes as input a public key pk, a value a, and a signature σ and outputs "accept" or "reject".

G, S, and V satisfy the following signature-correctness condition. For all (pk, sk) generated by G and all strings a, if $\sigma = S(\text{sk}, a)$, then $V(\text{pk}, a, \sigma) =$ "accept".

A set-membership proof is defined similarly. It consists of three functions, G', S', and V': 1) a randomized key generation algorithm G' takes as input a security parameter (e.g., the desired length of the output) and a set of elements, A, and generates a public/private key pair (pk', sk'); 2) a signing algorithm S' that takes as input a secret key sk', a set A, and an element of the set a, and computes a set membership proof π; and, 3) a verification algorithm V' which takes as input a public key pk', a value a, and a proof π and outputs *accept* or *reject*. Note that if a is not in A then S' outputs $\bot$.

G', S', and V' satisfy the following correctness condition. For all A and all (pk', sk') generated by G on A, and all strings $a \in A$, if $\pi = S'(\text{sk}', A, a)$, then $V'(\text{pk}', a, \pi) =$ "accept".

Definition: A set-membership proof scheme is (k, T, ϵ) secure against existential forgery if it also satisfies the following security requirement. An adversary is allowed to ask for public keys to be generated for sets of its choosing. The adversary is then allowed to see the signatures for k (set, set element) pairs where the pairs can be chosen by the adversary adaptively. No adversary running in time at most T can generate a (signature, set element, public key) triple that passes verification, except with probability at most ϵ.

The above definition of a set-membership proof scheme may be modified to include ancillary information about the set. That is, the signing algorithm may be modified to include this ancillary information about the set as input. If this is the case, the verification algorithm must be modified as well to include this ancillary information for proper verification.

A secure set-membership proof scheme can be constructed from a secure signature scheme and a hash function secure against second pre-image attacks (for random domain elements). The advantage of a set membership proof scheme over a signature scheme is that in practice for both the signer and the verifier, the expensive public key computations need only be done once and then cached for any given set. [8] This efficiency comes at the price of larger space requirements but we note that the size of the membership proofs can be made logarithmic in the cardinality of the set. An example of a set membership proof system is the combination of Merkle hash trees and public key signatures as in the example above.

An important property of a set-membership proof scheme to highlight is that the signer only needs to compute T and S once, regardless of how many set elements it will eventually compute membership proofs for. That is, the cost of one public key signature computation can be amortized over the cost of many set-membership proofs. Likewise, a verifier needs only to run the signature verification algorithm on one valid (τ, σ) pair. It can cache the positive result using τ as a key. Subsequent membership proofs with set tag τ require only the verifier to run E, which is not a public key algorithm. Thus, the cost of one public key signature verification can be amortized over the cost of many set-membership verifications. In subsequent sections, we analyze the amortization savings that can be realized in practice on real BGP data streams.

A.4 Route Attestation Tags

A attestation by an identity x about a string α is denoted $A(x; \alpha)$. An attestation is either a secure signature signed by the secret key of x or it is a membership proof of α by the identity x (using the secret key of x). We will denote an attestation by x about a string β *to an identity* y by $A(x; \beta : y)$. This is just an attestation $A(x; \alpha)$ with $\alpha = \beta : y$. Attestation may also have timestamps or expiration times. These may be used, in part, as anti-replay mechanisms. For purposes of exposition, for now we do not include timestamps in the notation. We defer discussion of the issue of of replay to the technical report.

---

[8]This is not transparent from the abstract description of a set membership proof scheme above. A formal description of a set membership proof scheme that explicitly breaks out the public key computations is cumbersome and omitted here for lack of space.

Definition: For a given route we define a *route attestation tag* or RAT, as follows. A RAT takes as an input a route $r = (b, p)$. $RAT(r)$ is a sequence of attestations defined recursively as follows.

$$RAT(r_i^{\leq}) = RAT(r_{i-1}^{\leq}), A(p_{i-1}; r_i^{\leq} : p_i)$$

for $i = 2, ..., |p|$. The base case is $RAT(r_1^{\leq})$. This is the origin authentication tag, or OAT, for ownership of the address block $r_0 = b$ by the AS with identifier p_1. The semantics of $OAT(b, a)$ were discussed extensively in [2]. Briefly, the $OAT(b, a)$ includes: a.) a chain of attestations from IANA to an organization O attesting to the fact that the ownership of the address block b has been delegated to O; b.) an attestation by IANA that it has assigned the AS identifier a to O; and c.) an attestation by O that it has assigned the address block b to AS a.

As an example, let $p = (a_1, a_2, a_3, a_4)$. Then
$$\begin{aligned} RAT(b; a_1, a_2, a_3, a_4) = & \ OAT(b, a_1), \\ & A(a_1; (b; a_1) : a_2), \\ & A(a_2; (b; a_1, a_2) : a_3), \\ & A(a_3; (b; a_1, a_2, a_3) : a_4) \end{aligned}$$

Note that the final attestation in $RAT(b; p)$ is by the second to last AS in the path, i.e., by AS $p_{|p|-1}$.

A RAT is valid only if all of the associated attestations are valid and the OAT is valid. Note that RATs as defined here are nearly identical to the definition of route attestations in defined in [18]. The only minor differences are the inclusion of the origin authentication tag and the slight generalization to allow both signatures and set-membership proofs in the individual router attestations. In the technical report we discuss the addition of the origin authentication tags to RATs.

We denote the concatenation of a route $r = (b; p)$ and an AS a by $r.a$, where this is just the route given by the pair $(b; p.a)$, i.e., the path of r extended by one hop to a.

Definition: A route $r = (b, p)$, and an accompanying $RAT(r.a')$, when received in an update over an eBGP session by a router in AS a is considered valid only if:

1. $a = a'$,
2. $p.a$ is almost simple,
3. the RAT of $r.a$ is valid, i.e., the pair $(r.a, RAT(r.a))$ validates, and
4. the route was received over an authenticated eBGP session with a router in AS a^* where a^* must equal the last AS in the AS path p, i.e., $a^* = p_{|p|}$.

As defined above, a router that announces its new best AS path for a given address block to all of its neighbors must send a slightly different attestation to each of its eBGP neighbors. That is, to announce the route r it must send $r, RAT(r.a)$ to an eBGP peer in AS a, and $r, RAT(r.a')$ to an eBGP peer in AS a' etc. At first glance this may seem unnecessary. However, different routers in the same AS may announce a different best AS path for the same prefix. If when advertising the route r, the router simply attested to the route up to and including its AS, it is easy to construct cases in which upstream routers can forge routes [31].

A similar reason argues for the requirement that the prefix be included in all of the attestations of a RAT. The alternative is to have the attestations in the RAT include only the AS path and to separately include the origin authentication tag for the prefix and origin AS. However, such a scheme allows for the following type of attack. Suppose a router in AS b receives routes for two different prefixes both originated by AS a, e.g, $(b; a.p.b)$ and $(b'; a.p'.b)$ and the origin authentication tags binding b and b' to a. If the attestations in the RATs contain the appropriate AS path prefixes but are not required to contain the address block, then the router in AS b can create RATs that will validate for routes it did not receive. In this example the router can create a valid RAT for $(b; a.p'.b)$ and $(b'; a.p.b)$, thus altering in an undetected fashion the routes for the prefixes b and b'.

Note that in order for a router in AS a to check the validity of $RAT(r.a)$, it is not sufficient for the router to simply have the certified value of the public key of its eBGP neighbor that sent it the route. The router must have the certified public keys of all of the ASes in order to check the attestations of each AS in the route. Here we assume a PKI provides each router with the certified public keys of all ASes. For a discussion of such a PKI see [41].

We now address the issue of the security guarantee provided by the RAT construction. Intuitively, we would like to say that as long as the attestation scheme used in a RAT is not existentially forgeable, then that RAT scheme is not existentially forgeable in the sense that an adversary cannot create a valid (route, RAT) pair that it has not previously seen. Unfortunately, it is not quite that simple. This is due to the fact that every AS, including malicious ones, are able to extract or extend valid $(r, RAT(r))$ pairs sent to them legitimately in several ways. For example, from a valid route attestation for r, it is easy to extract a valid route attestation tag for each prefix of r, i.e., $r_i^{\leq}$ for $i = 1, \cdots, |r|$. This follows directly from the recursive definition. As another example, if a router in AS a receives a valid pair $(r.a, RAT(r.a))$, then a (possibly different) router in a can compute a valid $RAT(r.a.a')$ for any neighboring AS a'. This is due to the fact that $RAT(r.a.a') = RAT(r.a), A(a; r.a : a')$ where $RAT(r.a)$ is given to AS a and $A(a; r.a : a')$ is an attestation by a itself. Moreover, since AS padding is allowed in BGP, a can form valid RATs for the form $RAT(r.a^i.a')$ for any neighboring AS a' and any $i \geq 1$, where a^i is a repeated i times. Let us call these extensions of a RAT *transit extensions*.

Below we will define all possible transit extensions of a given set of routes. Then we will show that if the adversary can compute a valid RAT for a route that is neither in the set of routes for which it has seen a valid RAT, nor in the set of its transit extensions for those routes, then the adversary must have computed an existential forgery of an attestation.

Let $\mathcal{P}$ be a set of AS paths. Since all "good" routers check whether a path is almost simple, assume without loss of generality that all the paths in $\mathcal{P}$ are almost simple. Denote the transit extensions of $\mathcal{P}$ by x as $TE(\mathcal{P}, x)$. We define it iteratively as follows. First, for each $p \in \mathcal{P}$ all of the prefixes of p are added to $TE(\mathcal{P}, x)$, including p itself. Now, for each $p \in TE(\mathcal{P}, x)$, except for those that contain x, add the set $p.\{x\}^*$ and the set $p.\{x\}^+.\mathcal{Q}_{p,x}$ to $TE(\mathcal{P}, x)$, where $\{x\}^* = \{x^i \,|\, i \geq 0\}$ and $\{x\}^+ = \{x^i \,|\, i \geq 1\}$. Here $\mathcal{Q}_{x,p}$ is $\mathcal{ASN}$ minus x and minus the ASes in p and is defined so that all of the extensions are almost simple paths. An example of the transit extension is given in our associated technical report [1].

Definition: A secure RAT is defined as follows. An adversarial AS x is given access to a RAT oracle. That is, x can query the RAT oracle on routes r of its choice in a dynamic fashion and receive $RAT(r)$ for each of its queries. Let $\mathcal{P}$ be the set of such routes. A RAT forgery by x is a valid $(r, RAT(r))$ pair for some r not in $TE(\mathcal{P}, x)$, the set of transit extensions of $\mathcal{P}$. A RAT is secure if no time bounded adversary with access to a RAT oracle can compute a RAT forgery except with negligible probability. This definition of security can be parameterized in the standard fashion by a time bound, a query bound, and a probability bound but we omit the details of this parameterization here. These definitions lead to the main security lemma for RATs.

Lemma: If AS x has a strategy for computing a RAT forgery then there is an efficient strategy for computing an attestation forgery.

A proof of the lemma is included in the technical report. The implication of the lemma is that if the attestations are secure as per the definitions above then the route attestation tag will be secure as per the definition above. The security lemma and proof can be easily modified to include security parameters. That is, the above lemma can be extended to give the security parameters of the RAT scheme as a function of the security parameters of the underlying attestation scheme.

Note that the adversarial model and proof of security are quite strong. They allow an adversarial AS x to see RATs for any routes of its choosing including, for example, routes that do not correspond to actual topology and routes that may have already transited x and continued for several more hops. If x is colluding with another AS, it may indeed be able to see the latter RATs. The security model protects against forgeries even with this type of collusion. It is better, of course, to overestimate the power of an adversary since it is always difficult to bound the information that a determined adversary can uncover. Even with this more powerful model, the security of RATs reduces to the security of the underlying attestations.

It is possible to capture the case of several ASes colluding with definitions and a security lemma similar to that above. However, the definitions are more complex and are omitted here. But, intuitively suppose a set of adversaries X is given valid RATs for a set of paths $\mathcal{P}$ of their choosing. The transit extensions of $\mathcal{P}$, $TE(\mathcal{P}, X)$, consist of all of the almost simple paths for which the adversaries can derive valid RATs from the valid RATs for the paths in $\mathcal{P}$. If the adversaries X can succeed in computing a valid (route,RAT) pair for a route not in $TE(\mathcal{P}, X)$ then they have succeeded in computing a RAT forgery. As long is it is computationally difficult to forge an attestation with non-negligible probability, it is computationally difficult for the adversaries X to compute a RAT forgery with non-negligible probability.

A.5 Constructions

In this section we describe the attestation schemes used by S-BGP and Nicol et al. In addition, we propose a different attestation scheme and several variants. S-BGP attestations include a validity interval $I = [t_s, t_e)$ and are denoted $A(a_1; r : a_2 | I)$ As noted before, the S-BGP attestations are implemented as a public key signature. That is, $A(a_1; r : a_2 | I) = I, \sigma$ where σ is the signature of the string $r : a_2 | I$ using the private key of a AS a_1.

Both the Nicol et al. scheme and our schemes are based on set-membership proofs. Nicol et al. make crucial use of the fact that BGP speaking routers do not continuously send BGP updates to their neighbors. Instead, BGP speaking routers group route updates into 30-second intervals, and only send these updates to their neighbors at the end of the 30-second interval. For a given router in AS a, define $\mathcal{R}_t$ to be the set of all tuples $(r : a')$ such that route r is sent by the given router in an eBGP update to a router in a' in the 30-second interval ending at time t. The routers in this scheme create set-membership proofs for the set $\mathcal{R}_t$ for each time t that ends an interval. As discussed previously, ancillary information can be included in a set-membership proof. In this case, both the time of the update t and the expiration time of the announcements t_e are included. Let π_α be the set-membership proof for $\alpha \in \mathcal{R}_t$. Then for each $\alpha \in \mathcal{R}_t$ the attestation $A(a; \alpha | [t, t_e))$ is simply $(\pi_\alpha, [t, t_e))$. Since the router is sending the attestations for each $\alpha \in \mathcal{R}_t$, this set of attestations can be more parsimonious than the collection of individual attestations (we omit details for brevity). Nonetheless, as (route, attestation) pairs for routes represented in $\mathcal{R}_t$ are forwarded downstream in the appropriate RAT for an extension of the route, the amortized length of the encoding of (π_α) will increase. This is because fewer and fewer of the attestations for elements of $\mathcal{R}_t$ will be included in downstream updates.

Our scheme is similar to Nicol et al. in that we also use set-membership proofs. However, the method with which we choose to aggregate updates into sets is different. Assume for now that a router in AS a knows in advance all of the routes it will send during time interval $I = [t_s, t_e]$. That is, let $\mathcal{T}_I$ be the set of all tuples $(r : a')$ where the route r was sent by the given router to a router in a' within the interval I. Within the interval I, when the router needs to compute the attestation $A(a; \beta | I)$, for $\beta \in \mathcal{T}_I$, it computes the se-membership proof π from β, $\mathcal{T}_I$, and ancillary information I. The attestation for β is (π, I). Of course, a router cannot know in advance all of the routes it will receive in an interval I. However, as we will show in subsequent sections, for BGP updates, the past is a fairly accurate predictor of the future. Thus the set $\mathcal{T}_I$ is an approximation based on past history of the routes that will be needed for updates in period I. When the router needs to send an attestation for a route not in $\mathcal{T}_I$, it simply computes an S-BGP attestation. If $\mathcal{T}_I$ is required to have a maximum size bound, as it must, then there are a variety of caching strategies for maintaining $\mathcal{T}_I$ from one interval to the next.

As we will see, for reasonably sized intervals I, the set $\mathcal{T}_I$ can get quite big. However, $\mathcal{T}_I$ can be partitioned into smaller sets in a number of ways, and then a set-membership proof scheme can be applied to each set. This affords a time-space tradeoff. For example, for all address blocks b in tuples in $\mathcal{T}_I$, let $\mathcal{T}_{b,I}$ be the tuples that have address block b. In what we denote the *prefix scheme*, the router creates set tags and set-tag signatures for each $\mathcal{T}_{b,I}$. And the attestations are membership proofs for the appropriate set and member of that set. In another variant, $\mathcal{T}_I$ is partitioned according to the origin AS. That is, we define $\mathcal{T}_{a,I}$ as the elements of $\mathcal{T}_I$ that share the same origin AS. The scheme based on this partition is denoted the *origin AS scheme*.

A final variant of our scheme allows the expiration time of an attestation to be different from the expiration time of the set-tag signature. For example, suppose we partition I into k subintervals. Let the set of intervals be $\mathcal{K}$. Using the origin AS scheme as an example, for each $\mathcal{T}_{a,I}$, the router creates a proof system for the set $\mathcal{T}_{a,I} \times \mathcal{K}$. Note that since the size of membership proofs can be made to be logarithmic in the size of sets, this only adds $\log |\mathcal{K}|$ to the length of the membership proofs. In this case, the output of an attestation is the same as above plus the particular subinterval used. That is I is used in the public key signature validation of σ and the subinterval is used to encode the leaf that the verifier must use.

Note that for every address block includes an empty path in $\mathcal{T}_I$. Within our setting, we consider a withdrawal of an address block to be denoted by a route advertisement of that address block with an empty path.

Replayer: Automatic Protocol Replay by Binary Analysis

James Newsome, David Brumley, Jason Franklin, Dawn Song*
Carnegie Mellon University
Pittsburgh, PA, USA
{jnewsome, dbrumley, jfranklin, dawnsong}@cmu.edu

ABSTRACT

We address the problem of replaying an application dialog between two hosts. The ability to accurately replay application dialogs is useful in many security-oriented applications, such as replaying an exploit for forensic analysis or demonstrating an exploit to a third party.

A central challenge in application dialog replay is that the dialog intended for the original host will likely not be accepted by another without modification. For example, the dialog may include or rely on state specific to the original host such as its hostname, a known cookie, *etc*. In such cases, a straight-forward byte-by-byte replay to a different host with a different state (e.g., different hostname) than the original observed dialog participant will likely fail. These state-dependent protocol fields must be updated to reflect the different state of the different host for replay to succeed.

We formally define the replay problem. We present a solution which makes novel use of program verification techniques such as theorem proving and weakest pre-condition. By employing these techniques, we create the first *sound* solution to the replay problem: replay succeeds whenever our approach yields an answer. Previous techniques, though useful, are based on unsound heuristics. We implement a prototype of our techniques called *Replayer*, which we use to demonstrate the viability of our approach.

Categories and Subject Descriptors

C.2 [**Computer Systems Organization**]: COMPUTER COMMUNICATION NETWORKS

---

*This material is based upon work supported by the National Science Foundation under Grant No. 0448452. Jason Franklin performed this research while on appointment as a U.S. Department of Homeland Security (DHS) Fellow under the DHS Scholarship and Fellowship Program, a program administered by the Oak Ridge Institute for Science and Education (ORISE) for DHS through an interagency agreement with the U.S Department of Energy (DOE). Any opinions, findings, and conclusions or recommendations expressed in this material are those of the author(s) and do not necessarily reflect the views of DHS, DOE, ORISE, or the National Science Foundation.

Permission to make digital or hard copies of all or part of this work for personal or classroom use is granted without fee provided that copies are not made or distributed for profit or commercial advantage and that copies bear this notice and the full citation on the first page. To copy otherwise, to republish, to post on servers or to redistribute to lists, requires prior specific permission and/or a fee.
CCS'06, October 30–November 3, 2006, Alexandria, Virginia, USA.
Copyright 2006 ACM 1-59593-518-5/06/0010 ...$5.00.

General Terms

Security

Keywords

application protocol replay, weakest pre-condition

1. INTRODUCTION

In many scenarios, it would be extremely useful to automatically *replay* an application dialog as seen by one of the participants. This problem is termed the *replay of application dialog* [8]. Scenarios which benefit from automatic application dialog replay include dynamically analyzing programs through repeated execution in unique environments, demonstrating observed software vulnerabilities to interested parties, forensics, and determining the range of software versions vulnerable to an exploit.

However, as shown in [8], automatic replay of an application dialog is a challenging task. A primary issue is that a successful dialog may rely on state specific to the participants, such as hostnames, shared cookies, *etc*. These state-specific protocol fields cause problems when replaying the dialog to a different host. Cui et. al. recently proposed using machine learning to identify certain fields such as IP addresses, host names, etc., and then modify them accordingly to replay the application dialog [8]. They show that their approach works for many interesting protocols. However, this approach is intrinsically heuristics-based and cannot guarantee the correctness of the replay.

In this paper, we first develop a formal definition for the replay problem. The formal definition provides new insight into the problem, and opens the door to using formal techniques to solve the problem instead of relying on heuristics.

Based on our formal definition, we design an approach to solve the application replay problem by making novel use of existing program verification techniques. At a high level, we first build a *symbolic formula* of how the original host processed the application dialog. We then use an off-the-shelf decision procedure to derive an input tailored to a different host from the symbolic formula. This approach provides a *sound* solution to the application protocol replay problem. Unlike previous, heuristic-based approaches, our approach is general, and can handle unanticipated types of state-dependent protocol fields.

For example, successfully replaying a dialog may require modifying parts of the application dialog specific to the original host. However, changing the trace may result in an inconsistent dialog, e.g., check-sums on protocol messages may be incorrect. Thus, automatically determining which parts of a dialog to change, automatically changing them, and subsequently making the messages consistent in an automated fashion is difficult. Our techniques can soundly handle such cases, while previous work required manually applying domain-specific knowledge.

We describe and evaluate our implementation of *Replayer*, a working prototype of our approach for automatic protocol replay. We use Replayer to solve several examples of the automatic protocol replay problem.

Specifically, this paper makes the following contributions:
- We provide the first formal definition of the general replay problem.
- We show how to make novel use of the concept of weakest pre-condition and theorem proving to solve the replay problem. Our adaptation of these techniques results in the first *sound* solution to the replay problem.
- Our approach is general and enables successful replay in cases that the previous approach could not handle. For example, our replay can handle protocol messages which include check-sums.
- We have implemented our approach in a system called *Replayer*. Our experiments demonstrate the viability of this approach.

2. PROBLEM STATEMENT

Example 1 Program P

1: session.cookie := counter
2: counter := counter + 1
3: send(session.cookie)
4: recv(request)
5: **if** (request.data.cookie != session.cookie) **then**
6:    fail()
7: **else if** (request.data.hostname != gethostname()) **then**
8:    fail()
9: **else if** (request.cksum != cksum(request.data)) **then**
10:    fail()
11: **else**
12:    process(request)
13: **end if**

Consider an application dialog between two hosts: an *initiator* and *host A* . The problem we address is how to *replay* the initiator's side of the network dialog to another party—*host B*. For non-trivial network protocols, it is insufficient to simply replay the exact input sent by the initiator. Certain protocol fields must be updated to reflect the state of host B for the replayed input to have the same effect on host B as it did on the host A.

Example 1 demonstrates three common types of protocol fields that must be updated to successfully replay a network dialog. The field checked on line 5 in the example is a cookie, which is an example of a *session*-specific protocol field. Cookies are opaque byte strings generated by one dialog participant, sent to the other, and then included in subsequent messages. They are used, for example, to identify a particular session or resource. Replaying a network dialog requires updating session-specific protocol fields to match the actual value that was sent by host B.

The field checked on line 7 of Example 1 is the host name of host B. This is an example of a *configuration*-specific protocol field. This type of protocol field is normally filled in using knowledge obtained by some out-of-band mechanism. Other examples of such state are user credentials, names of shared resources, etc. Replaying a network dialog requires updating configuration-specific protocol fields to reflect the configuration of host B.

The field checked on line 9 of Example 1 is a checksum over the rest of the request data, including the cookie and host-name

Term	Definition
S	Set of possible process states
$s_a, s_b, s_o, s_v \in S$	Initial state of host A, host B, observer, and verifier
$\sigma_a, \sigma_b, \sigma_o, \sigma_v \in S$	Post-state of host A, host B, observer, and verifier
I	Set of possible process inputs
$i_a, i_b, i_o, i_v \in I$	Input sent to host A, host B, observer, and verifier
$P : S \times I$	Program executed on host A, host B, observer, and verifier
Q	Set of possible post-conditions
$q \in Q$	Post-condition that is being satisfied
$\Phi : S \to Q$	Function to define a post-condition

Table 1: Terminology

fields. This is an example of a *consistency* protocol field. This type of protocol field is normally filled in using knowledge of the protocol. Another example of this type of protocol field is a length field—that is one specifying the number of bytes of one or more other protocol fields. If replay requires modifying other parts of the message to reflect host B's state, the corresponding consistency fields must be updated.

We next provide a formal and general definition of the protocol replay problem, which will lead to a formal and general solution.

2.1 Formal Definition

We define the problem of application protocol replay as follows. Let S represent the space of possible process states, and I represent the space of possible network inputs. A process with state $s \in S$ and input $i \in I$, executing a deterministic [1] program $P : S \times I \to S$, will result in a post-state $\sigma \in S$. The state s includes the *program counter* (the next instruction to be executed in the program), values of processor registers and memory, the state of the file system, *etc*. The input i is data written into the process via the network [2]. For simplicity, the final state σ also specifies the output produced during the execution.

A process on host A is executing the program P. While in initial state s_a, the process receives input i_a, and after continuing execution of the program, reaches a later state $\sigma_a \in S$, for which some *post-condition* $q \in Q$ is true, where Q is the set of all post-conditions defined as $Q = \{B | B : S \to \{T, F\}\}$. That is, $q(\sigma_a) = T$. Intuitively, the post-condition will be a function that is true if and only if the input has a "similar" effect on host B as it did on host A. We discuss how to set the post-condition later in this section.

We wish to provide an input to Host B, who is also running a process executing program P, such that it reaches a state σ_b that satisfies the post-condition. *I.e.*, such that $q(\sigma_b) = T$. However, the process being run by host B will be in a different initial state s_b. As a result, the input i_a seen by host A may *not* cause the post-condition to be satisfied. That is, it may be that $q(P(s_b, i_a)) = F$. In Example 1, the original input i_a will have incorrect values for the cookie and hostname protocol fields, assuming that state s_b specifies a different cookie and hostname. Depending on the exact post-condition q being used, i_a will likely not satisfy $q(P(s_b, i_a))$ as a result.

The protocol replay problem is to modify the previous input to

---

[1] In Section 6.5 we discuss how to deal with nondeterminism such as scheduling, other inputs, *etc*.

[2] We refer to i as being a single input for simplicity. However, i could refer to multiple network messages.

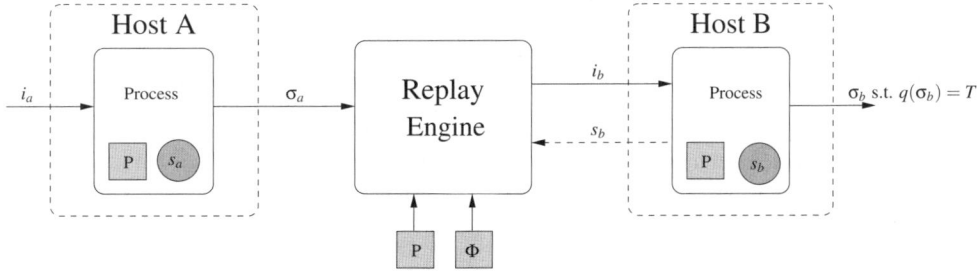

Figure 1: Our replay setting.

reach a σ_b s.t. $q(\sigma_b) = T$. More specifically, the protocol replay problem is given input i_a, a post-condition q such that $q(\sigma_a) = T$, and an initial state s_b, find a new input i_b such that $q(P(s_b, i_b)) = T$.[3]

Setting the post-condition Intuitively, the post-condition q is true if and only if the replayed dialog has a "similar" result on host B as it did on host A. Exactly what is meant by "similar" will vary, depending on the purpose of performing application protocol replay.

In our approach, we define a function $\Phi : S \to Q$. That is, Φ takes the final state $\sigma_a \in S$ and returns the post-condition $q \in Q$. This formulation is illustrated in Figure 1. Φ, in turn, is dictated by the specific goal of the entity performing replay. Naively, one might specify Φ to produce a q that is satisfied if and only if the final state σ is the exact final state that was reached by host A, σ_a. However, since the state potentially encapsulates not only every memory value of the process, but the state of the machine it is running on, such a post condition is likely to be unsatisfiable.

A generic replay application may specify Φ to produce q that is true if and only if σ_b includes the same output as σ_a. This type of Φ would be suitable for many applications, as it would result in the same observable behavior on host B as was seen on host A.

More specialized applications may use a more specialized Φ. For example, one of the applications for protocol replay is to allow an exploit to be replayed to verify that a vulnerability exists and/or to further study the vulnerability. For this application, Φ might be "q returns T if and only if the security property that was violated in σ_a is also violated in σ_b," with a security violation such as "instruction x overwrites return address at memory address y."

3. OUR APPROACH

In this section we describe the design of our solution to the application protocol replay problem. Throughout the remainder of this paper, we refer to the host that received the original input as the observer, and the host to whom we are attempting to replay that input as the verifier. Our solution is illustrated in Figure 2. We first create a symbolic formula from the program P and the post-condition q. The resulting formula relates the input and the initial state to the final program state. We then substitute the verifier's initial state into the formula, and use a *decision procedure* to derive i_v for replay to the verifier.

3.1 Creating Symbolic Formulas for Application Replay

To replay an application dialog, we must determine what conditions are necessary for a host to accept the dialog and terminate

---
[3]In the most general sense, we could also consider altering the program P or the host B's initial state s_b. While these approaches may be useful for some specialized applications, we do not address them in this work.

in the a final state satisfying the post-condition. The program itself calculates a function from the initial state and input to a final state. This calculation can be expressed as a *symbolic formula* over the program state space and the input. We can further refine the symbolic formula to include only those final states that satisfy the post-condition. For example, the symbolic formula would include clauses that check for self-consistency, the relation between session-specific protocol fields, and relationships on the configuration state.

There are multiple approaches for computing a symbolic formula that represents a host accepting the replayed dialog. For example, one popular approach for this sort of problem is forward symbolic execution. Forward symbolic execution entails "executing" the program on symbolic inputs, which results in a symbolic formula for that program path. If we symbolically execute all program paths, we arrive at a symbolic formula for the entire program. We could then augment the symbolic formula for the program such that the formula is satisfied iff the program would accept the replayed dialog.

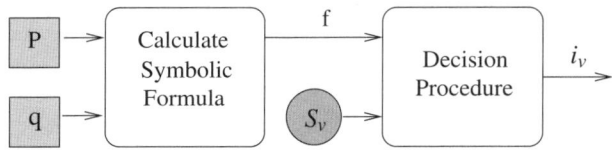

Figure 2: Our approach.

Our approach We take a different approach, where we compute the *weakest pre-condition* for a program P to terminate in a state satisfying post-condition q. A weakest pre-condition with respect to a post-condition q is a Boolean formula that characterizes the inputs and initial states which result in the program satisfying the post-condition. In our context, the weakest pre-condition formula characterizes the inputs and initial states that will cause the verifier to satisfy the desired post-condition. A satisfying assignment of values to variables in the weakest pre-condition produces an input that allows the verifier to reach the desired post-state.

The advantage of using the weakest pre-condition is that the weakest pre-condition formula is usually much smaller than that produced by forward symbolic execution. In Section 5 we show that we achieve much better performance using the weakest pre-condition than by using forward symbolic execution.

More formally, the weakest pre-condition $wp(P,q)$ characterizes all inputs to the program P that execution will result in a terminating state satisfying q. At a high level, $wp(P,q)$ is the weakest pre-condition on i_v which implies executing P on i_v from initial state s_v will terminate in a state satisfying q.

Computing the Weakest Pre-Condition We compute the weakest pre-condition on P by:

1. Translate P into the guarded command language (GCL). The GCL program, denoted P_g, is semantically equivilant to P, but much simpler for analysis.
2. Compute the weakest pre-condition $f = wp(P_g, q)$ in a syntax-directed fashion. The resulting formula f is a Boolean predicate over (verifier) program states and inputs.
3. Simplify the weakest pre-condition so it is more efficiently solved by the decision procedure in the next step.

Translating to GCL To calculate the weakest pre-condition for a program, we must first define how each instruction may affect program state. To simplify this task, we first translate the assembly instructions of P into a simplified language, called the *guarded command language* (GCL). The GCL has a relatively small number of distinct instructions, making it simpler to analyze. During this transformation, we also make all implicit modifications to the program state, such as processor status flags, explicit. The resulting program P_g is semantically equivalent to P, but can be reasoned about in a syntax-directed manner.

The GCL language constructs we use are shown in Table 2. Although GCL may look unimpressive, it is sufficiently expressive for reasoning about complex programs [9, 10, 13, 15] [4]. GCL is quite simple to understand. Statements in GCL mirror statements in assembly, e.g., store, load, assign, etc. Statements consist of a side-effect free rhs expression, and a lhs location to store the result. The lhs is always a variable name (i.e., a register) or memory location (both stack and heap locations are treated uniformly). $A;B$ denotes a sequence where statement A is executed, then statement B. $A \square B$ is a choice statement where either A is executed or B, and corresponds loosely to a conditional jump statement. **assume** e assumes a particular (side-effect free) expression is true, and is used to reason about conditional jump predicates. **skip** is a semantic no-op, and provided to make subsequent analysis simpler. $lhs := e$ denotes an assignment of the expression e to location lhs. **true** and **false** are the logical constants (and we also allow for normal Boolean operators such as negation ($\neg$)).

GCL is best demonstrated with a simple example. The statement:

$$\text{if } (x < 0) \text{ then } x := x - 1 \text{ else } x := x + 1;$$

is translated as:

$$(\textbf{assume } x < 0; x := x - 1) \square (\textbf{assume } \neg(x < 0); x := x + 1;)$$

System calls can be translated into a series of assignments to special variables. For example, input into the program via the `recv` system call can be written as a series of assignments to memory from $input_i$, for $i = 0$ to len, where len is the parameter passed to the system call specifying the maximum number of bytes read.

Likewise, data written into memory by other system calls, such as `gettimeofday`, reads from files or other sockets, *etc.*, can be represented as assignments to memory via specially named variables. The values of such variables can be considered part of the initial state s_v. We further discuss how to set these variables in Section 3.2.

Computing the weakest pre-condition We compute the weakest pre-condition for P_g in a syntax directed manner. The rules for computing the weakest pre-condition are shown in Table 2. Most rules are self-explanatory, e.g., to calculate the weakest pre-condition $wp(A; B, Q)$, we calculate $wp(A, wp(B, Q))$. Similarly $wp(\textbf{assume } e, Q) \equiv e \Rightarrow Q$.

---
[4]The GCL defines a few additional commands such as a **do-while** loop, which we do not use.

The one tricky rule is for assignment, i.e., calculating $wp(lhs := e, Q)$ where lhs is a variable name or memory reference and e is an expression. The rule $Q[lhs/e]$ specifies that all occurrences of lhs in the post-condition Q are substituted for e. If lhs is a variable, then we substitute all occurrences of lhs in Q for e. For example, $wp(j := i + 1, j < 3) \equiv i + 1 < 3$. However we must take into account any possible memory aliasing relationships when lhs is a memory reference. Consider computing $wp(mem[w] := e, mem[t] < 3)$. If $t = w$, then the resulting weakest pre-condition is $e < 3$. If $t \neq w$, then this statement has no effect and the weakest pre-condition is $mem[t] < 3$. Therefore, the weakest pre-condition for $wp(mem[w] := e, mem[t] < 3)$ is

$$(\text{if } w = t \text{ then } e \text{ else } mem[t]) < 3$$

The complete weakest pre-condition calculation for post-condition $x < 3$ for our previous example is:

$$wp((\textbf{assume } x < 0; x := x - 1) \square (\textbf{assume } \neg(x < 0); x := x + 1;),$$
$$x < 3)$$
$$\equiv (wp(\textbf{assume } x < 0; x := x - 1, x < 3)$$
$$\land wp(\textbf{assume } \neg(x < 0); x := x + 1, x < 3))$$
$$\equiv (x < 0 \Rightarrow x - 1 < 3) \land (\neg(x < 0) \Rightarrow x + 1 < 3)$$

3.2 Obtaining the verifier's initial state

To replay an input to the verifier, we must substitute the initial state s_v into the symbolic formula before we can use it to find a satisfying input. There are several ways this state may be obtained and represented in the symbolic formula, depending on the type of state to be provided, and the particular application scenario.

Example 2 GCL

1: session.cookie := counter;
2: counter := counter + 1;
3: SENT := session.cookie;
4: request := INPUT;
5: **assume**(request.data.cookie != session.cookie) $\Rightarrow$ fail()
   $\square$ **assume**(request.data.cookie = session.cookie) $\Rightarrow$
6: (**assume**(request.data.hostname != HOSTNAME) $\Rightarrow$ fail()
   $\square$ **assume**(request.data.hostname = HOSTNAME) $\Rightarrow$
7: (**assume**(request.cksum != cksum(request.data)) $\Rightarrow$ fail()
   $\square$ **assume**(request.cksum = cksum(request.data)) $\Rightarrow$
8: process(request))))

We use Example 2 as an illustrative example. Example 2 is the program from Example 1 translated into GCL. Notice that the `if` statements have been converted to `assume` statements and that the system calls corresponding to `send`, `recv`, and `gethostname` have been converted to assignments to or from the special variables SENT, INPUT, and HOSTNAME, respectively. This example is *without* computing the weakest pre-condition, for greater readability.

In this example there are two parts of the verifier's initial state that are needed to satisfy the `assume` statements, and therefore to produce a successful replay: the value of the cookie, and the value of the host name. We first show how the appropriate state can be obtained if we have *direct access* to the verifier. We then show how in many cases, including this one, the necessary state can be obtained even when direct access to the verifier is unavailable, using *a priori* knowledge, and knowledge inferred from the verifier's output.

A,B ∈ GCL stmt	::= A;B	GCL stmt	wp(stmt, Q)
	| **assume** *e* (*e* is an expression)	**assume** *e*	$e \Rightarrow Q$
	| *lhs* := *e* (*lhs* ∈ VARS $S \times I$)	*lhs* := *e*	$Q[lhs/e]$
	| A □ B	A; B	wp(A, wp(B,Q))
	| **skip**	A □ B	wp(A, Q) ∧ wp(B,Q)
	| **true** | **false**		

Table 2: The guarded command language (left), along with the corresponding weakest precondition predicate transformer (right).

3.2.1 Direct Access

The most straight-forward method of obtaining the verifier's state is to access the memory, registers, and system configuration directly. In this Example 2, the state of memory in s_v includes the value of *session.cookie*, allowing the corresponding protocol field to be correctly updated.

In this example, the host name is derived during execution, from a system call. Hence, the initial values of memory and registers are insufficient to find this part of the verifier's state. However, we can use direct access to the verifier to predict what the system call will return, and provide that value as part of the initial state, as an assignment to *HOSTNAME*. For many system calls, including this one, the return value can be found simply by executing that system call in a separate process on the verifier.

3.2.2 Non-direct access

Obtaining the initial state is more challenging if direct access to the verifier is unavailable. However, in many cases the necessary state can be automatically obtained by analyzing previous output of the verifier, and using *a priori* knowledge. In Example 2, we can obtain the value of the cookie using output sent by the verifier, and we can obtain the host name using *a priori* knowledge.

Inference from output   In Section 2, we described three types of protocol fields that must be updated for replay to be successful: session-specific fields, configuration-specific fields, and consistency fields. Session-specific fields can typically be updated using only state inferred from output of the verifier.

In Example 2, session.cookie is an example of such state. By inspection, clearly if we know the value sent by the verifier, SENT, we can infer the value of session.cookie. If we substitute the variable SENT with x, where x is the value actually sent, then the decision procedure will likewise be able to infer that session.cookie = x, and further that INPUT$_i$ = x, where i is the offset of the cookie field.

A minor caveat to this approach is that *SENT* will not appear directly in the weakest pre-condition. Therefore, to derive state from the output in this way, we must incorporate the weakest pre-condition of the corresponding output variables into the symbolic formula provided to the decision procedure.

A priori knowledge   Configuration-specific protocol fields sometimes cannot be derived directly from the output of the verifier, as is the case with the hostname field in Example 2. In this example, a client must have *a priori* knowledge of the host name, or be able to find it by some out-of-band means.

We can handle some of these cases by modeling system calls that read configuration state, as we do with the call to gethostname in Example 2. When such a variable appears in the symbolic formula, we can attempt to find its value by some out-of-band means. In this case, we could perform a reverse DNS lookup of the remote host, and use that data to provide an assignment to HOSTNAME in our symbolic formula.

As in the previous case, we must augment the symbolic formula with the weakest pre-condition of any such system calls we want to handle, because they will not appear directly in the weakest pre-condition of the post-condition.

3.3 Finding a satisfying input

The output of the weakest-precondition phase is a boolean formula f over program states $s \in S$ and input variables $i \in I$. We then assign values to the state variables s from the verifier's initial state s_v, as discussed in the previous section. We then use an off-the-shelf *decision procedure* to provide an example assignment i satisfying the weakest pre-condition. This i will result in a post-state that satisfies the post-condition.

The decision procedure provides such an assignment if one exists given initial state s_v. Otherwise, the decision procedure returns that the formula f with the state variables specified by s_v is not satisfiable.

If the decision procedure returns an assignment, the input i_v can be directly constructed using the assignments to the input variables. Again, because we are using sound techniques, such an input *will* satisfy the post condition.

In most cases, we will not be able to incorporate every possible execution path into the symbolic formula. We further discuss how we bound the program before computing the weakest pre-condition in Section 4.2.2. If the decision procedure is not able to find a satisfying assignment to the formula, it may still be satisfiable via an execution path not included in the formula. If desired we could backtrack and build a bounded program that includes additional execution paths, and try again. Note again that because there may be *infinite* paths in the full program, it is possible that we will never find a satisfying input regardless of how many paths we add to the bounded program. This is unsurprising, since in the general case one could easily construct a program and post-condition where finding a satisfying input can be reduced to deciding the halting problem.

Finally, there are of course cases in which *no* input will satisfy the post-condition given the initial state. For example, the verifier could be configured to not accept any incoming connection.

4. IMPLEMENTATION

We have implemented a proof-of-concept of our approach in a tool called *Replayer*. We describe the relevant implementation details in this section.

4.1 Trace recorder

We use Valgrind [20] to produce the execution trace T. Valgrind is an open source dynamic binary rewriting tool. Our choice was made for convenience, many other tools also produce traces [5, 19, 24]. We wrote a Valgrind plug-in that produces a log of the first address of each basic block executed.

4.2 Symbolic Formula Generator

We have built an analyzer that reads in the binary program P and the execution trace T, and outputs the corresponding weakest pre-

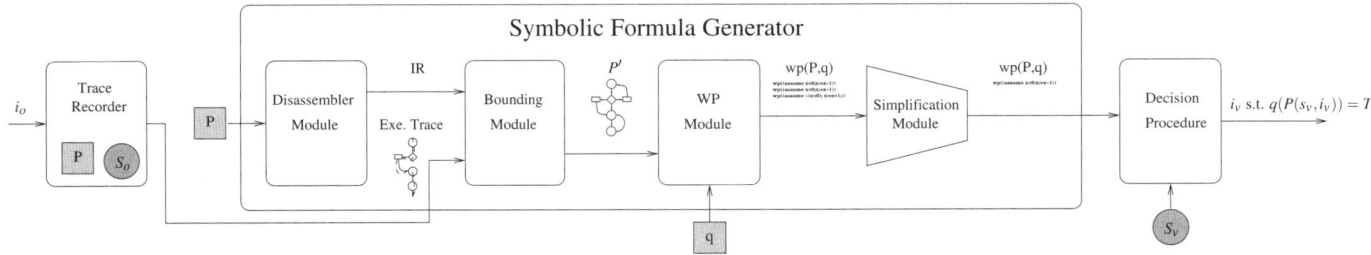

Figure 3: The Replayer design and implementation

condition formula. The code base for the analyzer is approximately 19,000 lines of C++ code. The code has several modules.

4.2.1 Disassembler module

The disassembler module is an extension of [6]. This module is responsible for disassembling the basic blocks logged in the trace and converting the assembly into an unambiguous intermediate representation. Translating to the intermediate representation, although straight-forward, requires us to explicitly model the effects of overflow, underflow, sign extension, etc. We simplified this task by first using Valgrind's libVEX to translate the assembly instructions to VEX—Valgrind's intermediate representation, thus saving us from having to specify every IA-32 instruction. However, since VEX is designed to generate efficient executable code rather than easily analyzable code, we transformed VEX to our own simpler intermediate representation. In particular, processor status flags in VEX are updated at run-time just before they are accessed. In our IR, status flags are instead updated explicitly.

One additional problem we must deal with is memory reads and writes, some of which may be unaligned. We call a memory read unaligned if it doesn't correspond to an atomic write. For example, consider a 32-bit write to memory locations 1-4. A subsequent read of byte 2 is unaligned. To address this, we post-process the assembly so all writes and reads are single bytes, e.g., a 32-bit write becomes 4 8-bit writes.

4.2.2 Bounding module

There are a large number of possible execution paths through most programs; infinitely many if the program does not halt. To reason about a program at all, we must convert the program to one that has a finite number of execution paths. A common technique for doing this is to bound the number of times each loop in the program executes, while adding an additional check to ensure that executions that would have executed a loop a greater number of times are not considered. Any reasoning over a program transformed in this way is still *sound*, but may not be *complete* since not all possible execution paths are considered.

We employ this technique, transforming the original program P into a modified program P'. Further, we use an execution trace T of the execution path followed by the observer as it processed the original input i_o to help determine *how* to bound the program.

We already know that a process executing P, with initial state s_o will reach a post-state that satisfies the post-condition (by definition), following the execution path specified in T. Intuitively, it is likely that the program can reach a post-state satisfying the post-condition by following an identical or similar execution path, even when starting in a different initial state s_v. Therefore, it makes sense to bound the program to execution paths similar to that in T. We call this *trace-guided bounding*.

In our implementation, trace-guided bounding bounds the program to follow the *exact* execution path from the trace T. This results in the most scalable, but the least complete solution. Therefore, it may be desirable to also consider *similar* execution paths. For example, one might consider allowing loops to be executed n greater or fewer times than they were executed in the trace T. One might also consider executions where alternate paths of "diamond" structures in the control flow graph are executed. For example, by inspection it makes little difference whether the if or the else clause is executed in if (input > 3) printf(``foo''); else printf(``bar'');.

In our current implementation, we bound the program to follow the exact path specified in T. We build a program consisting of the concatenation of each instruction executed in the trace T. All conditional and indirect jumps are replaced with direct jumps to the destination they jumped to in the trace T. To ensure soundness, we later add the corresponding condition from each jump (or calculation in the case of an indirect jump) to the post-condition, thus ensuring that we only consider inputs that would actually result in this execution path.

4.2.3 Weakest pre-condition module

The weakest pre-condition module translates the intermediate representation into the guarded command language from Table 2. The weakest pre-condition module implements the predicate transformers shown in Table 2. This code takes as input the GCL and an arbitrary post-condition, and outputs the corresponding weakest pre-condition formula.

4.2.4 Simplification module

After computing the weakest precondition, the model consists of a single, relatively large, formula. We perform a number of simplifications on this formula to reduce the size of the model. Smaller models are easier for the decision procedure (described next) to reason about.

We perform:

- Arithmetic simplification and re-association. For example, the expression $4 + esp + 4 < e$ would be simplified to $8 + esp < e$. Arithmetic simplification and re-association leaves arithmetic expressions as a sum-of-products.
- Boolean simplification. For example, many memory accesses are a constant offset from the stack pointer—esp on the IA-32 architecture. As a result, the conditions in many of the if-then-else clauses that test for memory aliasing are of the form $esp + x = esp + y$, where x and y are constants. Obviously, such a condition is true if and only if $x = y$, allowing us to simplify away the if-then-else.
- Common sub-expression elimination. For example, consider the formula $a + 2 < 3 \wedge a + 2 < 3$. Surprisingly, a decision procedure will consider the expression $a + 2 < 3$ twice. We eliminate common sub-expressions via a let binding, e.g.,

let $t = a+2 < 3$ in $t \wedge t$, which results in $a+2 < 3$ being considered only once by the decision procedure. Note that applying Boolean simplification will further reduce the formula to let $t = a+2 < 3$ in t.

We have found these simple optimizations to be extremely important. In many test cases simplification reduced time spent in the decision procedure by orders of magnitude. Our running example is simplified as:

$$(x < 0 \Rightarrow x < 4) \wedge (\neg(x < 0) \Rightarrow x < 2)$$

Note that the above formula could be further simplified, but we leave such simplification as future work.

4.3 Replay Engine

State Extractor For simplicity, we use the direct-access method of finding the verifier's state, as described in Section 3.2. The state extractor uses ptrace to read the state s_v of memory and registers of the process running program P at the point where P is waiting for input (*e.g.* blocked on a read system call).

The Replayer module substitutes the state s_v into the weakest pre-condition $wp(P,q)$, and runs the simplification module again. It then translates the weakest pre-condition into the appropriate syntax for a decision procedure.

Decision Procedure The decision procedure is a modular component, and we could in theory use any off-the-shelf product. We currently use STP [14], a decision procedure that specializes in modeling bit-vectors. After translation, the module asks the decision procedure for a satisfying assignment of values to input variables. The decision procedure will either output a satisfying input i_v for state s_v, or output that there does not exist such an input.

When a satisfying input is found, we replay the new input i_v and verify that the post-condition is satisfied. This step serves as a self-check of our implementation.

5. EVALUATION

We evaluate Replayer on several variations of Example 1. Each test is compiled as an unoptimized C program. We specify Φ as "execution reaches process(request)". In each test, we take the initial state at the point of the *recv* at line 4. We take the final state at the point where process(request) is called at line 12. Our measurements are performed on a machine running Ubuntu Linux version 5, with a Pentium 4 2.20 GHz CPU, and 1 GB of RAM.

When measuring performance, we consider two steps. The first step is to build and simplify the symbolic formula. This step needs to be performed only once to replay a particular observed input i_o. The model can then be used to replay the input to any number of verifiers. The second step is to substitute the initial state of a particular verifier into the model, performing any additional simplification of the model, and use the decision procedure to find a satisfying input. The second step must be performed each time the input is replayed to a verifier with a different initial state. We also provide the number of instructions in the executable. Note that the size of the trace can exceed the number of instructions as the instruction trace may include a single instruction multiple times, e.g., for when a loop is executed.

Simplification effectiveness Naturally, if we want to replay an input more than once, it pays to pre-process the symbolic formula as much as possible, to minimize the performance overhead of solving the symbolic formula. With this in mind, we compute the weakest pre-condition and simplify the formula as much as possible, rather than performing forward symbolic execution. Our first test is to measure the effectiveness of these steps.

We consider a version of Example 1 with only the test for the cookie field enabled. Here, the cookie field is a four-byte integer. Keep in mind that while this example is very simple at the source code level, the object code is significantly more complex.

In Table 3, we compare our performance using forward symbolic execution, computing the weakest pre-condition, and performing the weakest pre-condition with additional simplification. Our implementation of forward symbolic execution is to translate our GCL version of the program directly into the decision procedure's language. As expected, the formula resulting from computing the weakest pre-condition is much more efficiently solved by the decision procedure than the forward symbolic execution formula. The weakest pre-condition formula takes roughly one third the time for the decision procedure to solve.

Interestingly, our simplification module improves the performance of solving the final formula by two orders of magnitude. Further, the time to *generate* the formula in the first place is greatly reduced. This is because we perform simplification on the intermediate forms of the formula as we calculate the weakest pre-condition, reducing the complexity of computing the weakest pre-condition itself. Some of these simplification techniques coudl be built into decision procedures.

Updating a checksum We next enabled the check for the checksum, in addition to the check for the cookie. Hence, because the replayer must update the cookie, it must also update the checksum. In our program, the checksum is the integer addition of the data. We evaluate performance using a 16 byte checksum (4 integer additions), and an 80 byte checksum (20 integer additions). An advantage of our approach over machine-learning based approaches such as Roleplayer [8] is that we can replay protocols that have such relationships, without needing to know the exact algorithm ahead of time.

Table 4 shows the execution time for each Replayer step for when only the cookie check is enabled (same as in Table 3), for when the checksum is computed over 16 bytes of input, and for when the checksum is computed over 80 bytes of input. As one would expect, the checksum computation significantly increases the complexity of the generated formula. Replayer handles each of these cases in a reasonable amount of time, though further work will be needed to scale Replayer to larger formulas.

We have demonstrated the validity of our approach and of our implementation prototype. While further work is necessary for our approach to scale, we believe that our initial results are promising. We discuss some possible strategies to improve the scalability of our approach in Section 6.6.

6. DISCUSSION

6.1 Preserving similarity to the original input

Replayer is designed to soundly address the problem formulation given in Section 2. That is, Replayer produces an input i_v that will replay and reach the desired final state satisfying the desired post-condition. Whether or not the produced i_v resembles the original input i_o depends completely on the post-condition. Here, we discuss how the post-condition can be specified so that i_o and i_v are textually similar.

For example, suppose that the original input i_o is a message to an SMTP server that causes the SMTP server to send an email message. Suppose the post-condition only specifies that the input should trigger a message on the verifier host. Using our current approach, we would be likely to generate an i_v that *also* causes an SMTP server to send an email message, but the body of that message would likely be gibberish, rather than the contents of the

	Building the Formula (s)	Total time to solve formula (s)
Forward Symbolic Execution	.944	34.4
WP without simplification	12.2	11.3
WP with simplification	1.15	0.142

Table 3: Performance improvements from weakest pre-condition and simplifications

	Cookie only	Cookie + 16 byte checksum	Cookie + 80 byte checksum
Executable size (w and w/o glibc)	87188/329	87188/329	87188/329
IA-32 instructions in trace	35	84	229
1. Trace to GCL (s)	.901	.947	.892
2. Compute WP (s)	.205	7.92	355
Total time to build formula (1+2) (s)	1.106	8.867	356
3. Substitute and resimplify (s)	.029	.652	7.54
4. Translate to decision procedure (s)	.013	.142	2.02
5. Compute decision procedure (STP) (s)	0.10	.95	5.73
Total time to solve formula (3+4+5) (s)	.142	1.744	15.29

Table 4: Checksum replay performance.

message in i_o. In general, specifying a Φ that, for messages in any possible network protocol, generates a q that specifies that i_v is entirely semantically equivalent to i_o is impossible.

There are several techniques for creating a post-condition such that i_o and i_v are textually similar. Each technique *constraints* the input variables for non-state-dependent protocol fields to be the value that they took on in i_o, thus only allowing the decision procedure to select new values for the state-dependent protocol fields.

The challenge to accomplish this is that we would need to *identify* which parts of the input correspond to state-dependent protocol fields. There are several techniques we could use to do this. Note that *whichever* bytes of the input we choose to constrain, the system is still sound. However, every additional constraint risks causing the symbolic formula to become unsatisfiable. In particular, if we mistakenly constrain a state-dependent protocol field to its original value, of course we will be unable to find a satisfying input.

There are several techniques we can use to heuristically identify which bytes of the input do not correspond to state-dependent fields, and may hence be safely constrained. First, we can of course constrain any input bytes that do not appear at all in our symbolic formula, without reducing the completeness of the symbolic formula. Unfortunately, such bytes will not exist in most protocols. For example, in a text-based protocol, even input bytes that are otherwise ignored by the program will likely be constrained by the symbolic formula to not be NULL. Hence, this technique of identifying constrained bytes is not likely to be useful in practice.

A more promising technique is to identify the state-dependent fields, and consistency fields, and then constrain the rest of the input. For example:

- **Session-specific fields**, such as cookies, are characterized by the program sending data derived from some internal state (*e.g.*, a saved cookie), and later comparing that internal state to subsequent input.

- **Configuration-specific fields**, such as the name of a host, may be compared with data derived from another system call— *e.g.*, reading a configuration file. Another possibility is that the input data is passed as a *parameter* to a system call— *e.g.* to open a specified file.

- **Consistency fields**, such as a check-sum, result in an expression in the symbolic formula comparing a relatively small protocol field, *e.g.* a check-sum, with an expression involving a relatively large protocol field, *e.g.*, the rest of the request.

Another technique is to use an iterative greedy algorithm to determine which parts of the input can remain the same as i_o. First, we would use the decision procedure to find a satisfying input i_v as before. We would then attempt to constrain one of the input variables to have the same value as in i_o, and try again. If the decision procedure succeeds, we can continue to constrain that variable to have the same value as in i_o, otherwise we un-constrain it again. We can continue this process until no more bytes can be constrained without causing the decision procedure to fail.

The greedy part of this approach stems from the order to attempt to constrain the input bytes. Suppose the message has a consistency field such as a check-sum. While it may be possible to keep the consistency field value from the original input and allow the decision procedure to find new values for the data fields, it would be preferable to keep the data field values from the original input and derive a new consistent value for the consistency field. To encourage this property, we greedily prefer to constrain bytes that appears on the other side of a comparison operator as the fewest number of other input variables in the Boolean formula. Thus if we have $cksum = a + b + c + d$, we would prefer to constrain a through d before $cksum$.

We leave implementing and evaluating these techniques as future work. If successful, these techniques could greatly increase Replayer's ability to act as a generic protocol replay tool, by helping preserve any hidden semantics of the replayed message.

6.2 Generating an observer-independent protocol replay engine

In our current solution, we use the post-state σ_o and the execution trace T obtained by monitoring the *observer* as it processes the original input i_o. In some replay applications, this information may not be available. In particular, it would be useful to replay an input obtained from a logged network trace.

We may be able to build a replay system that can replay messages of a particular protocol, after observing and building symbolic formulas for most of the distinct types of messages in that protocol.

Assuming we are able to identify which fields are state-dependent using the techniques described in Section 6.1, we could constrain the non-state-dependent variables to the values of some other input i_2 to replay *that* input. To build a tool to replay any message of a particular protocol in this way, we would also need some mechanism to determine *which* of our symbolic formulas to employ to replay a particular input. One way of doing this is to build a *signature* of each type of message when generating the initial symbolic formulas. We can build these signatures using similar techniques to those described in [6]. That is, given the original observed σ_o and input i_o, we determine which parts of the input must remain the same to satisfy the formula. The bytes that cannot change will typically correspond to protocol keywords, which identify the type of message being sent. These protocol keywords can be built into a signature, which can later be used to identify subsequent inputs of the same message type.

6.3 Different program versions

In our current design, we assume that observer and the verifier are running the exact same program, P. In practice, we may be able to replay to a verifier that is running a different program P', if P' behaves as P in externally observable ways. This will often be the case for two slightly different versions of the same program, or perhaps even for two independently developed programs implementing the same protocol.

Direct access to the memory and registers of the verifier process are unlikely to be useful in supplying the initial state s_0, because P' will have a different memory and register allocation. However, if the necessary state can be inferred from the verifier's output and from a priori knowledge as described in Section 3.2, and if the program P' implements the same protocol specification as P, an input i_v derived from our symbolic formula is likely to result in the same *externally observable* behavior on the verifier as i_o did on the observer.

6.4 Complexity of finding a satisfying input

The general problem of finding an input that satisfies an arbitrary post-condition can easily be shown to be undecidable. *E.g.* the post-condition could be "Program outputs 1 iff $f(s_v, i_v)$ produces a program that halts." Naturally, we do not claim to be able to solve the problem for such post-conditions. In practice, we expect most useful post-conditions to be relatively simple, such as the examples given in Section 2.

Likewise, even a simple post-condition such as "execution reaches the same final *eip* as specified in σ_o" could be thwarted by a program that, for example, checks whether a cryptographic pseudo-random function computed on the input i_v is equal to the state s_v. However, for most programs, one would not expect the problem of finding an input to cause that program to reach a desired final state to intentionally be made into a hard problem. With few exceptions, programs are designed so that inputs can easily be constructed to cause a program to reach a desired state.

Note that handling common cases of cryptographic functions is not fundamentally difficult. For example, suppose that a protocol field of the input must include a correct cryptographic message authentication code (MAC) of the rest of the input. There is no need to "invert" the MAC to derive a correct input. The cryptographic key is part of the process state (or the process would not be able to verify the MAC itself). Hence, it is possible to derive a correct input by first satisfying other constraints on the input, and then performing a forward execution of the MAC computation code, which again is part of the program itself. We do not provide a general solution to automatically recognize and handle such cases in this work, though it would be straight-forward to include logic to recognize and handle calls into common cryptographic libraries. We believe a more general solution may be possible, which we leave for future work.

6.5 Non-determinism

Our approach assumes that the program behaves in a deterministic manner between the time the input i_v is read in, and the time that the final state s_v is reached. This is the case in many programs. In the presence of non-determinism not accounted for in our symbolic formula, our approach is no longer sound. However, it is likely that in many cases it would still work, as intuitively programs are constructed to *behave* in a deterministic manner based on their input, even when some non-determinism is present.

For example, consider a program that uses threads that are preemptively scheduled by the kernel. From the process's perspective, the scheduling of these threads is non-deterministic. However, assuming the absence of race conditions[5], the particular scheduling order of the threads has no effect on the final state reached by the program. Hence, we can compute the weakest pre-condition based on the scheduling order we observed in the original execution trace, and assume that the weakest pre-condition holds for other scheduling orders. The same reasoning applies to a program that is concurrently processing other requests, either via threads or via asynchronous I/O.

A potential challenge exists when the weakest pre-condition depends on the result of a system call made in between reading the input and reaching the post-condition. For example, the behavior of the program may depend on data read from a file on disk after receiving the replayed input. As we showed in Section 3.2, some system calls can be incorporated into our symbolic formula as part of the verifier's initial state. That is, if we can predict what data will be returned by a particular system call, we can substitute that data into the symbolic formula, thus removing the non-determinism. Hence, approach is still sound in these cases if we can accurately determine what these system calls will return. As discussed in Section 3.2, whether we can do this depends on the type of system call, and how much access we have to the verifier.

6.6 Insights into improving scalability, performance, and efficiency

Our approach was fundamentally motivated by the need for sound techniques for application replay. Our experience indicates that implementation details can make huge differences in efficiency, scalability, and performance. In particular, the availability of a compiler optimization will significantly (often exponentially) result in smaller and easier to prove formulas. We have found ourselves repeatedly implementing common compiler optimizations to get better performance. For example, without our simple algebraic simplifications, the 16 byte check-sum took over 80 minutes to compute the weakest pre-condition, instead of the current 7.9 seconds.

We believe further dramatic improvements are easily obtainable by implementing known compiler techniques. For example:

- We lack alias analysis, which often results in formulas (sometimes exponentially) larger than needed. During execution, many instructions are memory references, either to the stack for local variables spilled due to register contention, or to the heap. We take a strictly sound approach, where any two variables that may be aliased are considered aliases, resulting

---

[5]We do not attempt to address the problem of reproducing race conditions in this work. However, it may be possible to extend our approach to do so, for example by computing the weakest pre-condition over the scheduling algorithm as well.

in a `if-then-else` formula as described in Section 3.1. The `if-then-else` can potentially double the size of the program. Implementing x86 alias analysis such as value-set analysis [4] would significantly help.

- Further simplifications (e.g., common sub-expression elimination, more aggressive simplification, global value numbering, etc.). As mentioned in Section 4.2.4, we found that such simplifications can result in an order of magnitude speedup.

- We use a classical approach to calculating the weakest precondition. Flanagan and Saxe [13] propose a method that can exponentially reduce formula size.

7. RELATED WORK

The work closest to our own is that of Cui *et. al.* [8]. Cui *et. al.* develop a heuristic-based approach to automatically identify and update application fields in protocol dialog. A primary aim of their work is to decouple application semantics from the replay process. This decoupling requires the identification of application protocol dependent fields which must be modified for correct replay. The process by which different classes of fields are identified, modified, and replayed is a manual process based on the semantics of each field type. The accuracy of the developed techniques is not guaranteed and new heuristics must be developed to handle new classes of dynamic fields.

Several projects have addressed the problem of network traffic replay. They typically focus on the network or transport layers [1, 2, 7] or include application semantics through the manual development of application-level responders or plug-ins [21, 22, 26]. Our approach works at the application layer and utilizes the application itself for semantic information rather than a simplified version of the application encapsulated in a plug-in.

Replaying the execution of a program has been the focus of numerous projects [3, 11, 16, 17, 23, 25]. These projects typically focus on ensuring deterministic execution in the face of non-determinism or logging the precise sequence of instructions for debugging, intrusion analysis, or simple instruction-by-instruction replay. Most of these previous approaches simply log the non-deterministic event, for example, the order of memory accesses on a multiprocessor, or log the individual instructions executed. However, this is insufficient to correctly replay the execution of a program on a machine with different state. We incorporate the application and the machine state in order to soundly determine the input that would arrive at the same post condition.

We compute the weakest pre-condition using direct, classical techniques. More advanced techniques can be implemented and may resulting significant improvements. For instance, Flanagan and Saxe which can significantly reduce the size of the weakest pre-condition formula [13]. We adopt the standard technique of unrolling loops, which may lead to incompleteness, however, unrolling is not necessary if loop invariants can be provided. Others have explored automatically determining loop invariants, e.g., [12, 18].

8. CONCLUSION

We developed a solution to the protocol replay problem: that of replaying an application dialog observed by one host to another host. The challenge in protocol replay is that a dialog sent to one host will likely not be accepted by another. A simple byte-by-byte replay to a host that is in a different initial state than the original will fail. We developed a general and formal definition of the replay problem. The solution developed in this paper makes novel use of programming language techniques such as theorem proving and weakest pre-condition. By apply these techniques, we developed the first *sound* solution to the protocol replay problem. We implemented and evaluated a prototype of our replay system called *Replayer*. Our evaluation demonstrates both the viability and generality of our problem formulation and corresponding approach.

Acknowledgments

We would especially like to thank Cristian Cadar, David Dill, and Vijay Ganesh for their generous help with CVCL and STP. We would also like to thank Ivan Jager, Eric Li, Vern Paxson, and the anonymous reviewers for their helpful comments and suggestions during the preparation of this paper.

9. REFERENCES

[1] Cybertrace. http://www.cybertrace.com/ctids.html.

[2] Tcpreplay: Pcap editing and replay tools for *NIX. http://tcpreplay.sourceforge.net.

[3] David F. Bacon and Seth Copen Goldstein. Hardware-assisted replay of multiprocessor programs. In *Proceedings of the ACM/ONR Workshop on Parallel and Distributed Debugging*, May 1991.

[4] G. Balakrishnan and T. Reps. Analyzing memory accesses in x86 executables. In *Proc. Int. Conf. on Compiler Construction*, 2004.

[5] P Bosch, A Carloganu, and D Etiemble. Complete x86 instruction trace generation from hardware bus collect. In *23rd IEEE EUROMICRO Conference*, 1997.

[6] D. Brumley, J. Newsome, D. Song, H. Wang, and S. Jha. Towards automatic generation of vulnerability-based signatures. In *Proceedings of the IEEE Symposium on Security and Privacy (Oakland)*, 2006.

[7] Yu-Chung Cheng, Urs Hoelzle, Neal Cardwell, Stefan Savage, and Geoffrey M. Voelker. Monkey see, monkey do: A tool for tcp tracing and replaying. In *Proceedings of the 2004 USENIX Annual Technical Conference*, June 2004.

[8] Weidong Cui, Vern Paxson, Nicholas C. Weaver, and Randy H. Katz. Protocol-independent adaptive replay of application dialog. In *Proceedings of the 13th Annual Network and Distributed System Security Symposium*, February 2006.

[9] D.L. Detlefs, K. Rustan M. Leino, G. Nelson, and J.B. Saxe. Extended static checking. Technical Report 159, Compaq Systems Research Center, December 1998.

[10] E.W. Dijkstra. *A Discipline of Programming*. Prentice Hall, Englewood Cliffs, NJ, 1976.

[11] George W. Dunlap, Samuel T. King, Sukru Cinar, Murtaza Basrai, and Peter M. Chen. ReVirt: Enabling intrusion analysis through virtual-machine logging and replay. In *Proceedings of the 2002 Symposium on Operating Systems Design and Implementation (OSDI)*, December 2002.

[12] M. D. Ernest, J. Cockrell, W. G. Griswold, and D. Notkin. Dynamically discovering likely program invariants to support program evolution. *IEEE Transactions on Software Engineering*, 27(2), Feb 2001.

[13] C. Flanagan and J.B. Saxe. Avoiding exponential explosion: Generating compact verification conditions. In *Proceedings of the 28th ACM Symposium on the Principles of Programming Languages (POPL)*, 2001.

[14] Vijay Ganesh and David L. Dill. System description of STP. http://www.csl.sri.com/users/demoura/smt-comp/descriptions/stp.ps, August 2006.

[15] David Gries, editor. *Programming in the 1990's: An Introduction to the calculation of programs*. Springer Verlag, 1990.

[16] Samuel T. King, George W. Dunlap, and Peter M. Chen. Debugging operating systems with time-traveling virtual machines. In *Proceedings of the 2005 USENIX Annual Technical Conference*, April 2005.

[17] T. J. LeBlanc and J. M. Mellor-Crummey. Debugging parallel programs with instant replay. *IEEE Transactions on Computers*, 36(4):471–482, 1987.

[18] K. Rustan M. Leino and Francesco Logozzo. Loop invariants on demand. In *Asian Symposium on Programming Languages and Systems APLAS*, 2005.

[19] Chi-Keung Luk, Robert Cohn, Robert Muth, Harish Patil, Artur Klauser, Geoff Lowney, Steven Wallace, Vijay Janapa Reddi, and Kim Hazelwood. Pin: Building customized program analysis tools with dynamic instrumentation. In *Proc. of 2005 Programming Language Design and Implementation (PLDI) conference*, june 2005.

[20] Nicholas Nethercote and Julian Seward. Valgrind: A program supervision framework. In *Proceedings of the Third Workshop on Runtime Verification (RV'03)*, Boulder, Colorado, USA, July 2003.

[21] R. Pang, V. Yegneswaran, P. Barford, V. Paxson, and L. Peterson. Characteristics of internet background radiation. In *Proceedings of Internet Measurement Conference*, October 2004.

[22] Niels Provos. A virtual honeypot framework. In *Proceedings of the 13th USENIX Security Symposium*, August 2004.

[23] M. Russinovich and B. Cagswell. Replay for concurrent non-deterministic shared-memory applications. In *Proceedings of the 1996 Conference on Programming Language Design and Implementation*, May 1996.

[24] P. A. Sandon, Y.C. Liao, T.E. Cook, D.M. Schultz, and P Martin de Nicolas. Nstrace: A bus-driven instruction trace tool for powerpc microprocessors. *IBM Journal of Research and Development*, 41(3), 1997.

[25] S. Srinivasan, S. Kandula, C. Andrews, and Y. Zhou. Flashback: A light-weight rollback and deterministic replay extension for software debugging. In *Proceedings of the 2004 USENIX Annual Technical Conference*, June 2004.

[26] A. Turner. Flowreplay design notes. http://www.synfin.net/papers/flowreplay.pdf.

EXE: Automatically Generating Inputs of Death

Cristian Cadar, Vijay Ganesh, Peter M. Pawlowski, David L. Dill, Dawson R. Engler
Computer Systems Laboratory
Stanford University
Stanford, CA 94305, U.S.A
{cristic, vganesh, piotrek, dill, engler} @cs.stanford.edu

ABSTRACT

This paper presents EXE, an effective bug-finding tool that automatically generates inputs that crash real code. Instead of running code on manually or randomly constructed input, EXE runs it on symbolic input initially allowed to be "anything." As checked code runs, EXE tracks the constraints on each symbolic (i.e., input-derived) memory location. If a statement uses a symbolic value, EXE does not run it, but instead adds it as an input-constraint; all other statements run as usual. If code conditionally checks a symbolic expression, EXE forks execution, constraining the expression to be true on the true branch and false on the other. Because EXE reasons about all possible values on a path, it has much more power than a traditional runtime tool: (1) it can force execution down any feasible program path and (2) at dangerous operations (e.g., a pointer dereference), it detects if the current path constraints allow *any* value that causes a bug. When a path terminates or hits a bug, EXE automatically generates a test case by solving the current path constraints to find concrete values using its own co-designed constraint solver, STP. Because EXE's constraints have no approximations, feeding this concrete input to an uninstrumented version of the checked code will cause it to follow the same path and hit the same bug (assuming deterministic code).

EXE works well on real code, finding bugs along with inputs that trigger them in: the BSD and Linux packet filter implementations, the udhcpd DHCP server, the pcre regular expression library, and three Linux file systems.

Categories and Subject Descriptors

D.2.5 [**Software Engineering**]: Testing and Debugging—*Testing tools, Symbolic execution*

General Terms

Reliability, Languages

Keywords

Bug finding, test case generation, constraint solving, symbolic execution, dynamic analysis, attack generation.

1. INTRODUCTION

Attacker-exposed code is often a tangled mess of deeply-nested conditionals, labyrinthine call chains, huge amounts of code, and frequent, abusive use of casting and pointer operations. For safety, this code must exhaustively vet input received directly from potential attackers (such as system call parameters, network packets, even data from USB sticks). However, attempting to guard against all possible attacks adds significant code complexity and requires awareness of subtle issues such as arithmetic and buffer overflow conditions, which the historical record unequivocally shows programmers reason about poorly.

Currently, programmers check for such errors using a combination of code review, manual and random testing, dynamic tools, and static analysis. While helpful, these techniques have significant weaknesses. The code features described above make manual inspection even more challenging than usual. The number of possibilities makes manual testing far from exhaustive, and even less so when compounded by programmer's limited ability to reason about all these possibilities. While random "fuzz" testing [35] often finds interesting corner case errors, even a single equality conditional can derail it: satisfying a 32-bit equality in a branch condition requires correctly guessing one value out of four billion possibilities. Correctly getting a sequence of such conditions is hopeless. Dynamic tools require test cases to drive them, and thus have the same coverage problems as both random and manual testing. Finally, while static analysis benefits from full path coverage, the fact that it inspects rather than executes code means that it reasons poorly about bugs that depend on accurate value information (the exact value of an index or size of an object), pointers, and heap layout, among many others.

This paper describes EXE ("EXecution generated Executions"), an unusual but effective bug-finding tool built to deeply check real code. The main insight behind EXE is that code can *automatically* generate its own (potentially highly complex) test cases. Instead of running code on manually or randomly constructed input, EXE runs it on *symbolic* input that is initially allowed to be "anything." As checked code runs, if it tries to operate on symbolic (i.e., input-derived) expressions, EXE replaces the operation with its corresponding input-constraint; it runs all other operations as usual. When code conditionally checks a symbolic expression, EXE forks execution, constraining the expression to be true on the true branch and false on the other. When a path terminates or hits a bug, EXE automatically generates a test case that will run this path by solving the path's constraints for concrete values using its co-designed constraint solver, STP.

EXE amplifies the effect of running a single code path since the use of STP lets it reason about *all possible values* that the path could be run with, rather than a single set of concrete values from an individual test case. For instance, a dynamic memory checker such as Purify [30] only catches an out-of-bounds array access if the index (or pointer) has a specific concrete value that is out-of-bounds. In contrast, EXE identifies this bug if there is any possible input value on the given path that can cause an out-of-bounds access to the array. Similarly, for an arithmetic expression that uses symbolic data, EXE can solve the associated constraints for values that cause an overflow or a division/modulo by zero. Moreover, for an assert statement, EXE can reason about all possible input values on the given path that may cause the assert to fail. If the assert does not fail, then either (1) no input on this path can cause it to fail, (2) EXE does not have the full set of constraints, or (3) there is a bug in EXE.

The ability to automatically generate input to execute paths has several nice features. First, EXE can test any code path it wishes (and given enough time, exhaust all of them), thereby getting coverage out of practical reach from random or manual testing. Second, EXE generates actual attacks. This ability lets it show that external forces can exploit a bug, improving on static analysis, which often cannot distinguish minor errors from showstoppers. Third, the EXE user sees no false positives: re-running input on an uninstrumented copy of the checked code either verifies that it hits a bug or automatically discards it if not.

Careful co-design of EXE and STP has resulted in a system with several novel features. First, STP primitives let EXE build constraints for all C expressions with perfect accuracy, down to a single bit. (The one exception is floating-point, which STP does not handle.) EXE handles pointers, unions, bit-fields, casts, and aggressive bit-operations such as shifting, masking, and byte swapping. Because EXE is dynamic (it runs the checked code) it has access to all the information that a dynamic analysis has, and a static analysis typically does not. All non-symbolic (i.e., *concrete*) operations happen exactly as they would in uninstrumented code, and produce exactly the same values: when these values appear in constraints they are correct, not approximations. In our context, what this accuracy means is that if (1) EXE has the full set of constraints for a given path, (2) STP can produce a concrete solution from those constraints, and (3) the path is deterministic, then rerunning the checked system on these concrete values will force the program to follow the same exact path to the error or termination that generated this set of constraints.

In addition, STP provides the speed needed to make perfect accuracy useful. Aggressive customization makes STP often 100 times faster than more traditional constraint solvers while handling a broader class of examples. Crucially, STP efficiently reasons about constraints that refer to memory using symbolic pointer expressions, which presents more challenges than one may expect. For example, given a concrete pointer a and a symbolic variable i with the constraint $0 \leq i \leq n$, then the conditional expression if(a[i] == 10) is essentially equivalent to a big disjunction: if(a[0] == 10 || ... || a[n] == 10). Similarly, an assignment a[i] = 42 represents a potential assignment to any element in the array between 0 and n.

The result of these features is that EXE finds bugs in real code, and automatically generates concrete inputs to trigger them. It generates evil packet filters that exploit buffer overruns in the very mature and audited Berkeley Packet Filter (BPF) code as well as its Linux equivalent (§ 5.1). It generates packets that cause invalid memory reads and writes in the udhcpd DHCP server (§ 5.2), and bad regular expressions that compromise the pcre library (§ 5.3), previously audited for security holes. In prior work, it generated raw disk images that, when mounted by a Linux kernel, would crash it or cause a buffer overflow [46].

Both EXE and STP are contributions of this paper, which is organized as follows. We first give an overview of the entire system (§ 2), then describe STP and its key optimizations (§ 3), and do the same for EXE (§ 4). Finally, we present results (§ 5), discuss related work (§ 6), and conclude (§ 7).

2. EXE OVERVIEW

This section gives an overview of EXE. We illustrate EXE's main features by walking the reader through the simple code example in Figure 1. When EXE checks this code, it explores each of its three possible paths, and finds two errors: an illegal memory write (line 12) and a division by zero (line 16). Figure 2 gives a partial transcript of a checking run.

To check their code with EXE, programmers only need to mark which memory locations should be treated as holding *symbolic data* whose values are initially entirely unconstrained. These memory locations are typically the input to the program. In the example, the call make_symbolic(&i) (line 4) marks the four bytes associated with the 32-bit variable i as symbolic. They then compile their code using the EXE compiler, exe-cc, which instruments it using the CIL source-to-source translator [36]. This instrumented code is then compiled with a normal compiler (e.g., gcc), linked with the EXE runtime system to produce an executable (in Figure 2, ./a.out), and run.

As the program runs, EXE executes each feasible path, tracking all constraints. When a program path terminates, EXE calls STP to solve the path's constraints for concrete values. A path terminates when (1) it calls exit(), (2) it crashes, (3) an assertion fails, or (4) EXE detects an error. Constraint solutions are literally the concrete bit values for an input that will cause the given path to execute. When generated in response to an error, they provide a concrete attack that can be launched against the tested system.

The EXE compiler has three main jobs. First, it inserts checks around every assignment, expression, and branch in the tested program to determine if its operands are concrete or symbolic. An operand is defined to be concrete if and only if all its constituent bits are concrete. If all operands are concrete, the operation is executed just as in the uninstrumented program. If any operand is symbolic, the operation is not performed, but instead passed to the EXE runtime system, which adds it as a constraint for the current path. For the example's expression p = (char *)a + i * 4 (line 8), EXE checks if the operands a and i on the right hand side of the assignment are concrete. If so, it executes the expression, assigning the result to p. However, since i is symbolic, EXE instead adds the constraint that p equals $(char*)a + i * 4$. Note that because i can be one of four values ($0 \leq i \leq 3$), p simultaneously refers to four different locations $a[0]$, $a[1]$, $a[2]$ and $a[3]$. In addition, EXE treats memory as untyped bytes (§ 3.2) and thus does

```
 1 : #include <assert.h>
 2 : int main(void) {
 3 :   unsigned i, t, a[4] = { 1, 3, 5, 2 };
 4 :   make_symbolic(&i);
 5 :   if(i >= 4)
 6 :     exit(0);
 7 :   // cast + symbolic offset + symbolic mutation
 8 :   char *p = (char *)a + i * 4;
 9 :   *p = *p - 1; // Just modifies one byte!
10:
11:   // ERROR: EXE catches potential overflow i=2
12:   t = a[*p];
13:   // At this point i != 2.
14:
15:   // ERROR: EXE catches div by 0 when i = 0.
16:   t = t / a[i];
17:   // At this point: i != 0 && i != 2.
18:
19:   // EXE determines that neither assert fires.
20:   if(t == 2)
21:     assert(i == 1);
22:   else
23:     assert(i == 3);
24: }
```

Figure 1: A contrived, but complete C program (*simple.c*) that generates five test cases when run under EXE, two of which trigger errors (a memory overflow at line 12 and a division by zero at line 16). This example is used heavily throughout the paper. We assume it runs on a 32-bit little-endian machine.

```
% exe-cc simple.c
% ./a.out
% ls exe-last
    test1.forks  test2.out      test3.forks   test4.out
    test1.out    test2.ptr.err  test3.out     test5.forks
    test2.forks  test3.div.err  test4.forks   test5.out
% cat exe-last/test3.div.err
    ERROR: simple.c:16 Division/modulo by zero!
% cat exe-last/test3.out
    # concrete byte values:
    0 # i[0]
    0 # i[1]
    0 # i[2]
    0 # i[3]
% cat exe-last/test3.forks
    # take these choices to follow path
    0 # false branch (line 5)
    0 # false (implicit: pointer overflow check on line 9)
    1 # true (implicit: div-by-0 check on line 16)
% cat exe-last/test2.out
    # concrete byte values:
    2 # i[0]
    0 # i[1]
    0 # i[2]
    0 # i[3]
```

Figure 2: Transcript of compiling and running the C program shown in Figure 1.

not get confused by this (dubious) cast, nor the subsequent type-violating modification of a low-order byte at line 9.

Second, exe-cc inserts code to fork program execution when it reaches a symbolic branch point, so that it can explore each possibility. Consider the if-statement at line 5, if(i >= 4). Since i is symbolic, so is this expression. Thus, EXE forks execution (using the UNIX fork() system call) and on the true path asserts that $i \geq 4$ is true, and on the false path that it is not. Each time it adds a branch constraint, EXE queries STP to check that there exists at least one solution for the current path's constraints. If not, the path is impossible and EXE stops executing it. In our example, both branches are possible, so EXE explores both (though the true path exits immediately at line 6).

Third, exe-cc inserts code that calls to check if a symbolic expression could have any possible value that could cause either (1) a null or out-of-bounds memory reference or (2) a division or modulo by zero. If so, EXE forks execution and (1) on the true path asserts that the condition does occur, emits a test case, and terminates; (2) on the false path asserts that the condition does not occur and continues execution (to find more bugs). Extending EXE to support other checks is easy. If EXE has the entire set of constraints on such expressions and STP can solve them, then EXE detects if *any* input exists on that path that causes the error. Similarly, if the check passes, then no input exists that causes the error on that path — i.e., the path has been *verified* as safe under all possible input values.

These checks find two errors in our example. First, the symbolic index *p in the expression a[*p] (line 12) can cause an out-of-bounds error because *p can equal 4: the pointer p was computed using i with the constraint $0 \leq i < 4$ (line 8). Thus, $i = 2$ is legal, which means p can point to the low-order byte of a[2] (recall that each element of a has four bytes). The value of this byte is 4 after the subtraction at line 9. Since a[4] references an illegal location one past the end of a, EXE forks execution and on one path asserts that $i = 2$ and emits an error (test2.ptr.err) and a test case (test2.out), and on the other that $i \neq 2$ and continues.

Second, the symbolic expression t / a[i] (line 16) can generate a division by zero, which EXE detects by tracking and solving the constraints that (1) i can equal 0, 1, or 3 and (2) a[0] can equal 0 after the decrement at line 9. EXE again forks execution, emits an error (test3.div.err) and a test case (test3.out) and exits. The other path adds the constraint that $i \neq 0$ and continues.

Note, EXE automatically turns a programmer assert(e) on a symbolic expression *e* into a universal check of *e* simply because it tries to exhaust both paths of if-statements. If EXE determines that *e* can be false, it will go down the assertion's false path, hitting its error handling code. Further, if STP cannot find any such value, none exists on this path. In the example, EXE explores both branches at line 20, and proves that no input value exists that can cause either assert (line 21 and line 23) to fail. We leave working through this logic as an exercise for the more energetic reader. Even a cursory attempt should show the trickiness of manual reasoning about all-paths and all-values for even trivial code fragments. (We spent more time than we would like to admit puzzling over our own hand-crafted example and eventually gave up, resorting to using EXE to double-check our oft-wrong reasoning.)

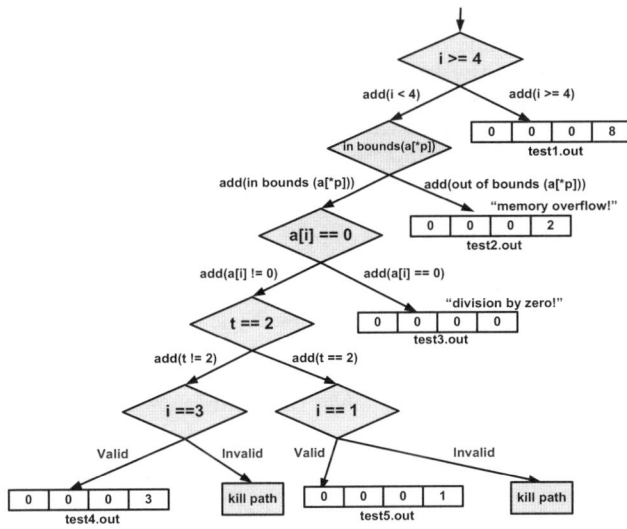

Figure 3: Execution for the simple C program in Figure 1: EXE generates five test cases, two of which are errors.

The paths followed by EXE are shown graphically in Figure 3. The branch points (both explicit and implicit) where EXE forks a new process are represented by rhombuses, and the test cases it generates by sequences of four bytes.

Mechanically, at each run of the instrumented code, EXE creates a new directory and, for each path, creates two files: one to hold the concrete bytes it generates, the other to hold the values for each decision (1 to take the true branch, 0 to take the false). The choice points enable easy replay of a single path for debugging. The values can either be read back by using a trivial driver (which EXE provides) or used completely separately from EXE.

In our example, the three paths and two errors lead to five pairs of files that hold (1) concrete byte values for i (these files have the suffix .out) and (2) the branch decisions for that path (suffix .forks). EXE creates a symbolic link exe-last pointing to the most recent output directory. The two errors are in .err files. If we look at the contents of the file for the division bug (test3.out), it shows that each byte of i is zero, which when concatenated in the right order and treated as an unsigned 32-bit quantity equals 0, as required. The branch decision states that we take the false branch at line 5, followed by the (implicit) false branch of the memory overflow check at line 9, and finally the (implicit) true branch of the division check at line 16. Similarly, the concrete values for the pointer error are byte 0 equals 2 and bytes 1, 2, 3 equal 0, which when concatenated yields the 32-bit value 2 as needed.

3. KEY FEATURES OF STP

This section gives a high-level overview of STP's key features, including the support it provides to EXE for accurately modeling memory. It then describes the optimizations STP performs, and shows experimental numbers evaluating their efficiency.

EXE's constraint solver is, more precisely, a *decision procedure* for bitvectors and arrays. Decision procedures are programs which determine the satisfiability of logical formulas that can express constraints relevant to software and hardware, and have been a mainstay of program verification for several decades. In the past, these decision procedures have been based on variations of Nelson and Oppen's *cooperating decision procedures framework* [37] for combining a collection of specialized decision procedures into a more comprehensive decision procedure capable of handling a more expressive logic than any of the specialized procedures can do individually.

The Nelson-Oppen approach has two downsides. Whenever a specialized decision procedure can infer that two expressions are equal, it must do so explicitly and communicate the equality to the other specialized decision procedures, which can be expensive. Worse, the framework tends to lead to a web of complex dependencies, which makes its code difficult to understand, tune, or get right. These problems hampered CVCL [6, 7], a state-of-the-art decision procedure that we implemented previously.

Our CVCL travails motivated us to simplify the design of STP by exploiting the extreme improvement in SAT solvers over the last decade. STP forgoes Nelson-Oppen contortions, and instead preprocesses the input through the application of mathematical and logical identities, and then eagerly translates constraints into a purely propositional logical formula that it feeds to an off-the-shelf SAT solver (we use MiniSAT [21]). As a result, the STP implementation has four times less code than CVCL, yet often runs orders of magnitude faster. STP is also more modular, because its pieces work in isolation. Modularity and simplicity help constraint solvers as they do everything else. In a sense, STP can be viewed as the result of applying the systems approach to constraint solving that has worked so well in the context of SAT: start simple, measure bottlenecks on real workloads, and tune to exactly these cases. STP was recently judged the co-winner of the QF_UFBV32 (32-bit bitvector) division of the SMTLIB competition [1] held as a satellite event of CAV 2006 [3].

Recently, several other decision procedures have been based on eager translation to SAT, including Saturn[45], UCLID[11], and Cogent[15]. Saturn is a static program analysis framework that translates C operations to SAT. It does not directly deal with arrays, so it avoids many interesting problems and optimizations. UCLID implements features such as arrays and arbitrary precision integer arithmetic, but does not focus on bitvector operations. Cogent is perhaps the most similar in architecture and purpose to STP. Judging from the published descriptions of these systems, STP's focus on optimizations for arrays is unique (and uniquely important for use with EXE). STP also has simplifications on word-level operations that are not discussed in the description of Cogent. (At this time, it is difficult to do side-by-side performance comparisons because of lack of common benchmarks and input syntax; Saturn, UCLID and Cogent also didn't participate in the SMTLIB competition.)

3.1 STP primitives

System code often treats memory as untyped bytes, and observes a single memory location in multiple ways. For example, by casting signed variables to unsigned, or (in the code we checked) treating an array of bytes as a network packet, inode, packet filter, etc. through pointer casting.

As a result, STP also views memory as untyped bytes. It provides only three data types: booleans, bitvectors, and arrays of bitvectors. A bitvector is a fixed-length sequence of bits. For example, 0010 is a constant, 4-bit bitvector representing the constant 2. With the exception of floating-point, which STP does not support, all C operators have a corresponding STP operator that can be used to impose constraints on bitvectors. STP implements all arithmetic operations (even non-linear operations such as multiplication, division and modulo), bitwise boolean operations, relational operations (less than, less than or equal, etc.), and multiplexers, which provide an "if-then-else" construct that is converted into a logical formula (similar to C's ternary operator). In addition, STP supports bit concatenation and bit extraction, features EXE makes extensive use of in order to translate untyped memory into properly-typed constraints.

STP implements its bitvector operations by translating them to operations on individual bits. There are two expression types: *terms*, which have bitvector values, and *formulas*, which have boolean values. If x and y are 32-bit bitvector values, $x + y$ is a term returning a 32-bit result, and $x + y < z$ is a formula. In the implementation, terms are converted into vectors of boolean formulas consisting entirely of single bit operations (AND, XOR, etc.). Each operation is converted in a fairly obvious way: for example, a 32-bit add is implemented as a ripple-carry adder. Formulas are converted into DAGs of single bit operations, where expressions with identical structure are represented uniquely (expression nodes are looked up in a hash table whenever they are created to see whether an identical node already exists). Simple boolean optimizations are applied as the nodes are created; for example, a call to create a node for AND(x, FALSE) will just return the FALSE node. The resulting boolean DAG is then converted to CNF by the standard method of naming intermediate nodes with new propositional variables.

3.2 Mapping C code to STP constraints

EXE represents each symbolic data block as an array of 8-bit bitvectors. The main advantage of using bitvectors is that they, like the C memory blocks that they represent, are essentially untyped. This property allows us to easily express constraints that refer to the same memory in different ways; each read of memory generates constraints based on the static type of the read (e.g., int, unsigned, etc.) but these types do not persist.

EXE uses STP to solve constraints on input as follows. First, it tracks what memory locations in the checked code hold symbolic values. Second, it translates expressions to bitvector based constraints. We discuss each step below.

Initially, there are no symbolic bytes in the checked code. When the user marks a byte-range, b, as symbolic, EXE calls into STP to create a corresponding, identically-sized array b_{sym}, and records in a table that b corresponds to b_{sym}. In Figure 1 (line 4), the call to make the 32-bit variable i symbolic causes EXE to allocate a bitvector array i_{sym} with four 8-bit elements and record that the concrete address of i (&i) corresponds to it.

As the program executes, the table mapping concrete bytes to STP bitvectors grows in exactly two cases:

1. v = e: where e is a symbolic expression (i.e., has at least one symbolic operand). EXE builds the symbolic expression e_{sym} representing e, and records that &v maps to it. Note that EXE does not allocate a new STP variable in this case but instead will substitute e_{sym} for v in subsequent constraints. EXE removes this mapping when v is overwritten with a concrete value or deallocated. In Figure 1 (line 8), EXE records the fact that p maps to the symbolic expression $(char*)a + i_{sym} * 4$ and substitutes any subsequent use of p's value with this expression. (Note that a is replaced by the actual base address of array a in the program.)

2. b[e]: where e is a symbolic expression and b is a concrete array. Since STP must reason about the set of values that b[e] could reference, EXE imports b into STP by allocating an identically-sized STP array b_{sym}, and initializing it to have the same (constant) contents as b. It then records that b maps to b_{sym} and removes this mapping only when the array is deallocated.

In Figure 1 (line 12), the array expression a[*p] causes EXE to allocate a_{sym}, a 16-element array of 8-bit bitvectors, and assert that:

$$a_{sym} = \{1, 0, 0, 0, 3, 0, 0, 0, 5, 0, 0, 0, 2, 0, 0, 0\}$$

Each expression e used in a symbolic operation is constructed in the following way. For each read of size n of a storage location l in e, EXE checks if l is concrete. If so, the read of l is replaced by its concrete value (i.e., a constant). Otherwise, EXE breaks down l into its corresponding bytes $b_0, \ldots, b_{n-1}$. It then builds a symbolic expression with the same size as l by concatenating each byte's (possibly symbolic) value. For each byte b_i it queries its data structures to check if b_i is symbolic. If not, it uses its current concrete value (an 8-bit constant), otherwise it looks up and uses its symbolic expression $(b_i)_{sym}$.

For example, in Figure 1 (line 8), EXE builds the symbolic expression corresponding to (char*)a + i*4 as follows. EXE determines that the first read of a is concrete and so replaces a with its concrete address (denoted a) represented as a 32-bit bitvector constant. It then determines that i is symbolic, and thus breaks it down into its four bytes, which are mapped to their corresponding STP bitvector array elements $i_{sym}[0]$, $i_{sym}[1]$, $i_{sym}[2]$, and $i_{sym}[3]$. Then, the four bitvectors are concatenated to obtain the expression $i_{sym}[3] @ i_{sym}[2] @ i_{sym}[1] @ i_{sym}[0]$ (where "@" denotes bitvector concatenation), which corresponds to the four-byte read of i. Finally, the constant 4 is replaced by the corresponding 32-bit bitvector constant 0...00000100. The resulting expression is

$$a + (i_{sym}[3] @ i_{sym}[2] @ i_{sym}[1] @ i_{sym}[0]) * 0...00000100$$

A limitation of STP is that it does not support pointers directly. EXE emulates symbolic pointer expressions by mapping them as an array reference at some offset. For each pointer p in the checked code, EXE tracks the data object to which p points by instrumenting all allocation and deallocation sites as well as all pointer arithmetic expressions (standard techniques developed by bounds-checking compilers [41]). For example, in Figure 1 (line 4), EXE records that p points to the data block a of size 16. Then, when EXE encounters a pointer dereference *p: (1) it looks up the block b to which pointer p refers; (2) looks up the corresponding STP array b_{sym} associated with b; and (3) computes the (possibly symbolic) offset of p from the base of the object it points to (i.e., o = p - b). EXE can then use the symbolic expression $b_{sym}[i_{sym} + o_{sym}]$ in symbolic constraints.

However, STP's lack of pointer support means that when EXE encounters a double-dereference **p of a symbolic pointer p it *concretizes* the first dereference (*p), fixing it to one of the possibly many storage locations it could refer to. (However, the result of **p can still be a symbolic expression.) This situation has rarely shown up in practice (see § 4.3), but we are working on removing it.

3.3 The key to speed: fast array constraints

The main bottleneck in STP when used in EXE is almost always reasoning about arrays. This subsection discusses STP's key array optimizations.

STP is an implementation of logic, so it is a purely functional language. The logic has one-dimensional arrays that are indexed by bitvectors and contain bitvectors. The operations on arrays are $read(A, i)$, which returns the value at location $A[i]$ where A is an array and i is an index expression of the correct type, and $write(A, i, v)$, which returns a new array with the same value as A at all indexes except i, where it has the value v. Array reads and writes can appear as subexpressions of an if-then-else construct, denoted by $ite(c, a, b)$, where c is the condition, a the then expression, and b the else expression.

STP eliminates array expressions by translating them to bitvector primitives (which it then translates to SAT). This is accomplished through two main transformations. The first, **read-over-write**, eliminates all $write(A, i, v)$ expressions:[1]

$$read(write(A, i, v), j) \Rightarrow ite(i = j, v, read(A, j))$$

The second eliminates all *read* expressions via a transformation mentioned in [11] that enforces the axiom that if two indexes i_s and i_t are the same, then $read(A, i_s)$ and $read(A, i_t)$ should return the same value. Mechanically, STP first replaces each occurrence of a read $read(A, i_j)$ with a new variable v_j, and then for each two terms i_s, i_t ever used to index into the same array A, it adds the *array axiom*:

$$i_s = i_t \Rightarrow v_s = v_t$$

For example, consider the formula:

$$(read(A, i_1) = e_1) \wedge (read(A, i_2) = e_2) \wedge (read(A, i_3) = e_3)$$

The transformed result would be:

$$(v_1 = e_1) \wedge (v_2 = e_2) \wedge (v_3 = e_3) \wedge (i_1 = i_2 \Rightarrow v_1 = v_2) \wedge$$

$$(i_1 = i_3 \Rightarrow v_1 = v_3) \wedge (i_2 = i_3 \Rightarrow v_2 = v_3)$$

Read elimination expands each formula by $n(n-1)/2$ nodes, where n is the number of syntactically distinct index expressions. Unfortunately, this blowup is lethal for arrays of a few thousand elements, which occur frequently in EXE. Fortunately, while finessing this problem appears hard in general, two optimizations we developed work well on the constraints generated by EXE.

The *array substitution optimization* reduces the number of array variables by substituting out all constraints of the form $read(A, c) = e$, where c is a constant and e does not contain another array read. Programs often index into arrays using constant indexes, so this is a case that occurs often in practice (see § 4.3). The optimization has two passes.

[1] Note that a *write* makes sense only inside a *read* node. A *write* node by itself has no effect, and can be ignored.

The first pass builds a substitution table with the left-hand-side of each such equation ($read(A, c)$) as the key and the right-hand-side (e) as the value, and then deletes the equation from the EXE query. The second pass over the expression replaces each occurrence of a key by the corresponding table entry. Note that for soundness, if we encounter a second equation whose left-hand-side is already in the table, the second equation is not deleted and the table is not changed. For our example, if we saw a subsequent equation $read(A, i_1) = e_4$ we would leave it; the second pass of the algorithm would rewrite it as $e_1 = e_4$.

The second optimization, *array-based refinement*, delays the translation of array *reads* with non-constant indexes, in effect introducing some laziness into STP's handling of arrays, in the hope of avoiding the $O(n^2)$ blowup from the read elimination transformation. Its main trick is to solve a less-expensive approximation of the formula, check the result in the original formula, and try again with a more accurate approximation if the result is incorrect.

Initially, all array read expressions are replaced by variables to yield an approximation of the original formula. The resulting logical formula is under-constrained, since it ignores the array axioms that require that array reads return the same values when indexes are the same. If the resulting under-constrained formula is not satisfiable, there is no solution for the original formula and STP returns unsatisfiable.

If, however, the SAT solver finds a solution to the under-constrained formula, then that solution is not guaranteed to be correct because it could violate one of the array axioms. For example, suppose STP is given the formula $(read(A, 0) = 0) \wedge (read(A, i) = 1)$. STP would first apply the substitution optimization by deleting the constraint $read(A, 0) = 0$ from the formula, and inserting the pair $(read(A, 0), 0))$ in the substitution table. Then, it would replace $read(A, i)$ by a new variable v_i, thus generating the under-constrained formula $v_i = 1$. Suppose STP finds the solution $i = 1$ and $v_i = 1$. STP then translates the solution to the variables of the original formula to get $(read(A, 0) = 0) \wedge (read(A, 1) = 1)$. This solution is satisfiable in the original formula as well, so STP terminates since it has found a true satisfying assignment.

However, suppose that STP finds the solution $i = 0$ and $v_i = 1$. Under this solution, the original formula evaluates to $(read(A, 0) = 0) \wedge (read(A, 0) = 1)$, which gives $0 = 1$. Hence, the solution to the under-constrained formula is not a solution to the original formula. When this happens, it must be because some array axiom was violated. STP adds array axioms to the formula and solves again until it gets a correct result. There are many policies for adding axioms, any of which is correct and will terminate so long as all of the axioms are added in the worst case. The current policy, which seems to work well, is to find an array index term for which at least one axiom is violated, then add all of the axioms involving that term. In our example, it will add the axiom $i = 0 \Rightarrow read(A, i) = read(A, 0)$. Then, the process of finding a satisfying assignment is repeated, by calling the SAT solver on the new under-constrained formula. The result must satisfy the newly added axioms, which the previous assignment violated, so the algorithm will not repeat assignments and will not violate previously added axioms. This process must terminate since there are only finitely many array axioms.

In the worst case, the algorithm will add all $n(n-1)/2$

Solver	Total Time	Timeouts
CVCL	60,366s	546
STP (no optimizations)	3,378s	36
STP (substitution)	1,216s	1
STP (refinement)	624s	1
STP (simplifications)	336s	0
STP (subst+refinement)	513s	1
STP (simplif+subst)	233s	0
STP (simplif+refinement)	220s	0
STP (all optimizations)	110s	0

Table 1: STP vs.CVCL. Queries time out (are aborted) after 60 seconds, which underestimates performance differences, since they could run for much longer. Using this conservative estimate, fully optimized STP is roughly 30X faster than the unoptimized version and 550X faster than CVCL and has no timeouts.

array axioms, at which time it is guaranteed to return a correct result because there are no more axioms it can violate. However, in practice, this loop will often terminate quickly because the formula can be proved unsatisfiable without all the array axioms, or because it luckily finds a true satisfying assignment without adding all the axioms.

Besides the above mentioned optimizations, STP implements several boolean and mathematical identities. These identities, or *simplifications*, also dramatically reduce the size of the input, before it is fed to the SAT solver. Some example identities include associativity and commutativity laws for addition and multiplication, distribution of multiplication by constants over addition, and combining like terms (e.g., $x + (-x)$ is simplified to 0).

All these optimizations have made it possible to deal with fairly large constant arrays when there are relatively few non-constant index expressions, which is sufficient to permit considerable progress in using EXE on real examples.

3.4 Measured performance

Table 1 gives experimental measurements for these optimizations. The experiment consists of running different versions of STP and our old solver, CVCL, over the performance regression suite we have built up of 8495 test cases taken from our test programs. The experiments for all solvers were run on a Pentium 4 machine at 3.2 GHz, with 2 GB of RAM and 512 KB of cache. The table gives the times taken by CVCL, baseline STP with no optimizations, STP with a subset of all optimizations enabled, and STP with full optimizations, i.e. substitution, array-based refinement, and simplifications. The third column shows the number of examples on which each solver timed out. The timeout was set at 60 seconds, and is added as penalty to the time taken by the solver (but in fact causes us to grossly underestimate the time taken by CVCL and earlier versions of STP since they could run for many minutes or even hours on some of the examples).

The baseline STP is nearly 20 times faster than CVCL, and more interestingly times out in far fewer cases. The fully optimized version of STP is about 30 times faster than the unoptimized version, almost 550 times faster than CVCL, and there are no timeouts.

4. EXE OPTIMIZATIONS

This section presents optimizations EXE uses and measures their effectiveness on five benchmarks. We first present two optimizations: caching constraints to avoid calling STP (§ 4.1), and removing irrelevant constraints from the queries EXE sends to STP (§ 4.2). We then measure the cumulative improvement of these optimizations, and provide an empirical feel for what symbolic execution looks like, including the time spent in various parts of EXE, and a description of the symbolic slice through the code (§ 4.3). Finally, we discuss and measure EXE's search heuristics (§ 4.4).

4.1 Constraint caching

EXE caches the result of satisfiability queries and constraint solutions in order to avoid calling STP when possible. This cache is managed by a server process so that multiple EXE processes (created by forking at each conditional) can coordinate. Before invoking STP on a query q, an EXE process prints q as a string, computes an MD4 cryptographic hash of this string, and sends this hash to the server. The server checks its persistent cache (a file) and if it gets a hit, returns the result. If not, the EXE process does a local STP query and then sends the $(hash, result)$ pair back to the server. Constraint solutions are cached in a similar way.

4.2 Constraint independence optimization

This section describes one of EXE's most important optimizations, *constraint independence*, which exploits the fact that we can often divide the set of constraints EXE tracks into multiple independent subsets of constraints. Two constraints are considered to be independent if they have disjoint sets of operands (i.e. disjoint sets of array reads).

For example, assume EXE tracks the following set of three constraints:
$(A[1] = A[2] + A[3]) \wedge (A[2] > A[4]) \wedge (A[7] = A[8])$
We can divide this set into two subsets of independent constraints
$$(A[1] = A[2] + A[3]) \wedge (A[2] > A[4])$$
and
$$A[7] = A[8]$$
and solve them separately.

Breaking a constraint into multiple independent subsets has two benefits. First, EXE can discard irrelevant constraints when it asks STP if a constraint c is satisfiable, with a corresponding decrease in cost. Instead of sending all the constraints collected so far to STP, EXE only sends the subset of constraints s_c to which c belongs, ignoring all other constraints. The worst case, when no irrelevant constraints are found, costs no more than the original query (omitting the small cost of computing the independent subsets).

Second, this optimization yields additional cache hits, since a given a subset of independent constraints may have appeared individually in previous runs. Conversely, including all constraints vastly increases the chance that at least one is different and so gets no cache hit. To illustrate, assume we have the following code fragment, which operates on two unconstrained symbolic arrays A and B:
```
if (A[i] > A[i+1]) {
   ...
}
if (B[j] + B[j-1] == B[j+1]) {
   ...
}
```

There are four paths through this code; EXE will thus create four processes. After forking and following each branch, EXE checks if the path is satisfiable. Without the constraint independence optimization, each of these four satisfiability queries will differ and miss in the cache. However, if the optimization is applied, some queries repeat. For example, when the second branch is reached, two of the four queries will be

$$(A[i] > A[i+1]) \land (B[j] + B[j-1] = B[j+1])$$

and

$$(A[i] \leq A[i+1]) \land (B[j] + B[j-1] = B[j+1])$$

which both devolve to

$$B[j] + B[j-1] = B[j+1]$$

since, in each query, the first constraint is unrelated to the last one, and its satisfiability was already determined when EXE reached the first branch.

Real programs often have many independent branches, which introduce many irrelevant constraints. These add up quickly. For example, assuming n consecutive independent branches (the example above is such an instance for $n = 2$), EXE will issue $2(2^n - 1)$ queries to STP (for each `if` statement, we issue two queries to check if both branches are possible). The optimization exponentially reduces this query count to $2n$ (two queries the first time we see each branch), since the rest of the time we find the result in the cache.

We compute the constraint independence subsets by constructing a graph G, whose nodes are the set of all array reads used in the given set of constraints. For the first example in the section, the set of nodes is $\{A[1], A[2], A[3], A[4], A[7], A[8]\}$. We add an edge between nodes n_i and n_j of G if and only if there exists a constraint c that contains both as operands. Once the graph G is constructed, we apply a standard algorithm to determine G's connected components. Finally, for each connected component, we create a corresponding independent subset of constraints by adding all the constraints that contain at least one of the nodes in that connected component. At the implementation level, we don't construct the graph G explicitly. Instead, we keep the nodes of G in a union-find structure [17], which we update each time we add a new constraint.

There are two additional issues that our algorithm has to take into account. First, an array read may contain a symbolic index. In this case, we are conservative, and merge all the elements of that array into a single subset.

The second issue relates to array writes. Since EXE and STP arrays are functional, each array read explicitly contains an ordered list of all array writes performed so far. Each array write is remembered as a pair consisting of the location that was updated, and the expression that was written to that location. When processing this list of array writes, we are again conservative, and merge all the expressions written into the array (the right hand side of each array write) into the subset of the original read. In addition, if any array write is performed at a symbolic index, we merge all the elements of the array into a single subset.

4.3 Experiments

We evaluate our optimizations on five benchmarks. These benchmarks consist of the three applications discussed in

	bpf	expat	pcre	tcpdump	udhcpd
Test cases	7333	360	866	2140	328
None	30.6	28.4	31.3	28.2	30.4
Caching	32.6	30.8	34.4	27.0	36.4
Independence	17.8	25.2	10.0	24.9	30.5
All	10.3	26.3	7.5	23.6	32.1
STP cost	6.9	24.6	2.8	22.4	23.1

Table 2: Optimization measurements, times in minutes. STP cost gives time spent in STP when all optimizations are enabled. Tables 3, 4, and 5 explore the fully optimized run (All) in more detail.

Section 5, `bpf`, `pcre`, and `udhcpd`, to which we added two more: `expat`, an XML parser library, and `tcpdump`, a tool for printing out the headers of packets on a network interface that match a boolean expression.

We run each benchmark under four versions of EXE: no optimization, caching only, independence only, and finally with both optimizations turned on. As a baseline, we run each benchmark for roughly 30 minutes using the unoptimized version of EXE, and record the number of test cases n that this run generates. We then run the other versions until they generate n test cases. All experiments are performed on a dual-core 3.2 GHz Intel Pentium D machine with 2 GB of RAM, and 2048 KB of cache.

Table 2 gives the number of test cases generated, as well as the runtime for each optimization combination. Full optimization ("All") significantly sped up two of five benchmarks: `bpf` by roughly a factor of three, and `pcre` by more than a factor of four. Both `tcpdump` and `expat` had marginal improvements (20% and 7% faster respectively), but `udhcpd` slows down by 5.6%. As the last row shows, with the exception of `pcre`, the time spent in STP represents by far the dominant cost of EXE checking.

Table 3 breaks down the full optimization run. As its first three rows show, caching without independence is not a win — its overhead (see Table 2) actually increases runtime for most applications, varying between 6.5% for `bpf` and 19.7% for `pcre`. With independence, the hit rate jumps sharply for both `bpf` and `pcre` (and, to a lesser extent, `tcpdump`), due to its removal of irrelevant constraints. The other two applications show no benefit from these optimizations — `udhcpd` has no independent constraints and `expat` has no cache hits. The average number of independent subsets (row 3) shows how interdependent our constraints are, varying from over 2,800 subsets for `expat` to only 1 (i.e., no independent constraints) for `udhcpd`.

The next three rows (4–6) measure the overhead spent in various parts of EXE. Reassuringly, the cost of independence is near zero. On the other hand, cache lookup overhead (row 5) is significant, due almost entirely to our naive implementation. On each cache lookup (§ 4.1), EXE prints the query as a string and then hashes it. As the table shows (row 6) the cost of printing the string dominates all other cache lookup overheads. Obviously, we plan to eliminate this inefficiency in the next version of the system.

Table 4 breaks down the queries sent to STP. The first three rows give the total number of: queries, constraints, and nodes. These last two numbers give a feel for query complexity: `bpf` is the easiest case (a small number of constraints, with roughly five nodes per constraint), whereas `udhcpd` is the worst with 688 nodes per constraint.

The next two rows give the number of non-linear con-

		bpf	expat	pcre	tcpdump	udhcpd
1	Cache hit rate	92.8%	0%	83%	35%	9.1%
2	Hit rate w/o independence	0.1%	0%	17.5%	12.6%	9.1%
3	Avg. # of independent subsets	19	2,824	122	13	1
4	Independence overhead	0m	0m	.1m	0m	0m
5	Cache lookup cost	1.1m	1.2m	1.9m	0.4m	2.1m
6	% of lookup spent printing	72%	96%	84%	90%	95%

Table 3: Optimization breakdown

		bpf	expat	pcre	tcpdump	udhcpd
1	# of queries (cache misses)	162,959	5,427	188,481	22,242	3,572
2	Total # of constraints	402,496	9,649,411	3,478,517	1,268,316	626,795
3	Total # of nodes	2,048,704	32,711,503	17,844,792	20,673,550	431,705,056
4	# non-linear constraints	3,758	10,679	95,623	343,312	508,096
5	% constraints non-linear	0.9%	0.1%	2.8%	27.1%	81.1%
6	Reads from symbolic array	405,501	11,788,264	3,757,238	1,619,843	3,855,965
7	% sym. array reads with sym. index	0.3%	0.3%	2.9%	7.8%	62.9%
8	Writes to symbolic array	62	2,310,903	706,214	0	0
9	% sym. array writes with sym. index	100%	0%	1.8%	0%	0%

Table 4: Dynamic counts from queries sent to STP.

straints (row 4) and their percentage (row 5) of the total constraints (from row 2). Non-linear constraints contain one or more non-linear operators — multiplication, division, or modulo — whose right hand side is not a constant power of two. In general, the more non-linear operations, the slower constraint solving gets, as the SAT circuits that STP constructs for these operations are expensive. For our benchmarks, only udhcpd has a large number of non-linear constraints, which translates into a large amount of time spent in STP.

The final four rows (6–9) give the number of reads and writes from and to symbolic data blocks, and the percentage of these that use symbolic indexes. While there are many array operations, with the exception of udhcpd, very few use symbolic indexes, which explains why the STP array substitution optimization (§ 3.3) was such a big win.

Table 5 gives more dynamic execution counts from the full optimization runs. The first row gives the number of bytes initially marked as symbolic; this represents the size of the symbolic filter and data in bpf, the size of the XML expression to be parsed in expat, the packet length in udhcpd and tcpdump, and the regular expression pattern length in pcre.

The next row (row 2) gives the total number of dynamic statements executed (assignments, branches, parameter and return value passing) across all paths executed by EXE, while the next (row 3) gives the percentage that are symbolic. For our benchmarks, this percentage varies from only 8.46% for expat to 41.70% for tcpdump. This numbers are encouraging and validate our approach of mixing concrete and symbolic execution, which lets us ignore a large amount of code in the programs we check.

The next three rows (4–6) look at symbolic branches, including the implicit branches EXE does for checking. Row 4 gives the total number of explicit symbolic branch points and row 5 the percentage of these branch points that had both branches feasible. (EXE pruned the other branches because the path's constraints were not satisfiable.) On our benchmarks, EXE was able to prune more than 80% of the branches it encountered, with the exception of udhcpd where it pruned (only) 47.18% of the branches. These results are reassuring for scalability – while the potential number of paths in the search space grows exponentially with the number of symbolic branches, the actual growth is much smaller: real code appears to have many dependencies between program points.

Row 6 measures the average number of symbolic branches (both implicit and explicit) per path. This number is large: ranging from around 38 up to 200 branches, which means that random guessing would have a hard time satisfying all the branches to get to the end of one path, much less the hundreds or thousands that EXE can systematically explore.

Row 7 gives the total number of times EXE performed a symbolic check. (In addition to these checks, EXE performs many more similar concrete checks.) Row 8 shows how many times EXE had to concretize a pointer because it encountered a symbolic dereference of a symbolic pointer (§ 3.2). This situation occurs in only one of our five benchmarks, tcpdump. Finally, row 9 shows that no uninstrumented functions were called with symbolic data as arguments.

4.4 Search heuristics

When EXE forks execution, it must pick which branch to follow first. By default, EXE uses depth-first search (DFS), picking randomly between the two branches. DFS keeps the current number of processes small (linear in the depth of the process chain), but works poorly in some cases. For example, if EXE encounters a loop with a symbolic variable as a bound, DFS can get "stuck" since it attempts to execute the loop as many times as possible, thus potentially taking a very long time to exit the loop.

In order to overcome this problem, we use search heuristics to drive the execution along "interesting" execution paths (e.g., that cover unexplored statements). After a `fork` call, each forked EXE process calls into a search server with a description of its current state (e.g., its current file, line number, and backtrace) and blocks until the server replies. The search server examines all blocked processes and picks the best one in terms of some heuristic that is more global than simply picking a random branch to follow. Our current heuristic uses a mixture of best-first and depth-first search. The search server picks the process blocked at the line of code run the fewest number of times and then runs this

		bpf	expat	pcre	tcpdump	udhcpd
1	Symbolic input size (bytes)	96	10	16	84	548
2	Total statements run (not unique)	298,195	41,345	423,182	40,097	15,258
3	% of statements symbolic	29.2%	8.5%	34.7%	41.7%	23.6% %
4	Explicit symbolic branch points	77,024	1,969	98,138	11,425	888
5	% with both branches feasible	11.3%	19.3%	0.9%	19.4%	52.8%
6	Avg. # symbolic branches per path	38.33	43.44	55.72	103.37	200.14
7	Symbolic checks	1,490	904	4,451	552	1,535
8	Pointer concretizations	0	0	0	73	0
9	Symbolic args. to uninstr. calls	0	0	0	0	0

Table 5: Dynamic counts from EXE execution runs.

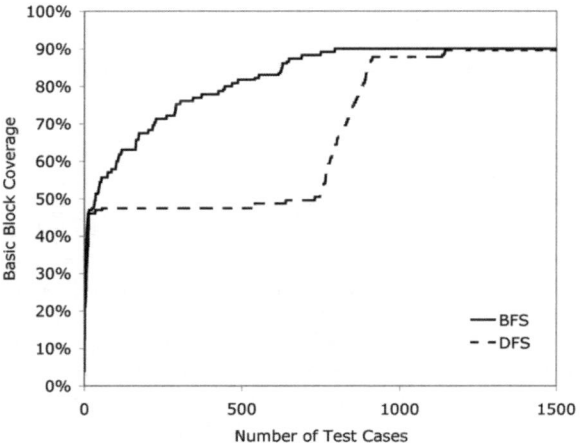

Figure 4: Best-first search vs. depth-first search.

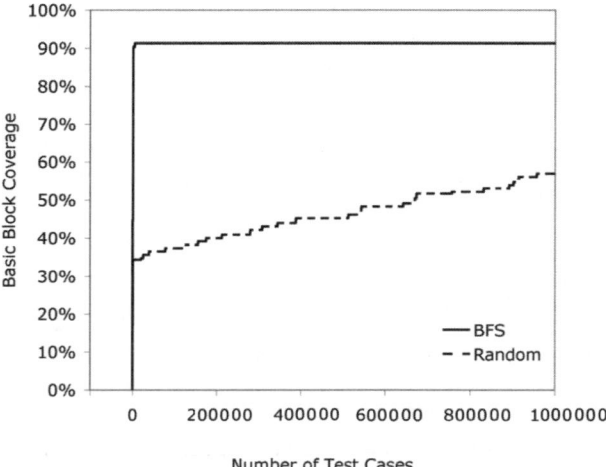

Figure 5: EXE with best-first search vs. random testing.

process (and its children) in a DFS manner for a while. It then picks another best-first candidate and iterates. This is just one of many possible heuristics, and the server is structured so that new heuristics are easy to plug in.

We experimentally evaluate our best-first search (BFS) heuristic in the context of one of our benchmarks, the Berkeley Packet Filter (BPF) (described in more detail in § 5.1). We start two separate executions of EXE, one using DFS and the other using BFS. We let both EXE executions run until they achieved full basic block coverage. Figure 4 compares BFS to DFS in terms of basic block coverage. (For visual clarity the graph only shows block coverage for the first 1500 test cases, as only a few blocks are missing from the coverage by these test cases.) BFS converges to full coverage more than twice as fast as DFS: 7,956 test cases versus 18,667. More precisely, EXE gets 91.74% block coverage, since there are several basic blocks in BPF that EXE cannot reach, such as dead code (e.g. the failure branch of asserts), or branches that do not depend on the input marked as symbolic.

Figure 5 then compares EXE against random testing also in terms of basic block coverage. We generate one million random test cases of the same size as those generated by EXE, and run these random test cases through a lightly-instrumented version of BPF that records basic block coverage. These test cases only cover 56.96% of the blocks in BPF; EXE achieves the same coverage in only 75 tests when using BFS. Even more strikingly, these million random test cases yield only 131 unique paths through the code, while each of EXE's test cases represents a unique path.

5. USING EXE TO FIND BUGS

This section presents three case studies that use EXE to find bugs in: (1) two packet filter implementations, (2) the udhcpd DHCP server, and (3) the pcre Perl compatible regular expressions library. We also summarize a previous effort of applying EXE to file system code.

5.1 Packet filters

Many operating systems allow programs to specify packet filters which describe the network packets they want to receive. Most packet filter implementations are variants of the Berkeley Packet Filter (BPF) system. BPF filters are written in a pseudo-assembly language, downloaded into the kernel, validated by the BPF system, and then applied to incoming packets. We used EXE to check the packet filter in both FreeBSD and Linux. FreeBSD uses BPF, while Linux uses a heavily modified version of it. EXE found two buffer overflows in the former and four errors in the latter. BPF is one particularly hard test of EXE — small, heavily-inspected and mature code, written by programmers known for their skill.

A filter is an array of instructions specifying an opcode (code), a possible memory offset to read or write (k), and several other fields. The BPF interpreter iterates over this filter, executing each opcode's corresponding action. This loop is the main source of vulnerabilities but is hard to test exhaustively (e.g., hitting all opcodes even once using random testing takes a long time).

We used a two-part checking process. First, we marked a fixed-sized array of filter instructions as symbolic and passed

```
s[0].code = BPF_STX; // also: (BPF_LDX|BPF_MEM)
s[0].k    = 0xfffffff0UL;
s[1].code = BPF_RET;
```

Figure 6: A BPF filter of death

```
// Code extracted from bpf_validate. Rejects
// filter if opcode's memory offset is more than
// BPF_MEMWORDS.
// Forgets to check opcodes LDX and STX!
if((BPF_CLASS(p->code) == BPF_ST
   || (BPF_CLASS(p->code) == BPF_LD &&
       (p->code & 0xe0) == BPF_MEM))
   && p->k >= BPF_MEMWORDS )
  return 0;
...
// Code extracted from bpf_filter: pc points to current
// instruction. Both cases can overflow mem[pc->k].
  case BPF_LDX|BPF_MEM:
    X = mem[pc->k]; continue;
  ...
  case BPF_STX:
    mem[pc->k] = X; continue;
```

Figure 7: The BPF code Figure 6's filter exploits.

it to the packet filter validation routine bpf_validate, which returns 1 if it considers a filter legal. For each valid filter, we then mark a fixed-size byte array (representing a packet) as symbolic and run the filter interpreter bpf_filter on the symbolic filter with the symbolic packet, thus checking the filter against all possible data packets of that length.

This checking illustrates one of EXE's interesting features: it turns interpreters into generators of the programs they can interpret. In our example, running the BPF interpreter on a symbolic filter causes it to generate all possible filters of that length, since each branch of the interpreter will fork execution, adding a constraint corresponding to the opcode it checked.

Figure 6 shows one of the two filters EXE found that cause buffer overflows in FreeBSD's BPF. The bug can occur when the opcode of a BPF instruction is either BPF_STX or BPF_LDX | BPF_MEM. As shown in Figure 7, bpf_validate forgets to bounds check the memory offset given by these instructions, as it does for instructions with opcodes BPF_ST or BPF_LD | BPF_MEM. This missing check means these instructions can write or read arbitrary offsets off the fixed-sized buffer mem, thus crashing the kernel or allowing a trivial exploit.

Linux had a trickier example. EXE found three filters that can crash the kernel because of an arithmetic overflow in a bounds check, shown in Figure 8. As with BPF,

```
// other filters that cause this error:
//     code = (BPF_LD|BPF_B|BPF_IND)
//     code = (BPF_LD|BPF_H|BPF_IND)
s[0].code = BPF_LD|BPF_B|BPF_ABS;
s[0].k    = 0x7fffffffUL;
s[1].code = BPF_RET;
s[1].k    = 0xfffffff0UL;
```

Figure 8: A Linux filter of death

```
static inline void *
skb_header_pointer(struct sk_buff *skb,
                   int offset, int len, void *buffer) {

  int hlen = skb_headlen(skb);

  // Memory overflow. offset=s[0].k; a filter
  // can make this value very large, causing
  // offset + len to overflow, trivially passing
  // the bounds check.
  if (offset + len <= hlen)
    return skb->data + offset;
```

Figure 9: The Linux code Figure 8's filter exploits.

the offset field (k) causes the problem. Here, the code to interpret BPF_LD instructions eventually calls the function skb_header_pointer, which computes an offset into a given packet's data and returns it. This routine is passed s[0].k as the offset parameter, and values 4 or 2 as the len parameter. It extracts the size of the current message header into hlen and checks that offset + len $\leq$ hlen. However, the filter can cause offset to be very large, which means the signed addition offset + len will overflow to a small value, passing the check, but then causing that very large offset value to be added to the message data pointer. This allows attackers to easily crash the machine. This error would be hard to find with random testing. Its occurrence in highly-visible, widely-used code, demonstrates that such tricky cases can empirically withstand repeated manual inspection.

5.2 A complete server: udhcpd

We also checked udhcpd-0.9.8, a clean, well-tested user-level DHCP server. We marked its input packet as symbolic, and then modified its network read call to return a packet of at most 548 bytes. After running udhcpd long enough to generate 596 test cases, EXE detected five different memory errors: four-byte read overflows at lines 213 and 214 in dhcpd.c and three similar errors at lines 79, 94, and 99 in options.c. These errors were not found when we tested the code using random testing. EXE generated packets to trigger all of these errors, one of which is shown in Figure 10. We confirmed these errors by rerunning the concrete error packets on an uninstrumented version of udhcpd while monitoring it with valgrind, a tool that dynamically checks for some types of memory corruption and storage leaks [38].

5.3 Perl Compatible Regular Expressions

The pcre library [39] is used by several popular open-source projects, including Apache, PHP, and Postfix. For speed, pcre provides a routine pcre_compile, which compiles a pattern string into a regular expression for later use. This routine has been the target of security advisories in the past [40].

We checked this routine by marking a null-terminated pattern string as symbolic and then passing it to pcre_compile. EXE quickly found a class of issues with this routine in a recent version of pcre (6.6). The function iterates over the provided pattern twice, first to do basic error checking and to estimate how much memory to allocate for the compiled pattern, and second to do actual compilation. The bugs

Offset	Hex value
0000	0000 0000 0000 0000 0000 0000 0000 0000
0010	0000 0000 0000 0000 0000 0000 5A00 0000
....	....
00F0	2100 00F9 0000 0000 0000 0000 0000 0000
....	....
01E0	0000 0000 0000 0000 0000 0000 2734 0000
01F0	0000 0000 0000 0000 0000 0000 0000 0000
0200	0000 0000 0000 0000 0000 0000 0000 3500
0210	030F 0000 0000 0000 0000 0000 0000 0000
0220	0032 0036

Figure 10: An EXE generated packet that causes an out-of-bounds read in udhcpd.

```
[^[\0^\0]\*-?]{\0              [\-\'[\0^\0]\']{\0
[\*-\'[\0^\0]\'-?]{\0          [\*-\'[\0^\0]\'-?]\0
[\*-\'[\0^\0]\'-?]\0           [\-\'[\0^\0]\'-]\0
(?#)\?[[[\0\0]\-]{\0           (?#)\?[[[\0\0]\-]\0
(?#)\?[[[\0\0]\[]\0            (?#)\?[:[[\0\0]\-]\0
(?#)\?[[[\0\0]\-]\0            (?#)\?[[[\0\0]\]\0
(?#)\?[[[\0\0][\0^\0]]\0       (?#)\?[[[\0\0][\0^\0]-]\0
(?#)\?[[[\0\0][\0^\0]\]\0      (?#)\?[=[[\0\0][\0^\0]\?]\0
```

Figure 11: EXE-generated regular expression patterns that cause out-of-bounds writes (leading to aborts in glibc on free) when passed as the first argument to pcre_compile.

found included overflowing reads in the check_posix_syntax helper function (pcre_compile.c:1361-1363), called during the first pass, as well as more dangerous overflowing reads and writes in the compile_regex and compile_branch helpers (illegal writes on pcre_compile.c lines 3400-3401 and 3515-3616), which are called during the compilation pass. While the first problem may appear to be an innocent read past the end of the buffer, it allows illegal expressions to enter the second pass, causing more serious issues. The substring "[\0^\0]" is especially dangerous because strings which end with this sequence will cause pcre to skip over both null characters and continue parsing unallocated or uninitialized memory. Figure 11 show a representative sample of EXE-generated patterns that trigger overflows in pcre, which in turn cause glibc aborts. The author of the library fixed the bug soon after being notified, and so the latest version of pcre as of this writing (6.7) does not exhibit this problem.

5.4 Generating disks of death

We previously used EXE to generate disk images for three file systems (ext2, ext3, and JFS) that when mounted would crash or compromise the Linux kernel [46]. At a high level, the checking worked as follows. We wrote a special device driver that returned symbolic blocks to its callers. We then compiled Linux using EXE and ran it as a user-level process (so fork would work) and invoked the mount system call, which caused the file system to read symbolic blocks, thereby driving checking.

We found bugs in all three file systems, demonstrating that EXE can handle complex systems code. Further, these errors would almost certainly be beyond the reach of random testing. For example, the Linux ext2 "read super block" routine has over forty if-statements to check the data associated with the super block. Any randomly-generated super block must satisfy these tests before it can reach even the next level of error checking, much less triggering the execution of "real code" that performs actual file system operations.

6. RELATED WORK

We described an initial, primitive version of EXE (then called EGT) in an invited workshop paper [13]. EGT did not support reads or writes of symbolic pointer expressions, symbolic arrays, bit-fields, casting, sign-extension, arithmetic overflow, and our symbolic checks. We also gave an overview of EXE in the file system checking paper [46] discussed in Section 5.4. That paper took EXE as a given and used it to find bugs. In contrast, both STP and EXE are contributions of this paper, which we describe in more detail as well as focus on a broader set of applications.

Simultaneously with our initial work [13], DART [27] also generated test cases from symbolic inputs. DART runs the tested unit code on random input and symbolically gathers constraints at decision points that use input values. Then, DART negates one of these symbolic constraints to generate the next test case. DART only handles integer constraints and devolves to random testing when pointer constraints are used, with the usual problems of missed paths.

The CUTE project [42] extends DART by tracking symbolic pointer constraints of the form: p = NULL, p ≠ NULL, p = q, or p ≠ q. In addition, CUTE tracks constraints formed by reading or writing symbolic memory at constant offsets (such as a field dereference p→field), but unlike EXE it cannot handle symbolic offsets. For example, the paper on CUTE shows that on the code snippet a[i] = 0; a[j] = 1; if (a[i] == 0) ERROR, CUTE fails to find the case when i equals j, which would have driven the code down both paths. In contrast to both DART and CUTE, EXE has completely accurate constraints on memory, and thus can (potentially) check code much more thoroughly.

CBMC is a bounded model checker for ANSI-C programs [14] designed to cross-check an ANSI C re-implementation of a circuit against its Verilog implementation. Unlike EXE, which uses a mixture of concrete and symbolic execution, CBMC runs code entirely symbolically. It takes (and requires) an entire, strictly-conforming ANSI C program, which it translates into constraints that are passed to a SAT solver. CBMC provides full support for C arithmetic and control operations, as well as reads and writes of symbolic memory. However, it has several serious limitations. First, it has a strongly-typed view of memory, which prevents it from checking code that accesses memory through pointers of different types. Second, because CBMC must translate the entire program to SAT, it can only check stand-alone programs that do not interact with the environment (e.g., by using system calls or even calling code for which there is no source). Both of these limits seem to prevent CBMC from checking the applications in this paper. Finally, CBMC unrolls all loops and recursive calls, which means that it may miss bugs that EXE can find and also that it may execute some symbolic loops more times than the current set of constraints allows.

Larson and Todd [34] present a system that dynamically tracks primitive constraints associated with "tainted" data (e.g., data that comes from untrusted sources such as network packets) and warns when the data could be used in a potentially dangerous way. They associate tainted integers with an upper and lower bound and tainted strings with their maximum length and whether the string is null-terminated. At potentially dangerous uses of inputs, such as array references or calls to the string library, they check whether the integer could be out of bounds, or if the string could violate the library function's contract. Thus, as EXE, this system can detect an error even if it did not actually occur during the program's concrete execution. However, their system lacks almost all of the symbolic power that EXE provides. Further, they cannot generate inputs to cause paths to be executed; users must provide test cases and they can only check paths covered by these test cases.

Static checking and static input generation. There has been much recent work on static bug finding, including better type systems [20, 25, 23], static analysis tools [25, 5, 18, 19, 24, 12, 43], and statically solving constraints to generate inputs that would cause execution to reach a specific program point or path [8, 28, 2, 4, 10]. The insides of these tools look dramatically different from EXE. An exception is Saturn [44], which expresses program properties as boolean constraints and models pointers and heap data down to the bit level. Dynamic analysis requires running code, static analysis does not. Thus, static tools often take less work to apply (just compile the source and skip what cannot be handled), can check all paths (rather than only executed ones), and can find bugs in code it cannot run (such as operating systems code). However, because EXE runs code, it can check much deeper properties, such as complex expressions in assertions, or properties that depend on accurate value information (the exact value of an index or size of an object), pointers, and heap layout, among many others. Further, unlike static analysis, EXE has no false positives. However, we view the two approaches as complementary: there is no reason not to use lightweight static techniques and then use EXE.

Software Model Checking. Model checkers have been used to find bugs in both the design and the implementation of software [31, 32, 9, 16, 5, 26, 47]. These approaches often require a lot of manual effort to build test harnesses. However, to some degree, the approaches are complementary to EXE: the tests EXE generates could be used to drive the model checked code, similar to the approach embraced by the Java PathFinder (JPF) project [33]. JPF combines model checking and symbolic execution to check applications that manipulate complex data structures written in Java. JPF differs from EXE in that it does not have support for untyped memory (not needed because Java is a strongly typed language) and does not support symbolic pointers.

Dynamic techniques for test and input generation. Past dynamic input generation work seem to focus on generating an input to follow a specific path, motivated by the problem of answering programmer queries as to whether control can reach a specific statement or not [22, 29]. EXE instead focuses on bug finding, in particular the problems of exhausting all input-controlled paths and universal checking, neither addressed by prior work.

7. CONCLUSION

We have presented EXE, which uses robust, bit-level accurate symbolic execution to find deep errors in code and automatically generate inputs that will hit these errors. A key aspect of EXE is its modeling of memory and its co-designed, fast constraint solver STP. We have applied EXE to a variety of real, tested programs where it was powerful enough to uncover subtle and surprising bugs.

Acknowledgments

We would like to thank Paul Twohey for his work on the regression suite, Martin Casado for providing us tcpdump in an easy to check form, and Suhabe Bugrara, Ted Kremenek, Darko Marinov, Adam Oliner, Ben Pfaff, and Paul Twohey for their valuable comments.

This research was supported by National Science Foundation (NSF) CAREER award CNS-0238570-001, Department of Homeland Security (DHS) grant FA8750-05-2-0142, NSF grant CCR-0121403, and a Junglee Corporation Stanford Graduate Fellowship.

8. REFERENCES

[1] SMTLIB competition. http://www.csl.sri.com/users/demoura/smt-comp, August 2006.

[2] T. Ball. A theory of predicate-complete test coverage and generation. In *Proceedings of the Third International Symposium on Formal Methods for Components and Objects*, Nov. 2004.

[3] T. Ball and R. B. Jones, editors. *Computer Aided Verification, 18th International Conference, CAV 2006, Seattle, WA, USA, August 17-20, 2006, Proceedings*, volume 4144 of *Lecture Notes in Computer Science*. Springer, 2006.

[4] T. Ball, R. Majumdar, T. Millstein, and S. K. Rajamani. Automatic predicate abstraction of C programs. In *PLDI '01: Proceedings of the ACM SIGPLAN 2001 conference on Programming language design and implementation*, pages 203–213. ACM Press, 2001.

[5] T. Ball and S. Rajamani. Automatically validating temporal safety properties of interfaces. In *SPIN 2001 Workshop on Model Checking of Software*, May 2001.

[6] C. Barrett and S. Berezin. CVC Lite: A new implementation of the cooperating validity checker. In R. Alur and D. A. Peled, editors, *CAV*, Lecture Notes in Computer Science. Springer, 2004.

[7] C. Barrett, S. Berezin, I. Shikanian, M. Chechik, A. Gurfinkel, and D. L. Dill. A practical approach to partial functions in CVC Lite. In *PDPAR'04 Workshop, Cork, Ireland*, July 2004.

[8] R. S. Boyer, B. Elspas, and K. N. Levitt. Select – a formal system for testing and debugging programs by symbolic execution. *ACM SIGPLAN Notices*, 10(6):234–45, June 1975.

[9] G. Brat, K. Havelund, S. Park, and W. Visser. Model checking programs. In *IEEE International Conference on Automated Software Engineering (ASE)*, 2000.

[10] D. Brumley, J. Newsome, D. Song, H. Wang, and S. Jha. Towards automatic generation of vulnerability-based signatures. In *Proceedings of the 2006 IEEE Symposium on Security and Privacy*, 2006.

[11] R. E. Bryant, S. K. Lahiri, and S. A. Seshia. Modeling and verifying systems using a logic of counter arithmetic with lambda expressions and uninterpreted functions. In E. Brinksma and K. G. Larsen, editors, *Proc. Computer-Aided Verification (CAV)*, pages 78–92. Springer-Verlaag, July 2002.

[12] W. Bush, J. Pincus, and D. Sielaff. A static analyzer for finding dynamic programming errors. *Software: Practice and Experience*, 30(7):775–802, 2000.

[13] C. Cadar and D. Engler. Execution generated test cases: How to make systems code crash itself. In *Proceedings of the 12th International SPIN Workshop on Model Checking of Software*, August 2005. A longer version of this paper appeared as Technical Report CSTR-2005-04, Computer Systems Laboratory, Stanford University.

[14] E. Clarke and D. Kroening. Hardware verification using ANSI-C programs as a reference. In *Proceedings of ASP-DAC 2003*, pages 308–311. IEEE Computer Society Press, January 2003.

[15] B. Cook, D. Kroening, and N. Sharygina. Cogent: Accurate theorem proving for program verification. In K. Etessami and S. K. Rajamani, editors, *Proceedings of CAV 2005*, volume 3576 of *Lecture Notes in Computer Science*, pages 296–300. Springer Verlag, 2005.

[16] J. Corbett, M. Dwyer, J. Hatcliff, S. Laubach, C. Pasareanu, Robby, and H. Zheng. Bandera: Extracting finite-state models from Java source code. In *ICSE 2000*, 2000.

[17] T. H. Cormen, C. E. Leiserson, R. L. Rivest, and C. Stein. *Introduction to Algorithms*. The MIT Electrical Engineering and Computer Science Series. MIT Press/McGraw Hill, 2001.

[18] SWAT: the Coverity software analysis toolset. http://coverity.com.

[19] M. Das, S. Lerner, and M. Seigle. Path-sensitive program verification in polynomial time. In *Proceedings of the ACM SIGPLAN 2002 Conference on Programming Language Design and Implementation*, Berlin, Germany, June 2002.

[20] R. DeLine and M. Fähndrich. Enforcing high-level protocols in low-level software. In *Proceedings of the ACM SIGPLAN 2001 Conference on Programming Language Design and Implementation*, June 2001.

[21] N. Een and N. Sorensson. An extensible SAT-solver. In *Proc. of the Sixth International Conference on Theory and Applications of Satisfiability Testing*, pages 78–92, May 2003.

[22] R. Ferguson and B. Korel. The chaining approach for software test data generation. *ACM Trans. Softw. Eng. Methodol.*, 5(1):63–86, 1996.

[23] C. Flanagan and S. N. Freund. Type-based race detection for Java. In *SIGPLAN Conference on Programming Language Design and Implementation*, pages 219–232, 2000.

[24] C. Flanagan, K. Leino, M. Lillibridge, G. Nelson, J. Saxe, and R. Stata. Extended static checking for Java. In *Proceedings of the ACM SIGPLAN 2002 Conference on Programming Language Design and Implementation*. ACM Press, 2002.

[25] J. Foster, T. Terauchi, and A. Aiken. Flow-sensitive type qualifiers. In *Proceedings of the ACM SIGPLAN 2002 Conference on Programming Language Design and Implementation*, June 2002.

[26] P. Godefroid. Model Checking for Programming Languages using VeriSoft. In *Proceedings of the 24th ACM Symposium on Principles of Programming Languages*, 1997.

[27] P. Godefroid, N. Klarlund, and K. Sen. DART: Directed automated random testing. In *Proceedings of the Conference on Programming Language Design and Implementation (PLDI)*, Chicago, IL USA, June 2005. ACM Press.

[28] A. Gotlieb, B. Botella, and M. Rueher. Automatic test data generation using constraint solving techniques. In *ISSTA '98: Proceedings of the 1998 ACM SIGSOFT international symposium on Software testing and analysis*, pages 53–62. ACM Press, 1998.

[29] N. Gupta, A. P. Mathur, and M. L. Soffa. Automated test data generation using an iterative relaxation method. In *SIGSOFT '98/FSE-6: Proceedings of the 6th ACM SIGSOFT International Symposium on Foundations of Software Engineering*, pages 231–244. ACM Press, 1998.

[30] R. Hastings and B. Joyce. Purify: Fast detection of memory leaks and access errors. In *Proceedings of the Winter USENIX Conference*, Dec. 1992.

[31] G. J. Holzmann. The model checker SPIN. *Software Engineering*, 23(5):279–295, 1997.

[32] G. J. Holzmann. From code to models. In *Proc. 2nd Int. Conf. on Applications of Concurrency to System Design*, pages 3–10, Newcastle upon Tyne, U.K., 2001.

[33] S. Khurshid, C. S. Pasareanu, and W. Visser. Generalized symbolic execution for model checking and testing. In *Proceedings of the Ninth International Conference on Tools and Algorithms for the Construction and Analysis of Systems*, 2003.

[34] E. Larson and T. Austin. High coverage detection of input-related security faults. In *Proceedings of the 12th USENIX Security Symposium, August 2003*.

[35] B. P. Miller, L. Fredriksen, and B. So. An empirical study of the reliability of UNIX utilities. *Communications of the Association for Computing Machinery*, 33(12):32–44, 1990.

[36] G. C. Necula, S. McPeak, S. Rahul, and W. Weimer. CIL: Intermediate language and tools for analysis and transformation of C programs. In *International Conference on Compiler Construction*, March 2002.

[37] G. Nelson and D. Oppen. Simplification by cooperating decision procedures. *ACM Transactions on Programming Languages and Systems*, 1(2):245–57, 1979.

[38] N. Nethercote and J. Seward. Valgrind: A program supervision framework. *Electronic Notes in Theoretical Computer Science*, 89(2), 2003.

[39] PCRE - Perl Compatible Regular Expressions. http://www.pcre.org/.

[40] PCRE Regular Expression Heap Overflow. US-CERT Cyber Security Bulletin SB05-334. http://www.us-cert.gov/cas/bulletins/SB05-334.html#pcre.

[41] O. Ruwase and M. S. Lam. A practical dynamic buffer overflow detector. In *Proceedings of the 11th Annual Network and Distributed System Security Symposium*, pages 159–169, 2004.

[42] K. Sen, D. Marinov, and G. Agha. CUTE: A concolic unit testing engine for C. In *In 5th joint meeting of the European Software Engineering Conference and ACM SIGSOFT Symposium on the Foundations of Software Engineering (ESEC/FSE'05)*, Sept. 2005.

[43] D. Wagner, J. Foster, E. Brewer, and A. Aiken. A first step towards automated detection of buffer overrun vulnerabilities. In *The 2000 Network and Distributed Systems Security Conference. San Diego, CA*, Feb. 2000.

[44] Y. Xie and A. Aiken. Scalable error detection using boolean satisfiability. In *Proceedings of the 32nd Annual Symposium on Principles of Programming Languages (POPL 2005)*, January 2005.

[45] Y. Xie and A. Aiken. Saturn: A SAT-based tool for bug detection. In K. Etessami and S. K. Rajamani, editors, *CAV*, volume 3576 of *Lecture Notes in Computer Science*, pages 139–143. Springer, 2005.

[46] J. Yang, C. Sar, P. Twohey, C. Cadar, and D. Engler. Automatically generating malicious disks using symbolic execution. In *Proceedings of the 2006 IEEE Symposium on Security and Privacy*, May 2006.

[47] J. Yang, P. Twohey, D. Engler, and M. Musuvathi. Using model checking to find serious file system errors. In *Symposium on Operating Systems Design and Implementation*, December 2004.

A Scalable Approach to Attack Graph Generation

Xinming Ou *
Purdue University
xou@cerias.purdue.edu

Wayne F. Boyer
Idaho National Laboratory
Wayne.Boyer@inl.gov

Miles A. McQueen
Idaho National Laboratory
Miles.McQueen@inl.gov

ABSTRACT

Attack graphs are important tools for analyzing security vulnerabilities in enterprise networks. Previous work on attack graphs has not provided an account of the scalability of the graph generating process, and there is often a lack of logical formalism in the representation of attack graphs, which results in the attack graph being difficult to use and understand by human beings. Pioneer work by Sheyner, et al. is the first attack-graph tool based on formal logical techniques, namely model-checking. However, when applied to moderate-sized networks, Sheyner's tool encountered a significant exponential explosion problem. This paper describes a new approach to represent and generate attack graphs. We propose logical attack graphs, which directly illustrate logical dependencies among attack goals and configuration information. A logical attack graph always has size polynomial to the network being analyzed. Our attack graph generation tool builds upon MulVAL, a network security analyzer based on logical programming. We demonstrate how to produce a derivation trace in the MulVAL logic-programming engine, and how to use the trace to generate a logical attack graph in quadratic time. We show experimental evidence that our logical attack graph generation algorithm is very efficient. We have generated logical attack graphs for fully connected networks of 1000 machines using a Pentium 4 CPU with 1GB of RAM.

Categories and Subject Descriptors

C.2.0 [**Computer-Communication Networks**]: General; K.6.5 [**Management of Computing and Information Systems**]: Security and Protection

General Terms

Security, Management

---

*As of August 14, 2006, Xinming Ou's affiliation is Kansas State University. This work was conducted when he was a post-doctoral research associate at Purdue University. We would like to thank Purdue University and CERIAS for supporting his work.

Copyright 2006 Association for Computing Machinery. ACM acknowledges that this contribution was authored or co-authored by an employee, contractor or affiliate of the U.S. Government. As such, the Government retains a nonexclusive, royalty-free right to publish or reproduce this article, or to allow others to do so, for Government purposes only.
CCS'06, October 30–November 3, 2006, Alexandria, Virginia, USA.
Copyright 2006 ACM 1-59593-518-5/06/0010 ...$5.00.

Keywords

Attack graphs, enterprise network security, logic-programming

1. INTRODUCTION

When analyzing the security of an enterprise network, it is important to consider multi-stage, multi-host attacks. A determined attacker is not likely to stop at the machine he first compromises, but can be expected to try to penetrate deeper into the network by jumping from one machine to another. For this reason, configuring an enterprise network securely is a daunting task for human beings. There are many potential interactions among multiple hosts and components in a network, such that the configuration of one machine will affect the security of others in the network. It is therefore important to design automatic tools that can analyze the configuration of an enterprise network and find potential security vulnerabilities. Such a tool will not be very useful if it cannot inform a system administrator with detailed information about the discovered problems. In particular, an *attack graph* that illustrates all possible multi-stage, multi-host attack paths is crucial for a system administrator to understand the nature of the threats and decide upon appropriate countermeasures.

Various kinds of attack graphs have been proposed for analyzing network security [8, 11, 1, 5, 2, 13]. Although some of them addressed the scalability problem [1, 5], none of the works has shown solid evidence that the graph generation tool can scale to an enterprise network with realistic sizes. In practice it is desirable to compute attack graphs for enterprise networks with 1000 to 10,000 hosts. Lippmann and Ingols gave a good overview on various attack graph tools in the past [3]. It shows that "although research has made significant progress in the past few years, no system has analyzed networks with more than 20 hosts, and computation for most approaches scales poorly and would be impractical for networks with more than even a few hundred hosts."

Besides the scalability problem, many of the existing attack graph tools adopt an ad-hoc way to represent input information and output graph data structures. The graph generation tools often required various auxiliary inputs in custom-designed data format, and the resulting attack graphs are often hard to comprehend and use by a human. These have made those attack graph tools difficult to use in practice.

The work by Sheyner, et al. [11] is the first formal treatment of attack graphs. Sheyner uses model checking to compute multi-stage, multi-host attack paths in a network. The state of the network is formally modeled as a collection of Boolean variables, representing configuration parameters and attacker's privileges. Attacker's actions are modeled as state-transition relations. The security property of the network is specified as a temporal formula, which can be automatically checked against the model by a model checker. Unlike a traditional model checker, which only outputs one counter example when the temporal formula is not satisfied, Sheyner's tool can output all counter examples in the form of a *scenario graph*. In the case of network security, the scenario graph is an attack graph illustrating all the multi-stage, multi-host attack paths that can potentially break a network's security property.

A formal, logic-based approach to attack graph generation, like the one by Sheyner, is advantageous compared to ad-hoc graph generation methods. Using mature logical techniques is less error-prone than custom-designed algorithms, especially for the complex problem of security analysis. A clear logical semantics for attack graphs also makes it easier to conduct further analysis based on the graph data structure. However, a logic-based attack graph tool must scale well with the size of the network to make it feasible to use in practice. When we tried to apply Sheyner's tool to analyze real world networks we found that the graph generation time and graph size were prohibitively large. For example, a network of only 10 hosts with 5 vulnerabilities per host takes about 15 minutes to generate and results in a graph of 10 million edges.

We observe that in Sheyner's attack graph, every node is a collection of Boolean variables encoding the *entire* network state at an attack stage. While the number of variables is polynomial in the size of the network, the possible number of states is exponential. Even though not all the states are reachable in the search, the potential state explosion associated with a model-checking based methodology makes it impractical except for small networks with very few vulnerabilities.

In this paper, we propose a new logic-based approach for representing and generating attack graphs. In our representation, a node in the graph is a logical statement. This logical statement does not encode the entire state of the network, but only some aspect of it. In some sense it can be viewed as one Boolean variable in the nodes of Sheyner's graph. The edges in the graph specify the causality relations between network configurations and an attacker's potential privileges. Intuitively, Sheyner's attack graph illustrates snapshots of attack steps, or "how the attack can happen", whereas our attack graph illustrates causes of the attacks, or "why the attack can happen". To differentiate these two kinds of attack graphs, we call the former "scenario attack graphs", and the latter "logical attack graphs".

One advantage of a logical attack graph is that it clearly specifies the causality relations between system configuration information and an attacker's potential privileges. In a scenario attack graph, one would have to delve into the Boolean variables and follow several steps upstream to identify what really causes the adverse situation that enables an attacker's action at a stage. In a logical attack graph, such causality would be specific as graph edges. From a logical attack graph, it is also possible to enumerate all possible attack scenarios by depth-first traversing. Most importantly, the size of a logical attack graph is always polynomial in the size of the network, whereas in the worst case a scenario attack graph's size could be exponential.

In the past people have proposed "exploit dependency graph" as a way to represent computer attacks [5, 2]. In an exploit dependency graph, the pre- and post-conditions for exploits are encoded as graph nodes and edges. Semantically this is equivalent to our logical attack graph. Our contribution is to formally represent such dependency relations in the form of logic, and generate attack graphs through automatic logic deduction, as opposed to a custom-designed graph search algorithm. We believe the logic-based approach has the advantage of better clarity and trustworthiness.

Our logical attack graphs require that the cause of an attacker's potential privileges be expressible as a propositional formula in terms of network configuration parameters. Attack conditions that cannot be expressed in propositional formulas cannot be captured by logical attack graphs. In our experience, we have not found any situations where this becomes a problem. After all, all computer attacks have a cause, and unsurprisingly those causes are almost always rooted in misconfigurations. For a security tool that aims at finding such misconfigurations, the semantics of logical attack graphs exactly suits that goal.

1.1 Security analysis based on logic programming

The work presented in this paper is based on the MulVAL project [6]. MulVAL is a reasoning system for automatically identifying security vulnerabilities in enterprise networks. The key idea behind MulVAL is that most configuration information can be represented as Datalog[1] tuples, and most attack techniques and OS security semantics can be specified using Datalog rules. Following is a Datalog rule for the remote exploit of a privilege-escalation vulnerability in a service program.

```
execCode(Attacker, Host, User) :-
  networkService(Host, Program,
                 Protocol, Port, User),
  vulExists(Host, VulID, Program,
            remoteExploit, privEscalation),
  netAccess(Attacker, Host, Protocol, Port).
```

Figure 1: **An interaction rule for remote exploit.**

Capitalized identifiers are variables in Datalog and can be instantiated with any concrete term during evaluation. This is a generic rule specifying the pre- and post-condition for this attack: if `Program` is running as `User` on `Host` as a service listening on `Protocol` and `Port`, it contains a remotely exploitable vulnerability whose impact is privilege escalation, and the attacker can access the service through the network, then the attacker can execute arbitrary code on the machine as `User`. Logically, ":-" can be replaced with a reversed "implies" connector. The rule can be viewed as a logical formula $\forall \Phi.Rule$, where Φ contains all the Prolog variables appearing in the rule.

Predicates such as `vulExists` and `networkService` are "primitive" and they represent configuration information reported by host and network scanners. Predicates such as `execCode` and `netAccess` are "derived" and they are computed from the configuration information by iteratively applying the interaction rules on the input. The architecture

---
[1] A syntactic subset of Prolog

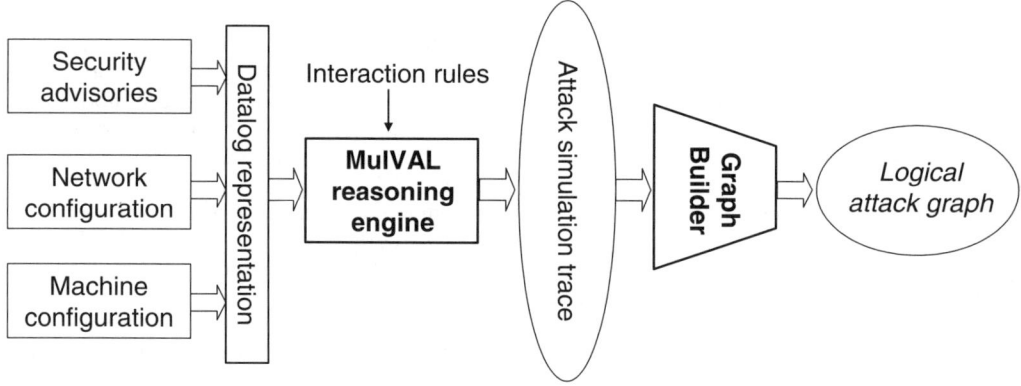

Figure 2: Logical Attack Graph Generator

of the Logical Attack Graph tool is illustrated in Figure 2. The MulVAL reasoning engine uses XSB [9], a Prolog system developed by StonyBrook, to evaluate the Datalog interaction rules on input facts. We modified the MulVAL engine so that the trace of the evaluation is recorded and sent to the graph builder, where the logical attack graph is output.

Our logical attack graph is in essence a derivation graph for Datalog programs. XSB, like other Prolog systems, does not provide functionalities for generating proofs of a successful query. The justifier program for XSB [7] can record execution trace for interactively retrieving reasons for a successful Prolog query, but does not provide the capability of constructing derivation graphs that illustrate all possible derivations. We use a similar approach to XSB's justifier program to record derivation steps when evaluating the MulVAL Datalog program. Those derivation steps are called *attack simulation trace*, which contains sufficient information to construct a logical attack graph.

In the following sections, we will review some of the related works. We will then give a formal definition of logical attack graphs, describe the algorithms for constructing them, and analyze the complexity of the algorithms. We have implemented the algorithm in a combination of Prolog and C++. Experimental results show that our logical attack graph tool is very efficient and can handle networks with thousands of machines.

2. RELATED WORK

Sheyner *et al.* uses model checking techniques to compute attack graphs [11]. We encountered significant scalability problems in applying this tool. One reason for the blow up is that there are many duplicate attack paths in the graph that differ only in the order in which independent attack steps are attempted. Partial-order reduction can remove such duplicate paths, but it has not been shown that the technique can significantly improve the scalability for attack graphs. Even after removing such duplicate paths, the resulting graphs could still be exponential. We also find it is hard to decode the meaning of the Boolean values in a node, and logical correlation among nodes is not always obvious.

Philips and Swiler developed a tool for generating attack graphs [8, 12] in 1998. Like the model checking approach, the nodes in their attack graphs represent the state of the network in the form of a collection of variables, and the edges represent an attacker's actions that change the state. Instead of using a model checker, Philips and Swiler developed a customized search engine to conduct the analysis. Like Sheyner's work, this state-based attack graph representation has inherent exponential problems, and such explosion was indeed reported by the authors. They hence used a technique similar to partial-order reduction to eliminate the duplicate attack paths that contributed to the explosion, but it is not clear from the paper how effective this method has been and no performance data was given.

Ammann, *et al.* also noticed the scalability problem in the model checking-based attack graph tool, and proposed a graph search-based algorithm, which was then used in the Topological Vulnerability Analysis tool [2]. They pointed out that for most computer attacks, one can assume the *monotonicity property*, where an attacker does not decrease his ability by launching attacks, and hence does not need to relinquish privileges he already gained. Under this assumption, an attacker's privileges always increase during the analysis. Since there are only a polynomial number of privileges an attacker can gain, the analysis algorithm will terminate in polynomial time. Our logical attack graph gives another perspective for this monotonicity property. We observe that most attacks, whether monotonic or non-monotonic, have rooted causes in configuration information. Thus, at an appropriate level all those attacks' preconditions can be specified using propositional formulas on configuration information. In some sense non-monotonic attacks can be treated as monotonic if one ignores the low-level details on how the attack can happen. For this reason simple Datalog rules can capture almost all kinds of attack conditions in a network. Ammann gave a theoretical upper bound for their algorithm as $O(|A|^2 \cdot |E|)$, where $|A|$ is the number of "attributes" (describing attack pre- and post-conditions) and $|E|$ is the number of "exploits". The paper stated that typically an exploit involves two hosts, yielding a quadratic number of concrete exploits. The paper did not discuss the number of attributes in terms of network size. We believe the attributes should include host connectivity information, thus the number of attributes is also quadratic in the number of hosts. This will give us an $O(N^6)$ complexity. It is certainly a very conservative estimation of the algorithm's complexity, but we could not find experimental data showing the performance of the proposed algorithm on large configuration settings. In this paper we will show an algorithm that has $O(N^2)$ complexity

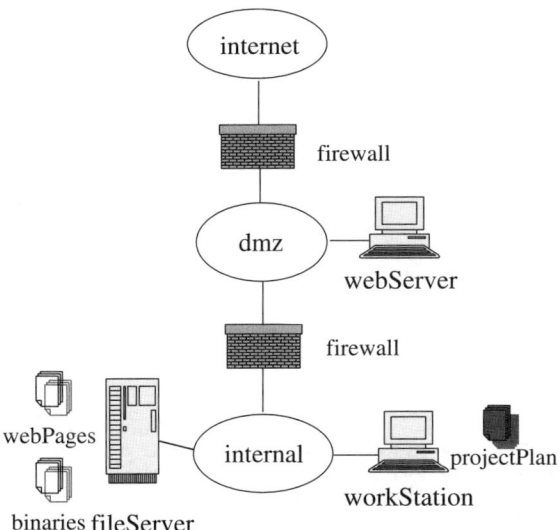

Figure 3: Example

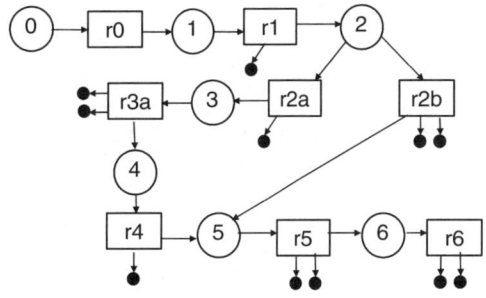

Figure 4: An example logical attack graph

under the assumption of constant table look-up time, and experimental results that demonstrate the worst-case running time of our graph generator grows between $O(N^2)$ and $O(N^3)$ for networks from a handful to a thousand machines.

Noel and Jajodia gave a good review on various representations of attack graphs in their work on using aggregation techniques to manage attack graph complexity [4]. Their work finally chose the "exploit-dependency graph". An exploit-dependency graph can be viewed as a logical attack graph and the two can be translated back and forth. However, we believe an attack-graph that explicitly uses predicates and logical connectives to represent security correlations in a network is better suited for rigorous security analysis and hardening. Noel's work focuses on aggregation techniques for exploit dependency graphs, and does not describe how the attack graphs can be built from configuration information or the scalability of the graph building process.

Schneier introduced the general idea of *attack trees* for representing security threats [10]. The logical attack graph presented in this paper is a special case of Schneier's attack tree. We apply the idea to the specific problem of enterprise network security, formally define the semantics of the attack graph in this context, and describe algorithms for automatically computing attack graphs from network and machine configuration information. We also show experimental evidence on the scalability of our approach.

Our logical attack graph toolkit was based on the MulVAL toolkit. The original MulVAL work [6] did not have the ability to compute complete attack graphs. Rather, separate attack paths can be output by using Prolog's meta-programming techniques. Even for a polynomial attack simulation process, the number of unique attack paths could be exponential in the worst case. And we indeed experienced such exponential blow-up when applying MulVAL's meta-programming based attack path generator to analyze a real network with 20 machines. The logical attack graph tool described in this paper has the ability of generating complete attack graphs for networks with thousands of machines.

3. LOGICAL ATTACK GRAPH

The example network in Figure 3 is directly borrowed from the MuLVAL paper [6]. Suppose the following potential attack paths are discovered after analyzing the configuration.

An attacker first compromises webServer by remotely exploiting vulnerability CVE-2002-0392 to get local access on the server. Since webServer is allowed to access fileServer through the NFS protocol, he can then try to modify data on the file server. There are two ways to achieve this. If there are vulnerabilities in the NFS service daemons, he can try to exploit them and get local access on the machine; or if the NFS export table is not set up appropriately, he can modify files on the server through the NFS protocol by using programs like NFS Shell[2]. Once he can modify files on the file server, the attacker can install a Trojan-horse program in the executable binaries on fileServer that are mounted by machine workStation. The attacker can now wait for an innocent user on workStation to execute it and obtain control on the machine.

The logical attack graph corresponding to the above scenarios is illustrated in Figure 4. A logical attack graph is a directed graph and can be represented in the form of a tree with possible cross links between nodes. Figure 5 shows the tree representation of the same attack graph. There are two kinds of nodes in the graph: a *derivation node* and a *fact node*. A derivation node is represented as a rectangle and a fact node is represented as a circle. There are also two kinds of fact nodes: a primitive fact node (represented as a solid small circle), and a derived fact node (represented as a circle with a number in it).

Every fact node in a logical graph is labeled with a logical statement in the form of a predicate applied to its arguments. The root node is the attack goal; in the example it is execCode(attacker,workStation,root), meaning "the attacker can execute arbitrary code as user root on machine workStation". Every derivation node is labeled with an interaction rule that is used for the derivation step. In the tree representation, every internal node is started with a node number in a bracket, followed by the node's label. A leaf node does not have a node number and is led by an empty square bracket.

The edges in the graph represent the "depends on" relation. A fact node is dependent on one or more derivation nodes, each of which represents an application of an inter-

---
[2]Downloadable from
http://www.deter.com/unix/software/nfsshell.c

```
<0>|--execCode(attacker,workStation,root)
   <r0>Rule5: Trojan horse installation
      <1>|--accessFile(attacker,workStation,write,/usr/local/share)
         <r1>Rule14: NFS semantics
            []-nfsMounted(workStation,/usr/local/share,fileServer,/export,read)
         <2>||--accessFile(attacker,fileServer,write,/export)
            <r2a>Rule10: execCode implies file access
               []-fileSystemACL(fileServer,root,write,/export)
            <3>|--execCode(attacker,fileServer,root)
               <r3>Rule3: remote exploit of a server program
                  []-networkServiceInfo(fileServer,mountd,rpc,100005,root)
                  []-vulExists(fileServer,CVE-2003-0252,mountd,
                              remoteExploit,privEscalation)
               <4>|--netAccess(attacker,fileServer,rpc,100005)
                  <r4>Rule6: multi-hop access
                     []-hacl(webServer,fileServer,rpc,100005)
                  <5>|--execCode(attacker,webServer,apache)
                     <r5>Rule3: remote exploit of a server program
                        []-networkServiceInfo(webServer,httpd,tcp,80,apache)
                        []-vulExists(webServer,CAN-2002-0392,httpd,
                                     remoteExploit,privEscalation)
                     <6>|--netAccess(attacker,webServer,tcp,80)
                        <r6>Rule7: direct network access
                           []-hacl(internet,webServer,tcp,80)
                           []-located(attacker,internet)
            <r2b>Rule15: NFS shell
               []-hacl(webServer,fileServer,rpc,100003)
               []-nfsExportInfo(fileServer,/export,write,webServer)
               |--execCode(attacker,webServer,apache)==>  <5>
```

Figure 5: An example logical attack graph, tree representation

action rule that yields the fact; A derivation node is dependent on one or more fact nodes, which together satisfy the preconditions of the rule. Thus a logical attack graph is a bipartite directed graph. The derivation nodes serve as a medium between a fact and its "reasons", i.e., how the fact becomes true. Since a fact may have different ways to become true, the derivation nodes directed from a fact node form a disjunction. A derivation node represents a successful application of an interaction rule, where all its preconditions are satisfied by its children. Thus the fact nodes directed from a derivation node form a conjunction.

For example, node 2 has two derivation nodes as its children: r2a and r2b (note that the tree representation uses || to signify that a fact node has more than one derivation). That is, there are two ways the attacker can modify files on fileServer. One way is to get root on file server by exploiting bug CVE-2003-0252 in the mountd program, and the other is to use the NFS Shell program. Both depend on the condition that an attacker already gained some access on webServer (node 5). In the tree representation, there is a cross link (==> <5>) pointing to node 5 in the second derivation branch.

A logical attack graph can be viewed as a derivation graph for a successful Datalog query. There may be many different ways to derive a fact in Datalog (corresponding to multiple paths to break into a network), thus we explicitly introduced the derivation node to represent one possible derivation step. Logically, a derivation node is an "and" node, where all its children are the arguments of a conjunction that derives the node; a derived fact node is an "or" node, where all its children represent different ways to derive them. A primitive fact node is a leaf node in the graph. It represents a piece of configuration information. Following is the formal definition of our logical attack graph.

Definition 1. $(N_r, N_p, N_d, E, \mathcal{L}, \mathcal{G})$ is a logical attack graph, where N_r, N_p and N_f are three sets of disjoint nodes in the graph, $E \subset (N_r \times (N_p \cup N_d)) \cup (N_d \times N_r)$, $\mathcal{L}$ is a mapping from a node to its label, and $\mathcal{G} \in N_d$ is the attacker's goal.

N_r, N_p and N_d are the sets of derivation nodes, primitive fact nodes, and derived fact nodes, respectively. A fact is primitive if it comes from the input to the MulVAL reasoning engine. A derived fact is the result of applying interaction rules iteratively on the input facts. The edges in a logical attack graph can only go from a derived fact node to a derivation node, or from a derivation node to a fact node. The labeling function maps a fact node to the fact it represents, and a derivation node to the rule that is used for the derivation. Formally, the semantics of a logical attack graph is defined as follows.

PROPERTY 1. *For every derivation node R, let P be R's parent node and C be the set of R's child nodes, then $(\wedge \mathcal{L}(C)) \Rightarrow \mathcal{L}(P)$ is an instantiation of interaction rule $\mathcal{L}(R)$.*

Here $\wedge$ is the conjunction operator. For example, the derivation node r3 is an application of the interaction rule shown in Figure 1. The free variables in the rule have all been instantiated with ground terms.

4. ALGORITHMS

We modified the MulVAL reasoning engine so that besides a "yes" or "no" answer, a Datalog query also records an attack simulation trace as a side effect of the evaluation. This is achieved by a source-to-source translation of MulVAL interaction rules. For example, for the interaction rule shown in Figure 1, it will be translated into the following form:

```
execCode(Attacker, Host, User) :-
  networkService(Host, Program,
                 Protocol, Port, User),
  vulExists(Host, VulID, Program,
            remoteExploit, privEscalation),
  netAccess(Attacker, Host, Protocol, Port),
  assert_trace(because(
  'remote exploit of a server program',
    execCode(Attacker, Host, User),
      [networkService(Host, Program,
                      Protocol, Port, User),
      vulExists(Host, VulID, Program,
                remoteExploit, privEscalation),
      netAccess(Attacker, Host,
                Protocol, Port)])).
```

In addition to the sub-goals in the original rule, a new sub-goal is added which calls the function assert_trace. When the evaluation of the rule succeeds, this function will record the successful derivation into a trace file. In essence this method for recording execution traces is similar to the one used by the "justifier" program in XSB [7]. The attack simulation trace has the following format.

Definition 2. Attack simulation trace.
TraceStep ::= **because**(*interactionRule, Fact, Conjunct*)
Fact ::= *predicate(list of constant)*
Conjunct ::= *[list of Fact]*

interactionRule is a string uniquely associated with a MulVAL interaction rule. A list is represented as a series of items separated by commas. The semantic meaning of a trace step is "*Conjunct* ⇒ *Fact* is an instantiation of *interactionRule*". It records the reason why a goal is true during Datalog evaluation.

In order to compute attack graphs that contain all possible attack paths, the logic engine must traverse all possible derivation paths and record trace steps in the process. The XSB logic engine used in MulVAL is a Prolog system that supports tabled execution [15]. Tabling is a form of memoization that can both save computation time and resolve cycles in the derivation. The added assert_trace predicate will be called whenever XSB successfully satisfies all the preconditions of an interaction rule. Under tabled execution, *all* possible answers to a query will be computed. Thus the logic engine will have traversed all possible derivation paths before returning.

A logical attack graph can be constructed from the trace step information. The algorithm is depicted in Figure 6. In simple words, every *TraceStep* term becomes a derivation node in the attack graph. The *Fact* field in the trace step becomes the node's parent and the *Conjunct* field becomes its children. The maximum number of iterations for the inner loop at line 7 is the same as the largest number of pre-conditions among all the interaction rules, which is constant for a fixed interaction rule set. Thus, if the look-up operation on line 4 and line 8 is constant time, the graph building algorithm takes time linear in the number of trace steps.

4.1 Loops in attack graphs

Even though the interaction rules contain cycles, the XSB logic engine can avoid entering an infinite loop through tabling. However, the resulting trace steps can still contain loops

Input: set $\mathcal{T}$ containing all the *TraceStep* terms, attacker's goal $\mathcal{G}$
Output: logical attack graph $(N_r, N_p, N_d, E, \mathcal{L}, \mathcal{G})$.

1. $N_r, N_p, N_d, E, \mathcal{L} \leftarrow \emptyset$
2. For each $t \in \mathcal{T}$ {
   let $t = \textbf{because}(interactionRule, Fact, Conjunct)$
3. Create a derivation node r
   $N_r \leftarrow N_r \cup \{r\}$
   $\mathcal{L} \leftarrow \mathcal{L} \cup \{r \rightarrow interactionRule\}$
4. Look up $n \in N_d$ such that $\mathcal{L}(n) = Fact$.
5. If such n does not exist {
   create a new fact node n
   $\mathcal{L} \leftarrow \mathcal{L} \cup \{n \rightarrow Fact\}$
   $N_d \leftarrow N_d \cup \{n\}$
   }
6. $E \leftarrow E \cup \{(n, r)\}$
7. For each fact f in *Conjunct* {
8. Look up fact node $c \in (N_p \cup N_d)$ such that $\mathcal{L}(c) = f$,
9. If such c does not exist {
   create a new fact node c
   $\mathcal{L} \leftarrow \mathcal{L} \cup \{c \rightarrow f\}$
   If f is primitive $\{ N_p \leftarrow N_p \cup \{c\} \}$
   else $\{ N_d \leftarrow N_d \cup \{c\} \}$
   }
10. $E \leftarrow E \cup \{(r, c)\}$
    }
}

Figure 6: Graph building algorithm

from those cyclic rules. For example, the following two interaction rules form a cycle.

```
Rule 'execCode implies file access':
accessFile(Attacker, Host, write, Path) :-
            execCode(Attacker, Host, root).

Rule 'Trojan horse installation':
execCode(Attacker, Host, root) :-
   accessFile(Attacker, Host, write, Path).
```

An attacker can write files on a machine if he can execute arbitrary code on the machine as root; conversely, if an attacker can modify files on a machine, he can install a Trojan horse on it and potentially become root as well. A standard Prolog system will enter an infinite loop when encountering such rules, even though the program has a well defined meaning under the least fixed point semantics. A tabled Prolog system such as XSB will not loop. However, if both rules' preconditions are satisfied, both of them will be evaluated, in which case the output trace file will contain trace steps that form a loop. For example,

```
because('execCode implies file access',
    accessFile(attacker,workStation,
                write,/usr/local/share),
    [execCode(attacker, workStation, root)]).

because('Trojan horse installation',
    execCode(attacker, workStation, root),
    [accessFile(attacker,workStation,
                write,/usr/local/share)]).
```

Such loops often times render useless information in the attack graph. For example, the above two trace steps would introduce a loop in the attack graph of Figure 4. We show the loop in Figure 7 (only relevant derived fact nodes are shown; the derivation nodes and primitive fact nodes are ignored).

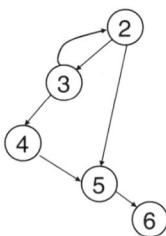

Figure 7: A loop in the example

Obviously, the back edge from node 3 to node 2 is meaningless, because the reason node 2 can be true is that node 3 is true in the first place. Such back edges should be eliminated from the graph. One may tend to think that a standard directed graph DFS algorithm that removes all the back edges will give us a correct DAG that represents all meaningful attack paths. However, this is not the case. Consider the two attack graphs in Figure 8. When starting from node 1, both (2, 3) and (3, 2) could be back edges, depending on the order in which the DFS algorithm traverses node 1's two child nodes. In case (a), only edge (3, 2) should be removed. If edge (2, 3) were removed, the attack path (1, 2, 3, 4) would be lost. Note that path (1, 3, 2) is not a valid derivation because it does not end in a leaf node. For case (b), neither (2, 3) nor (3, 2) should be removed, because then either attack path (1, 2, 3, 4) or (1, 3, 2, 5) would be lost.

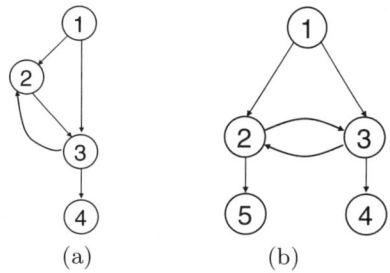

Figure 8: Two more loop examples

In general, an edge in an attack graph does not provide any useful information if it does not contribute to any valid logical derivation for attack goals. We call such edges "useless edges". To determine if an edge (u, v) is useless, we can remove u from the graph and test if v can still be derived in the graph. This can be done by a DFS search from v. If so the edge is not useless because there is a derivation path for v that does not involve u; otherwise the edge is useless and should be removed. The algorithm for finding all useless edges in an attack graph is at most quadratic in the size of the graph. We leave the implementation of this algorithm for future work. We also note that loops are not unique phenomena for logical attack graphs; they exist in other attack graph works as well, although we have not seen the problem addressed in the past.

5. COMPLEXITY ANALYSIS

The process of computing a logical attack graph consists of two stages. The first stage computes attack simulation traces through Datalog evaluation in XSB; the second stage builds attack graph data structures using the algorithm in Figure 6.

5.1 Complexity of computing attack trace

The generation of attack trace only introduces a constant-time overhead for every successful Datalog derivation. So the complexity of the first stage is the same as the complexity of evaluating the MulVAL Datalog program in XSB. The complexity of evaluating a fixed Datalog program against variable size inputs depends on the particular details of the program. The XSB documentation has some discussion on how to determine the complexity of evaluating a tabled Datalog program in XSB [14]. To make it easy to understand, let's consider the following Datalog interaction rule in MulVAL:

```
netAccess(Attacker, H2, Protocol, Port) :-
    execCode(Attacker, H1, _User),
    hacl(H1, H2, Protocol, Port).
```

The meaning of the rule is: if an attacker can become a local user on machine H1, and the network allows H1 to access H2 through Protocol and Port, then the attacker can access H2 through the protocol and port. This rule illustrates multi-hop network access in a network: an attacker can use a machine he controls as a stepping stone to compromise other machines.

When XSB evaluates this rule, it will first compute all possible machines an attacker can execute arbitrary code on (the first sub-goal), and then it will exhaustively search all H1 and H2's between which network access is possible (the second sub-goal). When all these tuples are computed, the goal predicate netAccess will be computed by matching the results of the two sub-goals. Pattern matching in XSB is very efficient due to the use of hash tables and tries. So the time spent is dominated by the number of intermediate tuples that need to be computed. The intermediate computation may of course invoke other interaction rules. In XSB's tabled execution, an invocation will compute *all* results of that goal, thus they can be reused later. One can think that all the rules are evaluated simultaneously in parallel with all possible instantiation of variables in their bodies. Each rule's evaluation time is determined by the number of different instantiations it needs to try. For a fixed Datalog program, the total running time is dominated by the rules that has the maximum number of different instantiations for the variables in its body.

THEOREM 1. *Evaluating MulVAL interaction rules against configuration tuples representing N hosts takes $O(N^2)$ derivation steps.*

PROOF. In MulVAL, the rule that has the most number of different body-variable instantiations happens to be the one we have shown above. In this rule two variables, H1 and H2, can be instantiated with every possible machine in the network; the other variables in the rule are not affected by the size of the network. Thus there are $O(N^2)$ possible instantiations for this rule. □

If the pattern matching in XSB is constant time, every derivation step needs constant time to finish and the overall running time for MulVAL Datalog evaluation will be quadratic. In our experiments we have seen a slightly higher growth than quadratic in the worst test case, due to increased time in pattern matching for large inputs.

Since every trace step was produced by one derivation step in Datalog evaluation, we have the following:

COROLLARY 1. *The number of trace-step terms produced in attack simulation is $O(N^2)$.*

5.2 Complexity of graph building

THEOREM 2. *The logical attack graph for a network with N machines has a size at most $O(N^2)$.*

PROOF. There is a one-to-one correspondence between *TraceStep* terms and derivation nodes. Let D be the number of trace steps, then there are D derivation nodes in the graph. If the maximum number of preconditions for an interaction rule is m, the number of edges in the graph is at most mD, and the maximum number of fact nodes is $mD+1$. By Corollary 1 we know D is $O(N^2)$, and so is $mD+1$. □

THEOREM 3. *The graph building algorithm in Figure 6 takes $O(\delta N^2)$ time to complete, where N is the number of hosts in the network, and δ is the maximum time spent in table look up at line 4 and 8 of the algorithm.*

PROOF. The loop in the graph building algorithm in Figure 6 goes through all the *TraceStep* terms. By Corollary 1, we know there are $O(N^2)$ such terms. In each iteration, the algorithm creates a derivation node for the *TraceStep* term and makes links from its parent and to its children. Every operation is constant time except for table look-up at line 4 and 8. □

Thus the time needed to construct the graph data structure is quadratic in the number of hosts, given a constant-time lookup table to store graph nodes. For our implementation we simply used the "map" container in C++'s standard library, which has $log(n)$ look up time. From Theorem 2 we know that the table size is $O(N^2)$. So $\delta = log(N^2)$ and the graph generation running time will be $O(N^2 log(N))$.

The next section shows performance data from our experiments.

6. EXPERIMENTAL RESULTS

Our Logical Attack Graph Generator was tested in the following environment. The CPU was a Pentium 4 3.2 GHz with 1GB of RAM, the operating system was Microsoft Windows XP Professional Version 2002 Service Pack 2, with XSB version 2.7.1.

The network configuration, machine configuration and vulnerability information were simulated for a variety of network sizes, topologies and vulnerability densities. The network configuration was simulated by the creation of a set of hacl(hostname1,hostname2,protocol,port) Datalog tuples as input to the MulVAl reasoning engine. Those tuples specify the allowed network traffic among machines in the network, including the attacker machine located on the Internet. The vulnerabilities were simulated by the creation of a set of vulExists and vulProperty Datalog tuples such that the same vulnerabilities exist on each of the simulated machines, and each vulnerability is a remote exploit of a service program running at a unique protocol and port number.

The "fully-connected" network topology simulation specified network accessibility of all protocols and ports between every pair of machines. The "star" topology was simulated to consist of one centralized machine (not the target machine) that has two-way accessibility of all protocols and ports to every other machine. The non-centralized machines, among which are the attacking and target machines, have no direct network access to any other machine. The "ring" topology was simulated with one machine (not the target machine) of the ring connected to the Internet, and all the other machines on the ring connected only to its two immediate neighbors with two-way access of all protocols and ports. The "partitioned" topology was simulated as two approximately equal sized fully-connected networks connected to each other only by one pair of machines (neither is the target machine), one on each sub-network. The only connection to the Internet is through a third machine located on the subnet that does not contain the target machine.

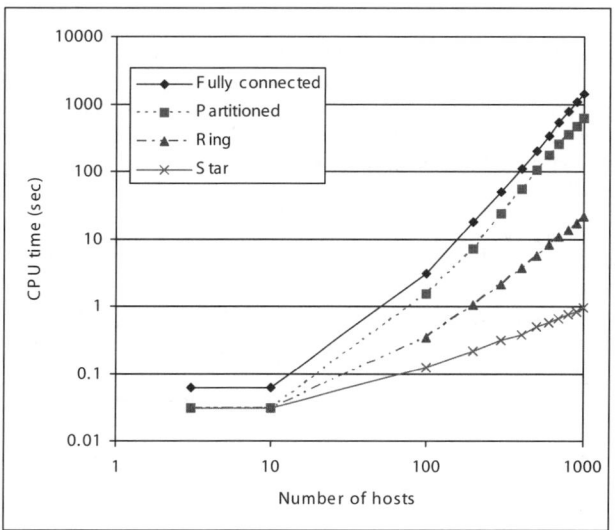

Figure 9: Graph generation CPU usage as a function of network size for several network topologies.

Figure 9 shows the graph generation CPU time for each of the simulated analysis problems of various sizes and topologies. The worst case is for a fully connected network. In this case the asymptotic CPU time is between $O(n^2)$ and $O(n^3)$, where n is the number of hosts. In the discussion of Section 4, we noted that ideally the complexity is $O(n^2)$, if table look-up takes constant time. However, our implementation uses the simple "map" template in C++ standard library and its look-up time depends on the size of the table. We believe after we replace it with a custom-designed hash table implementation the graph generation time will be near quadratic even for the worst case.

Figure 10 shows the memory usage as a function of network size for the same four network topologies. The worst case here is again a fully connected network, which has an asymptotic memory usage slightly lower than $O(n^2)$. In the

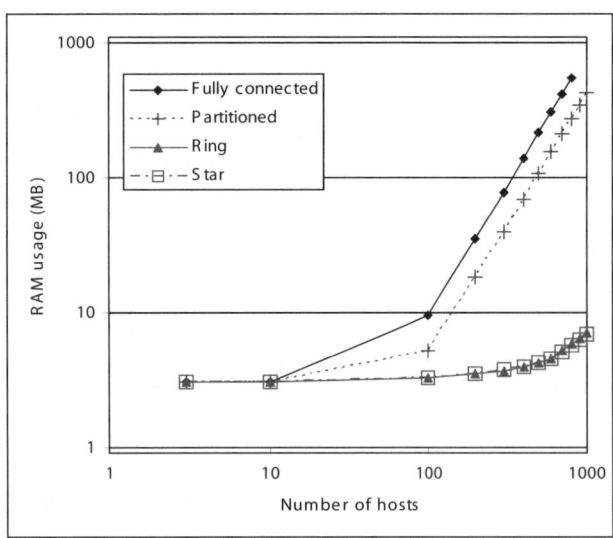

Figure 10: Graph generation memory usage as a function of network size for several network topologies.

two biggest cases (1000 host for fully-connected and partitioned network), we almost exhausted the 1GB memory on the test machine. The memory usage for the "star" and "ring" topology are not identical, although the difference is not visible on logarithmic scale.

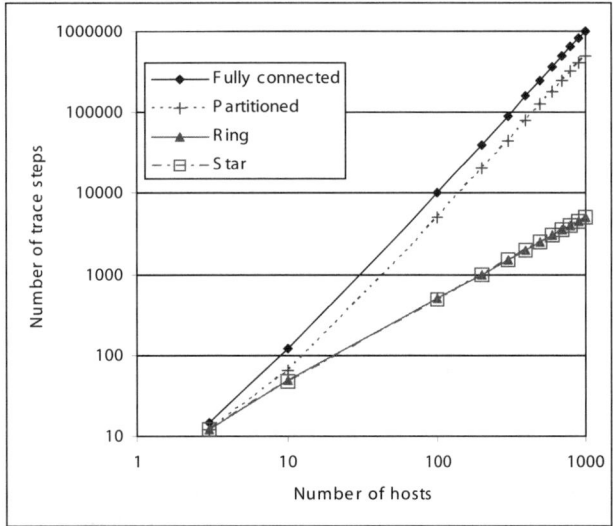

Figure 11: Number of trace-steps as a function of network size for various network topologies.

In Figure 11 the number of attack simulation trace steps, which is the input to the graph builder, is shown for the same set of test cases. For the worst case scenario, the number of trace steps is a quadratic function of the number of hosts. This verifies that Datalog evaluation in MulVAL reasoning engine takes $O(n^2)$ derivation steps to complete (Theorem 1).

Figure 12 shows that the number of derived fact nodes in the attack graph grows linearly with the size of the network.

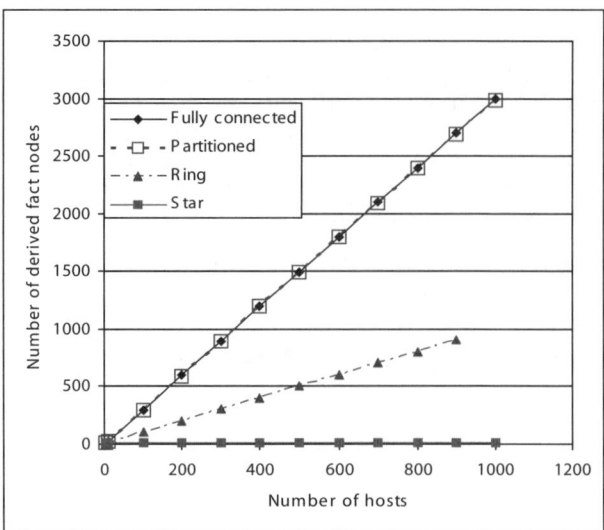

Figure 12: Number of derived fact nodes as a function of network size for various network topologies.

An interesting case is the one for the "star" topology, where the graph nodes remain constant regardless of the network size. This is because in that topology, the only attack path is from the attack machine to the hub, and then from the hub to the target machine.

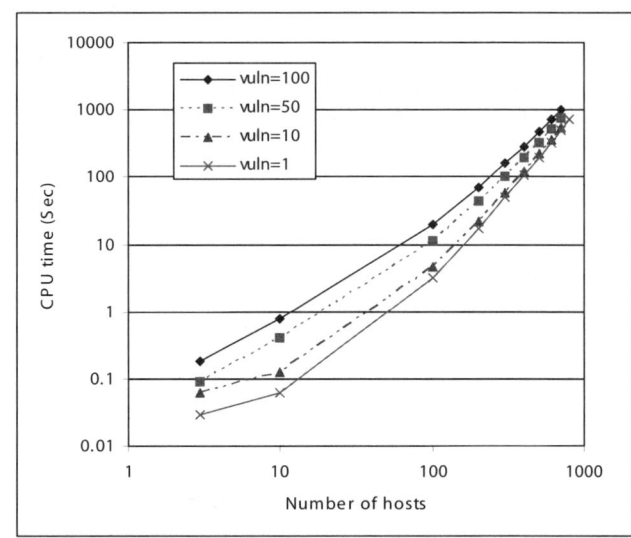

Figure 13: Graph generation CPU time for a fully connected network and number of vulnerabilities per host varying from 1 to 100.

In Figure 13 the attack graph generation CPU time is shown as a function of the network size for a fully connected network and for the number of vulnerabilities per host varied from 1 to 100. It shows that vulnerability density has a bigger impact when the network size is small. As the network size grows the CPU time is dominated by the number of machines, and thus vulnerability density has a less visible impact.

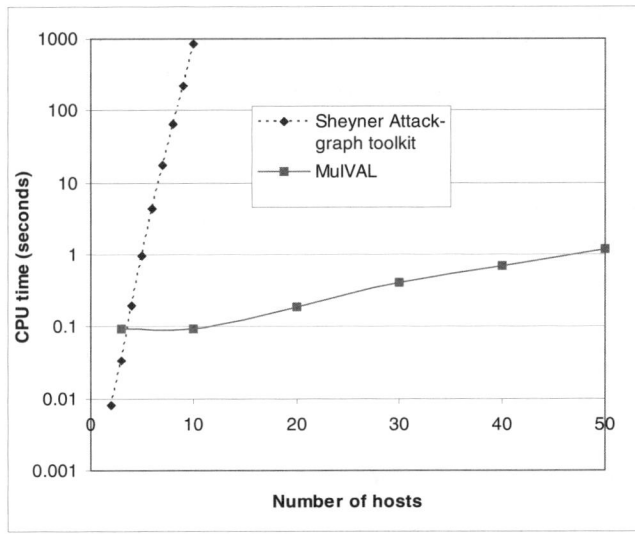

Figure 14: Graph generation CPU time compared to Sheyner attack graph toolkit. Fully connected network and 5 vulnerabilities per host.

Our graph builder was directly compared to the Sheyner attack graph toolkit by running both tools with equivalent input data. The Sheyner attack graph toolkit was tested on a Pentium III-M CPU, 256MB RAM, Fedora Core 1 LINUX operating system. Figure 14 is a comparison of graph builder CPU time for the case of a fully connected network and 5 vulnerabilities per host (note that only the Y axis is on logarithmic scale in this chart). From the diagram it is clear that the running time for Sheyner's tool grows exponentially. The growth trend for MulVAL is not obvious in this diagram because the running time is too short. But the difference between the two tools is obvious.

7. CONCLUSIONS

We have proposed a new approach to represent and generate attack graphs. Logical attack graphs directly illustrate logical dependencies among attack goals and configuration information and therefore have the significant advantage of improved clarity in guiding the user to an understanding of the causality relationship between system configuration and a successful attack. Our logical attack graph approach has dramatically improved scalability compared to previous approaches. We have shown that a logical attack graph has size polynomial to the network being analyzed. Our attack graph generation tool builds upon MulVAL, a network security analyzer based on logical programming. We have demonstrated how to produce a derivation trace in the MulVAL logic-programming engine, and how to use the trace to generate a logical attack graph in quadratic time. We have shown experimental evidence that our logical attack graph generation algorithm is very efficient.

8. REFERENCES

[1] P. Ammann, D. Wijesekera, and S. Kaushik. Scalable, graph-based network vulnerability analysis. In *Proceedings of 9th ACM Conference on Computer and Communications Security*, Washington, DC, November 2002.

[2] S. Jajodia, S. Noel, and B. O'Berry. Topological analysis of network attack vulnerability. In V. Kumar, J. Srivastava, and A. Lazarevic, editors, *Managing Cyber Threats: Issues, Approaches and Challanges*, chapter 5. Kluwer Academic Publisher, 2003.

[3] R. Lippmann and K. Ingols. An annotated review of past papers on attack graphs. Technical report, MIT Lincoln Laboratory, March 2005.

[4] S. Noel and S. Jajodia. Managing attack graph complexity through visual hierarchical aggregation. In *VizSEC/DMSEC '04: Proceedings of the 2004 ACM workshop on Visualization and data mining for computer security*, pages 109–118, New York, NY, USA, 2004. ACM Press.

[5] S. Noel, S. Jajodia, B. O'Berry, and M. Jacobs. Efficient minimum-cost network hardening via exploit dependency graphs. In *19th Annual Computer Security Applications Conference (ACSAC)*, December 2003.

[6] X. Ou, S. Govindavajhala, and A. W. Appel. MulVAL: A logic-based network security analyzer. In *14th USENIX Security Symposium*, Baltimore, MD, USA, August 2005.

[7] G. Pemmasani, H.-F. Guo, Y. Dong, C. Ramakrishnan, and I. Ramakrishnan. Online justification for tabled logic programs. In *The 7th International Symposium on Functional and Logic Programming*, April 2004.

[8] C. Phillips and L. P. Swiler. A graph-based system for network-vulnerability analysis. In *NSPW '98: Proceedings of the 1998 workshop on New security paradigms*, pages 71–79. ACM Press, 1998.

[9] P. Rao, K. F. Sagonas, T. Swift, D. S. Warren, and J. Freire. XSB: A system for efficiently computing well-founded semantics. In *Proceedings of the 4th International Conference on Logic Programming and Non-Monotonic Reasoning (LPNMR'97)*, pages 2–17, Dagstuhl, Germany, July 1997. Springer Verlag.

[10] B. Schneier. *Secrets & Lies: Digital Security in a Networked World*, chapter 21. John Wiley & Sons, 2000.

[11] O. Sheyner, J. Haines, S. Jha, R. Lippmann, and J. M. Wing. Automated generation and analysis of attack graphs. In *Proceedings of the 2002 IEEE Symposium on Security and Privacy*, pages 254–265, 2002.

[12] L. P. Swiler, C. Phillips, D. Ellis, and S. Chakerian. Computer-attack graph generation tool. In *DARPA Information Survivability Conference and Exposition (DISCEX II'01)*, volume 2, June 2001.

[13] T. Tidwell, R. Larson, K. Fitch, and J. Hale. Modeling Internet attacks. In *Proceedings of the 2001 IEEE Workshop on Information Assurance and Security*, West Point, NY, June 2001.

[14] D. S. Warren. *On the Complexity of Tabled Datalog Programs*. Department of Computer Science, SUNY @ Stony Brook, Stony Brook, NY 11794-4400, U.S.A., July 1999.

[15] D. S. Warren. *Programming in Tabled Prolog*. Department of Computer Science SUNY @ Stony Brook, July 1999.

Formal Specification and Verification of Data Separation in a Separation Kernel for an Embedded System

Constance L. Heitmeyer, Myla Archer, Elizabeth I. Leonard, and John McLean[*]
Information Technology Division
Naval Research Laboratory
Washington, DC 20375
{heitmeyer, archer, leonard, mclean}@itd.nrl.navy.mil

ABSTRACT

Although many algorithms, hardware designs, and security protocols have been formally verified, formal verification of the security of software is still rare. This is due in large part to the large size of software, which results in huge costs for verification. This paper describes a novel and practical approach to formally establishing the security of code. The approach begins with a well-defined set of security properties and, based on the properties, constructs a compact security model containing only information needed to reason about the properties. Our approach was formulated to provide evidence for a Common Criteria evaluation of an embedded software system which uses a separation kernel to enforce data separation. The paper describes 1) our approach to verifying the kernel code and 2) the artifacts used in the evaluation: a Top Level Specification (TLS) of the kernel behavior, a formal definition of data separation, a mechanized proof that the TLS enforces data separation, code annotated with pre- and postconditions and partitioned into three categories, and a formal demonstration that each category of code enforces data separation. Also presented is the formal argument that the code satisfies the TLS.

Categories and Subject Descriptors: D.2.4 [Software]: Software Engineering

General Terms: security, verification, languages, theory

Keywords: formal model, formal specification, theorem proving, separation kernel, code verification

1. INTRODUCTION

A critical objective of many military systems is to protect the confidentiality and integrity of sensitive information. Preventing unauthorized disclosure and modification of information is of enormous importance in military systems, since violations can jeopardize national security. Compelling evidence is required therefore that military systems satisfy their security requirements.

[*]This research was funded by the Office of Naval Research. C. Heitmeyer, M. Archer, and E. Leonard are members of the Software Engineering Section (Code 5546) of NRL's Center for High Assurance Computer Systems. J. McLean heads NRL's Information Technology Division (Code 5500).

This paper is authored by employees of the United States Government and is in the public domain.
CCS'06, October 30–November 3, 2006, Alexandria, Virginia, USA.
ACM 1-59593-518-5/06/0010.

A promising approach to demonstrating the security of code is formal verification, which has been successfully applied to algorithms, such as floating point division [27] and clock synchronization [32], and security protocols such as cryptographic protocols [25, 21]. However, most past efforts to verify security-critical software have been extremely expensive. One reason is that these efforts often built security models containing too much detail (see, for example, [12]) or tried to prove too many properties (see, for example, [37]). The result was that model building and property proving became prohibitively expensive.

A challenging problem therefore is how to make the verification of security-critical code affordable. This paper describes an approach to verifying the security of software that is both novel and practical. This approach was formulated in preparation for a Common Criteria evaluation of the security of a software-based embedded device called ED (Embedded Device). For the ED application, satisfying the Common Criteria required a formal proof of correspondence between a formal specification of ED's security functions and its required security properties *and* a demonstration that the code implementing ED satisfied the formal specification. ED, which processes data stored in different partitions of memory, is required to enforce a critical security property called *data separation*; for example, ED must ensure that data in one memory partition neither influences nor is influenced by data in another partition. To ensure that data separation is not violated, or if it is violated an exception occurs, the ED architecture includes a separation kernel [34], a tamper-proof, non-bypassable program that mediates every access to memory.

The task of our group was to provide evidence to the certifying authority that the ED separation kernel enforces data separation. The kernel code, which consists of over 3000 lines of C and assembly code, was annotated with pre- and postconditions in the style of Hoare and Floyd. To provide evidence that ED enforces data separation, we produced a Top Level Specification (TLS) of the separation-relevant behavior of the kernel, a formal statement of data separation, and a mechanized formal proof that the TLS satisfies data separation. Then, the annotated code was partitioned into three categories, each requiring a different proof strategy. Finally, the formal correspondence between the annotated code and the TLS was established for each category of code. Recently, five artifacts—the TLS, the formal statement of data separation, proofs that the TLS satisfies data separation, the organization of the annotated code into the three categories, and the documents showing correspondence of each category of code with the TLS—were presented along with the annotated code as evidence in a Common Criteria evaluation of ED.

This paper summarizes the process we followed in producing evidence for the Common Criteria evaluation of ED's separation kernel, describes each artifact developed during this process, and summarizes both the formal state machine model that underlies the TLS and the formal argument justifying our approach to establishing code conformance with the TLS. The paper makes two technical contributions. First, it describes a novel technique for partitioning the code into three different categories—namely, Event, Trusted, and Other Code—and for reasoning about the security of each category. Second, it describes a method for demonstrating the security of the code that is both original and practical. While the method combines a number of well-known techniques for specifying and reasoning about security—e.g., a state machine model represented both formally and in natural language, mechanized reasoning using PVS [35], and a demonstration of correspondence between the TLS and the annotated source code—which techniques to apply, how to apply them, and how to combine them was far from obvious and required significant discussion during the course of the project. Along the way, many alternative approaches and techniques were considered, and several were discarded. In our view, both the technique for partitioning the code and the method we formulated for proving that the code is secure should prove useful in future efforts to verify the security of software.

The paper is organized as follows. Section 2 reviews the notion of a separation kernel, summarizes the requirements of a Common Criteria evaluation, and presents some details of ED. Section 3 describes the process we followed to demonstrate data separation and describes the five artifacts that the process produced, including the three categories of code and how we proved that each category of code is secure. Section 4 presents the formal argument for demonstrating code conformance. Sections 5 and 6 discuss some lessons learned and describe topics requiring more research, i.e., the need for more powerful tool support. Section 7 describes related work. Finally, Section 8 presents some conclusions.

2. BACKGROUND
2.1 Separation Kernel

A *separation kernel* [34] mimics the separation of a system into a set of independent virtual machines by dividing the memory into partitions and restricting the flow of information between those partitions. Separation kernels are being developed for military applications requiring Multiple Independent Levels of Security (MILS) by commercial companies such as Wind River Systems, Green Hills Software, and LynuxWorks [5]. In a MILS environment, a separation kernel acts as a reference monitor [7], i.e., is non-bypassable, evaluatable, always invoked, and tamper proof.

2.2 Common Criteria

Seven international organizations established the Common Criteria to provide a single basis for evaluating the security of information technology products [3]. Associated with the Common Criteria are seven Evaluation Assurance Levels. EAL7, the highest assurance level, requires a formal specification of a product's security functions and its security model, and formal proof of correspondence between the two.

2.3 Embedded Device (ED)

The device of interest, ED, processes data in an embedded system whose memory has been divided into non-overlapping partitions. While at any given time the data stored and processed by ED in one memory partition is classified at a single security level, ED may later reconfigure that partition to store and process data at a different security level. Because it stores and processes data classified at different security levels, security violations by ED could cause significant damage. To prevent violations of data separation, e.g., the "leaking" of data from one memory partition to another, the ED design uses a separation kernel to mediate access to memory. By mediating every access, the kernel ensures that every memory access is authorized and that every transfer of data from one ED memory location to another is authorized. Any attempted memory access by ED that is unauthorized will cause an exception. Section 3.3 describes how TAME [9, 8], an interface to SRI's theorem prover PVS [35], was used to support the Common Criteria evaluation of ED's separation kernel.

3. CODE VERIFICATION PROCESS

The process followed in constructing the five ED artifacts consists of five steps. The process described below is an idealization of the actual process since, in any real-world process, one frequently returns to a former step to make corrections and add missing information. However, the sequence of steps that follows is a logical order for producing the various artifacts.

1. Formulate a Top Level Specification (TLS) of the kernel as a state machine model, using the style introduced in [22, 24].

2. Formally express the data separation property in terms of the inputs, state variables, and transitions defined in the state machine model that underlies the TLS.

3. Translate the TLS and the data separation property into the language of a mechanical prover, and prove formally that the TLS satisfies the data separation property.

4. Given a source code implementation of the kernel annotated with pre- and postconditions, partition the code into Event, Other, and Trusted Code, where, informally, Event Code is code corresponding to an event in the TLS that touches a Memory Area of Interest (defined below), Trusted Code is code that touches a Memory Area of Interest but is not Event Code, and Other Code is neither Event Code nor Trusted Code. Section 3.4 provides precise definitions of the three different code categories.

5. Demonstrate that the Event Code does not violate separation by constructing 1) a mapping from the Event Code to the TLS events and from the code states to the states in the TLS, and 2) a mapping from pre- and postconditions of the TLS events to pre- and postconditions that annotate the corresponding Event Code. Demonstrate separately that Trusted Code and Other Code do not violate data separation.

3.1 Top Level Specification

Major goals of the Top Level Specification (TLS) are to provide a precise, yet understandable description of the required external behavior of ED's separation kernel and to make explicit the assumptions on which the specification is based. To achieve this, the TLS represents the kernel as a state machine model using precise natural language. Such a natural language description was introduced in 1984 to describe the behavior of a secure military message system (MMS) [22, 24]. The advantage of natural language is that it enables stakeholders from differing backgrounds and with different objectives—the project manager, software developers, evaluators, and formal methods team—to communicate precisely about the required kernel behavior and helps ensure that misunderstandings are weeded out and issues resolved early in the verification process. Another goal of the TLS is to provide a formal context and precise vocabulary for defining data separation.

Like the secure MMS model, the state machine representing the behavior of the ED kernel is defined in terms of an input alphabet, a set of states, an initial state, and a transform relation describing the allowed state transitions. The input alphabet contains internal and external events, where an *internal event* can cause the kernel to invoke some process and an *external event* is performed by an external host. An example of an internal event is an event instructing ED to copy data from an input buffer associated with memory partition i to a data area in partition i. An example of an external event is the event occurring when an external host writes data into an input buffer assigned to partition i. The transform (also called the *next-state relation*) is defined on triples consisting of an event in the input alphabet, the current state, and the new state. Provided below are excerpts from the TLS as well as an example internal event. This event, Copy_Buf1In_Data1In_i, copies data from an input buffer for partition i into a data area in partition i.

Partitions, State Variables, Events and States. We assume the existence of $n \geq 1$ dedicated memory partitions and a single shared memory area. We also assume the existence of the following sets:

- V is a union of types, where each type is a non-empty set of values.

- R is a set of state variable names. For all r in R, $\mathrm{TY}(r) \subseteq V$ is the set of possible values of state variable r. $\mathcal{M}$ is a union of N non-overlapping memory areas, each represented by a state variable.

- $H = P \cup E$ is a set of M events, where each event is either an internal event in P or an external event in E.

A *system state* is a function mapping each state variable name r in R to a value. Formally, for all $r \in R$: $s(r) \in \mathrm{TY}(r)$.

Memory Areas. The N memory areas contain $N - 1$ Memory Areas of Interest, where $N - 1 = mn$, and m is the number of Memory Areas of Interest per partition. Informally, a Memory Area of Interest (MAI) is a memory area containing data whose leakage would violate data separation. The m MAIs for a partition i, $1 \leq i \leq n$, include partition i's input and output buffers and k data areas where data in partition i is stored and processed. The Nth memory area called G is the single shared memory area and contains all programs and data not residing in an MAI. The set $\mathcal{M}$ of all memory areas is defined as the union $A \cup \{G\}$, where $A = \{A_{i,j} \mid 1 \leq i \leq n \wedge 1 \leq j \leq m\}$ contains the mn MAIs. For all i, $1 \leq i \leq n$, $A_i = \{A_{i,j} \mid 1 \leq j \leq m\}$ is the set of memory areas for partition i. To guarantee that the memory areas of $\mathcal{M}$ are non-overlapping, the memory areas are required to be pairwise disjoint.

State Variables. The set of state variables[1] contained in R are

- a partition id c,

- the N memory areas in $\mathcal{M}$, and

- a set of n sanitization vectors $\mathcal{W}_\mathrm{D}[1], \ldots, \mathcal{W}_\mathrm{D}[n]$, each vector containing k elements.

The partition id c is 0 if no data processing in any partition is in progress, and i, $1 \leq i \leq n$, if data processing is in progress in partition i. (Data processing can occur in only one partition at a time.) For $1 \leq j \leq k$, the boolean value of the jth element $\mathcal{W}_\mathrm{D}^j[i]$ of the sanitization vector for partition i is *true* if the jth memory area of the ith partition has been sanitized and *false* otherwise. A sanitized memory area is modeled as having the value 0.

Events. The set of internal events $P \subset H$ is the union of n sets, $P_1, \ldots, P_n$, of *partition events*, one set for each partition i, and a singleton set Q; thus P is defined by $P = [\cup_{i=1}^{n} P_i] \cup Q$. Processing occurs on partition i when a sequence of events from P_i is processed. The sole member of Q is the event Other_NonParProc, an abstract internal event representing all internal events which invoke data processing in the shared message area G. One example of such an event is Assign_Val, which causes some value to be stored in G. The set of external events $E \subset H$ is defined by $E = E^{\mathrm{In}} \cup E^{\mathrm{Out}} \cup \{\texttt{Ext\_Ev\_Other}\}$, where $E^{\mathrm{In}} = \cup_{i=1}^{n} E_i^{\mathrm{In}}$ and $E^{\mathrm{Out}} = \cup_{i=1}^{n} E_i^{\mathrm{Out}}$. E_i^{In} is the set of external events writing into or clearing the input buffers assigned to partition i, and E_i^{Out} is the set of external events reading from or clearing the output buffers assigned to partition i. The event Ext_Ev_Other represents all other external events.

Partition Functions. Operations on data in partition i, for example, an operation copying data from one MAI in partition i to another MAI in i, are called 'partition functions.' For all i, $1 \leq i \leq n$, and for each internal event e in P_i, there exists a *partition function* Γ_e associated with e. Each function Γ_e computes a value stored in an MAI in A. For all $e \in P_i$, Γ_e has the signature $\Gamma_e : \mathrm{TY}(a_1) \to \mathrm{TY}(a_2)$, where a_1 and a_2 are MAIs in A_i. Thus, each function Γ_e, where e is an internal event in P_i, takes a single argument, the value stored in some MAI a_1, and uses that argument to compute a value to be stored in MAI a_2.

Access Control Matrix. Associated with the M events and the N memory areas is an M by N access control matrix AM, which indicates the read and write access that each internal event e in P (and its associated process) and each external event e in H has for each memory area a in $\mathcal{M}$. Each entry in the matrix is either null meaning no access, R for read access, W for write access, or RW for both read and write access. The left-most column of AM lists the events in H, and the headings of the remaining columns list the N memory areas in $\mathcal{M}$ as well as G.

For all i, j, $1 \leq i, j \leq n$, $i \neq j$, an event associated with partition i has null access to an MAI associated with partition j or to G; similarly, an event associated with j has null access to an MAI associated with i or to G. Moreover, the single event that invokes non-partition processing, namely, Other_NonParProc, has R and W access for G and access null for all other memory areas, i.e., the MAIs. Finally, the external events associated with partition i can only write into or read from input and output buffers associated with i.

System. A system is a state machine whose transitions from one state to the next are triggered by events. Formally, a system Σ is a 4-tuple $\Sigma = (H, S, s_0, T)$, where

- H is the set of events,

- S is the set of states,

- s_0 is the initial state, and

- T is the system transform, a partial function from $H \times S$ into S. T is partial because not all events are 'enabled' to be executed in the current state.

---

[1] By convention, state variable names may refer to the values of the variables.

Initial State. In the initial state s_0, the partition id c is 0; for all i, $1 \leq i \leq n$, the MAIs in A_i are 0; and each element of the sanitization vectors, $\mathcal{W}_D[1] \ldots \mathcal{W}_D[n]$, is *true*. Hence, in the initial state, no processing in any partition is authorized, only a non-partition process is authorized to execute, each element of every sanitization vector has the value *true*, and all MAIs have the value zero.

System Transform. The transform T is defined in terms of a set $\mathcal{R}$ of transform rules, $\mathcal{R} = \{ \mathcal{R}_e \mid e \in H \}$, where each *transform rule* $\mathcal{R}_e$ describes how an event e transforms a current state into a new state. The number of rules is M, one rule for each of the M events in H. No rule requires access privileges other than those defined by the access control matrix AM. The notation s and s' represents the current state and the new state. Given state s and state variable r, r's value in s is denoted by r_s. When an internal or external event e does not affect the value of any state variable r, when the precondition is not satisfied, or when the event e is not enabled, the value of r does not change from state s to state s', and the state variable r retains its current value, i.e., $r_s = r_{s'}$.

To denote that no state variable except those explicitly named changes, we write $\text{NOC}_{\hat{R}}$ (NO Change except to variables in $\hat{R}$), where $\hat{R} \subset R$. This includes the case where the ith element of a sanitization vector changes but no other vector elements change. For example, the postcondition $r_{s'} = x \land \text{NOC}_{\{r\}}$, where $x \in \text{TY}(r)$, is equivalent to $r_{s'} = x \land \forall \hat{r} \in R, \hat{r} \neq r : \hat{r}_{s'} = \hat{r}_s$.

Suppose s is a state in S, e is an event in H, and R is the set of state variables. Let $\texttt{pre}_e$ be a state predicate associated with e such that $\texttt{pre}_e$ evaluates to *true* if e has the potential to occur in state s and and *false* otherwise, and let $\texttt{post}_e$ be a predicate associated with e such that $\texttt{post}_e(s, s')$ holds whenever e occurs in state s and s' is a possible poststate of s when event e occurs in state s. Formally, the transform rule $\mathcal{R}_e$ in $\mathcal{R}$ is defined by

$$\mathcal{R}_e : \texttt{pre}_e(s) \Rightarrow \texttt{post}_e(s, s').$$

Whenever the result state of every event e is deterministic (which is true in the TLS), the assertion $\texttt{post}_e(s, s')$ defines the poststate $s' = T(e, s)$. To make T total on $H \times S$, the complete definition of T is written as

$$T(e, s) = \begin{cases} s' & \text{if } \texttt{pre}_e(s), \text{ where } \texttt{post}_e(s, s') \\ s & \text{otherwise.} \end{cases}$$

In the above definition, $\texttt{pre}_e(s)$ is not satisfied implies that e has no effect—i.e., essentially, did not occur.

Example of a Transform Rule. Consider an internal event, $e = \texttt{Copy\_Bfr1In\_Data1In\_}i$, which invokes a process copying data from partition i's Input Buffer 1, denoted B_i^1, into partition i's Data Area 1, denoted D_i^1. The transform rule for e is denoted $\mathcal{R}_{\texttt{Copy\_Bfr1In\_Data1In\_}i}$. Preconditions for e are (1) the partition id c equals i, and (2) the invoked process must have read access ('R') for partition i's Input Buffer 1 and write access ('W') for Data Area 1 in i. Postconditions for e are that (3) the element for Data Area 1 in i's sanitization vector becomes *false*, (4) a function of the value stored in i's Input Buffer 1 is written into i's Data Area 1, and (5) no other state variable changes. For all i, the rule $\mathcal{R}_e$ for event $e = \texttt{Copy\_Bfr1In\_Data1In\_}i$ is defined by

$\mathcal{R}_{\texttt{Copy\_Bfr1In\_Data1In\_}i}$:

$$c_s = i \land \qquad (1)$$
$$\text{AM}[e, a_1] = \texttt{R} \text{ and } \text{AM}[e, a_2] = \texttt{W} \qquad (2)$$
$$\text{where } a_1 = B_i^1 \text{ and } a_2 = D_i^1$$
$$\Rightarrow \quad \mathcal{W}_{D,s'}^1[i] = \textit{false} \land \qquad (3)$$
$$D_{i,s'}^1 = \Gamma_e(B_{i,s}^1) \land \qquad (4)$$
$$\text{NOC}_{\{\mathcal{W}_D^1[i], D_i^1\}} \qquad (5)$$

3.2 Security Policy: Data Separation

To operate securely, ED must enforce data separation. Informally, this means that ED must prevent data in a partition i from influencing or being influenced by 1) data in a partition j, where $i \neq j$, 2) an earlier configuration of partition i, or 3) data stored in G. Thus ED must prevent non-secure data flows. To demonstrate that the TLS enforces data separation, we proved that it satisfies five subproperties, namely, *No-Exfiltration*, *No-Infiltration*, *Temporal Separation*, *Separation of Control*, and *Kernel Integrity*. Below, each subproperty is defined formally using the notation introduced in Section 3.1.

3.2.1 No-Exfiltration Property

The No-Exfiltration Property states that data processing in any partition j cannot influence data stored outside the partition. This property is defined in terms of the set A_j (the MAIs of partition j); the entire memory $\mathcal{M}$; the internal events in P_j, which invoke data processing in j; and external events in $E_j^{\text{In}} \cup E_j^{\text{Out}}$, which affect data in j's input and output buffers.

Property 3.1 (No-Exfiltration) *Suppose that states s and s' are in state set S, event e is in H, memory area a is in $\mathcal{M}$, and j is a partition id, $1 \leq j \leq n$. Suppose further that $s' = T(e, s)$. If e is an event in $P_j \cup E_j^{\text{In}} \cup E_j^{\text{Out}}$ and $a_s \neq a_{s'}$, then a is in A_j.*

3.2.2 No-Infiltration Property

The No-Infiltration Property states that data processing in any partition i is not influenced by data outside that partition. It is defined in terms of the set A_i, which contains the MAIs of partition i.

Property 3.2 (No-Infiltration) *Suppose that states s_1, s_2, s_1', and s_2' are in S, event e is in H, and i is a partition id, $1 \leq i \leq n$. Suppose further that $s_1' = T(e, s_1)$ and $s_2' = T(e, s_2)$. If for all a in A_i: $a_{s_1} = a_{s_2}$, then for all a in A_i: $a_{s_1'} = a_{s_2'}$.*

3.2.3 Temporal Separation Property

The objective of this property is to guarantee that the k data areas in any partition i are clear when the system is not processing data in that partition, e.g., from the end of a processing thread in one partition to the start of a new processing thread in the same or a different partition.[2] Satisfying this property implies a second property—i.e., no data (e.g., Top Secret data) stored in the ith partition during one configuration of that partition can remain in any memory area of a later configuration (e.g., processing and storing data at the Unclassified level) of that same partition i. The set of states in which the system is not processing data stored in a partition is exactly the set of states in which the partition id c_s is 0. This fact can be used in stating the Temporal Separation Property.

Property 3.3 (Temporal Separation) *For all states s in S, for all i, $1 \leq i \leq n$, if the partition id c_s is 0, then the k data areas of partition i are clear, i.e., $D_{i,s}^1 = 0, \ldots, D_{i,s}^k = 0$.*

---

[2] The proof of this property depends on a constraint imposed by the transform rules on the partition id c: If c changes, it changes from 0 to non-zero or vice versa.

3.2.4 Separation of Control Property

This property states that when data processing is in progress on partition i, no data is being processed on partition j, $j \neq i$, until processing on partition i terminates. The property is defined in terms of the partition id c, which is i if processing is in progress on partition i, where $i > 0$, and 0 otherwise, and the set D_i of k data areas in partition i, $D_i = \{D_i^j \mid 1 \leq j \leq k\}$.

Property 3.4 (Separation of Control) *Suppose that states s and s' are in S, event e is in H, data area a is in $\mathcal{M}$, and j, where $1 \leq j \leq n$, is a partition id. Suppose further that $s' = T(e, s)$. If neither c_s nor $c_{s'}$ is j, then $a_s = a_{s'}$ for all $a \in D_j$.*

3.2.5 Kernel Integrity Property

The Kernel Integrity Property states that when data processing is in progress on partition i, the data stored on memory area G does not change. This property is defined in terms of G and the set P_i of events for partition i.

Property 3.5 (Kernel Integrity) *Suppose that states s and s' are in state set S, event e is in H, and i is a partition id, $1 \leq i \leq n$. Suppose further that $s' = T(e, s)$. If e is a partition event in P_i, then $G_{s'} = G_s$.*

3.3 Formal Verification

To formally verify that the TLS enforces data separation, the natural language formulation of the TLS was translated into TAME (Timed Automata Modeling Environment) [9, 8], a front-end to the mechanical prover PVS [28] which helps a user specify and reason formally about automata models. This translation requires the completion of a template to define the initial states, state transitions, input events, and other attributes of the state machine Σ. The TAME specification provides a machine version of the TLS that can be shown mechanically to satisfy the properties defined above.

After constructing the TAME specification of the TLS, we formulated two sets of TLS properties in TAME—invariant properties and other properties—that together formalize the five subproperties. Then, for each set of properties, we interactively constructed (TAME) proofs showing that the TAME specification satisfies each property. The scripts of these proofs, which are saved by PVS, can be rerun easily by the evaluators and serve as the formal proofs of data separation. One benefit of TAME is that the saved PVS proof scripts can be largely understood without rerunning them in PVS.

3.4 Partitioning the Code

To show formally that the ED separation kernel enforces data separation, we must prove that the kernel is a secure partial instantiation of the state machine Σ defined by the TLS. The formal verification described in Section 3.3 establishes formally that a strict instantiation of the TLS enforces data separation. A *partial instantiation* of the TLS is an implementation that contains fine-grain details which do not correspond to the state machine Σ defined in the TLS. A *secure partial instantiation* of the TLS is a partial instantiation of the TLS in which the fine-grain details that do not correspond to the TLS are benign. Section 4 contains the formal foundation for the proof that the code is a secure partial instantiation of the TLS.

The proof that the code for the ED kernel is a secure partial instantiation of the TLS is based on a demonstration that all kernel code falls into three major categories and one subcategory, with proofs that the code in each category satisfies certain properties. The categories are as follows:

1. *Event Code* is kernel code which implements a TLS internal event e in H and touches one or more MAIs. For each segment of Event Code, it is checked that

    (i) the concrete translation of the precondition in the TLS for the corresponding event e is satisfied at the point in the kernel code where the execution of the Event Code is initiated, and

    (ii) the concrete translation of the postcondition in the TLS for the corresponding event e is satisfied at the conclusion of Event Code execution.

2. *Trusted Code* is kernel code which touches MAIs but is not Event Code. This code does not correspond to behavior defined by the TLS and may have read and write access both to MAIs and to memory areas outside of the MAIs. It is validated either by a proof that the code does not permit any non-secure information flows or, in rare instances, by assumption. The TLS makes explicit any assumptions used in connection with Trusted Code and its behavior. The proofs for a given segment of Trusted Code characterize the entire functional behavior of that Trusted Code using Floyd-Hoare style assertions at the code level and show that no non-secure information flows can occur in that code.

3. *Other Code* is kernel code that is neither Event nor Trusted Code. More specifically, Other Code is kernel code which does not correspond to any behavior defined by the TLS and which has no access to any MAIs. Apple's Xcode development tool [2] was used to search the kernel code to locate all code segments with access to MAIs, i.e, code segments classified as Event or Trusted Code. This involved identifying all places in the kernel code where the MMU is reset and observing the permissions assigned. By observing the access granted for code segments categorized as Other Code, we can ensure that they have no access to any MAI.

    (a) A subset of Other Code, called *Verified Code*, is code with no access to MAIs which is still security-relevant because it performs functions necessary for the kernel to enforce data separation. These functions include setting up the MMU, establishing preconditions for Event Code, etc. Floyd-Hoare style assertions at the code level are used to prove that Verified Code correctly implements the required functions.

3.5 Demonstrating Code Conformance

Demonstrating code conformance requires the definition of two mappings. To establish correspondence between concrete states in the kernel code and abstract states in the TLS, a function α is defined that relates concrete states to abstract states by relating concrete entities (such as memory areas, code variables, and logical variables) at the code level to abstract state variables (such as MAIs and the partition id) in the TLS. For example, the actual physical addresses of the MAIs are mapped to their corresponding abstract state variables in the TLS. The map α also maps Event Code to events in the TLS. Another map Φ relates assertions at the abstract TLS level to assertions at the code level derived from the map α. See Section 4 for more details.

Using Φ to relate pre- and postconditions for an event in the TLS to derived pre- and postconditions for the corresponding Event Code, we next determine for each piece of Event Code sets of code-level pre- and postconditions that match the derived pre- and postconditions as closely as possible. Figure 1 shows the Event Code

```
{CopyDIn_partition_id : partition = partition_id}
{CopyDIn_priv :
        {(R, KER_INBUFFER_1_partition), (W, KER_PAR_DATA_STORAGE_1_partition)} ⊆ MMU}
{CopyDIn_value_data : ∀j.0 ≤ j < byte_length.
        A[j] = KER_INBUFFER_1_partition_START + j}
{CopyDIn_def_value_rest : ∀j.byte_length ≤ j < KER_PAR_DATA_STORAGE_1_partition_SIZE.
        B[j] = KER_PAR_DATA_STORAGE_1_partition_START + j}
{CopyDIn_local_inbuffer : buffer_in_start = KER_INBUFFER_1_partition_START}
{CopyDIn_local_datain : part_data_start = KER_PAR_DATA_STORAGE_1_partition_START}
```

```
        if (byte_length < (unsigned long)&__INBUFFER_SIZE)
{
        /* copy data from inbuffer 1 to partition      */
        /* part_data_start contains the starting address of */
        /* the memory area, buffer_in_start contains   */
        /* the starting address of the inbuffer        */
        /* kernel_memcopy is a copy routine whose functional correctness */
        /* has been verified using Floyd-Hoare assertions */
        kernel_memcopy(part_data_start, buffer_in_start, byte_length);
}
```

```
{CopyDIn_copy_size_datain : byte_length > KER_PAR_DATA_STORAGE_partition_SIZE → false}
{CopyDIn_copy_size_inbuffer : byte_length > KER_INBUFFER_1_partition_SIZE → false}
{CopyDIn_gamma_copy : ∀j.0 ≤ j < byte_length.
        KER_PAR_DATA_STORAGE_1_partition_START + j = A[j]}
{CopyDIn_gamma_rest : ∀j.byte_length ≤ j < KER_PAR_DATA_STORAGE_1_partition_SIZE.
        KER_PAR_DATA_STORAGE_1_partition_START + j = B[j]}
{CopyDIn_sanitize : part_data_sanitized_partition = false}
{CopyDIn_NOC : No concrete state variables have changed value except possibly
        KER_PAR_DATA_STORAGE_partition and part_data_sanitized_partition.}
```

Figure 1: Event Code and Code Level Assertions for Event Copy_Bfr1In_Data1In_i

corresponding to the Copy_Bfr1In_Data1In_i event in the TLS and the code level pre- and postconditions for this Event Code. In Figure 1, the top box contains the preconditions, then the indented Event Code is listed, and finally the bottom box contains the postconditions; each precondition and postcondition has the form {Assertion_Name : Assertion}. Generally, the match between assertions in the TLS and derived code-level assertions is not exact because auxiliary assertions are added 1) to express the correspondence between variables in the code and physical memory areas[3] (e.g., CopyDIn_local_datain), 2) to save values in memory areas as the values of logical variables (e.g., CopyDIn_value_data), and 3) to express error conditions that the TLS implicitly assumes to be impossible (e.g., CopyDIn_copy_size_datain).

After defining the desired sets of code-level pre- and postconditions, we check whether these assertions are among the assertions already proven in the annotated C code. The annotated C code refers to memory areas by indexing into arrays that define memory maps in the code, whereas the mapping α refers to memory areas by their actual physical addresses. Thus, to be equivalent to the desired assertions, the assertions in the annotated code frequently need dereferencing. For example, the annotated C code assertion §8.4, TLS2, is defined by

part_data_start =
(unsignedchar*)ker_rtime_mmu_map[partition].part_data_start,

which sets the variable part_data_start to the starting address of the data area in the partition by indexing into the real-time memory map in the code and selecting the part_data_start member of the structure corresponding to that array element. Dereferencing the index into the array and pointer into the structure yields the memory area KER_PAR_DATA_STORAGE_1_partition_START, the actual physical address of the partition data area, which stores the value used in the code-level precondition CopyDIn_local_datain.

In our initial attempt to match a pre- and postcondition in the annotated C code with each desired pre- and postcondition, four different outcomes were possible:

- The desired assertion exactly matched an assertion in the annotated code.

- The desired assertion exactly matched an assertion in the annotated code but dereferencing was required.

- The desired assertion was a close match with an assertion in the annotated code.

- No code assertion exactly or approximately matched the desired assertion.

We worked with the group annotating the C code to ensure that assertions corresponding to all desired pre- and postconditions were added to and verified on the code. (In general, it is sufficient to include strongest postconditions implying our derived assertions.) To show correspondence between the pre- and postconditions in the code and the TLS, two tables were created for each TLS event. Tables 1 and 2 are the correspondence tables for the pre- and postconditions for the TLS event $e =$ Copy_Bfr1In_Data1In_i defined in Section 3.1. In the tables, s and $s' = T(e, s)$ represent the abstract pre- and poststate; s_c, and s'_c represent the concrete pre- and poststate; and Φ, which is formally defined in Section 4, maps abstract predicates to corresponding concrete predicates.

In the tables, the first column contains the label of a desired code-level pre- or postcondition, the second column gives the location (section number and assertion label) of the corresponding assertion

---
[3]This facilitates Floyd-Hoare reasoning at the code level.

Table 1: Mapping Preconditions in the Code to Preconditions in the TLS

Precondition $\Phi(\text{pre}_e)(s_c)$ Desired in the Code	Assertion in Annotated Code	Precondition $\text{pre}_e(s)$ in the TLS	Ref. No.	Description
CopyDIn_partition_id	§8.4,P5	$c_s = i$	(1)	Partition id is i
CopyDIn_priv	§8.4,TLS1*	$\text{AM}(e, B_i^1) = \text{R}$ $\text{AM}(e, D_i^1) = \text{W}$	(2)	R access for Input Buffer 1, W access for Data Area 1
CopyDIn_value_data	§8.4,P4*	$B_{i,s}^1$	-	Value of data in Input Buffer 1
CopyDIn_def_value_rest	§8.4,TLS4	$D_{i,s}^1$	-	Value of Data Area 1
CopyDIn_local_inbuffer	§8.4, TLS3*	-	-	Local variable for Input Buffer 1
CopyDIn_local_datain	§8.4,TLS2*	-	-	Local variable for Data Area 1

Table 2: Mapping Postconditions in the Code to Postconditions in the TLS

Postcondition $\Phi(\text{post}_e)(s_c, s_c')$ Desired in the Code	Assertion in Annotated Code	Postcondition $\text{post}_e(s, s')$ in the TLS	Ref. No.	Description
CopyDIn_copy_size_datain	§8.4,R2*	-	-	Wrong size $\to$ Error return
CopyDIn_copy_size_inbuffer	§8.4, R3*	-	-	Wrong size $\to$ Error return
CopyDIn_gamma_copy	§8.4, R7*	$D_{i,s'}^1 = \Gamma(B_{i,s}^1)$	(4)	Copy to Data Area 1
CopyDIn_gamma_rest	§8.4,TLS6	-	-	Rem Data Area 1 unchged
CopyDIn_sanitize	§8.4,TLS5*	$\mathcal{W}_{D,s'}^1[i] = false$	(3)	Data Area 1 not sanitized
CopyDIn_NOC	By inspection	$\text{NOC}_{\{\mathcal{W}_D^1[i], D_i^1\}}$	(5)	No other change

in the annotated C code, the third column contains the corresponding pre- or postcondition (if any) in the TLS, the fourth column gives the reference number of the corresponding assertion in the transform rule, and the fifth column briefly describes the assertion. In cases where no corresponding assertion exists in the TLS, '-' appears in both the third and fourth columns. An asterisk '*' in the second column indicates that, for equivalence between the assertion in the annotated code and the desired code assertion to hold, the assertion in the annotated code requires dereferencing.

4. FORMAL FOUNDATIONS

This section adapts the classical theory of *refinement* [4], a technique for proving that a concrete state machine model satisfies (i.e., is a refinement of) an abstract state machine model, to show that the kernel code conforms to the behavior captured in the TLS. To begin, a function α is defined that maps each concrete state at the code level to a corresponding abstract state in the TLS state machine Σ by relating variables at the concrete code level to variables at the abstract TLS level. Variables at the concrete level include variables in the code, predicates defined on the code, logical history variables, and memory areas. Among the most important memory areas treated as concrete state variables are the data areas and the input and output buffers assigned to each partition, which are central to reasoning about possible information flows. Because each possible value of a concrete state variable can be represented by some possible value of the corresponding abstract state variable, the map α from concrete to abstract state variables induces a map $\alpha : S_c \to S_a$ from concrete to abstract states in the obvious way.[4] Once α is defined at the level of states in terms of state variables, the set E_c of Event Code code segments transferring data either to or from an MAI in the current partition is identified, and α is extended to map each code segment e_c in E_c to a corresponding internal event $e_a = \alpha(e_c)$ in the TLS.

The map α from concrete states to abstract states provides a means to take any assertion P_a about abstract states and derive a corresponding assertion $\Phi(P_a)$ about concrete states as follows:

$$\Phi(P_a)(s_c) \triangleq P_a(\alpha(s_c)),$$

where s_c is any state in S_c. Analogously, α can be used to derive an assertion $\Phi(P_a)(s_c^1, s_c^2)$ about a pair of concrete states from an assertion about a pair of abstract states as follows:

$$\Phi(P_a)(s_c^1, s_c^2) \triangleq P_a(\alpha(s_c^1), \alpha(s_c^2)).$$

The map Φ is used to relate preconditions and postconditions in the code to preconditions and postconditions in the TLS (see Figure 2). Note that preconditions (at both levels) apply only to one state. To capture the fact that an event changes only certain state variables (indicated at the abstract level using NOC), the postconditions are represented at both levels as predicates on two states.

To establish equivalence between the behavior of the kernel code and a subset of the behavior modeled in the TLS, it is sufficient to prove, in the simplest case, that for every e_c in E_c,

1. Whenever the concrete code segment e_c is ready to execute in state s_c, some concrete precondition Pre_{e_c} holds, where Pre_{e_c} implies $\Phi(\text{Pre}_{e_a})$, the concrete precondition derived from the abstract precondition for $e_a = \alpha(e_c)$;

2. Whenever the concrete precondition Pre_{e_c} holds for the current program state s_c, some concrete postcondition Post_{e_c} holds for the pair of program states $(s_c, e_c(s_c))$ immediately before and immediately after execution of e_c, where Post_{e_c} implies $\Phi(\text{Post}_{e_a})$, the concrete postcondition derived from the abstract postcondition for e_a;

3. The diagram in Figure 2 commutes when conditions 1 and 2 are satisfied and $\text{Pre}_{e_c}(s_c)$ holds.

Provided $\text{Post}_{e_a}(s_a, s_a') \equiv (s_a' = e_a(s_a))$ (as holds for post_e in the TLS transform described in Section 3.1), to establish condition 3, it is sufficient to prove that $\text{Pre}_{e_c}(s_c) \Rightarrow \Phi(\text{Pre}_{e_a})(s_c)$ and that $\text{Post}_{e_c}(s_c, e_c(s_c)) \Rightarrow \Phi(\text{Post}_{e_a})(s_c, e_c(s_c))$. Establishing conditions 1–3 guarantees that whenever the code segment e_c executes in the code, there is an enabled event e_a in the TLS that causes a transition from the abstract image s_a under α of the concrete prestate s_c at the code level into an abstract state $e_a(s_a)$ that is the abstract image under α of the concrete poststate $e_c(s_c)$ at the code level. More concisely, conditions 1, 2, and 3 imply that there exists an abstract transition that models the concrete transition.

---

[4]To distinguish abstract from concrete entities, this section tags abstract entities with an a and concrete entities with a c; for example, S_a represents the abstract states s and S_c the concrete states s.

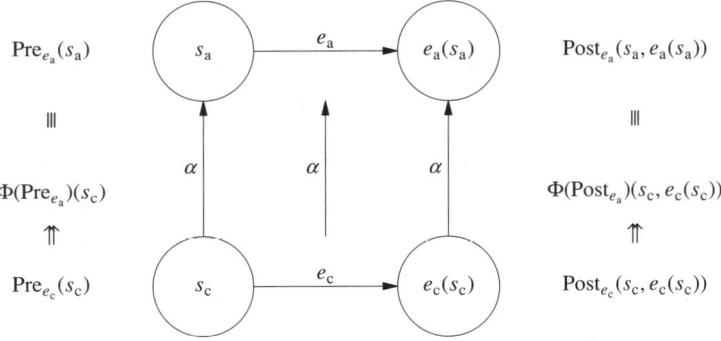

Figure 2: Relation between concrete and abstract transitions

The relation of Event Code segments to abstract events can be slightly more complex than shown in Figure 2. For example, in some cases, e_c may implement more than one event. However, these more complex cases can be handled similarly. When a concrete event implements n abstract events, for example, one looks for a partition $\text{Pre}_c \equiv \text{Pre}_c^1 \oplus ... \oplus \text{Pre}_c^n$ of the concrete precondition Pre_c such that when the ith part Pre_c^i holds, the code e_c implements the ith abstract event. Then, one establishes for each i a commutative diagram analogous to the diagram in Figure 2.

The argument that the kernel ensures data separation is based on relating executions of the kernel code to executions in the TLS. It begins by observing that α maps ED's initial state via α to an allowed initial state in the TLS. To support the remainder of the argument, the Event Code set E_c and the code-level map α are extended to cover the Other Code, and it is shown that the Trusted Code can be safely ignored. Most Event Code segments consist of a single program statement. In contrast, Other Code contains many lengthy code segments which simply manipulate local variables inside a function or procedure and do not map to any abstract event; such segments typically occur prior to an Event Code segment. We model these Other Code segments at the abstract level by a no−op ("do nothing") event implicitly included in the TLS.

Because every code segment in the Event or Other Code is modeled either by an abstract TLS event with concrete and abstract transitions related as in Figure 2 or by a no−op in the TLS, it follows that every execution of this part of the code corresponds to an execution in the TLS. Because parts of the Trusted Code have been verified and the remaining parts have been certified to cause no insecure information flows, modeling this code at the abstract level is unnecessary. Combining this reasoning with the additional assurance that α relates concrete data and buffer memory areas to abstract ones and thus models all information flows involving Memory Areas of Interest, it follows that all kernel behavior relevant to data separation at the concrete level is modeled at the abstract level. Thus, the Data Separation Property proved at the abstract level also holds at the concrete level.

5. LESSONS LEARNED

5.1 Software Design Decisions

Three software design decisions were critical in making code verification feasible. One major decision was to use a separation kernel, a single software module to mediate all memory accesses. A design that distributed the checking of memory accesses would have made the task of proving data separation much more difficult, if not impossible. A second critical decision was to keep the software simple. For example, once initiated, data processing in a partition was run to completion unless an exception occurred. In addition, ED's services were limited to the essential ones—the temptation to add new services late in development was resisted. A third critical decision was to enforce the "least privilege principle." For example, if a process only requires read access to a memory area, the kernel only grants read, and *not* read and write, access.

5.2 Top-Level Specification

One major challenge was to understand the required behavior of the separation kernel. Both scenarios and the SCR (Software Cost Reduction) tools [20, 19] were useful in validating and extending our understanding of the kernel behavior. To begin, we formulated several scenarios, i.e., sequences of events and how the kernel responded to those events. After specifying a state machine model of the kernel in SCR, we ran the scenarios through the SCR simulator. As expected, formulating the scenarios and running them through the simulator exposed gaps in our understanding. Both the scenarios and the questions raised were valuable in eliciting details of the required kernel behavior from ED's development team.

Keeping the size of the TLS small was critical for many reasons. It simplified communicating with the other stakeholders, changing the specification when the kernel behavior changed, translating the specification into TAME, and proving that the TLS enforced data separation.

The natural language representation of the TLS enabled stakeholders from differing backgrounds and with different objectives—the project manager, the software developers, and the evaluators—to communicate easily with the formal methods team about the kernel's required behavior. Discussion among these various stakeholders helped ensure that misunderstandings were avoided and issues resolved early in the certification process. This natural language representation of the TLS for ED contrasts with the representations used in many other formal specifications of secure systems, which are often expressed in specialized languages such as ACL2 (see, e.g, [18]). Moreover, any ambiguity inherent in the natural language representation was removed by translating the TLS into TAME, since the state machine semantics underlying TAME is expressed as a PVS theory.

5.3 Mechanized Verification

TAME's specification and proof support significantly simplified the verification effort. Translating the TLS into TAME required about three days. Because the number of memory areas is unspecified in the TLS, the overall memory content in the TLS had to be captured in TAME as a function from a set of memory areas to storable values. The higher order nature of PVS, which made this feasible, also contributed to the compactness of the TAME specification, which is only 368 lines long. In translating the TLS to TAME, the correspondence between entities in the natural language formulation and TAME entities was documented. Adjusting the TAME specification to reflect later changes in the TLS required less than three hours. Representing the five subproperties in TAME required about two hours.

About two weeks were needed to formally verify that the TLS enforces data separation. Adding and proving a new property (Kernel Integrity) suggested by an evaluator required under one hour. In proving the subproperties, a few days were needed to formulate an efficient proof approach. This exploration led to a new PVS strategy designed to simplify the proof guidance in the most complex proof. This strategy was useful in proving all five subproperties and has also been useful in other TAME applications. The proof of each subproperty completes in two minutes or less. Once the correct proof approach was identified, the time required to develop the proof scripts interactively in TAME was one day.

5.4 Showing Code Conformance

Two months were required to establish conformance between the TLS and the annotated C code. In the first month, we experimented with several different approaches for demonstrating conformance before the approach presented in this paper was selected. Once an approach was selected, the formal correspondence argument required one week. Three weeks were needed to construct the correspondence of Event Code to TLS events, i.e., developing the code level assertions necessary for the TLS pre- and postconditions to hold and locating the corresponding assertions in the annotated C code. One day was spent using the Xcode tool to locate all Event and Trusted Code and to verify that the permissions for the Other Code did not include access to MAIs. One week was needed to add the required assertions to the annotated code.

Our method for demonstrating code conformance relies on the notions of MAIs and Event Code. The extent to which our method can be extended to other applications depends on whether an analogous method of identifying Event Code (and Trusted Code) can be found. This is likely to be possible in other applications that must enforce data separation.

6. OPEN PROBLEMS
6.1 Code Annotations

For many years, researchers have recommended annotating code with pre- and postconditions and invariants (see, e.g., [26]). Code annotations are already used in practice. For example, software developers at Praxis annotate Spark programs with assertions and use tools to automatically check the validity of the assertions [11]. Moreover, at Microsoft, annotations are a mandated part of the software development process in the largest product groups [13]. However, manual annotation of source code with pre- and postconditions remains rare in the wider software development community because it is both tedious and error-prone. Hence, automated tools for checking code annotations would be extremely valuable. Even more valuable are tools that can construct pre- and postconditions automatically. One approach may be for a developer to generate some key pre- and postconditions. Given a small set of annotations, a tool could then generate additional annotations automatically.

6.2 A Code Conformance Proof Assistant

The semantic distance between the abstract TLS required for a Common Criteria evaluation and a low-level C program is huge. While the TLS describes the security-relevant program behavior in terms of sets, functions, and relations, the description of the behavior of a C program is in terms of low-level constructs, such as arrays, integers, and bits stored in registers and memory areas. Hence, automatic demonstration of conformance of low level C code to a TLS is unrealistic. A more realistic goal is a proof assistant with two inputs, a C program annotated with assertions and a TLS of the security-relevant functions of that program, for helping the user establish that the C program satisfies the TLS.

6.3 Automatic Code Generation

One promising way to obtain high assurance that an implementation satisfies critical security properties is to generate code automatically from a specification that has been proven to satisfy the properties. Automatic code generation is already feasible for some low-level specification languages such as Esterel [1]. While constructing efficient source code from more abstract specifications is possible for simple program constructs using simple data types (see, e.g., [31]), new research is needed to produce efficient code from specifications containing richer constructs and data types. Such technology should drastically reduce the effort required to produce efficient code and to increase assurance that the code satisfies critical security properties.

7. RELATED WORK

In the 1980s, the SCOMP [14], SeaView [23], LOCK [36], and Multinet Gateway [15] projects all applied formal methods to the specification and verification of systems. All developed TLSs and formal statements of the system security policies. For SCOMP, Multinet Gateway, and LOCK, the TLS was shown formally to satisfy the security policy. For SeaView, only two of 31 operations in the TLS were verified against the security policy model [37]. Conformance between the TLS and the SCOMP code was shown by constructing several mappings: English language to TLS, TLS to pseudo-code, and TLS to actual code [12]. The mapping was top down from the TLS to code; as a result, some code was unmapped. This approach is similar to our mapping of Event Code to the TLS, although the mapping is in the other direction. The LOCK project constructed mappings partially relating the TLS to the source code; specification-based testing provided additional evidence of correspondence. In Multinet Gateway, verification conditions were generated to show conformance between the specification and the code. If and how these conditions were discharged is unclear. Each project used tools to aid in specification and verification: SCOMP used HDM [30], Seaview used EHDM [33], and Multinet Gateway and LOCK used Gypsy [16]. More recently, in 2006, we formulated a second possible approach to software verification, based on TAME, which uses verified formal pseudo-code as "glue" relating a TLS to actual code [10].

In [18, 6], Greve, Wilding, and Vanfleet (GWV) present an ACL2 model for a generic separation kernel. In the model, a function describes the possible information flows between memory areas. This notion of flow is not as fine-grained as in our model, where access (with its possible information flows) is granted to each process only when it executes in a partition, thus providing least privilege in addition to separation. In the GWV approach, separation includes No-Exfiltration and No-Infiltration but not Temporal Separation, since the model does not allow reconfigurable partitions. How the GWV model was used to verify the AAMP7 microprocessor is described in [17, 29]. A traditional verification process was followed: build a formal security policy, an abstract and detailed model, and an implementation; prove that the abstract model satisfies the security policy; and show correspondence between the abstract and detailed models and between the detailed model and the implementation. Whether correctness was proven at either the detailed design level or code level is unclear.

8. CONCLUSIONS

This paper has introduced a novel and affordable approach for verifying security down to the source code level. The approach begins with a well-defined security policy, builds the minimal state machine model needed to prove that the model satisfies the policy, and proves, using a mechanical verifier, that the security model satisfies the policy. Once complete, the code is annotated with preconditions and postconditions and then partitioned into Event, Trusted, and Other Code. The final step is to 1) demonstrate conformance of the Event Code and the code pre- and postconditions with the internal events and pre- and postconditions of the TLS and 2) show that the Trusted Code and the Other Code are benign.

Tools such as model checkers and theorem provers are already available for verifying that a formal specification satisfies a security policy. A research challenge is to develop tools 1) for validating and constructing pre- and postconditions from source code, including C code, 2) to help show conformance of annotated code with a TLS, and 3) to automatically construct efficient, provably correct code from specifications. Research that addresses these three problems should significantly increase the affordability of constructing verified security-critical software.

9. ACKNOWLEDGMENTS

We acknowledge the monumental effort of the group who annotated the kernel code with pre- and postconditions and of the ED project leader, who had the foresight to include a separation kernel and to keep the design simple. Without the annotated code and solid design decisions, our effort would have been impossible. We also thank the members of the ED design team for answering questions about ED's operational behavior.

10. REFERENCES

[1] SCADE Tool Suite. Tools and documentation available at http://www.esterel-technologies.com/products/scade-suite.
[2] Xcode 2.1. Tool and documentation available at http://developer.apple.com/tools/xcode/index.html.
[3] Common criteria for information technology security evaluation, Parts 1–3. Tech. Report CCIMB-2004-01-001—003, Version 2.2, Rev. 256, Jan. 2004.
[4] M. Abadi and L. Lamport. The existence of refinement mappings. *Theoretical Computer Science*, 82(2):253–284, 1991.
[5] C. Adams. Keeping secrets in integrated avionics. *Aviation Today*, 2004.
[6] J. Alves-Foss and C. Taylor. An analysis of the GWV security policy. In *5th International Workshop on ACL2 Prover and Its Applications (ACL2-2004)*, 2004.
[7] J.P. Anderson. Computer security technology planning study. Technical Report ESD-TR-73-51, ESD/AFSC, Hanscom AFB, Bedford, MA, 1972.
[8] M. Archer, C. Heitmeyer, and E. Riccobene. Proving invariants of I/O automata with TAME. *Automated Software Engineering*, 9:201–232, 2002.
[9] M. Archer. TAME: Using PVS strategies for special-purpose theorem proving. *Annals of Mathematics and Artificial Intelligence*, 29(1-4):131–189, 2000.
[10] M. Archer and E. Leonard. Establishing high confidence in code implementations of algorithms using formal verification of pseudo-code. In *Proc., Third International Verification Workshop (VERIFY '06)*, Seattle, WA, August 2006.
[11] J. Barnes. *High Integrity Software: The SPARK Approach to Safety and Security*. Addison-Wesley, 2003.
[12] T.V. Benzel. Analysis of a kernel verification. In *Proceedings of the IEEE Security and Privacy Conference*, April 1984.
[13] M. Das. Formal specifications on industrial-strength code – From myth to reality. In *Proc., Computer-Aided Verification (CAV 2006)*, Seattle, WA, August 2006.
[14] L.J. Fraim. Secure office management system: The first commodity application on a trusted system. In *Proc., 1987 Fall Joint Computer Conference on Exploring Technology: Today and Tomorrow*, 1987.
[15] S. Gerhart, D. Craigen, and T. Ralston. Case study: Multinet Gateway System. *IEEE Software*, pages 37–39, 1994.
[16] D.I. Good. *Mechanical Proofs about Computer Programs*, chapter 3, pages 55–75. Prentice Hall, 1985.
[17] D. Greve, R. Richards, and M. Wilding. A summary of intrinsic partitioning verification. In *Fifth International Workshop on the ACL2 Prover and Its Applications (ACL2-2004)*, 2004.
[18] D. Greve, M. Wilding, and W.M. Vanfleet. A separation kernel formal security policy. In *Fourth International Workshop on the ACL2 Prover and Its Applications (ACL2-2003)*, July 2003.
[19] C. Heitmeyer, M. Archer, R. Bharadwaj, and R. Jeffords. Tools for constructing requirements specifications: The SCR toolset at the age of ten. *International Journal of Computer Systems Science and Engineering*, 1:19–35, 2005.
[20] C. L. Heitmeyer, R. D. Jeffords, and B. G. Labaw. Automated consistency checking of requirements specifications. *ACM Transactions on Software Engineering and Methodology*, 5(3):231–261, 1996.
[21] J. Jürjens. Sound methods and effective tools for model-based security engineering with UML. In *Proc., 27th International Conference on Software Engineering*, St. Louis, MO, 2005.
[22] C. E. Landwehr, C. L. Heitmeyer, and J. McLean. A security model for military message systems. *ACM Transactions on Computer Systems*, 2(3):198–222, August 1984.
[23] T.F. Lunt, D.E. Denning, R.R. Schell, M. Heckman, and W.R. Shockley. The SeaView security model. *IEEE Transactions on Software Engineering*, 16(6), 1990.
[24] J. McLean, C. Landwehr, and C. Heitmeyer. A formal statement of the military message system security model. In *Proc., 1984 IEEE Symposium on Security and Privacy*, pages 188–194, 1984.
[25] C. Meadows. Analysis of the internet key exchange protocol using the NRL protocol analyzer. In *Proc., IEEE Symposium on Security and Privacy*, Oakland, 1999.
[26] B. Meyer. Applying "design by contract". *IEEE Computer*, 25(10):40–51, 1992.
[27] J Strother Moore, Thomas W. Lynch, and Matt Kaufmann. A mechanically checked proof of the AMD5K86TM floating-point division program. *IEEE Transactions on Computers*, 7(9), 1998.
[28] S. Owre, J. Rushby, N. Shankar, and F. von Henke. Formal verification for fault-tolerant architectures: Prolegomena to the design of PVS. *IEEE Transactions on Software Eng.*, 21(2), 1995.
[29] R. Richards, D. Greve, M. Wilding, and W.M. Vanfleet. The common criteria, formal methods and ACL2. In *Fifth International Workshop, ACL2 Prover and Its Applications (ACL2-2004)*, 2004.
[30] L. Robinson. The HDM handbook, volume 1: The foundations of HDM, SRI project 4828. Tech. report, SRI International, Menlo Park, CA, June 1979.
[31] T. Rothamel, C. Heitmeyer, E. Leonard, and A. Liu. Generating optimized code from SCR specifications. In *Proc., ACM SIGPLAN/SIGBED Conf. on Languages, Compilers and Tools for Embedded Systems (LCTES 2006)*, Ottawa, Canada, June 2006.
[32] J. Rushby. A formally verified algorithm for clock synchronization under a hybrid fault model. In *13th ACM Symposium on Principles of Distributed Computing*, Los Angeles, CA, August 1994.
[33] J. Rushby, F. von Henke, and S. Owre. An introduction to formal specification and verification using EHDM. Technical Report CSL-91-2, SRI International, Menlo Park, CA, February 1991.
[34] J. Rushby. Design and verification of secure systems. In *Proc., Eighth ACM Symposium on Operating System Principles*, pages 12–21, December 1981.
[35] N. Shankar, S. Owre, J. M. Rushby, and D. W. J. Stringer-Calvert. PVS Prover Guide, Version 2.4. Technical report, Computer Science Lab, SRI Internat., Menlo Park, CA, November 2001.
[36] R. Smith. Cost profile of a highly assured, secure operating system. *ACM Transactions on Information and System Security*, 4(1):72–101, 2001.
[37] R.A. Whitehurst and T.F. Lunt. The SeaView verification. In *Proc., Computer Security Foundations Workshop II, 1989*, June 1989.

Beyond Separation of Duty: An Algebra for Specifying High-level Security Policies

Ninghui Li
ninghui@cs.purdue.edu

Qihua Wang
wangq@cs.purdue.edu

Center for Education and Research in Information Assurance and Security
and Department of Computer Science
Purdue University

ABSTRACT

A high-level security policy states an overall requirement for a sensitive task. One example of a high-level security policy is a separation of duty policy, which requires a sensitive task to be performed by a team of at least k users. It states a high-level requirement about the task without the need to refer to individual steps in the task. While extremely important and widely used, separation of duty policies state only quantity requirements and do not capture qualification requirements on users involved in the task. In this paper, we introduce a novel algebra that enables the specification of high-level policies that combine qualification requirements with quantity requirements motivated by separation of duty considerations. A high-level policy associates a task with a term in the algebra and requires that all sets of users that perform the task satisfy the term. We give the syntax and semantics of the algebra and study algebraic properties of its operators. We also study several computational problems related to the algebra.

Categories and Subject Descriptors

D.4.6 [**Operating Systems**]: Security and Protection—*Access controls*; K.6.5 [**Management of Computing and Information Systems**]: Security and Protection; F.2.2 [**Analysis of Algorithms and Problem Complexity**]: Nonnumerical Algorithms and Problems—*Complexity of proof procedures*

General Terms

Security, Theory, Languages

Keywords

Access Control, Separation of Duty, Policy Design

1. INTRODUCTION

Separation of Duty (SoD) is widely recognized as a fundamental principle in computer security [7, 18]. In its simplest form, the principle states that a sensitive task should be performed by two different users acting in cooperation. The concept of SoD has long existed before the information age; it has been widely used in, for example, the banking industry and the military, sometimes under the name "the two-man rule". More generally, an SoD policy requires the cooperation of at least k different users to complete the task. SoD has been identified as a high-level mechanism that is "at the heart of fraud and error control" [7]. An SoD policy is a high-level policy in the sense that it does not restrict which users are allowed to carry out the individual steps in a sensitive task, but rather states an overall requirement that must be satisfied by any set of users that together complete a task. In many situations, it is not enough to require only that k different users be involved in a sensitive task; there are also minimal qualification requirements for these users. For example, one may want to require users that are involved to be physicians, certified nurses, certified accountants, or directors of a company. Because a high-level SoD policy states only a quantity requirement and does not express such qualification requirements, existing work addresses this by specifying such requirements at individual steps of a task. For example, if a policy requires a manager and two clerks to be involved in a task, one may divide the task into three steps and require two clerks to each perform step 1 and step 3, and a manager to perform step 2.

Specifying such requirements at the lower level of steps, however, results in the loss of the following important advantages offered by a higher-level policy. First, as the specification abstracts away details of how a task is implemented, a higher-level policy is likely to be closer to organizational policy guidelines. It would thus be easier for administrators to specify and understand such policies. Second, a high-level policy can be specified even before the actual steps involved in a task are designed. In fact, a formal specification of task-level policies may help in the process of designing steps to implement the task. Third, a task-level policy is often more flexible than a corresponding step-level policy, which can be more restrictive than necessary. For example, to enforce a task-level policy that requires a manager and two clerks, a step-level policy may require a manager to execute a particular step, which is too restrictive. Finally, a higher-level policy specification allows flexibility in the choice of the mechanism for enforcing the policy. For example, one can use either static enforcement or dynamic enforcement. In static enforcement, one ensures that any set of users that together have enough permissions to perform the task satisfy the high-level policy. In dynamic enforcement, one records the history of who performs which steps in a task instance. When policies have to be associated with steps in a task, all the advantages discussed above are lost.

In this paper we introduce a novel algebra that enables the specification of high-level policies that combine qualification requirements with quantity requirements motivated by separation of duty considerations. A term in our algebra specifies a requirement on sets of users (we call these usersets). A high-level policy, rather than referring to the steps, simply associates a task with a term in the algebra. This policy requires that all sets of users that complete an instance of the task satisfy the term. Our algebra has four binary operators: $\sqcup, \sqcap, \odot, \otimes$, and two unary operators $\neg, +$. An SoD policy that requires 3 different users can be expressed using the term (All $\otimes$ All $\otimes$ All), where All is a keyword that refers to the set of all users. A policy that requires either a manager or two different clerks is expressed using the term (Manager $\sqcup$ (Clerk $\otimes$ Clerk)).

We define the syntax and semantics of terms in the algebra, and study the algebraic properties of the operators. We show that all four binary operators are commutative and associative. We also show that $\sqcap$ and $\sqcup$ distribute over each other and both $\odot$ and $\otimes$ distribute over $\sqcup$. The four binary operators result in 12 ordered pair of operators. For the eight pairs other than the four mentioned above, distributivity does not hold.

We then study the Term Satisfiability (TSAT) problem and the Userset-Term Satisfaction (UTS) problem. The TSAT problem asks whether a given term is satisfiable at all. We show that TSAT is **NP**-complete in general and remains **NP**-complete in certain sub-algebras with only a subset of the operators. We also identify a sub-algebra whose satisfiability problem is efficiently solvable.

The UTS problem asks whether a userset satisfies a term. We show that the UTS problem is **NP**-complete in general. To better understand the properties of the operators, we also study computational complexities for the UTS problem in all sub-algebras with only a subset of the operators. We identify syntactic restrictions so that even for terms with all six operators, as long as they satisfy these restrictions, UTS can be solved efficiently. We also present a heuristic algorithm for solving UTS in the general case, and show that for terms whose size is not very large, even if they do not satisfy the syntactic restrictions, UTS can be solved in reasonable amount of time.

Finally, some operators in our algebra are similar to the ones in regular expressions. A regular expression describes a set of strings, while a term in our algebra describes a set of sets. The relationships between the two are discussed.

The remainder of the paper is organized as follows. We introduce the syntax and semantics of the algebra in Section 2. We study the TSAT problem in Sections 3 and the UTS problem in Section 4. In Section 5, we discuss limitations of the algebra and extensions to it, as well as the relationship with regular expressions. We discuss related work in Section 6 and conclude with Section 7. Proofs not included in the main body are included in the appendices.

2. THE ALGEBRA

In this section, we introduce our algebra for expressing high-level security policies.

2.1 Syntax and Semantics

In our definition of the algebra, we use the notion of roles. We use a role to denote a set of users that have some common qualification or common job responsibility. We emphasize, however, that the algebra is not restricted to Role-Based Access Control (RBAC) systems [21]. In our algebra, a role is simply a named set of users. The notion of roles can be replaced by groups or user attributes. We use $\mathcal{U}$ to denote the set of all users, and $\mathcal{R}$ to denote the set of all roles.

Definition 1 (Terms in the Algebra). Terms in the algebra are defined as follows:

- An *atomic term* takes one of the following three forms: a role $r \in \mathcal{R}$, the keyword All, or a set $S \subseteq \mathcal{U}$ of users.

- An atomic term is a *term*; furthermore, if ϕ_1 and ϕ_2 are terms, then $\neg\phi_1, \phi_1^+, (\phi_1 \sqcup \phi_2), (\phi_1 \sqcap \phi_2), (\phi_1 \otimes \phi_2)$, and $(\phi_1 \odot \phi_2)$ are also terms, with the following restriction: For $\neg\phi_1$ or ϕ_1^+ to be a term, ϕ_1 must be a *unit term*, that is, it must not contain $^+, \otimes$, or $\odot$.

The unary operator $\neg$ has the highest priority, followed by the unary operator $+$, then by the four binary operators (namely $\sqcap, \sqcup, \odot, \otimes$), which have the same priority.

We now give several simple example terms to illustrate the intuition behind the operators in the algebra. The term "(Manager $\sqcap$ Accountant)" requires a user that is both a Manager and an Accountant. The term "(Manager $\sqcap \neg\{Alice, Bob\}$)" requires a user that is a manager, but is neither Alice nor Bob; here, the sub-term "$\neg\{Alice, Bob\}$" implements a blacklist. The term "(Physician $\sqcup$ Nurse)" requires a user that is either a Physician or a Nurse. The term "(Manager $\odot$ Clerk)" requires a user who is a Manager and a user who is a Clerk; however, when one user is both a Manager and a Clerk, that user by itself also satisfies the requirement. The term "((All $\otimes$ All) $\otimes$ All)" requires 3 different users. The keyword All allows us to refer to the set of all users. The term "Accountant$^+$" requires a set of one or more users, where each user in the set is an Accountant.

To formally assign meanings to terms, we need to first assign meanings to the roles used in the term. For this, we introduce the notion of configurations.

Definition 2 (Configurations). A *configuration* is given by a pair $\langle U, UR \rangle$, where U denotes the set of all users in the configuration, and $UR \subseteq U \times \mathcal{R}$ determines role memberships. When $(u, r) \in UR$, we say that u is a member of the role r.

Note that in configuration $\langle U, UR \rangle$, UR should not be confused with the user-role assignment relation UA in RBAC. When an RBAC system has both UA and a role hierarchy RH, the two relations UA and RH together determine UR.

Definition 3 (Satisfaction of a Term). Given a configuration $\langle U, UR \rangle$, we say that a userset X *satisfies* a term ϕ under $\langle U, UR \rangle$ if and only if one of the following holds[1]:

- The term ϕ is the keyword All, and X is a singleton set $\{u\}$ such that $u \in U$.

- The term ϕ is a role r, and X is a singleton set $\{u\}$ such that $(u, r) \in UR$.

- The term ϕ is a set S of users, and X is a singleton set $\{u\}$ such that $u \in S$.

- The term ϕ is of the form $\neg\phi_0$ where ϕ_0 is a unit term, and X is a singleton set that does not satisfy ϕ_0.

- The term ϕ is of the form ϕ_0^+ where ϕ_0 is a unit term, and X is a nonempty userset such that for every $u \in X$, $\{u\}$ satisfies ϕ_0.

- The term ϕ is of the form $(\phi_1 \sqcup \phi_2)$, and either X satisfies ϕ_1 or X satisfies ϕ_2.

---
[1] We sometimes say X satisfies ϕ, and omit "under $\langle U, UR \rangle$" when it is clear from the context.

- The term ϕ is of the form $(\phi_1 \sqcap \phi_2)$, and X satisfies both ϕ_1 and ϕ_2.

- The term ϕ is of the form $(\phi_1 \otimes \phi_2)$, and there exist usersets X_1 and X_2 such that $X_1 \cup X_2 = X$, $X_1 \cap X_2 = \emptyset$, X_1 satisfies ϕ_1, and X_2 satisfies ϕ_2.

- The term ϕ is of the form $(\phi_1 \odot \phi_2)$, and there exist usersets X_1 and X_2 such that $X_1 \cup X_2 = X$, X_1 satisfies ϕ_1, and X_2 satisfies ϕ_2. This differs from the definition for $\otimes$ in that it does not require $X_1 \cap X_2 = \emptyset$.

For example, given the term (Manager $\odot$ Clerk), and the configuration $\langle U = \{Alice, Bob, Carl\}, UR \rangle$, in which UR is such that: $U_{\text{Manager}} = \{Alice\}$ and $U_{\text{Clerk}} = \{Alice, Carl\}$, where $U_r = \{u \mid (u,r) \in UR\}$, we have $\{Alice\}$ satisfies the term, $\{Alice, Carl\}$ also satisfies the term, but $\{Alice, Bob\}$ and $\{Bob, Carl\}$ do not satisfy the term.

2.2 Examples

The following examples help illustrate the expressive power of the algebra.

- $\{Alice, Bob, Carl\} \otimes \{Alice, Bob, Carl\}$

  This term requires any two users out of the list of three.

- (Accountant $\sqcup$ Treasurer)$^+$

  This term requires that all participants must be either an Accountant or a Treasurer. But there is no restriction on the number of participants.

- ((Manager $\odot$ Accountant) $\otimes$ Treasurer)

  This term requires a Manager, an Accountant, and a Treasurer; the first two requirements can be satisfied by a single user.

- ((Physician $\sqcup$ Nurse) $\otimes$ (Manager $\sqcap \neg$Accountant))

  This term requires two different users, one of which is either a Physician or a Nurse, and the other is a Manager, but not an Accountant.

- ((Manager $\odot$ Accountant $\odot$ Treasurer) $\sqcap$ (Clerk $\sqcap \neg\{Alice, Bob\})^+$)

  This term requires a Manager, an Accountant and a Treasurer. In addition, everybody involved must be a Clerk and must not be *Alice* or *Bob*.

Definition 4 (Value of a term). Given a configuration $\langle U, UR \rangle$ and a term ϕ, we use $S_{\langle U, UR \rangle}(\phi)$ to denote the set of all usersets that satisfy ϕ under $\langle U, UR \rangle$, and call this the *value* of term ϕ under the configuration.

Consider the term ϕ = ((Manager $\odot$ Accountant $\odot$ Treasurer) $\sqcap$ (Clerk $\sqcap \neg\{Alice, Bob\})^+$) and the configuration $\langle U = \{Alice, Bob, Carl, Doris, Elaine, Frank\}, UR \rangle$, in which UR is such that:

U_{Manager} = $\{Alice, Doris, Elaine\}$
$U_{\text{Accountant}}$ = $\{Doris, Frank\}$
$U_{\text{Treasurer}}$ = $\{Bob, Carl, Doris\}$
U_{Clerk} = $\{Alice, Bob, Carl, Doris, Frank\}$

The sub-term (Clerk $\sqcap \neg\{Alice, Bob\})^+$ means that only subsets of $\{Carl, Doris, Frank\}$ may satisfy ϕ. We have

$S_{\langle U, UR \rangle}(\phi) = \{\{Doris\}, \{Carl, Doris\}, \{Doris, Frank\}, \{Carl, Doris, Frank\}\}$

That is, there are four usersets that satisfy the term ϕ.

2.3 Algebraic Properties

We now introduce the notion of equivalence among terms, which enables us to study the algebraic properties of the operators in the algebra.

Definition 5 (Term Equivalence). We say that two terms ϕ_1 and ϕ_2 are *equivalent* (denoted by $\phi_1 \equiv \phi_2$) when for every userset X and every configuration $\langle U, UR \rangle$, X satisfies ϕ_1 under $\langle U, UR \rangle$ if and only if X satisfies ϕ_2 under $\langle U, UR \rangle$. In other words, $\phi_1 \equiv \phi_2$ if and only if $\forall \langle U, UR \rangle \left[S_{\langle U, UR \rangle}(\phi_1) = S_{\langle U, UR \rangle}(\phi_2) \right]$.

Using a straightforward induction on the structure of terms, one can show that if $\phi_1 \equiv \phi_2$, then, for any term ϕ in which ϕ_1 occurs, let ϕ' be the term obtained by replacing in ϕ one or more occurrences of ϕ_1 with ϕ_2, we have $\phi \equiv \phi'$.

Theorem 1. *The operators have the following algebraic properties:*

1. *The operators $\sqcup, \sqcap, \odot, \otimes$ are commutative and associative. That is, for each* op $\in \{\sqcup, \sqcap, \odot, \otimes\}$, *and any terms ϕ_1, ϕ_2, and ϕ_3, we have $(\phi_1$ op $\phi_2) \equiv (\phi_2$ op $\phi_1)$ and $((\phi_1$ op $\phi_2)$ op $\phi_3) \equiv (\phi_1$ op $(\phi_2$ op $\phi_3))$.*

2. *The operators $\sqcup$ and $\sqcap$ distribute over each other. That is, $(\phi_1 \sqcup (\phi_2 \sqcap \phi_3)) \equiv ((\phi_1 \sqcup \phi_2) \sqcap (\phi_1 \sqcup \phi_3))$ and $(\phi_1 \sqcap (\phi_2 \sqcup \phi_3)) \equiv ((\phi_1 \sqcap \phi_2) \sqcup (\phi_1 \sqcap \phi_3))$.*

3. *The operator $\odot$ distributes over $\sqcup$. That is, $(\phi_1 \odot (\phi_2 \sqcup \phi_3)) \equiv ((\phi_1 \odot \phi_2) \sqcup (\phi_1 \odot \phi_3))$.*

4. *The operator $\otimes$ distributes over $\sqcup$. That is, $(\phi_1 \otimes (\phi_2 \sqcup \phi_3)) \equiv ((\phi_1 \otimes \phi_2) \sqcup (\phi_1 \otimes \phi_3))$.*

5. *No other ordered pair of binary operators have the distributive property. (There are 12 such pairs altogether; the four of them listed above have the distributive property.)*

6. $(\phi_1 \sqcap \phi_2)^+ \equiv (\phi_1^+ \sqcap \phi_2^+)$

7. *DeMorgan's Law:* $\neg(\phi_1 \sqcap \phi_2) \equiv (\neg\phi_1 \sqcup \neg\phi_2)$, $\neg(\phi_1 \sqcup \phi_2) \equiv (\neg\phi_1 \sqcap \neg\phi_2)$

See Appendix A for the proof of the above theorem, which also gives a counter example for each case that the distributive property does not hold.

Because of the associativity properties, in the rest of this paper we omit parentheses in a term when doing so does not cause any confusion.

We now describe some other facts about the operators, to further illustrate the operators and their relationships.

- Any userset that satisfies $(\phi_1 \sqcap \phi_2)$ also satisfies $(\phi_1 \sqcup \phi_2)$, but not the other way around.

- Any userset that satisfies $(\phi_1 \sqcap \phi_2)$ also satisfies $(\phi_1 \odot \phi_2)$, but not the other way around.

- Any userset that satisfies $(\phi_1 \otimes \phi_2)$ also satisfies $(\phi_1 \odot \phi_2)$, but not the other way around.

- Any userset that satisfies $\phi_1^+ \sqcup \phi_2^+$ also satisfies $(\phi_1 \sqcup \phi_2)^+$, but not the other way around.

  If X satisfies $(\phi_1^+ \sqcup \phi_2^+)$, then X satisfies either ϕ_1^+ or ϕ_2^+. Without loss of generality, assume that X satisfies ϕ_1^+. Then, for every $u \in X$, $\{u\}$ satisfies ϕ_1 and thus satisfies $(\phi_1 \sqcup \phi_2)$. Hence, X satisfies $(\phi_1 \sqcup \phi_2)^+$. For the other direction, if $\{u_1\}$ satisfies ϕ_1 but not ϕ_2, and $\{u_2\}$ satisfies ϕ_2 but not ϕ_1, then $\{u_1, u_2\}$ satisfies $(\phi_1 \sqcup \phi_2)^+$ but not $\phi_1^+ \sqcup \phi_2^+$.

2.4 Rationale of the Design of the Algebra

We now discuss the rationale for some of the design decisions for the algebra.

Monotonicity SoD policies satisfy the property of monotonicity; that is, if an SoD policy requires two users to perform a task, then having three or more users certainly satisfies this policy. Similarly, one may want a security algebra like ours to also satisfy the monotonicity property; that is, if a userset X satisfies a term ϕ, then any superset of X also satisfies ϕ. McLean [15] adopts this property in his security algebra for N-person policies.

Our algebra is designed to support both monotonic policies and policies that are not monotonic. for example, the term (Accountant $\otimes$ Accountant) can be satisfied only by a set of two users; a set that contains more than two users cannot satisfy the term. More generally, in Definition 3, term satisfaction is defined in such a way that every user in the userset is used to satisfy certain component of the term. No "extra" user is allowed.

We have considered a design having monotonicity property, in which we call the notion of satisfaction in Definition 3 "strict satisfaction" and define a userset X satisfies a term ϕ if and only if X contains a subset that strictly satisfies ϕ. The monotonicity property follows directly. We chose our current design over the one that has monotonicity because it is more expressive. Consider the following example. When one says that "a task requires two Accountants", this may mean one of the following three policies:

1. The task must be performed by a set of two users, both of who are Accountants. A group containing more (or less) than two people is not allowed.

2. The task must be performed by a set that contains two Accountants. In particular, a userset that contains two Accountants and a third user who is not an Accountant is allowed to perform the task.

3. The task must be performed by a set of two or more Accountants. In particular, a set of three Accountants can perform the task, but a set of two Accountants and one non-Accountant cannot. This ensures that everyone involved in the task has the qualification of an Accountant.

Policies 1 and 3 cannot be expressed using an algebra that has the monotonicity property. Suppose that one tries to use a term ϕ to express policy 1 (or policy 3) in an algebra that has the monotonicity property, then a set X of two Accountants satisfies ϕ. By monotonicity property, any superset of X also satisfies ϕ. This violates the intention of policies 1 and 3. More generally, a monotonic algebra cannot express policies that disqualify usersets that contain extra users, nor can it express security requirements in the form of "all involved users must satisfy certain requirements".

By dropping the monotonicity property, our algebra is able to express all the three policies. Policy 1 is expressed using the term (Accountant$\otimes$Accountant). Policy 2 is expressed using the term ((Accountant $\otimes$ Accountant) $\odot$ All$^+$). Note that the term All$^+$ can be satisfied by any nonempty userset. Policy 3 is expressed using the term (Accountant $\otimes$ Accountant$^+$).

Restrictions on "$\neg$" and "+" The syntax of our algebra (Definition 1) restricts that the two operators "$\neg$" and "+" be applied only to unit terms, i.e., those terms that do not contain $\odot$, $\otimes$, or $+$. The motivation for this design decision is the psychological acceptability principle [18]. We would like each operator to have a clear and intuitive meaning so that when one writes down a policy as a term, there is less chance for making mistakes and one is more confident that the term expresses the intended policy.

When $\neg$ is applied to a unit term, it expresses negative qualification about one user; this has a clear meaning; the term $\neg\phi_0$ means a user that does not satisfy ϕ_0. However, if $\neg$ is applied to a term that involves $\odot$, $\otimes$, or $+$; then the meaning becomes unclear. Consider the term $\neg$(Accountant $\odot$ Manager). Any userset of size three does not satisfy (Accountant $\odot$ Manager); therefore, it should satisfy $\neg$(Accountant $\odot$ Manager), even if every user in the userset is both an Accountant and a Manager. It is unclear to us what kind of real-world security policies such a term expresses.

The term ϕ_0^+, when ϕ_0 is a unit term, has a clear meaning; it means that every user must satisfy ϕ_0. The same term, when ϕ_0 involves operators such as $\odot$ and $\otimes$, has at least two possible meanings. One is to interpret $+$ as the closure operator of $\odot$, that is, a userset X satisfies ϕ_0^+ if and only if X can be divided into a number of (*possibly overlapping*) subsets such that each subset satisfies ϕ_0. The other is to interpret $+$ as the closure operator for $\otimes$, that is, a userset X satisfies ϕ_0^+ if and only if X can be divided into a number of *disjoint* subsets such that each subset satisfies ϕ_0. The two meanings coincide when ϕ_0 is a unit term. We could use two operators, one for each meaning, and allow them to be applied to non-unit terms. However, this adds complexity to the algebra and we have not seen the need for this. For simplicity and usability, we chose to allow $+$ only be applied on unit terms. The algebra can be extended to have two closure operators that can be applied to non-unit terms, if the need for them arises in the future.

2.5 Enforcing Policies Specified in the Algebra

To use the algebra to specify high-level security policies, the administrators first identify sensitive tasks and then for each sensitive task, specify a term such that every set of users that together perform the task must satisfy the term. For instance, a simple SoD policy that requires at least two users to perform the task can be specified as (All $\otimes$ All$^+$). Our algebra also allows the specification of more sophisticated policies based on user qualifications. In summary, a security policy is a pair $\langle task, \phi \rangle$, where ϕ is a term; it means that only usersets that satisfy ϕ can perform $task$.

Once a policy is specified, it needs to be enforced. Enforcement techniques for policies in this algebra can benefit from research in enforcement techniques for Separation of Duty policies [8, 13, 16, 20, 22]. We say that a policy $\langle task, \phi \rangle$ is monotonic if the term ϕ is monotonic, i.e., if a userset X satisfies ϕ, then any superset of X also satisfies ϕ. A monotonic policy can be enforced either statically or dynamically, whereas a policy that is not monotonic cannot be enforced statically.

To statically enforce a policy $\langle task, \phi \rangle$, one identifies the set of all permissions that are needed to perform the task, and requires that any userset such that users in the set together possess all these permissions satisfies the term ϕ. That is, one defines *static safety policies*, each of which takes the form $\mathsf{sp}\langle P, \phi \rangle$, where $P = \{p_1, \cdots, p_n\}$ is a set of permissions. Such a policy means that any userset such that all users in the set together have all permissions in P must satisfy ϕ. Note that if a userset has all permissions in P, then its superset also has all permissions in P; therefore, static enforcement is only for monotonic policies.

To dynamically enforce a policy $\langle task, \phi \rangle$, one identifies the steps in performing the task. And the system maintains a history of each instance of a task, which includes information on who has performed which step. For any task instance, one can compute the set of users (denoted as U_{past}) who have performed at least one step on the instance. Before a user u performs a step on the instance, the system checks to ensure that there exists a superset of

$U_{past} \cup \{u\}$ that can satisfy ϕ upon finishing all steps of the task. In particular, if u is about to perform the last step of the task instance, it is required by the policy that $U_{past} \cup \{u\}$ satisfies ϕ. An example on dynamic enforcement of policy $\langle task, \phi \rangle$ is given as follow:

A company has a high-level policy $\langle Purchase, (\texttt{Manager} \odot (\texttt{Clerk} \otimes \texttt{Clerk})) \rangle$ which states that a `Manager` and two `Clerks` are required to purchase supplies for the company. The task *Purchase* consists of three steps which are *MakeOrder*, *PrepareCheck* and *SignCheck*. Step-specific requirements state that *MakeOrder* must be performed by `Manager`, while *PrepareCheck* and *SignCheck* may be performed by either `Manager` or `Clerk`. Suppose *Alice* is a `Manager` who made an order and now tries to prepare a check for the order. If *Alice* is a `Clerk`, then the system will allow her to do so. The reason is that as long as a `Clerk` different from *Alice* (say *Bob*) signs the check later, $\{Alice, Bob\}$ satisfies the high-level requirement ($\texttt{Manager} \odot (\texttt{Clerk} \otimes \texttt{Clerk})$). If *Alice* is not a `Clerk`, then she is not allowed to sign the check she prepared. The reason is that no matter who signs the check in future, *Alice* plus that person cannot satisfy the high-level requirement ($\texttt{Manager} \odot (\texttt{Clerk} \otimes \texttt{Clerk})$). Note that if *Alice* performed both the first step and the second step, then she is precluded from performing the last step as two different `Clerks` are required to complete the task.

Dynamic enforcement can enforce both monotonic policies and policies that are not monotonic. It is also more flexible than static enforcement. Enforcement of high-level policies specified in the algebra generates many interesting open technical problems.

3. TWO TERM SATISFIABILITY PROBLEMS

Given the definitions of terms and term satisfaction, the following problems naturally arise.

The Term Satisfiability (TSAT) Problem: Given a term ϕ, determine whether there exists a configuration $\langle U, UR \rangle$ and a userset X such that X satisfies ϕ under $\langle U, UR \rangle$.

This asks whether the term ϕ is satisfiable at all. This provides a basic level of sanity check, as a term that cannot be satisfied in any configuration is probably not what a policy author intended.

The Term-Configuration Satisfiability (TCSAT) Problem: Given a term ϕ and a configuration $\langle U, UR \rangle$, determine whether there exists a userset X that satisfies ϕ under $\langle U, UR \rangle$.

This asks whether a term ϕ is satisfiable under a given configuration. This is useful when determining whether a term is meaningful in the current configuration.

The Userset-Term Satisfaction (UTS) Problem: Given a term ϕ, a configuration $\langle U, UR \rangle$, and a userset X, determine whether X satisfies ϕ under $\langle U, UR \rangle$.

This is probably the most fundamental problem related to the algebra. When an administrator specifies a policy that associates a sensitive task with a term; this means that every set of users that together perform an instance of the task must satisfy the term. To enforce this policy, one needs to check whether a userset satisfies the term and to forbid users in the set to finish the task if it does not satisfy the term. This requires solving the UTS problem.

In this section we will study TSAT and TCSAT. The UTS problem will be studied in Section 4.

3.1 The Term Satisfiability (TSAT) Problem

As the algebra supports negation, it is not surprising that unsatisfiable terms exist. A simple example of a term that is not satisfiable is $(r \sqcap \neg r)$. Another source of unsatisfiable terms is the use of explicit sets of users in a term. For example, the term $(\{Alice, Bob\} \sqcap \{Carl\})$ is not satisfiable. However, even if a term does not contain negation or explicit sets of users, it may still be unsatisfiable. An example of such a term is $\phi = (r_1 \sqcap (r_2 \otimes r_3))$, where r_1, r_2 and r_3 are roles. In the example, r_1 is satisfiable only by a singleton userset, and $(r_2 \otimes r_3)$ is satisfiable only by a userset of cardinality 2. Therefore, there does not exist any userset that satisfies ϕ.

In this section, we show that TSAT is **NP**-complete in general. We identify the source of intractability by identifying two special cases that are **NP**-hard. One special case involves the negation operator, and the other involves explicit sets of users. In the next section, we show that for terms that are free of negation and explicit sets of users, TSAT can be efficiently solved.

Lemma 2. *TSAT for unit terms using only roles and the operators* $\neg, \sqcap,$ *and* $\sqcup$ *is* **NP**-*hard.*

Lemma 3. *TSAT for terms using only usersets and the operators* $\sqcap, \sqcup,$ *and* $\odot$ *is* **NP**-*hard.*

Theorem 4. *TSAT is* **NP**-*complete.*

See Appendix B.1 for the proofs of Lemmas 2 and 3 and Theorem 4.

3.2 TSAT for the Sub-Algebra Free of Negation and Explicit Sets of Users

Lemmas 2 and 3 show that if a term involves negation or explicit sets of users, then determining whether it is satisfiable or not may be intractable. We now study the term satisfiability problem for terms that are free of explicit sets of users and negation. For convenience, we call such terms *UNF (Userset-and-Negation Free) terms*.

One property of UNF terms is that if a userset X satisfies a term ϕ under configuration $\langle U, UP \rangle$, then X also satisfies ϕ under $\langle U', UP' \rangle$, where $U \subseteq U'$ and $UP \subseteq UP'$. That is, because a UNF term does not have the negation operator, then if X satisfies the term under a configuration, enlarging the configuration will not make X fail to satisfy the term.

A key observation is that, in order to satisfy a term, a userset must be of certain size. For example, $(r_1 \odot (r_2 \otimes r_3))$ can be satisfied by a set of 2 or 3 users, but not by a set of 1 or 4 users. This observation leads us to introduce the notion of characteristic numbers of a UNF term.

Definition 6 (Characteristic Numbers). Given a UNF term ϕ and a positive integer k, we say that k is a *characteristic number* of ϕ when there exists a userset of size k that satisfies ϕ under some configuration. A term ϕ may have more than one characteristic numbers. We use $C(\phi)$ to denote the set of all characteristic numbers of ϕ and call it the *characteristic set* of ϕ.

It follows from the definition that a UNF term ϕ is satisfiable if and only if $C(\phi) \neq \emptyset$.

Theorem 5. *The characteristic set of a UNF term can be computed as follows:*

- $C(\textsf{All}) = C(r) = \{1\}$, *where r is a role*
- $C(\phi_1 \sqcup \phi_2) = C(\phi_1) \cup C(\phi_2)$

- $C(\phi_1 \sqcap \phi_2) = C(\phi_1) \cap C(\phi_2)$

- $C(\phi^+) = \{i \mid i \in [1, \infty)\}$, where ϕ is a unit term free of usersets and negations

- $C(\phi_1 \odot \phi_2) = \{i \mid \exists c_1 \in C(\phi_1) \, \exists c_2 \in C(\phi_2)$
  $[\, max(c_1, c_2) \leq i \leq c_1 + c_2 \,]\}$

- $C(\phi_1 \otimes \phi_2) = \{\, c_1 + c_2 \mid c_1 \in C(\phi_1) \,\wedge\, c_2 \in C(\phi_2) \,\}$

The proof for the theorem is in Appendix B.2. We now illustrate the computation of characteristic set according to the theorem using some examples.

- $C(\text{All} \otimes \text{All} \otimes \text{All}) = \{3\}$

- $C(\text{Manager} \odot \text{Accountant}) \otimes \text{Treasurer}) = \{2, 3\}$

  The term (Manager $\odot$ Accountant) can be satisfied by two users as well as by a single user who is both a Manager and an Accountant. An additional user is needed to satisfy Treasurer.

- $C((\text{Clerk} \sqcup \text{Accountant}) \otimes (\text{Clerk} \sqcap \text{Manager})) = \{2\}$

  Only one user is required for both (Clerk $\sqcup$ Accountant) and (Clerk $\sqcap$ Manager), and the $\otimes$ mandates that these users must be different from one another.

- $C((\text{Manager} \odot \text{Accountant} \odot \text{Treasurer}) \sqcap \text{Clerk}^+) = \{1, 2, 3\} \cap \{i \mid i \in [1, \infty)\} = \{1, 2, 3\}$

For the unsatisfiable term considered earlier in this section, namely $(r_1 \sqcap (r_2 \otimes r_3))$, we observe that $C(r_1 \sqcap (r_2 \otimes r_3)) = \{1\} \cap \{2\} = \emptyset$.

In Appendix B.3 we show that computing $C(\phi)$ using a straightforward algorithm that follows Theorem 5 takes at most quadratic time.

One can use $C(\phi)$ to determine whether ϕ satisfies some minimal SoD requirements. If the smallest characteristic number of a term is at least k, then we know that no $k-1$ users can satisfy the term.

We can extend the method of calculating the characteristic set stated in Theorem 5 to non-UNF terms as well, by defining $C(\neg \phi) = \{1\}$, where ϕ is a unit term, and $C(S) = 1$, where S is a set of users. But in that case, it is no longer true that for every integer $k \in C(\phi)$, there is a userset of size k that satisfies ϕ. For example, $C(\{Alice, Bob\} \sqcap \{Carl\}) = C(\{Alice, Bob\}) \cap C(\{Carl\}) = \{1\}$, even though the term $(\{Alice, Bob\} \sqcap \{Carl\})$ is not satisfiable. It remains true that for any userset X that satisfies a term ϕ, $|X| \in C(\phi)$. In other words, $C(\phi)$ gives an upperbound on the actual set of characteristic numbers of ϕ.

3.3 The Term-Configuration Satisfiability (TCSAT) Problem

We have discussed the TSAT problem, which asks whether a term is satisfiable at all. We now examine the TCSAT problem, which asks whether a term is satisfiable under a certain configuration. When a company comes up with a new security requirement for a task, it may want to know whether there exists a set of users that satisfies the new requirement and hence can perform the task under the current configuration of the company.

Observe that TCSAT is equivalent to TSAT for the terms using explicit sets of users but not roles or the keyword All. Given an instance of TCSAT, which consists of a term ϕ and a configuration $\langle U, UR \rangle$; one can replace each role (or the keyword All) in ϕ with the corresponding set of users in the configuration, which results in a new term ϕ'. In this case, ϕ' is independent of configuration, and ϕ is satisfiable under $\langle U, UR \rangle$ if and only if ϕ' is satisfiable. Therefore, it follows from Lemma 3 and Theorem 4 that TCSAT is **NP**-complete; this is stated in the following theorem.

Theorem 6. TCSAT *is* **NP**-*complete*.

4. THE USERSET-TERM SATISFACTION (UTS) PROBLEM

In Section 3, we have studied the problems of determining whether a term is satisfiable at all, as well as whether a term is satisfiable under a given configuration. In this section, we study the Userset-Term Satisfaction (UTS) Problem, which asks given a configuration $\langle U, UR \rangle$, a userset X, and a term ϕ, whether X satisfies ϕ under $\langle U, UR \rangle$. We will show that UTS in the most general case (i.e., arbitrary terms in which all operators are allowed) is **NP**-complete. In order to understand how the operators affect the computational complexity, we consider all sub-algebras in which only some subset of the six operators $\{\neg, +, \sqcap, \sqcup, \odot, \otimes\}$ is allowed. For example, UTS$\langle \neg, +, \sqcup, \sqcap \rangle$ denotes the sub-case of UTS where ϕ does not contain operators $\odot$ or $\otimes$, while UTS$\langle \otimes \rangle$ denotes the sub-case of UTS where $\otimes$ is the only kind of operator in ϕ. UTS$\langle \neg, +, \sqcup, \sqcap, \odot, \otimes \rangle$ denotes the general case. Observe that unlike in the case of TSAT, whether or not to allow explicit sets of users in a term does not affect the computational complexity of UTS, because a fixed configuration is given in UTS and one can always replace each role with the corresponding set of users.

Theorem 7. *The computational complexities for* UTS *and its subcases are given in Figure 1.*

The proof of Theorem 7 is done in three parts. First, in Appendix C.1, we prove that the five cases UTS$\langle \sqcup, \odot \rangle$, UTS$\langle \sqcap, \odot \rangle$, UTS$\langle \sqcup, \otimes \rangle$, UTS$\langle \sqcap, \otimes \rangle$, and UTS$\langle \odot, \otimes \rangle$ are **NP**-hard by reducing the **NP**-complete problems SET COVERING, DOMATIC NUMBER, and SET PACKING to them. Second, in Appendix C.2, we prove that the general case UTS$\langle \neg, +, \sqcup, \sqcap, \odot, \otimes \rangle$ is in **NP**. Finally, the tractable cases are discussed in Section 4.1, where we identify a wide class of syntactically restricted terms for which the UTS problem is tractable. The class of restricted terms subsumes all the cases listed as in **P** in Figure 1.

4.1 UTS is Tractable for Terms in Canonical Forms

From Figure 1, UTS is **NP**-complete in general in all but one sub-algebras that contain at least two binary operators; however, using any one binary operator by itself remains tractable. In this subsection, we show that if a term satisfies certain syntactic restrictions, then even if all operators appear in the term, one can still efficiently determine whether a userset satisfies the term.

Definition 7 (Canonical Forms for Terms). The canonical forms for terms are defined as follows:

- A term is *in level-1 canonical form* (called an 1CF term) if it is t or t^+, where t is a unit term. Recall that a unit term can use the operators $\neg$, $\sqcap$, and $\sqcup$. We call t the *base* of the 1CF term.

- A term is *in level-2 canonical form* (called a 2CF term) if it consists of one or more sub-terms that are 1CF terms, and (when there are two or more sub-terms) these sub-terms are connected only by the operator $\sqcap$.

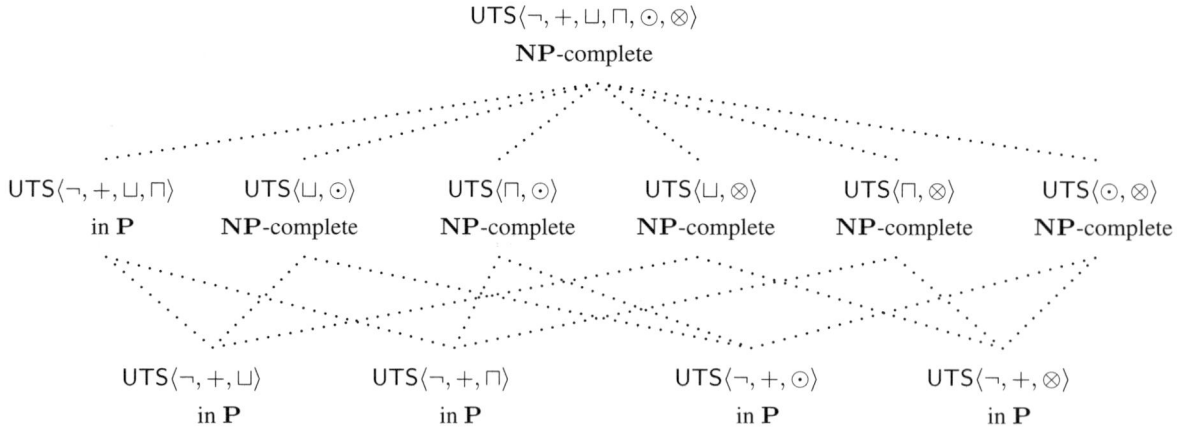

Figure 1: Various sub-cases of the Userset Term Satisfaction (UTS) problem and the corresponding time-complexity. Some sub-cases are omitted from the figure, as their time-complexities are implied from what are in the figure.

- A term is *in level-3 canonical form* (called a 3CF term) if it consists of one or more sub-terms that are 2CF terms, and (when there are two or more sub-terms) these sub-terms are connected only by the operator $\otimes$.

- A term is *in level-4 canonical form* (called a 4CF term) if it consists of one or more sub-terms that are 3CF terms, and these sub-terms are connected only by the operator $\odot$.

- A term is *in level-5 canonical form* (called a 5CF term) if it consists of one or more sub-terms that are 4CF terms, and these sub-terms are connected only by operators in the set $\{\sqcup, \sqcap\}$.

We say that a term is *in canonical form* if it is in level-5 canonical form. Observe that any term that is in level-i canonical form is also in level-$(i+1)$ canonical form for any $i \in [1, 4]$.

Theorem 8. *Given a term ϕ in canonical form, a set X of users, and a configuration $\langle U, UR \rangle$, checking whether X satisfies ϕ under $\langle U, UR \rangle$ can be done in polynomial time.*

PROOF. We first recall that, by definition, X satisfies $\phi_1 \sqcap \phi_2$ if and only if X satisfies both ϕ_1 and ϕ_2, and X satisfies $\phi_1 \sqcup \phi_2$ if and only if X satisfies either ϕ_1 or ϕ_2. Therefore, to determine whether X satisfies a 5CF term, one can first determine whether X satisfies each of the 4CF sub-terms, and then combine these results using logical conjunction and disjunction.

For a 1CF term ϕ, if it is a unit term, then it is straightforward to determine whether X satisfies ϕ, because a unit term can be satisfied only by a singleton set, and because of the definitions of $\sqcap$ and $\sqcup$. If ϕ is of the form t^+, where t is a unit term, then one just needs to determine whether each user in X satisfies t. Therefore, one can efficiently check whether X satisfies an 1CF term.

Given a 2CF term, if at least one sub-term is a unit term, then one can get an equivalent 1CF term by removing all occurrences of $^+$. For example, $(t_1 \sqcap t_2^+)$ is equivalent to $t_1 \sqcap t_2$. Given a 2CF term where all sub-terms have $^+$, from the algebraic property, it is equivalent a 1CF term. For example, $(t_1^+ \sqcap t_2^+)$ is equivalent to $(\phi_1 \sqcap \phi_2)^+$. Hence, any 2CF term can be revised to an equivalent 1CF term. We assume that the revision is performed whenever applicable so that we don't need to consider 2CF terms explicitly.

Given a 3CF term $P = (\phi_1 \otimes \cdots \otimes \phi_m)$, where each ϕ_i is an 1CF term. Let us first consider a special case that each ϕ_i is a unit term t_i. In this case, one can determine whether X satisfies ϕ_i by solving the following bipartite graph maximal matching problem. One constructs a bipartite graph such that one set of nodes consists of users in X and the other consists of the m unit terms $t_1, t_2, \cdots, t_m$; and there is an edge between $u \in X$ and t_i if and only if $\{u\}$ satisfies t_i. One then computes a maximal matching of the graph (which can be done in polynomial time); if it has size the same as $\max(|X|, m)$, then X satisfies P; otherwise, X does not satisfy P.

The case that a 3CF term contains $+$ is more complicated, so is the case for a 4CF term. Because of space limitation, we give the proof for the case of a 4CF term (which subsumes the case for a 3CF term) in Appendix C.3.

Terms in canonical form appear to be general enough to specify many high-level security policies in practice. We arrive at these canonical forms by excluding the intractable cases involving combinations of multiple operators in different ordering, and by studying how to handle each binary operator individually and examining how combinations of them can still be handled efficiently.

4.2 An Algorithm for UTS

We have shown that for terms that are in canonical forms, it is efficient to check whether a userset satisfies them or not. However, it is not necessary to restrict the use of the algebra to only such terms. Even if one writes policies that use terms that are not in canonical forms, these terms may not be in the pathological cases that lead to intractability, or they may not be very large. We now present a heuristic algorithm for UTS that works for all terms. We show through experimental results that this algorithm works well for instances of UTS with complicated terms and relative large usersets.

To determine whether a userset X satisfies a term ϕ under a configuration $\langle U, UR \rangle$, our algorithm first computes the syntax tree T of ϕ. Given the syntax tree, there are two approaches. The first one is top-down processing. One starts with X and the root of the syntax tree; if the root is the operator $\odot$, then for each subset $X_1 \subseteq X$, one recursively checks whether X_1 satisfies the left child, and if it does, one tries all $X_2 \subseteq X$ such that $X_1 \cup X_2 = X$ and checks whether X_2 satisfies the right child. Other operators can be handled similarly. The second approach is bottom-up processing. One starts with unit terms. For each unit term, one calculates all subsets of X that satisfy the term. One then goes bottom-up to calculate those for each node in the syntax tree.

Our algorithm combines top-down processing with bottom-up processing. It first performs bottom-up processing until encountering nodes such that bottom-up processing becomes too expensive. For example, if each user in a set Y can satisfy t, then node t^+ can be satisfied by the $2^{|Y|} - 1$ non-empty subsets of Y; so we avoid bottom-up processing for t^+. After the bottom-up phase, the algorithm performs top-down search. When the search encounters a node that has been bottom-up processed, it can perform a lookup. Our algorithm also includes several optimizations to improve top-down search. For example, it computes the characteristic set for each sub-term so as to prune the subsets of X that need to be checked; it also sorts sub-terms of ϕ according to the size of their characteristic sets so that sub-terms that are "harder to be satisfied" are processed first.

We prototyped both the above algorithm and the algorithm for terms in canonical forms, and have performed some experiments. Our prototypes are written in Java, and our experiments were carried out on a Workstation with a 3.2GHz Pentium 4 CPU and 512MB RAM. Some of our experimental results are presented in Table 1. As we can see in Table 1, UTS is efficiently solvable for terms in canonical form. Furthermore, our algorithm for canonical form scales well over the size of userset, as in Test 2 and Test 3, increasing userset size from 25 to 50 only results in 1 millisecond's increment on runtime. Finally, experimental data in Test 4 and Test 5 indicates that UTS can be solved quickly even for complicated terms that are not in canonical form.

Further improvements and optimizations can be made to our algorithms and prototypes; they are beyond the scope of this paper. Our goal here is to verify that UTS can be solved in reasonable amount of time for complicated terms, even though the problem is **NP**-complete in general.

5. DISCUSSIONS AND OPEN PROBLEMS

In this section we discuss a few small extensions to the algebra, the similarities and differences with regular expressions, and expressive power limitations of the algebra.

5.1 Extensions to the Algebra

In this paper, we have defined the basic operators in the algebra and examined their properties. We now discuss some additional operators that could be added to the algebra as syntactic sugars.

As discussed in Section 2.4, SoD policies are monotonic, as are policies in McLean's formulation of N-person policies [15]; our algebra supports both monotonic policies and policies that are not monotonic. To express a monotonic policy that requires a task to be performed by a userset that either satisfies a term ϕ or contains a subset that satisfies ϕ, one can use $(\phi \odot \text{All}^+)$. As monotonic policies may be quite common, we introduce a unary operator $\triangledown$ as a syntactic sugar. That is, $\triangledown \phi$ is defined to be $(\phi \odot \text{All}^+)$.

Besides monotonic policies, another type of policies mentioned in Section 2.4 is policies stating that every user involved in a task must satisfy certain requirement and there need to be at least a certain number of users involved. Let ϕ be a unit term that expresses the requirement. A policy that requires 2 or more users that satisfy ϕ can be expressed as $((\phi \otimes \phi) \odot \phi^+)$. To simplify the expression of these policies, we define ϕ^{2+} as a syntactic sugar for $((\phi \otimes \phi) \odot \phi^+)$. In general, ϕ^{k+} means that at least k ($k \geq 2$) users are required and every user involved must satisfy ϕ.

Similar to the above, ϕ^k is a syntactic sugar for a term using operator $\otimes$ to connect k unit terms ϕ. For instance, Accountant[3] is defined as (Accountant $\otimes$ Accountant $\otimes$ Accountant). More generally, ϕ^k states that exactly k users are required and every user involved must satisfy ϕ. Using ϕ^k rather than $(\phi \otimes \cdots \otimes \phi)$ explicitly states that all the k sub-terms connected together by $\otimes$ are the same. This makes the policy more succinct and easier to process.

5.2 Relationship with Regular Expressions

The syntax of terms in our algebra may remind readers of regular expressions. A regular expression is a string that describes or matches a set of strings, while a term in the algebra is a string that describes or matches a set of sets. Given an alphabet, a regular expression evaluates to *a set of strings*. Given a configuration, a term in our algebra evaluates to *a set of sets*. In the following, we compare our algebra with regular expressions.

For example, the regular expression "$a(b|c)[\hat{}abc]^+$" matches all strings that start with the letter a, followed by either b or c, and then by one or more symbols that are not in $\{a, b, c\}$. A term that is close in spirit to the regular expression is $\{a\} \otimes (\{b\} \sqcup \{c\}) \otimes (\neg\{a, b, c\})^+$, which is satisfied by all sets that contain a, either b or c, and one or more symbols that are not in $\{a, b, c\}$.

From the example, one can draw some analogies between the operators in regular expressions and the ones in our algebra. The operator $|$ in regular expressions is similar to $\sqcup$. Concatenation in regular expression may seem related to $\otimes$. One clear difference is that concatenation is order sensitive, whereas $\otimes$ is not, because a string is order sensitive but a set is not. A more subtle difference comes from the property that $\otimes$ requires the two sub-terms be satisfied by disjoint sets. For instance, $\{a\} \otimes \{a\}$ cannot be satisfied by any set. The usage of negation in regular expressions is similar to negation in the algebra; in both cases, negation can be applied only to an expression corresponds to a single element. In regular expression, the closure operator ($*$ or $+$) can be applied to arbitrary sub-expressions. Our algebra requires that repetition (using operator $^+$) can only be applied to unit terms. As we discussed in Section 2.4, since the algebra is proposed for security policy specification, we impose such restriction so as to clearly capture real-word security requirements. If the algebra is used in areas other than security policy specification, it is certainly possible to relax such restriction so that the algebra can define a wider range of sets. The remaining binary operators $\odot$ and $\sqcap$ have flavor of set intersection, which does not have counterparts in regular expressions.

Observe that determining whether a string satisfies a regular expression is in **NL**-complete, where **NL** stands for Nondeterministic Logarithmic-Space, and is contained in **P**. On the other hand, determining whether a userset satisfies a term is **NP**-complete, even if the term uses only $\sqcup$ and $\otimes$ or only $\sqcup$ and $\odot$. It appears that this increase in complexity is due to the unordered nature of sets. Checking a string against a regular expression can be performed from the beginning of a string to its end; on the other hand, there is no such order in checking a set against a term in the algebra.

As a fundamental tool for defining sets of strings, regular expression is used in many areas. Analogically, because our algebra is about the fundamental concept of defining sets of sets, we conjecture that, besides expression of security policies, the algebra could be used in other areas where set specification is desired.

5.3 Limitations of the Algebra's Expressive Power

It is well-known that using regular expression, one cannot express languages that require counting to an unbounded number; for example, one cannot express all strings over the alphabet $\{0, 1\}$ that contain the same number of a's as of b's.

Similarly, the algebra as defined in Section 2.1 cannot express a policy that requires equal number of users who are r_1 as those that are r_2. However, if we allow the application of $+$ to non-unit terms and define it as follows:

$$\phi^+ \stackrel{def}{=} \phi \sqcup (\phi \otimes \phi) \sqcup (\phi \otimes \phi \otimes \phi) \sqcup \cdots$$

	Term	In CF?	Userset Size	Runtime (ms)
1	$((r_1 \sqcap r_2) \otimes (r_1 \sqcup r_3) \otimes \neg r_4 \otimes r_5 \otimes \mathsf{All})$	Yes	5	< 1
2	$((((r_1 \sqcap r_2) \otimes r_3 \otimes (r_1 \sqcup r_5)^+) \odot (r_4 \otimes (\neg r_1)^+)))$	Yes	25	15
3	$((((r_1 \sqcap r_2) \otimes r_3 \otimes (r_1 \sqcup r_5)^+) \odot (r_4 \otimes (\neg r_1)^+)))$	Yes	50	16
4	$((((r_1 \otimes r_2) \sqcap (r_3 \odot r_5)) \odot (r_1 \sqcup r_3)) \otimes \mathsf{All}^+)$	No	15	16
5	$(((r_1 \sqcap r_2)^+ \odot r_3 \odot \mathsf{All}^+) \otimes (r_6 \sqcap r_1) \otimes ((r_2^+ \odot (\neg r_1 \otimes r_2)^+ \odot r_4) \sqcap (r_2 \sqcup \neg r_5)^+))$	No	50	219

Table 1: A table that shows the runtime of testing whether a userset satisfies a term.

then we can express the policy that requires equal number of users who are r_1 as those that are r_2 using the term

$$(r_1 \sqcap r_2)^+ \otimes ((r_1 \sqcap \neg r_2) \otimes (r_2 \sqcap \neg r_1))^+ \otimes (\neg r_1 \sqcap \neg r_2)^+.$$

Even with the extension, there are sets of usersets that cannot be expressed. For example, it seems unlikely that one can express a policy that requires that the number of users who are member of r_1 is the same as the square of the number of users who are r_2. Further discussions of expressive power and more general algebras are interesting future research topics and are beyond the scope of this paper.

6. RELATED WORK

The concept of SoD has long existed in the physical world, sometimes under the name "the two-man rule", for example, in the banking industry and the military. To our knowledge, in the information security literature the notion of SoD first appeared in Saltzer and Schroeder [18] under the name "separation of privilege." Clark and Wilson's commercial security policy for integrity [7] identified SoD along with well-formed transactions as two major mechanisms of fraud and error control. Nash and Poland [16] explained the difference between dynamic and static enforcement of SoD policies. In the former, a user may perform any step in a sensitive task provided that the user does not also perform another step on that data item. In the latter, users are constrained a-priori from performing certain steps.

Sandhu [19, 20] presented Transaction Control Expressions, a history-based mechanism for dynamically enforcing SoD policies. A transaction control expression associates each step in the transaction with a role. By default, the requirement is such that each step must be performed by a different user. One can also specify that two steps must be performed by the same user. In Transaction Control Expressions, user qualification requirements are associated with individual steps in a transaction, rather than a transaction as a whole.

There exists a wealth of literature [1, 2, 8, 10, 11, 12, 22, 23] on constraints in the context of RBAC. They either proposed and classified new kinds of constraints [10, 22] or proposed new languages for specifying sophisticated constraints [1, 2, 8, 12, 23]. Most of these constraints are motivated by SoD and are variants of role mutual exclusion constraints, which may declare two roles to be mutually exclusive so that no user can be a member of both roles.

There has also been recent interest in static and dynamic constraints to enforce separation of duty in workflow systems. Atluri and Huang [3] proposed an access control model for workflow environments, which supports temporal constraints. Bertino et al. [4] proposed a language for specifying static and dynamic constraints for separation of duty in role-based workflow systems. In these works, security requirements are associated with individual steps in the workflows.

McLean [15] introduced a framework that includes various mandatory access control models. These models differ in which users are allowed to change the security levels. They form a boolean algebra. McLean also looked at the issue of N-person policies, where a policy may allow multiple subjects acting together to perform some action. McLean adopted the monotonicity requirement in such N-person policies.

Several algebras have been proposed for combining security policies. These include the work by Bonatti et al. [5, 6], Wijesekera and Jajodia [24], Pincus and Wing [17]. These algebras are designed for purpose that are different from ours; therefore, they are quite different from our algebra. Each element in their algebra is a policy that specifies what subjects are allowed to access which resources, whereas each element in our algebra maps to a user.

The two operators $\odot$ and $\otimes$ in our algebra are taken from the RT family of role-based trust-management languages designed by Li et al. [14]. In [14], the notion of manifold roles was introduced, which are roles that have usersets, rather than individual users, as their members. The two operators $\otimes$ and $\odot$ are used to define manifold roles. This paper differs in that we propose to combine these two operators together with four other operators $\sqcup$, $\sqcap$, $\neg$, and $+$ (which are not in RT) in an algebra for specifying high-level security policies. In addition, we also study the algebraic properties of these operators, the satisfaction problems, and the term satisfiability problem related to the algebra.

7. SUMMARY

While separation of duty policies are extremely important and widely used, they state only quantity requirements and cannot capture qualification requirements on users involved in the task. We have introduced a novel algebra that enables the specification of high-level policies that combine qualification requirements with quantity requirements motivated by separation of duty considerations. A high-level policy associates a task with a term in the algebra and requires that all sets of users that perform the task satisfy the term. Specifying security policies at the task level has a number of advantages over the current approach of specifying such policies at the individual step level. Our algebra has two unary and four binary operators, and is expressive enough to specify a large number of diverse policies. We have also studied algebraic properties of these operators and several computational problems related to the algebra, including determining whether a term is satisfiable at all, determining whether a term is satisfiable under a given configuration, and determining whether a userset satisfies a term under a given configuration. As our algebra is about the general concept of sets of sets, we conjecture that it will prove to be useful in other contexts as well.

Acknowledgement

This work is supported by NSF CNS-0448204 (CAREER: Access Control Policy Verification Through Security Analysis And Insider

Threat Assessment), and by sponsors of CERIAS. We thank Mahesh V. Tripunitara for helpful discussions. We also thank the anonymous reviewers for their helpful comments.

8. REFERENCES

[1] G.-J. Ahn and R. S. Sandhu. The RSL99 language for role-based separation of duty constraints. In *Proceedings of the 4th Workshop on Role-Based Access Control*, pages 43–54, 1999.

[2] G.-J. Ahn and R. S. Sandhu. Role-based authorization constraints specification. *ACM Transactions on Information and System Security*, 3(4):207–226, Nov. 2000.

[3] V. Atluri and W. Huang. An authorization model for workflows. In *Proceedings of the 4th European Symposium on Research in Computer Security (ESORICS)*, pages 44–64, 1996.

[4] E. Bertino, E. Ferrari, and V. Atluri. The specification and enforcement of authorization constraints in workflow management systems. *ACM Transactions on Information and System Security*, 2(1):65–104, Feb. 1999.

[5] P. Bonatti, S. de Capitani di Vimercati, and P. Samarati. A modular approach to composing access control policies. In *Proceedings of the 7th ACM conference on Computer and Communications Security (CCS)*, pages 164–173, Nov. 2000.

[6] P. Bonatti, S. de Capitani di Vimercati, and P. Samarati. An algebra for composing access control policies. *ACM Transactions on Information and System Security (TISSEC)*, 5(1):1–35, Feb. 2002.

[7] D. D. Clark and D. R. Wilson. A comparison of commercial and military computer security policies. In *Proceedings of the 1987 IEEE Symposium on Security and Privacy*, pages 184–194. IEEE Computer Society Press, May 1987.

[8] J. Crampton. Specifying and enforcing constraints in role-based access control. In *Proceedings of the Eighth ACM Symposium on Access Control Models and Technologies (SACMAT 2003)*, pages 43–50, Como, Italy, June 2003.

[9] M. R. Garey and D. J. Johnson. *Computers And Intractability: A Guide to the Theory of NP-Completeness*. W.H. Freeman and Company, 1979.

[10] V. D. Gligor, S. I. Gavrila, and D. F. Ferraiolo. On the formal definition of separation-of-duty policies and their composition. In *Proceedings of IEEE Symposium on Research in Security and Privacy*, pages 172–183, May 1998.

[11] T. Jaeger. On the increasing importance of constraints. In *Proceedings of ACM Workshop on Role-Based Access Control*, pages 33–42, 1999.

[12] T. Jaeger and J. E. Tidswell. Practical safety in flexible access control models. *ACM Transactions on Information and System Security*, 4(2):158–190, May 2001.

[13] N. Li, Z. Bizri, and M. V. Tripunitara. On mutually-exclusive roles and separation of duty. In *Proceedings of the 11th ACM Conference on Computer and Communications Security (CCS-11)*, pages 42–51. ACM Press, Oct. 2004.

[14] N. Li, J. C. Mitchell, and W. H. Winsborough. Design of a role-based trust management framework. In *Proceedings of the 2002 IEEE Symposium on Security and Privacy*, pages 114–130. IEEE Computer Society Press, May 2002.

[15] J. McLean. The algebra of security. In *Proceedings of IEEE Symposium on Security and Privacy*, pages 2–7, Apr. 1988.

[16] M. J. Nash and K. R. Poland. Some conundrums concerning separation of duty. In *Proceedings of IEEE Symposium on Research in Security and Privacy*, pages 201–209, May 1990.

[17] J. Pincus and J. M. Wing. Towards an algebra for security policies (extended abstract). In *Proceedings of ICATPN 2005*, number 3536 in LNCS, pages 17–25. Springer, 2005.

[18] J. H. Saltzer and M. D. Schroeder. The protection of information in computer systems. *Proceedings of the IEEE*, 63(9):1278–1308, September 1975.

[19] R. Sandhu. Separation of duties in computerized information systems. In *Proceedings of the IFIP WG11.3 Workshop on Database Security*, Sept. 1990.

[20] R. S. Sandhu. Transaction control expressions for separation of duties. In *Proceedings of the Fourth Annual Computer Security Applications Conference (ACSAC'88)*, Dec. 1988.

[21] R. S. Sandhu, E. J. Coyne, H. L. Feinstein, and C. E. Youman. Role-based access control models. *IEEE Computer*, 29(2):38–47, February 1996.

[22] T. T. Simon and M. E. Zurko. Separation of duty in role-based environments. In *Proceedings of The 10th Computer Security Foundations Workshop*, pages 183–194. IEEE Computer Society Press, June 1997.

[23] J. Tidswell and T. Jaeger. An access control model for simplifying constraint expression. In *Proceedings of ACM Conference on Computer and Communications Security*, pages 154–163, 2000.

[24] D. Wijesekera and S. Jajodia. A propositional policy algebra for access control. *ACM Transactions on Information and Systems Security (TISSEC)*, 6(2):286–325, May 2003.

APPENDIX

A. PROOFS FOR THEOREMS IN SECTION 2

Proof for Theorem 1 on Algebraic Properties

1. The operators $\sqcup, \sqcap, \otimes, \odot$ are commutative and associative.
   This is straightforward from Definition 3.

2. The operator $\sqcup$ distributes over $\sqcap$.
   If a userset X satisfies $(\phi_1 \sqcup (\phi_2 \sqcap \phi_3))$, then either X satisfies ϕ_1, or X satisfies both ϕ_2 and ϕ_3. It follows that X satisfies $((\phi_1 \sqcup \phi_2) \sqcap (\phi_1 \sqcup \phi_3))$.
   If X satisfies $((\phi_1 \sqcup \phi_2) \sqcap (\phi_1 \sqcup \phi_3))$, then X satisfies $(\phi_1 \sqcup \phi_2)$ and $(\phi_1 \sqcup \phi_3)$. There are only two cases: (1) X satisfies ϕ_1; and (2) X satisfies both ϕ_2 and ϕ_3. In either case, X satisfies $(\phi_1 \sqcup (\phi_2 \sqcap \phi_3))$.
   The operator $\sqcap$ distributes over $\sqcup$.
   If X satisfies $(\phi_1 \sqcap (\phi_2 \sqcup \phi_3))$, then X satisfies both ϕ_1 and $(\phi_2 \sqcup \phi_3)$, which means X satisfies either ϕ_2 or ϕ_3. It follows that X satisfies $((\phi_1 \sqcap \phi_2) \sqcup (\phi_1 \sqcap \phi_3))$.
   If X satisfies $((\phi_1 \sqcap \phi_2) \sqcup (\phi_1 \sqcap \phi_3))$, then either (1) X satisfies $(\phi_1 \sqcap \phi_2)$ or (2) X satisfies $(\phi_1 \sqcap \phi_3)$. In both cases, X satisfies ϕ_1; furthermore, X satisfies either ϕ_2 or ϕ_3. It follows that X satisfies $(\phi_1 \sqcap (\phi_2 \sqcup \phi_3))$.

3. The operator $\odot$ distributes over $\sqcup$.
   If X satisfies $(\phi_1 \odot (\phi_2 \sqcup \phi_3))$, then there exist X_1 and X_2 such that $X_1 \cup X_2 = X$, X_1 satisfies ϕ_1, and X_2 satisfies $(\phi_2 \sqcup \phi_3)$. By Definition 3, X_2 satisfies ϕ_2 or satisfies ϕ_3. In the former case, X satisfies $(\phi_1 \odot \phi_2)$, which implies that X satisfies $((\phi_1 \odot \phi_2) \sqcup (\phi_1 \odot \phi_3))$, as desired. The argument is analogous if X_2 satisfies ϕ_3 but not ϕ_2.

If X satisfies $((\phi_1 \odot \phi_2) \sqcup (\phi_1 \odot \phi_3))$, then either X satisfies $(\phi_1 \odot \phi_2)$ or X satisfies $(\phi_1 \odot \phi_3)$. Without loss of generality, assume that X satisfies $(\phi_1 \odot \phi_2)$, then there exist X_1, X_2 such that $X_1 \cup X_2 = X$, X_1 satisfies ϕ_1 and X_2 satisfies ϕ_2. Therefore, X_2 satisfies $(\phi_2 \sqcup \phi_3)$, and consequently, X satisfies $(\phi_1 \odot (\phi_2 \sqcup \phi_3))$ as desired.

4. The operator $\otimes$ distributes over $\sqcup$.

   If X satisfies $(\phi_1 \otimes (\phi_2 \sqcup \phi_3))$, X can be partitioned into two disjoint sets X_1 and X_2 such that X_1 satisfies ϕ_1 and X_2 satisfies ϕ_2 or ϕ_3. In this case, by definition, X satisfies $(\phi_1 \otimes \phi_2)$ or $(\phi_1 \otimes \phi_3)$, which means X satisfies $((\phi_1 \otimes \phi_2) \sqcup (\phi_1 \otimes \phi_3))$.

   For the other direction, if X satisfies $((\phi_1 \otimes \phi_2) \sqcup (\phi_1 \otimes \phi_3))$, it satisfies either $(\phi_1 \otimes \phi_2)$ or $(\phi_1 \otimes \phi_3)$. Without loss of generality, assume that X satisfies $(\phi_1 \otimes \phi_2)$. Then, X can be partitioned into two disjoint sets X_1 and X_2 such that X_1 satisfies ϕ_1 and X_2 satisfies ϕ_2. By definition, X_2 satisfies $(\phi_2 \sqcup \phi_3)$. Therefore, X satisfies $(\phi_1 \otimes (\phi_2 \sqcup \phi_3))$.

5. No other ordered pair of operators have the distributive property.

   We show a counter example for each case. In the following, $U_r = \{u | (u, r) \in UR\}$.

   (a) The operator $\odot$ does not distribute over $\sqcap$.

   If X satisfies $(\phi_1 \odot (\phi_2 \sqcap \phi_3))$, then X also satisfies $((\phi_1 \odot \phi_2) \sqcap (\phi_1 \odot \phi_3))$.

   However, the other direction of implication does not hold. Counter example: Let $U_{r_1} = \{u_1, u_2\}$, $U_{r_2} = \{u_1\}$, and $U_{r_3} = \{u_2\}$, then $\{u_1, u_2\}$ satisfies $((r_1 \odot r_2) \sqcap (r_1 \odot r_3))$, but does not satisfy $(r_1 \odot (r_2 \sqcap r_3))$.

   (b) The operator $\sqcap$ does not distribute over $\odot$. Neither direction holds.

   Counter example: Let $U_{r_1} = U_{r_3} = \{u_1\}$ and $U_{r_2} = U_{r_4} = \{u_2\}$, let $\phi_1 = (r_1 \odot r_2)$, then $\{u_1, u_2\}$ satisfies $(\phi_1 \sqcap (r_3 \odot r_4))$, but does not satisfy $((\phi_1 \sqcap r_3) \odot (\phi_1 \sqcap r_4))$.

   Counter example: Let $U_{r_1} = \{u_1, u_2\}$, $U_{r_2} = \{u_1\}$, and $U_{r_3} = \{u_2\}$, then $\{u_1, u_2\}$ satisfies $((r_1 \sqcap r_2) \odot (r_1 \sqcap r_3))$, but does not satisfy $(r_1 \sqcap (r_2 \odot r_3))$.

   (c) The operator $\sqcup$ does not distribute over $\odot$.

   If X satisfies $(\phi_1 \sqcup (\phi_2 \odot \phi_3))$, then X satisfies $((\phi_1 \sqcup \phi_2) \odot (\phi_1 \sqcup \phi_3))$.

   However, the other direction of implication does not hold. Counter example: Let $U_{r_1} = \{u_1, u_2\}$, $U_{r_2} = \emptyset$ and $U_{r_3} = \emptyset$, then $\{u_1, u_2\}$ satisfies $((r_1 \sqcup r_2) \odot (r_1 \sqcup r_3))$, but does not strictly satisfy $(r_1 \sqcup (r_2 \odot r_3))$.

   (d) The operator $\sqcup$ does not distribute over $\otimes$. Neither direction holds.

   Counter example: Let $U_{r_1} = \{u_1, u_2\}$, $U_{r_2} = \emptyset$ and $U_{r_3} = \emptyset$, then $\{u_1, u_2\}$ satisfies $((r_1 \sqcup r_2) \otimes (r_1 \sqcup r_3))$, but does not satisfy $(r_1 \sqcup (r_2 \otimes r_3))$.

   Counter example: Let $U_{r_1} = \{u_1\}$, $U_{r_2} = \emptyset$ and $U_{r_3} = \emptyset$, then $\{u_1\}$ satisfies $(r_1 \sqcup (r_2 \otimes r_3))$, but does not satisfy $((r_1 \sqcup r_2) \otimes (r_1 \sqcup r_3))$.

   (e) The operator $\otimes$ does not distribute over $\sqcap$.

   If X satisfies $(\phi_1 \otimes (\phi_2 \sqcap \phi_3))$, then X satisfies $((\phi_1 \otimes \phi_2) \sqcap (\phi_1 \otimes \phi_3))$.

   However, the other direction of implication does not hold. Counter example: Let $U_{r_1} = \{u_1, u_2\}$, $U_{r_2} = \{u_1\}$ and $U_{r_3} = \{u_2\}$, then $\{u_1, u_2\}$ satisfies $((r_1 \otimes r_2) \sqcap (r_1 \otimes r_3))$, but does not satisfy $(r_1 \otimes (r_2 \sqcap r_3))$.

   (f) The operator $\sqcap$ does not distribute over $\otimes$. Neither direction holds.

   Counter example: Let $U_{r_1} = \{u_1, u_2\}$, $U_{r_2} = \{u_1\}$ and $U_{r_3} = \{u_2\}$, then $\{u_1, u_2\}$ satisfies $((r_1 \sqcap r_2) \otimes (r_1 \sqcap r_3))$, but does not satisfy $(r_1 \otimes (r_2 \sqcap r_3))$.

   Counter example: Let $U_{r_1} = U_{r_3} = \{u_1\}$ and $U_{r_2} = U_{r_4} = \{u_2\}$, and let $\phi_1 = (r_1 \odot r_2)$, then $\{u_1, u_2\}$ satisfies $(\phi_1 \sqcap (r_3 \otimes r_4))$, but does not satisfy $((\phi_1 \sqcap r_3) \otimes (\phi_1 \sqcap r_4))$.

   (g) The operator $\odot$ does not distribute over $\otimes$. Neither direction holds.

   Counter example: Let $U_{r_1} = \{u_1, u_4\}$, $U_{r_2} = \{u_2\}$ and $U_{r_3} = \{u_3\}$, then $\{u_1, u_2, u_3, u_4\}$ satisfies $((r_1 \odot r_2) \otimes (r_1 \odot r_3))$, but does not satisfies $(r_1 \odot (r_2 \otimes r_3))$.

   Counter example: Let $U_{r_1} = \{u_1\}$, $U_{r_2} = \{u_1\}$ and $U_{r_3} = \{u_2\}$, then $\{u_1, u_2\}$ satisfies $(r_1 \odot (r_2 \otimes r_3))$, but does not satisfy $((r_1 \odot r_2) \otimes (r_1 \odot r_3))$.

   (h) The operator $\otimes$ does not distribute over $\odot$.

   If X satisfies $(\phi_1 \otimes (\phi_2 \odot \phi_3))$, then X satisfies $((\phi_1 \otimes \phi_2) \odot (\phi_1 \otimes \phi_3))$.

   However, the other direction of implication does not hold. Counter example: Let $U_{r_1} = \{u_1, u_2\}$, $U_{r_2} = \{u_2\}$ and $U_{r_3} = \{u_1\}$, then $\{u_1, u_2\}$ satisfies $((r_1 \otimes r_2) \odot (r_1 \otimes r_3))$, but does not satisfy $(r_1 \otimes (r_2 \odot r_3))$.

6. $(\phi_1 \sqcap \phi_2)^+ \equiv (\phi_1^+ \sqcap \phi_2^+)$.

   If a userset X satisfies $(\phi_1 \sqcap \phi_2)^+$, then for every $u \in X$, $\{u\}$ satisfies $(\phi_1 \sqcap \phi_2)$ and thus satisfies ϕ_1 and ϕ_2. Hence, X satisfies ϕ_1^+ and ϕ_2^+, which means that X satisfies $(\phi_1^+ \sqcap \phi_2^+)$.
   If X satisfies $(\phi_1^+ \sqcap \phi_2^+)$, then X satisfies both ϕ_1^+ and ϕ_2^+. For every $u \in X$, $\{u\}$ satisfies both ϕ_1 and ϕ_2. Hence, X satisfies $(\phi_1 \sqcap \phi_2)^+$.

7. DeMorgan's Law: $\neg(\phi_1 \sqcap \phi_2) \equiv (\neg\phi_1 \sqcup \neg\phi_2)$, $\neg(\phi_1 \sqcup \phi_2) \equiv (\neg\phi_1 \sqcap \neg\phi_2)$

   The proof is straight forward by definition of $\neg, \sqcap$ and $\sqcup$.

B. PROOF FOR THEOREMS IN SECTION 3

In the following proofs, $(\text{op}_k \phi)$ denotes k copies of ϕ connected together by operator op and $(\text{op}_{i=1}^n r_i)$ denotes $(r_1 \text{ op} \cdots \text{op } r_n)$. Given $R = \{r_1, \cdots, r_m\}$, $(\text{op} R)$ denotes $(r_1 \text{ op} \cdots \text{op } r_m)$.

B.1 Proof for Lemma 2, Lemma 3, and Theorem 4

Proof for Lemma 2: TSAT with just role, $\neg$, $\sqcap$, and $\sqcup$ is NP-hard.

We reduce the **NP**-complete SAT problem to TSAT problem of terms consisting of just role, $\neg$, $\sqcap$, and $\sqcup$. Given a propositional logic formula e, let $\{v_1, \cdots, v_n\}$ be the set of propositional variables that appear in e. Construct a unit term ϕ by substituting every occurrence of v_i ($i \in [1, n]$) in e with atomic term r_i, every occurrence of $\neg v_i$ ($i \in [1, n]$) with $\neg r_i$, and replacing logical AND with $\sqcap$ and logical OR with $\sqcup$. By Definition 3, a term without $\odot, \otimes$ and $^+$ can be satisfied by singletons only. If ϕ is satisfiable, then there exists a configuration $\langle U, UR \rangle$ and a user u such that $\{u\}$ satisfies ϕ. We can then construct a truth assignment T in which v_i is TRUE if and only if $(u, r_i) \in UR$. It is clear that e evaluates to TRUE under T. Similarly, if there exists a truth assignment T such that e evaluates to TRUE under T, we can construct UR in which u is a member of r_i if and only if v_i is TRUE in T. In that case, $\{u\}$

satisfies ϕ under $\langle U, UR \rangle$. Therefore, e is satisfiable if and only if ϕ is satisfiable.

Proof for Lemma 3: TSAT with just userset, $\sqcap$, $\sqcup$, and $\odot$ is NP-hard.

We reduce the **NP**-complete SET COVERING problem to the TSAT problem of terms consisting of just sets of users, $\sqcap$, $\sqcup$, and $\odot$. In the set covering problem, we are given a finite set $U = \{u_1, \cdots, u_n\}$, a family $F = \{U_1, \cdots, U_m\}$ of subsets of U, and an integer k no larger than m, and we ask whether there are k sets in family F whose union is U.

We view each element in U as a user. For every $j \in [1, m]$ we construct a term $\phi_j = \bigodot\{\{u_i\} \mid u_i \in U_j\}$; that is, $\phi_j = \{u_{j_1}\} \odot \{u_{j_2}\} \odot \cdots \odot \{u_{j_x}\}$, where $U_j = \{u_{j_1}, u_{j_2}, \cdots, u_{j_x}\}$. It is clear that ϕ_i can only be satisfied by U_i. Finally, we construct a term $\phi = ((\bigodot_k (\bigsqcup_{i=1}^n \phi_i)) \sqcap (\bigodot_{i=1}^m \{u_i\}))$. Since $(\bigodot_{i=1}^m \{u_i\})$ can be satisfied only by U, U is the only userset that could satisfy ϕ.

We now demonstrate that ϕ is satisfiable if and only if there are no more than k sets in family F whose union is U. On the one hand, if ϕ is satisfiable, then it must be satisfied by U. In this case, U satisfies $(\bigodot_k (\bigsqcup_{i=1}^n \phi_i))$, which means that there exist k sets $U'_1, \cdots, U'_k$ such that $\bigcup_{i=1}^k U'_i = U$ and each U'_i satisfies $(\bigsqcup_{i=1}^n \phi_i)$. Since ϕ_i can be satisfied only by $U_i \in F$, we have $U'_j \in F$ for every $j \in [1, k]$. The answer to the SET COVERING problem is thus "yes". On the other hand, without loss of generality, assume that $\bigcup_{i=1}^k U_i = U$. We have, for any $i \in [1, k]$, U_i satisfies ϕ_i and thus satisfies $(\bigsqcup_{i=1}^n \phi_i)$. Therefore, U satisfies $(\bigodot_k (\bigsqcup_{i=1}^n \phi_i))$. Since U also satisfies $(\bigodot_{i=1}^m \{u_i\})$, U satisfies $((\bigodot_k (\bigsqcup_{i=1}^n \phi_i)) \sqcap (\bigodot_{i=1}^m \{u_i\}))$.

Proof for Theorem 4: TSAT is NP-complete.

Since we have already proved that certain subcases of TSAT is NP-hard, to prove the theorem, we just need to show that the problem is in **NP**.

To prove that the problem is in **NP**, we need to show that there exists a nondeterministic Turing machine M that is able to generate a userset X and a configuration $\langle U, UR \rangle$ and then check whether X satisfies ϕ under $\langle U, UR \rangle$ in polynomial time. In Lemma 15, we show that checking whether a userset satisfies a term under a given configuration is in **NP**. In other words, one can design a nondeterministic Turing machine N that checks whether a userset satisfies a term in polynomial time. Here, M is the same as N except that M nondeterministically generates userset X and configuration $\langle U, UR \rangle$ at the very beginning. It is obvious that the additional steps M taken can be done in polynomial time.

B.2 Proof for Theorem 5

- $C(\text{All}) = C(r) = \{1\}$ is straightforward.
- That $C(\phi_1 \sqcup \phi_2) = C(\phi_1) \cup C(\phi_2)$ follows from the definition of satisfaction (Definition 3).
- That $C(\phi_1 \sqcap \phi_2) = C(\phi_1) \cap C(\phi_2)$ follows from the definition of satisfaction (Definition 3).
- $C(\phi^+) = \{i \mid i \in [1, \infty)\}$: It follows from the computation of $C(\text{All}), C(r), C(\phi_1 \sqcup \phi_2), C(\phi_1 \sqcap \phi_2)$ that the characteristic set of any unit term ϕ free of usersets and negations is $\{1\}$. Given a configuration $\langle U, UR \rangle$ and a singleton $\{u_1\}$ such that $\{u_1\}$ satisfies ϕ, we can make $n-1$ copies of u_1 for any $n \geq 2$ so that the n users together satisfies ϕ^+. In other words, ϕ^+ may be satisfied by n users for any $n \geq 1$.
- $C(\phi_1 \odot \phi_2) = \{i \mid \exists c_1 \in C(\phi_1) \; \exists c_2 \in C(\phi_2) \; [max(c_1, c_2) \leq i \leq c_1 + c_2]\}$

Let X be a userset that satisfies $(\phi_1 \odot \phi_2)$. There exist X_1 and X_2 such that X_1 satisfies ϕ_1, X_2 satisfies ϕ_2, and $X_1 \cup X_2 = X$. By definition of characteristic set, there exist $c_1 \in C(\phi_1)$ and $c_2 \in C(\phi_2)$ such that $|X_1| = c_1$ and $|X_2| = c_2$. Hence, $max(c_1, c_2) \leq |X| \leq c_1 + c_2$.

Given $c_1 \in C(\phi_1)$ and $c_2 \in C(\phi_2)$, there exist X_1 and X_2 such that X_1 satisfies ϕ_1 under $\langle U_1, UR_1 \rangle$, X_2 satisfies ϕ_2 under $\langle U_2, UR_2 \rangle$, $|X_1| = c_1$ and $|X_2| = c_2$. For any integer $k \in [max(c_1, c_2), c_1 + c_2]$, we may name users in such a way that $|X_1 \cap X_2| = c_1 + c_2 - k$. In this case, $X = X_1 \cup X_2$ satisfies $(\phi_1 \odot \phi_2)$ under $\langle U_1 \cup U_2, UR_1 \cup UR_2 \rangle$ and $|X| = k$.

- $C(\phi_1 \otimes \phi_2) = \{c_1 + c_2 \mid c_1 \in C(\phi_1) \wedge c_2 \in C(\phi_2)\}$

A userset X satisfies $(\phi_1 \otimes \phi_2)$ if and only if there exist X_1 and X_2 such that $X_1 \cup X_2 = X$, $X_1 \cap X_2 = \emptyset$ and X_1, X_2 satisfy ϕ_1, ϕ_2 respectively. By definition of characteristic set, $|X_1| \in C(\phi_1)$ and $|X_2| \in C(\phi_2)$. Therefore, $|X| = (|X_1| + |X_2|) \in \{c_1 + c_2 \mid c_1 \in C(\phi_1) \wedge c_2 \in C(\phi_2)\}$.

On the other hand, given any $c_1 \in C(\phi_1)$ and $c_2 \in C(\phi_2)$, by definition of characteristic number, there exist X_1 and X_2 that satisfy ϕ_1 and ϕ_2 under $\langle U_1, UR_1 \rangle$ and $\langle U_1, UR_1 \rangle$ respectively, such that $|X_1| = c_1$ and $|X_2| = c_2$. Name the users in such a way that $X_1 \cap X_2 = \emptyset$. We have $X = X_1 \cup X_2$ satisfies $(\phi_1 \otimes \phi_2)$ under $\langle U_1 \cup U_2, UR_1 \cup UR_2 \rangle$, where $|X| = |X_1| + |X_2| = c_1 + c_2$.

B.3 Proof that computing the characteristic set takes quadratic time

A straightforward algorithm to compute $C(\phi)$ is to follow Theorem 5. We now show that this can be done in time quadratic to the size of ϕ, denoted by $|\phi|$, which is defined to be the number of occurrences of atomic terms in ϕ. Using induction on the structure of ϕ, it is easy to show that $|\phi|$ is equal to the number of binary operators in ϕ plus 1. We need the following lemma to prove this.

Lemma 9. $C(\phi)$ either is equal to a subset of $\{1, 2, \cdots, |\phi|\}$ or $W \cup \{i \mid i \in [|\phi|, \infty)\}$, where W is a subset of $\{1, 2, \cdots, |\phi|\}$.

PROOF. Proof by induction on the structure of term ϕ.

Based case: When ϕ is a unit term, $C(\phi) = \{1\}$ is a subset of $\{1, 2, \cdots, |\phi|\}$. Otherwise, when ϕ is in the form of ϕ_1^+, where ϕ_1 is a unit term, from Theorem 5, $C(\phi) = \{i \mid i \in [1, \infty)\} = W \cup \{i \mid i \in [|\phi|, \infty)\}$, where $W = \{1, 2, \cdots, |\phi|\}$.

Induction case: When ϕ is in the form of $(\phi_1 \text{ op } \phi_2)$, assume that the lemma holds for ϕ_1 and ϕ_2. Let W_1 denote a subset of $\{1, 2, \cdots, |\phi_1|\}$ and W_2 denote a subset of $\{1, 2, \cdots, |\phi_2|\}$. We have the following three cases:

Case 1: $C(\phi_1) = W_1$ and $C(\phi_2) = W_2$. Since $|\phi| = |\phi_1| + |\phi_2|$, it follows from Theorem 5 that $C(\phi) = W$, where W is a subset of $\{1, 2, \cdots, |\phi|\}$.

Case 2: Exactly one of $C(\phi_1)$ and $C(\phi_2)$ is an infinite set. Without loss of generality, assume that $C(\phi_1) = W_1$ and $C(\phi_2) = W_2 \cup \{i \mid i \in [|\phi_2|, \infty)\}$. We compute $C(\phi)$ according to op:

- op $= \sqcup$: $C(\phi) = C(\phi_1) \cup C(\phi_2) = W_1 \cup W_2 \cup \{i \mid i \in [|\phi_2|, \infty)\} = W_1 \cup W_2 \cup \{i \mid i \in [|\phi_2|, |\phi|]\} \cup \{i \mid i \in [|\phi|, \infty)\}$, in which $W_1 \cup W_2 \cup \{i \mid i \in [|\phi_2|, |\phi|]\}$ is a subset of $\{1, 2, \cdots, |\phi|\}$.
- op $= \sqcap$: $C(\phi) = C(\phi_1) \cap C(\phi_2)$ which is a subset of W_1.
- op $= \odot$: $C(\phi) = \{i \mid \exists c_1 \in W_1 \; \exists c_2 \in W_2 \; [max(c_1, c_2) \leq i \leq c_1 + c_2]\} \cup \{i \mid i \in [max(min(W_1), |\phi_2|), \infty)\} = \{i \mid \exists c_1 \in W_1 \; \exists c_2 \in W_2 \; [max(c_1, c_2) \leq i \leq c_1 + c_2]\} \cup \{i \mid i \in [max(min(W_1), |\phi_2|), |\phi|]\} \cup \{i \mid i \in [|\phi|, \infty)\}$. Note that $\{i \mid \exists c_1 \in W_1 \; \exists c_2 \in W_2 \; [max(c_1, c_2) \leq i \leq c_1 +$

$c_2]\} \cup \{i \mid i \in [max(min(W_1), |\phi_2|), |\phi|]\}$ is a subset of $\{1, 2, \cdots, |\phi|\}$.

- op $= \otimes$: $C(\phi) = \{c_1 + c_2 | c_1 \in W_1 \wedge (c_2 \in W_2 \vee c_2 \in [|\phi_2|, \infty))\} = \{c_1 + c_2 | c_1 \in W_1 \wedge c_2 \in W_2\} \cup \{i | i \in [min(W_1) + |\phi_2|, \infty)\} = \{c_1 + c_2 | c_1 \in W_1 \wedge c_2 \in W_2\} \cup \{i | i \in [min(W_1) + |\phi_2|, |\phi|]\} \cup \{i | i \in [|\phi|, \infty)\}$. Note that $\{c_1 + c_2 | c_1 \in W_1 \wedge c_2 \in W_2\} \cup \{i | i \in [min(W_1) + |\phi_2|, |\phi|]\}$ is a subset of $\{1, 2, \cdots, |\phi|\}$.

Case 3: Both $C(\phi_1)$ and $C(\phi_2)$ are infinite sets, where $C(\phi_1) = W_1 \cup \{i | i \in [|\phi_1|, \infty)\}$ and $C(\phi_2) = W_2 \cup \{i | i \in [|\phi_2|, \infty)\}$. The argument is similar to Case 2.

Given $C(\phi_1)$ and $C(\phi_2)$, Lemma 9 states that $C(\phi_i)$ contains at most $|\phi_i|$ ($i \in \{1, 2\}$) distinct numbers plus a consecutive numerical range, where the range may be treated as a unit during computation. Therefore, calculating $C(\phi_1 \text{ op } \phi_2)$ takes time at most linear in $|\phi_1| + |\phi_2|$. Thus, for each operator in ϕ, the algorithm takes time $O(|\phi|)$; therefore, it takes time at most quadratic in $|\phi|$ to calculate $C(\phi)$. Because ϕ is satisfiable if and only if $C(\phi) \neq \emptyset$, it follows that one can decide whether ϕ is satisfiable or not in time $O(|\phi|^2)$.

C. PROOFS FOR THEOREMS IN SECTION 4

In the following proofs, $(\text{op}_k \phi)$ denotes k copies of ϕ connected together by operator op and $(\text{op}_{i=1}^n r_i)$ denotes $(r_1 \text{ op} \cdots \text{op } r_n)$. Given $R = \{r_1, \cdots, r_m\}$, $(\text{op}R)$ denotes $(r_1 \text{ op} \cdots \text{op } r_m)$.

C.1 The five intractability subcases of UTS

Lemma 10. UTS $\langle \sqcup, \odot \rangle$ is **NP**-hard.

PROOF. We use a reduction from the **NP**-complete SET COVERING problem [9]. In the set covering problem, we are given a finite set $S = \{e_1, \cdots, e_n\}$, family of S's subsets $F = \{S_1, \cdots, S_m\}$, and an integer $k < m$, and we ask whether there are k sets in family F whose union is S. Given such an instance, our reduction maps each element in S to a user and to a role. We construct a configuration $\langle U, UR \rangle$ such that $U = \{u_1, \cdots, u_n\}$ and $UR = \{(u_i, r_i) \mid i \in [1, n]\}$, and a term $\phi = (\bigodot_k (\bigsqcup_{i=1}^m (\bigodot R_i)))$, where R_i is a set of roles such that $r_j \in R_i$ if and only if $e_j \in S_i$.

We now demonstrate that U satisfies ϕ under $\langle U, UR \rangle$ if and only if there exist k sets in family F whose union is S. On the one hand, if U satisfies ϕ, by definition, U has k subsets $U_1, \cdots, U_k$ such that $\bigcup_{i=1}^k U_i = U$ and every U_i satisfies $(\bigsqcup_{i=1}^m (\bigodot R_i))$. U_i satisfies $(\bigsqcup_{i=1}^m (\bigodot R_i))$ if and only if U_i satisfies a certain $(\bigodot R_{x_i})$, where $x_i \in [1, m]$. From the construction of R_{x_i}, U_i satisfies $(\bigodot R_{x_i})$ if and only if $U_i = \{u_a \mid e_a \in S_{x_i}\}$. Since $\bigcup_{i=1}^k U_i = U$, we have $\bigcup_{i=1}^k S_{x_i} = S$. The answer to the set covering problem is "yes". On the other hand, if k subsets in F cover S, without loss of generality, assume that $\bigcup_{i=1}^k S_i = S$. In this case, we divide U into k sets $U_1, \cdots, U_k$ such that $U_i = \{u_j \mid e_j \in S_i\}$. Since $\bigcup_{i=1}^k S_i = S$, we have $\bigcup_{i=1}^k U_i = U$. Furthermore, since $U_i = \{u_j \mid e_j \in S_i\}$, from the construction of R_i, we have U_i satisfies $(\bigodot R_i)$ for every $i \in [1, k]$. Therefore, U satisfies $\phi = (\bigodot_k (\bigsqcup_{i=1}^m (\bigodot R_i)))$.

Lemma 11. UTS $\langle \sqcap, \odot \rangle$ is **NP**-hard.

PROOF. We use a reduction from the **NP**-complete SET COVERING problem [9]. Given $S = \{e_1, \cdots, e_n\}$, a family of S's subsets $F = \{S_1, \cdots, S_m\}$, and an integer $k < m$, our reduction maps each element $e_j \in S$ to a role r_j and each subset $S_i \in F$ to a user u_i. We construct a configuration $\langle U, UR \rangle$ such that $U = \{u_1, \cdots, u_m\}$ and $UR = \{(u_i, r_j) \mid e_j \in S_i\}$, and a term $\phi = (((\bigodot_k \text{All}) \sqcap (\bigodot_{i=1}^n r_i)) \odot (\bigodot_m \text{All}))$.

We now demonstrate that U satisfies ϕ under $\langle U, UR \rangle$ if and only if there exist k sets in family F whose union is S. On one hand, assume U satisfies ϕ. Since $(\bigodot_m \text{All})$ can be satisfied by any nonempty userset with no more than m users, U always satisfies $(\bigodot_m \text{All})$ and it satisfies ϕ if and only if there is a subset U' of U such that U' satisfies $((\bigodot_k \text{All}) \sqcap (\bigodot_{i=1}^n r_i))$. U' satisfying $(\bigodot_k \text{All})$ indicates that $|U'| \leq k$, while U' satisfying $(\bigodot_{i=1}^n r_i)$ indicates that users in U' together have membership of all roles in $\{r_1, \cdots, r_n\}$. Without loss of generality, suppose $U' = \{u_1, \cdots, u_t\}$, where $t \leq k$. As $(u_i, r_j) \in UR$ if and only if $e_j \in S_i$, the union of $\{S_1, \cdots, S_t\}$ is S. The answer to the set covering problem is "yes". On the other hand, if k subsets in F cover S, without loss of generality, assume that $\bigcup_{i=1}^k S_i = S$. From the construction of UR, users $u_1, \cdots, u_k$ together have membership of all roles in $\{r_1, \cdots, r_n\}$. In this case, $\{u_1, \cdots, u_k\}$ satisfies $(\bigodot_{i=1}^n r_i)$. Also, $\{u_1, \cdots, u_k\}$ satisfies $(\bigodot_k \text{All})$. Hence, $\{u_1, \cdots, u_k\}$ satisfies $((\bigodot_k \text{All}) \sqcap (\bigodot_{i=1}^n r_i))$. $(\bigodot_m \text{All})$ is also satisfied by U. Therefore, U satisfies ϕ.

Lemma 12. UTS $\langle \odot, \otimes \rangle$ is **NP**-hard.

PROOF. We use a reduction from the **NP**-complete DOMATIC NUMBER problem [9]. Given a graph $G(V, E)$, the Domatic Number problem asks whether V can be partitioned into k disjoint nonempty sets $V_1, V_2, \cdots, V_k$, such that each V_i is a dominating set for G. V' is a dominating set for $G = (V, E)$ if for every node u in $V - V'$, there is a node v in V' such that $(u, v) \in E$.

Given a graph $G = (V, E)$ and a threshold k, let $U = \{u_1, u_2, \cdots, u_n\}$ and $R = \{r_1, r_2, \cdots, r_n\}$, where n is the number of nodes in V. Each user in U corresponds to a node in G, and $v(u_i)$ denotes the node corresponding to user u_i. $UR = \{(u_i, r_j) \mid i = j \text{ or } (v(u_i), v(u_j)) \in E\}$. Let $\phi = (\bigotimes_k (\bigodot_{i=1}^n r_i))$.

A dominating set in G corresponds to a set of users that together have membership of all the n roles. U satisfies ϕ under $\langle U, UR \rangle$ if and only if U can be divided into k pairwise disjoint sets, each of which has role membership of $r_1, r_2, \cdots, r_n$. Therefore, the answer to the Domatic Number problem is "yes" if and only if U satisfies ϕ under $\langle U, UR \rangle$.

Lemma 13. UTS $\langle \otimes, \sqcup \rangle$ is **NP**-hard.

PROOF. We use a reduction from the **NP**-complete SET PACKING problem [9], which asks, given a finite set $S = \{e_1, \cdots, e_n\}$, a family of S's subsets $F = \{S_1, \cdots, S_m\}$, and an integer k, whether there are k subsets in family F such that these k sets are pairwise disjoint. Without loss of generality, we assume that $S_i \not\subseteq S_j$ if $i \neq j$. (If $S_i \subseteq S_j$, one can remove S_j without affecting the answer.) Let $U = \{u_0, u_1, \cdots, u_n\}$, $R = \{r_1, \cdots, r_n\}$ and $UR = \{(u_i, r_i) \mid 1 \leq i \leq n\}$. In particular, u_0 is a user that is not assigned to any role. We then construct a term $\phi = ((\bigotimes_k (\bigsqcup_{i=1}^m (\bigotimes R_j))) \otimes \phi_{nonempty})$, where $R_j = \{r_i \mid e_i \in S_j\}$ and $\phi_{nonempty} = (\text{All} \sqcup (\text{All} \otimes \text{All}) \sqcup \cdots \sqcup (\bigotimes_m \text{All}))$.

We show that U satisfies ϕ under $\langle U, UR \rangle$ if and only if there are k pairwise disjoint sets in family F. As the only member of r_i is u_i, the only userset that satisfies $\phi_i = (\bigotimes R_j)$ is $U_j = \{u_i \mid e_i \in S_j\}$. Hence, a userset X satisfies $\phi' = (\bigsqcup_{i=1}^m \phi_i)$ if and only if X equals to some U_j.

Without loss of generality, assume that $S_1, \cdots, S_k$ are k pairwise disjoint sets. Then, $U_1, \cdots, U_k$ are k pairwise disjoint sets of users. U_1 satisfies ϕ_1, and thus satisfies ϕ'. Similarly, we have U_i satisfies ϕ' for every i from 1 to k. Furthermore, since $u_0 \notin U_i$

for any $i \in [1,k]$, we have $\bigcup_{i=1}^{k} U_i \subset U$. Hence, U can be divided into two nonempty subset $\bigcup_{i=1}^{k} U_i$ and $U' = U - \bigcup_{i=1}^{k} U_i$ such that $\bigcup_{i=1}^{k} U_i$ satisfies $(\bigotimes_k (\bigsqcup_{i=1}^{m} (\bigotimes R_j)))$ and U' satisfies $\phi_{nonempty}$. In other words, U satisfies ϕ.

On the other hand, suppose U satisfies ϕ. Then, U has a strict subset U' with $u_0 \notin U'$, such that U' can be divided into k pairwise disjoint sets $\hat{U}_1, \cdots, \hat{U}_k$, such that each $\hat{U}_i$ satisfies ϕ'. In order to satisfy ϕ', $\hat{U}_i$ must satisfy a certain ϕ_{a_i} and hence be equivalent to U_{a_i}, where $a_i \in [1,m]$. The assumption that $\hat{U}_1, \cdots, \hat{U}_k$ are pairwise disjoint indicates that $U_{a_1}, \cdots, U_{a_k}$ are also pairwise disjoint. Therefore, their corresponding sets $S_{a_1}, \cdots, S_{a_k}$ are pairwise disjoint. The answer to the Set Packing problem is "yes".

Lemma 14. UTS $\langle \sqcap, \otimes \rangle$ *is* **NP**-*hard*.

PROOF. We use a reduction from the NP-complete SET COVERING problem, which asks, given a family $F = \{S_1, \cdots, S_m\}$ of subsets of a finite set S and an integer k no larger than m, whether there are k sets in family F whose union is S.

Given $S = \{e_1, \cdots, e_n\}$ and a family of S's subsets $F = \{S_1, \cdots, S_m\}$, let $U = \{u_1, u_2, \cdots, u_m\}$, $R = \{r_1, r_2, \cdots, r_n\}$ and $UR = \{(u_i, r_j) \mid e_j \in S_i\}$. Let $\phi = ((\sqcap_{i=1}^{n} (r_i \otimes (\bigotimes_{k-1} \text{All}))) \otimes (\bigotimes_{m-k} \text{All}))$. We now demonstrate that U satisfies ϕ under $\langle U, UR \rangle$ if and only if there are k sets in family F whose union is S. Without loss of generality, assume that $k < m$.

Assume that U satisfies ϕ. Since $(\bigotimes_{m-k} \text{All})$ can be satisfied by any userset with $m - k$ users, U satisfies ϕ if and only if there is a size-k subset U' of U that satisfies $(r_i \otimes (\bigotimes_{k-1} \text{All}))$ for every i from 1 to n. This means that users in U' together have membership of all roles in $\{r_1, \cdots, r_n\}$. Suppose $U' = \{u_{a_1}, \cdots, u_{a_k}\}$, where $a_i \in [1, m]$. As $(u_i, r_j) \in UR$ if and only if $e_j \in S_i$, the union of $\{S_{a_1}, \cdots, S_{a_k}\}$ is S. The answer to the Set Covering problem is "yes".

On the other hand, without loss of generality, assume that $\bigcup_{i=1}^{k} S_i = S$. From the construction of UR, users $u_1, \cdots, u_k$ together have membership of $r_1, \cdots, r_n$. In this case, $\{u_1, \cdots, u_k\}$ satisfies $(r_i \otimes (\bigotimes_{k-1} \text{All}))$ for every i from 1 to n. Since $k < m$, $\{u_1, \cdots, u_k\}$ is a strict subset of U. Therefore, U can be divided into two nonempty subset $\{u_1, \cdots, u_k\}$ and $U - \{u_1, \cdots, u_k\}$ such that $\{u_1, \cdots, u_k\}$ satisfies $(\sqcap_{i=1}^{n} (r_i \otimes (\bigotimes_{k-1} \text{All})))$ and $U - \{u_1, \cdots, u_k\}$ satisfies $(\bigotimes_{m-k} \text{All})$. In other words, U satisfies ϕ.

C.2 Proof that UTS is in NP

Lemma 15. UTS $\langle \neg, +, \sqcup, \sqcap, \odot, \otimes \rangle$ *is in* **NP**.

PROOF. To determine whether a userset X satisfies a term ϕ under a configuration $\langle U, UR \rangle$, we first compute the syntax tree T of ϕ. When constructing T, an 1CF term (i.e., a term of the form ϕ or ϕ^+, where ϕ is a unit term, see Definition 7) is treated as a unit and is not further decomposed. In other words, the leaves in T correspond to sub-terms of ϕ that are 1CF terms and the inner nodes correspond to binary operators connecting these sub-terms. If X satisfies ϕ, then for each node in the tree, there exists a subset of X that satisfies the term rooted at that node, and the root of T corresponds to the set X. After these subsets are guessed and labeled with each node, verifying that they indeed satisfy the terms can be done efficiently. Verifying that a userset satisfies a 1CF term can be done efficiently. (See Proof of Theorem 8.) When the two children of a node are verified, checking that node is labeled correctly can also be done efficiently. Therefore, the problem is in **NP**.

C.3 The tractable cases

Lemma 16. UTS *for 4CF terms is in* **P**.

PROOF. Given a 4CF term $\phi = P_1 \odot \cdots \odot P_n$, where for each k such that $1 \leq k \leq n$, P_k is a 3CF term of the form $\phi_{k,1} \otimes \phi_{k,2} \otimes \cdots \otimes \phi_{k,m_k}$, and each $\phi_{k,j}$ is an 1CF term. Let $t_{k,j}$ be the base unit term in $\phi_{k,j}$. Let T_k be the multiset of base unit terms in P_k, that is, $T_k = \{t_{k,1}, t_{k,2}, \cdots, t_{k,m_k}\}$, and $|T_k| = m_k$.

Given a userset $X = \{u_1, \cdots, u_n\}$ and configuration $\langle U, UR \rangle$, we present an algorithm that determines whether X satisfies ϕ under $\langle U, UR \rangle$.

Step 1 The first step checks that each P_k is satisfied by some subset of X. For each k such that $1 \leq k \leq n$, do the following. Construct a bipartite graph $G(X, T_k)$, in which one partition consists of users in X and the other consists of all the $t_{k,j}$'s in T_k; and there is an edge between $u \in X$ and $t_{k,j}$ if and only if $\{u\}$ satisfies $t_{k,j}$. Compute a maximal matching of the graph $G(X, T_k)$, if the matching has size less than m_k, returns "no", as this means that X does not contain a subset that satisfies P_k; thus X does not satisfy ϕ.

Step 2 The second step checks that each user in X can be "consumed" by some unit term in ϕ. Let $G(A, B)$ denote the bipartite graph in which one partition, A, consists of users in X, and the other partition, B, consists of all the $t_{k,j}$'s in $T_1 \cup T_2 \cup \cdots \cup T_n$. Furthermore, for any unit term t that occurs as t^+ in ϕ, we make sure that B has at least $|X|$ copies of t, by adding additional copies of t if necessary. There is an edge between $u \in A$ and $t \in B$ if and only if $\{u\}$ satisfies t. Compute a maximal matching of the graph $G(X, T)$, if the matching has size less than $|X|$, returns "no".

Step 3 Return "yes".

It is not difficult to see that if the algorithm returns "no", then X does not satisfy ϕ. We now show that if the algorithm returns "yes", then X satisfies ϕ. If the algorithm returns "yes", then for each k, the graph $G(X, T_k)$ has a matching of size m_k; let X_k be the set of users involved in the matching, then X_k satisfies P_k. Let $X' = X_1 \cup X_2 \cup \cdots \cup X_n$. If $X' = X$, then clearly X satisfies ϕ. If $X' \subset X$, then find a user u in $X \setminus X'$, and do the following: Find the term t that is matched with u in the maximal matching computed in step 2. Such a term must exist, since the matching has size $|X|$. Without loss of generality, assume that t appears in P_1, and X_1 contains a user w that is matched with t; then change X_1 by replacing w with u. Clearly, the new X_1 still satisfies P_1. Compute X' again, and if $X' \subset X$, find another user in $X \setminus X'$ and repeat the above process. Note that X' will grow if w appears in some other X_k. Also observe that, the newly added matching between u and t will never be removed again in future, because no other user is matched with t in the maximal matching computed in step 2; as a result, u will always remain in X'. Therefore, after each step, one new user will be added to X' and will never be removed. After at most $|X|$ steps, we will have $X' = X$.

Computationally Sound Secrecy Proofs by Mechanized Flow Analysis

Michael Backes
Dept. of Computer Science
Saarland University
66041 Saarbrücken, Germany
backes@cs.uni-sb.de

Peeter Laud
Dept. of Mathematics and Computer Science
Tartu University
J. Liivi 2, 50409 Tartu, Estonia
peeter.laud@ut.ee

ABSTRACT

We present a novel approach for proving secrecy properties of security protocols by mechanized flow analysis. In contrast to existing tools for proving secrecy by abstract interpretation, our tool enjoys cryptographic soundness in the strong sense of blackbox reactive simulatability/UC which entails that secrecy properties proven by our tool are automatically guaranteed to hold for secure cryptographic implementations of the analyzed protocol, with respect to the more fine-grained cryptographic secrecy definitions and adversary models.

Our tool is capable of reasoning about a comprehensive language for expressing protocols, in particular handling symmetric encryption and asymmetric encryption, and it produces proofs for an unbounded number of sessions in the presence of an active adversary. We have implemented the tool and applied it to a number of common protocols from the literature.

Categories and Subject Descriptors: C.2.2 [Network protocols]: Protocol verification; D.3.3 [Language Constructs and Features]: Concurrent programming structures, Coroutines

General Terms: Security, Verification.

Keywords: Simulatability, Data flow analysis.

1. INTRODUCTION

Security proofs of cryptographic protocols are known to be difficult and the automation of such proofs has been studied soon after the first protocols were developed. From the start, the actual cryptographic operations in such proofs were idealized into so-called Dolev-Yao models, following [22, 23, 39], e.g., see [31, 47, 4, 38, 44, 13]. This idealization simplifies proof construction by freeing proofs from cryptographic details such as computational restrictions, probabilistic behavior, and error probabilities. Conducting secrecy proofs by typing based on these abstractions has shown to be a particularly salient technique as it allowed for elegant and fully automated proofs, often even for an unbounded number of sessions.

A type system was recently presented in [35] that combines the conciseness of language-based reasoning in Dolev-Yao models with strong computational soundness guarantees, i.e., if an abstract protocol typechecks then its cryptographic realization provably keeps the quantities handed to it by the protocol users (payload data) secret in the computational sense. Such computational soundness guarantees of abstract proofs have recently been identified as central for gaining trustworthy guarantees of security protocols: the computational model strives for stronger, more fine-grained security notions and furthermore considers a more realistic adversary that is allowed to perform arbitrary bitstring manipulations as long as they can be performed in probabilistic polynomial-time. However, despite being the first type system that allows for abstract, computationally sound reasoning under active attacks, the major drawback of [35] was that type inference was not considered. As a consequence, this work did not entail an automated procedure for analyzing secrecy aspects of cryptographic protocols with cryptographic soundness guarantees, which arguably is the central goal of unifying the advantages of both approaches.

We remedy this shortcoming by presenting a mechanized approach for soundly proving secrecy of payload data in cryptographic protocols by analysing the possible flows of data during the execution of the protocol. Our approach is capable of reasoning about a comprehensive language for expressing protocols, in particular handling symmetric encryption and asymmetric encryption, allows for more precise analyses compared with the type system of [35], is fully automated, and produces proofs for an unbounded number of sessions in the presence of an active adversary.

Our results (and the one of [35] as well) rely on a variant of the Dolev-Yao model of Backes, Pfitzmann, and Waidner, henceforth called the BPW model, which has been shown to be computationally sound in the strong sense of of blackbox reactive simulatability (BRSIM). The security notion of BRSIM means that one system (here, the cryptographic realization) can be plugged into arbitrary protocols instead of another system (here, the BPW model) [45, 18]. While first security proofs of several common protocols have been hand-crafted using the BPW model [8, 6], recent work has shown that the BPW model is accessible to theorem proving techniques as well [48]. Our work shows that soundly proving secrecy properties in a fully automated manner is possible using the BPW model, and it identifies cryptographically sound secrecy by typing as a promising direction for future work in general. In particular, our line complements the large number of existing works that aims at establishing computational soundness of Dolev-Yao models without considering secrecy by typing, cf. the section on related work for more details.

The analysis presented in this paper builds on the spi-calculus-style language, its deterministic semantics and the corresponding type system from [35] and is inspired by methods from control flow analysis. It works by collecting for each defined variable at each protocol point the possible shapes of terms that this variable may point to, including the possible creation points of the atomic subterms. The same information is also collected for channels between participants for encryption keys, thus yielding information which terms may be communicated over which channel, and which terms may be encrypted with which keys by honest participants, respectively. Finally, the same abstraction is also collected for terms that the adversary may learn during the run of the protocol.

There are a couple of noteworthy points. First, all inputs from the adversary are modeled using a single abstract value, thus freeing the analyser from the necessity to model every new term that the adversary may construct. Instead, we consider explicit rules for decomposing this abstract value, i.e., the adversary's input, which allows us to keep the description of the adversary's knowledge finite. Secondly, parts of the protocol statically following a public-key decryption are analysed twice — once assuming that the ciphertext was created by an honest participant and once assuming that it was created by the adversary. The distinction of these two cases (which was already present in [2] and also in [35]) is important for the precision of the analyser. Thirdly, we collect not only the possible values of variables but also relationships between them. Whenever certain operations restrict the set of possible values of some variables, we exploit these recorded relationships in order to restrict the set of values of related variables as well. This collection of relationships is reminiscent of shape analysis [51], although our task is considerably simpler here than a full shape analysis because we do not have destructive updates. We record the relationships between variables by collecting a set of constraints that their abstractions must satisfy.

Our prover (consisting of constraint generator and solver) has been implemented in O'Caml and can be downloaded at http://www.ut.ee/~peeter_l/research/brsiman.

1.1 Related Work

Early work on linking Dolev-Yao-style symbolic models and cryptography [5, 28, 32, 29] only considered passive attacks, and therefore cannot make general statements about protocols.

The security notion of BRSIM was first defined generally in [45], based on simulatability definitions for secure (one-step) function evaluation. It was extended in [46, 18], the latter with somewhat different details and called UC (universal composability), and has been widely applied to prove individual cryptographic systems secure and to derive general theoretical results. In particular, BRSIM/UC allows for plugging one system into arbitrary protocols instead of another system while retaining essentially arbitrary security properties [45, 18, 10].

A cryptographic justification of a Dolev-Yao model in the sense of BRSIM/UC was first given in [11] with extensions in [12, 9]. Later papers [40, 34, 19] considered to what extent restrictions to weaker security properties or less general protocol classes allow simplifications compared with [11]: Laud [34] has presented cryptographic foundations for a Dolev-Yao model of symmetric encryption but specific to certain confidentiality properties where the surrounding protocols are restricted to straight-line programs. Warinschi et al. [40, 20] have presented cryptographic underpinnings for a Dolev-Yao model of public-key encryption, yet for a restricted class of protocols and protocol properties that can be analyzed using this primitive. Baudet, Cortier, and Kremer [14] have established the soundness of specific classes of equational theories in a Dolev-Yao model under passive attacks. Canetti and Herzog [19] have shown that a Dolev-Yao-style symbolic analysis can be conducted using the framework of universal composability for a restricted class of protocols, namely mutual authentication and key exchange protocols with the additional constraint that the protocols must be expressible as loop-free programs using public-key encryption as their only cryptographic operation. We stress that none of these works build on type inference for proving secrecy properties of security protocols.

The work that comes closest to our work is the work of Laud [35] who designed a type system for proving secrecy aspects of security protocols based on the BPW model. He shows that if an abstract protocol typechecks in his system, then its cryptographic realization provably keeps the quantities handed to it by the protocol users secret in the computational sense. The proof of this fact exploits the BRSIM/UC soundness result of [11, 9, 10] for carrying over symbolic proofs of secrecy in the BPW model to the actual cryptographic realization, similar to the present paper. However, type inference has not been implemented yet in this paper so that the paper did not entail a mechanized procedure for soundly proving secrecy aspects of security protocols.

Efforts are also under way to formulate syntactic calculi with a probabilistic, polynomial-time semantics, including approaches based on process algebra [41, 37], security logics [30, 21] and cryptographic games [15]. In particular, Datta et al. [21] have proposed a promising logical deduction system to prove computational security properties. We are not aware of any implementations of these frameworks, except for Blanchet's [15], who has recently presented an automated tool for proving secrecy properties of security protocols based on transforming cryptographic games. This line of work is orthogonal to the work of justifying Dolev-Yao models, which offer a higher level of abstractions and thus much simpler proofs where applicable, so that proofs of larger systems can be automated.

Let us also mention some of the work in the area of type systems for cryptographic protocol analysis. The first type system of this kind was proposed by Abadi [1], which could be used for verifying the secrecy of payloads or nonces in the protocols using only symmetric encryption. This type system, as well as all the remaining ones that we describe work in the Dolev-Yao model. The type system was extended to cope with asymmetric encryption by Abadi and Blanchet [2]. Abadi and Blanchet [3] further generalized this type system to handle *generic* cryptographic primitives. The type system of Abadi has also been extended by Gordon and Jeffrey [25, 26, 27] to check for integrity properties. Finally, a static program analysis [33] and a type system [36] exist that work directly in the computational model, handling programs containing symmetric encryption, both for passive adversaries only.

Abstract interpretation, which is in most cases automatable using data flow analysis, has also been considered for the analysis of cryptographic protocols within the Dolev-Yao model. See for example [16] and the references contained therein.

1.2 Structure of the paper

We start by describing our (machine-based) execution model and the language used to program these machines for expressing security protocols in Sec. 2. We continue in Sec. 3 with the description of the analysis. In particular, we give the correctness theorem stating under which conditions the results of the abstract analysis entail computational security of a cryptographic protocol. Sec. 4 describes the implementation of our tool and its applicability to common security protocols from the literature. In Sec. 5 we give the main technical lemma, similar to subject reduction, used to prove the previously given correctness theorem.

2. EXECUTION MODEL

We use the same setup of a system as in [35]. In short, the BPW model (sometimes also called abstract cryptographic library in the following corresponding to the original title of [11]) for n honest users is implemented by a machine $\mathcal{TH}_n$ which has input ports $\mathsf{in}_{u_i}?$ to receive commands from the i-th user, output ports $\mathsf{out}_{u_i}!$ to return the results of commands and (handles of) received messages, ports $\mathsf{in}_a?$ and $\mathsf{out}_a!$ for the communication with the adversary, and a database of terms. The database records the structure of messages and the knowledge of messages by the parties (n users and the adversary). The users and the adversary access messages through *handles*, the transmission of messages involves the translation of handles. The possible commands are the construction, taking apart, and sending of messages. The protocol logic for the i-th user is implemented by a machine P_i that connects to the ports $\mathsf{in}_{u_i}?$ and $\mathsf{out}_{u_i}!$ and offers the ports $\mathsf{pin}_{u_i}?$ and $\mathsf{pout}_{u_i}!$ to the user through which it may send and receive data. An execution step of a machine P_i consists of receiving a message (either from $\mathcal{TH}_n$ or the user), performing some computations on the terms, and optionally sending a message. The machines P_i are programmed in a language resembling the spi-calculus, defined below.

$$
\begin{aligned}
e &::= n \mid \underline{\mathsf{keypair}^\ell} \mid \mathsf{store}(x) \\
&\mid x \mid \underline{\mathsf{retrieve}}(x) \mid \mathsf{list}(x_1,\ldots,x_k) \\
&\mid \underline{\mathsf{pubkey}}(x) \mid \underline{\mathsf{pubenc}^\ell}(x_\mathrm{k},x_\mathrm{t}) \\
&\mid \underline{\mathsf{privenc}^\ell}(x_\mathrm{k},x_\mathrm{t}) \mid \pi_i^j(x) \\
&\mid \underline{\mathsf{gen\_symenc\_key}}(i)^\ell \mid \underline{\mathsf{pubdec}}(x_\mathrm{k},x_\mathrm{t}) \\
&\mid \underline{\mathsf{privdec}}(x_\mathrm{k},x_\mathrm{t}) \mid \mathsf{gen\_nonce}^\ell
\end{aligned}
$$

$$
\begin{aligned}
SIP &::= \mathbf{receive}^\ell_c[x_\mathrm{p}](x) \\
IP &::= SIP \mid !SIP \\
I &::= IP.P \\
I^* &::= \mathbf{0} \mid I \mid I^* \\
P &::= I^* \mid \mathcal{II} \\
&\mid \mathbf{send}_c[x_\mathrm{p}](x).I^* \\
&\mid \mathbf{let}^\ell\ x := e\ \mathbf{in}\ P \\
&\qquad\qquad\qquad \mathbf{else}\ P' \\
&\mid \mathbf{if}^\ell\ x = x'\ \mathbf{then}\ P \\
&\qquad\qquad\qquad \mathbf{else}\ P'
\end{aligned}
$$

Here x-s are variables, e-s are expressions, I-s are input processes, P-s are output processes, and ℓ-s are labels for program points and expressions of interest. No label may occur twice in the protocol text, nor can a variable be defined twice or used before being defined. The language contains public-key and symmetric-key encryption as the cryptographic primitives (as well as nonces). A public and secret key pair is created by the expression keypair, the public key is extracted by pubkey. A *level i* is associated with each symmetric key to prevent encryption cycles (and make the proof relating $\mathcal{TH}_n$ and its concrete implementation go through); a symmetric key may only encrypt keys of lower level. The store- and retrieve-expressions are used to convert payloads (data that can be communicated with the user) to handles and back. The expression $\pi_i^j(x)$ extracts the i-th component from the list of length j pointed to by x. In $\mathbf{receive}_c[x_\mathrm{p}](x)$ and $\mathbf{send}_c[x_\mathrm{p}](x)$, the variable x is the message and x_p is the identity of the other party. The channel for the message is given by the constant *abstract channel c*. An abstract channel is used to group messages sent between protocol participants, as well as between the protocol user and participant (although the abstract channel does not alone determine the sender and the receiver of a message). Furthermore, the set of abstract channels **Chan** is partitioned into four parts, denoted $\mathbf{Chan}_\mathsf{x}$, where $\mathsf{x} \in \{\mathsf{s},\mathsf{a},\mathsf{i},\mathsf{u}\}$. If a message is sent on an abstract channel from $\mathbf{Chan}_\mathsf{s}$ [resp. $\mathbf{Chan}_\mathsf{a}$, $\mathbf{Chan}_\mathsf{i}$] then it means that the message travels between protocol participants over a secure (resp. authentic, insecure) channel. If a message is sent on an abstract channel from $\mathbf{Chan}_\mathsf{u}$ then it travels between the protocol user and the protocol participant (i.e. over one of the concrete channels pin_{u_i} or pout_{u_i}). The variables x and x_p are bound in a **receive**-statement. The variable x is also bound in the default-branch of a **let**-statement, but not in the else-branch, which is taken upon a failure of evaluating e.

The internal state of an inactive (i.e. not currently running) P_i consists of a list of input processes together with their execution environments, giving values to already defined variables. The "program" (or initial state) of each P_i is a list of input processes. An active P_i additionally contains the received message (together with the apparent sender and the name of the channel it was received on) and the currently running (output) process (together with its environment). When P_i receives the message, it is handed over to the first input process $(!)\mathbf{receive}_c[x_\mathrm{p}](x).P$ with matching channel name c in its list of processes. The variables x and x_p are bound to the message and the apparent sender and the process executes until it has become $\mathcal{II}$ or a list of input processes I^*. The value $\mathcal{II}$ means rejecting the message — the list of input processes of P_i is not changed, the currently executing process and its environment are discarded (thereby forgetting all references to any new terms that may have been created since receiving that message) and the message is handed over to the next input process with the matching channel name in the list of input processes of P_i. When a process accepts the message, it executes until it has become a list of input processes I^*. All processes in this list I^*, together with the environment of the output process, are put to the list of input processes of P_i instead of or in addition of (depending on the presence of replication) the original process. When no process accepts the message, it is simply lost.

Security

The security property we are interested in is the *secrecy of payloads* [10]. We also considered the same property in [35] and our treatment here does not differ from that. In short, we want the system implementing the protocol (consisting of the machines $\mathsf{P}_1,\ldots,\mathsf{P}_n$ and $\mathcal{TH}_n$) to retain the secrecy of any payloads *handed to it by the users* over the ports $\mathsf{pin}_{u_i}?$. The secrecy of payloads means that the user and the adversary together cannot figure out whether the system implementing the protocol is really computing with the values received from the user or with some other values. In the definition of payload secrecy, there is a scrambler / descrambler inserted between the user and the system implementing the protocol. If it is turned on, it replaces all messages from the user to the system (i.e. all payloads) with random values; and replaces these random values, if they are sent from the system to the user, with the values received from the user again. The user (together with the adversary) has to guess whether the scrambler / descrambler is turned on. If the user cannot guess then the payloads are kept secure in the system — they do not flow from the user through the system to the adversary. A precise definition can be found in [10] and a concise description in [35]. In [35] the following five properties were stated to be sufficient for the secrecy of payloads and for the simulatability of the machine $\mathcal{TH}_n$:

(I) the bit-strings that the machines P_i receive from the ports $\mathsf{pin}_{u_i}?$ do not affect the control flow of P_i, i.e. this data is not used in the **if**-statements;

(II) the machines P_i may pass the bit-strings received from the user to the cryptographic library only in store-commands;

(III) the terms resulting from these store-commands will not become available to the adversary, i.e. the adversary does not get handles for these terms.

(IV) symmetric keys of order i only encrypt terms of order less than i (note that symmetric keys created by the adversary have no order and are thereby not restricted by this condition);

(V) if a symmetric key unknown to the adversary (i.e. the adversary does not have a handle to it) is used for encryption then this key will never become known to the adversary.

The analysis presented in this paper verifies that these five properties hold.

A different secrecy property was also considered in [10] — the secrecy of keys generated by the system during the protocol run. We do not consider this property here, although a corresponding list of sufficient properties would not be difficult to fix (key secrecy can be considered a simpler property than payload secrecy); and these properties could also be verified by our analysis. However, this remains future work.

3. ANALYSIS

We set up a constraint system whose solutions upper-approximate the values flowing through the protocol. A constraint system consists of two main parts — constraint variables and the set of constraints. A constraint variable is just an identifier together with an associated domain (an upper semilattice) of possible values. A constraint is a statement of the form $E \leq C$ where C is a constraint variable and E is a monotone expression over constraint variables. A solution of a constraint system is a valuation, mapping each constraint variable to a value in its domain, such that all constraints are satisfied.

We will prove that any solution to our constraint system will be a safe approximation of any possible protocol run. To get the best precision, we are interested in the least solution. There are well-known methods for finding the least solutions for constraint systems with monotone constraints.

3.1 Abstract Domain

The possible values of protocol variables are abstracted by sets of the following abstract values AV. The sets of abstract values are used as domains for certain constraint variables below.

$$AV ::= AV_I \mid AV_H \mid \mathsf{seckey}(\ell, b) \qquad AV_I = \mathsf{X_P} \mid \mathsf{X_S}$$

$$\begin{aligned}
AV_H ::= &\ \mathsf{pubenc}(AV_H, AV_H, \ell, b) \mid \mathsf{nonce}(\ell, b) \\
&\mid \mathsf{symenc}(AV_H, AV_H, \ell, b) \mid \mathsf{AnyPubVal} \\
&\mid \mathsf{symkeyname}(\ell, b) \mid \mathsf{pubkey}(\ell, b) \\
&\mid (AV_H, \ldots, AV_H) \mid \mathsf{store}(AV_I) \\
&\mid \mathsf{symkey}(i, \ell, b)
\end{aligned} \tag{1}$$

Here AV_I contains the possible abstractions of payloads — they may be either public ($\mathsf{X_P}$) or secret ($\mathsf{X_S}$). The addresses of the communication partners (variable x_p in **send**- and **receive**-commands) are public. Data received from the protocol users are secret. The terms AV_H are the possible abstractions of terms in the database of $\mathcal{TH}_n$. They should be mostly self-descriptive. The arguments ℓ refer to program points (labels at expressions) where these values have been created. The arguments b also resemble program points — we have mentioned before that we analyse the parts of the protocol following a public-key decryption twice — once assuming that the ciphertext was generated by a protocol participant and once assuming that it was generated by the adversary. Hence, if n public-key decryptions occur before the program point ℓ then this point really counts as 2^n different program points for the analysis. If ℓ is a program point following n decryptions then b is a bit-string of length n where i-th bit records the assumed creator of the i-th decrypted ciphertext (1 — some honest participant; 0 — the adversary). We call b the *decryption context*.

The argument i in $\mathsf{symkey}(i, \ell, b)$ records the level of the symmetric key. The abstract value $\mathsf{symkeyname}(\ell, b)$ corresponds to the identities of the symmetric keys created at the program point ℓ (with the decryption context b). According to $\mathcal{TH}_n$, the adversary is able to find the identities of symmetric keys from the ciphertexts created with them. The abstract value $\mathsf{AnyPubVal}$ denotes any value that the adversary knows and may have constructed. All other AV_H denote values constructed by protocol participants. The secret decryption keys $\mathsf{seckey}(\ell, b)$ are not listed as a possible case for AV_H because $\mathcal{TH}_n$ puts severe restrictions on their use — they may only be used for decrypting ciphertexts; they cannot appear as subterms of more complex terms.

3.2 Constraint Variables

Given a protocol $\wp$ with its set of labels, we introduce the following constraint variables.

First, $\mathbf{S}_\ell^b$ for all statement labels ℓ occurring in the protocol (here we only consider the labels of **if**-, **let**- and **receive**-statements, not the labels occurring in expressions). Here b is a bit-string whose length equals the number asymmetric decryption operations that occur in the protocol before and including the point labeled with ℓ. Hence for a program point ℓ that is preceded by n asymmetric decryptions we have 2^n different variables $\mathbf{S}_\ell^b$.

Let $\mathbf{Var}_\ell^\circ$ be the set of variables defined before the protocol point ℓ. Let $\mathbf{Var}_\ell^\bullet$ be the union of $\mathbf{Var}_\ell^\circ$ with the set of protocol variables that are assigned a value at ℓ (depending on whether ℓ labels an **if**-, **let**- or **receive**-statement, this set has 0, 1 or 2 elements). The possible values for $\mathbf{S}_\ell^b$ are mappings from $\mathbf{Var}_\ell^\bullet$ to sets of abstract values AV. These mappings are ordered pointwise, with the sets of abstract values AV ordered by subset inclusion.

The variable $\mathbf{S}_\ell^b$ records the possible values of protocol variables after a successful completion of the operation at program point ℓ. A **let**- or **if**-statement is successful if the default-/true-branch was taken. A **receive**-statement is always successful.

Second, $\mathbf{R}_\ell^b$ for all statement labels ℓ occurring in the protocol and b having the same possible values as for $\mathbf{S}_\ell^b$ (and ordered the same way). These constraint variables are introduced to ease the presentation of the constraint system. Namely, the handling of a statement (if it succeeds) proceeds in two steps: first the constraints giving the abstraction(s) of the newly defined variable(s) are evaluated, followed by the evaluation of constraints describing the relationships between the values of different variables. The constraint variable $\mathbf{S}_\ell^b$ contains the result of these two steps. The constraint variable $\mathbf{R}_\ell^b$ contains the result of the first step only.

Third, $\mathbf{C}_c$ for all abstract channels $c \in \mathbf{Chan}_s \cup \mathbf{Chan}_a$ occurring in the protocol. The possible values of these variables are sets of abstract values AV_H, ordered by subset inclusion. These variables will record an abstraction of the possible messages sent over the abstract channel c.

Fourth, $\mathbf{P}$. This will record the values that the adversary knows. The possible values of this variable are sets of abstract values AV, ordered by subset inclusion.

Fifth, $\mathbf{E}_\ell^b$ for a label ℓ occurring at a key generation. This set records all abstract values that are encrypted with the key generated at ℓ for the preceding asymmetric decryption results described by b. The bit-string b has the same meaning as for the variables $\mathbf{S}_\ell^b$ (the point of interest is the occurrence of ℓ in the protocol). The possible values of these variables are sets of abstract values AV_H, ordered by subset inclusion.

Sixth, $\mathbf{L}_{\ell,\mathsf{true}}^b$ and $\mathbf{L}_{\ell,\mathsf{false}}^b$ for labels ℓ at *let*- and *while*-statements. They denote whether the true- (default-) and false-branch of the statement are alive or not. The bit-string has the same meaning as before. The possible values of these variables are false and true, ordered by $\mathsf{false} \leq \mathsf{true}$.

$$\text{store}(AV) \in \mathbf{P} \Rightarrow AV \in \mathbf{P}$$

$$(AV_1, \ldots, AV_j) \in \mathbf{P} \Rightarrow AV_i \in \mathbf{P}$$

$$\text{symenc}(AV_k, AV_t, \ell, b) \in \mathbf{P} \Rightarrow$$
$$(\exists AV' \in \mathbf{P} : AV_k \cong_\mathbf{P} AV') \Rightarrow AV_t \in \mathbf{P}$$

$$\text{pubenc}(\text{AnyPubVal}, AV_t, \ell, b) \in \mathbf{P} \Rightarrow AV_t \in \mathbf{P}$$

$$\text{pubenc}(AV_k, AV_t, \ell, b) \in \mathbf{P} \Rightarrow AV_k \in \mathbf{P}$$

$$\text{symenc}(\text{symkey}(i, \ell, b), AV_t, \ell', b') \in \mathbf{P} \Rightarrow$$
$$\text{symkeyname}(\ell, b) \in \mathbf{P}$$

$$\{\mathsf{X_P}, \text{AnyPubVal}\} \subseteq \mathbf{P}$$

Figure 1: Constraints describing the adversary's power

$$AV \in \mathbf{P} \Rightarrow AV \cong_\mathbf{P} \text{AnyPubVal}$$

$$\text{store}(\mathsf{X_P}) \cong_\mathbf{P} \text{AnyPubVal}$$

$$\big(\forall i : AV_i \cong_\mathbf{P} AV'_i\big) \wedge (AV'_1, \ldots, AV'_j) \cong_\mathbf{P} \text{AnyPubVal}$$
$$\Rightarrow (AV_1, \ldots, AV_j) \cong_\mathbf{P} \text{AnyPubVal}$$

$$(\text{AnyPubVal}, \ldots, \text{AnyPubVal}) \cong_\mathbf{P} \text{AnyPubVal}$$

$$AV_k \cong_\mathbf{P} AV'_k \wedge AV_t \cong_\mathbf{P} AV'_t \wedge \text{pubenc}(AV'_k, AV'_t, \ell, b) \in \mathbf{P}$$
$$\Rightarrow \text{pubenc}(AV_k, AV_t, \ell, b) \cong_\mathbf{P} \text{AnyPubVal}$$

$$AV_k \cong_\mathbf{P} AV'_k \wedge AV_t \cong_\mathbf{P} AV'_t \wedge \text{symenc}(AV'_k, AV'_t, \ell, b) \in \mathbf{P}$$
$$\Rightarrow \text{symenc}(AV_k, AV_t, \ell, b) \cong_\mathbf{P} \text{AnyPubVal}$$

$$\mathsf{X_S} \cong_\mathbf{P} \mathsf{X_P}$$

Figure 2: The "possibly equal" relation $\cong_\mathbf{P}$

3.3 Constraints

There are two sources for constraints — the protocol and the adversary. The first describes the movement of values during the computations performed by the protocol, while the second describes the capabilities of the adversary in decomposing messages. This second set of constraints, given in Fig. 1 is quite straightforward: The first two constraints are obvious — the adversary can retrieve stored payloads and decompose lists. The third constraint states that the adversary can decrypt a symmetric encryption if it has the key. The relation $\cong_\mathbf{P}$ relates two abstract values if the sets of terms they correspond to may intersect. Because the meaning of AnyPubVal depends on the adversary's knowledge, this relation must also depend on it. The relation $\cong_\mathbf{P}$ is the least reflexive, symmetric and structure-respecting relation on abstract values that satisfies the conditions given in Fig. 2. The fourth constraint for the adversary's capabilities states that if the public key used for public encryption may have been created by the adversary (which means that the secret key was also created by the adversary) then the adversary may find out the plaintext. The fifth and sixth constraints state that the adversary is capable of determining the identity of the key used to produce the ciphertext. For asymmetric encryption, this identity is the public key itself, while for symmetric encryption, it is the symkeyname. Finally, the public values $\mathsf{X_P}$ and AnyPubVal may be known to the adversary.

The set of constraints generated by an input or output process P is given by the mapping $\langle\!\langle\!\langle P \rangle\!\rangle\!\rangle$ that we are going to define below. For defining it, we also define the following mappings.

- $\langle\!\langle e \rangle\!\rangle(\mathbf{I}, b)$ gives the set of abstract values for the result of evaluating the expression e when the decryption context (after evaluating e) is b and the abstractions of already defined variables are given by the mapping $\mathbf{I}$.

- $\langle\!\langle e \rangle\!\rangle_\mathrm{s}(\mathbf{I})$ and $\langle\!\langle e \rangle\!\rangle_\mathrm{f}(\mathbf{I})$ give some necessary conditions for the evaluation of e to succeed or fail.

- $\langle\!\langle e \rangle\!\rangle_\varepsilon(\mathbf{I}, \mathbf{L})$ gives the set of constraints for the variables $\mathbf{E}^b_\ell$, as generated by e. Here $\mathbf{L}$ is a boolean showing whether this expression is live code.

- $\lfloor\!\lfloor x := e \rfloor\!\rfloor(b, \mathbf{X}, y)$, where $\mathbf{X}$ gives the abstractions of variables defined before the assignment $x := e$, and y is either x or a variable occurring in e gives a set of abstract values that certainly abstracts the value of y after the successful evaluation of e. The mapping $\lfloor\!\lfloor x := e \rfloor\!\rfloor$ is used to collect the relationships between values of variables.

The relationships between variables allow us to make the analysis more precise — they bound the abstractions of values of variables from above. When combining several of these upper bounds, we have to form their greatest lower bound. While the least upper bound of two sets of abstract values may be just the union of sets, the greatest lower bound cannot be simply the intersection because when two sets of abstract values $\mathbf{A}$ and $\mathbf{B}$ are both valid abstractions of some concrete value then we want their greatest lower bound $\mathbf{A} \dot{\cap} \mathbf{B}$ be a valid abstraction of that value as well. But certain concrete values may correspond to several different abstract values, for example a nonce that has become known to the adversary may occur in our abstractions either as $\text{nonce}(\ell, b)$ for some ℓ and b or as AnyPubVal.

For defining $\dot{\cap}$, we first define a *partial* binary operation $\sqcap$ on abstract values as the smallest (i.e. defined for as few arguments as possible) idempotent symmetric structure-preserving operation that satisfies $AV_H \sqcap \text{AnyPubVal} = AV_H$ for any abstract value AV_H defined in (1). Now we can just define $\mathbf{A} \dot{\cap} \mathbf{B} = \{AV \sqcap AV' \mid AV \in \mathbf{A}, AV' \in \mathbf{B}\}$. We also define $\mathbf{A} \dot{\subseteq} \mathbf{B}$ iff $\mathbf{A} \dot{\cap} \mathbf{B} = \mathbf{A}$.

The mappings $\langle\!\langle e \rangle\!\rangle$, $\langle\!\langle e \rangle\!\rangle_\mathrm{s}$, $\langle\!\langle e \rangle\!\rangle_\mathrm{f}$ and $\langle\!\langle e \rangle\!\rangle_\varepsilon$ are given in Figures 3 and 4. If we have left out the definition of $\langle\!\langle e \rangle\!\rangle_\mathrm{s}$ or $\langle\!\langle e \rangle\!\rangle_\mathrm{f}$ for some e then it is true. If we have left out the definition of $\langle\!\langle e \rangle\!\rangle_\varepsilon$ for some e then it is $\emptyset$. In Fig. 4, L_a [resp. L_s] denotes the set of all labels ℓ occurring in the protocol in the positions $\underline{\text{keypair}}^\ell$ [resp. $\underline{\text{gen\_symenc\_key}(i)}^\ell$]. In Fig. 3 we can see the special treatment of AnyPubVal — for example, payloads can be extracted from it and projections can be taken. The result is still a public value. During encryption, AnyPubVal may serve as the encryption key (of course, such ciphertexts can be decrypted by the adversary). During decryption, when the ciphertext is AnyPubVal, we use the variables $\mathbf{E}^b_\ell$ to determine the possible plaintexts.

The distinction between participant-generated and adversarially generated ciphertexts in public-key decryption can be seen in two definitions for $\langle\!\langle \underline{\text{pubdec}}(x_k, x_t) \rangle\!\rangle$. First of them assumes that the ciphertext is generated by some protocol participant, while the second assumes that the adversary is the source of the ciphertext. Both of these cases are also present in symmetric decryption, but they have been joined together, so that the analysis does not handle them separately.

The relationships between newly defined and existing variables are given by $\lfloor\!\lfloor x := e \rfloor\!\rfloor(b, \mathbf{X}, y)$, defined in Fig. 5. Recall that it gives for a variable y a set of abstract values that is guaranteed to

$\langle\!\langle n \rangle\!\rangle(\mathbf{I}, b) = \{\mathsf{X_P}\} \qquad \langle\!\langle x \rangle\!\rangle(\mathbf{I}, b) = \mathbf{I}(x)$

$\langle\!\langle \underline{\mathsf{keypair}}^\ell \rangle\!\rangle(\mathbf{I}, b) = \{\mathsf{seckey}(\ell, b)\}$

$\langle\!\langle \mathsf{store}(x) \rangle\!\rangle(\mathbf{I}, b) = \{\mathsf{store}(AV) \mid AV \in \mathbf{I}(x)\}$

$\langle\!\langle \mathsf{retrieve}(x) \rangle\!\rangle(\mathbf{I}, b) =$
$\quad \{AV \mid \mathsf{store}(AV) \in \mathbf{I}(x)\} \cup \{\mathsf{X_P} \mid \mathsf{AnyPubVal} \in \mathbf{I}(x)\}$

$\langle\!\langle \mathsf{list}(x_1, \ldots, x_k) \rangle\!\rangle(\mathbf{I}, b) = \{(AV_1, \ldots, AV_k) \mid AV_i \in \mathbf{I}(x_i)\}$

$\langle\!\langle \underline{\mathsf{gen\_symenc\_key}}(i)^\ell \rangle\!\rangle(\mathbf{I}, b) = \{\mathsf{symkey}(i, \ell, b)\}$

$\langle\!\langle \underline{\pi_i^j}(x) \rangle\!\rangle(\mathbf{I}, b) = \{AV_i \mid (AV_1, \ldots, AV_k) \in \mathbf{I}(x)\} \cup$
$\quad \{\mathsf{AnyPubVal} \mid \mathsf{AnyPubVal} \in \mathbf{I}(x)\}$

$\langle\!\langle \underline{\mathsf{pubkey}}(x) \rangle\!\rangle(\mathbf{I}, b) = \{\mathsf{pubkey}(\ell, b) \mid \mathsf{seckey}(\ell, b) \in \mathbf{I}(x)\}$

$\langle\!\langle \underline{\mathsf{gen\_nonce}} \rangle\!\rangle(\mathbf{I}, b) = \{\mathsf{nonce}(\ell, b)\}$

$\langle\!\langle \underline{\mathsf{pubenc}}^\ell(x_k, x_t) \rangle\!\rangle(\mathbf{I}, b) = \{\mathsf{pubenc}(AV_k, AV_t, \ell, b) \mid$
$\quad AV_k \in \mathbf{I}(x_k), AV_t \in \mathbf{I}(x_t), AV_k = \mathsf{pubkey}(\ldots)\} \cup$
$\quad \{\mathsf{pubenc}(\mathsf{AnyPubVal}, AV_t, \ell, b) \mid$
$\quad\quad \mathsf{AnyPubVal} \in \mathbf{I}(x_k), AV_t \in \mathbf{I}(x_t)\}$

$\langle\!\langle \underline{\mathsf{privenc}}^\ell(x_k, x_t) \rangle\!\rangle(\mathbf{I}, b) = \{\mathsf{symenc}(AV_k, AV_t, \ell, b) \mid$
$\quad AV_k \in \mathbf{I}(x_k), AV_t \in \mathbf{I}(x_t), AV_k = \mathsf{symkey}(\ldots)\} \cup$
$\quad \{\mathsf{symenc}(\mathsf{AnyPubVal}, AV_t, \ell, b) \mid$
$\quad\quad \mathsf{AnyPubVal} \in \mathbf{I}(x_k), AV_t \in \mathbf{I}(x_t)\}$

$\langle\!\langle \underline{\mathsf{privdec}}(x_k, x_t) \rangle\!\rangle(\mathbf{I}, b) =$
$\quad \{AV_p \mid \mathsf{symenc}(AV_k, AV_p, \ell', b') \in \mathbf{I}(x_t),$
$\quad\quad AV_k' \in \mathbf{I}(x_k), AV_k \cong_\mathbf{P} AV_k'\} \cup$
$\quad \{\mathsf{AnyPubVal} \mid \mathsf{AnyPubVal} \in \mathbf{I}(x_t), \mathbf{I}(x_k) \cap \mathbf{P} \neq \emptyset\} \cup$
$\quad \textit{if } \mathsf{AnyPubVal} \in \mathbf{I}(x_t) \textit{ then } \bigcup_{\mathsf{symkey}(i,\ell,b) \in \mathbf{I}(x_k)} \mathbf{E}_\ell^b \textit{ else } \emptyset$

$\langle\!\langle \underline{\mathsf{pubdec}}(x_k, x_t) \rangle\!\rangle(\mathbf{I}, b1) =$
$\quad \{AV_p \mid \mathsf{pubenc}(\mathsf{pubkey}(\ell'', b''), AV_p, \ell', b') \in \mathbf{I}(x_t),$
$\quad\quad \mathsf{seckey}(\ell'', b'') \in \mathbf{I}(x_k)\} \cup$
$\quad \textit{if } \mathsf{AnyPubVal} \in \mathbf{I}(x_t) \textit{ then } \bigcup_{\mathsf{seckey}(\ell, b) \in \mathbf{I}(x_k)} \mathbf{E}_\ell^b \textit{ else } \emptyset$

$\langle\!\langle \underline{\mathsf{pubdec}}(x_k, x_t) \rangle\!\rangle(\mathbf{I}, b0) = \{\mathsf{AnyPubVal} \mid \mathsf{AnyPubVal} \in \mathbf{I}(x_t)\}$

Figure 3: Abstract semantics of expressions: mapping $\langle\!\langle e \rangle\!\rangle$

$\langle\!\langle n \rangle\!\rangle_\mathrm{f}(\mathbf{I}) = \mathsf{false} \qquad \langle\!\langle x \rangle\!\rangle_\mathrm{f}(\mathbf{I}) = \mathsf{false}$

$\langle\!\langle \underline{\mathsf{keypair}}^\ell \rangle\!\rangle_\mathrm{f}(\mathbf{I}) = \mathsf{false} \qquad \langle\!\langle \mathsf{store}(x) \rangle\!\rangle_\mathrm{f}(\mathbf{I}) = \mathsf{false}$

$\langle\!\langle \mathsf{retrieve}(x) \rangle\!\rangle_\mathrm{s}(\mathbf{I}) =$
$\quad \mathsf{AnyPubVal} \in \mathbf{I}(x) \vee \exists AV : \mathsf{store}(AV) \in \mathbf{I}(x)$

$\langle\!\langle \mathsf{retrieve}(x) \rangle\!\rangle_\mathrm{f}(\mathbf{I}) = \exists AV \in \mathbf{I}(x) : AV \neq \mathsf{store}(\ldots)$

$\langle\!\langle \mathsf{list}(x_1, \ldots, x_k) \rangle\!\rangle_\mathrm{s}(\mathbf{I}) = \forall i : \exists AV \in \mathbf{I}(x_i) : AV \neq \mathsf{seckey}(\ldots)$

$\langle\!\langle \mathsf{list}(x_1, \ldots, x_k) \rangle\!\rangle_\mathrm{f}(\mathbf{I}) = \exists i : \exists AV \in \mathbf{I}(x_i) : AV = \mathsf{seckey}(\ldots)$

$\langle\!\langle \underline{\mathsf{gen\_symenc\_key}}(i)^\ell \rangle\!\rangle_\mathrm{f}(\mathbf{I}) = \mathsf{false}$

$\langle\!\langle \underline{\pi_i^j}(x) \rangle\!\rangle_\mathrm{s}(\mathbf{I}) = \mathsf{AnyPubVal} \in \mathbf{I}(x) \vee \exists (AV_i, \ldots, AV_j) \in \mathbf{I}(x)$

$\langle\!\langle \underline{\pi_i^j}(x) \rangle\!\rangle_\mathrm{f}(\mathbf{I}) = \exists AV \in \mathbf{I}(x) : AV \neq (AV_1, \ldots, AV_j)$

$\langle\!\langle \underline{\mathsf{pubkey}}(x) \rangle\!\rangle_\mathrm{s}(\mathbf{I}) = \exists \ell, b : \mathsf{seckey}(\ell, b) \in \mathbf{I}(x)$

$\langle\!\langle \underline{\mathsf{pubkey}}(x) \rangle\!\rangle_\mathrm{f}(\mathbf{I}) = \exists AV \in \mathbf{I}(x) : AV \neq \mathsf{seckey}(\ldots)$

$\langle\!\langle \underline{\mathsf{gen\_nonce}} \rangle\!\rangle_\mathrm{f}(\mathbf{I}) = \mathsf{false}$

$\langle\!\langle \underline{\mathsf{pubenc}}^\ell(x_k, x_t) \rangle\!\rangle_\mathrm{s}(\mathbf{I}) =$
$\quad \mathsf{AnyPubVal} \in \mathbf{I}(x_k) \vee \exists \ell', b' : \mathsf{pubkey}(\ell', b') \in \mathbf{I}(x_k)$

$\langle\!\langle \underline{\mathsf{pubenc}}^\ell(x_k, x_t) \rangle\!\rangle_\mathcal{E}(\mathbf{I}, \mathbf{L}) =$
$\quad \{\mathsf{pubkey}(\ell', b') \in \mathbf{I}(x_k) \wedge \mathbf{L} \Rightarrow \mathbf{I}(x_t) \subseteq \mathbf{E}_{\ell'}^{b'} \mid \ell' \in \mathsf{L}_a\}$

$\langle\!\langle \underline{\mathsf{privenc}}^\ell(x_k, x_t) \rangle\!\rangle_\mathrm{s}(\mathbf{I}) =$
$\quad \mathsf{AnyPubVal} \in \mathbf{I}(x_k) \vee \exists i, \ell', b' : \mathsf{symkey}(i, \ell', b') \in \mathbf{I}(x_k)$

$\langle\!\langle \underline{\mathsf{privenc}}^\ell(x_k, x_t) \rangle\!\rangle_\mathcal{E}(\mathbf{I}, \mathbf{L}) =$
$\quad \{\mathsf{symkey}(i, \ell', b') \in \mathbf{I}(x_k) \wedge \mathbf{L} \Rightarrow \mathbf{I}(x_t) \subseteq \mathbf{E}_{\ell'}^{b'} \mid \ell' \in \mathsf{L}_s\}$

$\langle\!\langle \underline{\mathsf{privdec}}(x_k, x_t) \rangle\!\rangle_\mathrm{s}(\mathbf{I}) =$
$\quad \mathsf{AnyPubVal} \in \mathbf{I}(x_k) \vee \exists i, \ell', b' : \mathsf{seckey}(\ell', b') \in \mathbf{I}(x_k)$

$\langle\!\langle \underline{\mathsf{pubdec}}(x_k, x_t) \rangle\!\rangle_\mathrm{s}(\mathbf{I}) = \exists \ell', b' : \mathsf{seckey}(\ell', b') \in \mathbf{I}(x_k)$

Figure 4: Abstract semantics of expressions: mappings $\langle\!\langle e \rangle\!\rangle_\mathrm{s}$, $\langle\!\langle e \rangle\!\rangle_\mathrm{f}$ and $\langle\!\langle e \rangle\!\rangle_\mathcal{E}$

abstract its concrete value. If the definition is missing for some $\lfloor\!\lfloor x := e \rfloor\!\rfloor(b, \mathbf{X}, y)$ in Fig. 5, it is equal to $\mathbf{X}(y)$ (i.e. no precision is gained). In Fig. 5, the message constructor C may be one of list, $\underline{\mathsf{pubenc}}^\ell$ or $\underline{\mathsf{privenc}}^\ell$. The corresponding abstract value constructor C is then either the list constructor, $\mathsf{pubenc}(\cdot, \cdot, \ell, b)$ or $\mathsf{privenc}(\cdot, \cdot, \ell, b)$.

The usage of $\lfloor\!\lfloor x := e \rfloor\!\rfloor$ may become clearer when we look at

$$\lfloor\!\lfloor x := \mathsf{C}(x_1,\ldots,x_k)\rfloor\!\rfloor(b,\mathbf{X},x) =$$
$$\{C(AV_1,\ldots,AV_k) \mid AV_i \in \mathbf{X}(x_i)\}$$

$$\lfloor\!\lfloor x := \mathsf{C}(x_1,\ldots,x_k)\rfloor\!\rfloor(b,\mathbf{X},x_i) =$$
$$\{AV_i \mid C(AV_1,\ldots,AV_k) \in \mathbf{X}(x)\}$$

$$\lfloor\!\lfloor x := \underline{\pi_i^j}(y)\rfloor\!\rfloor(b,\mathbf{X},x) = \{AV_i \mid (AV_1,\ldots,AV_k) \in \mathbf{X}(y)\}$$
$$\cup \{\mathsf{AnyPubVal} \mid \mathsf{AnyPubVal} \in \mathbf{X}(y)\}$$

$$\lfloor\!\lfloor x := \underline{\pi_i^j}(y)\rfloor\!\rfloor(b,\mathbf{X},y) =$$
$$\{(AV_1',\ldots,AV_{i-1}',AV_i \sqcap AV_i',AV_{i+1}',\ldots,AV_j') \mid$$
$$AV_i \in \mathbf{X}(x),(AV_1',\ldots,AV_j') \in \mathbf{X}(y)\}\cup$$
$$\{(\mathsf{AnyPubVal}^{i-1},AV_i,\mathsf{AnyPubVal}^{j-i}) \mid$$
$$AV_i \in \mathbf{X}(x),\mathsf{AnyPubVal} \in \mathbf{X}(y)\}$$

$$\lfloor\!\lfloor x := y\rfloor\!\rfloor(b,\mathbf{X},x) = \mathbf{X}(y)$$

$$\lfloor\!\lfloor x := y\rfloor\!\rfloor(b,\mathbf{X},y) = \mathbf{X}(x)$$

$$\lfloor\!\lfloor y := \underline{\mathsf{pubkey}}(x)\rfloor\!\rfloor(b,\mathbf{X},y) =$$
$$\{\mathsf{pubkey}(\ell,b') \mid \mathsf{seckey}(\ell,b') \in \mathbf{X}(x)\}$$

$$\lfloor\!\lfloor y := \underline{\mathsf{pubkey}}(x)\rfloor\!\rfloor(b,\mathbf{X},x) =$$
$$\{\mathsf{seckey}(\ell,b') \mid \mathsf{pubkey}(\ell,b') \in \mathbf{X}(x)\}$$

$$\lfloor\!\lfloor y := \underline{\mathsf{pubdec}}(x_k,x_t)\rfloor\!\rfloor(b1,\mathbf{X},y) = \textit{if } \mathsf{AnyPubVal} \in \mathbf{X}(x_k)$$
$$\textit{then } \mathbf{X}(y) \textit{ else} \bigcup_{\mathsf{seckey}(\ell',b') \in \mathbf{X}(x_k)} \mathbf{E}_{\ell'}^{b'}$$

$$\lfloor\!\lfloor y := \underline{\mathsf{privdec}}(x_k,x_t)\rfloor\!\rfloor(b,\mathbf{X},y) = \textit{if } \mathsf{AnyPubVal} \in \mathbf{X}(x_k)$$
$$\textit{then } \mathbf{X}(y) \textit{ else} \bigcup_{\mathsf{symkey}(i,\ell',b') \in \mathbf{X}(x_k)} \mathbf{E}_{\ell'}^{b'}$$

Figure 5: Constraints giving the relationships between values of variables

the constraints generated by processes. This generation is done by $\langle\!\langle\!\langle P\rangle\!\rangle\!\rangle(\mathbf{I},b,\mathbf{L},\mathcal{C})$ where $\mathbf{I}$ is a constraint variable describing the protocol state before the execution of the process P, the bitstring b is the current decryption context, the variable $\mathbf{L}$ denotes whether this process is alive, and $\mathcal{C}$ is a set of constraints of the form $\mathbf{X}(x) \dot\subseteq E$ where E is a monotone expression with respect to $\mathbf{X}$ that evaluates to a set of abstract values. The constraints in $\mathcal{C}$ relate the abstract values of variables that have been defined before the execution of P. If $\mathcal{C}$ contains a constraint $\mathbf{X}(x) \dot\subseteq E$ then the result of E is a suitable abstraction for the value of x.

Let $\mathbf{R}$ be a mapping from variables to sets of abstract values, and let $\mathcal{C}$ be a set of constraints in the form $\mathbf{X}(x) \dot\subseteq E$. We let $\mathfrak{L}(\mathbf{R},\mathcal{C})$ denote the greatest solution to the constraints $\mathcal{C}$ (with $\mathbf{X}$ as the variable) that is less than or equal to $\mathbf{R}$. The mapping $\langle\!\langle\!\langle P\rangle\!\rangle\!\rangle$ is given in Fig. 6.

Let us explain the generated constraints. For an assignment, the

If e is a not a public-key decryption then

$$\langle\!\langle\!\langle \mathbf{let}^\ell \; y := e \; \mathbf{in} \; P \; \mathbf{else} \; P'\rangle\!\rangle\!\rangle(\mathbf{I},b,\mathbf{L},\mathcal{C}) =$$
$$\textit{let } \mathcal{C}'\langle b\rangle = \mathcal{C} \cup \{\mathbf{X}(y) \dot\subseteq \lfloor\!\lfloor e\rfloor\!\rfloor(b,\mathbf{X},y) \mid y \in \mathbf{Var}_\ell^\bullet\} \textit{ in}$$
$$\{\mathbf{L} \wedge \langle\!\langle e\rangle\!\rangle_\mathsf{s}(\mathbf{I}) \Rightarrow \mathbf{L}_{\ell,\mathsf{true}}^b, \; \mathbf{L} \wedge \langle\!\langle e\rangle\!\rangle_\mathsf{f}(\mathbf{I}) \Rightarrow \mathbf{L}_{\ell,\mathsf{false}}^b,$$
$$\mathbf{L}_{\ell,\mathsf{true}}^b \Rightarrow \mathbf{R}_\ell^b \geq \mathbf{I}[y \mapsto \langle\!\langle e\rangle\!\rangle(\mathbf{I},b)], \; \mathbf{S}_\ell^b \geq \mathfrak{L}(\mathbf{R}_\ell^b,\mathcal{C}'\langle b\rangle)\} \cup$$
$$\langle\!\langle e\rangle\!\rangle_\mathcal{E}(\mathbf{I},\mathbf{L}_{\ell,\mathsf{true}}^b) \cup \langle\!\langle\!\langle P\rangle\!\rangle\!\rangle(\mathbf{S}_\ell^b,b,\mathbf{L}_{\ell,\mathsf{true}}^b,\mathcal{C}'\langle b\rangle) \cup$$
$$\langle\!\langle\!\langle P'\rangle\!\rangle\!\rangle(\mathbf{I},b,\mathbf{L}_{\ell,\mathsf{false}}^b,\mathcal{C})$$

If e is a public-key decryption then

$$\langle\!\langle\!\langle \mathbf{let}^\ell \; y := e \; \mathbf{in} \; P \; \mathbf{else} \; P'\rangle\!\rangle\!\rangle(\mathbf{I},b,\mathbf{L},\mathcal{C}) =$$
$$\textit{let } \mathcal{C}'\langle b\rangle = \mathcal{C} \cup \{\mathbf{X}(y) \dot\subseteq \lfloor\!\lfloor e\rfloor\!\rfloor(b,\mathbf{X},y) \mid y \in \mathbf{Var}_\ell^\bullet\} \textit{ in}$$
$$\{\mathbf{L} \wedge \langle\!\langle e\rangle\!\rangle_\mathsf{s}(\mathbf{I}) \Rightarrow \mathbf{L}_{\ell,\mathsf{true}}^b, \; \mathbf{L} \wedge \langle\!\langle e\rangle\!\rangle_\mathsf{f}(\mathbf{I}) \Rightarrow \mathbf{L}_{\ell,\mathsf{false}}^b,$$
$$\mathbf{L}_{\ell,\mathsf{true}}^b \Rightarrow \mathbf{R}_\ell^{b1} \geq \mathbf{I}[y \mapsto \langle\!\langle e\rangle\!\rangle(\mathbf{I},b1)], \; \mathbf{S}_\ell^{b1} \geq \mathfrak{L}(\mathbf{R}_\ell^{b1},\mathcal{C}'\langle b1\rangle),$$
$$\mathbf{L}_{\ell,\mathsf{true}}^b \Rightarrow \mathbf{R}_\ell^{b0} \geq \mathbf{I}[y \mapsto \langle\!\langle e\rangle\!\rangle(\mathbf{I},b0)], \; \mathbf{S}_\ell^{b0} \geq \mathfrak{L}(\mathbf{R}_\ell^{b0},\mathcal{C}'\langle b0\rangle)\} \cup$$
$$\langle\!\langle e\rangle\!\rangle_\mathcal{E}(\mathbf{I},\mathbf{L}_{\ell,\mathsf{true}}^b) \cup \langle\!\langle\!\langle P\rangle\!\rangle\!\rangle(\mathbf{S}_\ell^{b1},b1,\mathbf{L}_{\ell,\mathsf{true}}^b,\mathcal{C}'\langle b1\rangle) \cup$$
$$\langle\!\langle\!\langle P\rangle\!\rangle\!\rangle(\mathbf{S}_\ell^{b0},b0,\mathbf{L}_{\ell,\mathsf{true}}^b,\mathcal{C}'\langle b0\rangle) \cup \langle\!\langle\!\langle P'\rangle\!\rangle\!\rangle(\mathbf{I},b,\mathbf{L}_{\ell,\mathsf{false}}^b,\mathcal{C})$$

$$\langle\!\langle\!\langle \mathbf{if}^\ell \; x = x' \; \mathbf{then} \; P \; \mathbf{else} \; P'\rangle\!\rangle\!\rangle(\mathbf{I},b,\mathbf{L},\mathcal{C}) =$$
$$\textit{let } \mathcal{C}' = \mathcal{C} \cup \{\mathbf{X}(x) \dot\subseteq \mathbf{X}(x'),\mathbf{X}(x') \dot\subseteq \mathbf{X}(x)\} \textit{ in}$$
$$\{\mathbf{L} \wedge (\exists AV \in \mathbf{I}(x) \exists AV' \in \mathbf{I}(x') : AV \cong_\mathbf{P} AV') \Rightarrow \mathbf{L}_{\ell,\mathsf{true}}^b,$$
$$\mathbf{L} \Rightarrow \mathbf{L}_{\ell,\mathsf{false}}^b, \; \mathbf{R}_\ell^b \geq \mathbf{I}, \; \mathbf{S}_\ell^b \geq \mathfrak{L}(\mathbf{R}_\ell^b,\mathcal{C}')\} \cup$$
$$\langle\!\langle\!\langle P\rangle\!\rangle\!\rangle(\mathbf{S}_\ell^b,b,\mathbf{L}_{\ell,\mathsf{true}}^b,\mathcal{C}') \cup \langle\!\langle\!\langle P'\rangle\!\rangle\!\rangle(\mathbf{I},b,\mathbf{L}_{\ell,\mathsf{false}}^b,\mathcal{C})$$

$$\langle\!\langle\!\langle \mathbf{send}_c[x_\mathrm{p}](x).I^*\rangle\!\rangle\!\rangle(\mathbf{I},b,\mathbf{L},\mathcal{C}) = \langle\!\langle\!\langle I^*\rangle\!\rangle\!\rangle(\mathbf{I},b,\mathbf{L},\mathcal{C}) \cup$$
$$\{\mathbf{L} \Rightarrow \mathbf{I}(x_\mathrm{p}) \subseteq \mathbf{P},$$
$$\mathbf{L} \wedge (c \in \mathbf{Chan}_\mathsf{s} \cup \mathbf{Chan}_\mathsf{a}) \Rightarrow \mathbf{I}(x) \subseteq \mathbf{C}_c,$$
$$\mathbf{L} \wedge (c \in \mathbf{Chan}_\mathsf{a} \cup \mathbf{Chan}_\mathsf{i}) \Rightarrow \mathbf{I}(x) \subseteq \mathbf{P}\}$$

$$\langle\!\langle\!\langle (!)\mathbf{receive}_c^\ell[x_\mathrm{p}](x).P\rangle\!\rangle\!\rangle(\mathbf{I},b,\mathbf{L},\mathcal{C}) =$$
$$\{\mathbf{L} \Rightarrow \mathbf{L}_{b,\mathsf{true}}^\ell, \; \mathbf{S}_\ell^b \geq \mathbf{R}_\ell^b\} \cup \langle\!\langle\!\langle P\rangle\!\rangle\!\rangle(\mathbf{S}_\ell^b,b,\mathbf{L}_{\ell,\mathsf{true}}^b,\mathcal{C}) \cup$$
$$\begin{cases} \{\mathbf{L} \Rightarrow \mathbf{R}_\ell^b \geq \mathbf{I}[x \mapsto \mathbf{C}_c,x_\mathrm{p} \mapsto \{\mathsf{X}_\mathsf{P}\}]\}, \\ \qquad\qquad\qquad\qquad\qquad \textit{if } c \in \mathbf{Chan}_\mathsf{s} \cup \mathbf{Chan}_\mathsf{a} \\ \{\mathbf{L} \Rightarrow \mathbf{R}_\ell^b \geq \mathbf{I}[x \mapsto \{\mathsf{AnyPubVal}\},x_\mathrm{p} \mapsto \{\mathsf{X}_\mathsf{P}\}]\}, \\ \qquad\qquad\qquad\qquad\qquad \textit{if } c \in \mathbf{Chan}_\mathsf{i} \\ \{\mathbf{L} \Rightarrow \mathbf{R}_\ell^b \geq \mathbf{I}[x \mapsto \{\mathsf{X}_\mathsf{S}\},x_\mathrm{p} \mapsto \{\mathsf{X}_\mathsf{P}\}]\}, \\ \qquad\qquad\qquad\qquad\qquad \textit{if } c \in \mathbf{Chan}_\mathsf{u} \end{cases}$$

$$\langle\!\langle\!\langle I_1 \mid \cdots \mid I_n\rangle\!\rangle\!\rangle(\mathbf{I},b,\mathbf{L},\mathcal{C}) = \bigcup_{i=1}^n \langle\!\langle\!\langle I_i\rangle\!\rangle\!\rangle(\mathbf{I},b,\mathbf{L},\mathcal{C})$$

$$\langle\!\langle\!\langle \mathbf{0}\rangle\!\rangle\!\rangle(\mathbf{I},b,\mathbf{L},\mathcal{C}) = \langle\!\langle\!\langle \mathfrak{II}\rangle\!\rangle\!\rangle(\mathbf{I},b,\mathbf{L},\mathcal{C}) = \emptyset \; .$$

Figure 6: Constraints generated by processes

following constraints are generated. If the process is alive ($\mathbf{L}$ is true) and the expression e may succeed [resp. fail] then we demand that the boolean variable reflecting that — $\mathbf{L}_{\ell,\mathsf{true}}^b$ [resp. $\mathbf{L}_{\ell,\mathsf{false}}^b$] is true, too. If e may succeed and hence $\mathbf{L}_{\ell,\mathsf{true}}^b$ is true then we let the mapping $\mathbf{R}_\ell^b$ be (at least) the mapping $\mathbf{I}$, but additionally we fix the abstraction of the left-hand side y. Here this "at least" means "equal to" because there will be no other constraints for $\mathbf{R}_\ell^b$. If $\mathbf{L}_{\ell,\text{true}}^b$ is false then $\mathbf{R}_\ell^b$ has no constraints, hence it maps everything to $\emptyset$.

The set of constraints $\mathcal{C}'\langle b \rangle$ includes all the relationships between variables that are defined up to the successful execution of $y := e$. Note that the inequality signs in the constraints in $\mathcal{C}'\langle b \rangle$ has the opposite direction from the inequality signs in the constraints generated by $\langle\!\langle\!\langle P \rangle\!\rangle\!\rangle$. The constraint $\mathbf{S}_\ell^b \geq \mathfrak{L}(\mathbf{R}_\ell^b, \mathcal{C}'\langle b \rangle)$ states that $\mathbf{S}_\ell^b$ contains basically the same abstractions as $\mathbf{R}_\ell^b$, but all recorded relationships between variables have been taken into account. As this constraint is the only one for $\mathbf{S}_\ell^b$, the inequality $\mathbf{S}_\ell^b \leq \mathbf{R}_\ell^b$ always holds.

We also add the constraints for the variables $\mathbf{E}_{\ell'}^{b'}$ and we recursively invoke $\langle\!\langle\!\langle \cdot \rangle\!\rangle\!\rangle$ for the default- and the false-branch. The arguments for these recursive calls are also worth noting. We see that as we pass through the protocol, we collect the constraints expressing the relationships between variables. For the default-branch, $\mathbf{S}_\ell^b$ is the abstraction of the initial state, while for the false-branch, the same mapping $\mathbf{I}$ is used because no variable was assigned to if the evaluation of e failed. Also we collect no new relationships between variables if e fails (although it would be possible for some e, we have found that it does not change the precision of the analysis in practice).

If e is a public-key decryption then we have "two different default-branches", with decryption contexts $b1$ and $b0$. These two default-branches are reflected in the variables $\mathbf{R}_\ell^{b1}, \mathbf{S}_\ell^{b1}, \mathbf{R}_\ell^{b0}, \mathbf{S}_\ell^{b0}$, as well as in two invocations of $\langle\!\langle\!\langle \cdot \rangle\!\rangle\!\rangle$ for the default-branch P.

Consider now other cases for the process P. In an if-statement we check whether the abstractions of x and x' may intersect. We also add the equality of x and x' to the set of constraints $\mathcal{C}'$.

A send-command always succeeds, the intended recipient becomes known to the adversary and the message is recorded as occurring on the channel c and/or becoming known to the adversary.

In a receive-statement (no matter whether replicated or not), a message from the adversary is abstracted as AnyPubVal and a message from the user as a secret payload. The sender of the message is already known to the adversary, hence x_p is a public integer. No new constraints are added, hence the invocation of $\mathfrak{L}$ is not needed for $\mathbf{S}_\ell^b$.

If I_i is the program for the machine P_i then the set of constraints for the protocol is the union of $\langle\!\langle\!\langle I_i \rangle\!\rangle\!\rangle(\{\}, \varepsilon, \mathsf{true}, \emptyset)$ over all i, where $\{\}$ is the mapping with empty domain. Together with the constraints in Fig. 1, they make up the entire constraint system that we have to solve.

Suppose that we have solved the constraint system for some protocol $\wp$. Let ℓ be a label occurring in this protocol (labeling a subprocess P). Let $\mathbf{I}_\ell^b$ denote the variable $\mathbf{S}_{\ell'}^{b'}$ (or the empty mapping) that occurs as the first argument in the call $\langle\!\langle\!\langle P^\ell \rangle\!\rangle\!\rangle(\mathbf{I}, b, \mathbf{L}, \mathcal{C})$, invoked during the construction of constraints for $\wp$. That is, $\mathbf{I}_\ell^b$ gives the abstract values of variables before entering the subprocess labeled with ℓ in the context b.

The following theorem states how the security of a protocol can be established using our analysis.

THEOREM 1. *Let $\wp$ be a protocol and let Let $\mathbf{S}_\ell^b, \mathbf{E}_\ell^b, \mathbf{P}, \mathbf{C}_c, \mathbf{L}_{\ell,b}^b$ be such that the constraints given above are fulfilled. If the following conditions hold then the composition of machines $\mathcal{TH}_n$ and P_i ($1 \leq i \leq n$) preserves the secrecy of payloads, i.e., the payloads are cryptographically secret if $\mathcal{TH}_n$ is replaced by its cryptographic realization.*

(I) *If the protocol contains a statement of the form $\mathsf{if}^\ell\ x = x' \ldots$ then $\mathsf{X}_\mathsf{S} \notin \mathbf{I}_\ell^b(x)$, $\mathsf{X}_\mathsf{S} \notin \mathbf{I}_\ell^b(x')$, $\mathsf{store}(\mathsf{X}_\mathsf{S}) \notin \mathbf{I}_\ell^b(x)$ and $\mathsf{store}(\mathsf{X}_\mathsf{S}) \notin \mathbf{I}_\ell^b(x')$ for any b.*

(II) *If $\mathsf{X}_\mathsf{S} \in \mathbf{S}_\ell^b(x)$ for some b, x, and this x occurs as an argument to some operation where the abstract values at entry are given by $\mathbf{S}_\ell^b$, then this operation is store or a send to a user.*

(III) $\mathsf{X}_\mathsf{S} \notin \mathbf{P}$.

(IV) *If $\mathsf{AV} \in \mathbf{E}_\ell^b$ and a symm. key of order i is generated at ℓ then the order of AV is less than i.*

- *The order of $\mathsf{symkey}(i, \ell, b)$ is i. The order of a tuple is the maximum order of its members. The order of other abstract values is 0.*

(V) $\mathsf{symkey}(i, \ell, b) \notin \mathbf{P}$ *for any i, ℓ, b.*

4. IMPLEMENTATION

We find the (componentwise) least solution for the aforementioned collection of inequalities. The least fixed point is computed iteratively, using a version of the solver from [24], which is specifically tailored to systems of constraints. The computation might not terminate but we believe that this is not a problem for real protocols; in fact, we have never encountered this situation when we applied our tool to common protocols of the literature. The only case in which computation is not guaranteed to terminate is if the protocol is able to create terms of arbitrary complexity all by itself, without the help from the adversary. We also believe that the potentially exponential number of variables in the size of the protocol is not a problem in practice because protocol descriptions are short.

There are ways to deal with the divergence of the fixed-point computation (add suitable widenings). Also, we believe that the exponential number of sets can be represented in a more compact way, if necessary.

The value of $\mathfrak{L}(\mathbf{R}, \mathcal{C})$ is also computed iteratively, using the same solution method. The mapping $\mathbf{X}$ is initialized with $\mathbf{R}$ and the iteration proceeds downwards.

The constraint generator and solver have been implemented in O'Caml (version 3.09 was used to compile it to native code). We have tested the analyser on several protocols from the literature, namely Needham-Schroeder public key [42], its fix by Lowe [38], Otway-Rees [43], Yahalom, and its modification by Burrows et. al [17]. The goal of all these protocols is to exchange a symmetric key between two parties. We use our analysis to find out whether it is safe to use the exchanged key to protect secret payloads in transit over public networks. We have thus added an extra message at the end of the protocol sessions in all of these protocols. This extra message contains a secret payload encrypted under the freshly exchanged key. (Note again that we do not consider real-or-random secrecy of the keys in this paper, but secrecy of the (encrypted) payloads.) Our analysis considers all these protocols secure, except for the original (and indeed flawed) Needham-Schroeder protocol. If one allows old session keys to become known to the adversary, there may be attacks that are not discovered by our analyser since the BPW model does not consider the leakage of secret keys that have been used (this would cause a so-called commitment problem which makes a proof of computational soundness in the sense of BRSIM/UC impossible). Problems with the Needham-Schroeder-Lowe and the modified Yahalom protocol have been published [19, 49] but these problems do not affect payload secrecy properties — they do not allow an adversary to learn the new key or inject its own key which could then be used to learn information on the payload. The running times of the analyser on a computer with 1 GHz Intel Celeron processor and 256 MB of main memory are between one and eight seconds for these protocols with less than two seconds for Needham-Schroeder-Lowe and both versions of Yahalom.

5. CORRECTNESS OF THE ANALYSIS

Theorem 1 is a straightforward corollary of a lemma similar to subject reduction. We are going to give the statement of that lemma here, its proof can be found in the full version of this paper [7].

When arguing about the correctness, we need to distinguish public and secret payloads. Hence we change the semantics of the system a little bit and assume that each payload is labeled with its secrecy level. An integer received from the protocol user is labeled as secret; a constant integer or the apparent sender of some message is labeled as public. When the payloads are stored in the database of terms then the labels are stored as well. They are also retrieved together with labels. When two integers are compared their secrecy levels are ignored.

Let $\Re$ be a run of the protocol (a finite sequence of steps). Let $\mathcal{O}$ be the state of the database of $\mathcal{TH}_n$ at the end of this run. Let x be a protocol variable whose definition occurs under k replications. If the variable x has been assigned a value inside the i_1-th replica of the outermost replication, i_2-th replica of the next replication, etc., then we denote this value $\mathcal{O}(x, [i_1, \ldots, i_k])$. This value is either a handle to a term in $\mathcal{O}$, a secret integer, or a public integer. Similarly, if a program point ℓ inside k replications and (syntactically) preceded by n public-key decryptions then let $\mathcal{O}(\ell, [i_1, \ldots, i_k])$ be the n-bit string whose bits describe the source of n ciphertexts whose decryptions are visible at the point ℓ in the replica $[i_1, \ldots, i_k]$. Note that a ciphertext is a term in the database $\mathcal{O}$ and its creator is simply the first principal that had a handle to it.

For a set $\mathcal{T}$ of terms in the database $\mathcal{O}$ we let its *downwards closure* $\overline{\mathcal{T}}$ be the smallest set of terms containing $\mathcal{T}$ and being closed with respect to list projection, decryption with keys in $\overline{\mathcal{T}}$, and extracting the public keys and symmetric key names from ciphertexts. For an abstract value AV we define its semantics $[\![AV]\!]_\mathcal{O}$ with respect to the contents of the database $\mathcal{O}$. The semantics is either a set of terms in $\mathcal{O}$ or a set of payloads.

- $[\![\mathsf{X_P}]\!]_\mathcal{O} = \{\text{"public } n\text{"} \mid n \in \mathbb{N}\}$.
- $[\![\mathsf{X_S}]\!]_\mathcal{O} = \{\text{"secret } n\text{"} \mid n \in \mathbb{N}\}$.
- $[\![\mathsf{store}(AV)]\!]_\mathcal{O}$ is the set of all terms of type data in $\mathcal{O}$ whose argument belongs to $[\![AV]\!]_\mathcal{O}$.
- $[\![\mathsf{nonce}(\ell, b)]\!]_\mathcal{O}$ is the set of all terms of type nonce in $\mathcal{O}$ that are generated by the gen_nonce-expressions at the protocol point ℓ at replicas $[i_1, \ldots, i_k]$, such that $b = \mathcal{O}(\ell, [i_1, \ldots, i_k])$
- $[\![\mathsf{symkey}(i, \ell, b)]\!]_\mathcal{O}$, $[\![\mathsf{symkeyname}(\ell, b)]\!]_\mathcal{O}$, $[\![\mathsf{seckey}(\ell, b)]\!]_\mathcal{O}$, and $[\![\mathsf{pubkey}(\ell, b)]\!]_\mathcal{O}$ — defined the same way as $[\![\mathsf{nonce}(\ell, b)]\!]_\mathcal{O}$ (only replace nonce with skse, pkse, ske, or pke).
- $[\![(AV_1, \ldots, AV_j)]\!]_\mathcal{O}$ is the set of all terms of type list in $\mathcal{O}$ whose length is j and whose i-th component term ($1 \leq i \leq j$) belongs to $[\![AV_i]\!]_\mathcal{O}$.
- $[\![\mathsf{pubenc}(AV_k, AV_p, \ell, b)]\!]_\mathcal{O}$ is the set of all terms of type enc, such that
  - they have been created by the pubenc-expressions at the protocol point ℓ at replicas $[i_1, \ldots, i_k]$, such that $b = \mathcal{O}(\ell, [i_1, \ldots, i_k])$;
  - the term corresponding to the public key must belong to $[\![AV_k]\!]_\mathcal{O}$;
  - the term corresponding to the plaintext must belong to $[\![AV_p]\!]_\mathcal{O}$.
- $[\![\mathsf{symenc}(AV_k, AV_p, \ell, b)]\!]_\mathcal{O}$ is defined similarly, where enc is replaced with symenc.
- $[\![\mathsf{AnyPubVal}]\!]_\mathcal{O}$ is the downwards closure of the set of all terms that the adversary knows.

Let $\tilde{\mathbf{P}}$ be the largest set that $\tilde{\mathbf{P}} \subseteq \mathbf{P} \backslash \{\mathsf{AnyPubVal}\}$ and $(AV_1, \ldots, AV_j) \in \tilde{\mathbf{P}}$ implies $AV_i \in \tilde{\mathbf{P}}$ for $1 \leq i \leq j$. Informally, $\tilde{\mathbf{P}}$ is obtained from $\mathbf{P}$ by deleting AnyPubVal and also any abstract value that is a list, one of whose components (after flattening lists) is AnyPubVal. The set $\tilde{\mathbf{P}}$ is a better characterization than $\mathbf{P}$ for the set of terms created by honest participants and learned by the adversary.

Definition 1. A term T from the downwards closure of the set of all terms known to the adversary is *adversarially well-constructed* with respect to $\mathbf{P}$ if one of the following holds:

- $\exists AV \in \tilde{\mathbf{P}}$, such that $T \in [\![AV]\!]_\mathcal{O}$.
- All immediate subterms of T (the immediate subterm of a public key or a symmetric key name is the corresponding secret key) are known to the adversary and are also adversarially well-constructed. Also, if T is of type nonce, ske, enc, garbage, skse, symenc then T must have been constructed by the adversary.

That is, a term is adversarially well-constructed if the adversary knows how to construct this term from the terms that the analysis has found him to know.

LEMMA 2 (SUBJECT REDUCTION). *Let $\wp$ be a protocol and let Let $\mathbf{S}^b_\ell$, $\mathbf{E}^b_\ell$, $\mathbf{P}$, $\mathbf{C}_c$, $\mathbf{L}^b_{\ell,b}$ be such that the constraints given in Sec. 3.3 are fulfilled. Let $\Re$ be a run of the protocol $\wp$. Let $\mathcal{O}$ be the state of the database of $\mathcal{TH}_n$ at the end of $\Re$. The following claims hold.*

$\boxed{P}$ *If a term T is known to adversary then it is adversarially well-constructed wrt. $\mathbf{P}$.*

$\boxed{X}$ *If $\mathcal{O}(x, [i_1, \ldots, i_k]) = T$ (here T may be both a term or an immediate value) and the replica $[i_1, \ldots, i_k]$ passes through the point ℓ with the operation at ℓ succeeding and the value of x being defined, then there exists $AV \in \mathbf{S}^{\mathcal{O}(\ell, [i_1, \ldots, i_k])}_\ell(x)$, such that $T \in [\![AV]\!]_\mathcal{O}$.*

$\boxed{C}$ *If a term T is communicated over an abstract channel $c \in \mathbf{Chan}_s \cup \mathbf{Chan}_a$ then there exists $AV \in \mathbf{C}_c$, such that $T \in [\![AV]\!]_\mathcal{O}$.*

$\boxed{E}$ *If T_k is the term representing an asymmetric or symmetric key generated at the program point ℓ in the replica $[i_1, \ldots, i_k]$ and T_p is a term that occurs as the plaintext in an encryption where T_k is the key, then at least one of the following holds:*

- *there exists $AV \in \mathbf{E}^{\mathcal{O}(\ell, [i_1, \ldots, i_k])}_\ell$, such that $T_p \in [\![AV]\!]_\mathcal{O}$.*
- *T_k and T_p are both known to the adversary and the term representing encryption of T_p with T_k is constructed by the adversary.*

$\boxed{L}$ *If ℓ is a branching point in the protocol (a let- or if-statement) and if the B-branch was taken at the replica $[i_1, \ldots, i_k]$ (here B is either true/default or false), then $\mathbf{L}^{\mathcal{O}(\ell, [i_1, \ldots, i_k])}_{\ell, B} = \text{true}$.*

The lemma is proved by induction over the length of $\Re$.

6. ACKNOWLEDGMENTS

We thank the anonymous reviewers for their helpful comments. M. Backes was supported by the German Research Foundation (DFG) under grant 3194/1-1. P. Laud was supported by Estonian Science Foundation, grant #6095, and by EU Integrated Project AEOLUS (contract no. IST-15964). The constraint solver used in the implementation is from the project Goblin [50].

7. REFERENCES

[1] M. Abadi. Secrecy by Typing in Security Protocols. *Journal of the ACM*, 46(5):749–786, Sept. 1999.

[2] M. Abadi and B. Blanchet. Secrecy types for asymmetric communication. *Theoretical Computer Science*, 298(3):387–415, 2003.

[3] M. Abadi and B. Blanchet. Analyzing Security Protocols with Secrecy Types and Logic Programs. *Journal of the ACM*, 52(1):102–146, Jan. 2005.

[4] M. Abadi and A. D. Gordon. A calculus for cryptographic protocols: The spi calculus. In *Proc. 4th ACM CCS*, pages 36–47, 1997.

[5] M. Abadi and P. Rogaway. Reconciling two views of cryptography: The computational soundness of formal encryption. In *Proc. 1st IFIP TCS*, volume 1872 of *LNCS*, pages 3–22. Springer, 2000.

[6] M. Backes. A cryptographically sound Dolev-Yao style security proof of the Otway-Rees protocol. In *Proc. 9th ESORICS*, volume 3193 of *LNCS*, pages 89–108. Springer, 2004.

[7] M. Backes and P. Laud. Computationally Sound Secrecy Proofs by Mechanized Flow Analysis. Cryptology ePrint Archive: Report 2006/266, 10 Aug. 2006.

[8] M. Backes and B. Pfitzmann. A cryptographically sound security proof of the Needham-Schroeder-Lowe public-key protocol. *IEEE Journal on Selected Areas in Comm.*, 22(10):2075–2086, 2004.

[9] M. Backes and B. Pfitzmann. Symmetric encryption in a simulatable Dolev-Yao style cryptographic library. In *Proc. 17th IEEE CSFW*, pages 204–218, 2004.

[10] M. Backes and B. Pfitzmann. Relating symbolic and cryptographic secrecy. *IEEE Transactions on Dependable and Secure Computing*, 2(2):109–123, 2005.

[11] M. Backes, B. Pfitzmann, and M. Waidner. A composable cryptographic library with nested operations. In *Proc. 10th ACM CCS*, pages 220–230, 2003.

[12] M. Backes, B. Pfitzmann, and M. Waidner. Symmetric authentication within a simulatable cryptographic library. In *Proc. 8th ESORICS*, volume 2808 of *LNCS*, pages 271–290. Springer, 2003.

[13] D. Basin, S. Mödersheim, and L. Viganò. OFMC: A symbolic model checker for security protocols. *Intern. Journal of Information Security*, 2004.

[14] M. Baudet, V. Cortier, and S. Kremer. Computationally sound implementations of equational theories against passive adversaries. In *Proc. 32nd ICALP*, volume 3580 of *LNCS*, pages 652–663. Springer, 2005.

[15] B. Blanchet. A computationally sound mechanized prover for security protocols. In *Proc. 27th IEEE Symp. on Security & Privacy*, pages 140–154, 2006.

[16] C. Bodei, M. Buchholtz, P. Degano, F. Nielson, and H. R. Nielson. Static Validation of Security Protocols. *Journal of Computer Security*, 13(3):347–390, 2005.

[17] M. Burrows, M. Abadi, and R. Needham. A Logic of Authentication. *ACM Transactions on Computer Systems*, 8(1):18–36, Feb. 1990.

[18] R. Canetti. Universally composable security: A new paradigm for cryptographic protocols. In *Proc. 42nd IEEE FOCS*, pages 136–145, 2001.

[19] R. Canetti and J. Herzog. Universally composable symbolic analysis of mutual authentication and key exchange protocols. In *Proc. 3rd TCC*, pages 380–403. Springer, 2006.

[20] V. Cortier and B. Warinschi. Computationally sound, automated proofs for security protocols. In *Proc. 14th ESOP*, pages 157–171, 2005.

[21] A. Datta, A. Derek, J. Mitchell, V. Shmatikov, and M. Turuani. Probabilistic polynomial-time semantics for a protocol security logic. In *Proc. 32nd ICALP*, volume 3580 of *LNCS*, pages 16–29. Springer, 2005.

[22] D. Dolev and A. C. Yao. On the security of public key protocols. *IEEE Transactions on Information Theory*, 29(2):198–208, 1983.

[23] S. Even and O. Goldreich. On the security of multi-party ping-pong protocols. In *Proc. 24th IEEE FOCS*, pages 34–39, 1983.

[24] C. Fecht and H. Seidl. An Even Faster Solver for General Systems of Equations. In *Proc. 3rd SAS*, volume 1145 of *LNCS*, pages 189–204. Springer, 1996.

[25] A. D. Gordon and A. Jeffrey. Authenticity by Typing for Security Protocols. *Journal of Computer Security*, 11(4):451–520, 2003.

[26] A. D. Gordon and A. Jeffrey. Typing correspondence assertions for communication protocols. *Theoretical Computer Science*, 300(1-3):379–409, 7 May 2003.

[27] A. D. Gordon and A. Jeffrey. Types and effects for asymmetric cryptographic protocols. *Journal of Computer Security*, 12(3-4):435–483, 2004.

[28] J. D. Guttman, F. J. Thayer Fabrega, and L. Zuck. The faithfulness of abstract protocol analysis: Message authentication. In *Proc. 8th ACM CCS*, pages 186–195, 2001.

[29] J. Herzog, M. Liskov, and S. Micali. Plaintext awareness via key registration. In *Proc. CRYPTO 2003*, volume 2729 of *LNCS*, pages 548–564. Springer, 2003.

[30] R. Impagliazzo and B. M. Kapron. Logics for reasoning about cryptographic constructions. In *Proc. 44th IEEE FOCS*, pages 372–381, 2003.

[31] R. Kemmerer, C. Meadows, and J. Millen. Three systems for cryptographic protocol analysis. *Journal of Cryptology*, 7(2):79–130, 1994.

[32] P. Laud. Semantics and program analysis of computationally secure information flow. In *Proc. 10th ESOP*, pages 77–91, 2001.

[33] P. Laud. Handling Encryption in Analyses for Secure Information Flow. In *Proc. ESOP 2003*, volume 2618 of *LNCS*, pages 159–173. Springer, 2003.

[34] P. Laud. Symmetric encryption in automatic analyses for confidentiality against active adversaries. In *Proc. 25th IEEE Symp. on Security & Privacy*, pages 71–85, 2004.

[35] P. Laud. Secrecy types for a simulatable cryptographic library. In *Proc. 12th ACM CCS*, pages 26–35, 2005.

[36] P. Laud and V. Vene. A Type System for Computationally Secure Information Flow. In *Proc. 15th FCT*, volume 3623 of *LNCS*, pages 365–377. Springer, 2005.

[37] P. Lincoln, J. Mitchell, M. Mitchell, and A. Scedrov. A probabilistic poly-time framework for protocol analysis. In *Proc. 5th ACM CCS*, pages 112–121, 1998.

[38] G. Lowe. Breaking and fixing the Needham-Schroeder public-key protocol using FDR. In *Proc. 2nd TACAS*, volume 1055 of *LNCS*, pages 147–166. Springer, 1996.

[39] M. Merritt. *Cryptographic Protocols*. PhD thesis, Georgia Institute of Technology, 1983.

[40] D. Micciancio and B. Warinschi. Soundness of formal encryption in the presence of active adversaries. In *Proc. 1st TCC*, volume 2951 of *LNCS*, pages 133–151. Springer, 2004.

[41] J. Mitchell, M. Mitchell, and A. Scedrov. A linguistic characterization of bounded oracle computation and probabilistic polynomial time. In *Proc. 39th FOCS*, pages 725–733, 1998.

[42] R. M. Needham and M. D. Schroeder. Using Encryption for Authentication in Large Networks of Computers. *Communications of the ACM*, 21(12):993–999, Dec. 1978.

[43] D. Otway and O. Rees. Efficient and timely mutual authentication. *Operating Systems Review*, 21(1):8–10, 1987.

[44] L. Paulson. The inductive approach to verifying cryptographic protocols. *Journal of Cryptology*, 6(1):85–128, 1998.

[45] B. Pfitzmann and M. Waidner. Composition and integrity preservation of secure reactive systems. In *Proc. 7th ACM CCS*, pages 245–254, 2000.

[46] B. Pfitzmann and M. Waidner. A model for asynchronous reactive systems and its application to secure message transmission. In *Proc. 22nd IEEE Symp. on Security & Privacy*, pages 184–200, 2001. Extended version (with M, Backes) in IACR ePrint Report 2004/082.

[47] S. Schneider. Security properties and CSP. In *Proc. 17th IEEE Symp. on Security & Privacy*, pages 174–187, 1996.

[48] C. Sprenger, M. Backes, D. Basin, B. Pfitzmann, and M. Waidner. Cryptographically sound theorem proving. In *Proc. 19th IEEE CSFW*, 2006.

[49] P. F. Syverson. A Taxonomy of Replay Attacks. In *Proceedings of the 7th IEEE CSFW*, pages 187–191, 1994.

[50] V. Vojdani. Mitmelõimeliste C-programmide kraasimine analüsaatoriga Goblin (Linting multi-threaded C programs with the Goblin). Master's thesis, Tartu University, 2006.

[51] R. Wilhelm, S. Sagiv, and T. W. Reps. Shape Analysis. In *Proc. 9th CC*, volume 1781 of *LNCS*, pages 1–17. Springer, 2000.

Stateful Public-Key Cryptosystems: How to Encrypt with One 160-bit Exponentiation

Mihir Bellare
U. of California San Diego
Dept of Comp Sci & Eng
9500 Gilman Drive
La Jolla, CA 92093, USA
mihir@cs.ucsd.edu

Tadayoshi Kohno
U. of Washington
Dept of Comp Sci & Eng
AC101 Paul G. Allen Center
185 Stevens Way
Seattle, WA 98195-2350
yoshi@cs.washington.edu

Victor Shoup
New York University
Courant Institute
251 Mercer Street
New York, NY 10012, USA
shoup@cs.nyu.edu

ABSTRACT

We show how to significantly speed-up the encryption portion of some public-key cryptosystems by the simple expedient of allowing a sender to maintain state that is re-used across different encryptions. In particular we present stateful versions of the DHIES and Kurosawa-Desmedt schemes that each use only 1 exponentiation to encrypt, as opposed to 2 and 3 respectively in the original schemes, yielding the fastest discrete-log based public-key encryption schemes known in the random-oracle and standard models respectively. The schemes are proven to meet an appropriate extension of the standard definition of IND-CCA security that takes into account novel types of attacks possible in the stateful setting.

Categories and Subject Descriptors

C.2.0 [**General**]: Security and protection

General Terms

Security

Keywords

Cryptography, public-key encryption.

1. INTRODUCTION

The discrete-exponentiation computations underlying public key cryptography are expensive, in particular having a cost thousands of times that of blockcipher or hash function computations. This not only results in slowdown, but, on systems of limited computing power, can even be a barrier to adding public-key cryptography at all. Today, the cost of public-key cryptography is being felt even more acutely due to new resource constraints on emerging computing platforms: public-key cryptography operations are a severe drain on power and reduce battery life, which is the main limitation on mobile devices such as cell phones, PDAs, RFID chips and sensors. For these and more well-known reasons, there is an obvious interest in reducing the number of discrete exponentiations required for public-key cryptography. This is a domain where a 10% improvement would be very welcome and a 50% improvement would be dramatic.

Much work goes into improved algorithms, time-space tradeoffs (pre-computation) and faster implementations for discrete exponentiation. However these methods have been pushed pretty much as far as they can go, and bring only minor improvements at this point. Our approach is different. We propose to change the *model* for public-key encryption. The change is simple, namely that we allow senders to be stateful, maintaining information that they re-use across different encryptions. We will show that this is highly effective by presenting stateful encryption schemes whose encryption time beats that of all known (stateless) schemes by margins of 50% or more without impacting decryption time or requiring extra storage. We will show that these gains are not at the cost of assurance by providing security definitions and proofs to support our designs. We will also explain why stateful encryption is convenient to implement and deploy in systems so that the cost benefits obtained are realizable in practice. Let us now look at all this in some more detail.

BACKGROUND AND STATE OF THE ART. In the standard model of public-key encryption [27], the sender is stateless (i.e. memoryless), encrypting each message as a function (only) of the message, recipient's public key and freshly-chosen coins. Due to its being what is needed for the security of encryption-using applications [38], security against chosen-ciphertext attack (IND-CCA) [35, 23, 9] is the accepted security goal. We are interested in discrete-logarithm based, proven-IND-CCA schemes. (We discuss RSA later.) The most efficient such scheme is **DHIES** [2], an ElGamal-based scheme where encryption and decryption cost 2 and 1 exponentiations respectively.

THE NEW MODEL. In the new model we propose, the sender maintains state information. The encryption function applied by the sender takes not only the message, receiver's public-key and freshly-chosen coins, but also the current state, returning a ciphertext and a possibly updated state that replaces the previous one. Decryption is unchanged from a standard stateless system, and depends only on the ciphertext and secret key of the receiver. (The receiver is not stateful, and need not even know that the sender is.) Why

Scheme	Cost		Security	
	Enc	Dec	Assumption	RO?
DHIES [2]	2	1	Gap-DH	Yes
StDH	1	1	Gap-DH	Yes
KD [31]	3	1	DDH	No
StKD	1	1	DDH	No

Figure 1: DHIES and KD compared to our stateful variants StDH and StKD. We show the number of exponentiations for encryption and decryption as well as the assumptions for the proofs of security and whether or not it is in the RO model. Note all these schemes are hybrid, but for simplicity the table does not show the (negligible) costs and (standard) assumptions related to the symmetric components of the schemes.

---

this simple change to the model should result in (much) faster encryption may not be clear, but we will now see that it is so.

STATEFUL DH. We present a stateful variant of DHIES that drops the encryption cost to 1 exponentiation without increasing decryption cost. The idea is simple. Recall that the ElGamal encryption of message M under public key g^x has the form $(g^r, g^{xr} \cdot M)$ where r is the random coins chosen by the sender, anew for each encryption. It is natural to try to use (r, g^r) as the state, meaning re-use r across different encryptions so that g^r does not need to be computed each time, but this is clearly insecure, in particular because the resulting scheme is deterministic. However, if instead we derive a symmetric key from g^{xr} via a hash function, and send as ciphertext g^r together with a symmetric encryption of M under a (randomized, stateless) IND-CCA secure symmetric scheme, then the scheme can be shown to be secure. Note that symmetric schemes of the required type are easily and cheaply obtained [8, 11]: one can for example use an AES mode of operation like CBC in encrypt-then-mac combination with a secure MAC like HMAC [7] or CMAC [33]. An interesting feature of our scheme is the crucial use made of hybrid encryption, namely the combination of asymmetric and symmetric primitives.

STATEFUL KD. It is common in cryptography that the security proofs of the most efficient proven-secure schemes are in the random-oracle (RO) model [14]. This is the case for DHIES and its stateful variant StDH discussed above. Concerns about the difficulty of instantiating ROs [19, 5] have lead to interest in standard-model schemes. In this domain, the most efficient known (stateless) IND-CCA scheme is the Kurosawa-Desmedt [31] variant KD of the Cramer-Shoup [22] scheme, where encryption and decryption cost 3 and 1 exponentiations respectively. We present a stateful variant of KD which needs *just 1* exponentiation each for encryption and decryption, just as for StDH. Remarkably, not only is StKD the first non-RO scheme that is as efficient as RO ones, but also it is more efficient than any previous (stateless) schemes, whether with ROs or not!

INSTANTIATION. The preferred choice of group is an elliptic-curve one, where the discrete logarithm problem is already hard for 160-bits. This minimizes exponentiation time and ciphertext size.

VARIANTS. In StDH and StKD, encryption does not even modify the current state, which stays of constant size. If one is willing to modify and grow the state with encryption, further optimizations are possible, taking advantage of the fact that both StDH and StKD encrypt symmetrically under a key that is a deterministic function of the state and recipient public key. By caching this key in the state the first time it is computed, subsequent encryptions to the same recipient require only symmetric operations.

CRASH-ROBUSTNESS AND STATE-RESET. A sender-side system or application crash is a concern for stateful schemes because the current state would be lost. The danger this poses is exemplified with a typical stateful symmetric encryption scheme such as counter-mode with zero-initialized counter. If the current counter value is lost and encryption restarts with a re-initialized counter, privacy is compromised. Not so in our schemes, which are robust in the face of crashes. The sender can pick a new initial state (i.e. reset its state) and restart, and security is maintained. It is safe to pick a new state when a system reboots, and safe to have different concurrent applications each handling its own state. Resetting state does have a computational cost, but in our schemes this is exactly the difference in cost between the stateless and stateful versions. If you reset rarely (as we imagine will usually be the case) your cost is that of the stateful scheme. In the worst case that you reset for each and every encryption, you only return to the cost of the stateless scheme.

MAINTAINING STATE. Maintaining state is not difficult on a system, particularly given the robustness discussed above. Indeed, even though the current model of encryption is stateless at the mathematical level, systems will typically use state in implementing it, in the form of a seed from which the encryption algorithm's random choices are pseudorandomly generated. One issue is that privacy is lost if the state is compromised, but this can be addressed by not too infrequent (eg. once a day) state resets.

WHAT ABOUT PRE-COMPUTATION? The above-quoted factor of 2 or more performance improvement of our schemes compared to previous ones does not take into account pre-computation based speed-up. The bottom-line is that while pre-computation narrows the gap it does not eliminate it. Most importantly, the stateful schemes achieve the greater efficiency *without* any of the storage cost associated to pre-computation, which is attractive for platforms which are storage-limited. To elaborate, recall that exponentiation to a fixed base can be made a factor w faster by pre-computing and storing a table of $(160/w)2^w$ appropriate powers of the base. This can be applied to one of the 2 exponentiations underlying DHIES (the base is not fixed for the other one) and not at all to StDH (where the base is not fixed). If we set w, to, say, 5, the ratio of the cost of DHIES to the cost of StDH drops from 2 to 1.2. However, not only is a 20% improvement not to be sneezed at, but StDH achieves this without the need to store the $160 \cdot (160/5) \cdot 2^5 = 163,840$ bit table needed to speed-up DHIES, so that, overall, statefulness remains a win.

WHAT ABOUT RSA? Encryption is already fast in RSA because we can use a small encryption exponent like 3, but one needs a 1024-bit modulus to get the same security as elliptic-curve discrete-logarithm based schemes offer with 160-bit groups. The result is that RSA decryption, even though a single exponentiation, is slower than in the discrete-log based schemes. Also, ciphertexts are larger. Thus the discrete-log based systems are more attractive in many settings, particularly if encryption time is dropped as in our stateful schemes.

SECURITY PROOFS. In introducing a new model and schemes, there is the danger of having also introduced new security vulnerabilities. The high cost associated to software or hardware updates resulting from bugs in deployed cryptography means that we want to address this by providing pre-implementation security assurance. The most convincing form for this is a proof that the schemes meet an appropriate, well-defined notion of security. This is what we provide.

SECURITY DEFINITION. We begin with a threat-analysis that identifies paths for attack not covered by the classical definition of IND-CCA security for stateless schemes [35, 23, 9]. Briefly, there are two main issues. The first is that encryption now depends on a quantity not a priori known to the adversary, namely the sender's state, and so the ability of an adversary to see ciphertexts is no longer captured merely by giving it the public key as in [27, 35, 23]. We address this by giving the adversary an oracle for encryption under the sender's state. The second issue is that encryptions sent by one sender to different receivers, being computed using the same or related state, are not independent of each other, and so an adversary might be able to compromise privacy of messages encrypted under one public key by seeing ciphertexts of related messages encrypted by the same sender under a different, *maliciously chosen* public key. Thus we need to consider attacks in which the adversary is allowed to choose public keys for malicious receivers and see ciphertexts that a sender encrypted under these keys in an attempt to determine information about a ciphertext that the same sender encrypted to an honest recipient. A definition taking all these issues into account is provided in Section 3, and, above and below, when we refer to a stateful scheme being IND-CCA, we mean that it meets this definition.

PROOFS FOR OUR SCHEMES. We prove StDH is IND-CCA secure in the RO model assuming Gap-DH. (The Gap-DH assumption, due to Okamoto and Pointcheval [34], says that CDH remains hard even given an oracle for DDH.) We prove StKD is IND-CCA secure assuming, as in [31, 26], that the DDH (Decision Diffie-Hellman) problem is hard. In both cases we also make assumptions on the security of the symmetric encryption schemes used. These are slightly stronger than the ones made in the original papers [2, 26], but symmetric schemes satisfying these assumptions are easily and efficiently built via blockcipher modes of operation and MACs.

However there is one difference between what is proven about StDH and StKD. Recall that in our model the adversary can choose public keys and ask the sender to encrypt under them. For StKD, we only let the adversary provide public keys if it knows the corresponding secret key. (In the formal model, which we call the known secret key model and was first used in [6, 16], we simply require it to provide the secret key.) This reflects the assumption that the CA requires a proof of possession of a secret key corresponding to any public key it certifies. Removing this assumption without invoking ROs is an interesting open problem.

RELATED WORK. The multi-recipient encryption schemes of Kurosawa [30] and Bellare, Boldyreva and Staddon [6] allow a sender to batch-encrypt n messages to n different recipients at a cost lower than that of n separate encryptions. For example, their ElGamal based scheme uses for this task only $n+1$ exponentiations rather than the naive $2n$. However the recipients in a batch have to all be different, meaning have different public keys, and the sender must have all the public keys and messages together and process them as a batch. These schemes are not stateful. Stateful encryption is more general. Once an initial state is chosen, we can process any number of on-line encryption requests, these to any recipient, whether the same or different from previous ones. Yet we pay only 1 exponentiation per encryption. Our schemes use the randomness-recycling idea of [30, 6] but differ from theirs in the crucial use of hybrid encryption. Note that our stateful encryption schemes yield new, efficient multi-recipient encryption schemes, in particular, via StDH, the first one that does not use the known secret key model. (All the schemes of [30, 6] use this model.)

In the signcryption model [39, 4, 3] where both the sender and receiver have (certified) public keys, say g^r, g^x respectively, one can encrypt (and authenticate) cheaply under the implicitly shared key g^{rx}. In our setting however the sender does not need a public key or certificate. (And we are interested only in privacy.) This is the more pragmatic setting, reflecting a typical Internet SSL connection, where the server has a certificate but the client does not.

IBE-based approaches have been yielding efficient IND-CCA public-key encryption schemes in the standard model [20, 15, 29]. It is interesting to ask whether these schemes have stateful variants with lower encryption cost.

As noted in Figure 1, DHIES too can be proved under the Gap-DH assumption in the RO model. However, DHIES is also proved IND-CCA secure in [2] in the standard model under an assumption they call Oracle-DH (ODH). StDH can be proved under this assumption too, thereby avoiding the RO model. But ODH is a non-standard assumption mixing a hash function and number-theory.

Stateful encryption is well-known in the symmetric setting with schemes like counter mode encryption, analyzed in [8]. Our work extends this to the asymmetric setting.

2. NOTATION

A string means a binary one. If x, y are strings, then $|x|$ is the length of x. If S is a finite set, then $|S|$ is its cardinality, and $s \xleftarrow{\$} S$ denotes the operation of assigning to s an element of S chosen at random. If A is a randomized algorithm, then $y \xleftarrow{\$} A(x_1, \ldots)$ denotes the operation (experiment) of running A on inputs $x_1, \ldots$, and letting y denote the output.

3. STATEFUL ENCRYPTION SCHEMES

SYNTAX AND OPERATION. A *stateful public-key encryption scheme* StPE = (Setup, KG, PKCk, NwSt, Enc, Dec) is specified by six algorithms (all possibly randomized except the last) whose operation is illustrated in Figure 2. The setup algorithm Setup is run by an authority to produce system parameters sp. Any entity that wants to receive encrypted data can run the key-generation algorithm KG on input sp to get a public and secret key for itself. A sender maintains state st whose initial value is obtained by running the new state algorithm NwSt on input sp. At any time, the sender can apply the encryption algorithm to sp, a public key pk, message M and its current state st to get two out-

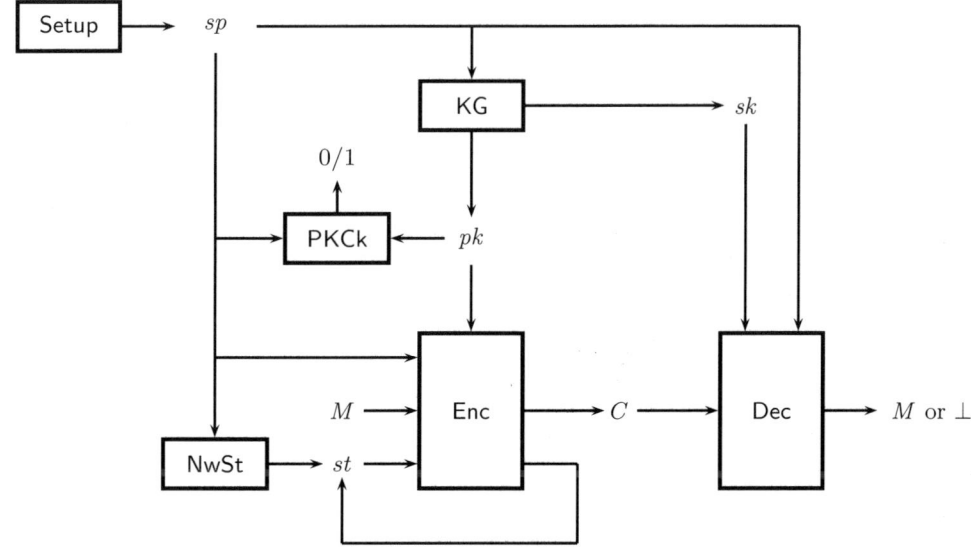

Figure 2: Components and operation of a stateful public-key encryption scheme.

puts. The first, C, is a ciphertext for the receiver whose public key is pk, and the second is an updated value for the state variable for the sender. We write this algorithmically as $(C, st) \xleftarrow{\$} \mathsf{Enc}(sp, pk, M, st)$. At any time, the sender can refresh its state by again running NwSt, which is important for robustness in the face of state-loss due to system failures as discussed in Section 1. The deterministic decryption algorithm Dec, run by the receiver, takes input sp, the receiver's secret key sk and a ciphertext C and returns either a message M or $\bot$ to indicate that the ciphertext is invalid.

The security of our schemes requires that the sender avoid encrypting unless the given public key passes some simple check, which in our case just involves testing that certain components of the key are members of the underlying group. Such a group-membership check could either be done by the encryption algorithm or by a CA as a condition to issuing a certificate for the public key. (Implicitly we are assuming that encryption is not done unless the sender has a valid certificate for the receiver's public key, just as for standard public-key encryption.) Which is better would depend on the cost or other things. In any case, since such a check could be placed in several places, we have represented it in the syntax via a separate public-key verification algorithm PKCk that given sp and a string pk returns a bit representing the outcome of the check on candidate public key pk. Note we are talking only about very simple checks, like whether components of the public key are members of the underlying group, not whether the supplier of the public key "knows" a corresponding secret key, which we will discuss below in the context of security. Simple checks like this (that things are group elements) are sometimes required for the security of standard schemes as well but not made explicit [18, 2].

Note that the sender uses the same state for communicating with different receivers who have different public keys, and the generation of a fresh state requires only sp and no information about potential receivers. In our main schemes, the encryption algorithm does not even modify the state, meaning that the output state is the same as the input one and in particular has constant size, but we will see variants

in which it does. Different senders maintain different states. Receivers are not stateful, and in particular there is thus no state synchronization issue.

We require a natural consistency condition, which says that for any sp generated by Setup, any $n \geq 1$, and any keys $(pk_1, sk_1), \ldots, (pk_n, sk_n)$ generated via $\mathsf{KG}(sp)$, if st_0 is an output of $\mathsf{NwSt}(sp)$ and (C_i, st_i) is an output of $\mathsf{Enc}(sp, pk_i, M_i, st_{i-1})$ for all $i \in \{1, \ldots, n\}$, then $\mathsf{Dec}(sk_i, C_i) = M_i$ for all $i \in \{1, \ldots, n\}$.

BASIC SECURITY MODEL. To define IND-CCA security of stateful public-key encryption scheme StPE = (Setup, KG, PKCk, NwSt, Enc, Dec) we consider a game played with an adversary A. The game begins with the initializations

$$sp \xleftarrow{\$} \mathsf{Setup} \,;\, (pk_1, sk_1) \xleftarrow{\$} \mathsf{KG}(sp) \,;\, c \xleftarrow{\$} \{0, 1\} \,;$$
$$n \leftarrow 1 \,;\, st \xleftarrow{\$} \mathsf{NwSt}(sp)$$

to generate system parameters, a key-pair for a (single) honest recipient, and a challenge bit c. Adversary A is then given input sp, pk_1 and allowed to query various oracles, after which it outputs a bit. Figure 3 shows what types of oracle queries A can make, what action the game takes in each case, and what response the game returns to A. The game is modeling a (single) stateful sender communicating with multiple receivers. Only one receiver, who in the game has index 1, is honest, having keys generated via KG. A controls the public keys of all other receivers: via $\mathrm{M}_\mathrm{K}\mathrm{B}_\mathrm{D}\mathrm{R}\mathrm{E}\mathrm{C}(pk)$, it can create a receiver with whose public key is any string pk of A's choice that passes the public-key verification test. (This models the fact that encryption will not be performed under a public key that does not pass the test, either because the CA would not certify it or because the encryption algorithm would refuse to do the test, depending on where we place the check in the scheme.) A can call LOR with arguments M_0, M_1 of its choice, with the restriction that $|M_0| = |M_1|$, to have the sender encrypt challenge message M_c, under the sender's current state and the public key of the honest receiver, and return the resulting ciphertext to A. In this game, we restrict A to at most one LOR query in its entire execution. (We will see below that this is wlog,

Oracle query	Action taken by game	Response
$\text{MkBdRec}(pk)$	If $\text{PKCk}(sp, pk) = 1$ then $n \leftarrow n+1$; $pk_n \leftarrow pk$	n
$\text{LOR}(M_0, M_1)$	$(C, st) \xleftarrow{\$} \text{Enc}(sp, pk_1, M_c, st)$	C
$\text{Encrypt}(i, M)$	$(C, st) \xleftarrow{\$} \text{Enc}(sp, pk_i, M, st)$	C
$\text{Decrypt}(C)$	$M \leftarrow \text{Dec}(sp, sk_1, C)$	M

Figure 3: Types of oracle queries adversary A can make, the action taken by the game in each case, and the response returned to A.

and it simplifies proofs.) A can call $\text{Encrypt}(i, M)$, with the restriction that $1 \leq i \leq n$, to have the sender encrypt M, under the sender's current state and the public key of receiver i, and return the ciphertext to A. (Encryptions to different users are not independent since they depend on a common state. This query is to capture the possibility that encryption under maliciously chosen public keys leaks state information compromising encryption under the honest public key.) The Decrypt oracle allows a chosen-ciphertext attack on the honest receiver, and of course A can call $\text{Decrypt}(C)$ only if C has not been previously returned in response to a LOR query. At the end of its execution, A must output a bit. Denoting it by d, the ind-cca-advantage of A is

$$\mathbf{Adv}_{\text{StPE}}^{\text{ind-cca}}(A) = 2 \cdot \Pr[d = c] - 1 \;.$$

The above reflects the *unknown secret key* (USK) model in which the CA is not assumed to require a proof of knowledge of the secret key corresponding to a registered public key. In the *known secret key* (KSK) model, where this assumption is made, the only change is that the MkBdRec oracle takes not one but two arguments, pk and sk, and the adversary is restricted to calling this oracle with pk, sk which are possible outputs of $\text{KG}(sp)$. (This way of abstracting out the proofs of knowledge begins with [6, 16] and leads to a simple and convenient model.) The action taken by the game is simply $n \leftarrow n+1$; $pk_n \leftarrow pk$, and the response returned is n. (The check as to whether $\text{PKCk}(sp, pk) = 1$ is now redundant and thus omitted.) We will generally refer to the notion of security as IND-CCA in either case, clarifying whether it is the USK or KSK model as necessary.

An extended security model and equivalence with the basic one. The above model is a simplified one. It turns out security in this model implies many extra security features not explicit in the model. For example, it implies that the sender can reset its state at any time and retain security, which as we have seen is important to running a stateful system in practice. It also implies that, even if the current state is compromised, encryption under a reset state remains secure, so that resets can be used to recover from compromise. The basic model allows only a single sender, single honest receiver, and single LOR query, but one should ask whether allowing all these to be multiple would give the adversary more power. It does not, even if the adversary is allowed to dynamically corrupt receivers and obtain their secret keys, and reset or expose sender's states. In the full version of this paper [10] we reflect all this. We define a richer, extended model which allows the extra capabilities we have discussed, and, via standard hybrid and guessing arguments, show that security in the extended model follows from security in the basic one. Specifically, the adversary's advantage changes by a factor that is at most polynomial in the number of oracle queries it makes. Given this, we use the basic model throughout the paper because doing so simplifies our proofs.

Random oracle model. For schemes in the random oracle (RO) model [14], as usual, any of the constituent algorithms may have access to a RO, and in the security experiment, the adversary gets access to it as well.

4. THE STATEFUL DH SCHEME

This section describes our stateful DH/ElGamal scheme StDH which is effectively a stateful version of the IND-CCA DHIES [2]. (Note ElGamal is only IND-CPA.)

Building blocks. We describe the building blocks used and assumptions made about them:

- A cyclic group $\mathbb{G}$ whose order is denoted m. Let $\text{Gen}(\mathbb{G})$ denote the set of generators of $\mathbb{G}$. We assume the Gap-CDH problem [34] is hard in $\mathbb{G}$, captured by defining $\mathbf{Adv}_{\mathbb{G}}^{\text{gap-dh}}(A_{\mathbb{G}})$, the gap-dh-advantage of an adversary $A_{\mathbb{G}}$, as the probability that $Z = g^{xr}$ under the experiment

$$g \xleftarrow{\$} \text{Gen}(\mathbb{G}) \;;\; r, x \xleftarrow{\$} \mathbb{Z}_m \;;\; Z \xleftarrow{\$} A_{\mathbb{G}}^{\text{DDH}_g(g^r, \cdot, \cdot)}(g, g^r, g^x) \;,$$

where the *restricted DDH oracle* $\text{DDH}_g(g^r, \cdot, \cdot)$, on input Y, W, returns 1 if $Y, W \in \mathbb{G}$ and $W = Y^r$, and 0 otherwise. This is weaker than a full DDH oracle because g^r is fixed, but the ensuing weaker assumption suffices for our results.

- A symmetric encryption scheme $\text{SE} = (\text{SEnc}, \text{SDec})$ with keylength k, specified by its (randomized) encryption algorithm SEnc and (deterministic) decryption algorithm SDec. It is assumed IND-CCA secure, captured by defining $\mathbf{Adv}_{\text{SE}}^{\text{ind-cca}}(A_{\text{SE}})$, the ind-cca-advantage of an adversary A_{SE}, as $2 \cdot \Pr[d = c] - 1$ in the experiment where K is chosen at random from $\{0,1\}^k$, then c is chosen at random from $\{0, 1\}$, and then

$$d \xleftarrow{\$} A_{\text{SE}}^{\text{SEnc}(K, \cdot), \text{LOR}(K, \cdot, \cdot, c), \text{SDec}(K, \cdot)} \;.$$

Above, $\text{LOR}(K, M_0, M_1, c)$ returns $C_s \xleftarrow{\$} \text{SEnc}(K, M_c)$. A_{SE} is allowed only one query to this left-or-right encryption oracle, consisting of a pair of equal-length messages, and is not allowed to query $\text{SDec}(K, \cdot)$ on a ciphertext previously returned by the LOR oracle. (Thus we are using the find-then-guess notion from [8], formulated in a different but equivalent way.)

Scheme. The Setup algorithm of the *stateful DH scheme* $\text{StDH} = (\text{Setup}, \text{KG}, \text{PKCk}, \text{NwSt}, \text{Enc}, \text{Dec})$ returns as system parameters a generator g chosen at random from $\text{Gen}(\mathbb{G})$. A receiver's secret and public keys, as created by $\text{KG}(g)$, are (x, X) and X respectively, where x is chosen at random from $\mathbb{Z}_m$ and $X = g^x$. $\text{PKCk}(g, X)$ simply verifies that

$X \in \mathbb{G}$. The new state algorithm $\mathsf{NwSt}(g)$ picks r at random from $\mathbb{Z}_m$ and returns $st = (r, R)$ where $R = g^r$. Given system parameters g, public key X assumed to be in $\mathbb{G}$, message M, state $st = (r, R)$, and access to random oracle $H \colon \mathbb{G}^3 \to \{0,1\}^k$, encryption algorithm Enc computes the k-bit key $K = H(R, X, X^r)$ and symmetric ciphertext $C_s \xleftarrow{\$} \mathsf{SEnc}(K, M)$, returns (R, C_s) as the ciphertext, and returns an unmodified state (r, R). Note encryption in this scheme does not change the state, and *all* the randomness for the encryption (without which the scheme cannot be secure) comes from the symmetric encryption scheme. Given g, secret key (x, X), ciphertext $C = (R, C_s)$ and oracle H, decryption algorithm Dec returns $\bot$ if $R \notin \mathbb{G}$ and otherwise returns $\mathsf{SDec}(H(R, X, R^x), C_s)$ (which may be $\bot$).

INSTANTIATIONS AND COST. The cost for public-key operations is 1 exponentiation each for encryption and decryption, namely half and the same, respectively as for standard (stateless) DH/ElGamal type schemes such as DHIES [2]. State initialization costs 1 exponentiation, but this is expected to be done rarely. (And in the worst case that it is done before each encryption, the total cost only returns to that of the stateless scheme.) The most suitable group to use is a 160-bit elliptic curve one, in which case the cost for PKCk, namely a group membership test, is cheap enough to neglect. Note one can use a group with an efficiently computable pairing, in which case the assumption drops to just CDH since DDH is easy in such groups. (Interestingly, in this case, the pairing is used only by the algorithms in the proof of security and does not impact the performance of the scheme.) One can also use a group of integers modulo a prime or subgroup thereof, but there seems no point since the group will need to be larger, slowing everything down, and the check could cost an exponentiation. Ciphertext expansion (length of ciphertext minus length of plaintext) is 160 bits plus the expansion from the symmetric encryption.

For a symmetric IND-CCA scheme, the easiest is to use an encrypt-then-mac generic composition [11] with an AES mode of operation like CBC and a MAC like CBC-MAC or HMAC [7]. Using a dedicated authenticated encryption mode like OCB [36] will halve the cost. In both of these cases, we get a 128+80 bit expansion for the symmetric part. The expansion can be reduced at the cost of more block cipher operations by the schemes of [24]. But in any case, the symmetric costs are negligible compared to the asymmetric ones unless the message is very long.

The symmetric scheme needed in DHIES is slightly weaker than the one here because it only needs to be one-time IND-CCA. The computational cost of upgrading to IND-CCA is however negligible. But we do add 128 bits to the ciphertext.

VARIANT. The symmetric key $H(R, X, X^r)$ could be computed only the first time the sender communicates with the receiver whose public key is X and then "cached" as part of the state, reducing the cost of further encryptions to this receiver to merely symmetric operations. Note in this scheme the encryption algorithm does modify the state.

SECURITY OF STATEFUL DH. The following implies that if the Gap-DH problem is hard in $\mathbb{G}$ and SE is IND-CCA secure then Stateful DH meets the notion of IND-CCA security for stateful schemes we defined previously, even in the USK model.

THEOREM 4.1. *Let* StDH *be the stateful DH public-key encryption scheme associated to cyclic group $\mathbb{G}$ and symmetric encryption scheme* SE. *Let A be an ind-cca-adversary against* StDH *in the USK model. Then there exists a gap-dh-adversary $A_\mathbb{G}$ against $\mathbb{G}$ and a ind-cca-adversary A_{SE} against* SE *such that*

$$\mathbf{Adv}^{\text{ind-cca}}_{\mathsf{StDH}}(A) \leq 2 \cdot \mathbf{Adv}^{\text{gap-dh}}_{\mathbb{G}}(A_\mathbb{G}) + \mathbf{Adv}^{\text{ind-cca}}_{\mathsf{SE}}(A_{\mathsf{SE}}) \,.$$

Furthermore the running times of $A_\mathbb{G}, A_{\mathsf{SE}}$ are that of A plus some overhead, in the first case for some computations of SEnc, SDec, *and in the second for a number of exponentiations equal to the number of random-oracle queries of A.* ∎

PROOF THEOREM 4.1. We use game-playing in the style of [12]. We first recall some of the notation and conventions from the latter used in the games of Figure 4. When adversary A interacts with game $G \in \{G_0, G_1\}$, first the **Initialize** procedure is executed. Then A is executed, its queries answered by the corresponding game procedures. The output bit d of A is the input to the **Finalize** procedure which then produces the game output, denoted G^A. The games use arrays $K[\cdot,\cdot], H[\cdot,\cdot,\cdot]$ assumed to be initially everywhere undefined, and a test of the form (not $K[S, X]$) returns true if $K[S, X]$ is undefined and false otherwise.

We assume wlog that A never repeats a query to the random oracle. We first claim the following:

$$\begin{aligned}
&\mathbf{Adv}^{\text{ind-cca}}_{\mathsf{StDH}}(A) \\
&= 2\Pr[G_0^A \Rightarrow 1] - 1 \qquad\qquad\qquad (1) \\
&= 2 \cdot (\Pr[G_0^A \Rightarrow 1] - \Pr[G_1^A \Rightarrow 1]) \\
&\quad + 2 \cdot \Pr[G_1^A \Rightarrow 1] - 1 \qquad\qquad (2) \\
&\leq 2 \cdot \Pr[G_0^A \text{ sets bad}] + 2 \cdot \Pr[G_1^A \Rightarrow 1] - 1 \,. \quad (3)
\end{aligned}$$

To conclude the proof, we will design $A_\mathbb{G}$ and A_{SE} so that

$$\begin{aligned}
\Pr[G_0^A \text{ sets bad}] &\leq \mathbf{Adv}^{\text{gap-dh}}_{\mathbb{G}}(A_\mathbb{G}) \qquad (4)\\
2 \cdot \Pr[G_1^A \Rightarrow 1] - 1 &\leq \mathbf{Adv}^{\text{ind-cca}}_{\mathsf{SE}}(A_{\mathsf{SE}}) \,. \qquad (5)
\end{aligned}$$

The theorem follows from (3), (4) and (5). We will now justify the above claims.

To justify (1), we claim that game G_0 perfectly mimics the game defining the ind-cca-advantage of A. Game G_0 begins by picking the secret and public key for recipient 1, the challenge bit c, and an initial state (r, R) for the sender. For any public key X, where either $X = X_1$ or A queried MKBDREC(X), the game maintains as $K[R, X]$ a k-bit key that plays the role of $H(R, X, X^r)$. This value is picked ahead of time as needed to answer queries to the ENCRYPT, LOR and DECRYPT oracles. Note that neither x_1 nor r are used in replying to these queries. In replying to random-oracle queries, the game "patches" H to return $K[R, X]$ to query R, X, X^r, and returns a random value if the query does not have this form. The patching code is broken into cases for $X = X_1$ and $X \neq X_1$, but since line 509 is included in G_0, the result is to always patch as we have just said. Game G_0 sets the flag bad when A makes the "crucial" random-oracle query R, X_1, X_1^r, the response to which should be the symmetric key $K[R, X_1]$ under which the challenge message is symmetrically encrypted. However, the flag does not affect the oracle responses. It also saves X_1^r as D^* in this case. Note that in responding to H queries, the value r is used, at line 502.

Games G_0, G_1 are identical-until-bad as defined in [12], meaning differ only in statements that follow the setting of the flag bad to true. The Fundamental Lemma of Game Playing [12] thus says that $\Pr[G_0^A \Rightarrow 1] - \Pr[G_1^A \Rightarrow 1] \leq \Pr[G_0^A \text{ sets bad}]$. This justifies (3).

procedure Initialize	**procedure** DECRYPT(C)
001 $x_1 \xleftarrow{\$} \mathbb{Z}_m$; $r \xleftarrow{\$} \mathbb{Z}_m$	401 Parse C as (S, C_s)
002 $X_1 \leftarrow g^{x_1}$; $R \leftarrow g^r$	402 If $S \notin \mathbb{G}$ then reply $\bot$
003 $c \xleftarrow{\$} \{0,1\}$	403 If (not $K[S, X_1]$) then $K[S, X_1] \xleftarrow{\$} \{0,1\}^k$
004 $n \leftarrow 1$	404 Reply $\mathsf{SDec}(K[S, X_1], C_s)$
005 $K[R, X_1] \xleftarrow{\$} \{0,1\}^k$	
	procedure $H(S, X, D)$
procedure MKBDREC(X)	501 $H[S, X, D] \xleftarrow{\$} \{0,1\}^k$
101 If $X \in \mathbb{G}$ then	502 If $S = R$ and $D = X^r$ then
102 $\quad n \leftarrow n+1$; $X_n \leftarrow X$	503 $\quad$ If $X \neq X_1$ then
103 $\quad$ If (not $K[R, X]$) then	504 $\quad\quad$ If (not $K[R, X]$) then $K[R, X] \xleftarrow{\$} \{0,1\}^k$
104 $\quad\quad K[R, X] \xleftarrow{\$} \{0,1\}^k$	505 $\quad\quad H[R, X, D] \leftarrow K[R, X]$
105 Reply n	506 $\quad$ Else
	507 $\quad\quad D^* \leftarrow D$
procedure ENCRYPT(i, M)	508 $\quad\quad$ bad $\leftarrow$ true
201 $C_s \xleftarrow{\$} \mathsf{SEnc}(K[R, X_i], M)$	509 $\quad\quad \boxed{H[R, X_1, D] \leftarrow K[R, X_1]}$
202 Reply (R, C_s)	510 Reply $H[S, X, D]$
procedure LOR(M_0, M_1)	**procedure** Finalize(d)
301 $C_s \xleftarrow{\$} \mathsf{SEnc}(K[R, X_1], M_c)$	601 If $c = d$ then return 1
302 Reply (R, C_s)	602 Else return 0

Figure 4: Game G_0 includes the boxed statement, while game G_1 does not.

$A_\mathbb{G}$ has as input g and group elements that we (suggestively) denote as R, X_1. It also has access to the restricted DDH oracle $\mathrm{DDH}_g(R, \cdot, \cdot)$. It begins by executing lines 003, 004 of **Initialize**, so that quantities $R, X_1, c, n, K[R, X_1]$ corresponding to the ones in G_0 are defined. (But $A_\mathbb{G}$ does not know r, x_1.) Now it runs A on inputs g, X_1, replying to A's oracle queries as per the procedures of G_0. We argue that it can do this. This is easy for MKBDREC, ENCRYPT, LOR and DECRYPT queries because the corresponding procedures have already been written to not use r or x_1 and can thus be executed as shown. But $A_\mathbb{G}$ can also execute the procedure to reply to H queries because it can call its $\mathrm{DDH}_g(R, \cdot, \cdot)$ oracle on inputs X, D to implement the line 502 test as to whether $D = X^r$. When A halts, $A_\mathbb{G}$ ignores its output bit d. If the value D^* is defined (meaning if line 507 was executed) then $A_\mathbb{G}$ outputs D^*, and else it outputs some default value, like $1 \in \mathbb{G}$. If R, X_1 are random group elements then $A_\mathbb{G}$ perfectly mimics game G_0. D^* is defined (as X_1^r) iff bad is set to true. We have justified (4).

In Game G_1, the reply to query $H(R, X_1, X_1^r)$ will not be the "correct" value $K[R, X_1]$ but rather a random, unrelated value. Adversary A_SE takes advantage of this. It begins by executing lines 001, 002 and 004 of **Initialize**. It has oracles that we suggestively denote by $\mathsf{SEnc}(K[R, X_1], \cdot)$, $\mathsf{LOR}(K[R, X_1], \cdot, \cdot, c)$, $\mathsf{SDec}(K[R, X_1], \cdot)$, so that we are identifying the key underlying these oracles with $K[R, X_1]$ and the challenge bit c with that of Game G_1, so that quantities $R, X_1, c, n, K[R, X_1]$ corresponding to the ones in G_1 are defined. (But $A_\mathbb{G}$ does not know $K[R, X_1]$ or c.) Now it runs A on inputs g, X_1, replying to A's oracle queries as per the procedures of G_0. We argue that it can do this. For MKBDREC, it can execute the procedure as shown. For ENCRYPT, it uses its $\mathsf{SEnc}(K[R, X_1], \cdot)$ oracle. For LOR, it uses its $\mathsf{LOR}(K[R, X_1], \cdot, \cdot, c)$ oracle. For DECRYPT, to simulate line 404, it uses its $\mathsf{SDec}(K[R, X_1], \cdot)$ oracle if $S = R$ and otherwise runs the code shown in that line. It can execute G_1's procedure to reply to H queries as shown because it does have r and also because line 509 is omitted. When A halts with an output bit d, A_SE also outputs d and halts. If $K[R, X_1]$ is a random k-bit string and c is a random bit then A_SE has perfectly simulated G_1, and thus $\Pr[c = d]$, in the game defining the ind-cca-advantage of A_SE, equals $\Pr[G_1^A \Rightarrow 1]$. We have justified (5). $\square$

5. THE STATEFUL KD SCHEME

This section describes our stateful version StKD of the Kurosawa-Desmedt [31] variant of the Cramer-Shoup [22] scheme.

BUILDING BLOCKS. We describe the building blocks used and assumptions made about them:

- A cyclic group $\mathbb{G}$ whose order is denoted m. We assume m is a prime. Let $\mathbb{G}^* = \mathbb{G} - \{1\}$ denote the set of generators of $\mathbb{G}$. We assume the DDH problem is hard in $\mathbb{G}$, captured by defining $\mathbf{Adv}_\mathbb{G}^{\mathrm{ddh}}(A_\mathbb{G})$, the ddh-advantage of an adversary $A_\mathbb{G}$, as

$$\Pr[A_\mathbb{G}(g_1, g_2, g_1^r, g_2^r) = 1] - \Pr[A_\mathbb{G}(g_1, g_2, g_1^r, g_2^s) = 1] ,$$

where $g_1, g_2 \xleftarrow{\$} \mathbb{G}^*$ and $r, s \xleftarrow{\$} \mathbb{Z}_m$.

- A symmetric encryption scheme $\mathsf{SE} = (\mathsf{SEnc}, \mathsf{SDec})$ with keylength k. It is assumed IND-CPA secure [8], captured by defining $\mathbf{Adv}_\mathsf{SE}^{\mathrm{ind\text{-}cpa}}(A_\mathsf{SE})$, the ind-cpa-advantage of an adversary A_SE, as $2 \cdot \Pr[d = c] - 1$ in the experiment

$$K \xleftarrow{\$} \{0,1\}^k \;;\; c \xleftarrow{\$} \{0,1\} \;;\; d \xleftarrow{\$} A_\mathsf{SE}^{\mathsf{SEnc}(K, \cdot), \mathsf{LOR}(K, \cdot, \cdot, c)} ,$$

where the LOR oracle is as in Section 4. We also assume integrity of ciphertexts (INT-CTXT) [28, 11], captured by defining $\mathbf{Adv}_\mathsf{SE}^{\mathrm{int\text{-}ctxt}}(A_\mathsf{SE})$, the int-ctxt-advantage of an adversary A_SE, as

$$\Pr\left[A_\mathsf{SE}^{\mathsf{SEnc}(K, \cdot), \mathsf{SDec}(K, \cdot)} \text{ forges} \;:\; K \xleftarrow{\$} \{0,1\}^k \right] ,$$

where A_SE forges if it makes a query C to $\mathsf{SDec}(K, \cdot)$ such that the latter does not return $\bot$ but C was not previously returned by $\mathsf{SEnc}(K,\cdot)$. Note IND-CPA+INT-CTXT is (strictly) stronger than IND-CCA [11], but schemes with this property are easily obtained via the encrypt-then-mac construction [11].

- A family $H\colon \{0,1\}^k \times \mathbb{G} \times \mathbb{G} \to \mathbb{Z}_m$ of hash functions in which $H(K,\cdot,\cdot)\colon \mathbb{G} \times \mathbb{G} \to \mathbb{Z}_m$ for each k-bit key K. It is assumed universal one-way [32], also called TCR [13], captured by defining $\mathbf{Adv}_H^{\mathrm{tcr}}(A_H)$, the tcr-advantage of an adversary A_H, as

$$\Pr[\, H(K,S_1,T_1) = H(K,S_2,T_2) \wedge (S_1,T_1) \neq (S_2,T_2)\,]$$

in the experiment where we first run A_H to get (S_1,T_1), only then pick key K at random, given K to A_H, and have it now provide (S_2,T_2) [32]. Note TCR is a weaker requirement than collision-reistance, so that, in particular, any practical collision-resistant function can be used.

- A key-derivation function $F\colon \mathbb{G} \to \{0,1\}^k$. We assume its output on a random input is computationally indistinguishable from a random k-bit string, captured by defining the kdf-advantage of an adversary A_F as

$$\mathbf{Adv}_F^{\mathrm{kdf}}(A_F) = \Pr[\,A_F(F(S)) = 1\,] - \Pr[\,A_F(K) = 1\,]\,,$$

where $S \xleftarrow{\$} \mathbb{G}$ and $K \xleftarrow{\$} \{0,1\}^k$. Such functions are easily built out of cryptographic hash functions.

The assumptions are the same as made in [26] for the KD scheme except that we need a randomized symmetric encryption scheme whereas a one-time scheme sufficed in [26]. The cost of this upgrade is however negligible.

SCHEME. The Setup algorithm of the *stateful KD scheme* StKD = (Setup, KG, PKCk, NwSt, Enc, Dec) returns as system parameters a pair g_1, g_2 of generators chosen at random from $\mathbb{G}^*$ and a random k-bit key K_H for H. A receiver's secret and public keys, as created by $\mathsf{KG}(g_1,g_2)$, are (x_1,x_2,y_1,y_2) and X,Y respectively, where x_1,x_2,y_1,y_2 are chosen at random from $\mathbb{Z}_m$ and $X = g_1^{x_1}g_2^{x_2}$ and $Y = g_1^{y_1}g_2^{y_2}$. $\mathsf{PKCk}((g_1,g_2,K_H),(X,Y))$ does nothing, simply returning 1, since the conditions put on keys by the KSK model already imply that any public key arising in the system will be a possible output of $\mathsf{KG}((g_1,g_2,K_H))$ so that, in particular it will consist of a pair of group elements. The new state algorithm $\mathsf{NwSt}((g_1,g_2,K_H))$ picks r at random from $\mathbb{Z}_m$ and returns $st = (r,R_1,R_2,\alpha)$ where $R_1 = g_1^r$, $R_2 = g_2^r$ and $\alpha = H(K_H,R_1,R_2)$. Given system parameters g_1, g_2, K_H, public key (X,Y) assumed to consist of a pair of group elements, message M and state $st = (r,R_1,R_2,\alpha)$, encryption algorithm Enc sets $Z = X^r Y^{r\alpha}$, $K = F(Z)$, computes the symmetric ciphertext $C_s \xleftarrow{\$} \mathsf{SEnc}(K,M)$, returns (R_1,R_2,C_s) as the ciphertext, and returns an unmodified state (r,R_1,R_2,α). (As with StDH, encryption does not change the state, and all the randomness for the encryption comes from the symmetric encryption scheme.) Given system parameters (g_1,g_2,K_H), secret key (x_1,x_2,y_1,y_2) and ciphertext $C = (R_1,R_2,C_s)$, decryption algorithm Dec returns $\bot$ if R_1 or R_2 are not in $\mathbb{G}$ and otherwise sets $\alpha = H(K_H,R_1,R_2)$, $Z = R_1^{x_1+y_1\alpha} R_2^{x_2+y_2\alpha}$ and $K = F(Z)$, and returns $\mathsf{SDec}(K,C_s)$ (which may be $\bot$).

INSTANTIATIONS AND COST. Since a multi-exponentiation $a,b,A,B \mapsto A^a B^b$ has essentially the same cost as an exponentiation, the cost for public-key operations is 1 exponentiation each for encryption and decryption, namely one-third and the same, respectively as for KD, and the same as for StDH. State initialization costs 2 exponentiations, but this is expected to be done rarely, and in the worst case that it is done before each encryption, the total cost only returns to that of KD. The symmetric primitives used can be instantiated in practice via block ciphers and cryptographic hash functions.

VARIANT. As with StDH, the StKD scheme has the nice feature that the symmetric key K computed by the encryption algorithm is a deterministic function of the system parameters, state and public key. Thus it could be computed only the first time the sender communicates with a particular receiver and then cached in the state, reducing the cost of further encryptions to this receiver to merely symmetric operations.

SECURITY. While the proof of security of this scheme does not rely on the random oracle model, we unfortunately have to make the assumption that we are in the KSK model. Recall this captures the assumption that the CA issues a certificate for a public key only upon receiving a proof of knowledge of the corresponding secret key. In practice, this would likely be implemented by having receivers run proofs of knowledge of representations of X and Y with respect to g_1 and g_2 (see [21, 17]) as part of the certification protocol. When a sender encrypts under a public key, it suffices that it simply verifies the certificate, as usual. Note that some security practitioners have long advocated a "proof of possession of secret key" when certifying a public key. The current application and those in [6, 16] are some that appear to benefit from this.

THEOREM 5.1. *Let $\mathbb{G}$ be a group of prime order m in which the DDH problem is hard. Let $\mathsf{SE} = (\mathsf{SEnc}, \mathsf{SDec})$ be an IND-CPA+INT-CTXT secure symmetric encryption scheme with keylength k. Let $H\colon \{0,1\}^k \times \mathbb{G} \times \mathbb{G} \to \mathbb{Z}_m$ be a family of TCR hash functions. Let $F\colon \mathbb{G} \to \{0,1\}^k$ be a secure key derivation function as defined above. Then the StKD scheme associated to these primitives is IND-CCA secure in the stateful KSK model.* ∎

Theorem 5.1 can be proved by slightly modifying the proof of security of the ordinary KD scheme given in [26]. We sketch this at a very high level, but the reader should be able to easily fill in the details by consulting [26].

PROOF SKETCH. Let A be an adversary attacking StKD in our stateful KSK model. Let $st = (r^*, R_1^*, R_2^*, \alpha^*)$ denote the sender's state, which is generated at the beginning of the game, and remains fixed throughout, and let X^*, Y^* denote the public-key of the honest receiver (namely, receiver 1), and $(x_1^*, x_2^*, y_1^*, y_2^*)$ its corresponding secret key. We also define $Z^* = (X^*)^{r^*} (Y^*)^{r^* \alpha^*}$ and $K^* = F(Z^*)$. For a given ciphertext $C = (R_1, R_2, C_s)$, we call (R_1, R_2) the *preamble*, and C_s the *body*. We call (R_1^*, R_2^*) the *target preamble*. We say a preamble (R_1, R_2) is *valid* if $\log_{g_1}(R_1) = \log_{g_2}(R_2)$.

We may assume A makes no ENCRYPT queries except those of the form ENCRYPT$(1,\cdot)$. This is because when the adversary registers a public key (X,Y) for a "dishonest receiver" (namely one whose index is $i > 1$), he must, according to the KSK model, present a corresponding secret key (x_1,x_2,y_1,y_2) such that $X = g_1^{x_1}g_2^{x_2}$ and $Y = g_1^{y_1}g_2^{y_2}$. Because he knows this secret key, he can encrypt any message M he likes with respect to the public key (X,Y) (and the given state) by computing $Z \leftarrow (R_1^*)^{x_1+y_1\alpha^*}(R_2^*)^{x_2+y_2\alpha^*}$, $K \leftarrow F(Z)$, $C_s \xleftarrow{\$} \mathsf{SEnc}(K,M)$. This is where we make critical use of the fact that all users must register private keys (or at least present a proof of knowledge to the CA). Note that to make this work, the adversary may need to make a preliminary query of the form ENCRYPT(1, ·), just to obtain the target preamble. With this simplification, note that the ENCRYPT and LOR oracles encrypt messages using the same symmetric key K^*.

We further assume the adversary does not submit to the DECRYPT oracle a ciphertext that was previously output by the LOR or ENCRYPT oracles. This is justified, since (a) by definition, the adversary is not allowed to get an output of LOR decrypted, and (b) the adversary already "knows" the decryption of any ciphertext output by ENCRYPT.

We now have an attack game which is much closer to the standard (non-stateful) IND-CCA attack game. The only difference is that the same state is used to encrypt several messages, instead of just one. In the proof of [26], only a single message was encrypted using a given state, and because of that, a weaker assumption on the symmetric encryption scheme, namely that it was IND-CPA+INT-CTXT in a single encryption attack, sufficed. Having strengthened the assumption to IND-CPA+INT-CTXT in a multiple encryption attack, it is really just an exercise to modify the proof in [26] appropriately.

We give an outline of this proof here. The proof is a sequence of games [37, 12], Game 0, Game 1, etc. Game 0 is the original attack game (with the above simplifications). We denote by S_i the event that A guesses the hidden bit in Game i. Our goal is to show that $\Pr[S_0] \approx 1/2$.

Game 1. We modify the decryption oracle so that it rejects all ciphertexts whose preambles do not match the target preamble, yet hash to the same value. The TCR assumption on H implies that $\Pr[S_1] \approx \Pr[S_0]$.

Game 2. We modify the way that the challenger computes Z^* (and hence K^*): as
$$Z^* \leftarrow (R_1^*)^{x_1^* + y_1^* \alpha} (R_2^*)^{x_2^* + y_2^* \alpha^*}.$$
This is purely a conceptual change, so $\Pr[S_2] = \Pr[S_1]$.

Game 3. We now replace the target preamble by a randomly chosen, invalid preamble. The DDH assumption implies that $\Pr[S_3] \approx \Pr[S_2]$.

Game 4. We now modify the decryption oracle, so that it rejects any ciphertext whose preamble does not match the target preamble, and is invalid (note that to implement this efficiently, we now assume the challenger knows $\log_{g_1}(g_2)$). Let E_4 be the event that some ciphertext is rejected using this new rejection rule that would not have been otherwise rejected. We have $|\Pr[S_4] - \Pr[S_3]| \le \Pr[E_4]$. We want to show that $\Pr[E_4] \approx 0$. To do this, we augment Game 4 slightly: we have the challenger choose at random one particular decryption query, called the *chosen decryption query*, and define E_4' to be the event that that particular ciphertext was rejected using the new rule. If Q is a bound on the number of decryption queries, then $\Pr[E_4] \le Q \Pr[E_4']$, and so it will suffice to show that $\Pr[E_4'] \approx 0$. We will establish this later, in Game $6'$.

Game 5. We make two changes. First, we replace Z^* by a completely random group element. This is done in a consistent fashion, so that not only the ENCRYPT and LOR oracles use this value of Z^*, but so does the decryption oracle on ciphertexts whose preamble matches the target preamble. Second, in processing the chosen decryption query, if the preamble is different from the target preamble, does not hash to the same value as the target preamble, and is invalid, then either the symmetric encryption algorithm will reject the ciphertext, or if not, the new rejection rule introduced in Game 4 will reject it. For the purposes of determining if the new rejection rule is needed, a random group element $\tilde{Z}$ is chosen, and the body of the ciphertext is decrypted using the symmetric key $\tilde{K} = F(\tilde{Z})$: if this decryption does not result in $\bot$, we say "the new rejection rule applies"; in any case, the decryption oracle returns $\bot$ to the adversary.

Let E_5' denote the event that the new rejection rule applies in Game 5. We have $\Pr[S_5] = \Pr[S_4]$ and $\Pr[E_5'] = \Pr[E_4']$. This follows from the usual 4-wise independence argument, as in the original Cramer-Shoup analysis: the joint distribution of all the relevant random variables defining these events is identical. The proof now forks in two different directions, making two different modifications to Game 5.

Game 6. Now we replace K^* be a random k-bit key. This is done consistently throughout the game: in the LOR oracle, the ENCRYPT oracle, and the DECRYPT oracle (for those ciphertexts whose preamble matches the target preamble). Under the assumption that F is a good key derivation function, we have $\Pr[S_6] \approx \Pr[S_5]$. Moreover, note that K^* is essentially used in this game as a key in a standard symmetric chosen ciphertext attack, and under the security assumption for SE, it follows that $\Pr[S_6] \approx 1/2$.

Game 6'. We now modify Game 5 in a different way. Instead of computing $\tilde{K}$ as $F(\tilde{Z})$, we simply choose $\tilde{K}$ as a random k-bit key. Define $E_{6'}'$ to be the event that the new rejection rule applies in Game 6. Under the assumption that F is a good key derivation function, we have $\Pr[E_{6'}'] \approx \Pr[E_5']$. Moreover, under the assumption that SE provides ciphertext integrity, we must have $\Pr[E_{6'}'] \approx 0$. It follows now that $\Pr[E_4'] = \Pr[E_5'] \approx \Pr[E_{6'}'] \approx 0$, and hence $\Pr[S_4] \approx \Pr[S_3]$.

Tracing through the logic, we see that $\Pr[S_0] \approx \Pr[S_6] \approx 1/2$, which proves the theorem.

Note that from the proof, we see that SE does not need to provide full ciphertext integrity. The properties needed are that it is IND-CCA secure, and that for any randomly chosen key, and any adversarially chosen ciphertext, the ciphertext is rejected with overwhelming probability (in this game, the adversary has no access to an encryption oracle).

6. ACKNOWLEDGMENTS

Bellare was supported in part by NSF grant CNS-0524765 and a gift from Intel Corporation. Work done while Kohno was at UCSD, supported in part by Bellare's grants. Shoup was supported in part by NSF grant CCR-0310297.

7. REFERENCES

[1] M. Abe, R. Gennaro, K. Kurosawa and V. Shoup. Tag-KEM/DEM: A New Framework for Hybrid Encryption and a New Analysis of Kurosawa Desmedt KEM. *EUROCRYPT '05*, LNCS 3494, Springer-Verlag.

[2] M. Abdalla, M. Bellare and P. Rogaway. The Oracle Diffie-Hellman Assumptions and an Analysis of DHIES. *CT-RSA '01*, LNCS 2020, Springer-Verlag.

[3] J. An. Authenticated Encryption in the Public-Key Setting: Security Notions and Analyses. Cryptology ePrint Archive: Report 2001/079.

[4] J. An, Y. Dodis and T. Rabin. On the Security of Joint Signature and Encryption. *EUROCRYPT '02*, LNCS 2332, Springer-Verlag.

[5] M. Bellare, A. Boldyreva and A. Palacio. An Uninstantiable Random-Oracle-Model Scheme for a Hybrid-Encryption Problem. *EUROCRYPT '04*, LNCS 3027, Springer-Verlag.

[6] M. Bellare, A. Boldyreva and J. Staddon. Multi-Recipient Encryption Schemes: Security Notions and Randomness Re-Use. *PKC '03*, LNCS 2567, Springer-Verlag.

[7] M. Bellare, R. Canetti and H. Krawczyk. Keying hash functions for message authentication. *CRYPTO '96*, LNCS 1109, Springer-Verlag.

[8] M. Bellare, A. Desai, E. Jokipii and P. Rogaway. A Concrete Security Treatment of Symmetric Encryption: Analysis of the DES Modes of Operation. *Proc. 38th FOCS*, IEEE, 1997.

[9] M. Bellare, A. Desai, D. Pointcheval and P. Rogaway. Relations among notions of security for public-key encryption schemes. *CRYPTO '98*, LNCS 1462, Springer-Verlag.

[10] M. Bellare, T. Kohno and V. Shoup. Stateful Public-Key Cryptosystems: How to Encrypt with One 160-bit Exponentiation. Full version of this paper. Cryptology ePrint Archive: Report 2006/267.

[11] M. Bellare and C. Namprempre. Authenticated Encryption: Relations among notions and analysis of the generic composition paradigm. *ASIACRYPT '00*, LNCS 1976, Springer-Verlag.

[12] M. Bellare and P. Rogaway. Code-Based Game-Playing Proofs and the Security of Triple Encryption. *EUROCRYPT '06*, LNCS 4004, Springer-Verlag.

[13] M. Bellare and P. Rogaway. Collision-Resistant Hashing: Towards Making UOWHFs Practical. *CRYPTO '97*, LNCS 1294, Springer-Verlag.

[14] M. Bellare and P. Rogaway. Random oracles are practical: A paradigm for designing efficient protocols. *Proceedings of the 1st Annual Conference on Computer and Communications Security*, ACM, 1993.

[15] D. Boneh and J. Katz. Improved Efficiency for CCA-Secure Cryptosystems Built Using Identity Based Encryption. *CT-RSA '05*, LNCS 3376, Springer-Verlag.

[16] A. Boldyreva. Threshold Signatures, Multisignatures and Blind Signatures Based on the Gap Diffie-Hellman Group Signature Scheme. *PKC '03*, LNCS 2567, Springer-Verlag.

[17] S. Brands. Untraceable off-line cash in wallets with observers. *CRYPTO '93*, LNCS 773, Springer-Verlag.

[18] M. Burmester and Y. Desmedt. Remarks on soundness of proofs. *Electronics Letters*, 25(22), 1509–1511, 1989.

[19] R. Canetti, O. Goldreich and S. Halevi. The random oracle methodology, revisited. *Proc. 30th STOC*, ACM, 1998.

[20] R. Canetti, S. Halevi and J. Katz. Chosen-Ciphertext Security from Identity-Based Encryption. *EUROCRYPT '04*, LNCS 3027, Springer-Verlag.

[21] D. Chaum, J. Evertse and J. van de Graaf. An improved protocol for demonstrating possession of discrete logarithms and some generalizations. *EUROCRYPT '87*, LNCS 304, Springer-Verlag.

[22] R. Cramer and V. Shoup. Design and analysis of practical public-key encryption schemes secure against adaptive chosen ciphertext attack. *SIAM J. Comp.*, 33(1), 167–226, 2003.

[23] D. Dolev, C. Dwork and M. Naor. Non-Malleable Cryptography. *SIAM J. Comp.*, 30(2), 391–437, 2000.

[24] A. Desai. New Paradigms for Constructing Symmetric Encryption Schemes Secure against Chosen Ciphertext Attack. *CRYPTO '00*, LNCS 1880, Springer-Verlag.

[25] T. ElGamal. A public key cryptosystem and signature scheme based on discrete logarithms. *IEEE Trans. Info. Theory*, 31(4), 469–472, 1985.

[26] R. Gennaro and V. Shoup. A note on an encryption scheme of Kurosawa and Desmedt. Cryptology ePrint Archive: Report 2004/194.

[27] S. Goldwasser and S. Micali. Probabilistic Encryption. *J. Comput. & Sys. Sci.*, 28, 270–299, 1984.

[28] J. Katz and M. Yung. Unforgeable Encryption and Chosen Ciphertext Secure Modes of Operation. *FSE '00*, LNCS 1978, Springer-Verlag.

[29] E. Kiltz. Chosen-Ciphertext Security from Tag-Based Encryption. *Theory of Cryptography – TCC '06*, LNCS 3876, Springer-Verlag.

[30] K. Kurosawa. Multi-Recipient Public-Key Encryption with Shortened Ciphertext. *PKC '02*, LNCS 2274, Springer-Verlag.

[31] K. Kurosawa and Y. Desmedt. A New Paradigm of Hybrid Encryption Scheme. *CRYPTO '04*, LNCS 3152, Springer-Verlag.

[32] M. Naor and M. Yung. Universal one-way hash functions and their cryptographic applications. *Proc. 21st STOC*, ACM, 1989.

[33] NIST. Recommendation for Block Cipher Modes of Operation: The CMAC Mode for Authentication. Document SP 800-38B, May 2005.

[34] T. Okamoto and D. Pointcheval. The Gap Problems: A New Class of Problems for the Security of Cryptographic Schemes. *PKC '01*, LNCS 1992, Springer-Verlag.

[35] C. Rackoff and D. Simon. Non-interactive zero knowledge proof of knowledge and chosen ciphertext attack. *CRYPTO '91*, LNCS 576, Springer-Verlag.

[36] P. Rogaway, M. Bellare and J. Black. OCB: A block-cipher mode of operation for efficient authenticated encryption. *ACM Trans. Info. Sys. Sec.*, 6(3), 365–403, 2003.

[37] V. Shoup. Sequences of Games: A Tool for Taming Complexity in Security Proofs. Cryptology ePrint Archive: Report 2004/332.

[38] V. Shoup. Why chosen ciphertext security matters. IBM Research Report RZ 3076, November 1998.

[39] Y. Zheng. Digital Signcryption or How to Achieve Cost(Signature & Encryption) $\ll$ Cost(Signature) + Cost(Encryption). *CRYPTO '97*, LNCS 1294, Springer-Verlag.

Multi-Signatures in the Plain Public-Key Model and a General Forking Lemma

Mihir Bellare
University of California San Diego
Department of Computer Science & Engineering
9500 Gilman Drive
La Jolla, California 92093, USA
mihir@cs.ucsd.edu

Gregory Neven
Katholieke Universiteit Leuven
B-3001 Heverlee, Belgium;
and Ecole Normale Supérieure
75005 Paris, France
Gregory.Neven@esat.kuleuven.be

ABSTRACT

A multi-signature scheme enables a group of signers to produce a compact, joint signature on a common document, and has many potential uses. However, existing schemes impose key setup or PKI requirements that make them impractical, such as requiring a dedicated, distributed key generation protocol amongst potential signers, or assuming strong, concurrent zero-knowledge proofs of knowledge of secret keys done to the CA at key registration. These requirements limit the use of the schemes. We provide a new scheme that is proven secure in the plain public-key model, meaning requires nothing more than that each signer has a (certified) public key. Furthermore, the important simplification in key management achieved is not at the cost of efficiency or assurance: our scheme matches or surpasses known ones in terms of signing time, verification time and signature size, and is proven secure in the random-oracle model under a standard (not bilinear map related) assumption. The proof is based on a simplified and general Forking Lemma that may be of independent interest.

Categories and Subject Descriptors

C.2.0 [**General**]: Security and protection

General Terms

Security

Keywords

Cryptography, digital signatures, multi-signatures, Forking Lemma.

1. INTRODUCTION

Consider entities $1, \ldots, N$, each having a public key and corresponding secret key. A multi-signature (MS) scheme allows any subset $L \subseteq \{1, \ldots, N\}$ of them, at any time, to engage in an interactive protocol whose output is a joint signature on a message m of their choice. Verification can be done by any external party given just L, m, the purported multi-signature σ, and the public keys of all signers in L. Such a system could be useful for contract signing, co-signing, or distribution of a certificate authority.

A trivial way to implement a multi-signature scheme is to let the multi-signature σ of message m be the list $(\sigma_i : i \in L)$ where σ_i is i's signature on m. This multi-signature is however large, in particular of size proportional to the number $|L|$ of signers. There are several important practical reasons for which this is costly and undesirable. For example, on wireless devices such as PDAs, cell phones, RFID chips and sensors, battery life is the main limitation. Communicating even one bit of data uses significantly more power than executing one 32-bit instruction [4]. Reducing the number of bits to communicate saves power and is important to increase battery life. Also, in many settings, communication is not reliable, and so the fewer the number of bits one has to communicate, the better. For such reasons, we want multi-signature schemes that are non-trivial, meaning the size of the multi-signature is about the same as that of a single ordinary signature and in particular not proportional to the number $|L|$ of signers.

ROGUE-KEY ATTACKS. The early literature of MS schemes [25, 20, 28, 24, 27, 32, 34, 35] features numerous attacks breaking proposed schemes. In most cases, this was due to weaknesses related to key setup, in particular the ability to mount a *rogue-key attack*. In such an attack, an adversary who is a group member (insider) chooses its public key as a function of that of honest users in such a way that it can then easily forge multi-signatures. Although they might at first hearing sound far-fetched, rogue-key attacks are in fact possible to mount in practice and are a real threat. When, eventually, precise definitions [31] and proven secure schemes [31, 11, 29] emerged, they obviously paid a lot of attention to key setup. These schemes were, happily, proven secure against rogue-key attacks, but, unhappily, at the cost of complexity and expense in the scheme, or using unrealistic and burdensome assumptions on the public-key infrastructure (PKI), as we will now explain in detail.

DRAWBACKS OF PREVIOUS SCHEMES. The MS-MOR [31] scheme requires, as a pre-processing step, that the set of *potential* signers engages in an interactive key generation protocol that provides to each a public and secret key. The purpose of the protocol is to ensure that no dishonest player can choose its public key as a function of public keys of honest players. This type of *dedicated key generation* is however not practical for several reasons. First, it means the group of potential signers is restricted to being static: it must be decided and fixed before signing can start, and new signers cannot be added later. However, in practice the potential set of signers is dynamic and may not be known ahead of time, and we would like to be able to add potential signers at will. Second, the key generation protocol of [31] is expensive and results in large, complex public keys. In particular, the public key of each signer has a size that depends on the number N of potential signers. (However, once keys are established, multi-signature generation and verification are actually attractively cheap.)

The drawback of the **MS-Bo** [11] and **MS-LOSSW** [29] MS schemes is that the security model makes the *knowledge of secret key* (KOSK) assumption. There is no dedicated key generation, but, when the adversary, mounting a rogue-key attack, provides a public key of its choice for a group member, it is required, in the model, to provide also a matching secret key. Of course "real" adversaries would not do any such thing, so what does this mean? It is explained by the authors as modeling the assumption that a user provides the certification authority (CA) with a proof of knowledge of its secret key before the CA certifies the corresponding public key. However, it is not that simple. The current implementation of these proofs is represented by standards PKCS#10 [38] —used by VeriSign— and RFC 4210/11 [3, 41]. Here the proof is implemented by having the user send the CA a signature, under the public key it is attempting to get certified, of some message that includes the public key and the user identity. While such methods might intuitively seem to prove knowledge of the secret key, they do *not* suffice to realize the abstract model of [11, 29] in which the attacker actually hands the challenger the secret keys. In particular, not only does this type of proof of possession not suffice to prove secure the **MS-Bo** and **MS-LOSSW** schemes, but there are actually attacks against these schemes if such proofs of possession are used [9]. (That is, if we attempt to drop the KOSK assumption and substitute these proofs of possession instead.) To obtain proofs of possession sufficient to implement the KOSK assumption, it appears one should use zero-knowledge (ZK) proofs of knowledge (POKs) that meet strong, formal extractability requirements [5]. However, that is not all. Since in practice we would expect that the users may register keys at any time, concurrently with other users or with executions of the multi-signature protocol, one would require ZK POKs extractable under such concurrent conditions. This eliminates many standard protocols, including standard POKs of discrete logarithms. Still, it is true that protocols with the desired properties do seem to exist. For example, in the random-oracle model one could use [17] or, in the standard model, non-interactive ZK POKs [14]. But the first is not cheap and the second is prohibitive, and even if one were willing, adding them to the PKI requires modifying the client and CA functioning and software, which is difficult, costly and preferably avoided. Also, one is still left with the task of actually formally justifying a claim that this would implement the abstract KOSK model of [11, 29]. (We are not making such a claim here.) However, the main reason this route is impractical is simply that CAs do not right now implement such POKs. If, today, some corporation or person wishes to implement a MS scheme, it seems unlikely that VeriSign is going to oblige them by suddenly changing their offerings to include appropriate POKs of secret keys at registration.

OUR MS SCHEME. In summary, the most significant practical obstacle to multi-signatures at present is the key-setup requirements or assumptions of previous works. Our contribution is to remove this obstacle by presenting a multi-signature scheme in the *plain public-key model*. This means that, with regard to key setup, nothing more is required than in *any* usage of public-key cryptography, namely that any potential signer has a public key. There is no dedicated key generation protocol. A signer is *not* assumed to have proved knowledge of its secret key to the CA, but only to have a standard certificate. Yet, security against rogue-key attacks is proved *without* the KOSK assumption.

To elaborate, in our setting, the group of potential signers is dynamic: anyone possessing a (certified) public key can join at any time. In our security model, the adversary can corrupt a signer and choose its public key as a function of those of other (honest) signers. It is not required to supply the challenger with a matching secret key, meaning we prove security even when the adversary does not know the secret key underlying a public key it makes for itself. The fact that we do not need to assume any kind of proof-of-knowledge of the secret key performed to the CA at the time a public key is registered and a certificate is obtained reduces the demands on the PKI and allows our protocols to be implemented within the current PKI. CAs need not take any special actions or change their functioning or software. Indeed, a CA does not need to even know that a key is to be used in our multi-signature scheme; it can be treated like any other key.

EFFICIENCY AND OTHER SCHEME ATTRIBUTES. The significant gains in key setup achieved by our MS scheme are not at the cost of performance. As Table 1 indicates, our scheme, denoted **MS-BN**, compares favorably with previous ones in terms of signing time, verifying time, signature size and other attributes. We now discuss the information in the table in a little more depth.

MS-BN is based on the Schnorr [42] scheme and is proven secure in the random-oracle (RO) model [10] assuming hardness of the standard discrete logarithm problem. (**MS-Bo** and **MS-MOR** also use the RO model, but **MS-LOSSW** does not.) **MS-BN** allows secure concurrent executions of signing protocols by different subsets of the set of potential signers, which is important because applications on the Internet are inherently placed in a concurrent execution environment. Concurrent signing is explicitly disallowed in [31]. Security of our scheme is proved even when the adversary can control the scheduling and mount rogue-key attacks, yet without the KOSK assumption.

Unlike [11, 29], we do not use pairings (bilinear maps), which not only results, for us, in greater efficiency and ease of implementation, but also means we do not rely on the relatively new and untested hardness assumptions related to pairing-based cryptography. Verification in **MS-BN** is cheaper than in **MS-Bo**, and both signing and verification are cheaper in **MS-BN** than in **MS-LOSSW**. We have included the system parameter size in the table mainly to note that this is very large (25,920 bits) for **MS-LOSSW**, unlike for any other scheme. Our signatures are 320 bits as opposed to the 160 of **MS-Bo** because the latter uses the

Scheme	Sign	Verify	|sig|	|pk|	|par|	Key setup	Assump
MS-MOR [31]	1 exp	1 exp	$2 \cdot 160$	$[3 + 2\lg(N)] \cdot 160$	0	dedicated key-reg	DL
MS-Bo [11]	1 exp	2 pr	160	$6 \cdot 160$	$6 \cdot 160$	KOSK model	(co)CDH
MS-LOSSW [29]	3 exp	2 pr	$7 \cdot 160$	$6 \cdot 160$	$162 \cdot 160$	KOSK model	(co)CDH
MS-BN	1 exp	1 exp	$2 \cdot 160$	160	160	plain pk model	DL

Table 1: MS scheme comparisons. For each scheme (the last is ours) we show the computational cost of signing (per signer), the computational cost of verification of a multi-signature, the size of a multi-signature, the size of the public key of an individual signer, the size of the system parameters common to all signers, the type of key-setup, and the assumption used to prove security. All sizes are in bits. By "exp" we mean an exponentiation. (Some of the exponentiations are actually multi-exponentiations, but these have the same cost as single exponentiations.) N is the total number of signers in the system. By "pr" we mean a pairing, whose cost estimate is 6–20 exponentiations. We assume we work over a 160-bit elliptic-curve (EC) group for the DL-based schemes. For the coCDH-based schemes we assume an asymmetric pairing, that is, e: $\mathbb{G}_1 \times \mathbb{G}_2 \to \mathbb{G}_T$ with $\mathbb{G}_1 \neq \mathbb{G}_2$ (this to make the signatures as short as possible) and an isomorphism $\psi\colon \mathbb{G}_2 \to \mathbb{G}_1$ (this to make the proofs go through) with group-element representation sizes in $\mathbb{G}_1, \mathbb{G}_2$ and $\mathbb{G}_T$ being, respectively, 160-bits, $6 \cdot 160$ bits and $6 \cdot 160$ bits, which is what is needed, in this asymmetric-with-isomorphism setting, to provide the 1024-bit RSA level of security achieved by 160-bit EC groups [18].

pairing-based BSL [13] short signature scheme, but our gain in verification time (by a factor of 12-40) more than compensates. The public keys in our scheme are shorter than in any other scheme, an important benefit in case they have to be transmitted with the multi-signature. Additionally, of course, MS-BN is in the plain public-key model. Our efficiency estimates do not take security into account, meaning that all schemes are not necessarily compared at the same level of security. We do this as we do not think our analyses are tight (that is, the real security is better) and thus comparing at the same level would be misleading for practice.

NEW FORKING LEMMA. Our proof of security of the MS-BN scheme relies on a generalization of the Forking Lemma of [40] that may be of independent interest. The original Forking Lemma of Pointcheval and Stern [40] applies to signature schemes obtained from three-move identification schemes via the Fiat-Shamir [16] transform in the random oracle model. Roughly it says that in an *expected* $O(1/\epsilon)$ repeated executions of a forger A with success probability ϵ, one can find two accepting conversations that agree in the first prover move but not the verifier challenge, leading, via the special soundness property of Σ protocols, to recovery of the secret key and hence a proof of security of the signature scheme in the RO model. This lemma has been important in proving security of signature schemes via the rewinding technique. However, the lemma seems hard (if not impossible) to apply in situations like ours where we are not dealing with a regular signature scheme but a MS scheme. Indeed, in the past, variants of the lemma had to be formulated and proved for different types of signatures. (For example, a version for blind signatures is in [40], and one for ring signatures in [22]. Another variant is in [39].) The statement of our Forking Lemma (cf. Lemma 1), in contrast to previous ones, makes no mention of signatures or even, for that matter, random oracles. Rather it asserts a simple lower bound on the probability that two executions of an arbitrary algorithm on certain (related) inputs both accept. This statement, we feel, distills the probabilistic essence of the Forking Lemma and divorces it from any particular application context. (In our view, the Forking Lemma is something purely probabilistic, not about signatures. Previous Forking Lemmas mixed these things up.) In this form, it can be applied not only to prove security of regular signature schemes but also, as we show, to prove security of schemes like ours where the setting is more complex. Our Forking Lemma also provides worst-case rather than expected-time guarantees on the constructed algorithm, in contrast to [40]. We feel this meshes better with standard assumptions. (In using the Lemma of [40], you need to assume, say, hardness of discrete-logarithm computations against expected-time adversaries. This assumption may be true, but is not the standard one.) Our Forking Lemma can be viewed as an extension of the Reset Lemma of [8], and our proof, which uses the techniques of the latter, is simpler than that of [40].

RELATION TO AGGREGATE SIGNATURES. A natural thought is that multi-signatures are a special case of aggregate signatures, and we know the latter have been implemented without the KOSK assumption [12], so doesn't this yield multi-signatures in the plain public-key model? Let us explore this.

Suppose signer i has produced a BSL signature [13] σ_i on a message m_i $(i = 1, \ldots, n)$. The procedure of [12] aggregates $\sigma_1, \ldots, \sigma_n$ into a single, aggregate signature σ. Multi-signatures is simply the special case where $m_i = m$ is a common message for all $i \in L$. Now, [12] do prove security without the KOSK assumption, but for this *need to assume that the messages* $m_1, \ldots, m_n$ *are distinct*. So the multi-signature case is exactly the one where they do *not* have security without the KOSK assumption. In fact, in this case, there actually is a rogue-key attack on the scheme. Indeed, this MS scheme is exactly MS-Bo, which we know is not secure against rogue-key attack without the KOSK assumption.

However, [12] also suggest a workaround to the message distinctness assumption. Have each signer prepend its public key to its message before signing, so that the individual signatures now are (in the MS case) on the *enhanced messages* $pk_1 \| m, \ldots, pk_l \| m$. Now, security is guaranteed as long as the enhanced messages are distinct. However, this is not enough for security against rogue key attack in our plain public-key model. If an attacker, playing the role of signer 2, sets its public key pk_2 to equal the public key pk_1 of honest user 1 (an easy task) and outputs some forgery on some message m for some group including signers 1, 2, then we have a situation where two enhanced messages are the same, so the result of [12] does not apply. To fix this one can use the analysis of [6] that shows the scheme is secure even if enhanced messages are not distinct, and then we do obtain a secure MS scheme. However, verification of a multi-signature for n signers costs $n+1$ pairings, making it substantially less efficient than all the other schemes we have discussed, where verification time does not depend on the number of signers in the group.

Another potential route to multi-signatures is via sequential aggregate signatures [30]. These can be built from trapdoor permutation families in which there is a single domain underlying the entire family [30], but in fact there seems to be no example of such a family. RSA does not have the desired property. The authors build some RSA-based schemes directly, another one can be constructed using techniques from [21]. Some limitations of the schemes of [30] are lifted in [6]. However, a sequential-aggregate based MS scheme will require a number of communication rounds proportional to the number n of signers involved in a signature, as well as n applications of the trapdoor function to verify, while all previous protocols, including ours, are constant-round and have constant verification cost.

2. NOTATION

Let $\mathbb{N} = \{1, 2, 3, \ldots\}$. A string means a binary one. The empty string is denoted ε. If x, y are strings, then $|x|$ is the length of x. If $x_1, x_2, \ldots$ are objects then $x_1 \| x_2 \| \ldots$ denotes an encoding of them as strings from which the constituent objects are easily recoverable. If S is a (multi)set, then $|S|$ is its cardinality, $s \xleftarrow{\$} S$ denotes the operation of assigning to s an element of S chosen at random, and $\langle S \rangle$ is a unique encoding of S as a string. If A is a randomized algorithm, then $\mathsf{A}(x_1, \ldots; \rho)$ denotes its output on inputs $x_1, \ldots$ and coins ρ, while $y \xleftarrow{\$} \mathsf{A}(x_1, \ldots)$ means that we choose ρ at random and let $y = \mathsf{A}(x_1, \ldots; \rho)$.

3. A GENERAL FORKING LEMMA

Here we state and prove our Forking Lemma here that we will use later to prove the security of our multi-signature scheme. Our Forking Lemma, unlike that of Pointcheval and Stern [40], makes no mention of signature schemes or random oracles, but rather concentrates on the output behavior of an algorithm when run twice on related inputs. This makes it easily applicable in contexts other than standard signature schemes, and separates the probabilistic analysis of the rewinding from the actual simulation in the security proof, allowing for more modular (and hence easier to verify) proofs. In the following, think of x as a public key and $h_1, \ldots, h_q$ as replies to queries to a random oracle.

LEMMA 1. [**General Forking Lemma**] *Fix an integer $q \geq 1$ and a set H of size $h \geq 2$. Let A be a randomized algorithm that on input $x, h_1, \ldots, h_q$ returns a pair, the first element of which is an integer in the range $0, \ldots, q$ and the second element of which we refer to as a side output. Let IG be a randomized algorithm that we call the input generator. The accepting probability of A, denoted acc, is defined as the probability that $J \geq 1$ in the experiment*

$$x \xleftarrow{\$} \mathsf{IG} \; ; \; h_1, \ldots, h_q \xleftarrow{\$} H \; ; \; (J, \sigma) \xleftarrow{\$} \mathsf{A}(x, h_1, \ldots, h_q) \; .$$

The forking algorithm $\mathsf{F_A}$ associated to A is the randomized algorithm that takes input x proceeds as follows:

Algorithm $\mathsf{F_A}(x)$
   Pick coins ρ for A at random
   $h_1, \ldots, h_q \xleftarrow{\$} H$
   $(I, \sigma) \leftarrow \mathsf{A}(x, h_1, \ldots, h_q; \rho)$
   If $I = 0$ then return $(0, \varepsilon, \varepsilon)$
   $h'_I, \ldots, h'_q \xleftarrow{\$} H$
   $(I', \sigma') \leftarrow \mathsf{A}(x, h_1, \ldots, h_{I-1}, h'_I, \ldots, h'_q; \rho)$
   If $(I = I'$ and $h_I \neq h'_I)$ then return $(1, \sigma, \sigma')$
   Else return $(0, \varepsilon, \varepsilon)$.

Let
$$\mathrm{frk} = \Pr\left[b = 1 \; : \; x \xleftarrow{\$} \mathsf{IG} \; ; \; (b, \sigma, \sigma') \xleftarrow{\$} \mathsf{F_A}(x) \right] \; .$$
Then
$$\mathrm{frk} \geq \mathrm{acc} \cdot \left(\frac{\mathrm{acc}}{q} - \frac{1}{h} \right) \; . \tag{1}$$
Alternatively,
$$\mathrm{acc} \leq \frac{q}{h} + \sqrt{q \cdot \mathrm{frk}} \; . \quad \blacksquare \tag{2}$$

We proceed to the proof. We first recall two sublemmas that we will use in the proof. The proofs are skipped here but provided in the full version of our paper [7] for completeness. The first is a standard fact which one can derive from Jensen's inequality, or as a consequence of the fact that the variance of any random variable is non-negative:

LEMMA 2. *Let X be real-valued random variable. Then $\mathbf{E}\left[X^2\right] \geq \mathbf{E}[X]^2$.* $\blacksquare$

The next lemma is actually a consequence of the above although it appears in [2] with a different proof:

LEMMA 3. *Suppose $q \geq 1$ is an integer, and $x_1, \ldots, x_q \geq 0$ are real numbers. Then*
$$\sum_{i=1}^{q} x_i^2 \geq \frac{1}{q} \cdot \left(\sum_{i=1}^{q} x_i \right)^2 \; . \quad \blacksquare$$

PROOF OF LEMMA 1. We first prove (1) and then show that it implies (2). For any input x let $\mathrm{acc}(x)$ denote the probability that $J \geq 1$ in the experiment
$$h_1, \ldots, h_q \xleftarrow{\$} H \; ; \; (J, \sigma) \xleftarrow{\$} \mathsf{A}(x, h_1, \ldots, h_q) \; .$$
Also let
$$\mathrm{frk}(x) = \Pr\left[b = 1 \; : \; (b, \sigma, \sigma') \xleftarrow{\$} \mathsf{F_A}(x) \right] \; .$$
We claim that for all x,
$$\mathrm{frk}(x) \geq \mathrm{acc}(x) \cdot \left(\frac{\mathrm{acc}(x)}{q} - \frac{1}{h} \right) \; . \tag{3}$$
Then, with the expectation taken over $x \xleftarrow{\$} \mathsf{IG}$, we have
$$\begin{aligned}
\mathrm{frk} &= \mathbf{E}[\mathrm{frk}(x)] \geq \mathbf{E}\left[\mathrm{acc}(x) \cdot \left(\frac{\mathrm{acc}(x)}{q} - \frac{1}{h} \right) \right] \\
&= \frac{\mathbf{E}[\mathrm{acc}(x)^2]}{q} - \frac{\mathbf{E}[\mathrm{acc}(x)]}{h} \\
&\geq \frac{\mathbf{E}[\mathrm{acc}(x)]^2}{q} - \frac{\mathbf{E}[\mathrm{acc}(x)]}{h} = \mathrm{acc} \cdot \left(\frac{\mathrm{acc}}{q} - \frac{1}{h} \right) \; .
\end{aligned}$$
This establishes (1). Above, we used (3), Lemma 2 and also the fact that $\mathbf{E}[\mathrm{acc}(x)] = \mathrm{acc}$.

We proceed to the proof of (3). For any input x, with probabilities taken over the coin tosses of F_A we have

$$\begin{aligned}
\mathrm{frk}(x) &= \Pr\left[\, I = I' \wedge I \geq 1 \wedge h_I \neq h'_I \,\right] \\
&\geq \Pr\left[\, I = I' \wedge I \geq 1 \,\right] - \Pr\left[\, I \geq 1 \wedge h_I = h'_I \,\right] \\
&= \Pr\left[\, I = I' \wedge I \geq 1 \,\right] - \frac{\Pr\left[\, I \geq 1 \,\right]}{h} \\
&= \Pr\left[\, I = I' \wedge I \geq 1 \,\right] - \frac{\mathrm{acc}(x)}{h}\,.
\end{aligned}$$

It remains to show that $\Pr\left[\, I = I' \wedge I \geq 1 \,\right] \geq \mathrm{acc}(x)^2/q$. Let $\mathcal{R}$ denote the set from which A draws its coins at random. For each $i \in \{1, \ldots, q\}$ let $X_i \colon \mathcal{R} \times H^{i-1} \to [0,1]$ be defined by setting $X_i(\rho, h_1, \ldots, h_{i-1})$ to

$$\Pr\left[\, J = i \;:\; h_i, \ldots, h_q \stackrel{\$}{\leftarrow} H\,;\, (J, \sigma) \leftarrow \mathsf{A}(x, h_1, \ldots, h_q; \rho) \,\right]$$

for all $\rho \in \mathcal{R}$ and $h_1, \ldots, h_{i-1} \in H$. Regard X_i as a random variable over the uniform distribution on its domain. Then

$$\begin{aligned}
&\Pr\left[\, I = I' \wedge I \geq 1 \,\right] \\
&= \sum_{i=1}^{q} \Pr\left[\, I = i \wedge I' = i \,\right] \\
&= \sum_{i=1}^{q} \Pr\left[\, I = i \,\right] \cdot \Pr\left[\, I' = i \mid I = i \,\right] \\
&= \sum_{i=1}^{q} \sum_{\rho, h_1, \ldots, h_{i-1}} X_i(\rho, h_1, \ldots, h_{i-1})^2 \cdot \frac{1}{|\mathcal{R}| \cdot |H|^{i-1}} \\
&= \sum_{i=1}^{q} \mathbf{E}\left[X_i^2\right] \;\geq\; \sum_{i=1}^{q} \mathbf{E}\left[X_i\right]^2\,,
\end{aligned}$$

where in the last step we used Lemma 2. Now let $x_i = \mathbf{E}[X_i]$ for $i \in \{1, \ldots, q\}$, and apply Lemma 3. We get

$$\sum_{i=1}^{q} \mathbf{E}[X_i]^2 \;\geq\; \frac{1}{q} \cdot \left(\sum_{i=1}^{q} \mathbf{E}[X_i]\right)^2 \;=\; \frac{1}{q} \cdot \mathrm{acc}(x)^2\,.$$

This completes the proof of (3) and thus of (1). We now show how to obtain (2). Using (1) we have

$$\left(\mathrm{acc} - \frac{q}{2h}\right)^2 \;=\; \mathrm{acc}^2 - \frac{q}{h} \cdot \mathrm{acc} + \frac{q^2}{4h^2} \;\leq\; q \cdot \mathrm{frk} + \frac{q^2}{4h^2}\,.$$

Taking the square root of both sides, and using the fact that $\sqrt{a+b} \leq \sqrt{a} + \sqrt{b}$ for any real numbers $a, b \geq 0$, we get

$$\mathrm{acc} - \frac{q}{2h} \;\leq\; \sqrt{q \cdot \mathrm{frk}} + \sqrt{\frac{q^2}{4h^2}} \;=\; \frac{q}{2h} + \sqrt{q \cdot \mathrm{frk}}\,.$$

Re-arranging terms yields (2). □

4. MULTI-SIGNATURES

Here we provide our definitions for multi-signatures with security in the plain public-key model.

THE MODEL. Consider a group of signers signing the same message m, each having as input its own public and secret key as well as a list of the public keys of the other signers. The signers want to interact in a protocol which eventually outputs a compact signature σ that represents the signature of each individual signer on the message m. We assume the signers are connected to each other via point-to-point links over which they can send messages. We do not assume these links are secure (that is, they are neither private nor authenticated) and we do not assume a broadcast primitive. The signers interact for some number of rounds. In each round, view a signer as receiving a (but not necessarily the same) message from every other signer, performing some computation, and then sending a message to every other signer, except that in the first round the "received message" is the party's input (and so is not really received) and in the last round the "sent message" is a local output (and so is not really sent). The local output is either $\perp$ to indicate failure or is the compact signature σ. Instances of the protocol may be executed concurrently, with one signer possibly participating in several concurrent instances at the same time.

In describing protocols, we will have each signer assign indices $1, \ldots, n$ to the signers, with itself being signer 1. We clarify that these are local references to the cosigners participating in this protocol instance. (That is, each signer in this protocol instance chooses its own indexing, so that signer 3 on my list and your list may not be the same. Think of the index a signer gives to its cosigners as locally identifying the link over which they communicate.) These indices have no certified relationship with the public keys. In particular, they are not identities.

Formally a multi-signature scheme $\mathsf{MS} = (\mathsf{Pg}, \mathsf{Kg}, \mathsf{Sign}, \mathsf{Vf})$ consists of four algorithms. A central authority runs the parameter generation algorithm Pg to generate the system-wide parameters par. Each signer independently generates its own public and private key pair via $(pk, sk) \stackrel{\$}{\leftarrow} \mathsf{Kg}(par)$. We stress that this is a non-interactive process that can be performed by any signer at any given time. New signers can join the system at will, and need not engage in expensive protocols with a CA or with other signers to prove knowledge of the corresponding secret key before participating in signing protocols. The Sign algorithm represents the signing protocol as indicated above. The verification algorithm Vf takes as input a multiset of public keys $L = \{pk_1, \ldots, pk_n\}$, the message m and a candidate signature σ, and outputs 1 if σ is a valid signature for L and m, or outputs 0 otherwise. (Because users may let their keys depend on those of other users, we explicitly allow them to be the same by modeling L as a multiset.) We have the obvious correctness requirement, namely that if a group of signers begin their interaction with public keys $L = \{pk_1, \ldots, pk_n\}$ and message m, and all signers follow the protocol (meaning, perform their computations according to Sign) then all have local output σ such that Vf returns 1 on input L, σ.

When describing protocols we will not specify the algorithms directly but instead describe them informally by saying what parties receive, compute and send in each round.

SECURITY. The notion of security requires that it be infeasible to forge multi-signatures involving at least one honest signer. As in previous works [31, 12, 29] we can in fact assume there is a single honest signer. Our adversary will be viewed as having effectively corrupted all other signers. It can choose their public keys as it likes, even as a function of the public key of the honest signer, and can interact arbitrarily with the honest signer in any number of concurrent instances before outputting its forgery attempt. In somewhat more detail, we consider the following three-phase game associated to multi-signature scheme $\mathsf{MS} = (\mathsf{Pg}, \mathsf{Kg}, \mathsf{Sign}, \mathsf{Vf})$ and adversary (forger) F:

Setup: The game chooses system-wide parameters $par \stackrel{\$}{\leftarrow} \mathsf{Pg}$ and a key-pair $(pk^*, sk^*) \stackrel{\$}{\leftarrow} \mathsf{Kg}(par)$ for the for the "target" (honest) signer. The target public key pk^* is given as input to the forger F.

Attack: F can start a protocol instance with the honest signer by providing the latter with a message m and a multiset $L = \{pk_1, \ldots, pk_n\}$ of purported cosigners, where pk^* occurs in L at least once. It can choose these public keys as it wishes, including as a function of pk^* and previous protocol flows. In interacting with the honest signer, F will play the role of all signers in L other than one instance of pk^*, sending messages to, and receiving messages from, the honest signer. The forger can schedule an arbitrary number of protocol instances concurrently, interacting with "clones" of the honest signer, where each clone maintains its own state and uses its own coins but all use the keys pk^*, sk^* and follow the protocol (meaning use algorithm Sign) to compute their responses to received messages. When the honest signer terminates then its local output (whether $\bot$ or a compact signature) is returned to F.

Forgery: At the end of its execution, F outputs a multiset $L = \{pk_1, \ldots, pk_n\}$, a message m and a forged signature σ. The forger is said to win the game if $\mathsf{Vf}(L, m, \sigma) = 1$, $pk^* \in L$ and F never initiated a signing protocol with L, m.

The advantage of algorithm F in breaking MS, denoted as $\mathbf{Adv}_{\mathsf{MS}}^{\mathsf{uf\text{-}cma}}(\mathsf{F})$, is defined as the probability that F wins the above game, where the probability is taken over the coin tosses of the forger, the honest signer, and the setup phase. We say that a forger F (t, q_S, N, ϵ)-breaks MS if F runs in time at most t, F initiates at most q_S signing protocols with the honest signer, the number of public keys in the multiset L involved in any signing query or in the forgery is at most N, and $\mathbf{Adv}_{\mathsf{MS}}^{\mathsf{uf\text{-}cma}}(\mathsf{F}) \geq \epsilon$. The scheme MS is said to be (t, q_S, N, ϵ)-secure if no forger (t, q_S, N, ϵ)-breaks it. In the random oracle model, the Sign and Vf algorithms, as well as the adversary, additionally have access to a random oracle $\mathsf{H} : \{0,1\}^* \to D$, where D is a set possibly depending on the system parameters. The additional parameter q_H denotes the maximum number of F's random oracle queries. (If there is more than one random oracle, we mean the total number of queries to all random oracles.) We say that F $(t, q_S, q_H, N, \epsilon)$-breaks MS in the random oracle model, and that MS is $(t, q_S, q_H, N, \epsilon)$-secure in the random oracle model.

DISCUSSION. Note that the game described above does not require the adversary to fix the public keys of all signers in the system at the beginning of the game (as is required by the notions of [31, 11]), or to submit the secret keys of all corrupt signers to a special certification oracle (as is required by the notions of [11, 29]). Rather, our model allows the adversary to dynamically add new signers to the system, using arbitrary public keys that may depend on the target public key or on previous signing interactions. It thereby avoids the KOSK assumption and does not presume expensive proof of knowledge protocols to be performed with the CA. This security notion reflects a real-world system with the desirable features that new signers can join on-the-fly using self-generated keys, and that existing (external) CA infrastructure can be reused for the certification of these keys.

INTERACTIVE AGGREGATE SIGNATURES. Although multi-signatures are presented as being about a bunch of signers signing a single common message, one can consider something more general where each party has its own message. Let us call this an *interactive aggregate signature* (IAS). Here party i has message M_i. The parties start knowing all messages and each other's public keys, interact, and finally produce an aggregate signature that is supposed to validate that party i signed M_i for all i involved. An IAS scheme can be viewed either as a generalization of a sequential aggregate signature scheme [30] where the interaction between signers is arbitrary rather than sequential, or as an extension of a MS scheme where each signer has a different message. IAS schemes potentially have more applications than MS schemes. However, we observe that IAS and MS are equivalent in the sense that any scheme for one is easily turned into a scheme for the other. Indeed, we can implement IAS given a MS scheme by using as message in the latter the tuple of all messages and public keys of the former. On the other hand, obviously, we an implement an MS scheme given an IAS scheme by setting the messages of all parties to the single message of the MS scheme. For this reason we do not explicitly consider IAS schemes further, but it is worthwhile to note that that the single-message restriction of an MS scheme is not really limiting. We also think IAS schemes are interesting in that they unify aggregate and multi-signatures, both of which are special cases of IASs.

5. OUR MULTI-SIGNATURE SCHEME

Our scheme is based on Schnorr's signature scheme [42]. Let $\mathbb{G}$ be a cyclic group of prime order p, and let g be a generator of $\mathbb{G}$. Recall that a Schnorr signature of a message m under public key $X \in \mathbb{G}$ is a pair $(R, s) \in \mathbb{G} \times \mathbb{Z}_p$ such that $g^s = RX^c$ in $\mathbb{G}$, where c is the response to a random oracle query on $R\|m$. A first idea to aggregate signatures may be to let a signature under keys $L = \{X_1, \ldots, X_n\}$ of message m be a pair (R, s) such that $g^s = R \prod_{i=1}^n X_i^c$, where $c = \mathsf{H}(R\|\langle L\rangle\|m)$ is determined by a random oracle. Without restrictions on key generation however, this type of scheme is vulnerable to a well-known attack [24, 27, 32, 31] where a corrupt signer chooses $x_1 \xleftarrow{\$} \mathbb{Z}_p$ and sets its public key $X_1 \leftarrow g^{x_1} \cdot \prod_{i=2}^n X_i^{-1}$. This way, x_1 essentially becomes the "secret key" for the entire group of signers $L = \{X_1, \ldots, X_n\}$: signer 1 can by himself sign any message m in name of the entire group L by choosing $r \xleftarrow{\$} \mathbb{Z}_p$ and computing $(R = g^r, s = cx_1 + r \bmod p)$ where $c = \mathsf{H}(R\|\langle L\rangle\|m)$.

We counteract this attack by using a different value c_i in the exponent of each public key X_i, so that the verification equation becomes $g^s = R\prod_{i=1}^n X_i^{c_i}$, where the values for c_i are determined by independent random oracle queries $c_i = \mathsf{H}(X_i\|R\|\langle L\rangle\|m)$. With the help of our general Forking Lemma (see Lemma 1), we succeed in extracting from any forger the discrete logarithm of the target public key. The way we apply the Forking Lemma is particularly interesting because we need certain random oracle responses to be the same in both executions of the adversary, even though the corresponding queries may not occur until *after* the fork.

A second problem is that, in order to respond to the forger's signature queries, the simulator needs to know the value of R before the forger does so that it can program the random oracle. The value of R is typically computed as the product of individual shares of R chosen by each signer. Unless the target signer is the last to reveal his share (which we cannot assume to always be the case), the forger knows R before the simulator does, enabling the forger to perform a random oracle query involving R and thereby to prevent the simulator from programming this entry later on. We overcome this problem by letting signers first "commit" to their share of R through an additional random oracle query. The simulator, who sees all random oracle queries, can therefore look up the individual shares of R before the forger can, and can thus correctly program the random oracle.

We now proceed to describe the scheme and analyze its security. We begin by recalling some necessary definitions.

THE DISCRETE LOGARITHM ASSUMPTION. Let $\mathbb{G}$ be a multiplicative group of prime order p, and let $\mathbb{G}^* = \mathbb{G} \setminus \{1\}$. The advantage of algorithm A in solving the discrete logarithm problem in $\mathbb{G}$ is defined as

$$\mathbf{Adv}_{\mathbb{G}}^{\mathrm{dlog}}(\mathsf{A}) = \Pr\left[g^x = y \mid g \xleftarrow{\$} \mathbb{G}^*;\ y \xleftarrow{\$} \mathbb{G};\ x \xleftarrow{\$} \mathsf{A}(y)\right].$$

We say that A (t, ϵ)-solves the discrete logarithm problem in $\mathbb{G}$ if it runs in time at most t and $\mathbf{Adv}_{\mathbb{G}}^{\mathrm{dlog}}(\mathsf{A}) \geq \epsilon$, and we say that the discrete logarithm problem in $\mathbb{G}$ is (t, ϵ)-hard if no algorithm A (t, ϵ)-solves it.

THE SCHEME. Let $k = \lfloor \log_2 p \rfloor$, let $l_0, l_1 \in \mathbb{N}$, and let $\mathsf{H}_0 : \{0,1\}^* \to \{0,1\}^{l_0}$ and $\mathsf{H}_1 : \{0,1\}^* \to \{0,1\}^{l_1}$ be random oracles. To these, we associate the multi-signature scheme $\mathsf{MS\text{-}BN} = (\mathsf{Pg}, \mathsf{Kg}, \mathsf{Sign}, \mathsf{Vf})$ as follows:

Parameter generation. A trusted center chooses a random generator $g \xleftarrow{\$} \mathbb{G}^*$ and publishes $(\mathbb{G}, p, g)$ as system-wide parameters.

Key generation. Each signer runs the Kg algorithm to generate a random secret key $x \xleftarrow{\$} \mathbb{Z}_p$ and the corresponding public key $X \leftarrow g^x$.

Signing. Let X_1 and x_1 be the public and private key of a signer, let m be the message to be signed, and let $X_2, \ldots, X_n$ be the public keys of all other cosigners. We recall that the indices $1, \ldots, n$ are merely local references to cosigners, defined by one signer within one protocol instance. These indices are not tied to public keys in a global way, and in particular are not unique identities of signers. The communication proceeds in a number of rounds, where in each round each signer receives a message from every other signer, performs some local computation, and sends a message to every other signer.

Round 1:
- Local input: x_1, $L = \{X_1, \ldots, X_n\}$, m
- Computation: Choose $r_1 \xleftarrow{\$} \mathbb{Z}_p$, compute $R_1 \leftarrow g^{r_1}$ and query $t_1 \leftarrow \mathsf{H}_0(R_1)$.
- Send to signer i: t_1

Round 2:
- Receive from signer i: t_i
- Send to signer i: R_1

Round 3:
- Receive from signer i: R_i
- Computation: For all $2 \leq i \leq n$, check that $t_i = \mathsf{H}_0(R_i)$. Abort the protocol with local output $\perp$ if this is not the case; otherwise, compute $R \leftarrow \prod_{i=1}^n R_i$, query $c_1 \leftarrow \mathsf{H}_1(X_1 \| R \| \langle L \rangle \| m)$ where $\langle L \rangle$ is a unique encoding of L (e.g. the sequence of keys in lexicographic order), and compute $s_1 \leftarrow x_1 c_1 + r_1 \mod p$.
- Send to signer i: s_1

Round 4:
- Receive from signer i: s_i
- Computation: $s \leftarrow \sum_{i=1}^n s_i \mod p$
- Local output: the signature $\sigma = (R, s)$

Verification. Given a multiset of public keys $L = \{X_1, \ldots, X_n\}$, message m and signature $\sigma = (R, s)$, the verifier computes $c_i \leftarrow \mathsf{H}_1(X_i \| R \| \langle L \rangle \| m)$ for all $1 \leq i \leq n$. He accepts the signature if $g^s = R \prod_{i=1}^n X_i^{c_i}$, and rejects otherwise.

EFFICIENCY. An overview comparing the efficiency of our scheme to that of other (provably secure) multi-signature schemes is given in Table 1. Compared to the MS-MOR scheme [31], our MS-BN scheme avoids the expensive interactive key generation protocol, while offering considerably shorter public keys and maintaining the same signature length and signing/verification costs. Moreover, our scheme allows for concurrent signing sessions at the cost of one extra round of interaction (and the computation of some hash values). Compared to the MS-Bo scheme [11], our scheme has about twice the signature size, yet offers faster verification. Our scheme beats the MS-LOSSW scheme [29] in all costs. Moreover, our scheme has the advantage over MS-Bo and MS-LOSSW of not relying on pairings to be defined over the underlying group. On the other hand, the signing protocol in our scheme has more rounds of interaction than in the other schemes.

We note that when we motivated communication reduction in the Introduction (in particular for mobile devices), we were referring to the size of the signature, not to the communication cost of the protocol that computes the signature. This is because it is the signatures that are frequently communicated. In fact the protocol may take place over a a different, cheaper medium, yet produce signatures (e.g. certificates) that are placed on mobile devices and then frequently transmitted by these devices.

SECURITY. The following theorem implies that the MS-BN scheme meets our definition of security in the plain public-key model (i.e., without the KOSK assumption).

THEOREM 4. *If there exists a $(t, q_\mathsf{S}, q_\mathsf{H}, N, \epsilon)$-forger F in the random-oracle model against the MS scheme MS-BN described above, then there exists an algorithm B that (t', ϵ')-breaks the discrete logarithm problem in $\mathbb{G}$, where*

$$\begin{aligned}\epsilon' &\geq \frac{\epsilon^2}{q_\mathsf{H} + q_\mathsf{S}} - \frac{2q_\mathsf{H} + 16N^2 q_\mathsf{S}}{2^{l_0}} - \frac{8N q_\mathsf{S}}{2^k} - \frac{1}{2^{l_1}}, \quad (4)\\ t' &= 2t + q_\mathsf{S} t_{\exp} + O((q_\mathsf{S} + q_\mathsf{H})(1 + q_\mathsf{H} + N q_\mathsf{S}))\end{aligned}$$

and $t_{\exp}$ is the time of an exponentiation in $\mathbb{G}$. ∎

PROOF OF THEOREM 4. The idea of the proof is to use our Forking Lemma to obtain from the forger F two forgeries (R, s) and (R', s') satisfying

$$g^s = R \prod_{i=1}^n X_i^{c_i} \quad \text{and} \quad g^{s'} = R \prod_{i=1}^n X_i^{c_i'},$$

such that $c_i = c_i'$ if X_i is the target public key X^*, and $c_i \neq c_i'$ for all other keys. We can then extract the discrete logarithm of X^* by dividing the two equations above. Special care must be taken however in responding F's random oracle queries so that the above relations between c_i and c_i' are ensured. In particular, F may not ask the queries defining c_i and c_i' until *after* the fork, where the two executions of F have already diverged. We overcome this by fixing the response values to these queries *before* the fork, and by recognizing the queries when they actually occur after the fork. It is mainly due to the modularity of our simplified Forking Lemma that the complexity of the proof is kept manageable.

We are now ready to present the actual proof. Given a $(t, q_S, q_H, N, \epsilon)$-forger F, consider the following algorithm A. On inputs $g \in \mathbb{G}^*$, $X^* \in \mathbb{G}$ and $h_1, \ldots, h_{q_H+q_S} \in \{0,1\}^{l_1}$, algorithm A runs the forger F on input system parameters $par = (\mathbb{G}, p, g)$ and target public key $pk^* = X^*$. Algorithm A initializes counters ctr_1, ctr_2 to zero, and maintains initially empty associative arrays $T_0[\cdot], T_1[\cdot, \cdot], T_2[\cdot]$. It assigns $T_2[X^*] \leftarrow 0$ and answers F's oracle queries as follows. Tables T_0 and T_1 are used to simulate random oracles H_0 and H_1, respectively, while T_2 assigns a unique index $1 \le i \le q_H + Nq_S$ to each public key X occurring either as a cosigner's public key in one of F's signature queries, or as the first item in the argument of one of F's queries to H_1. Algorithm A assigns index 0 to the target public key X^* by setting $T_2[X^*] \leftarrow 0$. It responds to F's oracle queries as follows:

- $H_0(R)$: If $T_0[R]$ is undefined, then A chooses $T_0[R] \stackrel{\$}{\leftarrow} \{0,1\}^{l_0}$. It returns $T_0[R]$ to F.

- $H_1(X\|Q)$: If $T_2[X]$ is undefined then A increases ctr_2 and sets $T_2[X] \leftarrow ctr_2$. Let $i = T_2[X]$. If $T_1[i, Q]$ has not yet been defined, then A immediately assigns random values to *all* entries $T_1[j, Q]$ for $1 \le j \le q_H + Nq_S$, increases ctr_1 and assigns $T_0[0, Q] \leftarrow h_{ctr_1}$. (If the argument of the query cannot be parsed as $X\|Q$, then A simply returns a random element of $\{0,1\}^{l_1}$, preserving consistency if the same query is asked again.)

- Signing query with public keys L and message m: If $X^* \notin L$ then algorithm A returns $\perp$ to F; otherwise, it parses the elements of L as $\{X_1 = X^*, X_2, \ldots, X_n\}$ and continues as follows. First, it checks for all $2 \le i \le n$ whether $T_2[X_i]$ has already been defined; it increases ctr_2 and sets $T_2[X_i] \leftarrow ctr_2$ if not. Then, A increases counter ctr_1 and sets $c_1 \leftarrow h_{ctr_1}$. It chooses $s_1 \stackrel{\$}{\leftarrow} \mathbb{Z}_p$, computes $R_1 \leftarrow g^{s_1} X_1^{-c_1}$ and sends $t_1 = H_0(R_1)$ to all cosigners.

After receiving $t_2, \ldots, t_n$ from F (who's playing the role of the cosigners), A searches in table T_0 for the values R_i so that $t_i = T_0[R_i]$. If no such R_i can be found for some $2 \le i \le n$, then A sets a flag $alert \leftarrow \mathtt{true}$ and sends R_1 to all cosigners. If more than one such value is found for some R_i, then it sets $bad_1 \leftarrow \mathtt{true}$, aborts the execution of F and halts with output $(0, \varepsilon)$. Otherwise, A computes $R \leftarrow \prod_{i=1}^n R_i$ and checks whether $T_1[0, Q]$ has already been defined for $Q = R\|\langle L \rangle\|m$. If so, it sets $bad_2 \leftarrow \mathtt{true}$, aborts the execution of F and halts with output $(0, \varepsilon)$. If not, it sets $T_1[0, Q] \leftarrow c_1$, chooses $T_1[i, Q] \stackrel{\$}{\leftarrow} \{0,1\}^{l_1}$ for all $1 \le i \le q_H + Nq_S$, and sends R_1 to all cosigners.

If, after receiving $R_2, \ldots, R_n$, there exists an index $1 \le i \le n$ such that $H_0(R_i) \ne t_i$, then A stops the signing protocol returning $\perp$. If $alert = \mathtt{true}$ while $H_0(R_i) = t_i$ for all $1 \le i \le n$, then it sets $bad_3 \leftarrow \mathtt{true}$, aborts the execution of F and halts outputting $(0, \varepsilon)$. Otherwise, it sends s_1 to all cosigners.

After receiving $s_2, \ldots, s_n$, A computes $s \leftarrow \sum_{i=1}^n s_i \mod p$ and returns (R, s) as the signature.

Eventually, F outputs a forged signature (R, s) together with multiset $L = \{X_1, \ldots, X_n\}$ and message m. Algorithm A first performs additional queries $H_1(X_i\|R\|\langle L \rangle\|m)$ for $1 \le i \le n$, thereby making sure that $T_2[X_i]$ is defined. Let $1 \le J \le q_H + q_S$ be the index such that $T_1[0, R\|\langle L \rangle\|m] = h_J$, and let n^* be the number of times that X^* occurs in L. If F's forgery is valid, algorithm A halts returning $(J, (R, h_J, s, n^*))$; if not, it halts returning $(0, \varepsilon)$.

Let $\Pr[bad_i]$ denote the probability of the event that flag bad_i gets set to $\mathtt{true}$. Consider set $H = \{0, 1\}^{l_1}$ and input generator IG that returns random elements $g, X^* \stackrel{\$}{\leftarrow} \mathbb{G}$. We bound the accepting probability acc of A with respect to these, as defined in Lemma 1, as follows:

$$\begin{aligned} \mathrm{acc} &\ge \epsilon - \Pr[bad_1] - \Pr[bad_2] - \Pr[bad_3] \\ &\ge \epsilon - \frac{(q_H + Nq_S + 1)^2}{2^{l_0+1}} \\ &\quad - \sum_{i=1}^{q_S} \left(\frac{q_H + Nq_S}{2^k} + \frac{q_H + q_S}{2^k} + \frac{N}{2^{l_0}} \right) \\ &\ge \epsilon - \frac{(q_H + Nq_S + 1)^2}{2^{l_0}} - \frac{2q_S(q_H + Nq_S)}{2^k} . \end{aligned}$$

We clarify how the bounds in the second inequality were obtained. If at some point in the execution of F two values $R_i \ne R'_i$ are found such that $t_i = H_0(R_i) = H_0(R'_i)$, then clearly at least one collision must have occurred in H_0. However, all response values of H_0 are chosen uniformly at random from $\{0,1\}^{l_0}$, and since there are at most $q_H + Nq_S$ queries to H_0, the probability that at least one collision occurs is at most $((q_H + Nq_S)(q_H + Nq_S + 1)/2)/2^{l_0} \le (q_H + Nq_S + 1)^2/2^{l_0}$.

To cause bad_2 to be set to $\mathtt{true}$ during the i-th signing query, we distinguish between the case that $H_0(R_1)$ was previously queried by the forger, and the case that it wasn't. In the first case, F probably knows R and may have deliberately queried $H_1(X\|R\|\langle L \rangle\|m)$ for some X. But since R_1 was chosen by A independently from F's view at the beginning of the signing protocol, the probability that F queried $H_0(R_1)$ is at most $(q_H + Nq_S)/p \le (q_H + Nq_S)/2^k$. In the second case, F's view is completely independent of R_1, and hence of R. The probability that R occurred by chance in a previous query to H_1 or was set by A in one of the $i - 1$ previous signature simulations is at most $(q_H + q_S)/p \le (q_H + q_S)/2^k$.

Lastly, in order to set $bad_3 = \mathtt{true}$, F must have predicted the value of $H_0(R_i)$ for at least one $1 \le i \le n$, which it can do with probability at most $N/2^{l_0}$. The third inequality follows from simple rearranging of terms after assuming (without loss of generality) that $q_H, q_S, N > 0$.

Now consider an algorithm B that on input X^* runs the forking algorithm $F_A(X^*)$, which with probability frk returns $(1, (R, h, s, n^*), (R', h', s', n'^*))$ with $h \ne h'$. These forgeries are such that

$$g^s = R \prod_{i=1}^n X_i^{c_i} \quad \text{and} \quad g^{s'} = R' \prod_{i=1}^{n'} X_i'^{c_i'}$$

where $L = \{X_1, \ldots, X_n\}$ and m are the public keys and the message involved in F's forgery and $c_i = H_1(X_i\|R\|\langle L \rangle\|m)$ are the relevant random oracle responses from the first run. Let $I^* \subseteq \{1, \ldots, n\}$ be the set of indices such that $X_i = X^*$. Variables $L', X'_1, \ldots, X'_{n'}, m', c'_1, \ldots, c'_{n'}, I'^*$ are defined analogously for the second run of F. We will show later that, due to the way that A simulates F's environment, it must hold that $n = n'$, that $L = L'$, that $I^* = I'^*$, that $n^* = n'^*$, that $c_i = c'_i$ for $i \notin I^*$, and that $c_i = h$ and $c'_i = h'$ for $i \in I^*$. Dividing the two equations above then gives

$$g^{s-s'} = \prod_{i \in I^*}(X^*)^{h-h'} = (X^*)^{n^*(h-h')},$$

so that B can compute the discrete logarithm of X^* as $(s - s')/(n^*(h - h')) \bmod p$. The probability that algorithm B succeeds in doing so is given by

$$\begin{aligned}\epsilon' &\geq \text{frk} \\ &\geq \frac{\text{acc}^2}{q_H + q_S} - \frac{1}{2^{l_1}} \\ &\geq \frac{\epsilon^2}{q_H + q_S} - \frac{2(q_H + Nq_S + 1)^2}{(q_H + q_S) \cdot 2^{l_0}} - \frac{4q_S(q_H + Nq_S)}{(q_H + q_S) \cdot 2^k} - \frac{1}{2^{l_1}} \\ &\geq \frac{\epsilon^2}{q_H + q_S} - \frac{2q_H + 16N^2 q_S}{2^{l_0}} - \frac{8Nq_S}{2^k} - \frac{1}{2^{l_1}},\end{aligned}$$

where again we assume without loss of generality that $q_H, q_S, N > 0$. The theorem follows.

We still have to argue why the equalities between all the variables in both runs of A hold. In the case that F returned $(1, (R, h, s, n^*), (R, h', s', n'^*))$, let J be the index that A returned after both executions by F_A. In A's first execution, $h_J = h$ is assigned to $T_1[0, R\|\langle L\rangle\|m] = H_1(X^*\|R\|\langle L\rangle\|m)$ at the moment when F makes its first query $H_1(X\|R\|\langle L\rangle\|m)$ for *some* public key X (so not necessarily X^*), where $L = \{X_1, \ldots, X_n\}$. Likewise, in the second run, $h'_J = h'$ is assigned to $T_1[0, R'\|\langle L'\rangle\|m']$ when F queries $H_1(X'\|R'\|\langle L'\rangle\|m')$ for some public key X', where $L' = \{X'_1, \ldots, X'\}$. Up to the point of this hash query, however, the environments of F provided by A in the first and the second run are identical, because A uses the same inputs, random tape and values $h_1, \ldots, h_{J-1}$ to generate F's inputs, random tape and oracle responses. Therefore, the two executions of F are identical up to this point, and in particular the arguments of both hash queries must be the same, implying that $R = R'$, $L = L'$, $n = n'$, $X_i = X'_i$ and $m_i = m'_i$ for $1 \leq i \leq n$. Moreover, the entries for $X_1, \ldots, X_n$ in T_2 are assigned *at the latest* when parsing the arguments of this hash query, causing the values of $T_2[X_i]$ to be the same in both runs as well. The forger's other queries $H_1(X_i\|R\|\langle L\rangle\|m)$ may not occur until much later, but the response values $T_1[T_2[X_i], R\|\langle L\rangle\|m]$ for these queries are chosen *before* the fork, and hence are the same in both runs as well. Therefore, it holds that $c_i = c'_i$ for all $1 \leq i \leq n$, that $c_i = h_J = h$ for $i \in I$, and that $c'_i = h'_J = h'$ for $i \in I$, which concludes the proof.

The running time t' of B is twice that of A plus the time needed to compute the discrete logarithm $(s - s')/(n^*(h - h')) \bmod p$. The running time of A is the running time t of F plus the time needed to answer $q_H + Nq_S$ random oracle queries and q_S signature queries. We assume that exponentiations in $\mathbb{G}$ take time $t_{\exp}$, and all other operations take unit time. Each random oracle query may cause A to perform $O(1 + q_H + Nq_S)$ unit-time operations. Each signature query involves one multi-exponentiation in $\mathbb{G}$ and $O(1 + q_H + Nq_S)$ unit-time operations. Therefore, we have $t' = 2t + q_S t_{\exp} + O((q_S + q_H)(1 + q_H + Nq_S))$. □

REDUCTION TIGHTNESS. The reduction presented above is tighter than that of the MS-MOR scheme [31], but is still not tight, as can be seen from (4). In comparison, the security proof of the MS-MOR scheme requires two applications of the forking technique (once to extract the secret keys of corrupt players, and once to obtain two forgeries) and $q_H \cdot q_S$ rewindings (to simulate signing protocols) of the forger, yielding a considerable loss in the tightness of the security reduction. The pairing-based MS-Bo and MS-LOSSW schemes do not have tight security reductions either.

6. FURTHER RESULTS

Our scheme is based on the Schnorr signature scheme [42], but our techniques can be applied to other schemes following the Fiat-Shamir paradigm as well. However, unlike the case of standard signatures [1], a three-move identification scheme does not automatically give rise to a multi-signature scheme: compression of signatures requires a special homomorphism to exist on conversation transcripts. A number of three-move identification schemes in the literature turn out to have such a homomorphism though. In particular, one can obtain efficient multi-signature schemes based on discrete logarithms from [36], based on RSA from [19], based on factoring from [16, 15, 33, 37], and based on pairings from [23]. In the full version of this paper [7], we also adapt ideas from Katz and Wang [26] to construct a scheme with a tight security reduction to the decisional Diffie-Hellman problem, at the cost of a slight increase in signing and verification time.

Acknowledgments

Mihir Bellare was supported by NSF grant CNS-0524765 and a gift from Intel Corporation. Gregory Neven is a Postdoctoral Fellow of the Research Foundation - Flanders (FWO-Vlaanderen), and was supported in part by the Concerted Research Action (GOA) Ambiorics 2005/11 of the Flemish Government and in part by the European Commission through the IST Programme under Contract IST-2002-507932 ECRYPT.

7. REFERENCES

[1] M. Abdalla, J. H. An, M. Bellare, and C. Namprempre. From identification to signatures via the Fiat-Shamir transform: Minimizing assumptions for security and forward-security. *EUROCRYPT 2002*, LNCS 2332, Springer-Verlag.

[2] M. Abdalla and L. Reyzin. A new forward-secure digital signature scheme. *ASIACRYPT 2000*, LNCS 1976, Springer-Verlag.

[3] C. Adams, S. Farrell, T. Kause, and T. Monen. Internet X.509 Public Key Infrastructure Certificate Management Protocol (CMP). Internet Engineering Task Force RFC 4210, 2005.

[4] K. Barr and K. Asanovic. Energy aware lossless data compression. *MobiSys 2003*, ACM Press.

[5] M. Bellare and O. Goldreich. On defining proofs of knowledge. *CRYPTO 1992*, LNCS 740, Springer-Verlag.

[6] M. Bellare, C. Namprempre, and G. Neven. Unrestricted aggregate signatures. *Cryptology ePrint Archive*, Report 2006/285, 2006.

[7] M. Bellare and G. Neven. New multi-signatures and a general forking lemma. Full version of this paper, available from http://www.cs.ucsd.edu/users/mihir, 2006.

[8] M. Bellare and A. Palacio. GQ and Schnorr identification schemes: Proofs of security against impersonation under active and concurrent attacks. *CRYPTO 2002*, LNCS 2442, Springer-Verlag.

[9] M. Bellare, T. Ristenpart, and S. Yilek. Work in progress, 2006.

[10] M. Bellare and P. Rogaway. Random oracles are practical: A paradigm for designing efficient protocols. *ACM CCS 1993*, ACM Press.

[11] A. Boldyreva. Threshold signatures, multisignatures and blind signatures based on the gap-Diffie-Hellman-group signature scheme. *PKC 2003*, LNCS 2567, Springer-Verlag.

[12] D. Boneh, C. Gentry, B. Lynn, and H. Shacham. Aggregate and verifiably encrypted signatures from bilinear maps. *EUROCRYPT 2003*, LNCS 2656, Springer-Verlag.

[13] D. Boneh, H. Shacham, and B. Lynn. Short signatures from the Weil pairing. *ASIACRYPT 2001*, LNCS 2248, Springer-Verlag.

[14] A. De Santis and G. Persiano. Zero-knowledge proofs of knowledge without interaction. *FOCS 1992*, IEEE Computer Society Press.

[15] U. Feige, A. Fiat, and A. Shamir. Zero knowledge proofs of identity. *Journal of Cryptology*, 1(2):77–94, 1988.

[16] A. Fiat and A. Shamir. How to prove yourself: Practical solutions to identification and signature problems. *CRYPTO 1986*, LNCS 263, Springer-Verlag.

[17] M. Fischlin. Communication-efficient non-interactive proofs of knowledge with online extractors. *CRYPTO 2005*, LNCS 3621, Springer-Verlag.

[18] S. D. Galbraith, K. G. Paterson, and N. P. Smart. Pairings for cryptographers. *Cryptology ePrint Archive*, Report 2006/165, 2006.

[19] L. C. Guillou and J.-J. Quisquater. A "paradoxical" indentity-based signature scheme resulting from zero-knowledge. *CRYPTO 1988*, LNCS 403, Springer-Verlag.

[20] L. Harn. Group-oriented (t, n) threshold digital signature scheme and digital multisignature. *IEE Proceedings – Computers and Digital Techniques*, 141(5):307–313, 1994.

[21] R. Hayashi, T. Okamoto, and K. Tanaka. An RSA family of trap-door permutations with a common domain and its applications. *PKC 2004*, LNCS 2947, Springer-Verlag.

[22] J. Herranz and G. Sáez. Forking lemmas for ring signature schemes. *INDOCRYPT 2003*, LNCS 2947, Springer-Verlag.

[23] F. Hess. Efficient identity based signature schemes based on pairings. *SAC 2002*, LNCS 2595, Springer-Verlag.

[24] P. Horster, M. Michels, and H. Petersen. Meta-multisignatures schemes based on the discrete logarithm problem. *IFIP/SEC 1995*. Chapman & Hall.

[25] K. Itakura and K. Nakamura. A public-key cryptosystem suitable for digital multisignatures. *NEC Research & Development*, 71:1–8, 1983.

[26] J. Katz and N. Wáng. Efficiency improvements for signature schemes with tight security reductions. *ACM CCS 2003*, ACM Press.

[27] S. K. Langford. Weakness in some threshold cryptosystems. *CRYPTO 1996*, LNCS 1109, Springer-Verlag.

[28] C.-M. Li, T. Hwang, and N.-Y. Lee. Threshold-multisignature schemes where suspected forgery implies traceability of adversarial shareholders. *EUROCRYPT 1994*, LNCS 950, Springer-Verlag.

[29] S. Lu, R. Ostrovsky, A. Sahai, H. Shacham, and B. Waters. Sequential aggregate signatures and multisignatures without random oracles. *EUROCRYPT 2006*, LNCS 4004, Springer-Verlag.

[30] A. Lysyanskaya, S. Micali, L. Reyzin, and H. Shacham. Sequential aggregate signatures from trapdoor permutations. *EUROCRYPT 2004*, LNCS 3027, Springer-Verlag.

[31] S. Micali, K. Ohta, and L. Reyzin. Accountable-subgroup multisignatures. *ACM CCS 2001*, ACM Press.

[32] M. Michels and P. Horster. On the risk of disruption in several multiparty signature schemes. *ASIACRYPT 1996*, LNCS 1163, Springer-Verlag.

[33] K. Ohta and T. Okamoto. A modification of the Fiat-Shamir scheme. *CRYPTO 1988*, LNCS 403, Springer-Verlag.

[34] K. Ohta and T. Okamoto. A digital multisignature scheme based on the Fiat-Shamir scheme. *ASIACRYPT 1991*, LNCS 739, Springer-Verlag.

[35] K. Ohta and T. Okamoto. Multi-signature schemes secure against active insider attacks. *IEICE Transactions on Fundamentals of Electronics Communications and Computer Sciences*, E82-A(1):21–31, 1999.

[36] T. Okamoto. Provably secure and practical identification schemes and corresponding signature schemes. *CRYPTO 1992*, LNCS 1751, Springer-Verlag.

[37] H. Ong and C.-P. Schnorr. Fast signature generation with a Fiat Shamir–like scheme. *EUROCRYPT 1990*, LNCS 473, Springer-Verlag.

[38] PKCS #10: Certification request syntax standard. RSA Data Security, Inc., 2000.

[39] D. Pointcheval, E. Brickell, S. Vaudenay, and M. Yung. Design validations for discrete logarithm based signature schemes. *PKC 2000*, LNCS 1751, Springer-Verlag.

[40] D. Pointcheval and J. Stern. Security arguments for digital signatures and blind signatures. *Journal of Cryptology*, 13(3):361–396, 2000.

[41] J. Schaad. *Internet X.509 Public Key Infrastructure Certificate Request Message Format*, Internet Engineering Task Force RFC 4211, 2005.

[42] C.-P. Schnorr. Efficient signature generation by smart cards. *Journal of Cryptology*, 4(3):161–174, 1991.

Deniable Authentication and Key Exchange*

Mario Di Raimondo
Dipartimento di Matematica ed Informatica
Università di Catania, Italy
diraimondo@dmi.unict.it

Rosario Gennaro, Hugo Krawczyk
IBM T.J.Watson Research Center, USA
rosario@watson.ibm.com,
hugo@ee.technion.ac.il

ABSTRACT

We extend the definitional work of Dwork, Naor and Sahai from deniable authentication to deniable key-exchange protocols. We then use these definitions to prove the deniability features of SKEME and SIGMA, two natural and efficient protocols which serve as basis for the Internet Key Exchange (IKE) protocol.

SKEME is an encryption-based protocol for which we prove full deniability based on the plaintext awareness of the underlying encryption scheme. Interestingly SKEME's deniability is possibly the first "natural" application which essentially requires plaintext awareness (until now this notion has been mainly used as a tool for proving chosen-ciphertext security).

SIGMA, on the other hand, uses non-repudiable signatures for authentication and hence cannot be proven to be fully deniable. Yet we are able to prove a weaker, but meaningful, "partial deniability" property: a party may not be able to deny that it was "alive" at some point in time but can fully deny the contents of its communications and the identity of its interlocutors.

We remark that the deniability of SKEME and SIGMA holds in a *concurrent* setting and does not essentially rely on the random oracle model.

Categories and Subject Descriptors

C.2.2 [**Computer-Communication Networks**]: Network Protocols – *Applications*; D.4.6 [**Operating Systems**]: Security and Protection – *Authentication*

General Terms

Security, Theory

Keywords

Authentication, deniability, key exchange

---
*A full version of this paper appears on the IACR Eprint Archive at http://eprint.iacr.org/2006/280

1. INTRODUCTION

Privacy of communications has been the main object of study in cryptography for centuries. Its classical goal: prevent unauthorized parties from accessing secret or confidential information. In this setting the focus is on establishing authenticated and secret communication (a.k.a. "secure channels") with authorized peers. The intent is to defend the communications from an "unauthorized third party" in the form of an eavesdropper or active attacker; there is no attempt at preventing an authorized peer from disclosing information it receives legitimately. Today, with the transfer of our personal, social, economic and political lives to digital form, privacy has become a much wider and central notion. Modern cryptography recognized these issues early on through anonymity-related notions [13, 14], mix networks [12], undeniable signatures [15], private information retrieval [17], and more. More recently, we have seen a huge increase in the treatment of broader privacy issues in the crypto/security community (e.g. [36, 24, 16]). This paper focuses on an essential aspect of privacy: the deniability of every-day digital communications.

Deniable communication has always been a central concern in personal and business communications, with off-the-record communication serving as an essential social and political tool. Given that much of these interactions now happen over digital media (email, instant messaging, web transactions, virtual private networks) it is of central importance to provide these communications with "off-the-record" or deniability capabilities: the author or sender of a message should be able to deny, e.g., in court, that he or she sent that message. (Needless to say, there are special applications where non-repudiation is essential, but this is not the case for most of our communications.) One of the challenges of deniability in the digital world is that deniability is at odds, at least without careful design, with remote authentication. That is, Alice needs to be able to get a digital proof that she is talking to Bob but that proof should not leave any trace that will convince a third party that Bob talked (or said something specific) to Alice. This should be the case even if Alice herself is trying to prove the existence of the conversation to such third party! Thus, while in the traditional "secure channels" setting one is not defending against misbehavior by the (authorized) peer to the communication, in the deniability setting the potential attackers include the authorized (identified and authenticated) peer.

The first to formally treat the deniable authentication problem were Dwork, Naor and Sahai in [25], followed by a series of papers including [39, 32, 21]. On the more applied front, a practical (and widespread) protocol that set "plausible deniability" as a desirable property (though, not as an essential goal) is the IKE protocol [29]; in particular, this motivated the design of the SKEME protocol [34] that became part of IKEv1. More recently, [7, 22] treat explicitly off-the-record communications in the setting of instant-messaging communications.

Our present paper is motivated by the above works. One missing link in these works is the formal treatment of deniability for key-exchange (KE) protocols. Note that when using symmetric shared keys to authenticate/encrypt information then deniability is easy to achieve at least *as long as the secret key cannot be tied, via third-party verifiable proofs, to the identities of the peers*. However, when the symmetric keys are established via a KE protocol, which in turn uses public key techniques for authentication (as it is common in today's communications), then the weak link for deniability becomes this KE protocol. If its authentication mechanisms leave a "proof of communication" then deniability is lost.

The following example is useful to illustrate how a KE protocol can indeed leave such a "proof of communication". Here is a simple 3-message variant of an ISO Diffie-Hellman KE protocol [30, 9]:
(1) $A \to B: g^x$
(2) $B \to A: g^y, sig_B(g^x, g^y, id_A)$
(3) $A \to B: sig_A(g^y, g^x, id_B)$

The protocol makes use of digital signatures as the most natural (and scalable) tool for remote authentication. In addition, as part of the essential features of a secure KE protocol the identities of the communicating parties (i.e., id_A and id_B) are tied to the exchanged key via these signatures. However, by signing the peer's identity each of the parties of the protocol leaves an undeniable proof of communication between these parties. Note that in this case the non-deniability of the protocol is particularly serious: not only can A prove that B talked to her, but even an eavesdropper that obtains these signatures will be able to provide such a proof. Unfortunately, if one omits these identities in the signatures the protocol becomes completely insecure [26].

This example serves to stress the conflict between authentication and deniability; and, in the case of KE, the conflict between deniability and the binding between identities and the exchanged key so central to KE security. One of our central contributions is in showing that carefully designed protocols may provide for sound KE security as well as for significant levels of deniability. Fortunately, we show this to be the case for some well-known and practical KE protocols.

DEFINING DENIABILITY. The notion of *deniability* in public key authentication is formalized by Dwork, Naor and Sahai [25] using the *simulation* paradigm. Informally, a protocol is deniable if the view of any receiver (or verifier) can be simulated by a machine that does not know the secret key of the sender (or prover). The idea behind this natural and appealing definition is that the transcript of the protocol owned by the receiver, cannot be used to trace this conversation back to a specific sender, since the receiver could have produced it via the simulator machine.

Thus, the basic property of a deniable protocol follows the notion of *zero-knowledge* [27]. However, for deniability one needs a stronger simulator than in the case of ZK: while in ZK the simulator is basically a "mental experiment", for deniability, as pointed out by Pass [39], the simulator must be a real machine that works in the real world (ruling out some typical ZK simulators in the common reference string (CRS) and random oracle models). Another challenge in dealing with ZK-based proofs is that those usually make use of "rewinding". As pointed out in [25], this technique significantly limits the applicability of the proof in a real-life concurrent-executions setting.

DENIABILITY FOR KEY EXCHANGE PROTOCOLS. In a KE protocol, two parties engage in a protocol whose result is a (session) key K which only the two of them know, and they are assured to be sharing K with each other. They will use K to encrypt and authenticate messages in the session, using a symmetric-key authentication mechanism that is deniable *provided that the key cannot be traced to either party*.

Thus, the most important component for the deniability of electronic communications is the deniability of KE protocols. If the parties can deny having exchanged a key with the other party, then the rest of the communication can also be denied.

OUR CONTRIBUTIONS. After recalling the definition of deniable authentication from [25] we propose a definition of deniable key exchange protocols, which still adheres to the simulation paradigm. The extension is not trivial as we are moving to a protocol which outputs a secret value (rather than the simple accept/reject bit of an authentication protocol). In particular KE deniability requires the simulation not only of the entire transcript, but of the output key as well, since the key will be passed to an arbitrary security protocol after the exchange phase is completed.

We then study the deniability of SKEME [34] and SIGMA [35], two practical KE protocols which form the cryptographic core of IKE, the Internet Key Exchange of IPSEC [29, 33].

SKEME uses encryption to authenticate the parties. For SKEME we show the strong form of deniability guaranteed by our definition. The analysis has several interesting features. We first abstract out a basic message authentication mechanism from the key exchange protocol. This authentication scheme was proven unforgeable in [1] if the encryption scheme used inside is CCA2-secure [23]. Our first result is to show that CCA2-security is *not* sufficient for deniability by showing that there is a CCA2-secure encryption scheme for which the above protocol is not deniable. We then show that deniability holds if the encryption scheme is *plaintext-aware* (either in the standard model [4, 20] or in the random oracle model [6, 2]). Interestingly this makes deniability one of the first applications to essentially require plaintext awareness; so far plaintext awareness was mainly used as a tool to prove CCA2 security.

The case of SIGMA is more involved. Since SIGMA uses signature-based authentication, deniability (defined as full simulation) cannot be achieved. However, in spite of its use of signatures (and in contrast to the above ISO example) SIGMA offers some valuable deniability properties. We capture these properties via a modified notion of *partial deniability* for key-exchange protocols, in which a party can deny the identity of the parties he or she exchanges keys with, as well as the content of the subsequent communications protected by those keys. We show the 4-message variant of SIGMA (known as SIGMA-R) to be partially deniable, under a "key awareness" notion, which can be plausibly assumed to hold for "natural" MAC and hashing functions (and formally shown to hold in the random oracle model).

CONCURRENCY An important feature is that our proof of deniability for SKEME and SIGMA holds in a *concurrent* setting, where the adversary can open and schedule sessions in an arbitrary way [25]. This is of utmost importance for practical applications run in an open network like the Internet. In addition, our notions and proofs, while meaningful in the random oracle model, do *not* essentially depend on it.

RELATED WORK. As we said earlier deniable authentication has been studied from both the theoretical [25, 39, 32, 21] and the practical [34, 29, 8, 38, 7, 22] points of view. For the case of key exchange protocols, where both parties have registered public keys there are other approaches to deniable authentication, such as : *Designated Verifier Proofs* [31] and *Ring Signatures* [40]. Our approach is different in that our goal is to prove deniability for real-life protocols used in practice (which do not use the above tools).

Formalisms of KE security has been extensively studied [1, 41, 9, 11] yet none of those studies considered deniability explicitly and/or formally. Informal treatment of deniability issues for KE protocols can be found in [34, 7, 8, 38, 22].

2. DENIABLE KEY EXCHANGE

Our presentation and definition of the notions of deniability in this section follows essentially the approach and definition from Dwork, Naor and Sahai [25]. We first recall their definition of deniable authentication that we use as the basis for formalizing deniability of key exchange protocols. (Throughout the paper we will use the standard polynomial-time complexity notions, such as indistinguishability, negligible probabilities, etc.; in particular, all algorithms and machines are probabilistic polynomial-time.)

Deniable Authentication. We assume the author is familiar with the notion of a secure (unforgeable) message authentication protocol from [25]. Intuitively an authentication protocol is deniable if a (possibly dishonest) receiver cannot convince a third party (let's call it, the *judge*) that a given sender S authenticated a given message m. This notion was formalized in [25] using a zero-knowledge formalism. The idea is that an authentication protocol is deniable if the receiver's *view* of the protocol can be *simulated* by an efficient machine (called the *simulator*) that does *not* know the secret key of the sender S. In other words, the receiver interacting with the simulator obtains views with the same distribution than when interacting with the real sender S. Thus, when the receiver (the attacker in this setting) brings such a view to the judge, this view will not be a convincing evidence of the interaction with S since the same view could have been generated by the receiver alone by running the simulator.

Consider an adversary $\mathcal{M}$ (for "malicious") acting as a receiver on input pk. The adversary may also have some auxiliary input aux taken from a distribution AUX of such inputs. This auxiliary input models some extra information that the adversary might have gathered in some other form; for example, if $\mathcal{M}$ has been eavesdropping on correctly executed protocols between other parties and S, AUX will consist of legal transcripts of runs of the authentication protocol.

The adversary $\mathcal{M}$ starts an arbitrary number of executions of the authentication protocol with S with public key pk, choosing the input messages for these executions. These executions can be arbitrarily scheduled and interleaved by $\mathcal{M}$. The adversary's *view* of this interaction is then defined as the transcript of the full interaction between $\mathcal{M}$ and S, together with the internal coin tosses of $\mathcal{M}$. We denote this as $\text{View}_{\mathcal{M}}^{S}(\text{pk}, aux)$.

DEFINITION 1. *[25] We say that (AKG,S,R) is concurrently deniable with respect to the class AUX of auxiliary inputs if for any adversary $\mathcal{M}$, acting as the receiver on input pk and any auxiliary input $aux \in AUX$, there exists a simulator $SIM_{\mathcal{M}}^{S}$ that, running on the same inputs, produces a simulated view which is indistinguishable from the real one.*

We stress that the the simulator has *all* the same inputs as $\mathcal{M}$, including its random coins (alternatively assume that the simulator provides these coins to $\mathcal{M}$).

Remark (*Off-line vs. on-line judges.*) In the above definition, the real and simulated views are required to be indistinguishable: i.e. no efficient machine (a *distinguisher*) can tell them apart. In the deniability context (also in the case of KE protocols) the *distinguisher* represents the role of the *judge* which needs to decide if the transcript presented by $\mathcal{M}$ corresponds to a real execution of the protocol with S or to a simulated view of such run. This distinguisher is presented with the transcript as well as with the inputs of $\mathcal{M}$ including the auxiliary input aux (which are also the inputs on which SIM is run). Hence, this formulation corresponds to the situation in which the transcript is generated without the judge intervention, i.e., the judge is invoked *a posteriori* to decide if the message m was really authenticated by S or not. One could also contemplate a stronger setting in which the judge is allowed to interact with the adversary *before* the authentication protocol takes place or even *during* the run of the protocol between $\mathcal{M}$ and S. In the latter case, there is little one can do to provide deniability with respect to this "on-line" judge since he is a direct witness of the conversation between S and R. In the case of interaction between the judge and adversary $\mathcal{M}$ before the (alleged) run between $\mathcal{M}$ and S, but not during the run, the above definition per se is not sufficient to ensure deniability. Yet, some protocols will achieve some form of deniability also in this case. We will not formalize this stronger notion here but when presenting specific protocols we will discuss the extent to which they achieve this stronger notion.

Deniable Key Exchange. We now extend the above definition of deniability to the setting of (authenticated) key-exchange (KE) protocols. We first present a simplified definition of a KE protocol (for formal definitions of KE security see [5, 41, 9, 10]). Then we define the deniability property in a concurrent-execution setting.

In a key exchange protocol, two parties, say A and B, are associated with public keys pk_A and pk_B respectively, for which they each own the matching secret key sk_A and sk_B. We assume that public/secret keys are generated according to a key generation algorithm KG which is part of the specification of the KE protocol, and these are used in the authentication steps of the KE protocol. The protocol specifies the interaction between A and B (one acting as "initiator" and the other as "responder") and its result is either a (session) key K or "error". The basic security requirement in a KE protocol is that if A outputs a session key K and associates it to peer B then the only party that may possibly know K is B; and if B outputs the same session key then it associates it to peer A. Note however that this security guarantee is provided only for sessions (i.e., runs of the KE protocol) in which both peers are uncorrupted.

In great *contrast*, the deniability guarantee of a KE session is most relevant when one of the peers is dishonest. The goal is to prevent either A or B from proving to a judge that they exchanged a key with a specific party, and to prevent a proof of what the contents of a communication protected with that key were. Once again we model deniability via simulation along the lines of Definition 1 but with some important differences, arising mainly from the fact that not only is the KE protocol itself that needs deniability but also the communications that use the resultant session key.

Let Σ be a key-exchange protocol defined by a key generation algorithm KG and interactive machines Σ_I, Σ_R specifying the roles of the (honest) initiator (the party who sends the first message) and responder, respectively. Both Σ_I and Σ_R run on input their own secret and public keys and, typically, also on input the identity and public key of a specified peer. Each run of the KE protocol by a party is called a session. Upon completion, the protocol outputs either error (e.g., an authentication operation failed) or outputs a session key.

Consider an adversary $\mathcal{M}$ which runs on input an arbitrary number of public keys $\vec{pk} = (pk_1, \ldots, pk_\ell)$, randomly chosen according to KG and associated with the honest users in the network. $\mathcal{M}$ also has an auxiliary input aux as in Definition 1. The adversary initiates an arbitrary number of executions of Σ with the honest parties, some as an initiator, others as a responder. The executions are concurrent, i.e. scheduled and interleaved arbitrarily by the adversary. The view of $\mathcal{M}$ consists of its internal coin tosses, the transcript of the entire interaction *and the session keys computed in all the protocols in which $\mathcal{M}$ participated* (if the session does not complete, the session key is defined as an error value). We denote this view as $\text{View}_\mathcal{M}(\vec{pk}, aux)$.

The definition below follows the traditional approach of simulation of the adversary's view. However, it is important to highlight an element in this definition that differentiates it essentially from deniability in the message authentication setting of the previous subsection: the inclusion of the computed secret key to the adversary's view. Recall that the goal of deniability in a KE protocol is not only to prevent a (adversarial) party $\mathcal{M}$ from proving that another (honest) party B talked to $\mathcal{M}$ but also to prevent $\mathcal{M}$ from proving to a third party the contents of a communication in which B participated. Since KE protocols are typically run in order to agree on a session key which is later used to authenticate further communication, *it is essential that not only the communication during the KE session be simulatable but also the value of the session key should be part of the output of the simulation*. In this case, when an attacker brings to a "judge" evidence against B in the form of a key exchange or subsequent authenticated information, the simulatability of the exchange and the resultant key will make this evidence worthless.

DEFINITION 2. *We say that* (KG, Σ_I, Σ_R) *is a* concurrently deniable *key exchange protocol with respect to the class AUX of auxiliary inputs if for any adversary $\mathcal{M}$, for any input of public keys $\vec{pk} = (pk_1, \ldots, pk_\ell)$ and any auxiliary input $aux \in AUX$, there exists a simulator $SIM_\mathcal{M}$ that, running on the same inputs as $\mathcal{M}$, produces a simulated view which is indistinguishable from the real view of $\mathcal{M}$.*

Remarks on the definition. First note that impersonation is not the issue here: when $\mathcal{M}$ is interacting with B she is not trying to impersonate A but rather, she is trying to obtain a proof that she herself interacted with B and established a key with him. The goal of deniability is to prevent $\mathcal{M}$ from proving to a third party that this was the case. Thus, when $\mathcal{M}$ interacts with B we can assume (wlog) that she will do so using a public key $pk_\mathcal{M}$ (which may or may not be associated with $\mathcal{M}$'s identity). Indeed, since our definition guarantees that even when the attacker $\mathcal{M}$ runs the key generation algorithm to generate a public key (thus knowing the corresponding secret key) she cannot prove that B talked to her, then $\mathcal{M}$ can certainly not be able to prove that B talked to any *other* party A (in particular, this implies deniability with respect to eavesdroppers). In addition, while $\mathcal{M}$ may decide to reveal her secret key $sk_\mathcal{M}$ to help in proving that B talked to her, she does not have to do so (actually $\mathcal{M}$ may use a public key for which she does not know a corresponding private key!).

3. DENIABILITY OF SKEME

In this section we prove the deniability of the key exchange protocol SKEME [34] which uses public key encryption as a means of authentication. First we abstract out the authentication protocol which is at the core of SKEME and prove that it is deniable if the encryption scheme used is plaintext-aware according to the definition in [4]. Then we use this result to prove the deniability of the SKEME key exchange protocol[1]. We also show that chosen-ciphertext security is not sufficient to prove deniability. In the following we denote an encryption scheme $\mathcal{E} = (\text{gen}, \text{enc}, \text{dec})$ which are the key generation, encryption and decryption algorithms, respectively. We assume the reader is familiar with the notions of security against adaptive chosen-ciphertext attack [23] and plaintext-awareness [4], though we informally recall them below.

3.1 Encryption-based Deniable Authentication

We first study the authentication protocol based on public-key encryption which is at the core of SKEME [34] (related protocols are studied in [23, 25]).

Let pk be the sender's S public key and sk the related secret key, for a public key encryption scheme $\mathcal{E}$. The authentication protocol $\lambda_\mathcal{E}$ on input a message m works as follows:

1. The receiver R chooses a random key k and encrypts it for S as $c = \text{enc}(pk, k)$. R sends c to S.

2. S decrypts $k' = \text{dec}(sk, c)$ (if the decryption fails, the sender chooses a random key k'). S computes $t = \text{MAC}_{k'}(m)$ and sends t to R.

3. R accepts if $t = \text{MAC}_k(m)$.

The receiver's belief that S is really authenticating m comes from the fact that she is the only one able to decrypt k. Indeed this authentication protocol is proven secure in [1] if $\mathcal{E}$ is secure against adaptive chosen-ciphertext attack.

The protocol $\lambda_\mathcal{E}$ is perfectly deniable against an honest receiver, since the simulator SIM_R can easily produce valid

---
[1]Interestingly, this appears to be the first "essential" application of plaintext-awareness; before, this notion was used mainly as a tool to prove chosen-ciphertext security.

transcripts on his own. What happens against a dishonest receiver $\mathcal{M}$? To formally prove the deniability we need to create a simulator $SIM_\mathcal{M}$ that is able to produce valid transcripts interacting with the malicious receiver $\mathcal{M}$. When the simulator receives the encrypted challenge $c = \mathsf{enc}(\mathsf{pk}, k)$, how are we going to simulate the reply $t = \mathsf{MAC}_k(m)$? The simulator doesn't know the sender's secret key sk!

The scheme can be made deniable by adding a challenge-response sub-protocol (in a way similar to the authentication protocol in Dwork et al. [25]). Basically, instead of replying with the MAC t, the sender replies with a commitment to t. Upon receiving such commitment, the receiver reveals the key k encrypted in the first message. If this key equals the one that he decrypted, the sender opens the commitment, revealing the MAC t.

This variant is still unforgeable assuming that $\mathcal{E}$ is secure against adaptive chosen ciphertext attack. Deniability follows from a standard black-box ZK simulation, which however includes a "rewinding" step for the malicious receiver. This rewinding step limits the applicability of the protocol in a concurrent setting: in this case the proof of security guarantees deniability only if a logarithmic (in the security parameter) number of sessions are concurrently executed [25].

We are interested in the deniability of the original protocol $\lambda_\mathcal{E}$, which is the authentication core of SKEME. Our first result is to show security against adaptive chosen ciphertext attack is *not* sufficient to make $\lambda_\mathcal{E}$ is not deniable.

THEOREM 3. *There exists a encryption scheme $\mathcal{E}$, which is secure against adaptive chosen ciphertext attack and such that $\lambda_\mathcal{E}$ is not deniable.*

In contrast we show next that *concurrent deniability* for $\lambda_\mathcal{E}$ can be proven by making a stronger assumption on the underlying encryption scheme, namely *(PA-2) plaintext awareness* [4].

Intuitively, we say that $\mathcal{E}$ is *PA-1 plaintext-aware* if for any adversary **C** that on input a public key pk outputs a valid ciphertext c there is a "companion" machine **C*** that outputs the matching plaintext. Think of **C*** as the *alter ego* of **C**: the definition basically implies that if **C** produces a valid ciphertext it must "know" the corresponding plaintext. A strengthening of this notion, PA-2, accommodates the fact that an attacker may have access to a set of ciphertexts computed under public key pk but not produced by the attacker itself; for example, these ciphertexts could have been generated by other, honest, parties in the system communicating with the owner of this public key. Hence, the definition of PA-1 is strengthened so that the above adversary **C** is also given as input a list of valid ciphertexts for which it does not know the corresponding plaintexts, (thus modeling the fact that the attacker may copy ciphertexts from the network and replay them). In a PA-2 scheme, and the companion machine **C*** is defined to yield the corresponding plaintext only for valid ciphertexts produced by **C** which are not in the above list. The resultant notion is called *PA-2 plaintext awareness*.

When proving the deniability property of protocol $\lambda_\mathcal{E}$, the PA-1 notion is therefore too weak to represent the common situation in which multiple copies of the protocol are run concurrently since in this case the attacker does have access to valid ciphertexts created by other parties. In particular, in the case of protocol $\lambda_\mathcal{E}$ the attacker may "replay" such ciphertexts in a communication with the owner of pk without knowing the encrypted plaintext. Our result, therefore, depends on the encryption function being PA-2. In this case we will use the auxiliary input AUX to represent a list of valid transcripts of protocol $\lambda_\mathcal{E}$ (with pk as the sender's public key) gathered by the attacker in the network.

We denote by $TR(\mathsf{pk})$ the set of legal transcripts of the protocol $\lambda_\mathcal{E}$ under public key pk (i.e., those transcripts that contain legal ciphertexts under pk) and let the auxiliary input aux, with respect to which the deniability of $\lambda_\mathcal{E}$ is established, be a list of transcripts sampled from $TR(\mathsf{pk})$ and for which the adversary $\mathcal{M}$ has no information on the plaintexts (and randomness) used to generate the ciphertexts included in these transcripts. In other words, we assume these to be transcripts generated by *honest parties* interacting with S.

THEOREM 4. *If the encryption scheme $\mathcal{E}$ is PA-2 plaintext-aware and semantically secure, then the protocol $\lambda_\mathcal{E}$ is concurrently deniable with respect to the class of auxiliary inputs $TR(\mathsf{pk})$.*

3.2 The SKEME Key Exchange Protocol

The SKEME key exchange from [34] is described in Figure 1. It consists of two parallel executions of the authentication protocol $\lambda_\mathcal{E}$ applied to the messages (g^x, g^y) and (g^y, g^x) where A and B act as the sender in one and the receiver in the other. Yet, deniability for SKEME does not immediately follow, from the simulatability of $\lambda_\mathcal{E}$. Recall that for a key exchange, we must simulate not only the transcript, but also the output key.

THEOREM 5. *If $\mathcal{E}$ is a PA-2 and IND-CPA secure encryption scheme, then SKEME is a concurrently deniable key exchange protocol.*

Notice that Theorem 5 holds for the regular case in which the judge (or distinguisher) is not present during the run of the KE protocol nor it provides inputs to the protocol, but rather is presented with a transcript *a posteriori*. The full version discusses how SKEME can be made deniable in the case in which the adversary cooperates with the judge before the protocol takes place.

3.3 On Plaintext-Aware Encryptions

Above we used the notion of plaintext-awareness in the standard model (i.e. without random oracle) from [4]. Dent in [20] shows that the Cramer-Shoup scheme [18] is PA-2 plaintext-aware under a non-black-box type of assumption known as "Knowledge of exponent assumption" (KEA1) [19, 28, 3], which we recall below.

Let G be a cyclic group of prime order q, and g, h both both generators of G. We assume that for every algorithm $\mathcal{M}$ that on input G, q, g, h outputs (y, z) where $y = g^x$ and $z = h^x$ for some $x \in Z_q$, there exists an algorithm $\mathcal{M}^*$ which outputs x. In other words, the only way for $\mathcal{M}$ to output a pair of elements y, z that have the same discrete log x with respect to two different basis g, h is to know x.

RANDOM ORACLE MODEL SCHEMES. Typical instantiations of SKEME, such as in IKE, use encryption schemes like OAEP which are plaintext-aware in the random oracle model. It is not hard to see that our proof of deniability will also hold for such schemes. Indeed the basic tool to simulate the authentication protocol $\lambda_\mathcal{E}$ is a "plaintext extractor" for correct ciphertexts, which is guaranteed to exist by the definition of plaintext-awareness in the random oracle model.

Note that the "programmability feature" of the random oracle is *not* used in the above argument (only the ability to "see" where the adversary queries the oracle.) Thus our simulator is a valid deniability simulator (a simulator that "programs" the random oracle, does not guarantee deniability [39]).

4. PARTIAL DENIABILITY OF SIGMA

The SIGMA key-exchange protocol was proposed in [35] and proven secure in [11]. It forms the cryptographic basis for the Internet Key Exchange (IKE) protocol (specifically, the signature-based mode in IKEv1 [29] and the public key mode in IKEv2 [33]). Its basic goal is to provide a secure Diffie-Hellman exchange authenticated using digital signatures.

While deniability was not an explicit design goal of SIGMA (in IKEv1 deniability was offered via the encryption-based mode based on SKEME), we show that the protocol provides some significant level of deniability even though it does not achieve the full deniability of SKEME.

The 4-round SIGMA protocol is presented in Figure 2: each party signs its own DH exponential as well as the peer's exponential and computes a MAC on its own identity and a value 0/1 depending on whether the party the initiator or responder; this serves to prevent reflection attacks[2]. We stress that if one adds a differentiator, such as this bit 0/1, to the signatures then the proof of deniability (for the responder) as presented below will not work. Indeed, the proof uses the fact that the signatures produced by the parties do not include evidence of whether the signature was produced in the role of initiator and responder.

Another aspect of SIGMA that will be of importance in our analysis in Section 4.2 relates to the way the MAC key K_m and session key K_s are derived from the DH value g^{xy}. This key derivation function (KDF) typically consists of three components: (i) computing the DH value g^{xy}, (ii) extracting a key K from g^{xy} via a hashing operation (implemented via SHA-1, a universal hashing scheme, etc.), (iii) using a PRF with key K to derive the two values K_m, K_s (e.g., the first is set to $PRF_K(0)$ and the latter to $PRF_K(1)$). We do not specify the way these components are implemented but in the analysis we will need to assume certain properties of these functions. Finally, we point out that in some variations of SIGMA, the third and fourth message are encrypted (to provide *identity protection*): the deniability of the protocol is preserved (the proof is just a straightforward adaptation of the proof of Theorem 7).

4.1 Partial Deniability

The challenge in creating a deniable key-exchange protocol that uses digital signatures is that the sole fact that a signature was generated, even if on random inputs, may provide significant information (e.g., that the signer was "alive" or active). Yet, even in this case the range of deniability may vary widely: from no deniability in the case of a protocol that signs the peer's identity (as the ISO protocol discussed in the introduction) to a protocol that only signs self-generated random information. Here we investigate the position of SIGMA in this "deniability range". We'll see that the provision of not signing the peer's identity (but rather MACing one's own identity), needed to achieve identity protection, also provides the basis for deniability.

Let's consider first the role of the initiator, or A. The core observation is that A will sign g^y irrespective of who generated it. For example, consider an attacker $\mathcal{M}$ that encodes in the exponent y (used to generate the value g^y sent to A) its own identity (e.g., choosing y to be a signature by this attacker). The fact that A will sign g^y says nothing about whether A talked or not to $\mathcal{M}$ (or even if she was willing to talk to $\mathcal{M}$) since A will sign g^y regardless of who sent it. In other words, A's transcripts are *peer-independent*. The case of the responder B is similar to the above and its transcript is also "peer-independent"[3]. This peer-independence property provides a limited, yet meaningful and significant form of deniability.

To formalize it we resort to a weaker notion of deniability for message authentication introduced by Dwork, Naor and Sahai [25]. This definition from [25] still follows the simulation paradigm but allows the simulator to interact with an oracle representing the real sender (or prover) on random messages rather than the input one. This captures the fact that whatever is learned from the authentication protocol is *independent* of the authenticated message. Or in other words, the attacker can prove that he talked to a specific party, but can't prove which message the party authenticated.

Roughly speaking, we will say that a key-exchange protocol is *partially deniable for the initiator* if the runs of the (honest) initiator A with a given responder B are indistinguishable from runs with any other responder B'. In the formal definition we provide the initiator's simulator SIM_I, simulating an initiator A and acting on input (B, pk_B), with one of two oracles: an oracle that acts as the real initiator A running with peer B' where $B' \neq B$, or an oracle that acts as A running as the responder with peer B'. Similarly, we define partial deniability for the responder. Then we will say that the protocol is *partially deniable* if it is deniable for the initiator and responder. This is formalized in Definition 6 below.

NOTATION. Let Σ be a key-exchange protocol with interactive machines Σ_I, Σ_R defining the roles of the (honest) initiator and responder, respectively. We use the symbol $\Sigma_I^C(D, pk_D)$ to denote the interactive machine implementing a honest party C as the initiator in a run of protocol Σ with peer D and peer's public key pk_D. Implicit in this notation is that Σ_I^C has as input the secret and public keys of C. The responder machine $\Sigma_R^C(D, pk_D)$ is defined analogously.

DEFINITION 6. *Let $\Sigma = (\Sigma_I, \Sigma_R)$ be a key-exchange protocol. We say that Σ_I is partially deniable with respect to an I-oracle (resp. R-oracle) if for any adversary $\mathcal{M}$ and any (honest) party C, the interaction between C as initiator with $\mathcal{M}$ as responder can be simulated (as in Definition 2) by a simulator SIM_I that is given oracle access to $\Sigma_I^C(D, pk_D)$ (resp. $\Sigma_R^C(D, pk_D)$) where pk_D is a public key chosen independently of $\mathcal{M}$'s public key $pk_{\mathcal{M}}$.*

---

[2] Alternatively, as in the case of IKE, initiator and responder can use different MAC keys (both derived from g^{xy}) in generating their MAC values. The deniability of the protocol is preserved with either technique.

[3] However, we will see later that the fact that B sends its signature on (g^y, g^x) *after* verifying the sender's identity may provide some extra information under special circumstances.

We say that Σ_I is partially deniable *if it is partially deniable with respect to an I-oracle or with respect to an R-oracle. The definition of Σ_R being partially deniable is similar. Finally, we say that $\Sigma = (\Sigma_I, \Sigma_R)$ is partially deniable if both Σ_I and Σ_R are partially deniable.*

CONCURRENCY. For ease of presentation the above definition is formulated in terms of a single "stand-alone" execution of the protocol. Adding full concurrency to the definition is straightforward (see Definition 2). More importantly, we note that the proof of partial deniability of SIGMA (Theorem 7) avoids rewinding, and thus holds in the concurrent model.

AUXILIARY INPUT. The treatment of auxiliary input, omitted in the above simplified definition, is identical to the case of Definition 2; in particular, the proof of Theorem 7 below holds with respect to any auxiliary input. What is important to stress is that the SIGMA protocol has the salient property of being (partially) deniable even if the judge provides DH values to the attacker to be used in the protocol and for which the attacker does not know the exponent (see the discussion on off-line/on-line judges in Section 2). Indeed, when $\mathcal{M}$ acts as an initiator then receiving g^x (but not x) from a third party does not allow $\mathcal{M}$ to produce the correct third protocol message and hence the execution is aborted without the responder generating the authentication information in message 4. In the case that $\mathcal{M}$ acts as responder then a third-party provided g^y makes no difference since the initiator authenticates g^y independently of the peer (and without "inspecting" the g^y value itself).

4.2 Deniability Analysis of SIGMA

We now give an informal overview of the proof that SIGMA (as depicted in Figure 2) is partially deniable as in Definition 6. The main difficulties in the simulation of the protocol are (i) the generation of the signatures on behalf of the simulated parties, and (ii) the computation by the simulator of the session key K_s and the MAC key K_m. Note that producing K_s is a necessary condition for satisfying the definition of deniability and partial deniability (Definitions 2 and 6) while computing (or learning) K_m is a necessary condition to simulate the generation and verification of the MAC values exchanged in the protocol.

For point (i) the simulator will use the real parties as oracles (as allowed by the definition of partial deniability) to produce the signatures. Specifically, the simulation of a given party C acting as initiator will use an oracle to the real party C acting as initiator. The simulation of C as responder will also use an oracle to the real party C acting as initiator (rather than as a responder – see discussion in Section 4.3).

For point (ii), since each of the exponentials g^x, g^y are chosen by either the simulator's oracles or by the adversary then the corresponding exponents x, y are not known to the simulator who thus cannot easily compute the values g^{xy}, K_m or K_s. We solve this problem as follows. If the attacker itself cannot compute its own MAC values then the simulator will be able to succeed without computing or verifying these MAC values (in this case, the simulated party would abort before having to compute the MAC values). However, if the attacker can compute the MAC then, intuitively, the simulator (which has non-black-box access to the attacker) can learn these values as well.

However, what we really need is for the simulator to learn not just the MAC values but the keys K_m and K_s. In most natural/practical implementations of the key derivation function of SIGMA one will have the property that for the adversary to compute the right MAC values, it has to compute the right MAC key and to do so it has to compute the PRF key from which also K_s is computed. This, however, is not a necessary condition that follows just from the regular definitions of MAC and PRF: one may be able to construct artificial functions where the attacker succeeds in computing its own correct MAC value without necessarily having to compute the key K_m (note that the key K_m does not have to be random as the attacker can influence it via the choice of the DH exponential). Moreover, one can envision a situation where the attacker can prove that it could have not possibly known how to compute the MAC produced by its peer to an exchange in which case the exchange (and possibly the following communication) may not be deniable. Therefore, our proof of SIGMA will apply to KDF construction where the above artificial situation does not arise. We formalize this via the following "key awareness" assumption.

The Key-Awareness Assumption for SIGMA's KDF. *We say that a KDF procedure for SIGMA has the Key-Awareness property if for all deniability attackers $\mathcal{M}$ against the protocol the following holds. If on exchange of DH values X, Y an attacker $\mathcal{M}$ computes a correct value $\mathsf{MAC}_{K_m}(t)$, for some input t, where K_m is the MAC key derived by the KDF from X, Y, then there is an "extractor machine" $\mathcal{M}'$ that on the same inputs of $\mathcal{M}$ outputs the keys K_m and K_s.*[4]

In other words, there are no "shortcuts" available for the attacker in computing the MAC without going through the key derivation steps and explicitly computing the MAC key K_m as well as the companion key K_s. We note that the above very informal definition of Key-Awareness can be formalized in ways similar to other non-black box "extraction assumptions" such as the "knowledge of exponent" and plaintext awareness assumptions discussed in Section 3.

As we said earlier we expect most natural implementations of key derivation in SIGMA to have the above property, in particular when the hashing, PRF and MAC are implemented using HMAC or CBC-MAC using strong hash and block-cipher functions as in the implementation of SIGMA in IKE. Of course, we cannot prove this, but we have to explicitly assume it.

Next we show that this assumption holds when the hash function H used to derive the PRF key $K = H(g^{xy})$ is modelled as a random oracle; in this case, the whole key derivation has the key awareness property. Namely, we show that the simulator can extract the correct keys K_m and K_s from the run of the attacker, provided the latter computes correct MAC values. (Interestingly, while intuitive the following argument contains some unexpected subtleties.)

Due to the randomness of H, if the attacker $\mathcal{M}$ does not compute g^{xy} then the key $K = H(g^{xy})$ (and thus K_m) is indistinguishable from random for $\mathcal{M}$. In this case, the attacker cannot possibly compute the correct value of a MAC under K_m. Thus, the simulator (which has access to the or-

---

[4]In the case in which the protocol uses additional keys, such as directional MAC keys, encryption keys, etc, then we assume that the extractor returns all these keys; alternatively, if all these keys are derived from a single PRF key K then it suffices that extractor returns this key K.

acle H) can check the inputs to H provided by $\mathcal{M}$ and hence learn K, and with it both K_m and K_s. There is however a problem in this argument. The simulator will not be able to compute, in general, the value g^{xy} by itself; so how will it know which of the inputs to H was the real value g^{xy}? (In particular, $\mathcal{M}$ may use a key K, or K_m, derived from an output of H on a point different than g^{xy}). The solution to this problem uses the fact that the simulator SIM (shown in the proof of Theorem 7) will have an example of a *correct* value of a MAC, computed under the *correct* key K_m, produced by a real player! Thus, in this case, SIM proceeds as follows. It considers each output of H as a candidate PRF key K^* and derives from it two candidate keys K_s^* and K_m^*. Now, SIM, uses the candidate key K_m^* to verify the MAC value received from the real player. If the verification succeeds then SIM assumes K_m^* to be the correct MAC key K_m (and K_s^* to be the correct session key). It is easy to see that a wrong candidate key K_m^* will succeed in verifying the real MAC with negligible probability. Indeed, since the distribution of wrong K_m^* keys is uniform (and independent from K_m) then the latter probability is at most as the probability to forge a MAC and hence negligible. (Clearly, if two independent random keys have a high probability of producing, or verifying, the same MAC value then one can forge MAC values computed under a secret random key by simply re-computing the MAC value under another random and independent key.)

Another subtlety in the above argument is what is meant by the "right value" of g^{xy}. What happens if the attacker chooses, say, a value Y (instead of g^y) not in the subgroup generated by g (as in the Lim-Lee attacks [37])? In this case the value g^{xy} is not even well-defined. However, since in this case we will have that the value $X = g^x$ was chosen by a real (and honest) player then the value Y^x is well defined and it is this value that we actually refer to as the "right g^{xy}" (conversely, when X is the value chosen by the attacker and $Y = g^y$ is chosen by the honest player then the "real value of g^{xy}" refers to X^y).

NOTE. We note that the above simulation requires no rewinding and hence it does not break the concurrency of the deniability property. Also worth noting is that as pointed out by Pass [39] one has to be careful when arguing deniability using the random oracle model. Specifically, one cannot use in such an argument the so called "programmability feature" of random oracles. We note that this property is *not* used in the above argument. (We only use the ability to "see" where the adversary queries the oracle.)

Summarizing, we can prove the following Theorem (details of the proof are in the full version):

THEOREM 7. *The SIGMA protocol from Figure 2 with a key-aware key derivation is partially deniable according to Definition 6. (In particular, this is the case if one models the hashing of g^{xy} in SIGMA as a random oracle.)*

4.3 Discussion on Partial Deniability

It is important to understand the "real-life" semantics of our notion of partial deniability. The idea behind our definition is that the transcript available to the adversary $\mathcal{M}$ could have been produced by a party C when interacting with any other party D (this is what we refer to as the "peer independence" property). Thus C can deny having been involved with $\mathcal{M}$. A point to notice is the role of C during this "claimed" interaction with D. The simpler case is when C acts as initiator: in this case the information generated by C is totally independent of the peer's identity and hence interaction with any specific peer can be denied by the initiator. In the case that C runs as responder, the "peer independence" property can be formally proven provided that C also runs as initiator in *other* instances of the protocol. In other words, the authentication information created by C as responder is simulatable from the information created by C as initiator. How common is it in real-life protocols that parties run as both initiators and responders? Clearly, this depends on the application. For example, in applications of SIGMA to Instant Messaging (as in [7, 22]) the common case is that end-points to the IM service act as both responders and initiators. In contrast, in a typical client-server configuration of IPSec, parties will run as either initiators or responders but not both (fortunately, in these cases clients usually run as initiators so they are fully protected by deniability). However, as we see next, SIGMA provides some level of deniability also in cases where parties act exclusively as responders and not as initiators.

DENIABILITY FOR PARTIES WHO ONLY ACT AS RESPONDERS. For a party, B, that only acts as a responder (never as an initiator), SIGMA still provides some form of deniability whose significance may depend on the application. Indeed, note that there is nothing explicit in the authentication information sent by the responder that ties it to a specific peer. The problem, however, is that B will send this information only after verifying the identity of the peer. Let's consider an example. Say, Charlie, acting as initiator, sends Bob g^x where x encodes Charlie's signature on some value. Bob signs this g^x and later Charlie brings to court Bob's signature and the value x showing that this value was generated by him (Charlie). In itself this proves nothing about Bob having *knowingly* communicated with Charlie: g^x could have been sent by David, a party with whom Bob was willing to talk. In other words, David could have been collaborating with Charlie in "framing" Bob (or maybe Charlie just broke into David's computer). In this sense, Bob enjoys deniability even though it acts as responder-only in SIGMA. On the other hand, one can imagine cases where additional "circumstantial evidence" may make it harder for Bob to deny interacting with Charlie, especially since Bob needs to convince that Charlie was using someone with whom Bob was willing to communicate for mounting the attack.

The above considerations apply also to the case of an implementation of SIGMA that adds information under the signatures that make the signatures produced by C as initiator distinguishable from those generated by C as responder (the addition of such information is not needed for the security of SIGMA and is certainly not recommended in any setting where deniability is significant).

DENIABILITY OF SIGMA-I. We remark that the 3-message variant of SIGMA, called SIGMA-I [35], is partially deniable, according to Definition 6, *only for the responder*. In this case it is the responder's behavior which is peer-independent since his signature is produced before seeing the identity of the initiator. The initiator, Alice, in SIGMA-I is in a position similar to the responder in the 4-rounds version of SIGMA (Fig. 2, also referred SIGMA-R) since she signs after seeing the identity of the other peer. Unlike the SIGMA-R case, however, the initiator Alice may not be able to claim that her signature on (g^x, g^y) was performed in her role of responder. Indeed, a malicious responder could choose his DH value dependent on Alice's DH value, and this is never the case when Alice acts as responder. In other words, the signatures of initiators and responders can be made distinguishable by a dishonest peer in SIGMA-I, something that is not possible in SIGMA-R. Thus SIGMA-R may be somewhat preferable in the deniability setting.

FINAL NOTE. The deniability limitations of SIGMA follow from the use of signatures (essential to the protocol) and the fact that these signatures are applied to a peer-provided value. The latter issue can be avoided if one replaces the peer's DH value under the signature with a freshness value not chosen by the peer, such as a non-repeating counter or a timestamp; however, these values are seldom available, or secure enough, in practical settings. As said earlier, a more essential limitation of partial deniability is that having a signature of a party, even on random information, is sufficient proof to show that the party was "alive". Moreover, it is easy to see that in the case of SIGMA an attacker can encode into its DH value, information that will allow to prove not only that a party was alive but that it was alive after certain time or event (for example, the attacker can encode into the DH value the hashing of today's New York Times). Leaving a proof of such information seems unavoidable in any protocol that needs to include some "freshness guarantee" inside a signature to prevent its replay.

5. REFERENCES

[1] M. Bellare, R. Canetti and H. Krawczyk. *A Modular Approach to the Design and Analysis of Authentication and Key Exchange Protocols.* STOC '98, 419–428, ACM 1998.

[2] M. Bellare, A. Desai, D. Pointcheval and P. Rogaway. *Relations among Notions of Security for Public-Key Encryption Schemes.* CRYPTO '98, LNCS 1462, 26–45, Springer 1998.

[3] M. Bellare and A. Palacio. *The Knowledge of Exponent Assumptions and 3-Round Zero-Knowledge Protocols.* CRYPTO '04, LNCS 3152, 273–289, Springer 2004.

[4] M. Bellare and A. Palacio. *Towards Plaintext-Aware Public-Key Encryption without Random Oracles.* ASIACRYPT '04, LNCS 3329, 48–62, Springer 2004.

[5] M. Bellare and P. Rogaway. *Entity authentication and key distribution.* CRYPTO '93, LNCS 773, 232–249, Springer 1994.

[6] M. Bellare and P. Rogaway. *Optimal Asymmetric Encryption.* EUROCRYPT '94, LNCS 950, 92–111, Springer 1994.

[7] N. Borisov, I. Goldberg and E. Brewer. *Off-the-Record Communication, or, Why Not To Use PGP.* ACM WPES'04, 77–84, ACM 2004.

[8] C. Boyd, W. Mao and K. Paterson. *Key Agreement using Statically Keyed Authenticators.* ACNS 2004, LNCS 3089, 248–262, Springer 2004.

[9] R. Canetti and H. Krawczyk. *Analysis of Key-Exchange Protocols and Their Use for Building Secure Channels.* EUROCRYPT '01, LNCS 2045, 453–474, Springer 2001.

[10] R. Canetti and H. Krawczyk. *Universally Composable Notions of Key Exchange and Secure Channels.* EUROCRYPT '02, LNCS 2332, 337–351, Springer 2002. Full version at eprint.iacr.org/2002/059.

[11] R. Canetti and H. Krawczyk. *Security Analysis of IKE's Signature-based Key-Exchange Protocol.* CRYPTO '02, LNCS 2442, 143–161, Springer 2002.

[12] D. Chaum. *Untraceable Electronic Mail, Return Addresses, and Digital Pseudonyms.* Communications of the ACM, 24(2), February 1981.

[13] D. Chaum. *Blind Signatures for Untraceable Payments.* CRYPTO '82, 199–203, Plenum 1982.

[14] D. Chaum. *Security Without Identification: Transaction Systems to Make Big Brother Obsolete.* Communications of the ACM, 28(10):1030–1044, October 1985.

[15] D. Chaum and H. van Antwerpen. *Undeniable Signatures.* CRYPTO '89, LNCS 435, 212–226, Springer 1990.

[16] S. Chawla, C. Dwork, F. McSherry, A. Smith and H. Wee. *Toward Privacy in Public Databases.* TCC'05, LNCS 3378, 363–385, Springer 2005.

[17] B. Chor, O. Goldreich, E. Kushilevitz and M. Sudan. *Private Information Retrieval.* FOCS'95, 41–50, IEEE 1995.

[18] R. Cramer and V. Shoup. *A Practical Public-Key Cryptosystem Secure Against Adaptive Chosen Ciphertexts Attacks.* CRYPTO '98, LNCS 1462, 13–25, Springer 1998.

[19] I. Damgard. *Towards Practical Public Key Systems Secure Against Chosen Ciphertext Attacks.* CRYPTO '91, LNCS 576, 445–456, Springer 1992.

[20] A. Dent. *Cramer-Shoup is Plaintext-Aware in the Standard Model.* EUROCRYPT '06, LNCS 4004, 289–307, Springer 2006.

[21] M. Di Raimondo and R. Gennaro. *New Approaches for Deniable Authentication.* ACM CCS '05, 112–121, ACM 2005.

[22] M. Di Raimondo, R. Gennaro and H. Krawczyk. *Secure Off-the-Record Messaging.* ACM WPES'05, 81–89, ACM Press, 2005.

[23] D. Dolev, C. Dwork and M. Naor. *Non-Malleable Cryptography*, SIAM J. Comp., 30(2):391–437, April 2000.

[24] C. Dwork and K. Nissim. *Privacy-Preserving Datamining on Vertically Partitioned Databases.* CRYPTO '04, LNCS 3152, 528–544, Springer 2004.

[25] C. Dwork, M. Naor and A. Sahai. *Concurrent Zero-Knowledge.* J. ACM 51(6): 851-898 (2004).

[26] W. Diffie, P. Van Oorschot and M. Wiener. *Authentication and Authenticate Key Exchange.* Designs, Codes and Cryptography, no. 2, 107–125, 1992.

[27] S. Goldwasser, S. Micali, and C. Rackoff. *The Knowledge Complexity of Interactive Proof-systems.* SIAM J. Comp., 18(1):186–208, February 1989.

[28] S. Hada and T. Tanaka. *On the Existence of 3-Round Zero-Knowledge Protocols.* CRYPTO '98, LNCS 1462, 408–423, Springer 1998.

[29] D. Harkins and D. Carrel, eds. *The Internet Key Exchange (IKE).* RFC 2409, November 1998.

[30] ISO/IEC IS 9798-3, "Entity authentication mechanisms — Part 3: Entity authentication using asymmetric techniques", 1993.

[31] M. Jakobsson, K. Sako and R. Impagliazzo. *Designated Verifier Proofs and Their Applications.* EUROCRYPT '96, LNCS 1070, 143–154, Springer 1996.

[32] J. Katz, *Efficient and Non-Malleable Proofs of Plaintext Knowledge and Applications.* EUROCRYPT '03, LNCS 2656, 211–228, Springer 2003.

[33] C. Kaufman, ed., Internet Key Exchange (IKEv2) Protocol, draft-ietf-ipsec-ikev2-17.txt, September 2004 (pending RFC).

[34] H. Krawczyk. *SKEME: a versatile secure key exchange mechanism for Internet.* IEEE SNDSS '96, 114–127, IEEE Press 1996.

[35] H. Krawczyk. *SIGMA: The 'SiGn-and-MAc' Approach to Authenticated Diffie-Hellman and Its Use in the IKE Protocols.* CRYPTO '03, LNCS 2729, 400–425, Springer 2003. Available at www.research.ibm.com/security/sigma.ps

[36] Y. Lindell and B. Pinkas. *Privacy Preserving Data Mining.* J.of Cryptology, 15(3):177–206, Springer 2002.

[37] C.H. Lim and P.J. Lee. *A Key Recovery Attack on Discrete Log-based Schemes Using a Prime Order Subgroup.* CRYPTO '97, LNCS 1294, 249–263, Springer 1997.

[38] W. Mao and K.G. Paterson. *On the Plausible Deniability Feature of Internet Protocols.* Manuscript.

[39] R. Pass. *On Deniability in the Common Reference String and Random Oracle Model.* CRYPTO '03, LNCS 2729, 316–337, Springer 2003.

[40] R. Rivest, A. Shamir and Y. Tauman. *How to Leak a Secret.* ASIACRYPT '01, LNCS 2248, 552–565, Springer 2001.

[41] V. Shoup. *On Formal Models for Secure Key Exchange.* IBM Research Report RZ 3120, April 1999.

SKEME

Public Input: $\mathsf{pk}_A, \mathsf{pk}_B$ encryption public keys of A and B respectively.
Secret Input of A: sk_A; Secret Input of B: sk_B

$\qquad$ A $\qquad\qquad\qquad\qquad\qquad\qquad\qquad\qquad$ B

x, k_A at random
$c_A = \mathsf{enc}(\mathsf{pk}_B, k_A)$ $\xrightarrow{\quad g^x,\ c_A \quad}$

$\qquad\qquad\qquad\qquad\qquad\qquad\qquad\qquad$ y, k_B at random and
$\qquad\qquad\qquad\qquad\qquad\qquad\qquad\qquad$ $k'_A = \mathsf{dec}(\mathsf{sk}_B, c_A)$,
$\xleftarrow{\quad g^y,\ c_B,\ t_B \quad}$ $\qquad t_B = \mathsf{MAC}_{k'_A}(g^y, g^x)$,
$\qquad\qquad\qquad\qquad\qquad\qquad\qquad\qquad$ $c_B = \mathsf{enc}(\mathsf{pk}_A, k_B)$

$t_B \stackrel{?}{=} \mathsf{MAC}_{k_A}(g^y, g^x)$ and
$k'_B = \mathsf{dec}(\mathsf{sk}_A, c_B)$, $\xrightarrow{\quad t_A \quad}$ $t_A \stackrel{?}{=} \mathsf{MAC}_{k_B}(g^x, g^y)$
$t_A = \mathsf{MAC}_{k'_B}(g^x, g^y)$

Output: Shared key $K = PRF_{k_A}(g^{xy}) \oplus PRF_{k_B}(g^{xy})$

Figure 1: The three rounds version of SKEME

The SIGMA Protocol

Public Input: $\mathsf{pk}_A, \mathsf{pk}_B$ signature public keys of A and B respectively.
Secret Input of A: sk_A; Secret Input of B: sk_B

$\qquad$ A $\qquad\qquad\qquad\qquad\qquad\qquad\qquad\qquad$ B

x at random $\xrightarrow{\quad g^x \quad}$

$\xleftarrow{\quad g^y \quad}$ $\qquad y$ at random

$K_m, K_s \leftarrow KDF(g^{xy})$
$t_A = \mathsf{MAC}_{K_m}(0, \mathsf{A})$ $\xrightarrow{\quad \mathsf{A}, \sigma_A, t_A \quad}$
$\sigma_A = \mathsf{Sig}_{\mathsf{sk}_A}(g^x, g^y)$

$\qquad\qquad\qquad\qquad\qquad\qquad\qquad\qquad$ $K_m, K_s \leftarrow KDF(g^{xy})$
If $\mathsf{Ver}_{\mathsf{pk}_B}((g^y, g^x), \sigma_B) = 1$ $\qquad\qquad\qquad$ If $\mathsf{Ver}_{\mathsf{pk}_A}((g^x, g^y), \sigma_A) = 1$
and $t_B = \mathsf{MAC}_{K_m}(1, \mathsf{B})$ $\xleftarrow{\quad \mathsf{B}, \sigma_B, t_B \quad}$ and $t_A = \mathsf{MAC}_{K_m}(0, \mathsf{A})$
then output session key K_s $\qquad\qquad\qquad\qquad$ then $\sigma_B = \mathsf{Sig}_{\mathsf{sk}_B}(g^y, g^x)$
else output error $\qquad\qquad\qquad\qquad\qquad\qquad$ and $t_B = \mathsf{MAC}_{K_m}(1, \mathsf{B})$
$\qquad\qquad\qquad\qquad\qquad\qquad\qquad\qquad$ output session key K_s
$\qquad\qquad\qquad\qquad\qquad\qquad\qquad\qquad$ else $\sigma_B = t_B = \mathsf{error}$

Figure 2: The four-round version of SIGMA

Secure Function Evaluation with Ordered Binary Decision Diagrams

Louis Kruger and Somesh Jha
University of Wisconsin-Madison
[lpkruger,jha]@cs.wisc.edu

Eu-Jin Goh and Dan Boneh
Stanford University
[eujin, dabo]@cs.stanford.edu

ABSTRACT

Privacy-preserving protocols allow multiple parties with private inputs to perform joint computation while preserving the privacy of their respective inputs. An important cryptographic primitive for designing privacy-preserving protocols is secure function evaluation (SFE). The classic solution for SFE by Yao uses a gate representation of the function that the two parties want to jointly compute. Fairplay is a system that implements the classic solution for SFE. In this paper, we present a new protocol for SFE that uses a graph-based representation of the function. Specifically we use the graph-based representation called ordered binary decision diagrams (OBDDs). For a large number of Boolean functions, OBDDs are more succinct than the gate-based representation. Preliminary experimental results based on a prototype implementation shows that for several functions, our protocol results in a smaller bandwidth than Fairplay. For example, for the classic millionaire's problem, our new protocol results in a approximately 45% bandwidth reduction over Fairplay. Therefore, our protocols will be particularly useful for applications for environments with limited bandwidth, such as applications for wireless and sensor networks.

Categories and Subject Descriptors

C.2.0 [**Computer-Communication Networks**]: Security and Protection

General Terms

Security, Algorithms, Theory

Keywords

Binary Decision Diagrams, Secure Function Evaluation

1. INTRODUCTION

The ease and transparency of information flow on the Internet has heightened concerns of personal privacy [10, 25].

Permission to make digital or hard copies of all or part of this work for personal or classroom use is granted without fee provided that copies are not made or distributed for profit or commercial advantage and that copies bear this notice and the full citation on the first page. To copy otherwise, to republish, to post on servers or to redistribute to lists, requires prior specific permission and/or a fee.
CCS'06, October 30–November 3, 2006, Alexandria, Virginia, USA.
Copyright 2006 ACM 1-59593-518-5/06/0010 ...$5.00.

Various Internet activities, such as Web surfing, email, and other services leak sensitive information. As a result, there has been interest in developing technologies [9, 13, 24] and protocols to address these concerns. In particular, privacy-preserving protocols [11, 12, 18, 20] that allow multiple parties to perform joint computations without revealing their private inputs have been the subject of much interest. Our focus in this paper is on two party privacy-preserving protocols.

One of the fundamental cryptographic primitives for designing privacy-preserving protocols is *secure function evaluation (SFE)*. A protocol for SFE enables two parties A and B with inputs x and y respectively to jointly compute a function $f(x,y)$ while preserving the privacy of the two parties (i.e., at the end of the protocol, party A only knows its input x and the value of the function $f(x,y)$, and similarly for B). Yao showed that for a polynomial-time computable function f, there exists a SFE protocol that executes in polynomial time [15, 27] (details about this protocol can be found in Goldreich's book [14, Chapter 7]). Yao's classic solution for SFE has been used to design privacy-preserving protocols for various applications [1]. The importance of Yao's protocol spurred researchers to design a compiler that takes a description of the function f and emits code corresponding to Yao's protocol for secure evaluation of f. Such compilers, for example Fairplay [22], enable wider applicability of SFE. MacKenzie *et al.* [21] implemented a compiler for generating secure two-party protocols for a restricted but important class of functions, which is particularly suited for applications where the secret key is protected using threshold cryptography. For most applications, the classic protocol for SFE is quite expensive, which has led researchers to develop more efficient privacy-preserving protocols for specific problems [11, 12, 18, 20].

In the classic SFE protocol, the function f is represented as circuit comprised of gates. Fairplay uses this circuit representation of f. *Ordered Binary Decision Diagrams (OBDDs)* are a graph-based representation of Boolean functions that have been used in a variety of applications in computer-aided design, including symbolic model checking (a technique for verifying designs), verification of combinational logic, and verification of finite-state concurrent systems [3, 7]. OBDDs can be readily extended to represent functions with arbitrary domains and ranges.

Given an OBDD representation of the function to be jointly computed by the two parties, Yao's protocol can be directly used by first converting the OBDD into a circuit. Converting an OBDD to a circuit, however, incurs a blow-up in the number of gates required. To empirically measure this blowup, we implemented a compiler that takes an OBDD and converts it into a circuit description that can be used in Fairplay. On the average, this conversion from OBDD to circuit resulted in a increase in size by a factor of 10. Details of this experiment can be found in Section 4.

In this paper, we present a SFE algorithm that directly uses an OBDD representation of the function f that the two parties want to jointly compute. The advantage of using an OBDD representation over the gate-representation is that OBDDs are more succinct for certain widely used classes of functions than the gate representation. For example, among other functions, our results show the OBDD representation is more efficient than the gate representation for 8-bit AND, 8-bit addition, and the millionaire's and billionaire's problems [27]. As a result, our protocol has reduced bandwidth consumption over the classic Yao protocol implemented in Fairplay. Because processor speeds have increased at a more rapid pace than bandwidth availability over the past years, network bandwidth is likely to be the bottleneck for a number of applications. In particular, our protocols are especially useful for applications operating over networks with limited bandwidth, such as wireless and sensor networks. Furthermore, we have empirically confirmed this statement by implementing our protocol and comparing it with Fairplay.

This paper makes the following contributions:

- We present a SFE protocol that uses the OBDD representation of the function to be jointly computed by two parties. Our new protocol along with the correctness proof is provided in Section 3.

- Experimental results based upon a prototype implementation of our protocol demonstrate that for certain functions, our implementation results in a smaller encrypted circuit than Fairplay. For example, for the classic millionaire's problem, our implementation reduces the bandwidth by approximately 45% over Fairplay. Our implementation and experimental results are described in Section 4.

In summary, this paper presents a new SFE protocol that uses the OBDD representation. The OBDD representation is more efficient for several practical functions of interest. For other functions, the circuit description (and therefore FairPlay) will be more efficient. This paper presents a generic alternative to Boolean circuits that can be used when appropriate.

2. ORDERED BINARY DECISION DIAGRAMS (OBDDS)

Ordered binary decision diagrams (OBDDs) are a canonical representation for Boolean formulas [3]. They are often substantially more compact than traditional normal forms, such as conjunctive normal form (CNF) and disjunctive normal form (DNF), and they can be manipulated efficiently. Therefore, they are widely used for a variety of applications in computer-aided design, including symbolic model checking, verification of combinational logic, and verification of finite-state concurrent systems [7]. A detailed discussion of OBDDs can be found in Bryant's seminal article [3].

Given a Boolean function $f(x_1, x_2, \cdots, x_n)$ of n variables $x_1, \cdots, x_n$ and a total ordering on the n variables, the OBDD for f, denoted by $OBDD(f)$, is a rooted, directed acyclic graph (DAG) with two types of vertices: *terminal* and *nonterminal* vertices. $OBDD(f)$ also has the following components:

- Each vertex v has a level, denoted by $level(v)$, between 0 and n. There is a distinguished vertex called *root* whose level is 0.

- Each nonterminal vertex v is labeled by a variable $var(v) \in \{x_1, \cdots, x_n\}$ and has two successors, $low(v)$ and $high(v)$. Each terminal vertex is labeled with either 0 or 1. There are only two terminal vertices in an OBDD. Moreover, the labeling of vertices respects the total ordering $<$ on the variables, i.e., if u has a nonterminal successor v, then $var(u) < var(v)$.

Given an assignment $\mathcal{A} = \langle x_1 \leftarrow b_1, \cdots, x_n \leftarrow b_n \rangle$ to the variables $x_1, \cdots, x_n$ the value of the Boolean function $f(b_1, \cdots, b_n)$ can be found by starting at the root and following the path where the edges on the path are labeled with $b_1, \cdots, b_n$. OBDDs can also be used to represent functions with finite range and domain. Let g be a function of n Boolean variables with output that can be encoded by k Boolean variables. The function g can be represented as an array of k OBDDs where the i-th OBDD represents the Boolean function corresponding to the i-th output bit of g. For the rest of the paper we will assume that the function f is a Boolean function, but our protocols can be easily extended for the case of functions with a finite range. We will illustrate OBDDs with an example.

EXAMPLE 1. *Figure 1 shows the OBDD for the function $f(x_1, x_2, x_3, x_4) = (x_1 = x_2) \wedge (x_3 = x_4)$ of four variables x_1, x_2, x_3, x_4 with the total ordering $x_1 < x_2 < x_3 < x_4$.[1] Notice that the ordering of the labels on the vertices on any path from the root to the terminals of the OBDD corresponds to the total ordering of the Boolean variables. Consider the assignment $\langle x_1 \leftarrow 1, x_2 \leftarrow 1, x_3 \leftarrow 0, x_4 \leftarrow 0 \rangle$. In the OBDD shown in Figure 1, if we start at the root and follow the edges corresponding to the assignment, we end up at the terminal vertex labeled with 1. Therefore, the value of $f(1, 1, 0, 0)$ is 1.*

One of the advantages of OBDDs is that they can be manipulated efficiently, i.e., given OBDDs for f and g, OBDDs for $f \wedge g$, $f \vee g$, and $\neg f$ can be computed efficiently. We now describe an operation called *restriction*, which is used in our protocol. Given a n variable Boolean function $f(x_1, x_2, \cdots, x_n)$ and a Boolean value b, $f \mid_{x_i \leftarrow b}$ is a Boolean function of $n-1$ variables $x_1, \cdots, x_{i-1}, x_{i+1}, \cdots, x_n$ defined as follows:
$f \mid_{x_i \leftarrow b} (x_1, \cdots, x_{i-1}, x_{i+1}, \cdots, x_n)$ is equal to $f(x_1, \cdots, x_{i-1}, b, x_{i+1}, \cdots, x_n)$. Essentially $f \mid_{x_i \leftarrow b}$ is the function obtained by substituting the value b for the variable x_i in the function f. Given the OBDD for f, the $OBDD$ for $f \mid_{x_i \leftarrow b}$ can be efficiently computed [3, Section 4]. The restriction operation can be extended to multiple variables in a straightforward manner, e.g., $f \mid_{x_i \leftarrow b, x_j \leftarrow b'}$ can be computed as $(f \mid_{x_i \leftarrow b}) \mid_{x_j \leftarrow b'}$. We explain the algorithm using our example; the reader is referred to [3] for details. Consider the function $f(x_1, x_2, x_3, x_4)$ described in example 1.

---
[1] OBDDs are sensitive to variable ordering, e.g., with the ordering $x_1 < x_3 < x_2 < x_4$ the OBDD for $(x_1 = x_2) \wedge (x_3 = x_4)$ has 11 nodes.

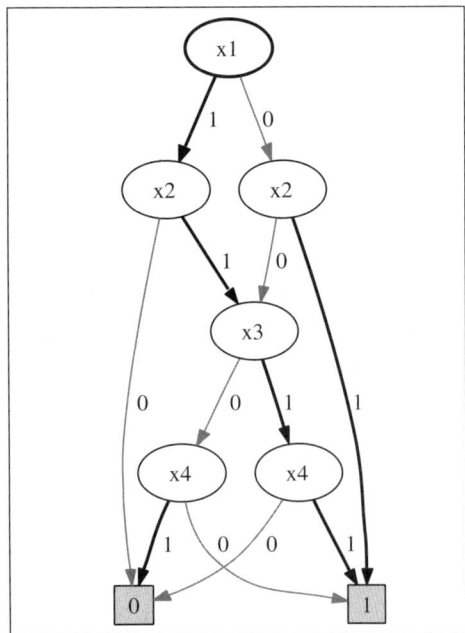

Figure 1: OBDD for the function $f(x_1, x_2, x_3, x_4) = (x_1 = x_2) \wedge (x_3 = x_4)$.

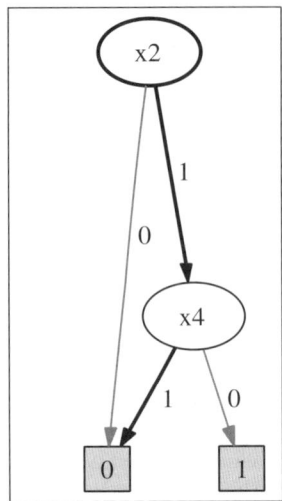

Figure 2: OBDD for the restriction $f|_{x_1 \leftarrow 1, x_3 \leftarrow 0}$ where $f(x_1, x_2, x_3, x_4) = (x_1 = x_2) \wedge (x_3 = x_4)$.

The OBDD corresponding to $f|_{x_1 \leftarrow 1, x_3 \leftarrow 0}$ is shown in Figure 2. Since $x_1 \leftarrow 1$, the root of OBDD ($f|_{x_1 \leftarrow 1, x_3 \leftarrow 0}$) is the left vertex labeled with x_2. Consider the two vertices v_1 and v_2 labeled with x_2. If v_1 has an edge that points to the vertex labeled with x_3, then that edge is changed to point to the right vertex labeled with x_4 (because this is the vertex reached if x_3 is equal to 1). Notice that in the reduced OBDD shown in Figure 2 the vertices that are labeled with x_1 and x_3 have been eliminated.

3. TWO PARTY SFE WITH OBDDS

For our protocols, we require a symmetric encryption scheme with two easily attained special properties [19], which are (1) *elusive range*: an encryption under one key is in the range of an encryption with a different key with negligible probability, and (2) *efficiently verifiable range*: given a key, a user can efficiently verify that a ciphertext is in the range of that key. These properties are required so that the receiver of the garbled OBDD can correctly decrypt nodes in the OBDD. The formal definition of these properties by Lindell and Pinkas [19] is provided with the proofs. An example of a symmetric key encryption scheme that fulfills these properties is $E_k(m) = (r, f_k(r) \oplus m \| 0^n)$, where $f : \{0,1\}^n \times \{0,1\}^n \to \{0,1\}^{2n}$ is a pseudo-random function and $r \xleftarrow{R} \{0,1\}^n$ is a n-bit random sequence. Unless stated otherwise, all symmetric key encryption schemes in this paper, besides being semantically secure [14, Chapter 5], also require these two properties.

Our protocol also uses a 1-out-of-2 oblivious transfer (denoted OT_1^2) protocol. A 1-out-of-k oblivious transfer OT_1^k is a protocol that lets Bob obtain one of k secrets held by Alice, without Alice learning which secret Bob obtains.

We now give the protocol for securely computing an OBDD between two parties where each party holds a part of the input. Assume f is a Boolean function $f(x_1, x_2, \cdots, x_n)$ of n Boolean variables $x_1, x_2, \cdots, x_n$. Let $OBDD(f)$ denote the OBDD for f with the ordering $x_1 < x_2 < \cdots < x_n$. We describe the protocol in stages. Protocol 1 described in Section 3.1 assumes that Alice holds inputs corresponding to the first k variables, and Bob has the inputs corresponding to last $n - k$ variables $x_{k+1}, \ldots, x_n$. Protocol 2 described in Section 3.2 allows arbitrary sharing of inputs, and it uses the restriction operation on OBDDs described earlier to reduce the bandwidth requirement of the protocol.

3.1 Protocol 1

For this protocol, we assume that Alice holds the inputs $(i_1, \ldots, i_k)$ corresponding to the first k variables $x_1, \ldots, x_k$, and Bob has the inputs $(i_{k+1}, \ldots, i_n)$ corresponding to last $n - k$ variables $x_{k+1}, \ldots, x_n$. In our protocol, Alice and Bob want to compute $f(x_1, \cdots, x_n)$ on their inputs using the OBDD for f. As the outcome, Bob learns $f(i_1, \ldots, i_n)$. This protocol is described in Figure 3.

One of the difficulties in developing and proving the protocol is that OBDDs allow skipping of levels. For example, Figure 4(a) shows the OBDD for the Boolean function $x_1 \wedge x_2$. Assume that the vertex at level 0 is labeled with x_1 and vertices at level 1 are labeled with x_2. Suppose Alice owns variable x_1 and Bob owns variable x_2. If Alice's input is 1, then Bob follows one more edge than if Alice's input were 0, which allows Bob to determine Alice's input. Compare this to Figure 4(b), where a dummy vertex is added so that, regardless of Alice's input, Bob has to follow the same number of edges. In this case, Bob learns nothing about Alice's input. Before Alice garbles the OBDDs, she adds dummy vertices so that each path from the root to a terminal node has the same number of edges. Alice adds dummy nodes whenever $OBDD(f)$ allows Bob to skip levels when evaluating the OBDD on his share of the input. Recall that the 0-successor and 1-successor of v is denoted by $low(v)$

Input: Both parties' inputs include the $OBDD(f)$ for the Boolean function $f(x_1, x_2, \cdots, x_n)$ with the ordering $x_1 < x_2 < \cdots < x_n$. Furthermore, Alice holds the inputs $(i_1, \ldots, i_k)$ corresponding to the first k variables $x_1, \ldots, x_k$, and Bob has the inputs $(i_{k+1}, \ldots, i_n)$.

1. Alice performs the following steps:

    (a) She traverses the $OBDD(f)$ using her input $(i_1, \cdots, i_k)$, which results in a node v_{init} at level k.

    (b) She uniformly and independently at random creates $(n-k)$ pairs of secrets $(s_1^0, s_1^1), \cdots, (s_{n-k}^0, s_{n-k}^1)$. In addition, for each node v in the $OBDD(f)$ whose level is between k and $n-1$, Alice also creates a secret s_v.

    (c) She assigns a uniformly random label to each node whose level is between k and n. We refer to the randomly assigned label of node v using the notation $label(v)$.

    (d) Next, Alice augments $OBDD(f)$ with some number of dummy nodes (to ensure that Bob always traverses $n-k$ nodes in his phase of the protocol).

    (e) Alice garbles all nodes whose level is between k and $n-1$ in the following manner. Let v be a node in $OBDD(f)$ such $k \leq level(v) \leq n-1$ and define $level(v) = \ell$. The encryption of node v, denoted by $E^{(v)}$, is a label and a randomly ordered ciphertext pair

    $$\left(label(v) \ , \ E_{s_v \oplus s^0_{\ell-k+1}}(label(low(v)) \parallel s_{low(v)}) \ , \ E_{s_v \oplus s^1_{\ell-k+1}}(label(high(v)) \parallel s_{high(v)}) \right) \ ,$$

    where the labels are pre-pended to the secret with a separator symbol and the order of the ciphertexts is determined by a fair coin flip. Roughly speaking, the secrets corresponding to the 0-successor and 1-successor of node v are encrypted with the secret corresponding to v and its level.

    Note that dummy nodes have the same structure as normal nodes, except that the ciphertext pair contain encryptions of the same message since dummy nodes have the same 0 and 1-successors. Provided the encryption scheme is semantically secure, this poses no problem since the keys are chosen uniformly at random.

    Lastly, there are two terminal nodes of the form $(b, label(t_b))$ for $b = 0$ or 1. Recall that $OBDD(f)$ has two terminal nodes, denoted as 0 and 1, that are at level n.

    (f) Once Alice is done encrypting, she sends to Bob the encryption of all nodes whose level is between k and n and the secret $s_{v_{init}}$ corresponding to node v_{init} at level k. We called this the garbled OBDD.

2. Bob performs the following steps:

    (a) He engages in $n-k$ 1-out-of-2 oblivious transfers to obtain the secrets corresponding to his input. For example, if his input i_j is 0, then he obtains the (level) secret s^0_{j-k}; otherwise, he obtains the secret s^1_{j-k}.

    (b) Now Bob is ready to start his computation. Suppose $i_{k+1} = 0$. With s_1^0 and $s_{v_{init}}$, he decrypts both ciphertexts in $E^{(v_{init})}$ and decides which gives the correct result by using the verifiable range property of the encryption scheme. Bob now has both $s_{low(v)}$ (the secret corresponding to the 0-successor of v_{init}) and $label(low(v))$ (which tells Bob which encrypted node is used to evaluate his next input). Continuing this way, Bob eventually obtains a label corresponding to one of the terminal nodes, which determines the result of the OBDD on the shared inputs. Bob sends this result to Alice.

Figure 3: Protocol 1.

and $high(v)$. For example, if node n at level j has node n''' at level $j+3$ as its 0-successor, then Alice inserts two dummy nodes n' and n'' at level $j+1$ and $j+2$ respectively. The 0-successor of node n is changed to n', both 0 and 1-successors of n' are set to n'', and both 0 and 1-successors of n'' are set to n'''.

We prove correctness and security in the case of semi-honest parties. In the semi-honest model, both parties are assumed to perform computations and send messages according to their prescribed actions in the protocol. They may also record whatever they see during the protocol (i.e. their own input and randomness, and the messages they receive). We refer readers to Goldreich's book [14] for the complete definitions. Claim 1 proves that the protocol shown in Figure 3 is correct; that is, Bob computes the function $f(i_1, \cdots, i_n)$. Claim 2 proves that protocol 1 is secure in the semi-honest model; that is, at the end of protocol 1, Alice only knows its input and the value of the function $f(i_1, \cdots, i_n)$, and similarly for Bob. For our proofs, we require the definition of elusive range and efficiently verifiable range from Lindell and Pinkas [19].

DEFINITION 1. *Let (G,E,D) be a symmetric key encryption scheme with key-generation, encryption, and decryption algorithms. Denote the range of the scheme by* $\mathsf{Range}_n(k) = \{E_k(x)\}_{x \in \{0,1\}^n}$. *We say that*

1. *(G,E,D) has an **elusive range** if for every probabilistic poly-time machine A, every polynomial $p(\cdot)$, and all sufficiently large n, $\Pr_{k \leftarrow \mathsf{G}(1^n)}[A(1^n) \in \mathsf{Range}_n(k)] < 1/p(n)$.*

2. *(G,E,D) has an **efficiently verifiable range** if there exists a probabilistic poly-time machine M such that $M(1^n, k, c) = 1$ if and only if $c \in \mathsf{Range}_n(k)$.*

CLAIM 1. *If the encryption scheme has an elusive range and the oblivious transfer protocol is secure, then Protocol 1 is correct for semi-honest Alice and Bob.*

Proof: First we show that every node in the garbled OBDD sent to Bob can be evaluated correctly. For an encrypted node v, let c_0 and c_1 be the first and second ciphertext term. Suppose k is the key that Bob obtains to decrypt the ciphertext in node v. Because the encryption scheme is elusive and all the keys used in the garbled OBDD are chosen uniformly and independently at random, it follows

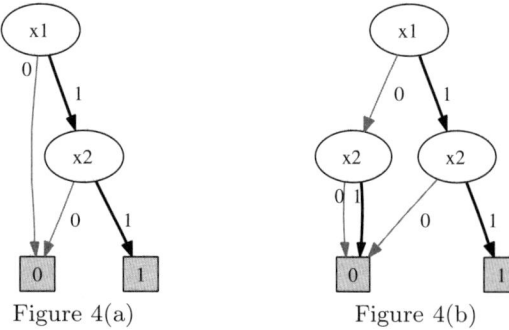

Figure 4(a) Figure 4(b)

Figure 4: Adding dummy vertices

immediately that, except with negligible probability, only one ciphertext in an encrypted node decrypts correctly; that is, either $c_0 \in \mathsf{Range}(k)$ or $c_1 \in \mathsf{Range}(k)$ but not both. Hence, except with negligible probability, every node in the garbled OBDD can be evaluated correctly.

By induction, we show that the correct key is obtained for every node input and output. The base case is the key $s_{v_{init}}$ and the keys $s_{k+1}^{i_{k+1}}, \ldots, s_n^{i_n}$ (corresponding to Bob's input) obtained from Alice through the $n-k$ oblivious transfers. The inductive step at node v assumes that the correct label (pointing Bob to node v) and node key s_v was obtained in the previous step. Furthermore, the correct level key $s_l^{i_l}$ was obtained by executing an oblivious transfer protocol. Since every node in the garbled OBDD can be evaluated correctly and the garbled OBDD is built by a semi-honest Alice, Bob obtains the correct label and key output at node v, which concludes the inductive step. We can conclude that the entire garbled OBDD can be evaluated to give the correct result.[2] ∎

CLAIM 2. *If the encryption scheme is semantically secure and has an efficiently verifiable elusive range, and the oblivious transfer protocol is secure, then Protocol 1 is secure against semi-honest Alice and Bob.*

Proof of this claim is tedious and is given in Appendix A.

3.2 Protocol 2

The protocol presented in this section allows both Alice and Bob to possess arbitrary input sets instead of assuming that Alice (and also Bob) holds either the first k or the last $n-k$ input variables. In this new protocol, before garbling the OBDD, Alice can reduce the size of the OBDD by eliminating vertices whose labels correspond to Alice's inputs. Let $f(x_1, \cdots, x_n)$ be the function to be computed and X_A denotes the inputs of Alice. Alice first computes the OBDD for the restriction $f\mid_{X_A}$ of f for the variables in its input set X_A. Alice then encrypts the reduced OBDD and sends it to Bob. During the restriction operation, all vertices whose labels correspond to Bob's input should be included. If this is not the case, then there is a risk that different restrictions could produce different numbers of nodes, which would leak information to Bob about Alice's inputs. For example, in

---
[2]Note that the negligible probability of error during decryption can be removed by Alice first checking that every encrypted node decrypts correctly before sending the garbled OBDD to Bob.

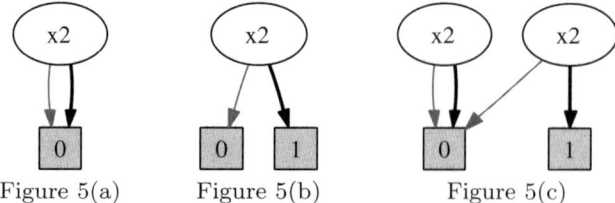

Figure 5(a) Figure 5(b) Figure 5(c)

Figure 5: OBDD restriction

> **Input:** Both parties' inputs include the $OBDD(f)$ for the Boolean function $f(x_1, x_2, \cdots, x_n)$ with the ordering $x_1 < x_2 < \cdots < x_n$. Furthermore, Alice holds the inputs for the variables in the set X_A and Bob holds the inputs for the variables in the set $X_B = \{x_1, \cdots, x_n\} - X_A$.
>
> 1. Alice performs the following steps:
>   (a) Alice computes the OBDD $\mathcal{O}_A$ for the function $f\mid_{X_A}$. This is the restriction operation described in Section 2.
>   (b) Alice encrypts the OBDD for the function $f\mid_{X_A}$ and sends it to Bob. This step is exactly the same as for Protocol 1 described in Figure 3. Alice also sends the secret corresponding to the root of the OBDD $\mathcal{O}_A$.
> 2. The computation for Bob is exactly the same as that for Protocol 1.

Figure 6: Protocol 2.

Figure 5(a), where Alice's value is 0, there are only 2 nodes, but in Figure 5(b), there are 3 nodes. Figure 5(c) shows the result of retaining extra vertices, in which case there are 4 nodes regardless of Alice's inputs. The description of this protocol is given in Figure 6. The proof of correctness of this protocol is almost identical to the one presented in Section 3.1. Example 2 shows an execution of this protocol on a small function.

EXAMPLE 2. *Assume that Alice and Bob want to compute $f(x_1, x_2) = x_1 \wedge x_2$, where Alice has input x_1 and Bob has input x_2, or in other words $X_A = \{x_1\}$ and $X_B = \{x_2\}$. Assume that $x_1 = 0$ and $x_2 = 1$. $OBDD(f)$ with dummy nodes is shown in figure 4(b). Alice computes the OBDD for the function $f\mid_{x_1 \leftarrow 0}$, which results in a structure shown in Figure 5(c). Let the two nonterminal nodes in Figure 5(c) be v_1 and v_2. First, Alice generates 2 secrets s_{v_1} and s_{v_2}, which are assigned to the nodes v_1 and v_2, respectively. Alice also generates random labels for the four nodes in Figure 5(c), and generates a pair of secrets (s_1^0, s_1^1). The garbled OBDD corresponding to Figure 5(c) is shown below (terminal nodes are shown as 0 and 1 and lab denotes label).*

$(lab(v_1), E_{s_{v_1} \oplus s_1^0}(lab(0) \parallel s_0), E_{s_{v_1} \oplus s_1^1}(lab(0) \parallel s_0))$
$(lab(v_2), E_{s_{v_2} \oplus s_1^0}(lab(0) \parallel s_0), E_{s_{v_2} \oplus s_1^1}(lab(1) \parallel s_1))$
$(0, lab(0))$
$(1, lab(1))$

Alice reveals the secret s_{v_1} corresponding node v_1. Alice and Bob engage in a 1-out-of-2 (OT_1^2) protocol and Bob obtains the secret s_1^1 (recall that $x_2 = 1$). Bob can now decrypt the second component of the first entry of the garbled OBDD and obtain $label(0) \parallel s_0$, and Bob can infer that the output is 0.

CLAIM 3. *If the encryption scheme has an elusive range and the oblivious transfer protocol is secure, then Protocol 2 is correct for semi-honest Alice and Bob.*

Proof: The proof of this claim is exactly same as the proof of Claim 1. One has to assume that the restriction operation used by Alice is correct. ∎

CLAIM 4. *If the encryption scheme is semantically secure and has an efficiently verifiable elusive range, and the oblivious transfer protocol is secure, then Protocol 2 is secure against semi-honest Alice and Bob.*

Proof of this claim is tedious and is given in Appendix A.

4. IMPLEMENTATION AND EXPERIMENTAL RESULTS

In this section we describe various components of our implementation (called SFE-OBDD), which is based on protocol 2 described in Section 3.2. We also present experimental results comparing the performance of our implementation against Fairplay. The following major conclusions can be drawn from our experimental investigation:

- The restriction operation used in protocol 2 can significantly reduce the size of the OBDD, which can lead to reduced bandwidth while executing the protocol.

- Our OBDD protocol outperforms Fairplay circuit in terms of bandwidth in most functions in our benchmark. However, some functions which have inefficient OBDD representations can perform far worse.

- Execution times of our implementation and Fairplay are dominated by the oblivious-transfer protocol. Since the oblivious-transfer component of our protocol and the protocol used by Fairplay is the same, with respect to execution times we did not observe as much improvement as in bandwidth.

- Converting an OBDD into a circuit results in a blowup in size. We implemented a reverse compiler that takes an OBDD and converts it into a circuit description, which can be used in Fairplay. Typically this conversion from OBDD to circuit resulted in a blowup in size by a factor of $5-10$, depending on post-conversion optimizations. However, if the original circuit is particularly inefficient, it is possible for some gain to be achieved due to the canonical representation property of OBDDs.

4.1 Implementation

Our implementation consists of the following components:

1. An implementation of protocol 2 as described in Section 3.2.

2. Fairplay uses the *secure hardware definition language (SHDL)* to describe circuits. Because we wanted to compare the performance of our protocol with the Yao circuit protocol on identical functions, we implemented an OBDD compiler that takes as input a file describing a function in SHDL, and produces the corresponding OBDD. Note that both the SHDL used by Fairplay and the BDD representation originate from the same high level SFDL description, this means that the OBDD and circuit are evaluating the same functions.

For the cryptographic primitives we use exactly the same implementation as Fairplay. We use the 1-out-of-2 (OT_1^2) proposed by Noar and Pinkas [23], and the encryption function is $E_k(m)$ was $SHA - 1(k) \oplus (m \parallel 0^n)$.

Our OBDD compiler allows us to directly compare the efficiency of our implementation to Fairplay. The OBDD compiler takes as input a file containing an SHDL description and produces a file containing the description of the corresponding OBDD. This file can then be used as an input to the SFE protocol. Our compiler uses the JavaBDD [16], and BuDDy [5] libraries, which provide functions to construct and manipulate OBDDs. It is well known that the size of an OBDD can be sensitive to the ordering of variables [3]. In some cases, variable ordering can make the difference between a OBDD that is linear versus exponential in the number of variables. Our SHDL to OBDD compiler allows the user to specify a particular variable ordering, which is useful if the user has domain knowledge about the function. If this is not practical, the compiler includes an optimizer that attempts to automatically find a variable ordering that yields an efficient OBDD, making use of heuristic functions built into the BuDDy library. Although in general finding the optimal variable ordering is NP-hard [2], we have found that in practice the optimizer can find good orderings for various functions we considered.

4.2 Experimental Results

We used various functions, some of which are included in the Fairplay distribution, as test cases to perform a comparison of the Fairplay protocol with our OBDD-based SFE protocol. The description of the functions are given in Figure 7. Each function was evaluated at several word sizes to evaluate scalability.

For each function, Figure 8 shows the sizes of the OBDDs and the corresponding circuit used by Fairplay. The size of an OBDD is the number of vertices in it. The size of a Fairplay circuit with n_1 gates of arity 1, n_2 gates of arity 2, and n_3 gates of arity 3 was computed as $2 \times n_1 + 4 \times n_2 + 8 \times n_3$ (this represents the number of entries in the truth table for the circuit). For the OBDDs, we show the sizes of the original OBDDs (in column marked as **Original**), with the dummy nodes added (in column marked as **Full**), and after Alice has performed the restriction operation on OBDDs with dummy nodes (in column marked as **Res**). Recall that dummy nodes are added so that regardless of Alice's inputs Bob has to follow the same number of edges. In protocol 2 Alice computes the OBDD for restriction on its input of the function to be jointly computed for the variable. These operations are described in detail in Section 3. Two observations can be made from Figure 8.

- Restriction can significantly reduce the size of the OBDD. For example, for the function Mil16 restriction reduces the size of the OBDD by more than half.

- Notice that for all functions except MUL8, MUL16, KDS4, KDS8, and KDS16 the size of the OBDD after restriction is smaller than the size of the circuit used in Fairplay. This suggests the choice of when to use our system over Fairplay depends on the function to be computed.

Our experimental results were obtained using a pair of machines connected on a local 100-megabit network. The

And	This is a circuit that computes the bitwise AND of two N bit numbers. Alice has N inputs, Bob has N inputs, and there are N bits of output.
Add	This is a circuit that computes the addition of two N bit numbers. Alice has N inputs, Bob has N inputs, and there are N bits of output. The high bit is discarded.
Eq	This is a simple equality comparator of two N-bit numbers. There is one bit of output.
Mul	This is a circuit that computes the unsigned multiplication of two N bit numbers. Alice has N inputs, Bob has N inputs, and there are N bits of output. The output is modulo 2^N. We were unable to test N=16 because of exponential blowup in the BDD
KDS	This is a circuit that implements a simply-keyed database lookup. Alice supplies N key/value pairs, and Bob supplies a key. The output is the value of that key, or 0 if the key is not found. The keys are $\log_2(N)$ bits, and the data are 24 bits. We were unable to test N=16 because of exponential blowup in the BDD.
Mil	This is the millionaire's problem. Alice and Bob each have and N bit integer as inputs, and there is 1 bit of output indicating if Alice's input is larger than Bob's.
Parity	Alice and Bob each have N-bits of input. They want to jointly compute the parity of their combined input bits. There is one bit of output.

Figure 7: Description of the functions used in our experiments. Each function was tested with N=4, N=8, and N=16 except where indicated

	BDD			FairPlay
	Original	Full	Res	
Add4	32	40	22	56
Add8	72	96	54	128
Add16	152	208	118	272
And4	14	18	10	24
And8	26	34	18	48
And16	50	66	34	96
Eq4	18	24	11	102
Eq8	27	41	18	230
Eq16	51	81	34	486
KDS4	416	578	466	356
KDS8	4084	7149	6283	780
KDS16	*	*	*	2244
MUL4	54	75	28	114
MUL8	1685	1800	1087	586
MUL16	*	*	*	2682
Mil4	24	34	22	52
Mil8	46	70	40	116
Mil16	94	150	90	244
parity4	18	18	10	30
parity8	34	34	18	62
parity16	66	66	34	126

Figure 8: Size of the OBDDs and the circuit used in Fairplay for functions shown in Figure 7. Values labeled "*" could not be converted to OBDDs because of exponential blowup.

machines were configured with 3.0Ghz Intel Pentium4 processors, 1 gigabyte of memory, and the Centos Linux 4.0 operating system using a modified Linux 2.6.9 kernel. For each function shown in Figure 7 we executed our OBDD-based and Fairplay code on a Sun Microsystems Java 1.5.0_04 JVM. Alice was run on one machine, and Bob on the other. For each execution, we measured the network bandwidth used (number of bytes transferred between Alice and Bob) and the execution time. The number reported for each trial is the average of three trials. Figure 9 shows the size of the garbled OBDD and garbled circuit in bytes and the network bandwidth for our implementation and Fairplay. Recall that the garbled OBDD is the structure that Alice sends to Bob to evaluate. With respect to network bandwidth our implementation outperformed Fairplay for seven out of the nine functions. We have implemented a reverse compiler that takes as input an OBDD, and outputs an SHDL description of a boolean circuit to evaluate the OBDD. This is performed via a straightforward transformation that takes each node in the OBDD and produces corresponding 3-input MUX gate in the boolean circuit. Then, an optimization pass is run using the same techniques described in [22]. The column labeled as "Converted Fairplay" in Figure 9 shows the size of the encrypted circuits produced by running the FairPlay protocol on the converted BDDs. Note that in a few cases, the converted circuit is actually more efficient than the corresponding Fairplay circuit. This occurs because FairPlay is not guaranteed to produce an optimal circuit from the function description. However, it is clear that our protocol that directly uses OBDD is much more efficient than the protocol produced by the reverse compiler.

Figures 10 and 11 show the execution times for SFE-OBDD and Fairplay. The elapsed execution times (EET) are shown in the last column. Columns 2-5 show the breakdown by sub-task, which are IPCG (initializations, parsing, and garbling), CC (circuit communication, Alice sending the garbled structure to Bob), OT (Oblivious Transfer, Bob obtaining secrets corresponding to its input), and EV (circuit evaluation, Bob evaluating the garbled structure). These sub-tasks were also used by the Fairplay paper [22]. In general, because the time for OT dominates the execution time, we only observe moderate improvement in SFE-BDD over Fairplay for execution times.

Function	Size in bytes.			Bandwidth in bytes	
	SFE-OBDD	Fairplay	Converted Fairplay	SFE-OBDD	Fairplay
Add4	970	1915	5382	3604	4684
Add8	1979	4214	11408	6590	9000
Add16	3992	8821	23442	12557	17645
And4	739	1080	1866	3373	3849
And8	1206	2153	3140	5813	6938
And16	2134	4299	5546	10696	13117
Eq4	582	2977	2690	3214	5716
Eq8	828	6527	3918	5434	11240
Eq16	1324	13626	7012	9892	22295
KDS4	14966	12248	65946	51219	13984
KDS8	185282	25608	682333	261077	27838
MUL4	1108	3286	8354	3739	6052
MUL8	32206	15706	288317	36814	20493
Mil4	892	1662	4278	3524	4399
Mil8	1400	3542	8248	6012	8256
Mil16	2790	7306	16859	11356	15972
parity4	577	1092	3237	3209	3830
parity8	828	2181	6172	5438	6897
parity16	1324	4359	11983	9889	13033

Figure 9: Size in bytes of the garbled OBDD, garbled circuit, and garbled circuit using the reverse compiler. Network bandwidth in bytes.

Fn	IPCG	CC	OT	Eval	EET
Add4	13.75%	5.62%	79.69%	0.94%	0.32
Add8	13.08%	3.80%	82.07%	1.05%	0.47
Add16	11.39%	2.36%	84.42%	1.83%	0.76
And4	12.42%	5.73%	81.21%	0.64%	0.31
And8	9.38%	4.02%	85.94%	0.67%	0.45
And16	7.44%	2.75%	89.39%	0.41%	0.73
Eq4	12.66%	5.38%	81.33%	0.63%	0.32
Eq8	9.33%	3.90%	86.12%	0.65%	0.46
Eq16	8.97%	2.48%	88.14%	0.41%	0.72
KDS4	4.79%	0.95%	93.86%	0.40%	2.52
KDS8	10.91%	1.69%	87.13%	0.27%	5.50
MUL4	14.77%	5.23%	79.38%	0.62%	0.33
MUL8	31.80%	4.11%	63.45%	0.63%	0.63
Mil4	13.44%	5.62%	80.00%	0.94%	0.32
Mil8	11.37%	3.86%	84.12%	0.64%	0.47
Mil16	10.55%	3.03%	85.88%	0.53%	0.76
parity4	10.67%	4.78%	83.71%	0.84%	0.36
parity8	9.37%	3.92%	86.06%	0.65%	0.46
parity16	8.55%	2.62%	88.41%	0.41%	0.72

Figure 10: Elapsed execution time (EET) in seconds and their breakdowns into sub-tasks for SFE-OBDD.

Fn	IPCG	CC	OT	Eval	EET
Add4	17.65%	19.00%	63.12%	0.23%	0.44
Add8	16.13%	15.96%	67.38%	0.53%	0.56
Add16	10.74%	9.67%	79.12%	0.48%	0.84
And4	11.92%	20.44%	67.40%	0.24%	0.41
And8	11.78%	16.07%	71.96%	0.19%	0.54
And16	9.78%	6.35%	83.48%	0.38%	0.79
Eq4	19.51%	11.85%	68.15%	0.49%	0.41
Eq8	15.44%	16.52%	67.68%	0.36%	0.56
Eq16	13.23%	9.84%	76.70%	0.23%	0.85
KDS4	33.33%	12.75%	53.33%	0.58%	0.34
KDS8	35.98%	11.92%	51.43%	0.66%	0.45
MUL4	21.89%	2.16%	75.41%	0.54%	0.37
MUL8	21.75%	7.99%	69.89%	0.37%	0.54
Mil4	38.27%	8.26%	53.28%	0.19%	0.53
Mil8	16.98%	9.25%	73.40%	0.38%	0.53
Mil16	18.78%	9.01%	71.99%	0.22%	0.92
parity4	20.78%	11.25%	67.73%	0.24%	0.41
parity8	49.55%	0.39%	49.94%	0.13%	0.77
parity16	14.63%	1.15%	83.97%	0.25%	0.79

Figure 11: Elapsed execution time (EET) in seconds and their breakdowns into sub-tasks for Fairplay.

5. FUTURE WORK

There are other optimizations to OBDDs that we have not explored in this paper. For example, adding negated edges to OBDDs can result in smaller structures for some Boolean functions.[3] Incorporating these optimizations in our protocol while preserving privacy is a direction for future work. There are several other OBDD-like representations developed by the computer-aided design and computer aided verification research communities, such as Binary Moment Diagrams (BMDs) [4] and Hybrid Decision Diagrams (HDDs) [8]. For a certain class of functions, these representations are more succinct than OBDDs. For example, BMDs can efficiently represent integer multiplication, which cannot be represented efficiently at the bit-level with OBDDs. Extending our protocol for these representations is an important direction of future research. Our vision is to provide an option for all these representations in our system so that a user can choose the representation that is suitable for the problem.

OBDDs have been used for a variety of applications, such efficient filtering in publish-subscribe systems [6], program analysis [26], and planning [17]. In the future we will investigate whether our protocol can be extended to design privacy-preserving algorithms for these applications.

6. REFERENCES

[1] Gagan Aggarwal, Nina Mishra, and Benny Pinkas. Secure computation of the k-th ranked element. In Christian Cachin and Jan Camenisch, editors, *Proceedings of Eurocrypt 2004*, volume 3027 of *LNCS*, pages 40–55. Springer-Verlag, May 2004.

[2] Beate Bollig and Ingo Wegener. Improving the variable ordering of OBDDs is NP-complete. *IEEE Transactions on Computers*, 45(9), September 1996.

[3] Randal E. Bryant. Graph-based algorithms for boolean function manipulation. *IEEE Transactions on Computers*, 35(8):677–691, 1986.

[4] Randal E. Bryant and Yirng-An Chen. Verification of arithmetic circuits with binary moment diagrams. In *Proceedings of the 32nd Conference on Design Automation (DAC)*, 1995.

[5] Buddy. http://sourceforge.net/projects/buddy.

[6] A. Campailla, S. Chaki, E. M. Clarke, S. Jha, and H. Veith. Efficient filtering in publish-subscribe systems using binary decision diagrams. In *Proceedings of the 23rd International Conference on Software Engineering (ICSE)*, 2001.

[7] E.M. Clarke, O. Grumberg, and D.A. Peled. *Model Checking*. The MIT Press, 2000.

[8] E.M. Clarke, M. Khaira, and X. Zhao. Hybrid decision diagrams: Overcoming the limitations of MTBDDs and BMDs. In *Proceedings of the International Conference on Computer-Aided Design (ICCAD)*, 1995.

[9] Lorrie Cranor, Marc Langheinrich, Massimo Marchiori, Martin Presler-Marshall, and Joseph Reagle. *The Platform for Privacy Preferences 1.0 (P3P1.0) Specification*. W3C Recommendation, 16 April 2002.

[10] Lorrie Faith Cranor. Internet privacy. *Communications of the ACM*, 42(2):28–38, 1999.

[11] J. Feigenbaum, B. Pinkas, R. Ryger, and F. Saint-Jean. Secure computation of surveys. In *2004 EU Workshop on Secure Multiparty Protocols (SMP)*, 2004.

[12] Michael Freedman, Kobbi Nissim, and Benny Pinkas. Efficient private matching and set intersection. In Christian Cachin and Jan Camenisch, editors, *Proceedings of Eurocrypt 2004*, volume 3027 of *LNCS*, pages 1–19. Springer-Verlag, May 2004.

[13] Ian Goldberg, David Wagner, and Eric Brewer. Privacy-enhancing technologies for the internet. In *Proc. of 42nd IEEE Spring COMPCON*. IEEE Computer Society Press, February 1997.

[14] O. Goldreich. *The Foundations of Cryptography — Volume 2*. Cambridge University Press, 2004.

[15] O. Goldreich, S. Micali, and A. Wigderson. How to play any mental game – a completeness theorem for protocols with honest majority. In *19th STOC*, pages 218–229, 1987.

[16] Javabdd - java binary decision diagram library. http://javabdd.sourceforge.net/.

[17] R. M. Jensen and M. M. Veloso. Obdd-based universal planning for multiple synchronized agents in non-deterministic domains. In *Proceedings of the Fifth International Conference on Artificial Intelligence Planning Systems (AIPS)*, 2000.

[18] Y. Lindell and B. Pinkas. Privacy preserving data mining. *Journal of Cryptology*, 15(3), 2002.

[19] Yehuda Lindell and Benny Pinkas. A proof of Yao's protocol for secure two-party computation. Cryptology ePrint Archive, Report 2004/175, 2004. http://eprint.iacr.org/2004/175.

[20] B. Pinkas M. Naor and R. Sumner. Privacy preserving auctions and mechanism design. In *Proceedings of the 1st ACM conf. on Electronic Commerce*, 1999.

[21] Philip D. MacKenzie, Alina Oprea, and Michael K. Reiter. Automatic generation of two-party computations. In *Proceedings of ACM Conference on Computer and Communications Security*, 2003.

[22] D. Malkhi, N. Nisan, B. Pinkas, and Y. Sella. Fairplay — a secure two-party computation system. In *Proceedings of the 13th Usenix Security Symposium*, San Diego, CA, USA, August 2004.

[23] Moni Naor and Benny Pinkas. Efficient oblivious transfer protocols. In *Proceedings of the Twelfth Annual Symposium on Discrete Algorithms (SODA)*, pages 448–457, 2001.

[24] D. M. Rind, I. S. Kohane, P. Szolovits, C. Safran, H. C. Chueh, and G. O. Barnett. Maintaining the confidentiality of medical records shared over the internet and the world wide web. *Annals of Internal Medicine*, 127(2), July 1997.

[25] Joseph Turow. Americans and online privacy: The system is broken. Technical report, Annenberg Public Policy Center, June 2003.

[26] J. Whaley and M. S. Lam. Cloning-based context-sensitive pointer alias analysis using binary decision diagrams. In *Proceedings of the ACM SIGPLAN 2004 Conference on Programming Language Design and Implementation (PLDI)*, 2004.

---

[3] Following a negated edge in an OBDD flips the value of the result.

[27] A.C. Yao. How to generate and exchange secrets. In *Proceedings of the 27th IEEE Symposium on Foundations of Computer Science*, 1986.

APPENDIX
A. PROOF OF CORRECTNESS

Proof: [*Proof of Claim 2*] Intuitively, Bob's security follows directly from the security of the 1-out-of-2 oblivious transfer protocol he uses to obtain the secrets corresponding to his input. Alice's security follows from both the security of the oblivious transfer protocol (allowing Bob to only obtain only one key per node) and the semantic security of the encryption scheme (which allows Bob to only decrypt one entry in each node). We now flesh out the details by providing a simulation proof from Alice and Bob's view of the protocol. Let x and y be the inputs of Alice and Bob respectively and Π be a protocol for secure-function evaluation of $f(x,y)$. Let $\text{VIEW}_A^\Pi(x,y)$ and $\text{VIEW}_B^\Pi(x,y)$ be the view of Alice and Bob for the run of the protocol Π on input x and y (view of a party consists of its input, output, and all messages it receives during the execution of the protocol). In a simulation proof one needs to show two probabilistic polynomial-time algorithms S_A and S_B such that $S_A(x, f(x,y))$ and $S_B(y, f(x,y))$ are computationally indistinguishable from $\text{VIEW}_A^\Pi(x,y)$ and $\text{VIEW}_B^\Pi(x,y)$, respectively. For a precise definition of a simulation proof the reader should refer to [14, Chapter 7].

We first consider the case where Alice is corrupt. Alice's view in an execution of Protocol 1 consists of her view of the oblivious transfer protocol executions and the output of the function from Bob at the end. We now build a simulator that simulates Alice's view given access only to her input and output. Because the oblivious transfer protocol is secure, there exists a simulator that can simulate the transcript of Alice's view of the oblivious transfer protocol without knowing Bob's input. On input $(i_1, \ldots, i_k, f(i_1, \ldots, i_n))$, the simulator first simulates Alice's view of all $n-k$ executions of the oblivious transfer protocol by repeatedly running the oblivious transfer protocol simulator. Using a standard hybrid argument on the transcripts of all $n-k$ executions oblivious transfer protocols, we see that if the oblivious transfer protocol is secure, then the distributions of the simulated and real combined transcripts of all $n-k$ oblivious transfer executions are indistinguishable with non-negligible probability.

Finally, the simulator writes $f(i_1, \ldots, i_n)$ on the transcript of Alice's view. We now show that the distribution of the output f is indistinguishable (except with negligible probability) from the real output, which amounts to showing that Bob outputs $f(i_1, \ldots, i_n))$ correctly on a real interaction. By the security of the oblivious transfer protocol, Bob is provided with the correct keys corresponding to its input during each execution of the oblivious transfer protocol. Applying claim 1, it follows immediately that, except with negligible probability, Alice obtains the correct output from Bob, except with negligible probability. Therefore, the distribution of the simulated transcript is indistinguishable, except with negligible probability, from a real transcript, concluding the case when Alice is corrupt.

We now consider the case when Bob is corrupt. Given $(i_{k+1}, \ldots, i_n, f(i_1, \ldots, i_n))$, the simulator $\mathcal{S}$ must simulate both a garbled OBDD that Bob can use to correctly compute $f(i_1, \ldots, i_n)$, and Bob's view of the $n-k$ executions of the oblivious transfer protocol. We first show how $\mathcal{S}$ simulates the $n-k$ oblivious transfer protocol executions. As in the previous case, because the oblivious transfer protocol is secure, there exists a simulator that can simulate the transcript of Bob's view of the oblivious transfer protocol without knowing Alice's input. Therefore, $\mathcal{S}$ simulates Bob's view of all $n-k$ executions of the oblivious transfer protocol by running the oblivious transfer protocol simulator $n-k$ times. Using a standard hybrid argument on the transcripts of the oblivious transfer protocols, we see that if the oblivious transfer protocol is secure, then the distributions of the simulated and real transcripts of all $n-k$ oblivious transfer executions are indistinguishable except with negligible probability.

We now show how $\mathcal{S}$ builds a garbled OBDD that Bob can use to successfully compute $f(i_1, \ldots, i_n)$. Since $\mathcal{S}$ does not know $i_1, \ldots, i_k$, it cannot generate the garbled OBDD according to the protocol instructions. Instead, $\mathcal{S}$ generates a garbled OBDD that always evaluates to $f(i_1, \ldots, i_n)$ regardless of the keys used. Such a garbled OBDD is built by first generating a chain of $n-k$ garbled nodes $n_{k+1}, \ldots, n_n$ such that Bob's computation starts at n_{k+1} and proceeds along the chain through n_{k+2} and so on, before ending at node n_n; note that there is one such node for every level from $k+1$ to n. To ensure the computation always proceeds along this chain, *both* ciphertexts in garbled nodes $n_{k+1}, \ldots, n_{n-1}$ are encryptions (under different keys) of the same label-key message such that the label points to the next node along the chain and the node key combined with the level key allows successful decryption of that node; for example, simulated node n_j for $k+1 \leq j \leq n-1$ has the form

$$\left(label(n_j), \, E_{s_{n_j} \oplus s_l^0}(label(n_{j+1}) \| s_{n_{j+1}}) \right.$$
$$\left. E_{s_{n_j} \oplus s_l^1}(label(n_{j+1}) \| s_{n_{j+1}}) \right) \, .$$

Node n_n is the terminal node and it is set to $f(i_1, \ldots, i_n)$. Once $n_{k+1}, \ldots, n_n$ is generated, the simulator generates a number of "fake" nodes so that the simulated garbled OBDD contains the correct number of nodes; this number can be determined from $OBDD(f)$. Fake nodes are nodes whose ciphertext pair contain encryptions under different keys of the same label-key message; in a fake node, the label, the keys used to encrypt the ciphertext pair, and the label-key message encrypted in the ciphertext pair are chosen randomly.

All that remains is to show that the distribution of the simulated garbled OBDD is indistinguishable from that of a real garbled OBDD. We do this by using a standard hybrid argument over the nodes in the garbled OBDD. Specifically, we run hybrid experiments with garbled OBDDs where real nodes are replaced by simulated nodes. We define the hybrid distributions such that $H_0(i_1, \ldots, i_n)$ contains the real garbled OBDD and $H_B(i_1, \ldots, i_n)$ contains the simulated garbled OBDD where B is the number of non-dummy nodes in the real garbled OBDD. We do not need to consider dummy nodes in our hybrid experiment OBDDs because dummy nodes have the same distribution as the simulated fake nodes and do not affect our argument.

We now define the hybrid garbled OBDD in experiment $H_i(i_1, \ldots, i_n)$; the difficulty here is that the hybrid OBDD contains both real and simulated nodes but must still allow Bob to correctly compute $f(i_1, \ldots, i_n)$. First, we traverse the real garbled OBDD and label a node as active if it is used by Bob in the process of evaluating the OBDD and inactive otherwise. Note that there will be only $n - k$ active nodes. Next, we order the nodes in the garbled OBDD by their level with level $j + 1$ nodes placed ahead of level $j + 2$ and so on; within the same level, nodes are ordered arbitrarily. The hybrid OBDD is defined as follows: first take the real garbled OBDD and replace the first i non-dummy nodes as follows: inactive nodes are replaced with simulated fake nodes. An active node at level j is altered by replacing its current ciphertext pair with two encryptions of the label-key message corresponding to the next active node at level $j + 1$. These replacement ciphertexts are created with the keys used to create the original ciphertext pair. Note that the distribution of this altered active node is identical to that of the simulated node n_j in the node chain described above. It is easy to see that a garbled OBDD built with this definition has the same distribution as 1) a real garbled OBDD when $i = 0$ (i.e. for H_0), and 2) a simulated garbled OBDD when $i = B$ (i.e. for H_B).

We are now ready to show that the distribution of the simulated garbled OBDD is indistinguishable from that of a real garbled OBDD; that is, we will show that $\{H_0(i_1, \ldots, i_n)\} = \{H_B(i_1, \ldots, i_n)\}$. Suppose to the contrary that the distributions are distinguishable; that is, there exists a poly-time distinguisher $\mathcal{D}$ that

$$|\Pr[\mathcal{D}(H_0(i_1, \ldots, i_n)) = 1] - \Pr[\mathcal{D}(H_B(i_1, \ldots, i_n)) = 1]| > 1/p$$

for some polynomial p. Then there exists a j such that

$$|\Pr[\mathcal{D}(H_{j-1}(i_1, \ldots, i_n)) = 1] - \Pr[\mathcal{D}(H_j(i_1, \ldots, i_n)) = 1]| > 1/pB$$

Using $\mathcal{D}$, we now build an adversary that breaks the semantic security of the encryption scheme used to encrypt the garbled nodes. Recall in a semantic security game, the adversary sends two messages m_0, m_1 to the challenger and receives the encryption of m_b for $b = \{0, 1\}$; the adversary's goal is to determine b. Let n^j be the jth node and we denote its two ciphertext terms as c_0^j and c_1^j. Note that node n^j in the hybrid OBDD in distribution H_{j-1}^* is a real garbled node, whereas the same node for distribution H_j^* is a simulated garbled node; specifically, c_0^j and c_1^j in distribution H_{j-1}^* are encryptions of different label-key messages, whereas they are encryptions of the same label-key message in distribution H_j^*. We exploit this fact to build the adversary $\mathcal{A}$ that breaks semantic security of the encryption scheme.

First, $\mathcal{A}$ creates the hybrid garbled OBDD corresponding to the distribution $H_{j-1}^*(i_1, \ldots, i_n)$. One of the two ciphertexts c_0^j and c_1^j in node n^j is an encryption of the label and key for the active node n^{j+1}, whereas the other ciphertext is an encryption of the label and key for an inactive node. Let ℓ_0 be the label-key message encrypted in c_0^j and ℓ_1 be that encrypted in c_1^j. Next, $\mathcal{A}$ sends ℓ_0 and ℓ_1 to the semantic security challenger and receives c^*, which is an encryption of either ℓ_0 or ℓ_1. Without loss of generality, let ℓ_0 be the label-key message (contained in ciphertext c_0^j) that leads to the next active node. $\mathcal{A}$ replaces c_1^j with c^* in node n^j in the garbled OBDD that it built in the first step, and then feeds the altered OBDD together with the other required inputs to the hybrid distinguisher $\mathcal{D}$. Note that c^* cannot be decrypted with the node and level keys for node n^j. This fact, however, does not prevent the garbled OBDD from being evaluated correctly because c^* replaces c_1^j, which contains the label and key to an inactive node, and would not be successfully decrypted while evaluating the garbled OBDD on the inputs $(i_1, \ldots, i_n)$.

$\mathcal{D}$ eventually outputs a result stating that the input is of distribution H_{j-1}^* or H_j^*. If $\mathcal{D}$ outputs that the input is of distribution H_{j-1}^*, then $\mathcal{A}$ outputs that c^* is an encryption of ℓ_1; otherwise $\mathcal{A}$ outputs that c^* is an encryption of ℓ_0. Notice that if c^* is an encryption of ℓ_0, then both ciphertexts in node n^j are encryptions of the same label-message, and the input to $\mathcal{D}$ has distribution H_j^*. Similarly, if c^* is an encryption of ℓ_1, then the input to $\mathcal{D}$ has distribution H_{j-1}^*. Since $\mathcal{D}$ distinguishes between H_{j-1}^* and H_j^* with non-negligible probability, we see that $\mathcal{A}$ wins the semantic security game with non-negligible advantage. Since we assume that the encryption scheme is semantically secure, this implication is a contradiction, and there is no such distinguisher $\mathcal{D}$ that distinguishes between $H_0(i_1, \ldots, i_n)$ and $H_B(i_1, \ldots, i_n)$; that is, the distribution of the simulated garbled OBDD is indistinguishable from that of a real garbled OBDD. Therefore, Bob's simulated view is indistinguishable, except with negligible probability, to the real view, concluding the proof. ∎

Proof: [*Proof Sketch for Claim 4*] The full proof is very similar for that in Claim 2, and we provide only a sketch. The case when Alice is corrupt is identical to that in Claim 2. We briefly discuss why the proof is also almost identical in the case when Bob is corrupt. The main observation is that the simulator $\mathcal{S}$ builds the simulated garbled OBDD in the same way as in Claim 2 because there is no difference in how Protocol 2 requires Bob to traverse the garbled nodes given the same level keys. Therefore, the hybrid distributions are defined the same way and the rest of the proof follows. ∎

Author Index

Aiello, W. 298
Akritidis, P. 221
Anagnostakis, K. G. 221
Antonatos, S. 221
Archer, M. 346
Ateniese, G. 288
Atluri, V. 144
Backes, M. 370
Bellare, M. 380, 390
Berger, Y. 245
Boneh, D. 211, 410
Boyen, X. 191
Boyer, W. F. 336
Brainard, J. 168
Brumley, D. 311
Butler, K. 298
Cadar, C. 322
Camenisch, J. 201
Chen, H. 278
Choi, J. Y. 37
Curtmola, R. 79
De Santis, A. 288
Di Raimondo, M. 400
Dill, D. L. 322
Eisner, J. 235
Engler, D. R. 322
Ferrara, A. L. 288
Fogla, P. 59
Franklin, J. 311
Ganesh, V. 322
Garay, J. 79
Gennaro, R. 400
Goh, E.-J. 410
Golle, P. 69
Goyal, V. 89
Heitmeyer, C. L. 346
Hohenberger, S. 201
Irwin, K. 134
Jajodia, S. 6

Jha, S. 47, 410
Juels, A. 168
Kamara, S. 79
Karlof, C. 154
Kil, C. 37
Kohlweiss, M. 201
Kohno, T. 380
Krawczyk, H. 400
Kruger, L. 410
Lam, V. T. 221
Laud, P. 370
Lee, A. J. 124
Lee, W. 59
Leonard, E. I. 346
Levine, B. N. 255
Li, N. 113, 356
Li, Z. 37
Liberatore, M. 255
Liu, A. 264
Lysyanskaya, A. 201
Mason, J. 235
Masucci, B. 288
McDaniel, P. 99, 298
McLean, J. 346
McQueen, M. A. 336
McSherry, F. 69
Meyerovich, M. 201
Miller, B. P. 47
Mironov, I. 69
Murdoch, S. J. 27
Nambiar, A. 17
Neumann, P. G. 1
Neven, G. 390
Newsome, J. 311
Ning, P. 264
Ostrovsky, R. 79
Ou, X. 336
Pandey, O. 89
Pawlowski, P. M. 322

Perrig, A. 278
Pirretti, M. 99
Reiter, M. K. 37
Rivest, R. L. 168
Rubin, S. 47
Sahai, A. 89
Setia, S. 6
Shacham, H. 191
Shankar, U. 154
Shen, E. 191
Shoup, V. 380
Song, D. 278, 311
Stubblefield, A. 235
Sun, K. 264
Szydlo, M. 168
Traynor, P. 99
Tripunitara, M. V. 113
Vaidya, J. 144
Wang, C. 264
Wang, H. 179
Wang, Q. 113, 356
Wang, XF. 37
Warner, J. 144
Waters, B. 89, 99, 191, 211
Waters, K. 235
Winsborough, W. H. 134
Winslett, M. 124
Wool, A. 245
Wright, M. 17
Xie, M. 179
Xu, J. 37
Yeredor, A. 245
Yin, H. 179
Yu, T. 134
Yung, M. 168
Zhou, Y. 264
Zhu, B. 6

NOTES

NOTES

NOTES